KT-385-643
T078796

Useful Information to Remember

The stronger the acid, the more readily it gives up a proton. (2.1)

The stronger the acid, the weaker is its conjugate base. (2.1)

The stronger the acid, the smaller is its pK_a value. (2.2)

Acid strength depends on the stability of the base formed when the acid gives up its proton—the more stable the base, the stronger is its conjugate acid. (2.6)

Going across a row in the periodic table: the more electronegative the atom to which the hydrogen is attached, the stronger is the acid. Electronegativity: $F > O > N > C$ (2.6)

Going down a column in the periodic table: the larger the atom to which the hydrogen is attached, the stronger is the acid. (2.6)

A compound exists primarily in its acid form (with its proton) in a solution that is more acidic than its pK_a value and in its basic form (without its proton) in a solution that is more basic than its pK_a value. (2.7)

Alkene stability: the fewer hydrogens bonded to the sp^2 carbons, the more stable is the alkene. (4.6)

Alkyl substituents stabilize alkenes and carbocations. (4.6, 5.2)

Electron-rich atoms or molecules are attracted to electron-deficient atoms or molecules; that is, nucleophiles are attracted to electrophiles. (4.7)

Curved arrows show the flow of electrons; they are drawn from an electron-rich center to an electron-deficient center. (4.7)

Carbocation stability: $3° >$ benzyl ~ allyl ~ $2° > 1° >$ methyl > vinyl (5.2)

In an electrophilic addition reaction, the electrophile adds to the sp^2 carbon bonded to the greater number of hydrogens. (5.3)

A reduction reaction increases the number of C—H bonds. (5.12)

Electronegativity: $sp > sp^2 > sp^3$ (5.13)

A chiral molecule has a nonsuperimposable mirror image. (6.4)

A compound with one or more asymmetric centers will be optically active except if it is a meso compound. (6.10)

The greater the predicted stability of a resonance contributor, the more it contributes to the structure of the hybrid. (7.5)

The delocalization energy tells how much more stable a compound with delocalized electrons is than it would be if its electrons were localized. (7.6)

Conjugated dienes are more stable than isolated dienes. (7.7)

To be aromatic, a compound must be cyclic and planar and have an uninterrupted cloud of π electrons. The π cloud must contain an odd number of pairs of π electrons. (8.1)

All substituents that activate a benzene ring toward electrophilic aromatic substitution are ortho/para directors. The halogens are also ortho/para directors. All substituents (except the halogens) that deactivate a benzene ring toward electrophilic aromatic substitution are meta directors. (8.15)

The weaker the base, the better it is as a leaving group. (9.3)

The major product of an elimination reaction is the most stable alkene. (9.8)

The weaker the base, the more readily it can be displaced. (10.2)

An oxidation decreases the number of C—H bonds. (10.4)

A carboxylic acid derivative will undergo a nucleophilic acyl substitution reaction provided that the newly added group in the tetrahedral intermediate is not a much weaker base than the group that was attached to the acyl group in the reactant. (11.5)

The relative reactivity of carboxylic acid derivatives: acyl chlorides > esters ~ carboxylic acids > amides (11.6)

In an acid-catalyzed reaction, the acid protonates the atom in the reactant with the greatest electron density. (11.9)

The relative reactivity of carbonyl compounds: acyl chlorides > aldehydes > ketones > esters (12.2)

Stronger bonds show absorption bands at larger wavenumbers. (14.8)

Electron withdrawal causes NMR signals to appear at higher frequencies (at larger δ values). (14.18)

Interest Boxes

Essential Organic Chemistry

Second Edition

Paula Yurkanis Bruice

University of California, Santa Barbara

Prentice Hall

Boston Columbus Indianapolis New York San Francisco Upper Saddle River
Amsterdam Cape Town Dubai London Madrid Milan Munich Paris Montreal Toronto
Delhi Mexico City Sao Paulo Sydney Hong Kong Seoul Singapore Taipei Tokyo

Acquisitions Editor: Dawn Giovanniello
Assistant Editor: Jessica Neumann
Editor in Chief, Science: Nicole Folchetti
Marketing Manager: Elizabeth Averbeck
Editorial Assistant: Kristen Wallerius
Marketing Assistant: Keri Parcells
Managing Editor, Chemistry and Geosciences: Gina M. Cheselka
Project Manager, Science: Beth Sweeten
Senior Operations Supervisor: Alan Fischer
Art Director: Suzanne Behnke
Associate Media Producer: Kristin Mayo
Art Editor: Connie Long
Art Studio: Imagineering
Photo Researcher: Yvonne Gerin
Spectra: Reproduced by permission of Aldrich Chemical Co.
Interior Designer: Joseph Sengotta
Cover Designer: Suzanne Behnke
Cover Photo: Biwa Inc/Photonica/Getty Images
Production Services/Composition: Preparé, Inc.

To Meghan, Kenton, and Alec
with love and immense respect;
and to Tom, my best friend

Pearson Education, Inc.
Upper Saddle River, NJ 07458

Printed in the United States of America
10 9 8 7 6 5 4 3 2 1

ISBN-10: 0-321-64416-6
ISBN-13: 978-0-321-64416-9

Brief Contents

Contents

Preface

TO THE INSTRUCTOR

My guiding principle in writing this book has been to present organic chemistry as an exciting and vitally important science. Too many students look upon organic chemistry as a necessary evil, a course they have to take for reasons that to them are unknown. Most likely, they come to organic chemistry without having the opportunity to study a science that is not a collection of individual topics, but one that unfolds and grows, allowing them to use what they learn at the beginning of the course to explain and predict what follows. To counter the impression that the study of organic chemistry consists primarily of memorizing a diverse collection of molecules and reactions, this book is organized around shared features and unifying concepts, and emphasizes principles that can be applied again and again. I want students to learn how to apply what they have learned to a new setting, reasoning their way to a solution rather than memorizing a multitude of facts.

I also want students to see that organic chemistry is integral to their daily lives. Consequently, there are about 100 interest boxes sprinkled throughout the book. These boxes are designed to show students the relevance of organic chemistry to medicine (e.g., dissolving sutures, cholesterol and heart disease, artificial blood), to agriculture (e.g., resisting herbicides, acid rain, pesticides: natural and synthetic), to nutrition (e.g., trans fats, basal metabolic rate, omega fatty acids), and to our shared life on this planet (e.g., fossil fuels, measuring toxicity, biodegradable polymers).

Many students taking organic chemistry are interested in the biological sciences. The last few chapters of the book focus on the organic chemistry of compounds found in the biological world. The material in these chapters is not necessarily the same as what students will see if they take a course in biochemistry. Instead, the students will see how the principles that govern the reactions of organic compounds, for example, electron delocalization, leaving-group tendency, electrophilicity, and nucleophilicity, also govern the reactions of compounds found in biological systems. For example, DNA has T's instead of the U's found in RNA because of imine hydrolysis; the phosphorylation of glucose is an S_N2 reaction; NADH is a hydride donor just like $NaBH_4$ and $LiAlH_4$; the metabolism of a fatty acid involves a reverse Claisen condensation; and so on.

Some hard choices had to be made in deciding what was "essential" to this audience of students. The choices were made with three goals in mind: students should understand how and why organic compounds react the way they do, they should experience the fun and challenge of designing simple syntheses, and they should learn in earlier chapters those reactions they will encounter again in later chapters that focus on bioorganic topics. When I wrote the chapter on spectroscopy, I did not want students to be overwhelmed by a topic that they may never revisit in their lives, but I did want them to enjoy being able to interpret some simple spectra. This chapter was written so that it can be covered at any time during the course; there is a table of functional groups inside the back cover of the book for those who teach spectroscopy before all the functional groups have been introduced.

I hope you find this edition even more appealing to your students than the first edition has been. I am eager to hear your comments, bearing in mind that positive comments are the most fun, but critical comments are the most useful.

PEDAGOGICAL FEATURES

Problems, Solved Problems, and Problem-Solving Strategies

The book contains lots of problems. The answers (and explanations, when needed) are in the accompanying *Study Guide and Solutions Manual*, which I authored to ensure consistency in language with the text. The *Study Guide and Solutions Manual* also contains exercises on "electron pushing" and on "drawing resonance contributors." I have found these exercises to help my students feel comfortable

▼ Problem-Solving Strategy

PROBLEM-SOLVING STRATEGY

Write the number of hydrogens attached to each of the indicated carbon atoms in the following compound:

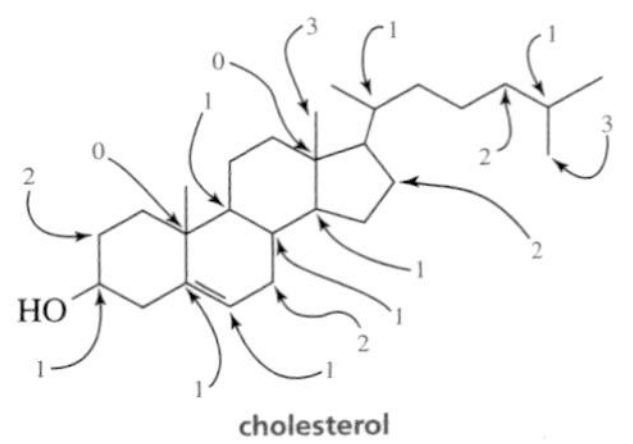

cholesterol

All the carbon atoms in the compound are neutral, so each needs to be bonded to four atoms. Thus if the carbon has only one bond that is showing, it must be attached to three hydrogens that are not shown; if the carbon has two bonds that are showing, it must be attached to two hydrogens that are not shown, and so on.

Now continue on to Problem 9.

◀ End-of-chapter problems

PROBLEMS

21. For each of the following compounds, draw the form in which it will predominate at pH = 3, pH = 6, pH = 10, and pH = 14:

a. CH_3COOH pK_a = 4.8 b. $CH_3CH_2\overset{+}{N}H_3$ pK_a = 11.0 c. CF_3CH_2OH pK_a = 12.4

22. Write the products of the following acid–base reactions, and indicate whether reactants or products are favored at equilibrium (use the pK_a values that are given in Section 2.3):

a. $CH_3\overset{O}{\overset{\|}{C}}OH + CH_3O^- \rightleftharpoons$ c. $CH_3\overset{O}{\overset{\|}{C}}OH + CH_3NH_2 \rightleftharpoons$

b. $CH_3CH_2OH + {}^-NH_2 \rightleftharpoons$ d. $CH_3CH_2OH + HCl \rightleftharpoons$

Biographical sketch ▶

BIOGRAPHY

Linus Carl Pauling (1901–1994) *was born in Portland, Oregon. A friend's home*

with topics that should be easy but somehow perplex even the best of them unless they have sufficient practice. The problems within each chapter are primarily drill problems that allow students to test themselves on material just covered before moving on to the next section. Selected problems are accompanied by worked-out solutions to provide insight into problem-solving techniques. Short answers provided at the end of the book for problems marked with a diamond give students immediate feedback concerning their mastery of a skill or concept. Most chapters also contain at least one **Problem-Solving Strategy**, a feature that teaches students how to approach various kinds of problems. For example, the problem-solving strategy in Chapter 9 teaches students how to determine whether a reaction will be more apt to take place by an S_N1 or an S_N2 pathway. Each Problem-Solving Strategy is followed by an exercise giving the student an opportunity to use the problem-solving skill just learned.

The **end-of-chapter problems** vary in difficulty. Frequently, they begin with drill problems that integrate material from the entire chapter, requiring the student to think in terms of all the material in the chapter rather than focusing on individual sections. The problems become more challenging as the student proceeds, often reinforcing concepts from previous chapters. The net result for the student is a progressive building of both problem-solving ability and confidence.

▼ Interest boxes

RECYCLING SYMBOLS

When plastics are recycled, the various types must be separated from one another. To aid in the separation, many states require manufacturers to place a recycling symbol on their products to indicate the type of plastic it is. You are probably familiar with these symbols, which are often embossed on the bottom of plastic containers. The symbols consist of three arrows around one of seven numbers; an abbreviation below the symbol indicates the type of polymer from which the container is made. The lower the number in the middle of the symbol, the greater is the ease with which the material can be recycled: 1 (PET) stands for poly(ethylene terephthalate), 2 (HDPE) for high-density polyethylene, 3 (V) for poly(vinyl chloride), 4 (LDPE) for low-density polyethylene, 5 (PP) for polypropylene, 6 (PS) for polystyrene, and 7 for all other plastics.

Summaries and Voice Boxes to Help the Student

Each chapter concludes with a Summary to help students synthesize the key points. Chapters that cover reactions conclude with a Summary of Reactions. Annotations (voice boxes) help students focus on points being discussed.

- **End-of-Chater Summaries** review the major concepts of the chapter in a concise, narrative format.
- **Summary of Reactions** sections list reactions covered in the chapter for review. Cross–references make it easy to locate the sections covering specific reaction types.

Margin Notes and Boxed Material to Engage the Student

Margin notes and biographical sketches appear throughout the text. The **margin notes** encapsulate key points that students should remember, and **biographical sketches** give students some appreciation of the history of chemistry and the people who contributed to that history.

Interest boxes connect chemistry to real life (discussing, for example, why Dalmatians are the only mammals that excrete uric acid, why life is based on carbon instead of silicon, how fleas can be controlled, and why SAMe is a product prominently displayed in health food stores).

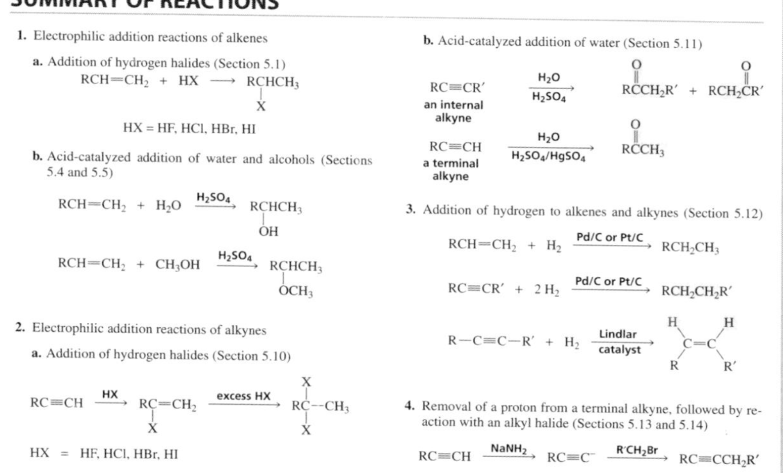
SUMMARY OF REACTIONS

1. Electrophilic addition reactions of alkenes

a. Addition of hydrogen halides (Section 5.1)

$RCH{=}CH_2 + HX \longrightarrow RCHCH_3$ (with X on the CH)

HX = HF, HCl, HBr, HI

b. Acid-catalyzed addition of water and alcohols (Sections 5.4 and 5.5)

$RCH{=}CH_2 + H_2O \xrightarrow{H_2SO_4} RCHCH_3$ (with OH on the CH)

$RCH{=}CH_2 + CH_3OH \xrightarrow{H_2SO_4} RCHCH_3$ (with OCH_3 on the CH)

2. Electrophilic addition reactions of alkynes

a. Addition of hydrogen halides (Section 5.10)

$RC{\equiv}CH \xrightarrow{HX} RC(X){=}CH_2 \xrightarrow{\text{excess HX}} RCX_2{-}CH_3$

HX = HF, HCl, HBr, HI

b. Acid-catalyzed addition of water (Section 5.11)

$RC{\equiv}CR'$ (an internal alkyne) $\xrightarrow[H_2SO_4]{H_2O} RCOCH_2R' + RCH_2COR'$

$RC{\equiv}CH$ (a terminal alkyne) $\xrightarrow[H_2SO_4/HgSO_4]{H_2O} RCOCH_3$

3. Addition of hydrogen to alkenes and alkynes (Section 5.12)

$RCH{=}CH_2 + H_2 \xrightarrow{\text{Pd/C or Pt/C}} RCH_2CH_3$

$RC{\equiv}CR' + 2\,H_2 \xrightarrow{\text{Pd/C or Pt/C}} RCH_2CH_2R'$

$R{-}C{\equiv}C{-}R' + H_2 \xrightarrow{\text{Lindlar catalyst}}$ cis-$RHC{=}CHR'$

4. Removal of a proton from a terminal alkyne, followed by reaction with an alkyl halide (Sections 5.13 and 5.14)

$RC{\equiv}CH \xrightarrow{NaNH_2} RC{\equiv}C^- \xrightarrow{R'CH_2Br} RC{\equiv}CCH_2R'$

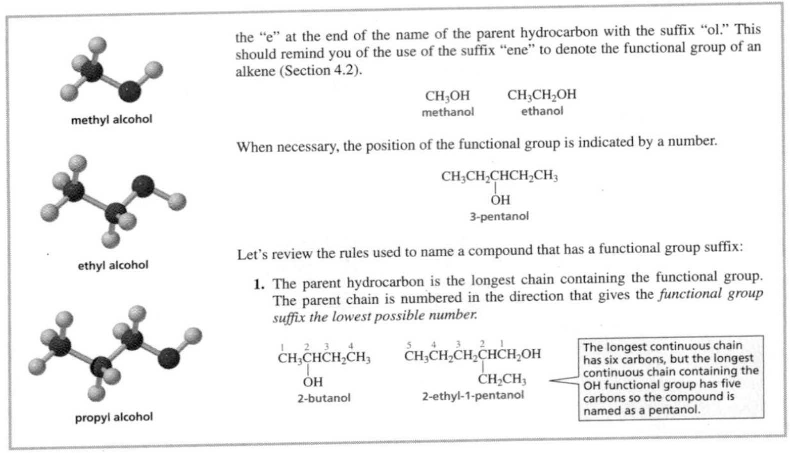

methyl alcohol
ethyl alcohol
propyl alcohol
the "e" at the end of the name of the parent hydrocarbon with the suffix "ol." This should remind you of the use of the suffix "ene" to denote the functional group of an alkene (Section 4.2).
CH3OH methanol
CH3CH2OH ethanol
When necessary, the position of the functional group is indicated by a number.
CH3CH2CHCH2CH3 OH
3-pentanol
Let's review the rules used to name a compound that has a functional group suffix:
1. The parent hydrocarbon is the longest chain containing the functional group. The parent chain is numbered in the direction that gives the functional group suffix the lowest possible number.
CH3CHCH2CH3 OH
2-butanol
CH3CH2CH2CHCH2OH CH2CH3
2-ethyl-1-pentanol
The longest continuous chain has six carbons, but the longest continuous chain containing the OH functional group has five carbons so the compound is named as a pentanol.

▲ Art program

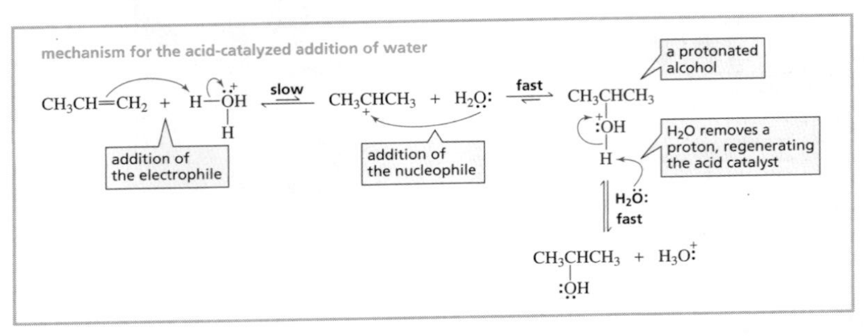

▲ Mechanism

Art Program: Rich in Three-Dimensional, Computer-Generated Structures

This edition continues to present energy-minimized, **three-dimensional structures** throughout the text that give students an appreciation of the three-dimensional shapes of organic molecules. Color in the illustration program is used not simply for show but to highlight and organize the information. Colors are employed in consistent ways (for example, mechanism arrows are always red), but there is no need for a student to memorize a color palette.

Companion Website

WWW icons in the margins identify 3-D molecules, movies, and interactive animations on the Companion Website (*http://www.chemplace.com*) that are pertinent to the material being discussed. For a description of the Website's content, see The Chemistry Place for Essential Organic Chemistry under Resources for Students.

Changes to This Edition

Responses from instructors and students have led to adjustments in the coverage and distribution of certain topics and encouraged expansion of the book's most successful pedagogical features.

Content and Organization

The chapter on isomers and stereochemistry has been moved closer to the front of the book. The chapter on radicals has been deleted; however, the discussion of radicals in biological systems is retained and is incorporated with the radical chemistry found in Chapter 5. A discussion of catalysis has been added to Chapter 5, and mass spectrometry has been added to the chapter on "Determining the Structures of Organic Compounds." There are new sections that will allow students to apply what they have learned about reactivity to synthetic problems: for example, using substitution reactions to synthesize organic compounds, and using orientation effects to synthesize disubstituted benzenes. There is also new material on solvent effects, and the reactions of carbohydrates in basic solutions. Mechanistic details are expanded to facilitate students' understanding. Lastly, much of the book has been rewritten with the goal of making the material easier to understand.

Pedagogical Elements

The mechanisms in this edition are introduced by bold green titles and are presented in the form of bulleted steps to make the individual steps stand out more clearly. The mechanisms are not boxed; students should see mechanisms as being central to the understanding of the discipline and not something that is set aside or that they can come back to later. This edition also includes more voice boxes to aid student learning, more interest boxes, and many new problems.

LIST OF RESOURCES

For Students

Study Guide and Solutions Manual (0321592581) by Paula Yurkanis Bruice. This *Study Guide and Solutions Manual* contains complete and detailed explanations of the solutions to the problems in the text. It also contains exercises on "drawing curved arrows" and on "drawing resonance contributors." Each chapter contains a Practice Test, with the answers found at the back of the book.

The Chemistry Place for Essential Organic Chemistry, 2e (*http://www.chemplace.com*) Built to complement *Essential Organic Chemistry* as part of an integrated course package and identified by WWW icons in the margins, the easy-to-use Companion Website features the following modules for each chapter:

- **Interactive Tutorial and Animation Galleries** highlight central concepts and illustrate key mechanisms.

- **Molecule Galleries** feature hundreds of 3-D molecular models of compounds. Students can rotate and compare models, change their representation, and examine electrostatic potential map surfaces—a unique feature of learning organic chemistry on the Web.

- **Practice Exercises and Quizzes** allow students to test their understanding of the material. Each question includes a hint with a cross-reference to a text reading and detailed feedback. Although I had never been a fan of multiple-choice questions, the quality of these questions, written by a team of authors, has changed my mind.

Molecular Modeling Workbook (0131410407) Features SpartanView™ and SpartanBuild™ software. This workbook includes a software tutorial and numerous challenging exercises students can tackle to solve problems involving structure building and analysis using the tools included in the two pieces of Spartan software. Available free when packaged with the text. Contact your Pearson Prentice Hall representative for details.

Prentice Hall Molecular Model Kit (02055081363) This best-selling model kit allows students to build space-filling and ball-and-stick models of common organic molecules. It allows accurate depiction of double and triple bonds, including heteroatomic molecules (which some model kits cannot handle well).

Prentice Hall Framework Molecular Model Kit (0133300765) This model kit allows students to build scale models with precise interatomic distances and bond angles. This is the most accurate model kit available.

For Instructors

Online Test Item File (0321602668) by Debbie Beard, Mississippi State University. Includes a selection of more than 1200 multiple-choice, short answer, and essay test questions.

Online Instructor Resource Center (0321633466) This resource features almost all the art from the text, including tables, in PowerPoint™ format. Also included are Lecture PowerPoints, Classroom Response PowerPoints and Problem PowerPoints for each chapter.

ACKNOWLEDGMENTS

I am enormously grateful to the following reviewers who made this book a reality. The value of their work cannot be overstated.

Second Edition Reviewers
Deborah Booth, *University of Southern Mississippi*
Paul Buonora, *California State University–Long Beach*
Tom Chang, *Utah State University*
Dana Chatellier, *University of Delaware*
Amy Deveau, *University of New England*
J. Brent Friesen, *Dominican University*
Anne Gorden, *Auburn University*
Christine Hermann, *University of Radford*
Scott Lewis, *James Madison University*
Cynthia McGowan, *Merrimack College*
Keith Mead, *Mississippi State University*
Amy Pollock, *Michigan State University*

Manuscript Reviewers
Ardeshir Azadnia, *Michigan State University*
Debbie Beard, *Mississippi State University*
J. Phillip Bowen, *University of North Carolina–Greensboro*
Tim Burch, *Milwaukee Area Technical College*
Dana Chatellier, *University of Delaware*
Michelle Chatellier, *University of Delaware*
Long Chiang, *University of Massachusetts–Lowell*
Jan Dekker, *Reedley College*
Olga Dolgounitcheva, *Kansas State University*
John Droske, *University of Wisconsin–Stevens Point*
Eric Enholm, *University of Florida*
Gregory Friestad, *University of Vermont*
Wesley Fritz, *College of Dupage*
Robert Gooden, *Southern University*
Michael Groziak, *California State University–Hayward*
Steve Holmgren, *Montana State University*
Robert Hudson, *University of Western Ontario*
Richard Johnson, *University of New Hampshire*
Alan Kennan, *Colorado State University*
Spencer Knapp, *Rutgers University*
Mike Nuckols, *North Carolina State University*
Ed Parish, *Auburn University*
Mark W. Peczuh, *University of Connecticut*
Suzanne Purrington, *North Carolina State University*
Charles Rose, *University of Nevada–Reno*
Preet Saluja, *Triton College*
Joseph Sloop, *United States Military Academy*
Robert Swindell, *University of Arkansas*

Amar Tung, *Lincoln University*
Kraig Wheeler, *Delaware State University*
Randy Winchester, *Grand Valley State University*
Mark Workentin, *University of Western Ontario*

Focus Group Attendees
Ardeshir Azadina, *Michigan State University*
Gregory L. Baker, *Michigan State University*
Jay Brown, *Southwest Minnesota State University*
Jerry Easdon, *College of Ozarks*
Nancy Gardner, *California State University–Long Beach*
Cyril Parkanyi, *Florida Atlantic University*
Bob Swindell, *University of Arkansas*
Kathleen Trahanovsky, *Iowa State University*

Accuracy Reviewer
Malcolm Forbes, *University of North Carolina*

I am deeply grateful to my editor, Dawn Giovanniello, whose creative talents are extraordinary. I am also grateful to my assistant editor, Jessica Neumann, who was always ready to do whatever was needed to make this book the best that it could be. I want to thank the other talented and dedicated people at Pearson Prentice Hall who played important roles in the development of the book. Gina Cheselka, Managing Editor, kept the project on track and managed an infinite number of critical details; Elizabeth Averbeck, Marketing Manager, brought the book to the attention of the global community of organic chemistry instructors; Kristin Mayo developed and managed the media program; Suzanne Behnke, Art Director, created this edition's striking cover; Connie Long produced the new art; Dave Theisen, National Sales Director, has been, as always, a champion, resource, and friend in the field. I am also grateful to the compositor, Rosaria Cassinese, who kept me on track—with more patience than I deserved—during the production process.

I particularly want to thank the many wonderful and talented students who have taught me more than they will ever know. And I want to thank my children, from whom I may have learned the most.

To make this book as user friendly as possible, I would appreciate any comments that will help me achieve this goal in future editions. If you find sections that should be clarified or expanded, please let me know. Finally, I am enormously grateful to Malcolm Forbes of the University of North Carolina, who painstakingly combed the book looking for errors. Any that remain are my responsibility; if you find any, please send me a quick e-mail so that they can be corrected in future printings.

Paula Yurkanis Bruice
University of California, Santa Barbara
pybruice@chem.ucsb.edu

TO THE STUDENT

Welcome to organic chemistry! You are about to embark on an exciting journey. This book has been written with students like you in mind—those who are encountering the subject for the first time. The journey into organic chemistry can be an exciting one. The book's central goal is to make this journey both stimulating and enjoyable by helping you understand the central principles of the subject and asking you to apply them as you progress through the pages.

You should start by familiarizing yourself with the book. The inside front and back covers display information you may want to refer to often during the course. Chapter Summaries and Reaction Summaries at each chapter's end provide helpful reminders of the material you should understand after studying the chapter. The Glossary at the end of the book can be a very useful study aid. The molecular models and electrostatic potential maps that you will find throughout the book are provided to give you an appreciation of what molecules look like in three dimensions and to show how charge is distributed within a molecule. Think of the margin notes as the author's opportunity to inject reminders of facts that are important to remember. Be sure to read them.

Work all the problems within each chapter. These are drill problems that allow you to check whether you have mastered the skills and concepts the chapter is teaching. Some of them (or parts of them) are solved for you in the text. Short answers to some of the others—those marked with a diamond—are provided at the end of the book. Do not overlook the Problem-Solving Strategies that are also sprinkled throughout the text; they provide practical suggestions on the best way to approach important types of problems.

In addition to the within-chapter problems, work as many end-of-chapter problems as you can. The more problems you work, the more comfortable you will be with the subject matter and the better prepared you will be for the material in subsequent chapters. Do not let any problem frustrate you. If you cannot figure out the answer in a reasonable amount of time, turn to the *Study Guide and Solutions Manual* to learn how you should have approached the problem. Later on, go back and try to work the problem on your own again. Be sure to visit the Companion Website (*http://www.chemplace.com*) to try out some of the study tools such as the Student Tutorials, Molecule Galleries, and Exercise Sets and Quizzes.

The most important advice to remember (and follow) in studying organic chemistry is DO NOT FALL BEHIND! The individual steps to learning organic chemistry are quite simple; each by itself is relatively easy to master. But they are numerous, and the subject can quickly become overwhelming if you do not keep up.

Before many of its theories and mechanisms were figured out, organic chemistry was a discipline that could be mastered only through memorization. Fortunately, that is no longer true. You will find many unifying principles that will allow you to use what you have learned in one situation to predict what will happen in other situations. So, as you read the book and study your notes, always try to understand why each chemical event or behavior happens. For example, when the reasons behind reactivity are understood, most reactions can be predicted. Approaching the course with the misconception that to succeed you must memorize hundreds of unrelated reactions could be your downfall. There is simply too much material to memorize. Understanding and reasoning, not memorization, provides the necessary foundation on which to lay subsequent learning. Nevertheless, from time to time, some memorization will be required: some fundamental rules have to be memorized, and you will need to memorize the common names of a number of organic compounds. But the latter should not be a problem; after all, your friends have common names that you have been able to learn.

Good luck in your study. I hope you enjoy your course in organic chemistry and learn to appreciate the logic of this fascinating discipline. If you have any comments about the book or any suggestions for improving it, I would love to hear from you. Remember, positive comments are the most fun, but negative comments are the most useful.

Paula Yurkanis Bruice
pybruice@chem.ucsb.edu

About the Author

Paula Bruice with Zeus and Abigail

Paula Yurkanis Bruice was raised primarily in Massachusetts. After graduating from the Girls' Latin School in Boston, she received an A.B. from Mount Holyoke College and a Ph.D. in chemistry from the University of Virginia. She was awarded an NIH postdoctoral fellowship for study in the Department of Biochemistry at the University of Virginia Medical School and held a postdoctoral appointment in the Department of Pharmacology at Yale Medical School.

She has been a member of the faculty at the University of California, Santa Barbara, since 1972, where she has received the Associated Students Teacher of the Year Award, the Academic Senate Distinguished Teaching Award, two Mortar Board Professor of the Year Awards, and the UCSB Alumni Association Teaching Award. Her research interests center on the mechanism and catalysis of organic reactions, particularly those of biological significance. Paula has a daughter and a son who are physicians and a son who is a lawyer. Her main hobbies are reading mystery/suspense novels and enjoying her pets (two dogs, two cats, and two parrots).

Electronic Structure and Covalent Bonding

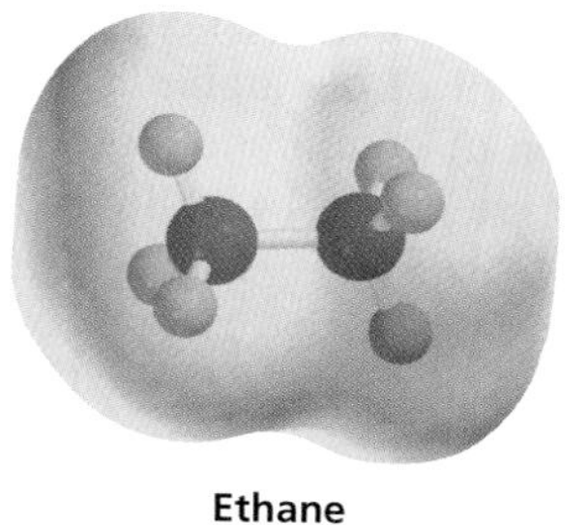

Ethane

Ethene

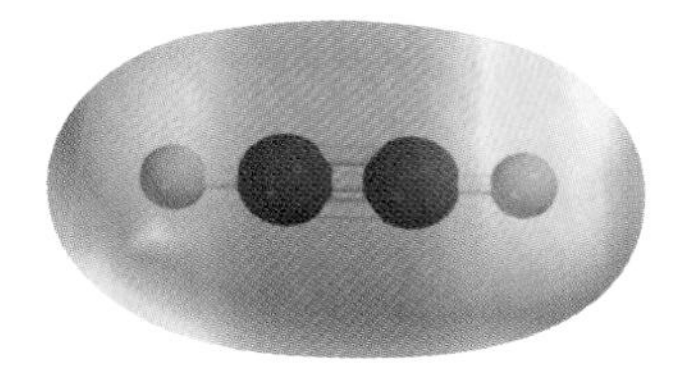

Ethyne

To stay alive, early humans must have been able to tell the difference between two kinds of materials in their world. "You can live on roots and berries," they might have said, "but you can't live on dirt. You can stay warm by burning tree branches, but you can't burn rocks."

By the early eighteenth century, scientists thought they had grasped the nature of that difference. Compounds derived from living sources were believed to contain an unmeasurable vital force—the essence of life. Because they came from organisms, they were called "organic" compounds. Compounds derived from minerals—those lacking that vital force—were "inorganic."

Because chemists could not create life in the laboratory, they assumed they could not create compounds that had a vital force. Since this was their mind-set, you can imagine how surprised chemists were in 1828 when Friedrich Wöhler produced urea—a compound known to be excreted by mammals—by heating ammonium cyanate, an inorganic mineral.

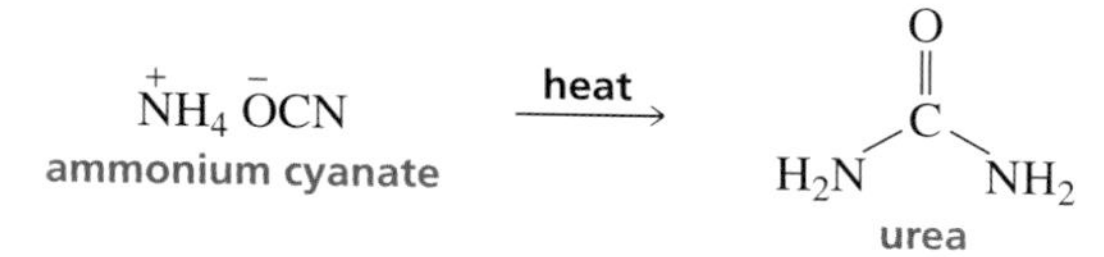

For the first time, an "organic" compound had been obtained from something other than a living organism and certainly without the aid of any kind of vital force. Clearly, chemists needed a new definition for "organic compounds." **Organic compounds** are now defined as *compounds that contain carbon*.

Why is an entire branch of chemistry devoted to the study of carbon-containing compounds? We study organic chemistry because just about all of the molecules that

German chemist **Friedrich Wöhler (1800–1882)** *began his professional life as a physician and later became a professor of chemistry at the University of Göttingen. Wöhler codiscovered the fact that two different chemicals could have the same molecular formula. He also developed methods of purifying aluminum—at the time, the most expensive metal on Earth—and beryllium.*

make life possible—proteins, enzymes, vitamins, lipids, carbohydrates, and nucleic acids—contain carbon; thus, the chemical reactions that take place in living systems, including our own bodies, are reactions of organic compounds. Most of the compounds found in nature—those we rely on for food, medicine, clothing (cotton, wool, silk), and energy (natural gas, petroleum)—are organic compounds as well.

Organic compounds are not, however, limited to those found in nature. Chemists have learned to synthesize millions of organic compounds never found in nature, including synthetic fabrics, plastics, synthetic rubber, medicines, and even things like photographic film and Super Glue. Many of these synthetic compounds prevent shortages of naturally occurring products. For example, it has been estimated that if synthetic materials were not available for clothing, all of the arable land in the United States would have to be used for the production of cotton and wool just to provide enough material to clothe us. Currently, there are about 16 million known organic compounds, and many more are possible.

What makes carbon so special? Why are there so many carbon-containing compounds? The answer lies in carbon's position in the periodic table. Carbon is in the center of the second row of elements. The atoms to the left of carbon have a tendency to give up electrons, whereas the atoms to the right have a tendency to accept electrons (Section 1.3).

the second row of the periodic table

Because carbon is in the middle, it neither readily gives up nor readily accepts electrons. Instead, it shares electrons. Carbon can share electrons with several different kinds of atoms, and it can also share electrons with other carbon atoms. Consequently, carbon is able to form millions of stable compounds with a wide range of chemical properties simply by sharing electrons.

When we study organic chemistry, we study how organic compounds react. When an organic compound reacts, some existing bonds break and some new bonds form. Bonds form when two atoms share electrons, and bonds break when two atoms no longer share electrons. How readily a bond forms and how easily it breaks depend on the particular electrons that are shared, which, in turn, depend on the atoms to which the electrons belong. So if we are going to start our study of organic chemistry at the beginning, we must start with an understanding of the structure of an atom—what electrons an atom has and where they are located.

NATURAL VERSUS SYNTHETIC

It is a popular belief that natural substances—those made in nature—are superior to synthetic ones—those made in the laboratory. Yet when a chemist synthesizes a compound, such as penicillin, it is exactly the same in all respects as the compound synthesized in nature. Sometimes chemists can improve on nature. For example, chemists have synthesized analogs of morphine—compounds with structures similar to but not identical to that of morphine—that have painkilling effects like morphine but, unlike morphine, are not habit forming. Chemists have synthesized analogs of penicillin that do not produce the allergic responses that a significant fraction of the population experiences from naturally produced penicillin, or that do not have the bacterial resistance of the naturally produced antibiotic.

A field of poppies growing in Afghanistan. Commercial morphine is obtained from opium, the juice obtained from this species of poppy.

1.1 THE STRUCTURE OF AN ATOM

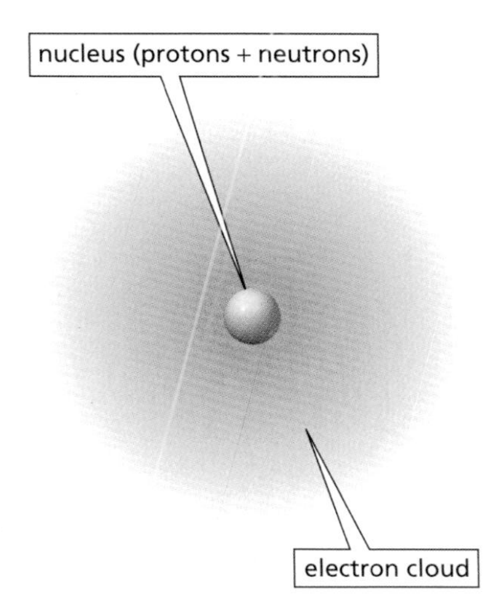

An atom consists of a tiny dense nucleus surrounded by electrons that are spread throughout a relatively large volume of space around the nucleus. The nucleus contains *positively charged protons* and *neutral neutrons*, so it is positively charged. The electrons are *negatively charged*. Because the amount of positive charge on a proton equals the amount of negative charge on an electron, a neutral atom has an equal number of protons and electrons. Atoms can gain electrons and thereby become negatively charged, or they can lose electrons and become positively charged. However, the number of protons in an atom does not change.

Protons and neutrons have approximately the same mass and are about 1800 times more massive than an electron. This means that most of the *mass* of an atom is in its nucleus. However, most of the *volume* of an atom is occupied by its electrons, and that is where our focus will be because it is the electrons that form chemical bonds.

The **atomic number** of an atom equals the *number of protons* in its nucleus. The atomic number is also the number of electrons that surround the nucleus of a neutral atom. For example, the atomic number of carbon is 6, which means that a neutral carbon atom has six protons and six electrons.

The **mass number** of an atom is the *sum of its protons and neutrons*. All carbon atoms have the same atomic number because they all have the same number of protons. They do not all have the same mass number because they do not all have the same number of neutrons. For example, 98.89% of naturally occurring carbon atoms have six neutrons—giving them a mass number of 12—and 1.11% have seven neutrons—giving them a mass number of 13. These two different kinds of carbon atoms (^{12}C and ^{13}C) are called isotopes. **Isotopes** have the same atomic number (that is, the same number of protons), but different mass numbers because they have different numbers of neutrons.

Naturally occurring carbon also contains a trace amount of ^{14}C, which has six protons and eight neutrons. This isotope of carbon is radioactive, decaying with a half-life of 5730 years. (The *half-life* is the time it takes for one-half of the nuclei to decay.) As long as a plant or an animal is alive, it takes in as much ^{14}C as it excretes or exhales. When it dies, it no longer takes in ^{14}C, so the ^{14}C in the organism slowly decreases. Therefore, the age of an organic substance can be determined by its ^{14}C content.

The **atomic weight** (or **atomic mass**) of a naturally occurring element is the *average mass of its atoms*. For example, carbon has an atomic weight of 12.011 atomic mass units. The **molecular weight** of a compound is the *sum of the atomic weights* of all the atoms in the molecule.

PROBLEM 1 ♦

Oxygen has three isotopes with mass numbers of 16, 17, and 18. The atomic number of oxygen is eight. How many protons and neutrons does each of the isotopes have?

1.2 HOW THE ELECTRONS IN AN ATOM ARE DISTRIBUTED

The electrons in an atom can be thought of as occupying a set of shells that surround the nucleus. The way in which the electrons are distributed in these shells is based on a theory developed by Einstein. The first shell is the smallest and the one closest to the nucleus; the second shell is larger and extends farther from the nucleus; and the third and higher numbered shells extend even farther out. Each shell consists of subshells known as **atomic orbitals**. The first shell has only an *s* atomic orbital; the second shell consists of *s* and *p* atomic orbitals; and the third shell consists of *s*, *p*, and *d* atomic orbitals (Table 1.1).

Table 1.1 Distribution of Electrons in the First Three Shells That Surround the Nucleus

	First shell	Second shell	Third shell
Atomic orbitals	*s*	*s*, *p*	*s*, *p*, *d*
Number of atomic orbitals	1	1, 3	1, 3, 5
Maximum number of electrons	2	8	18

Each shell contains one *s* orbital. The second and higher shells—in addition to their *s* orbital—each contain three *p* orbitals. The three *p* orbitals have the same energy. The third and higher shells—in addition to their *s* and *p* orbitals—also contain five *d* orbitals. Because an orbital can contain no more than two electrons (see below), the first shell, with only one atomic orbital, can contain no more than two electrons. The second shell, with four atomic orbitals—one *s* and three *p*—can have a total of eight electrons. Eighteen electrons can occupy the nine atomic orbitals—one *s*, three *p*, and five *d*—of the third shell.

An important point to remember is that *the closer the atomic orbital is to the nucleus, the lower is its energy*. Because the *s* orbital in the first shell (called a 1*s* orbital) is closer to the nucleus than is the *s* orbital in the second shell (called a 2*s* orbital), the 1*s* orbital is lower in energy. Comparing orbitals in the same shell, we see that an *s* orbital is lower in energy than a *p* orbital, and a *p* orbital is lower in energy than a *d* orbital.

The closer the orbital is to the nucleus, the lower is its energy.

Relative energies of atomic orbitals: $1s < 2s < 2p < 3s < 3p < 3d$

The **electronic configuration** of an atom describes what orbitals the electrons occupy. The following three rules are used to determine an atom's electronic configuration:

1. An electron always goes into the available orbital with the lowest energy.
2. No more than two electrons can occupy each orbital, and the two electrons must be of opposite spin. (Notice in Table 1.2 that spin in one direction is designated by ↑, and spin in the opposite direction by ↓.)

From these first two rules, we can assign electrons to atomic orbitals for atoms that contain one, two, three, four, or five electrons. The single electron of a hydrogen atom occupies a 1*s* orbital, the second electron of a helium atom fills the 1*s* orbital, the third electron of a lithium atom occupies a 2*s* orbital, the fourth electron of a beryllium atom fills the 2*s* orbital, and the fifth electron of a boron atom occupies one of the 2*p* orbitals. (The subscripts *x*, *y*, and *z* distinguish the three 2*p* orbitals.) Because the three

Table 1.2 The Electronic Configurations of the Smallest Atoms

Atom	Name of element	Atomic number	1*s*	2*s*	$2p_x$	$2p_y$	$2p_z$	3*s*
H	Hydrogen	1	↑					
He	Helium	2	↑↓					
Li	Lithium	3	↑↓	↑				
Be	Beryllium	4	↑↓	↑↓				
B	Boron	5	↑↓	↑↓	↑			
C	Carbon	6	↑↓	↑↓	↑	↑		
N	Nitrogen	7	↑↓	↑↓	↑	↑	↑	
O	Oxygen	8	↑↓	↑↓	↑↓	↑	↑	
F	Fluorine	9	↑↓	↑↓	↑↓	↑↓	↑	
Ne	Neon	10	↑↓	↑↓	↑↓	↑↓	↑↓	
Na	Sodium	11	↑↓	↑↓	↑↓	↑↓	↑↓	↑

ALBERT EINSTEIN

Albert Einstein (1879–1955) was born in Germany. When he was in high school, his father's business failed and his family moved to Milan, Italy. Although Einstein wanted to join his family in Italy, he had to stay behind because German law required compulsory military service after high school. To help him, his high school mathematics teacher wrote a letter saying that Einstein could have a nervous breakdown without his family and also that there was nothing left to teach him. Eventually, Einstein was asked to leave the school because of his disruptive behavior. Popular folklore says he left because of poor grades in Latin and Greek, but his grades in those subjects were fine.

Einstein was visiting the United States when Hitler came to power, so he accepted a position at the Institute for Advanced Study in Princeton, becoming a U.S. citizen in 1940. Although a lifelong pacifist, he wrote a letter to President Roosevelt warning of ominous advances in German nuclear research. This led to the creation of the Manhattan Project, which developed the atomic bomb and tested it in New Mexico in 1945.

$2p$ orbitals have the same energy, the electron can be put into any one of them. Before we can continue to atoms containing six or more electrons, we need the third rule:

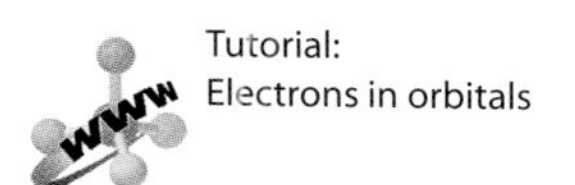

Tutorial: Electrons in orbitals

3. When there are two or more orbitals with the same energy, an electron will occupy an empty orbital before it will pair up with another electron.

The sixth electron of a carbon atom, therefore, goes into an empty $2p$ orbital, rather than pairing up with the electron already occupying a $2p$ orbital (Table 1.2). There is one more empty $2p$ orbital, so that is where the seventh electron of a nitrogen atom goes. The eighth electron of an oxygen atom pairs up with an electron occupying a $2p$ orbital rather than going into a higher energy $3s$ orbital.

Electrons in inner shells (those below the outermost shell) are called **core electrons**. Electrons in the outermost shell are called **valence electrons**. Carbon, for example, has two core electrons and four valence electrons (Table 1.2).

Lithium and sodium each have one valence electron. Elements in the same column of the periodic table have the same number of valence electrons. Because the number of valence electrons is the major factor determining an element's chemical properties, elements in the same column of the **periodic table** have similar chemical properties. (You can find a periodic table inside the back cover of this book.) Thus, the chemical behavior of an element depends on its electronic configuration.

Shown is a bronze sculpture of **Einstein** *on the grounds of the National Academy of Sciences in Washington, D.C. It measures 21 feet from the top of the head to the tip of the feet and weighs 7000 pounds. In his left hand, Einstein holds the mathematical equations that represent his three most important contributions to science: the photoelectric effect, the equivalency of energy and matter, and the theory of relativity. At his feet is a map of the sky.*

PROBLEM 2 ◆

How many valence electrons do the following atoms have?

a. carbon **b.** nitrogen **c.** oxygen **d.** fluorine

PROBLEM 3 ◆

Table 1.2 shows that lithium and sodium each have one valence electron. Find potassium (K) in the periodic table and predict how many valence electrons it has.

PROBLEM-SOLVING STRATEGY

Write the ground-state electronic configuration for chlorine.

The periodic table in the back of the book shows that chlorine has 17 electrons. Now we need to assign the electrons to orbitals using the rules that determine an atom's electronic configuration. Two electrons are in the $1s$ orbital, two are in the $2s$ orbital, six are in the $2p$ orbitals, and two are in the $3s$ orbital. This accounts for 12 of the 17 electrons. The remaining five electrons are in the $3p$ orbitals. Therefore, the electronic configuration is written as: $1s^2, 2s^2, 2p^6, 3s^2, 3p^5$

Now continue on to Problem 4.

PROBLEM 4 ♦

a. Compare the ground-state electronic configurations for nitrogen and phosphorus and look at their relative positions in the periodic table. How many valence electrons does each have? How many core electrons does each have?
b. Compare the ground-state electronic configurations for oxygen and sulfur and look at their relative positions in the periodic table. How many valence electrons does each have? How many core electrons does each have?

PROBLEM 5 ♦

How many valence electrons do chlorine, bromine, and iodine have?

1.3 IONIC AND COVALENT BONDS

BIOGRAPHY

American chemist **Gilbert Newton Lewis (1875–1946)** *was born in Weymouth, Massachusetts, and received a Ph.D. from Harvard in 1899. He was the first person to prepare "heavy water," which has deuterium atoms in place of the usual hydrogen atoms (D_2O versus H_2O). Because heavy water can be used as a moderator of neutrons, it became important in the development of the atomic bomb. Lewis started his career as a professor at the Massachusetts Institute of Technology and joined the faculty at the University of California, Berkeley, in 1912.*

In trying to explain why atoms form bonds, G. N. Lewis proposed that *an atom is most stable if its outer shell is either filled or contains eight electrons and it has no electrons of higher energy*. According to Lewis's theory, an atom will give up, accept, or share electrons in order to achieve a filled outer shell or an outer shell that contains eight electrons. This theory has come to be called the **octet rule** (even though the filled outer shell of hydrogen has only two electrons).

Lithium (Li) has a single electron in its 2*s* orbital. If it loses this electron, the lithium atom ends up with a filled outer shell—a stable configuration. Lithium, therefore, loses an electron relatively easily. Sodium (Na) has a single electron in its 3*s* orbital, so it too loses an electron easily.

When we draw the electrons around an atom, as in the following equations, core electrons are not shown; only valence electrons are shown because only valence electrons are used in bonding. Each valence electron is shown as a dot. Notice that when the single valence electron of lithium or sodium is removed, the resulting atom—now called an ion—carries a positive charge.

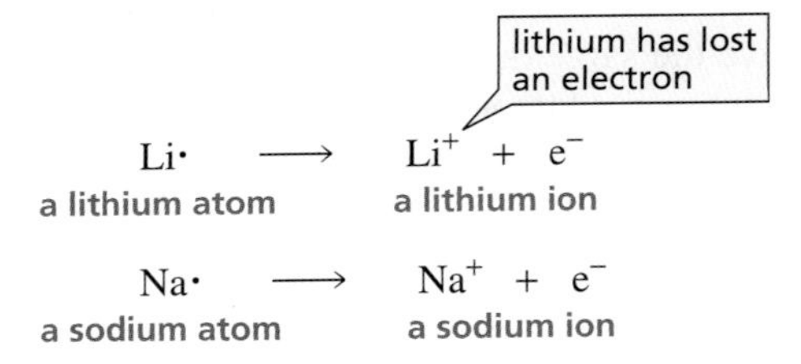

Fluorine and chlorine each have seven valence electrons (Table 1.2 and Problem 5). Consequently, they readily acquire an electron in order to have an outer shell of eight electrons.

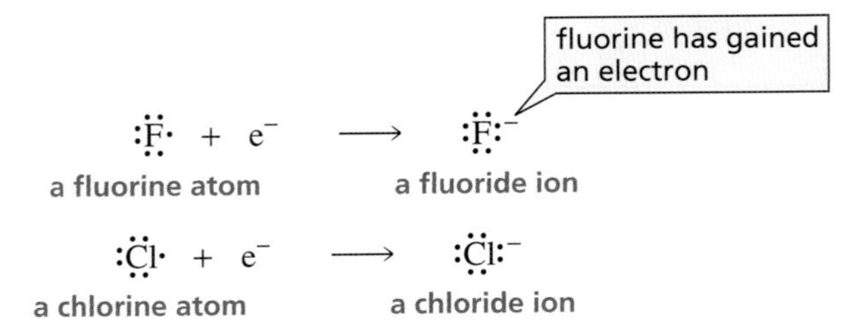

Ionic Bonds Are Formed by the Attraction Between Ions of Opposite Charge

We have just seen that sodium gives up an electron easily and chlorine readily acquires an electron. Therefore, when sodium metal and chlorine gas are mixed, each sodium atom transfers an electron to a chlorine atom, and crystalline sodium chloride (table salt) is formed as a result. The positively charged sodium ions and negatively charged chloride ions are independent species held together by the attraction of opposite charges (Figure 1.1). A **bond** is an attractive force between two ions or between two atoms. A bond that results from the attraction between ions of opposite charge is called an **ionic bond**.

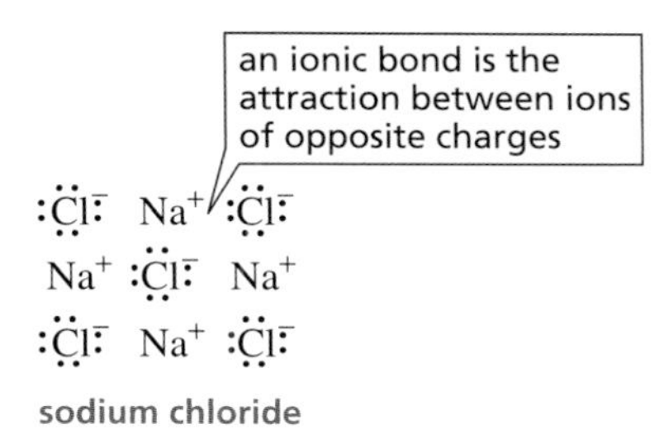

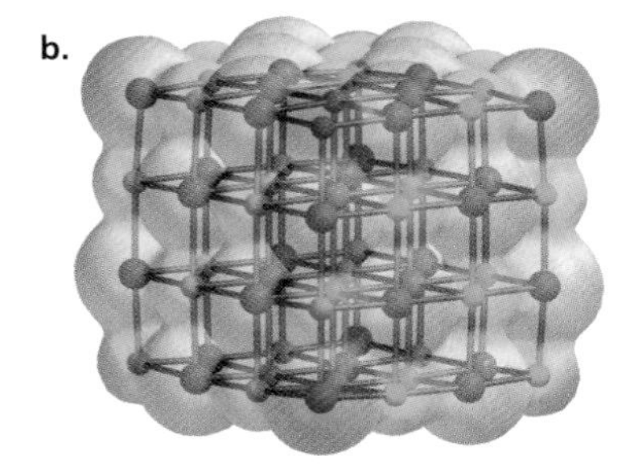

◀ **Figure 1.1**
(a) Crystalline sodium chloride.
(b) The electron-rich chloride ions are red, and the electron-poor sodium ions are blue. Each chloride ion is surrounded by six sodium ions, and each sodium ion is surrounded by six chloride ions. Ignore the sticks holding the balls together; they are there only to keep the model from falling apart.

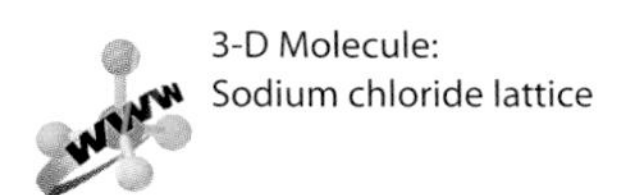

Covalent Bonds Are Formed by Sharing Electrons

Instead of giving up or acquiring electrons, an atom can achieve a filled outer shell (or an outer shell of eight electrons) by sharing electrons. For example, two fluorine atoms can each attain a filled second shell by sharing their unpaired valence electrons. A bond formed as a result of *sharing electrons* is called a **covalent bond**.

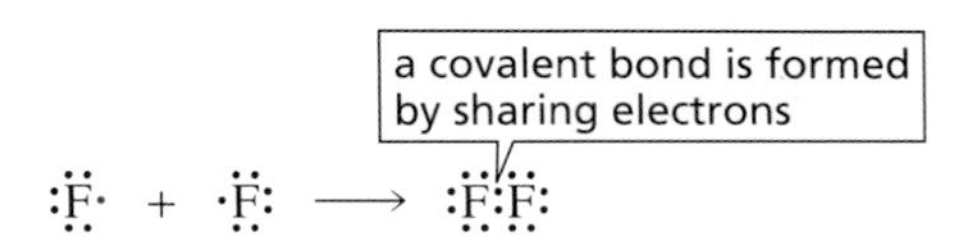

Two hydrogen atoms can form a covalent bond by sharing electrons. As a result of covalent bonding, each hydrogen acquires a stable, filled first shell.

$$\mathrm{H\cdot} + \mathrm{\cdot H} \longrightarrow \mathrm{H{:}H}$$

Similarly, hydrogen and chlorine can form a covalent bond by sharing electrons. In doing so, hydrogen fills its only shell and chlorine achieves an outer shell of eight electrons.

$$\mathrm{H\cdot} + \mathrm{\cdot\ddot{\underset{..}{Cl}}{:}} \longrightarrow \mathrm{H{:}\ddot{\underset{..}{Cl}}{:}}$$

A hydrogen atom can achieve a completely empty shell by losing an electron. Loss of its sole electron results in a positively charged **hydrogen ion**. A positively charged hydrogen ion is called a **proton** because when a hydrogen atom loses its valence electron, only the hydrogen nucleus—which consists of a single proton—remains.

A hydrogen atom can achieve a filled outer shell by gaining an electron, thereby forming a negatively charged hydrogen ion, called a **hydride ion.**

$$\underset{\text{a hydrogen atom}}{\mathrm{H}\cdot} \longrightarrow \underset{\text{a proton}}{\mathrm{H}^+} + e^-$$

$$\underset{\text{a hydrogen atom}}{\mathrm{H}\cdot + e^-} \longrightarrow \underset{\text{a hydride ion}}{\mathrm{H}:^-}$$

Because oxygen has six valence electrons, it needs to form two covalent bonds to achieve a filled outer shell. Nitrogen, with five valence electrons, must form three covalent bonds, and carbon, with four valence electrons, must form four covalent bonds to achieve a filled outer shell. Notice that all the atoms in water, ammonia, and methane have filled outer shells.

$$2\,\mathrm{H}\cdot + \cdot\ddot{\mathrm{O}}: \longrightarrow \mathrm{H}:\ddot{\mathrm{O}}:\mathrm{H}$$

water — oxygen has formed 2 covalent bonds

$$3\,\mathrm{H}\cdot + \cdot\dot{\mathrm{N}}\cdot \longrightarrow \mathrm{H}:\ddot{\mathrm{N}}:\mathrm{H}\ (\mathrm{H})$$

ammonia — nitrogen has formed 3 covalent bonds

$$4\,\mathrm{H}\cdot + \cdot\dot{\mathrm{C}}\cdot \longrightarrow \mathrm{H}:\mathrm{C}(\mathrm{H})(\mathrm{H}):\mathrm{H}$$

methane — carbon has formed 4 covalent bonds

Nonpolar Covalent Bonds and Polar Covalent Bonds

The atoms that share the bonding electrons in the F—F or the H—H covalent bond are identical. Therefore, they share the electrons equally; that is, each electron spends as much time in the vicinity of one atom as in the other. Such a bond is called a **nonpolar covalent bond**.

In contrast, the bonding electrons in hydrogen chloride, water, and ammonia are more attracted to one atom than to another because the atoms that share the electrons in these molecules are different and have different electronegativities. **Electronegativity** is a measure of the ability of an atom to pull the bonding electrons toward itself. The electronegativities of some of the elements are shown in Table 1.3. The

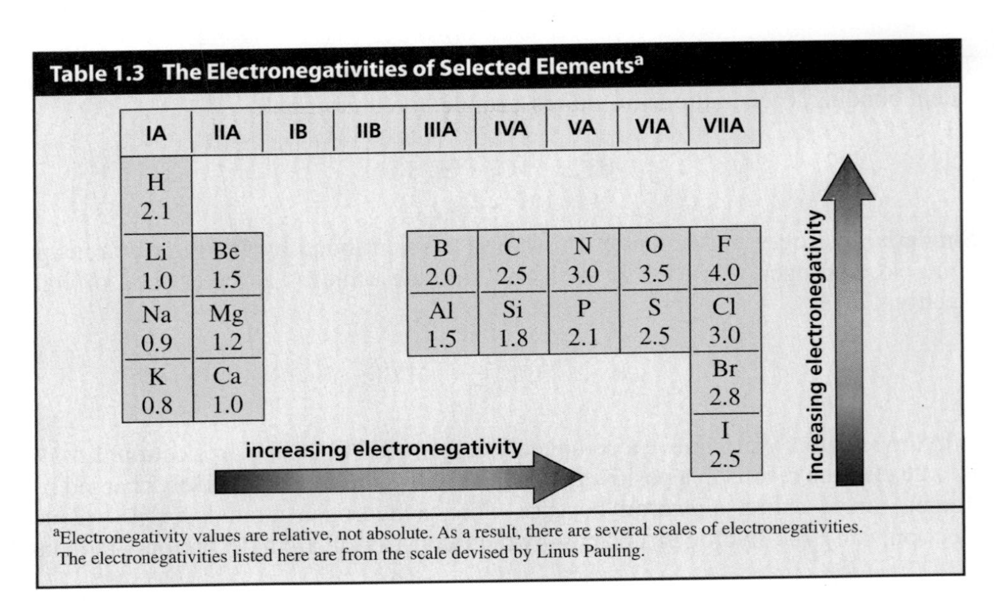

Table 1.3 The Electronegativities of Selected Elements[a]

IA	IIA	IB	IIB	IIIA	IVA	VA	VIA	VIIA
H 2.1								
Li 1.0	Be 1.5			B 2.0	C 2.5	N 3.0	O 3.5	F 4.0
Na 0.9	Mg 1.2			Al 1.5	Si 1.8	P 2.1	S 2.5	Cl 3.0
K 0.8	Ca 1.0							Br 2.8
								I 2.5

[a]Electronegativity values are relative, not absolute. As a result, there are several scales of electronegativities. The electronegativities listed here are from the scale devised by Linus Pauling.

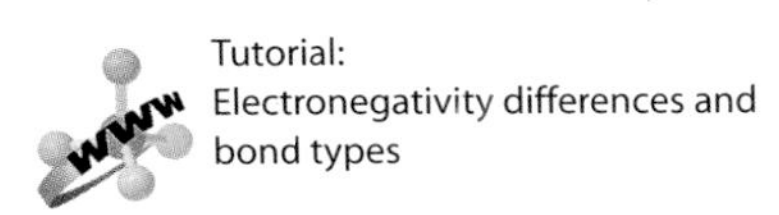

Tutorial:
Electronegativity differences and bond types

bonding electrons in hydrogen chloride, water, and ammonia are more attracted to the atom with the greater electronegativity. A **polar covalent bond** is a covalent bond between atoms of different electronegativities. Notice that electronegativity increases from left to right across a row of the periodic table or going up any of the columns.

A polar covalent bond has a slight positive charge on one end and a slight negative charge on the other. Polarity in a covalent bond is indicated by the symbols $\delta+$ and $\delta-$, which denote partial positive and partial negative charges, respectively. The negative end of the bond is the end that has the more electronegative atom. The greater the difference in electronegativity between the bonded atoms, the more polar the bond will be. (Notice that a pair of shared electrons can be shown as a line between two atoms.)

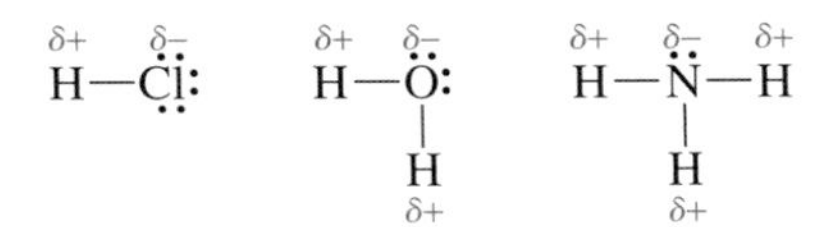

You can think of ionic bonds and nonpolar covalent bonds as being at the opposite ends of a continuum of bond types. At one end is an ionic bond, a bond in which there is no sharing of electrons. At the other end is a nonpolar covalent bond, a bond in which the electrons are shared equally. Polar covalent bonds fall somewhere in between, and the greater the difference in electronegativity between the atoms forming the bond, the closer the bond is to the ionic end of the continuum. C—H bonds are relatively nonpolar, because carbon and hydrogen have similar electronegativities (electronegativity difference = 0.4; see Table 1.3). N—H bonds are relatively polar (electronegativity difference = 0.9), but not as polar as O—H bonds (electronegativity difference = 1.4). Even closer to the ionic end of the continuum is the bond between sodium and chloride ions (electronegativity difference = 2.1), but sodium chloride is not as ionic as potassium fluoride (electronegativity difference = 3.2).

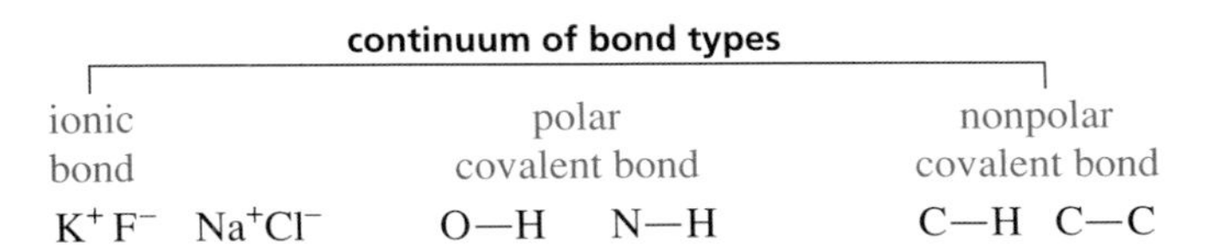

PROBLEM 6 ◆

Which bond is more polar?

a. H—CH_3 or Cl—CH_3

b. H—OH or H—H

c. H—Cl or H—F

d. Cl—Cl or Cl—CH_3

PROBLEM 7 ◆

Which of the following has

a. the most polar bond?

b. the least polar bond?

NaI LiBr Cl_2 KCl

PROBLEM 8 ◆

Use the symbols $\delta+$ and $\delta-$ to show the direction of polarity of the indicated bond in each of the following compounds (for example, $H_3\overset{\delta+}{C}-\overset{\delta-}{O}H$)

a. HO—H

b. H_3C—NH_2

c. HO—Br

d. I—Cl

Electrostatic potential maps (often simply called potential maps) are models that show how charge is distributed in the molecule under the map. The potential maps for LiH, H_2, and HF are shown below.

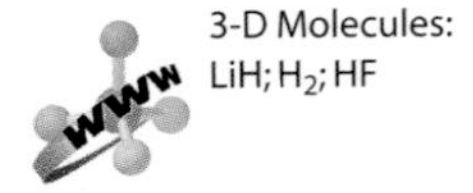

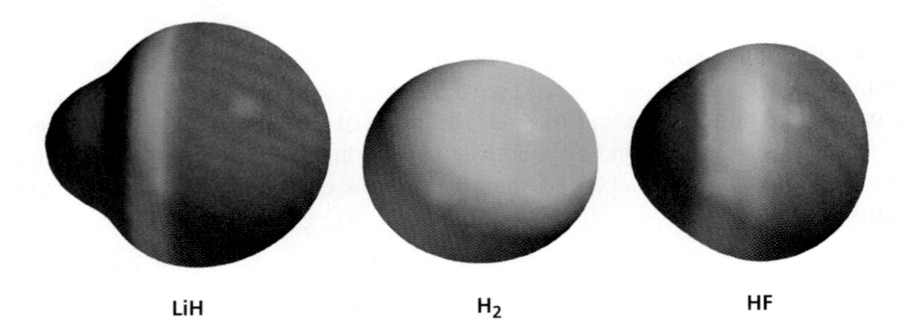

The colors indicate the distribution of charge in the molecule: red signifies electron-rich areas (negative charge); blue signifies electron-deficient areas (positive charge); and green signifies no charge. For example, the potential map for LiH indicates that the hydrogen atom is more negatively charged than the lithium atom. By comparing the three maps, we can tell that the hydrogen in LiH is more negatively charged than a hydrogen in H_2, and the hydrogen in HF is more positively charged than a hydrogen in H_2.

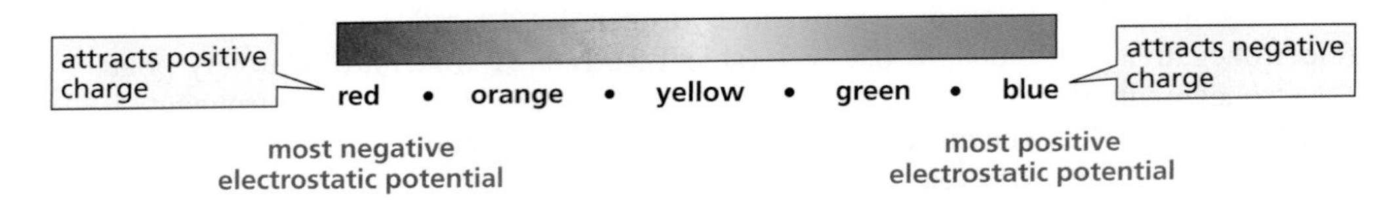

PROBLEM 9 ♦

After examining the potential maps for LiH, HF, and H_2, answer the following questions:

a. Which compounds are polar?

b. Which compound has the most positively charged hydrogen?

1.4 HOW THE STRUCTURE OF A COMPOUND IS REPRESENTED

First we will see how compounds are drawn using Lewis structures. Then we will look at the representations of structures that are used more commonly for organic compounds.

Lewis Structures

The chemical symbols we have been using, in which the valence electrons are represented as dots, are called **Lewis structures** (or electron dot structures) after G. N. Lewis (Section 1.3). Lewis structures are useful because they show us which atoms are bonded together and tell us whether any atoms *possess lone-pair electrons* or have a *formal charge*, two concepts we describe below. The Lewis structures for H_2O, H_3O^+, HO^-, and H_2O_2 are

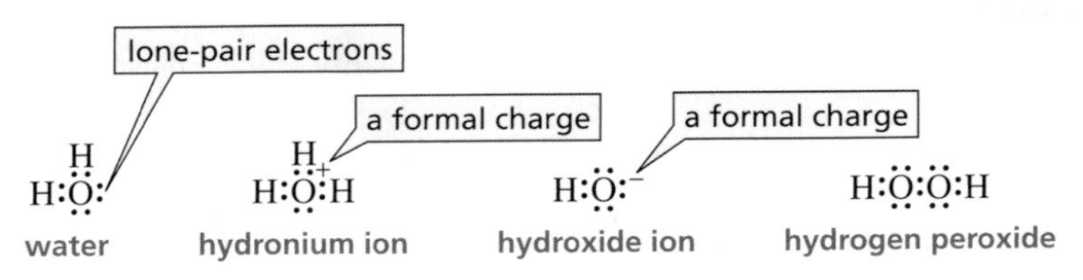

When you draw a Lewis structure, make sure that hydrogen atoms are surrounded by no more than two electrons and that C, O, N, and halogen (F, Cl, Br, I) atoms are surrounded by no more than eight electrons, in accordance with the octet rule. Valence electrons not used in bonding are called **nonbonding electrons, lone-pair electrons**, or simply **lone pairs**.

Once you have all the atoms and the electrons in place, you must examine each atom to see whether a formal charge should be assigned to it. A **formal charge** is the *difference* between the number of valence electrons an atom has when it is not bonded to any other atoms and the number of electrons it "owns" when it is bonded. An atom "owns" all of its lone-pair electrons and half of its bonding (shared) electrons.

Movie: Formal charge

formal charge = number of valence electrons − (number of lone-pair electrons + 1/2 number of bonding electrons)

For example, an oxygen atom has six valence electrons (Table 1.2). In water (H_2O), oxygen "owns" six electrons (four lone-pair electrons and half of the four bonding electrons). Because the number of electrons it "owns" is equal to the number of its valence electrons ($6 - 6 = 0$), the oxygen atom in water does not have a formal charge. The oxygen atom in the hydronium ion (H_3O^+) "owns" five electrons: two lone-pair electrons plus three (half of six) bonding electrons. Because the number of electrons it "owns" is one less than the number of its valence electrons, its formal charge is $+1$ ($6 - 5 = 1$). The oxygen atom in hydroxide ion HO^- "owns" seven electrons: six lone-pair electrons plus one (half of two) bonding electron. Because it "owns" one more electron than the number of its valence electrons, its formal charge is -1 ($6 - 7 = -1$).

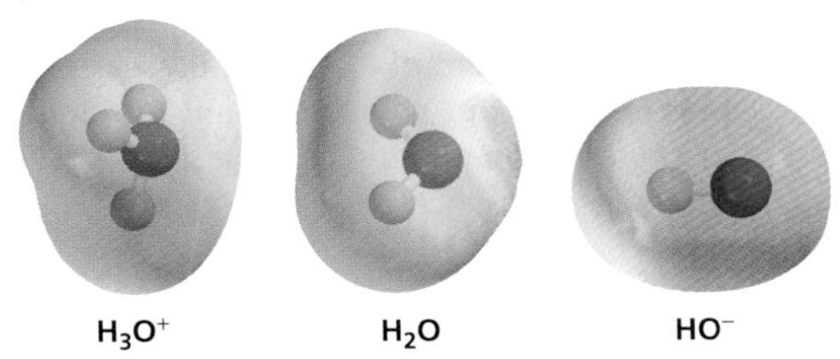

PROBLEM 10 ◆

The formal charge does not necessarily indicate that the atom has greater or less electron density than other atoms in the molecule without formal charges. You can see this by examining the potential maps for H_2O, H_3O^+, and HO^-.

a. Which atom bears the formal negative charge in the hydroxide ion?
b. Which atom has the greater electron density in the hydroxide ion?
c. Which atom bears the formal positive charge in the hydronium ion?
d. Which atom has the least electron density in the hydronium ion?

Nitrogen has five valence electrons (Table 1.2). Prove to yourself that the appropriate formal charges have been assigned to the nitrogen atoms in the following Lewis structures:

H:N̈:H Ḧ	H H:N̈:H (+) Ḧ	H:N̈:⁻ Ḧ	H:N̈:N̈:H Ḧ Ḧ
ammonia	ammonium ion	amide anion	hydrazine

Carbon has four valence electrons. Take a moment to understand why the carbon atoms in the following Lewis structures have the indicated formal charges:

H H:C̈:H Ḧ	H H:C̈⁺ Ḧ	H H:C̈:⁻ Ḧ	H H:C̈· Ḧ	H H H:C̈:C̈:H Ḧ Ḧ
methane	methyl cation a carbocation	methyl anion a carbanion	methyl radical	ethane

A species containing a positively charged carbon atom is called a **carbocation**, and a species containing a negatively charged carbon atom is called a **carbanion**. (Recall that a *cation* is a positively charged ion and an *anion* is a negatively charged ion.) A species containing an atom with a single unpaired electron is called a **radical** (often called a **free radical**).

Hydrogen has one valence electron, and each halogen (F, Cl, Br, I) has seven valence electrons, so the following species have the indicated formal charges:

H^+	$H:^-$	$H\cdot$	$:\ddot{Br}:^-$	$:\ddot{Br}\cdot$	$:\ddot{Br}:\ddot{Br}:$	$:\ddot{Cl}:\ddot{Cl}:$
hydrogen ion	hydride ion	hydrogen atom a radical	bromide ion	bromine atom a radical	bromine	chlorine

PROBLEM 11 ◆

Give each atom the appropriate formal charge:

a. $CH_3-\ddot{O}(H)-CH_3$ b. $H-\ddot{C}(H)-H$ c. $CH_3-N(CH_3)(CH_3)-CH_3$ d. $H-N(H)(H)-B(H)(H)-H$

In studying the molecules in this section, notice that when the atoms do not bear a formal charge or an unpaired electron, hydrogen and the halogens always have *one* covalent bond, oxygen always has *two* covalent bonds, nitrogen always has *three* covalent bonds, and carbon has *four* covalent bonds. Atoms that have more bonds or fewer bonds than the number required for a neutral atom will have either a formal charge or an unpaired electron. These numbers are very important to remember when you are first drawing structures of organic compounds because they provide a quick way to recognize when you have made a mistake.

$H-$	$:\ddot{F}-$ $:\ddot{Cl}-$ $:\ddot{I}-$ $:\ddot{Br}-$	$:\ddot{O}-$ (with one more bond)	$-\ddot{N}-$ (with one more bond)	$-C-$ (with two more bonds)
one bond	one bond	two bonds	three bonds	four bonds

In the following Lewis structures, notice that each atom has a filled outer shell. Also notice that since none of the molecules has a formal charge or an unpaired electron, C forms four bonds, N forms three bonds, O forms two bonds, and H and Br each form one bond.

$H_3C:\ddot{Br}:$ $\quad H_3C:\ddot{O}:H \quad H_3C:\ddot{O}:CH_3 \quad H_3C:\ddot{N}H_2 \quad H_3C:\ddot{N}(H):CH_3$

Because a pair of shared electrons can be shown as a line between two atoms, compare the preceding structures with the following ones:

$H_3C-\ddot{Br}: \quad H_3C-\ddot{O}-H \quad H_3C-\ddot{O}-CH_3 \quad H_3C-\ddot{N}-H(H) \quad H_3C-\ddot{N}(H)-CH_3$

PROBLEM 12 ◆

Draw the Lewis structure for each of the following:

a. $CH_3\overset{+}{N}H_3$ b. $^-C_2H_5$ c. NaOH d. NH_4Cl

Kekulé Structures

In **Kekulé structures**, the bonding electrons are drawn as lines and the lone-pair electrons are usually left out entirely, unless they are needed to draw attention to some chemical property of the molecule. (Although lone-pair electrons are not shown, you should remember that neutral nitrogen, oxygen, and halogen atoms always have them: one pair in the case of nitrogen, two pairs in the case of oxygen, and three pairs in the case of a halogen.)

```
   H           H              H     H          H               H       H
   |           |              |     |          |               |       |
H—C—Br     H—C—O—H      H—C—O—C—H     H—C—N—H     H—C—N—C—H
   |           |              |     |          |   |           |   |   |
   H           H              H     H          H   H           H   H   H
```

Condensed Structures

Frequently, structures are simplified by omitting some (or all) of the covalent bonds and listing atoms bonded to a particular carbon (or nitrogen or oxygen) next to it with subscripts as necessary. These structures are called **condensed structures**. Compare the following examples with the Kekulé structures shown above:

CH_3Br CH_3OH CH_3OCH_3 CH_3NH_2 CH_3NHCH_3

You can find more examples of condensed structures and the conventions commonly used to create them in Table 1.4.

PROBLEM 13 ◆

Draw the lone-pair electrons that are not shown in the following structures:

a. $CH_3CH_2NH_2$ **c.** CH_3CH_2OH **e.** CH_3CH_2Cl

b. CH_3NHCH_3 **d.** CH_3OCH_3 **f.** $HONH_2$

PROBLEM 14 ◆

Draw condensed structures for the compounds represented by the following models (black = C, white = H, red = O, blue = N, green = Cl):

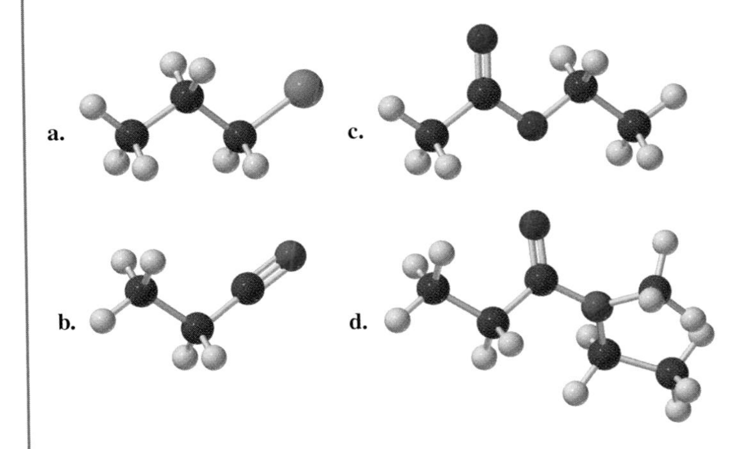

PROBLEM 15 ◆

Which of the atoms in the molecular models in Problem 14 have

a. three lone pairs? **b.** two lone pairs? **c.** one lone pair? **d.** no lone pairs?

Table 1.4 Kekulé and Condensed Structures

Kekulé structure	Condensed structures

Atoms bonded to a carbon are shown to the right of the carbon. Atoms other than H can be shown hanging from the carbon.

```
    H   H   H   H   H   H
    |   |   |   |   |   |
H — C — C — C — C — C — C — H
    |   |   |   |   |   |
    H   Br  H   H   Cl  H
```

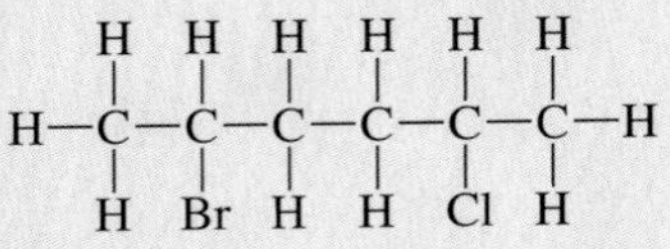

$CH_3CHBrCH_2CH_2CHClCH_3$ or

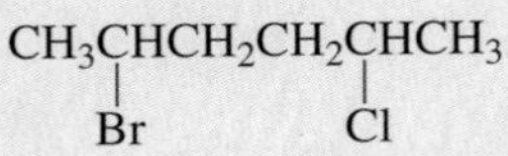

```
CH3CHCH2CH2CHCH3
   |        |
   Br       Cl
```

Repeating CH_2 groups can be shown in parentheses.

```
    H   H   H   H   H   H
    |   |   |   |   |   |
H — C — C — C — C — C — C — H
    |   |   |   |   |   |
    H   H   H   H   H   H
```

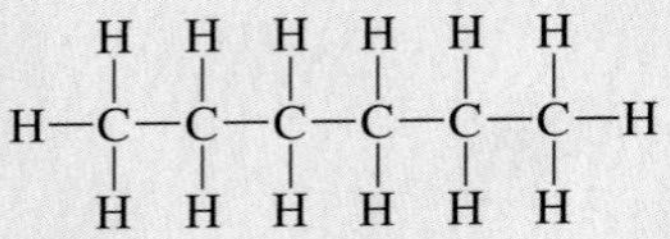

$CH_3CH_2CH_2CH_2CH_2CH_3$ or $CH_3(CH_2)_4CH_3$

Groups bonded to a carbon can be shown (in parentheses) to the right of the carbon, or hanging from the carbon.

```
    H   H   H     H   H   H
    |   |   |     |   |   |
H — C — C — C  —  C — C — C — H
    |   |   |     |   |   |
    H   H   CH3   H   OH  H
```

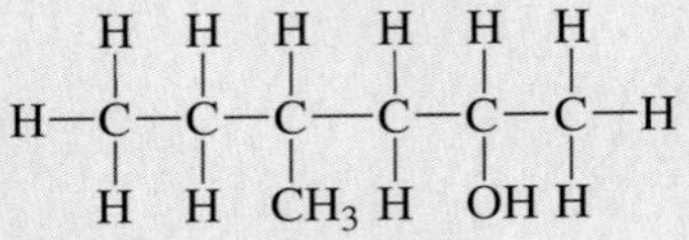

$CH_3CH_2CH(CH_3)CH_2CH(OH)CH_3$ or

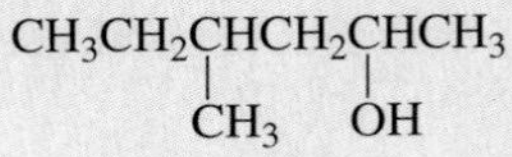

```
CH3CH2CHCH2CHCH3
       |    |
       CH3  OH
```

Groups bonded to the far-right carbon are not put in parentheses.

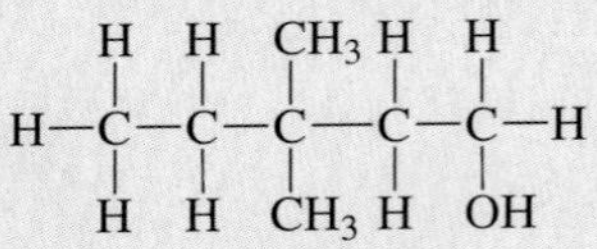

```
    H   H   CH3   H   H
    |   |   |     |   |
H — C — C — C  —  C — C — H
    |   |   |     |   |
    H   H   CH3   H   OH
```

$CH_3CH_2C(CH_3)_2CH_2CH_2OH$ or

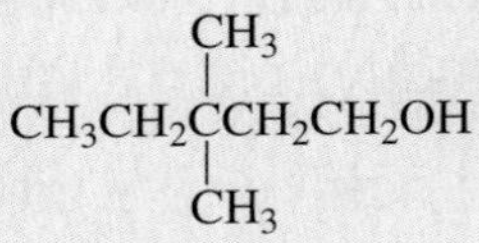

```
        CH3
        |
CH3CH2CCH2CH2OH
        |
        CH3
```

Two or more identical groups considered bonded to the "first" atom on the left can be shown (in parentheses) to the left of that atom, or hanging from the atom.

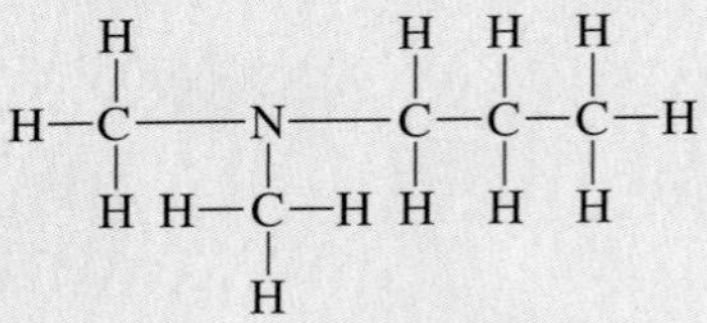

```
    H                 H   H   H
    |                 |   |   |
H — C  ——  N  ——  C — C — C — H
    |      |          |   |   |
    H  H — C — H      H   H   H
           |
           H
```

$(CH_3)_2NCH_2CH_2CH_3$ or

```
CH3NCH2CH2CH3
   |
   CH3
```

```
    H      H          H   H   H
    |      |          |   |   |
H — C  ——  C  ——  C — C — C — H
    |      |          |   |   |
    H  H — C — H      H   H   H
           |
           H
```

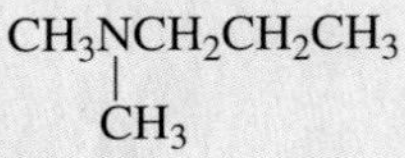

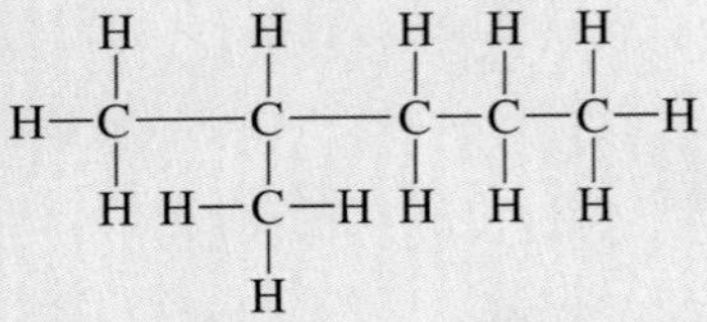

$(CH_3)_2CHCH_2CH_2CH_3$ or

```
CH3CHCH2CH2CH3
   |
   CH3
```

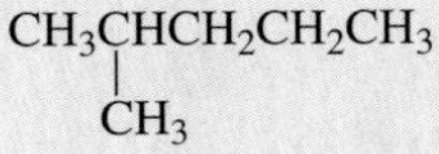

PROBLEM 16

Which of the following formulas are not possible for an organic compound?

C_2H_6 C_2H_7 C_3H_9 C_3H_8 C_4H_{10}

PROBLEM 17

a. Draw two Lewis structures for C_2H_6O.
b. Draw three Lewis structures for C_3H_8O.
(*Hint:* The two Lewis structures in part a are **constitutional isomers**; they have the same atoms, but differ in the way the atoms are connected; see page 143. The three Lewis structures in part b are also constitutional isomers.)

PROBLEM 18

Expand the following condensed structures to show the covalent bonds and lone pairs:

a. $CH_3NH(CH_2)_2CH_3$
b. $(CH_3)_2CHCl$
c. $(CH_3)_3COH$
d. $(CH_3)_3C(CH_2)_3CH(CH_3)_2$

1.5 ATOMIC ORBITALS

We have seen that electrons are distributed into different atomic orbitals (Table 1.2). An **orbital** is a three-dimensional region around the nucleus where an electron is most likely to be found. But what does an orbital look like? The *s* orbital is a sphere with the nucleus at its center. Thus, when we say that an electron occupies a 1*s* orbital, we mean that there is a greater than 90% probability that the electron is in the space defined by the sphere.

An orbital tells us the volume of space around the nucleus where an electron is most likely to be found.

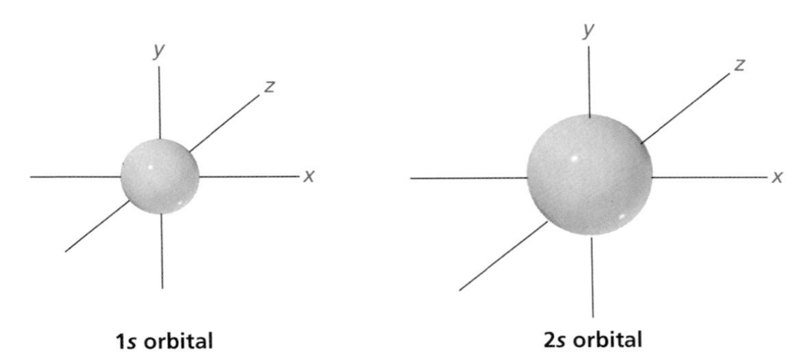

1*s* orbital   2*s* orbital

Because the second shell lies farther from the nucleus than the first shell (Section 1.2), the average distance from the nucleus is greater for an electron in a 2*s* orbital than for an electron in a 1*s* orbital. A 2*s* orbital, therefore, is represented by a larger sphere. Because of the greater size of the 2*s* orbital, its average electron density is less than the average electron density of a 1*s* orbital.

Unlike *s* orbitals, which resemble spheres, a *p* orbital has two lobes. Generally, the lobes are depicted as teardrop-shaped, but computer-generated representations reveal that they are shaped more like doorknobs. In Section 1.2, we saw that the second and higher numbered shells each contain three *p* orbitals, and the three *p* orbitals have the same energy. The p_x orbital is symmetrical about the *x*-axis, the p_y orbital is symmetrical about the *y*-axis, and the p_z orbital is symmetrical about the *z*-axis. This means that each *p* orbital is perpendicular to the other two *p* orbitals. The energy of a 2*p* orbital is greater than that of a 2*s* orbital because the average location of an electron in a 2*p* orbital is farther away from the nucleus.

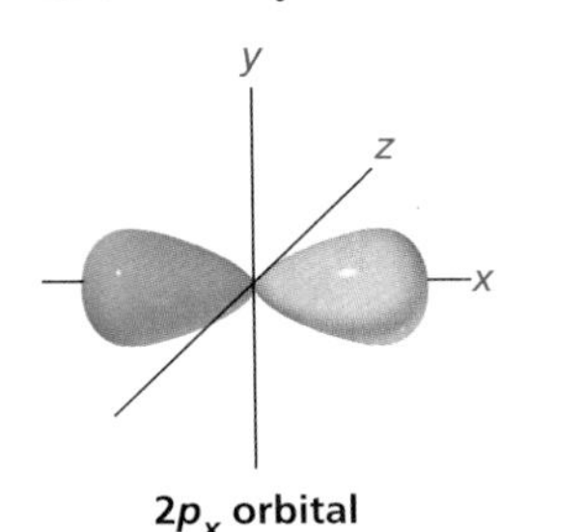

$2p_x$ orbital

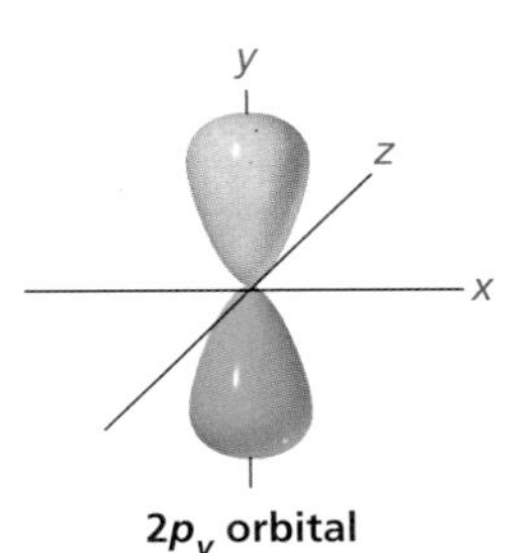

$2p_y$ orbital

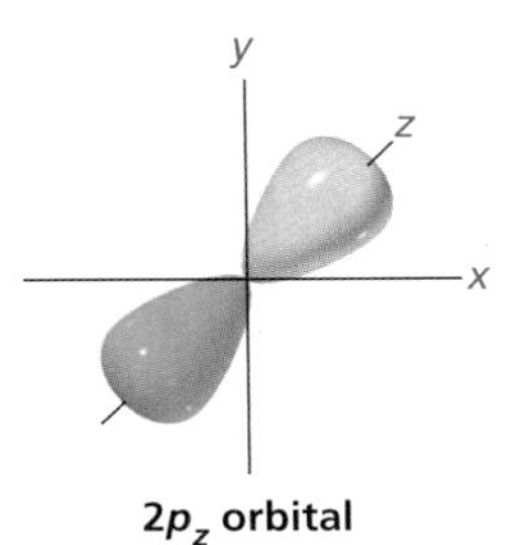

$2p_z$ orbital

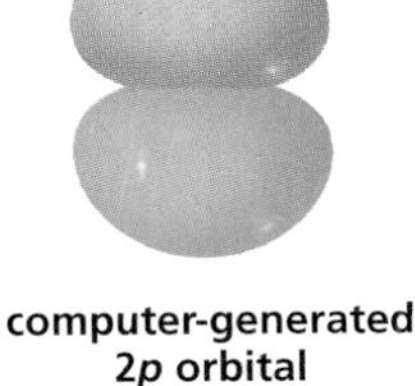

computer-generated 2*p* orbital

1.6 HOW ATOMS FORM COVALENT BONDS

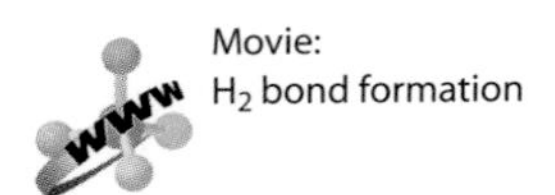

Movie: H_2 bond formation

How do atoms form covalent bonds in order to form molecules? Let's look first at the bonding in a hydrogen molecule (H_2). The covalent bond is formed when the 1*s* orbital of one hydrogen atom overlaps the 1*s* orbital of a second hydrogen atom. The covalent bond that is formed when the two orbitals overlap is called a **sigma (σ) bond**.

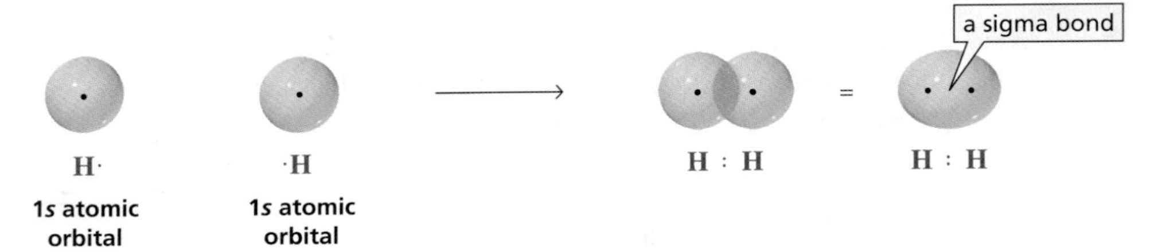

Why do atoms form covalent bonds? As the two orbitals start to overlap to form the covalent bond, energy is released (and stability increases) because the electron in each atom is attracted both to its own nucleus and to the positively charged nucleus of the other atom (Figure 1.2). Thus atoms form covalent bonds because the covalently bonded atoms are more stable than the individual atoms. The attraction of the negatively charged electrons for the positively charged nuclei is what holds the atoms together. The more the orbitals overlap, the more the energy decreases until the atoms are so close together that their positively charged nuclei start to repel each other. This repulsion causes a large increase in energy. Maximum stability (that is, minimum energy) is achieved when the nuclei are a certain distance apart. This distance is the **bond length** of the new covalent bond. The length of the H—H bond is 0.74 Å.

As Figure 1.2 shows, energy is released when a covalent bond forms. When the H—H bond forms, 105 kcal/mol or 439 kJ/mol of energy is released (1 kcal = 4.184 kJ).* Breaking the bond requires precisely the same amount of energy. Thus, the **bond strength**—also called the **bond dissociation energy**—is the energy required to break the bond, or the energy released when the bond is formed. Every covalent bond has a characteristic bond length and bond strength.

Maximum stability corresponds to minimum energy.

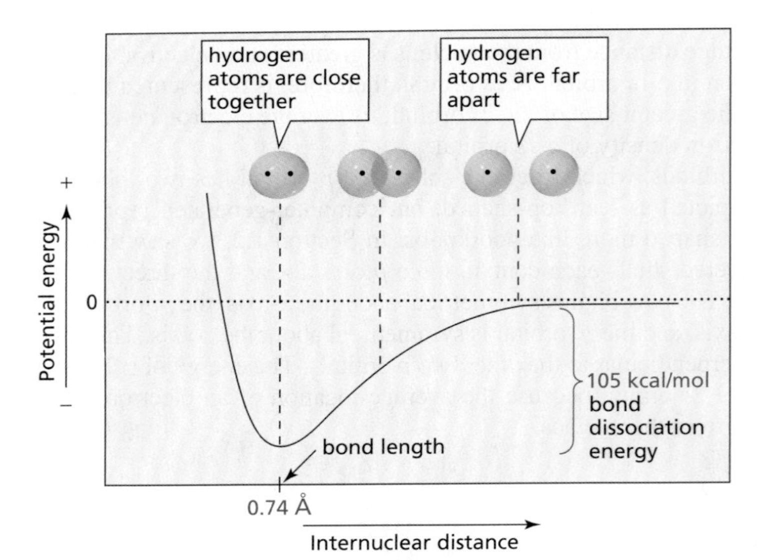

▶ **Figure 1.2**
The change in energy that occurs as two 1*s* atomic orbitals approach each other. The internuclear distance at minimum energy is the length of the H—H covalent bond.

*Joules are the Système International (SI) units for energy, although many chemists use calories. We will use both in this book.

1.7 HOW SINGLE BONDS ARE FORMED IN ORGANIC COMPOUNDS

We will begin the discussion of bonding in organic compounds by looking at the bonding in methane, a compound with only one carbon atom. Then we will examine the bonding in ethane, a compound with two carbons attached by a carbon–carbon *single bond.*

The Bonds in Methane

Methane (CH_4) has four covalent C—H bonds. Because all four bonds have the same length and all the bond angles are the same (109.5°), we can conclude that the four C—H bonds in methane are identical. Four different ways to represent a methane molecule are shown here.

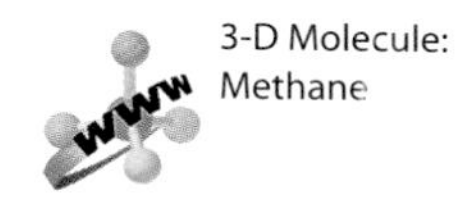

3-D Molecule: Methane

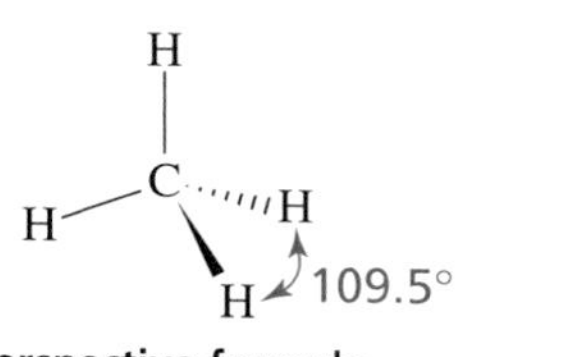

perspective formula of methane

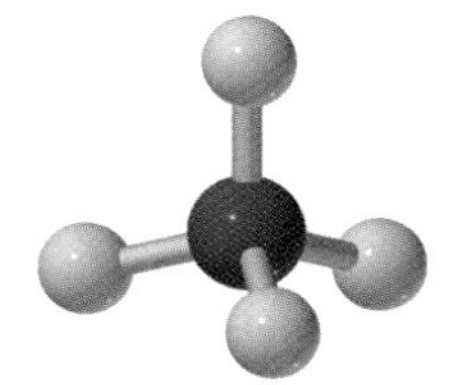
ball-and-stick model of methane

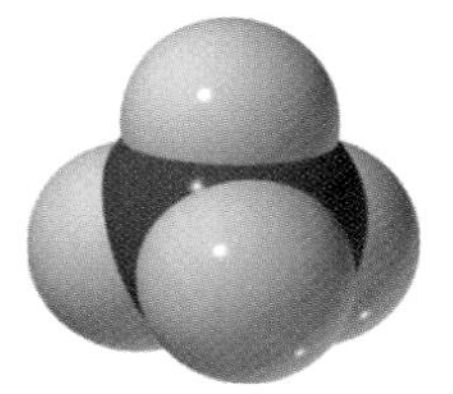
space-filling model of methane

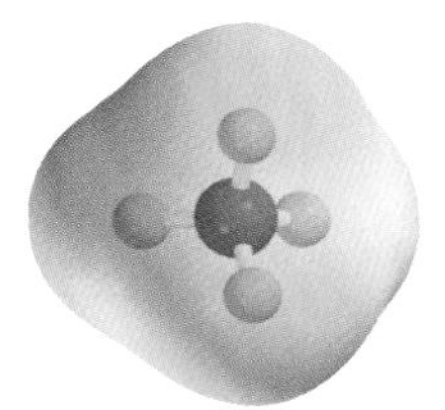
electrostatic potential map for methane

In a **perspective formula**, bonds in the plane of the paper are drawn as solid lines, bonds protruding out of the plane of the paper toward the viewer are drawn as solid wedges, and those protruding back from the plane of the paper away from the viewer are drawn as hatched wedges.

The potential map of methane shows that neither carbon nor hydrogen carries much of a charge: there are neither red areas, representing partially negatively charged atoms, nor blue areas, representing partially positively charged atoms. (Compare this map with the potential map for water on page 11). The absence of partially charged atoms is due to the similar electronegativities of carbon and hydrogen, which cause them to share their bonding electrons relatively equally. Methane is therefore a **nonpolar molecule**.

You may be surprised to learn that carbon forms four covalent bonds since you know that carbon has only two unpaired electrons in its electronic configuration (Table 1.2). But if carbon formed only two covalent bonds, it would not complete its octet. We therefore need to come up with an explanation that accounts for carbon's forming four covalent bonds.

If one of the electrons in carbon's 2*s* orbital were promoted into the empty 2*p* orbital, the new electronic configuration would have four unpaired electrons; thus, four covalent bonds could be formed.

If carbon used an *s* orbital and three *p* orbitals to form these four bonds, the bond formed with the *s* orbital would be different from the three bonds formed with *p* orbitals. What could account for the fact that the four C—H bonds in methane are identical if they are made using one *s* and three *p* orbitals? The answer is that carbon uses *hybrid orbitals*.

Hybrid orbitals are mixed orbitals that result from combining atomic orbitals. The concept of combining orbitals was first proposed by Linus Pauling in 1931. If the one

Linus Carl Pauling (1901–1994) *was born in Portland, Oregon. A friend's home chemistry laboratory sparked Pauling's early interest in science. He received a Ph.D. from the California Institute of Technology and remained there for most of his academic career. He received the Nobel Prize in chemistry in 1954 for his work on molecular structure. Like Einstein, Pauling was a pacifist, winning the 1962 Nobel Peace Prize for his work on behalf of nuclear disarmament.*

s and three *p* orbitals of the second shell are all combined and then apportioned into four equal orbitals, each of the four resulting orbitals will be one part *s* and three parts *p*. Therefore, each orbital has 25% *s* character and 75% *p* character. This type of mixed orbital is called an sp^3 (stated "*s*-*p*-three" not "*s*-*p*-cubed") orbital. (The superscript 3 means that three *p* orbitals were mixed with one *s* orbital to form the hybrid orbitals.) Each of the four sp^3 orbitals has the same energy.

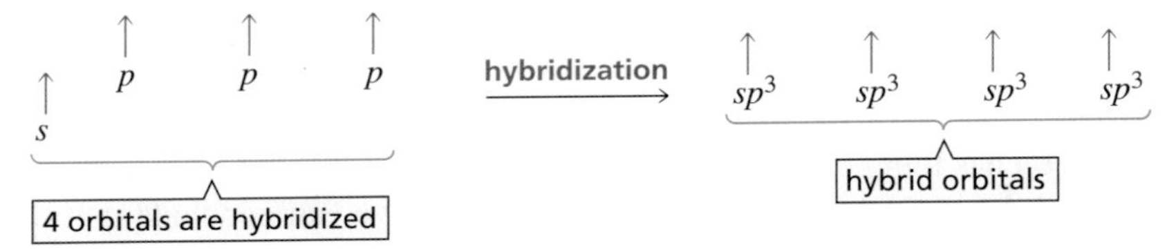

Like a *p* orbital, an sp^3 orbital has two lobes. Unlike the lobes of a *p* orbital, the two lobes of an sp^3 orbital are not the same size (Figure 1.3). The larger lobe of the sp^3 orbital is used in covalent bond formation.

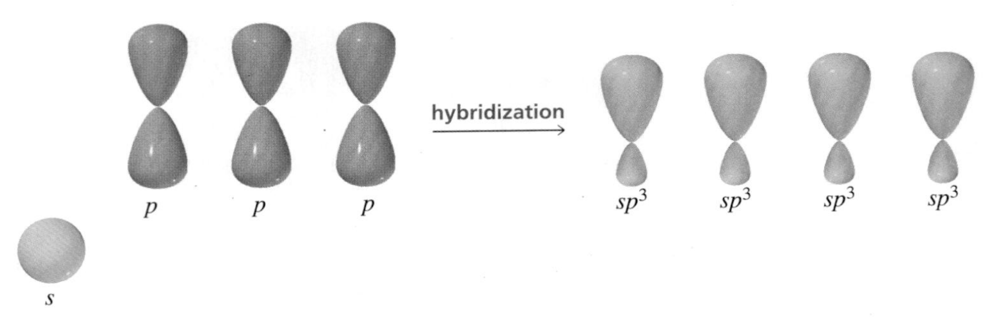

▶ Figure 1.3
An *s* orbital and three *p* orbitals hybridize to form four sp^3 orbitals. An sp^3 orbital is more stable than a *p* orbital, but not as stable as an *s* orbital.

The four sp^3 orbitals adopt a spatial arrangement that keeps them as far away from each other as possible (Figure 1.4a). They do this because electrons repel each other, and moving the orbitals as far from each other as possible minimizes the repulsion. When four orbitals move as far from each other as possible, they point toward the corners of a regular tetrahedron (a pyramid with four faces, each an equilateral triangle). Each of the four C—H bonds in methane is formed from overlap of an sp^3 orbital of carbon with the *s* orbital of a hydrogen (Figure 1.4b). This explains why the four C—H bonds are identical.

▶ Figure 1.4
(a) The four sp^3 orbitals are directed toward the corners of a tetrahedron, causing each bond angle to be 109.5°. (b) An orbital picture of methane, showing the overlap of each sp^3 orbital of carbon with the *s* orbital of a hydrogen. (For clarity, the smaller lobes of the sp^3 orbitals are not shown.)

The angle between any two bonds that point from the center to the corners of a tetrahedron are 109.5°. The bond angles in methane therefore are 109.5°. This is called a **tetrahedral bond angle**. A carbon, such as the one in methane, that forms covalent bonds using four equivalent sp^3 orbitals is called a **tetrahedral carbon**.

Electron pairs stay as far from each other as possible.

Hybrid orbitals may appear to have been contrived just to make things fit—and that is exactly the case. Nevertheless, they give us a very good picture of the bonding in organic compounds.

> **Note to the student**
> It is important to understand what molecules look like in three dimensions. Therefore be sure to visit the textbook's website (*http://www.chemplace.com*) and look at the three-dimensional representations of molecules that can be found in the molecule gallery prepared for each chapter.

The Bonds in Ethane

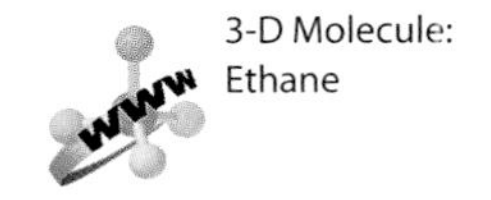

The two carbon atoms in ethane (CH_3CH_3) are tetrahedral. Each carbon uses four sp^3 orbitals to form four covalent bonds:

```
    H   H
    |   |
H — C — C — H
    |   |
    H   H
```

ethane

One sp^3 orbital of one carbon overlaps an sp^3 orbital of the other carbon to form the C—C bond. Each of the remaining three sp^3 orbitals of each carbon overlaps the s orbital of a hydrogen to form a C—H bond. Thus, the C—C bond is formed by sp^3–sp^3 overlap, and each C—H bond is formed by sp^3–s overlap (Figure 1.5).

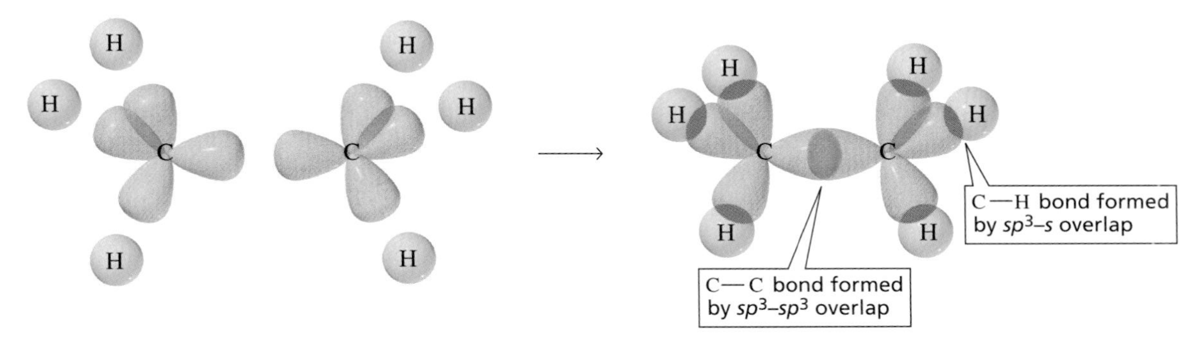

▲ **Figure 1.5**
An orbital picture of ethane. The C—C bond is formed by sp^3–sp^3 overlap, and each C—H bond is formed by sp^3–s overlap. (The smaller lobes of the sp^3 orbitals are not shown.)

Each of the bond angles in ethane is nearly the tetrahedral bond angle of 109.5°, and the length of the C—C bond is 1.54 Å. Ethane, like methane, is a nonpolar molecule.

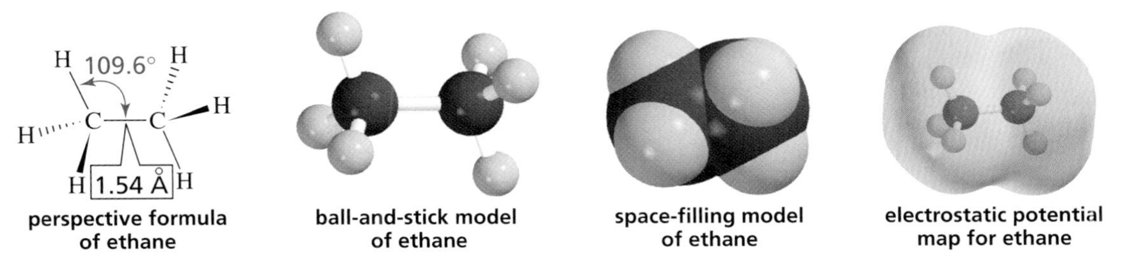

perspective formula of ethane | **ball-and-stick model of ethane** | **space-filling model of ethane** | **electrostatic potential map for ethane**

All the bonds in methane and ethane are sigma (σ) bonds. We will see that all **single bonds** found in organic compounds are *sigma bonds*.

PROBLEM 19 ♦

What orbitals are used to form the 10 covalent bonds in propane ($CH_3CH_2CH_3$)?

1.8 HOW A DOUBLE BOND IS FORMED: THE BONDS IN ETHENE

Each of the carbon atoms in ethene (also called ethylene) forms four bonds, but each is bonded to only three atoms:

```
H       H
 \     /
  C = C
 /     \
H       H
```

ethene
ethylene

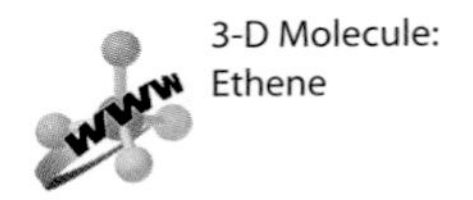

To bond to three atoms, each carbon hybridizes three atomic orbitals: an *s* orbital and two of the *p* orbitals). Because three orbitals are hybridized, three hybrid orbitals are formed. These are called sp^2 orbitals. After hybridization, each carbon atom has three identical sp^2 orbitals and one *p* orbital:

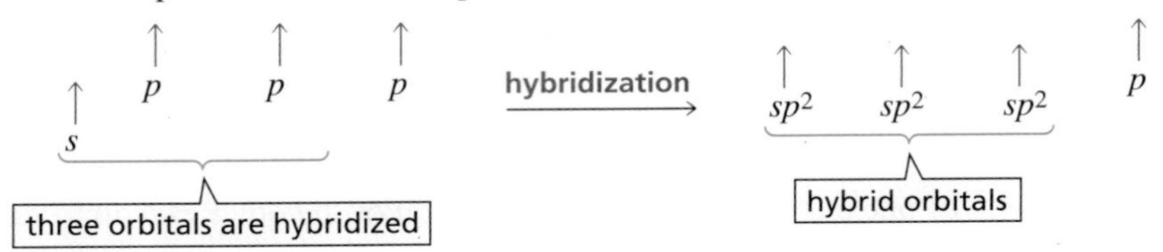

The axes of the three sp^2 orbitals lie in a plane (Figure 1.6a). To minimize electron repulsion, the three orbitals need to get as far from each other as possible. Therefore the bond angles are all close to 120°. The unhybridized *p* orbital is perpendicular to the plane defined by the axes of the sp^2 orbitals (Figure 1.6b).

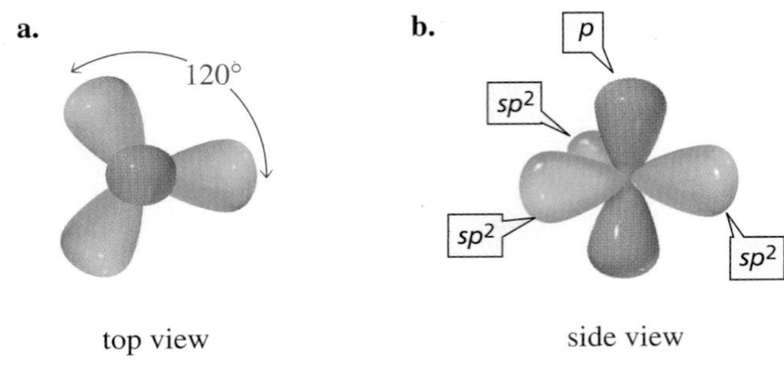

▶ Figure 1.6
(a) The three sp^2 orbitals lie in a plane.
(b) The unhybridized *p* orbital is perpendicular to the plane. (The smaller lobes of the sp^2 orbitals are not shown.)

The carbons in ethene are held together by two bonds. Two bonds connecting two atoms is called a **double bond**. The two carbon–carbon bonds in the double bond are not identical. One of them results from the overlap of an sp^2 orbital of one carbon with an sp^2 orbital of the other carbon; this is a sigma (σ) bond. Each carbon uses its other two sp^2 orbitals to overlap the *s* orbital of a hydrogen to form the C—H bonds (Figure 1.7a). The second carbon–carbon bond results from side-to-side overlap of the two unhybridized *p* orbitals (Figure 1.7b). Side-to-side overlap of *p* orbitals forms a **pi (π) bond**. Thus, one of the bonds in a double bond is a σ bond, and the other is a π bond. All the C—H bonds are σ bonds.

Side-to-side overlap of two *p* atomic orbitals forms a π bond.

The two *p* orbitals that overlap to form the π bond must be parallel to each other for maximum overlap to occur. This forces the triangle formed by one carbon and two hydrogens to lie in the same plane as the triangle formed by the other carbon and two

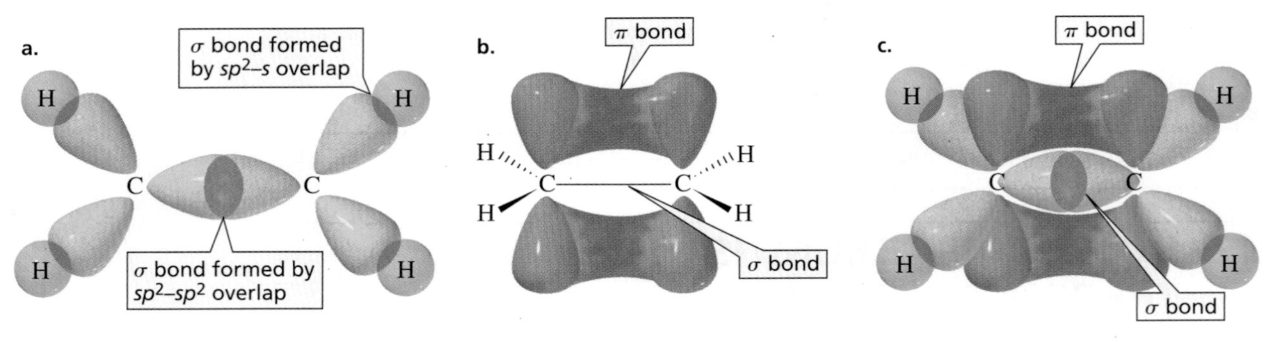

▲ Figure 1.7
(a) One C—C bond in ethene is a σ bond formed by sp^2–sp^2 overlap, and the C—H bonds are formed by sp^2–*s* overlap.
(b) The second C—C bond is a π bond formed by side-to-side overlap of a *p* orbital of one carbon with a *p* orbital of the other carbon.
(c) There is an accumulation of electron density above and below the plane containing the two carbons and four hydrogens.

hydrogens. As a result, all six atoms of ethene lie in the same plane, and the electrons in the *p* orbitals occupy a volume of space above and below the plane (Figure 1.7c). The potential map for ethene shows that it is a nonpolar molecule with a slight accumulation of negative charge (the pale orange area) above the two carbons. (If you could turn the potential map over to show the hidden side, a similar accumulation of negative charge would be found there.)

H 121.7° H
C=C 116.6°
H 1.33 Å H

a double bond consists of one σ bond and one π bond

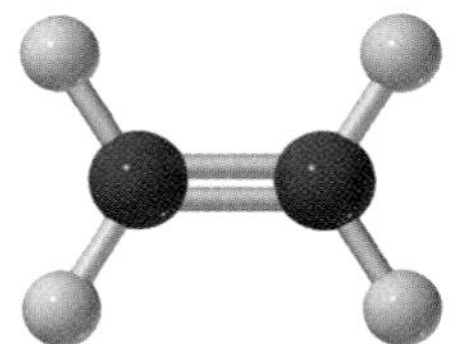

ball-and-stick model of ethene

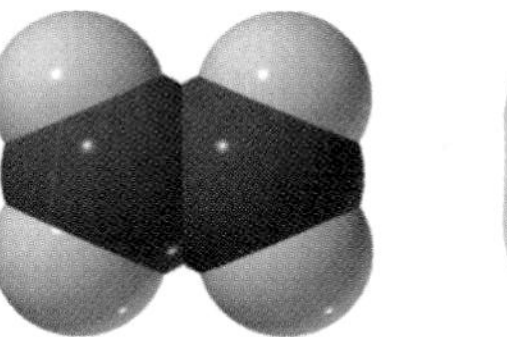

space-filling model of ethene

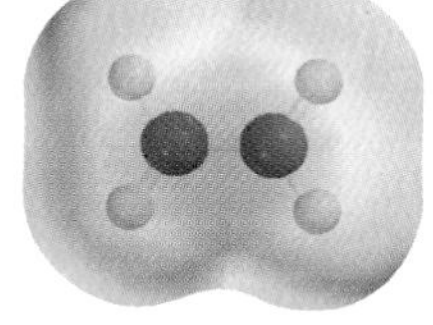

electrostatic potential map for ethene

Four electrons hold the carbons together in a carbon–carbon double bond; only two electrons hold the carbons together in a carbon–carbon single bond. This means that a carbon–carbon double bond is stronger and shorter than a carbon–carbon single bond.

DIAMOND AND GRAPHITE: SUBSTANCES THAT CONTAIN ONLY CARBON ATOMS

Diamond is the hardest of all substances. Graphite, in contrast, is a slippery, soft solid most familiar to us as the "lead" in pencils. Both materials, in spite of their very different physical properties, contain only carbon atoms. The two substances differ solely in the nature of the bonds holding the carbon atoms together. Diamond consists of a rigid three-dimensional network of atoms, with each carbon bonded to four other carbons via sp^3 orbitals. The carbon atoms in graphite, on the other hand, are sp^2 hybridized, so each bonds to only three other carbons. This planar arrangement causes the atoms in graphite to lie in flat, layered sheets that can shear off of neighboring sheets. When you write with a pencil, sheets of carbon atoms shear off to leave a thin trail of graphite.

1.9 HOW A TRIPLE BOND IS FORMED: THE BONDS IN ETHYNE

The carbon atoms in ethyne (also called acetylene) are each bonded to only two atoms—a hydrogen and another carbon:

H—C≡C—H
ethyne
acetylene

Because each carbon forms covalent bonds with two atoms, only two orbitals (an *s* and a *p*) are hybridized. Two identical *sp* orbitals result. Each carbon atom in ethyne, therefore, has two hybridized *sp* orbitals and two unhybridized *p* orbitals (Figure 1.8).

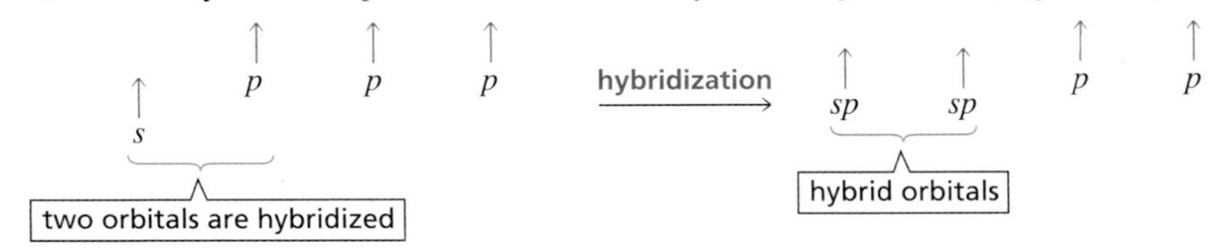

To minimize electron repulsion, the two *sp* orbitals point in opposite directions (Figure 1.8).

The carbons in ethyne are held together by three bonds. Three bonds connecting two atoms is called a **triple bond**. One of the *sp* orbitals of one carbon in ethyne overlaps an *sp* orbital of the other carbon to form a carbon–carbon σ bond. The other *sp* orbital of each carbon overlaps the *s* orbital of a hydrogen to form a C—H σ bond

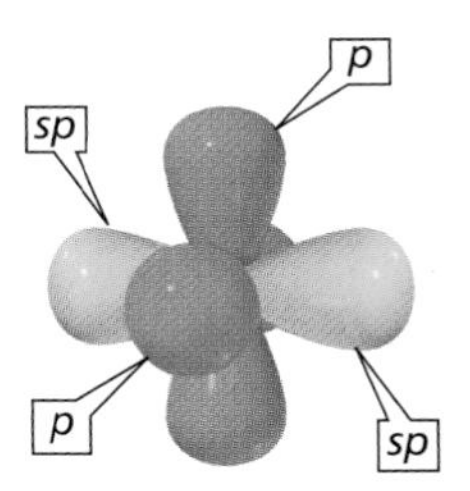

▲ **Figure 1.8**
The two *sp* orbitals are oriented 180° away from each other, perpendicular to the two unhybridized *p* orbitals. (The smaller lobes of the *sp* orbitals are not shown.)

(Figure 1.9a). Because the two *sp* orbitals point in opposite directions, the bond angles are 180°. The two unhybridized *p* orbitals are perpendicular to each other, and both are perpendicular to the *sp* orbitals. Each of the unhybridized *p* orbitals engages in side-to-side overlap with a parallel *p* orbital on the other carbon, with the result that two π bonds are formed (Figure 1.9b).

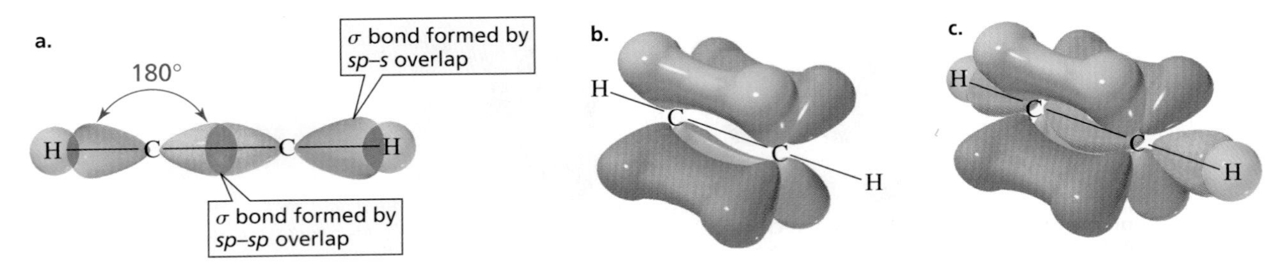

▲ **Figure 1.9**
(a) The C—C σ bond in ethyne is formed by *sp–sp* overlap, and the C—H bonds are formed by *sp–s* overlap. The carbon atoms and the atoms bonded to them are in a straight line.
(b) The two π bonds are formed by side-to-side overlap of the *p* orbitals of one carbon with the *p* orbitals of the other carbon.
(c) The triple bond has an electron-dense region above and below and in front of and in back of the internuclear axis of the molecule.

A **triple bond** therefore consists of one σ bond and two π bonds. Because the two unhybridized *p* orbitals on each carbon are perpendicular to each other, there is a region of high electron density above and below, *and* in front of and in back of, the internuclear axis of the molecule (Figure 1.9c). The potential map for ethyne shows that negative charge accumulates in a cylinder that wraps around the egg-shaped molecule.

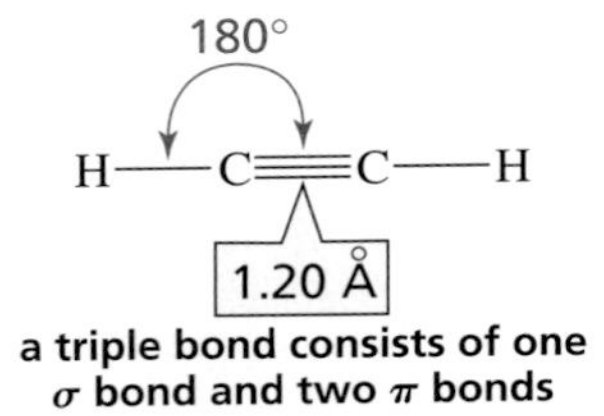

a triple bond consists of one σ bond and two π bonds

ball-and-stick model of ethyne

space-filling model of ethyne

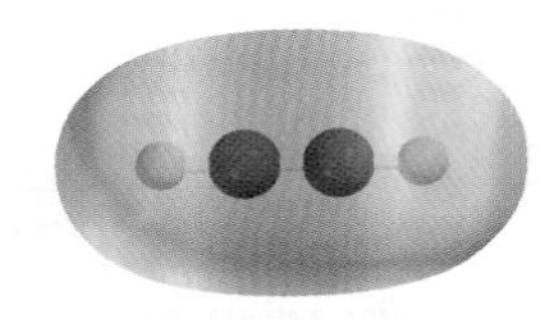

electrostatic potential map for ethyne

Because the two carbon atoms in a triple bond are held together by six electrons, a triple bond is stronger and shorter than a double bond.

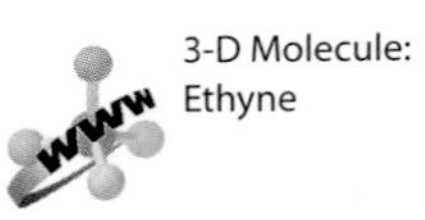

3-D Molecule: Ethyne

PROBLEM 20 ***SOLVED***

For each of the following species:

a. Draw its Lewis structure.
b. Describe the orbitals used by each carbon atom in bonding and indicate the approximate bond angles.
 1. HCOH **2.** CCl_4 **3.** HCN

Solution to 20a Because HCOH is neutral, we know that each H forms one bond, the oxygen forms two bonds, and the carbon forms four bonds. Our first attempt at a Lewis structure (drawing the atoms in the order given by the Kekulé structure) shows that carbon is the only atom that does not form the needed number of bonds.

H—C—O—H

If we place a double bond between the carbon and the oxygen and move an H, all the atoms end up with the correct number of bonds. All the C and H atoms have filled outer shells but lone-pair electrons need to be added to give the oxygen atom a filled outer shell. When we check to see if any atom needs to be assigned a formal charge, we find that none of them does.

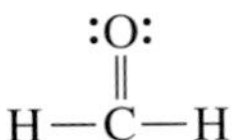

Solution to 20b Because the carbon atom forms a double bond, we know that carbon uses sp^2 orbitals (as it does in ethene) to bond to the two hydrogens and the oxygen. It uses its "leftover" p orbital to form the second bond to oxygen. Because carbon is sp^2 hybridized, the bond angles are approximately 120°.

120° 120° 120°

1.10 BONDING IN THE METHYL CATION, THE METHYL RADICAL, AND THE METHYL ANION

Not all carbon atoms form four bonds. A carbon with a positive charge, a negative charge, or an unpaired electron forms only three bonds. Now we will see what orbitals carbon uses when it forms three bonds.

The Methyl Cation ($^+CH_3$)

The positively charged carbon in the methyl cation is bonded to three atoms, so it hybridizes three orbitals—an s orbital and two p orbitals. Therefore, it forms its three covalent bonds using sp^2 orbitals. Its unhybridized p orbital remains empty. The positively charged carbon and the three atoms bonded to it lie in a plane. The p orbital stands perpendicular to the plane.

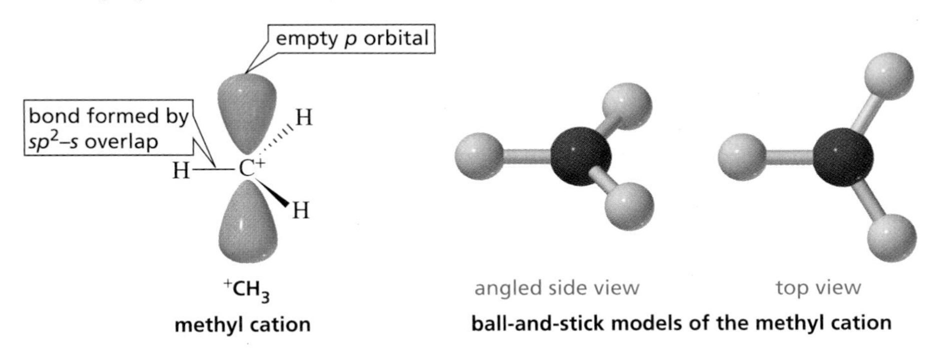

$^+CH_3$
methyl cation

ball-and-stick models of the methyl cation

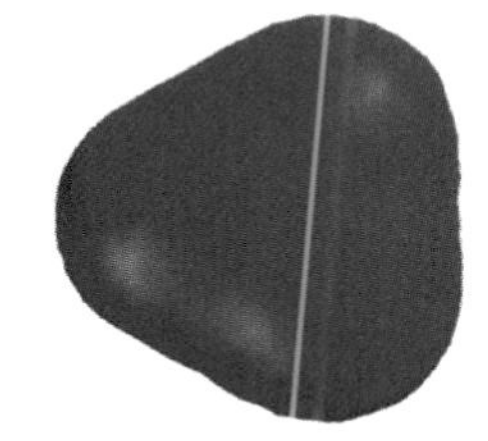

electrostatic potential map for the methyl cation

The Methyl Radical ($\cdot CH_3$)

The carbon atom in the methyl radical is also sp^2 hybridized. The methyl radical differs by one unpaired electron from the methyl cation. That electron is in the p orbital. Notice the similarity in the ball-and-stick models of the methyl cation and the methyl radical. The potential maps, however, are quite different because of the additional electron in the methyl radical.

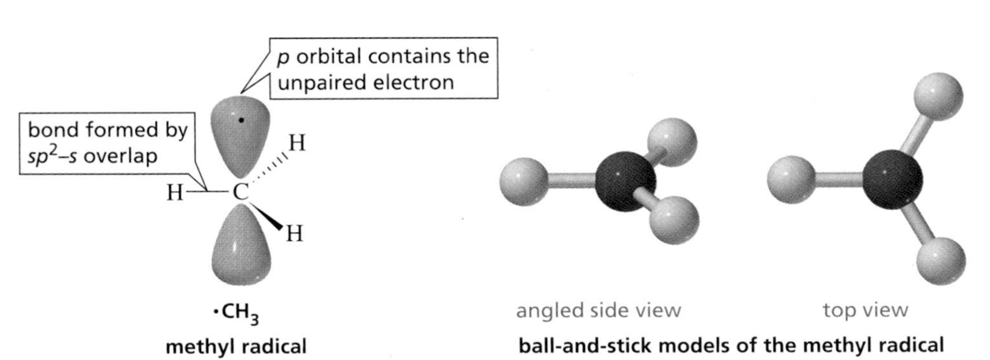

$\cdot CH_3$
methyl radical

ball-and-stick models of the methyl radical

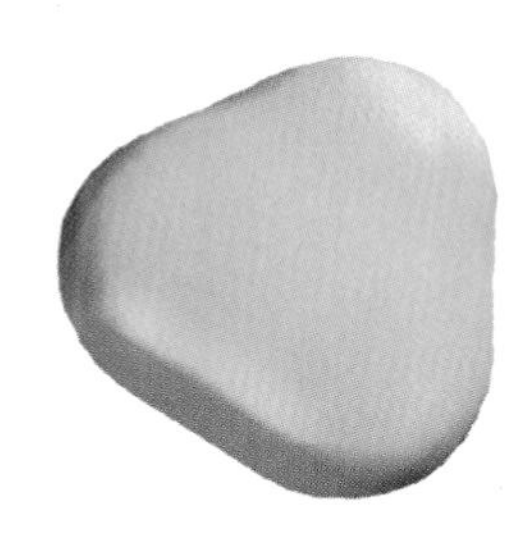

electrostatic potential map for the methyl radical

The Methyl Anion ($\ddot{:}CH_3$)

The negatively charged carbon in the methyl anion has three pairs of bonding electrons and one lone pair. The four pairs of electrons are farthest apart when the four orbitals containing the bonding and lone-pair electrons point toward the corners of a tetrahedron. In other words, a negatively charged carbon is sp^3 hybridized. In the methyl anion, three of carbon's sp^3 orbitals each overlap the s orbital of a hydrogen, and the fourth sp^3 orbital holds the lone pair.

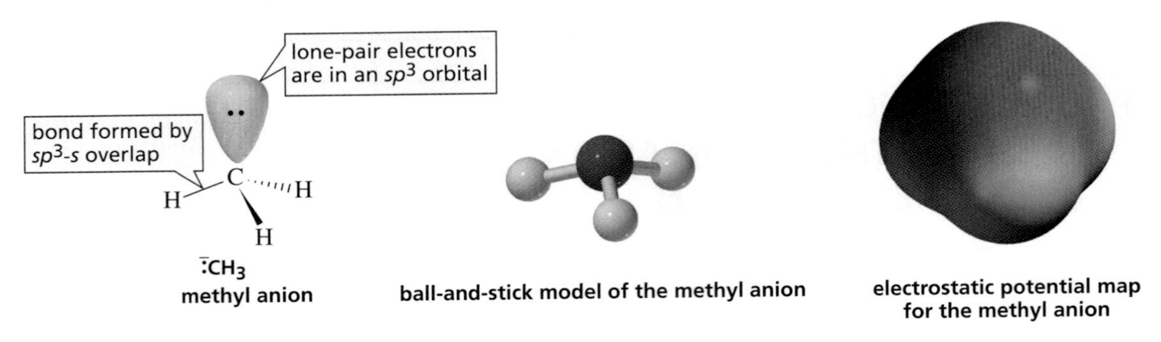

Take a moment to compare the potential maps for the methyl cation, the methyl radical, and the methyl anion.

1.11 THE BONDS IN WATER

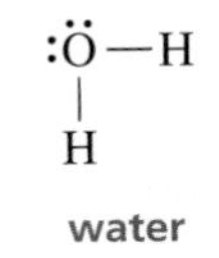

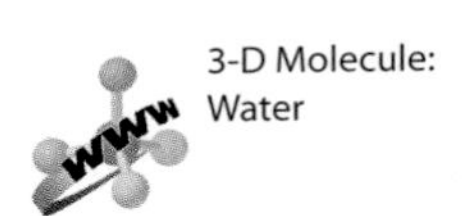

The bond angles in a molecule indicate which orbitals are used in bond formation.

The oxygen atom in water (H_2O) forms two covalent bonds and has two lone pairs.

Because the electronic configuration of oxygen shows that it has two unpaired electrons (Table 1.2), oxygen does not need to promote an electron to form the number (two) of covalent bonds required to complete its octet. If we assume that oxygen uses p orbitals to form the two O—H bonds, as predicted by oxygen's electronic configuration, we would expect a bond angle of about 90° because the two p orbitals are at right angles to each other. However, the experimentally observed bond angle is 104.5°. In addition, we would expect the lone pairs to be chemically different because one pair would be in an s orbital and the other would be in a p orbital. The lone pairs, however, are known to be identical.

To explain the observed bond angle and the fact that the lone pairs are identical, oxygen must use hybrid orbitals to form covalent bonds—just as carbon does. The s orbital and the three p orbitals must hybridize to produce four identical sp^3 orbitals.

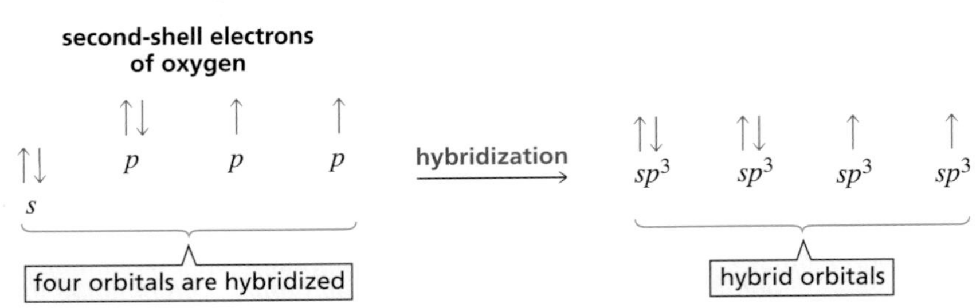

Each of the two O—H bonds is formed by the overlap of an sp^3 orbital of oxygen with the s orbital of a hydrogen. A lone pair occupies each of the two remaining sp^3 orbitals.

The bond angle in water (104.5°) is a little smaller than the bond angle in methane (109.5°) presumably because each of the lone pairs is held by only one nucleus, which makes a lone pair more diffuse than a bonding pair that is held by two nuclei and is therefore relatively confined between them. Consequently, there is more electron repulsion between lone pairs, causing the O—H bonds to squeeze closer together, thereby decreasing the bond angle.

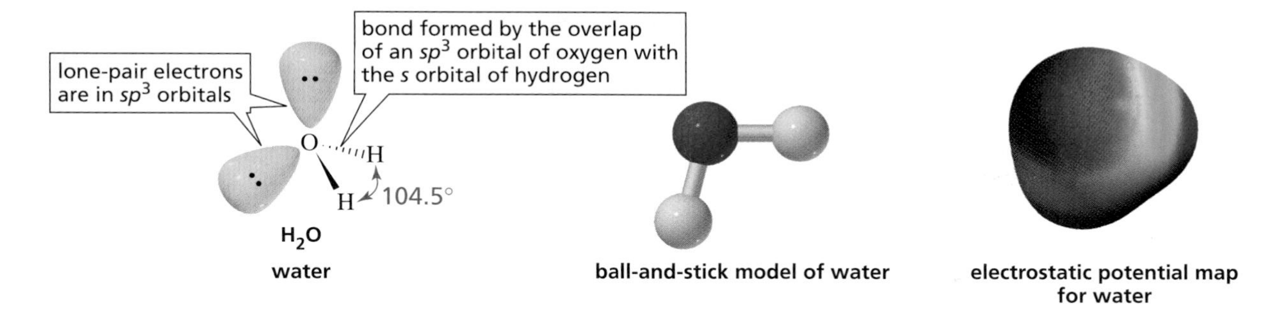

H_2O
water

ball-and-stick model of water

electrostatic potential map for water

Compare the potential map for water with that for methane. Water is a polar molecule; methane is nonpolar.

PROBLEM 21 ♦

The bond angles in H_3O^+ are greater than __________ and less than __________.

WATER—A UNIQUE COMPOUND

Water is the most abundant compound found in living organisms. Its unique properties have allowed life to originate and evolve. Its high heat of fusion (the heat required to convert a solid to a liquid) protects organisms from freezing at low temperatures because a lot of heat must be removed from water to freeze it. Its high heat capacity (the heat required to raise the temperature of a substance a given amount) minimizes temperature changes in organisms, and its high heat of vaporization (the heat required to convert a liquid to a gas) allows animals to cool themselves with a minimal loss of body fluid. Because liquid water is denser than ice, ice formed on the surface of water floats and insulates the water below. That is why oceans and lakes don't freeze from the bottom up. It is also why plants and aquatic animals can survive when the ocean or lake they live in freezes.

1.12 THE BONDS IN AMMONIA AND IN THE AMMONIUM ION

The nitrogen atom in ammonia (NH_3) forms three covalent bonds and has one lone pair.

H—N̈—H
|
H
ammonia

Because nitrogen has three unpaired electrons in its electronic configuration (Table 1.2), it can form three covalent bonds without having to promote an electron. The experimentally observed bond angles in NH_3 are 107.3°, indicating that nitrogen also uses hybrid orbitals when it forms covalent bonds. Like carbon and oxygen, the one s and three p orbitals of the second shell of nitrogen hybridize to form four identical sp^3 orbitals:

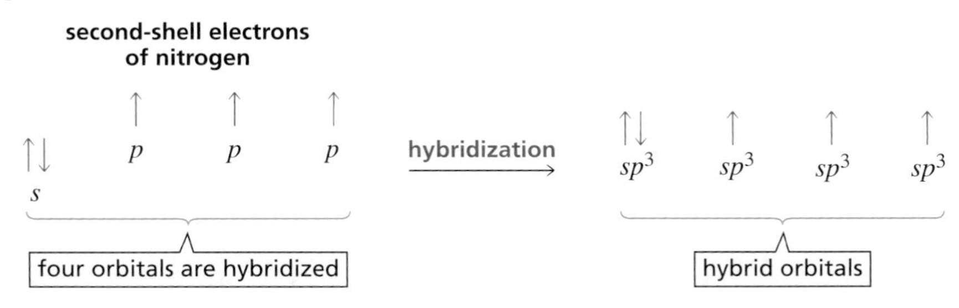

Each of the N—H bonds in NH_3 is formed by the overlap of an sp^3 orbital of nitrogen with the s orbital of a hydrogen. The single lone pair occupies an sp^3 orbital. The bond angle (107.3°) is smaller than the tetrahedral bond angle (109.5°) because of the relatively diffuse lone pair. Notice that the bond angles in NH_3 (107.3°) are larger than the bond angles in H_2O (104.5°) because nitrogen has only one lone pair, whereas oxygen has two lone pairs.

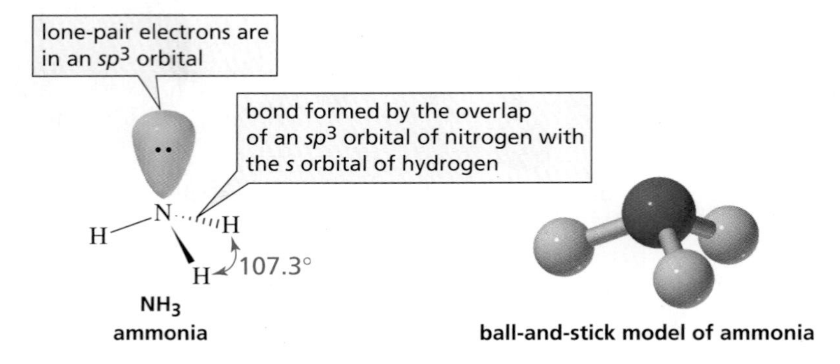

NH_3
ammonia

ball-and-stick model of ammonia

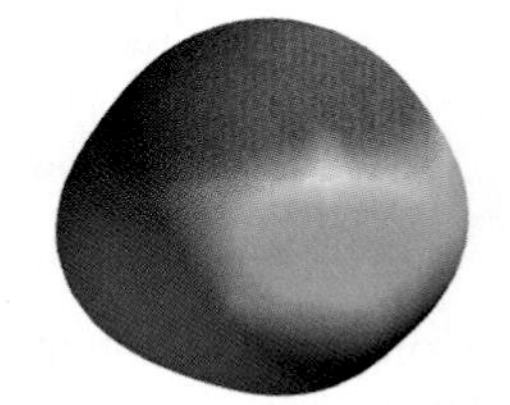

electrostatic potential map for ammonia

3-D Molecule: Ammonia

Because the ammonium ion ($^+NH_4$) has four identical N—H bonds and no lone pairs, all the bond angles are 109.5°, just like the bond angles in methane.

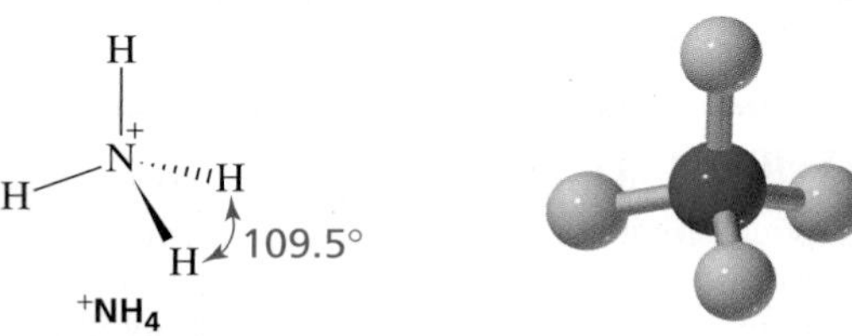

$^+NH_4$
ammonium ion

ball-and-stick model of the ammonium ion

electrostatic potential map for the ammonium ion

PROBLEM 22 ♦

According to the potential map for the ammonium ion, which atom(s) has (have) the least electron density?

PROBLEM 23 ♦

Compare the potential maps for methane, ammonia, and water. Which is the most polar molecule? Which is the least polar?

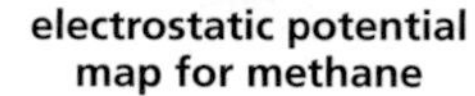

electrostatic potential map for methane

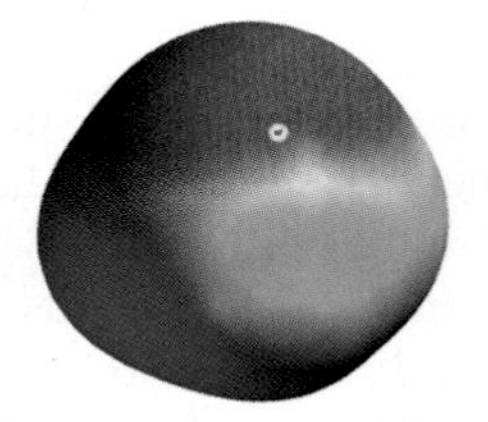

electrostatic potential map for ammonia

electrostatic potential map for water

PROBLEM 24 ♦

Predict the approximate bond angles in the methyl carbanion.

1.13 THE BOND IN A HYDROGEN HALIDE

Fluorine, chlorine, bromine, and iodine are known as the halogens; HF, HCl, HBr, and HI are called hydrogen halides. Bond angles will not help us determine the orbitals that form the hydrogen halide bond, as they did with other molecules, because a hydrogen halide has only one bond and therefore no bond angle. We do know, however, that a halogen's three lone pairs are identical and that lone pairs position themselves to minimize electron repulsion (Section 1.7). Both of these facts suggest that a halogen's three lone pairs are in sp^3 orbitals. Therefore, we will assume that a hydrogen–halogen bond is formed by the overlap of an sp^3 orbital of the halogen with the *s* orbital of hydrogen.

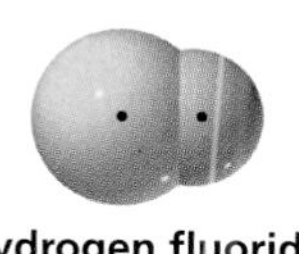

hydrogen fluoride

hydrogen chloride

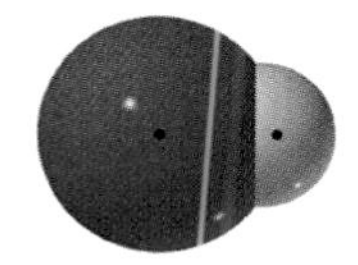

hydrogen bromide

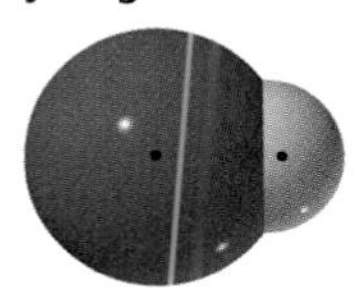

hydrogen iodide

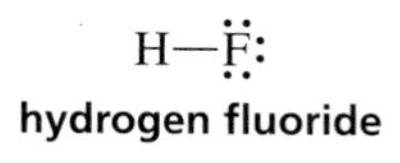

hydrogen fluoride

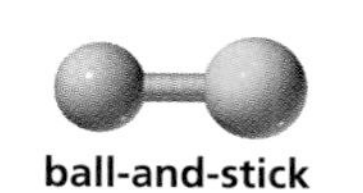

ball-and-stick model of hydrogen fluoride

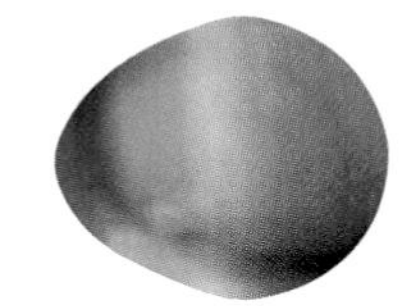

electrostatic potential map for hydrogen fluoride

In the case of fluorine, the sp^3 orbital used in bond formation belongs to the second shell of electrons. In chlorine, the sp^3 orbital belongs to the third shell of electrons. Because the average distance from the nucleus is greater for an electron in the third shell than for an electron in the second shell, the average electron density is less in a $3sp^3$ orbital than in a $2sp^3$ orbital. This means that the electron density in the region where the *s* orbital of hydrogen overlaps the sp^3 orbital of the halogen decreases as the size of the halogen increases (Figure 1.10). Therefore, the hydrogen–halogen bond becomes longer and weaker as the size (atomic weight) of the halogen increases (Table 1.5).

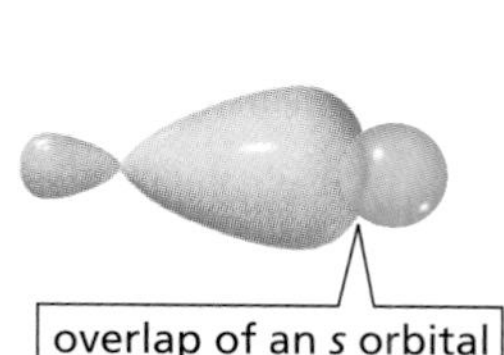

◀ **Figure 1.10**
There is greater electron density in the region of overlap of an *s* orbital with a $2sp^3$ orbital than in the region of overlap of an *s* orbital with a $3sp^3$ orbital.

The greater the electron density in the region of orbital overlap, the stronger is the bond.

The shorter the bond, the stronger it is.

Table 1.5 Hydrogen–Halogen Bond Lengths and Bond Strengths

Hydrogen halide	Bond length (Å)	Bond strength kcal/mol	Bond strength kJ/mol
H—F	0.917	136	571
H—Cl	1.2746	103	432
H—Br	1.4145	87	366
H—I	1.6090	71	298

PROBLEM 25 ♦

a. Predict the relative lengths and strengths of the bonds in Cl_2 and Br_2.
b. Predict the relative lengths and strengths of the bonds in HF, HCl, and HBr.

PROBLEM 26 ♦

a. Which bond is longer? **b.** Which bond is stronger?

1. C—Cl or C—Br **2.** C—C or C—H **3.** H—Cl or H—H

1.14 SUMMARY: HYBRIDIZATION, BOND LENGTHS, BOND STRENGTHS, AND BOND ANGLES

All single bonds found in organic compounds are sigma bonds.

The hybridization of a C, O, or N is $sp^{(3 \text{ minus the number of } \pi \text{ bonds})}$

All single bonds are σ bonds. All double bonds are composed of one σ bond and one π bond. All triple bonds are composed of one σ bond and two π bonds. The easiest way to determine the hybridization of a carbon, oxygen, or nitrogen atom is to look at the number of π bonds it forms: if it forms no π bonds, it is sp^3 hybridized; if it forms one π bond, it is sp^2 hybridized; if it forms two π bonds, it is sp hybridized. The exceptions are carbocations and carbon radicals, which are sp^2 hybridized—not because they form a π bond, but because they have an empty or a half-filled p orbital (Section 1.10).

$CH_3-\ddot{N}H_2$: sp^3, sp^3

$(CH_3)_2C=\ddot{N}-\ddot{N}H_2$: sp^3, sp^2, sp^2, sp^3

$CH_3-C\equiv N:$: sp^3, sp, sp

$CH_3-\ddot{O}H$: sp^3, sp^3

$CH_3C(=\ddot{O})CH_3$: sp^3, sp^2, sp^3; O: sp^2

$:\ddot{O}=C=\ddot{O}:$: sp^2, sp, sp^2

In comparing the lengths and strengths of carbon–carbon single, double, and triple bonds, we see that the more bonds holding two carbon atoms together, the shorter and stronger is the carbon–carbon bond (Table 1.6): Triple bonds are shorter and stronger than double bonds, which are shorter and stronger than single bonds.

A double bond (a σ bond plus a π bond) is stronger than a single bond (a σ bond), but it is not twice as strong. We can conclude, therefore, that a π bond is weaker than a σ bond.

A π bond is weaker than a σ bond.

You may wonder how an electron "knows" what orbital it should go into. In fact, electrons know nothing about orbitals. They simply arrange themselves around atoms in the most stable manner possible. It is chemists who use the concept of orbitals to explain this arrangement.

PROBLEM 27 ♦

Which of the bonds of a carbon–carbon double bond has more effective orbital–orbital overlap, the σ bond or the π bond?

Table 1.6 Comparison of the Bond Angles and the Lengths and Strengths of the Carbon–Carbon Bonds in Ethane, Ethene, and Ethyne

Molecule	Hybridization of carbon	Bond angles	Length of C—C bond (Å)	Strength of C—C bond (kcal/mol)	(kJ/mol)
H_3C-CH_3 ethane	sp^3	109.5°	1.54	90	377
$H_2C=CH_2$ ethene	sp^2	120°	1.33	174	728
$H-C\equiv C-H$ ethyne	sp	180°	1.20	231	967

PROBLEM 28 ◆

a. What is the hybridization of each of the carbon atoms in the following compound?

$$\underset{\displaystyle CH_3}{CH_3\underset{|}{C}HCH{=}CHCH_2C{\equiv}CCH_3}$$

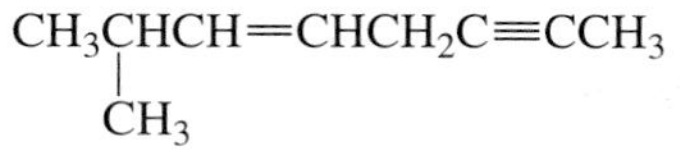

b. What is the hybridization of each of the carbon, oxygen, and nitrogen atoms in the following compounds?

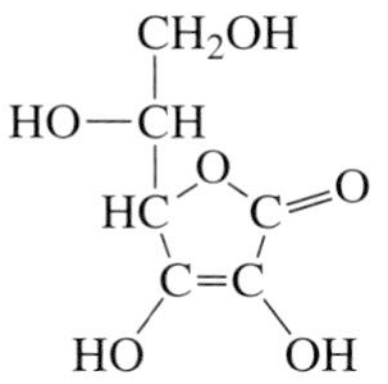

vitamin C

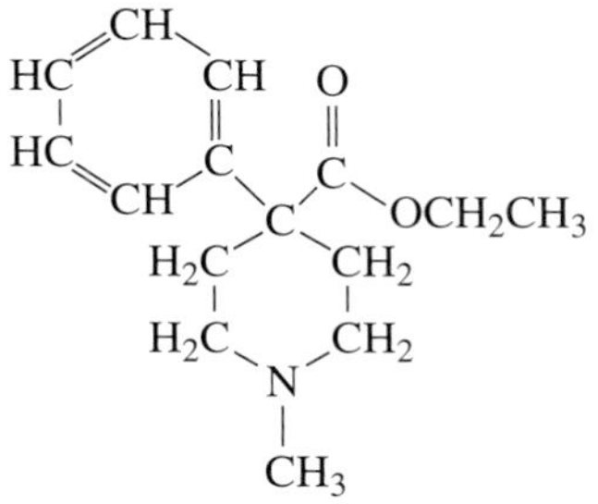

demerol
an analgesic

PROBLEM-SOLVING STRATEGY

Predict the approximate bond angle of the C—N—H bond in $(CH_3)_2NH$.

First we need to determine the hybridization of the central atom. Because the nitrogen atom forms only single bonds, we know it is sp^3 hybridized. Next we look to see if there are lone pairs that will affect the bond angle. A neutral nitrogen has one lone pair. Based on these observations, we can predict that the C—N—H bond angle will be about 107.3°, the same as the H—N—H bond angle in NH_3, another compound with a neutral sp^3 hybridized nitrogen.

Now continue on to Problem 29.

PROBLEM 29 ◆

Predict the approximate bond angles:

a. the C—N—C bond angle in $(CH_3)_2\overset{+}{N}H_2$
b. the C—N—C bond angle in $CH_3CH_2NH_2$
c. the H—C—N bond angle in $(CH_3)_2NH$
d. the C—O—C bond angle in CH_3OCH_3

PROBLEM 30

Describe the orbitals used in bonding and the approximate bond angles in the following compounds. (*Hint:* See Table 1.6.)

a. CH_3OH **b.** $HONH_2$ **c.** HCOOH **d.** N_2

SUMMARY

Organic compounds are compounds that contain carbon. The **atomic number** of an atom is the number of protons in its nucleus. The **mass number** of an atom is the sum of its protons and neutrons. **Isotopes** have the same atomic number but different mass numbers.

An **atomic orbital** indicates where there is a high probability of finding an electron. The closer the atomic orbital is to the nucleus, the lower is its energy. Electrons are assigned to atomic orbitals according to three rules: an electron goes into the available orbital with the lowest energy; no more than two electrons can be in an orbital; and an electron will occupy an empty orbital before pairing up with an electron in an orbital with the same energy.

The **octet rule** states that an atom will give up, accept, or share electrons in order to fill its outer shell or attain an outer shell with eight electrons. The **electronic configuration** of an atom describes the orbitals occupied by the atom's electrons. Electrons in inner shells are called **core electrons**; electrons in the outermost shell are called **valence electrons**. **Lone-pair electrons** are valence electrons that are not used in bonding. A **bond** formed as a result of the attraction of opposite charges is called an **ionic bond**; a bond formed as a result of sharing electrons is called a **covalent bond**. A **polar covalent bond** is a covalent bond between atoms with different **electronegativities**.

Lewis structures indicate which atoms are bonded together and show **lone-pair electrons** and **formal charges**. A **carbocation** has a positively charged carbon, a **carbanion** has a negatively charged carbon, and a **radical** has an unpaired electron.

Bond strength is measured by the **bond dissociation energy**. A σ bond is stronger than a π bond. All **single bonds** in organic compounds are **sigma (σ) bonds**. A **double bond** consists of one σ bond and one π bond, and a **triple bond** consists of one σ bond and two π bonds. Carbon–carbon triple bonds are shorter and stronger than carbon–carbon double bonds, which are shorter and stronger than carbon–carbon single bonds. To form four bonds, carbon promotes an electron from a 2*s* to a 2*p* orbital. C, N, O, and the halogens form bonds using **hybrid orbitals**. The **hybridization** of C, N, or O depends on the number of π bonds the atom forms: no π bonds means that the atom is ***sp*^{3} hybridized**, one π bond indicates that it is ***sp*^{2} hybridized**, and two π bonds signifies that it is ***sp* hybridized**. Exceptions are carbocations and carbon radicals, which are ***sp*^{2} hybridized**. Bonding and lone-pair electrons around an atom are positioned as far apart as possible.

PROBLEMS

31. Draw a Lewis structure for each of the following species:
a. H_2CO_3 **b.** CO_3^{2-} **c.** H_2CO **d.** CH_3NH_2 **e.** CO_2 **f.** N_2H_4

32. How many valence electrons do the following atoms have?
a. carbon and silicon **b.** nitrogen and phosphorus **c.** neon and argon **d.** magnesium and calcium

33. For each of the following species, give the hybridization of the central atom and indicate the bond angles:
a. NH_3 **b.** $^+NH_4$ **c.** $^-CH_3$ **d.** $C(CH_3)_4$ **e.** $\cdot CH_3$ **f.** $^+CH_3$ **g.** HCN **h.** H_3O^+

34. Use the symbols $\delta+$ and $\delta-$ to show the direction of polarity of the indicated bond in each of the following compounds:
a. F—Br **b.** H_3C—Cl **c.** H_3C—MgBr **d.** H_2N—OH

35. Draw the condensed structure of a compound that contains only carbon and hydrogen atoms and that has
a. three *sp*^{3} carbons.
b. one *sp*^{3} carbon and two *sp*^{2} carbons.
c. two *sp*^{3} carbons and two *sp* carbons.

36. Predict the approximate bond angles:
a. the H—C—O bond angle in CH_3OH
b. the C—O—H bond angle in CH_3OH
c. the H—C—H bond angle in $H_2C{=}O$
d. the C—C—N bond angle in $CH_3C{\equiv}N$

37. Give each atom the appropriate formal charge:
a. $\mathrm{H}{:}\ddot{\underset{\cdot\cdot}{\mathrm{O}}}{:}$ **b.** $\mathrm{H}{:}\ddot{\underset{\cdot\cdot}{\mathrm{O}}}\cdot$ **c.** $\mathrm{H}{-}\ddot{\underset{\cdot\cdot}{\mathrm{N}}}{-}\mathrm{H}$ **d.** $\mathrm{H}{-}\ddot{\mathrm{C}}{-}\mathrm{H}$

38. Write the electronic configuration for the following species (carbon's electronic configuration is written as $1s^2\,2s^2\,2p^2$):
a. Ca **b.** Ca^{2+} **c.** Ar **d.** Mg^{2+}

39. Only one of the following formulas describes a compound that exists. Fix the other formulas so they also describe compounds that exist.
a. $CH_3CH_3CH_3$ **c.** $(CH_3)_2CCH_3$ **e.** $CH_3CH_2CH_2$
b. CH_5 **d.** $(CH_3)_2CHCH_2CH_3$ **f.** $CH_3CHCH_2CH_3$

40. List the bonds in order of decreasing polarity (that is, list the most polar bond first).
a. C—O, C—F, C—N **b.** C—Cl, C—I, C—Br **c.** H—O, H—N, H—C **d.** C—H, C—C, C—N

41. Write the Kekulé structure for each of the following compounds:

a. CH_3CHO
b. CH_3OCH_3
c. CH_3COOH
d. $(CH_3)_3COH$
e. $CH_3CH(OH)CH_2CN$
f. $(CH_3)_2CHCH(CH_3)CH_2C(CH_3)_3$

42. Assign the missing formal charges.

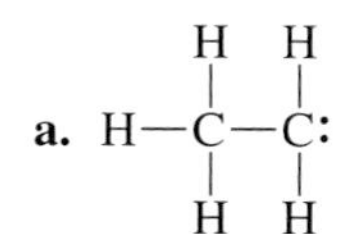

a. H—C—C: (each C bearing two H)

b. H—C—C (each C bearing two H)

c. H—C—C· (each C bearing two H)

d. H—C—C—C—H (with H on each carbon; :Ö: on the central carbon)

43. Draw the missing lone pairs and assign the missing formal charges.

a. H—C—O—H (C bearing two H)

b. H—C—O—H (C bearing two H; O bearing H)

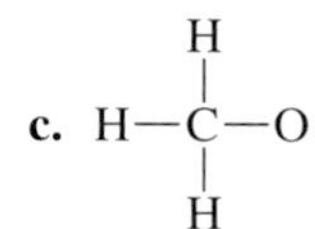

c. H—C—O (C bearing two H)

d. H—C—N—H (C bearing two H; N bearing H)

44. Account for the difference in the shape and color of the potential maps for ammonia and the ammonium ion in Section 1.12.

45. What is the hybridization of the indicated atom in each of the following compounds?

a. $CH_3CH{=}CH_2$
b. $CH_3\overset{\overset{\displaystyle O}{\|}}{C}CH_3$
c. CH_3CH_2OH
d. $CH_3C{\equiv}N$
e. $CH_3CH{=}NCH_3$
f. $CH_3OCH_2CH_3$

46. **a.** Which of the indicated bonds in each compound is shorter?
b. Indicate the hybridization of the C, O, and N atoms in each of the compounds.

1. $CH_3CH{=}CHC{\equiv}CH$ **2.** $CH_3\overset{\overset{\displaystyle O}{\|}}{C}CH_2{-}OH$ **3.** $CH_3NH{-}CH_2CH_2N{=}CHCH_3$

47. Which of the following have a tetrahedral geometry?

H_2O H_3O^+ $^+CH_3$ NH_3 $^+NH_4$ $^-CH_3$

48. Do the two sp^2 hybridized carbons and the two indicated atoms lie in the same plane?

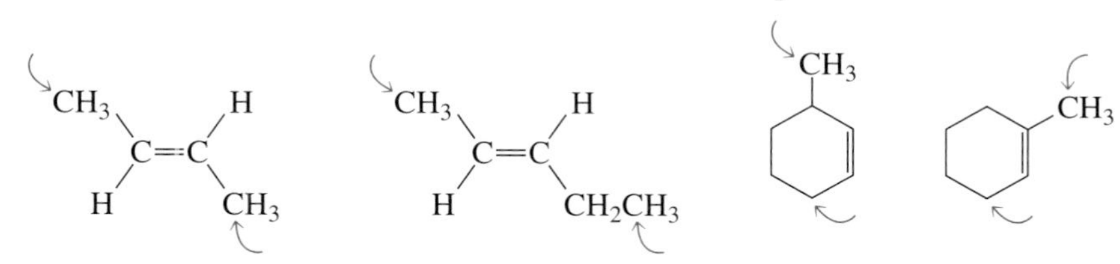

49. For each of the following compounds, indicate the hybridization of each carbon atom and give the approximate values of all the bond angles:

a. $CH_3C{\equiv}CH$ **b.** $CH_3CH{=}CH_2$ **c.** $CH_3CH_2CH_3$ **d.** $CH_2{=}CH{-}CH{=}CH_2$

50. Sodium methoxide (CH_3ONa) has both ionic and covalent bonds. Which bond is ionic? How many covalent bonds does it have?

51. **a.** Why is a H—H bond (0.74 Å) shorter than a C—C bond (1.54 Å)?
b. Predict the length of a C—H bond.

52. Explain why the following compound is not stable:

(six-membered ring containing a C≡C triple bond; the other four ring carbons each bear two H)

CHAPTER 2

Acids and Bases

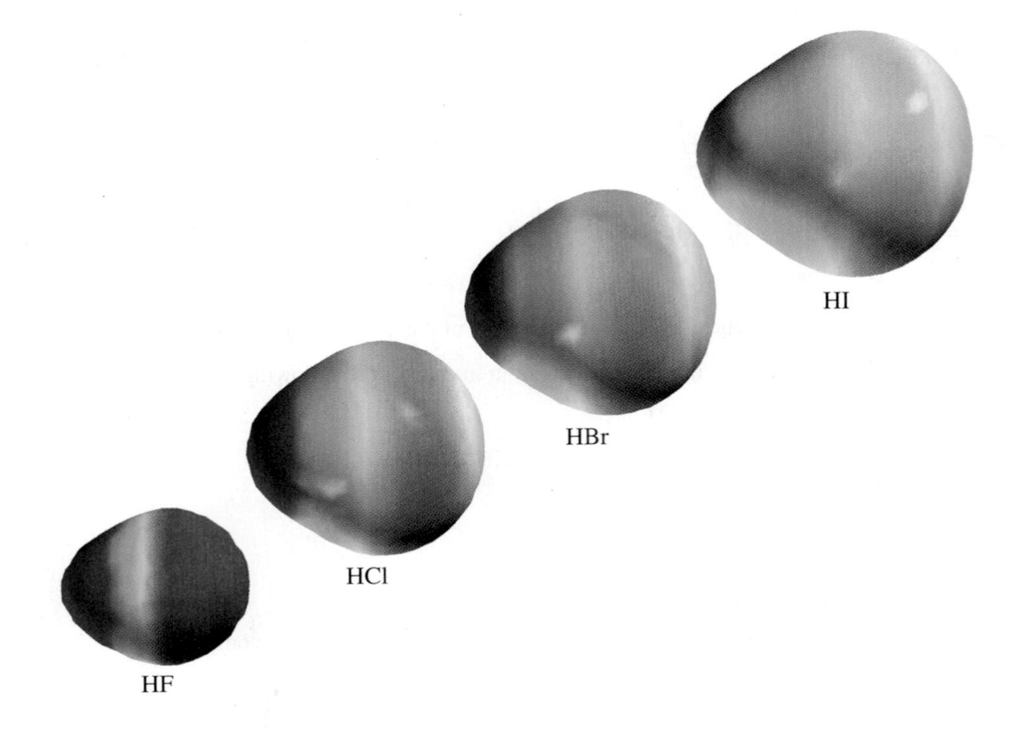

Early chemists called any compound that tasted sour an acid (from *acidus*, Latin for "sour"). Some familiar acids are citric acid (found in lemons and other citrus fruits), acetic acid (found in vinegar), and hydrochloric acid (found in stomach acid—the sour taste associated with vomiting). Compounds that neutralize acids were called bases, or alkaline compounds. Glass cleaners and solutions designed to unclog drains are familiar alkaline solutions.

2.1 AN INTRODUCTION TO ACIDS AND BASES

Brønsted and Lowry defined an **acid** as a species that donates a proton, and a **base** as a species that accepts a proton. (Remember that positively charged hydrogen ions are called protons.) In the reaction shown below, hydrogen chloride (HCl) is an acid because it donates a proton to water, and water is a base because it accepts a proton from HCl. Water can accept a proton because it has two lone pairs, either of which can form a covalent bond with a proton. In the reverse reaction, H_3O^+ is an acid because it donates a proton to Cl^-, and Cl^- is a base because it accepts a proton from H_3O^+. The reaction of an acid with a base is called an **acid–base reaction**. Both an acid and a base must be present in an acid–base reaction, because an acid cannot donate a proton unless a base is present to accept it.

$$\underset{\text{an acid}}{H\ddot{\underset{\cdot\cdot}{Cl}}:} + \underset{\text{a base}}{H_2\ddot{O}:} \rightleftharpoons \underset{\text{a base}}{:\ddot{\underset{\cdot\cdot}{Cl}}:^-} + \underset{\text{an acid}}{H_3\ddot{O}^+}$$

Notice that according to the Brønsted–Lowry definitions, *any species that has a hydrogen can potentially act as an acid, and any compound that has a lone pair can potentially act as a base.*

When a compound loses a proton, the resulting species is called its **conjugate base**. Thus, Cl^- is the conjugate base of HCl, and H_2O is the conjugate base of H_3O^+. When a compound accepts a proton, the resulting species is called its **conjugate acid**. Thus, HCl is the conjugate acid of Cl^-, and H_3O^+ is the conjugate acid of H_2O.

In a reaction between ammonia and water, ammonia (NH_3) is a base because it accepts a proton, and water is an acid because it donates a proton. Thus, HO^- is the conjugate base of H_2O, and $^+NH_4$ is the conjugate acid of NH_3. In the reverse reaction, ammonium ion ($^+NH_4$) is an acid because it donates a proton, and hydroxide ion (HO^-) is a base because it accepts a proton.

$$\underset{\text{a base}}{\ddot{N}H_3} + \underset{\text{an acid}}{H_2\ddot{O}:} \rightleftharpoons \underset{\text{an acid}}{^+NH_4} + \underset{\text{a base}}{H\ddot{O}:^-}$$

Notice that water can behave as either an acid or a base. It can behave as an acid because it has a proton that it can donate, but it can also behave as a base because it has a lone pair that can accept a proton. In Section 2.2, we will see how we can predict that water acts as a base in the first reaction and as an acid in the second reaction.

Acidity is a measure of the tendency of a compound to give up a proton. **Basicity** is a measure of a compound's affinity for a proton. A strong acid is one that has a strong tendency to give up its proton. This means that its conjugate base must be weak because it has little affinity for the proton. A weak acid has little tendency to give up its proton, indicating that its conjugate base is strong because it has a high affinity for the proton. Thus, the following important relationship exists between an acid and its conjugate base: *the stronger the acid, the weaker is its conjugate base*. For example, since HBr is a stronger acid than HCl, we know that Br^- is a weaker base than Cl^-.

BIOGRAPHY

Born in Denmark, **Johannes Nicolaus Brønsted (1879–1947)** *studied engineering before he switched to chemistry. He was a professor of chemistry at the University of Copenhagen. During World War II, he became known for his anti-Nazi position, and as a consequence was elected to the Danish parliament in 1947. He died before he could take his seat.*

The stronger the acid, the more readily it gives up a proton.

The stronger the acid, the weaker is its conjugate base.

PROBLEM 1 ◆

a. Draw the conjugate acid of each of the following:
   1. NH_3 2. Cl^- 3. HO^- 4. H_2O

b. Draw the conjugate base of each of the following:
   1. NH_3 2. HBr 3. HNO_3 4. H_2O

PROBLEM 2

a. Write an equation showing CH_3OH reacting as an acid with NH_3 and an equation showing CH_3OH reacting as a base with HCl.

b. Write an equation showing NH_3 reacting as an acid with HO^- and an equation showing NH_3 reacting as a base with HBr.

BIOGRAPHY

Thomas M. Lowry (1874–1936) *was born in England, the son of an army chaplain. He earned a Ph.D. at Central Technical College, London (now Imperial College). He was head of chemistry at Westminster Training College and, later, at Guy's Hospital in London. In 1920, he became a professor of chemistry at Cambridge University.*

2.2 pK_a AND pH

When a strong acid such as hydrogen chloride is dissolved in water, almost all the molecules dissociate (break into ions), which means that the *products* are favored at equilibrium—the equilibrium lies to the right. When a much weaker acid, such as acetic acid, is dissolved in water, very few molecules dissociate, so the *reactants* are favored at equilibrium—the equilibrium lies to the left. Two half-headed arrows are used to designate equilibrium reactions. A longer arrow is drawn toward the species favored at equilibrium.

$$\underset{\text{hydrogen chloride}}{H\ddot{C}l:} + H_2\ddot{O}: \rightleftharpoons H_3\ddot{O}^+ + :\ddot{C}l:^-$$

$$\underset{\text{acetic acid}}{CH_3C(=O)\ddot{O}H} + H_2\ddot{O}: \rightleftharpoons H_3\ddot{O}^+ + CH_3C(=O)\ddot{O}:^-$$

The degree to which an acid (HA) dissociates in an aqueous solution is indicated by the **acid dissociation constant, K_a**. Brackets are used to indicate the concentration in moles/liter, that is, the molarity (M).

$$K_a = \frac{[H_3O^+][A^-]}{[HA]}$$

The larger the acid dissociation constant, the stronger is the acid—that is, the greater is its tendency to give up a proton. Hydrogen chloride, with an acid dissociation constant of 10^7, is a stronger acid than acetic acid, with an acid dissociation constant of only 1.74×10^{-5}. For convenience, the strength of an acid is generally indicated by its **pK_a** value rather than by its K_a value, where

$$pK_a = -\log K_a$$

The pK_a of hydrogen chloride is −7, and the pK_a of acetic acid, a much weaker acid, is 4.76. Notice that the stronger the acid, the smaller is its pK_a value.

The stronger the acid, the smaller is its pK_a value.

very strong acids	$pK_a < 1$
moderately strong acids	$pK_a = 1–3$
weak acids	$pK_a = 3–5$
very weak acids	$pK_a = 5–15$
extremely weak acids	$pK_a > 15$

The concentration of positively charged hydrogen ions in the solution is indicated by **pH**. The concentration can be written as $[H^+]$ or, because a hydrogen ion in water is solvated, as $[H_3O^+]$.

$$pH = -\log[H_3O^+]$$

The lower the pH, the more acidic is the solution. Acidic solutions have pH values less than 7; basic solutions have pH values greater than 7. The pH values of some commonly encountered solutions are shown in the margin. The pH of a solution can be changed simply by adding acid or base to the solution. Do not confuse pH and pK_a: the pH scale is used to describe the acidity of a *solution*; the pK_a is characteristic of a particular *compound*, much like a melting point or a boiling point—it indicates the tendency of the compound to give up its proton.

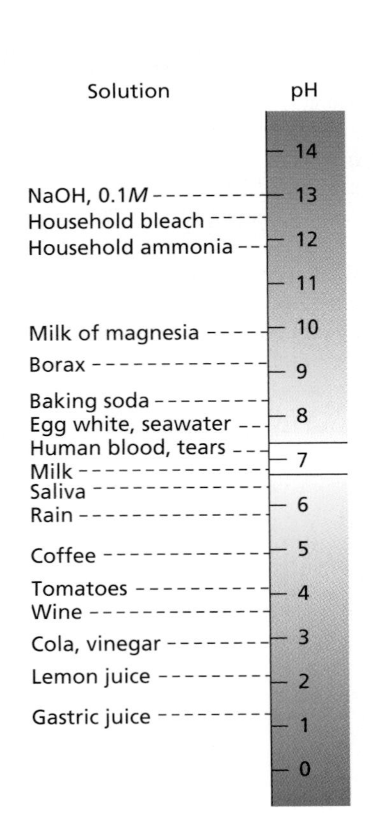

a. Which is a stronger acid, one with a pK_a of 5.2 or one with a pK_a of 5.8?
b. Which is a stronger acid, one with an acid dissociation constant of 3.4×10^{-3} or one with an acid dissociation constant of 2.1×10^{-4}?

PROBLEM-SOLVING STRATEGY

Vitamin C has a pK_a value of 4.17. What is its K_a value?

You will need a calculator to answer this question. Remembering that $pK_a = -\log K_a$:

1. Enter the pK_a value on your calculator.
2. Multiply it by −1.
3. Determine the inverse log by pressing the key labeled 10^x.

You should find that vitamin C has a K_a value of 6.76×10^{-5}.

Now continue on to Problem 4.

PROBLEM 4 ◆

Butyric acid, the compound responsible for the unpleasant odor and taste of sour milk, has a pK_a value of 4.82. What is its K_a value? Is it a stronger or a weaker acid than vitamin C?

ACID RAIN

Rain is mildly acidic (pH = 5.5) because when the CO_2 in the air reacts with water, a weak acid—carbonic acid (pK_a = 6.4)—is formed.

$$CO_2 + H_2O \rightleftharpoons \underset{\text{carbonic acid}}{H_2CO_3}$$

In some parts of the world, rain has been found to be much more acidic—with pH values as low as 4.3. Acid rain is formed when sulfur dioxide and nitrogen oxides are produced, because when these gases react with water, strong acids—sulfuric acid (pK_a = −5.0) and nitric acid (pK_a = −1.3)—are formed. Burning fossil fuels for the generation of electric power is the factor most responsible for forming these acid-producing gases.

Acid rain has many deleterious effects. It can destroy aquatic life in lakes and streams; it can make soil so acidic that crops cannot grow; and it can cause the deterioration of paint and building materials, including monuments and statues that are part of our cultural heritage. Marble—a form of calcium carbonate—decays because acid reacts with CO_3^{2-} to form carbonic acid, which decomposes to CO_2 and H_2O, the reverse of the reaction shown to the left.

$$CO_3^{2-} \overset{H^+}{\rightleftharpoons} HCO_3^- \overset{H^+}{\rightleftharpoons} H_2CO_3 \rightleftharpoons CO_2 + H_2O$$

photo taken in 1935

photo taken in 1994

Statue of George Washington in Washington Square Park in Greenwich Village, New York.

PROBLEM 5

Antacids are compounds that neutralize stomach acid. Write the equations that show how Milk of Magnesia, Alka-Seltzer, and Tums remove excess acid.

a. Milk of Magnesia: $Mg(OH)_2$

b. Alka-Seltzer: $KHCO_3$ and $NaHCO_3$

c. Tums: $CaCO_3$

PROBLEM 6 ♦

Are the following body fluids acidic or basic?

a. bile (pH = 8.4) **b.** urine (pH = 5.9) **c.** spinal fluid (pH = 7.4)

2.3 ORGANIC ACIDS AND BASES

The most common organic acids are carboxylic acids—compounds that have a COOH group. Acetic acid and formic acid are examples of carboxylic acids. Carboxylic acids have pK_a values ranging from about 3 to 5. (They are weak acids.) The pK_a values of a wide variety of organic compounds are given in Appendix II.

H_3C–C(=O)–OH
acetic acid
pK_a = 4.76

H–C(=O)–OH
formic acid
pK_a = 3.75

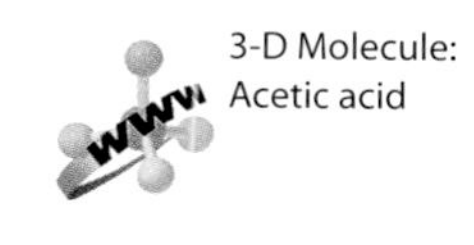

3-D Molecule: Acetic acid

The carboxyl group of a carboxylic acid can be represented in several different ways.

–C(=O)–OH
a carboxyl group

—COOH —CO_2H

carboxyl groups are frequently shown in abbreviated forms

Alcohols—compounds that have an OH group—are much weaker acids than carboxylic acids, with pK_a values close to 16. Methyl alcohol and ethyl alcohol are examples of alcohols.

CH_3OH
methyl alcohol
pK_a = 15.5

CH_3CH_2OH
ethyl alcohol
pK_a = 15.9

We have seen that water can behave both as an acid and as a base. An alcohol behaves similarly; it can behave as an acid and donate a proton, or as a base and accept a proton.

a curved arrow indicates where the electrons start from and where they end up

$$CH_3\ddot{O}-H + H-\ddot{O}:^- \rightleftharpoons CH_3\ddot{O}:^- + H-\ddot{O}-H$$

an acid

$$CH_3\ddot{O}-H + H-\ddot{O}^+(H)-H \rightleftharpoons CH_3\ddot{O}^+(H)-H + H-\ddot{O}-H$$

a base

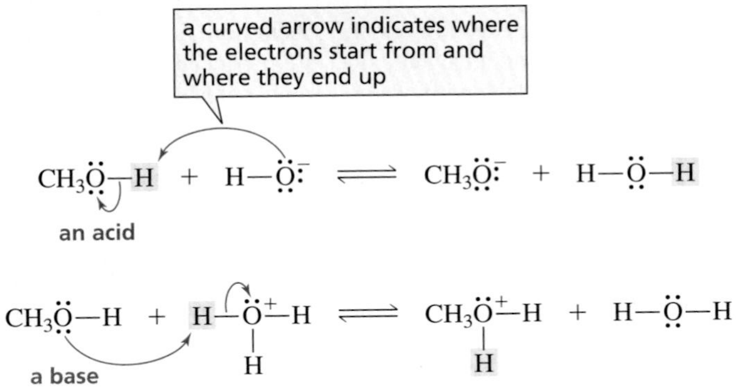

A carboxylic acid also can behave as an acid and donate a proton, or as a base and accept a proton.

$$CH_3C(=O)\ddot{O}-H + H-\ddot{O}:^- \rightleftharpoons CH_3C(=O)\ddot{O}:^- + H-\ddot{O}-H$$

an acid

$$CH_3C(=O)\ddot{O}-H + H-\ddot{O}^+(H)-H \rightleftharpoons CH_3C(=\ddot{O}^+-H)O-H + H-\ddot{O}-H$$

a base

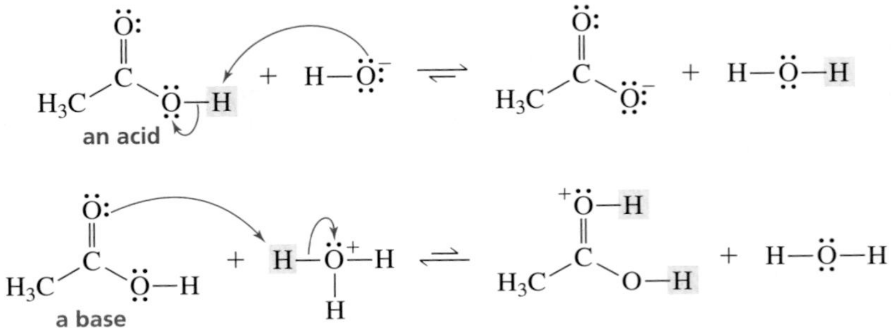

A *protonated* compound is a compound that has gained an additional proton. Protonated alcohols and protonated carboxylic acids are very strong acids. For example, protonated methyl alcohol has a pK_a of −2.5, protonated ethyl alcohol has a pK_a of −2.4, and protonated acetic acid has a pK_a of −6.1. Notice that the sp^2 oxygen of the carboxylic acid is the one that is protonated. We will see why this is so in Section 11.9.

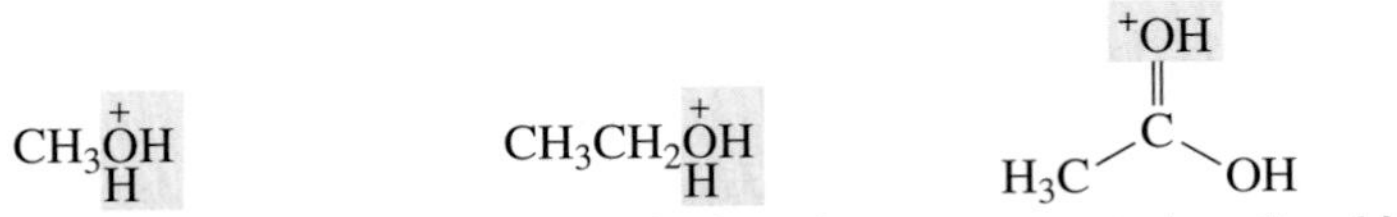

$CH_3\overset{+}{O}H_2$
protonated methanol
pK_a = −2.5

$CH_3CH_2\overset{+}{O}H_2$
protonated ethanol
pK_a = −2.4

$CH_3C(=\overset{+}{O}H)OH$
protonated acetic acid
pK_a = −6.1

A compound with an NH_2 group is an amine. An amine can behave as an acid and donate a proton, or as a base and accept a proton.

$$CH_3\ddot{N}H(H) + H-\ddot{O}:^- \rightleftharpoons CH_3\ddot{N}H^- + H-\ddot{O}-H$$

an acid

$$CH_3\ddot{N}H(H) + H-\ddot{O}^+(H)-H \rightleftharpoons CH_3\overset{+}{N}H(H)(H) + H-\ddot{O}-H$$

a base

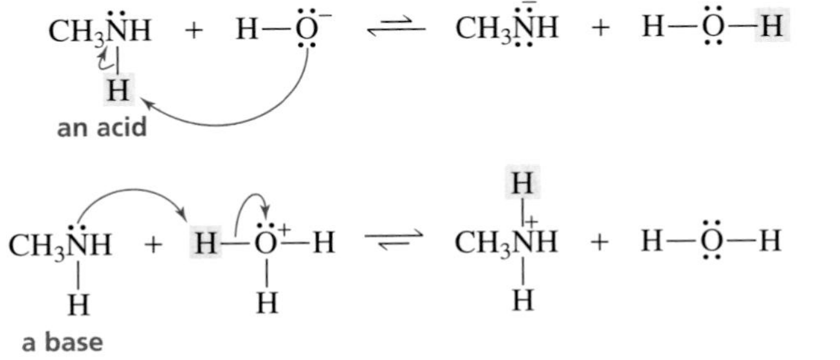

Amines, however, have such high pK_a values that they rarely behave as acids. Ammonia also has a high pK_a value.

CH_3NH_2
methylamine
$pK_a = 40$

NH_3
ammonia
$pK_a = 36$

Amines are much more likely to act as bases. In fact, amines are the most common organic bases. Instead of talking about the strength of a base in terms of its pK_b value, it is easier to talk about the strength of its conjugate acid as indicated by its pK_a value, remembering that the stronger the acid, the weaker is its conjugate base. For example, protonated methylamine is a stronger acid than protonated ethylamine, which means that methylamine is a weaker base than ethylamine. Notice that the pK_a values of protonated amines are about 10 to 11.

$CH_3\overset{+}{N}H_3$
protonated methylamine
$pK_a = 10.7$

$CH_3CH_2\overset{+}{N}H_3$
protonated ethylamine
$pK_a = 11.0$

It is important to know the approximate pK_a values of the various classes of compounds we have discussed. An easy way to remember them is in units of five, as shown in Table 2.1. (R is used when the particular carboxylic acid, alcohol, or amine is not specified.) Protonated alcohols, protonated carboxylic acids, and protonated water have pK_a values less than 0, carboxylic acids have pK_a values of about 5, protonated amines have pK_a values of about 10, and alcohols and water have pK_a values of about 15. These values are also listed inside the back cover of this book for easy reference.

Be sure to learn the approximate pK_a values given in Table 2.1.

Table 2.1 Approximate pK_a Values

$pK_a < 0$	$pK_a \sim 5$	$pK_a \sim 10$	$pK_a \sim 15$
$R\overset{+}{O}H_2$ a protonated alcohol	$RCOOH$ (R–C(=O)–OH) a carboxylic acid	$R\overset{+}{N}H_3$ a protonated amine	ROH an alcohol
$RC(=\overset{+}{O}H)OH$ a protonated carboxylic acid			H_2O water
H_3O^+ protonated water			

PROBLEM 7 ◆

a. Which is a stronger base, CH_3COO^- or $HCOO^-$? (The pK_a of CH_3COOH is 4.8; the pK_a of $HCOOH$ is 3.8.)

b. Which is a stronger base, HO^- or $^-NH_2$? (The pK_a of H_2O is 15.7; the pK_a of NH_3 is 36.)

c. Which is a stronger base, H_2O or CH_3OH? (The pK_a of H_3O^+ is −1.7; the pK_a of $CH_3\overset{+}{O}H_2$ is −2.5.)

PROBLEM 8 ◆

Using the pK_a values in Section 2.3, rank the following species in order of decreasing base strength (that is, list the strongest base first):

CH_3NH_2 CH_3NH^- CH_3OH CH_3O^- $CH_3\overset{O}{\overset{\|}{C}}O^-$

2.4 HOW TO PREDICT THE OUTCOME OF AN ACID–BASE REACTION

Now let's see how we can predict that water will act as a base in the first reaction in Section 2.1 and as an acid in the second reaction. To determine which of the two reactants of the first reaction will be the acid, we need to compare their pK_a values: the pK_a of hydrogen chloride is −7, and the pK_a of water is 15.7. Because hydrogen chloride is the stronger acid, it will donate a proton to water. Water, therefore, is a base in this reaction. When we compare the pK_a values of the two reactants of the second reaction, we see that the pK_a of ammonia is 36 and the pK_a of water is 15.7. In this case, water is the stronger acid, so it donates a proton to ammonia. Water, therefore, is an acid in this reaction.

PROBLEM 9 ♦

Using the pK_a values in Section 2.3, predict the products of the following reaction:

$$CH_3NH_2 + CH_3OH \rightleftharpoons$$

2.5 HOW TO DETERMINE THE POSITION OF EQUILIBRIUM

To determine the position of equilibrium for an acid–base reaction (that is, whether reactants or products are favored at equilibrium), we need to compare the pK_a value of the acid on the left of the arrow with the pK_a value of the acid on the right of the arrow. The equilibrium favors *reaction* of the stronger acid and *formation* of the weaker acid. Thus, the equilibrium lies away from the stronger acid and toward the weaker acid. Products, therefore, are favored in the first reaction, and reactants are favored in the second reaction. Notice that the stronger acid has the weaker (more stable) conjugate base, so the equilibrium favors formation of the more stable species.

Strong reacts to form weak.

$$\underset{\substack{\text{stronger acid}\\ \text{p}K_a = 4.8}}{CH_3COOH} + \underset{\text{stronger base}}{NH_3} \rightleftharpoons \underset{\text{weaker base}}{CH_3COO^-} + \underset{\substack{\text{weaker acid}\\ \text{p}K_a = 9.4}}{^+NH_4}$$

$$\underset{\substack{\text{weaker acid}\\ \text{p}K_a = 15.9}}{CH_3CH_2OH} + \underset{\text{weaker base}}{CH_3NH_2} \rightleftharpoons \underset{\text{stronger base}}{CH_3CH_2O^-} + \underset{\substack{\text{stronger acid}\\ \text{p}K_a = 10.7}}{CH_3\overset{+}{N}H_3}$$

PROBLEM 10

a. For each of the acid–base reactions in Section 2.3, compare the pK_a values of the acids on either side of the equilibrium arrows and convince yourself that the position of equilibrium is in the direction indicated. (The pK_a values you need can be found in Section 2.3 or in Problem 7.)

b. Do the same thing for the equilibria in Section 2.1. (The pK_a of $^+NH_4$ is 9.4.)

2.6 HOW THE STRUCTURE OF AN ACID AFFECTS ITS pK_a

The weaker the base, the stronger is its conjugate acid.

Stable bases are weak bases.

The more stable the base, the stronger is its conjugate acid.

The strength of an acid is determined by the stability of the conjugate base that is formed when the acid gives up its proton: the more stable the base, the stronger is its conjugate acid. A stable base is a base that readily bears the electrons it formerly shared with a proton. In other words, stable bases are weak bases—they don't share their electrons well. That is why we can say, *the weaker the base, the stronger is its conjugate acid* or, *the more stable the base, the stronger is its conjugate acid.*

Two factors that affect the stability of a base are its *size* and its *electronegativity*. The elements in the second row of the periodic table are all about the same size, but they have very different electronegativities, which increase across the row from left to right. Therefore, of the atoms shown, carbon is the least electronegative and fluorine is the most electronegative.

relative electronegativities: C < N < O < F (most electronegative)

If we look at the acids formed by attaching hydrogens to these elements, we see that the most acidic compound is the one that has its hydrogen attached to the most electronegative atom. Thus, HF is the strongest acid and methane is the weakest acid (Table 2.2).

relative acidities: CH_4 < NH_3 < H_2O < HF (strongest acid)

If we look at the stabilities of the conjugate bases of these acids, we find that they too increase from left to right because the more electronegative the atom, the better it can bear its negative charge. Thus, we see that the strongest acid has the most stable conjugate base.

relative stabilities: $^-CH_3$ < $^-NH_2$ < HO^- < F^- (most stable)

We therefore can conclude that *when the atoms are similar in size, the strongest acid will have its hydrogen attached to the most electronegative atom.*

When atoms are similar in size, the strongest acid will have its hydrogen attached to the most electronegative atom.

The effect that the electronegativity of the atom bonded to a hydrogen has on the acidity of that hydrogen can be appreciated when the pK_a values of alcohols and amines are compared. Because oxygen is more electronegative than nitrogen, an alcohol is more acidic than an amine.

CH_3OH	CH_3NH_2
methyl alcohol	methylamine
pK_a = 15.5	pK_a = 40

Similarly, a protonated alcohol is more acidic than a protonated amine.

$CH_3\overset{+}{O}H_2$	$CH_3\overset{+}{N}H_3$
protonated methyl alcohol	protonated methylamine
pK_a = −2.5	pK_a = 10.7

Table 2.2 The pK_a Values of Some Simple Acids

CH_4	NH_3	H_2O	HF
pK_a = ~60	pK_a = 36	pK_a = 15.7	pK_a = 3.2
		H_2S	HCl
		pK_a = 7.0	pK_a = −7
			HBr
			pK_a = −9
			HI
			pK_a = −10

When atoms are very different in size, the strongest acid will have its hydrogen attached to the largest atom.

In comparing atoms that are very different in size, the *size* of the atom is much more important than its *electronegativity* in determining how well it bears its negative charge. For example, as we proceed down a column in the periodic table, the atoms get larger and their electronegativity decreases. However the stability of the bases increases down the column, so the strength of their conjugate acid *increases*. Thus, HI is the strongest acid of the hydrogen halides, even though iodine is the least electronegative of the halogens. Therefore, *when atoms are very different in size, the strongest acid will have its hydrogen attached to the largest atom.*

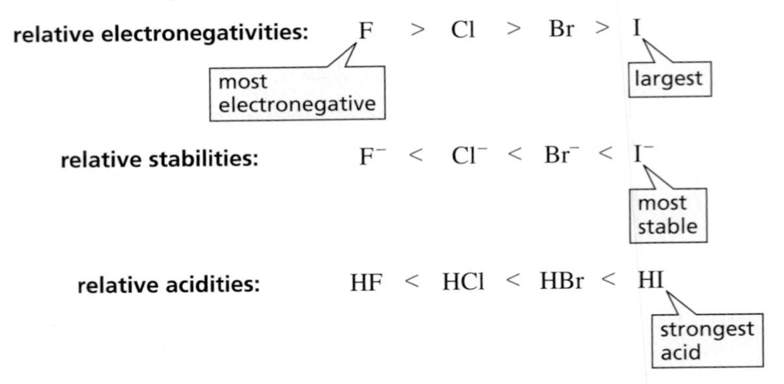

Why does the size of an atom have such a significant effect on the stability of the base that it more than overcomes the difference in electronegativity? The valence electrons of F^- are in a $2sp^3$ orbital, the valence electrons of Cl^- are in a $3sp^3$ orbital, those of Br^- are in a $4sp^3$ orbital, and those of I^- are in a $5sp^3$ orbital. The volume of space occupied by a $3sp^3$ orbital is significantly greater than the volume of space occupied by a $2sp^3$ orbital because a $3sp^3$ orbital extends out farther from the nucleus. Because its negative charge is spread over a larger volume of space, Cl^- is more stable than F^-.

Thus, as the halide ion increases in size, its stability increases because its negative charge is spread over a larger volume of space (its electron density decreases). Therefore, HI is the strongest acid of the hydrogen halides because I^- is the most stable halide ion, even though iodine is the least electronegative of the halogens (Table 2.2). The potential maps illustrate the large difference in size of the halogens:

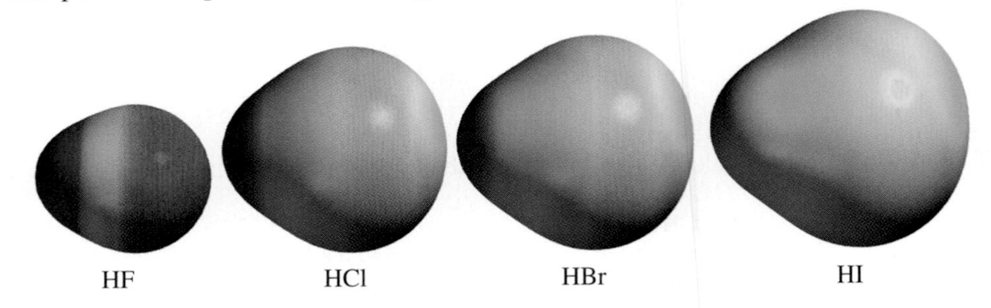

PROBLEM 11 ♦

For each of the following pairs, indicate which is the stronger acid:

a. HCl or HBr

b. $CH_3CH_2CH_2\overset{+}{N}H_3$ or $CH_3CH_2CH_2\overset{+}{O}H_2$

c.

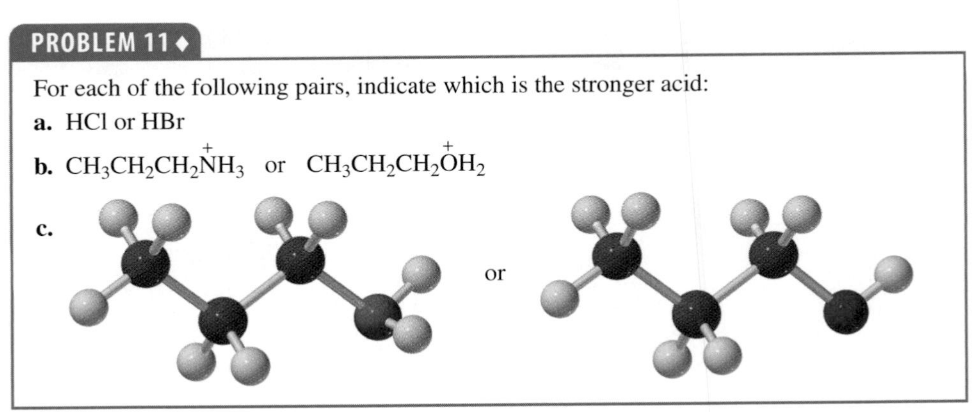

PROBLEM 12 ◆

a. Which of the halide ions (F^-, Cl^-, Br^-, I^-) is the strongest base?
b. Which is the weakest base?

PROBLEM 13 ◆

a. Which is more electronegative, oxygen or sulfur?
b. Which is a stronger acid, H_2O or H_2S?
c. Which is a stronger acid, CH_3OH or CH_3SH?
d. Which of the following is a stronger acid?

$H_3C{-}C({=}O){-}OH$ or $H_3C{-}C({=}O){-}SH$

PROBLEM 14 ◆

For each of the following pairs, indicate which is the stronger base:

a. H_2O or HO^- **b.** H_2O or NH_3 **c.** $CH_3C(=O)O^-$ or CH_3O^- **d.** CH_3O^- or CH_3S^-

2.7 HOW pH AFFECTS THE STRUCTURE OF AN ORGANIC COMPOUND

Whether an acid will lose a proton in an aqueous solution depends on both the pK_a of the acid and the pH of the solution. *A compound will exist primarily in its acidic form (with its proton) in solutions that are more acidic than the* pK_a *value of the group that undergoes dissociation. It will exist primarily in its basic form (without its proton) in solutions that are more basic than the* pK_a *value of the group that undergoes dissociation.* When the pH of a solution equals the pK_a value of the group that undergoes dissociation, the concentration of the compound in its acidic form will equal the concentration of the compound in its basic form.

A compound will exist primarily in its acidic form if the pH of the solution is less than the compound's pK_a.

A compound will exist primarily in its basic form if the pH of the solution is greater than the compound's pK_a.

acidic form		basic form
$RCOOH$	$\rightleftharpoons$	$RCOO^- + H^+$
$R\overset{+}{N}H_3$	$\rightleftharpoons$	$RNH_2 + H^+$

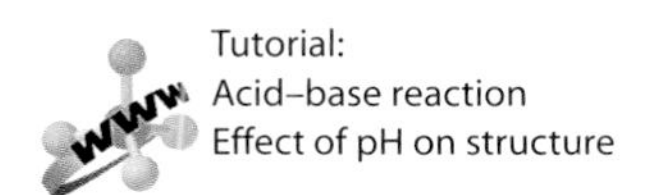

Tutorial:
Acid–base reaction
Effect of pH on structure

PROBLEM-SOLVING STRATEGY

Write the form in which the following compounds will predominate in a solution with a pH = 5.5:

a. CH_3CH_2OH (pK_a = 15.9) **c.** $CH_3\overset{+}{N}H_3$ (pK_a = 11.0)

b. $CH_3CH_2\overset{+}{O}H_2$ (pK_a = −2.5)

To answer this kind of question, we need to compare the pH of the solution with the pK_a value of the compound's dissociable proton.

a. The pH of the solution is more acidic (5.5) than the pK_a value of CH_3CH_2OH (15.9). Therefore, the compound will exist primarily as CH_3CH_2OH (with its proton).

b. The pH of the solution is more basic (5.5) than the pK_a value of $CH_3CH_2\overset{+}{O}H_2$ (−2.5). Therefore, the compound will exist primarily as CH_3CH_2OH (without its proton).

c. The pH of the solution is more acidic (5.5) than the pK_a value of $CH_3\overset{+}{N}H_3$ (10.7). Therefore, the compound will exist primarily as $CH_3\overset{+}{N}H_3$ (with its proton).

Now continue on to Problem 15.

PROBLEM 15 ♦

For each of the following compounds, shown in their acidic forms, write the form that will predominate in a solution with a pH = 5.5:

a. CH_3COOH (pK_a = 4.76)
b. $CH_3CH_2\overset{+}{N}H_3$ (pK_a = 11.0)
c. H_3O^+ (pK_a = −1.7)
d. HBr (pK_a = −9)
e. $^+NH_4$ (pK_a = 9.4)
f. $HC{\equiv}N$ (pK_a = 9.1)
g. HNO_2 (pK_a = 3.4)
h. HNO_3 (pK_a = −1.3)

PROBLEM 16 ♦ *SOLVED*

a. Indicate whether a carboxylic acid (RCOOH) with a pK_a of 4.5 will have more charged molecules or more neutral molecules in a solution with the following pH value:

1. pH = 1
2. pH = 3
3. pH = 5
4. pH = 7
5. pH = 10
6. pH = 13

b. Answer the same question for a protonated amine ($R\overset{+}{N}H_3$) with a pK_a of 9.

c. Answer the same question for an alcohol (ROH) with a pK_a of 15.

Solution to 16 a1. First determine whether more molecules will be in the acidic form or the basic form; if the pH is less than the pK_a, more molecules will be in the acidic form; if the pH is greater than the pK_a, more molecules will be in the basic form. For 16a1, the pH = 1 and the pK_a = 4.5, so more molecules will be in the acidic form. Now determine whether the acidic form is charged or neutral. The acidic form of a carboxylic acid is neutral, so there will be more neutral molecules in the solution.

PROBLEM 17 ♦

A naturally occurring amino acid such as alanine has both a carboxylic acid group and an amine group. The pK_a values of the two groups are shown.

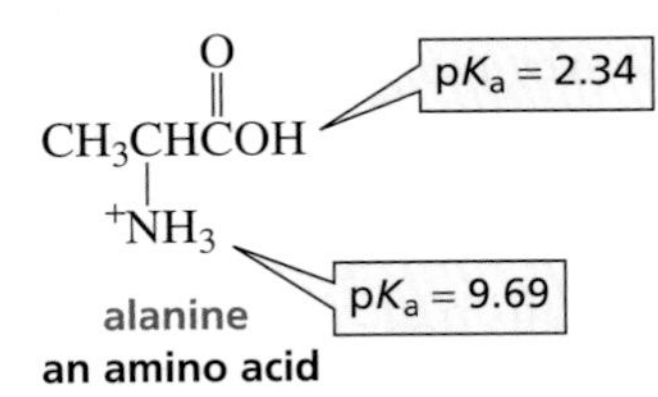

alanine
an amino acid

a. Write the structure of alanine in a solution at physiological pH (pH = 7.3).
b. Is there a pH at which alanine will be neutral (neither group will have a charge)?

2.8 BUFFER SOLUTIONS

A solution containing a weak acid (HA) and its conjugate base (A^-) is called a **buffer solution**. A buffer solution will maintain nearly constant pH when small amounts of acid or base are added to it, because the weak acid can donate a proton to any HO^- added to the solution, and its conjugate base can accept any H^+ that is added to the solution.

can donate an H^+ to HO^-

$$HA + HO^- \longrightarrow A^- + H_2O$$

$$A^- + H_3O^+ \longrightarrow HA + H_2O$$

can accept an H^+ from H_3O^+

BLOOD IS A BUFFERED SOLUTION

Blood is the fluid that transports oxygen to all the cells of the human body. The normal pH of human blood is 7.3 to 7.4. Death will result if this pH decreases to a value less than ~ 6.8 or increases to a value greater than ~ 8.0 for even a few seconds.

Oxygen is carried to cells by a protein in the blood called hemoglobin (HbH^+). When hemoglobin binds O_2, hemoglobin loses a proton, which would make the blood more acidic if it did not contain a buffer to maintain its pH.

$$HbH^+ + O_2 \rightleftharpoons HbO_2 + H^+$$

A carbonic acid/bicarbonate (H_2CO_3/HCO_3^-) buffer controls the pH of blood. An important feature of this buffer is that carbonic acid decomposes to CO_2 and H_2O:

$$CO_2 + H_2O \rightleftharpoons \underset{\text{carbonic acid}}{H_2CO_3} \rightleftharpoons \underset{\text{bicarbonate}}{HCO_3^-} + H^+$$

During exercise our metabolism speeds up, producing large amounts of CO_2. The increased concentration of CO_2 shifts the equilibrium between carbonic acid and bicarbonate to the right, which increases the concentration of H^+. Significant amounts of lactic acid are also produced during exercise, and this further increases the concentration of H^+. Receptors in the brain respond to the increased concentration of H^+ by triggering a reflex that increases the rate of breathing. Hemoglobin then releases more oxygen to the cells, and more CO_2 is eliminated by exhalation. Both processes decrease the concentration of H^+ in the blood by shifting both equilibria to the left.

Thus, any disorder that decreases the rate and depth of ventilation, such as emphysema, will decrease the pH of the blood—a condition called acidosis. In contrast, any excessive increase in the rate and depth of ventilation, such as hyperventilation due to anxiety, will increase the pH of blood—a condition called alkalosis.

PROBLEM 18

Write the equation that shows how a buffer made by dissolving CH_3COOH and $CH_3COO^-Na^+$ in water prevents the pH of a solution from changing when

a. a small amount of H^+ is added to the solution.
b. a small amount of HO^- is added to the solution.

2.9 LEWIS ACIDS AND BASES

In 1923, G. N. Lewis (page 6) offered new definitions for the terms *acid* and *base*. He defined an acid as a species that *accepts a share in an electron pair* and a base as a species that *donates a share in an electron pair*. All proton-donating acids fit the Lewis definition because all proton-donating acids lose a proton and the proton accepts a share in an electron pair.

Lewis acid: need two from you.

Lewis base: have pair, will share.

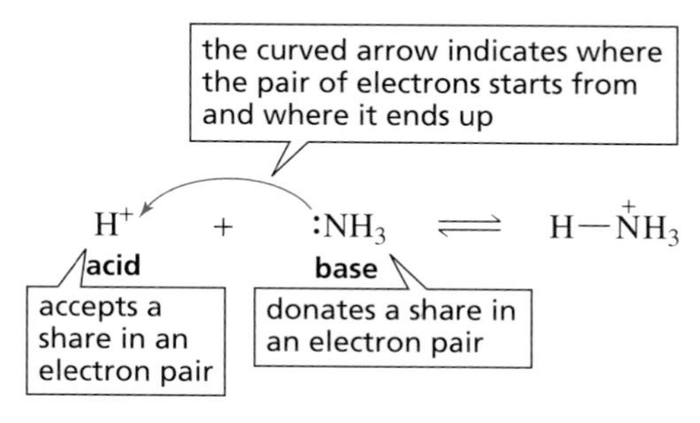

Lewis acids, however, are not limited to compounds that donate protons. According to the Lewis definition, compounds such as aluminum trichloride ($AlCl_3$) and borane (BH_3) are acids because they have unfilled valence orbitals and thus can accept a share in an electron pair. These compounds react with a compound that has a lone pair, just as a proton reacts with ammonia. Thus, the Lewis definition of an acid includes all proton-donating compounds and some additional compounds that do not have protons. Throughout this text, the term *acid* is used to mean a proton-donating acid, and the term **Lewis acid** is used to refer to a non-proton-donating acid such as $AlCl_3$ or BH_3.

All bases are **Lewis bases** because they all have a pair of electrons that they can share, either with an atom such as aluminum or boron or with a proton.

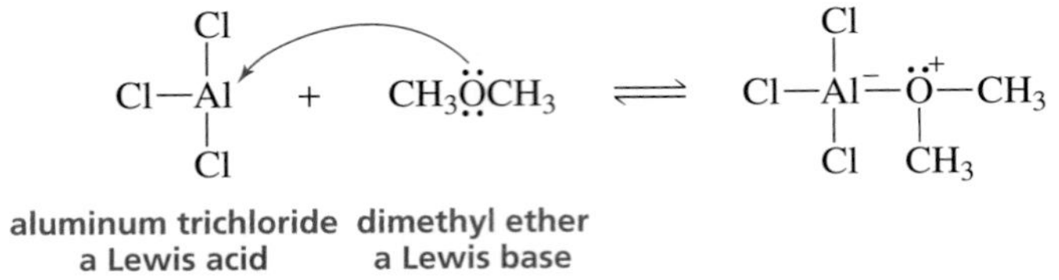

$$AlCl_3 + CH_3\ddot{O}CH_3 \rightleftharpoons Cl_3\overset{-}{Al}-\overset{+}{\ddot{O}}(CH_3)-CH_3$$

aluminum trichloride — a Lewis acid; dimethyl ether — a Lewis base

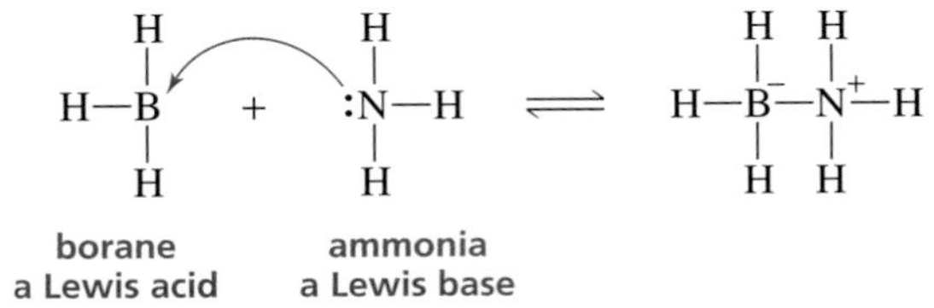

$$BH_3 + :NH_3 \rightleftharpoons H_3\overset{-}{B}-\overset{+}{N}H_3$$

borane — a Lewis acid; ammonia — a Lewis base

PROBLEM 19

Write the products of the following reactions using arrows to show where the pair of electrons starts and where it ends up:

a. $ZnCl_2 + CH_3\ddot{O}H \rightleftharpoons$

b. $FeBr_3 + :\ddot{Br}:^- \rightleftharpoons$

c. $AlCl_3 + :\ddot{Cl}:^- \rightleftharpoons$

PROBLEM 20

Write the products formed from the reaction of each of the following species with HO^-:

a. CH_3OH
b. $^+NH_4$
c. $CH_3\overset{+}{N}H_3$
d. BF_3
e. $^+CH_3$
f. $FeBr_3$
g. $AlCl_3$
h. CH_3COOH

SUMMARY

An **acid** is a species that donates a proton; a **base** is a species that accepts a proton. A **Lewis acid** is a species that accepts a share in an electron pair; a **Lewis base** is a species that donates a share in an electron pair.

Acidity is a measure of the tendency of a compound to give up a proton. **Basicity** is a measure of a compound's affinity for a proton. The stronger the acid, the weaker is its conjugate base. The strength of an acid is given by the **acid dissociation constant** (K_a). Approximate pK_a values are as follows: protonated alcohols, protonated carboxylic acids, protonated water < 0; carboxylic acids ~ 5; protonated amines ~ 10; alcohols and water ~ 15. The **pH** of a solution indicates the concentration of positively charged hydrogen ions in the solution. In **acid–base reactions**, the equilibrium favors reaction of the stronger acid and formation of the weaker acid.

The strength of an acid is determined by the stability of its conjugate base: the more stable the base, the stronger is its conjugate acid. When atoms are similar in size, the strongest acid will have its hydrogen attached to the most electronegative atom. When atoms are very different in size, the strongest acid will have its hydrogen attached to the largest atom.

A compound exists primarily in its acidic form in solutions more acidic than its pK_a value and primarily in its basic form in solutions more basic than its pK_a value. A **buffer solution** contains both a weak acid and its conjugate base.

PROBLEMS

21. For each of the following compounds, draw the form in which it will predominate at pH = 3, pH = 6, pH = 10, and pH = 14:

a. CH_3COOH pK_a = 4.8 **b.** $CH_3CH_2\overset{+}{N}H_3$ pK_a = 11.0 **c.** CF_3CH_2OH pK_a = 12.4

22. Write the products of the following acid–base reactions, and indicate whether reactants or products are favored at equilibrium (use the pK_a values that are given in Section 2.3):

a. $CH_3\overset{O}{\overset{\|}{C}}OH + CH_3O^- \rightleftharpoons$

b. $CH_3CH_2OH + {}^-NH_2 \rightleftharpoons$

c. $CH_3\overset{O}{\overset{\|}{C}}OH + CH_3NH_2 \rightleftharpoons$

d. $CH_3CH_2OH + HCl \rightleftharpoons$

23. a. Which of the following is the strongest acid?
b. Which is the weakest acid?
c. Which acid has the strongest conjugate base?

1. nitrous acid (HNO_2), $K_a = 4.0 \times 10^{-4}$
2. nitric acid (HNO_3), $K_a = 22$
3. bicarbonate (HCO_3^-), $K_a = 6.3 \times 10^{-11}$
4. hydrogen cyanide (HCN), $K_a = 7.9 \times 10^{-10}$
5. formic acid (HCOOH), $K_a = 2.0 \times 10^{-4}$

24. Which is the stronger base?

a. HS^- or HO^- **b.** CH_3O^- or $CH_3\bar{N}H$ **c.** CH_3OH or CH_3O^- **d.** Cl^- or Br^-

25. Locate the three nitrogen atoms in the electrostatic potential map of histamine, the compound that causes the symptoms associated with the common cold and allergic responses. Which of the two nitrogen atoms in the ring is the most basic?

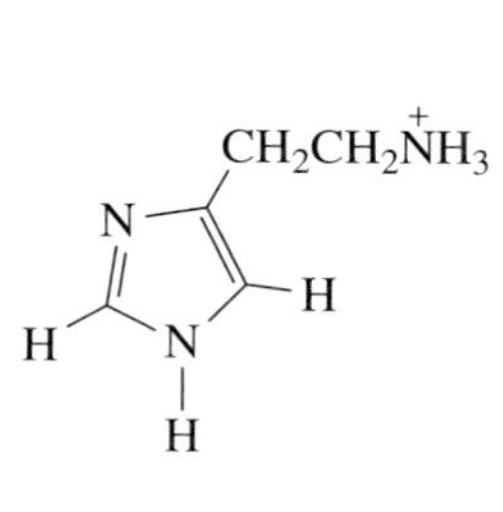

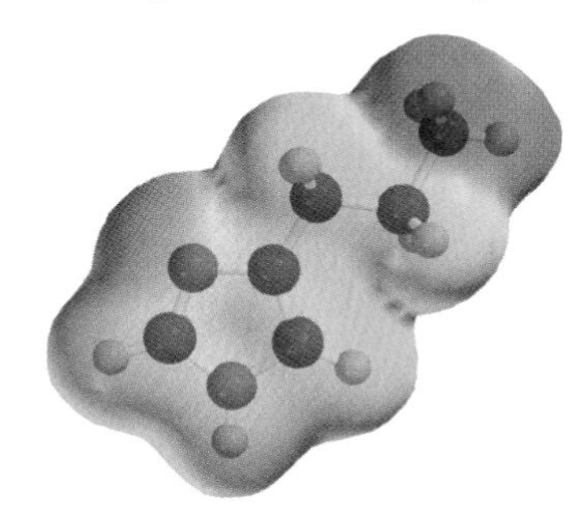

histamine

26. Using the table of pK_a values given in Appendix II, answer the following:
a. Which is the most acidic organic compound in the table?
b. Which is the least acidic organic compound in the table?
c. Which is the most acidic carboxylic acid in the table?

27. As long as the pH is greater than __________, more than 50% of a protonated amine with a pK_a of 10.4 will be in its neutral, nonprotonated form.

28. a. List the following carboxylic acids in order of decreasing acidity:

1. $CH_3CH_2CH_2COOH$ $K_a = 1.52 \times 10^{-5}$
2. $CH_3CH_2\underset{Cl}{\underset{|}{C}}HCOOH$ $K_a = 1.39 \times 10^{-3}$
3. $ClCH_2CH_2CH_2COOH$ $K_a = 2.96 \times 10^{-5}$
4. $CH_3\underset{Cl}{\underset{|}{C}}HCH_2COOH$ $K_a = 8.9 \times 10^{-5}$

b. How does the presence of an electronegative substituent such as Cl affect the acidity of a carboxylic acid?
c. How does the location of the substituent affect the acidity of a carboxylic acid?
d. Why does the electronegative substituent affect the acidity of the carboxylic acid?

29. Explain the difference in the pK_a values of the following compounds:

CH_3COOH	ICH_2COOH	$BrCH_2COOH$	$ClCH_2COOH$	FCH_2COOH
$pK_a = 4.76$	$pK_a = 3.15$	$pK_a = 2.86$	$pK_a = 2.81$	$pK_a = 2.66$

30. a. List the following alcohols in order of decreasing acidity:

CCl_3CH_2OH	CH_2ClCH_2OH	$CHCl_2CH_2OH$
$K_a = 5.75 \times 10^{-13}$	$K_a = 1.29 \times 10^{-13}$	$K_a = 4.90 \times 10^{-13}$

b. Explain their relative acidities.

31. Ethyne has a pK_a value of 25, water has a pK_a value of 15.7, and ammonia (NH_3) has a pK_a value of 36. Draw the equation, showing equilibrium arrows that indicate whether reactants or products are favored, for the reaction of ethyne with

a. HO^-
b. $^-NH_2$
c. Which would be a better base to use if you wanted to remove a proton from ethyne, HO^- or $^-NH_2$?

32. For each of the following pairs of reactions, indicate which one has the more favorable equilibrium constant (that is, which one most favors products):

a. $CH_3CH_2OH + NH_3 \rightleftharpoons CH_3CH_2O^- + {}^+NH_4$
or
$CH_3OH + NH_3 \rightleftharpoons CH_3O^- + {}^+NH_4$

b. $CH_3CH_2OH + NH_3 \rightleftharpoons CH_3CH_2O^- + {}^+NH_4$
or
$CH_3CH_2OH + CH_3NH_2 \rightleftharpoons CH_3CH_2O^- + CH_3\overset{+}{N}H_3$

33. Carbonic acid has a pK_a of 6.1 at physiological temperature. Is the carbonic acid/bicarbonate buffer system that maintains the pH of the blood at 7.3 better at neutralizing excess acid or excess base?

34. Water and diethyl ether are immiscible liquids. Charged compounds dissolve in water, and uncharged compounds dissolve in ether (Section 3.7). $C_6H_{11}COOH$ has a pK_a of 4.8 and $C_6H_{11}\overset{+}{N}H_3$ has a pK_a of 10.7.

a. What pH would you make the water layer in order to cause both compounds to dissolve in it?
b. What pH would you make the water layer in order to cause the acid to dissolve in the water layer and the amine to dissolve in the ether layer?
c. What pH would you make the water layer in order to cause the acid to dissolve in the ether layer and the amine to dissolve in the water layer?

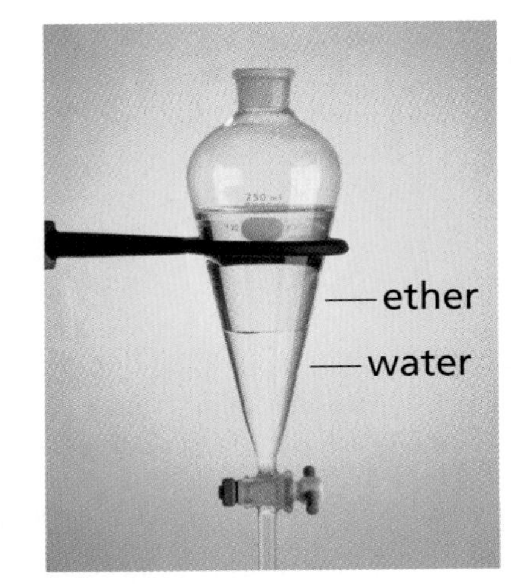

35. How could you separate a mixture of the following compounds? The reagents available to you are water, ether, 1.0 M HCl, and 1.0 M NaOH. (*Hint:* See Problem 34.)

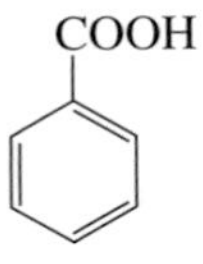

COOH (benzene ring)	$^+NH_3Cl^-$ (benzene ring)	OH (benzene ring)	Cl (benzene ring)	$^+NH_3Cl^-$ (cyclohexane ring)
$pK_a = 4.17$	$pK_a = 4.60$	$pK_a = 9.95$		$pK_a = 10.66$

CHAPTER 3

An Introduction to Organic Compounds

Nomenclature, Physical Properties, and Representation of Structure

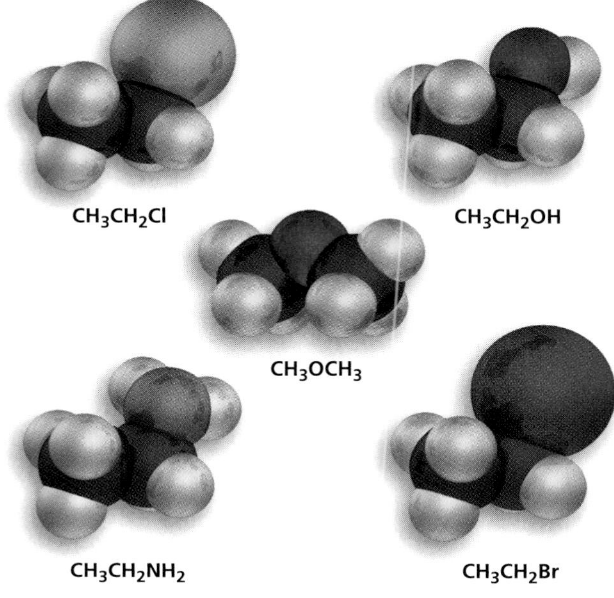

If we are going to talk about organic compounds, we need to know how to name them. First, we will learn how *alkanes* are named because their names form the basis for the names of almost all organic compounds. **Alkanes** are composed of only carbon atoms and hydrogen atoms and contain *only single bonds*. Compounds that contain only carbon and hydrogen are called **hydrocarbons**. Thus, alkanes are hydrocarbons.

Alkanes in which the carbons form a continuous chain with no branches are called **straight-chain alkanes**. The names of several straight-chain alkanes are given in Table 3.1.

Table 3.1 Nomenclature and Physical Properties of Some Straight-Chain Alkanes

Number of carbons	Molecular formula	Name	Condensed structure	Boiling point (°C)	Melting point (°C)	Density[a] (g/mL)
1	CH_4	methane	CH_4	−167.7	−182.5	
2	C_2H_6	ethane	CH_3CH_3	−88.6	−183.3	
3	C_3H_8	propane	$CH_3CH_2CH_3$	−42.1	−187.7	0.5005
4	C_4H_{10}	butane	$CH_3CH_2CH_2CH_3$	−0.5	−138.3	0.5787
5	C_5H_{12}	pentane	$CH_3(CH_2)_3CH_3$	36.1	−129.8	0.5572
6	C_6H_{14}	hexane	$CH_3(CH_2)_4CH_3$	68.7	−95.3	0.6603
7	C_7H_{16}	heptane	$CH_3(CH_2)_5CH_3$	98.4	−90.6	0.6837
8	C_8H_{18}	octane	$CH_3(CH_2)_6CH_3$	125.7	−56.8	0.7026
9	C_9H_{20}	nonane	$CH_3(CH_2)_7CH_3$	150.8	−53.5	0.7177
10	$C_{10}H_{22}$	decane	$CH_3(CH_2)_8CH_3$	174.0	−29.7	0.7299

[a]Density is temperature dependent. The densities given are those determined at 20 °C.

If you look at the relative numbers of carbon and hydrogen atoms in the alkanes listed in Table 3.1, you will see that the general molecular formula for an alkane is C_nH_{2n+2}, where n is any integer. So, if an alkane has one carbon atom, it must have four hydrogen atoms; if it has two carbon atoms, it must have six hydrogen atoms.

We have seen that carbon forms four covalent bonds and hydrogen forms only one covalent bond (Section 1.4). This means that there is only one possible structure for an alkane with molecular formula CH_4 (methane) and only one structure for an alkane with molecular formula C_2H_6 (ethane). We examined the structures of these two compounds in Section 1.7. There is also only one possible structure for an alkane with molecular formula C_3H_8 (propane).

name	Kekulé structure	condensed structure	ball-and-stick model
methane		CH_4	
ethane		CH_3CH_3	
propane		$CH_3CH_2CH_3$	
butane		$CH_3CH_2CH_2CH_3$	

```
methane:
      H
      |
   H—C—H
      |
      H

ethane:
      H  H
      |  |
   H—C—C—H
      |  |
      H  H

propane:
      H  H  H
      |  |  |
   H—C—C—C—H
      |  |  |
      H  H  H

butane:
      H  H  H  H
      |  |  |  |
   H—C—C—C—C—H
      |  |  |  |
      H  H  H  H
```

3-D Molecules: Methane; Ethane; Propane; Butane

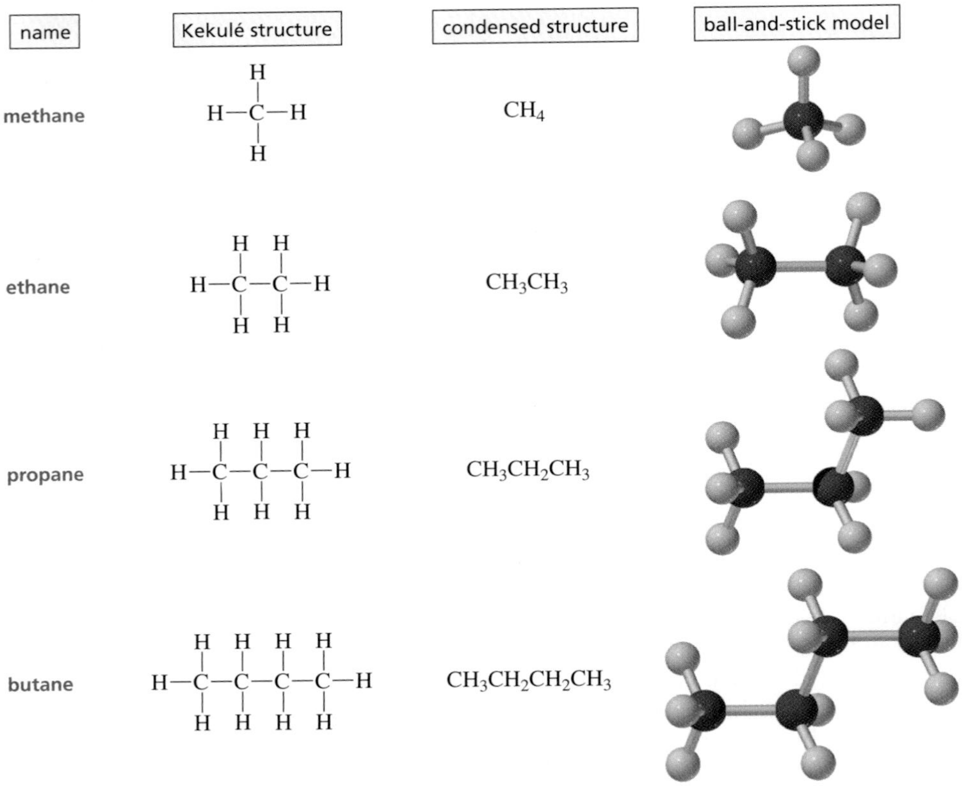

As the number of carbons in an alkane increases beyond three, the number of possible structures increases. There are two possible structures for an alkane with molecular formula C_4H_{10}. In addition to butane—a straight-chain alkane—there is a branched butane called isobutane. Both of these structures fulfill the requirement that each carbon forms four bonds and each hydrogen forms only one bond.

Compounds such as butane and isobutane that have the same molecular formula but differ in the order in which the atoms are connected are called **constitutional isomers**—their molecules have different constitutions. In fact, isobutane got its name because it is an "iso"mer of butane. The structural unit consisting of a carbon bonded to a hydrogen and two CH_3 groups—that occurs in isobutane—has come to be called "iso." Thus, the name isobutane tells you that the compound is a four-carbon alkane with an iso structural unit.

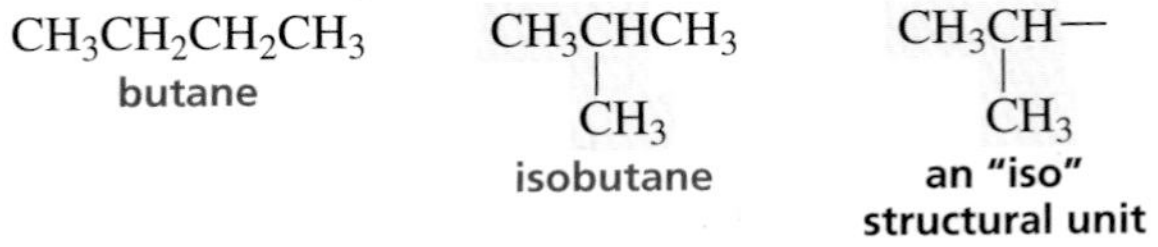

```
CH3CH2CH2CH3      CH3CHCH3        CH3CH—
                     |               |
                    CH3             CH3
```

$CH_3CH_2CH_2CH_3$ — butane

CH_3CHCH_3 with CH_3 on the middle carbon — isobutane

$CH_3CH—$ with CH_3 on the CH carbon — an "iso" structural unit

There are three alkanes with molecular formula C_5H_{12}. You have already learned to name two of them. Pentane is the straight-chain alkane. Isopentane, as its name indicates, has an iso structural unit and five carbon atoms. We cannot name the other branched-chain alkane without defining a name for a new structural unit. (For now, ignore the names written in blue.)

$CH_3CH_2CH_2CH_2CH_3$
pentane

$CH_3CH(CH_3)CH_2CH_3$
isopentane

$CH_3C(CH_3)_2CH_3$
2,2-dimethylpropane

There are five constitutional isomers with molecular formula C_6H_{14}. Again, we are able to name only two of them, unless we define new structural units.

	$CH_3CH_2CH_2CH_2CH_2CH_3$	$CH_3CH(CH_3)CH_2CH_2CH_3$	$CH_3C(CH_3)_2CH_2CH_3$
common name:	hexane	isohexane	
systematic name:	hexane	2-methylpentane	2,2-dimethylbutane

$CH_3CH_2CH(CH_3)CH_2CH_3$
3-methylpentane

$CH_3CH(CH_3)-CH(CH_3)CH_3$
2,3-dimethylbutane

The number of constitutional isomers increases rapidly as the number of carbons in an alkane increases. For example, there are 75 alkanes with molecular formula $C_{10}H_{22}$ and 4347 alkanes with molecular formula $C_{15}H_{32}$. To avoid having to memorize the names of thousands of structural units, chemists have devised rules for creating systematic names that describe the compound's structure. That way, only the rules have to be learned. Because the name describes the structure, these rules also make it possible to deduce the structure of a compound from its name.

FOSSIL FUELS

Alkanes are widespread both on Earth and on other planets. The atmospheres of Jupiter, Saturn, Uranus, and Neptune contain large quantities of methane (CH_4), the smallest alkane, an odorless and flammable gas. In fact, the blue colors of Uranus and Neptune are the result of methane in their atmospheres. Alkanes on Earth are found in natural gas and petroleum, which are formed by the decomposition of plant and animal material that has been buried for long periods in the Earth's crust, where oxygen is scarce. Natural gas and petroleum, therefore, are known as *fossil fuels.*

Natural gas is approximately 75% methane. The remaining 25% is composed of other small alkanes such as ethane, propane, and butane. In the 1950s, natural gas replaced coal as the main energy source for domestic and industrial heating in the United States.

Petroleum is a complex mixture of alkanes that can be separated into fractions by distillation. The fraction that boils off at the lowest temperature (hydrocarbons containing three and four carbons) is a gas that can be liquefied under pressure. This gas is used as a fuel for cigarette lighters, camp stoves, and barbecues. The fraction that boils at somewhat higher temperatures (hydrocarbons containing 5 to 11 carbons) is gasoline; the next fraction (9 to 16 carbons) includes kerosene and jet fuel. The fraction with 15 to 25 carbons is used for heating oil and diesel oil, and the highest boiling fraction is used for lubricants and greases. After distillation, a nonvolatile residue called asphalt or tar is left behind.

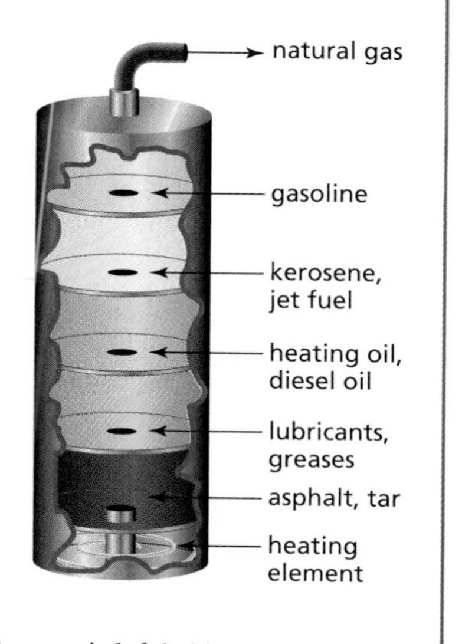

This method of nomenclature is called **systematic nomenclature**. It is also called **IUPAC nomenclature** because it was designed by a commission of the International Union of Pure and Applied Chemistry (abbreviated IUPAC and pronounced "eye-you-pack"). Names such as isobutane—nonsystematic names—are called **common names** and are shown in red in this text. The systematic (IUPAC) names are shown in blue. Before we can understand how a systematic name for an alkane is constructed, we must learn how to name alkyl substituents.

A PROBLEMATIC ENERGY SOURCE

Modern society faces three major problems as a consequence of our dependence on fossil fuels for energy. First, these fuels are a nonrenewable resource and the world's supply is continually decreasing. Second, a group of Middle Eastern and South American countries controls a large portion of the world's supply of petroleum. These countries have formed a cartel known as the *Organization of Petroleum Exporting Countries (OPEC)*, which controls both the supply and the price of crude oil. Political instability in any OPEC country can seriously affect the world oil supply. Third, burning fossil fuels increases the concentrations of CO_2 in the atmosphere; burning coal increases the concentration of both CO_2 and SO_2. Scientists have established experimentally that atmospheric SO_2 causes "acid rain," which represents a threat to the Earth's plants and, therefore, to our food and oxygen supplies (Section 2.2).

Since 1958, the concentration of atmospheric CO_2 at Mauna Loa, Hawaii has been measured periodically. The concentration has increased 20% since the first measurements were taken, causing scientists to predict an increase in the Earth's temperature as a result of the absorption of infrared radiation by CO_2 (the *greenhouse effect*). A steady increase in the temperature of the Earth would have devastating consequences, including the formation of new deserts, massive crop failure, decreasing polar ice sheets, and the melting of glaciers with a concomitant rise in sea level. Clearly, what we need is a renewable, nonpolitical, nonpolluting, and economically affordable source of energy.

3.1 HOW ALKYL SUBSTITUENTS ARE NAMED

Removing a hydrogen from an alkane results in an **alkyl substituent** (or an alkyl group). Alkyl substituents are named by replacing the "ane" ending of the alkane with "yl." The letter "R" is used to indicate any alkyl group.

$CH_3—$ a methyl group

$CH_3CH_2—$ an ethyl group

$CH_3CH_2CH_2—$ a propyl group

$CH_3CH_2CH_2CH_2—$ a butyl group

$CH_3CH_2CH_2CH_2CH_2—$ a pentyl group

R— any alkyl group

If a hydrogen of an alkane is replaced by an OH, the compound becomes an **alcohol**; if it is replaced by an NH_2, the compound becomes an **amine**; if it is replaced by a halogen, the compound becomes an **alkyl halide**; and if it is replaced by an OR, the compound becomes an ether.

methyl alcohol

R—OH an alcohol

$R—NH_2$ an amine

R—X an alkyl halide (X = F, Cl, Br, or I)

R—O—R an ether

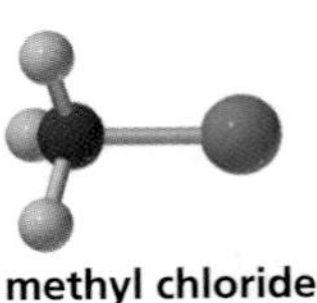
methyl chloride

The alkyl group name followed by the name of the class of the compound (alcohol, amine, and so on) yields the common name of the compound. The two alkyl groups in ethers are stated in alphabetical order. The following examples show how alkyl group names are used to build common names:

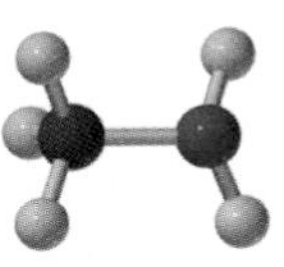
methylamine

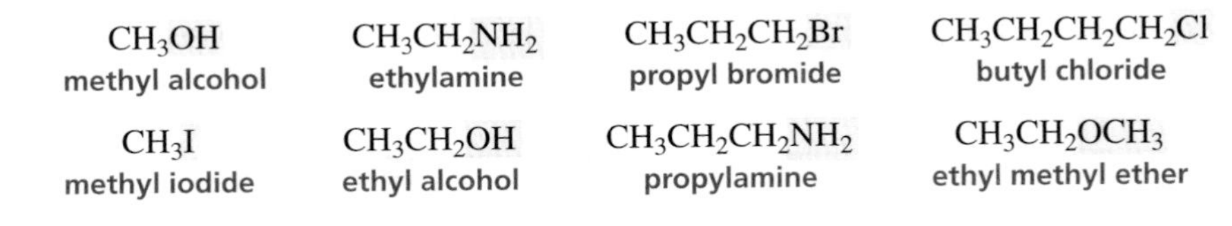
CH_3OH methyl alcohol

$CH_3CH_2NH_2$ ethylamine

$CH_3CH_2CH_2Br$ propyl bromide

$CH_3CH_2CH_2CH_2Cl$ butyl chloride

CH_3I methyl iodide

CH_3CH_2OH ethyl alcohol

$CH_3CH_2CH_2NH_2$ propylamine

$CH_3CH_2OCH_3$ ethyl methyl ether

Notice that there is a space between the name of the alkyl group and the name of the class of compound, except in the case of amines where the entire name is written as one word.

PROBLEM 1 ♦

Name the following:

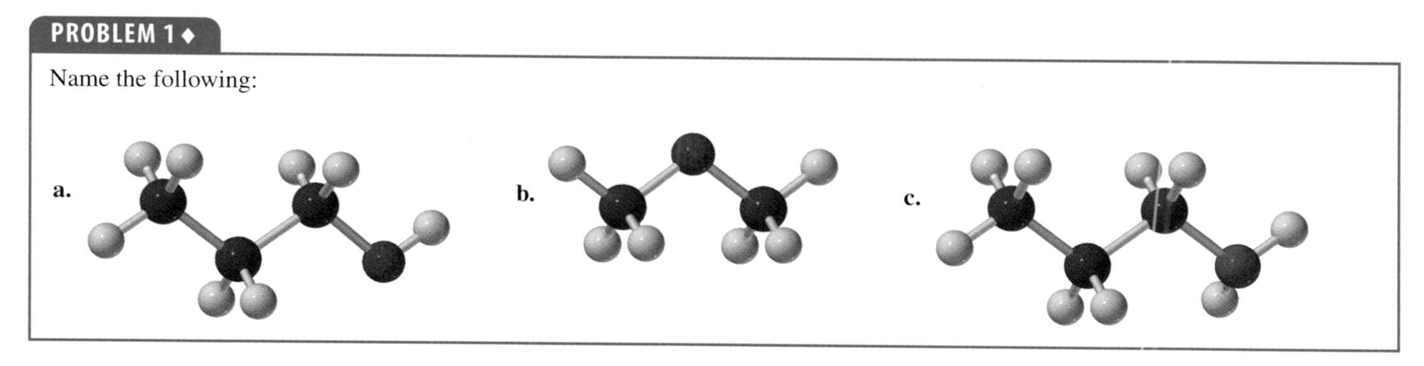

Two alkyl groups—the propyl group and the isopropyl group—have three carbons. A propyl group is obtained when a hydrogen is removed from a *primary carbon* of propane. A **primary carbon** is a carbon bonded to only one other carbon. An isopropyl group is obtained when a hydrogen is removed from the *secondary carbon* of propane. A **secondary carbon** is a carbon bonded to two other carbons. Notice that an isopropyl group, as its name indicates, has its three carbons arranged as an iso structural unit.

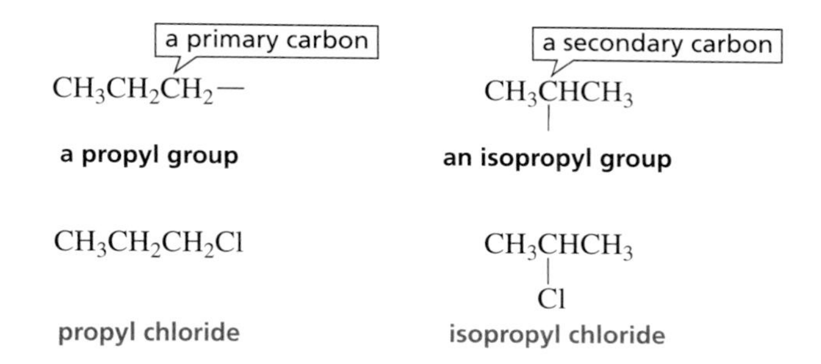

Molecular structures can be drawn in different ways. Isopropyl chloride, for example, is drawn here in two ways. Both representations depict the same compound, although, at first glance, the two-dimensional representations appear to be different: the methyl groups are placed at opposite ends in one structure and at right angles in the other. The structures are identical, however, because carbon is tetrahedral. The four groups bonded to the central carbon—a hydrogen, a chlorine, and two methyl groups—point to the corners of a tetrahedron. (If you visit the Molecule Gallery in Chapter 3 of the website (*www.chemplace.com*), you will be able to actually rotate isopropyl chloride to prove that the two molecules are identical.)

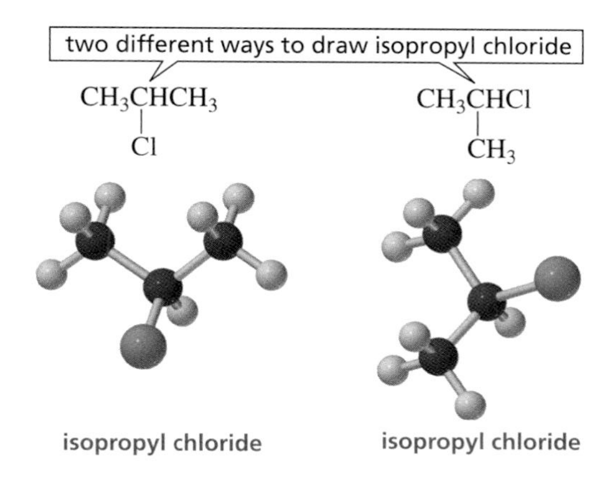

Mentally rotate the model on the right clockwise until it looks the same as the model on the left.

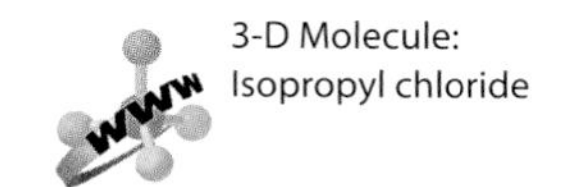

3-D Molecule:
Isopropyl chloride

A primary carbon is bonded to one carbon, a secondary carbon is bonded to two carbons, and a tertiary carbon is bonded to three carbons.

There are four alkyl groups that contain four carbons. Two of them, the butyl and isobutyl groups, have a hydrogen removed from a primary carbon. A *sec*-butyl group has a hydrogen removed from a secondary carbon (*sec*-, often abbreviated *s*-, stands for secondary), and a *tert*-butyl group has a hydrogen removed from a tertiary carbon (*tert*-, sometimes abbreviated *t*-, stands for tertiary). A **tertiary carbon** is a carbon that is bonded to three other carbons. Notice that the isobutyl group is the only group with an iso structural unit.

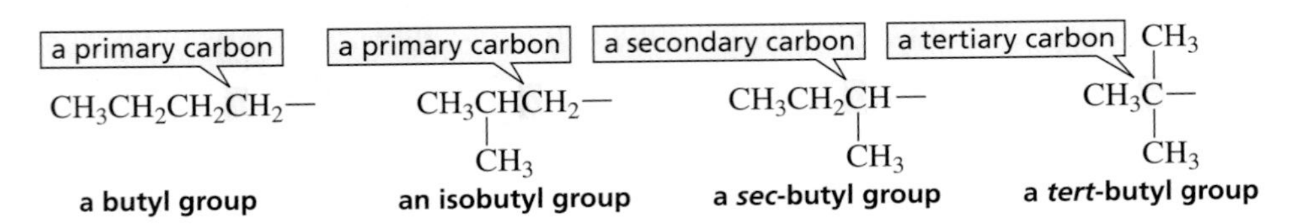

a butyl group **an isobutyl group** **a *sec*-butyl group** **a *tert*-butyl group**

The names of straight-chain alkyl groups sometimes have the prefix "*n*" (for "normal"), to emphasize that its carbons are in an unbranched chain. If the name does not have a prefix such as "*n*" or "iso," we assume that the carbons are in an unbranched chain.

$CH_3CH_2CH_2CH_2Br$ — **butyl bromide** or ***n*-butyl bromide**

$CH_3CH_2CH_2CH_2CH_2F$ — **pentyl fluoride** or ***n*-pentyl fluoride**

A chemical name must specify one compound only. The prefix "*sec*," therefore, can be used only with *sec*-butyl. The name "*sec*-pentyl," for example, cannot be used because pentane has two different secondary carbons, which means that removing a hydrogen from a secondary carbon of pentane produces one of two different alkyl groups, depending on which hydrogen is removed. As a result, "*sec*-pentyl chloride" would specify two different alkyl chlorides, and, therefore, it is not a correct name.

Both alkyl halides have five carbon atoms with a chlorine attached to a secondary carbon, so both compounds would be named *sec*-pentyl chloride.

$CH_3CH(Cl)CH_2CH_2CH_3$ $CH_3CH_2CH(Cl)CH_2CH_3$

If you examine the following structures, you will see that whenever the prefix "iso" is used, the iso structural unit will be at one end of the molecule and any group replacing a hydrogen will be at the other end:

$CH_3CH(CH_3)CH_2CH_2OH$ — **isopentyl alcohol**

$CH_3CH(CH_3)CH_2CH_2CH_2Cl$ — **isohexyl chloride**

$CH_3CH(CH_3)CH_2NH_2$ — **isobutylamine**

$CH_3CH(CH_3)CH_2Br$ — **isobutyl bromide**

$CH_3CH(CH_3)CH_2CH_2OH$ — **isopentyl alcohol**

$CH_3CH(CH_3)Br$ — **isopropyl bromide**

Tutorials: Common names of alkyl groups

Alkyl group names are used so frequently that you should learn them. Some of the most common alkyl group names are compiled in Table 3.2 for your convenience.

Table 3.2 Names of Some Alkyl Groups

methyl	CH_3-	isobutyl	$CH_3\underset{\displaystyle CH_3}{\underset{\vert}{C}}HCH_2-$	pentyl	$CH_3CH_2CH_2CH_2CH_2-$
ethyl	CH_3CH_2-			isopentyl	$CH_3\underset{\displaystyle CH_3}{\underset{\vert}{C}}HCH_2CH_2-$
propyl	$CH_3CH_2CH_2-$	*sec*-butyl	$CH_3CH_2\underset{\displaystyle CH_3}{\underset{\vert}{C}}H-$		
isopropyl	$CH_3\underset{\displaystyle CH_3}{\underset{\vert}{C}}H-$			hexyl	$CH_3CH_2CH_2CH_2CH_2CH_2-$
				isohexyl	$CH_3\underset{\displaystyle CH_3}{\underset{\vert}{C}}HCH_2CH_2CH_2-$
butyl	$CH_3CH_2CH_2CH_2-$	*tert*-butyl	$CH_3\overset{\displaystyle CH_3}{\overset{\vert}{\underset{\displaystyle CH_3}{\underset{\vert}{C}}}}-$		

PROBLEM 2

Draw the structures and name the four constitutional isomers with molecular formula C_4H_9Br.

PROBLEM 3 ◆

Write a structure for each of the following:

a. isopropyl alcohol
b. isopentyl fluoride
c. ethyl propyl ether
d. *sec*-butyl iodide
e. *tert*-butylamine
f. *n*-octyl bromide

PROBLEM 4 ◆

Name the following:

a. $CH_3OCH_2CH_3$
b. $CH_3OCH_2CH_2CH_3$
c. $CH_3CH_2\underset{\displaystyle CH_3}{\underset{\vert}{C}}HNH_2$
d. $CH_3CH_2CH_2CH_2OH$
e. $CH_3\underset{\displaystyle CH_3}{\underset{\vert}{C}}HCH_2Br$
f. $CH_3CH_2\underset{\displaystyle CH_3}{\underset{\vert}{C}}HCl$

PROBLEM 5 ◆

Draw the structure and give the systematic name of a compound with molecular formula C_5H_{12} that has

a. one tertiary carbon.
b. three secondary carbons.
c. no secondary or tertiary carbons.

3.2 THE NOMENCLATURE OF ALKANES

The systematic name of an alkane is obtained using the following rules:

1. Determine the number of carbons in the *longest continuous carbon chain*. This chain is called the **parent hydrocarbon**. The name that indicates the number of carbons in the parent hydrocarbon becomes the alkane's "last name." For example, a parent hydrocarbon with eight carbons would be called *octane*. The longest continuous chain is not always a straight chain; sometimes you have to "turn a corner" to obtain the longest continuous chain.

First, determine the number of carbons in the longest continuous chain.

$\overset{8}{C}H_3\overset{7}{C}H_2\overset{6}{C}H_2\overset{5}{C}H_2\overset{4}{C}H(CH_3)\overset{3}{C}H_2\overset{2}{C}H_2\overset{1}{C}H_3$ — 4-methyloctane

$\overset{8}{C}H_3\overset{7}{C}H_2\overset{6}{C}H_2\overset{5}{C}H_2\overset{4}{C}H(\overset{3}{C}H_2\overset{2}{C}H_2\overset{1}{C}H_3)CH_2CH_3$ — 4-ethyloctane

two different alkanes with an eight-carbon parent hydrocarbon

2. The name of any alkyl substituent that hangs off the parent hydrocarbon is placed in front of the name of the parent hydrocarbon, together with a number to designate the carbon to which the alkyl substituent is attached. The carbons in the parent hydrocarbon are numbered in the direction that gives the substituent as low a number as possible. The substituent's name and the name of the parent hydrocarbon are joined in one word, preceded by a hyphen that connects the number to the substituent's name.

Number the chain in the direction that gives the substituent the lower number.

$\overset{1}{C}H_3\overset{2}{C}H(CH_3)\overset{3}{C}H_2\overset{4}{C}H_2\overset{5}{C}H_3$ — 2-methylpentane

$\overset{6}{C}H_3\overset{5}{C}H_2\overset{4}{C}H_2\overset{3}{C}H(CH_2CH_3)\overset{2}{C}H_2\overset{1}{C}H_3$ — 3-ethylhexane

Notice that only systematic names contain numbers; common names never contain numbers.

Numbers are used only for systematic names, never for common names.

$CH_3CH(CH_3)CH_2CH_2CH_3$

common name: isohexane
systematic name: 2-methylpentane

3. If more than one substituent is attached to the parent hydrocarbon, the chain is numbered in the direction that will produce a name containing the lowest of the possible numbers. The substituents are listed in alphabetical (not numerical) order, with each substituent preceded by the appropriate number. In the following example, the correct name (5-ethyl-3-methyloctane) contains a 3 as its lowest number, whereas the incorrect name (4-ethyl-6-methyloctane) contains a 4 as its lowest number:

Substituents are listed in alphabetical order.

$CH_3CH_2CH(CH_3)CH_2CH(CH_2CH_3)CH_2CH_2CH_3$

5-ethyl-3-methyloctane
not
4-ethyl-6-methyloctane
because 3 < 4

If two or more substituents are the same, the prefixes "di," "tri," and "tetra" are used to indicate how many identical substituents the compound has. The numbers indicating the locations of the identical substituents are listed together, separated by commas. Notice that there must be as many numbers in a name as there are substituents. The prefixes "di," "tri," "tetra," "*sec*," and "*tert*" are ignored in alphabetizing substituent groups, but the prefixes "iso" and "cyclo" ("cyclo" is introduced in Section 2.3) are not ignored.

A number and a word are separated by a hyphen; numbers are separated by a comma.

di, tri, tetra, *sec*, and *tert* are ignored in alphabetizing.

iso and cyclo are not ignored in alphabetizing.

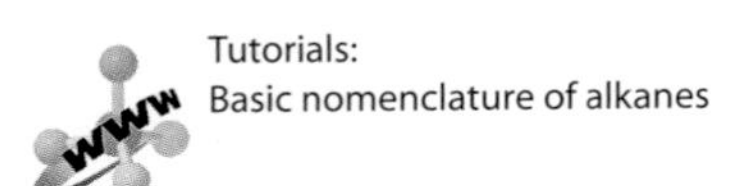

Tutorials: Basic nomenclature of alkanes

$CH_3CH_2CH(CH_3)CH_2CH(CH_3)CH_3$ — 2,4-dimethylhexane

$CH_3CH_2C(CH_2CH_3)(CH_3)CH_2CH_2CH(CH_3)CH_3$ — 5-ethyl-2,5-dimethylheptane

$CH_3CH_2C(CH_2CH_3)_2CH_2CH_2CH(CH_2CH_3)CH(CH_3)CH_2CH_2CH_3$ — 3,3,6-triethyl-7-methyldecane

$CH_3CH_2CH_2CH(CH_3CHCH_3)CH_2CH_2CH(CH_3)CH_3$ — 5-isopropyl-2-methyloctane

4. When numbering in either direction leads to the same lowest number for one of the substituents, the chain is numbered in the direction that gives the lowest possible number to one of the remaining substituents.

```
    CH3
    |
CH3CCH2CHCH3
    |    |
    CH3  CH3
```
2,2,4-trimethylpentane
not
2,4,4-trimethylpentane
because 2 < 4

```
           CH3  CH2CH3
           |    |
CH3CH2CHCHCH2CHCH2CH3
         |
         CH3
```
6-ethyl-3,4-dimethyloctane
not
3-ethyl-5,6-dimethyloctane
because 4 < 5

5. If the same substituent numbers are obtained in both directions, the first group stated receives the lower number.

Only if the same set of numbers is obtained in both directions does the first group stated get the lower number.

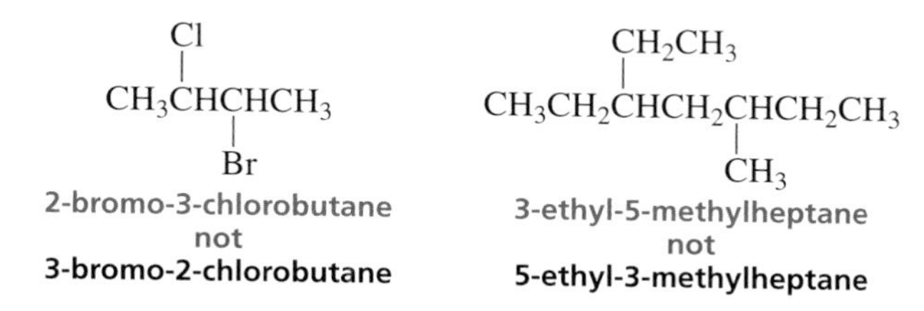

6. If a compound has two or more chains of the same length, the parent hydrocarbon is the chain with the greatest number of substituents.

In the case of two hydrocarbon chains with the same number of carbons, choose the one with the most substituents.

```
      3   4    5   6
CH3CH2CHCH2CH2CH3
       |
      2CHCH3
       |
      1CH3
```
3-ethyl-2-methylhexane **(two substituents)**

```
 1  2   3  4   5   6
CH3CH2CHCH2CH2CH3
       |
       CHCH3
       |
       CH3
```
not
3-isopropylhexane (one substituent)

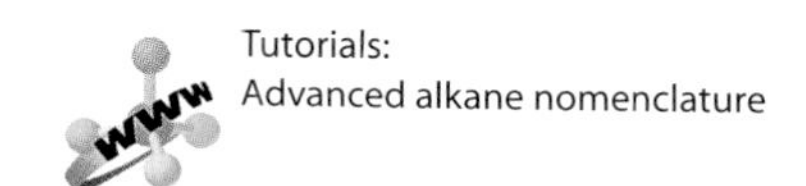

OCTANE NUMBER

When poor fuels are used in an engine, combustion can be initiated before the spark plug fires. A pinging or knocking may then be heard in the running engine. As the quality of the fuel improves, the engine is less likely to knock. The quality of a fuel is indicated by its octane number. Straight-chain hydrocarbons have low octane numbers and make poor fuels. Heptane, for example, with an arbitrarily assigned octane number of 0, causes engines to knock badly. Branched-chain alkanes have more hydrogens on primary carbons. These are the C—H bonds that require the most energy to break and, therefore, make combustion more difficult to initiate, thereby reducing knocking. 2,2,4-Trimethylpentane, for example, does not cause knocking and has arbitrarily been assigned an octane number of 100.

```
CH3CH2CH2CH2CH2CH2CH3
```
heptane
octane number = 0

```
    CH3  CH3
    |    |
CH3CCH2CHCH3
    |
    CH3
```
2,2,4-trimethylpentane
octane number = 100

The octane number of a gasoline is determined by comparing its knocking with the knocking of mixtures of heptane and 2,2,4-trimethylpentane. The octane number given to the gasoline corresponds to the percent of 2,2,4-trimethylpentane in the matching mixture. Thus, a gasoline with an octane rating of 91 has the same "knocking" property as a mixture of 91% 2,2,4-trimethylpentane and 9% heptane. The term *octane number* originated from the fact that 2,2,4-trimethylpentane contains eight carbons.

Two methods are used to determine the octane number. The Research Octane Number (RON) is obtained using a test engine with a variable compression ratio; the Motor Octane Number (MON), which is about 8–10 points less than the RON, uses a similar test engine but with a preheated fuel mixture and a higher engine speed. In Europe and Australia, the octane number on the gas pump is the RON; in Canada and the United States, it is the average of the RON and MON. Therefore, the octane number of a gasoline in the United States would be 4–5 points lower than the same gasoline in Europe.

PROBLEM 6 ♦

Draw the structure for each of the following:

a. 2,3-dimethylhexane
b. 4-isopropyl-2,4,5-trimethylheptane
c. 2,2-dimethyl-4-propyloctane
d. 4-isobutyl-2,5-dimethyloctane

PROBLEM 7 *SOLVED*

a. Draw the 18 constitutional isomers with molecular formula C_8H_{18}.
b. Give each isomer its systematic name.
c. Which isomers contain an isopropyl group?
d. Which isomers contain a *sec*-butyl group?
e. Which isomers contain a *tert*-butyl group?

Solution to 7a Start with the isomer with an eight-carbon continuous chain. Then draw isomers with a seven-carbon continuous chain plus one methyl group. Next, draw isomers with a six-carbon continuous chain plus two methyl groups or one ethyl group. Then draw isomers with a five-carbon continuous chain plus three methyl groups or one methyl group and one ethyl group. Finally, draw a four-carbon continuous chain with four methyl groups. (You will be able to tell whether you have drawn duplicate structures by your answers to 7b, because if two structures have the same systematic name, they represent the same compound.)

PROBLEM 8 ♦

Give the systematic name for each of the following:

a. 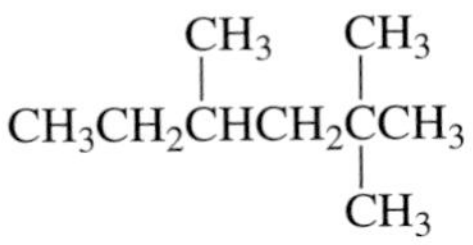
CH_3 (on C-3) CH_3 (on C-5)
$CH_3CH_2CHCH_2CCH_3$
CH_3 (on C-5)

b. $CH_3CH_2C(CH_3)_3$

c. $CH_3CH_2C(CH_2CH_3)_2CH_2CH_2CH_3$

d. 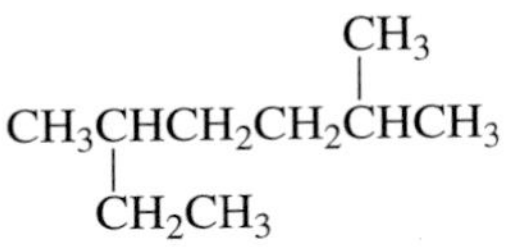
CH_3 (on C-6)
$CH_3CHCH_2CH_2CHCH_3$
CH_2CH_3 (on C-2)

e. 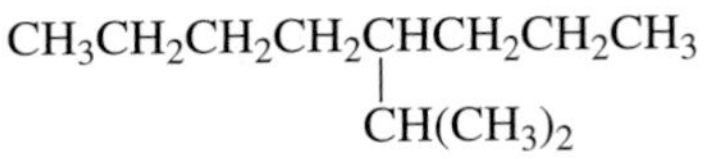
$CH_3CH_2CH_2CH_2CHCH_2CH_2CH_3$
$CH(CH_3)_2$ (on C-5)

f. $CH_3C(CH_3)_2CH(CH_3)CH(CH_2CH_3)_2$

3.3 THE NOMENCLATURE OF CYCLOALKANES • SKELETAL STRUCTURES

Cycloalkanes are alkanes with their carbon atoms arranged in a ring. Because of the ring, a cycloalkane has two fewer hydrogens than a noncyclic alkane with the same number of carbons. This means that the general molecular formula for a cycloalkane is C_nH_{2n}. Cycloalkanes are named by adding the prefix "cyclo" to the alkane name that signifies the number of carbon atoms in the ring.

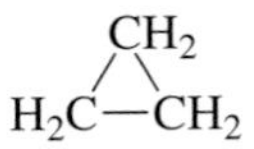

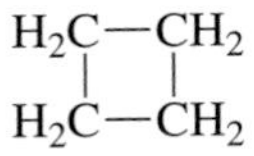

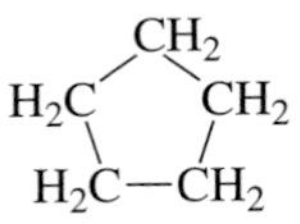

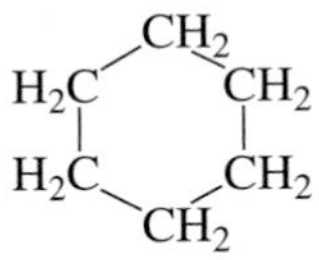

Cycloalkanes are almost always written as **skeletal structures**. Skeletal structures show the carbon–carbon bonds as lines, but do not show the carbons or the hydrogens bonded to carbons. Atoms other than carbon are shown, and hydrogens bonded to atoms other than carbon are shown. Each vertex in a skeletal structure represents a carbon, and each carbon is understood to be bonded to the appropriate number of hydrogens to give the carbon four bonds.

cyclopropane cyclobutane cyclopentane cyclohexane

Noncyclic molecules can also be represented by skeletal structures. In skeletal structures of noncyclic molecules, the carbon chains are represented by zigzag lines. Again, each vertex represents a carbon, and carbons are assumed to be present where a line begins or ends.

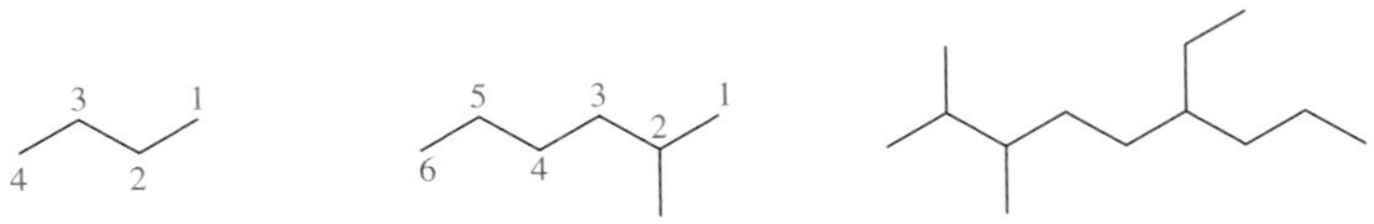

butane 2-methylhexane 6-ethyl-2,3-dimethylnonane

The rules for naming cycloalkanes resemble the rules for naming noncyclic alkanes:

1. In a cycloalkane with an attached alkyl substituent, the ring is the parent hydrocarbon. There is no need to number the position of a single substituent on a ring.

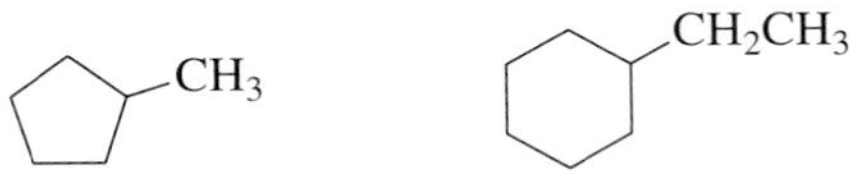

If there is only one substituent on a ring, do not give that substituent a number.

CH_3 CH_2CH_3

methylcyclopentane ethylcyclohexane

2. If the ring has two different substituents, they are stated in *alphabetical order* and the number 1 position is given to the substituent stated first.

CH_3 $CH_2CH_2CH_3$ CH_3 CH_3

1-methyl-2-propylcyclopentane 1,3-dimethylcyclohexane

PROBLEM-SOLVING STRATEGY

Write the number of hydrogens attached to each of the indicated carbons in the following compound:

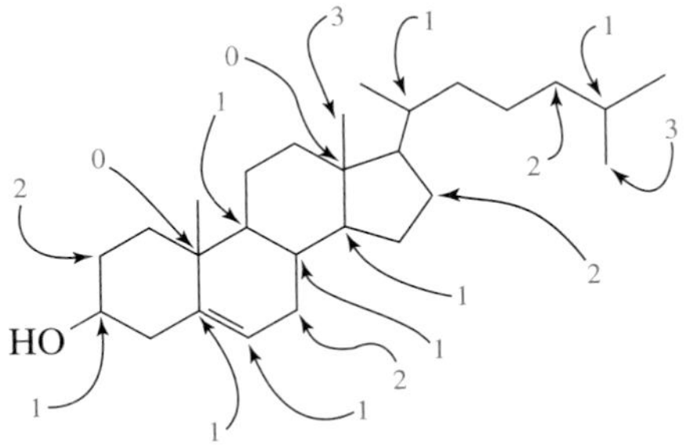

HO

cholesterol

All the carbons in the compound are neutral, so each needs to be bonded to four atoms. Thus if the carbon has only one bond that is showing, it must be attached to three hydrogens that are not shown; if the carbon has two bonds that are showing, it must be attached to two hydrogens that are not shown, and so on.

Now continue on to Problem 9.

PROBLEM 9

Write the number of hydrogens attached to each of the indicated carbons in the following compound:

HO O OH N

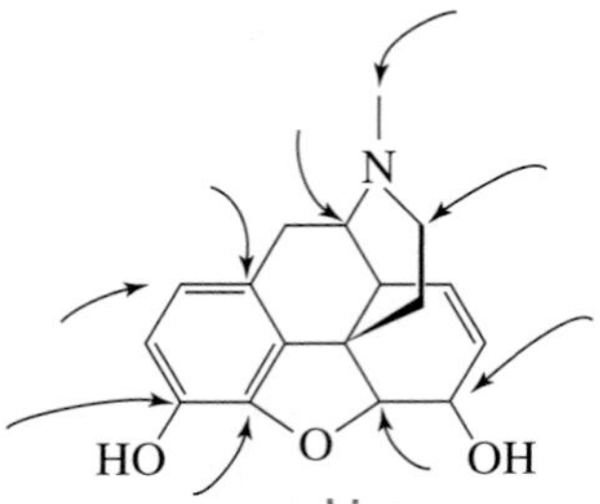

morphine

PROBLEM 10 ◆

Convert the following condensed structures into skeletal structures:

a. $CH_3CH_2CH_2CH_2CH_2CH_2OH$

b. $CH_3CH_2CH(CH_3)CH_2CH(CH_3)CH_2CH_3$

c. $CH_3CH(CH_3)CH_2CH_2CH(Br)CH_3$

d. $CH_3CH_2CH_2CH_2OCH_3$

PROBLEM 11 ◆

The molecular formula for ethyl alcohol (CH_3CH_2OH) is C_2H_6O. What is the molecular formula for the following compounds?

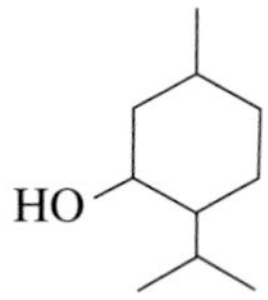

HO

menthol
found in peppermint oil

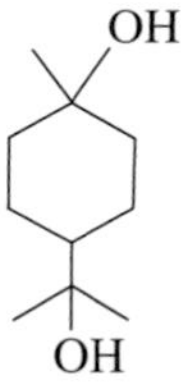

OH

OH

terpin hydrate
a common constituent of cough medicine

PROBLEM 12 ◆

Give the systematic name for each of the following:

a. 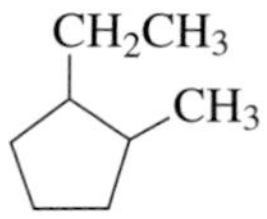 CH_2CH_3 CH_3

b. 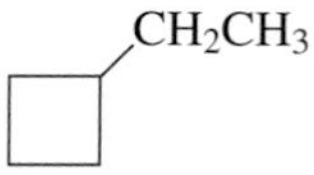 CH_2CH_3

c.

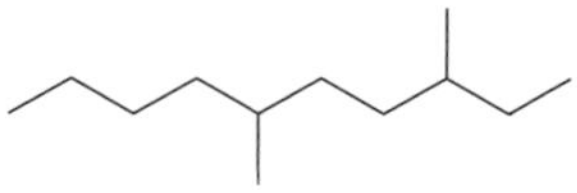

d.

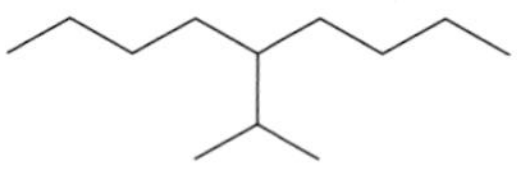

3.4 THE NOMENCLATURE OF ALKYL HALIDES

An alkyl halide is a compound in which a hydrogen of an alkane has been replaced by a halogen. The lone-pair electrons on the halogen are generally not shown unless they are needed to draw your attention to some chemical property of the atom. The common names of alkyl halides consist of the name of the alkyl group, followed by the name of the halogen—with the "ine" ending of the halogen name replaced by "ide" (that is, fluoride, chloride, bromide, iodide).

	CH_3Cl	CH_3CH_2F	CH_3CHICH_3 (I on C2)	$CH_3CH_2CHBrCH_3$ (Br on C2)
common name:	methyl chloride	ethyl fluoride	isopropyl iodide	*sec*-butyl bromide
systematic name:	chloromethane	fluoroethane	2-iodopropane	2-bromobutane

In the IUPAC system, alkyl halides are named as substituted alkanes. The prefixes for the halogens end with "o" (that is, "fluoro," "chloro," "bromo," "iodo"). Notice that although a name must specify only one compound, a compound can have more than one name.

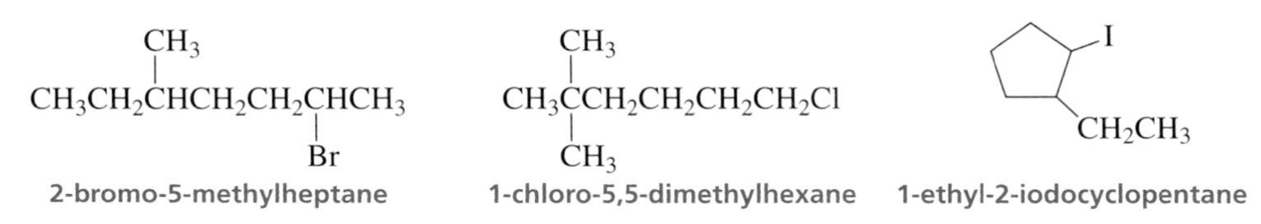

2-bromo-5-methylheptane 1-chloro-5,5-dimethylhexane 1-ethyl-2-iodocyclopentane

CH_3F
methyl fluoride

CH_3Cl
methyl chloride

CH_3Br
methyl bromide

CH_3I
methyl iodide

A compound can have more than one name, but a name must specify only one compound.

PROBLEM-SOLVING STRATEGY

Do the following structures represent the same compound or different compounds?

$CH_3CHCH_2CH_2CH_3$ (Cl on C2) and $CH_3CH_2CH_2CHCl$ (CH_3 on the CHCl carbon)

The easiest way to answer this question is to determine the systematic names of the compounds. If they have the same systematic name, they are identical compounds; if they do not have the same systematic name, they are different compounds. Both structures are named 2-chloropentane; therefore, they represent the same compound.

$\overset{1}{C}H_3\overset{2}{C}H\overset{3}{C}H_2\overset{4}{C}H_2\overset{5}{C}H_3$ (Cl on C2) and $\overset{5}{C}H_3\overset{4}{C}H_2\overset{3}{C}H_2\overset{2}{C}HCl$ (${}^{1}CH_3$ on C2)

2-chloropentane 2-chloropentane

Now continue on to Problem 13.

PROBLEM 13 ◆

Do the following structures represent the same compound or different compounds?

a. $CH_3CH_2CHCH_2CH_2Br$ (CH_3 on C3) and $CH_3CHCH_2CH_2Br$ (CH_2CH_3 on C2)

b. cyclohexane with CH_3 and I on adjacent carbons and cyclohexane with I and CH_3 on the same carbon

PROBLEM 14 ◆

Give two names for each of the following:

a. $CH_3CH_2CHCH_3$ with Cl on C-2

b. $CH_3CHCH_2CH_2CH_2Cl$ with CH_3 on C-2

c. Bromocyclohexane (cyclohexane ring with Br)

d. CH_3CHCH_3 with F on C-2

3.5 THE CLASSIFICATION OF ALKYL HALIDES, ALCOHOLS, AND AMINES

The number of alkyl groups attached to the carbon to which the halogen is bonded determines whether an alkyl halide is primary, secondary, or tertiary.

Alkyl halides are classified as *primary*, *secondary*, or *tertiary*, depending on the carbon to which the halogen is attached. **Primary alkyl halides** have a halogen bonded to a primary carbon, **secondary alkyl halides** have a halogen bonded to a secondary carbon, and **tertiary alkyl halides** have a halogen bonded to a tertiary carbon (Section 3.1).

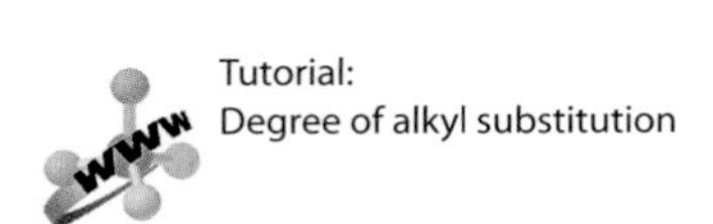

Tutorial: Degree of alkyl substitution

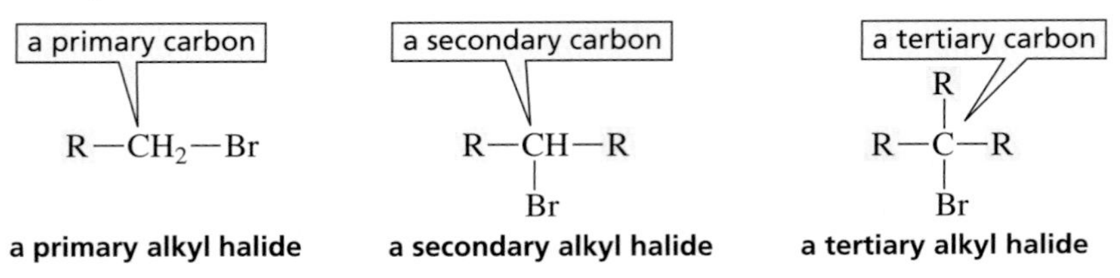

The number of alkyl groups attached to the carbon to which the OH group is bonded determines whether an alcohol is primary, secondary, or tertiary.

3-D Molecules: *n*-Butyl alcohol; *sec*-Butyl alcohol; *tert*-Butyl alcohol

Alcohols are classified in the same way.

$R—CH_2—OH$ — **a primary alcohol**

$R_2CH—OH$ (R—CH—OH with R above) — **a secondary alcohol**

$R_3C—OH$ (R—C—OH with R above and below) — **a tertiary alcohol**

The number of alkyl groups attached to the nitrogen determines whether an amine is primary, secondary, or tertiary.

There are also *primary*, *secondary*, or *tertiary amines*; but in the case of amines, the terms have different meanings. The classification refers to how many alkyl groups are bonded to the nitrogen. **Primary amines** have one alkyl group bonded to the nitrogen, **secondary amines** have two, and **tertiary amines** have three. The common name of an amine consists of the names of all the alkyl groups bonded to the nitrogen, in alphabetical order, followed by "amine."

$R—NH_2$ — **a primary amine**

R—NH with R above — **a secondary amine**

R—N—R with R above — **a tertiary amine**

$CH_3NCH_2CH_2CH_3$ with CH_2CH_3 on N — **ethylmethylpropylamine** **a tertiary amine**

BAD-SMELLING COMPOUNDS

Amines are associated with some of nature's unpleasant odors. Amines with relatively small alkyl groups, for example, have a fishy smell. Thus, fermented shark, a traditional dish in Iceland, smells exactly like triethylamine. The amines putrescine and cadaverine are poisonous compounds formed when amino acids are degraded. Because the body excretes them in the quickest way possible, their odors may be detected in both urine and breath. These compounds are also responsible for the odor of decaying flesh.

$H_2N(CH_2)_4NH_2$ — **putrescine**

$H_2N(CH_2)_5NH_2$ — **cadaverine**

PROBLEM 15 ♦

Tell whether the following are primary, secondary, or tertiary:

a. $CH_3{-}C(CH_3)_2{-}Br$ **b.** $CH_3{-}C(CH_3)_2{-}OH$ **c.** $CH_3{-}C(CH_3)_2{-}NH_2$

PROBLEM 16 ♦

Name the following amines and tell whether they are primary, secondary, or tertiary:

a. $CH_3NHCH_2CH_2CH_3$ **c.** $CH_3CH_2NHCH_2CH_3$

b. $CH_3N(CH_3)CH_3$ **d.** $CH_3N(CH_3)CH_2CH_2CH_2CH_3$

PROBLEM 17 ♦

Draw the structures for **a–c** by substituting a chlorine for a hydrogen of methylcyclohexane:

a. a primary alkyl halide
b. a tertiary alkyl halide
c. three secondary alkyl halides

3.6 THE STRUCTURES OF ALKYL HALIDES, ALCOHOLS, ETHERS, AND AMINES

The classes of compounds we have been looking at in this chapter have structural resemblances to the simpler compounds introduced in Chapter 1. Let's begin by looking at alkyl halides and their resemblance to alkanes. Both classes of compounds have the same geometry; the only difference is that the C—X bond (where X denotes a halogen) of the alkyl halide has replaced a C—H bond of an alkane. The C—X bond of an alkyl halide is formed from the overlap of an sp^3 orbital of carbon with an sp^3 orbital of the halogen (Section 1.13). Fluorine uses a $2sp^3$ orbital, chlorine a $3sp^3$ orbital, bromine a $4sp^3$ orbital, and iodine a $5sp^3$ orbital. Because the electron density of the orbital decreases with increasing volume, the C—X bond becomes longer and weaker as the size of the halogen increases (Table 3.3). Notice that this is the same trend shown by the H—X bond (Table 1.5, page 27).

Now consider the geometry of the oxygen in an alcohol, which is the same as the geometry of the oxygen in water (Section 1.11). In fact, an alcohol molecule can be thought of structurally as a water molecule with an alkyl group in place of one of the hydrogens. The oxygen atom in an alcohol is sp^3 hybridized, as it is in water. One of the four sp^3 orbitals of oxygen overlaps an sp^3 orbital of a carbon, one overlaps the s orbital of a hydrogen, and the other two each contain a lone pair.

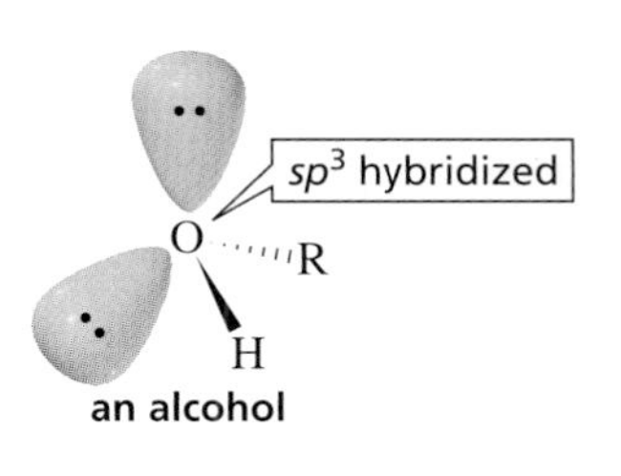

an alcohol

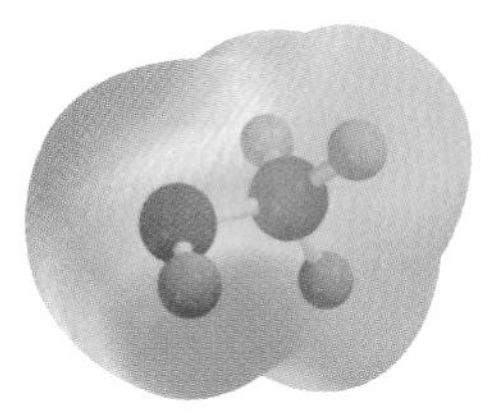
electrostatic potential map for methyl alcohol

Table 3.3 Carbon–Halogen Bond Lengths and Bond Strengths

	Orbital interactions	Bond lengths	Bond strength kcal/mol	Bond strength kJ/mol
$H_3C—F$		1.39 Å	108	451
$H_3C—Cl$		1.78 Å	84	350
$H_3C—Br$		1.93 Å	70	294
$H_3C—I$		2.14 Å	57	239

The oxygen of an ether also has the same geometry as the oxygen in water. An ether molecule can be thought of structurally as a water molecule with alkyl groups in place of both hydrogens.

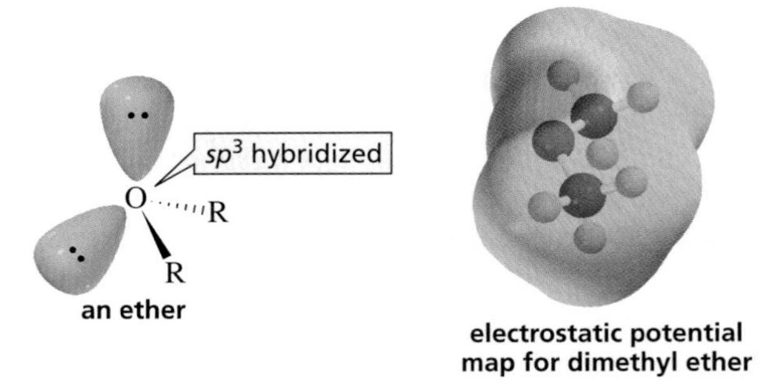

an ether

electrostatic potential map for dimethyl ether

The nitrogen of an amine has the same geometry as the nitrogen in ammonia (Section 1.12). It is sp^3 hybridized as in ammonia, with one, two, or three of the hydrogens replaced by alkyl groups. Remember that the number of hydrogens replaced by alkyl groups determines whether the amine is primary, secondary, or tertiary (Section 3.5).

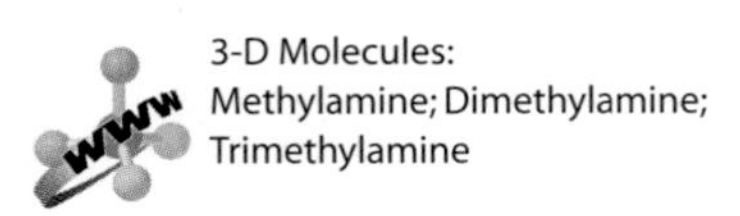

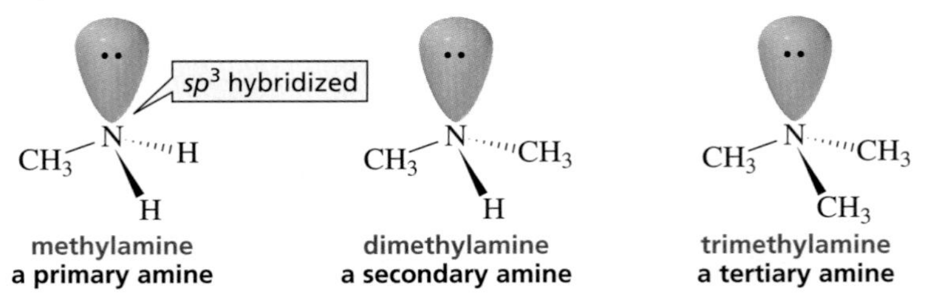

methylamine
a primary amine

dimethylamine
a secondary amine

trimethylamine
a tertiary amine

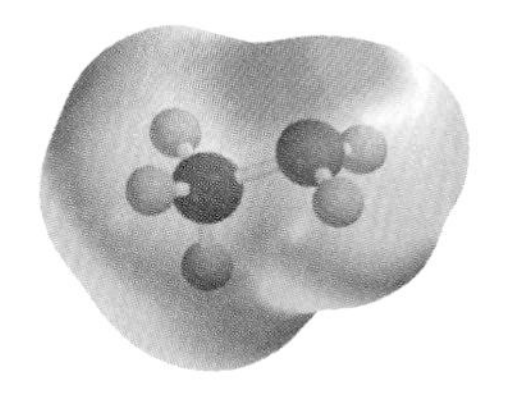
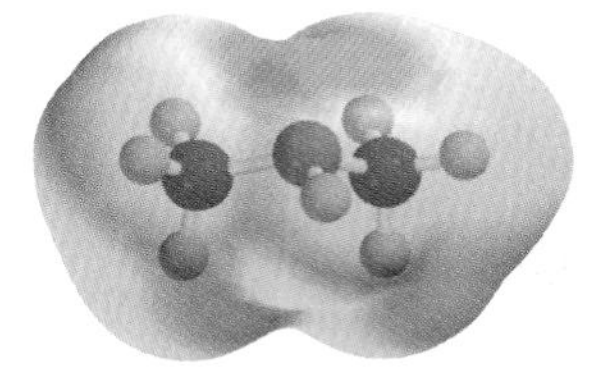
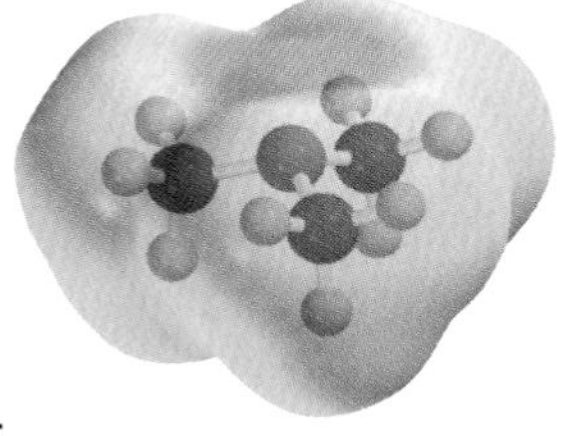

electrostatic potential maps for

methylamine dimethylamine trimethylamine

PROBLEM 18 ◆

Predict the approximate size of the following bond angles. (*Hint*: See Sections 1.11 and 1.12.)

a. the C—O—C bond angle in an ether
b. the C—N—C bond angle in a secondary amine
c. the C—O—H bond angle in an alcohol

3.7 THE PHYSICAL PROPERTIES OF ALKANES, ALKYL HALIDES, ALCOHOLS, ETHERS, AND AMINES

Boiling Points

The **boiling point (bp)** of a compound is the temperature at which the liquid form becomes a gas (vaporizes). In order for a compound to vaporize, the forces that hold the individual molecules close to each other in the liquid must be overcome. This means that the boiling point of a compound depends on the strength of the attractive forces between the individual molecules. If the molecules are held together by strong forces, a lot of energy will be needed to pull the molecules away from each other and the compound will have a high boiling point. In contrast, if the molecules are held together by weak forces, only a small amount of energy will be needed to pull the molecules away from each other and the compound will have a low boiling point.

The attractive forces between alkane molecules are relatively weak. Alkanes contain only carbon and hydrogen atoms, and the electronegativities of carbon and hydrogen are similar. As a result, the bonds in alkanes are nonpolar—there are no significant partial charges on any of the atoms. Therefore, alkanes are neutral (nonpolar) molecules. The nonpolar nature of alkanes gives them their oily feel.

However, it is only the average charge distribution over the alkane molecule that is neutral. The electrons are moving continuously, and at any instant the electron density on one side of the molecule can be slightly greater than that on the other side, causing the molecule to have a temporary dipole. A molecule with a dipole has a negative end and a positive end.

A temporary dipole in one molecule can induce a temporary dipole in a nearby molecule. As a result, the (temporarily) negative side of one molecule ends up adjacent to the (temporarily) positive side of another, as shown in Figure 3.1. Because the dipoles in the molecules are induced, the interactions between the molecules are called **induced-dipole–induced-dipole interactions**. The molecules of an alkane are held together by these induced-dipole–induced-dipole interactions, also known as **van der Waals forces**. Van der Waals forces are the weakest of all the intermolecular attractions.

The magnitude of the van der Waals forces that hold alkane molecules together depends on the area of contact between the molecules. The greater the area of contact, the stronger are the van der Waals forces and the greater the amount of energy needed to overcome them. If you look at the alkanes in Table 3.1 on page 47, you will see that their boiling points increase as their size increases. This relationship holds because

BIOGRAPHY

Johannes Diderik van der Waals (1837–1923) *was a Dutch physicist. He was born in Leiden, the son of a carpenter, and was largely self-taught when he entered the University of Leiden, where he earned a Ph.D. He was a professor of physics at the University of Amsterdam from 1877 to 1907. He won the 1910 Nobel Prize for his research on the gaseous and liquid states of matter.*

▶ **Figure 3.1**
Van der Waals forces are induced-dipole–induced-dipole interactions.

each additional methylene (CH_2) group increases the area of contact between the molecules. The four smallest alkanes have boiling points below room temperature (which is about 25 °C), so they exist as gases at room temperature.

Because the strength of the van der Waals forces depends on the area of contact between the molecules, branching in a compound lowers the compound's boiling point by reducing the area of contact. If you think of pentane, an unbranched alkane, as a cigar and think of 2,2-dimethylpropane as a tennis ball, you can see that branching decreases the area of contact between molecules: two cigars make contact over a greater area than do two tennis balls. Thus, if two alkanes have the same molecular weight, the more highly branched alkane will have a lower boiling point.

$CH_3CH_2CH_2CH_2CH_3$	$CH_3CH(CH_3)CH_2CH_3$	$CH_3C(CH_3)_2CH_3$
pentane bp = 36.1 °C	2-methylbutane bp = 27.9 °C	2,2-dimethylpropane bp = 9.5 °C

PROBLEM 19 ♦

What is the smallest alkane that is a liquid at room temperature?

The boiling points of a series of ethers, alkyl halides, alcohols, or amines also increase with increasing molecular weight because of the increase in van der Waals forces. (See Appendix I.) The boiling points of these compounds, however, are also affected by the polar character of the C—Z bond (where Z denotes N, O, F, Cl, or Br). The C—Z bond is polar because nitrogen, oxygen, and the halogens are more electronegative than the carbon to which they are attached.

$$R{-}\overset{\delta+}{C}{-}\overset{\delta-}{Z} \qquad Z = N, O, F, Cl, \text{ or } Br$$

Molecules with polar bonds are attracted to one another because they can align themselves in such a way that the positive end of one molecule is adjacent to the negative end of another. These attractive forces, called **dipole–dipole interactions**, are stronger than van der Waals forces, but not as strong as ionic or covalent bonds.

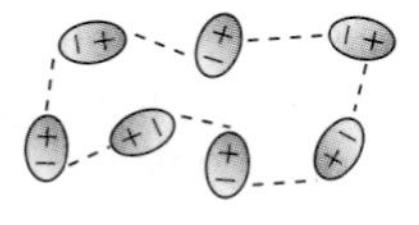

More extensive tables of physical properties can be found in Appendix I.

Ethers generally have higher boiling points than alkanes of comparable molecular weight because both van der Waals forces and dipole–dipole interactions must be overcome for an ether to boil (Table 3.4).

cyclopentane	tetrahydrofuran
bp = 49.3 °C	bp = 65 °C

Table 3.4 Comparative Boiling Points (°C)

Alkanes	Ethers	Alcohols	Amines
$CH_3CH_2CH_3$ **−42.1**	CH_3OCH_3 **−23.7**	CH_3CH_2OH **78**	$CH_3CH_2NH_2$ **16.6**
$CH_3CH_2CH_2CH_3$ **−0.5**	$CH_3OCH_2CH_3$ **10.8**	$CH_3CH_2CH_2OH$ **97.4**	$CH_3CH_2CH_2NH_2$ **47.8**
$CH_3CH_2CH_2CH_2CH_3$ **36.1**	$CH_3CH_2OCH_2CH_3$ **34.5**	$CH_3CH_2CH_2CH_2OH$ **117.3**	$CH_3CH_2CH_2CH_2NH_2$ **77.8**

As the table shows, alcohols have much higher boiling points than alkanes or ethers with similar molecular weights because, in addition to van der Waals forces and the dipole–dipole interactions of the C—O bond, alcohols can form **hydrogen bonds**. A hydrogen bond is a special kind of dipole–dipole interaction that occurs between a hydrogen that is bonded to an oxygen, a nitrogen, or a fluorine and the lone-pair electrons of an oxygen, nitrogen, or fluorine in another molecule.

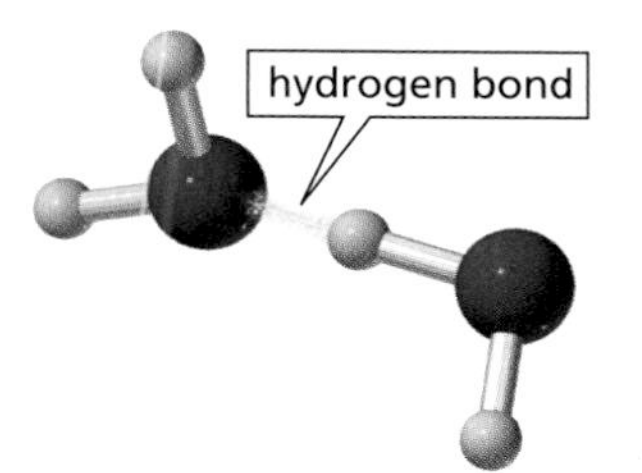

hydrogen bonding in water

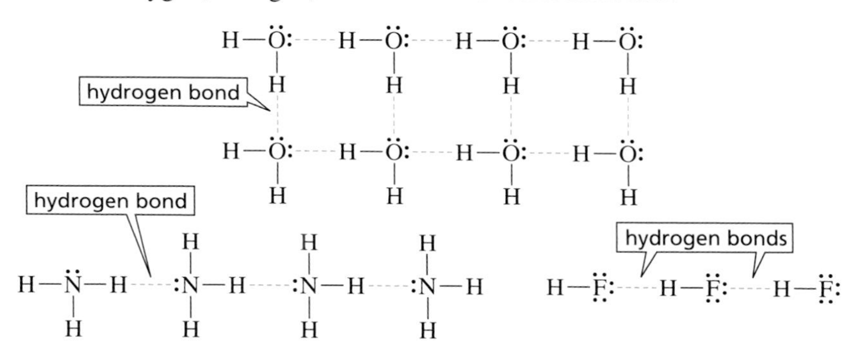

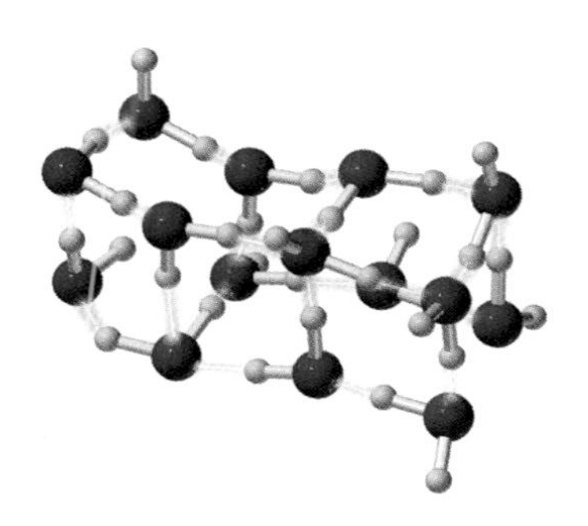

A hydrogen bond is stronger than other dipole–dipole interactions. The extra energy required to break the hydrogen bonds is why alcohols have much higher boiling points than alkanes or ethers with similar molecular weights.

The boiling point of water illustrates the dramatic effect that hydrogen bonding has on boiling points. Water has a molecular weight of 18 and a boiling point of 100 °C. The alkane nearest in size is methane, with a molecular weight of 16. Methane boils at − 167.7 °C.

Primary and secondary amines also form hydrogen bonds, so these amines have higher boiling points than alkanes with similar molecular weights. Nitrogen is not as electronegative as oxygen, however, which means that the hydrogen bonds between amine molecules are weaker than the hydrogen bonds between alcohol molecules. An amine, therefore, has a lower boiling point than an alcohol with a similar molecular weight (Table 3.4).

Because primary amines have two N—H bonds, hydrogen bonding is more significant in primary amines than in secondary amines. Tertiary amines cannot form hydrogen bonds between their own molecules because they do not have a hydrogen attached to the nitrogen. Consequently, when we compare amines with the same molecular weight and similar structures, we find that a primary amine has a higher boiling point than a secondary amine and a secondary amine has a higher boiling point than a tertiary amine.

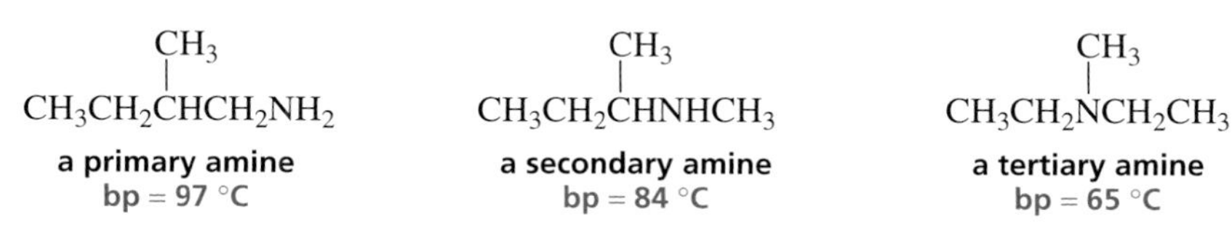

PROBLEM 20 ♦

a. Which is longer, an O—H hydrogen bond or an O—H covalent bond?
b. Which is stronger?

PROBLEM-SOLVING STRATEGY

a. Which of the following compounds will form hydrogen bonds between its molecules?
1. $CH_3CH_2CH_2OH$ **2.** $CH_3CH_2CH_2F$ **3.** $CH_3OCH_2CH_3$

b. Which of these compounds will form hydrogen bonds with a solvent such as ethanol?

In solving this type of question, start by defining the kind of compound that will do what is being asked.

a. A hydrogen bond forms when a hydrogen attached to an O, N, or F of one molecule interacts with a lone pair on an O, N, or F of another molecule. Therefore, a compound that will form hydrogen bonds with itself must have a hydrogen bonded to an O, N, or F. Only compound 1 will be able to form hydrogen bonds with itself.

b. A solvent such as ethanol has a hydrogen attached to an O, so it will be able to form hydrogen bonds with a compound that has a lone pair on an O, N, or F. All three compounds will be able to form hydrogen bonds with ethanol.

Now continue on to Problem 21.

PROBLEM 21 ♦

a. Which of the following will form hydrogen bonds between its molecules?
1. $CH_3CH_2CH_2COOH$
2. $CH_3CH_2N(CH_3)_2$
3. $CH_3CH_2CH_2CH_2Br$
4. $CH_3CH_2CH_2NHCH_3$
5. $CH_3CH_2OCH_2CH_2OH$
6. $CH_3CH_2CH_2CH_2F$

b. Which of the preceding compounds form hydrogen bonds with a solvent such as ethanol?

PROBLEM 22 ♦

List the following compounds in order of decreasing boiling point:

OH HO OH OH OH OH NH$_2$

Both van der Waals forces and dipole–dipole interactions must be overcome for an alkyl halide to boil. Moreover, as the halogen atom increases in size, the size of its electron cloud increases, and the larger the electron cloud, the stronger are the van der Waals interactions. Therefore, an alkyl fluoride has a lower boiling point than an alkyl chloride with the same alkyl group. Similarly, alkyl chlorides have lower boiling points than alkyl bromides, which have lower boiling points than alkyl iodides (Table 3.5).

Table 3.5 Comparative Boiling Points of Alkanes and Alkyl Halides (°C)

	Y				
	H	F	Cl	Br	I
$CH_3—Y$	−161.7	−78.4	−24.2	3.6	42.4
$CH_3CH_2—Y$	−88.6	−37.7	12.3	38.4	72.3
$CH_3CH_2CH_2—Y$	−42.1	−2.5	46.6	71.0	102.5
$CH_3CH_2CH_2CH_2—Y$	−0.5	32.5	78.4	101.6	130.5
$CH_3CH_2CH_2CH_2CH_2—Y$	36.1	62.8	107.8	129.6	157.0

DRUGS BINDING TO THEIR RECEPTORS

Many drugs exert their physiological effects by binding to a specific binding site called a receptor. A drug binds to a receptor using the same kinds of bonding interactions—van der Waals interactions, dipole–dipole interactions, hydrogen bonding—that molecules use to bind to each other. The most important factor in the interaction between a drug and its receptor is a snug fit. Therefore, drugs with similar shapes often have similar physiological effects. Salicylic acid has been used for the relief of fever and arthritic pain since 500 B.C. In 1897, acetylsalicylic acid (aspirin) was found to be a more potent anti-inflammatory agent and less irritating to the stomach; it became commercially available in 1899.

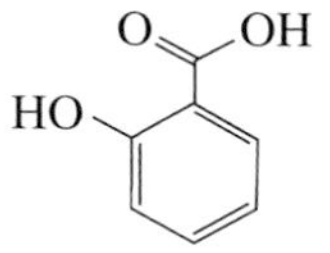

salicylic acid

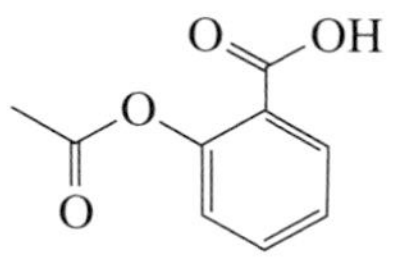

acetylsalicylic acid

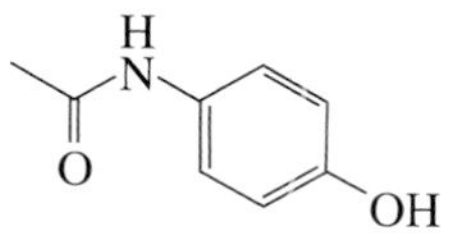

acetaminophen
Tylenol®

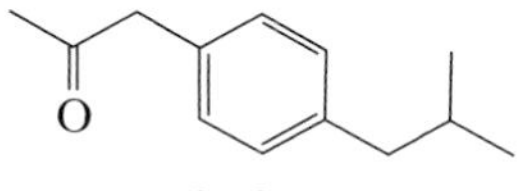

ibufenac

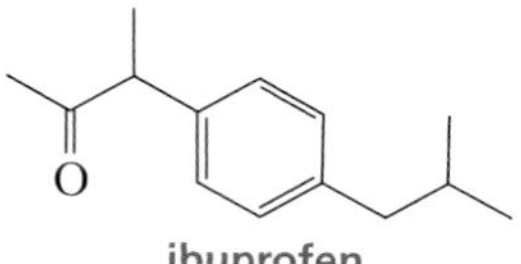

ibuprofen
Advil®

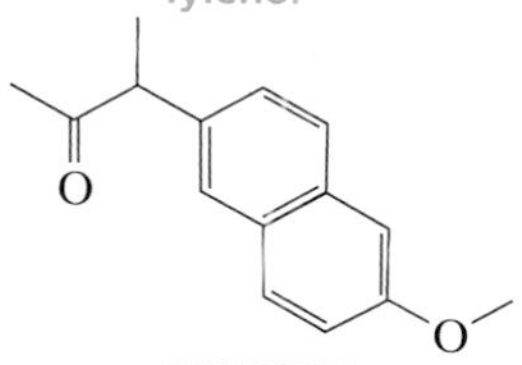

naproxen
Aleve®

Acetaminophen (Tylenol) was introduced in 1955. It became a widely used drug because it causes no gastric irritation. However, its effective dose is not far from its toxic dose. Subsequently, ibufenac emerged; adding a methyl group to ibufenac produced ibuprofen (Advil), a much safer drug. Naproxen (Aleve), which has twice the potency of ibuprofen, was introduced in 1976.

PROBLEM 23

List the compounds in each set in order of decreasing boiling point:

a. $CH_3CH_2CH_2CH_2CH_2CH_2Br$ $CH_3CH_2CH_2CH_2Br$ $CH_3CH_2CH_2CH_2CH_2Br$

b. $CH_3\underset{\displaystyle CH_3}{\underset{|}{C}H}CH_2CH_2CH_2CH_2CH_3$ $\quad$ $CH_3\underset{\displaystyle CH_3}{\underset{|}{\overset{\displaystyle CH_3}{\overset{|}{C}}}}-\underset{\displaystyle CH_3}{\underset{|}{\overset{\displaystyle CH_3}{\overset{|}{C}}}}CH_3$

$CH_3CH_2CH_2CH_2CH_2CH_2CH_2CH_3$ $\quad$ $CH_3CH_2CH_2CH_2CH_2CH_2CH_2CH_2CH_3$

c. $CH_3CH_2CH_2CH_2CH_3$ $CH_3CH_2CH_2CH_2OH$ $CH_3CH_2CH_2CH_2Cl$
$CH_3CH_2CH_2CH_2CH_2OH$

Melting Points

The **melting point (mp)** of a compound is the temperature at which its solid form is converted into a liquid. If you examine the melting points of the alkanes listed in Table 3.1, you will see that they increase (with a few exceptions) as the molecular weight increases (Figure 3.2). The increase in melting point is less regular than the increase in boiling point because *packing* influences the melting point of a compound. **Packing** is a property that determines how well the individual molecules in a solid fit together in a crystal lattice. The tighter the fit, the more energy is required to break the lattice and melt the compound.

PROBLEM 24♦

Use Figure 3.2 to answer this question. Which pack more tightly, the molecules of an alkane with an even number of carbons or the molecules with an odd number of carbons?

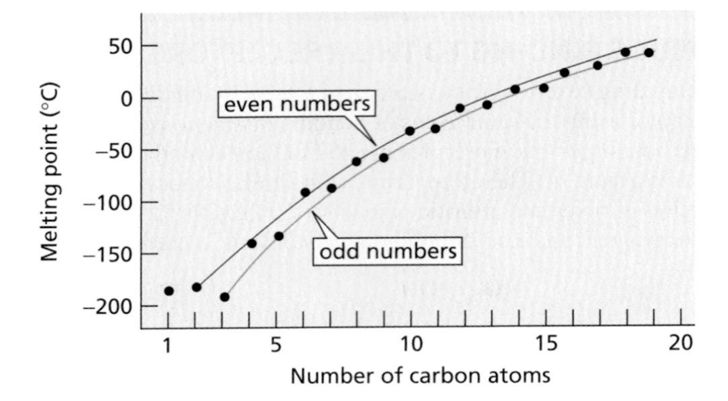

▶ **Figure 3.2**
Melting points of straight-chain alkanes.

Solubility

"Like" dissolves "like."

The general rule that governs **solubility** is "like dissolves like." In other words, *polar compounds dissolve in polar solvents, and nonpolar compounds dissolve in nonpolar solvents*. The reason "polar dissolves polar" is that a polar solvent, such as water, has partial charges that can interact with the partial charges on a polar compound. The negative poles of the solvent molecules surround the positive pole of the polar *solute*, and the positive poles of the solvent molecules surround the negative pole of the polar *solute*. A **solute** is a molecule or an ion dissolved in a solvent. Clustering of the solvent molecules around the solute molecules separates solute molecules from each other, which is what makes them dissolve. The interaction between solvent molecules and solute molecules is called **solvation**.

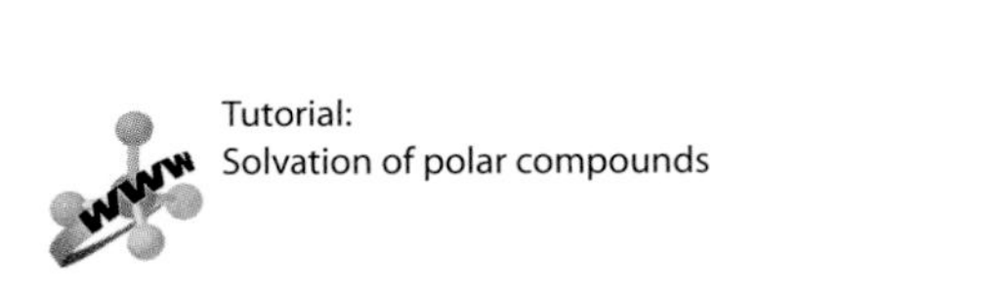

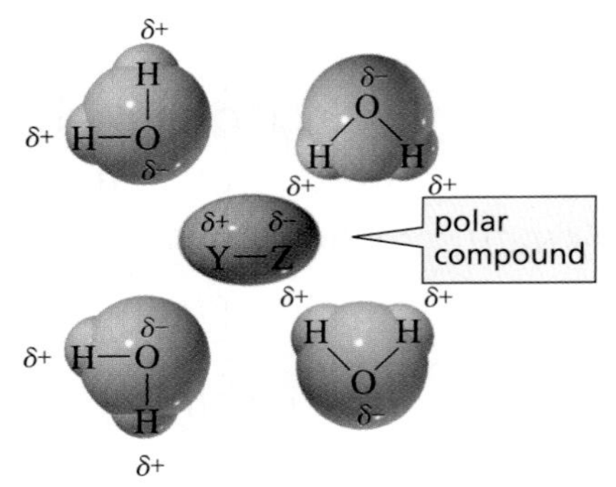

solvation of a polar compound by water

Because nonpolar compounds have no partial charges, polar solvents are not attracted to them. In order for a nonpolar molecule to dissolve in a polar solvent such as water, the nonpolar molecule would have to push the water molecules apart, disrupting their hydrogen bonding. Hydrogen bonding is strong enough to exclude the nonpolar compound. In contrast, nonpolar solutes dissolve in nonpolar solvents because the van der Waals interactions between solvent and solute molecules are about the same as between solvent–solvent and solute–solute molecules.

Alkanes are nonpolar, which causes them to be soluble in **nonpolar solvents** and insoluble in polar solvents such as water. The densities of alkanes (Table 3.1) increase with increasing molecular weight, but even a 30-carbon alkane is less dense than water. This means that a mixture of an alkane and water will separate into two distinct layers, with the less dense alkane floating on top. The Alaskan oil spill of 1989, the Persian Gulf spill of 1991, and the even larger spill off the northwest coast of Spain in 2002 are large-scale examples of this phenomenon. (Crude oil is primarily a mixture of alkanes.)

Oil from a 70,000-ton oil spill in 1996 off the coast of Wales.

An alcohol has both a nonpolar alkyl group and a polar OH group. So is an alcohol molecule nonpolar or polar? Is it soluble in a nonpolar solvent, or is it soluble in water? The answer depends on the size of the alkyl group. As the alkyl group increases in size, becoming a more significant fraction of the alcohol molecule, the compound becomes less and less soluble in water. In other words, the molecule becomes more and more like an alkane. Groups consisting of four carbons tend to straddle the dividing line at room temperature: alcohols with fewer than four carbons are soluble in water, but alcohols with more than four carbons are insoluble in water. In other words, an OH group can drag about three or four carbons into solution in water.

The four-carbon dividing line is only an approximate guide because the solubility of an alcohol also depends on the structure of the alkyl group. Alcohols with branched alkyl groups are more soluble in water than alcohols with nonbranched alkyl groups with the same number of carbons, because branching minimizes the contact surface of the nonpolar portion of the molecule. So *tert*-butyl alcohol is more soluble than *n*-butyl alcohol in water.

Similarly, the oxygen atom of an ether can drag only about three carbons into water (Table 3.6). We have already seen (photo on page 46) that diethyl ether—an ether with four carbons—is not soluble in water.

Table 3.6 Solubilities of Ethers in Water

2 C's	CH_3OCH_3	soluble
3 C's	$CH_3OCH_2CH_3$	soluble
4 C's	$CH_3CH_2OCH_2CH_3$	slightly soluble (10 g/100 g H_2O)
5 C's	$CH_3CH_2OCH_2CH_2CH_3$	minimally soluble (1.0 g/100 g H_2O)
6 C's	$CH_3CH_2CH_2OCH_2CH_2CH_3$	insoluble (0.25 g/100 g H_2O)

Low-molecular-weight amines are soluble in water because amines can form hydrogen bonds with water. Comparing amines with the same number of carbons, we find that primary amines are more soluble than secondary amines because primary amines have two hydrogens that can engage in hydrogen bonding. Tertiary amines, like primary and secondary amines, have lone-pair electrons that can accept hydrogen bonds, but unlike primary and secondary amines, tertiary amines do not have hydrogens to donate for hydrogen bonds. Tertiary amines, therefore, are less soluble in water than are secondary amines with the same number of carbons.

Alkyl halides have some polar character, but only alkyl fluorides have an atom that can form a hydrogen bond with water. This means that alkyl fluorides are the most water soluble of the alkyl halides. The other alkyl halides are less soluble in water than ethers or alcohols with the same number of carbons (Table 3.7).

Table 3.7 Solubilities of Alkyl Halides in Water

CH_3F very soluble	CH_3Cl soluble	CH_3Br slightly soluble	CH_3I slightly soluble
CH_3CH_2F soluble	CH_3CH_2Cl slightly soluble	CH_3CH_2Br slightly soluble	CH_3CH_2I slightly soluble
$CH_3CH_2CH_2F$ slightly soluble	$CH_3CH_2CH_2Cl$ slightly soluble	$CH_3CH_2CH_2Br$ slightly soluble	$CH_3CH_2CH_2I$ slightly soluble
$CH_3CH_2CH_2CH_2F$ insoluble	$CH_3CH_2CH_2CH_2Cl$ insoluble	$CH_3CH_2CH_2CH_2Br$ insoluble	$CH_3CH_2CH_2CH_2I$ insoluble

PROBLEM 25 ♦

Rank the following groups of compounds in order of decreasing solubility in water:

a. $CH_3CH_2CH_2OH$ $CH_3CH_2CH_2CH_2Cl$ $CH_3CH_2CH_2CH_2OH$ $HOCH_2CH_2CH_2OH$

b. CH_3 NH_2 OH

PROBLEM 26 ♦

In which solvent would cyclohexane have the lowest solubility: 1-pentanol, diethyl ether, ethanol, or hexane?

PROBLEM 27 ♦

The effectiveness of a barbiturate as a sedative is related to its ability to penetrate the nonpolar membrane of a cell. Which of the following barbiturates would you expect to be the more effective sedative?

CH_3CH_2 O $CH_3(CH_2)_4CH_2$ NH O N H O
hexethal

CH_3CH_2 O CH_3CH_2 NH O N H O
barbital

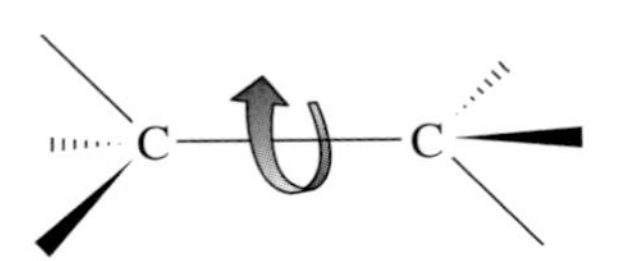

▲ **Figure 3.3**
Rotation about a carbon–carbon bond can occur without changing the amount of orbital overlap.

3.8 ROTATION OCCURS ABOUT CARBON–CARBON SINGLE BONDS

We have seen that a carbon–carbon single bond (a σ bond) is formed when an sp^3 orbital of one carbon overlaps an sp^3 orbital of a second carbon (Section 1.7). Figure 3.3 shows that rotation about a carbon–carbon single bond can occur without any change in the amount of orbital overlap. The different spatial arrangements of the atoms that result from rotation about a single bond are called **conformations**.

When rotation occurs about the carbon–carbon bond of ethane, two extreme conformations result—a *staggered conformation* and an *eclipsed conformation*. An infinite number of conformations between these two extremes are also possible.

H_3C-CH_3
ethane

Newman projections

$\overset{60^\circ}{\rightleftharpoons}$

a staggered conformation for rotation about the C—C bond in ethane

an eclipsed conformation for rotation about the C—C bond in ethane

Our drawings of molecules are two-dimensional attempts to communicate three-dimensional structures. Chemists commonly use *Newman projections* to represent the three-dimensional spatial arrangements resulting from rotation about a σ bond. A **Newman projection** assumes the viewer is looking along the longitudinal axis of a particular C—C bond. The carbon in front is represented by a point (where three lines are seen to intersect), and the carbon at the back is represented by a circle. The three lines emanating from each of the carbons represent its other three bonds.

A **staggered conformation** is more stable, and therefore lower in energy, than an **eclipsed conformation**. Because of this energy difference, rotation about a carbon–carbon single bond is not completely free. The eclipsed conformer is higher in energy, so an energy barrier must be overcome when rotation about the C—C bond occurs (Figure 3.4). However, the barrier in ethane is small enough (2.9 kcal/mol or 12 kJ/mol) to allow continuous rotation. A molecule's conformation changes from staggered to eclipsed millions of times per second at room temperature. Because of this continuous interconversion, the conformations cannot be separated from each other.

Figure 3.4 shows the potential energies of all the conformations of ethane obtained during one complete 360° rotation. Notice that the *staggered conformations* are at energy minima, whereas the *eclipsed conformations* are at energy maxima. Conformations at energy minima are often called **conformers**.

BIOGRAPHY

Melvin S. Newman (1908–1993) *was born in New York. He received a Ph.D. from Yale University in 1932 and was a professor of chemistry at Ohio State University from 1936 to 1973. He first suggested his technique for drawing organic molecules in 1952.*

A staggered conformation is more stable than an eclipsed conformation.

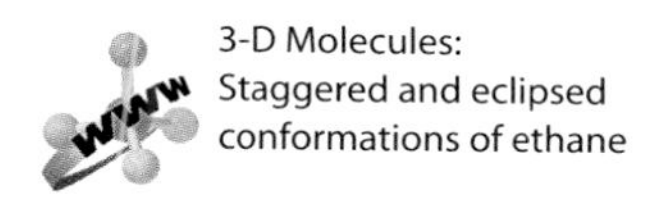

3-D Molecules: Staggered and eclipsed conformations of ethane

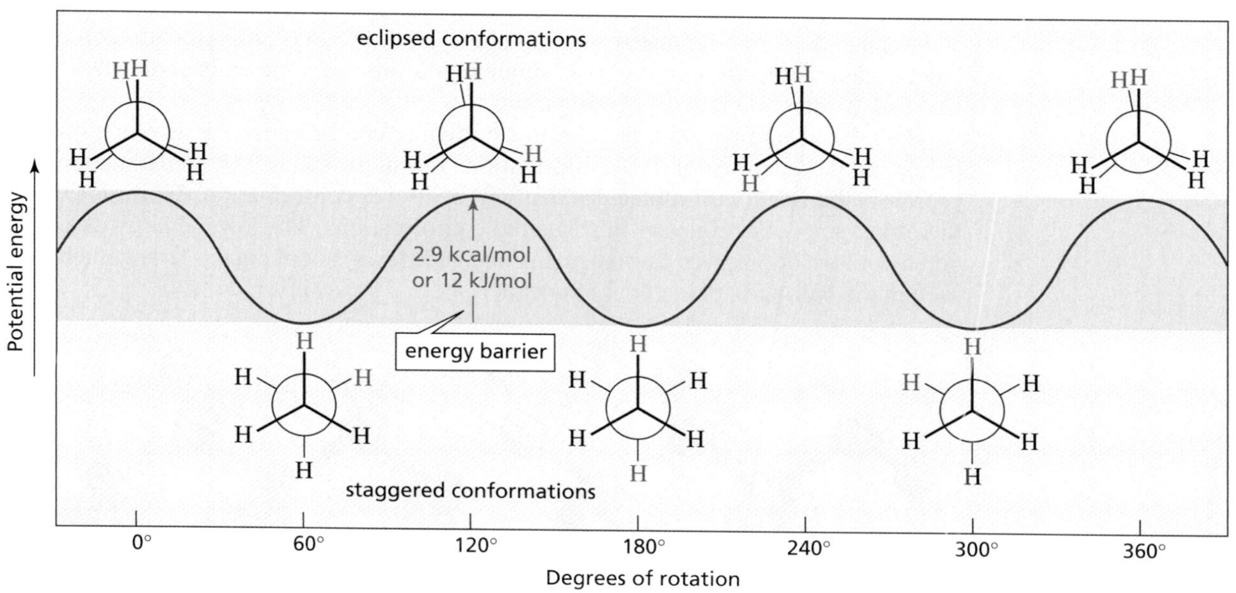

▲ **Figure 3.4**
The potential energy of ethane as a function of the angle of rotation about the carbon–carbon bond.

Butane has three carbon–carbon single bonds, and rotation can occur about each of them.

$\overset{1}{C}H_3-\overset{2}{C}H_2-\overset{3}{C}H_2-\overset{4}{C}H_3$

butane

the C-1—C-2 bond

the C-2—C-3 bond

the C-3—C-4 bond

ball-and-stick model of butane

The following Newman projections show staggered and eclipsed conformations that result from rotation about the C-2—C-3 bond.

Movie: Potential energy of butane conformations

0°

A B C D E F A

The three staggered conformations do not have the same energy. Conformation D, in which the two methyl groups are as far apart as possible, is more stable than the other two staggered conformations (B and F) because of steric strain. **Steric strain** is the strain (that is, the extra energy) experienced by a molecule when atoms or groups are too close to one another, causing their electron clouds to repel each other. In general, steric strain in molecules increases as the size of the interacting atoms or groups increases.

The eclipsed conformations resulting from rotation about the C-2—C-3 bond in butane also have different energies. The eclipsed conformation in which the two methyl groups are closest to each other (A) is less stable than the eclipsed conformations in which they are farther apart (C and E).

Because there is continuous rotation about all the C—C single bonds in a molecule, organic molecules with C—C single bonds are not static balls and sticks—they have many interconvertible conformations.

The relative number of molecules in a particular conformations at any one time depends on the stability of the conformation: the more stable the conformation, the greater is the fraction of molecules that will be in that conformation. Most molecules, therefore, at any one time are in staggered conformations. The lower energy of a staggered conformation gives carbon chains the tendency to adopt zigzag arrangements, as seen in the ball-and-stick model of decane.

3-D Molecule: Decane

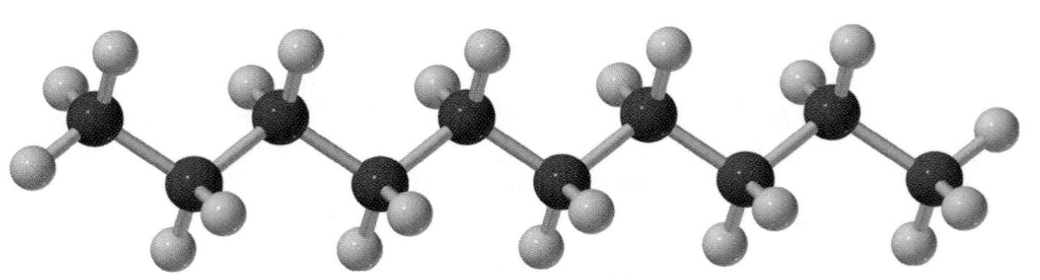

ball-and-stick model of decane

PROBLEM 28

a. Draw the three staggered conformations of butane for rotation about the C-1 — C-2 bond. (The carbon in the foreground in a Newman projection should have the lower number.)
b. Do the three staggered conformations have the same energy?
c. Do the three eclipsed conformations have the same energy?

PROBLEM 29

a. Draw the most stable conformation of pentane for rotation about the C-2 — C-3 bond.
b. Draw the least stable conformation of pentane for rotation about the C-2 — C-3 bond.

3.9 SOME CYCLOALKANES HAVE ANGLE STRAIN

Early chemists observed that cyclic compounds found in nature generally have five- or six-membered rings. This observation suggests that compounds with five- and six-membered rings are more stable than compounds with three- or four-membered rings.

We know that, ideally, an sp^3 carbon has bond angles of 109.5°. However, cyclopropane has bond angles of 60°, the bond angle associated with a planar three-membered ring. The bond angles, therefore, are 49.5° less than the ideal sp^3 bond angle of 109.5°. This deviation of the bond angle from the ideal bond angle causes strain called **angle strain**. Figure 3.5 shows that angle strain results from less effective orbital overlap, because the overlapping orbitals cannot point directly at each other. The less effective orbital overlap causes the C — C bonds of cyclopropane to be weaker than normal C — C bonds.

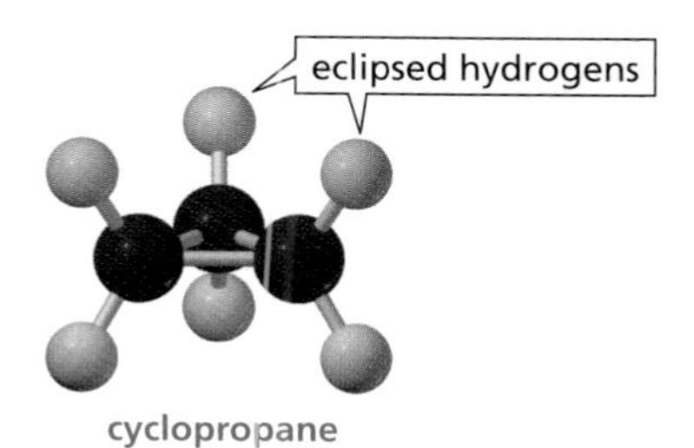

cyclopropane

3-D Molecules: Cyclopropane; Cyclobutane; Cyclopentane

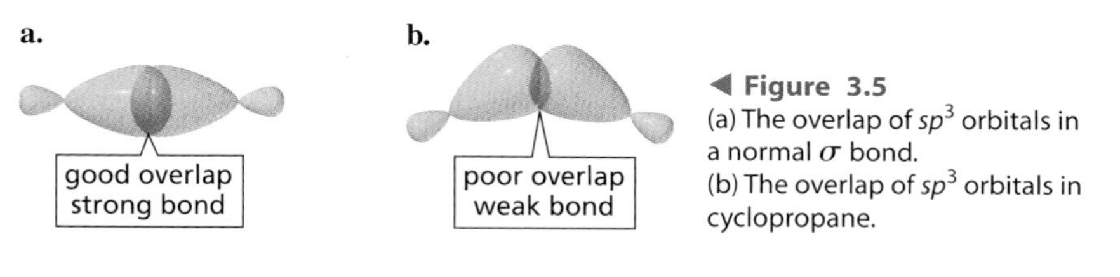

◀ **Figure 3.5**
(a) The overlap of sp^3 orbitals in a normal σ bond.
(b) The overlap of sp^3 orbitals in cyclopropane.

The bond angles in planar cyclobutane would have to be compressed from 109.5° to 90°, the bond angle associated with a planar four-membered ring. Planar cyclobutane would then be expected to have less angle strain than cyclopropane because the bond angles in cyclobutane are only 19.5° away from the ideal bond angle.

Although planar cyclobutane would have less angle strain than cyclopropane, it would have eight pairs of eclipsed hydrogens, compared with the six pairs in cyclopropane. Because of the eclipsed hydrogens, a cyclobutane molecule is not planar—one of its CH_2 groups is bent away from the plane defined by the other three carbons. Although bent cyclobutane has more angle strain than planar cyclobutane, the increase in angle strain is more than compensated by the decrease in eclipsed hydrogens.

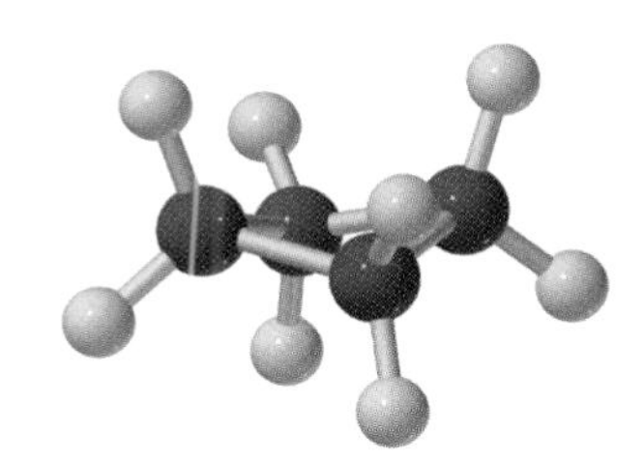
cyclobutane

If cyclopentane were planar, it would have essentially no angle strain (its bond angles would be 108°), but it would have 10 pairs of eclipsed hydrogens. So cyclopentane puckers, allowing the hydrogens to become nearly staggered. In the process, however, the compound acquires some angle strain.

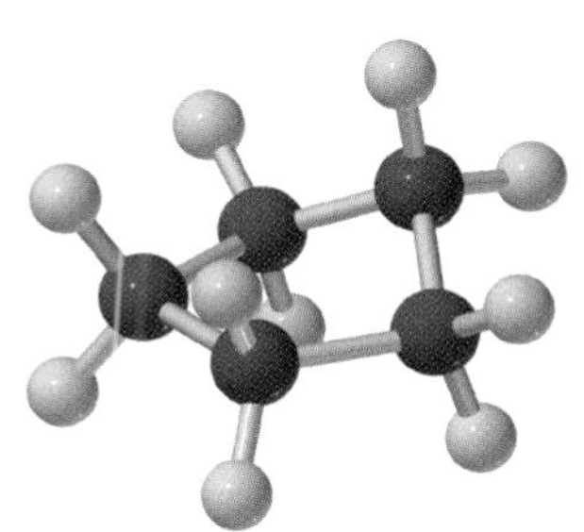
cyclopentane

3.10 CONFORMERS OF CYCLOHEXANE

The cyclic compounds most commonly found in nature contain six-membered rings because rings of that size can exist in a conformation—called a **chair conformer**—that is almost completely free of strain. All the bond angles in a *chair conformer* are 111°, which is very close to the ideal tetrahedral bond angle of 109.5°, and all the adjacent bonds are staggered (Figure 3.6).

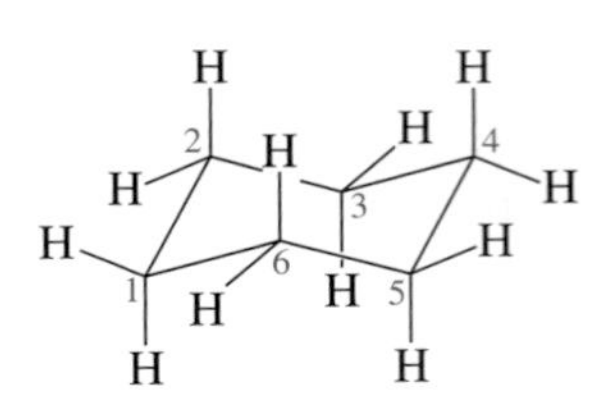

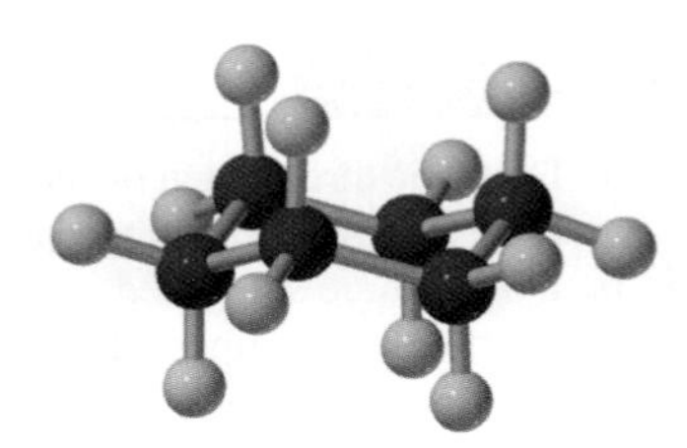

▶ **Figure 3.6**
The chair conformer of cyclohexane, a Newman projection of the chair conformer, and a ball-and-stick model showing that all the bonds are staggered.

The chair conformer is such an important conformer that you should learn how to draw it:

1. Draw two parallel lines of the same length, slanted upward and beginning at the same height.

2. Connect the tops of the lines with a V whose left-hand side is slightly longer than the right-hand side. Connect the bottoms of the lines with an inverted V. (The bottom-left and top-right lines should be parallel; the top-left and bottom-right lines should also be parallel.) This completes the framework of the six-membered ring.

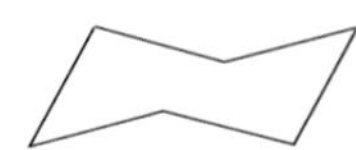

3. Each carbon has an axial bond and an equatorial bond. The **axial bonds** (red lines) are vertical and alternate above and below the ring. The axial bond on one of the uppermost carbons is up, the next is down, the next is up, and so on.

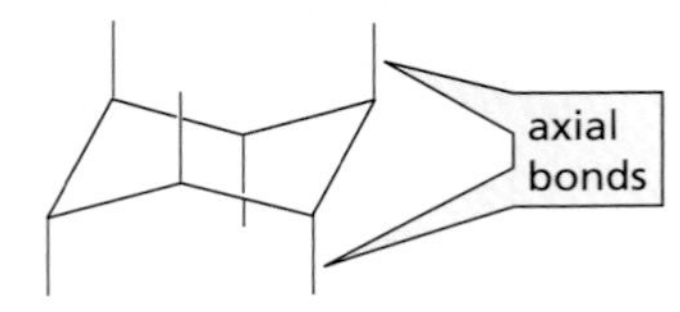

4. The **equatorial bonds** (red lines with blue balls) point outward from the ring. Because the bond angles are greater than 90°, the equatorial bonds are on a slant. If the axial bond points up, the equatorial bond on the same carbon is on a downward slant. If the axial bond points down, the equatorial bond on the same carbon is on an upward slant.

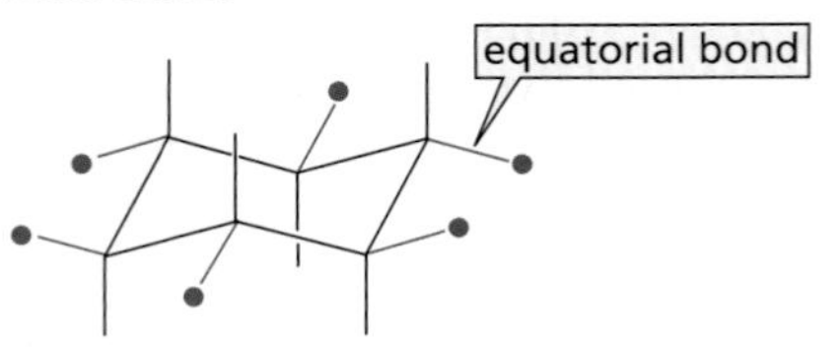

Notice that each equatorial bond is parallel to two ring bonds (two carbons over).

Remember that cyclohexane is viewed on edge. The lower bonds of the ring are in front, and the upper bonds of the ring are in back.

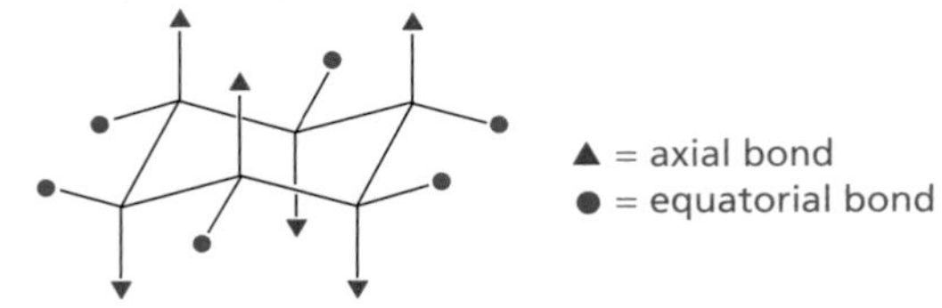

PROBLEM 30

Draw 1,1-dichlorocyclohexane and label the axial and equatorial bonds that connect the chloro substituents to the ring.

Cyclohexane rapidly interconverts between two stable chair conformers because of the ease of rotation about its C—C bonds. This interconversion is called ring flip; at room temperature, cyclohexane undergoes 10^5 ring flips per second. When the two chair conformers interconvert, bonds that are equatorial in one chair conformer become axial in the other chair conformer and vice versa (Figure 3.7).

Bonds that are equatorial in one chair conformer are axial in the other chair conformer.

pull this carbon down

push this carbon up

ring flip

◀ **Figure 3.7**
The bonds that are axial in one chair conformer are equatorial in the other chair conformer. The bonds that are equatorial in one chair conformer are axial in the other chair conformer.

To convert from one chair conformer to the other, the bottommost carbon must be pushed up and what was previously the topmost carbon must be pulled down. The conformations that cyclohexane can assume when interconverting from one chair conformer to the other are shown in Figure 3.8. Because the chair conformers are the most stable of the conformers, at any instant, more molecules of cyclohexane are in chair conformers than in any other conformer. For every 10,000 molecules of cyclohexane in a chair conformer, there is no more than one molecule in the next most stable conformer, the twist-boat.

Go to the website to see three-dimensional representations of the conformers of cyclohexane.

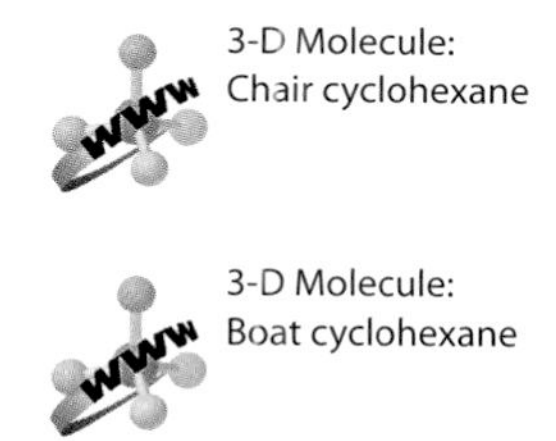

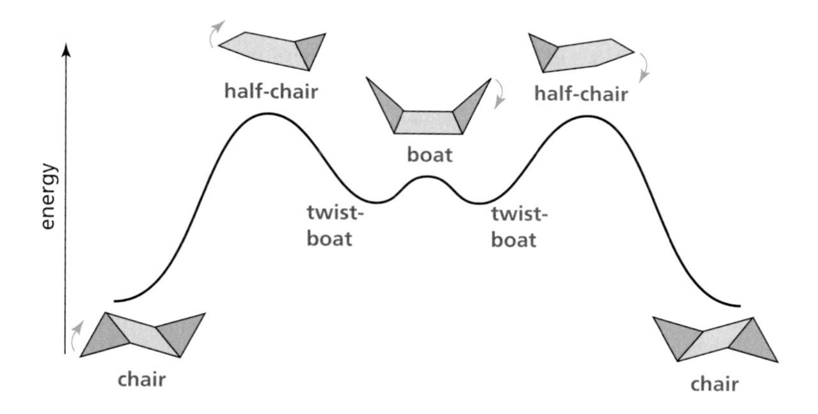

◀ **Figure 3.8**
The conformers of cyclohexane—and their relative energies—as one chair conformer interconverts to the other chair conformer.

3.11 CONFORMERS OF MONOSUBSTITUTED CYCLOHEXANES

Unlike cyclohexane, which has two equivalent chair conformers, the two chair conformers of a monosubstituted cyclohexane such as methylcyclohexane are not equivalent. The methyl substituent is in an equatorial position in one conformer and in an axial position in the other (Figure 3.9), because, as we have just seen, hydrogens (or substituents) that are equatorial in one chair conformer are axial in the other (Figure 3.7).

Because the three axial bonds on the same side of the ring are parallel to each other, any axial substituent will be relatively close to the axial substituents on the other two carbons. Therefore, the chair conformer with the methyl substituent in an equatorial position is more stable than the chair conformer with the methyl substituent in an axial

Build a model of methylcyclohexane, and convert it from one chair conformer to the other.

the methyl group is in an equatorial position

CH_3 ring flip $\rightleftharpoons$

the methyl group is in an axial position

CH_3

more stable chair conformer **less stable chair conformer**

▶ **Figure 3.9**
A substituent is in an equatorial position in one chair conformer and in an axial position in the other. The conformer with the substituent in the equatorial position is more stable.

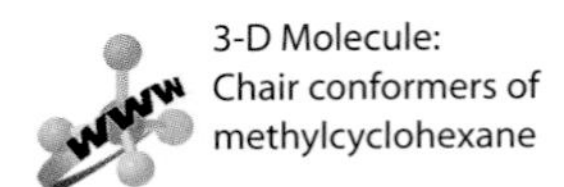

3-D Molecule: Chair conformers of methylcyclohexane

position, because a substituent has more room and, therefore, fewer steric interactions when it is in an equatorial position.

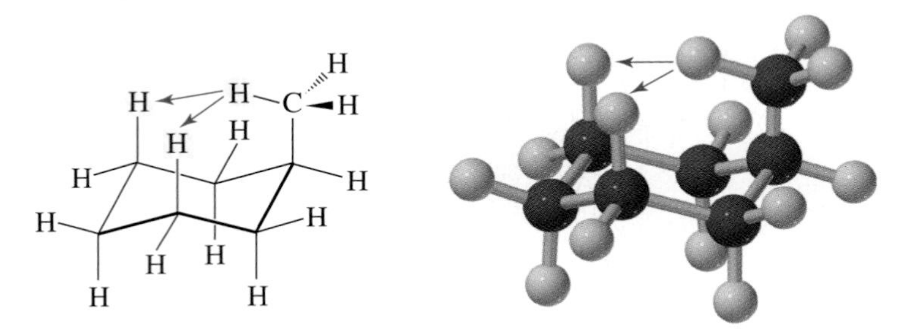

The larger the substituent on a cyclohexane ring, the more the equatorial-substituted conformer will be favored.

Because of the difference in stability of the two chair conformers, a sample of methylcyclohexane or any other substituted cyclohexane will at any point in time contain more chair conformers with the substituent in the equatorial position than with the substituent in the axial position.

PROBLEM 31 ◆

At any one time, would you expect there to be more conformers with the substituent in the equatorial position in a sample of ethylcyclohexane or in a sample of isopropylcyclohexane?

3.12 CONFORMERS OF DISUBSTITUTED CYCLOHEXANES

The cis isomer has its substituents on the same side of the ring.

The trans isomer has its substituents on opposite sides of the ring.

If a cyclohexane ring has substituents on two different carbons, we must take both substituents into account when predicting which of the two chair conformers is more stable. Let's use 1,4-dimethylcyclohexane as an example. First of all, note that there are two different dimethylcyclohexanes. One has both methyl substituents on the *same side* of the cyclohexane ring (they both point downward); it is called the **cis isomer** (*cis* is Latin for "on this side"). The other has the two methyl substituents on *opposite sides* of the ring (one points upward and one points downward); it is called the **trans isomer** (*trans* is Latin for "across").

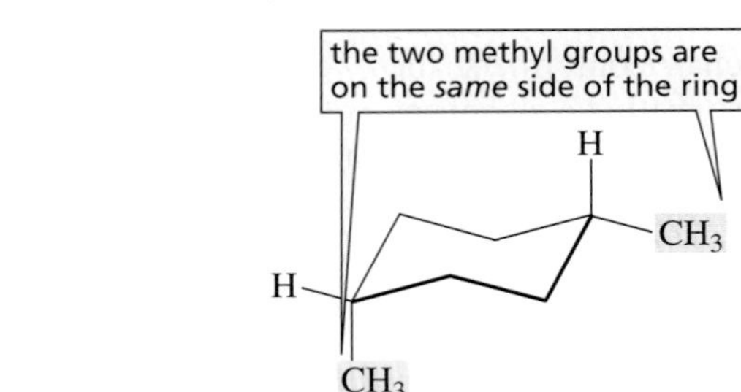

cis-1,4-dimethylcyclohexane

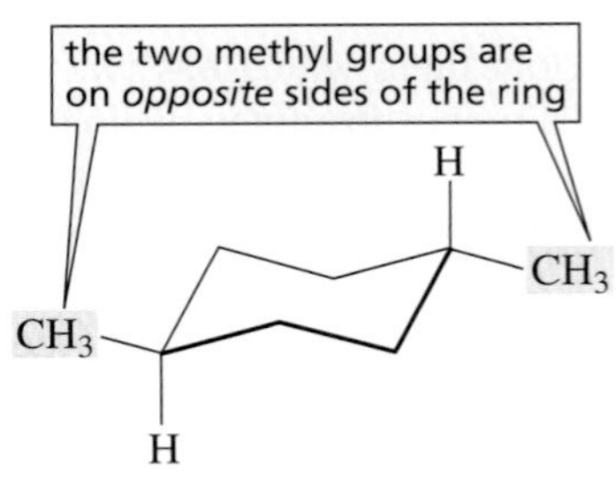

trans-1,4-dimethylcyclohexane

cis-1,4-Dimethylcyclohexane and *trans*-1,4-dimethylcyclohexane are examples of **cis–trans isomers** or **geometric isomers: cis–trans isomers** are compounds containing the same atoms, and the atoms are linked in the same order, but the atoms exhibit two different spatial arrangements. The cis and trans isomers are different compounds with different melting and boiling points. They can, therefore, be separated from one another.

PROBLEM-SOLVING STRATEGY

Is the conformer of 1,2-dimethylcyclohexane with one methyl group in an equatorial position and the other in an axial position the cis isomer or the trans isomer?

H
CH_3
H
CH_3

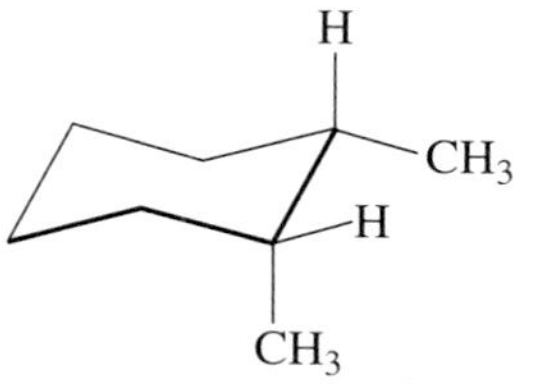

Is this the cis isomer or the trans isomer?

To solve this kind of problem, we need to determine whether the two substituents are on the same side of the ring (cis) or on opposite sides of the ring (trans). If the bonds bearing the substituents both point upward or both point downward, the compound is the cis isomer; if one bond points upward and the other downward, the compound is the trans isomer. Because the conformer in question has both methyl groups attached to downward-pointing bonds, it is the cis isomer.

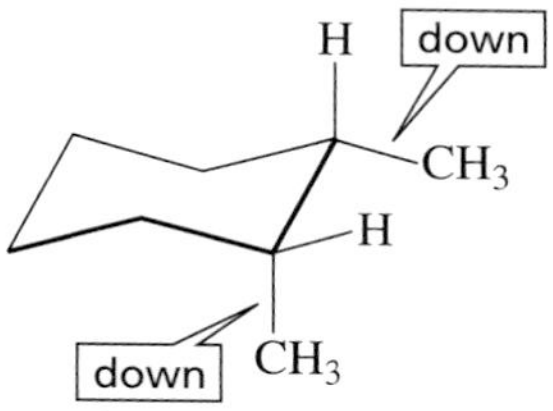

H down CH_3 H down CH_3

the cis isomer

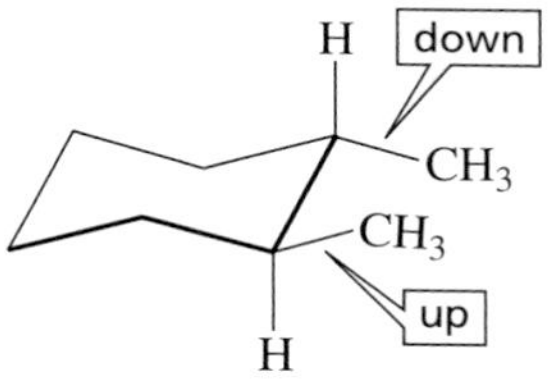

H down CH_3 CH_3 up H

the trans isomer

Now continue on to Problem 32.

PROBLEM 32 ♦

Determine whether each of the following is a cis isomer or a trans isomer:

a. H H Br Cl

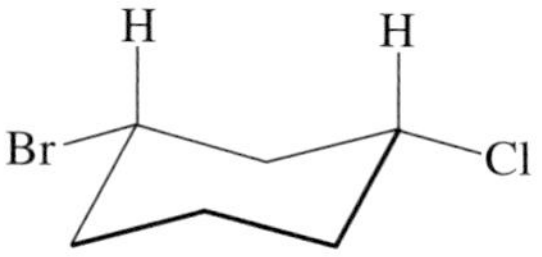

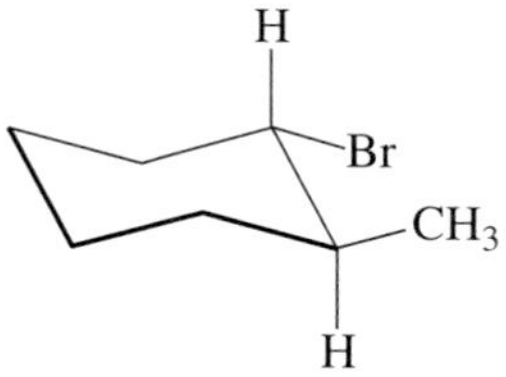

b. Br H CH_3 H

c. H Br CH_3 H

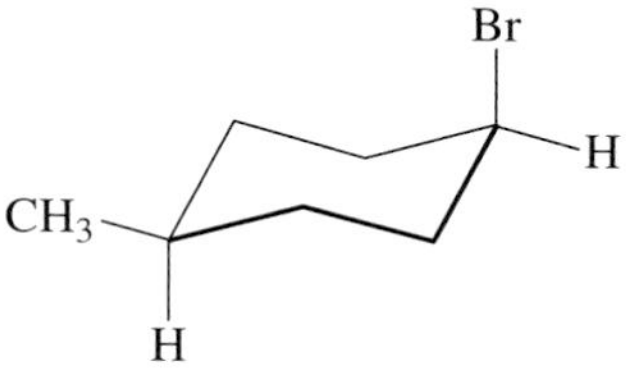

d. H Cl CH_3 H

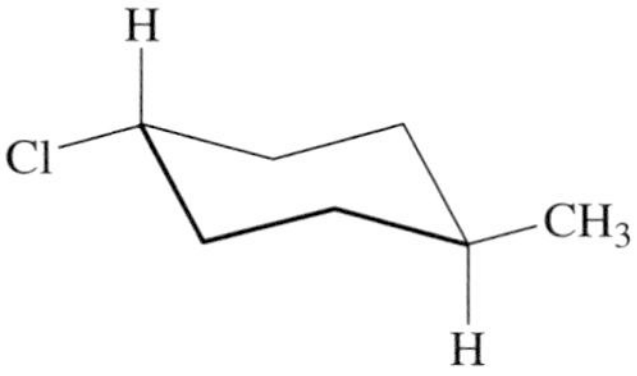

Every compound with a cyclohexane ring has two chair conformers; thus, the cis isomer and the trans isomer of a disubstituted cyclohexane each has two chair conformers. Let's compare the structures of the two chair conformers of *cis*-1,4-dimethylcyclohexane to see if we can predict any difference in their stabilities. The conformer shown on the left has one methyl group in an equatorial position and one methyl group in an axial position. The conformer shown on the right also has one methyl group in an equatorial position and one methyl group in an axial position. Therefore, both chair conformers are equally stable.

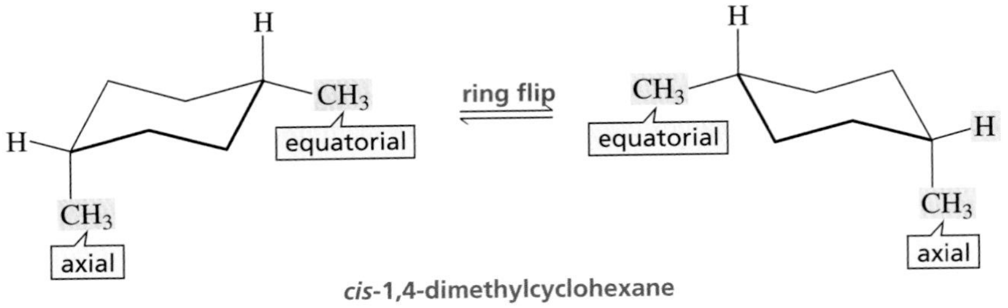

H ... CH_3 equatorial ... ring flip ... CH_3 equatorial ... H
CH_3 axial ... CH_3 axial

cis-1,4-dimethylcyclohexane

In contrast, the two chair conformers of *trans*-1,4-dimethylcyclohexane have different stabilities because one has both methyl substituents in equatorial positions and the other has both methyl groups in axial positions. The conformer with both methyl groups in equatorial positions is more stable.

CH_3 equatorial ... CH_3 equatorial ... ring flip ... axial CH_3 ... CH_3 axial

more stable ... less stable

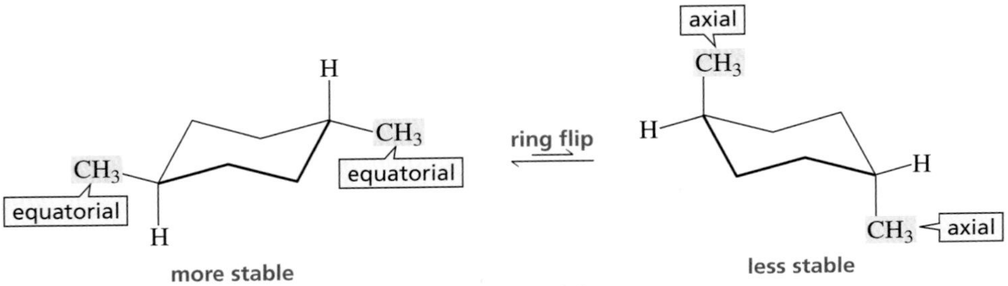

trans-1,4-dimethylcyclohexane

Now let's look at the geometric isomers of 1-*tert*-butyl-3-methylcyclohexane. Both substituents of the cis isomer are in equatorial positions in one conformer and in axial positions in the other conformer. The conformer with both substituents in equatorial positions is more stable.

equatorial CH_3 ... equatorial ... ring flip ... CH_3 axial ... axial

more stable ... less stable

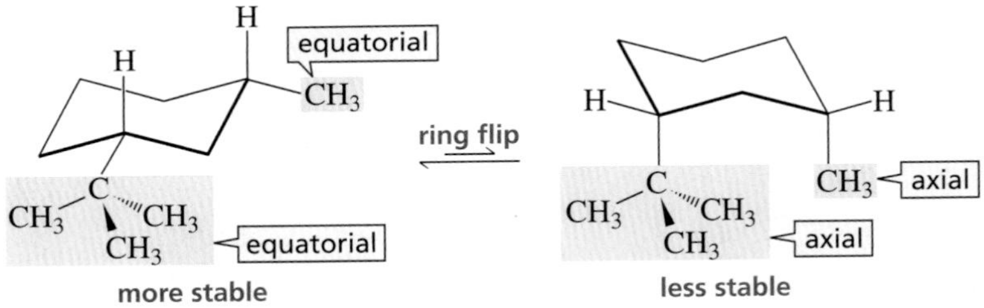

cis-1-*tert*-butyl-3-methylcyclohexane

Both conformers of the trans isomer have one substituent in an equatorial position and the other in an axial position. Because the *tert*-butyl group is larger than the methyl group, the conformer with the *tert*-butyl group in the equatorial position, where there is more room for a substituent, is more stable.

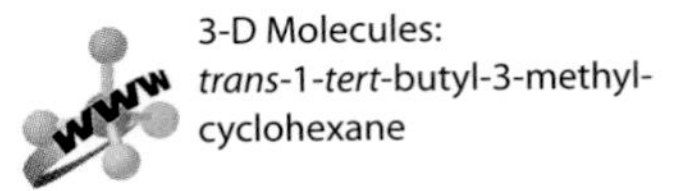

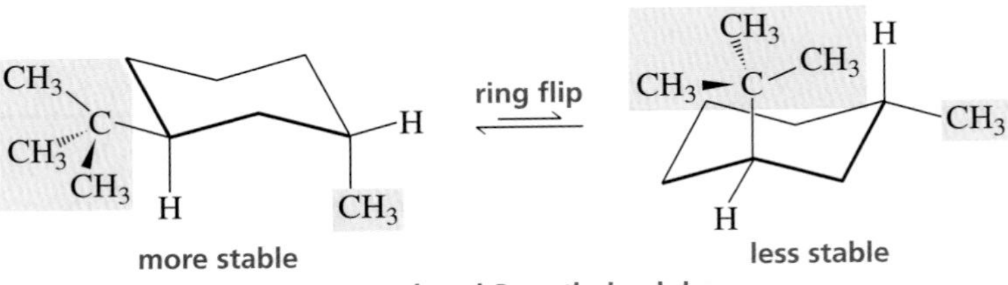

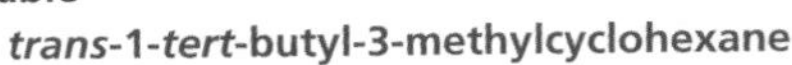
trans-1-*tert*-butyl-3-methylcyclohexane

PROBLEM 33 *SOLVED*

a. Draw the more stable chair conformer of *cis*-1-ethyl-2-methylcyclohexane.
b. Draw the more stable conformer of *trans*-1-ethyl-2-methylcyclohexane.
c. Which is more stable, *cis*-1-ethyl-2-methylcyclohexane or *trans*-1-ethyl-2-methylcyclohexane?

Solution to 33a If the two substituents of a 1,2-disubstituted cyclohexane are to be on the same side of the ring, one must be in an equatorial position and the other must be in an axial position. The more stable chair conformer is the one in which the larger of the two substituents (the ethyl group) is in the equatorial position.

3.13 FUSED CYCLOHEXANE RINGS

When two cyclohexane rings are fused together, the second ring can be considered to be a pair of substituents bonded to the first ring. As with any disubstituted cyclohexane, the two substituents can be either cis or trans. The trans isomer (one substituent bond points upward and the other downward) has both substituents in the equatorial position. The cis isomer has one substituent in the equatorial position and one in the axial position. **Trans-fused** rings, therefore, are more stable than **cis-fused** rings.

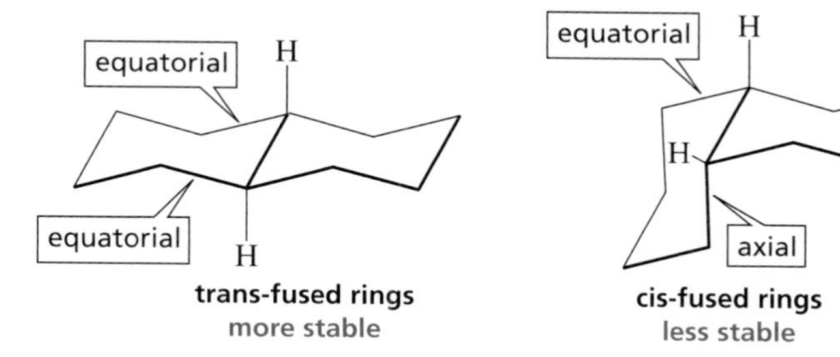

trans-fused rings
more stable

cis-fused rings
less stable

Tutorials:
Steroids

Hormones are chemical messengers—organic compounds synthesized in glands and delivered by the bloodstream to target tissues in order to stimulate or inhibit some process. Many hormones are **steroids**. The four rings in steroids are designated A, B, C, and D. The B, C, and D rings are all trans fused, and in most naturally occurring steroids, the A and B rings are also trans fused.

A B C D

the steroid ring system

A B C D

all the rings are trans fused

The most abundant member of the steroid family in animals is **cholesterol**, the precursor of all other steroids. Cholesterol is an important component of cell membranes. Its ring structure makes it more rigid than other membrane components (Section 20.5).

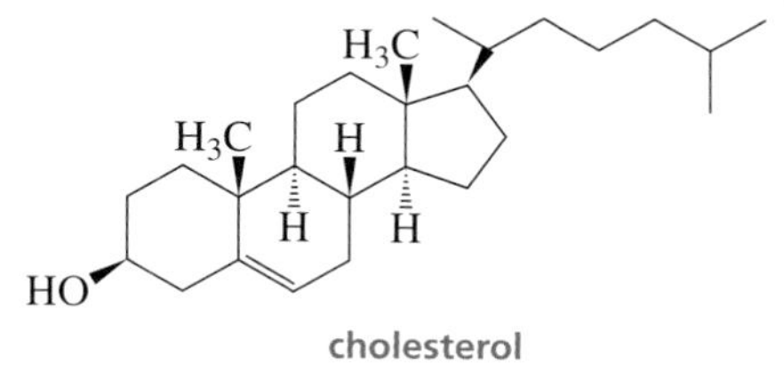

cholesterol

BIOGRAPHY

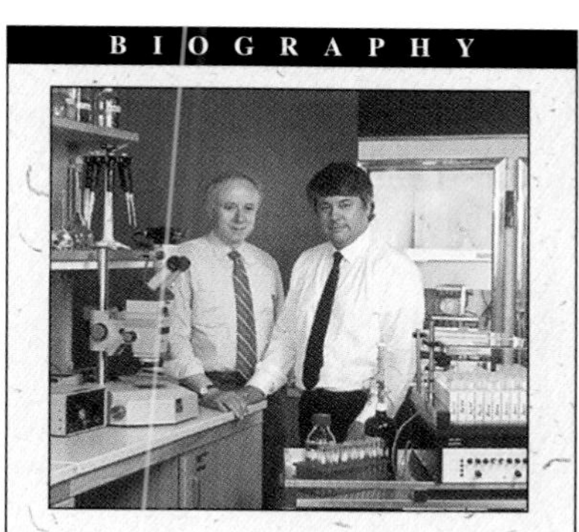

Michael S. Brown *and* **Joseph Leonard Goldstein** *shared the 1985 Nobel Prize in physiology or medicine for their work on the regulation of cholesterol metabolism and the treatment of disease caused by elevated cholesterol levels in the blood. Brown was born in New York in 1941, Goldstein in South Carolina in 1940. They are both professors of medicine at the University of Texas Southwestern Medical Center.*

CHOLESTEROL AND HEART DISEASE

Cholesterol is probably the best-known steroid because of the widely publicized correlation between cholesterol levels in the blood and heart disease. Cholesterol is synthesized in the liver and is also found in almost all body tissues. Cholesterol is found in many foods, but we do not require it in our diet because the body can synthesize all we need. A diet high in cholesterol can lead to high levels of cholesterol in the bloodstream, and the excess can accumulate on the walls of arteries, restricting the flow of blood. This disease of the circulatory system is known as *atherosclerosis* and is a primary cause of heart disease. Cholesterol travels through the bloodstream packaged in particles that are classified according to their density. Low-density lipoprotein (LDL) particles transport cholesterol from the liver to other tissues. Receptors on the surfaces of cells bind LDL particles, allowing them to be brought into the cell so that it can use the cholesterol. High-density lipoprotein (HDL) is a cholesterol scavenger, removing cholesterol from the surfaces of membranes and delivering it back to the liver, where it is converted into bile acids. LDL is the so-called bad cholesterol, whereas HDL is the "good" cholesterol. The more cholesterol we eat, the less the body synthesizes. But this does not mean that dietary cholesterol has no effect on the total amount of cholesterol in the bloodstream, because dietary cholesterol also inhibits the synthesis of the LDL receptors. So the more cholesterol we eat, the less the body synthesizes, but also, the less the body can get rid of by bringing it into target cells.

CLINICAL TREATMENT OF HIGH CHOLESTEROL

Statins are the newest class of cholesterol-reducing drugs. Statins reduce serum cholesterol levels by inhibiting the enzyme that catalyzes the formation of a compound needed for the synthesis of cholesterol. As a consequence of diminished cholesterol synthesis in the liver, the liver forms more LDL receptors—the receptors that help clear LDL from the bloodstream. Studies show that for every 10% that cholesterol is reduced, deaths from coronary heart disease are reduced by 15% and total death risk is reduced by 11%.

Lovastatin and simvastatin are natural statins used clinically under the trade names Mevacor and Zocor. Atorvastatin (Lipitor), a synthetic statin, is now the most popular statin. It has greater potency and a longer half-life than natural statins have, because its metabolites are as active as the parent drug in reducing cholesterol levels. Therefore, smaller doses of the drug may be administered. In addition, Lipitor is less polar than lovastatin and simvastatin, so it has a greater tendency to remain in the endoplasmic reticulum of the liver cells, where it is needed. Lipitor was the second most widely prescribed drug in the United States in 2004 and the first in 2006 (Section 21.0).

lovastatin
Mevacor®

simvastatin
Zocor®

atorvastatin
Lipitor®

SUMMARY

Alkanes are **hydrocarbons** that contain only single bonds. Their general molecular formula is C_nH_{2n+2}. **Constitutional isomers** have the same molecular formula, but their atoms are linked differently. Alkanes are named by determining the number of carbons in their **parent hydrocarbon**—the longest continuous chain. **Substituents** are listed in alphabetical order, with a number to designate their position on the chain.

Systematic names can contain numbers; **common names** never do. A compound can have more than one name, but a name must specify only one compound.

Whether alkyl halides or alcohols are **primary**, **secondary**, or **tertiary** depends on whether the X (halogen) or OH group is bonded to a primary, secondary, or tertiary carbon. A **primary carbon** is bonded to one carbon, a **secondary carbon** is bonded to two carbons, and a **tertiary carbon** is bonded to three carbons. Whether amines are **primary**, **secondary**, or **tertiary** depends on the number of alkyl groups bonded to the nitrogen.

The oxygen of an alcohol or an ether has the same geometry it has in water; the nitrogen of an amine has the same geometry it has in ammonia. The greater the attractive forces between molecules—**van der Waals forces, dipole–dipole interactions, hydrogen bonds**—the higher is the **boiling point** of the compound. A **hydrogen bond** is an interaction between a hydrogen bonded to an O, N, or F and a lone pair of an O, N, or F in another molecule. The boiling point of straight-chain compounds increases with increasing molecular weight. Branching lowers the boiling point.

Polar compounds dissolve in **polar solvents**, and **nonpolar compounds** dissolve in **nonpolar solvents**. The interaction between a solvent and a molecule or an ion dissolved in that solvent is called **solvation**. The oxygen of an alcohol or an ether can drag three or four carbons into solution in water.

Rotation about a C—C bond results in two extreme **conformations**, staggered and eclipsed, that rapidly interconvert. A **staggered conformation** is more stable than an **eclipsed conformation**.

Five- and six-membered rings are more stable than three- and four-membered rings because of the **angle strain** that results when bond angles deviate from the ideal bond angle of 109.5°. In a process called **ring flip**, cyclohexane rapidly interconverts between two stable chair conformers. **Bonds** that are **axial** in one chair conformer are **equatorial** in the other and vice versa. The chair conformer with a substituent in the equatorial position is more stable, because there is more room, and hence less **steric strain**, in an equatorial position. In the case of disubstituted cyclohexanes, the more stable conformer will have its larger substituent in the equatorial position. A **cis isomer** has its two substituents on the same side of the ring; a **trans isomer** has its substituents on opposite sides of the ring. Cis and trans isomers are called **cis–trans isomers** or **geometric isomers**. Cis and trans isomers are different compounds; each isomer has two chair conformers.

PROBLEMS

34. Write a structural formula for each of the following:

a. *sec*-butyl *tert*-butyl ether
b. isoheptyl alcohol
c. *sec*-butylamine
d. 4-*tert*-butylheptane
e. 1,1-dimethylcyclohexane
f. 4,5-diisopropylnonane
g. triethylamine
h. cyclopentylcyclohexane
i. 3,4-dimethyloctane
j. 5,5-dibromo-2-methyloctane

35. Give the systematic name for each of the following:

a. 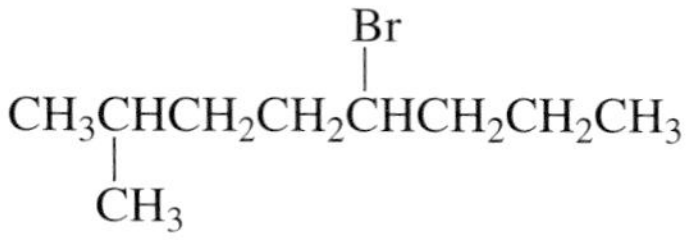

b. $(CH_3)_3CCH_2CH_2CH_2CH(CH_3)_2$

c. 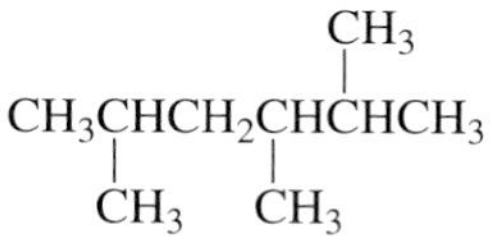

d. $(CH_3CH_2)_4C$

e. 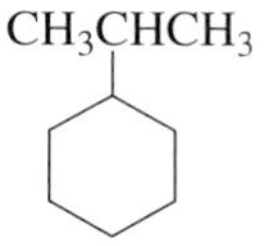

f. 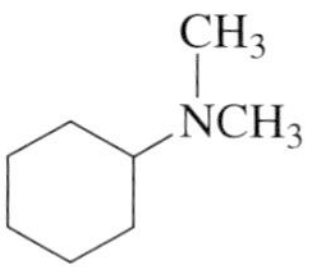

36. a. How many primary carbons does the following compound have?

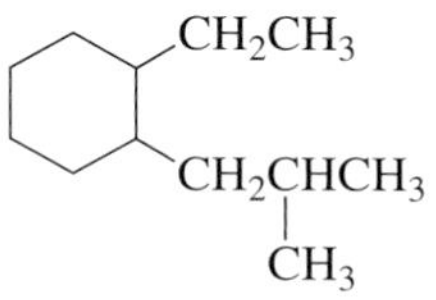

b. How many secondary carbons does it have?
c. How many tertiary carbons does it have?

37. Draw the structure and give the systematic name of a compound with a molecular formula C_7H_{14} that has

a. only one tertiary carbon
b. only two secondary carbons

38. Which of the following conformations of isobutyl chloride is the most stable?

Newman projection 1: front CH_3 (top), H, H; back H, CH_3, Cl (bottom)

Newman projection 2: front CH_3 (top), H, H; back Cl, CH_3, H (bottom)

Newman projection 3: front Cl (top), H, H; back H_3C, CH_3, H (bottom)

39. Name the following amines and tell whether they are primary, secondary, or tertiary:

a. $CH_3CH_2CH_2NCH_2CH_3$ (with CH_2CH_3 on N)

b. $CH_3CHCH_2NHCHCH_2CH_3$ (with CH_3 on C-2 and CH_3 on the CH after NH)

c. $CH_3CH_2CH_2NHCH_2CH_2CHCH_3$ (with CH_3 on the CH)

d. 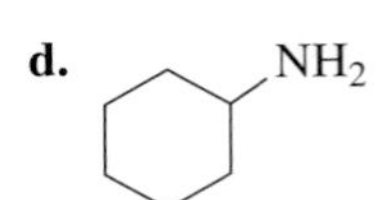

40. Draw the structural formula of an alkane that has
- **a.** six carbons, all secondary
- **b.** eight carbons and only primary hydrogens
- **c.** seven carbons with two isopropyl groups

41. Name each of the following:

a. $CH_3CH_2CH_2OCH_2CH_3$

b. $CH_3CHCH_2CH_2CH_2OH$ (with CH_3 on C-2)

c. $CH_3CH_2CHCH_3$ (with NH_2 on the CH)

d. $CH_3CH_2CHCH_3$ (with Cl on the CH)

e. $CH_3CHCH_2CH_2CH_3$ (with CH_3 on the CH)

f. CH_3CBr (with CH_3 and CH_2CH_3 on the C)

g. cyclohexane ring with OH

h. cyclopentane ring with Br

i. CH_3CHNH_2 (with CH_3 on the CH)

j. $CH_3CH_2CH(CH_3)NHCH_2CH_3$

42. Which of the following pairs of compounds has
- **a.** the higher boiling point: 1-bromopentane or 1-bromohexane?
- **b.** the higher boiling point: pentyl chloride or isopentyl chloride?
- **c.** the greater solubility in water: butyl alcohol or pentyl alcohol?
- **d.** the higher boiling point: hexyl alcohol or methyl pentyl ether?
- **e.** the higher melting point: hexane or isohexane?
- **f.** the higher boiling point: 1-chloropentane or pentyl alcohol?

43. Draw 1,2,3,4,5,6-hexamethylcyclohexane with
- **a.** all the methyl groups in axial positions.
- **b.** all the methyl groups in equatorial positions.

44. Ansaid and Motrin belong to the group of drugs known as nonsteroidal anti-inflammatory drugs (NSAIDs). Both are only slightly soluble in water, but one is a little more soluble than the other. Which of the drugs has the greater solubility in water?

Ansaid®: biphenyl bearing F, with $-CHCOOH$ (with CH_3 on the CH)

Motrin®: CH_3CHCH_2- (with CH_3 on the CH) on a benzene ring, with $-CHCOOH$ (with CH_3 on the CH)

45. Al Kane was given the structural formulas of several compounds and was asked to give them systematic names. How many did Al name correctly? Correct those that are misnamed.
- **a.** 3-isopropyloctane
- **b.** 2,2-dimethyl-4-ethylheptane
- **c.** isopentyl bromide
- **d.** 3,3-dichlorooctane
- **e.** 5-ethyl-2-methylhexane
- **f.** 2-methyl-2-isopropylheptane

46. Which of the following is the least stable?

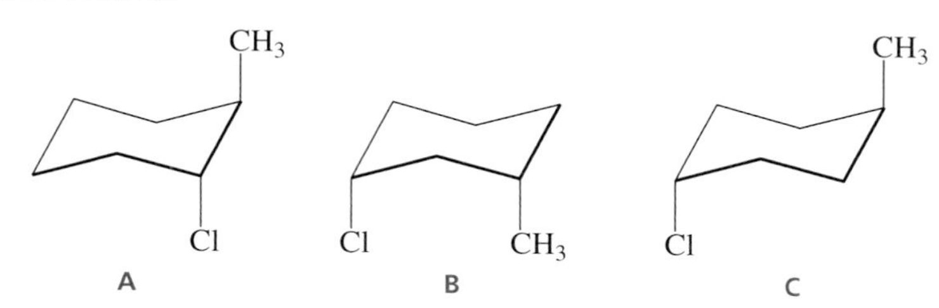

47. Give systematic names for all the alkanes with molecular formula C_7H_{16} that do not have any secondary hydrogens.

48. Draw skeletal structures for the following:
a. 5-ethyl-2-methyloctane
b. 1,3-dimethylcyclohexane
c. propylcyclopentane
d. 2,3,3,4-tetramethylheptane

49. Which of the following pairs of compounds has
a. the higher boiling point: 1-bromopentane or 1-chloropentane?
b. the higher boiling point: diethyl ether or butyl alcohol?
c. the greater density: heptane or octane?
d. the higher boiling point: isopentyl alcohol or isopentylamine?
e. the higher boiling point: hexylamine or dipropylamine?

50. Why are alcohols of lower molecular weight more soluble in water than those of higher molecular weight?

51. For rotation about the C-3—C-4 bond of 2-methylhexane:
a. Draw the Newman projection of the most stable conformation.
b. Draw the Newman projection of the least stable conformation.
c. About which other carbon–carbon bonds may rotation occur?
d. How many of the carbon–carbon bonds in the compound have staggered conformations that are all equally stable?

52. Which of the following structures represents a cis isomer?

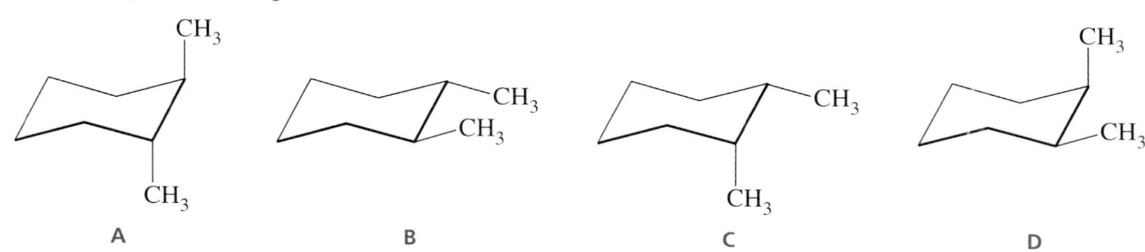

53. Draw all the isomers that have molecular formula $C_5H_{11}Br$. (*Hint:* There are eight of them.)
a. Give the systematic name for each of the isomers.
b. How many of the isomers are primary alkyl halides?
c. How many of the isomers are secondary alkyl halides?
d. How many of the isomers are tertiary alkyl halides?

54. Give the systematic name for each of the following:

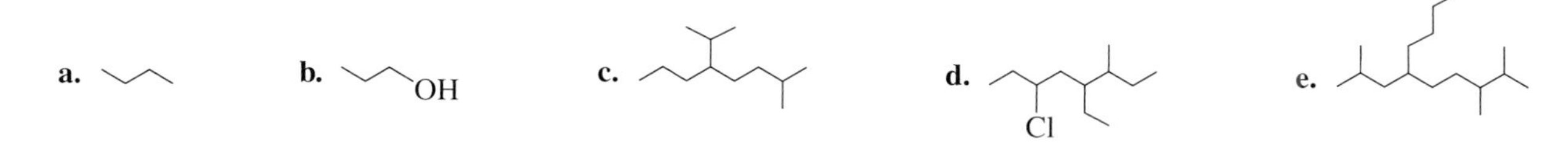

55. Draw the two chair conformers for each of the following, and indicate which conformer is more stable:
a. *cis*-1-ethyl-3-methylcyclohexane
b. *trans*-1-ethyl-2-isopropylcyclohexane
c. *trans*-1-ethyl-2-methylcyclohexane
d. *trans*-1-ethyl-3-methylcyclohexane
e. *cis*-1-ethyl-3-isopropylcyclohexane
f. *cis*-1-ethyl-4-isopropylcyclohexane

56. Explain why
a. H_2O has a higher boiling point than CH_3OH (65 °C).
b. H_2O has a higher boiling point than NH_3 (−33 °C).
c. H_2O has a higher boiling point than HF (20 °C).

57. How many ethers have molecular formula $C_5H_{12}O$? Draw their structures and name them.

58. Draw the most stable conformer of the following molecule:

CH_3 (wedge), H_3C (hash), CH_3 (hash) on cyclohexane ring

59. Give the systematic name for each of the following:

a. $CH_3CH_2CHCH_2CHCH_2CH_3$ with CH_3 on C-5 and $CHCH_3$ (bearing CH_3) on C-3

b. $CH_3CHCHCH_2CH_2CH_2Cl$ with CH_2CH_3 on C-2 and Cl on C-3

60. Which of the following can be used to verify that carbon is tetrahedral?
a. Methyl bromide does not have constitutional isomers.
b. Tetrachloromethane does not have a dipole.
c. Dibromomethane does not have constitutional isomers.

61. The most stable form of glucose (blood sugar) is a six-membered ring chair conformer with its five substituents all in equatorial positions. Draw the most stable form of glucose by putting the OH groups on the appropriate bonds in the chair conformer.

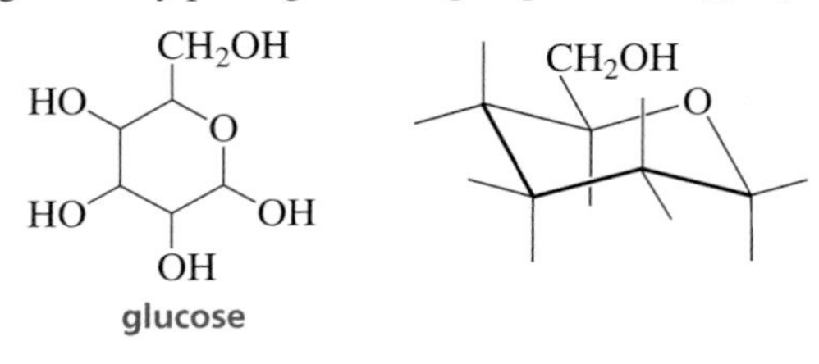

glucose

62. Draw the nine isomeric heptanes and name each isomer.

63. Draw the most stable conformer of 1,2,3,4,5,6-hexachlorocyclohexane.

64. a. Draw all the staggered and eclipsed conformations that result from rotation about the C-2—C-3 bond of pentane.
b. Draw a potential-energy diagram for rotation about the C-2—C-3 bond of pentane through 360°, starting with the least stable conformation.

65. Using Newman projections, draw the most stable conformation for the following:
a. 3-methylpentane, considering rotation about the C-2—C-3 bond
b. 3-methylhexane, considering rotation about the C-3—C-4 bond

66. For each of the following disubstituted cyclohexanes, indicate whether the substituents in its two chair conformers would be both equatorial in one chair conformer and both axial in the other *or* one equatorial and one axial in each of the chair conformers:

a. *cis*-1,2-
b. *trans*-1,2-
c. *cis*-1,3-
d. *trans*-1,3-
e. *cis*-1,4-
f. *trans*-1,4-

67. Which will have a higher percentage of the diequatorial-substituted conformer, compared with the diaxial-substituted conformer: *trans*-1,4-dimethylcyclohexane or *cis*-1-*tert*-butyl-3-methylcyclohexane?

Alkenes

Structure, Nomenclature, Stability, and an Introduction to Reactivity

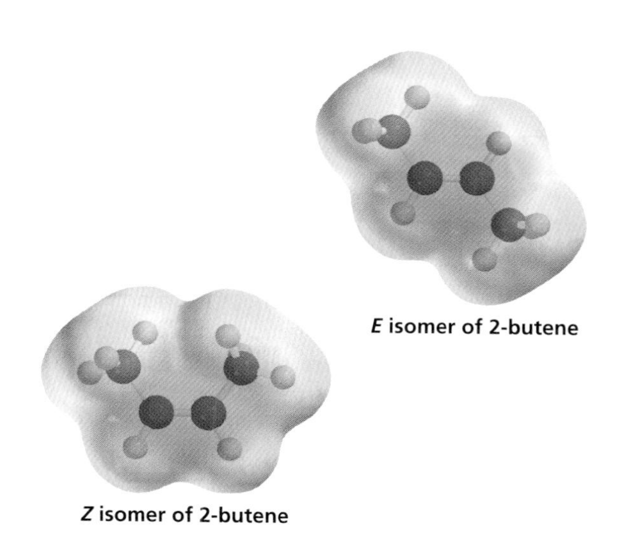

In Chapter 3, we saw that alkanes are hydrocarbons that contain only carbon–carbon *single* bonds. Hydrocarbons that contain a carbon–carbon *double* bond are called **alkenes**. Alkenes play many important roles in biology. Ethene, for example, is a plant hormone—a compound that controls growth and other changes in the plant's tissues. Among other things, ethene affects seed germination, flower maturation, and fruit ripening. Many of the flavors and fragrances produced by certain plants also belong to the alkene family.

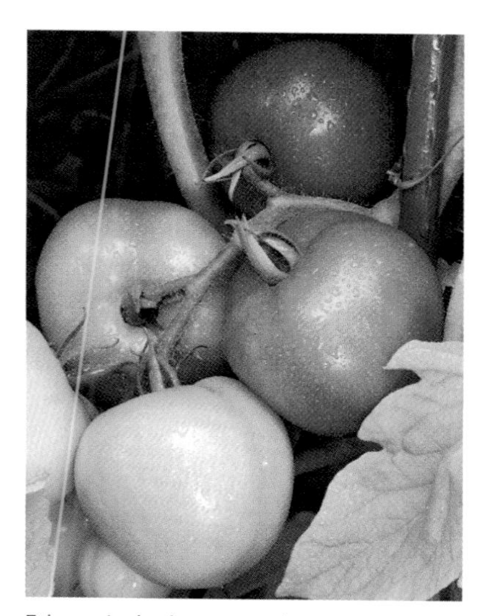

Ethene is the hormone that causes tomatoes to ripen.

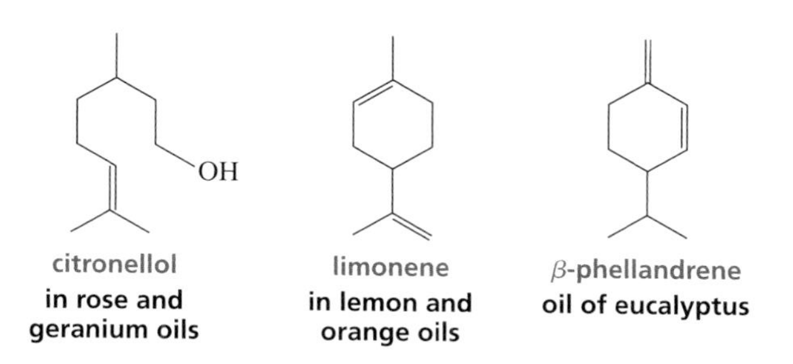

We will begin our study of alkenes by looking at how they are named, their structures, and their relative stabilities. Then we will examine a reaction of an alkene, paying close attention to the steps by which it occurs and the energy changes that accompany them. You will see that some of the discussion in this chapter revolves around concepts with which you are familiar, while some of the information is new and will broaden the foundation of knowledge you will be building on in subsequent chapters.

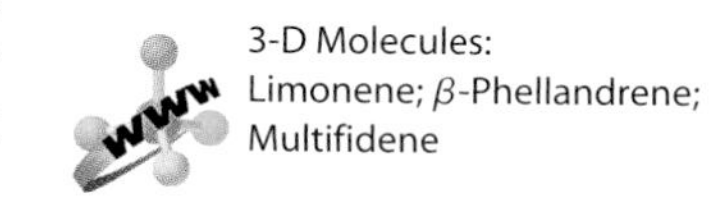

3-D Molecules: Limonene; β-Phellandrene; Multifidene

PHEROMONES

Insects communicate by releasing **pheromones**—chemical substances that other insects of the same species detect with their antennae. Many of the sex, alarm, and trail pheromones are alkenes. Interfering with an insect's ability to send or receive chemical signals is an environmentally safe way to control insect populations. For example, traps containing synthetic sex attractants have been used to capture such crop-destroying insects as the gypsy moth and the boll weevil.

muscalure
sex attractant of the housefly

multifidene
sex attractant of brown algae

4.1 MOLECULAR FORMULAS

We have seen that the general molecular formula for a noncyclic alkane is C_nH_{2n+2} (Section 3.0). We have also seen that the general molecular formula for a cyclic alkane is C_nH_{2n} because the cyclic structure reduces the number of hydrogens by two (Section 3.3).

The general molecular formula for a *noncyclic alkene* is also C_nH_{2n} because, as a result of the double bond, an alkene has two fewer hydrogens than an alkane with the same number of carbons. Thus, the general molecular formula for a *cyclic alkene* must be C_nH_{2n-2}. We can, therefore, make the following statement: *the general molecular formula for a hydrocarbon is* C_nH_{2n+2} *minus two hydrogens for every* π *bond or ring in the molecule*.

The general molecular formula for a hydrocarbon is C_nH_{2n+2}, minus two hydrogens for every π bond or ring present in the molecule.

$CH_3CH_2CH_2CH_2CH_3$
an alkane
C_5H_{12}
C_nH_{2n+2}

$CH_3CH_2CH_2CH{=}CH_2$
an alkene
C_5H_{10}
C_nH_{2n}

a cyclic alkane
C_5H_{10}
C_nH_{2n}

a cyclic alkene
C_5H_8
C_nH_{2n-2}

Because alkanes contain the maximum number of C—H bonds possible—that is, they are saturated with hydrogen—they are called **saturated hydrocarbons**. In contrast, alkenes are called **unsaturated hydrocarbons** because they have fewer than the maximum number of hydrogens.

$CH_3CH_2CH_2CH_3$
a saturated hydrocarbon

$CH_3CH{=}CHCH_3$
an unsaturated hydrocarbon

PROBLEM 1 ♦ SOLVED

Determine the molecular formula for each of the following:

a. a 5-carbon hydrocarbon with one π bond and one ring.
b. a 4-carbon hydrocarbon with two π bonds and no rings.
c. a 10-carbon hydrocarbon with one π bond and two rings.

Solution to 1a For a 5-carbon hydrocarbon with no π bonds and no rings, $C_nN_{2n+2} = C_5H_{12}$. A 5-carbon hydrocarbon with one π bond and one ring has four fewer hydrogens, because two hydrogens are subtracted for every π bond or ring present in the molecule. The molecular formula, therefore, is C_5H_8.

PROBLEM 2◆ SOLVED

Determine the total number of double bonds and/or rings for hydrocarbons with the following molecular formulas:

a. $C_{10}H_{16}$ **b.** $C_{20}H_{34}$ **c.** C_4H_8

Solution to 2a For a 10-carbon hydrocarbon with no π bonds and no rings, $C_nN_{2n+2} = C_{10}H_{22}$. A 10-carbon compound with molecular formula $C_{10}H_{16}$ has six fewer hydrogens. Therefore, it has a total of three double bonds and/or rings.

4.2 THE NOMENCLATURE OF ALKENES

The **functional group** is the center of reactivity in an organic molecule. In an alkene, the double bond is the functional group. The IUPAC system uses a suffix to denote certain functional groups. The systematic name of an alkene, for example, is obtained by replacing the *ane* at the end of the parent hydrocarbon's name with the suffix *ene*. Thus, a two-carbon alkene is called ethene, and a three-carbon alkene is called propene. Ethene also is frequently called by its common name: ethylene.

	$H_2C{=}CH_2$	$CH_3CH{=}CH_2$	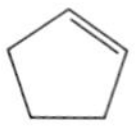 (cyclopentene ring)	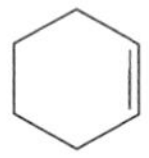 (cyclohexene ring)
systematic name:	ethene	propene	cyclopentene	cyclohexene
common name:	ethylene	propylene		

The following rules are used to name a compound with a functional group suffix:

1. The longest continuous chain containing the functional group (in this case, the carbon–carbon double bond) is numbered in a direction that gives the functional group suffix the lowest possible number. The position of the double bond is indicated by a number immediately preceding the name of the alkene. For example, 1-butene signifies that the double bond is between the first and second carbons of butene; 2-hexene signifies that the double bond is between the second and third carbons of hexene. (The four alkene names shown above do not need a number because there is no ambiguity.)

Number the longest continuous chain containing the functional group in the direction that gives the functional group suffix the lowest possible number.

$\overset{4}{C}H_3\overset{3}{C}H_2\overset{2}{C}H{=}\overset{1}{C}H_2$ — 1-butene

$\overset{1}{C}H_3\overset{2}{C}H{=}\overset{3}{C}H\overset{4}{C}H_3$ — 2-butene

$\overset{1}{C}H_3\overset{2}{C}H{=}\overset{3}{C}H\overset{4}{C}H_2\overset{5}{C}H_2\overset{6}{C}H_3$ — 2-hexene

Notice that 1-butene does not have a common name. You might be tempted to call it "butylene," which is analogous to "propylene" for propene, but butylene is not an appropriate name. A name must be unambiguous, and "butylene" could signify either 1-butene or 2-butene.

2. For a compound with two double bonds, the suffix is "diene."

$\overset{1}{C}H_2{=}\overset{2}{C}H{-}\overset{3}{C}H_2{-}\overset{4}{C}H{=}\overset{5}{C}H_2$ — 1,4-pentadiene

$\overset{1}{C}H_3\overset{2}{C}H{=}\overset{3}{C}H{-}\overset{4}{C}H{=}\overset{5}{C}H\overset{6}{C}H_2\overset{7}{C}H_3$ — 2,4-heptadiene

$\overset{5}{C}H_3\overset{4}{C}H{=}\overset{3}{C}H{-}\overset{2}{C}H{=}\overset{1}{C}H_2$ — 1,3-pentadiene

3. The name of a substituent is placed in front of the name of the longest continuous chain that contains the functional group, together with a number to designate the carbon to which the substituent is attached. Notice that *when a compound's name contains both a functional group suffix and a substituent, the functional group suffix gets the lowest possible number.*

When there is only a substituent, the substituent gets the lowest possible number.

When there is only a functional group suffix, the functional group suffix gets the lowest possible number.

When there is both a functional group suffix and a substituent, the functional group suffix gets the lowest possible number.

$\overset{1}{C}H_3\overset{2}{C}H{=}\overset{3}{C}H\overset{4}{C}H(CH_3)\overset{5}{C}H_3$ — 4-methyl-2-pentene

$CH_3\overset{3}{C}(\overset{2}{C}H_2\overset{1}{C}H_3){=}\overset{4}{C}H\overset{5}{C}H_2\overset{6}{C}H_2\overset{7}{C}H_3$ — 3-methyl-3-heptene

Substituents are cited in alphabetical order.

4. If a chain has more than one substituent, the substituents are listed in alphabetical order, using the same rules for alphabetizing discussed in Section 3.2. Then the appropriate number is assigned to each substituent.

3,6-dimethyl-3-octene

$$\underset{7\ \ 6\ \ 5\ \ 4\ \ 3\ \ 2\ \ 1}{CH_3CH_2\overset{Br}{C}H\overset{Cl}{C}HCH_2CH{=}CH_2}$$

5-bromo-4-chloro-1-heptene

A substituent receives the lowest possible number only if there is no functional group suffix or if the same number for the functional group suffix is obtained in both directions.

5. If counting in either direction results in the same number for the alkene functional group suffix, the correct name is the one containing the lowest substituent number.

$$CH_3CH_2CH_2\underset{CH_3}{C}{=}CHCH_2\underset{CH_3}{C}HCH_3$$

2,5-dimethyl-4-octene
not
4,7-dimethyl-4-octene
because 2 < 4

$$CH_3\underset{Br}{C}HCH{=}\underset{CH_3}{C}CH_2CH_3$$

2-bromo-4-methyl-3-hexene
not
5-bromo-3-methyl-3-hexene
because 2 < 3

6. A number is not needed to denote the position of the double bond in a cyclic alkene because the ring is always numbered so that the double bond is between carbons 1 and 2. To assign numbers to any substituents, count around the ring in the direction (clockwise or counterclockwise) that puts the lowest number into the name.

3-ethylcyclopentene   4,5-dimethylcyclohexene   4-ethyl-3-methylcyclohexene

7. Numbers are needed to denote the positions of the double bonds if the ring has two double bonds.

1,3-cyclohexadiene   1,4-cyclohexadiene   2-methyl-1,3-cyclopentadiene

Remember that the name of a substituent is placed *before* the name of the parent hydrocarbon, and the functional group suffix is placed *after* the name of the parent hydrocarbon.

substituent—parent hydrocarbon—functional group suffix

The sp^2 carbons of an alkene are called **vinylic carbons**. An sp^3 carbon that is adjacent to a vinylic carbon is called an **allylic carbon**.

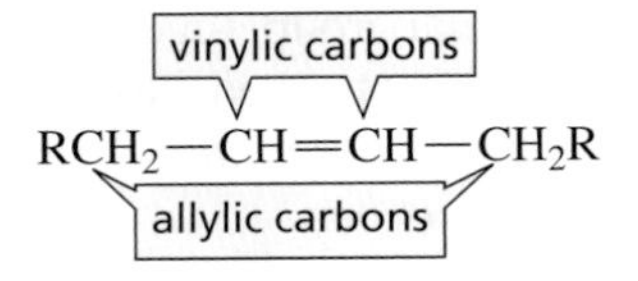

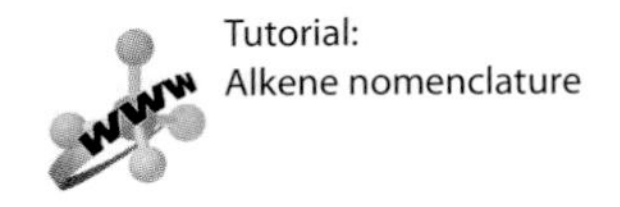

Tutorial:
Alkene nomenclature

Two groups containing a carbon–carbon double bond are used in common names—the **vinyl group** and the **allyl group**. The vinyl group is the smallest possible group containing a vinylic carbon; the allyl group is the smallest possible group containing an allylic carbon. When "vinyl" or "allyl" is used in a name, the substituent must be attached to the vinylic or allylic carbon, respectively.

	$H_2C{=}CH{-}$	$H_2C{=}CHCH_2{-}$
	the vinyl group	**the allyl group**
	$H_2C{=}CHCl$	$H_2C{=}CHCH_2Br$
systematic name:	chloroethene	3-bromopropene
common name:	vinyl chloride	allyl bromide

PROBLEM 3

Draw the structure for each of the following:

a. 3,3-dimethylcyclopentene
b. 6-bromo-2,3-dimethyl-2-hexene
c. ethyl vinyl ether
d. allyl alcohol

PROBLEM 4 ◆

Give the systematic name for each of the following:

a. $CH_3CHCH{=}CHCH_3$ (with CH_3 on C-2)

b. $CH_3CH_2C{=}CCHCH_3$ (with CH_3 on C-3, CH_3 on C-4, Cl on C-5)

c. $BrCH_2CH_2CH{=}CCH_3$ (with CH_2CH_3 on C-4)

d. H_3C–cyclohexene–CH_3

4.3 THE STRUCTURE OF ALKENES

The structure of the smallest alkene (ethene) was described in Section 1.8. Other alkenes have similar structures. Each double-bonded carbon of an alkene has three sp^2 orbitals that lie in a plane with angles of 120°. Each of these sp^2 orbitals overlaps an orbital of another atom to form a σ bond. Thus, one of the carbon–carbon bonds in a double bond is a σ bond, formed by the overlap of an sp^2 orbital of one carbon with an sp^2 orbital of the other carbon. The second carbon–carbon bond in the double bond (the π bond) is formed by side-to-side overlap of the remaining p orbital of one of the sp^2 carbons with the remaining p orbital of the other sp^2 carbon. Because three points determine a plane, each sp^2 carbon and the two atoms singly bonded to it lie in a plane. In order to achieve maximum orbital–orbital overlap, the two p orbitals must be parallel to each other. Therefore, all six atoms of the double-bond system are in the same plane.

$$(H_3C)_2C{=}C(CH_3)_2$$

the six carbon atoms are in the same plane

It is important to remember that *the π bond represents the cloud of electrons that is above and below the plane defined by the two sp^2 carbons and the four atoms bonded to them.*

p orbitals overlap to form a π bond

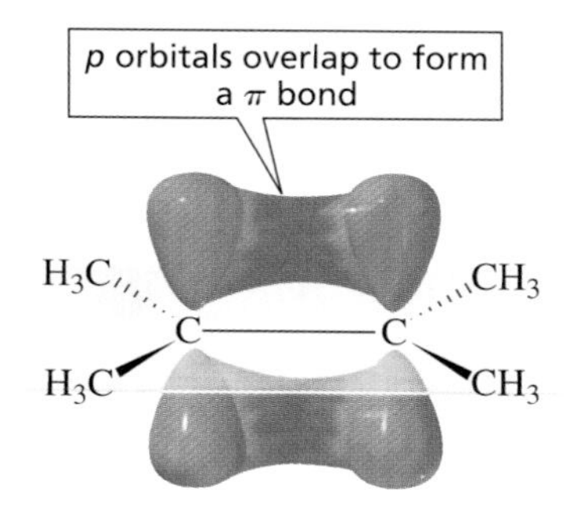

3-D Molecules: 2,3-Dimethyl-2-butene

PROBLEM 5◆ SOLVED

For each of the following compounds, tell how many of its carbon atoms lie in the same plane:

a. (cyclohexene with CH_3) b. (cyclohexene with CH_3) c. (cyclohexene with H_3C) d. (cyclohexene with CH_3, CH_3)

Solution to 5a The two sp^2 carbons (blue dots) and the carbons bonded to each of the sp^2 carbons (red dots) lie in the same plane. Therefore, five carbons lie in the same plane.

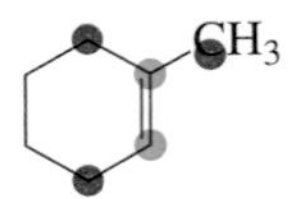
CH3

4.4 ALKENES CAN HAVE CIS–TRANS ISOMERS

We have just seen that the two p orbitals forming the π bond must be parallel to achieve maximum overlap. Therefore, rotation about a double bond does not readily occur. If rotation were to occur, the two p orbitals would cease to overlap, and the π bond would break (Figure 4.1).

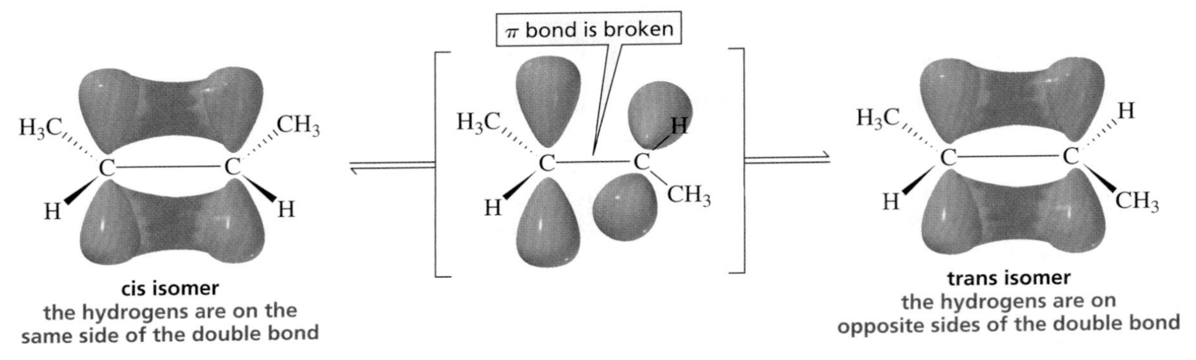
π bond is broken
H3C
CH3
C
H
cis isomer
the hydrogens are on the same side of the double bond
trans isomer
the hydrogens are on opposite sides of the double bond

▲ **Figure 4.1**
Rotation about the carbon–carbon double bond would break the π bond.

Because of the high-energy barrier to rotation about a double bond, an alkene such as 2-butene can exist in two distinct forms: the hydrogens bonded to the sp^2 carbons can be on the same side of the double bond or on opposite sides of the double bond.

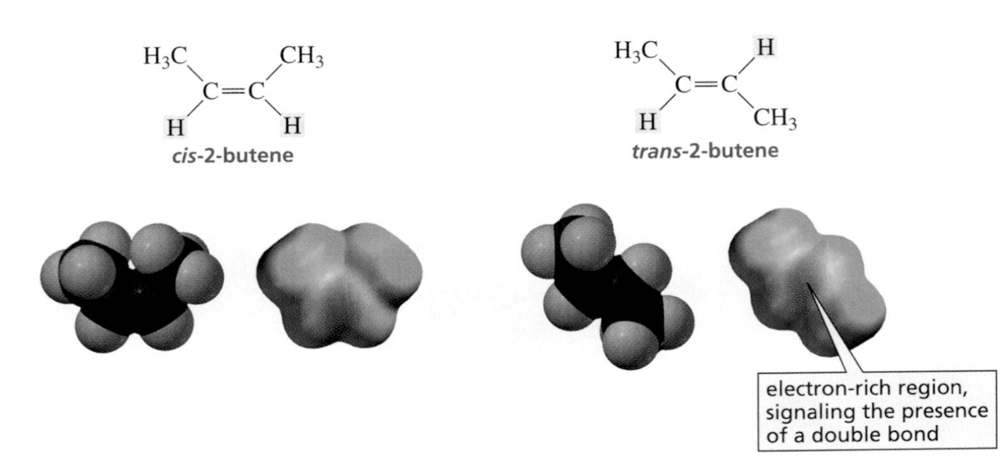
H3C
CH3
C=C
H
cis-2-butene
trans-2-butene
electron-rich region, signaling the presence of a double bond

3-D Molecules: *cis*-2-Butene; *trans*-2-Butene

The isomer with the hydrogens on the same side of the double bond is called the **cis isomer**, and the isomer with the hydrogens on opposite sides of the double bond is called the **trans isomer**. A pair of isomers such as *cis*-2-butene and *trans*-2-butene is called **cis–trans isomers** or **geometric isomers**. This should remind you of the cis–trans isomers of disubstituted cyclohexanes you encountered in Section 3.12— the cis isomer had its substituents on the same side of the ring, and the trans isomer had its substituents on opposite sides of the ring.

If one of the sp^2 carbons of the double bond is attached to two identical substituents, there is only one possible structure for the alkene. In other words, cis and trans isomers are not possible for an alkene that has identical substituents attached to one of the sp^2 carbons.

cis and trans isomers are not possible for these compounds because two substituents on an sp^2 carbon are the same

$(H)_2C{=}C(CH_3)Cl$ $\quad$ $CH_3CH_2(H)C{=}C(CH_3)_2$

3-D Molecules: 2-Methyl-2-pentene

The cis and trans isomers can be interconverted only when the molecule absorbs sufficient heat or light energy to cause the π bond to break, because once the π bond is broken, rotation can occur easily about the remaining σ bond (Section 3.8).

$$\underset{cis\text{-2-pentene}}{\text{H}_3\text{C(H)C}{=}\text{C(H)CH}_2\text{CH}_3} \underset{h\nu}{\overset{>180\ ^\circ\text{C}}{\rightleftharpoons}} \underset{trans\text{-2-pentene}}{\text{H}_3\text{C(H)C}{=}\text{C(H)CH}_2\text{CH}_3}$$

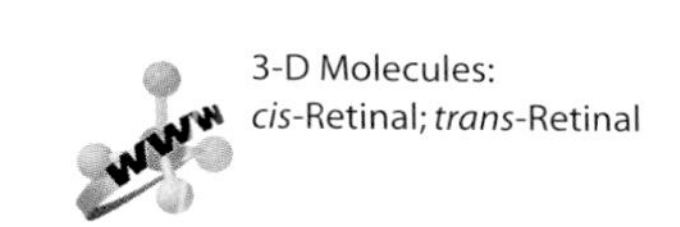

CIS–TRANS INTERCONVERSION IN VISION

When rhodopsin absorbs light, a double bond interconverts between the cis and trans forms. This process plays an important role in vision.

cis double bond — rhodopsin (N—opsin) $\xrightarrow{\text{light}}$ trans double bond (N—opsin)

OPSIN — cis form

OPSIN — trans form

PROBLEM 6 ♦

a. Which of the following compounds can exist as cis–trans isomers?
b. For those compounds, draw and label the cis and trans isomers.

1. $CH_3CH{=}CHCH_2CH_3$

2. $CH_3CH{=}C(CH_3)CH_3$

3. $CH_3CH{=}CHCH_3$

4. $CH_3CH_2CH{=}CH_2$

4.5 NAMING ALKENES USING THE *E,Z* SYSTEM

As long as each of the sp^2 carbons of an alkene is bonded to only one substituent, we can use the terms *cis* and *trans* to designate the structure of the alkene: *if the hydrogens are on the same side of the double bond, it is the cis isomer; if they are on opposite sides of the double bond, it is the trans isomer*. But how do we designate the isomers of a compound such as 1-bromo-2-chloropropene?

Br, H on one C; Cl, CH_3 on the other (Cl same side as Br) — Br, H on one C; CH_3, Cl on the other (CH_3 same side as Br)

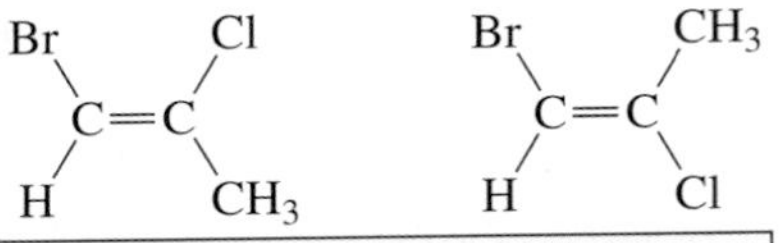

Which isomer is cis and which is trans?

For a compound such as 1-bromo-2-chloropropene, the cis-trans system of nomenclature cannot be used because there are four different groups on the two sp^2 carbons. The *E,Z* system of nomenclature was devised for such compounds.

To name an isomer by the *E,Z* system, we first determine the relative priorities of the two groups bonded to one of the sp^2 carbons and then the relative priorities of the two groups bonded to the other sp^2 carbon. (The rules for assigning relative priorities are explained below.) If the two high-priority groups (one from each carbon) are on the same side of the double bond, the isomer has the *Z* configuration (*Z* is for *zusammen*, German for "together"). If the high-priority groups are on opposite sides of the double bond, the isomer has the *E* configuration (*E* is for *entgegen*, German for "opposite").

The *Z* isomer has the high-priority groups on the same side.

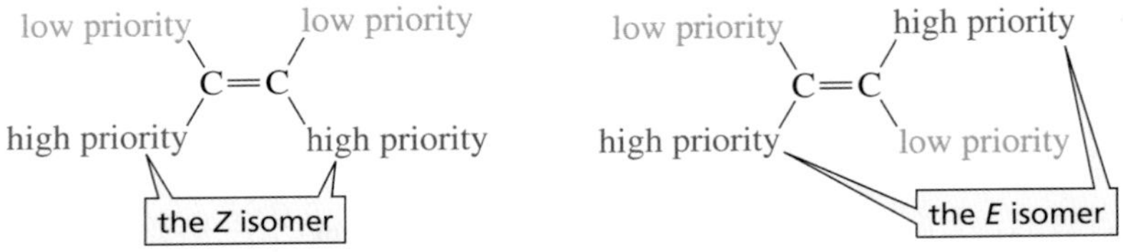

low priority / high priority on one C; low priority / high priority on the other (high priorities on same side) — the *Z* isomer

low priority / high priority on one C; high priority / low priority on the other (high priorities on opposite sides) — the *E* isomer

The relative priorities of the two groups bonded to an sp^2 carbon are determined using the following rules:

The greater the atomic number of the atom bonded to the sp^2 carbon, the higher is the priority of the substituent.

1. The relative priorities of the two groups depend on the atomic numbers of the two atoms bonded directly to the sp^2 carbon. The greater the atomic number, the higher the priority. For example, in the following compounds, one of the sp^2 carbons is bonded to a Br and to an H:

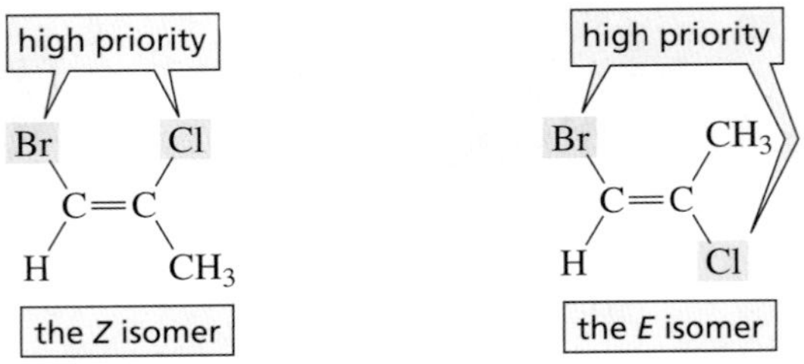

high priority: Br and Cl on same side (Br, H; Cl, CH_3) — the *Z* isomer

high priority: Br and Cl on opposite sides (Br, H; CH_3, Cl) — the *E* isomer

Br has a greater atomic number than H, so **Br** has a higher priority than **H**. The other sp^2 carbon is bonded to a Cl and to a C. Chlorine has the greater atomic number, so **Cl** has a higher priority than **C**. (Notice that you use the atomic number of C, not the mass of the CH_3 group, because the priorities are based on the atomic numbers of atoms, *not* on the masses of groups.) The isomer on the left has the high-priority groups (Br and Cl) on the same side of the double bond, so it is the **Z isomer**. (Zee groups are on Zee Zame Zide.) The isomer on the right has the high-priority groups on opposite sides of the double bond, so it is the ***E* isomer**.

2. If the two groups bonded to an sp^2 carbon start with the same atom (there is a tie), you then move outward from the point of attachment and consider the atomic numbers of the atoms that are attached to the "tied" atoms. For example, in the following compounds, both atoms bonded the sp^2 carbon on the left are C's (in a CH_2Cl group and a CH_2CH_2Cl group), so there is a tie.

If the atoms attached to an sp^2 carbon are the same, the atoms attached to the "tied" atoms are compared; the one with the greatest atomic number belongs to the group with the higher priority.

```
             CH3                          CH3
              |                            |
ClCH2CH2     CHCH3          ClCH2         CHCH3
       \    /                    \       /
        C=C                       C=C
       /    \                    /    \
  ClCH2      CH2OH       ClCH2CH2      CH2OH
```

the *Z* isomer   the *E* isomer

The C of the CH_2Cl group is bonded to **Cl**, **H**, **H**, and the C of the CH_2CH_2Cl group is bonded to **C**, **H**, **H**. Cl has a greater atomic number than C, so the CH_2Cl group has the higher priority. Both atoms bonded to the other sp^2 carbon are C's (from a CH_2OH group and a $CH(CH_3)_2$ group), so there is a tie on that side as well. The C of the CH_2OH group is bonded to **O**, **H**, and **H**, and the C of the $CH(CH_3)_2$ group is bonded to **C**, **C**, and **H**. Of these six atoms, O has the greatest atomic number, so $\mathbf{CH_2OH}$ has a higher priority than $\mathbf{CH(CH_3)_2}$. (Notice that you do not add the atomic numbers; you consider the single atom with the greatest atomic number.) The *E* and *Z* isomers are as shown above.

3. If an atom is doubly bonded to another atom, the priority system treats it as if it were singly bonded to two of those atoms. If an atom is triply bonded to another atom, the priority system treats it as if it were singly bonded to three of those atoms. For example, one of the sp^2 carbons in the following pair of isomers is bonded to a CH_2CH_2OH group and to a $CH_2C{\equiv}CH$ group:

If an atom is doubly bonded to another atom, treat it as if it were singly bonded to two of those atoms.

If an atom is triply bonded to another atom, treat it as if it were singly bonded to three of those atoms.

```
HOCH2CH2     CH=CH2        HOCH2CH2     CH2CH3
       \    /                     \    /
        C=C                        C=C
       /    \                     /    \
HC≡CCH2      CH2CH3        HC≡CCH2      CH=CH2
```

the *Z* isomer   the *E* isomer

Because the atoms immediately bonded to the sp^2 carbon on the left are both bonded to **C**, **H**, **H**, we ignore them and turn our attention to the groups attached to them. One of these is CH_2OH and the other is $C{\equiv}CH$. The triple-bonded C is considered to be bonded to **C**, **C**, and **C**; the other C is bonded to **O**, **H**, and **H**. Of the six atoms, O has the greatest atomic number, so $\mathbf{CH_2OH}$ has a higher priority than $\mathbf{C{\equiv}CH}$. Both atoms bonded to the other sp^2 carbon are C's, so they are tied. The first carbon of the $\mathbf{CH_2CH_3}$ group is bonded to **C**, **H**, and **H**. The first carbon of the $\mathbf{CH{=}CH_2}$ group is bonded to an H and doubly bonded to a C. Therefore, it is considered to be bonded to **H**, **C**, and **C**. One C cancels in each of the two groups, leaving H and H in the CH_2CH_3 group and C and H in the $CH{=}CH_2$ group. C has a greater atomic number than H, so $\mathbf{CH{=}CH_2}$ has a higher priority than $\mathbf{CH_2CH_3}$.

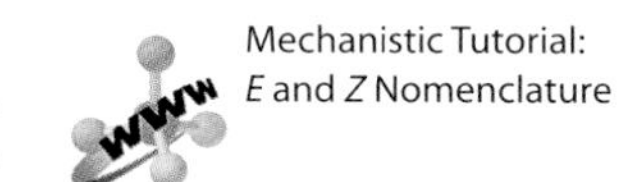

PROBLEM 7 ◆

Assign relative priorities to each set of substituents:

a. Br, —I, —OH, —CH_3

b. —CH_2CH_2OH, —OH, —CH_2Cl, —$CH{=}CH_2$

PROBLEM 8

Draw and label the *E* and *Z* isomers for each of the following:

a. $CH_3CH_2CH{=}CHCH_3$

b. $CH_3CH_2C(Cl){=}CHCH_2CH_3$

c. $CH_3CH_2C(CH_2CH_2CH_2CH_3){=}C(CH(CH_3)_2)CH_2Cl$

PROBLEM 9

Indicate whether each of the following is an *E* isomer or a *Z* isomer (black = carbon, white = hydrogen, green = chlorine, and red = oxygen):

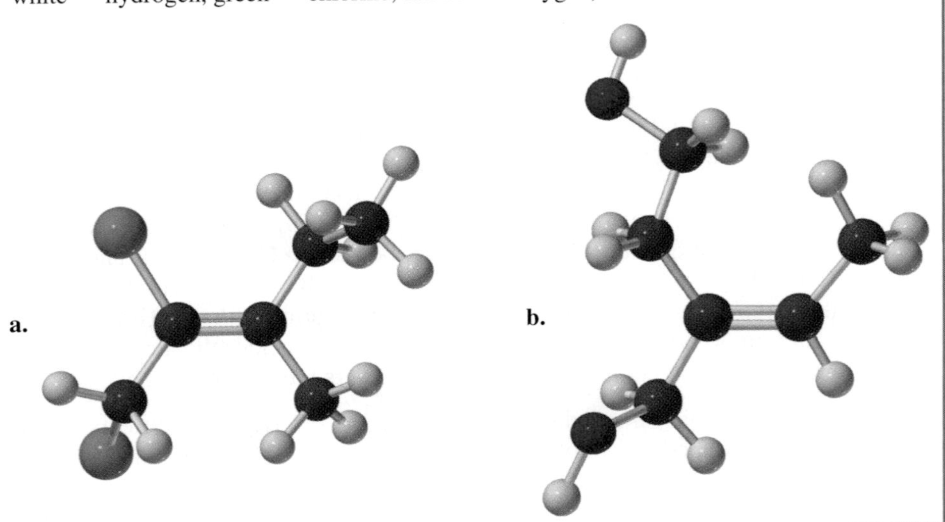

PROBLEM-SOLVING STRATEGY

Draw the structure of (*E*)-1-bromo-2-methyl-2-butene.

First draw the compound without specifying the isomer so you can see what groups are bonded to the sp^2 carbons. Now determine the relative priorities of the two groups on each of the sp^2 carbons.

$$BrCH_2C(CH_3){=}CHCH_3$$

The sp^2 carbon on the right is attached to a CH_3 and to an H; CH_3 has the higher priority. The other sp^2 carbon is attached to a CH_3 and to CH_2Br; CH_2Br has the higher priority. To get the *E* isomer, draw the compound with the two high-priority groups on opposite sides of the double bond.

$$\begin{matrix} BrCH_2 & & H \\ & C{=}C & \\ H_3C & & CH_3 \end{matrix}$$

Now continue on to Problem 10.

PROBLEM 10 ◆

Draw the structure of (*Z*)-3-isopropyl-2-heptene.

4.6 THE RELATIVE STABILITIES OF ALKENES

Alkyl substituents bonded to the sp^2 carbons of an alkene have a stabilizing effect on the alkene.

We can, therefore, make the following statement: *the more alkyl substituents bonded to the sp^2 carbons of an alkene, the greater is its stability*. (Some students find it easier to look at the number of hydrogens bonded to the sp^2 carbons. In terms of hydrogens, the statement is: *the fewer hydrogens bonded to the sp^2 carbons of an alkene, the greater is its stability.*)

The fewer hydrogens bonded to the sp^2 carbons of an alkene, the more stable it is.

relative stabilities of alkyl-substituted alkenes

most stable $R_2C{=}CR_2$ > $R_2C{=}CHR$ > $R_2C{=}CH_2$ > $RHC{=}CH_2$ least stable

PROBLEM 11

a. Which of the following compounds is the most stable?

[three isomeric diethylcyclohexenes, each with two CH_2CH_3 groups]

b. Which is the least stable?

Both *cis*-2-butene and *trans*-2-butene have two alkyl groups bonded to their sp^2 carbons. The trans isomer, in which the large substituents are farther apart, is more stable than the cis isomer, in which the large substituents are closer together.

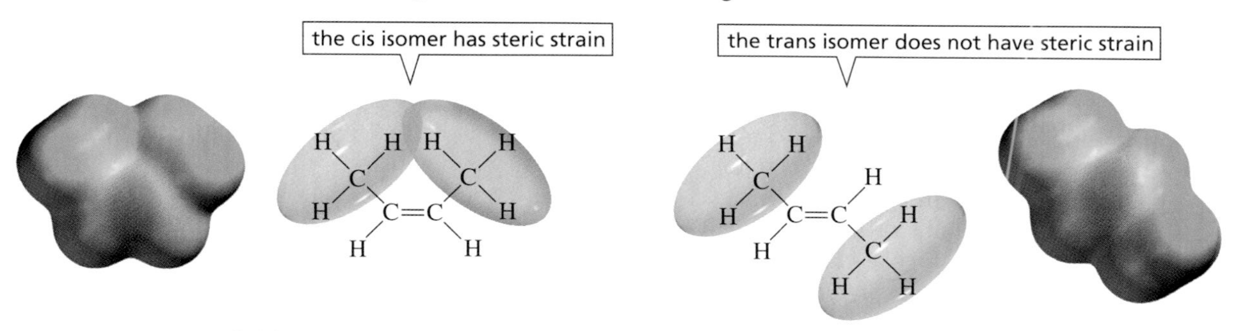
the cis isomer has steric strain
the trans isomer does not have steric strain
H
C
C=C

cis-2-butene *trans*-2-butene

When the large substituents are on the same side of the molecule, their electron clouds can interfere with each other, causing steric strain in the molecule and making it less stable (Section 3.8). When the large substituents are on opposite sides of the molecule, their electron clouds cannot interact so the molecule has less steric strain and is, therefore, more stable.

PROBLEM 12 ◆

Rank the following compounds in order of decreasing stability:

trans-3-hexene, *cis*-3-hexene, *cis*-2,5-dimethyl-3-hexene, *cis*-3,4-dimethyl-3-hexene

4.7 HOW ALKENES REACT • CURVED ARROWS SHOW THE BONDS THAT BREAK AND THE BONDS THAT FORM

There are many millions of organic compounds. If you had to memorize how each of them reacts, studying organic chemistry would not be a very pleasant experience. Fortunately, organic compounds can be divided into families, and all the members of a family react in similar ways. What makes learning organic chemistry even easier is that there are only a few rules that govern the reactivity of each family.

The family that an organic compound belongs to is determined by its functional group. The **functional group** is the center of reactivity of a molecule. You will find a table of common functional groups inside the back cover of this book. You are already familiar with the functional group of an alkene: the carbon–carbon double bond. All compounds with a carbon–carbon double bond react in similar ways, whether the compound is a small molecule like ethene or a large molecule like cholesterol.

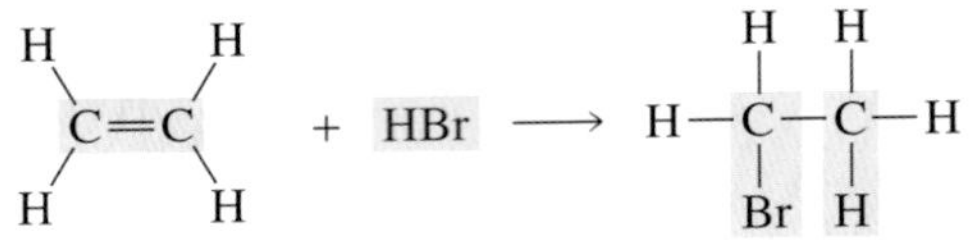

$$H_2C{=}CH_2 + HBr \longrightarrow H{-}\underset{Br}{\overset{H}{C}}{-}\underset{H}{\overset{H}{C}}{-}H$$

ethene

cholesterol + HBr ⟶ (product with Br and H added across the double bond)

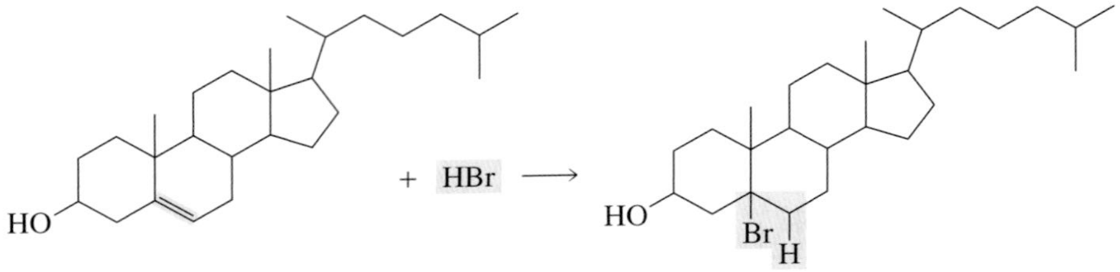

cholesterol

It is not sufficient to know that a compound with a carbon–carbon double bond reacts with HBr to form a product in which the H and Br atoms have taken the place of the π bond; we need to understand *why* the compound reacts with HBr. In each chapter that discusses the reactivity of a particular functional group, we will see how the nature of the functional group allows us to predict the kind of reactions it will undergo. Then, when you are confronted with a reaction you have never seen before, knowing how the structure of the molecule affects its reactivity will help you predict the products of the reaction.

In essence, organic chemistry is all about the interaction between electron-rich atoms or molecules and electron-deficient atoms or molecules. These are the forces that make chemical reactions happen. From this observation follows a very important rule for predicting the reactivity of organic compounds: *electron-rich atoms or molecules are attracted to electron-deficient atoms or molecules.* Each time you study a new functional group, remember that the reactions it undergoes can be explained by this very simple rule.

Electron-rich atoms or molecules are attracted to electron-deficient atoms or molecules.

Therefore, to understand how a functional group reacts, you must first learn to recognize electron-deficient and electron-rich atoms and molecules. An electron-deficient atom or molecule is called an **electrophile**. Literally, "electrophile" means "electron loving" (*phile* is the Greek suffix for "loving"). An electrophile looks for electrons.

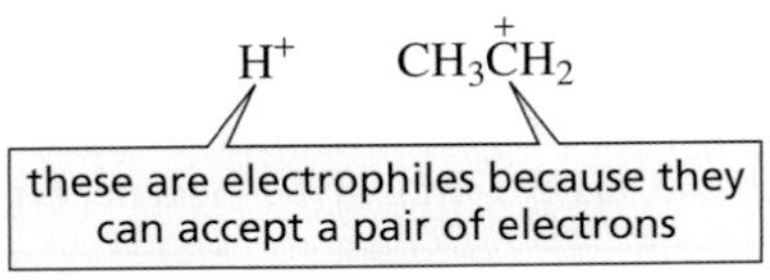

H^+ $\quad CH_3\overset{+}{C}H_2$

these are electrophiles because they can accept a pair of electrons

An electron-rich atom or molecule is called a **nucleophile**. A nucleophile has a pair of electrons it can share. Because a nucleophile has electrons to share and an electrophile is seeking electrons, it should not be surprising that they attract each other. Thus, the preceding rule can be restated as *a nucleophile reacts with an electrophile.*

A nucleophile reacts with an electrophile.

$$H\ddot{\underset{..}{O}}:^- \quad :\ddot{\underset{..}{Cl}}:^- \quad CH_3\ddot{N}H_2 \quad H_2\ddot{O}:$$

these are nucleophiles because they have a pair of electrons to share

PROBLEM 13 ♦

Identify the electrophile and the nucleophile in each of the following acid–base reactions:

a. $AlCl_3 + :NH_3 \rightleftharpoons Cl_3\bar{Al}-\overset{+}{N}H_3$

b. $H-\ddot{\underset{..}{Br}}: + H\ddot{\underset{..}{O}}:^- \rightleftharpoons :\ddot{\underset{..}{Br}}:^- + H_2\ddot{O}:$

We have seen that the π bond of an alkene consists of a cloud of electrons above and below the plane defined by the sp^2 carbons and the four atoms bonded to them. As a result of this cloud of electrons, an alkene is an electron-rich molecule—it is a nucleophile. (Notice the relatively electron-rich pale orange area in the electrostatic potential maps for *cis*- and *trans*-2-butene in Section 4.4.) We have also seen that a π bond is weaker than a σ bond (Section 1.14). The π bond, therefore, is the bond that is most easily broken when an alkene undergoes a reaction. For these reasons, we can predict that an alkene will react with an electrophile, and, in the process, the π bond will break. So if a reagent such as hydrogen bromide is added to an alkene, the alkene (a nucleophile) will react with the partially positively charged hydrogen (an electrophile) of hydrogen bromide and a carbocation will be formed. In the second step of the reaction, the positively charged carbocation (an electrophile) will react with the negatively charged bromide ion (a nucleophile) to form an alkyl halide.

$$CH_3CH{=}CHCH_3 + \overset{\delta+}{H}-\overset{\delta-}{Br} \longrightarrow CH_3\overset{+}{C}H-\underset{H}{\underset{|}{C}}HCH_3 + Br^- \longrightarrow CH_3\underset{Br}{\underset{|}{C}}H-\underset{H}{\underset{|}{C}}HCH_3$$

a carbocation — **2-bromobutane, an alkyl halide**

This step-by-step description of the process by which reactants (alkene + HBr) are changed into products (alkyl halide) is called the **mechanism of the reaction**. To help us understand a mechanism, curved arrows are drawn to show how the electrons move as new covalent bonds are formed and existing covalent bonds are broken. These are called "curved" arrows to distinguish them from the "straight" arrow used to link reactants with products in a chemical reaction. Each curved arrow represents the simultaneous movement of two electrons *from an electron-rich center* (at the tail of the arrow) and *toward an electron-deficient center* (at the point of the arrow). In this way, the curved arrows show which bonds are broken and which bonds are formed.

Curved arrows are used to show the bonds that break and the bonds that form; they are drawn from an electron-rich center to an electron-deficient center.

$$CH_3CH{=}CHCH_3 + \overset{\delta+}{H}-\overset{\delta-}{\ddot{\underset{..}{Br}}}: \longrightarrow CH_3\overset{+}{C}H-\underset{H}{\underset{|}{C}}HCH_3 + :\ddot{\underset{..}{Br}}:^-$$

π bond has broken

new σ bond has formed

For the reaction of 2-butene with HBr, an arrow is drawn to show that the two electrons of the π bond of the alkene are attracted to the partially positively charged hydrogen of HBr. The hydrogen is not immediately free to accept this pair of electrons because it is already bonded to a bromine, and hydrogen can be bonded to only one

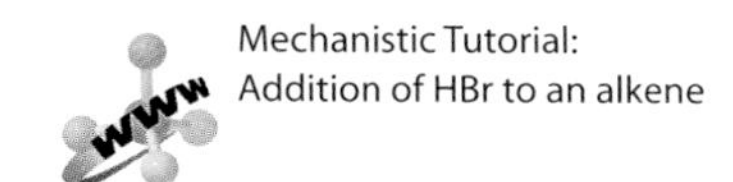

atom at a time (Section 1.4). However, as the π electrons of the alkene move toward the hydrogen, the H—Br bond breaks, with bromine keeping the bonding electrons. Notice that the π electrons are pulled away from one carbon but remain attached to the other. Thus, the two electrons that formerly formed the π bond now form a σ bond between carbon and the hydrogen from HBr. The product of this first step in the reaction is a carbocation because the sp^2 carbon that did not form the new bond with hydrogen has lost a share in an electron pair (the electrons of the π bond) and is, therefore, positively charged.

In the second step of the reaction, a lone pair on the negatively charged bromide ion forms a bond with the positively charged carbon of the carbocation. Notice that in both steps of the reaction *an electrophile reacts with a nucleophile.*

$$CH_3\overset{+}{C}H-CH(H)CH_3 + :\ddot{Br}:^- \longrightarrow CH_3CH(Br)-CH(H)CH_3$$

new σ bond

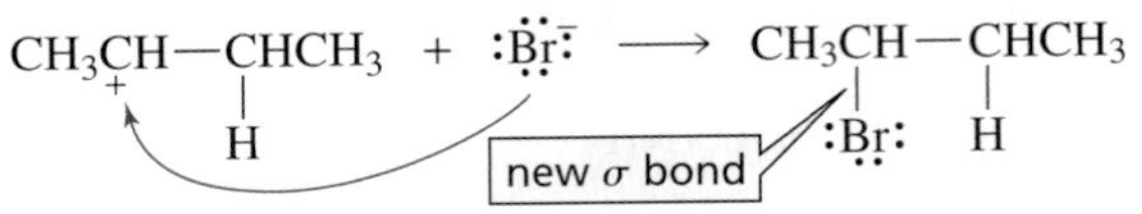

Solely from the knowledge that an electrophile reacts with a nucleophile and a π bond is the weakest bond in an alkene, we have been able to predict that the product of the reaction of 2-butene and HBr is 2-bromobutane. The overall reaction involves the *addition* of 1 mole of HBr to 1 mole of the alkene. The reaction, therefore, is called an **addition reaction**. Because the first step of the reaction is the addition of an electrophile (H^+) to the alkene, the reaction is more precisely called an **electrophilic addition reaction**. *Electrophilic addition reactions are the characteristic reactions of alkenes.*

At this point, you may think that it would be easier just to memorize the fact that 2-bromobutane is the product of the reaction, without trying to understand the mechanism that explains why 2-bromobutane is the product. Keep in mind, however, that you will soon be encountering a great number of reactions, and you will not be able to memorize them all. If you strive to understand the mechanism of each reaction, however, the unifying principles of organic chemistry will soon be clear to you, making mastery of the material much easier and a lot more fun.

It will be helpful to do the exercise on drawing curved arrows in the Study Guide/Solution Manual (Special Topic III).

PROBLEM 14 ◆

Which of the following are electrophiles, and which are nucleophiles?

H^- CH_3O^- $CH_3C{\equiv}CH$ $CH_3\overset{+}{C}HCH_3$ NH_3

A FEW WORDS ABOUT CURVED ARROWS

1. Draw a curved arrow so that it points in the direction of electron flow and never away from the flow. This means that an arrow will always be drawn away from a negative charge and/or toward a positive charge. An arrow is used to show both the bond that forms and the bond that breaks.

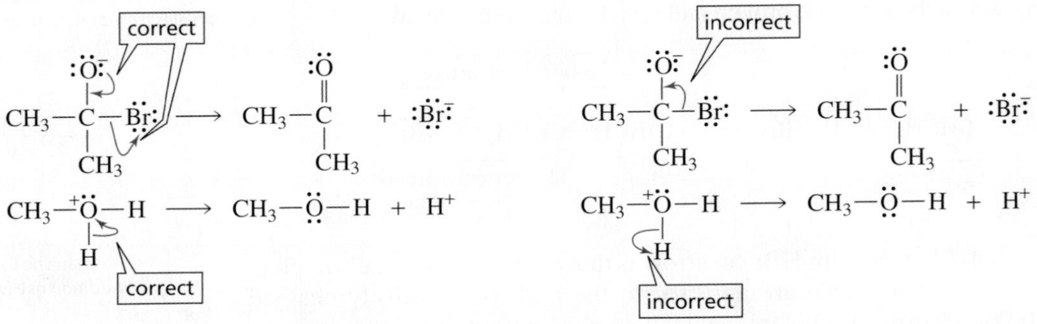

correct

$$CH_3-C(:\ddot{O}:^-)(CH_3)-\ddot{Br}: \longrightarrow CH_3-C(=\ddot{O})CH_3 + :\ddot{Br}:^-$$

incorrect

$$CH_3-C(:\ddot{O}:^-)(CH_3)-\ddot{Br}: \longrightarrow CH_3-C(=\ddot{O})CH_3 + :\ddot{Br}:^-$$

$$CH_3-\overset{+}{\ddot{O}}(H)-H \longrightarrow CH_3-\ddot{O}-H + H^+$$

correct

$$CH_3-\overset{+}{\ddot{O}}(H)-H \longrightarrow CH_3-\ddot{O}-H + H^+$$

incorrect

2. Curved arrows are meant to indicate the movement of electrons. Never use a curved arrow to indicate the movement of an atom. For example, do not use an arrow as a lasso to remove the proton, as shown here:

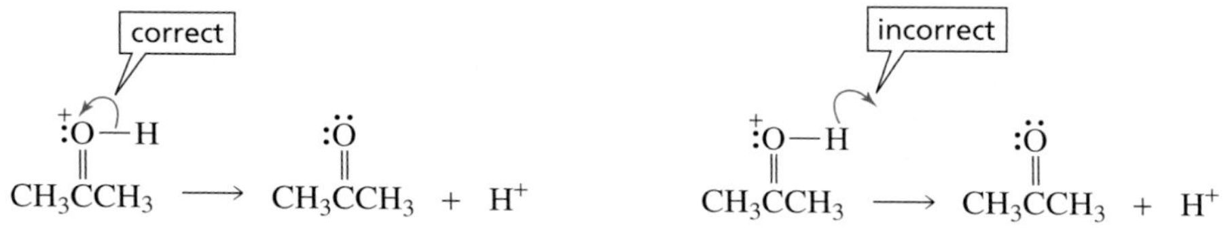

correct: $CH_3C(=\overset{+}{O}H)CH_3 \longrightarrow CH_3C(=O)CH_3 + H^+$

incorrect: $CH_3C(=\overset{+}{O}H)CH_3 \longrightarrow CH_3C(=O)CH_3 + H^+$

3. A head of a curved arrow always points at an atom or at a bond. Never draw the arrow head pointing out into space.

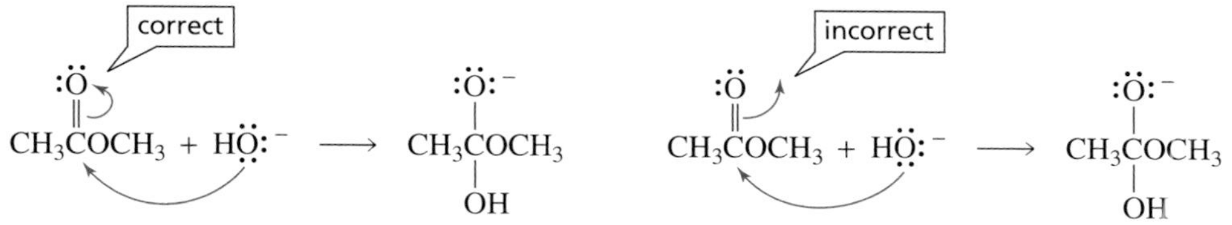

correct: $CH_3C(=O)OCH_3 + HO^- \longrightarrow CH_3C(O^-)(OH)OCH_3$

incorrect: $CH_3C(=O)OCH_3 + HO^- \longrightarrow CH_3C(O^-)(OH)OCH_3$

4. The arrow always starts at the electron source. In the following example, the arrow starts at the electron-rich π bond, not at a carbon atom:

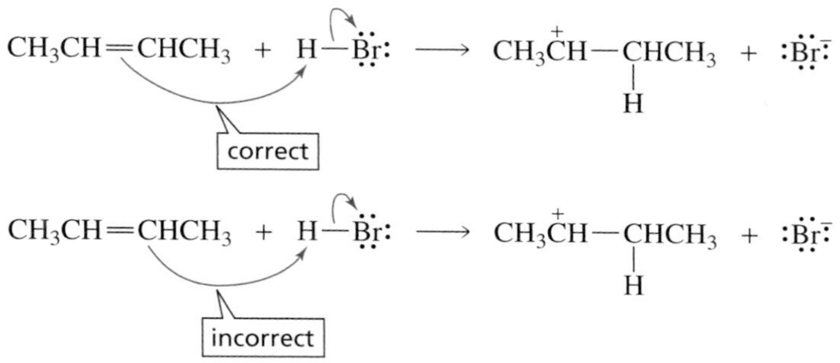

correct: $CH_3CH{=}CHCH_3 + H{-}Br \longrightarrow CH_3\overset{+}{C}H{-}CH(H)CH_3 + Br^-$

incorrect: $CH_3CH{=}CHCH_3 + H{-}Br \longrightarrow CH_3\overset{+}{C}H{-}CH(H)CH_3 + Br^-$

PROBLEM 15

Use curved arrows to show the movement of electrons in each of the following reaction steps. (*Hint:* Look at the starting material and look at the products, then draw the arrows.)

a. 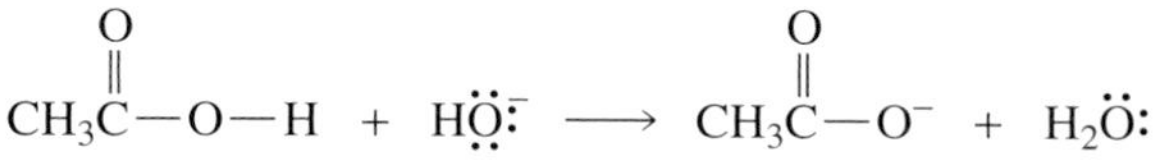 $CH_3C(=O){-}O{-}H + HO^- \longrightarrow CH_3C(=O){-}O^- + H_2O$

b. 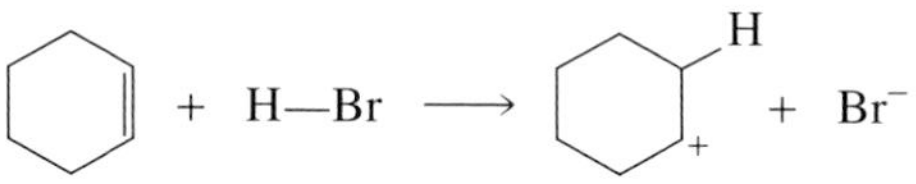 cyclohexene + H—Br ⟶ cyclohexyl cation (H added) + Br^-

c. 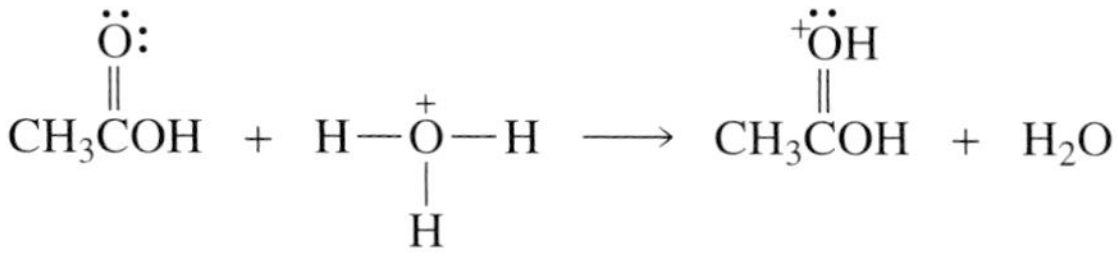 $CH_3C(=O)OH + H{-}\overset{+}{O}(H){-}H \longrightarrow CH_3C(={}^{+}OH)OH + H_2O$

d. 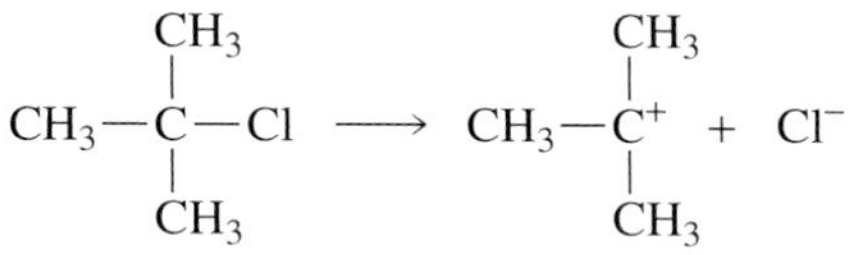 $CH_3{-}C(CH_3)_2{-}Cl \longrightarrow CH_3{-}C^+(CH_3)_2 + Cl^-$

PROBLEM 16

For reactions a-c in Problem 15, indicate which reactant is the nucleophile and which is the electrophile.

4.8 A REACTION COORDINATE DIAGRAM DESCRIBES THE ENERGY CHANGES THAT TAKE PLACE DURING A REACTION

We have just seen that the addition of HBr to 2-butene is a two-step process (Section 4.7). In each step, the reactants pass through a *transition state* as they are converted into products. The structure of the **transition state** for each of the steps is shown below in brackets; it lies somewhere between the structure of the reactants and the structure of the products. Notice that the bonds that break and the bonds that form during the course of the reaction are partially broken and partially formed in the transition state, as indicated by dashed lines. Similarly, atoms that either become charged or lose their charge during the course of the reaction are partially charged in the transition state. Transition states are always shown in brackets with a double-dagger superscript.

$$CH_3CH{=}CHCH_3 + H{-}Br \longrightarrow \left[\begin{array}{c} CH_3\overset{\delta+}{C}H{\cdots}CHCH_3 \\ \vdots \\ H \\ \vdots \\ {}^{\delta-}Br \end{array} \right]^{\ddagger} \longrightarrow CH_3\underset{+}{C}HCH_2CH_3 + Br^-$$

transition state

$$CH_3\underset{+}{C}HCH_2CH_3 + Br^- \longrightarrow \left[\begin{array}{c} CH_3\overset{\delta+}{C}HCH_2CH_3 \\ \vdots \\ {}^{\delta-}Br \end{array} \right]^{\ddagger} \longrightarrow \begin{array}{c} CH_3CHCH_2CH_3 \\ | \\ Br \end{array}$$

transition state

The energy changes that take place in each step of the reaction can be described by a **reaction coordinate diagram** (Figure 4.2). In a reaction coordinate diagram, the total energy of all species is plotted against the progress of the reaction. A reaction progresses from left to right as written in the chemical equation, so the energy of the reactants is plotted on the left-hand side of the x-axis and the energy of the products is plotted on the right-hand side.

Figure 4.2a shows that, in the first step of the reaction, the alkene is converted into a carbocation that is less stable than the reactants. Remember that *the more stable the species, the lower is its energy*. Because the product of the first step is less stable than the reactants, we know that this step consumes energy. We see that as the carbocation is formed, the reaction passes through the transition state. Notice that the transition state is a *maximum* energy state on the reaction coordinate diagram.

The more stable the species, the lower is its energy.

The carbocation reacts in the second step with the bromide ion to form the final product (Figure 4.2b). Because the product is more stable than the reactants, we know that this step releases energy.

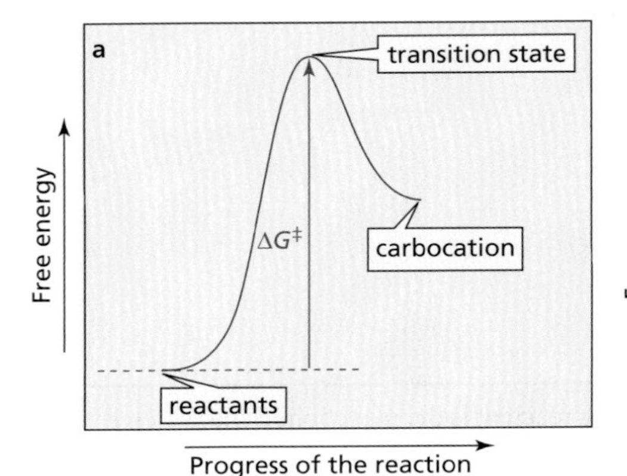

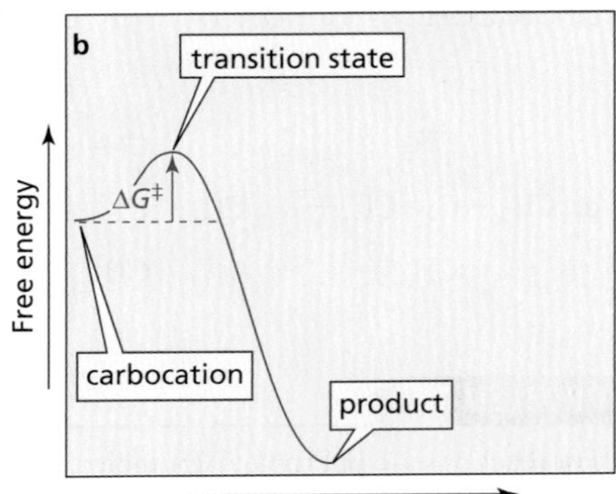

▶ **Figure 4.2**
The reaction coordinate diagrams for the two steps in the addition of HBr to 2-butene:
(a) the first step;
(b) the second step.

A chemical species that is a product of one step of a reaction and a reactant for the next step is called an **intermediate**. Thus, the carbocation is an intermediate. Although the carbocation is more stable than either of the transition states, it is still too unstable to be isolated. Do not confuse transition states with intermediates: *transition states have partially formed bonds, whereas intermediates have fully formed bonds.*

Transition states have partially formed bonds. Intermediates have fully formed bonds.

Because the product of the first step is the reactant of the second step, we can hook the two reaction coordinate diagrams together to obtain the reaction coordinate diagram for the overall reaction (Figure 4.3).

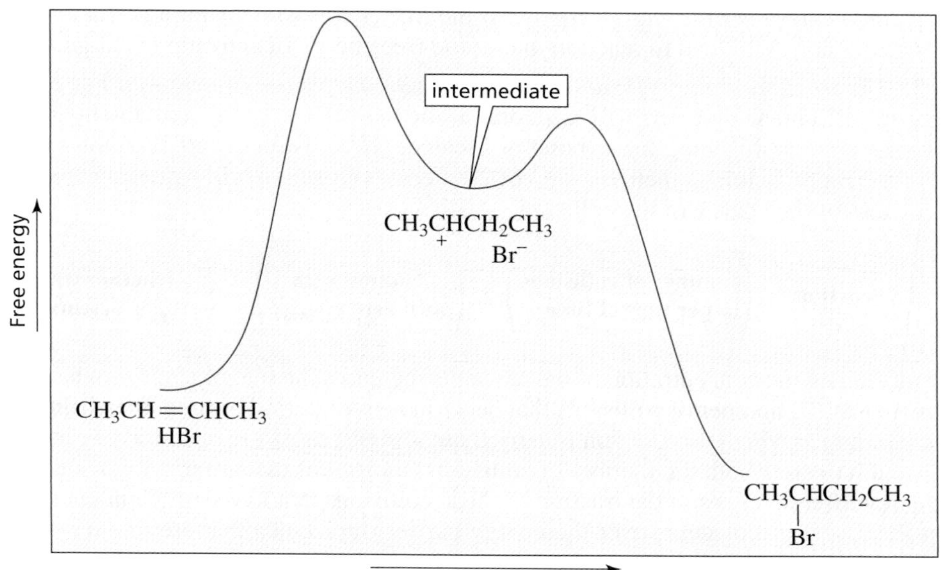

◀ **Figure 4.3**
The reaction coordinate diagram for the addition of HBr to 2-butene.

A reaction is over when the system reaches equilibrium. The relative concentrations of products and reactants at equilibrium depend on their relative stabilities: *the more stable the compound, the greater is its concentration at equilibrium*. Thus, if the products are more stable (have a lower free energy) than the reactants, there will be a higher concentration of products than reactants at equilibrium. On the other hand, if the reactants are more stable than the products, there will be a higher concentration of reactants than products at equilibrium. We see that the free energy of the final products in Figure 4.3 is lower than the free energy of the initial reactants. Therefore, we know that there will be more products than reactants when the reaction has reached equilibrium. A reaction that leads to a higher concentration of products compared with the concentration of reactants is called a **favorable reaction**.

The more stable the compound, the greater is its concentration at equilibrium.

How fast a reaction occurs is indicated by the energy "hill" that must be climbed for the reactants to be converted into products. The higher the energy barrier, the slower is the reaction. The energy barrier is called the **free energy of activation**. The free energy of activation for each step is indicated by $\Delta G^{\ddagger}$ in Figure 4.2. It is the difference between the free energy of the transition state and the free energy of the reactants:

The higher the energy barrier, the slower is the reaction.

$$\Delta G^{\ddagger} = \textbf{(free energy of the transition state)} - \textbf{(free energy of the reactants)}$$

We can see from the reaction coordinate diagram that the free energy of activation for the first step of the reaction is greater than the free energy of activation for the second step. In other words, the first step of the reaction is slower than the second step. This is what we would expect, considering that the molecules in the first step of this reaction must collide with sufficient energy to break covalent bonds, whereas no bonds are broken in the second step.

If a reaction has two or more steps, the step that has its transition state *at the highest point on the reaction coordinate* is called the **rate-determining step**. The rate-determining step controls the overall rate of the reaction because the overall rate of a reaction such as that shown in Figure 4.3 cannot exceed the rate of the rate-determining step. In Figure 4.3, the rate-determining step is the first step—the addition of the electrophile (the proton) to the alkene.

What determines how fast a reaction occurs? The rate of a reaction depends on the following factors:

1. *The number of collisions that take place between the reacting molecules in a given period of time.* The greater the number of collisions, the faster is the reaction.
2. *The fraction of the collisions that occur with sufficient energy to get the reacting molecules over the energy barrier.* If the free energy of activation is small, more collisions will lead to reaction than if the free energy of activation is large.
3. *The fraction of the collisions that occur with the proper orientation.* For example, 2-butene and HBr will react only if the molecules collide with the hydrogen of HBr approaching the π bond of 2-butene. If collision occurs with the hydrogen approaching a methyl group of 2-butene, no reaction will take place, regardless of the energy of the collision.

$$\textbf{rate of a reaction} = \left(\begin{array}{c}\textbf{number of collisions}\\\textbf{per unit of time}\end{array}\right) \times \left(\begin{array}{c}\textbf{fraction with}\\\textbf{sufficient energy}\end{array}\right) \times \left(\begin{array}{c}\textbf{fraction with}\\\textbf{proper orientation}\end{array}\right)$$

Increasing the concentration of the reactants increases the rate of a reaction because it increases the number of collisions that occur in a given period of time. Increasing the temperature at which the reaction is carried out also increases the rate of a reaction because it increases both the number of collisions (molecules that are moving faster collide more frequently) and the fraction of those collisions that have sufficient energy to get the reacting molecules over the energy barrier (molecules that are moving faster collide with greater energy).

PROBLEM 17

Draw a reaction coordinate diagram for

a. a fast reaction with products that are more stable than the reactants.
b. a slow reaction with products that are more stable than the reactants.
c. a slow reaction with products that are less stable than the reactants.

PROBLEM 18 ◆

Given the following reaction coordinate diagram for the reaction of A to give D, answer the following questions:

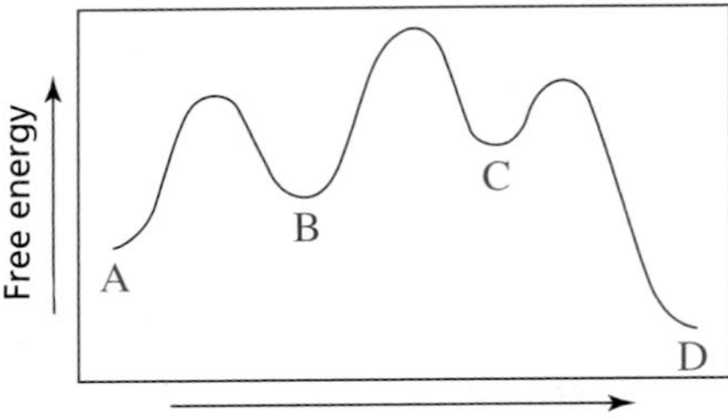

a. How many intermediates are there?
b. Which intermediate is the most stable?
c. How many transition states are there?
d. Which transition state is the most stable?
e. Which is more stable, reactants or products?
f. What is the fastest step in the reaction?
g. What is the reactant of the rate-determining step?
h. Is the overall reaction favorable?

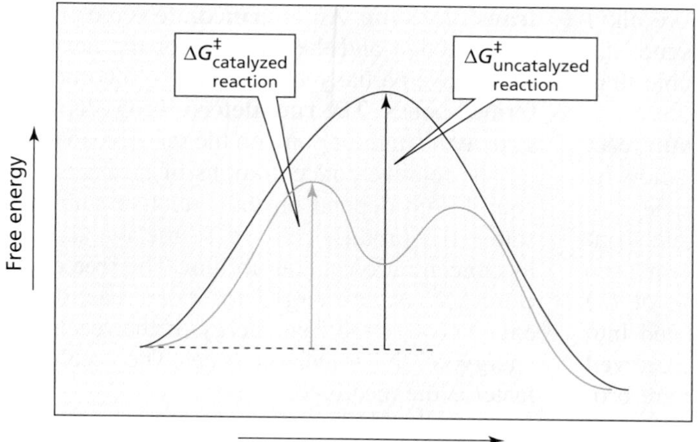

◀ **Figure 4.4**
The reaction coordinate diagrams for an uncatalyzed reaction (black) and for a catalyzed reaction. The catalyzed reaction (green) takes place by an alternative pathway with a lower "energy hill."

The rate of a reaction can also be increased by adding a catalyst to the reaction mixture. A catalyst gives the reactants a new pathway to follow—one with a smaller $\Delta G^{\ddagger}$ (Figure 4.4). Thus, a **catalyst** increases the rate of a reaction by decreasing the energy barrier that has to be overcome in the process of converting the reactants into products. A catalyst must participate in the reaction if it is going to make it go faster, but it is not consumed or changed during the reaction. Because the catalyst is not used up, only a small amount of it is needed to catalyze the reaction (typically 1 to 10% of the number of moles of reactant).

A catalyst gives the reagents a new pathway with a lower "energy hill."

Notice that the stability of the reactants and products is the same in both the catalyzed and uncatalyzed reactions. In other words, the catalyst does not change the relative concentrations of products and reactants when the system reaches equilibrium. Thus, the catalyst does not change the *amount* of product formed; it changes only the *rate* at which it is formed.

PESTICIDES: NATURAL AND SYNTHETIC

Long before chemists learned how to create compounds that would protect plants from predators, plants were doing the job themselves. Plants had every incentive to synthesize pesticides; when you cannot run, you need to find another way to protect yourself. But which pesticides are more harmful, those synthesized by chemists or those synthesized by plants? Unfortunately, we do not know the answer because although federal laws require all human-made pesticides to be tested for any cancer-causing effects, they do not require testing of plant-made pesticides. Besides, risk evaluations of chemicals are usually done on rats, and something that is carcinogenic in a rat may not be carcinogenic in a human. Furthermore, when rats are tested, they are exposed to much greater concentrations of the chemical than would be experienced by a human, and some chemicals are only harmful at high doses. For example, we all need sodium chloride for survival, but high concentrations are poisonous; and although we associate alfalfa sprouts with healthy eating, monkeys fed very large amounts of alfalfa sprouts develop an immune system disorder.

SUMMARY

Alkenes are hydrocarbons that contain a double bond. The double bond is the **functional group** or center of reactivity of the alkene. The **functional group suffix** of an alkene is *ene*. The general molecular formula for a hydrocarbon is C_nH_{2n+2}, minus two hydrogens for every π bond or ring in the molecule. Because alkenes contain fewer than the maximum number of hydrogens, they are called **unsaturated hydrocarbons**.

Rotation about the double bond is restricted, so an alkene can exist as **cis–trans isomers**. The **cis isomer** has its hydrogens on the same side of the double bond; the **trans isomer** has its hydrogens on opposite sides of the double bond. The ***Z* isomer** has the high-priority groups on the same side of the double bond; the ***E* isomer** has the high-priority groups on opposite sides of the double bond. The relative priorities depend on the atomic numbers of the atoms bonded directly to the sp^2 carbon. The more alkyl substituents bonded to the sp^2 carbons of an alkene, the greater is its stability. **Trans alkenes** are more stable than **cis alkenes** because of steric strain.

All compounds with a particular **functional group** react similarly. Due to the cloud of electrons above and below its π bond, an alkene is an electron-rich species (a **nucleophile**). Nucleophiles are attracted to electron-deficient species, called **electrophiles**. Alkenes undergo **electrophilic addition reactions**. The description of the step-by-step process by which reactants are changed into products is called the **mechanism of the reaction**. **Curved arrows** show which bonds are formed and which are broken in a reaction.

A **reaction coordinate diagram** shows the energy changes that take place in a reaction. The more stable the species, the lower is its energy. As reactants are converted into products, a reaction passes through a maximum energy **transition state**. An **intermediate** is the product of one step of a reaction and the reactant for the next step. Transition states have partially formed bonds; intermediates have fully formed bonds. The **rate-determining step** has its transition state at the highest point on the reaction coordinate.

The relative concentrations of products and reactants at equilibrium depend on their relative stabilities. The more stable the product relative to the reactant, the greater is its concentration at equilibrium. The **free energy of activation**, $\Delta G^{\ddagger}$, is the energy barrier of a reaction. It is the difference between the free energy of the reactants and the free energy of the transition state. The smaller the $\Delta G^{\ddagger}$, the faster is the reaction.

A **catalyst** increases the rate of a reaction but is not consumed or changed in the reaction. It changes the *rate* at which a product is formed by providing a pathway with a smaller $\Delta G^{\ddagger}$, but it does not change the *amount* of product formed.

PROBLEMS

19. Give the systematic name for each of the following:

a. CH_3 **b.** CH_2CH_3 **c.** CH_3

20. Squalene, a hydrocarbon with molecular formula $C_{30}H_{50}$, is obtained from shark liver. (*Squalus* is Latin for "shark.") If squalene is a noncyclic compound, how many π bonds does it have?

21. Draw and label the *E* and *Z* isomers for each of the following:

a. $CH_3CH_2C(CH_2CH_2CH_2CH_3){=}C(CH(CH_3)_2)CH_2Cl$

b. $HOCH_2CH_2C(CH{=}O){=}C(C(CH_3)_3)C{\equiv}CH$

22. For each of the following pairs, indicate which member is more stable:

a. $CH_3C(CH_3){=}CHCH_2CH_3$ or $CH_3CH{=}CHCH(CH_3)CH_3$

b. CH_3 or CH_3

23. a. Give the structures and the systematic names for all alkenes with molecular formula C_4H_8, ignoring cis–trans isomers. (*Hint:* There are three.)

b. Which of the compounds have *E* and *Z* isomers?

24. Draw the structure for each of the following:

a. (*Z*)-1,3,5-tribromo-2-pentene
b. (*Z*)-3-methyl-2-heptene
c. (*E*)-1,2-dibromo-3-isopropyl-2-hexene
d. vinyl bromide
e. 1,2-dimethylcyclopentene
f. diallylamine

25. Determine the total number of double bonds and/or rings for a hydrocarbon with molecular formula:

a. $C_{12}H_{20}$ **b.** $C_{40}H_{56}$

26. Name the following:

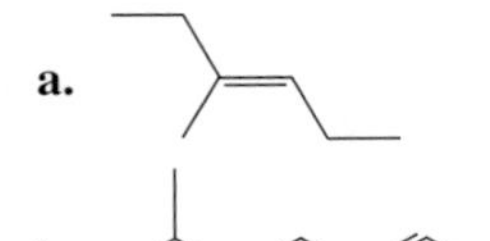

a. **c.** Br **e.**

b. **d.** **f.**

27. Draw curved arrows to show the flow of electrons responsible for the conversion of the reactants into the products:

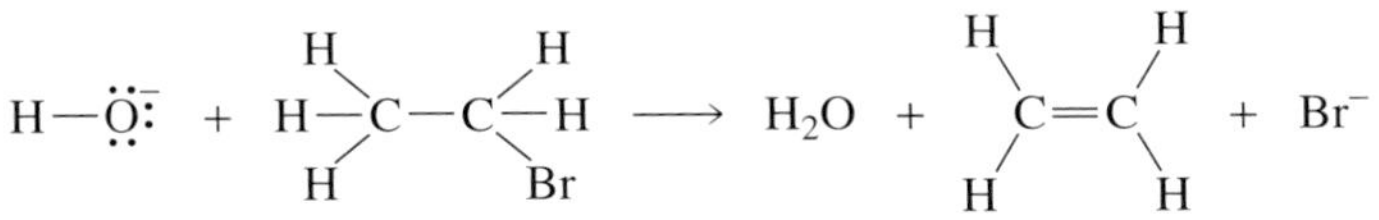

$$H-\ddot{\underset{..}{O}}:^- + H-\underset{H}{\overset{H}{C}}-\underset{Br}{\overset{H}{C}}-H \longrightarrow H_2O + \underset{H}{\overset{H}{}}C{=}C\underset{H}{\overset{H}{}} + Br^-$$

28. Draw three alkenes with molecular formula C_5H_{10} that do not have cis–trans isomers.

29. Tell whether each of the following has the *E* or the *Z* configuration:

a. 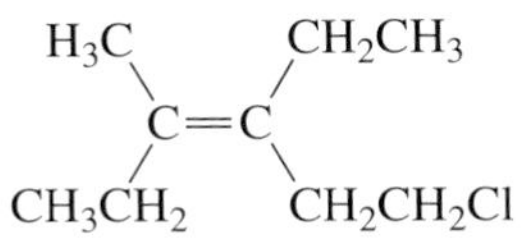 $(H_3C)(CH_3CH_2)C{=}C(CH_2CH_3)(CH_2CH_2Cl)$

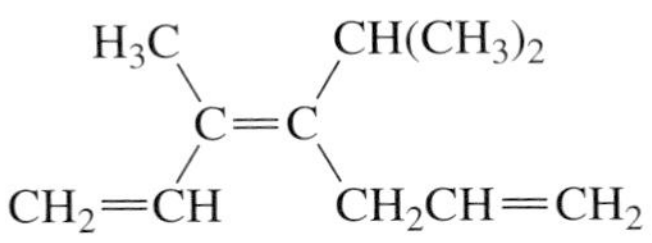

b. $(H_3C)(CH_2{=}CH)C{=}C(CH(CH_3)_2)(CH_2CH{=}CH_2)$

c. $(H_3C)(Br)C{=}C(CH_2Br)(CH_2CH_2CH_2CH_3)$

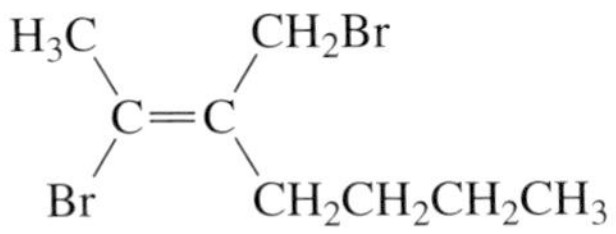

d. 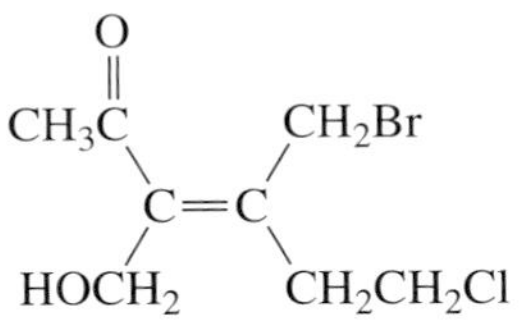 $(CH_3C(=O))(HOCH_2)C{=}C(CH_2Br)(CH_2CH_2Cl)$

30. Which of the following compounds is the most stable? Which is the least stable?

3,4-dimethyl-2-hexene; 2,3-dimethyl-2-hexene; 4,5-dimethyl-2-hexene

31. Assign relative priorities to each set of substituents:

a. $-CH_2CH_2CH_3$ $-CH(CH_3)_2$ $-CH{=}CH_2$ $-CH_3$
b. $-CH_2NH_2$ $-NH_2$ $-OH$ $-CH_2OH$
c. $-C({=}O)CH_3$ $-CH{=}CH_2$ $-Cl$ $-C{\equiv}N$

32. Determine the molecular formula for each of the following:
a. a 5-carbon hydrocarbon with two π bonds and no rings
b. an 8-carbon hydrocarbon with three π bonds and one ring

33. Give the systematic name for each of the following:

a. 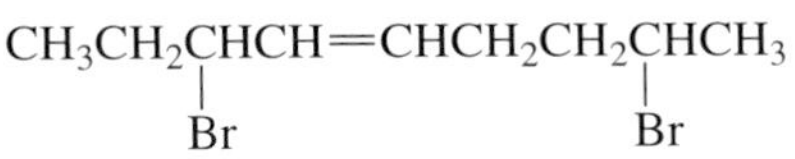 $CH_3CH_2CH(Br)CH{=}CHCH_2CH_2CH(Br)CH_3$

b. 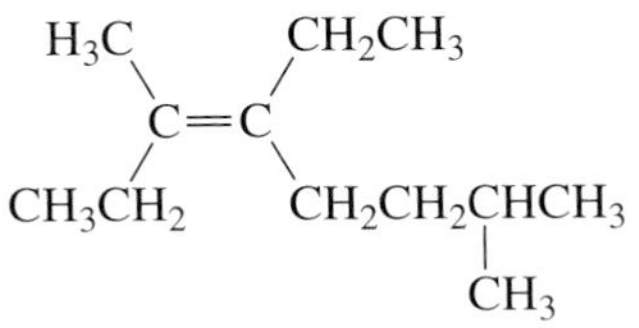 $(H_3C)(CH_3CH_2)C{=}C(CH_2CH_3)(CH_2CH_2CH(CH_3)CH_3)$

c. cyclopentene ring bearing CH_3 and CH_3 substituents

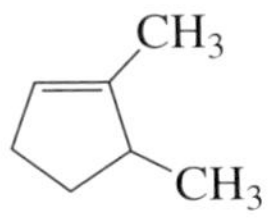

d. 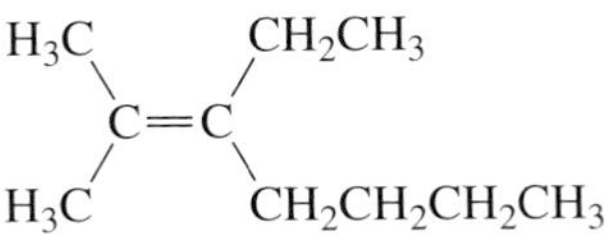 $(H_3C)(H_3C)C{=}C(CH_2CH_3)(CH_2CH_2CH_2CH_3)$

34. Draw a reaction coordinate diagram for a two-step reaction in which the products of the first step are less stable than the reactants, the reactants of the second step are less stable than the products of the second step, the final products are less stable than the initial reactants, and the second step is the rate-determining step. Label the reactants, products, intermediates, and transition states.

35. Molly Kule was a lab technician who was asked by her supervisor to add names to the labels on a collection of alkenes that showed only structures on the labels. How many did Molly get right? Correct the incorrect names.

a. 3-pentene
b. 2-octene
c. 2-vinylpentane
d. 1-ethyl-1-pentene
e. 5-ethylcyclohexene
f. 5-chloro-3-hexene
g. 5-bromo-2-pentene
h. (*E*)-2-methyl-1-hexene
i. 2-methylcyclopentene
j. 2-ethyl-2-butene

36. Determine the number of double bonds and/or π bonds and then draw possible structures for compounds with the following molecular formulas:

a. C_3H_6 b. C_3H_4 c. C_4H_6

37. Draw a reaction coordinate diagram for the following reaction in which C is the most stable and B is the least stable of the three species and the transition state going from A to B is more stable than the transition state going from B to C:

$$A \underset{k_{-1}}{\overset{k_1}{\rightleftharpoons}} B \underset{k_{-2}}{\overset{k_2}{\rightleftharpoons}} C$$

a. How many intermediates are there?
b. How many transition states are there?
c. Which step has the greater rate constant in the forward direction?
d. Which step has the greater rate constant in the reverse direction?
e. Of the four steps, which has the greatest rate constant?
f. Which is the rate-determining step in the forward direction?
g. Which is the rate-determining step in the reverse direction?

38. The rate constant for a reaction can be increased by __________ the stability of the reactant or by __________ the stability of the transition state.

39. α-Farnesene is a compound found in the waxy coating of apple skins. To complete its systematic name, include the *E* or *Z* designation after the number indicating the location of the double bond.

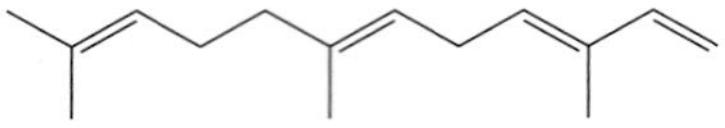

α-farnesene
3,7,11-trimethyl-(1,3?,6?,10)-dodecatetraene

40. Give the structures and the systematic names for all alkenes with molecular formula C_6H_{12}, ignoring cis–trans isomers. (*Hint:* There are 13.)

a. Which of the compounds have *E* and *Z* isomers? **b.** Which of the compounds is the most stable?

41. Tamoxifen slows the growth of some breast tumors by binding to estrogen receptors. Is tamoxifen an *E* or a *Z* isomer?

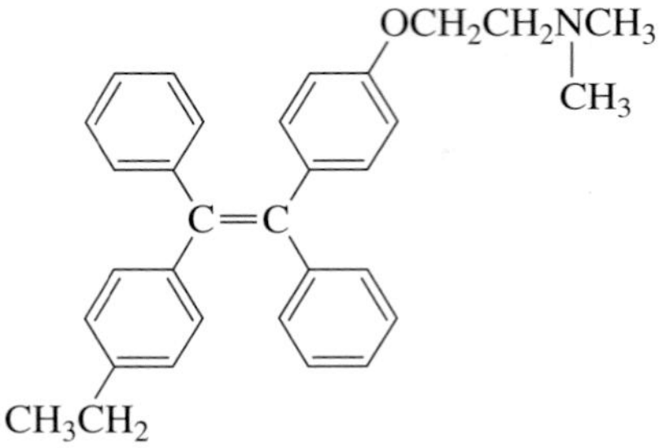

$OCH_2CH_2NCH_3$
CH_3
C=C
CH_3CH_2

tamoxifen

The Reactions of Alkenes and Alkynes

An Introduction to Multistep Synthesis

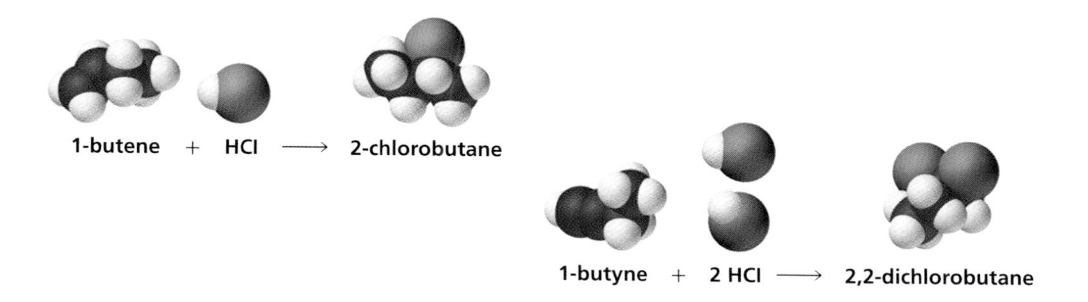

We will start this chapter by looking at the reactions that alkenes undergo. You will see that all the reactions take place by similar mechanisms. As you study each reaction, look for the feature that all alkene reactions have in common: *the relatively loosely held* π *electrons of the carbon–carbon double bond are attracted to an electrophile. Thus, each reaction starts with the addition of an electrophile to one of the* sp^2 *carbons of the alkene and concludes with the addition of a nucleophile to the other* sp^2 *carbon.* The end result is that the π bond breaks because it is weaker than the σ bond (Section 1.14), and the sp^2 carbons form new σ bonds with the electrophile and the nucleophile.

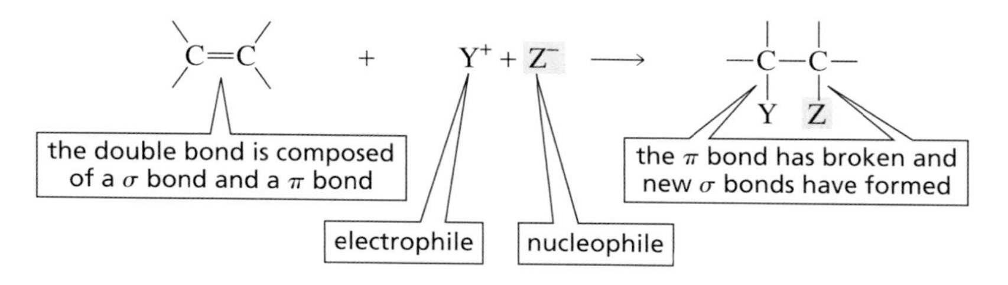

Because both the electrophile and the nucleophile add to the double bond, and the electrophile is the *first* species that adds, this characteristic reaction of alkenes is called an **electrophilic addition reaction.**

This reactivity makes alkenes an important class of organic compounds because they can be used to synthesize a wide variety of other compounds. For example, we will see that alkyl halides, alcohols, ethers, and alkanes all can be synthesized from alkenes by electrophilic addition reactions. The particular product obtained depends only on the *electrophile* and the *nucleophile* used in the addition reaction.

5.1 THE ADDITION OF A HYDROGEN HALIDE TO AN ALKENE

We have seen that an alkene undergoes an **electrophilic addition reaction** with a hydrogen halide (HF, HCl, HBr, or HI); the proton is the electrophile that adds to one of the sp^2 carbons, and the halide ion is the nucleophile that adds to the other sp^2 carbon. Therefore, the product of the reaction is an alkyl halide (Section 4.7).

$$(CH_3)_2C{=}C(CH_3)_2 + HBr \longrightarrow CH_3\underset{CH_3}{CH}-\underset{Br}{\overset{CH_3}{C}}CH_3$$

2,3-dimethyl-2-butene — 2-bromo-2,3-dimethylbutane

cyclohexene + HI ⟶ iodocyclohexane

We looked at the **mechanism of this reaction** in Section 4.7.

mechanism for the addition of a hydrogen halide

$$CH_3CH{=}CHCH_3 + H{-}\ddot{\underset{..}{Cl}}: \xrightarrow{\text{slow}} CH_3\overset{+}{C}HCH_2CH_3 + :\ddot{\underset{..}{Cl}}:^- \xrightarrow{\text{fast}} CH_3\underset{Cl}{CH}CH_2CH_3$$

addition of the electrophile — addition of the nucleophile

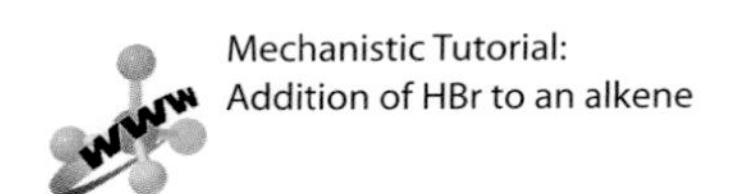

Mechanistic Tutorial:
Addition of HBr to an alkene

- The first step of the reaction is a relatively slow addition of the proton (an electrophile) to the alkene (a nucleophile) to form a carbocation intermediate.
- In the second step, the positively charged carbocation intermediate (an electrophile) reacts rapidly with the negatively charged chloride ion (a nucleophile).

PROBLEM 1

Write the mechanism for the reaction of cyclohexene with HI (the second reaction shown above).

Because the alkenes in the two preceding reactions have the same substituents on both sp^2 carbons, it is easy to predict the product of the reaction: the electrophile (H^+) adds to one of the sp^2 carbons, and the nucleophile (X^-) adds to the other sp^2 carbon. It doesn't matter which sp^2 carbon the electrophile attaches to, because the same product will be obtained in either case.

But what happens if the alkene does not have the same substituents on both of the sp^2 carbons? Which sp^2 carbon gets the hydrogen? For example, does the addition of HCl to 2-methylpropene produce *tert*-butyl chloride or isobutyl chloride?

$$CH_3\overset{CH_3}{C}{=}CH_2 + HCl \longrightarrow CH_3\underset{Cl}{\overset{CH_3}{C}}CH_3 \quad \text{or} \quad CH_3\overset{CH_3}{CH}CH_2Cl$$

2-methylpropene — 2-chloro-2-methylpropane (*tert*-butyl chloride) — 1-chloro-2-methylpropane (isobutyl chloride)

To answer this question, we need to carry out the reaction, isolate the products, and identify them. When we do, we find that the only product of the reaction is *tert*-butyl chloride. Now we need to find out why that compound is the product of the reaction so we can use this knowledge to predict the products of other alkene reactions. To do this, we need to look at the mechanism of the reaction.

The first step of the reaction—the addition of H^+ to an sp^2 carbon to form either the *tert*-butyl cation or the isobutyl cation—is the rate-determining step (Section 4.8). If there is any difference in the rate of formation of these two carbocations, then the one that is formed faster will be the predominant product of the first step. Moreover, because carbocation formation is rate determining, the particular carbocation that is formed in the first step determines the final product of the reaction. That is, if the *tert*-butyl cation is formed, it will react rapidly with Cl^- to form *tert*-butyl chloride. On the other hand, if the isobutyl cation is formed, it will react rapidly with Cl^- to form isobutyl chloride. Since the only product of the reaction is *tert*-butyl chloride, we know that the *tert*-butyl cation is formed much faster than the isobutyl cation.

The sp^2 carbon that does *not* become attached to the proton is the carbon that is positively charged in the carbocation.

$$CH_3C(CH_3){=}CH_2 + H{-}Cl \longrightarrow CH_3\overset{+}{C}(CH_3)CH_3 + :\ddot{Cl}:^- \longrightarrow CH_3C(CH_3)(Cl)CH_3$$

tert-butyl cation → *tert*-butyl chloride, only product formed

$$CH_3C(CH_3){=}CH_2 + H{-}Cl \not\longrightarrow CH_3CH(CH_3)\overset{+}{C}H_2 + :\ddot{Cl}:^- \longrightarrow CH_3CH(CH_3)CH_2Cl$$

isobutyl cation → isobutyl chloride, not formed

The question now is, why is the *tert*-butyl cation formed faster than the isobutyl cation? To answer this question, we need to take a look at the factors that affect the stability of carbocations and thus the ease with which they are formed.

5.2 CARBOCATION STABILITY DEPENDS ON THE NUMBER OF ALKYL SUBSTITUENTS ATTACHED TO THE POSITIVELY CHARGED CARBON

Carbocations are classified according to the number of alkyl substituents that are bonded to the positively charged carbon: a **primary carbocation** has one alkyl substituent, a **secondary carbocation** has two, and a **tertiary carbocation** has three. The stability of a carbocation increases as the number of alkyl substituents bonded to the positively charged carbon increases. Thus, tertiary carbocations are more stable than secondary carbocations, and secondary carbocations are more stable than primary carbocations.

Carbocation stability: 3° > 2° > 1°

relative stabilities of carbocations

$$R_3C^+ > R_2HC^+ > RH_2C^+ > H_3C^+$$

most stable: a tertiary carbocation > a secondary carbocation > a primary carbocation > methyl cation: least stable

The reason for this pattern of stability is that alkyl groups bonded to the positively charged carbon decrease the concentration of positive charge on the carbon since they can donate electrons through the σ bond better than hydrogens can; decreasing the concentration of positive charge makes the carbocation more stable. Notice that the blue (representing positive charge in these electrostatic potential maps, Section 1.3) is most intense for the least stable methyl cation and is least intense for the most stable *tert*-butyl cation.

The greater the number of alkyl substituents bonded to the positively charged carbon, the more stable the carbocation is.

Alkyl substituents stabilize both alkenes *and* carbocations.

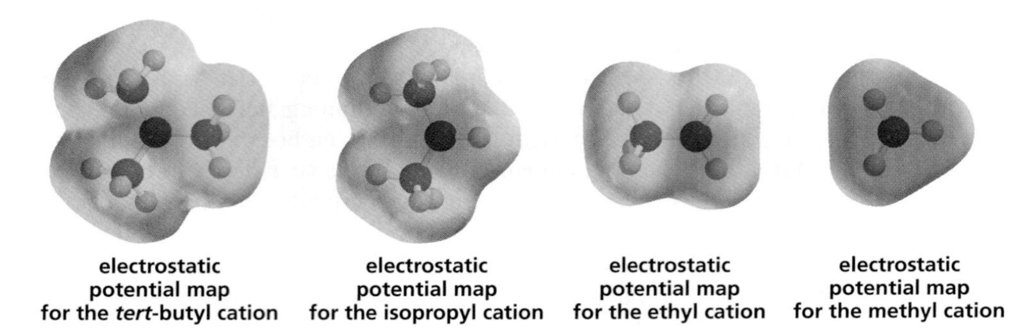

PROBLEM 2 ◆

Which is more stable, a methyl cation or an ethyl cation?

PROBLEM 3 ◆

List the following carbocations in order of decreasing stability:

$CH_3CH_2\overset{CH_3}{\underset{+}{C}}CH_3$ $\qquad CH_3CH_2\underset{+}{C}HCH_3$ $\qquad CH_3CH_2CH_2\overset{+}{C}H_2$

Now we can understand why the *tert*-butyl cation is formed faster than the isobutyl cation when 2-methylpropene reacts with HCl. We know that the *tert*-butyl cation (a tertiary carbocation) is more stable than the isobutyl cation (a primary carbocation). The same factors that stabilize the positively charged carbocation also stabilize the transition state for its formation because the transition state has a partial positive charge (Section 4.8). Therefore, the transition state leading to the *tert*-butyl cation is more stable (that is, lower in energy) than the transition state leading to the isobutyl cation (Figure 5.1).

We have seen that the rate of a reaction is determined by the free energy of activation ($\Delta G^\ddagger$), which is the difference between the free energy of the transition state and the free energy of the reactant: the more stable the transition state, the smaller is the free energy of activation, and therefore, the faster is the reaction (Section 4.8). Thus, the *tert*-butyl cation will be formed faster than the isobutyl cation.

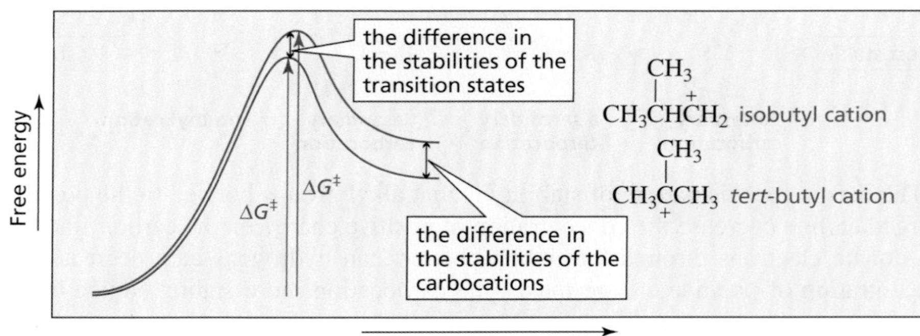

▶ **Figure 5.1**
The reaction coordinate diagram for the addition of H^+ to 2-methylpropene to form the primary isobutyl cation and the tertiary *tert*-butyl cation.

5.3 ELECTROPHILIC ADDITION REACTIONS ARE REGIOSELECTIVE

We have just seen that the major product of an electrophilic addition reaction is the one obtained by adding the electrophile to the sp^2 carbon that results in the formation of the more stable carbocation. For example, when propene reacts with HCl, the proton can add to the number-1 carbon (C-1) to form a secondary carbocation, or it can add to the number-2 carbon (C-2) to form a primary carbocation. The secondary carbocation is formed more rapidly because it is more stable than the primary carbocation. (Primary carbocations are so unstable that they form only with great difficulty.) The product of the reaction, therefore, is 2-chloropropane.

$$CH_3\overset{2}{C}H{=}\overset{1}{C}H_2 \xrightarrow{HCl} CH_3\overset{+}{C}HCH_3 + Cl^- \longrightarrow CH_3CH(Cl)CH_3$$

a secondary carbocation; 2-chloropropane

$$CH_3CH{=}CH_2 \xrightarrow{HCl} \not\rightarrow CH_3CH_2\overset{+}{C}H_2$$

a primary carbocation

The major product obtained from the addition of HI to 2-methyl-2-butene is 2-iodo-2-methylbutane; only a small amount of 2-iodo-3-methylbutane is obtained. The major product obtained from the addition of HBr to 1-methylcyclohexene is 1-bromo-1-methylcyclohexane. In both cases, the more stable tertiary carbocation is formed more rapidly than the less stable secondary carbocation, so the major product of each reaction is the one that results from forming the tertiary carbocation.

$$CH_3CH{=}C(CH_3)CH_3 + HI \longrightarrow CH_3CH_2C(CH_3)(I)CH_3 + CH_3CH(I)CH(CH_3)CH_3$$

2-methyl-2-butene; 2-iodo-2-methylbutane major product; 2-iodo-3-methylbutane minor product

1-methylcyclohexene + HBr ⟶ 1-bromo-1-methylcyclohexane (major product) + 1-bromo-2-methylcyclohexane (minor product)

The two different products of each of these reactions are *constitutional isomers*. We saw at the beginning of Chapter 3 that **constitutional isomers** have the same molecular formula, but differ in how their atoms are connected. A reaction (such as either of those just shown) in which two or more constitutional isomers could be obtained as products, but one of them predominates, is called a **regioselective reaction**.

Regioselectivity is the preferential formation of one constitutional isomer over another.

The addition of HBr to 2-pentene is not regioselective. Because the addition of H^+ to either of the sp^2 carbons produces a secondary carbocation, both carbocation intermediates have the same stability, so both will be formed equally easily. Thus, approximately equal amounts of the two alkyl halides will be formed.

$$CH_3CH{=}CHCH_2CH_3 + HBr \longrightarrow CH_3CH(Br)CH_2CH_2CH_3 + CH_3CH_2CH(Br)CH_2CH_3$$

2-pentene; 2-bromopentane 50%; 3-bromopentane 50%

BIOGRAPHY

Vladimir Vasilevich Markovnikov (1837–1904) *was born in Russia, the son of an army officer. He was a professor of chemistry at Kazan, Odessa, and Moscow Universities.*

The electrophile adds to the sp^2 carbon that is bonded to the greater number of hydrogens.

From the alkene reactions we have seen so far, we can devise a rule that applies to *all* alkene electrophilic addition reactions: *the* **electrophile** *adds to the* sp^2 *carbon that is bonded to the greater number of hydrogens.* Vladimir Markovnikov was the first to recognize that in the addition of a hydrogen halide to an alkene, the H^+ adds to the sp^2 carbon that is bonded to the greater number of hydrogens. As a result, this rule is often referred to as **Markovnikov's rule**.

This rule is simply a quick way to determine the relative stabilities of the intermediates that could be formed in the rate-determining step. You will get the same answer, whether you identify the major product of an electrophilic addition reaction by using the rule or whether you identify it by determining relative carbocation stabilities. In the following reaction, for example, H^+ is the electrophile:

$$CH_3CH_2\overset{2}{C}H{=}\overset{1}{C}H_2 + HCl \longrightarrow CH_3CH_2\underset{}{\overset{Cl}{\overset{|}{C}}}HCH_3$$

We can say that H^+ adds preferentially to C-1 because C-1 is bonded to two hydrogens, whereas C-2 is bonded to only one hydrogen. Or we can say that H^+ adds to C-1 because that results in the formation of a secondary carbocation, which is more stable than the primary carbocation that would have to be formed if H^+ added to C-2.

PROBLEM 4 ♦

What would be the major product obtained from the addition of HBr to each of the following compounds?

a. $CH_3CH_2CH{=}CH_2$

b. $CH_3CH{=}\overset{CH_3}{\overset{|}{C}}CH_3$

c. 1-methylcyclopentene (CH_3 on a cyclopentene ring double-bond carbon)

d. $CH_2{=}\overset{CH_3}{\overset{|}{C}}CH_2CH_2CH_3$

e. methylenecyclohexane (cyclohexane ring $={CH_2}$)

f. $CH_3CH{=}CHCH_3$

PROBLEM-SOLVING STRATEGY

a. What alkene should be used to synthesize 3-bromohexane?

$$? + HBr \longrightarrow CH_3CH_2\underset{Br}{\underset{|}{C}}HCH_2CH_2CH_3$$

3-bromohexane

The best way to answer this kind of question is to begin by listing all the alkenes that could be used. Because you want to synthesize an alkyl halide that has a bromo substituent at the 3-position, the alkene should have an sp^2 carbon at that position. Two alkenes fit the description: 2-hexene and 3-hexene.

$CH_3CH{=}CHCH_2CH_2CH_3$ — **2-hexene**

$CH_3CH_2CH{=}CHCH_2CH_3$ — **3-hexene**

Because there are two possibilities, we next need to determine whether there is any advantage to using one over the other. The addition of H^+ to 2-hexene can form two different carbocations, but they are both secondary carbocations. Because they have the same stability, approximately equal amounts of each will be formed. Therefore, half of the product will be the desired 3-bromohexane and half will be 2-bromohexane.

$$CH_3CH{=}CHCH_2CH_2CH_3 \xrightarrow{HBr} CH_3CH_2\overset{+}{C}HCH_2CH_2CH_3 + Br^- \longrightarrow CH_3CH_2CH(Br)CH_2CH_2CH_3$$

2-hexene → secondary carbocation → 3-bromohexane

$$CH_3CH{=}CHCH_2CH_2CH_3 \xrightarrow{HBr} CH_3\overset{+}{C}HCH_2CH_2CH_2CH_3 + Br^- \longrightarrow CH_3CH(Br)CH_2CH_2CH_2CH_3$$

2-hexene → secondary carbocation → 2-bromohexane

The addition of H^+ to either of the sp^2 carbons of 3-hexene, on the other hand, forms the same carbocation because the alkene is symmetrical. Therefore, all of the product will be the desired 3-bromohexane.

$$CH_3CH_2CH{=}CHCH_2CH_3 \xrightarrow{HBr} CH_3CH_2\overset{+}{C}HCH_2CH_2CH_3 + Br^- \longrightarrow CH_3CH_2CH(Br)CH_2CH_2CH_3$$

3-hexene → only one carbocation is formed → 3-bromohexane

Because all the alkyl halide formed from 3-hexene is 3-bromohexane, but only half the alkyl halide formed from 2-hexene is 3-bromohexane, 3-hexene is the best alkene to use to prepare 3-bromohexane.

b. What alkene should be used to synthesize 2-bromopentane?

$$? \quad + \quad HBr \longrightarrow CH_3CH(Br)CH_2CH_2CH_3$$

2-bromopentane

Either 1-pentene or 2-pentene could be used because both have an sp^2 carbon at the C-2 position.

$CH_2{=}CHCH_2CH_2CH_3$ (1-pentene) $\qquad CH_3CH{=}CHCH_2CH_3$ (2-pentene)

When H^+ adds to 1-pentene, one of the carbocations that could be formed is secondary and the other is primary. A secondary carbocation is more stable than a primary carbocation, which is so unstable that none will be formed. Thus, 2-bromopentane will be the only product of the reaction.

$$CH_2{=}CHCH_2CH_2CH_3 \xrightarrow{HBr} CH_3\overset{+}{C}HCH_2CH_2CH_3 + Br^- \longrightarrow CH_3CH(Br)CH_2CH_2CH_3$$

1-pentene → 2-bromopentane

$$CH_2{=}CHCH_2CH_2CH_3 \xrightarrow{HBr} \not\rightarrow \overset{+}{C}H_2CH_2CH_2CH_2CH_3$$

When H^+ adds to 2-pentene, on the other hand, each of the two carbocations that can be formed is secondary. Both are equally stable, so they will be formed in approximately equal amounts. Thus, only about half of the product of the reaction will be the desired 2-bromopentane. The other half will be 3-bromopentane.

$$CH_3CH{=}CHCH_2CH_3 \xrightarrow{HBr} CH_3\overset{+}{C}HCH_2CH_2CH_3 + Br^- \longrightarrow CH_3CH(Br)CH_2CH_2CH_3$$

2-pentene → 2-bromopentane

$$CH_3CH{=}CHCH_2CH_3 \xrightarrow{HBr} CH_3CH_2\overset{+}{C}HCH_2CH_3 + Br^- \longrightarrow CH_3CH_2CH(Br)CH_2CH_3$$

2-pentene → 3-bromopentane

Because all the alkyl halide formed from 1-pentene is 2-bromopentane, but only half the alkyl halide formed from 2-pentene is 2-bromopentane, 1-pentene is the best alkene to use to prepare 2-bromopentane.

Now continue on to Problem 5.

PROBLEM 5 ◆

What alkene should be used to synthesize each of the following alkyl bromides?

a. $CH_3C(CH_3)(Br)CH_3$

b. cyclohexyl–$CH_2CH(Br)CH_3$

c. cyclohexyl–$C(CH_3)(Br)CH_3$

d. 1-bromo-1-ethylcyclohexane (CH_2CH_3, Br)

5.4 THE ADDITION OF WATER TO AN ALKENE

An alkene does not react with water because there is no electrophile present to start a reaction by adding to the nucleophilic alkene. The O—H bonds of water are too strong—water is too weakly acidic—to allow the hydrogen to act as an electrophile for this reaction.

$$CH_3CH{=}CH_2 + H_2O \longrightarrow \text{no reaction}$$

If an acid is added to the solution (the acid most often used is H_2SO_4), then the outcome is much different: a reaction will occur because the acid provides an electrophile (H^+). The product of the reaction is an alcohol. The addition of water to a molecule is called **hydration**, so we can say that an alkene will be *hydrated* in the presence of water and acid.

$$CH_3CH{=}CH_2 + H_2O \xrightleftharpoons{H_2SO_4} CH_3CH(OH)-CH_2(H)$$

2-propanol
an alcohol

Because H_2SO_4 is a strong acid, it dissociates almost completely in water (Section 2.2). The acid that participates in the reaction, therefore, is most apt to be a hydronium ion.

$$H_2SO_4 + H_2O \rightleftharpoons H_3O^+ + HSO_4^-$$

hydronium ion

The acid is a catalyst—it increases the *rate* at which a product is formed, but it does not affect the *amount* of product formed. Because the catalyst in the hydration of an alkene is an acid, hydration is an **acid-catalyzed reaction**.

Notice that the first two steps of *the mechanism for the acid-catalyzed addition of water to an alkene* are essentially the same as the two steps of *the mechanism for the addition of a hydrogen halide to an alkene*:

mechanism for the acid-catalyzed addition of water

$$CH_3CH{=}CH_2 + H-\overset{+}{\ddot{O}}H(H) \xrightleftharpoons{\text{slow}} CH_3\overset{+}{C}HCH_3 + H_2\ddot{O}: \xrightleftharpoons{\text{fast}} CH_3CH(\overset{+}{:OH}H)CH_3$$

addition of the electrophile

addition of the nucleophile

a protonated alcohol

H_2O removes a proton, regenerating the acid catalyst

$H_2\ddot{O}$: fast

$$CH_3CH(\ddot{O}H)CH_3 + H_3\overset{+}{O}:$$

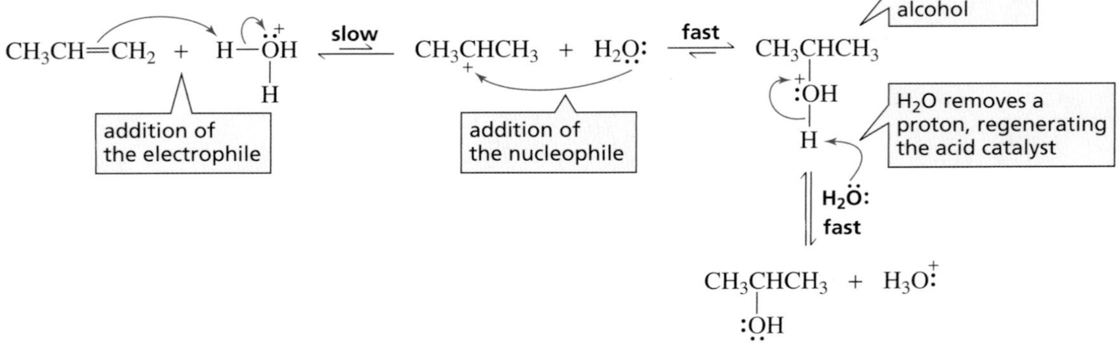

- The electrophile (H^+) adds to the sp^2 carbon that is bonded to the greater number of hydrogens.
- The nucleophile (H_2O) adds to the carbocation, forming a protonated alcohol.
- The protonated alcohol is a very strong acid (Section 2.2), so it loses a proton. The final product of the addition reaction is an alcohol; the regenerated acid catalyst returns to the reaction mixture.

Movies: The First Step to Hydration of an Alkene; The Second Step to Hydration of an Alkene; The Third Step to Hydration of an Alkene.

Mechanistic Tutorial: Addition of water to an alkene

As we saw in Section 4.7, the addition of the electrophile to the alkene is relatively slow, and the subsequent addition of the nucleophile to the carbocation occurs rapidly. The reaction of the carbocation with a nucleophile is so fast, in fact, that the carbocation combines with whatever nucleophile it collides with: note that there are two nucleophiles in solution—water and the conjugate base of the acid (HSO_4^-) that was used to start the reaction. Because the concentration of water is much greater than the concentration of the conjugate base, the carbocation is much more likely to collide with water. The product of the collision is a protonated alcohol. Notice that the catalyst is not consumed or changed during the reaction.

Do not memorize the products of alkene addition reactions. Instead, for each reaction, ask yourself, "What is the electrophile?" and "What nucleophile is present in the greatest concentration?"

PROBLEM 6 ◆

Give the major product obtained from the acid-catalyzed hydration of each of the following alkenes:

a. $CH_3CH_2CH_2CH{=}CH_2$

b. (cyclohexene)

c. $CH_3CH_2CH_2CH{=}CHCH_3$

d. (cyclohexane)$=CH_2$

5.5 THE ADDITION OF AN ALCOHOL TO AN ALKENE

Alcohols react with alkenes in the same way that water does, so this reaction too requires an acid catalyst. The product of the reaction is an ether.

$$CH_3CH{=}CH_2 + CH_3OH \xrightleftharpoons{H_2SO_4} \underset{\substack{|\\OCH_3}}{CH_3CH}{-}\underset{\substack{|\\H}}{CH_2}$$

2-methoxypropane
an ether

The mechanism for the acid-catalyzed addition of an alcohol is essentially the same as the mechanism for the acid-catalyzed addition of water. The only difference is that the nucleophile is ROH instead of HOH.

mechanism for the acid-catalyzed addition of an alcohol

$$CH_3CH{=}CH_2 + H{-}\overset{+}{\ddot{O}}(H)CH_3 \xrightleftharpoons{slow} CH_3\overset{+}{C}HCH_3 + CH_3\ddot{\underset{..}{O}}H \xrightleftharpoons{fast} CH_3CHCH_3 \text{ with } {-}\overset{+}{\ddot{O}}CH_3(H)$$

$$\xrightleftharpoons[fast]{CH_3\ddot{O}H} \underset{\substack{|\\:\ddot{O}CH_3}}{CH_3CHCH_3} + CH_3\overset{+}{\ddot{O}}H(H)$$

PROBLEM 7

a. Give the major product of each of the following reactions:

1. $CH_3\overset{\overset{\large CH_3}{|}}{C}{=}CH_2 \; + \; HCl \longrightarrow$

2. $CH_3\overset{\overset{\large CH_3}{|}}{C}{=}CH_2 \; + \; HBr \longrightarrow$

3. $CH_3\overset{\overset{\large CH_3}{|}}{C}{=}CH_2 \; + \; H_2O \xrightarrow{H_2SO_4}$

4. $CH_3\overset{\overset{\large CH_3}{|}}{C}{=}CH_2 \; + \; CH_3OH \xrightarrow{H_2SO_4}$

b. What do all the reactions have in common?
c. How do all the reactions differ?

Synthetic Tutorial: Addition of HBr to an alkene

Synthetic Tutorial: Addition of water to an alkene

Synthetic Tutorial: Addition of alcohol to an alkene

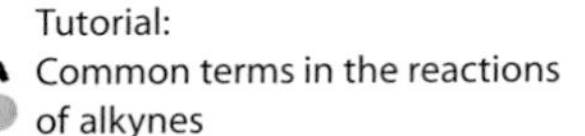
Tutorial: Common terms in the reactions of alkynes

PROBLEM 8

How could the following compounds be prepared, using an alkene as one of the starting materials?

a. 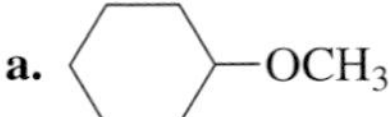

b. $CH_3O\overset{\overset{\large CH_3}{|}}{\underset{\underset{\large CH_3}{|}}{C}}CH_3$

c. $CH_3CH_2O\underset{\underset{\large CH_3}{|}}{C}HCH_2CH_3$

d. 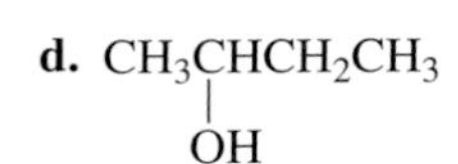
$CH_3\underset{\underset{\large OH}{|}}{C}HCH_2CH_3$

PROBLEM 9 ◆

When chemists write reactions, they show reaction conditions, such as the solvent, the temperature, and any required catalyst above or below the arrow.

$$CH_2{=}CHCH_2CH_3 \; + \; H_2O \xrightarrow{H_2SO_4} CH_3\underset{\underset{\large OH}{|}}{C}HCH_2CH_3$$

Sometimes reactions are written by placing only the organic (carbon-containing) reagent on the left-hand side of the arrow; the other reagents are written above or below the arrow.

$$CH_2{=}CHCH_2CH_3 \xrightarrow[H_2O]{H_2SO_4} CH_3\underset{\underset{\large OH}{|}}{C}HCH_2CH_3$$

There are two nucleophiles in each of the following reactions. For each reaction, explain why there is a greater concentration of one nucleophile than the other. What will be the major product of each reaction?

a. $CH_3CH{=}CHCH_3 \; + \; H_2O \xrightarrow{H_2SO_4}$ **b.** $CH_3CH{=}CHCH_3 \xrightarrow[CH_3OH]{H_2SO_4}$

PROBLEM 10

Give the major product obtained from the reaction of HBr with each of the following:

a. (methylenecyclohexane, CH_2)

b. $CH_3\underset{\underset{\large CH_3}{|}}{C}HCH_2CH{=}CH_2$

c. 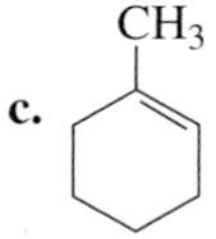
(CH_3)

d. (CH_3)

5.6 AN INTRODUCTION TO ALKYNES

An **alkyne** is a hydrocarbon that contains a carbon–carbon triple bond. Because of its triple bond, an alkyne has four fewer hydrogens than an alkane with the same number of carbons. Therefore, while the general molecular formula for an alkene is C_nH_{2n+2}, the general molecular formula for a noncyclic alkyne is C_nH_{2n-2}.

The few drugs in clinical use that contain alkyne functional groups are not naturally occurring compounds; they exist only because chemists have been able to synthesize them. Their brand names are shown below, in green. Brand names are always capitalized; only the company that holds the patent for a product can use the product's brand name for commercial purposes (Section 21.1).

Parsal®
Sinovial® 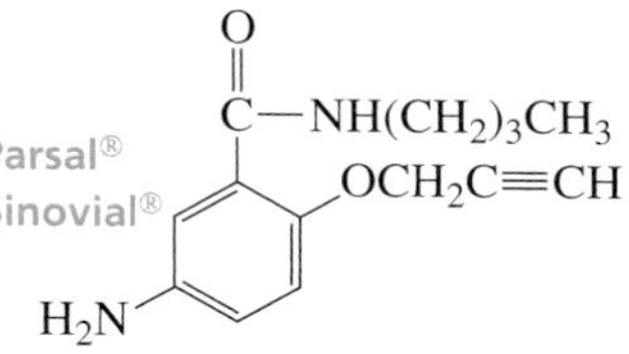
parsalmide
an analgesic

Eudatin®
Supirdyl® 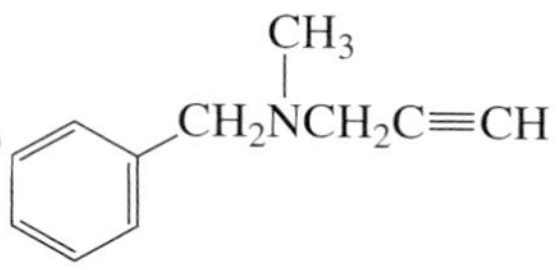
pargyline
an antihypertensive

Norquen®
Ovastol® 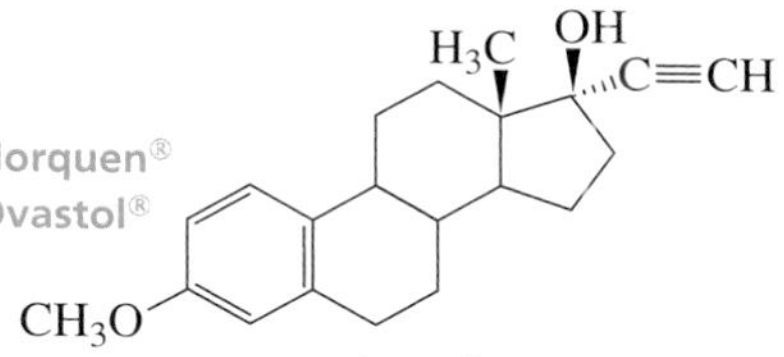
mestranol
a component in oral contraceptives

NATURALLY OCCURRING ALKYNES

There are only a few naturally occurring alkynes. Examples include capillin, which has fungicidal activity, and ichthyothereol, a convulsant used by the indigenous people of the Amazon for poisoned arrowheads. A class of naturally occurring compounds called enediynes has been found to have powerful antibiotic and anticancer properties. These compounds all have a nine- or ten-membered ring that contains two triple bonds separated by a double bond. Some enediynes are currently being tested in clinical trials (Section 21.10).

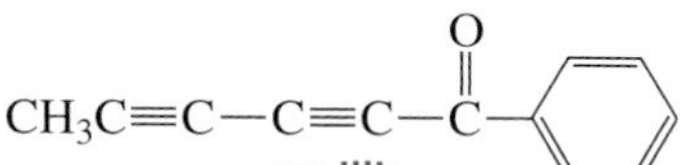

capillin

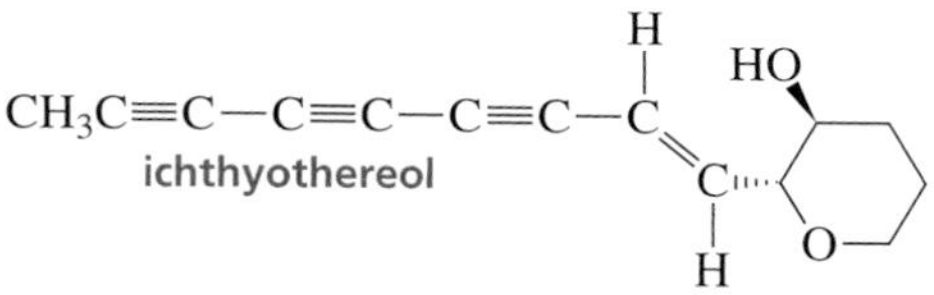

ichthyothereol

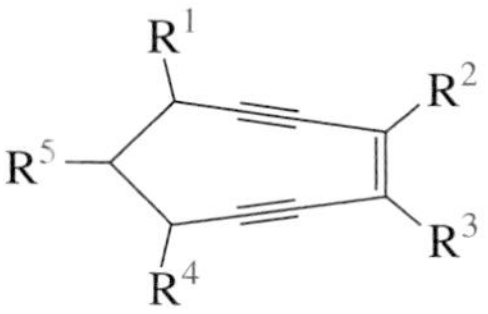

an enediyne

PROBLEM 11 ♦

What is the general molecular formula for a cyclic alkyne?

PROBLEM 12 ♦

What is the molecular formula for a cyclic hydrocarbon with 14 carbons and two triple bonds?

5.7 THE NOMENCLATURE OF ALKYNES

The systematic name of an alkyne is obtained by replacing the "ane" ending of the alkane name with "yne." Analogous to the way compounds with other functional groups are named, the longest continuous chain containing the carbon–carbon triple bond is numbered in the direction that gives the alkyne functional group suffix as low a number as possible. If the triple bond is at the end of the chain, the alkyne is classified as a **terminal alkyne**. Alkynes with triple bonds located elsewhere along the chain are called **internal alkynes**. For example, 1-butyne is a terminal alkyne, whereas 2-pentyne is an internal alkyne.

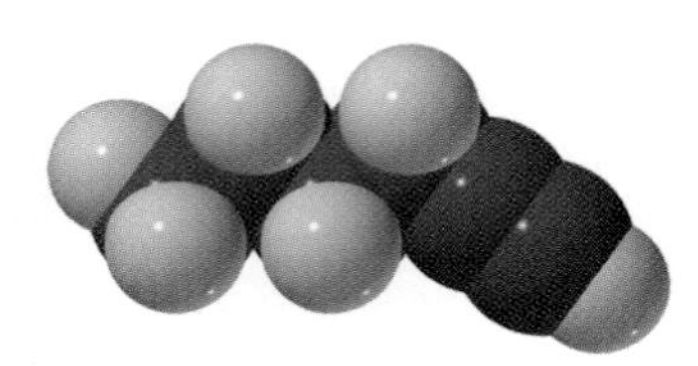

1-hexyne
a terminal alkyne

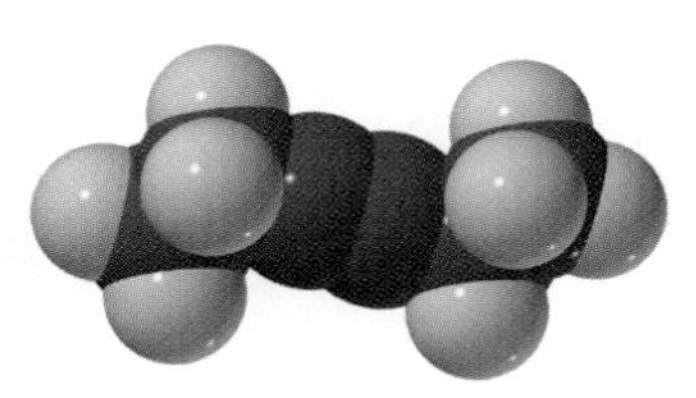

3-hexyne
an internal alkyne

	$HC{\equiv}CH$	$\overset{4}{C}H_3\overset{3}{C}H_2\overset{2}{C}{\equiv}\overset{1}{C}H$	$\overset{1}{C}H_3\overset{2}{C}{\equiv}\overset{3}{C}\overset{4}{C}H_2\overset{5}{C}H_3$	$\overset{5}{C}H_2\overset{6}{C}H_3$ on C4 of $CH_3\overset{4}{C}H\overset{3}{C}{\equiv}\overset{2}{C}\overset{1}{C}H_3$
systematic:	ethyne	1-butyne	2-pentyne	4-methyl-2-hexyne
common:	acetylene	a terminal alkyne	an internal alkyne	

If counting from either direction leads to the same number for the functional group suffix, the correct systematic name is the one that contains the lowest substituent number. If the compound contains more than one substituent, the substituents are listed in alphabetical order.

$$\underset{6}{C}H_3\underset{5}{C}H(CH_3)\underset{4}{C}{\equiv}\underset{3}{C}\underset{2}{C}H_2\underset{1}{C}H_2Br$$

1-bromo-5-methyl-3-hexyne
***not* 6-bromo-2-methyl-3-hexyne**
because 1 < 2

$$\underset{1}{C}H_3\underset{2}{C}H(Cl)\underset{3}{C}H(Br)\underset{4}{C}{\equiv}\underset{5}{C}\underset{6}{C}H_2\underset{7}{C}H_2\underset{8}{C}H_3$$

3-bromo-2-chloro-4-octyne
***not* 6-bromo-7-chloro-4-octyne**
because 2 < 6

PROBLEM 13 ◆

Draw the structure for each of the following:

a. 1-chloro-3-hexyne
b. 4-bromo-2-pentyne
c. 4,4-dimethyl-1-pentyne

PROBLEM 14 ◆

Name the following:

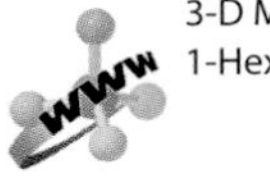

3-D Molecules:
1-Hexyne; 3-Hexyne

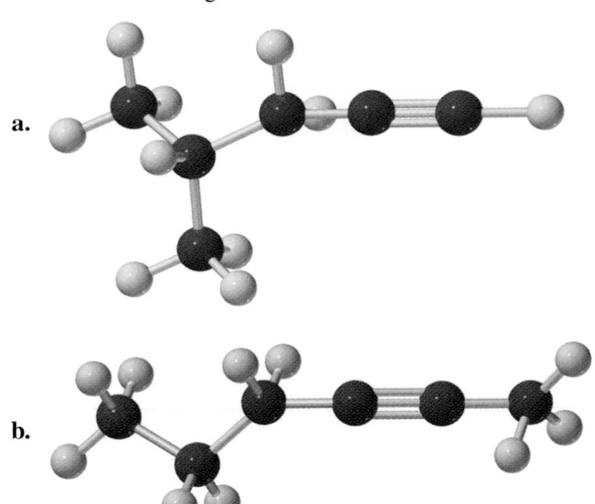

PROBLEM 15

Draw the structures and give the systematic names for the seven alkynes with molecular formula C_6H_{10}.

PROBLEM 16 ◆

Give the systematic name for each of the following:

a. $BrCH_2CH_2C{\equiv}CCH_3$

b. $CH_3CH_2CH(Br)C{\equiv}CCH_2CH(Cl)CH_3$ (Br on C3, Cl on C7)

c. $CH_3CH_2CH(CH_3)C{\equiv}CCH_2CH_3$

d. $CH_3CH_2CH(CH_2CH_2CH_3)C{\equiv}CH$

5.8 THE STRUCTURE OF ALKYNES

The structure of ethyne was discussed in Section 1.9. We saw that each carbon is *sp* hybridized, so each has two *sp* orbitals and two *p* orbitals. One *sp* orbital overlaps the *s* orbital of a hydrogen, and the other overlaps an *sp* orbital of the other carbon. Because the *sp* orbitals are oriented as far from each other as possible to minimize electron repulsion, ethyne is a linear molecule with bond angles of 180°.

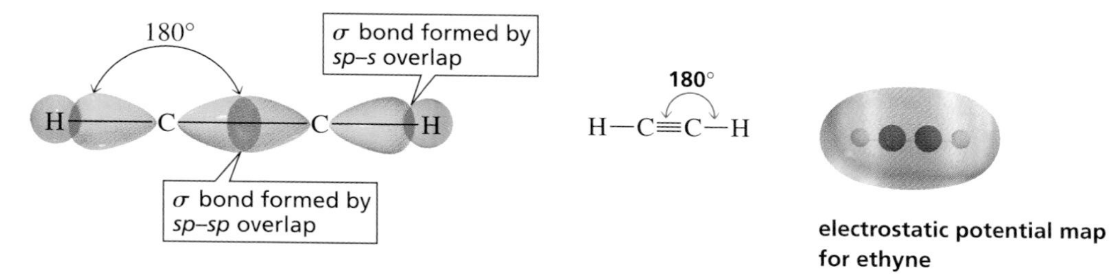

The two remaining *p* orbitals on each carbon are oriented at right angles to one another and to the *sp* orbitals (Figure 5.2). Each of the two *p* orbitals on one carbon overlaps the parallel *p* orbital on the other carbon to form two π bonds. One pair of overlapping *p* orbitals results in a cloud of electrons above and below the σ bond, and the other pair results in a cloud of electrons in front of and behind the σ bond. The electrostatic potential map of ethyne shows that the end result can be thought of as a cylinder of electrons wrapped around the σ bond.

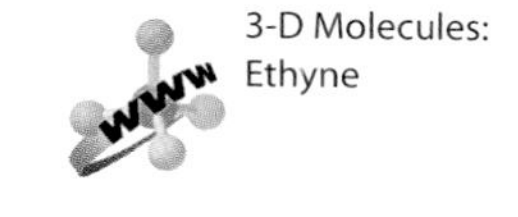

A triple bond is composed of a σ bond and two π bonds.

a.

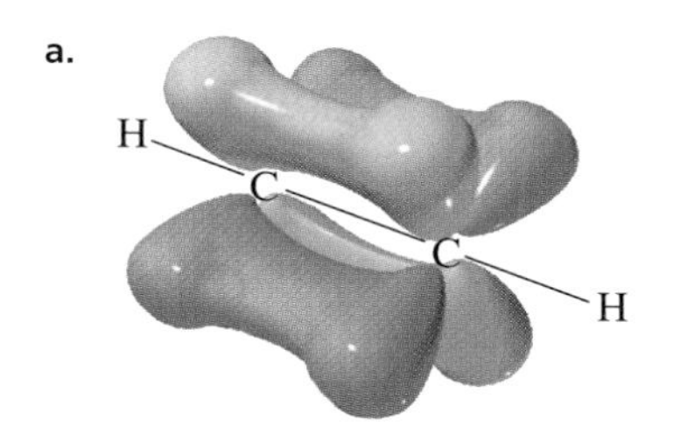

b.

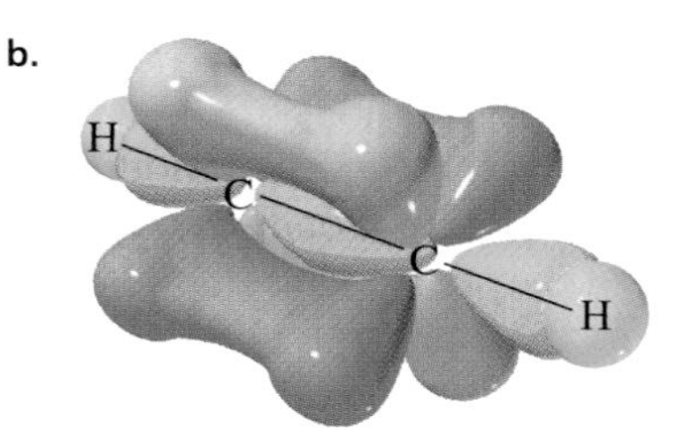

◀ **Figure 5.2**
(a) Each of the two π bonds of a triple bond is formed by side-to-side overlap of a *p* orbital of one carbon with a parallel *p* orbital of the adjacent carbon.
(b) A triple bond consists of a σ bond formed by *sp–sp* overlap (yellow) and two π bonds formed by *p–p* overlap (blue and purple).

PROBLEM 17 ◆

What orbitals are used to form the carbon–carbon σ bond between the highlighted carbons?

a. $CH_3CH{=}CHCH_3$ d. $CH_3C{\equiv}CCH_3$ g. $CH_3CH{=}CHCH_2CH_3$

b. $CH_3CH{=}CHCH_3$ e. $CH_3C{\equiv}CCH_3$ h. $CH_3C{\equiv}CCH_2CH_3$

c. $CH_3CH{=}C{=}CH_2$ f. $CH_2{=}CHCH{=}CH_2$ i. $CH_2{=}CHC{\equiv}CH$

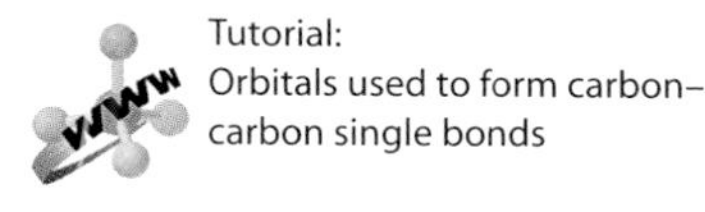

Tutorial:
Orbitals used to form carbon–carbon single bonds

5.9 THE PHYSICAL PROPERTIES OF UNSATURATED HYDROCARBONS

All hydrocarbons have similar physical properties. In other words, alkenes and alkynes have physical properties similar to those of alkanes (Section 3.7). All are insoluble in water and soluble in nonpolar solvents such as hexane. They are less dense than water and, like any other series of compounds, have boiling points that increase with increasing molecular weight (see Table 3.1 on page 47). Alkynes are more linear than alkenes, causing alkynes to have stronger van der Waals interactions. As a result, an alkyne has a higher boiling point than an alkene containing the same number of carbons (see Appendix I).

5.10 THE ADDITION OF A HYDROGEN HALIDE TO AN ALKYNE

With a cloud of electrons completely surrounding the σ bond, an alkyne is an electron-rich molecule. In other words, it is a nucleophile, and consequently it will react with an electrophile. Thus alkynes, like alkenes, undergo electrophilic addition reactions, and the same reagents that add to alkenes also add to alkynes. Moreover, the mechanism for electrophilic addition to an alkyne is the same as the mechanism for electrophilic addition to an alkene. For example, compare the mechanism for the addition of a hydrogen halide to an alkene shown in Section 5.1 with the mechanism for the addition of a hydrogen halide to an alkyne shown below.

mechanism for the addition of a hydrogen halide

$$CH_3C{\equiv}CCH_3 + H{-}\ddot{\underset{..}{Cl}}: \xrightarrow{\text{slow}} CH_3\overset{+}{C}{=}CHCH_3 + :\ddot{\underset{..}{Cl}}:^- \xrightarrow{\text{fast}} CH_3\overset{Cl}{\overset{|}{C}}{=}CHCH_3$$

a nucleophile (alkyne); an electrophile (HCl); an electrophile (carbocation); a nucleophile (chloride ion)

Tutorial: Addition of HCl to an alkyne

- The relatively weak π bond breaks because the π electrons are attracted to the electrophilic proton.
- The positively charged carbocation intermediate reacts rapidly with the negatively charged chloride ion.

The addition reactions of alkynes, however, have a feature that alkenes do not have: because the product of the addition of an electrophilic reagent to an alkyne is an alkene, a second electrophilic addition reaction can occur if excess hydrogen halide is present. In the second addition reaction, the electrophile (H^+) adds to the sp^2 carbon bonded to the greater number of hydrogens—as predicted by the rule that governs electrophilic addition reactions (Section 5.3).

$$CH_3C{\equiv}CCH_3 \xrightarrow{\text{HCl}} CH_3\overset{Cl}{\overset{|}{C}}{=}CHCH_3 \xrightarrow{\text{HCl}} CH_3\underset{\underset{Cl}{|}}{\overset{\overset{Cl}{|}}{C}}CH_2CH_3$$

a second electrophilic addition reaction occurs

The electrophile adds to the *sp* carbon of a terminal alkyne that is bonded to the hydrogen.

If the alkyne is a *terminal* alkyne, the H^+ will add to the *sp* carbon bonded to the hydrogen, because the *secondary* vinylic cation that results is more stable than the *primary* vinylic cation that would be formed if the H^+ added to the other *sp* carbon. (Recall that alkyl groups stabilize positively charged carbon atoms; see Section 5.2.)

$$\underset{\text{1-butyne}}{CH_3CH_2C{\equiv}CH} \xrightarrow{\text{HBr}} CH_3CH_2\overset{+}{C}{=}\overset{H}{\overset{|}{C}}H + Br^- \longrightarrow \underset{\text{2-bromo-1-butene}}{CH_3CH_2\overset{Br}{\overset{|}{C}}{=}\overset{H}{\overset{|}{C}}H}$$

the electrophile adds here

2-bromo-1-butene: a halo-substituted alkene

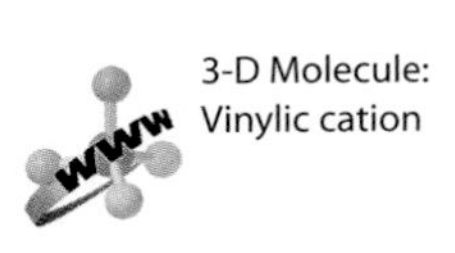

3-D Molecule: Vinylic cation

more stable $CH_3CH_2\overset{+}{C}{=}CH_2$ a secondary vinylic cation

$CH_3CH_2CH{=}\overset{+}{C}H$ less stable a primary vinylic cation

A second addition reaction will take place if excess hydrogen halide is present. Once again, the electrophile (H^+) adds to the sp^2 carbon that is bonded to the greater number of hydrogens.

electrophile adds here

$$\underset{\text{2-bromo-1-butene}}{CH_3CH_2\overset{\overset{\displaystyle Br}{|}}{C}{=}CH_2} \xrightarrow{HBr} \underset{\text{2,2-dibromobutane}}{CH_3CH_2\underset{\underset{\displaystyle Br}{|}}{\overset{\overset{\displaystyle Br}{|}}{C}}CH_3}$$

Addition of a hydrogen halide to an *internal* alkyne forms two products, because the initial addition of the proton can occur with equal ease to either of the *sp* carbons.

$$\underset{\text{2-pentyne}}{CH_3CH_2C{\equiv}CCH_3} + \underset{\text{excess}}{HCl} \longrightarrow \underset{\text{2,2-dichloropentane}}{CH_3CH_2CH_2\underset{\underset{\displaystyle Cl}{|}}{\overset{\overset{\displaystyle Cl}{|}}{C}}CH_3} + \underset{\text{3,3-dichloropentane}}{CH_3CH_2\underset{\underset{\displaystyle Cl}{|}}{\overset{\overset{\displaystyle Cl}{|}}{C}}CH_2CH_3}$$

Note, however, that if the same group is attached to each of the *sp* carbons of the internal alkyne, only one product will be obtained.

$$\underset{\text{3-hexyne}}{CH_3CH_2C{\equiv}CCH_2CH_3} + \underset{\text{excess}}{HBr} \longrightarrow \underset{\text{3,3-dibromohexane}}{CH_3CH_2CH_2\underset{\underset{\displaystyle Br}{|}}{\overset{\overset{\displaystyle Br}{|}}{C}}CH_2CH_3}$$

PROBLEM 18 ◆

Give the major product of each of the following reactions:

a. $HC{\equiv}CCH_3 \xrightarrow{HBr}$

b. $HC{\equiv}CCH_3 \xrightarrow[HBr]{\text{excess}}$

c. $CH_3C{\equiv}CCH_3 \xrightarrow[HBr]{\text{excess}}$

d. $CH_3C{\equiv}CCH_2CH_3 \xrightarrow[HBr]{\text{excess}}$

5.11 THE ADDITION OF WATER TO AN ALKYNE

In Section 5.4, we saw that alkenes undergo the acid-catalyzed addition of water. The product of the reaction is an alcohol.

$$\underset{\substack{\text{1-butene}\\ \text{an alkene}}}{CH_3CH_2CH{=}CH_2} + H_2O \xrightarrow{H_2SO_4} \underset{\substack{sec\text{-butyl alcohol}\\ \text{an alcohol}}}{CH_3CH_2\underset{\underset{\displaystyle OH}{|}}{CH}{-}\underset{\underset{\displaystyle H}{|}}{CH_2}}$$

Alkynes also undergo the acid-catalyzed addition of water. The initial product of the reaction is an *enol*. An **enol** is a compound with a carbon–carbon double bond and an OH group bonded to one of the sp^2 carbons. (The ending "ene" signifies the double bond, and "ol" the OH group; when the two are joined, the second *e* of "ene" is dropped to avoid two consecutive vowels, but the word is pronounced as if the *e* were still there: "ene-ol.")

$$CH_3C{\equiv}CCH_3 + H_2O \xrightarrow{H_2SO_4} \underset{\text{an enol}}{CH_3\overset{\overset{\displaystyle OH}{|}}{C}{=}CHCH_3} \rightleftharpoons \underset{\text{a ketone}}{CH_3\overset{\overset{\displaystyle O}{\|}}{C}{-}CH_2CH_3}$$

Addition of water to an alkyne forms a ketone.

The enol immediately rearranges to a *ketone*, a compound with the general structure shown below. A carbon doubly bonded to an oxygen is called a **carbonyl** ("car-bo-nil") **group**; a **ketone** ("key-tone") is a compound that has two alkyl groups bonded to a carbonyl group.

$$\overset{\text{O}}{\overset{\|}{\text{C}}} \qquad\qquad \text{R}\overset{\text{O}}{\overset{\|}{\text{C}}}\text{R}$$

a carbonyl group   a ketone

A ketone and an enol differ only in the location of a double bond and a hydrogen. The ketone and enol are called **keto–enol tautomers**. **Tautomers** ("taw-toe-mers") are isomers that are in rapid equilibrium. Interconversion of the tautomers is called **tautomerization**. Because the keto tautomer is usually more stable than the enol tautomer, it predominates at equilibrium.

$$\text{RCH}_2\text{—}\overset{\text{O}}{\overset{\|}{\text{C}}}\text{—R} \rightleftharpoons \text{RCH}{=}\overset{\text{OH}}{\overset{|}{\text{C}}}\text{—R}$$

keto tautomer   enol tautomer

tautomerization

The addition of water to an internal alkyne that has the same group attached to each of the *sp* carbons forms a single ketone as a product.

$$CH_3CH_2C{\equiv}CCH_2CH_3 + H_2O \xrightarrow{H_2SO_4} CH_3CH_2\overset{\text{O}}{\overset{\|}{\text{C}}}CH_2CH_2CH_3$$

If the two groups are not identical, two ketones are formed because the initial addition of the proton can occur to either of the *sp* carbons.

$$CH_3C{\equiv}CCH_2CH_3 + H_2O \xrightarrow{H_2SO_4} CH_3\overset{\text{O}}{\overset{\|}{\text{C}}}CH_2CH_2CH_3 + CH_3CH_2\overset{\text{O}}{\overset{\|}{\text{C}}}CH_2CH_3$$

Terminal alkynes are less reactive than internal alkynes toward the addition of water. Terminal alkynes will add water if mercuric ion (Hg^{2+}) is added to the acidic mixture. The mercuric ion is a catalyst—it increases the rate of the addition reaction.

$$CH_3CH_2C{\equiv}CH + H_2O \xrightarrow[HgSO_4]{H_2SO_4} CH_3CH_2\overset{\text{OH}}{\overset{|}{\text{C}}}{=}CH_2 \rightleftharpoons CH_3CH_2\overset{\text{O}}{\overset{\|}{\text{C}}}\text{—}CH_3$$

an enol   a ketone

PROBLEM 19 ◆

What ketones would be formed from the acid-catalyzed addition of water to 3-heptyne?

PROBLEM 20 ◆

Which alkyne would be the best reagent to use for the synthesis of each of the following ketones?

a. $CH_3\overset{\text{O}}{\overset{\|}{\text{C}}}CH_3$ **b.** $CH_3CH_2\overset{\text{O}}{\overset{\|}{\text{C}}}CH_2CH_2CH_3$ **c.** $CH_3\overset{\text{O}}{\overset{\|}{\text{C}}}\text{—}$cyclohexyl

PROBLEM 21 ♦

Draw the enol tautomers for the following ketone:

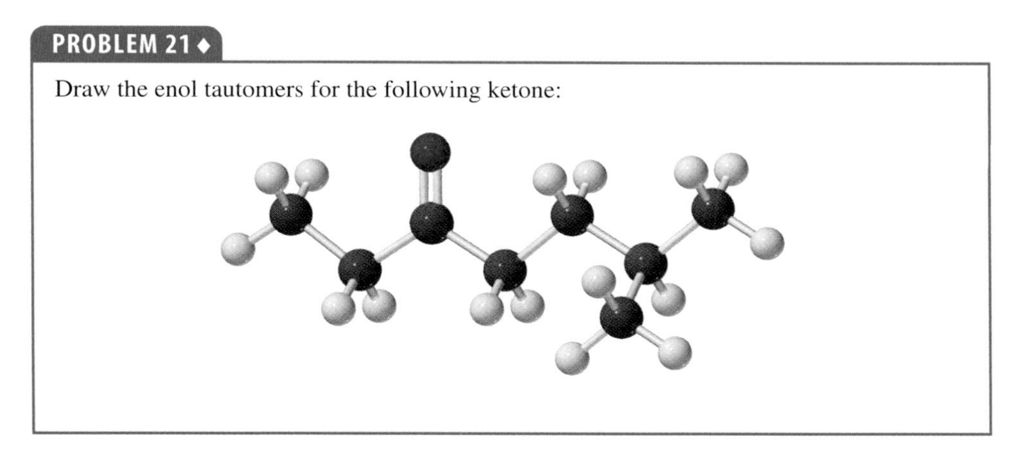

5.12 THE ADDITION OF HYDROGEN TO ALKENES AND ALKYNES

In the presence of a metal catalyst such as platinum or palladium, hydrogen (H_2) adds to the double bond of an alkene to form an alkane. Without the catalyst, the energy barrier to the reaction is enormous because the H—H bond is so strong. The catalyst decreases the energy of activation by breaking the H—H bond (Section 4.8). Platinum and palladium are used in a finely divided state adsorbed on charcoal (Pt/C, Pd/C).

$$\underset{\text{2-methylpropene}}{CH_3\overset{\overset{CH_3}{|}}{C}{=}CH_2} + H_2 \xrightarrow{\text{Pd/C}} \underset{\text{2-methylpropane}}{CH_3\overset{\overset{CH_3}{|}}{C}HCH_3}$$

$$\underset{\text{cyclohexene}}{\text{(cyclohexene)}} + H_2 \xrightarrow{\text{Pt/C}} \underset{\text{cyclohexane}}{\text{(cyclohexane)}}$$

The addition of hydrogen is called **hydrogenation**. Because the preceding reactions require a catalyst, they are **catalytic hydrogenations**. A reaction that increases the number of C—H bonds in a compound is called a **reduction reaction**. Thus, catalytic hydrogenation is a reduction reaction.

A reduction reaction increases the number of C—H bonds.

The details of the mechanism of catalytic hydrogenation are not completely understood. We know that hydrogen is adsorbed on the surface of the metal and that all the bond-breaking and bond-forming events occur on the surface of the metal. As the alkane product is formed, it diffuses away from the metal surface (Figure 5.3).

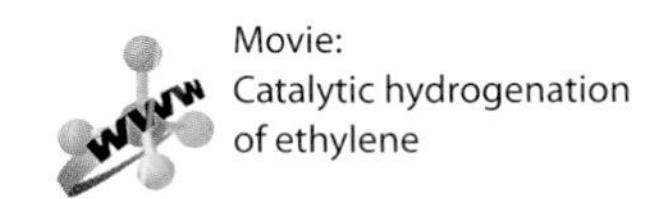
www

Movie: Catalytic hydrogenation of ethylene

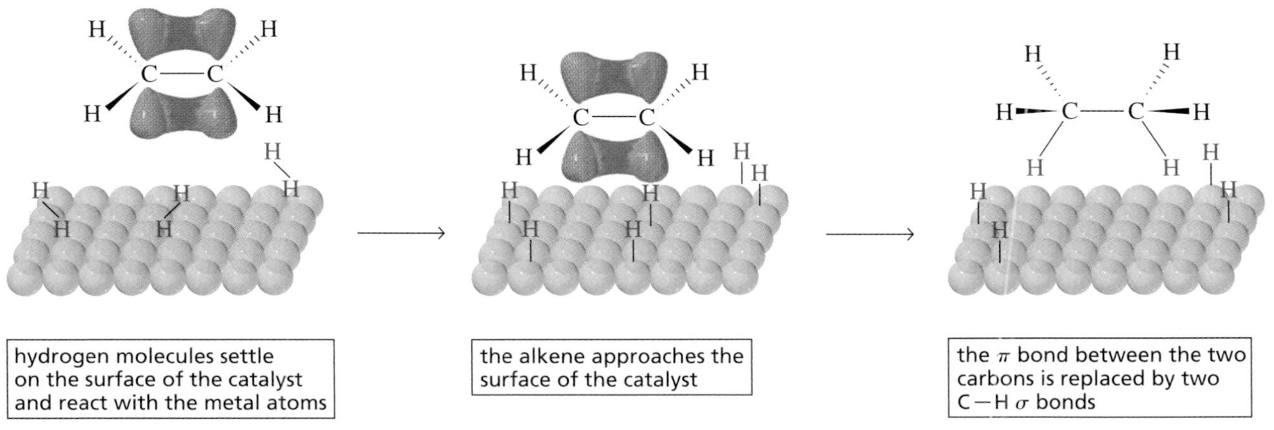
hydrogen molecules settle on the surface of the catalyst and react with the metal atoms
the alkene approaches the surface of the catalyst
the π bond between the two carbons is replaced by two C—H σ bonds

▲ **Figure 5.3**
Catalytic hydrogenation of an alkene.

BIOGRAPHY

Herbert H. M. Lindlar *was born in Switzerland in 1909 and received a Ph.D. from the University of Bern. He worked at Hoffmann–La Roche and Co. in Basel, Switzerland, and he authored many patents. His last patent was a procedure for isolating the carbohydrate xylose from the waste produced in paper mills.*

Hydrogen adds to an alkyne in the presence of a metal catalyst such as palladium or platinum in the same manner that it adds to an alkene. The initial product is an alkene, but it is difficult to stop the reaction at that stage because of hydrogen's strong tendency to add to alkenes in the presence of these efficient metal catalysts. The product of the hydrogenation reaction, therefore, is an alkane.

$$\underset{\text{alkyne}}{CH_3CH_2C{\equiv}CH} \xrightarrow[\text{Pt/C}]{H_2} \underset{\text{alkene}}{CH_3CH_2CH{=}CH_2} \xrightarrow[\text{Pt/C}]{H_2} \underset{\text{alkane}}{CH_3CH_2CH_2CH_3}$$

an alkyne is converted to an alkane

The reaction can be stopped at the alkene stage if a "poisoned" (partially deactivated) metal catalyst is used. The most commonly used partially deactivated metal catalyst is called **Lindlar catalyst**.

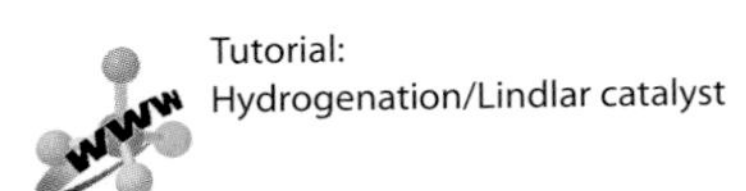

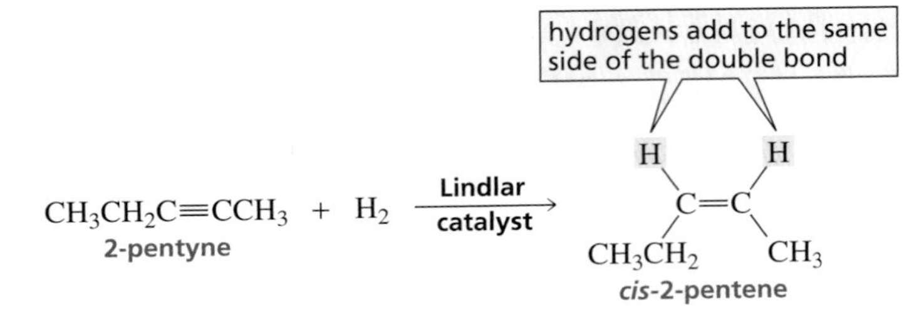

Because the alkyne sits on the surface of the metal catalyst and the hydrogens are delivered to the triple bond from the surface of the catalyst, both hydrogens are delivered to the same side of the triple bond. Therefore, the addition of hydrogen to an internal alkyne in the presence of Lindlar catalyst forms a **cis alkene**.

TRANS FATS

Fats and oils contain long-chain unbranched carboxylic acids (called fatty acids) with carbon–carbon double bonds in a cis configuration. Fats are solids at room temperature, whereas oils are liquids at room temperature because they contain more double bonds; the double bonds prevent the molecules from packing tightly together (Section 19.1).

COOH

linoleic acid
an 18-carbon fatty acid with two cis double bonds

Some or all of the double bonds in oils can be reduced by catalytic hydrogenation. For example, margarine and shortening are prepared by hydrogenating vegetable oils such as soybean oil and safflower oil until they have the desired creamy, solid consistency.

All the double bonds in naturally occurring fats and oils have the cis configuration. The heat used in the hydrogenation process breaks the π bond of these double bonds. Sometimes, instead of being hydrogenated, a double bond reforms; if the sigma bond rotates while the π bond is broken, the double bond can reform in the trans configuration (Section 4.4), forming what is known as a trans fat.

One reason trans fats are a health concern is that they do not have the same shape as natural cis fats but are able to take the place of cis fats in cell membranes, thereby affecting the ability of the membrane to control the flow of molecules into and out of our cells.

COOH

oleic acid
an 18-carbon fatty acid with one cis double bond
before being heated

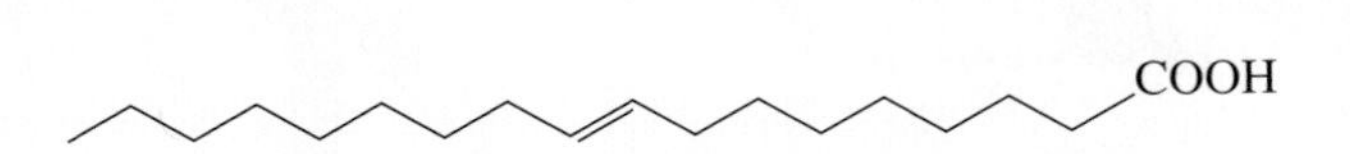

an 18-carbon fatty acid with one trans double bond
after being heated

PROBLEM-SOLVING STRATEGY

What alkene would you use if you wanted to synthesize methylcyclohexane?

You need to choose an alkene that has the same number of carbons, attached in the same way, as those in the desired product. Several alkenes could be used for this synthesis, because the double bond can be located anywhere in the molecule.

CH_2 (methylenecyclohexane) or CH_3 (1-methylcyclohexene) or CH_3 (3-methylcyclohexene) or CH_3 (4-methylcyclohexene) $\xrightarrow[\text{Pd/C}]{H_2}$ CH_3 (methylcyclohexane)

methylcyclohexane

Now continue on to answer the questions in Problem 22.

PROBLEM 22 ♦

What reagents would you use if you wanted to synthesize
a. *cis*-2-butene?
b. 1-hexene?

PROBLEM 23 ♦

How many different alkenes can be hydrogenated to form:
a. butane?
b. pentane?
c. methylcyclopentane?

5.13 A HYDROGEN BONDED TO AN *sp* CARBON IS ACIDIC

Carbon forms nonpolar covalent bonds with hydrogen because carbon and hydrogen, with similar electronegativities, share their bonding electrons almost equally. However, all carbon atoms do not have the same electronegativity. An *sp* carbon is more electronegative than an sp^2 carbon, which is more electronegative than an sp^3 carbon.

An *sp* carbon is more electronegative than an sp^2 carbon, which is more electronegative than an sp^3 carbon.

relative electronegativities of carbon atoms

most electronegative $sp > sp^2 > sp^3$ least electronegative

Because the most acidic compound is the one with the hydrogen attached to the most electronegative atom (when the atoms are the same size; see Section 2.6), ethyne is a stronger acid than ethene, and ethene is a stronger acid than ethane. (Recall that the stronger the acid, the lower its pK_a.)

$HC \equiv CH$	$H_2C{=}CH_2$	CH_3CH_3
ethyne	ethene	ethane
$pK_a = 25$	$pK_a = 44$	$pK_a > 60$

In order to remove a proton from an acid (in a reaction that strongly favors products), the base that removes the proton must be stronger than the base that is generated as a result of removing the proton (Section 2.5). In other words, you must start with a stronger base than the base that will be formed. Because NH_3 is a weaker acid (pK_a = 36) than a terminal alkyne ($pK_a = 25$), an amide ion ($^-NH_2$) is a stronger base than the

The stronger the acid, the weaker its conjugate base.

To remove a proton from an acid in a reaction that favors products, the base that removes the proton must be stronger than the base that is formed.

carbanion—called an **acetylide ion**—that is formed when a hydrogen is removed from the *sp* carbon of a terminal alkyne. (Remember, the stronger the acid, the weaker its conjugate base.) Therefore, an amide ion can be used to remove a proton from a terminal alkyne to form an acetylide ion.

$$\underset{\boxed{\text{stronger acid}}}{RC{\equiv}CH} + \underset{\substack{\text{amide ion} \\ \boxed{\text{stronger base}}}}{{}^{-}NH_2} \rightleftharpoons \underset{\substack{\text{acetylide ion} \\ \textbf{weaker base}}}{RC{\equiv}C^{-}} + \underset{\textbf{weaker acid}}{NH_3}$$

An amide ion cannot remove a hydrogen bonded to an sp^2 or an sp^3 carbon. Only a hydrogen bonded to an *sp* carbon is sufficiently acidic to be removed by an amide ion. Consequently, a hydrogen bonded to an *sp* carbon sometimes is referred to as an "acidic" hydrogen. The "acidic" property of terminal alkynes is one way their reactivity differs from that of alkenes. Be careful not to misinterpret what is meant when we say that a hydrogen bonded to an *sp* carbon is "acidic." It is more acidic than most other carbon-bound hydrogens, but it is much less acidic than a hydrogen of a water molecule, and water is only a very weakly acidic compound ($pK_a = 15.7$).

relative acid strengths

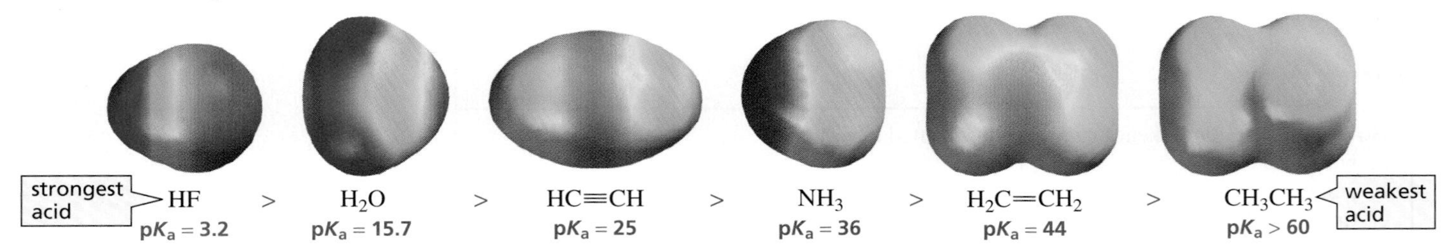

PROBLEM 24 ♦

Explain why sodium amide cannot be used to form a carbanion from an alkane in a reaction that favors products.

PROBLEM-SOLVING STRATEGY

a. List the following compounds in order of decreasing acidity:

$$CH_3CH_2\overset{+}{N}H_3 \qquad CH_3CH{=}\overset{+}{N}H_2 \qquad CH_3CH{\equiv}\overset{+}{N}H$$

To compare the acidities of a group of compounds, first look at how the compounds differ. These three compounds differ in the hybridization of the nitrogen to which the acidic hydrogen is attached. Now recall what you know about hybridization and acidity. You know that hybridization of an atom affects its electronegativity (*sp* is more electronegative than sp^2, and sp^2 is more electronegative than sp^3), and you know that the more electronegative the atom to which a hydrogen is attached, the more acidic is the hydrogen. Now you can answer the question.

$$\textbf{relative acidities} \quad CH_3C{\equiv}\overset{+}{N}H > CH_3CH{=}\overset{+}{N}H_2 > CH_3CH_2\overset{+}{N}H_3$$

b. Draw the conjugate bases of the above compounds and list them in order of decreasing basicity. First remove a proton from each acid to obtain the structures of the conjugate bases. The stronger the acid, the weaker is its conjugate base, so using the relative acid strengths obtained in part a, we find that the order of basicity is.

$$\textbf{relative basicities} \quad CH_3CH_2NH_2 > CH_3CH{=}NH > CH_3C{\equiv}N$$

Now continue on to Problem 25.

PROBLEM 25 ♦

List the following species in order of decreasing basicity:

a. $CH_3CH_2CH{=}\bar{C}H$ $\quad CH_3CH_2C{\equiv}C^-$ $\quad CH_3CH_2CH_2\bar{C}H_2$

b. $CH_3CH_2O^-$ $\quad F^-$ $\quad CH_3C{\equiv}C^-$ $\quad {}^-NH_2$

PROBLEM 26 *SOLVED*

Which carbocation in each of the following pairs is more stable?

a. $CH_3\overset{+}{C}H_2$ or $H_2C{=}\overset{+}{C}H$ $\qquad$ **b.** $H_2C{=}\overset{+}{C}H$ or $HC{\equiv}\overset{+}{C}$

Solution to 26a A double-bonded carbon is more electronegative than a single-bonded carbon. Therefore, a double-bonded carbon with a positive charge would be less stable than a single-bonded carbon with a positive charge. Therefore, the ethyl carbocation is more stable.

5.14 SYNTHESIS USING ACETYLIDE IONS

Reactions that form carbon–carbon bonds are important in the synthesis of organic compounds because, without such reactions, we could not convert molecules with small carbon skeletons into molecules with larger carbon skeletons. Instead, the product of a reaction would always have the same number of carbons as the starting material.

One reaction that forms a carbon–carbon bond is the reaction of an acetylide ion with an alkyl halide. Only primary alkyl halides or methyl halides should be used in this reaction.

$$\underset{\text{an acetylide ion}}{CH_3CH_2C{\equiv}C^-} + \underset{\text{an alkyl halide}}{CH_3CH_2CH_2Br} \longrightarrow \underset{\text{3-heptyne}}{CH_3CH_2C{\equiv}CCH_2CH_2CH_3} + Br^-$$

The mechanism of this reaction is well understood. Bromine is more electronegative than carbon, and as a result, the electrons in the C—Br bond are not shared equally by the two atoms. There is a partial positive charge on carbon and a partial negative charge on bromine. The negatively charged acetylide ion (a nucleophile) is attracted to the partially positively charged carbon (an electrophile) of the alkyl halide. As the electrons of the acetylide ion approach the carbon to form the new C—C bond, they push out the bromine and its bonding electrons because carbon can bond to no more than four atoms at a time.

$$CH_3CH_2C{\equiv}\ddot{C}^- + CH_3CH_2\overset{\delta+}{C}H_2{-}\overset{\delta-}{Br} \longrightarrow CH_3CH_2C{\equiv}CCH_2CH_2CH_3 + Br^-$$

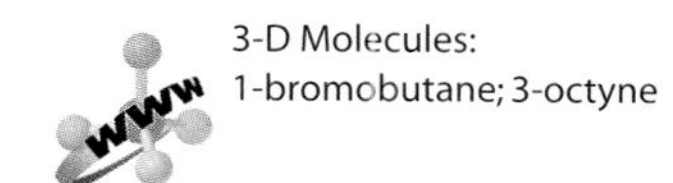

We can convert terminal alkynes into internal alkynes of any desired chain length simply by choosing an alkyl halide with the appropriate structure. Just count the number of carbons in the terminal alkyne and the number of carbons in the product to see how many carbons are needed in the alkyl halide.

$$\underset{\text{1-pentyne}}{CH_3CH_2CH_2C{\equiv}CH} \xrightarrow{NaNH_2} CH_3CH_2CH_2C{\equiv}C^- \xrightarrow{CH_3CH_2Br} \underset{\text{3-heptyne}}{CH_3CH_2CH_2C{\equiv}CCH_2CH_3}$$

PROBLEM 27 SOLVED

A chemist wants to synthesize 3-heptyne but cannot find any 1-pentyne, the starting material used in the synthesis shown on page 127. How else can 3-heptyne be synthesized?

Solution The *sp* carbons of 3-heptyne are bonded to an ethyl group and to a propyl group. Therefore, to produce 3-heptyne, the acetylide ion of 1-pentyne can react with an ethyl halide (as on page 127), or the acetylide ion of 1-butyne can react with a propyl halide. Since 1-pentyne is not available, the chemist should use 1-butyne and a propyl halide.

$$CH_3CH_2C{\equiv}C^- + CH_3CH_2CH_2Br \longrightarrow \underset{\text{3-heptyne}}{CH_3CH_2C{\equiv}CCH_2CH_2CH_3} + Br^-$$

5.15 AN INTRODUCTION TO MULTISTEP SYNTHESIS

Synthetic chemists consider time, cost, and yield in designing syntheses. In the interest of time, a well-designed synthesis will consist of as few steps (sequential reactions) as possible, and each of those steps will be a reaction that is easy to carry out. If two chemists in a pharmaceutical company were each asked to prepare a new drug, and one synthesized the drug in 3 simple steps while the other used 20 difficult steps, which chemist would not get a raise? The costs of the starting materials are also taken into consideration; the more reactant needed to synthesize one gram of product, the more expensive it is to produce. Moreover, each step in the synthesis should provide the greatest possible yield of the desired product. Sometimes a synthesis involving several steps is preferred because the starting materials are inexpensive, the reactions are easy to carry out, and the yield of each step is high. Such a synthesis is better than one with fewer steps if those steps require expensive starting materials and reactions that are more difficult to run or give lower yields. At this point in your chemical education, however, you are not yet familiar with the costs of different chemicals or the difficulty of carrying out specific reactions. So, for the time being, when you design a synthesis, just focus on finding the route with the fewest steps.

The following examples will give you an idea of the type of thinking required for designing a successful synthesis.

Example 1. Starting with 1-butyne, how could you make the ketone shown below? You can use any organic and inorganic reagents.

$$\underset{\text{1-butyne}}{CH_3CH_2C{\equiv}CH} \xrightarrow{?} CH_3CH_2\overset{O}{\overset{\|}{C}}CH_2CH_2CH_3$$

Many chemists find that the easiest way to design a synthesis is to work backward. Instead of looking at the starting material and deciding how to do the first step of the synthesis, look at the product and decide how to do the last step. The product is a ketone. At this point, the only reaction you know that forms a ketone is the addition of water (in the presence of an acid catalyst) to an alkyne. If the alkyne used in the reaction has identical substituents on both *sp* carbons, only one ketone will be obtained. Thus, 3-hexyne is the best alkyne to use for the synthesis of the desired ketone.

$$\underset{\text{3-hexyne}}{CH_3CH_2C{\equiv}CCH_2CH_3} \xrightarrow[H_2SO_4]{H_2O} CH_3CH_2\overset{OH}{\overset{|}{C}}{=}CHCH_2CH_3 \rightleftharpoons CH_3CH_2\overset{O}{\overset{\|}{C}}CH_2CH_2CH_3$$

3-Hexyne can be obtained from the four-carbon starting material by removing the proton from its *sp* carbon, followed by reaction with a two-carbon alkyl halide. (The numbers 1 and 2 in front of the reagents above and below the reaction arrow indicate two sequential reactions; the second reagent is not added until the reaction with the first reagent is completely over.)

$$\underset{\text{1-butyne}}{CH_3CH_2C{\equiv}CH} \xrightarrow[\text{2. } CH_3CH_2Br]{\text{1. } NaNH_2} \underset{\text{3-hexyne}}{CH_3CH_2C{\equiv}CCH_2CH_3}$$

Thus, the synthetic scheme for the synthesis of the desired ketone is given by

$$CH_3CH_2C{\equiv}CH \xrightarrow[\text{2. } CH_3CH_2Br]{\text{1. } NaNH_2} CH_3CH_2C{\equiv}CCH_2CH_3 \xrightarrow[H_2SO_4]{H_2O} CH_3CH_2\overset{O}{\overset{\|}{C}}CH_2CH_2CH_3$$

Example 2. Starting with ethyne, how could you make 2-bromopentane?

$$\underset{\text{ethyne}}{HC{\equiv}CH} \xrightarrow{?} \underset{\text{2-bromopentane}}{CH_3CH_2CH_2\underset{Br}{\underset{|}{C}}HCH_3}$$

The desired product can be prepared from 1-pentene, which can be prepared from 1-pentyne. 1-Pentyne can be prepared from ethyne and an alkyl halide with three carbons.

$$HC{\equiv}CH \xrightarrow[\text{2. } CH_3CH_2CH_2Br]{\text{1. } NaNH_2} CH_3CH_2CH_2C{\equiv}CH \xrightarrow[\text{Lindlar catalyst}]{H_2} CH_3CH_2CH_2CH{=}CH_2 \xrightarrow{HBr} CH_3CH_2CH_2\underset{Br}{\underset{|}{C}}HCH_3$$

Example 3. How could you prepare 3,3-dibromohexane from reagents that contain no more than two carbons?

$$\text{reagents with no more than 2 carbons} \xrightarrow{?} \underset{\text{3,3-dibromohexane}}{CH_3CH_2\overset{Br}{\overset{|}{\underset{Br}{\underset{|}{C}}}}CH_2CH_2CH_3}$$

The desired product can be prepared from an alkyne and excess HBr. 3-Hexyne is the alkyne of choice, because it will form one dibromide, whereas 2-hexyne would form two different dibromides (3,3-dibromohexane and 2,2-dibromohexane) because it is not a symmetrical alkyne. 3-Hexyne can be prepared from 1-butyne and ethyl bromide, and 1-butyne can be prepared from ethyne and ethyl bromide.

$$HC{\equiv}CH \xrightarrow[\text{2. } CH_3CH_2Br]{\text{1. } NaNH_2} CH_3CH_2C{\equiv}CH \xrightarrow[\text{2. } CH_3CH_2Br]{\text{1. } NaNH_2} CH_3CH_2C{\equiv}CCH_2CH_3 \xrightarrow{\text{excess HBr}} CH_3CH_2\overset{Br}{\overset{|}{\underset{Br}{\underset{|}{C}}}}CH_2CH_2CH_3$$

PROBLEM 28

Starting with acetylene, how could the following compounds be synthesized?

a. $CH_3CH_2CH_2C{\equiv}CH$

b. $CH_3CH_2CH_2\overset{O}{\overset{\|}{C}}CH_3$

c. $CH_3CH{=}CH_2$

d. $CH_3\underset{Br}{\underset{|}{C}}HCH_3$

e. cis-$CH_3CH{=}CHCH_3$ (both CH_3 groups on the same side of the C=C; H and H on the same side)

f. $CH_3\overset{Cl}{\overset{|}{\underset{Cl}{\underset{|}{C}}}}CH_3$

5.16 SYNTHETIC POLYMERS

A **polymer** is a large molecule made by linking together repeating units of small molecules called **monomers**. The process of linking them together is called **polymerization**.

$$n\text{M} \xrightarrow{\text{polymerization}} \text{—M—M—M—M—M—M—M—M—M—}$$

monomers → polymer

ethylene monomers → polyethylene

Polymers can be divided into two broad groups: **synthetic polymers** and **biopolymers**. Synthetic polymers are synthesized by scientists, whereas biopolymers are synthesized by organisms. Examples of biopolymers are DNA, which is the storage molecule for genetic information—the molecule that determines whether a fertilized egg becomes a human or a honeybee; RNA and proteins, the molecules that induce biochemical transformations; and polysaccharides, which store energy, act as recognition sites on cell surfaces, and also function as structural materials. The structures and properties of these biopolymers are presented in other chapters. Here, we will explore synthetic polymers.

Probably no group of synthetic compounds is more important to modern life than synthetic polymers. Some synthetic polymers resemble natural substances, but most are quite different from materials found in nature. Such diverse products as photographic film, compact discs, food wrap, artificial joints, Super Glue, toys, plastic bottles, weather stripping, automobile body parts, and shoe soles are made of synthetic polymers. More than 2.5×10^{13} kilograms of synthetic polymers are produced in the United States each year, and approximately 30,000 polymer patents are currently in force. We can expect scientists to develop many more new materials in the years to come.

Synthetic polymers can be divided into two major classes, depending on their method of preparation. Here we will look at *chain-growth polymers*. The second major class of polymers, *step-growth polymers*, are discussed in Section 11.14.

Chain-growth polymers are made by **chain reactions**—the addition of monomers to the end of a growing chain. The monomers used most commonly in chain-growth polymerization are ethylene and substituted ethylenes. Polystyrene—used for disposable food containers, insulation, and toothbrush handles, among other things—is an example of a chain-growth polymer. As its name suggests, the monomer used to form polystyrene is a substituted ethylene called styrene. Polystyrene can be pumped full of air to produce the material known as Styrofoam.

$$CH_2{=}CH(C_6H_5) \quad CH_2{=}CH(C_6H_5) \quad CH_2{=}CH(C_6H_5) \longrightarrow \text{—}CH_2\text{—}CH(C_6H_5)\text{—}[CH_2\text{—}CH(C_6H_5)]_n\text{—}CH_2\text{—}CH(C_6H_5)\text{—}$$

styrene → polystyrene, a chain-growth polymer (repeating unit in brackets)

Some of the many polymers synthesized by chain-growth polymerization are listed in Table 5.1.

Table 5.1 Some Important Chain-Growth Polymers and Their Uses

Monomer	Repeating unit	Polymer name	Uses
$CH_2{=}CH_2$	$-CH_2-CH_2-$	polyethylene	film, toys, bottles, plastic bags
$CH_2{=}CH(Cl)$	$-CH_2-CH(Cl)-$	poly(vinyl chloride)	"squeeze" bottles, pipe, siding, flooring
$CH_2{=}CH-CH_3$	$-CH_2-CH(CH_3)-$	polypropylene	molded caps, margarine tubs, indoor/outdoor carpeting, upholstery
$CH_2{=}CH(C_6H_5)$	$-CH_2-CH(C_6H_5)-$	polystyrene	packaging, toys, clear cups, egg cartons, hot drink cups
$CF_2{=}CF_2$	$-CF_2-CF_2-$	poly(tetrafluoroethylene) Teflon	nonsticking surfaces, liners, cable insulation
$CH_2{=}CH(C{\equiv}N)$	$-CH_2-CH(C{\equiv}N)-$	poly(acrylonitrile) Orlon, Acrilan	rugs, blankets, yarn, apparel, simulated fur
$CH_2{=}C(CH_3)-COCH_3$ (with C=O)	$-CH_2-C(CH_3)(COCH_3)-$ (with C=O)	poly(methyl methacrylate) Plexiglas, Lucite	lighting fixtures, signs, solar panels, skylights
$CH_2{=}CH-OCCH_3$ (with C=O)	$-CH_2-CH(OCCH_3)-$ (with C=O)	poly(vinyl acetate)	latex paints, adhesives

The two most common mechanisms for chain-growth polymerization are cationic polymerization and radical polymerization. Each of these mechanisms has three distinct phases; *initiation steps* that start the polymerization, *propagation steps* that allow the polymer chain to grow, and *termination* steps that stop the growth of the chain.

Chain reactions have initiation, propagation, and termination steps.

Cationic Polymerization

In cationic polymerization, the initiator is an electrophile that adds to the alkene monomer, causing it to become a cation. The initiator most often used in cationic polymerization is a proton, generated from the reaction of BF_3 with water (Section 2.9); because boron does not have a complete octet, it accepts a share in an electron pair from the oxygen of water.

BF_3 (F—B(F)—F): incomplete octet

initiation steps

$$F_3B + H_2\ddot{O}: \rightleftharpoons F_3\bar{B}{:}\overset{+}{O}H_2 \rightleftharpoons F_3\bar{B}{:}\ddot{O}H + H^+$$

$$H^+ + CH_2{=}C(CH_3)_2 \longrightarrow CH_3\overset{+}{C}(CH_3)_2$$

the alkene monomer reacts with an electrophile

- The cation (an electrophile) formed in the initiation steps reacts with a second monomer, forming a new cation that reacts in turn with a third monomer. These are called **propagation steps** because they propagate the chain reaction. The cation is now at the end of the unit that was most recently added to the end of the chain. This is called the **propagating site**.

propagation steps

$$(CH_3)_2\overset{+}{C}CH_3 + CH_2{=}C(CH_3)_2 \longrightarrow CH_3C(CH_3)_2CH_2\overset{+}{C}(CH_3)_2$$

$$CH_3C(CH_3)_2CH_2\overset{+}{C}(CH_3)_2 + CH_2{=}C(CH_3)_2 \longrightarrow CH_3C(CH_3)_2CH_2C(CH_3)_2CH_2\overset{+}{C}(CH_3)_2$$

propagating sites

As each subsequent monomer adds to the chain, the new positively charged propagating site is at the end of the last unit added. This process is repeated over and over. Hundreds or even thousands of alkene monomers can be added one at a time to the growing chain. Notice that in both the chain-initiating and chain-propagating steps, the rule governing electrophilic addition reactions is followed: the electrophile adds to the carbon bonded to the greater number of hydrogens (Section 5.3).

- Eventually, the chain reaction stops because the propagating sites are destroyed. A propagating site is destroyed when it reacts with a nucleophile. This is called a **termination step**.

termination step

$$CH_3C(CH_3)_2{-}[CH_2C(CH_3)_2]_n{-}CH_2\overset{+}{C}(CH_3)_2 \xrightarrow{Nu^-} CH_3C(CH_3)_2{-}[CH_2C(CH_3)_2]_n{-}CH_2C(CH_3)_2{-}Nu$$

DESIGNING A POLYMER

A polymer used for making contact lenses must be sufficiently hydrophilic (water-loving) to allow lubrication of the eye. Such a polymer, therefore, has many OH groups.

$$-CH_2-\underset{C(=O)CH_3}{CH}-CH_2-\underset{C(=O)CH_2CH_2OH}{CH}-CH_2-\underset{C(=O)CH_3}{CH}-CH_2-\underset{C(=O)CH_2CH_2OH}{CH}-$$

polymer used to make contact lenses

PROBLEM 29

Draw a short segment of the polymer that would be formed from cationic polymerization of methyl vinyl ether with H^+ as the initiator.

Radical Polymerization

In radical polymerization, the initiator is a species that breaks into radicals. Most radical initiators have an O—O bond because such a bond easily breaks in a way that allows each of the atoms that formed the bond to retain one of the bonding electrons. Each of the radicals that is formed seeks an electron to complete its octet. A radical can obtain an electron by adding to the electron-rich π bond of the alkene, thereby forming a new radical. The curved arrows that we have previously seen have arrowheads with two barbs because they represent the movement of two electrons. Notice that the arrowheads in the mechanism shown below have only one barb because they represent the movement of only one electron.

An arrowhead with two barbs signifies the movement of two electrons.

An arrowhead with one barb signifies the movement of one electron.

- Radical polymerization has two initiation steps; one creates radicals, and the other forms the radical that propagates the chain reaction. Because the radical is seeking an electron, a radical is an electrophile and, like other electrophiles, it adds to the sp^2 carbon bonded to the greater number of hydrogens.

initiation steps

$$\underset{\textbf{a radical initiator}}{RO{-}OR} \xrightarrow{\Delta} \underset{\textbf{radicals}}{2\ RO\cdot}$$

a radical is a reactive species

$$RO\cdot + CH_2{=}\underset{\displaystyle Z}{\overset{}{CH}} \longrightarrow RO{-}CH_2\underset{\displaystyle Z}{\dot{C}H}$$

the alkene monomer reacts with a radical

- The radical adds to another alkene monomer, converting it into a radical. This radical reacts with another monomer, adding a new subunit that propagates the chain reaction. Notice that when a radical is used to initiate polymerization, the propagating sites are also radicals.

propagation steps

propagating sites

$$RO{-}CH_2\underset{\displaystyle Z}{\dot{C}H} + CH_2{=}\underset{\displaystyle Z}{CH} \longrightarrow RO{-}CH_2\underset{\displaystyle Z}{CH}CH_2\underset{\displaystyle Z}{\dot{C}H}$$

$$RO{-}CH_2\underset{\displaystyle Z}{CH}CH_2\underset{\displaystyle Z}{\dot{C}H} + CH_2{=}\underset{\displaystyle Z}{CH} \longrightarrow RO{-}CH_2\underset{\displaystyle Z}{CH}CH_2\underset{\displaystyle Z}{CH}CH_2\underset{\displaystyle Z}{\dot{C}H}$$

- The chain reaction stops when the propagating sites are destroyed. A propagating site is destroyed when it reacts with a species (XY) that allows it to pair up its electron.

termination step

$$-CH_2{-}\left[CH_2\underset{\displaystyle Z}{CH}\right]_n{-}CH_2\underset{\displaystyle Z}{\dot{C}H} + X{-}Y \longrightarrow -CH_2{-}\left[CH_2\underset{\displaystyle Z}{CH}\right]_n{-}CH_2\underset{\displaystyle Z}{CH}X + Y\cdot$$

RECYCLING SYMBOLS

When plastics are recycled, the various types must be separated from one another. To aid in the separation, many states require manufacturers to place a recycling symbol on their products to indicate the type of plastic it is. You are probably familiar with these symbols, which are often embossed on the bottom of plastic containers. The symbols consist of three arrows around one of seven numbers; an abbreviation below the symbol indicates the type of polymer from which the container is made. The lower the number in the middle of the symbol, the greater is the ease with which the material can be recycled: 1 (PET) stands for poly(ethylene terephthalate), 2 (HDPE) for high-density polyethylene, 3 (V) for poly(vinyl chloride), 4 (LDPE) for low-density polyethylene, 5 (PP) for polypropylene, 6 (PS) for polystyrene, and 7 for all other plastics.

PROBLEM 30 ◆

What monomer would you use to form each of the following polymers?

a. $-CH_2CHCH_2CHCH_2CHCH_2CHCH_2CH-$ (each CH bearing a Cl substituent)

b. $-CH_2CCH_2CCH_2CCH_2CCH_2CCH_2C-$ (each C bearing a CH_3 group and a $C{=}O$ group bonded to O–CH_3)

c. $-CF_2CF_2CF_2CF_2CF_2CF_2CF_2CF_2CF_2CF_2-$

PROBLEM 31

Show the mechanism for the formation of a segment of poly(vinyl chloride) containing three units of vinyl chloride and initiated by HO·.

Branching of the Polymer Chain

If the propagating site removes a hydrogen atom from the polymer chain, a branch can grow off the chain at that point.

$$-CH_2CH_2CH_2\dot{C}H_2 \; + \; -CH_2CH_2\underset{|}{\overset{H}{C}}HCH_2CH_2CH_2-$$

$$\downarrow$$

$$-CH_2CH_2CH_2CH_2\text{–}H \; + \; -CH_2CH_2\dot{C}HCH_2CH_2CH_2- \xrightarrow{CH_2=CH_2} -CH_2CH_2\underset{\dot{C}H_2CH_2-}{C}HCH_2CH_2CH_2-$$

Removing a hydrogen atom from a carbon near the end of a chain leads to short branches, whereas removing a hydrogen atom from a carbon near the middle of a chain results in long branches. Short branches are more likely to be formed than long ones because the ends of the chain are more accessible.

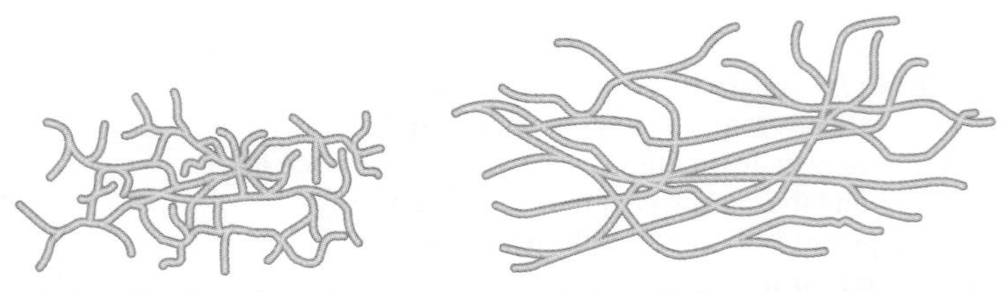

chain with short branches   chain with long branches

Branching greatly affects the physical properties of the polymer. Unbranched chains can pack together more closely than branched chains can. Consequently, linear polyethylene (known as high-density polyethylene) is a relatively hard plastic, used for the production of such things as artificial hip joints, whereas branched polyethylene (low-density polyethylene) is a much more flexible polymer, used for trash bags and dry-cleaning bags.

PROBLEM 32 ◆

Polyethylene can be used for the production of beach chairs as well as beach balls. Which of these items is made from more highly branched polyethylene?

5.17 RADICALS IN BIOLOGICAL SYSTEMS

Radicals are extremely reactive species. Fats and oils react with radicals to form compounds with strong odors that are responsible for the unpleasant taste and smell associated with sour milk and rancid butter. The molecules that form cell membranes can undergo this same reaction (Section 19.4). Radical reactions in biological systems also have been implicated in the aging process.

Clearly, unwanted radicals in biological systems must be destroyed before radical reactions have an opportunity to damage cells. Compounds known as **radical inhibitors** destroy radicals by creating compounds with only paired electrons. Hydroquinone is an example of a radical inhibitor. Two unwanted radicals each obtain an electron by removing a hydrogen atom from hydroquinone; rearrangement of the electrons in the product forms quinone, a compound in which all the electrons are paired.

hydroquinone + R· (reactive radical) ⟶ semiquinone + RH; semiquinone + R· ⟶ quinone + RH

Vitamin C and vitamin E are radical inhibitors present in biological systems. Vitamin C (also called ascorbic acid) is a water-soluble compound that traps radicals formed in the aqueous environment of the cell and in blood plasma. Vitamin E is a water-insoluble (therefore, fat-soluble) compound that traps radicals formed in nonpolar membranes. Why one vitamin functions in aqueous environments and the other in nonaqueous environments is apparent from their structures and electrostatic potential maps, which show that vitamin C is a relatively polar compound, whereas vitamin E is nonpolar.

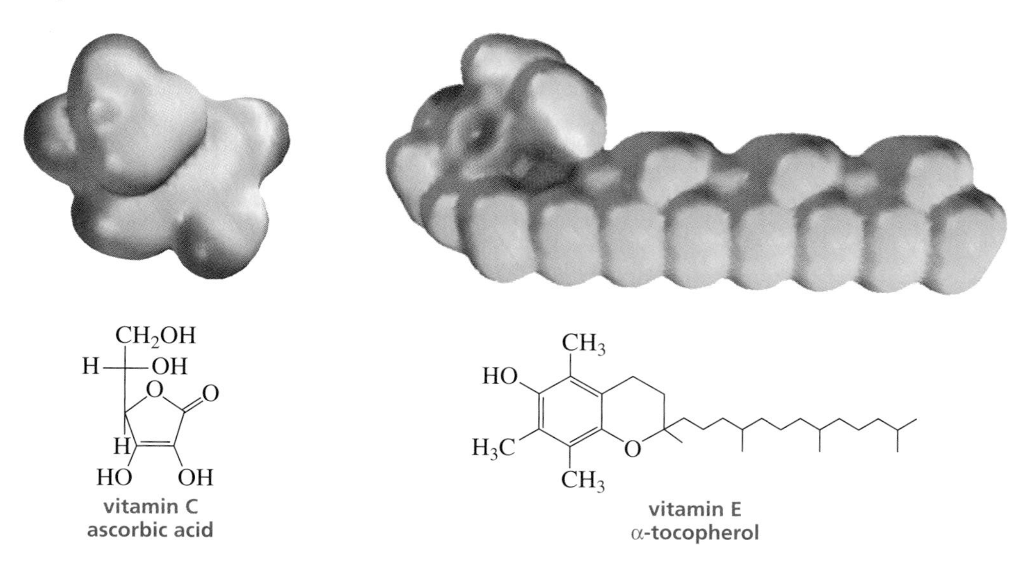

vitamin C
ascorbic acid

vitamin E
α-tocopherol

FOOD PRESERVATIVES

Radical inhibitors that are present in food are known as *preservatives* or *antioxidants*. They preserve food by destroying radicals, thereby preventing undesirable radical reactions. BHA and BHT are synthetic preservatives that are added to many packaged foods. Vitamin E is a naturally occurring preservative found in vegetable oil.

[Structures: butylated hydroxyanisole (BHA), two isomers: OH, $C(CH_3)_3$, OCH_3 substituted benzene rings; butylated hydroxytoluene (BHT): OH, two $C(CH_3)_3$, CH_3 substituted benzene ring]

butylated hydroxyanisole
BHA

butylated hydroxytoluene
BHT

food preservatives

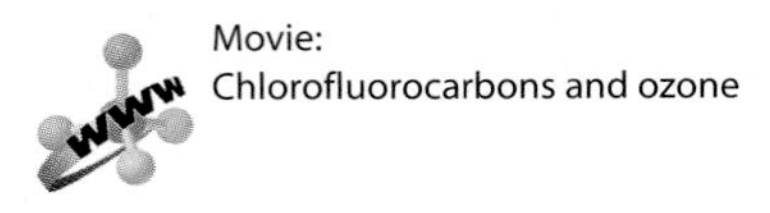

Movie:
Chlorofluorocarbons and ozone

RADICALS AND STRATOSPHERIC OZONE

Ozone (O_3), a major constituent of smog, is a health hazard at ground level. In the stratosphere, however, a layer of ozone shields the Earth from harmful solar radiation. The ozone layer is thinnest at the equator and densest toward the poles, with the greatest concentrations lying between 12 and 15 miles above the Earth's surface. Ozone is formed in the atmosphere from the interaction of molecular oxygen with very short wavelength ultraviolet light ($h\nu$).

$$O_2 \xrightarrow{h\nu} O + O$$

$$O + O_2 \longrightarrow \underset{\text{ozone}}{O_3}$$

The stratospheric ozone layer acts as a filter for biologically harmful ultraviolet radiation that otherwise would reach the surface of the Earth. Among other effects, high-energy short-wavelength ultraviolet light can damage DNA in skin cells, causing mutations that trigger skin cancer. We owe our very existence to this protective ozone layer. According to current theories of evolution, life could not have developed on land in the absence of this ozone layer. Instead, most if not all living things would have had to remain in the ocean, where water screens out the harmful ultraviolet radiation.

Since about 1985, scientists have noted a precipitous drop in stratospheric ozone over Antarctica. This area of ozone depletion, dubbed the "ozone hole," is unprecedented in the history of ozone observations. Scientists subsequently noted a similar decrease in ozone over Arctic regions; then, in 1988, they detected a depletion of ozone over the United States for the first time. Three years later, scientists determined that the rate of ozone depletion was two to three times faster than originally anticipated. Many in the scientific community blame recently observed increases in cataracts and skin cancer as well as diminished plant growth on the ultraviolet radiation that has penetrated the reduced ozone layer. Some predict that erosion of the protective ozone layer will cause an additional 200,000 deaths from skin cancer over the next 50 years.

Strong circumstantial evidence implicates synthetic chlorofluorocarbons (CFCs)—alkanes in which all the hydrogens have been replaced by fluorine and chlorine, such as $CFCl_3$ and CF_2Cl_2—as a major cause of ozone depletion. These gases, known commercially as Freons, have been used extensively as cooling fluids in refrigerators and air conditioners. They were also once widely used as propellants in aerosol spray cans (deodorant, hair spray, and so on) because of their odorless, nontoxic, and nonflammable properties, and, being chemically inert, they do not react with the contents of the can. Such use now, however, has been banned.

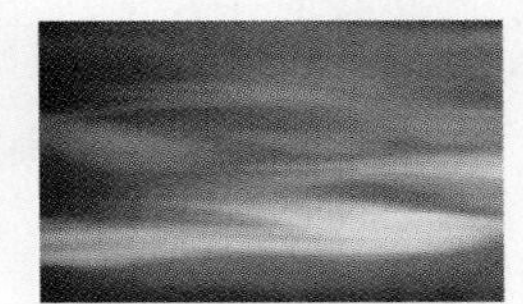

Polar stratospheric clouds increase the rate of ozone destruction. These clouds form over Antarctica during the cold winter months. Ozone depletion in the Arctic is less severe because the temperature generally does not get cold enough for the polar stratospheric clouds to form there.

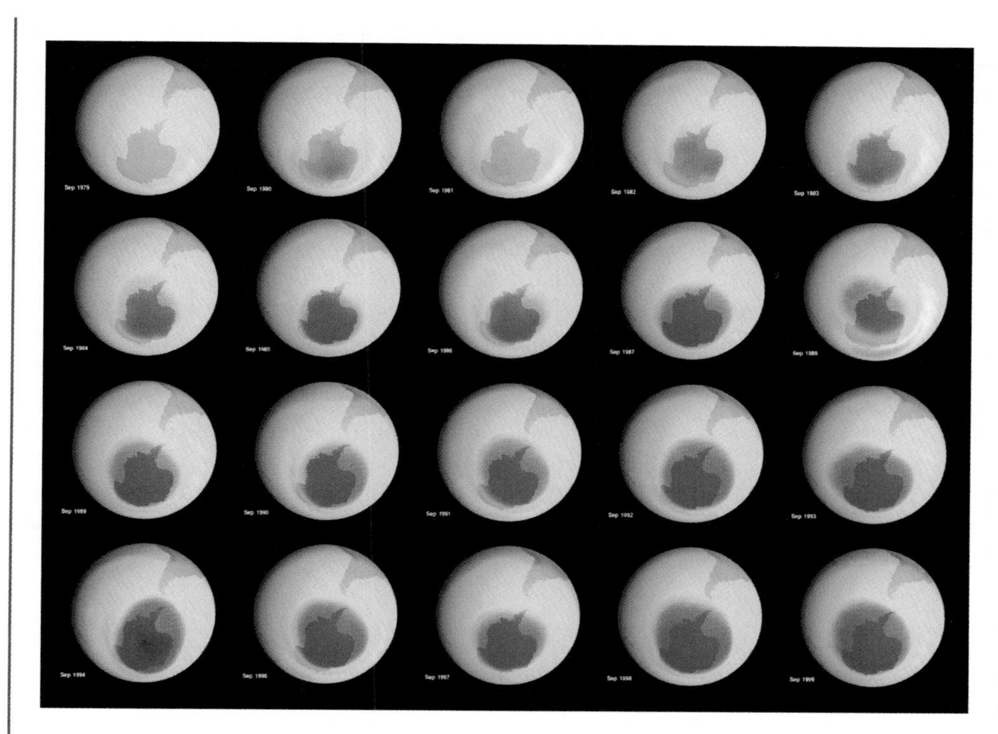

Growth of the Antarctic ozone hole, located mostly over the continent of Antarctica, since 1979. The images were made from data supplied by total ozone-mapping spectrometers (TOMSs). The color scale depicts the total ozone values in Dobson units, with the lowest ozone densities represented by dark blue.

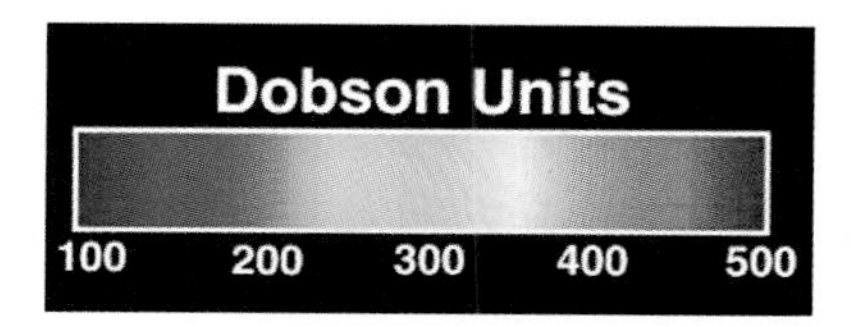

Chlorofluorocarbons remain very stable in the atmosphere until they reach the stratosphere. There they encounter wavelengths of ultraviolet light that cause the carbon–chlorine bond to break, generating chlorine radicals.

$$\mathrm{F{-}CCl_2{-}F} \xrightarrow{h\nu} \mathrm{F{-}\dot{C}Cl{-}F} + \mathrm{Cl\cdot}$$

The chlorine radicals are the ozone-removing agents. They react with ozone to form chlorine monoxide radicals and oxygen (O_2). The chlorine monoxide radicals then react with ozone to form chlorine dioxide, which dissociates to regenerate a chlorine radical. These three steps—two of which each destroy an ozone molecule—are repeated over and over. It has been calculated that each chlorine atom destroys 100,000 ozone molecules!

$$\mathrm{Cl\cdot} + \mathrm{O_3} \longrightarrow \mathrm{ClO\cdot} + \mathrm{O_2}$$

$$\mathrm{ClO\cdot} + \mathrm{O_3} \longrightarrow \mathrm{\cdot ClO_2} + \mathrm{O_2}$$

$$\mathrm{\cdot ClO_2} \longrightarrow \mathrm{Cl\cdot} + \mathrm{O_2}$$

SUMMARY

Alkenes undergo **electrophilic addition reactions**. These start with the addition of an electrophile to one of the sp^2 carbons and conclude with the addition of a nucleophile to the other sp^2 carbon. In all electrophilic addition reactions, the *electrophile* adds to the sp^2 carbon bonded to the greater number of hydrogens. **Regioselectivity** is the preferential formation of one **constitutional isomer** over another.

Regioselectivity results from the fact that the addition of hydrogen halides and the acid-catalyzed addition of water and alcohols form the most stable **carbocation intermediate**; **tertiary carbocations** are more stable than **secondary carbocations**, which are more stable than **primary carbocations**. We have now seen that alkyl groups stabilize both alkenes and carbocations.

An **alkyne** is a hydrocarbon that contains a carbon–carbon triple bond. The functional group suffix of an alkyne is "yne." A **terminal alkyne** has the triple bond at the end of the chain; an **internal alkyne** has the triple bond located elsewhere along the chain.

Alkynes, like alkenes, undergo electrophilic addition reactions. The same reagents that add to alkenes add to alkynes. Electrophilic addition to a *terminal* alkyne is regioselective; the *electrophile* adds to the *sp* carbon that is bonded to the hydrogen. If excess reagent is available, alkynes undergo a second addition reaction with hydrogen halides because the product of the first reaction is an alkene.

When an alkyne undergoes the acid-catalyzed addition of water, the product of the reaction is an enol, which immediately rearranges to a ketone. A **ketone** is a compound that has two alkyl groups bonded to a **carbonyl** (C=O) **group**. The ketone and enol are called **keto–enol tautomers**; they differ in the location of a double bond and a hydrogen. Interconversion of the tautomers is called **tautomerization**. The keto tautomer predominates at equilibrium. Terminal alkynes add water if mercuric ion is added to the acidic mixture.

Hydrogen adds to alkenes and alkynes in the presence of a metal catalyst (Pd/C or Pt/C) to form an alkane. The addition of H_2 to a compound is called **hydrogenation**. A **hydrogenation** reaction is a **reduction** reaction because the product has more C—H bonds than the reactant. Addition of hydrogen to an internal alkyne in the presence of Lindlar catalyst forms a *cis alkene*.

The electronegativities of carbon atoms decrease in the order: $sp > sp^2 > sp^3$. Ethyne is, therefore, a stronger acid than ethene, and ethene is a stronger acid than ethane. An amide ion can remove a hydrogen bonded to an *sp* carbon of a terminal alkyne because it is a stronger base than the **acetylide ion** that is formed. An acetylide ion reacts with a methyl halide or a primary alkyl halide to form an internal alkyne.

A **polymer** is a giant molecule made by linking together repeating units of small molecules called **monomers**. **Chain-growth polymers** are formed by chain reactions with **initiation**, **propagation**, and **termination steps**. Polymerization of alkenes can be initiated by electrophiles and by radicals.

SUMMARY OF REACTIONS

1. Electrophilic addition reactions of alkenes

a. Addition of hydrogen halides (Section 5.1)

$$RCH{=}CH_2 + HX \longrightarrow RCH(X)CH_3$$

HX = HF, HCl, HBr, HI

b. Acid-catalyzed addition of water and alcohols (Sections 5.4 and 5.5)

$$RCH{=}CH_2 + H_2O \xrightarrow{H_2SO_4} RCH(OH)CH_3$$

$$RCH{=}CH_2 + CH_3OH \xrightarrow{H_2SO_4} RCH(OCH_3)CH_3$$

2. Electrophilic addition reactions of alkynes

a. Addition of hydrogen halides (Section 5.10)

$$RC{\equiv}CH \xrightarrow{HX} RC(X){=}CH_2 \xrightarrow{\text{excess HX}} RC(X)_2{-}CH_3$$

HX = HF, HCl, HBr, HI

b. Acid-catalyzed addition of water (Section 5.11); R′ means that R and R′ do not have to be the same alkyl group.

$$\underset{\text{an internal alkyne}}{RC{\equiv}CR'} \xrightarrow[H_2SO_4]{H_2O} RC(=O)CH_2R' + RCH_2C(=O)R'$$

$$\underset{\text{a terminal alkyne}}{RC{\equiv}CH} \xrightarrow[H_2SO_4/HgSO_4]{H_2O} RC(=O)CH_3$$

3. Addition of hydrogen to alkenes and alkynes (Section 5.12)

$$RCH{=}CH_2 + H_2 \xrightarrow{\text{Pd/C or Pt/C}} RCH_2CH_3$$

$$RC{\equiv}CR' + 2\,H_2 \xrightarrow{\text{Pd/C or Pt/C}} RCH_2CH_2R'$$

$$R{-}C{\equiv}C{-}R' + H_2 \xrightarrow{\text{Lindlar catalyst}} \text{(H)(R)C=C(H)(R')}$$

4. Removal of a proton from a terminal alkyne, followed by reaction with an alkyl halide (Sections 5.13 and 5.14)

$$RC{\equiv}CH \xrightarrow{NaNH_2} RC{\equiv}C^- \xrightarrow{R'CH_2Br} RC{\equiv}CCH_2R'$$

PROBLEMS

33. Identify the electrophile and the nucleophile in each of the following reaction steps. Then draw curved arrows to illustrate the bond-making and bond-breaking processes.

a. 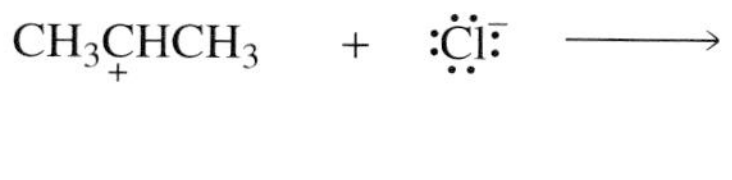 $CH_3\overset{+}{C}HCH_3$ + $:\ddot{Cl}:^-$ ⟶ $CH_3CH(:\ddot{Cl}:)CH_3$

b. 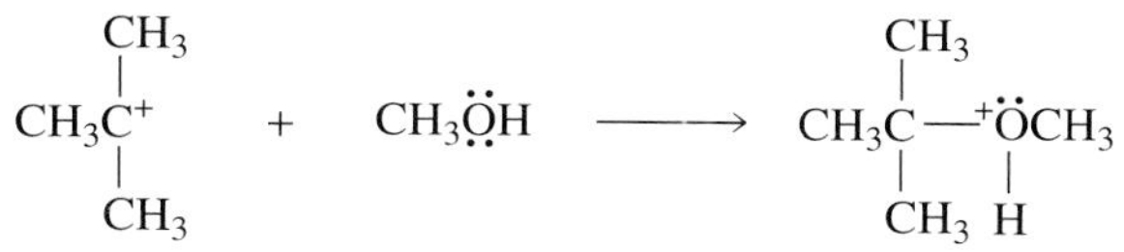 $CH_3C^+(CH_3)_2$ + $CH_3\ddot{O}H$ ⟶ $CH_3C(CH_3)_2—\overset{+}{\ddot{O}}(H)CH_3$

34. What will be the major product of the reaction of 2-methyl-2-butene with each of the following reagents?

a. HBr
b. HI
c. H_2, Pd/C
d. H_2O + trace H_2SO_4
e. CH_3OH + trace H_2SO_4
f. CH_3CH_2OH + trace H_2SO_4

35. Give the major product of each of the following reactions:

a. 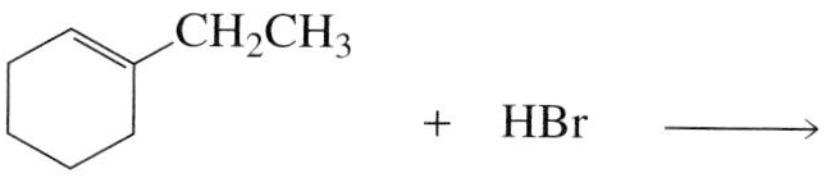 (cyclohexene with CH_2CH_3 substituent) + HBr ⟶

b. 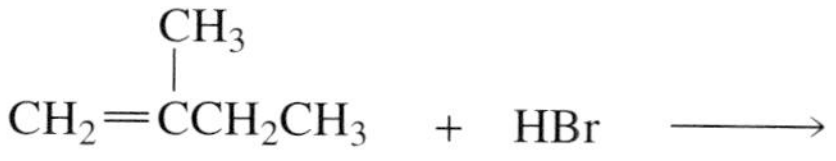 $CH_2{=}C(CH_3)CH_2CH_3$ + HBr ⟶

c. (cyclohexane with $=CHCH_3$) + HCl ⟶

d. 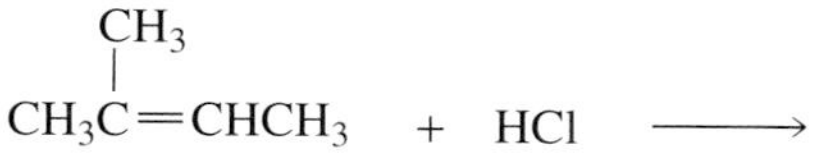 $CH_3C(CH_3){=}CHCH_3$ + HCl ⟶

36. Draw curved arrows to show the flow of electrons responsible for the conversion of reactants into products.

a. 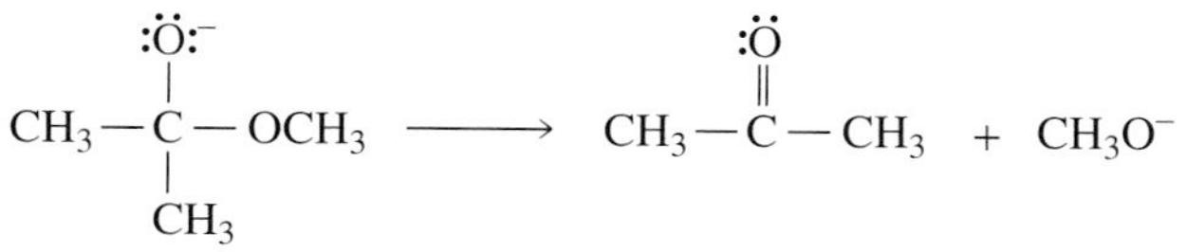 $CH_3—C(:\ddot{O}:^-)(CH_3)—OCH_3$ ⟶ $CH_3—C(=\ddot{O}:)—CH_3$ + CH_3O^-

b. $CH_3C{\equiv}C—H$ + $:\ddot{N}H_2^-$ ⟶ $CH_3C{\equiv}C^-$ + $\ddot{N}H_3$

c. $CH_3CH_2—Br$ + $CH_3\ddot{O}:^-$ ⟶ $CH_3CH_2—\ddot{O}CH_3$ + Br^-

37. Give the reagents that would be required to carry out the following syntheses:

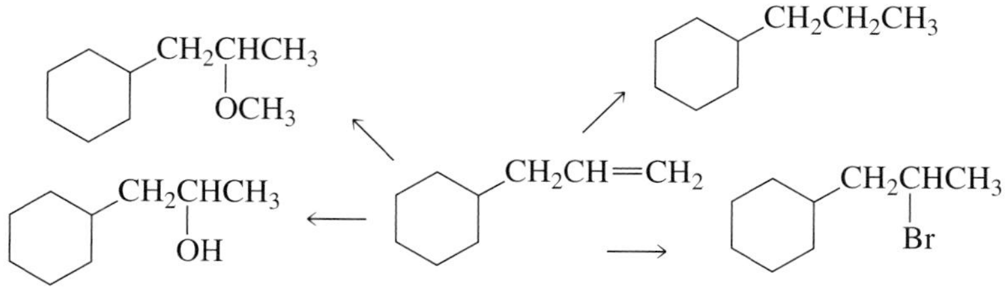

Starting material: cyclohexyl–$CH_2CH{=}CH_2$, converted to:
- cyclohexyl–$CH_2CH(OCH_3)CH_3$
- cyclohexyl–$CH_2CH_2CH_2CH_3$
- cyclohexyl–$CH_2CH(OH)CH_3$
- cyclohexyl–$CH_2CH(Br)CH_3$

38. Draw all the enol tautomers for each of the ketones in Problem 20.

39. What ketones are formed when the following alkyne undergoes the acid-catalyzed addition of water?

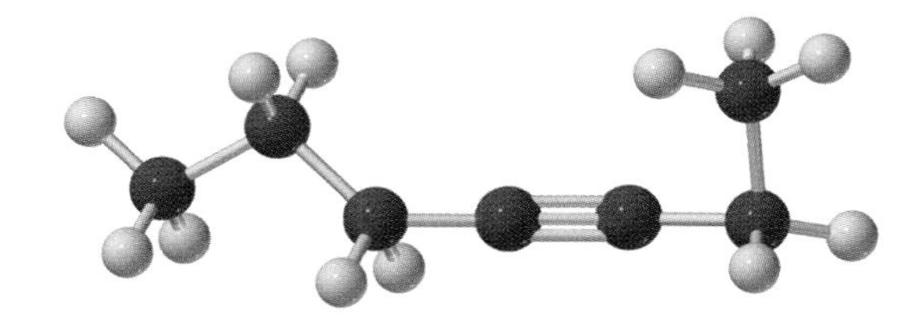

40. Give the major product of each of the following reactions:

a. 1-methylcyclohexene $\xrightarrow{HCl}$ **b.** 1-methylcyclohexene $\xrightarrow{H_2O}$ **c.** 1-methylcyclohexene $\xrightarrow[H_2O]{H_2SO_4}$ **d.** 1-methylcyclohexene $\xrightarrow{HBr}$ **e.** 1-methylcyclohexene $\xrightarrow[CH_3OH]{H_2SO_4}$

41. For each of the following pairs, indicate which member is more stable:

a. $CH_3\overset{+}{C}(CH_3)CH_3$ or $CH_3\overset{+}{C}HCH_2CH_3$ **b.** $CH_3CH_2\overset{+}{C}H_2$ or $CH_3\overset{+}{C}HCH_3$ **c.** $CH_3\overset{+}{C}H_2$ or $CH_2{=}\overset{+}{C}H$

42. Using an alkene and any other reagents, how would you prepare the following compounds?

a. cyclohexane

b. $CH_3CH_2CH_2CH(Cl)CH_3$

c. cyclohexyl–$CH_2CH(OH)CH_3$

43. Identify the two alkenes that react with HBr to give 1-bromo-1-methylcyclohexane.

44. The second-order rate constant (in units of $M^{-1}s^{-1}$) for acid-catalyzed hydration at 25 °C is given for each of the following alkenes:

$(H_3C)(H)C{=}CH_2$ — 4.95×10^{-8}

$(H_3C)(H)C{=}C(CH_3)(H)$ (cis) — 8.32×10^{-8}

$(H_3C)(H)C{=}C(H)(CH_3)$ (trans) — 3.51×10^{-8}

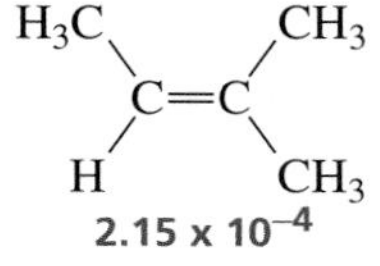
$(H_3C)(H)C{=}C(CH_3)_2$ — 2.15×10^{-4}

$(H_3C)_2C{=}C(CH_3)_2$ — 3.42×10^{-4}

a. Why does (*Z*)-2-butene react faster than (*E*)-2-butene?
b. Why does 2-methyl-2-butene react faster than (*Z*)-2-butene?
c. Why does 2,3-dimethyl-2-butene react faster than 2-methyl-2-butene?

45. a. Propose a mechanism for the following reaction (remember to use curved arrows when showing a mechanism):

$$CH_3CH_2CH{=}CH_2 + CH_3OH \xrightarrow{H_2SO_4} CH_3CH_2CH(OCH_3)CH_3$$

b. Which step is the rate-determining step?
c. What is the electrophile in the first step?
d. What is the nucleophile in the first step?
e. What is the electrophile in the second step?
f. What is the nucleophile in the second step?

46. The pK_a of protonated ethyl alcohol is −2.4 and the pK_a of ethyl alcohol is 15.9. Therefore, as long as the pH of the solution is greater than ________ and less than ________, more than 50% of ethyl alcohol will be in its neutral, nonprotonated form. (*Hint:* See Section 2.7.)

47. a. How many alkenes could you treat with H_2, Pt/C in order to prepare methylcyclopentane?
b. Which of the alkenes is the most stable?

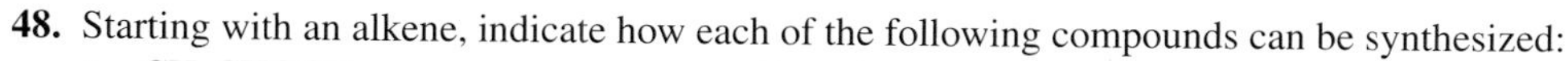

48. Starting with an alkene, indicate how each of the following compounds can be synthesized:

a. 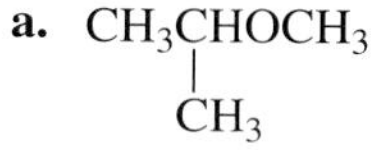
CH_3CHOCH_3 with CH_3 on C-2

b. 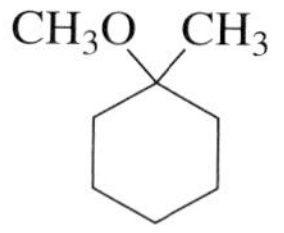 CH_3O and CH_3 on the same carbon of a cyclohexane ring

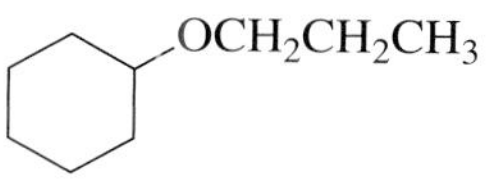

c. 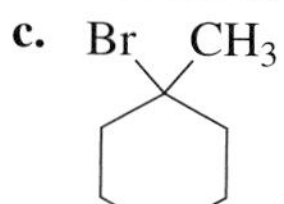
Br and CH_3 on the same carbon of a cyclohexane ring

d. $CH_3CHCH_2CH_3$ with OCH_2CH_3 on C-2

e. cyclohexane with $OCH_2CH_2CH_3$

f. $CH_3CCH_2CH_3$ with CH_3 and OH on C-2

49. Draw a structure for each of the following:

a. 2-hexyne
b. 5-ethyl-3-octyne
c. 1-bromo-1-pentyne
d. 5,6-dimethyl-2-heptyne

50. Give the major product obtained from the reaction of each of the following with excess HCl:

a. $CH_3CH_2C{\equiv}CH$ **b.** $CH_3CH_2C{\equiv}CCH_2CH_3$ **c.** $CH_3CH_2C{\equiv}CCH_2CH_2CH_3$

51. Give the systematic name for each of the following:

a. $CH_3C{\equiv}CCH_2CHCH_3$ with Br on the CH

b. $CH_3C{\equiv}CCH_2CHCH_3$ with $CH_2CH_2CH_3$ on the CH

c. $CH_3C{\equiv}CCH_2CCH_3$ with two CH_3 groups on the C

d. $CH_3CHCH_2C{\equiv}CCHCH_3$ with Cl on C-2 and CH_3 on C-6

52. Identify the electrophile and the nucleophile in each of the following reaction steps. Then draw curved arrows to illustrate the bond-making and bond-breaking processes.

a. $CH_3CH_2\overset{+}{C}{=}CH_2 + :\ddot{Cl}:^- \longrightarrow CH_3CH_2C(Cl){=}CH_2$

b. $CH_3C{\equiv}CH + H{-}Br \longrightarrow CH_3\overset{+}{C}{=}CH_2 + Br^-$

c. $CH_3C{\equiv}C{-}H + {}^-:\ddot{N}H_2 \longrightarrow CH_3C{\equiv}C:^- + \ddot{N}H_3$

53. Al Kyne was given the structural formulas of several compounds and was asked to give them systematic names. How many did Al name correctly? Correct those that are misnamed.

a. 4-ethyl-2-pentyne **b.** 1-bromo-4-heptyne **c.** 2-methyl-3-hexyne **d.** 3-pentyne

54. Draw the structures and give the common and systematic names for alkynes with molecular formula C_7H_{12}.

55. What reagents would you use for the following syntheses?

a. (*Z*)-3-hexene from 3-hexyne **b.** hexane from 3-hexyne

56. What is the molecular formula of a hydrocarbon that has 1 triple bond, 2 double bonds, 1 ring, and 32 carbons?

57. What will be the major product of the reaction of 1 mol of propyne with each of the following reagents?

a. HBr (1 mol)
b. HBr (2 mol)
c. aqueous H_2SO_4, $HgSO_4$
d. excess H_2, Pt/C
e. H_2/Lindlar catalyst
f. sodium amide
g. product of Problem 57f followed by 1-chloropentane

58. Answer Problem 57, using 2-butyne as the starting material instead of propyne.

59. What reagents could be used to carry out the following syntheses?

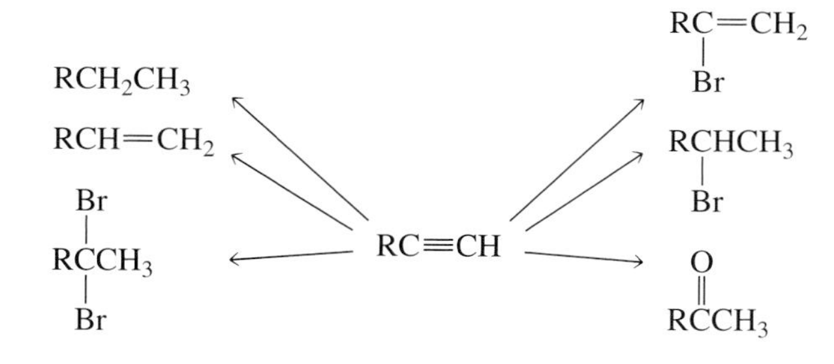

60. a. Starting with 5-methyl-2-hexyne, how could you prepare the following compound?

$CH_3CH_2CH(OH)CH_2CH(CH_3)CH_3$

b. What other alcohol would also be obtained?

61. How many of the following names are correct? Correct the incorrect names.

a. 4-heptyne
b. 2-ethyl-3-hexyne
c. 4-chloro-2-pentyne
d. 2,3-dimethyl-5-octyne
e. 4,4-dimethyl-2-pentyne
f. 2,5-dimethyl-3-hexyne

62. Which of the following pairs are keto–enol tautomers?

a. $CH_3CH(OH)CH_3$ and $CH_3C(=O)CH_3$

b. $CH_3CH_2CH_2C(OH){=}CH_2$ and $CH_3CH_2CH_2C(=O)CH_3$

c. $CH_3CH_2CH_2CH{=}CHOH$ and $CH_3CH_2CH_2C(=O)CH_3$

63. Using ethyne as the starting material, how can the following compounds be prepared?

a. $CH_3C(=O)CH_3$

b.

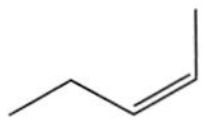

c.

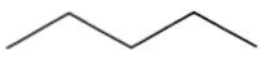

64. Draw the keto tautomer for each of the following:

a. $CH_3CH{=}C(OH)CH_3$

b. $CH_3CH_2CH_2C(OH){=}CH_2$

c. —OH

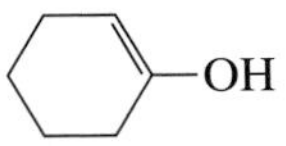

d. 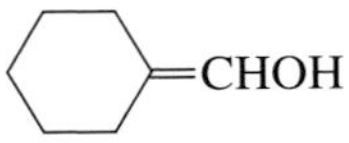 =CHOH

65. Show how each of the following compounds could be prepared using the given starting material, any necessary inorganic reagents, and any necessary organic compound that has no more than four carbon atoms:

a. $HC{\equiv}CH \longrightarrow CH_3CH_2CH_2CH_2C(=O)CH_3$

b. $HC{\equiv}CH \longrightarrow CH_3CH_2CH(Br)CH_3$

c. $HC{\equiv}CH \longrightarrow CH_3CH_2CH_2CH(OH)CH_3$

d. 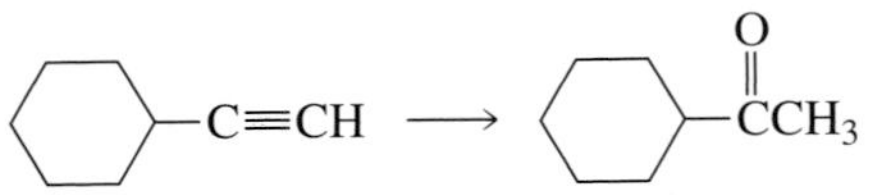 $-C{\equiv}CH \longrightarrow -C(=O)CH_3$

66. Any base whose conjugate acid has a pK_a greater than ________ can remove a proton from a terminal alkyne to form an acetylide ion in a reaction that favors products.

67. Dr. Polly Meher was planning to synthesize 3-octyne by adding 1-bromobutane to the product obtained from the reaction of 1-butyne with sodium amide. Unfortunately, however, she had forgotten to order 1-butyne. How else can she prepare 3-octyne?

68. Draw short segments of the polymers obtained from the following monomers:

a. $CH_2{=}CHF$
b. $CH_2{=}CHCO_2H$

69. Draw the structure of the monomer or monomers used to synthesize the following polymers:

a. $-[CH_2CH(CH_2CH_3)]_n-$

b. 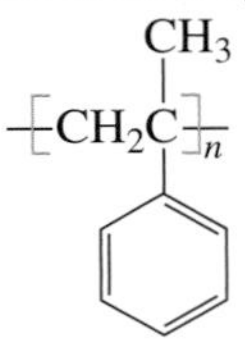 $-[CH_2C(CH_3)(C_6H_5)]_n-$

70. Draw short segments of the polymer obtained from 1-pentene, using H^+ as an initiator (obtained from the reaction of $BF_3 + H_2O$).

71. In cationic polymerization, H^+ is the initiator, obtained from the reaction of $BF_3 + H_2O$. Why is HCl not used as the source of H^+?

Isomers and Stereochemistry

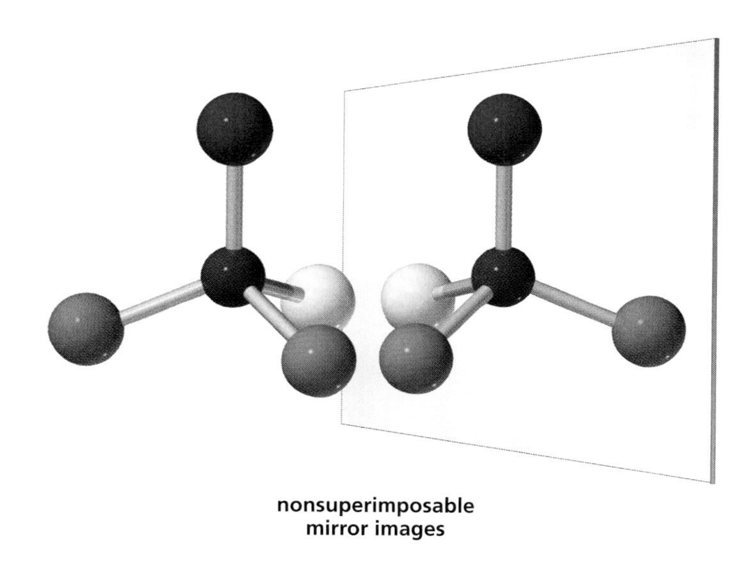

nonsuperimposable mirror images

Compounds that have the same molecular formula but do not have identical structures are called **isomers**. Isomers fall into two main classes: *constitutional isomers* and *stereoisomers*. **Constitutional isomers** differ in the way their atoms are connected (Section 3.0). For example, ethyl alcohol and dimethyl ether are constitutional isomers with molecular formula C_2H_6O. The oxygen in ethyl alcohol is bonded to a carbon and to a hydrogen, whereas the oxygen in dimethyl ether is bonded to two carbons.

constitutional isomers

CH_3CH_2OH (ethyl alcohol) and CH_3OCH_3 (dimethyl ether)

$CH_3CH_2CH_2CH_2Cl$ (1-chlorobutane) and $CH_3CH_2CH(Cl)CH_3$ (2-chlorobutane)

Unlike the atoms in constitutional isomers, the atoms in *stereoisomers* are connected in the same way. **Stereoisomers** differ in the way their atoms are arranged in space. There are two kinds of stereoisomers: **cis–trans isomers** and isomers that contain **asymmetric centers**.

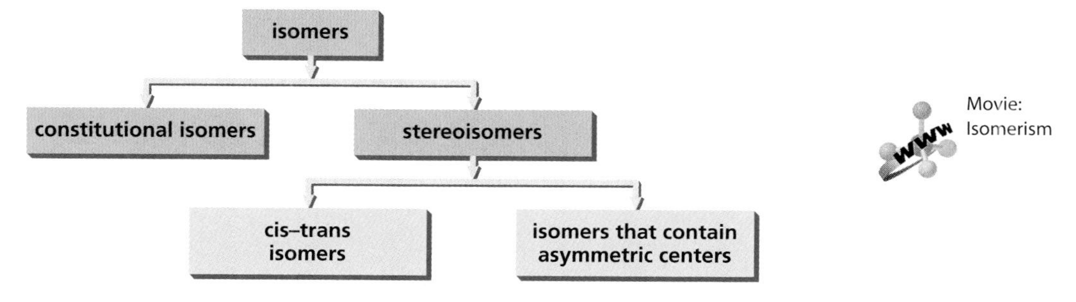

Movie: Isomerism

PROBLEM 1 ♦

a. Draw three constitutional isomers with molecular formula C_3H_8O.
b. How many constitutional isomers can you draw for $C_4H_{10}O$?

6.1 CIS–TRANS ISOMERS RESULT FROM RESTRICTED ROTATION

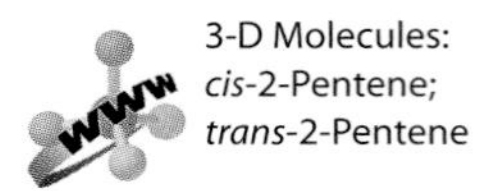

3-D Molecules: *cis*-2-Pentene; *trans*-2-Pentene

Cis–trans isomers result from restricted rotation. Restricted rotation can be caused either by a *double bond* or by a *cyclic structure*. We have seen that because of the restricted rotation about its carbon–carbon double bond, an alkene such as 2-pentene can exist as cis and trans isomers (Section 4.4). The **cis isomer** has the hydrogens on the *same side* of the double bond, and the **trans isomer** has the hydrogens on *opposite sides* of the double bond.

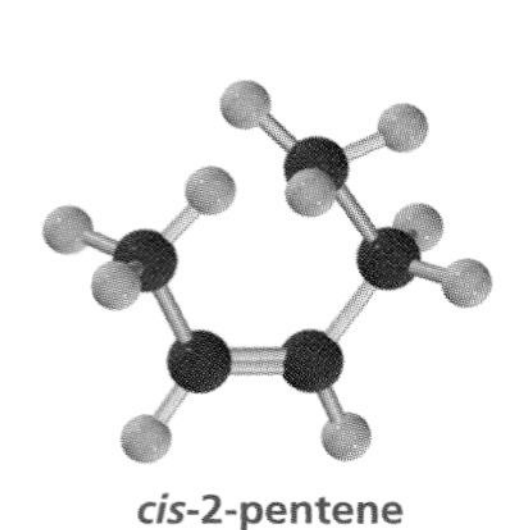

cis-2-pentene

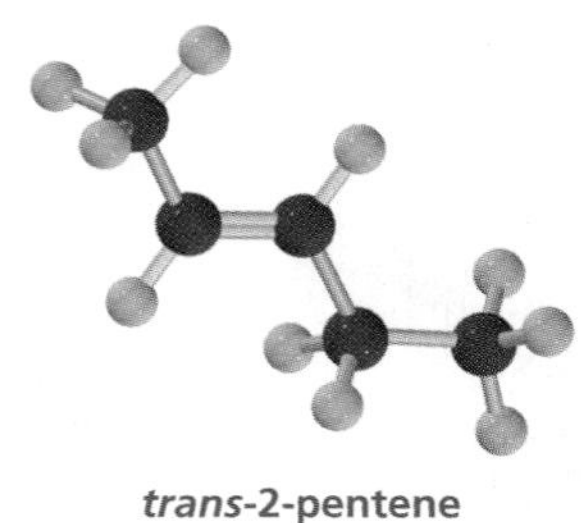

trans-2-pentene

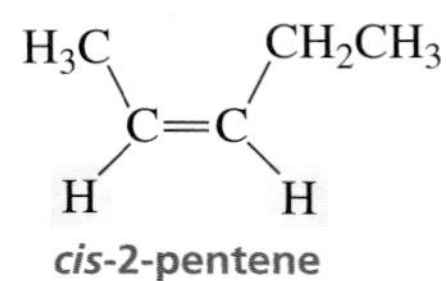

cis-2-pentene

H3C H C=C H CH2CH3

trans-2-pentene

As a result of restricted rotation about the bonds in a ring, cyclic compounds also can have cis and trans isomers (Section 3.12). The cis isomer has the hydrogens on the same side of the ring, whereas the trans isomer has the hydrogens on opposite sides of the ring.

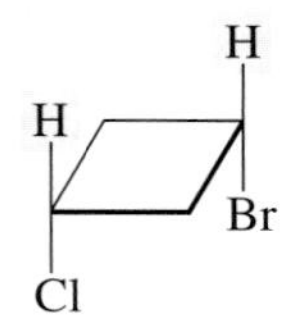

cis-1-bromo-3-chlorocyclobutane

Br H H Cl

trans-1-bromo-3-chlorocyclobutane

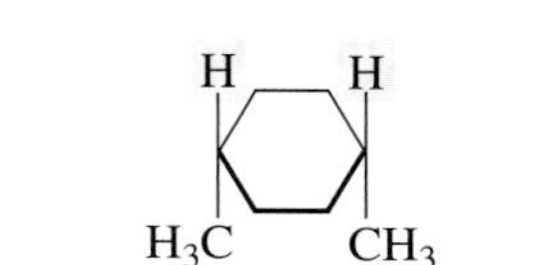

cis-1,4-dimethylcyclohexane

H CH3 H3C H

trans-1,4-dimethylcyclohexane

PROBLEM 2

Draw the cis and trans isomers for the following:

a. 1-ethyl-3-methylcyclobutane
b. 3-hexene
c. 1-bromo-4-chlorocyclohexane
d. 2-methyl-3-heptene

6.2 A CHIRAL OBJECT HAS A NONSUPERIMPOSABLE MIRROR IMAGE

Why can't you put your right shoe on your left foot? Why can't you put your right glove on your left hand? It is because hands, feet, gloves, and shoes have right-handed and left-handed forms. An object with a right-handed and a left-handed form is said to be **chiral** (ky-ral), a word derived from the Greek word *cheir*, which means "hand."

A chiral object has a *nonsuperimposable mirror image*. In other words, its mirror image is not the same as an image of the object itself. A hand is chiral because when you look at your right hand in a mirror, you see not a right hand but a left hand (Figure 6.1). In contrast, a chair is *not* chiral; the reflection of the chair in the mirror looks the same as the chair itself. Objects that are not chiral are said to be **achiral**. An achiral object has a *superimposable mirror image*. Some other achiral objects are a fork, a glass, and a balloon (assuming they are unadorned).

chiral objects

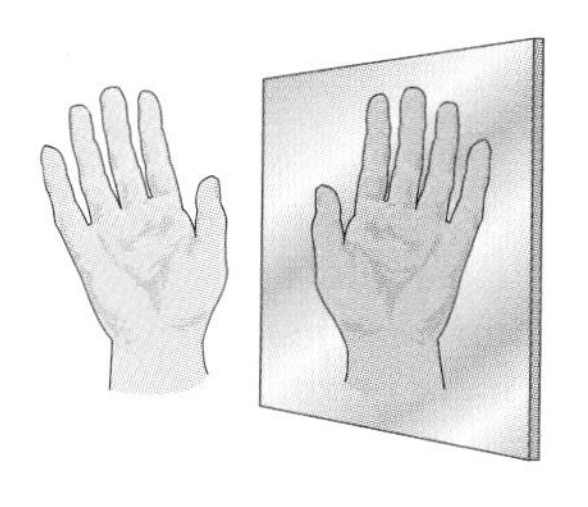
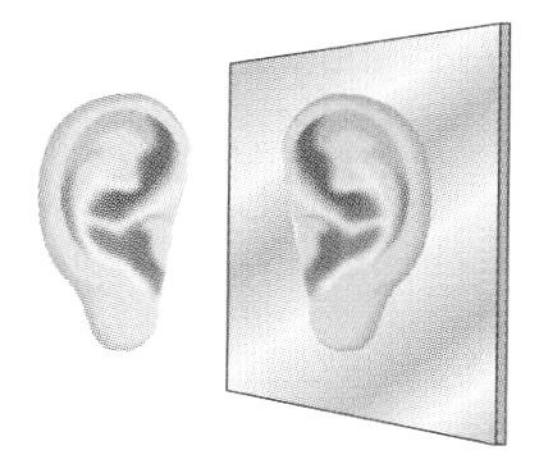
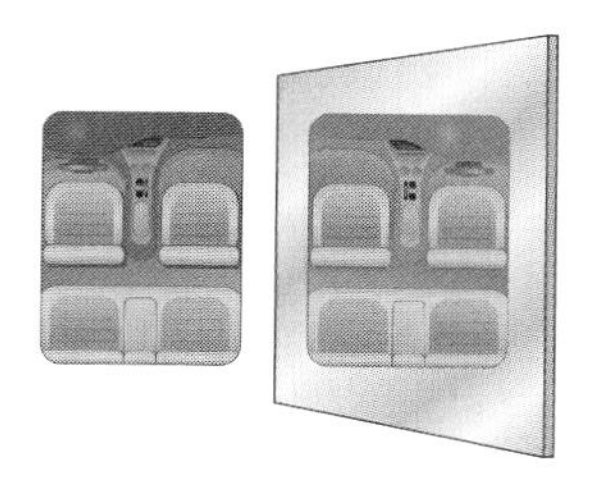

achiral objects

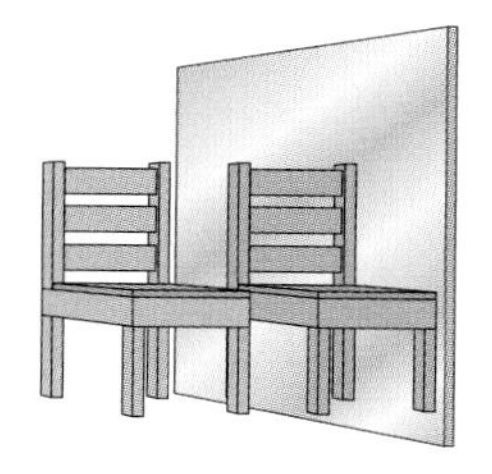

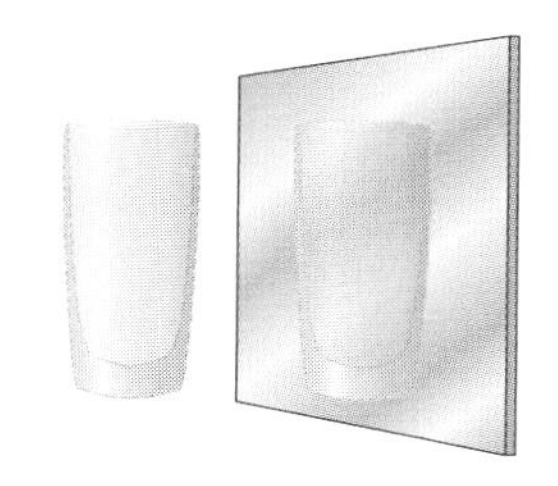

◀ **Figure 6.1**
Using a mirror to test for chirality.
A chiral object is not the same as its mirror image—they are nonsuperimposable.
An achiral object is the same as its mirror image—they are superimposable.

PROBLEM 3 ♦

a. Name five capital letters that are chiral. **b.** Name five capital letters that are achiral.

6.3 AN ASYMMETRIC CENTER IS THE CAUSE OF CHIRALITY IN A MOLECULE

A molecule with one asymmetric center is chiral.

Objects are not the only things that can be chiral. Molecules can be chiral, too. The usual cause of chirality in a molecule is an *asymmetric center*.

An **asymmetric center** is an atom bonded to four different groups. Each of the compounds shown below has an asymmetric center indicated by a star. For example, the starred carbon in 4-octanol is an asymmetric center because it is bonded to four different groups (H, OH, $CH_2CH_2CH_3$, and $CH_2CH_2CH_2CH_3$). Notice that the atoms immediately bonded to the asymmetric center are not necessarily different from one another; the propyl and butyl groups are different even though the point at which they differ is several atoms away from the asymmetric center. The starred carbon in 2,4-dimethylhexane is an asymmetric center because it is bonded to four different groups (methyl, ethyl, isobutyl, and hydrogen).

$$CH_3CH_2CH_2\overset{*}{C}HCH_2CH_2CH_2CH_3 \quad (\text{OH on starred C})$$
4-octanol

$$CH_3\overset{*}{C}HCH_2CH_3 \quad (\text{Br on starred C})$$
2-bromobutane

$$CH_3CH(CH_3)CH_2\overset{*}{C}H(CH_3)CH_2CH_3$$
2,4-dimethylhexane

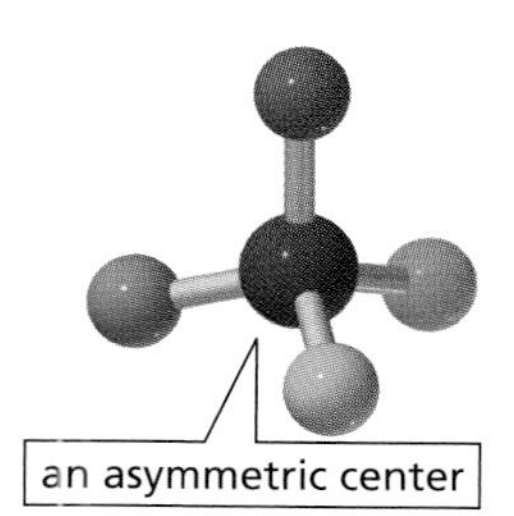

PROBLEM 4 ♦

Which of the following have an asymmetric center?

a. $CH_3CH_2CH(Cl)CH_3$

b. $CH_3CH_2CH(CH_3)CH_3$

c. $CH_3CH_2C(CH_3)(Br)CH_2CH_2CH_3$

d. CH_3CH_2OH

e. $CH_3CH_2CH(Br)CH_2CH_3$

f. $CH_2{=}CHCH(NH_2)CH_3$

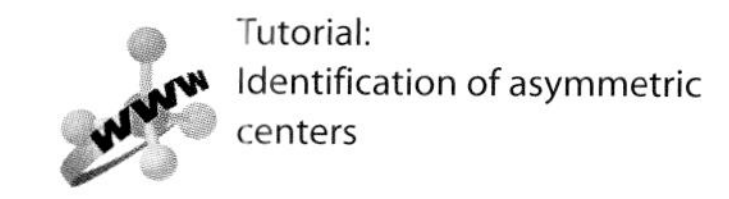

Tutorial:
Identification of asymmetric centers

PROBLEM 5 SOLVED

Tetracycline is called a broad-spectrum antibiotic because it is active against a wide variety of bacteria. How many asymmetric centers does tetracycline have?

Solution Only sp^3 carbons can be asymmetric centers, because an asymmetric center must have four different groups attached to it. Therefore, we start by locating all the sp^3 carbons in tetracycline. (They are numbered in red.) Tetracycline has nine sp^3 carbons. Four of them (numbers 1, 2, 5, and 8) are not asymmetric centers because they are not bonded to four different groups. Tetracycline, therefore, has five asymmetric centers.

tetracycline

6.4 ISOMERS WITH ONE ASYMMETRIC CENTER

A compound with one asymmetric center, such as 2-bromobutane, can exist as two stereoisomers. The two stereoisomers are analogous to a left and a right hand. If we imagine a mirror between the two stereoisomers, we can see they are mirror images of each other. Moreover, they are nonsuperimposable mirror images and thus different molecules.

$CH_3\overset{*}{C}HCH_2CH_3$ (with Br on C*)

2-bromobutane

mirror

the two stereoisomers of 2-bromobutane
enantiomers

Convince yourself that the two 2-bromobutane isomers are not identical by building ball-and-stick models; use four different-colored balls to represent the four different groups bonded to the asymmetric center. Try to superimpose them.

Movie: Nonsuperimposable mirror image

Molecules that are nonsuperimposable mirror images of each other are called **enantiomers**. Thus, the two stereoisomers of 2-bromobutane are enantiomers. A molecule that has a *nonsuperimposable* mirror image, like an object that has a nonsuperimposable mirror image, is **chiral**. Therefore, each member of a pair of enantiomers is chiral. Notice that chirality is a property of an entire object or an entire molecule.

A chiral molecule has a nonsuperimposable mirror image.

An achiral molecule has a superimposable mirror image.

A molecule that has a *superimposable* mirror image, like an object that has a superimposable mirror image, is **achiral**. To see that the achiral molecule below is superimposable on its mirror image (that they are identical molecules), mentally rotate it clockwise.

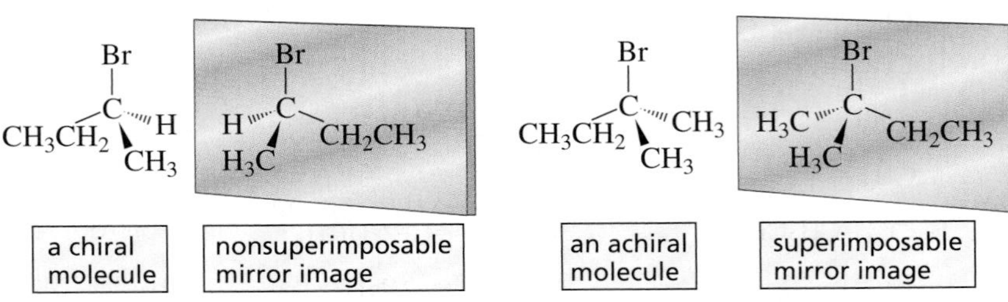

a chiral molecule | nonsuperimposable mirror image — enantiomers

an achiral molecule | superimposable mirror image — identical molecules

PROBLEM 6 ◆

Which of the compounds in Problem 4 can exist as enantiomers?

6.5 HOW TO DRAW ENANTIOMERS

Chemists draw enantiomers using *perspective formulas*. A **perspective formula** shows two of the bonds to the asymmetric center in the plane of the paper, one bond as a solid wedge protruding forward out of the paper, and the fourth bond as a hatched wedge extending behind the paper. The solid wedge and the hatched wedge must be adjacent to one another. When you draw the first enantiomer, the four groups bonded to the asymmetric center can be placed around it in any order. You then draw the second enantiomer by drawing the mirror image of the first.

A solid wedge represents a bond that extends out of the plane of the paper toward the viewer.

A hatched wedge represents a bond that points back from the plane of the paper away from the viewer.

When you draw a perspective formula, make certain that the two bonds in the plane of the paper are adjacent to one another; neither the solid wedge nor the hatched wedge should be drawn between them.

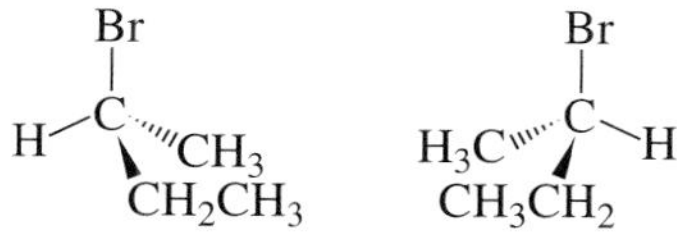

Br, H, C, CH_3, CH_2CH_3 — Br, H_3C, C, H, CH_3CH_2

perspective formulas of the enantiomers of 2-bromobutane

PROBLEM 7 ◆

Draw the enantiomers of each of the following compounds using perspective formulas:

a. $CH_3CH(Br)CH_2OH$ b. $ClCH_2CH_2CH(CH_3)CH_2CH_3$ c. $CH_3CH(CH_3)CH(OH)CH_3$

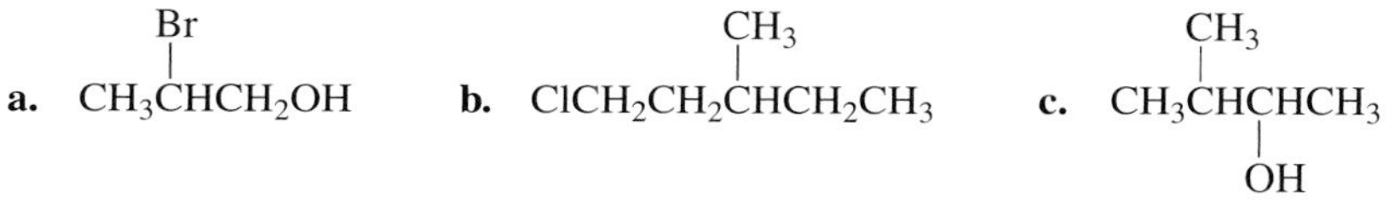

PROBLEM 8 *SOLVED*

Do the following structures represent identical molecules or a pair of enantiomers?

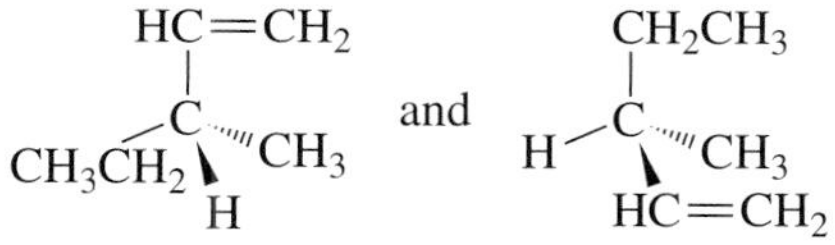

$HC{=}CH_2$, C, CH_3CH_2, CH_3, H and CH_2CH_3, H, C, CH_3, $HC{=}CH_2$

Solution Interchanging two atoms or groups attached to an asymmetric center of a perspective formula produces an enantiomer. Interchanging two atoms or groups a second time brings you back to the original molecule. Because groups have to be interchanged twice to get from one structure to the other, the two structures represent identical molecules.

$HC{=}CH_2$, C, CH_3CH_2, CH_3, H $\xrightarrow[\text{vinyl and H}]{\text{interchange}}$ H, C, CH_3CH_2, CH_3, $HC{=}CH_2$ $\xrightarrow[\text{ethyl and H}]{\text{interchange}}$ CH_2CH_3, H, C, CH_3, $HC{=}CH_2$

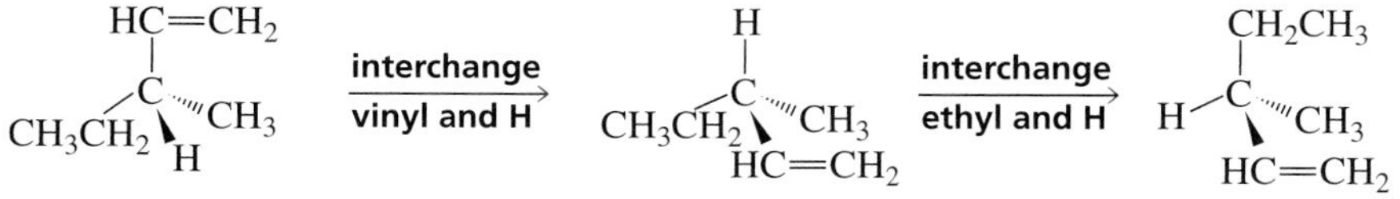

In Section 6.6 you will learn another way to determine if two structures represent identical molecules or enantiomers.

6.6 NAMING ENANTIOMERS BY THE *R,S* SYSTEM

How do we name the different stereoisomers of a compound like 2-bromobutane so that we know which one we are talking about? We need a system of nomenclature that indicates the arrangement of the atoms or groups about the asymmetric center. Chemists use the letters *R* and *S* for this purpose. For any pair of enantiomers with one asymmetric center, one member will have the ***R* configuration** and the other will have the ***S* configuration**.

Let's first look at how we can determine the configuration of a compound if we have a three-dimensional model.

1. **Rank the groups (or atoms) bonded to the asymmetric center in order of priority.** The atomic numbers of the atoms directly attached to the asymmetric center determine the relative priorities. The higher the atomic number, the higher the priority. This should remind you of the way that relative priorities are determined for *E* and *Z* isomers (Section 4.5), because the system of priorities was originally devised for the *R,S* system of nomenclature and was later adopted for the *E,Z* system.

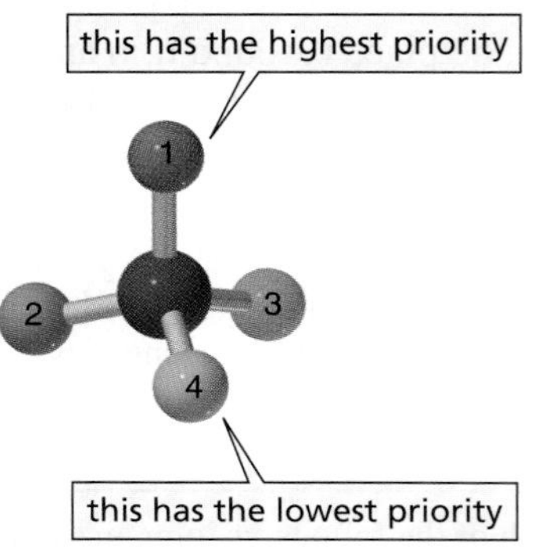

The molecule is oriented so the group with the lowest priority points away from the viewer. If an arrow drawn from the highest priority group to the next highest priority group points clockwise, the molecule has the *R* configuration.

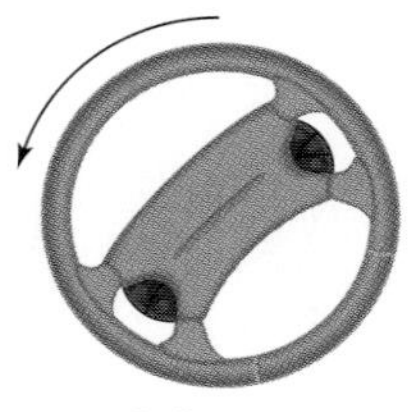

left turn

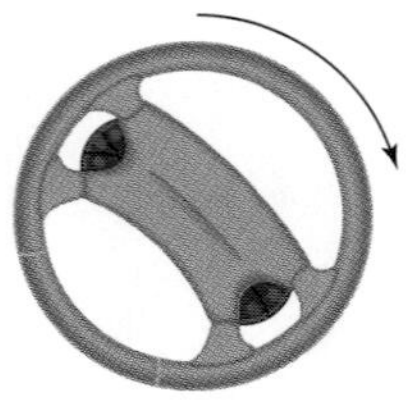

right turn

2. **Orient the molecule so that the group (or atom) with the lowest priority (4) is directed away from you. Then draw an imaginary arrow from the group (or atom) with the highest priority (1) to the group (or atom) with the next highest priority (2).** If the arrow points clockwise, the asymmetric center has the *R* configuration (*R* is for *rectus*, which is Latin for "right"). If the arrow points counterclockwise, the asymmetric center has the *S* configuration (*S* is for *sinister*, which is Latin for "left").

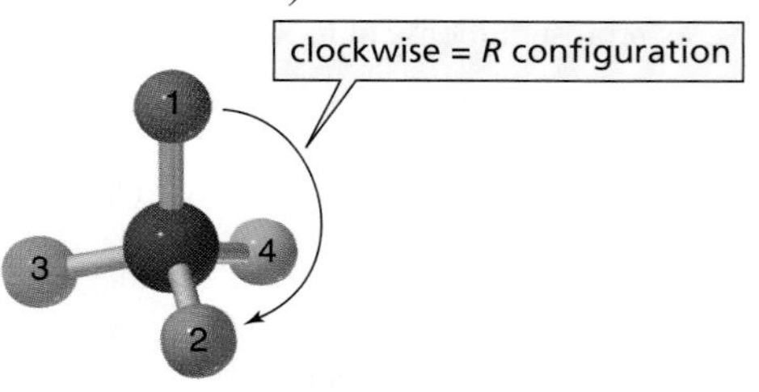

If you forget which direction corresponds to which configuration, imagine driving a car and turning the steering wheel clockwise to make a right turn or counterclockwise to make a left turn.

PROBLEM 9 ♦

Which of the following molecular models are identical?

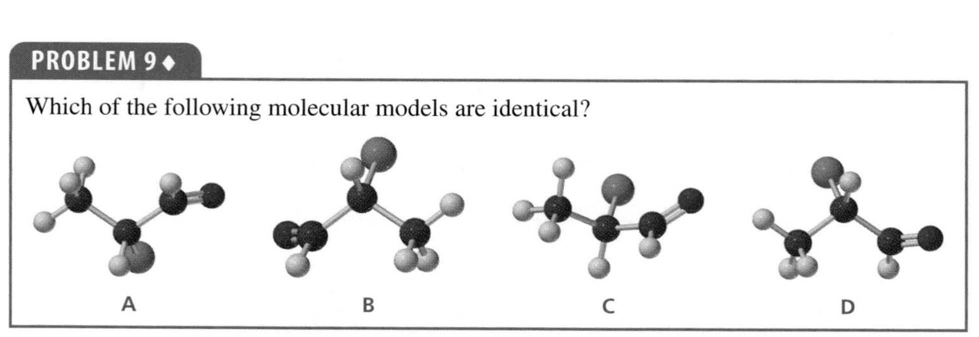

If you are able to easily visualize spatial relationships, the above two rules are all you need to determine whether the asymmetric center of a molecule written on a two-dimensional piece of paper has the *R* or the *S* configuration. Mentally rotate the molecule so that the group (or atom) with the lowest priority (4) is directed away from you, then draw an imaginary arrow from the group (or atom) with the highest priority to the group (or atom) with the next highest priority.

If you have trouble visualizing spatial relationships and you don't have access to a model, the following set of instructions will allow you to determine the configuration about an asymmetric center without having to rotate the molecule mentally. We will use the enantiomers of 2-bromobutane as an example.

the enantiomers of 2-bromobutane

1. **Rank the groups (or atoms) that are bonded to the asymmetric center in order of priority.** In our example, bromine has the highest priority (1), the ethyl group has the second highest priority (2), the methyl group is next (3), and hydrogen has the lowest priority (4). (Revisit Section 4.5 if you don't understand how these priorities are assigned.)

2. **If the group (or atom) with the lowest priority is bonded by a hatched wedge**, draw an arrow from the group (or atom) with the highest priority (1) to the group (or atom) with the second highest priority (2). If the arrow points clockwise, the compound has the *R* configuration, and if it points counterclockwise, the compound has the *S* configuration.

Clockwise specifies *R* if the lowest priority substituent is on a hatched wedge.

Counterclockwise specifies *S* if the lowest priority substituent is on a hatched wedge.

the group with the lowest priority is bonded by a hatched wedge

(*S*)-2-bromobutane (*R*)-2-bromobutane

3-D Molecules: (*R*)-2-Bromobutane; (*S*)-2-Bromobutane

3. **If the group with the lowest priority (4) is NOT bonded by a hatched wedge**, interchange two groups, so group 4 is bonded by a hatched wedge. Then proceed as in step 2: draw an arrow from the group (or atom) with the highest priority (1) to the group (or atom) with the second highest priority (2). Because you have interchanged two groups, you are determining the configuration of the *enantiomer* of the original molecule. So if the arrow points clockwise, the enantiomer (with the interchanged groups) has the *R* configuration, which means the original molecule has the *S* configuration. On the other hand, if the arrow points counterclockwise, the enantiomer (with the interchanged groups) has the *S* configuration, which means the original molecule has the *R* configuration.

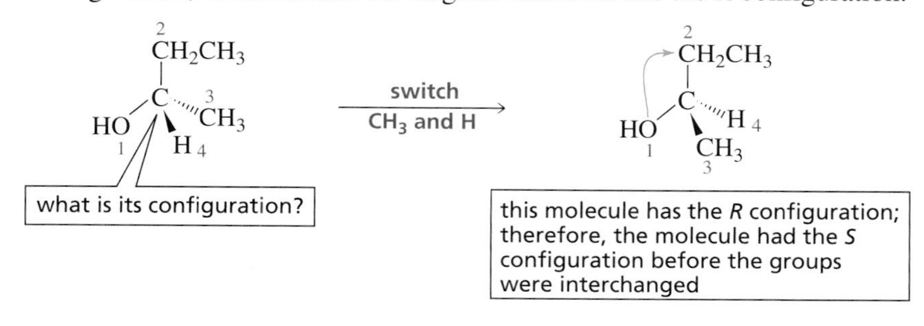
switch
CH3 and H
what is its configuration?
this molecule has the R configuration; therefore, the molecule had the S configuration before the groups were interchanged

4. In drawing the arrow from group 1 to group 2, you can draw past the group with the lowest priority (4), but never draw past the group with the next lowest priority (3).

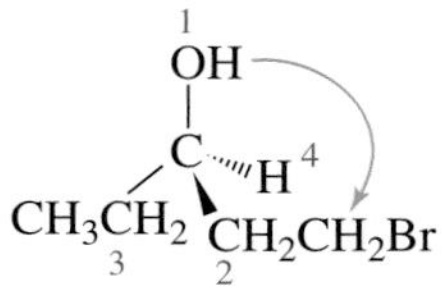

OH (1), H (4), CH_3CH_2 (3), CH_2CH_2Br (2)

(*R*)-1-bromo-3-pentanol

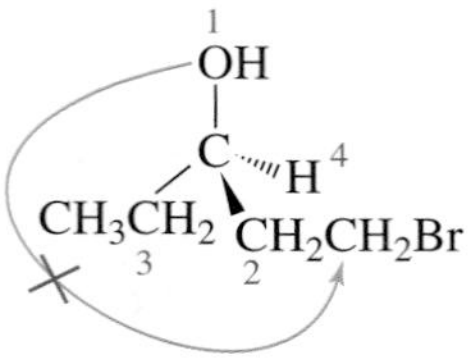

PROBLEM 10 ◆

Assign relative priorities to the following groups:

a. $-CH_2OH$ $-CH_3$ $-CH_2CH_2OH$ $-H$

b. $-CH{=}O$ $-OH$ $-CH_3$ $-CH_2OH$

c. $-CH(CH_3)_2$ $-CH_2CH_2Br$ $-Cl$ $-CH_2CH_2CH_2Br$

d. $-CH{=}CH_2$ $-CH_2CH_3$ $-C{\equiv}CH$ $-CH_3$

PROBLEM 11 ◆

Indicate whether each of the following structures has the *R* or the *S* configuration:

a. 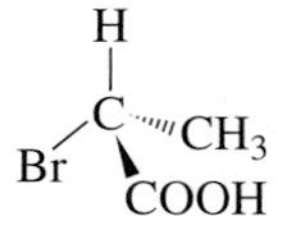 H, Br, CH_3, COOH

b.  H, OH

PROBLEM 12 ◆

Name the following:

a. 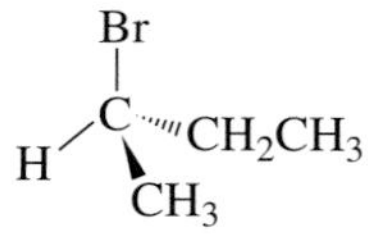 Br, H, CH_2CH_3, CH_3

b. 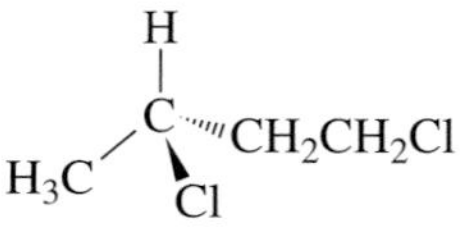 H, H_3C, CH_2CH_2Cl, Cl

PROBLEM-SOLVING STRATEGY

Do the following structures represent identical molecules or a pair of enantiomers?

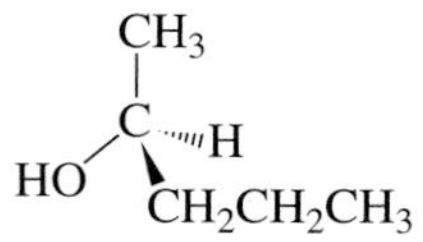

CH_3, HO, H, $CH_2CH_2CH_3$ and OH, $CH_3CH_2CH_2$, CH_3, H

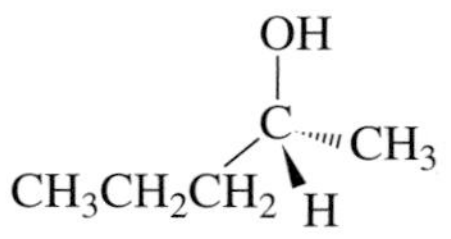

The easiest way to find out whether two structures are enantiomers or identical molecules is to determine their configurations. If one has the *R* configuration and the other has the *S* configuration, they are enantiomers. If they both have the *R* configuration or both have the *S* configuration, they are identical molecules. Because the structure on the left has the *S* configuration and the structure on the right has the *R* configuration, we know that they represent a pair of enantiomers.

Now continue on to Problem 13.

PROBLEM 13 ◆

Do the following structures represent identical molecules or a pair of enantiomers?

a. 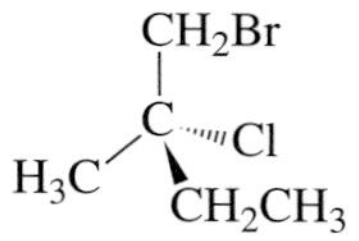 CH_2Br, H_3C, Cl, CH_2CH_3 and Cl, CH_3CH_2, CH_3, CH_2Br

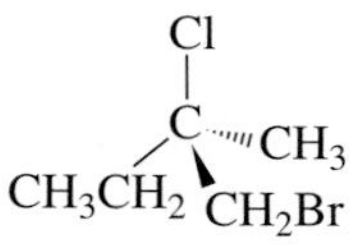

b. 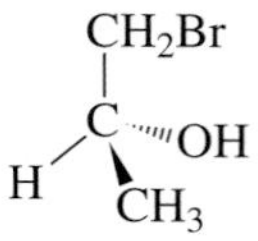 CH_2Br, H, OH, CH_3 and H, HO, CH_3, CH_2Br

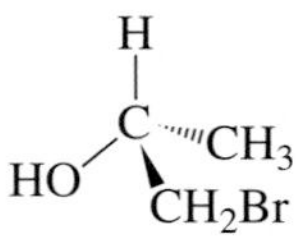

PROBLEM-SOLVING STRATEGY

(*S*)-Alanine is a naturally occurring amino acid. Draw its structure using a perspective formula.

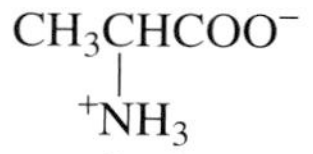

CH_3CHCOO^- with $^+NH_3$ on the central carbon

alanine

First draw the bonds about the asymmetric center. Remember that the solid wedge and the hatched wedge must be adjacent to one another.

Put the group with the lowest priority on the hatched wedge. Put the group with the highest priority on any remaining bond.

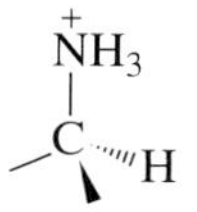

Because you have been asked to draw the *S* enantiomer, draw an arrow counterclockwise from the group with the highest priority to the next available bond; put the group with the next highest priority on that bond.

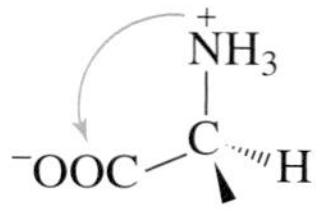

Put the remaining substituent on the last available bond.

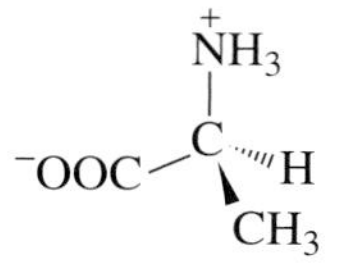

Now continue on to Problem 14.

PROBLEM 14

Draw a perspective formula for each of the following:

a. (*S*)-2-chlorobutane

b. (*R*)-1,2-dibromobutane

6.7 CHIRAL COMPOUNDS ARE OPTICALLY ACTIVE

Enantiomers share many of the same properties; they have the same boiling points, the same melting points, and the same solubilities. In fact, all the physical properties of enantiomers are the same except those that stem from how the groups bonded to the asymmetric center are arranged in space. One property that enantiomers do not share is the way they interact with *plane-polarized light*.

Normal light, such as that coming from a light bulb or the sun, consists of rays that oscillate in all directions. In contrast, all the rays in a beam of **plane-polarized light**

BIOGRAPHY

Born in France, **Jean-Baptiste Biot (1774–1862)** *was imprisoned for taking part in a street riot during the French Revolution. He became a professor of mathematics at the University of Beauvais and later a professor of physics at the Collège de France. He was awarded the Legion of Honor by Louis XVIII. (Also see page 159.)*

(or simply polarized light) oscillate in a single plane. Plane-polarized light is produced by passing normal light through a polarizer.

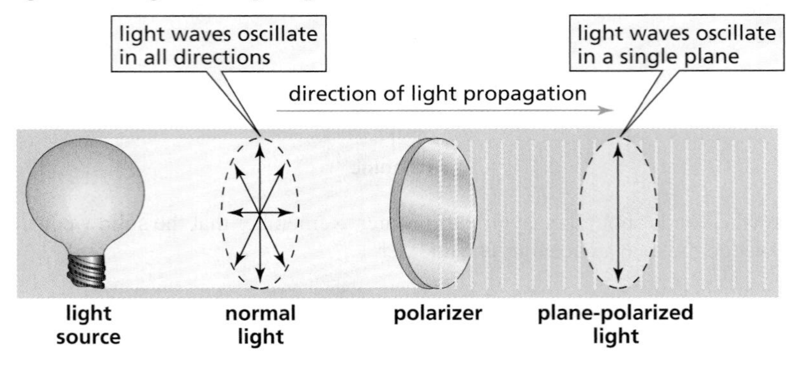

You can experience the effect of a polarized lens by wearing a pair of polarized sunglasses. Polarized sunglasses allow only light oscillating in a single plane to pass through them, which is why they block reflections (glare) more effectively than nonpolarized sunglasses do.

In 1815, the physicist Jean-Baptiste Biot discovered that certain naturally occurring organic substances, such as camphor and oil of turpentine, are able to rotate the plane of polarization of polarized light. He noted that some compounds rotated the plane clockwise and others counterclockwise, while some did not rotate the plane of polarization at all. He predicted that the ability to rotate the plane of polarization was attributable to some asymmetry in the molecules. It was later determined that the molecular asymmetry was associated with compounds having one or more asymmetric centers.

When polarized light passes through a solution of achiral molecules, the light emerges from the solution with its plane of polarization unchanged. *An achiral compound does not rotate the plane of polarization.*

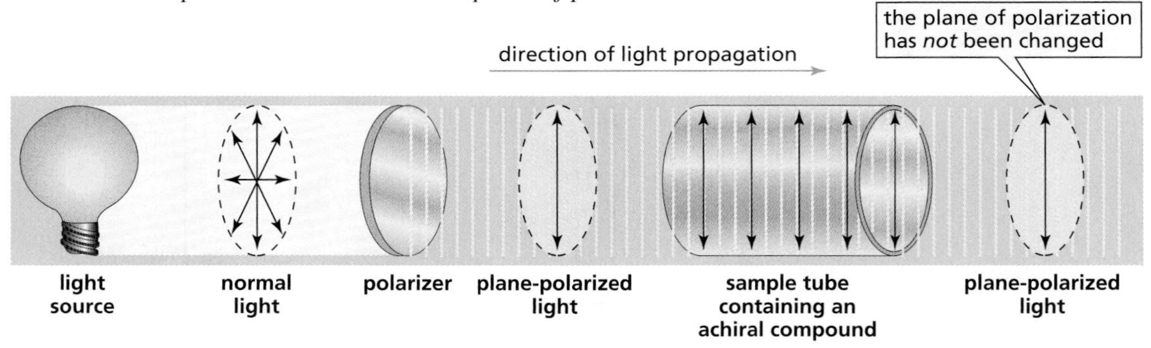

However, when polarized light passes through a solution of a chiral compound, the light emerges with its plane of polarization changed. Thus, *a chiral compound rotates the plane of polarization.* A chiral compound can rotate the plane of polarization clockwise or counterclockwise. If one enantiomer rotates the plane of polarization clockwise, its mirror image will rotate the plane of polarization exactly the same amount counterclockwise.

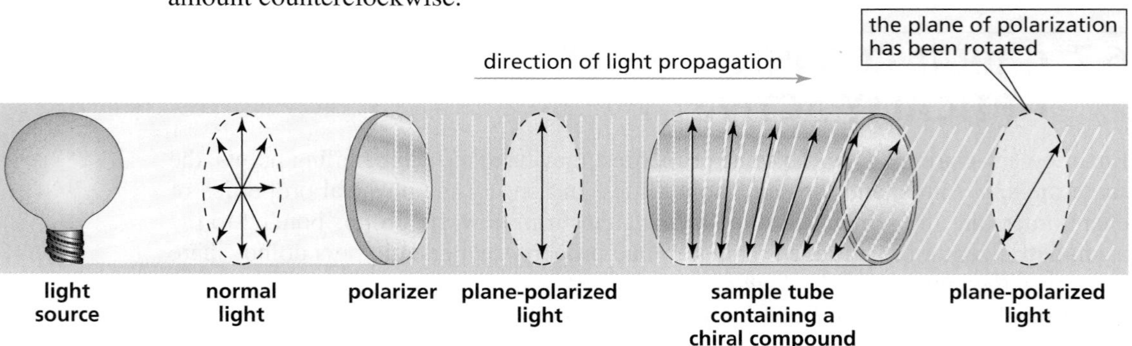

A compound that rotates the plane of polarization is said to be **optically active**. In other words, chiral compounds are optically active and achiral compounds are **optically inactive**.

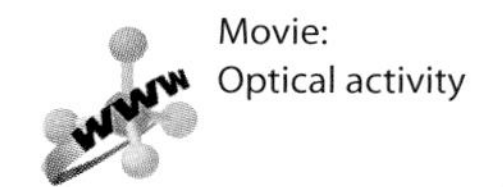

Movie: Optical activity

If an optically active compound rotates the plane of polarization clockwise, the compound is said to be **dextrorotatory**, indicated in the compound's name by the prefix (+). If it rotates the plane of polarization counterclockwise, it said to be **levorotatory**, indicated by (−).

When light is filtered through two polarized lenses at a 90° angle to one another, none of the light passes through.

Do not confuse (+) and (−) with *R* and *S*. The (+) and (−) symbols indicate the direction an optically active compound rotates the plane of polarization, whereas *R* and *S* indicate the arrangement of the groups about an asymmetric center. Some compounds with the *R* configuration are (+) and some are (−).

Some molecules with the *R* configuration are (+) and some are (−). Likewise, some molecules with the *S* configuration are (+) and some are (−).

We can tell by looking at the structure of a compound whether it has the *R* or the *S* configuration, but the only way we can tell whether a compound is dextrorotatory (+) or levorotatory (−) is to put the compound in a polarimeter, an instrument that measures the direction and the amount the plane of polarization is rotated (see Section 6.8). For example, (*S*)-lactic acid and (*S*)-sodium lactate both have an *S* configuration, but (*S*)-lactic acid has been found to be dextrorotatory, whereas (*S*)-sodium lactate is levorotatory. When we know the direction an optically active compound rotates the plane of polarization, we can incorporate (+) or (−) into its name.

CH_3 / C / HO, H, COOH — (*S*)-(+)-lactic acid

CH_3 / C / HO, H, COO^-Na^+ — (*S*)-(−)-sodium lactate

PROBLEM 15 ♦

a. Is (*R*)-lactic acid dextrorotatory or levorotatory?
b. Is (*R*)-sodium lactate dextrorotatory or levorotatory?

PROBLEM 16 ♦ *SOLVED*

What is the configuration of the following compounds?

a. (−)-glyceraldehyde
b. (−)-glyceric acid
c. (+)-isoserine
d. (+)-lactic acid

HC=O / C / HO, H, CH_2OH — (+)-glyceraldehyde

COOH / C / HO, H, CH_2OH — (−)-glyceric acid

COOH / C / HO, H, $CH_2\overset{+}{N}H_3$ — (+)-isoserine

COOH / C / HO, H, CH_3 — (−)-lactic acid

Solution to 16a We know that (+)-glyceraldehyde has the *R* configuration because the group with the lowest priority is on the hatched wedge and the arrow drawn from the OH group to the HC=O group is clockwise. Therefore, (−)-glyceraldehyde has the *S* configuration.

6.8 HOW SPECIFIC ROTATION IS MEASURED

The direction and amount an optically active compound rotates the plane of polarization can be measured with an instrument called a **polarimeter** (Figure 6.2). In a polarimeter, light passes through the polarizer and emerges as polarized light. The polarized light passes through a sample tube. If the tube is empty, or is filled with an optically inactive material, the light emerges from it with the plane of polarization unchanged. The light then passes through an analyzer, which is a second polarizer mounted on an eyepiece with a dial marked in degrees. Before the start of the experiment, the user looks through

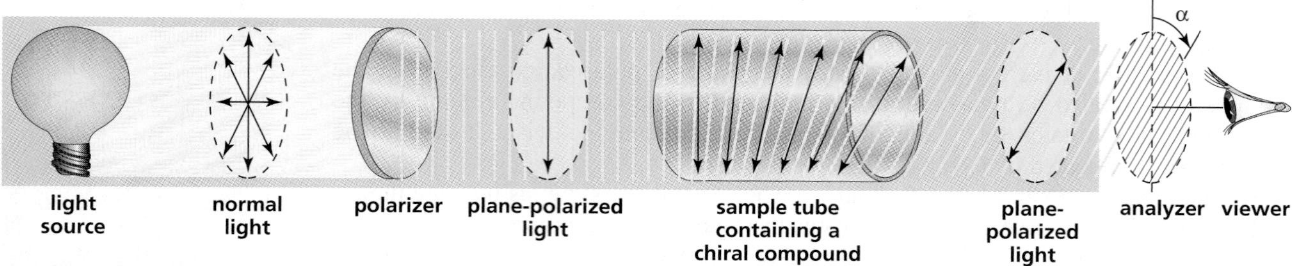

▲ **Figure 6.2**
A schematic drawing of a polarimeter.

the eyepiece and rotates the analyzer until he or she sees total darkness. At this point, the analyzer is at a right angle to the first polarizer, so no light passes through. This analyzer setting corresponds to zero rotation.

The sample to be measured is then placed in the sample tube. If the sample is optically active, it will rotate the plane of polarization. The analyzer, therefore, will no longer block all the light, so some light reaches the user's eye. Now the user rotates the analyzer again until no light passes through. The degree to which the analyzer is rotated can be read from the dial. This value, which is measured in degrees, is called the **observed rotation** (α).

Each optically active compound has a characteristic specific rotation. The **specific rotation** can be calculated from the observed rotation using the following formula:

$$[\alpha]_{\lambda}^{T} = \frac{\alpha}{l \times c}$$

where $[\alpha]$ is the specific rotation; T is temperature in °C; λ is the wavelength of the incident light (when the sodium D-line is used, λ is indicated as D); α is the observed rotation; l is the length of the sample tube in decimeters; and c is the concentration of the sample in grams per milliliter of solution.

If one enantiomer has a specific rotation of +5.75, the specific rotation of the other enantiomer must be −5.75, because the mirror-image molecule rotates the plane of polarization the same amount but in the opposite direction. The specific rotations of some common compounds are shown in Table 6.1.*

Table 6.1 The Specific Rotation of Some Naturally Occurring Compounds

Compound	Specific rotation
Cholesterol	−31.5
Cocaine	−16
Codeine	−136
Morphine	−132
Penicillin V	+233
Progesterone (female sex hormone)	+172
Sucrose (table sugar)	+66.5
Testosterone (male sex hormone)	+109

CH_2OH, H_3C, H, CH_2CH_3
(R)-2-methyl-1-butanol
$[\alpha]_{D}^{20\,°C} = +5.75$

CH_2OH, H, CH_3, CH_2CH_3
(S)-2-methyl-1-butanol
$[\alpha]_{D}^{20\,°C} = -5.75$

PROBLEM 17 ◆

The observed rotation of 2.0 g of a compound in 50 mL of solution in a polarimeter tube 20-cm long is +13.4°. What is the specific rotation of the compound?

A mixture of equal amounts of two enantiomers—such as (*R*)-(−)-lactic acid and (*S*)-(+)-lactic acid—is called a **racemic mixture**. Racemic mixtures do not rotate the plane of polarization. They are optically inactive because for every molecule in a racemic mixture that rotates the plane of polarization in one direction, there is a mirror-image molecule that rotates the plane in the opposite direction. As a result, the light

* Unlike observed rotation, which is measured in degrees, specific rotation has units of 10^{-1} deg cm^2 g^{-1}. In this book, values of specific rotation will be given without units.

emerges from a racemic mixture with its plane of polarization unchanged. The symbol (±) is used to specify a racemic mixture. Thus, (±)-2-bromobutane indicates a mixture of (+)-2-bromobutane and an equal amount of (−)-2-bromobutane.

PROBLEM 18 ◆

(*S*)-(+)-Monosodium glutamate (MSG) is a flavor enhancer used in many foods. Some people have an allergic reaction to MSG (headache, chest pain, and an overall feeling of weakness). "Fast food" often contains substantial amounts of MSG, which is widely used in Chinese food as well. MSG has a specific rotation of +24.

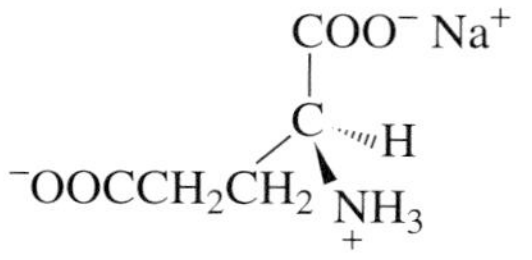

$$^{-}OOCCH_2CH_2-C(COO^{-}Na^{+})(H)(\overset{+}{N}H_3)$$

(*S*)-(+)-monosodium glutamate

a. What is the specific rotation of (*R*)-(−)-monosodium glutamate?
b. What is the specific rotation of a racemic mixture of monosodium glutamate?

PROBLEM 19 ◆

Naproxen, a nonsteroidal anti-inflammatory drug that is the active ingredient in Aleve, has a specific rotation of +66. Does naproxen have the *R* or the *S* configuration?

6.9 ISOMERS WITH MORE THAN ONE ASYMMETRIC CENTER

Many organic compounds have more than one asymmetric center. The more asymmetric centers a compound has, the more stereoisomers it can have. If we know the number of asymmetric centers, we can calculate the maximum number of stereoisomers for that compound: *a compound can have a maximum of* 2^n *stereoisomers, where* n *equals the number of asymmetric centers.* For example, the amino acid threonine has two asymmetric centers. Therefore, it can have a maximum of four ($2^2 = 4$) stereoisomers.

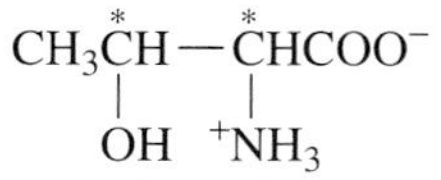

$$CH_3\overset{*}{C}H(OH)-\overset{*}{C}H(\overset{+}{N}H_3)COO^{-}$$

threonine

The four stereoisomers of threonine consist of two pairs of enantiomers. Stereoisomers **1** and **2** are nonsuperimposable mirror images. They, therefore, are enantiomers. Stereoisomers **3** and **4** are also enantiomers. Stereoisomers **1** and **3** are not identical, and they are not mirror images. Such stereoisomers are called **diastereomers**. *Diastereomers are stereoisomers that are not enantiomers.* Stereoisomers **1** and **4**, **2** and **3**, and **2** and **4** are also pairs of diastereomers. Notice that the configuration of one of the asymmetric centers is the same in both of a pair of diastereomers, but the configuration of the other asymmetric center is different.

Diastereomers are stereoisomers that are not enantiomers.

1	2	3	4
COO^-	COO^-	COO^-	COO^-
H–C–$\overset{+}{N}H_3$	$H_3\overset{+}{N}$–C–H	H–C–$\overset{+}{N}H_3$	$H_3\overset{+}{N}$–C–H
H–C–OH	HO–C–H	HO–C–H	H–C–OH
CH_3	CH_3	CH_3	CH_3

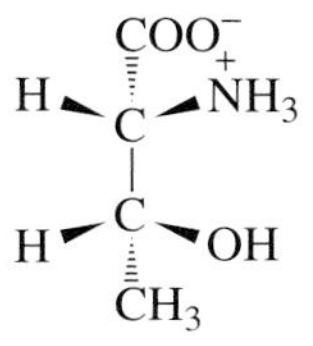

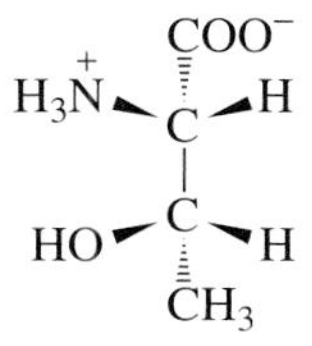

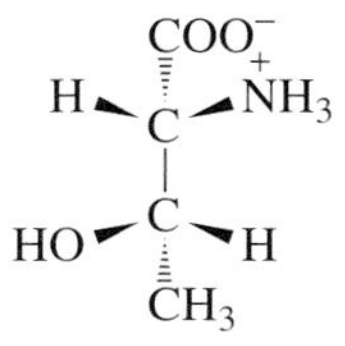

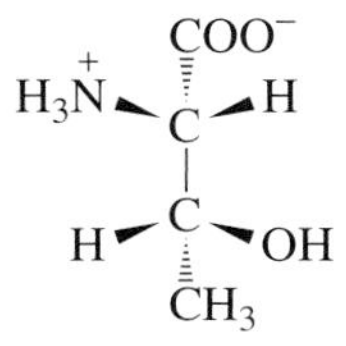

Enantiomers have *identical physical properties* (except for the way they interact with polarized light) and *identical chemical properties*, so they react at the same rate with an achiral reagent. Diastereomers have *different physical properties*, meaning different melting points, different boiling points, different solubilities, different specific rotations, and so on, and *different chemical properties*, so they react with an achiral reagent at different rates.

3-D Molecules:
(2*S*,3*R*)-3-Chloro-2-butanol;
(2*R*,3*S*)-3-Chloro-2-butanol;
(2*S*,3*S*)-3-Chloro-2-butanol;
(2*R*,3*R*)-3-Chloro-2-butanol

PROBLEM 20 ♦

a. Stereoisomers with two asymmetric centers are called _____ if the configuration of both asymmetric centers in one stereoisomer is the opposite of the configuration of the asymmetric centers in the other stereoisomer.
b. Stereoisomers with two asymmetric centers are called _____ if the configuration of both asymmetric centers in one stereoisomer is the same as the configuration of the asymmetric centers in the other stereoisomer.
c. Stereoisomers with two asymmetric centers are called _____ if one of the asymmetric centers has the same configuration in both stereoisomers and the other asymmetric center has the opposite configuration in the two stereoisomers.

PROBLEM 21 ♦

a. How many asymmetric centers does cholesterol have?
b. What is the maximum number of stereoisomers that cholesterol can have? (Only one of these is found in nature.)

H_3C H_3C CH_3 H_3C H CH_3 H H HO

cholesterol

PROBLEM 22

Draw the stereoisomers of the following amino acids. Indicate pairs of enantiomers and pairs of diastereomers.

$CH_3CH(CH_3)CH_2-CH(^+NH_3)COO^-$
leucine

$CH_3CH_2CH(CH_3)-CH(^+NH_3)COO^-$
isoleucine

PROBLEM 23 ♦ SOLVED

Indicate whether the following pairs of compounds are identical or are enantiomers, diastereomers, or constitutional isomers:

a. Cl, CH_3 and Cl, CH_3
c. Cl, CH_3 and CH_3, Cl
b. Cl, CH_3 and Cl, CH_3
d. Cl, CH_3 and Cl, CH_3

Solution to 23a The configuration of one of the asymmetric centers (the one bonded to Cl) is the same in both compounds; the configuration of the other asymmetric center (the one bonded to CH_3) is different in the two compounds. The two compounds, therefore, are diastereomers.

6.10 MESO COMPOUNDS HAVE ASYMMETRIC CENTERS BUT ARE OPTICALLY INACTIVE

In the examples we have just seen, the compounds with two asymmetric centers had four stereoisomers. However, some compounds with two asymmetric centers have only three stereoisomers. This is why we emphasized in Section 6.9 that the *maximum* number of stereoisomers a compound with n asymmetric centers can have is 2^n, instead of stating that a compound with n asymmetric centers has 2^n stereoisomers.

An example of a compound with two asymmetric centers that has only three stereoisomers is 2,3-dibromobutane.

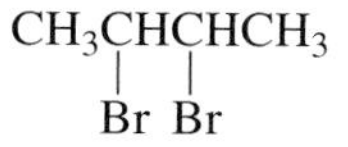

$CH_3CH(Br)CH(Br)CH_3$

2,3-dibromobutane

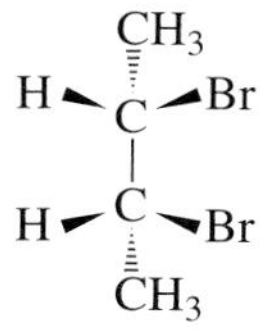

1 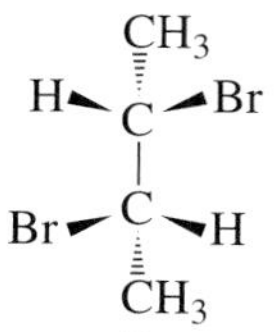 2 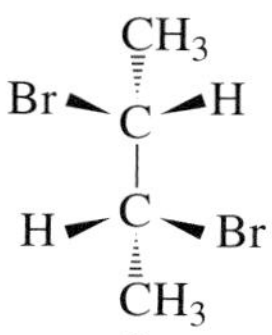 3

The "missing" isomer is the mirror image of **1**, because **1** and its mirror image are the same molecule. You can see that **1** and its mirror image are identical if you rotate the mirror image 180°.

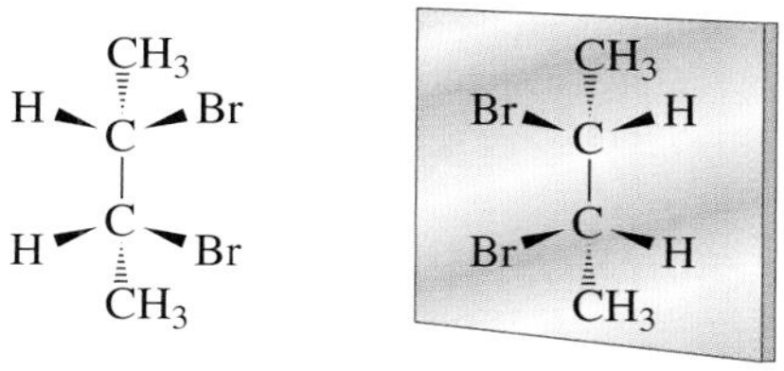

superimposable mirror image

Stereoisomer **1** is called a *meso compound.* Even though a meso (mee-zo) compound has asymmetric centers, it is achiral. A meso compound does not rotate plane-polarized light because it is superimposable on its mirror image.

A meso compound is achiral.

A **meso compound** can be recognized by the fact that it has two (or more) asymmetric centers and a plane of symmetry. A **plane of symmetry** cuts the molecule in half so that one half is the mirror image of the other. A molecule with a plane of symmetry does not have an enantiomer; it is achiral. Compare stereoisomer **1**, which has a plane of symmetry and thus no enantiomer, with stereoisomer **2**, which does not have a plane of symmetry and therefore *does* have an enantiomer.

A meso compound has two or more asymmetric centers and a plane of symmetry.

If a compound has a plane of symmetry, it is achiral.

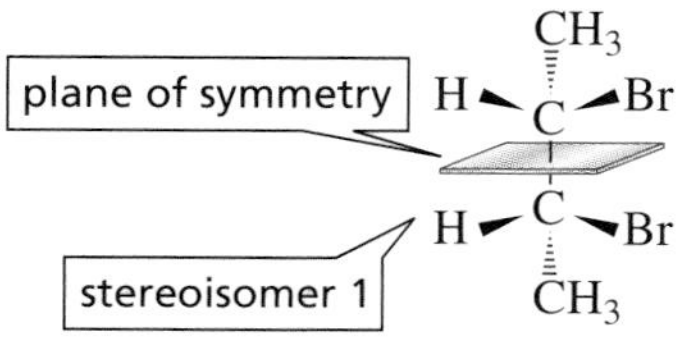

plane of symmetry

stereoisomer 1

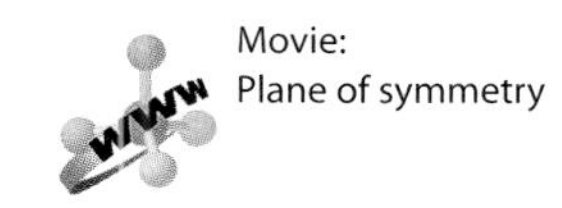
Movie: Plane of symmetry

It is easy to recognize when a compound with two asymmetric centers has a stereoisomer that is a meso compound—the four atoms or groups bonded to one asymmetric center are identical to the four atoms or groups bonded to the other asymmetric center. *A compound with the same four atoms or groups bonded to two different asymmetric centers will have three stereoisomers: one will be a meso compound, and the other two will be enantiomers.*

If a compound with two asymmetric centers has the same four groups bonded to each of the asymmetric centers, one of its stereoisomers will be a meso compound.

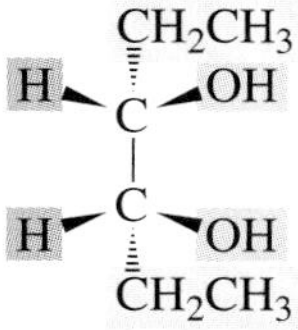

meso compound 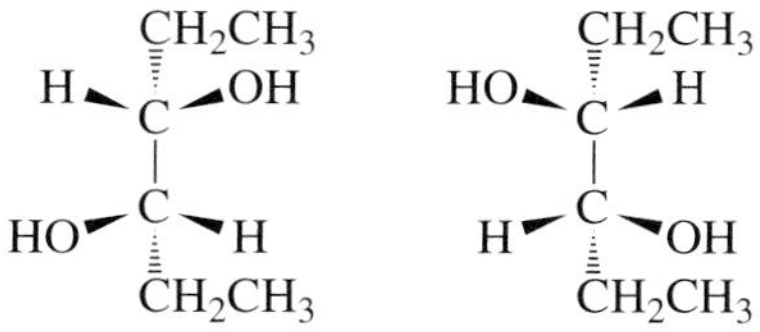 **enantiomers**

Tartaric acid has three stereoisomers because each of its two asymmetric centers has the same set of four substituents.

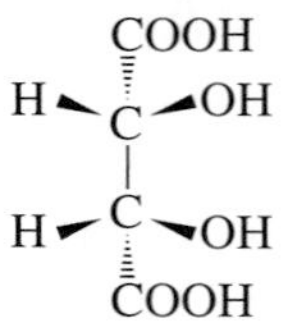

COOH
H–C–OH
H–C–OH
COOH

meso compound

COOH | COOH
H–C–OH | HO–C–H
HO–C–H | H–C–OH
COOH | COOH

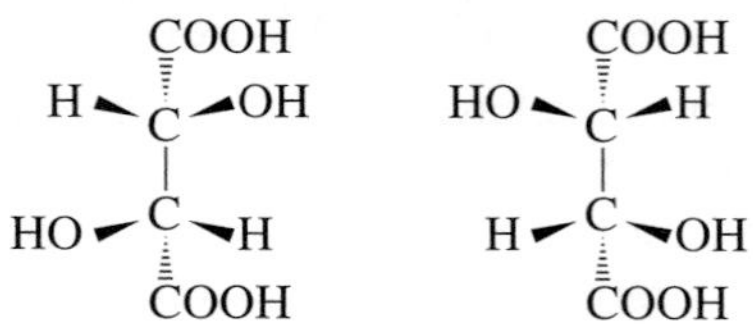

enantiomers

The physical properties of the three stereoisomers of tartaric acid are listed in Table 6.2. The meso compound and either of the enantiomers are diastereomers. Notice that the physical properties of the enantiomers are the same, whereas the physical properties of the diastereomers are different.

PROBLEM-SOLVING STRATEGY

Which of the following has a stereoisomer that is a meso compound?

2,3-dimethylbutane
A

2-bromo-3-methylpentane
C

3,4-dimethylhexane
B

3,4-diethylhexane
D

Check each compound to see if it has the necessary requirements to have a stereoisomer that is a meso compound. That is, does it have two asymmetric centers, and, if so, do they each have the same four substituents attached to them?

Compounds A and D do *not* have a stereoisomer that is a meso compound because they do not have any asymmetric centers.

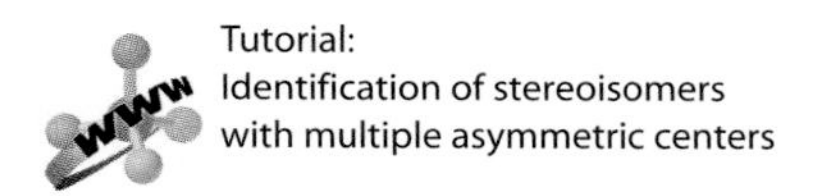
Tutorial:
Identification of stereoisomers with multiple asymmetric centers

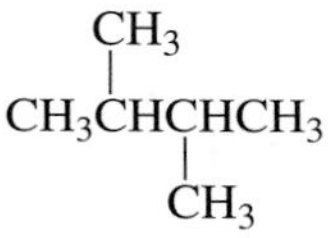

$CH_3CH(CH_3)CH(CH_3)CH_3$
A

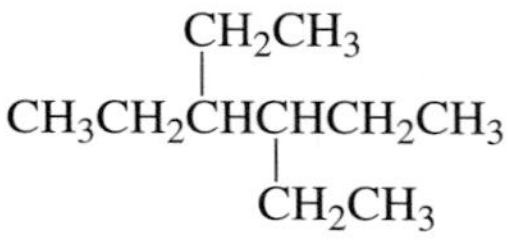

$CH_3CH_2CH(CH_2CH_3)CH(CH_2CH_3)CH_2CH_3$
D

Compound C has two asymmetric centers. It does *not* have a stereoisomer that is a meso compound because each of the asymmetric centers is *not* bonded to the same four substituents.

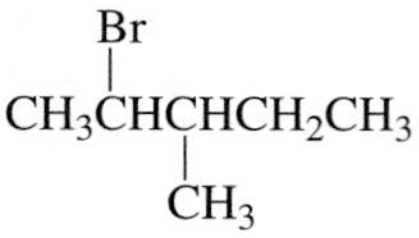

$CH_3CH(Br)CH(CH_3)CH_2CH_3$
C

Compound B has two asymmetric centers, and each asymmetric center is bonded to the same four atoms or groups. Therefore, compound B has a stereoisomer that is a meso compound.

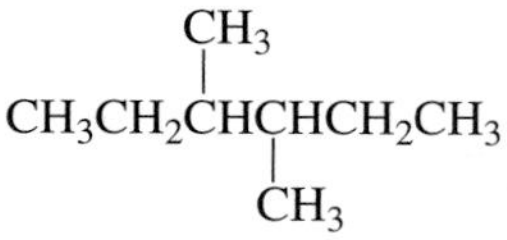

$CH_3CH_2CH(CH_3)CH(CH_3)CH_2CH_3$
B

The stereoisomer that is the meso compound is the one with a plane of symmetry.

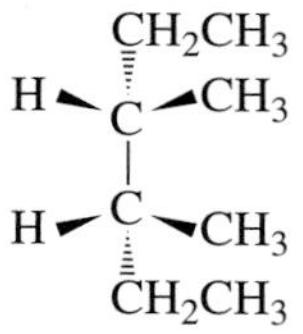

CH_2CH_3
H–C–CH_3
H–C–CH_3
CH_2CH_3

Now continue on to Problem 24.

Table 6.2 The Physical Properties of the Stereoisomers of Tartaric Acid

	Melting point, °C	$[\alpha]_D^{25\,°C}$	Solubility, g/100 g H_2O at 15 °C
(2*R*,3*R*)-(−)-Tartaric acid	171	+11.98	139
(2*S*,3*S*)-(−)-Tartaric acid	171	−11.98	139
(2*R*,3*S*)-Tartaric acid (meso)	140	0°	125
(±)-Tartaric acid	206	0°	

PROBLEM 24 ♦

Which of the following has a stereoisomer that is a meso compound?

2,4-dibromohexane	2,4-dibromopentane	2,4-dimethylpentane
A	B	C

PROBLEM 25

Draw all the stereoisomers for each of the following:

a. 1-bromo-2-methylbutane
b. 3-chloro-3-methylpentane
c. 2,3-dichloropentane
d. 1,3-dichloropentane

6.11 HOW ENANTIOMERS CAN BE SEPARATED

Enantiomers cannot be separated by the usual separation techniques such as distillation or crystallization because their identical boiling points and solubilities cause them to distill or crystallize simultaneously. Louis Pasteur was the first to succeed in separating a pair of enantiomers. While working with crystals of sodium ammonium tartrate, he noted that the crystals were not identical—some were "right-handed" and some were "left-handed." After painstakingly separating the two kinds of crystals with a pair of tweezers, he found that a solution of the right-handed crystals rotated the plane of polarization of polarized light clockwise, whereas a solution of the left-handed crystals rotated the plane of polarization counterclockwise.

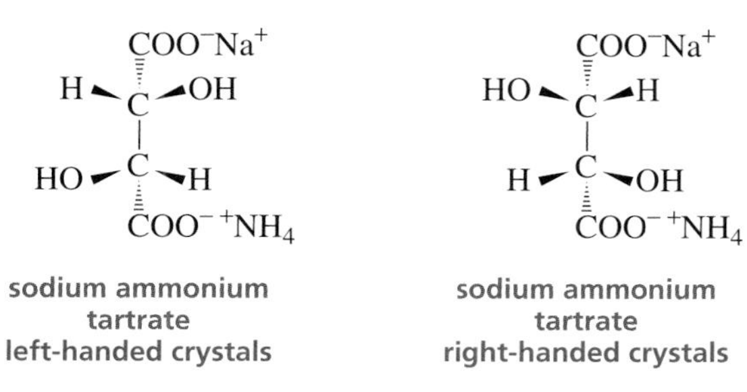

sodium ammonium tartrate left-handed crystals

sodium ammonium tartrate right-handed crystals

Pasteur, only 26 years old at the time and unknown in scientific circles, was concerned about the accuracy of his observations because a few years earlier, the well-known German organic chemist Eilhardt Mitscherlich had reported that crystals of sodium ammonium tartrate were all identical. Pasteur immediately reported his findings to Jean-Baptiste Biot (page 152) and repeated the experiment with Biot present. Biot was convinced that Pasteur had successfully separated the enantiomers of sodium ammonium tartrate. Pasteur's experiment also gave rise to a new chemical term. Tartaric acid is obtained from grapes, so it was also called racemic acid (*racemus* is Latin for "a bunch of grapes"). That is why a mixture of equal amounts of enantiomers came to be known as a **racemic mixture** (Section 6.8).

BIOGRAPHY

The French chemist and microbiologist **Louis Pasteur (1822–1895)** *was the first to demonstrate that microbes cause specific diseases. Asked by the French wine industry to find out why wine often went sour while aging, he showed that the microorganisms that cause grape juice to ferment, producing wine, also cause wine to become sour. Gently heating the wine after fermentation, a process called pasteurization, kills the organisms so that they cannot sour the wine.*

Crystals of potassium hydrogen tartrate, a naturally occurring salt found in wines. Most fruits produce citric acid, but grapes produce large quantities of tartaric acid instead.

BIOGRAPHY

Eilhardt Mitscherlich (1794–1863), *a German chemist, studied medicine so he could travel to Asia in order to satisfy his interest in Oriental languages. He later became fascinated by chemistry. He was a professor of chemistry at the University of Berlin and wrote a successful chemistry textbook that was published in 1829.*

Later, chemists recognized how lucky Pasteur had been. Sodium ammonium tartrate forms asymmetric crystals only under the precise conditions that Pasteur happened to employ. Under other conditions, the symmetrical crystals that had fooled Mitscherlich are formed. But to quote Pasteur, "Chance favors the prepared mind."

Separating enantiomers by hand, as Pasteur did, is not a universally useful method because few compounds form asymmetric crystals. Fortunately enantiomers can now be separated relatively easily by a technique called **chromatography**. In this method, the mixture to be separated is dissolved in a solvent, and the solution is passed through a column packed with a chiral material that tends to absorb organic compounds. The two enantiomers can be expected to move through the column at different rates because they will have different affinities for the chiral material—just as a right hand prefers a right-handed glove to a left-handed glove—so one enantiomer will emerge from the column before the other.

6.12 RECEPTORS

A **receptor** is a protein that binds a particular molecule. Because a receptor is chiral, it will bind one enantiomer better than the other. In Figure 6.3, the receptor binds the *R* enantiomer, but it does not bind the *S* enantiomer.

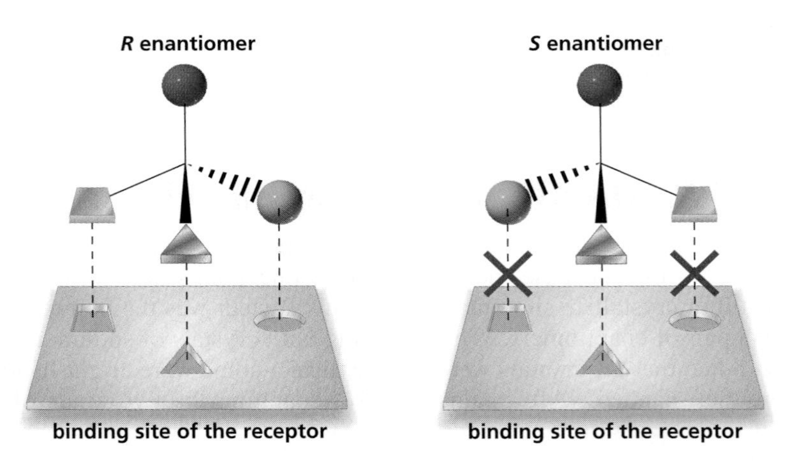

▶ **Figure 6.3**
A schematic diagram showing why only one enantiomer is bound by a receptor. One enantiomer fits into the binding site and one does not.

The fact that a receptor typically recognizes only one enantiomer causes enantiomers to have different physiological properties. For example, receptors located on the exteriors of nerve cells in the nose are able to perceive and differentiate the estimated 10,000 smells to which they are exposed. The reason that (*R*)-(−)-carvone (found in spearmint oil) and (*S*)-(+)-carvone (the main constituent of caraway seed oil) have such different odors is that each enantiomer fits into a different receptor.

(*R*)-(−)-carvone	(*S*)-(+)-carvone
spearmint oil	caraway seed oil
$[\alpha]_D^{20\,°C} = -62.5$	$[\alpha]_D^{20\,°C} = +62.5$

Frances O. Kelsey receives a medal from John F. Kennedy for preventing the sale of thalidomide *(see page 161).*

Many drugs exert their physiological activity by binding to cell-surface receptors. If the drug has an asymmetric center, the receptor can bind one of the enantiomers preferentially. Thus, enantiomers of a drug can have the same physiological activities, different degrees of the same activity, or very different activities, depending on the drug.

PROBLEM 26 ♦

Limonene exists as two different stereoisomers. The *R* enantiomer is found in oranges, and the *S* enantiomer is found in lemons. Which of the following is found in oranges?

(+)-limonene (−)-limonene

THE ENANTIOMERS OF THALIDOMIDE

Thalidomide was developed in then West Germany and was first marketed in 1957 for insomnia and morning sickness. At that time it was available in more than 40 countries but had not been approved for use in the United States because Frances O. Kelsey, a physician for the Federal Food and Drug Administration (FDA), had insisted upon additional tests. (See the box on Drug Safety on page 559.)

The dextrorotatory isomer has stronger sedative properties, but the commercial drug was a racemic mixture. No one knew that the levorotatory isomer was highly teratogenic—causes horrible birth defects—until women who had been given the drug during the first three months of pregnancy gave birth to babies with a wide variety of defects, deformed limbs being the most common. About 10,000 children were damaged by the drug.

It was eventually determined that the dextrorotatory isomer also has mild teratogenic activity and that each of the enantiomers can racemize (interconvert) in the body. Thus, it is not clear whether the birth defects would have been less severe if the women had been given the dextrorotatory isomer only. Because thalidomide was found to damage fast-growing cells in the developing fetus, it has recently has been approved—with restrictions—as an anticancer drug to discourage the production of cancer cells.

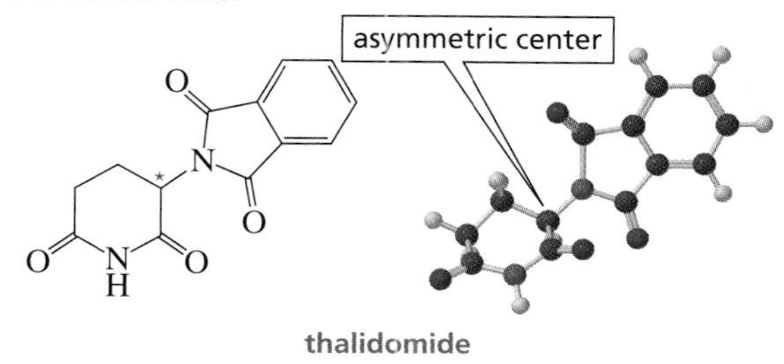

thalidomide

CHIRAL DRUGS

Until relatively recently, most drugs have been marketed as racemic mixtures because of the high cost of separating enantiomers. In 1992, however, the Food and Drug Administration (FDA) issued a policy statement encouraging drug companies to use recent advances in separation techniques to develop single-enantiomer drugs. Now, one-third of all drugs sold are single enantiomers.

If a drug is sold as a racemate, the FDA requires that both enantiomers be tested, since the enantiomers can have similar or very different properties. Examples are numerous. The *S* isomer of Prozac, an antidepressant, is better at blocking serotonin; but it is used up faster than the *R* isomer. Testing has shown that the anesthetic (*S*)-(+)-ketamine is four times more potent than (*R*)-(−)-ketamine and the disturbing side effects are apparently only associated with the (*R*)-(−)-enantiomer. Only the *S* isomer of the beta-blocker propranolol shows activity; the *R* isomer is inactive. The activity of ibuprofen, the popular analgesic marketed as Advil, Nuprin, and Motrin, resides primarily in the (*S*)-(+)-enantiomer. Heroin addicts can be maintained with (−)-α-acetylmethadol for a 72-hour period compared to 24 hours with racemic methadone. This means less frequent visits to the clinic; a single dose can get an addict through an entire weekend.

Prescribing a single enantiomer prevents the patient from having to metabolize the less potent enantiomer and decreases the chance of unwanted drug interactions. Drugs that could not be given as racemates because of the toxicity of one of the enantiomers can now be used. For example, (*S*)-penicillamine can be used to treat Wilson's disease even though (*R*)-penicillamine causes blindness.

6.13 THE STEREOCHEMISTRY OF REACTIONS

When we looked at the electrophilic addition reactions that alkenes undergo (Chapter 5), we examined the step-by-step process by which each reaction occurs (the mechanism of the reaction), and we determined what products are formed. However, we did not consider the stereochemistry of the reactions.

Stereochemistry is the field of chemistry that deals with the structures of molecules in three dimensions. When we study the stereochemistry of a reaction, we are concerned with the following question: if the product of a reaction can exist as two or more stereoisomers, does the reaction produce a single stereoisomer or all possible stereoisomers?

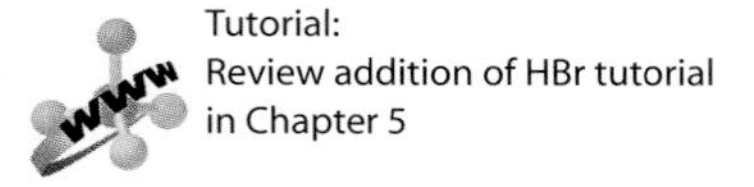

Tutorial:
Review addition of HBr tutorial in Chapter 5

The Stereochemistry of Electrophilic Addition Reactions of Alkenes

We have seen that when an alkene reacts with an electrophilic reagent such as HBr, the major product of the addition reaction is the one obtained by adding the electrophile (H^+) to the sp^2 carbon bonded to the greater number of hydrogens and adding the nucleophile (Br^-) to the other sp^2 carbon (Section 5.3). For example, the major product obtained from the reaction of propene with HBr is 2-bromopropane. This particular product does not have stereoisomers because it does not have an asymmetric center. Therefore, we do not have to be concerned with the stereochemistry of this reaction.

$$\underset{\text{propene}}{CH_3CH{=}CH_2} \xrightarrow{\text{HBr}} CH_3\overset{+}{C}HCH_3 \;\; Br^- \longrightarrow \underset{\substack{\text{2-bromopropane} \\ \text{major product}}}{CH_3CH(Br)CH_3}$$

If, however, the reaction creates a product with an asymmetric center, we need to know which stereoisomers are formed. For example, the reaction of HBr with 1-butene forms 2-bromobutane, a compound with an asymmetric center. But what is the configuration of the product? Do we get the *R* enantiomer, the *S* enantiomer, or both?

$$\underset{\text{1-butene}}{CH_3CH_2CH{=}CH_2} \xrightarrow{\text{HBr}} CH_3CH_2\overset{+}{C}HCH_3 \;\; Br^- \longrightarrow \underset{\text{2-bromobutane}}{CH_3CH_2CH(Br)CH_3}$$

asymmetric center

When a reactant that does not have an asymmetric center undergoes a reaction that forms a product with *one* asymmetric center, the product will be a racemic mixture.

When a reactant that does not have an asymmetric center undergoes a reaction that forms a product with *one* asymmetric center, the product will always be a racemic mixture. For example, the reaction of 1-butene with HBr forms identical amounts of (*R*)-2-bromobutane and (*S*)-2-bromobutane.

We can see why a racemic mixture is formed if we examine the structure of the carbocation formed in the first step of the reaction. The positively charged carbon is sp^2 hybridized, so the three atoms to which it is bonded lie in a plane (Section 1.10). When the bromide ion approaches the positively charged carbon from above the plane, one enantiomer is formed, but when it approaches from below the plane, the other enantiomer is formed. Because the bromide ion has equal access to both sides of the plane, identical amounts of the *R* and *S* enantiomers are obtained from the reaction.

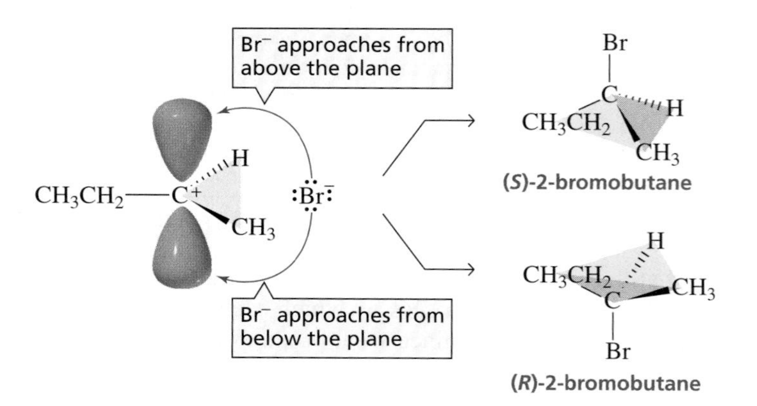

PROBLEM 27

What stereoisomers are obtained from each of the following reactions?

a. $CH_3CH_2CH_2CH{=}CH_2 + HCl \longrightarrow$

b. $(H)(CH_3CH_2)C{=}C(H)(CH_2CH_3) + H_2O \xrightarrow{H_2SO_4}$

c. $(H_3C)(H_3C)C{=}C(CH_3)(CH_2CH_2CH_3) + H_2 \xrightarrow{Pt/C}$

d. $(H_3C)(H_3C)C{=}C(CH_3)(H) + HBr \longrightarrow$

The Stereochemistry of Hydrogen Addition

We have seen that in catalytic hydrogenation, both hydrogens add to the same side of the double bond (Section 5.12). Thus, when the reactant is an alkyne, the product is a cis alkene.

$$CH_3C{\equiv}CCH_3 + H_2 \xrightarrow{\text{Lindlar catalyst}} (H)(H_3C)C{=}C(H)(CH_3)$$

cis-2-butene

When the reactant is a cyclic alkene, both hydrogens add to the same side of the ring, so only the cis isomer is formed.

$$\text{1,2-dimethylcyclohexene} + H_2 \xrightarrow{Pt/C} \text{cis-1,2-dimethylcyclohexane (H, H same side; } H_3C\text{, } CH_3\text{)}$$

PROBLEM 28 ♦

Give the configuration of the products obtained from the following reactions:

a. *trans*-2-butene $+ H_2O \xrightarrow{H_2SO_4}$

b. *cis*-3-hexene $+ HBr \longrightarrow$

c. 1,2-dimethylcyclopentene (H_3C, CH_3) $\xrightarrow[Pt/C]{H_2}$

6.14 THE STEREOCHEMISTRY OF ENZYME-CATALYZED REACTIONS

The chemistry associated with living organisms is called **biochemistry**. When you study biochemistry, you study the structures and functions of the molecules found in the biological world and the reactions involved in the synthesis and degradation of these molecules. Because the compounds in living organisms are organic compounds, it is not surprising that many of the reactions encountered in organic chemistry also occur in biological systems.

BIOGRAPHY

For studies on the stereochemistry of enzyme-catalyzed reactions, **Sir John Cornforth** *received the Nobel Prize in chemistry in 1975. Born in Australia in 1917, Conforth studied at the University of Sydney and received a Ph.D. from Oxford. His major research was carried out in laboratories at Britain's Medical Research Council and at Shell Research Ltd. He was declared "Australian of the Year" in 1975 and was knighted in 1977.*

BIOGRAPHY

Frank H. Westheimer (1912–2007) *did fundamental work on the stereochemistry of enzyme-catalyzed reactions. Born in Baltimore, he received his graduate training at Harvard University, served on the faculty of the University of Chicago, and subsequently returned to Harvard as a professor of chemistry.*

An achiral reagent reacts identically with both enantiomers. A sock, which is achiral, fits on either foot.

A chiral reagent reacts differently with each enantiomer. A shoe, which is chiral, fits on only one foot.

Reactions that occur in biological systems are catalyzed by proteins called **enzymes**. When an enzyme catalyzes a reaction that forms a product with an asymmetric center, *only one stereoisomer is formed*. For example, the enzyme known as fumarase catalyzes the addition of water to fumarate, thereby forming malate, a compound with an asymmetric center.

$$\underset{\text{fumarate}}{\text{(H)(}^-\text{OOC)C=C(COO}^-\text{)(H)}} + H_2O \xrightarrow{\text{fumarase}} \underset{\text{malate}}{^-OOCCH_2\overset{\text{asymmetric center}}{C}H(OH)COO^-}$$

But the reaction forms only (*S*)-malate; the *R* enantiomer is not formed.

$$\underset{(S)\text{-malate}}{^-OOCCH_2\text{–}C(COO^-)(H)(OH)}$$

Why does a reaction that is not catalyzed by an enzyme form a racemic mixture (Section 6.13), whereas an enzyme-catalyzed reaction forms only one stereoisomer? It is because an enzyme is chiral. Its chiral binding site restricts delivery of reagents to only one side of the reactant's functional group. Consequently, only one stereoisomer is formed.

If the reactant can exist as stereoisomers, an enzyme typically will catalyze the reaction of only one of the stereoisomers, because an enzyme (like a receptor) will bind only the stereoisomer whose substituents are in the correct positions to interact with the substituents in the chiral binding site (Figure 6.3). Other stereoisomers do not have substituents in the proper positions, so they cannot bind efficiently to the enzyme. For example, fumarase catalyzes the addition of water to fumarate (the trans isomer) but not to maleate (the cis isomer).

$$\underset{\text{maleate}}{\text{(}^-\text{OOC)(H)C=C(COO}^-\text{)(H)}} + H_2O \xrightarrow{\text{fumarase}} \text{no reaction}$$

An enzyme's behavior can be likened to a right-handed glove, which fits only the right hand: it forms only one stereoisomer and it reacts with only one stereoisomer.

PROBLEM 29 ♦

a. What would be the product of the reaction of fumarate and H_2O if an acid were used as a catalyst instead of fumarase?
b. What would be the product of the reaction of maleate and H_2O if an acid were used as a catalyst instead of fumarase?

SUMMARY

Stereochemistry is the field of chemistry that deals with the structures of molecules in three dimensions. Compounds that have the same molecular formula but are not identical are called **isomers**; they fall into two classes: constitutional isomers and stereoisomers. **Constitutional isomers** differ in the way their atoms are connected. **Stereoisomers** differ in the way their atoms are arranged in space. There are two kinds of stereoisomers: **cis–trans isomers** and isomers that contain **asymmetric centers**.

A **chiral** molecule has a nonsuperimposable mirror image; an **achiral** molecule has a superimposable mirror image. The feature that is most often the cause of chirality is an asymmetric center. An **asymmetric center** is an atom bonded to four different atoms or groups.

Nonsuperimposable mirror-image molecules are called **enantiomers**. **Diastereomers** are stereoisomers that are not enantiomers. Enantiomers have identical physical and chemical properties; diastereomers have different physical and chemical properties. An achiral reagent reacts identically with both enantiomers; a chiral reagent reacts differently with each enantiomer. A mixture of equal amounts of two enantiomers is called a **racemic mixture**.

The letters ***R*** and ***S*** indicate the **configuration** about an asymmetric center. If one stereoisomer has the *R* configuration and the other has the *S* configuration, they are enantiomers; if they both have the *R* configuration or both have the *S* configuration, they are identical.

Chiral compounds are **optically active**, meaning that they rotate the plane of polarization of polarized light; achiral compounds are **optically inactive**. If one enantiomer rotates the plane of polarization clockwise (+), its mirror image will rotate it the same amount counterclockwise (−). Each optically active compound has a characteristic **specific rotation**. A **racemic mixture** is optically inactive. A **meso compound** has two or more asymmetric centers and a plane of symmetry; it is an achiral molecule. A compound with the same four groups bonded to two different asymmetric centers will have three stereoisomers, a meso compound, and a pair of enantiomers.

When a reactant that does not have an asymmetric center forms a product with one asymmetric center, the product will be a racemic mixture. Catalytic hydrogenation adds both hydrogens to the same side of the reactant. An enzyme-catalyzed reaction forms only one stereoisomer; an enzyme typically catalyzes the reaction of only one stereoisomer.

PROBLEMS

30. Which of the following have an asymmetric center?

$CHBr_2Cl$ CH_2FCl CH_3CHCl_2 $CHFBrCl$ $CH_3CH_2CHClCH_3$

31. Disregarding stereoisomers, give the structures of all compounds with molecular formula C_5H_{10}. Which ones can exist as stereoisomers?

32. Draw all possible stereoisomers for each of the following compounds. Indicate if no stereoisomers are possible.

a. 2-bromo-4-methylpentane
b. 2-bromo-4-chloropentane
c. 3-heptene
d. 1-bromo-4-methylcyclohexane
e. 1-bromo-3-chlorocyclobutane
f. 2-iodopentane
g. 3,3-dimethylpentane
h. 3-chloro-1-butene

33. Name the following compounds using *R,S* and *E,Z* (Section 4.5) designations where necessary:

a. 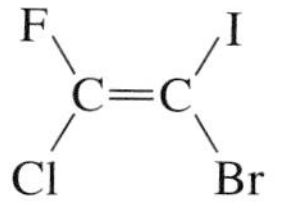 F, I, C=C, Cl, Br
b. 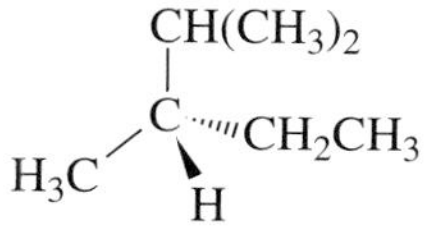 $CH(CH_3)_2$, C, CH_2CH_3, H_3C, H
c. 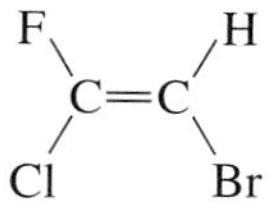 F, H, C=C, Cl, Br
d. 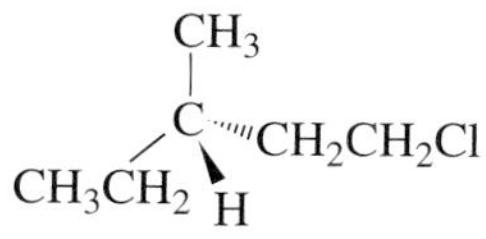 CH_3, C, CH_2CH_2Cl, CH_3CH_2, H

34. Mevacor is used clinically to lower serum cholesterol levels. How many asymmetric centers does it have?

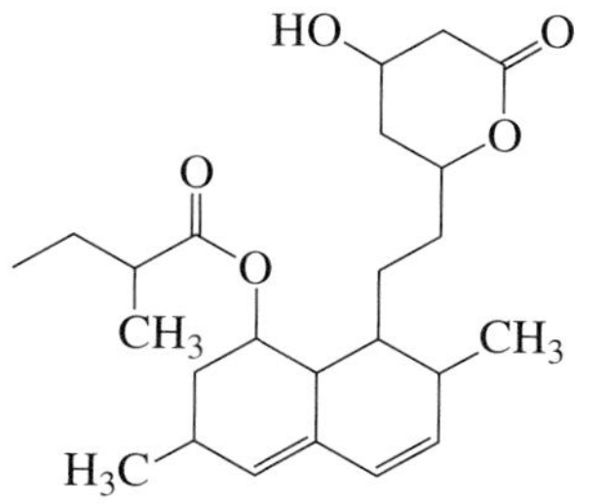

HO, O, O, O, O, CH_3, CH_3, H_3C

Mevacor®

35. Draw a diastereomer of the following:

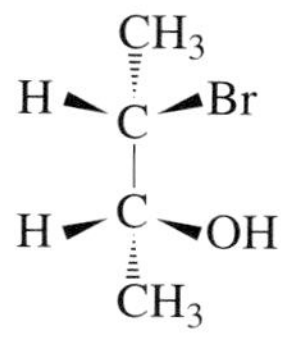

CH_3, H, C, Br, H, C, OH, CH_3

36. Indicate whether the following pairs of structures represent identical compounds, enantiomers, diastereomers, or constitutional isomers:

a. 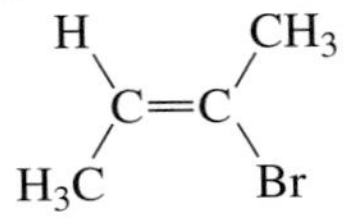 and

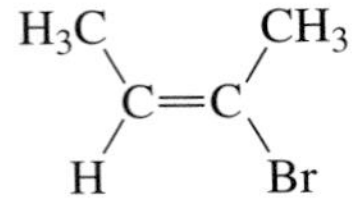

b. 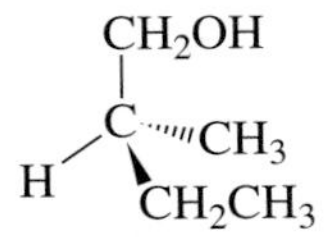 and

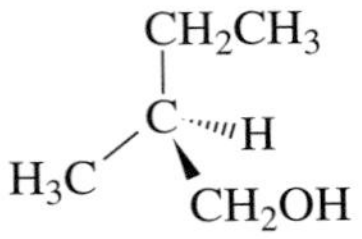

c. 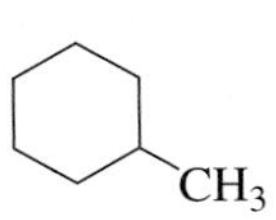 and

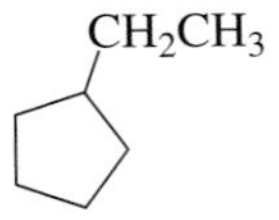

d. 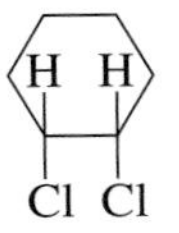 and

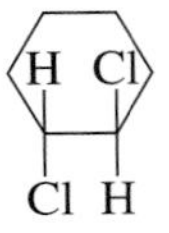

e. 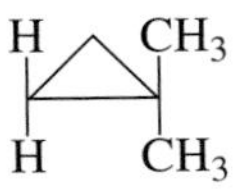 and

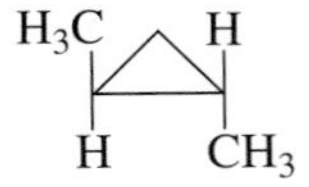

f. 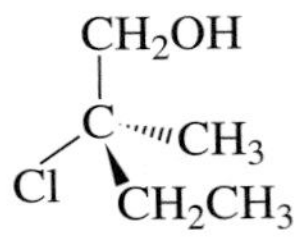 and

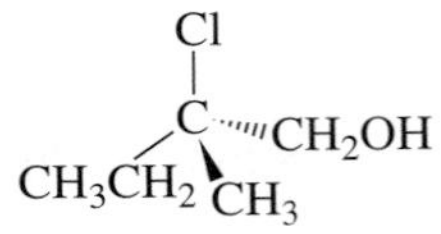

g. 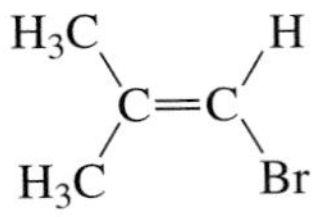 and

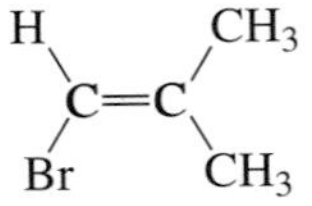

h. 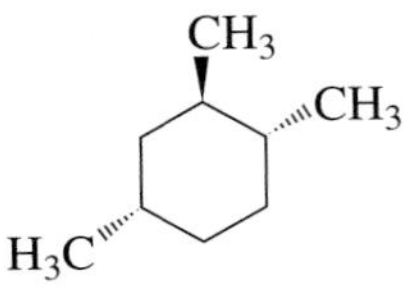 and

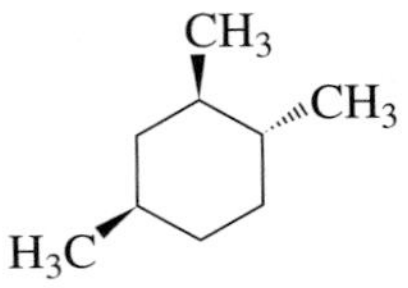

37. a. Give the product(s) that would be obtained from the reaction of *cis*-2-butene with each of the following reagents. If the products can exist as stereoisomers, show which stereoisomers are formed.

1. HCl **2.** H_2O + acid **3.** H_2, Pt/C **4.** CH_3OH + acid

b. How would the product(s) differ if the reactant had been *trans*-2-butene?

38. Which of the following compounds have an achiral stereoisomer?

a. 2,3-dichlorobutane **b.** 2,3-dichloropentane **c.** 2,4-dibromopentane **d.** 2,3-dibromopentane

39. Draw the stereoisomers of 2,4-dichlorohexane. Indicate pairs of enantiomers and pairs of diastereomers.

40. Give the products of the following reactions. If the products can exist as stereoisomers, show which stereoisomers are formed.

a. 1-butene + HCl **b.** *cis*-2-pentene + HCl **c.** 1-ethylcyclohexene + H_2O + acid

41. A solution of an unknown compound (3.0 g of the compound in 20 mL of solution), when placed in a polarimeter tube 2.0 dm long, was found to rotate the plane of polarization of polarized light 1.8° in a counterclockwise direction. What is the specific rotation of the compound?

42. Butaclamol is a potent antipsychotic that has been used clinically in the treatment of schizophrenia. How many asymmetric centers does it have?

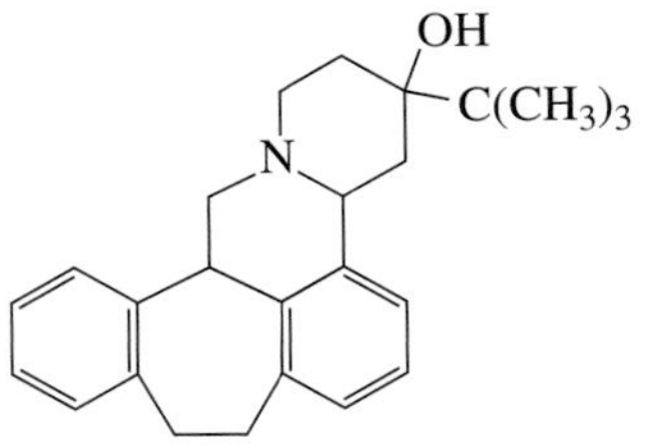

Butaclamol®

43. Which of the following objects are chiral?

a. a mug with DAD written on one side
b. a mug with MOM written on one side
c. a mug with DAD written opposite the handle
d. a mug with MOM written opposite the handle
e. an automobile
f. a wheelbarrow
g. a nail
h. a screw

44. Explain how *R* and *S* are related to (+) and (−).

45. Draw structures for each of the following:

a. (*S*)-1-bromo-1-chlorobutane **b.** two achiral isomers of 3,4,5-trimethylheptane

46. The stereoisomer of naproxen that is the active ingredient in Aleve and in several other over-the-counter nonsteroidal anti-inflammatory drugs is shown below. Is the active ingredient (*R*)-naproxen or (*S*)-naproxen?

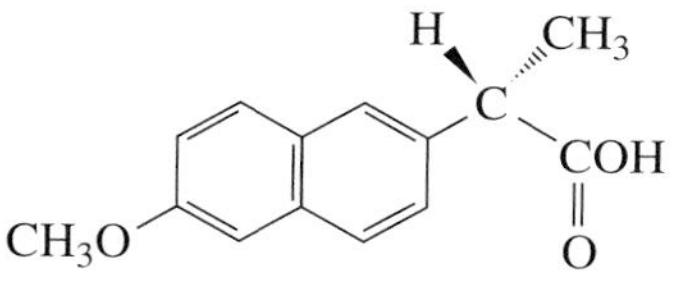

(?)-naproxen

47. Draw all the stereoisomers for each of the following:
a. 1-chloro-3-methylpentane **b.** 2,4-dichloroheptane **c.** 3,4-dichlorohexane

48. Indicate the configuration of the asymmetric centers in the following:

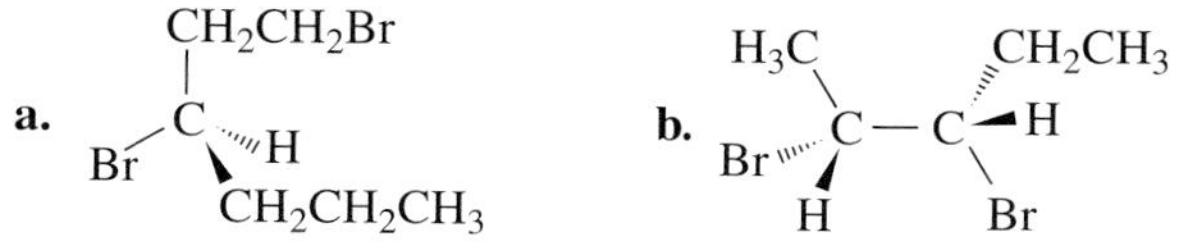

a. **b.**

49. Citrate synthase, one of the enzymes in the series of enzyme-catalyzed reactions known as the citric acid cycle (Section 18.7), catalyzes the synthesis of citric acid from oxaloacetic acid and acetyl-CoA. If the synthesis is carried out with acetyl-CoA that has radioactive carbon (^{14}C) in the indicated position, the isomer shown here is obtained. (Note: ^{14}C has a higher priority than ^{12}C.)

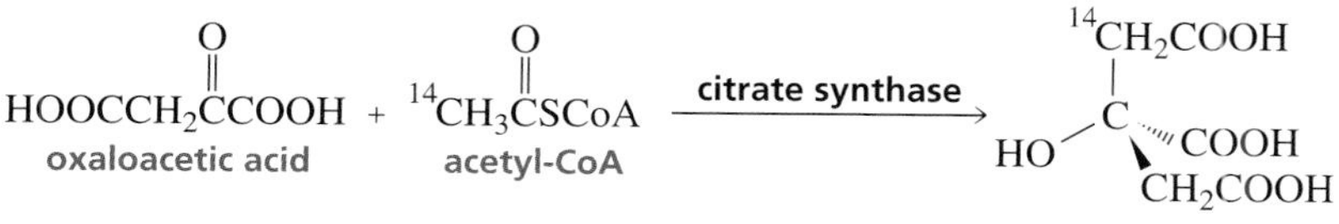

HOOCCH$_2$CCOOH (oxaloacetic acid) + ^{14}CH$_3$CSCoA (acetyl-CoA) —citrate synthase→ citric acid

a. Which stereoisomer of citric acid is synthesized, *R* or *S*?
b. Why is the other stereoisomer not obtained?
c. If the acetyl-CoA used in the synthesis does not contain ^{14}C, will the product of the reaction be chiral or achiral?

50. Chloramphenicol is a broad-spectrum antibiotic that is particularly useful against typhoid fever. What is the configuration of each of the asymmetric centers?

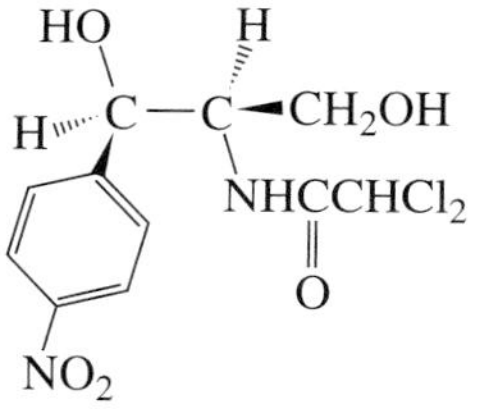

chloramphenicol

51. For many centuries, the Chinese have used extracts from a group of herbs known as ephedra to treat asthma. Chemists have been able to isolate a compound from these herbs, which they named ephedrine, a potent dilator of air passages in the lungs.

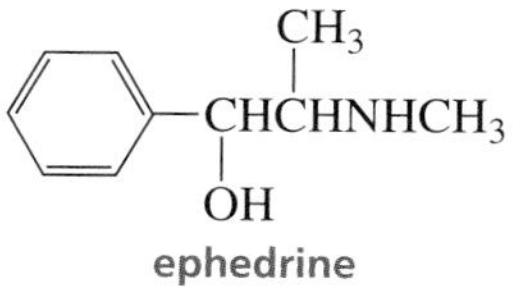

ephedrine

a. How many stereoisomers are possible for ephedrine?
b. The stereoisomer shown here is the one that is pharmacologically active. What is the configuration of each of the asymmetric centers?

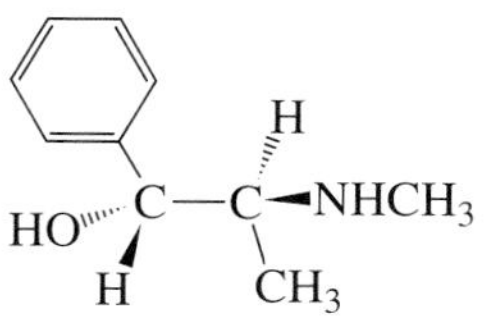

52. Indicate whether the following pairs of structures represent identical compounds or enantiomers:

a. 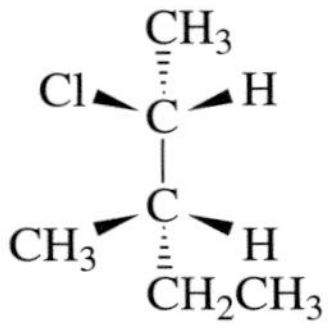 and 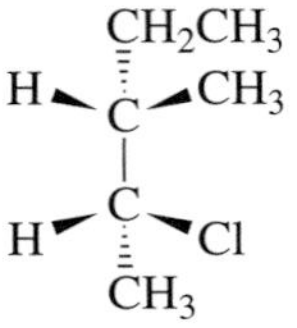 **b.** 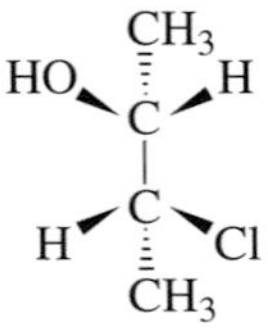 and

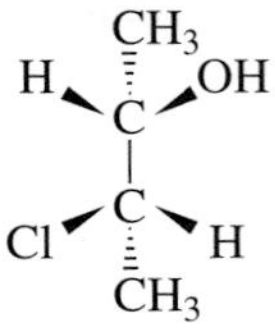

53. The following compound has only one asymmetric center. Why, then, does it have four stereoisomers?

$$CH_3CH{=}CHCHCH_3 \text{ (OH on C-4)}$$

54. Two stereoisomers are obtained from the reaction of HBr with (*S*)-4-bromo-1-pentene. One of the stereoisomers is optically active, and the other is not. Give the structures of the stereoisomers, indicating their configurations, and explain the difference in the optical properties.

55. Indicate whether the following pairs of structures represent identical compounds, enantiomers, diastereomers, or constitutional isomers:

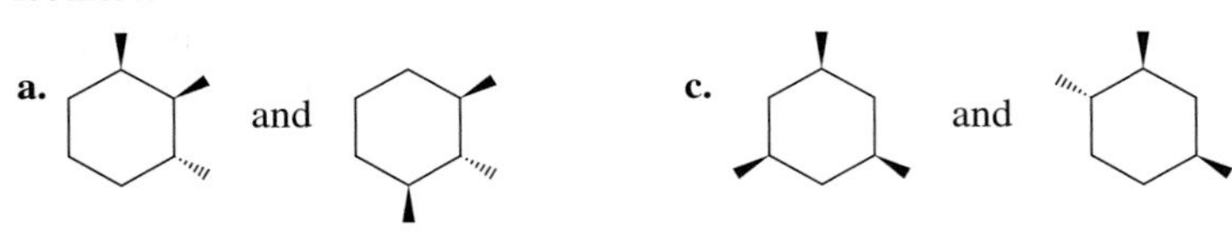

b. 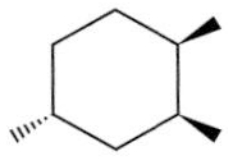 and **d.** 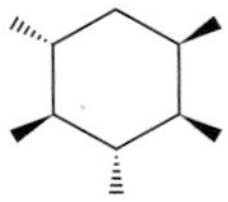 and

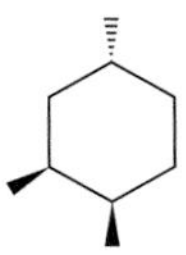

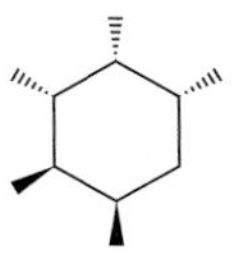

CHAPTER 7

Delocalized Electrons and Their Effect on Stability, Reactivity, and pK_a

Ultraviolet and Visible Spectroscopy

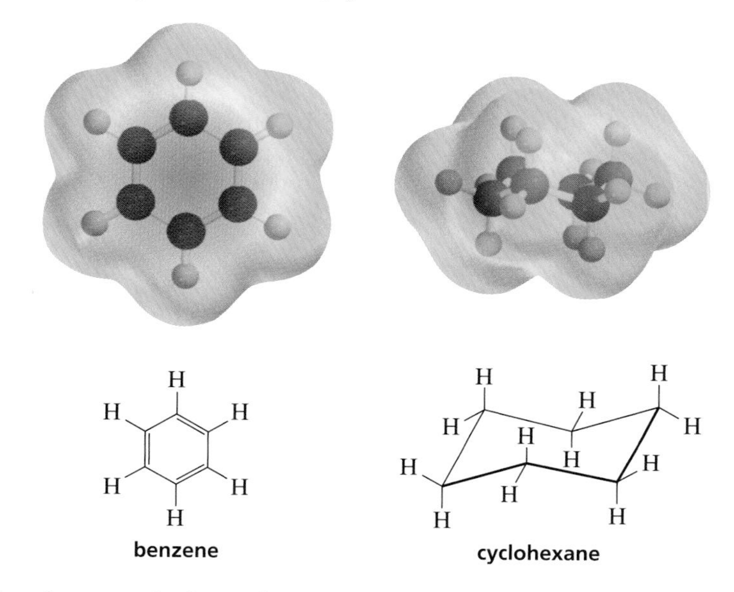

Electrons that are restricted to a particular region are called **localized electrons**. Localized electrons either belong to a single atom or are confined to a bond between two atoms.

$CH_3-\ddot{N}H_2$ — localized electrons

$CH_3-CH=CH_2$ — localized electrons

Not all electrons are confined to a single atom or bond. Many organic compounds contain *delocalized* electrons. **Delocalized electrons** neither belong to a single atom nor are confined to a bond between two atoms, but are shared by three or more atoms.

7.1 THE STRUCTURE OF BENZENE

To understand the concept of delocalized electrons, we will first look at the structure of benzene. Because they did not know about delocalized electrons, benzene's structure puzzled early organic chemists. They knew that benzene had a molecular formula of C_6H_6. They also knew that when a different atom is substituted for any one of the hydrogens of benzene, only one product is obtained and when the substituted product undergoes a second substitution, three products are obtained.

$$C_6H_6 \xrightarrow[\text{with an X}]{\text{replace a hydrogen}} C_6H_5X \xrightarrow[\text{with an X}]{\text{replace a hydrogen}} C_6H_4X_2 + C_6H_4X_2 + C_6H_4X_2$$

C_6H_5X: one monosubstituted compound

$C_6H_4X_2 + C_6H_4X_2 + C_6H_4X_2$: three disubstituted compounds

For every *two* hydrogens that are missing from the general molecular formula C_nH_{2n+2}, a hydrocarbon has either a π bond or a ring.

What kind of structure might we predict for benzene if we knew only what the early chemists knew? The molecular formula (C_6H_6) tells us that benzene has eight fewer hydrogens than a noncyclic alkane with six carbons ($C_nH_{2n+2} = C_6H_{14}$); thus, the number of rings and π bonds in benzene must total four (Section 4.1). Because only one product is obtained regardless of which of the hydrogens is replaced with another atom, we know that all the hydrogens must be identical. Two structures that fit these requirements are shown here.

$CH_3C{\equiv}C{-}C{\equiv}CCH_3$

shorter double bond

longer single bond

Neither of these structures is consistent with the observation that three compounds are formed if a second hydrogen is replaced with another atom. The noncyclic structure yields two disubstituted products.

$$CH_3C{\equiv}C{-}C{\equiv}CCH_3 \xrightarrow[\text{with Br's}]{\text{replace 2 H's}} CH_3C{\equiv}C{-}C{\equiv}CCHBr_2 \quad \text{and} \quad BrCH_2C{\equiv}C{-}C{\equiv}CCH_2Br$$

The cyclic structure, with alternating single and slightly shorter double bonds, yields four disubstituted products—a 1,3-disubstituted product, a 1,4-disubstituted product, and two 1,2-disubstituted products—because the two substituents can be placed either on two adjacent carbons joined by a single bond or on two adjacent carbons joined by a double bond.

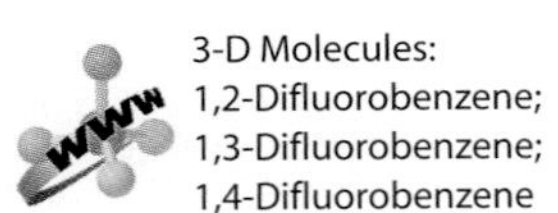

3-D Molecules: 1,2-Difluorobenzene; 1,3-Difluorobenzene; 1,4-Difluorobenzene

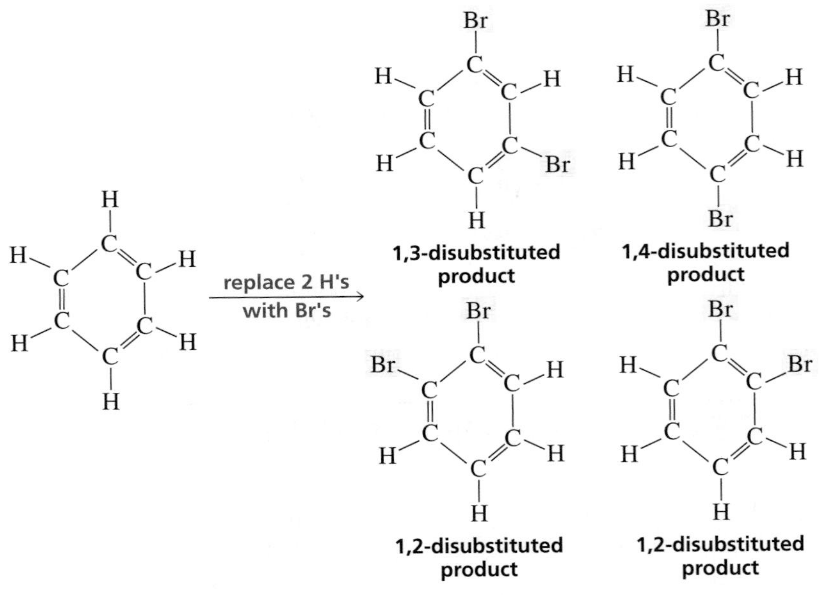

In 1865, the German chemist Friedrich Kekulé suggested a way of resolving this dilemma. He proposed that benzene was not a single compound, but a mixture of two compounds in rapid equilibrium.

rapid
equilibrium

Kekulé structures of benzene

Kekulé's proposal explained why only three disubstituted products are obtained when a monosubstituted benzene undergoes a second substitution. According to Kekulé, there actually *are* four disubstituted products, but the two 1,2-disubstituted products interconvert too rapidly to be separated from each other.

Br Br $\underset{\text{equilibrium}}{\overset{\text{rapid}}{\rightleftharpoons}}$ Br Br

Controversy over the structure of benzene continued until the 1930s, when modern analytical techniques showed that *benzene is a planar molecule and the six carbon–carbon bonds have the same length.* The length of each carbon–carbon bond is 1.39 Å, which is shorter than a carbon–carbon single bond (1.54 Å) but longer than a carbon–carbon double bond (1.33 Å; Section 1.14). In other words, benzene does not have alternating single and double bonds.

If the carbon–carbon bonds all have the same length, they must also have the same number of electrons between the carbon atoms. This can be so, however, only if the π electrons of benzene are delocalized around the ring, rather than each pair of π electrons being localized between two carbon atoms. To better understand the concept of delocalized electrons, we will now take a close look at the bonding in benzene.

KEKULÉ'S DREAM

Friedrich August Kekulé von Stradonitz (1829–1896) was born in Germany. He entered the University of Giessen to study architecture, but switched to chemistry after taking a course in the subject. He was a professor of chemistry at the University of Heidelberg, at the University of Ghent in Belgium, and then at the University of Bonn. In 1890, he gave an extemporaneous speech at the twenty-fifth-anniversary celebration of his first paper on the cyclic structure of benzene. In this speech, he claimed that he had arrived at the Kekulé structures as a result of dozing off in front of a fire while working on a textbook. He dreamed of chains of carbon atoms twisting and turning in a snakelike motion, when suddenly the head of one snake seized hold of its own tail and formed a spinning ring. Emperor William II of Germany made Kekulé a nobleman in 1895. This allowed him to add "von Stradonitz" to his name. Kekulé's students received three of the first five Nobel Prizes in chemistry.

Friedrich August Kekulé von Stradonitz

7.2 BENZENE HAS DELOCALIZED ELECTRONS

Each of benzene's six carbon atoms is sp^2 hybridized. An sp^2 carbon has bond angles of 120°—identical to the size of the angles of a planar hexagon. Thus, benzene is a planar molecule. Each of the carbons in benzene uses two sp^2 orbitals to bond to two other carbons; the third sp^2 orbital of each carbon overlaps the s orbital of a hydrogen (Figure 7.1a). Each carbon also has a p orbital at right angles to the sp^2 orbitals (Section 1.8). Because benzene is planar, the six p orbitals are parallel (Figure 7.1b). The p orbitals are close enough for side-to-side overlap, so each p orbital overlaps the p orbitals on *both* adjacent carbons. As a result, the overlapping p orbitals form a continuous doughnut-shaped cloud of electrons above, and another doughnut-shaped cloud of electrons below, the plane of the benzene ring (Figure 7.1c). The electrostatic potential map (Figure 7.1d) shows that all the carbon–carbon bonds have the same electron density.

a. b. c. d.

an sp^2 orbital

an s orbital

▲ **Figure 7.1**
(a) The carbon–carbon and carbon–hydrogen σ bonds in benzene.
(b) The p orbital on each carbon of benzene can overlap with two adjacent p orbitals.
(c) The clouds of π electrons above and below the plane of the benzene ring.
(d) The electrostatic potential map for benzene.

3-D Molecule: Benzene

Each of the six π electrons, therefore, is localized neither on a single carbon nor in a bond between two carbons (as in an alkene). Instead, each π electron is shared by all six carbons. In other words, six π electrons are delocalized—they roam freely within the doughnut-shaped clouds that lie over and under the ring of carbon atoms. Benzene is often drawn as a hexagon containing either dashed lines or a circle to symbolize the six delocalized π electrons.

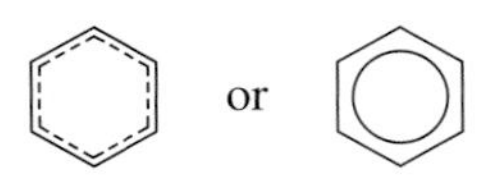

This type of representation makes it clear that there are no double bonds in benzene. We see now that Kekulé's structure was very nearly correct. The actual structure of benzene is a Kekulé structure with delocalized electrons. Be sure not to confuse planar benzene with chair-shaped cyclohexane (see the structures on page 169).

7.3 RESONANCE CONTRIBUTORS AND THE RESONANCE HYBRID

A disadvantage of using dashed lines (or a circle) to represent delocalized electrons is that those lines do not tell us how many π electrons they represent. For example, the dashed lines inside the hexagon indicate that the π electrons are shared equally by all six carbons of benzene and that all the carbon–carbon bonds have the same length, but they do not show how many π electrons are in the ring. Consequently, chemists prefer to use structures that portray the electrons as localized (and show the numbers of electrons), even though the electrons in the compound's actual structure are delocalized. The *approximate* structure with localized electrons is called a **resonance contributor**. The *actual* structure with delocalized electrons is called a **resonance hybrid**. Notice that it is easy to see that there are six π electrons in the ring of each resonance contributor.

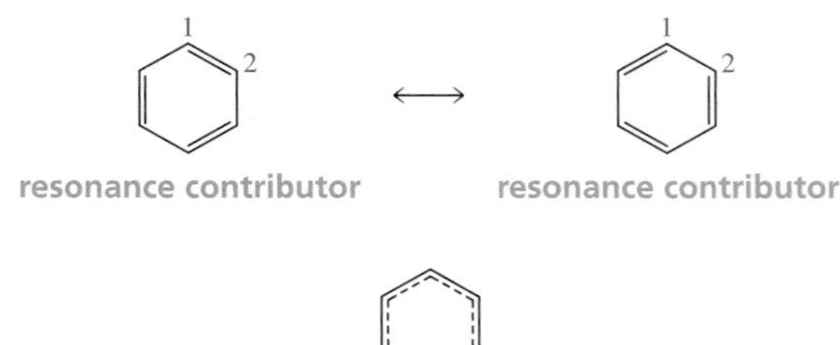

Resonance contributors are shown with a double-headed arrow between them. The double-headed arrow does *not* mean that the structures are in equilibrium with one another. Rather, it indicates that the actual structure lies somewhere between the structures of the resonance contributors. Resonance structures are merely a convenient way to show the π electrons; they do not depict any real electron distribution.

Electron delocalization is shown by double-headed arrows ($\longleftrightarrow$). Equilibrium is shown by two arrows pointing in opposite directions ($\rightleftharpoons$).

The following analogy illustrates the difference between resonance contributors and the resonance hybrid. Imagine that you are trying to describe to a friend what a rhinoceros looks like. You might tell your friend that a rhinoceros looks like a cross between a unicorn and a dragon. Like resonance contributors, the unicorn and the dragon do not really exist. Furthermore, like resonance contributors, they are not in equilibrium: a rhinoceros does not change back and forth between the two forms, looking like a unicorn one instant and a dragon the next. The unicorn and the dragon are simply ways to represent what the actual structure—the rhinoceros—looks like. *Resonance contributors, like unicorns and dragons, are imaginary, not real. Only the resonance hybrid, like the rhinoceros, is real.*

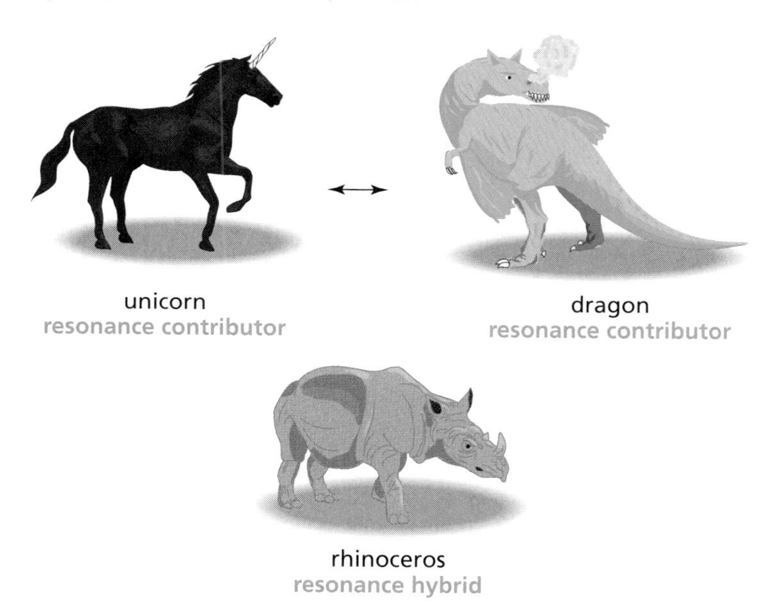

Electron delocalization is most effective if all the atoms sharing the delocalized electrons lie in or close to the same plane, so that their *p* orbitals can maximally overlap. For example, cyclooctatetraene is not planar; its sp^2 carbons have bond angles of 120°, whereas a planar eight-member ring would have bond angles of 135°. Because the ring is not planar, a *p* orbital can overlap with one adjacent *p* orbital, but it can have little or no overlap with the other adjacent *p* orbital. Therefore, the eight π electrons are not delocalized over the entire cyclooctatetraene ring, and its carbon–carbon bonds do not all have the same length.

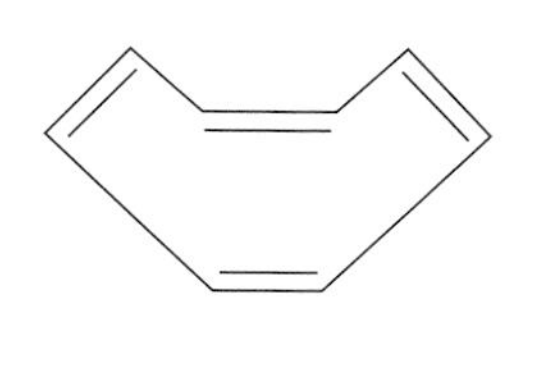

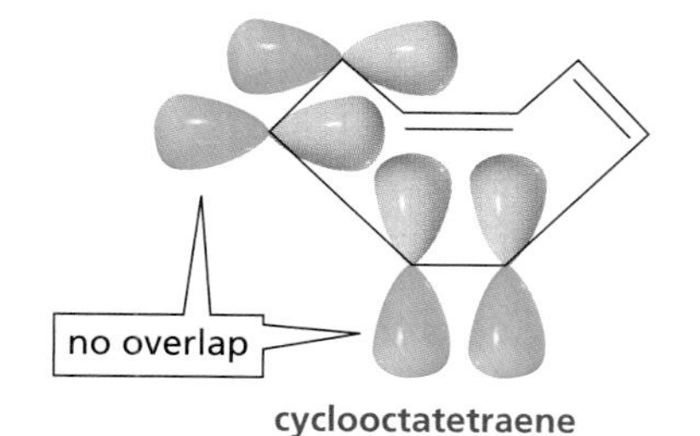

cyclooctatetraene

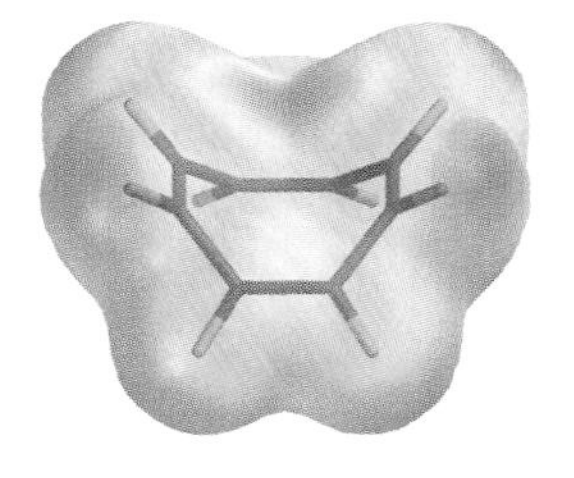

7.4 HOW TO DRAW RESONANCE CONTRIBUTORS

We have seen that an organic compound with delocalized electrons is generally represented as a structure with localized electrons to let us know how many π electrons are present in the molecule. For example, nitroethane is represented as having a nitrogen–oxygen double bond and a nitrogen–oxygen single bond.

$$CH_3CH_2-\overset{+}{N}(=\ddot{O}:)-\ddot{O}:^-$$

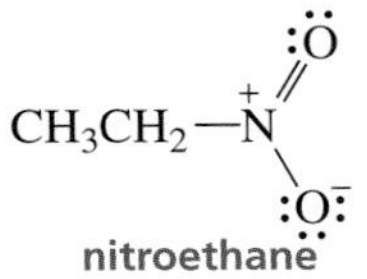

nitroethane

However, the two nitrogen–oxygen bonds in nitroethane are actually identical; they each have the same bond length. A more accurate description of the molecule's structure is obtained by drawing the two resonance contributors. Both resonance contributors show the compound with a nitrogen–oxygen double bond and a nitrogen–oxygen single bond, but they indicate that the electrons are delocalized by depicting the double bond in one contributor as a single bond in the other.

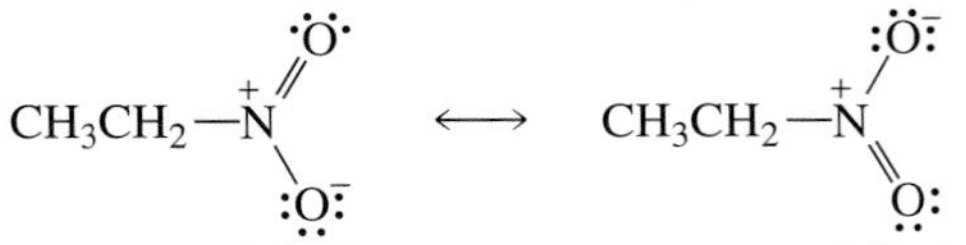

$$CH_3CH_2-\overset{+}{N}(=\ddot{O}:)-\ddot{O}:^- \longleftrightarrow CH_3CH_2-\overset{+}{N}(-\ddot{O}:^-)=\ddot{O}:$$

resonance contributor resonance contributor

Delocalized electrons result from a *p* orbital overlapping the *p* orbitals of more than one adjacent atom.

The resonance hybrid, in contrast, shows that the *p* orbital of nitrogen overlaps the *p* orbital of each oxygen. In other words, it shows that the two π electrons are shared by three atoms. The resonance hybrid also shows that the two nitrogen–oxygen bonds are identical and that the negative charge is shared equally by both oxygen atoms. Thus, we need to visualize and mentally average both resonance contributors to appreciate what the actual molecule—the resonance hybrid—looks like.

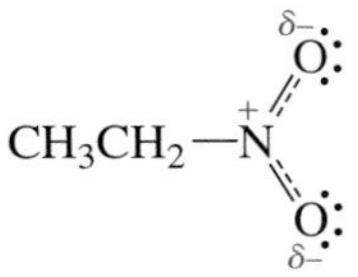

$$CH_3CH_2-\overset{+}{N}(\cdots\overset{\delta-}{\ddot{O}:})\cdots\underset{\delta-}{\ddot{O}:}$$

resonance hybrid

Rules for Drawing Resonance Contributors

To draw a set of resonance contributors, we first draw a Lewis structure (Section 1.4) for the molecule—it becomes our first resonance contributor—and then we move the electrons, following the rules listed below, to generate the next resonance contributor.

1. Only electrons move. Atoms never move.
2. Only π electrons (electrons in π bonds) and nonbonding electrons can move; σ electrons never move.
3. The total number of electrons in the molecule does not change. Therefore, each of the resonance contributors for a particular compound must have the same net charge. If one has a net charge of zero, all the others must also have net charges of zero. (A net charge of zero does not necessarily mean that there is no charge on any of the atoms: a molecule with a positive charge on one atom and a negative charge on another atom has a net charge of zero.)

To draw resonance contributors, move only π electrons or lone-pair electrons toward an sp^2 carbon.

Notice, as you study the following resonance contributors and practice drawing them yourself, that the electrons (π electrons or lone pairs) are always moved toward an sp^2 carbon. Remember that an sp^2 carbon is either a positively charged carbon or a double-bonded carbon (Sections 1.8 and 1.10). Electrons cannot be moved toward an sp^3 carbon because an sp^3 carbon has a complete octet so it cannot accommodate any more electrons.

The following carbocation has delocalized electrons. To draw its resonance contributor, we *move the π electrons toward an* sp^2 *carbon*. A curved arrow can help you decide how to draw the second contributor. Remember that the tail of the curved arrow shows where the electrons start from and the head of the arrow shows where the electrons are going. The resonance hybrid shows that the positive charge is shared by two carbons.

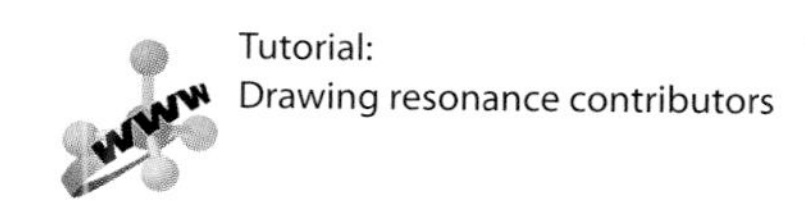

an sp^2 carbon

$$CH_3CH{=}CH{-}\overset{+}{C}HCH_3 \longleftrightarrow CH_3\overset{+}{C}H{-}CH{=}CHCH_3$$

resonance contributors

$$CH_3\overset{\delta+}{C}H{\overset{\cdots}{=}}CH{\overset{\cdots}{=}}\overset{\delta+}{C}HCH_3$$

resonance hybrid

Let's compare this carbocation with a similar compound in which all the electrons are localized. The π electrons in the compound shown below cannot move because the carbon they would move toward is an sp^3 carbon; sp^3 *carbons cannot accept electrons.*

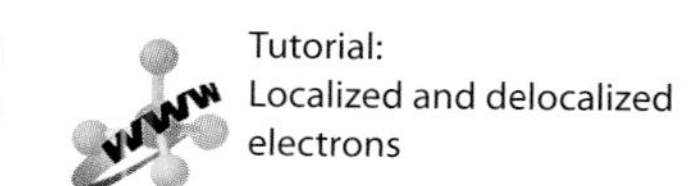

an sp^3 carbon cannot accept electrons

$$CH_2{=}CH{-}CH_2\overset{+}{C}HCH_3$$

localized electrons

In the next example, *π electrons again move toward* a positive charge. The resonance hybrid shows that the positive charge is shared by three carbons.

an sp^2 carbon

$$CH_3CH{=}CH{-}CH{=}CH{-}\overset{+}{C}H_2 \longleftrightarrow CH_3CH{=}CH{-}\overset{+}{C}H{-}CH{=}CH_2 \longleftrightarrow CH_3\overset{+}{C}H{-}CH{=}CH{-}CH{=}CH_2$$

resonance contributors

$$CH_3\overset{\delta+}{C}H{\overset{\cdots}{=}}CH{\overset{\cdots}{=}}\overset{\delta+}{C}H{\overset{\cdots}{=}}CH{\overset{\cdots}{=}}\overset{\delta+}{C}H_2$$

resonance hybrid

The resonance contributor for the next compound is obtained by *moving lone-pair electrons toward an* sp^2 *carbon*. The sp^2 carbon can accommodate the new electrons by breaking a π bond. The lone-pair electrons in the example, on the right below, are not delocalized because they would have to move toward an sp^3 carbon.

$$R{-}C({=}\ddot{O}{:}){-}\ddot{N}H_2 \longleftrightarrow R{-}C({-}{:}\ddot{O}{:}^-){=}\overset{+}{N}H_2$$

lone-pair electrons

an sp^2 carbon

resonance contributors

$$R{-}C({\overset{\cdots}{=}}O^{\delta-}){\overset{\cdots}{=}}\overset{\delta+}{N}H_2$$

resonance hybrid

$$R{-}C({=}O){-}CH_2{-}\ddot{N}H_2$$

an sp^3 carbon cannot accept electrons

For additional practice drawing resonance contributors, see Special Topics II in the Study Guide and Solutions Manual.

The following resonance contributors are obtained by moving π electrons toward an sp^2 carbon. Notice that the electrons move toward (not away from) the most electronegative atom (the oxygen).

$$CH_3\overset{\overset{\large\ddot{O}:}{\|}}{C}-CH=CH_2 \longleftrightarrow CH_3\overset{\overset{:\ddot{O}:^-}{|}}{C}=CH-\overset{+}{C}H_2$$

$$CH_3\overset{\overset{\ddot{O}:^{\delta-}}{\|}}{C}\text{=-}CH\text{=-}\overset{\delta+}{C}H_2$$

resonance hybrid

The only time you move electrons away from the most electronegative atom in order to arrive at a resonance contributor is when that is the only way electrons can be moved. In other words, movement of electrons away from the most electronegative atom is better than no movement at all, because electron delocalization makes a molecule more stable (as you will see in Section 7.5).

$$CH_2=CH-\ddot{O}CH_3 \longleftrightarrow \bar{\ddot{C}}H_2-CH=\overset{+}{\ddot{O}}CH_3$$

Radicals can also have delocalized electrons. The resonance contributors are obtained by moving single electrons toward sp^2 carbons.

$$CH_3-CH=CH-\dot{C}H_2 \longleftrightarrow CH_3-\dot{C}H-CH=CH_3$$

resonance contributors

$$CH_3-\overset{\delta\cdot}{C}H\text{=-}CH\text{=-}\overset{\delta\cdot}{C}H_2$$

resonance hybrid

PEPTIDE BONDS

Every third bond in a protein is a peptide bond. A resonance contributor can be drawn for a peptide bond by moving the lone pair on nitrogen toward the sp^2 carbon.

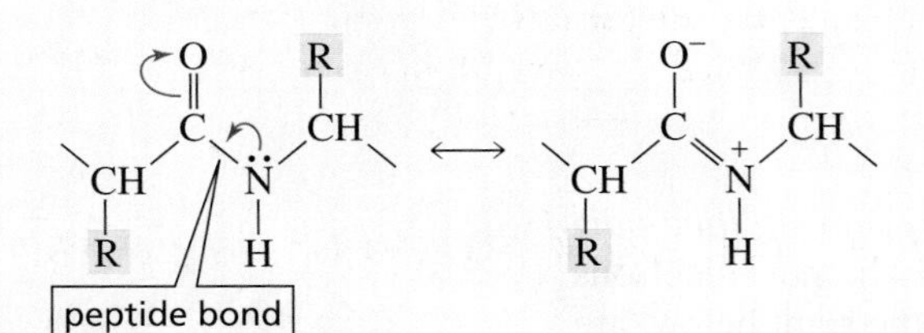

Because of the partial double bond character of the peptide bond, the carbon and nitrogen atoms and the two atoms bonded to each are held rigidly in a plane, as represented below by the blue and green boxes (Section 4.3). This planarity affects the way proteins fold, so it has important implications for the three-dimensional shape of these biological molecules (Section 16.10).

a segment of a protein

PROBLEM 1 ♦

Which of the following compounds have delocalized electrons?

a. 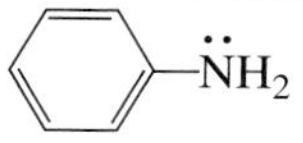 (benzene ring)–$\ddot{N}H_2$

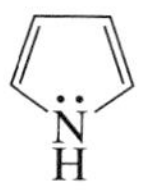

b. 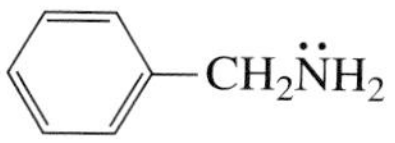 (benzene ring)–$CH_2\ddot{N}H_2$

c. (ring with $\ddot{O}$)

d. $CH_2{=}CHCH_2CH{=}CH_2$

e. (pyrrole ring, $\ddot{N}$–H)

f. $CH_3CH{=}CHCH{=}CH\overset{+}{C}H_2$

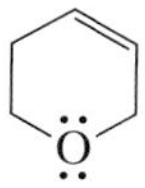

g. 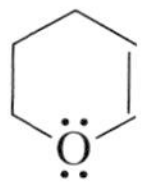 (ring with $\ddot{O}$)

h. $CH_3CH_2\ddot{N}HCH_2CH{=}CH_2$

PROBLEM 2

For the compounds in Problem 1 that have delocalized electrons, draw their resonance contributors.

7.5 THE PREDICTED STABILITIES OF RESONANCE CONTRIBUTORS

All resonance contributors do not necessarily contribute equally to the resonance hybrid. The degree to which each resonance contributor contributes depends on its predicted stability. Because resonance contributors are not real, their stabilities cannot be measured. Therefore, the stabilities of resonance contributors have to be predicted based on molecular features found in real molecules. *The greater the predicted stability of the resonance contributor, the more it contributes to the structure of the resonance hybrid; and the more it contributes to the structure of the resonance hybrid, the more similar the contributor is to the real molecule*. The examples that follow illustrate these points.

The greater the predicted stability of the resonance contributor, the more it contributes to the structure of the resonance hybrid.

The two resonance contributors for a carboxylic acid are shown below, labeled **A** and **B**. Structure **B** has two features that make it less stable than structure **A**: one of its oxygen atoms has a positive charge—not a comfortable situation for an electronegative atom—and the structure has separated charges. A molecule with **separated charges** has a positive charge and a negative charge that can be neutralized by the movement of electrons. Resonance contributors with separated charges are relatively unstable (relatively high in energy) because energy is required to keep the opposite charges separated. Structure **A**, therefore, is predicted to be more stable than structure **B**. Consequently, **A** makes a greater contribution to the resonance hybrid, so the resonance hybrid looks more like **A** than like **B**.

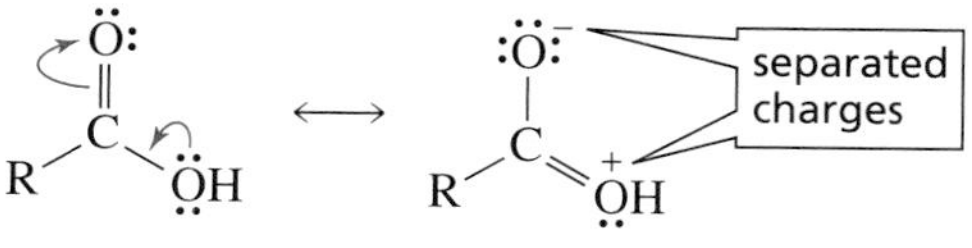

separated charges

A ⟷ B

a carboxylic acid

The two resonance contributors for a carboxylate ion are shown next.

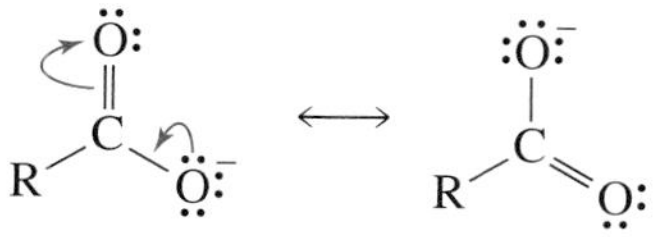

C ⟷ D

a carboxylate ion

Structures **C** and **D** are predicted to be equally stable and therefore are expected to contribute equally to the resonance hybrid.

Let's now see which of the two resonance contributors shown below has a greater predicted stability. Structure **E** has a negative charge on carbon, whereas structure **F** has a negative charge on oxygen. Oxygen is more electronegative than carbon, so oxygen can better accommodate the negative charge. Consequently, structure **F** is predicted to be more stable than structure **E**. The resonance hybrid, therefore, more closely resembles **F**; that is, the resonance hybrid has a greater concentration of negative charge on oxygen than on carbon.

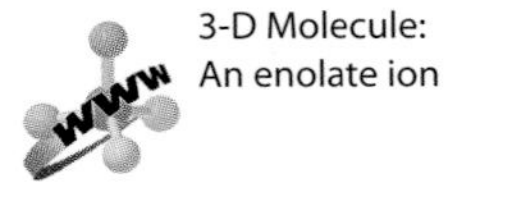

3-D Molecule: An enolate ion

$$\underset{\mathbf{E}}{R{-}\overset{\overset{:\ddot{O}}{\|}}{C}{-}\ddot{C}^{-}HCH_3} \longleftrightarrow \underset{\mathbf{F}}{R{-}\overset{\overset{:\ddot{O}:^{-}}{|}}{C}{=}CHCH_3}$$

PROBLEM 3 *SOLVED*

Draw resonance contributors for each of the following species, and rank the contributors in order of decreasing contribution to the hybrid:

a. $CH_3\overset{+}{C}(CH_3){-}CH{=}CHCH_3$ **b.** (cyclohexanone enolate: carbanion ring carbon adjacent to $C{=}\ddot{O}:$) **c.** $CH_3\overset{+}{C}H{-}CH{=}CHCH_3$

Solution to 3a Structure **A** is more stable than structure **B** because the positive charge is on a tertiary carbon in **A** and on a secondary carbon in **B**. (Recall that tertiary carbocations are more stable than secondary carbocations: Section 4.4)

$$\underset{\mathbf{A}}{CH_3\overset{+}{C}(CH_3){-}CH{=}CHCH_3} \longleftrightarrow \underset{\mathbf{B}}{CH_3C(CH_3){=}CH{-}\overset{+}{C}HCH_3}$$

PROBLEM 4 ◆

Draw the resonance hybrid for each of the species in Problem 3.

7.6 DELOCALIZATION ENERGY IS THE ADDITIONAL STABILITY DELOCALIZED ELECTRONS GIVE TO A COMPOUND

Delocalization energy indicates how much more stable a compound with delocalized electrons is than it would be if its electrons were localized.

The resonance hybrid is more stable than any of its resonance contributors is predicted to be.

Delocalized electrons stabilize a compound. The extra stability a compound gains from having delocalized electrons is called **delocalization energy**. Electron delocalization is also called resonance, so delocalization energy is also called **resonance energy**. Knowing that delocalized electrons increase the stability of a molecule, we can conclude that *a resonance hybrid is more stable than any of its resonance contributors is predicted to be.*

The delocalization energy associated with a compound that has delocalized electrons depends on the number *and* predicted stability of the resonance contributors: *the greater the number of relatively stable resonance contributors, the greater is the delocalization energy.* For example, the delocalization energy of a carboxylate ion with two relatively stable resonance contributors is significantly greater than the delocalization of a carboxylic acid with only one relatively stable resonance contributor.

$RCOOH \rightleftharpoons RCOO^- + H^+$

relatively stable ⟷ relatively unstable

relatively stable ⟷ relatively stable

resonance contributors of a carboxylic acid

resonance contributors of a carboxylate ion

The greater the number of relatively stable resonance contributors, the greater is the delocalization energy.

The more nearly equivalent the resonance contributors are in structure, the greater the delocalization energy. The species shown below is particularly stable because it has three equivalent resonance contributors.

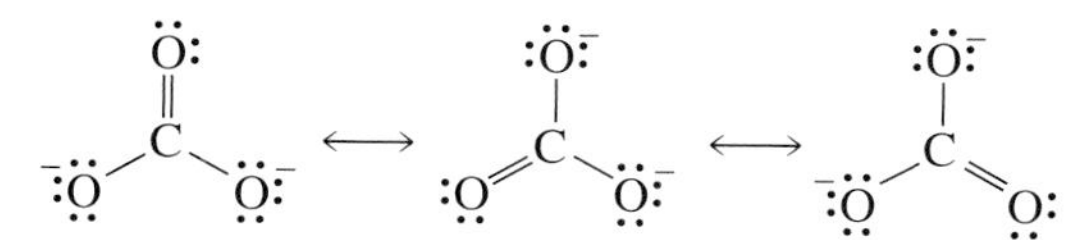

The more nearly equivalent the resonance contributors, the greater is the delocalization energy.

PROBLEM 5 ◆

a. Predict the relative bond lengths of the three carbon–oxygen bonds in CO_3^{2-}.
b. What would you expect the charge to be on each oxygen atom?

PROBLEM 6 ◆

Rank the following species in order of decreasing delocalization energy:

$HCOO^-$ $\quad$ $^-O{-}C({=}O){-}O^-$ $\quad$ $HCOOH$

PROBLEM 7 ◆

Which species has the greater delocalization energy?

$CH_2{=}CH{-}CH{=}CH_2$ or $H_3C{-}C({=}O){-}O^-$

7.7 EXAMPLES THAT ILLUSTRATE HOW DELOCALIZED ELECTRONS AFFECT STABILITY

Dienes are hydrocarbons with two double bonds. **Isolated dienes** have isolated double bonds; **isolated double bonds** are separated by more than one single bond. **Conjugated dienes** have conjugated double bonds; **conjugated double bonds** are separated by one single bond.

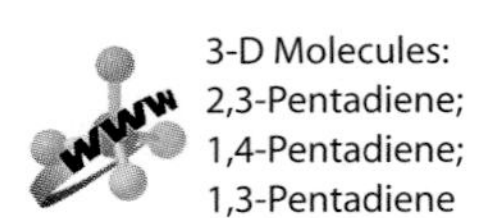

3-D Molecules: 2,3-Pentadiene; 1,4-Pentadiene; 1,3-Pentadiene

double bonds are separated by more than one single bond

$CH_2{=}CH{-}CH_2{-}CH{=}CH_2$
an isolated diene

double bonds are separated by one single bond

$CH_3CH{=}CH{-}CH{=}CHCH_3$
a conjugated diene

The π electrons in each of the double bonds of an isolated diene are *localized* between two carbons. In contrast, the π electrons in a conjugated diene are *delocalized*. As we discovered in Section 7.6, electron delocalization stabilizes a molecule. Therefore, conjugated dienes are more stable than isolated dienes. (Notice that because the compound does not have an electronegative atom that would determine the direction in which the electrons move, they can move both to the left and to the right.)

Electron delocalization stabilizes a molecule.

$$\overset{-}{C}H_2{-}CH{=}CH{-}\overset{+}{C}H_2 \longleftrightarrow CH_2{=}CH{-}CH{=}CH_2 \longleftrightarrow \overset{+}{C}H_2{-}CH{=}CH{-}\overset{-}{C}H_2$$

resonance contributors

delocalized electrons

$CH_2{\overset{\cdots}{=}}CH{\overset{\cdots}{=}}CH{\overset{\cdots}{=}}CH_2$

resonance hybrid

DELOCALIZED ELECTRONS IN VISION

We have seen that the interconversion of a double bond in rhodopsin between the cis and trans forms plays an important role in vision (Section 4.4). In the first step of this process, a π bond breaks to form a diradical. The energy required to break this bond depends on the stability of the diradical. Stabilization of the diradical by electron delocalization makes it easier to break the π bond.

N—opsin
rhodopsin

$\longrightarrow$

a diradical

N—opsin
rhodopsin

PROBLEM 8 ♦

Draw the resonance contributors for the following radical:

$$CH_2{=}CH{-}\dot{C}H{-}CH{=}CH_2$$

PROBLEM 9 ♦

How many carbon atoms share the unpaired electrons in rhodopsin?

An **allylic cation** has a positive charge on an allylic carbon (a carbon adjacent to an sp^2 carbon of an alkene). A **benzylic cation** has a positive charge on a benzylic carbon (a carbon adjacent to an sp^2 carbon of a benzene ring).

an allylic carbon

$CH_2{=}CH\overset{+}{C}HR$

an allylic cation

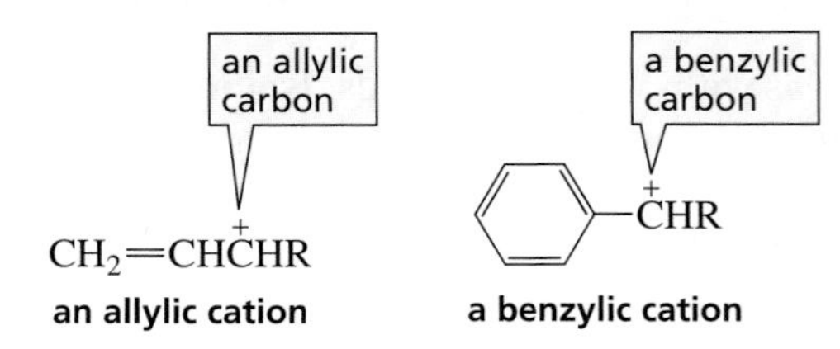

a benzylic cation

An allylic cation has two resonance contributors. The positive charge is not localized on a single carbon, but is shared by two carbons.

$$\mathrm{RCH{=}CH{-}\overset{+}{C}H_2 \longleftrightarrow R\overset{+}{C}H{-}CH{=}CH_2}$$

an allylic cation

A benzylic cation has five resonance contributors. Notice that the positive charge is shared by four carbons.

a benzylic cation

Because primary allylic and benzylic cations have delocalized electrons, they are more stable than other *primary* carbocations with localized electrons. (Indeed, they have about the same stability as secondary carbocations.) We can add these cations to the list of carbocations whose relative stabilities were shown in Section 5.2.

Allylic and benzylic cations are more stable than other primary carbocations.

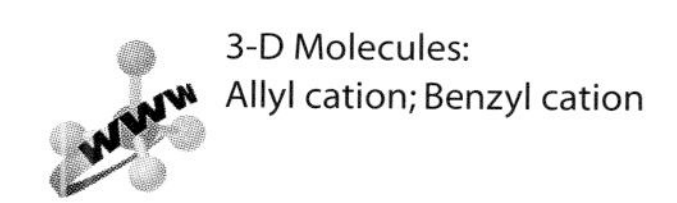

relative stabilities of carbocations

most stable → a tertiary carbocation > a primary benzylic cation ≈ a primary allylic cation ($\mathrm{CH_2{=}CH\overset{+}{C}H_2}$) ≈ a secondary carbocation > a primary carbocation > methyl cation > vinyl cation ($\mathrm{CH_2{=}\overset{+}{C}H}$) ← least stable

Not all allylic and benzylic cations have the same stability. Just as a tertiary carbocation is more stable than a secondary carbocation, a tertiary allylic cation is more stable than a secondary allylic cation, which in turn is more stable than a primary allylic cation. Similarly, a tertiary benzylic cation is more stable than a secondary benzylic cation, which is more stable than a primary benzylic cation.

relative stabilities

most stable → tertiary allylic cation > secondary allylic cation > primary allylic cation

most stable → tertiary benzylic cation > secondary benzylic cation > primary benzylic cation

Notice that it is the *primary* allylic and *primary* benzylic cations that have about the same stability as *secondary* carbocations. Secondary benzylic and allylic cations, as well as tertiary benzylic and allylic cations, are even more stable than primary benzylic and allylic cations.

PROBLEM-SOLVING STRATEGY

Which carbocation is more stable?

$$CH_3CH{=}CH{-}\overset{+}{C}H_2 \quad \text{or} \quad CH_3\overset{CH_3}{\overset{|}{C}}{=}CH{-}\overset{+}{C}H_2$$

Start by drawing the resonance contributors for each carbocation. Then compare their predicted stabilities.

$$CH_3CH{=}CH{-}\overset{+}{C}H_2 \longleftrightarrow CH_3\overset{+}{C}H{-}CH{=}CH_2 \qquad CH_3\overset{CH_3}{\overset{|}{C}}{=}CH{-}\overset{+}{C}H_2 \longleftrightarrow CH_3\underset{+}{\overset{CH_3}{\overset{|}{C}}}{-}CH{=}CH_2$$

Each carbocation has two resonance contributors. The positive charge of the carbocation on the left is shared by a primary allylic carbon and a secondary allylic carbon. The positive charge of the carbocation on the right is shared by a primary allylic carbon and a tertiary allylic carbon. Because a tertiary allylic carbon is more stable than a secondary allylic carbon, the carbocation on the right is more stable.

Now continue on to Problem 10.

PROBLEM 10 ♦

Which carbocation in each of the following pairs is more stable?

a. $CH_3CH{=}CH\overset{+}{C}HCH_3$ or $CH_3CH{=}CH\overset{+}{C}H_2$

b. 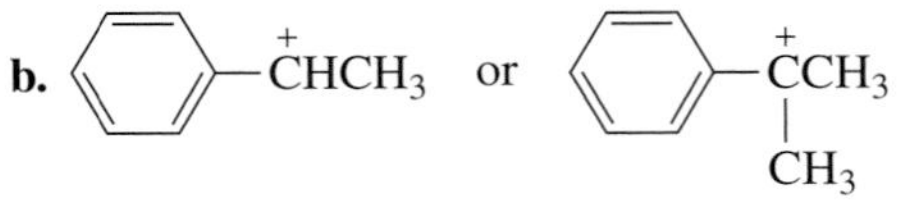 $C_6H_5{-}\overset{+}{C}HCH_3$ or $C_6H_5{-}\overset{+}{C}(CH_3)CH_3$

PROBLEM 11 ♦

Which species is more stable?

a. 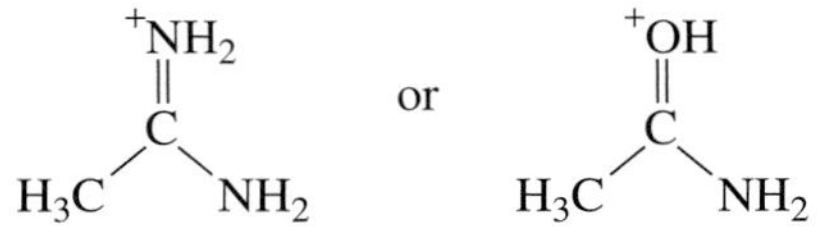 $H_3C{-}C({=}\overset{+}{N}H_2){-}NH_2$ or $H_3C{-}C({=}\overset{+}{O}H){-}NH_2$

b. $CH_3\overset{O^-}{\overset{|}{C}}HCH{=}CH_2$ or $CH_3\overset{O^-}{\overset{|}{C}}{=}CHCH_3$

7.8 EXAMPLES THAT ILLUSTRATE HOW DELOCALIZED ELECTRONS CAN AFFECT THE PRODUCT OF A REACTION

Our ability to predict the correct product of an organic reaction often depends on recognizing when organic molecules have delocalized electrons. For example, in the following reaction, both sp^2 carbons of the alkene are bonded to the same number of hydrogens:

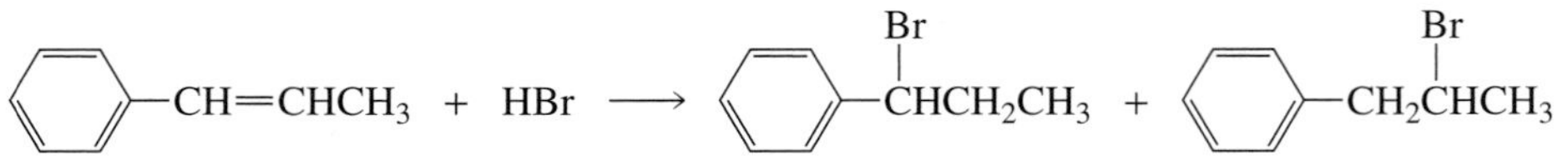

$$C_6H_5{-}CH{=}CHCH_3 + HBr \longrightarrow \underset{100\%}{C_6H_5{-}\overset{Br}{\overset{|}{C}}HCH_2CH_3} + \underset{0\%}{C_6H_5{-}CH_2\overset{Br}{\overset{|}{C}}HCH_3}$$

Therefore, the rule that tells us to add the electrophile to the sp^2 carbon bonded to the greater number of hydrogens predicts that approximately equal amounts of the two products will be formed. When the reaction is carried out, however, only one of the products is obtained. (In Chapter 8, you will see why the double bonds in the benzene ring do not undergo electrophilic addition reactions.)

The rule leads us to an incorrect prediction of the reaction product because it does not take electron delocalization into consideration. It presumes that both carbocation intermediates are equally stable since they are both secondary carbocations. The rule does not take into account the fact that one intermediate is a secondary carbocation and the other is a secondary benzylic cation. Because the secondary benzylic cation is more stable (this is, it is stabilized by electron delocalization), it is formed more readily. The difference in the rates of formation of the two carbocations is sufficient to cause only one to be obtained.

$C_6H_5\overset{+}{C}HCH_2CH_3$ — **a secondary benzylic cation**, more stable

$C_6H_5CH_2\overset{+}{C}HCH_3$ — **a secondary carbocation**

PROBLEM 12 *SOLVED*

Predict the sites on each of the following compounds where protonation can occur:

a. $CH_3CH{=}CHOCH_3 + H^+$

b. 1-cyclohexenyl–N-piperidine $+ H^+$

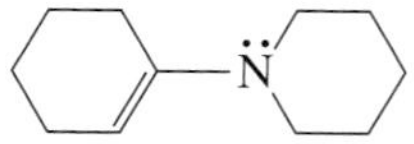

Solution to 12a The resonance contributors show that two sites can be protonated: the lone pair on oxygen and the lone pair on carbon.

$$CH_3CH{=}CH{-}\ddot{O}CH_3 \longleftrightarrow CH_3\overset{-}{\ddot{C}}H{-}CH{=}\overset{+}{O}CH_3$$

resonance contributors

sites of protonation: $CH_3CH{=}CH\ddot{O}CH_3$

Let's now compare the products formed when *isolated dienes* (dienes that have only localized electrons) undergo electrophilic addition reactions to the products formed when conjugated dienes (dienes that have delocalized electrons) undergo the same reactions.

Reactions of Isolated Dienes

The reactions of *isolated dienes* are like the reactions of alkenes. If an excess of the electrophilic reagent is present, two independent addition reactions will occur. In each reaction, *the electrophile adds to the sp^2 carbon that is bonded to the greater number of hydrogens*.

$$\underset{\text{1,5-hexadiene, an isolated diene}}{CH_2{=}CHCH_2CH_2CH{=}CH_2} + \underset{\text{excess}}{HBr} \longrightarrow CH_3\underset{Br}{C}HCH_2CH_2\underset{Br}{C}HCH_3$$

The reaction proceeds exactly as we would predict from our knowledge of the mechanism for the reaction of alkenes with electrophilic reagents.

mechanism for the reaction of an isolated diene with excess HBr

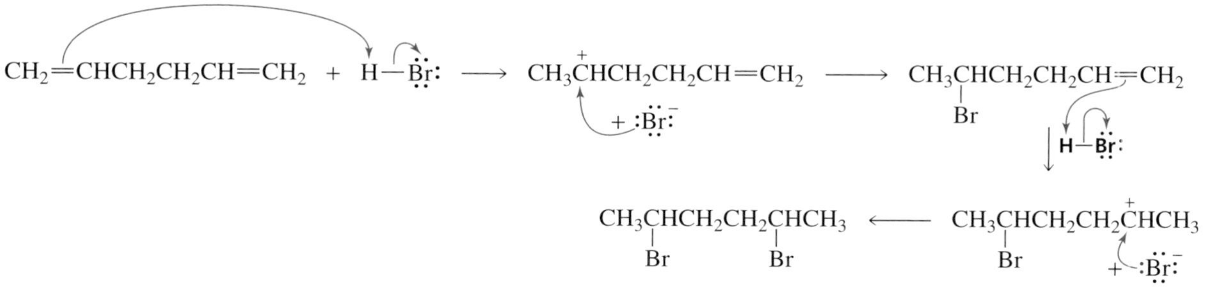

$$CH_2{=}CHCH_2CH_2CH{=}CH_2 + H{-}\ddot{B}r{:} \longrightarrow CH_3\overset{+}{C}HCH_2CH_2CH{=}CH_2 + {:}\ddot{B}r{:}^- \longrightarrow CH_3\underset{Br}{C}HCH_2CH_2CH{=}CH_2$$

$$CH_3\underset{Br}{C}HCH_2CH_2CH{=}CH_2 + H{-}\ddot{B}r{:} \longrightarrow CH_3\underset{Br}{C}HCH_2CH_2\overset{+}{C}HCH_3 + {:}\ddot{B}r{:}^- \longrightarrow CH_3\underset{Br}{C}HCH_2CH_2\underset{Br}{C}HCH_3$$

- The electrophile (H^+) adds to the electron-rich double bond in a manner that produces the more stable carbocation (Section 5.3).
- The bromide ion adds to the carbocation.
- Because there is an excess of the electrophilic reagent, there is enough reagent to add to the other double bond: the H^+ adds in a manner that produces the more stable carbocation.
- The bromide ion adds to the carbocation.

If there is only enough electrophilic reagent to add to one of the double bonds, it will add preferentially to the more reactive double bond. For example, in the reaction of 2-methyl-1,5-hexadiene with HCl, addition of HCl to the double bond on the left forms a secondary carbocation, whereas addition of HCl to the double bond on the right forms a tertiary carbocation. Because the transition state leading to formation of a tertiary carbocation is more stable than that leading to formation of a secondary carbocation, the tertiary carbocation is formed faster (Section 5.3). Therefore, in the presence of a limited amount of HCl, the major product of the reaction will be 5-chloro-5-methyl-1-hexene.

$$\underset{\substack{\textbf{2-methyl-1,5-hexadiene}\\ \textbf{1 mol}}}{CH_2{=}CHCH_2CH_2\overset{\overset{\displaystyle CH_3}{|}}{C}{=}CH_2} + \underset{\textbf{1 mol}}{HCl} \longrightarrow \underset{\substack{\textbf{5-chloro-5-methyl-1-hexene}\\ \textbf{major product}}}{CH_2{=}CHCH_2CH_2\overset{\overset{\displaystyle CH_3}{|}}{\underset{\underset{\displaystyle Cl}{|}}{C}}CH_3}$$

PROBLEM 13 ◆

Give the major product of each of the following reactions, assuming that one equivalent of each reagent is used in every reaction.

a. $CH_2{=}CHCH_2CH_2CH{=}\overset{\overset{\displaystyle CH_3}{|}}{C}CH_3 \xrightarrow{\text{HBr}}$

b. (1-methyl-1,3-cycloheptadiene structure) $\xrightarrow{\text{HCl}}$

PROBLEM 14 ◆

Which of the double bonds in zingiberene, the compound responsible for the odor of ginger, is the most reactive in an electrophilic addition reaction?

zingiberene

Reactions of Conjugated Dienes

When a diene with *conjugated double bonds*, such as 1,3-butadiene, reacts with a limited amount of electrophilic reagent so that addition can occur at only one of the double bonds, two addition products are formed. One is a **1,2-addition** product, which is the result of addition at the 1- and 2-positions. The other is a **1,4-addition** product, the result of addition at the 1- and 4-positions.

$$\underset{\substack{\textbf{1,3-butadiene}\\ \textbf{1 mol}}}{\overset{1}{C}H_2{=}\overset{2}{C}H{-}\overset{3}{C}H{=}\overset{4}{C}H_2} + \underset{\textbf{1 mol}}{HBr} \longrightarrow \underset{\substack{\textbf{3-bromo-1-butene}\\ \textbf{1,2-addition product}}}{\overset{1}{C}H_3\underset{\underset{\displaystyle Br}{|}}{\overset{2}{C}H}{-}CH{=}CH_2} + \underset{\substack{\textbf{1-bromo-2-butene}\\ \textbf{1,4-addition product}}}{\overset{1}{C}H_3{-}CH{=}CH{-}\underset{\underset{\displaystyle Br}{|}}{\overset{4}{C}H_2}}$$

On the basis of your knowledge of how electrophilic reagents add to double bonds, you would expect the 1,2-addition product to form. However, that the 1,4-addition product also forms may be surprising because not only did the reagent not add to adjacent carbons, but a double bond has changed its position. The double bond in the 1,4-product is between the 2- and 3-positions, whereas the reactant had a single bond in this position.

When we talk about addition at the 1- and 2-positions or at the 1- and 4-positions, the numbers refer to the four carbons of the conjugated system. Thus, the carbon in the 1-position is one of the sp^2 carbons at the end of the conjugated system—it is not necessarily the first carbon in the molecule.

$$R-\overset{1}{C}H=\overset{2}{C}H-\overset{3}{C}H=\overset{4}{C}H-R$$

the conjugated system

For example, take a minute to look at the products formed in the following reaction.

$$CH_3CH=CH-CH=CHCH_3 \xrightarrow{HBr} CH_3CH_2-\underset{Br}{CH}-CH=CHCH_3 + CH_3CH_2-CH=CH-\underset{Br}{CH}CH_3$$

2,4-hexadiene → 4-bromo-2-hexene (**1,2-addition product**) + 2-bromo-3-hexene (**1,4-addition product**)

To understand why an electrophilic addition reaction to a conjugated diene forms both 1,2-addition and 1,4-addition products, we must look at the mechanism for the reaction.

mechanism for the reaction of a conjugated diene with HBr

$$CH_2=CH-CH=CH_2 + H-\ddot{Br}: \longrightarrow CH_3-\overset{+}{C}H-CH=CH_2 \longleftrightarrow CH_3-CH=CH-\overset{+}{C}H_2$$

1,3-butadiene; $+ :\ddot{Br}:^-$ an allylic cation $+ :\ddot{Br}:^-$

1,3-butadiene does not form (crossed arrow):

$$\overset{+}{C}H_2-CH_2-CH=CH_2$$

a primary carbocation

$$\downarrow$$

$$CH_3-\underset{Br}{CH}-CH=CH_2 + CH_3-CH=CH-\underset{Br}{CH_2}$$

3-bromo-1-butene (**1,2-addition product**) + 1-bromo-2-butene (**1,4-addition product**)

- The proton adds to C-1, forming an allylic cation. The π electrons of the allylic cation are delocalized; therefore, the positive charge is shared by two carbons.
- The resonance contributors of the allylic cation show that the positive charge on the carbocation is shared by C-2 and C-4. Consequently, the bromide ion can attack either C-2 or C-4 to form the 1,2-addition product or the 1,4-addition product, respectively.

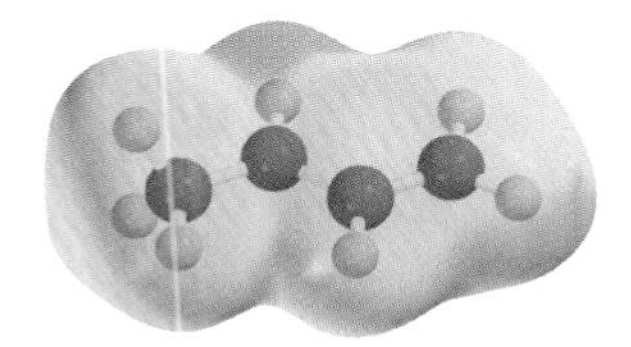

$$CH_3-\overset{\delta+}{C}H\text{---}CH\text{---}\overset{\delta+}{C}H_2$$

The first step in all electrophilic additions to conjugated dienes is addition of the electrophile to one of the sp^2 carbons at the end of the conjugated system.

Notice that in the first step of the reaction, adding to C-1 is the same as adding to C-4 because 1,3-butadiene is symmetrical. The proton does not add to C-2 or C-3 because doing so would form a primary carbocation. The π electrons of the primary carbocation would be localized; thus, it would not be as stable as the allylic cation that has delocalized π electrons. Therefore, the first step in all electrophilic additions to conjugated dienes is addition of the electrophile to one of the sp^2 carbons at the end of the conjugated system.

PROBLEM 15 ♦

What diene would you expect to be more stable, 2,4-heptadiene or 2,5-heptadiene?

PROBLEM 16 ◆

Give the products of the following reactions, assuming that one equivalent of each reagent is used in each reaction:

a. $CH_3CH{=}C(CH_3){-}C(CH_3){=}CHCH_3 \xrightarrow{HBr}$

b. 1,2-dimethylcyclohexa-1,3-diene $\xrightarrow{HBr}$

7.9 EXAMPLES THAT ILLUSTRATE HOW DELOCALIZED ELECTRONS CAN AFFECT pK_a

We have seen that a carboxylic acid is a much stronger acid than an alcohol. For example, the pK_a of acetic acid is 4.76, whereas the pK_a of ethanol is 15.9 (Section 2.3). We know, then, that the conjugate base of a carboxylic acid is considerably weaker and, therefore, more stable than the conjugate base of an alcohol. (Recall that the stronger the acid, the weaker and more stable is its conjugate base.)

$H_3C{-}C({=}O){-}OH$ acetic acid pK_a = 4.76 $\quad$ CH_3CH_2OH ethanol pK_a = 15.9

The factor most responsible for the increased stability of the carboxylate ion is its *greater delocalization energy* relative to that of its conjugate acid. The carboxylate ion has greater delocalization energy, because the ion has two equivalent resonance contributors that are predicted to be relatively stable, whereas the carboxylic acid has only one (Section 7.5). Therefore, loss of a proton from a carboxylic acid is accompanied by increased delocalization energy—in other words, an increase in stability.

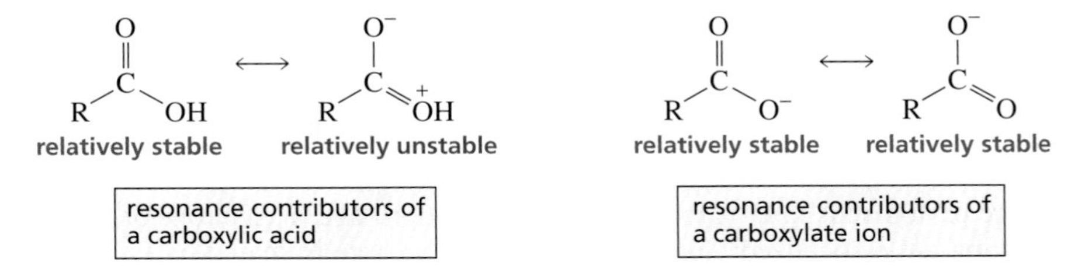

resonance contributors of a carboxylic acid

resonance contributors of a carboxylate ion

3-D Molecule: Acetate ion; Ethoxide ion

In contrast, all the electrons in an alcohol, such as ethanol, and its conjugate base are localized, so loss of a proton from an alcohol is not accompanied by an increase in delocalization energy.

$$CH_3CH_2\ddot{O}H \rightleftharpoons CH_3CH_2\ddot{O}\!:^- + H^+$$

ethanol

Phenol, a compound in which an OH group is bonded to a benzene ring, is a stronger acid than an alcohol such as ethanol or cyclohexanol.

phenol pK_a = 10 $\quad$ cyclohexanol pK_a = 16 $\quad$ CH_3CH_2OH ethanol pK_a = 16

While both phenol and the phenolate ion have delocalized electrons, the delocalization energy of the phenolate ion is greater than that of phenol because three of phenol's resonance contributors have separated charges and a positive charge on an electronegative oxygen. The loss of a proton from phenol, therefore, is accompanied by an increase in delocalization energy. In contrast, neither cyclohexanol nor its conjugate base has delocalized electrons, so loss of a proton is not accompanied by an increase in delocalization energy.

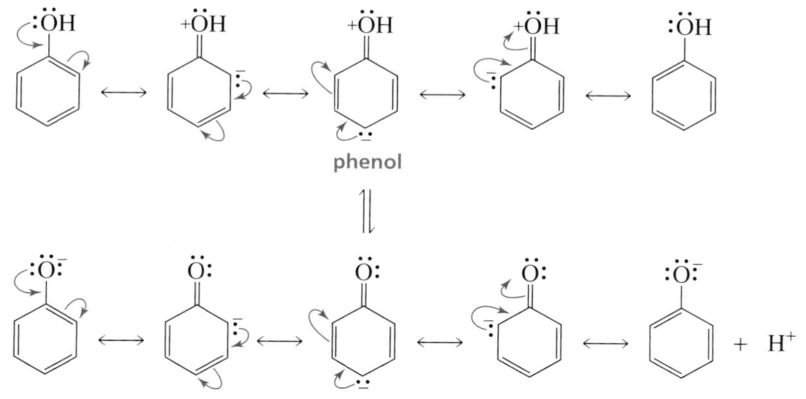

phenol

phenolate ion

Tutorial:
Acidity and electron delocalization

Protonated aniline is a much stronger acid than protonated cyclohexylamine.

protonated aniline
pK_a = 4.60

protonated cyclohexylamine
pK_a = 11.2

The nitrogen atom of protonated aniline lacks a lone pair that can be delocalized. When protonated aniline loses a proton, however, the lone pair that formerly held the proton can be delocalized. Loss of a proton, therefore, is accompanied by an increase in delocalization energy.

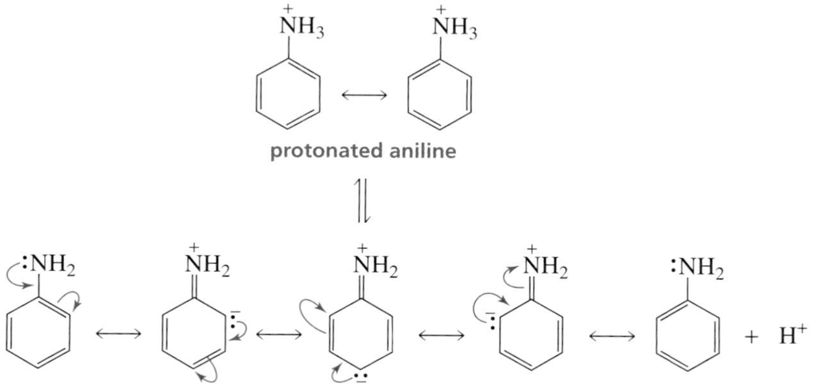

protonated aniline

aniline

An amine such as cyclohexylamine has no delocalized electrons either in the protonated form or in the unprotonated form to stabilize it.

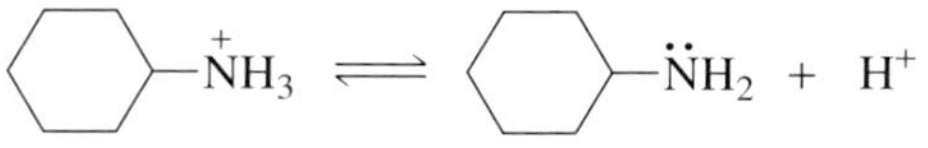

protonated cyclohexylamine

cyclohexylamine

We can now add phenol and protonated aniline to the classes of organic compounds whose approximate pK_a values you should know (Table 7.1). They are also listed inside the back cover for easy reference.

Table 7.1 Approximate pK_a Values

$pK_a < 0$	$pK_a \approx 5$	$pK_a \approx 10$	$pK_a \approx 15$
$R\overset{+}{O}H$ (with H)	$RCOH$ (C=O)	$R\overset{+}{N}H_3$	ROH
^{+}OH $\parallel$ $RCOH$	$C_6H_5\overset{+}{N}H_3$	C_6H_5OH	H_2O
H_3O^+			

PROBLEM 17 ◆ SOLVED

Which is a stronger acid?

a. $CH_3CH_2CH_2OH$ or $CH_3CH{=}CHOH$

b. $HC(=O)CH_2OH$ or $H_3CC(=O)OH$

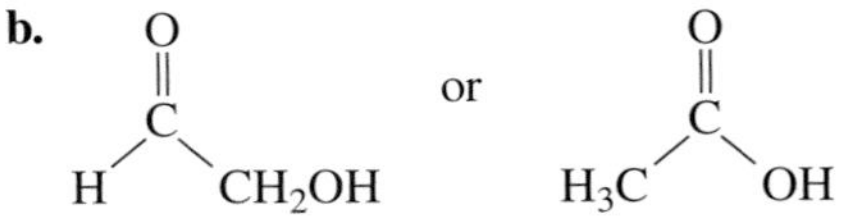

c. $CH_3CH{=}CHCH_2OH$ or $CH_3CH{=}CHOH$

d. $CH_3CH_2CH_2\overset{+}{N}H_3$ or $CH_3CH{=}CH\overset{+}{N}H_3$

Solution to 17a The conjugate base of propyl alcohol does not have delocalized electrons, but the conjugate base of allyl alcohol does. Because delocalized electrons stabilize a compound, the conjugate base of allyl alcohol is the weaker of the two bases. Therefore, allyl alcohol is a stronger acid than propyl alcohol, because the weaker the base, the stronger is its conjugate acid.

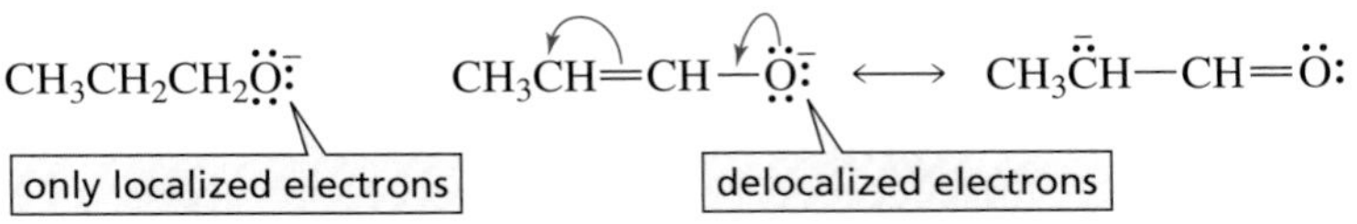

$$CH_3CH_2CH_2\ddot{\underset{..}{O}}{:}^- \qquad CH_3CH{=}CH{-}\ddot{\underset{..}{O}}{:}^- \longleftrightarrow CH_3\overset{-}{\ddot{C}}H{-}CH{=}\ddot{O}{:}$$

only localized electrons

delocalized electrons

PROBLEM 18 ◆

Which is a stronger base?

a. ethylamine or aniline

b. ethylamine or ethoxide ion ($CH_3CH_2O^-$)

c. phenolate ion or ethoxide ion

PROBLEM 19 ◆

Rank the following compounds in order of decreasing acid strength:

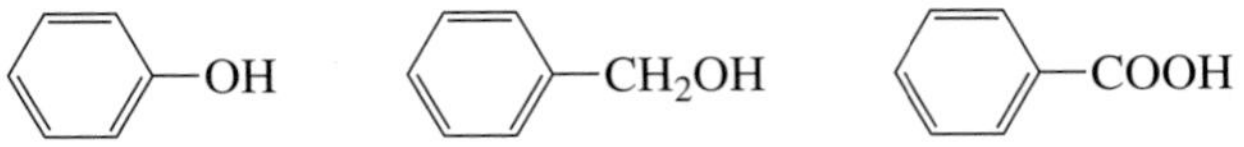

$C_6H_5{-}OH \qquad C_6H_5{-}CH_2OH \qquad C_6H_5{-}COOH$

7.10 ULTRAVIOLET AND VISIBLE SPECTROSCOPY

UV/Vis spectroscopy provides information about compounds that have conjugated double bonds. Ultraviolet (UV) light has wavelengths ranging from 180 to 400 nm (nanometers); visible (Vis) light has wavelengths ranging from 400 to 780 nm. The shorter the wavelength, the greater is the energy of the radiation. Ultraviolet light, therefore, has greater energy than visible light. If a compound absorbs **ultraviolet light**, a UV spectrum is obtained; if it absorbs **visible light**, a visible spectrum is obtained.

The shorter the wavelength, the greater is the energy of the radiation.

The UV spectrum of acetone is shown in Figure 7.2. The λ_{max} (stated as "lambda max") is the wavelength corresponding to the highest point of the absorption band. The UV spectrum shows that acetone has a λ_{max} = 195 nm.

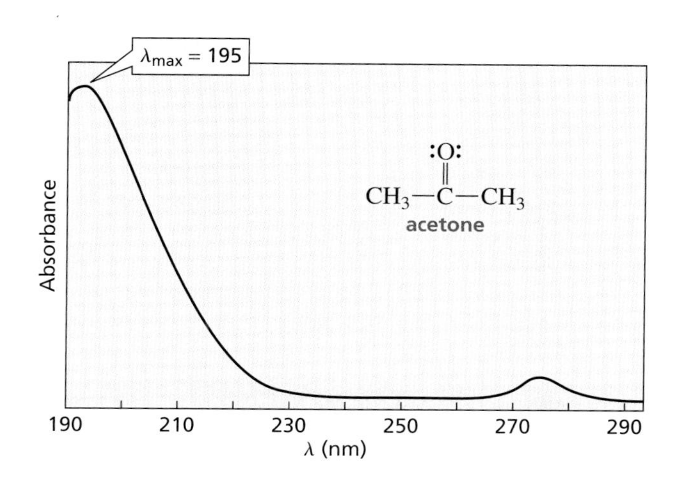

◀ **Figure 7.2**
The UV spectrum of acetone.

ULTRAVIOLET LIGHT AND SUNSCREENS

Exposure to ultraviolet light stimulates specialized cells in the skin to produce a black pigment known as melanin, which causes the skin to look tan. Melanin absorbs UV light, so it protects our bodies from the harmful effects of the sun. If more UV light reaches the skin than the melanin can absorb, the light will burn the skin and cause photochemical reactions that can result in skin cancer. UV-A is the lowest-energy UV light (315 to 400 nm) and does the least biological damage. Fortunately, most of the more dangerous, higher-energy UV light, UV-B (290 to 315 nm) and UV-C (180 to 290 nm), is filtered out by the ozone layer in the stratosphere. That is why we need to worry about the apparent thinning of the ozone layer (Section 5.17).

H_2N–C$_6$H$_4$–COOH
para-aminobenzoic acid
PABA

$(CH_3)_2N$–C$_6$H$_4$–$COOCH_2CH(CH_2CH_3)CH_2CH_2CH_2CH_3$
2-ethylhexyl 4-(dimethylamino)benzoate
Padimate O

CH_3O–C$_6$H$_4$–CH=CH–C(=O)$OCH_2CH(CH_2CH_3)CH_2CH_2CH_2CH_3$
2-ethylhexyl (*E*)-3-(4-methoxyphenyl)-2-propenoate
Giv Tan F

Applying a sunscreen can protect skin against UV light. The amount of protection provided by a particular sunscreen is indicated by its SPF (sun protection factor). The higher the SPF, the greater the protection. Some sunscreens contain an inorganic component, such as zinc oxide, that reflects the light as it reaches the skin. Others contain a compound that absorbs UV light. PABA was the first commercially available UV-absorbing sunscreen. PABA absorbs UV-B light, but is not very soluble in oily skin lotions. Less polar compounds, such as Padimate O, are now commonly used. Recent research has shown that sunscreens that absorb only UV-B light do not give adequate protection against skin cancer; both UV-A and UV-B protection are needed. Giv Tan F absorbs both UV-B and UV-A light, so it gives better protection.

7.11 THE λ_{max} INCREASES AS THE NUMBER OF CONJUGATED DOUBLE BONDS INCREASES

The more conjugated double bonds that a compound has, the longer is the wavelength at which the λ_{max} occurs. For example, methyl vinyl ketone has two conjugated double bonds, whereas acetone has only one; hence, the λ_{max} for methyl vinyl ketone is at 219 nm—a longer wavelength than the λ_{max} for acetone (195 nm).

3-D Molecule: Methyl vinyl ketone

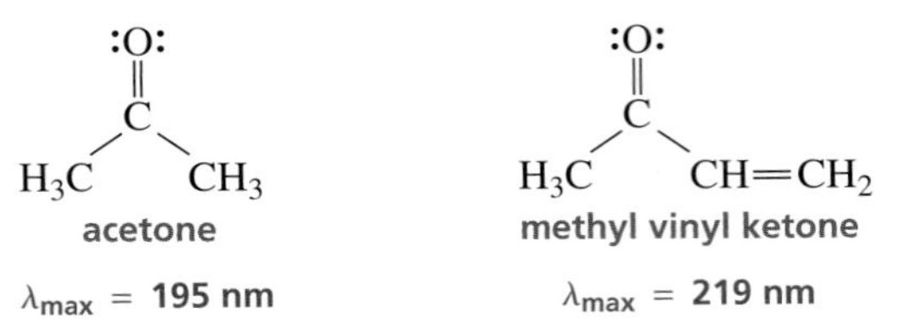

The λ_{max} increases as the number of conjugated double bonds increases.

The λ_{max} values for several conjugated dienes are shown in Table 7.2. Notice that the λ_{max} increases as the number of conjugated double bonds increases. Thus, the λ_{max} value can be used to estimate the number of conjugated double bonds in the compound.

Table 7.2 λ_{max} Values for Ethene and Conjugated Polyenes

Compound	λ_{max} (nm)
$H_2C{=}CH_2$	165
(2 conjugated double bonds)	217
(3 conjugated double bonds)	256
(4 conjugated double bonds)	290
(5 conjugated double bonds)	334
(6 conjugated double bonds)	364

If a compound has enough conjugated double bonds, it will absorb visible light (λ_{max} > 400 nm), and the compound will be colored. Thus, β-carotene, a precursor of vitamin A—found in carrots, apricots, and sweet potatoes—is an orange substance. Lycopene—found in tomatoes, watermelon, and pink grapefruit—is red.

β-carotene
λ_{max} = 455 nm

lycopene
λ_{max} = 474 nm

PROBLEM 20◆

Rank the compounds in order of decreasing λ_{max}:

benzene; biphenyl; Ph—CH=CH—Ph; Ph—CH=CH$_2$

7.12 A COMPOUND THAT ABSORBS VISIBLE LIGHT IS COLORED

White light is a mixture of all wavelengths of visible light. If any of these wavelengths are removed from white light, the remaining light is colored. Therefore, a compound that absorbs visible light is colored, and its color depends on the color of the wavelengths of the absorbed light. The wavelengths that the compound does not absorb are reflected back to the viewer, producing the color the viewer sees.

The relationship between the wavelengths of the light that a substance absorbs and the substance's observed color is shown in Table 7.3. Notice that two absorption bands are necessary to produce green. Most colored compounds have fairly broad absorption bands, although vivid colors have narrow absorption bands. The human eye is able to distinguish more than a million different shades of color!

Lycopene, β-carotene, and anthocyanins are found in the leaves of trees, but their characteristic colors are usually obscured by the green color of chlorophyll (Section 8.3). In the fall, when chlorophyll degrades, the other colors become apparent.

Table 7.3 Dependence of the Color Observed on the Wavelength of Light Absorbed

Wavelengths absorbed (nm)	Observed color
380–460	yellow
380–500	orange
440–560	red
480–610	purple
540–650	blue
380–420 and 610–700	green

Azobenzenes (benzene rings connected by an N=N bond) have an extended conjugated system that causes them to absorb light from the visible region of the spectrum. Some substituted azobenzenes, such as the two shown below, are used commercially as dyes. Varying the extent of conjugation and the substituents attached to the conjugated system creates a large number of different colors. Notice that the only difference between butter yellow and methyl orange is an $SO_3^-Na^+$ group. When margarine was first produced, it was colored with butter yellow to make it look more like butter. (White margarine would not have been very appetizing.) This dye was abandoned after it was found to be carcinogenic. β-Carotene is now used to color margarine (page 190). Methyl orange is a commonly used acid–base indicator (see Problem 47).

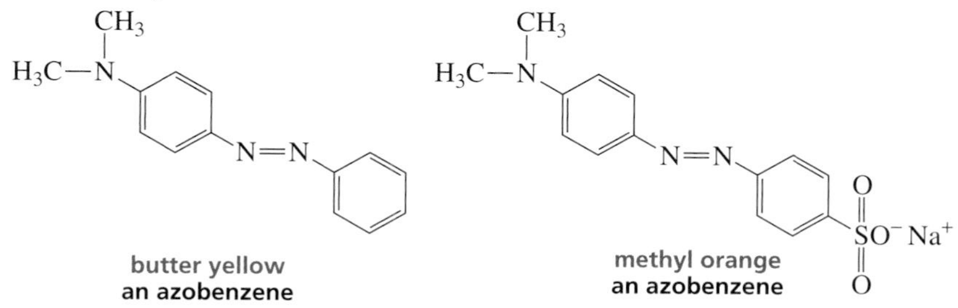

The lone-pair electrons on oxygen and nitrogen in the compounds shown below are available to interact with the π electron cloud of the ring; such an interaction increases the λ_{max}. Because the anilinium ion does not have a lone pair, its λ_{max} is similar to that of benzene.

	benzene	phenol	phenolate ion	aniline	anilinium ion
		:ÖH	:Ö:⁻	$\ddot{N}H_2$	$\overset{+}{N}H_3$
λ_{max} =	255 nm	270 nm	287 nm	280 nm	254 nm

ANTHOCYANINS: A COLORFUL CLASS OF COMPOUNDS

A class of highly conjugated compounds called anthocyanins is responsible for the red, purple, and blue colors of many flowers (poppies, peonies, cornflowers), fruits (cranberries, rhubarb, strawberries, blueberries), and vegetables (beets, radishes, red cabbage).

In a neutral or basic solution, the monocyclic fragment (on the right-hand side of the anthocyanin) is not conjugated with the rest of the molecule, so the anthocyanin does not absorb visible light and is therefore a colorless compound. In an acidic environment, however, the OH group becomes protonated and water is eliminated. Loss of water results in the third ring's becoming conjugated with the rest of the molecule. As a result of the number of conjugated double bonds, the anthocyanin absorbs visible light with wavelengths between 480 and 550 nm. The exact wavelength of light absorbed depends on the substituents (R and R′) on the anthocyanin. Thus, the flower, fruit, or vegetable appears red, purple, or blue, depending on what the R groups are. You can see this color change if you alter the pH of cranberry juice so that it is no longer acidic.

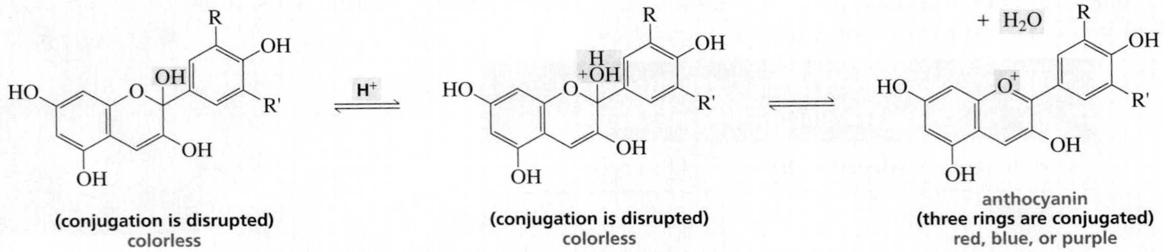

(conjugation is disrupted)
colorless

$\xrightleftharpoons{H^+}$

(conjugation is disrupted)
colorless

$\rightleftharpoons$ anthocyanin $+ H_2O$

anthocyanin
(three rings are conjugated)
red, blue, or purple

R = H, OH, or OCH_3
R′ = H, OH, or OCH_3

PROBLEM 21 ♦

a. At pH = 7, one of the ions shown below is purple and the other is blue. Which is which?

b. What would be the difference in the colors of the compounds at pH = 3?

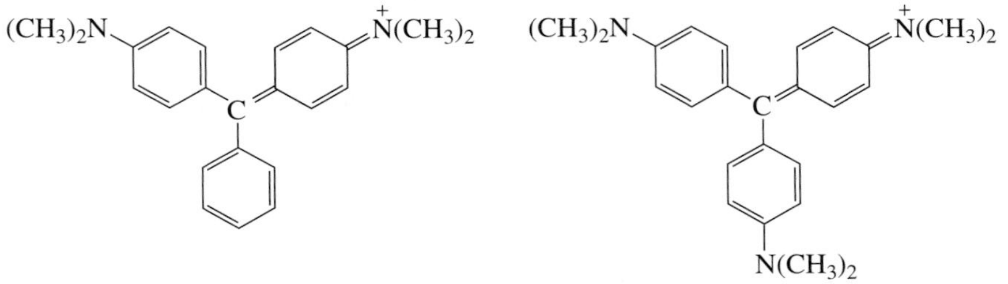

SUMMARY

Localized electrons belong to a single atom or are confined to a bond between two atoms. **Delocalized electrons** are shared by more than two atoms; they result when a *p* orbital overlaps the *p* orbitals of more than one adjacent atom. Electron delocalization occurs only if all the atoms sharing the delocalized electrons lie in or close to the same plane.

Each of benzene's six carbon atoms is sp^2 hybridized, with bond angles of 120°. A *p* orbital of each carbon overlaps the *p* orbitals of both adjacent carbons. The six π electrons are shared by all six carbons. Thus, benzene is a planar molecule with six delocalized π electrons.

Chemists use **resonance contributors**—structures with localized electrons—to approximate the actual structure of a compound that has delocalized electrons: the **resonance hybrid**. To draw resonance contributors, move only π electrons or nonbonding electrons toward an sp^2 carbon. The total number of electrons and the numbers of paired and unpaired electrons do not change.

The greater the **predicted stability** of the resonance contributor, the more it contributes to the hybrid and the more similar its structure is to the real molecule. The extra stability a compound gains from having delocalized electrons is called **delocalization energy**. It tells us how much more stable a compound with delocalized electrons is than it would be if its electrons were localized. The greater the number of relatively stable resonance contributors and the more nearly equivalent they are, the greater is the delocalization energy of the compound.

Because electron delocalization increases the stability of a compound, allylic and benzylic cations are more stable, since they have delocalized electrons, than similarly substituted carbocations with only localized electrons. Delocalized electrons can affect the pK_a of a compound; a carboxylic acid and a phenol are more acidic than an alcohol such as ethanol, and a protonated aniline is more acidic than a protonated amine because in each case, the loss of a proton is accompanied by an increase in electron delocalization.

Conjugated double bonds are separated by one single bond. **Isolated double bonds** are separated by more than one single bond. Because dienes with conjugated double bonds have delocalized electrons, they are more stable than dienes with isolated double bonds.

An isolated diene, like an alkene, undergoes only 1,2-addition. If there is only enough electrophilic reagent to add to one of the double bonds, it will add preferentially to the one that forms the more stable carbocation. A conjugated diene reacts with a limited amount of electrophilic reagent to form a **1,2-addition product** and a **1,4-addition product**. The first step is addition of the electrophile to one of the sp^2 carbons at the end of the conjugated system.

Ultraviolet and **visible (UV/Vis) spectroscopy** provide information about compounds with conjugated double bonds. UV light is higher in energy than visible light; the shorter the wavelength, the greater is its energy. The more conjugated double bonds there are in a compound, the greater is its λ_{max} value.

SUMMARY OF REACTIONS

1. In the presence of excess electrophilic reagent, both double bonds of a *diene with isolated double bonds* will undergo electrophilic addition.

$$CH_2{=}CHCH_2CH_2\overset{CH_3}{\overset{|}{C}}{=}CH_2 + \underset{\text{excess}}{HBr} \longrightarrow CH_3\underset{Br}{\underset{|}{C}}HCH_2CH_2\underset{Br}{\underset{|}{\overset{CH_3}{\overset{|}{C}}}}CH_3$$

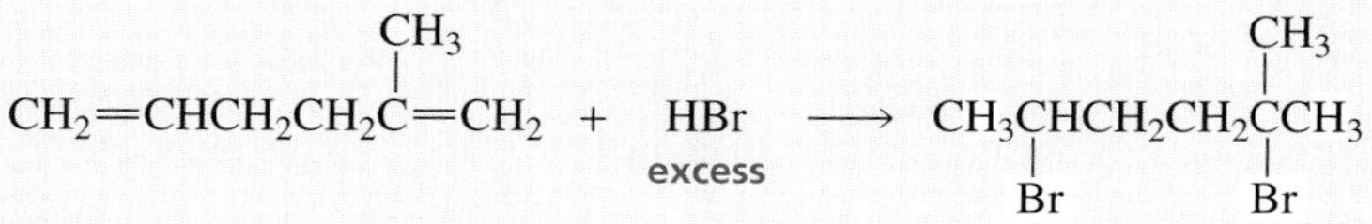

In the presence of only one equivalent of an electrophilic reagent, only the most reactive double bond of an isolated diene will undergo electrophilic addition (Section 7.8).

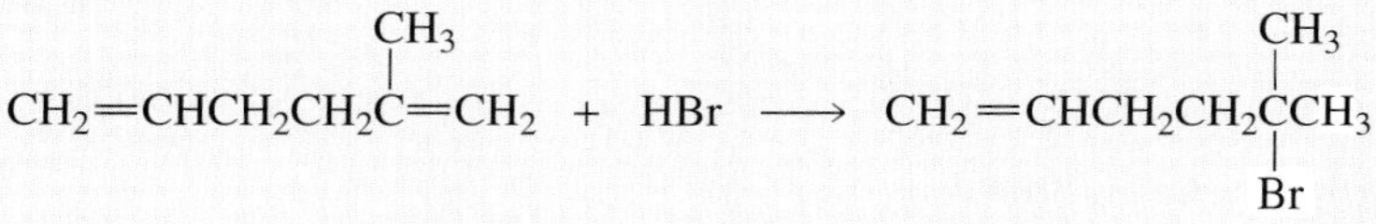

$$CH_2{=}CHCH_2CH_2\overset{CH_3}{\overset{|}{C}}{=}CH_2 + HBr \longrightarrow CH_2{=}CHCH_2CH_2\underset{Br}{\underset{|}{\overset{CH_3}{\overset{|}{C}}}}CH_3$$

2. *Conjugated dienes* undergo 1,2- and 1,4-addition with one equivalent of an electrophilic reagent (Section 7.8).

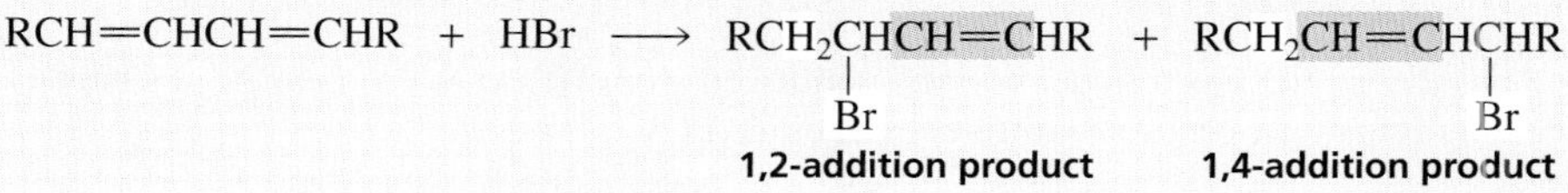

$$RCH{=}CHCH{=}CHR + HBr \longrightarrow \underset{\text{1,2-addition product}}{RCH_2\underset{Br}{\underset{|}{C}}HCH{=}CHR} + \underset{\text{1,4-addition product}}{RCH_2CH{=}CH\underset{Br}{\underset{|}{C}}HR}$$

PROBLEMS

22. Which of the following have delocalized electrons?

a. 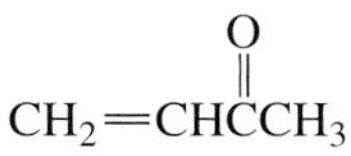 $CH_2{=}CH\overset{O}{\overset{\|}{C}}CH_3$

b. [cyclohexene]

c. $CH_3\underset{+}{\overset{CH_3}{\overset{|}{C}}}CH_2CH{=}CH_2$

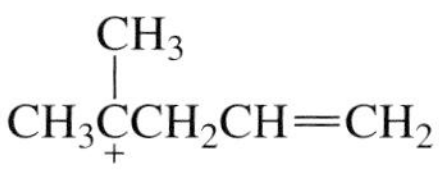

d. $CH_3CH{=}CHOCH_2CH_3$

e. $CH_2{=}CHCH_2CH{=}CH_2$

f. 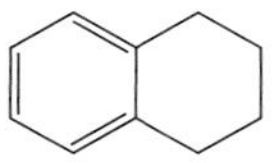 [bicyclic structure: benzene ring fused to cyclohexane ring]

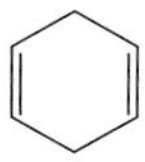

g. $CH_3CH_2NHCH_2CH{=}CHCH_3$

h. $CH_3CH_2NHCH{=}CHCH_3$

i. $CH_3CH_2\underset{+}{C}HCH{=}CH_2$

j. 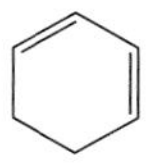 [1,3-cyclohexadiene]

23. Draw resonance contributors for the following ions:

a. 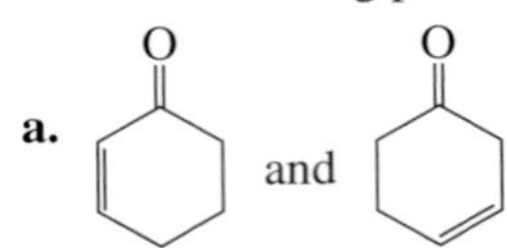

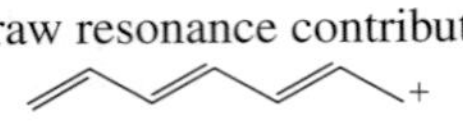

b.

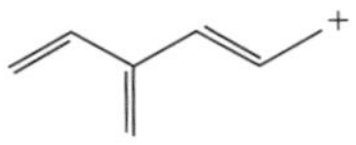

24. Are the following pairs of structures resonance contributors or different compounds?

a. and

b. 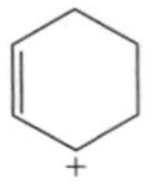 and

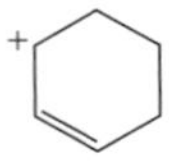

c. $CH_3\overset{O}{\overset{\|}{C}}CH_2CH_3$ and $CH_3\overset{OH}{\overset{|}{C}}{=}CHCH_3$

d. $CH_3\overset{+}{C}HCH{=}CHCH_3$ and $CH_3CH{=}CHCH_2\overset{+}{C}H_2$

25. a. Draw resonance contributors for the following species. Indicate which are major contributors and which are minor contributors to the resonance hybrid.

1. $CH_3CH{=}CHOCH_3$
2. 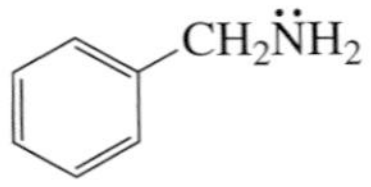 $CH_2\ddot{N}H_2$
3. $CH_3\overset{-}{\ddot{C}}HCH{=}NCH_3$
4. $CH_3CH{=}CH\overset{+}{C}H_2$
5. 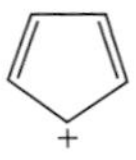
6. 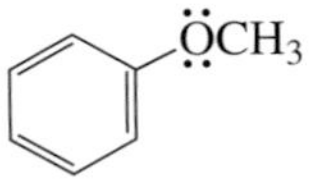 $\ddot{O}CH_3$

b. Do any of the species have resonance contributors that all contribute equally to the resonance hybrid?

26. Which resonance contributor makes the greater contribution to the resonance hybrid?

a. 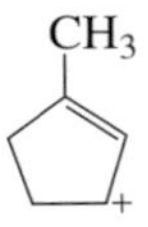 CH_3 or CH_3

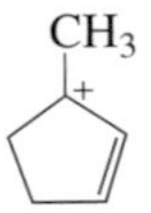

b. 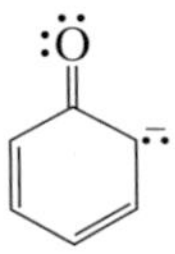 :Ö or :Ö:$^-$

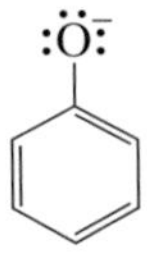

c. $CH_3\overset{+}{C}HCH{=}CH_2$ or $CH_3CH{=}CH\overset{+}{C}H_2$

27. a. Which oxygen atom has the greater electron density?

$CH_3\overset{O}{\overset{\|}{C}}OCH_3$

b. Which compound has the greater electron density on its nitrogen atom? (Revisit Problem 25 in Chapter 2. Now you can understand why the two ring nitrogens have different basicities.)

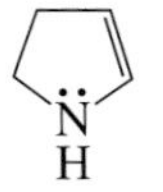

N H or N H

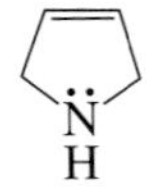

c. Which compound has the greater electron density on its oxygen atom?

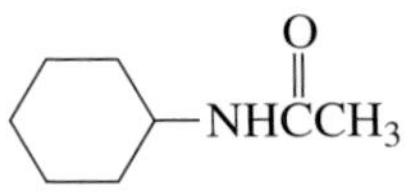

$-NH\overset{O}{\overset{\|}{C}}CH_3$ or $-NH\overset{O}{\overset{\|}{C}}CH_3$

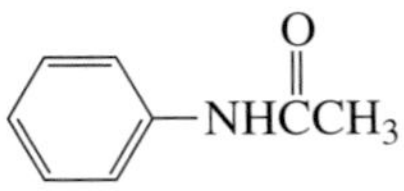

28. Which can lose a proton more readily, a methyl group bonded to cyclohexane or a methyl group bonded to benzene? Explain your answer.

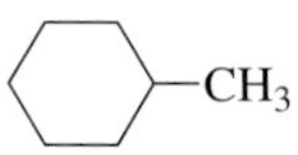

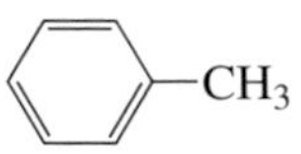

$-CH_3$ $-CH_3$

29. Rank the compounds in order of decreasing λ_{max}:

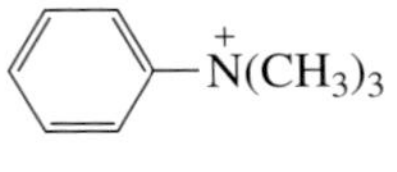

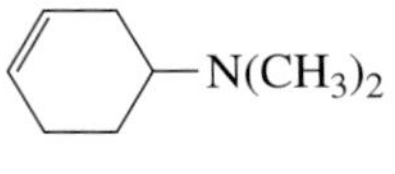

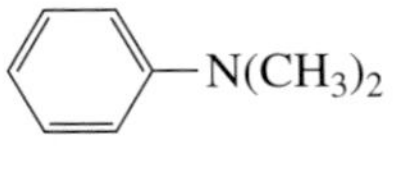

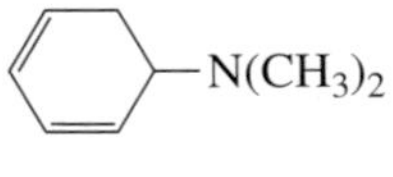

$-\overset{+}{N}(CH_3)_3$ $-N(CH_3)_2$ $-N(CH_3)_2$ $-N(CH_3)_2$

30. Rank the following compounds in order of decreasing acidity:

31. Which species in each pair is more stable?

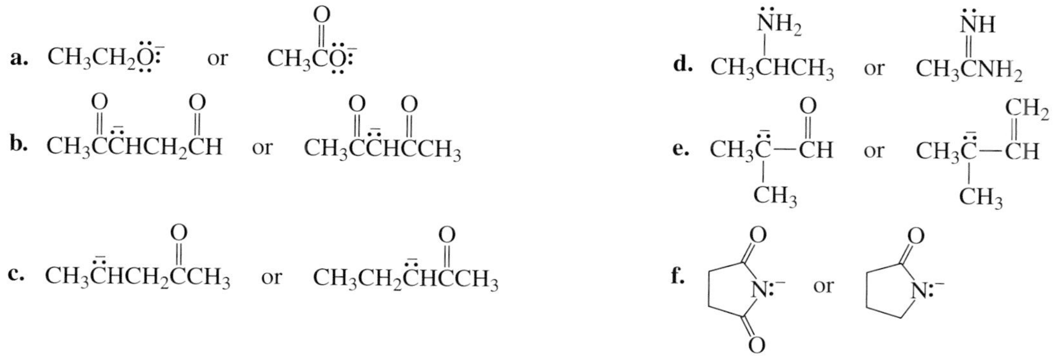

a. $CH_3CH_2\ddot{O}{:}^-$ or $CH_3C(=O)\ddot{O}{:}^-$

b. $CH_3C(=O)\bar{C}HCH_2CH(=O)$ or $CH_3C(=O)\bar{C}HC(=O)CH_3$

c. $CH_3\bar{C}HCH_2C(=O)CH_3$ or $CH_3CH_2\bar{C}HC(=O)CH_3$

d. $CH_3CH(\ddot{N}H_2)CH_3$ or $CH_3C(=\ddot{N}H)NH_2$

e. $CH_3\bar{C}(CH_3)-CH=O$ or $CH_3\bar{C}(CH_3)-CH=CH_2$

f. N:⁻ (ring with two C=O) or N:⁻ (ring with one C=O)

32. Which species in each of the pairs in Problem 31 is the stronger base?

33. Draw resonance contributors for the following species. Indicate which are major contributors to the resonance hybrid and which are minor contributors.

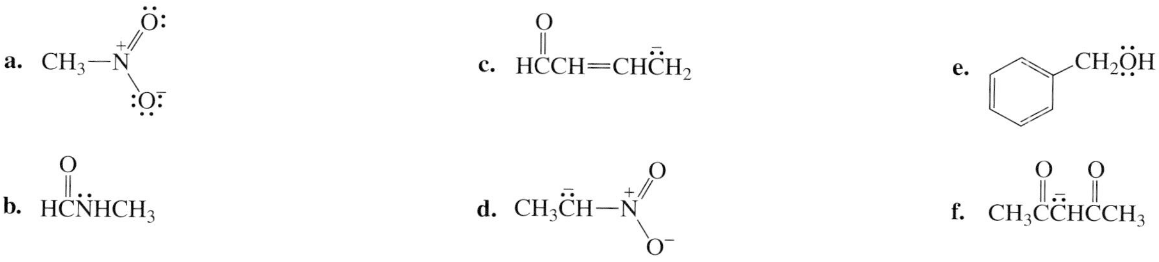

a. $CH_3-\overset{+}{N}(=\ddot{O}{:})\ddot{O}{:}^-$

b. $HC(=O)\ddot{N}HCH_3$

c. $HC(=O)CH=CH\ddot{\bar{C}}H_2$

d. $CH_3\ddot{\bar{C}}H-\overset{+}{N}(=O)O^-$

e. $C_6H_5CH_2\ddot{O}H$

f. $CH_3C(=O)\ddot{\bar{C}}HC(=O)CH_3$

34. Rank the following compounds in order of decreasing acidity of the indicated hydrogen:

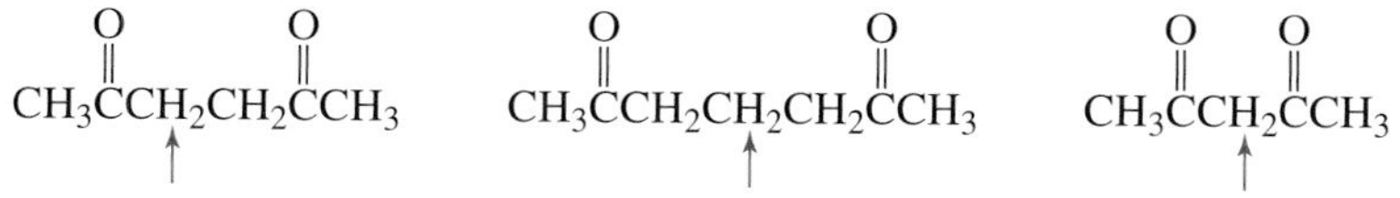

$CH_3C(=O)CH_2CH_2C(=O)CH_3$ $\quad$ $CH_3C(=O)CH_2CH_2CH_2C(=O)CH_3$ $\quad$ $CH_3C(=O)CH_2C(=O)CH_3$

35. Draw resonance contributors for each of the following species, and rank the resonance contributors in order of decreasing contribution to the hybrid:

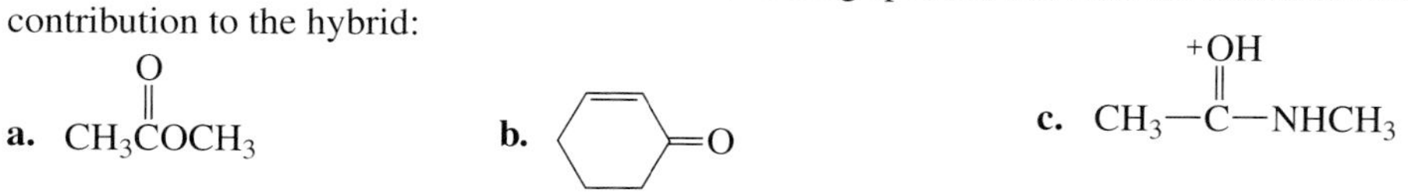

a. $CH_3C(=O)OCH_3$

b. (cyclohexenone)

c. $CH_3-C(=\overset{+}{O}H)-NHCH_3$

36. Draw the resonance hybrid for each of the species in Problem 35.

37. Name the following dienes and rank them in order of increasing stability. (Alkyl groups stabilize dienes in the same way that they stabilize alkenes; see Section 4.6.)

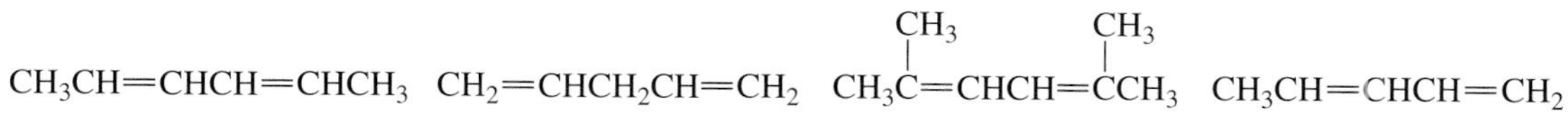

$CH_3CH=CHCH=CHCH_3$ $\quad$ $CH_2=CHCH_2CH=CH_2$ $\quad$ $CH_3C(CH_3)=CHCH=C(CH_3)CH_3$ $\quad$ $CH_3CH=CHCH=CH_2$

38. Which carbocation in each of the following pairs is more stable?

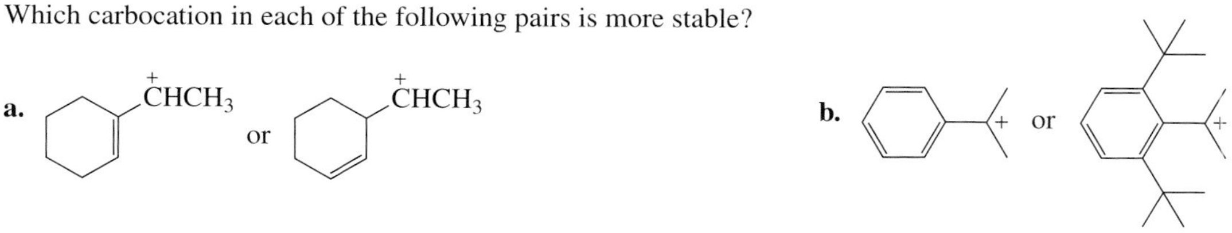

a. (cyclohexenyl)$\overset{+}{C}HCH_3$ or (cyclohexenyl)$\overset{+}{C}HCH_3$

b. (phenyl carbocation) or (di-tert-butylphenyl carbocation)

39. a. How many linear dienes have molecular formula C_6H_{10}? (Disregard cis–trans isomers.)
b. How many of the linear dienes in part a are conjugated dienes?
c. How many are isolated dienes?

40. What products would be obtained from the reaction of 1,3,5-hexatriene with one equivalent of HBr?

41. Give the major product of each of the following reactions, assuming the presence of one equivalent of each reagent:

a. 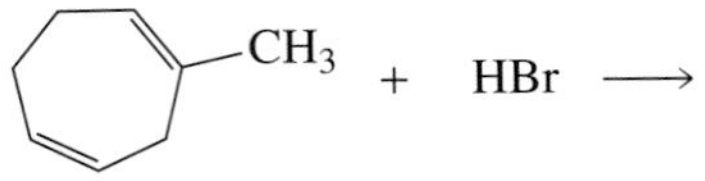 CH_3 + HBr ⟶

b. 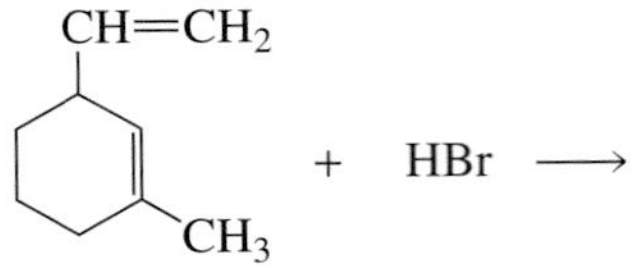 $CH{=}CH_2$, CH_3 + HBr ⟶

42. How could you use UV spectroscopy to distinguish between the compounds in each of the following pairs?

a. 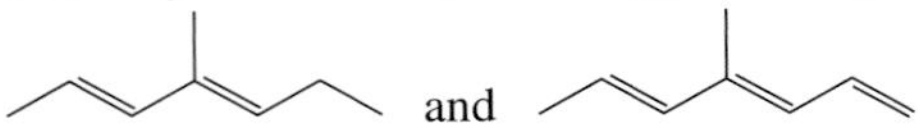 and

b. 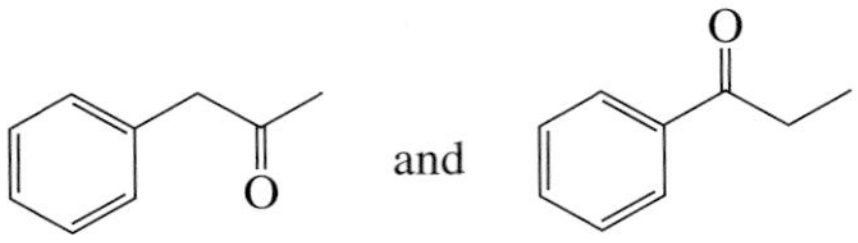 O and O

43. Give all the products of the following reaction:

Cl, + + HO^- ⟶

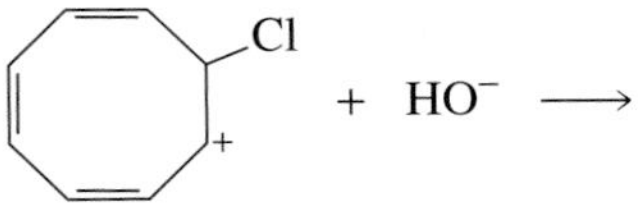

44. Some credit card slips have a top sheet of "carbonless paper" that transfers an imprint of your signature to a sheet lying underneath. The paper contains tiny capsules filled with the colorless compound shown below:

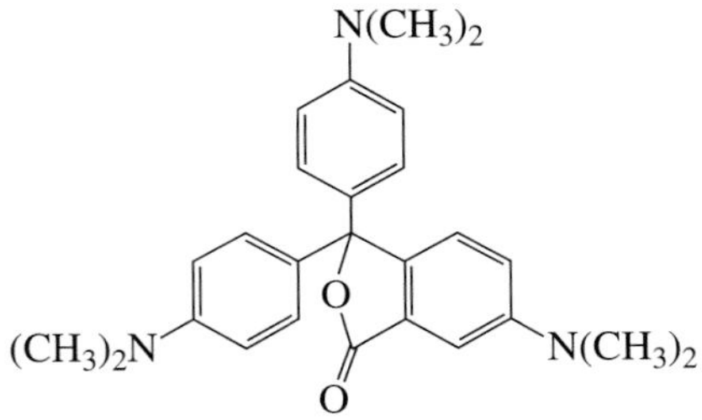

$N(CH_3)_2$, O, $(CH_3)_2N$, $N(CH_3)_2$, O

When you press on the paper, the capsules burst and the colorless compound comes into contact with the acid-treated bottom sheet, forming a highly colored compound. What is the structure of the colored compound?

45. a. How could each of the following compounds be prepared from a hydrocarbon in a single step?

1. 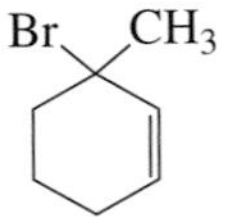 Br, CH_3

2. 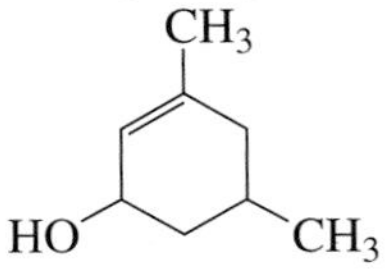 CH_3, HO, CH_3

b. What other organic compound would be obtained from each synthesis?

46. Give the major products obtained from the reaction of one equivalent of HCl with the following:
a. 2,3-dimethyl-1,3-pentadiene **b.** 2,4-dimethyl-1,3-pentadiene

47. a. Methyl orange (whose structure is given in Section 7.12) is an acid–base indicator. In solutions of pH > 4, it is red; in solutions of pH < 4, it is yellow. Account for the change in color.
b. Phenolphthalein, which is another indicator, exhibits a much more dramatic color change. In solutions of pH < 8.5, it is colorless; in solutions of pH > 8.5, it is deep red-purple. Account for the change in color.

O, O, OH, OH

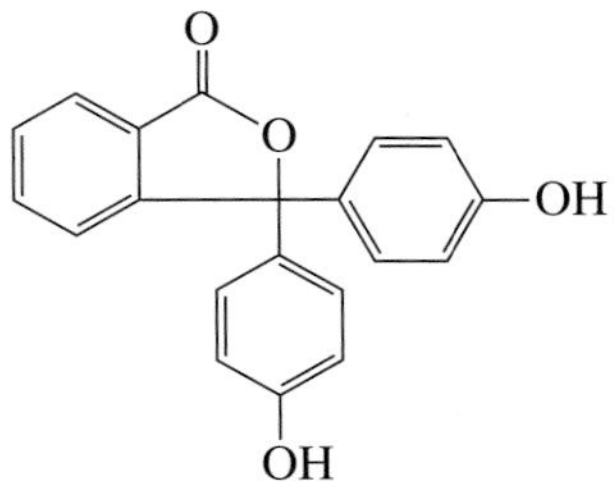

phenolphthalein

Aromaticity

Reactions of Benzene and Substituted Benzenes

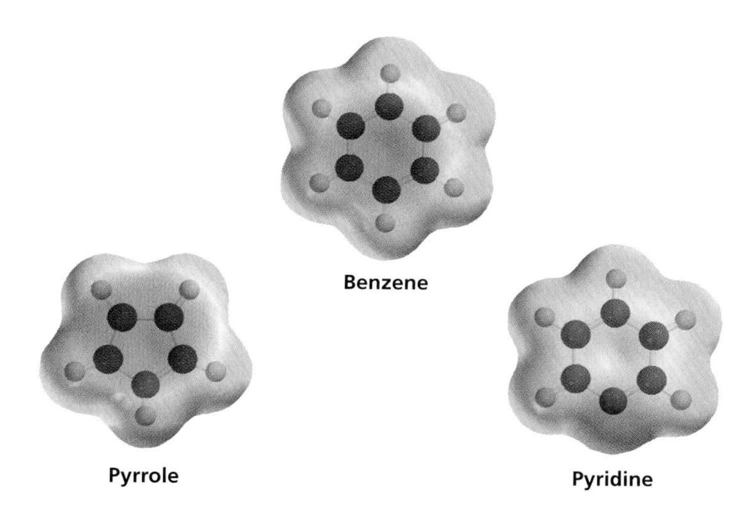

The compound we know as benzene was first isolated in 1825 by Michael Faraday. He extracted it from the liquid residue obtained after heating whale oil under pressure to produce the gas then being used to illuminate buildings in London.

Many substituted benzenes are found in nature. A few that have physiological activity are shown here:

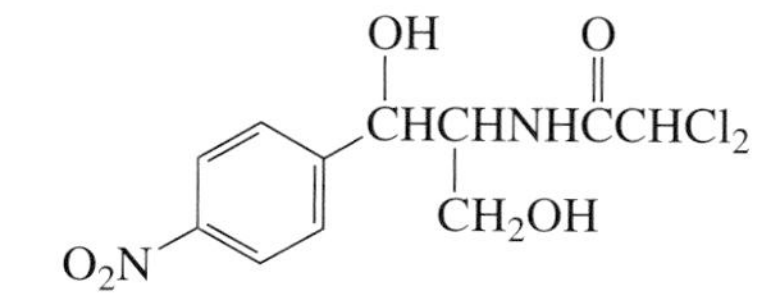

chloramphenicol
an antibiotic that is particularly effective against typhoid fever

adrenaline
epinephrine

ephedrine
a bronchodilator

mescaline
active ingredient of the peyote cactus

BIOGRAPHY

Michael Faraday (1791–1867) *was born in England, a son of a blacksmith. At the age of 14, he was apprenticed to a bookbinder and educated himself by reading the books that he bound. He became an assistant to Sir Humphry Davy at the Royal Institution of Great Britain in 1812, where Faraday taught himself chemistry. In 1825, he became director of a laboratory there and, in 1833, a professor of chemistry. He is best known for his work on electricity and magnetism.*

fenfluramine

phentermine

Many other physiologically active substituted benzenes are not found in nature, but exist because chemists have synthesized them. The now-banned diet drug fen-phen is a mixture of two synthetic substituted benzenes: fenfluramine and phentermine. Two other substituted benzenes, BHA and BHT, are preservatives found in a wide variety of packaged foods (Section 5.17).

When naturally occurring compounds are found to have desirable physiological activities, chemists will synthesize structurally similar compounds with the hope of developing them into useful products. For example, chemists have synthesized compounds with structures similar to adrenaline, producing amphetamine, a central nervous system stimulant, and the closely related methamphetamine, which are both used clinically as appetite suppressants. Methamphetamine, known as "speed," is also made and sold illegally because of its rapid and intense psychological effects. These compounds represent just a few of the many substituted benzenes that have been synthesized for commercial use by the chemical and pharmaceutical industries. The physical properties of several substituted benzenes are given in Appendix I.

amphetamine

methamphetamine
"speed"

acetylsalicylic acid
aspirin

hexachlorophene
a disinfectant

Compounds like benzene that have relatively few hydrogens in relation to the number of carbons, are typically found in oils produced by trees and other plants. Early chemists called such compounds **aromatic compounds** because of their pleasing fragrances. In this way, they were distinguished from **aliphatic compounds**, which have higher hydrogen-to-carbon ratios. Today, chemists use the word "aromatic" to signify certain kinds of chemical structures. We will now look at the features that cause a compound to be classified as aromatic.

MEASURING TOXICITY

Agent Orange, a defoliant widely used during the Vietnam War, is a mixture of two synthetic substituted benzenes: 2,4-D and 2,4,5-T. Dioxin (TCDD), a contaminant formed during the manufacture of Agent Orange, has been implicated as the causative agent of the various symptoms suffered by those exposed to this compound during the war.

2,4-dichlorophenoxyacetic acid
2,4-D

2,4,5-trichlorophenoxyacetic acid
2,4,5-T

2,3,7,8-tetrachlorodibenzo[*b*,*e*][1,4]dioxin
TCDD

The toxicity of a compound is indicated by its LD_{50} value—the dosage found to kill 50% of the test animals exposed to it. Dioxin, with an LD_{50} value of 0.0006 mg/kg for guinea pigs, is an extremely toxic compound. Compare this with the LD_{50} values of some well-known but far less toxic poisons: 0.96 mg/kg for strychnine and 15 mg/kg for both arsenic trioxide and sodium cyanide. One of the most toxic agents known is the botulism toxin, with an LD_{50} value of about 1×10^{-8} mg/kg.

8.1 THE TWO CRITERIA FOR AROMATICITY

In Chapter 7, we saw that benzene is a planar, cyclic compound with a cyclic cloud of delocalized electrons above and below the plane of the ring. (See Figure 7.1 on page 172.) Because its π electrons are delocalized, all the C—C bonds in benzene have the same length.

Benzene is a particularly stable compound because its **delocalization energy**—the extra stability it gains from having delocalized electrons—is unusually large (Section 7.6). Compounds with unusually large delocalization energies are called **aromatic compounds**.

Aromatic compounds are particularly stable.

How can we tell whether a compound is aromatic by looking at its structure? In other words, what structural features do aromatic compounds have in common?

To be classified as aromatic, a compound must meet both of the following criteria:

1. *It must have an uninterrupted cyclic cloud of π electrons* (called a π cloud) *above and below the plane of the molecule.* Let's look a little more closely at what this means:

   For the π cloud to be cyclic, *the molecule must be cyclic.*
   For the π cloud to be uninterrupted, *every atom in the ring must have a* p *orbital.*
   For the π cloud to form, each p orbital must overlap with the p orbitals on either side of it. Therefore, *the molecule must be planar.*

Tutorial: Aromaticity

2. *The π cloud must contain an odd number of pairs of π electrons.*

   Therefore, benzene is an aromatic compound because it is cyclic and planar, every carbon in the ring has a p orbital, and the π cloud contains *three* pairs of π electrons.

For a compound to be aromatic, it must be cyclic and planar, and have an uninterrupted cloud of π electrons. The π cloud must contain an odd number of pairs of π electrons.

8.2 APPLYING THE CRITERIA FOR AROMATICITY

Cyclobutadiene has two pairs of π electrons, and cyclooctatetraene has four pairs of π electrons. Unlike benzene, these compounds are *not* aromatic because they have an *even* number of pairs of π electrons. There is an additional reason why cyclooctatetraene is not aromatic—it is not planar, it is tub-shaped (Section 7.3). Because cyclobutadiene and cyclooctatetraene are not aromatic, they do not have the unusual stability of aromatic compounds.

3-D Molecules: Cyclobutadiene; Benzene; Cyclooctatetraene

cyclobutadiene

benzene cyclooctatetraene

Now let's look at some other compounds and determine whether they are aromatic. Cyclopentadiene is not aromatic because it does not have an uninterrupted ring of p orbital-bearing atoms. One of its ring atoms is sp^3 hybridized, and only sp^2 and sp carbons have p orbitals. Therefore, cyclopentadiene does not fulfill the first criterion for aromaticity.

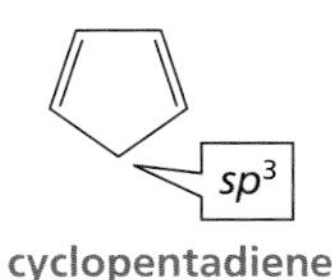

cyclopentadiene

cyclopentadienyl cation

cyclopentadienyl anion

The cyclopentadienyl cation also *is not* aromatic because, although it is planar and has an uninterrupted ring of *p* orbital-bearing atoms, its π cloud has *two* (an even number) pairs of π electrons. The cyclopentadienyl anion *is* aromatic: it is planar and has an uninterrupted ring of *p* orbital-bearing atoms, and its π cloud contains *three* (an odd number) pairs of π electrons.

When drawing resonance contributors, remember that only electrons move; atoms never move.

Notice that the negatively charged carbon in the cyclopentadienyl anion is sp^2 hybridized, because if it were sp^3 hybridized, the ion would not be aromatic. The resonance hybrid shows that all the carbons in the cyclopentadienyl anion are equivalent. Each carbon has exactly one-fifth of the negative charge associated with the anion.

resonance contributors of the cyclopentadienyl anion

$\delta-$ $\delta-$ $\delta-$ $\delta-$ $\delta-$

resonance hybrid

The criteria that determine whether a monocyclic hydrocarbon compound is aromatic can also be used to determine whether a polycyclic hydrocarbon compound is

BUCKYBALLS

Diamond and graphite are two familiar forms of pure carbon (Section 1.8). A third form was discovered unexpectedly in 1985, while scientists were conducting experiments designed to understand how long-chain molecules are formed in outer space. R. E. Smalley, R. F. Curl, Jr., and H. W. Kroto shared the 1996 Nobel Prize in chemistry for discovering this new form of carbon. They named the substance buckminsterfullerene (often shortened to fullerene) because its structure reminded them of the geodesic domes popularized by R. Buckminster Fuller, an American architect and philosopher. Its nickname is "buckyball."

A geodesic dome

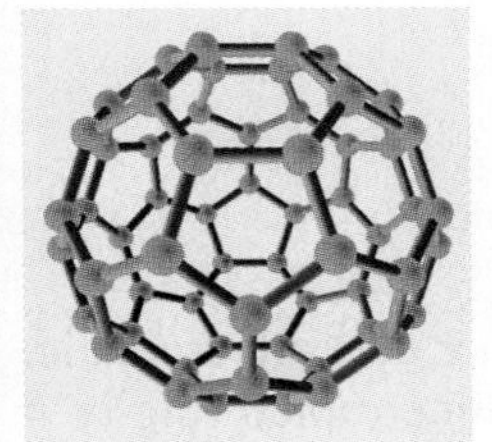

Consisting of a hollow cluster of 60 carbons, fullerene is the most symmetrical large molecule known. Like graphite, fullerene has only sp^2 carbons, but instead of being arranged in layers, the carbons are arranged in rings that fit together like the seams of a soccer ball. Each molecule has 32 interlocking rings (20 hexagons and 12 pentagons). At first glance, fullerene would appear to be aromatic because of its benzene-like rings. However, the curvature of the ball prevents the molecule from fulfilling the first criterion for aromaticity—that it be planar.

Buckyballs have extraordinary chemical and physical properties. For example, they are exceedingly rugged, as shown by their ability to survive the extreme temperatures of outer space. Because they are essentially hollow cages, they can be manipulated to make materials never before known. For example, when a buckyball is "doped" by inserting potassium or cesium into its cavity, it becomes an excellent organic superconductor. These molecules are now being studied for use in many other applications, including the development of new polymers, new catalysts, and new drug delivery systems. The discovery of buckyballs is a strong reminder of the technological advances that can be achieved as a result of basic research.

Richard E. Smalley (1943–2005) *was born in Akron, Ohio. He received a B.S. from the University of Michigan and a Ph.D. from Princeton University. He was a professor of chemistry at Rice University.*

Robert F. Curl, Jr., *was born in Texas in 1933. He received a B.A. from Rice University and a Ph.D. from the University of California, Berkeley. He is a professor of chemistry at Rice University.*

Sir Harold W. Kroto *was born in 1939 in England and was a professor of chemistry at the University of Sussex. Now he is a professor of chemistry at Florida State University.*

aromatic. Naphthalene (five pairs of π electrons), phenanthrene (seven pairs of π electrons), and chrysene (nine pairs of π electrons) are aromatic.

naphthalene

phenanthrene

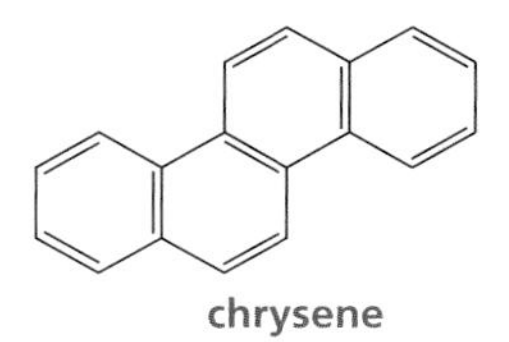
chrysene

3-D Molecules: Phenanthrene; Naphthalene

PROBLEM 1

a. Draw arrows to show the movement of electrons in going from one resonance contributor to the next in the cyclopentadienyl anion.
b. How many ring atoms share the negative charge?

PROBLEM 2 ◆

The pK_a of cyclopentadiene is 15, whereas the pK_a of cyclopentane is > 60. Explain the difference in pK_a values. (*Hint:* Compare the stabilities of the conjugate bases.)

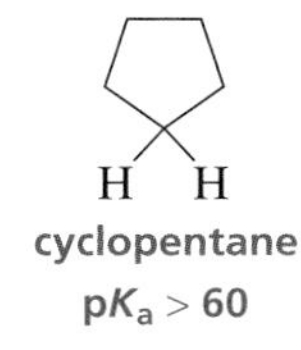

cyclopentane
pK_a > 60

PROBLEM 3 ◆

Which of the following are aromatic? Explain your choice.

cycloheptatriene

cycloheptatrienyl cation

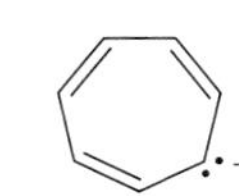
cycloheptatrienyl anion

PROBLEM 4 ◆

Which of the following are aromatic?

a. 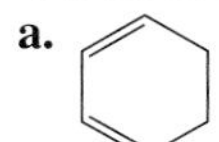

b. 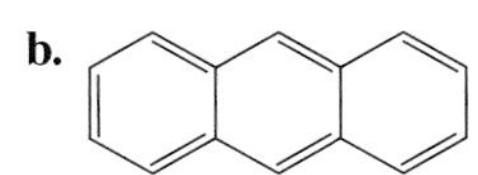

c. 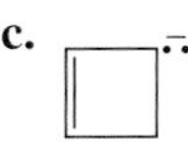

d. $CH_2{=}CHCH{=}CHCH{=}CH_2$

PROBLEM 5 ◆ SOLVED

a. How many monobromonaphthalenes are there?
b. How many monobromophenanthrenes are there?

3-D Molecules: 1-Chloronaphthalene; 2-Chloronaphthalene

Solution to 5a There are two monobromonaphthalenes. Substitution cannot occur at either of the carbons shared by both rings, because those carbons are not bonded to a hydrogen. Naphthalene is a flat molecule, so substitution for a hydrogen at any other carbon will result in one of the two compounds shown.

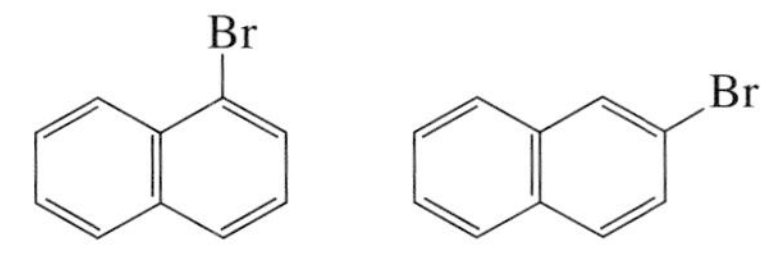

8.3 AROMATIC HETEROCYCLIC COMPOUNDS

A compound does not have to be a hydrocarbon to be aromatic. Many *heterocyclic compounds* are aromatic. A **heterocyclic compound** is a cyclic compound in which one or more of the ring atoms is an atom other than carbon. A ring atom that is not carbon is called a **heteroatom**. The name comes from the Greek word *heteros*, which means "different." The most common heteroatoms are N, O, and S.

heterocyclic compounds

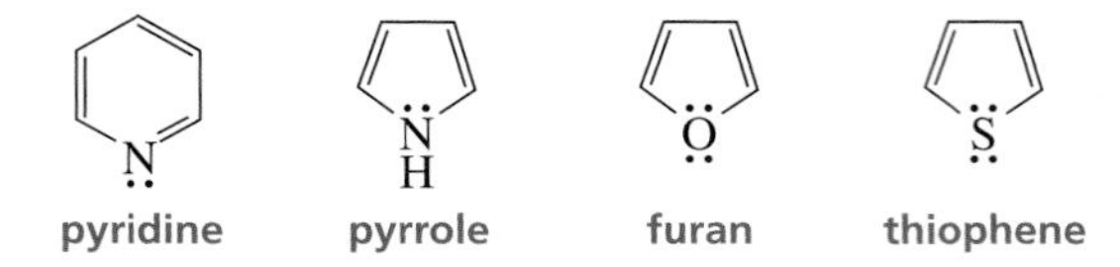
N
N
H
O
S
pyridine
pyrrole
furan
thiophene

Pyridine is an aromatic heterocyclic compound. Each of the six ring atoms of pyridine is sp^2 hybridized, which means that each has a p orbital; and the molecule contains three pairs of π electrons. (Do not be confused by the lone-pair electrons on the nitrogen; they are not π electrons.) Because nitrogen is sp^2 hybridized, it has three sp^2 orbitals and a p orbital. The p orbital is used to form the π bond. Two of nitrogen's sp^2 orbitals overlap the sp^2 orbitals of adjacent carbons, and nitrogen's third sp^2 orbital contains the lone pair.

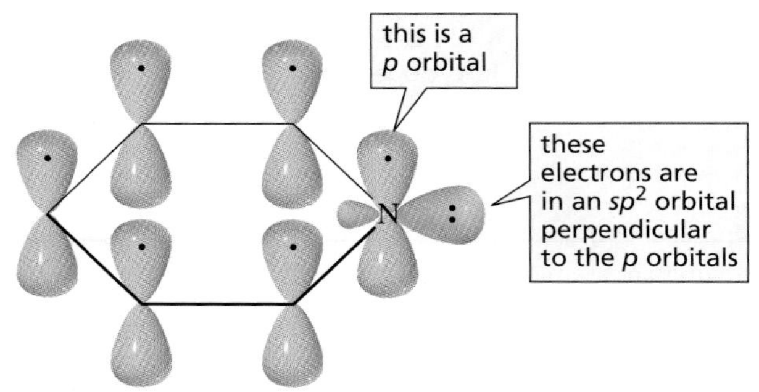
this is a p orbital
these electrons are in an sp² orbital perpendicular to the p orbitals
N

orbital structure of pyridine

It is not immediately apparent that the electrons represented as lone-pair electrons on the nitrogen atom of pyrrole are π electrons. The resonance contributors, however, show that the nitrogen is sp^2 hybridized and uses its three sp^2 orbitals to bond to two carbons and one hydrogen.

resonance contributors of pyrrole

The lone-pair electrons, therefore, must be in a p orbital that overlaps the p orbitals of adjacent carbons, forming a π bond; thus, they are π electrons. Pyrrole, therefore, has three pairs of π electrons and is aromatic.

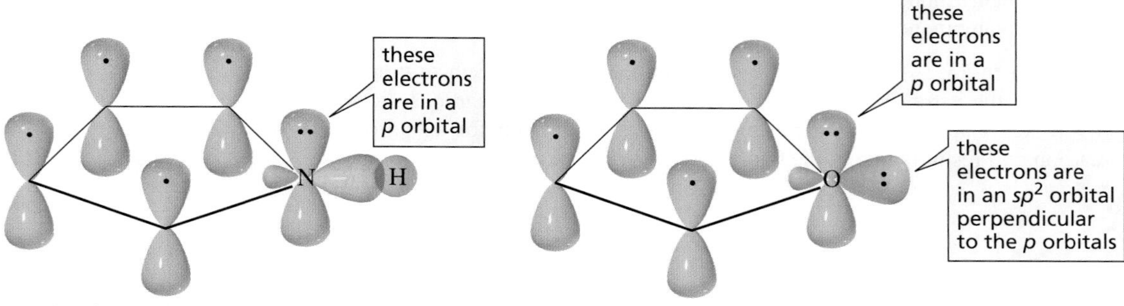
these electrons are in a p orbital
N
H
these electrons are in a p orbital
these electrons are in an sp² orbital perpendicular to the p orbitals
O

orbital structure of pyrrole

orbital structure of furan

Similarly, furan and thiophene are stable aromatic compounds. Both the oxygen in furan and the sulfur in thiophene are sp^2 hybridized and have one lone pair in an sp^2 orbital. The second lone pair of each compound is in a p orbital that overlaps the p orbitals of adjacent carbons, forming a π bond. Thus, they are π electrons.

resonance contributors of furan

Quinoline, indole, imidazole, purine, and pyrimidine are other examples of heterocyclic aromatic compounds.

quinoline indole imidazole purine pyrimidine

PROBLEM 6

a. Draw arrows to show the movement of electrons in going from one resonance contributor to the next in pyrrole.
b. How many ring atoms share the negative charge?

PROBLEM 7 ♦

Refer to the electrostatic potential maps on page 197 to answer the following questions:
a. Why is the bottom part of the electrostatic potential map of pyrrole blue?
b. Why is the bottom part of the electrostatic potential map of pyridine red?
c. Why is the center of the electrostatic potential map of benzene more red than the center of the electrostatic potential map of pyridine?

HEME AND CHLOROPHYLL

Heme, which is found in hemoglobin and myoglobin, contains an iron ion (Fe^{2+}) ligated by the four nitrogens of a **porphyrin ring system**. **Ligation** is the sharing of lone-pair electrons with a metal ion.

heme

a porphyrin ring system

The iron atoms in hemoglobin and myoglobin, in addition to being ligated to the four nitrogens of the porphyrin ring, are also ligated to a protein (globin) component, and the sixth ligand is oxygen or carbon dioxide. Carbon monoxide is about the same size and shape as O_2, but CO binds more tightly than O_2 to Fe^{2+}. Consequently, breathing carbon monoxide can be fatal because it binds to hemoglobin more tightly than oxygen does and interferes with the transport of oxygen through the bloodstream.

The extensive conjugated system of heme gives blood its characteristic red color. Concentrations as low as 1×10^{-8} M can be detected by UV spectroscopy (Section 7.12).

The ring system in chlorophyll *a*, the substance responsible for the green color of plants, is similar to a porphyrin ring

(continued)

system. The metal atom in chlorophyll *a* is magnesium (Mg^{2+}). Vitamin B_{12} also has a ring system similar to a porphyrin ring system, but in this case, the metal ion is cobalt (Co^{3+}) (Section 17.10).

chlorophyll *a*

Chlorophyll a is the pigment that makes plants look green. This highly conjugated compound absorbs purple light, which causes plants to reflect green light (Section 7.12).

PORPHYRIN, BILIRUBIN, AND JAUNDICE

The average human breaks down about 6 g of hemoglobin each day. The protein portion (globin) and the iron are reutilized, but the porphyrin ring is broken down, being reduced first to biliverdin, a green compound, and then to bilirubin, a yellow compound. If more bilirubin is formed than can be excreted by the liver, it accumulates in the blood. When its concentration there reaches a certain level, it diffuses into the tissues, giving them a yellow appearance. This condition is known as jaundice.

8.4 THE NOMENCLATURE OF MONOSUBSTITUTED BENZENES

Some monosubstituted benzenes are named simply by attaching "benzene" to the name of the substituent.

bromobenzene	chlorobenzene	nitrobenzene **used as a solvent in shoe polish**	ethylbenzene

Some monosubstituted benzenes have names that incorporate the name of the substituent. Unfortunately, such names have to be memorized.

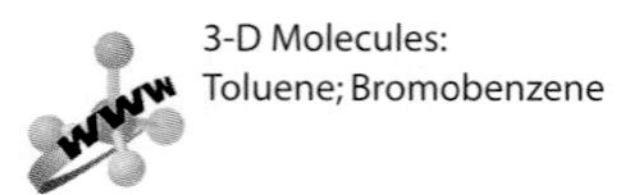

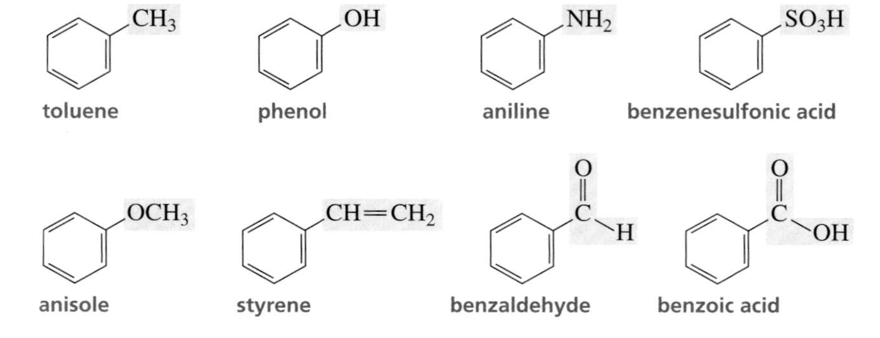

When a benzene ring is a substituent, it is called a **phenyl group**. A benzene ring with a methylene group is called a **benzyl group**.

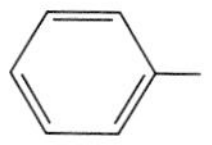

$-CH_2-$

a phenyl group 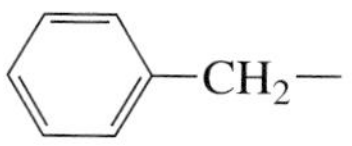 a benzyl group

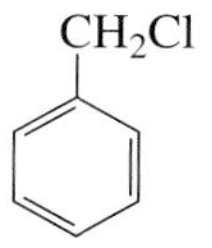

CH_2Cl

chloromethylbenzene
benzyl chloride

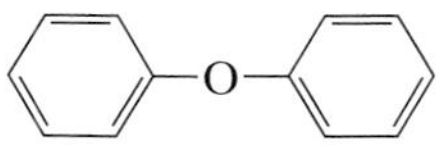

O

diphenyl ether

$-CH_2OCH_2-$

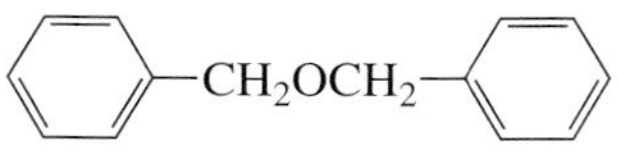

dibenzyl ether

An **aryl group** (Ar) is the general term for either a phenyl group or a substituted phenyl group, just as an alkyl group (R) is the general term for a group derived from an alkane. In other words, ArOH could be used to designate any of the following phenols:

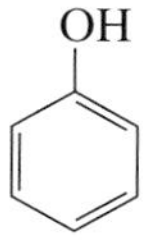

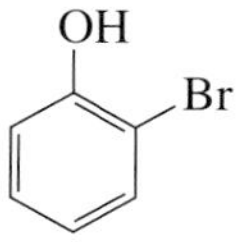

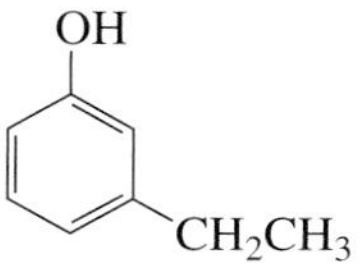

OH OH Br OH CH_2CH_3

THE TOXICITY OF BENZENE

Benzene, which has been widely used in chemical synthesis and has been frequently used as a solvent, is a toxic substance. The major adverse effects of chronic exposure are seen in the central nervous system and bone marrow; it causes leukemia and aplastic anemia. A higher than average incidence of leukemia, for example, has been found in industrial workers with long-term exposure to as little as 1 ppm benzene in the atmosphere. Toluene has replaced benzene as a solvent because, although it too is a central nervous system depressant, it does not cause leukemia or aplastic anemia. "Glue sniffing," a highly dangerous activity, produces narcotic central nervous system effects because glue contains toluene.

PROBLEM 8 ♦

Draw the structure of each of the following:

a. 2-phenylhexane **b.** benzyl alcohol **c.** 3-benzylpentane

8.5 HOW BENZENE REACTS

Aromatic compounds such as benzene undergo **electrophilic aromatic substitution reactions**: an electrophile substitutes for one of the hydrogens attached to the benzene ring.

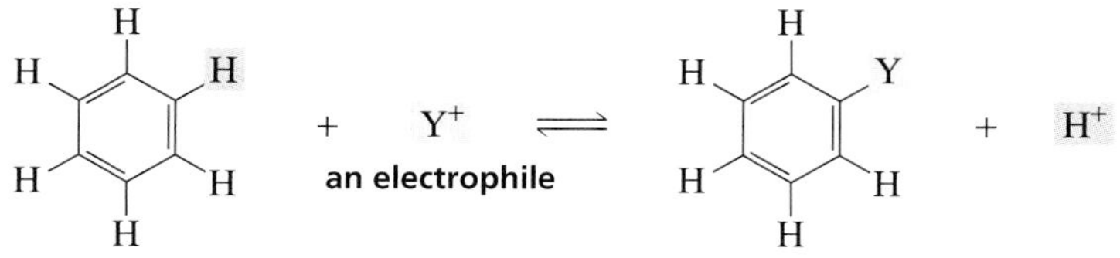

$$C_6H_6 + Y^+ \rightleftharpoons C_6H_5Y + H^+$$

an electrophile

Now let's look at why this substitution reaction occurs. The cloud of π electrons above and below the plane of its ring causes benzene to be a nucleophile. It will, therefore, react with an electrophile (Y^+). When an electrophile attaches itself to a benzene ring, a carbocation intermediate is formed.

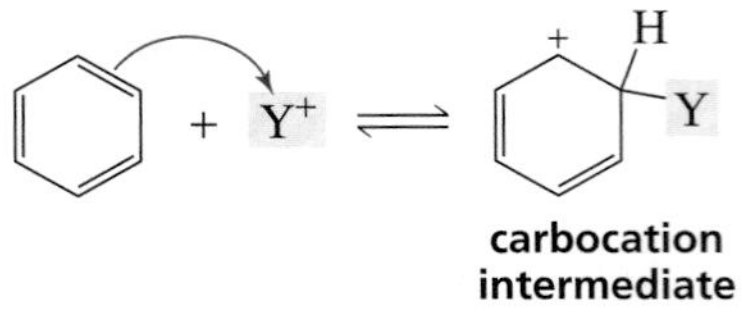

$$\text{benzene} + Y^+ \rightleftharpoons \text{carbocation intermediate}$$

carbocation intermediate

This description should remind you of the first step in an electrophilic addition reaction of an alkene: the alkene reacts with an electrophile and forms a carbocation intermediate (Section 5.1). In the second step of an electrophilic addition reaction of an alkene, the carbocation reacts with a nucleophile (Z^-) to form an addition product.

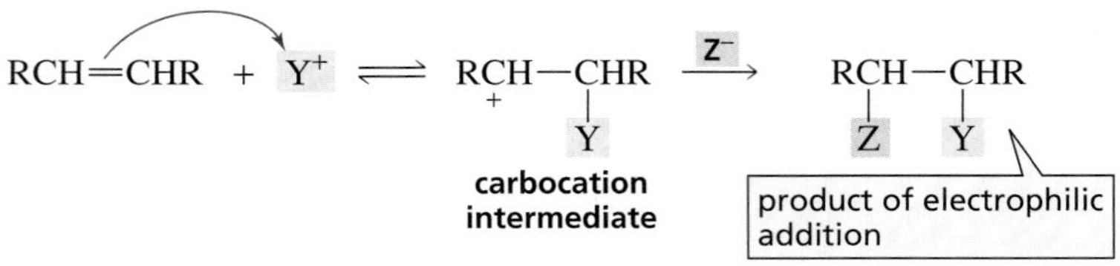

$$RCH{=}CHR + Y^+ \rightleftharpoons R\overset{+}{C}H{-}CHR(Y) \xrightarrow{Z^-} RCH(Z){-}CHR(Y)$$

carbocation intermediate

product of electrophilic addition

If the carbocation intermediate formed from the reaction of benzene with an electrophile were to react similarly with a nucleophile (depicted as path *b* in Figure 8.1), the product of the reaction would not be aromatic. But if the carbocation instead were to lose a proton from the site of electrophilic attack (depicted as path *a* in Figure 8.1), the aromaticity of the benzene ring would be restored.

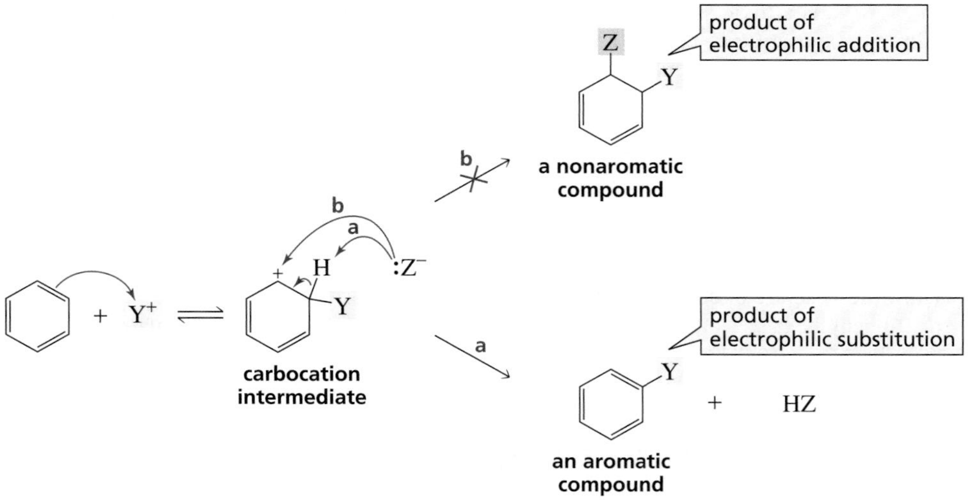

product of electrophilic addition

a nonaromatic compound

carbocation intermediate

product of electrophilic substitution

$+ \quad HZ$

an aromatic compound

▶ **Figure 8.1**
Reaction of benzene with an electrophile. Because the aromatic product is more stable, the reaction proceeds as (a) an electrophilic substitution reaction rather than (b) an electrophilic addition reaction.

Because the aromatic substitution product is much more stable than the nonaromatic addition product (Figure 8.2), benzene undergoes *electrophilic substitution reactions* that preserve aromaticity rather than *electrophilic addition reactions*—the reactions characteristic of alkenes—that would destroy aromaticity. The substitution reaction is more accurately called an **electrophilic aromatic substitution reaction**, since the electrophile substitutes for a hydrogen of an aromatic compound.

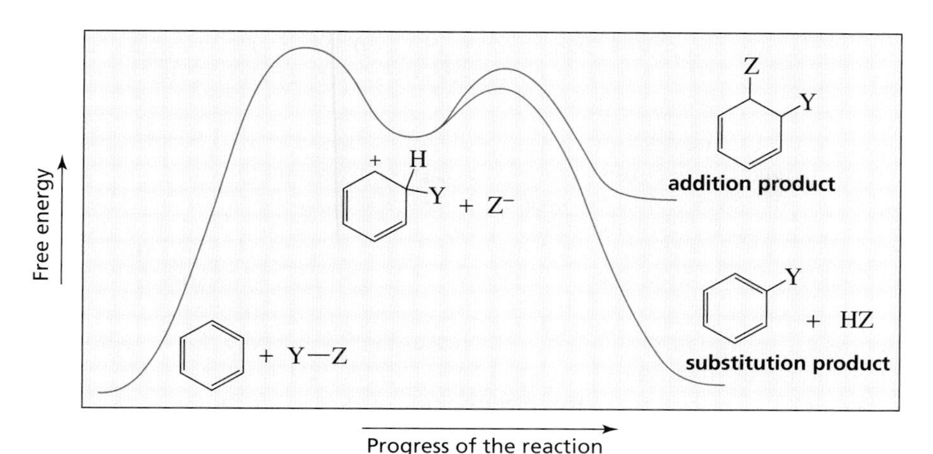

◀ **Figure 8.2**
Reaction coordinate diagrams for electrophilic substitution of benzene and electrophilic addition to benzene.

8.6 THE GENERAL MECHANISM FOR ELECTROPHILIC AROMATIC SUBSTITUTION REACTIONS

In an electrophilic aromatic substitution reaction, an electrophile becomes attached to a ring carbon, and an H^+ comes off the same ring carbon.

an electrophilic aromatic substitution reaction

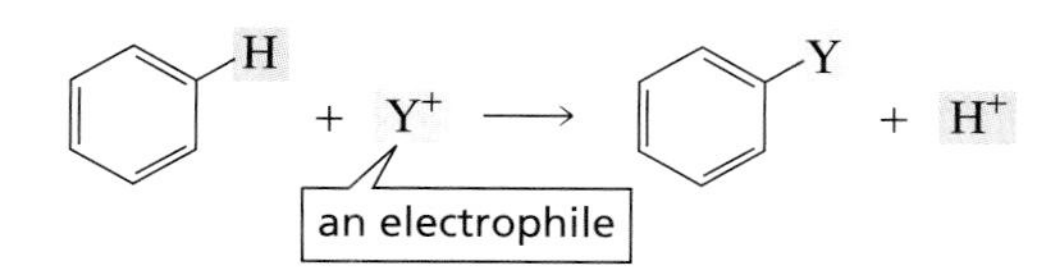

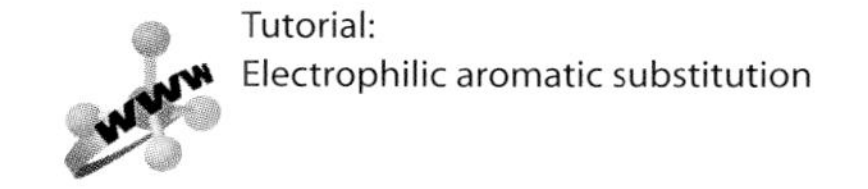

The following are the five most common electrophilic aromatic substitution reactions:

1. **Halogenation:** A bromine (Br) or a chlorine (Cl) substitutes for a hydrogen.
2. **Nitration:** A nitro (NO_2) group substitutes for a hydrogen.
3. **Sulfonation:** A sulfonic acid (SO_3H) group substitutes for a hydrogen.
4. **Friedel–Crafts acylation:** An acyl (RC=O) group substitutes for a hydrogen.
5. **Friedel–Crafts alkylation:** An alkyl (R) group substitutes for a hydrogen.

All of these electrophilic aromatic substitution reactions take place by the same two-step mechanism.

general mechanism for electrophilic aromatic substitution

the proton is removed from the carbon that has formed the new bond with the electrophile

slow

fast

$+ HB^+$

:B

a base in the reaction mixture

- Benzene (a nucleophile) reacts with an electrophile (Y^+), forming a carbocation intermediate. The structure of the carbocation intermediate can be approximated by three resonance contributors.
- A base (:B) in the reaction mixture pulls off a proton from the carbocation intermediate, and the electrons that held the proton move into the ring to reestablish its aromaticity. Notice that *the proton is always removed from the carbon that has formed the new bond with the electrophile.*

The first step is relatively slow and consumes energy because an aromatic compound is being converted into a much less stable nonaromatic carbocation intermediate (Figure 8.2). The second step is fast and releases energy because this step restores the stability-enhancing aromaticity.

In an electrophilic aromatic substitution reaction, an electrophile (Y^+) is put on a ring carbon, and the H^+ comes off the same ring carbon.

We will look at each of these five electrophilic aromatic substitution reactions individually. As you study them, notice that they differ only in how the electrophile (Y^+) needed to start the reaction is generated. Once the electrophile is formed, all five reactions follow the same two-step mechanism for electrophilic aromatic substitution.

8.7 HALOGENATION OF BENZENE

The bromination or chlorination of benzene requires a Lewis acid catalyst such as ferric bromide or ferric chloride. Recall that a *Lewis acid* is a compound that accepts a share in an electron pair (Section 2.9).

bromination

$$C_6H_6 + Br_2 \xrightarrow{FeBr_3} C_6H_5Br + HBr$$

bromobenzene

chlorination

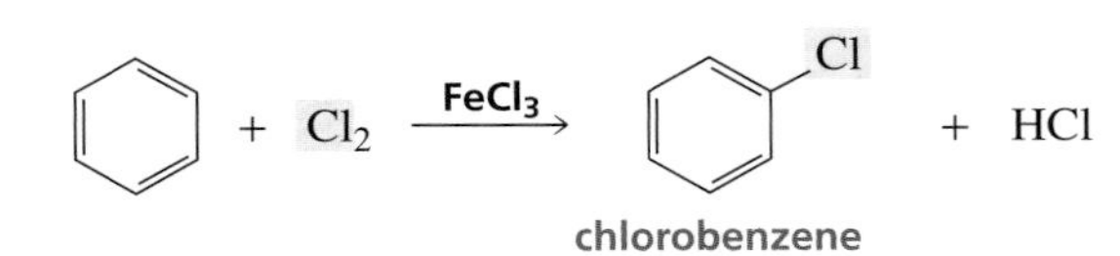

Chlorobenzene

Donating a lone pair to the Lewis acid weakens the Br—Br (or Cl—Cl) bond, which makes Br_2 (or Cl_2) a better electrophile.

$$:\ddot{Br}-\ddot{Br}: + FeBr_3 \longrightarrow :\ddot{Br}-\overset{+}{\ddot{Br}}-\overset{-}{Fe}Br_3$$

an electrophile — a better electrophile

The mechanism for bromination is shown below. For the sake of clarity, only one of the three resonance contributors of the carbocation intermediate is shown in this and subsequent mechanisms for electrophilic aromatic substitution reactions. Bear in mind, however, that each carbocation intermediate actually has the three resonance contributors shown in Section 8.6.

mechanism for bromination

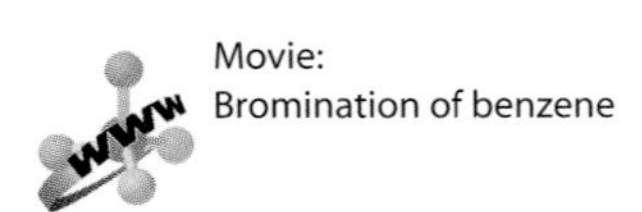

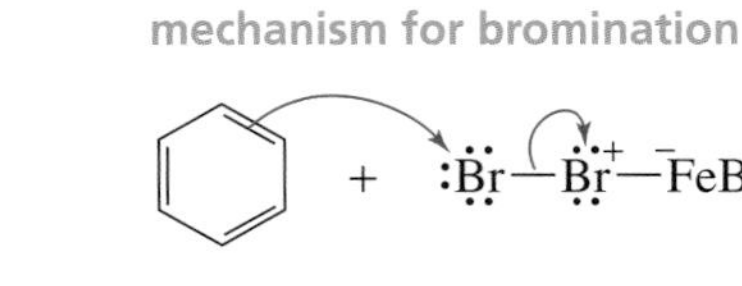

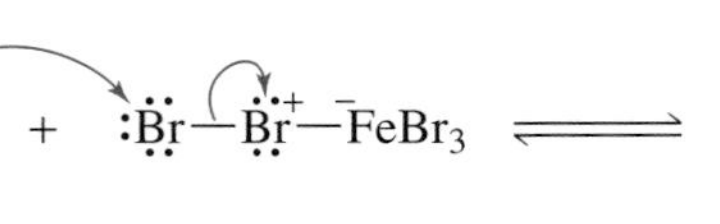

- The electrophile attaches to the benzene ring.
- A base (:B) from the reaction mixture (such as $^-FeBr_4$ or the solvent) removes a proton from the carbocation intermediate, thereby reforming the aromatic ring.

Chlorination of benzene takes place by the same mechanism as bromination.

mechanism for chlorination

$$\text{C}_6\text{H}_6 + :\ddot{\text{Cl}}-\ddot{\text{Cl}}^{+}-\bar{\text{F}}\text{eCl}_3 \rightleftharpoons \text{(carbocation with H and Cl)} + \bar{\text{F}}\text{eCl}_4 \xrightarrow{:\text{B}} \text{C}_6\text{H}_5\text{Cl} + \text{HB}^+$$

THYROXINE

Thyroxine, which is a hormone produced by the thyroid gland, regulates the metabolic rate in the body, causing an increase in the rate at which fats, carbohydrates, and proteins are metabolized. Without thyroxine, development in the young comes to a halt. Humans obtain thyroxine from tyrosine (an amino acid) and iodine. The thyroid gland is the only part of the body that uses iodine, which we acquire primarily from the iodized salt in our diet. An enzyme called iodoperoxidase converts the I^- we ingest to I^+, the electrophile needed to place an iodo substituent on the benzene ring.

HO–C₆H₄–$CH_2CH(^{+}NH_3)CO^-$ (C=O)

tyrosine
an amino acid

HO–C₆H₂I₂–O–C₆H₂I₂–$CH_2CH(^{+}NH_3)CO^-$ (C=O)

thyroxine

Chronically low levels of thyroxine cause enlargement of the thyroid gland, a condition known as goiter. Low thyroxine levels can be corrected by taking thyroxine orally. Synthroid, the most popular brand of thyroxine, is the sixth most-prescribed drug in the United States (see Table 21.1 on page 553).

8.8 NITRATION OF BENZENE

The nitration of benzene with nitric acid requires sulfuric acid as a catalyst.

nitration

$$\text{C}_6\text{H}_6 + \text{HNO}_3 \xrightarrow{\text{H}_2\text{SO}_4} \text{C}_6\text{H}_5\text{NO}_2 + \text{H}_2\text{O}$$

nitrobenzene

$HO-\overset{+}{N}(=O)-O^-$
nitric acid

To generate the necessary electrophile, sulfuric acid protonates nitric acid. Protonated nitric acid loses water to form a nitronium ion, the electrophile required for nitration.

$$\text{H}\ddot{\text{O}}-\text{NO}_2 + \text{H}-\text{OSO}_3\text{H} \rightleftharpoons \text{H}\overset{\text{H}}{\underset{+}{\text{O}}}-\text{NO}_2 + \text{HSO}_4^- \rightleftharpoons {}^{+}\text{NO}_2 + \text{H}_2\ddot{\text{O}}:$$

nitric acid — nitronium ion

$O=\overset{+}{N}=O$
nitronium ion

The mechanism for nitration is the same as the mechanisms described in Section 8.7.

mechanism for nitration

$$\text{C}_6\text{H}_6 + {}^{+}\text{NO}_2 \rightleftharpoons \text{(carbocation with H and NO}_2\text{)} \xrightarrow{:\text{B}} \text{C}_6\text{H}_5\text{NO}_2 + \text{HB}^+$$

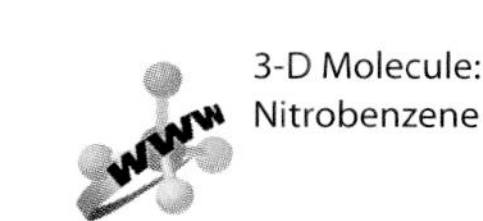

3-D Molecule: Nitrobenzene

- The electrophile attaches to the benzene ring.
- A base (:B) from the reaction mixture (for example, H_2O, HSO_4^-, or solvent) removes a proton from the carbocation intermediate, thereby reforming the aromatic ring.

8.9 SULFONATION OF BENZENE

Fuming sulfuric acid (a solution of SO_3 in sulfuric acid) or concentrated sulfuric acid is used to sulfonate aromatic rings.

sulfonation

$$\text{benzene} + H_2SO_4 \overset{\Delta}{\rightleftharpoons} \text{C}_6\text{H}_5\text{–}SO_3H + H_2O$$

benzenesulfonic acid

Take a minute to note the similarities in the mechanisms for forming the $^{+}SO_3H$ electrophile for sulfonation and the $^{+}NO_2$ electrophile for nitration.

$$HO{-}SO_3H + H{-}OSO_3H \rightleftharpoons H\overset{+}{O}(H){-}SO_3H + HSO_4^- \rightleftharpoons {}^{+}SO_3H + H_2O$$

sulfuric acid — sulfonium ion

The mechanism for sulfonation is the same as the other mechanisms we have seen for electrophilic aromatic substitution.

mechanism for sulfonation

$$\text{benzene} + {}^{+}SO_3H \rightleftharpoons \text{carbocation intermediate (H, }SO_3H\text{)} \xrightarrow{:B} \text{C}_6\text{H}_5\text{–}SO_3H + HB^+$$

- The electrophile attaches to the benzene ring.
- A base (:B) from the reaction mixture removes a proton from the carbocation intermediate, thereby reforming the aromatic ring.

BIOGRAPHY

Charles Friedel (1832–1899), *born in Strasbourg, France, was a professor of chemistry and director of research at the Sorbonne. At one point, his interest in mineralogy led him to attempt to make synthetic diamonds. He met James Crafts when they both were doing research at L'Ecòle de Médicine in Paris. They collaborated scientifically for most of their lives, discovering the Friedel–Crafts reactions in Friedel's laboratory in 1877.*

8.10 FRIEDEL–CRAFTS ACYLATION OF BENZENE

Two electrophilic substitution reactions bear the names of chemists Charles Friedel and James Crafts. *Friedel–Crafts acylation* places an acyl group on a benzene ring, and *Friedel–Crafts alkylation* places an alkyl group on a benzene ring.

$$R{-}C(=O){-} \qquad\qquad R{-}$$

an acyl group — an alkyl group

An acyl chloride is used to generate the electrophile for a **Friedel–Crafts acylation. An acyl chloride** has a Cl in place of the OH group of a carboxylic acid.

Friedel–Crafts acylation

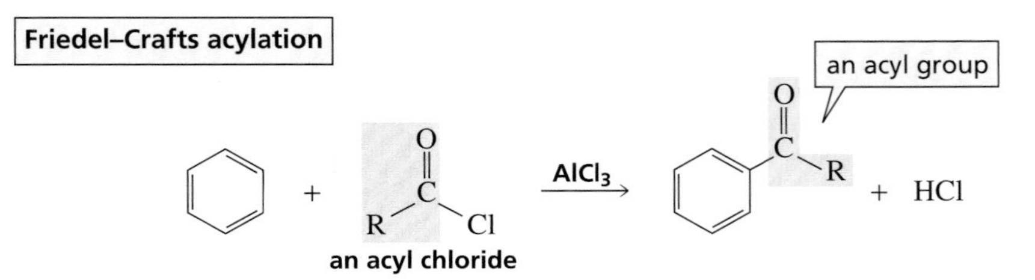

The electrophile (an acylium ion) is formed by the reaction of the acyl chloride with $AlCl_3$, a Lewis acid.

$$\text{RCOCl} + AlCl_3 \longrightarrow R{-}\overset{+}{C}{=}\ddot{O}{:} + {}^{-}AlCl_4$$

an acyl chloride   an acylium ion

The mechanism for Friedel–Crafts acylation is shown below.

mechanism for Friedel–Crafts acylation

$$C_6H_6 + R{-}\overset{+}{C}{=}\ddot{O}{:} \rightleftharpoons \text{(carbocation intermediate with H and :B)} \longrightarrow C_6H_5COR + HB^+$$

- The electrophile attaches to the benzene ring.
- A base (:B) from the reaction mixture removes a proton from the carbocation intermediate, thereby reforming the aromatic ring.

BIOGRAPHY

James Mason Crafts (1839–1917) *was born in Boston, the son of a woolen-goods manufacturer. He graduated from Harvard in 1858 and was a professor of chemistry at Cornell University and at the Massachusetts Institute of Technology. He was president of MIT from 1897 to 1900, when he was forced to retire because of poor health.*

PROBLEM 9

Write the mechanism for the following reaction:

$$C_6H_6 + H_3C{-}CO{-}Cl \xrightarrow{AlCl_3} C_6H_5{-}CO{-}CH_3 + HCl$$

8.11 FRIEDEL–CRAFTS ALKYLATION OF BENZENE

Friedel–Crafts alkylation places an alkyl group on a benzene ring.

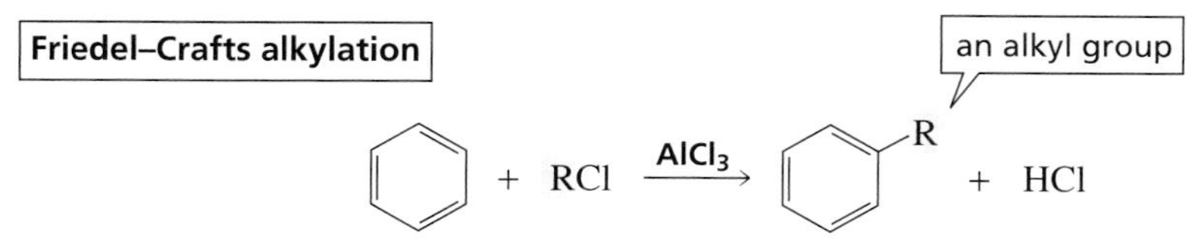

The electrophile in this reaction is a carbocation that is formed from the reaction of an alkyl halide with $AlCl_3$. Alkyl fluorides, alkyl chlorides, alkyl bromides, and alkyl iodides can all be used.

$$R{-}\ddot{\underset{\cdot\cdot}{Cl}}{:} + AlCl_3 \longrightarrow R^+ + {}^{-}AlCl_4$$

an alkyl halide   a carbocation

The mechanism for Friedel-Crafts alkylation is shown below.

mechanism for Friedel–Crafts alkylation

$$C_6H_6 + R^+ \longrightarrow \text{(carbocation intermediate with H and :B)} \longrightarrow C_6H_5R + HB^+$$

- The electrophile (R^+) attaches to the benzene ring.
- A base (:B) from the reaction mixture removes a proton from the carbocation intermediate, thereby reforming the aromatic ring.

In addition to reacting with carbocations generated from alkyl halides, benzene can react with carbocations generated from the reaction of an alkene with an acid (Section 5.1).

alkylation of benzene by an alkene

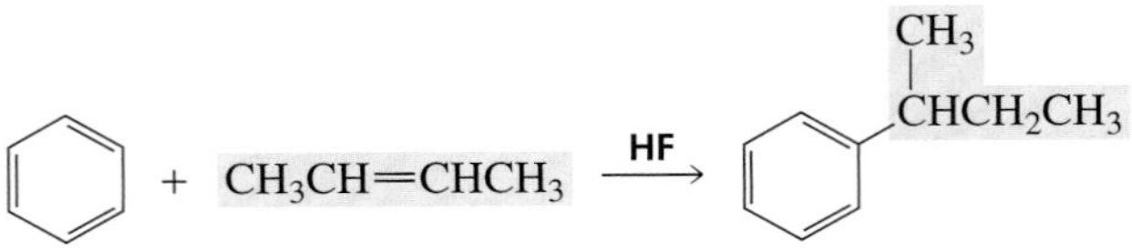

$$\text{C}_6\text{H}_6 + CH_3CH{=}CHCH_3 \xrightarrow{\text{HF}} \text{C}_6\text{H}_5\text{–}\underset{\substack{|\\ \text{CH}_3}}{\text{CH}}CH_2CH_3$$

sec-**butylbenzene**

PROBLEM 10

Write the mechanism for the alkylation of benzene by 2-butene + HF.

PROBLEM 11 ♦

What would be the product of a Friedel–Crafts alkylation reaction using the following alkyl chlorides?

a. CH_3CH_2Cl **b.** $CH_3CH_2CHCH_3$ (with Cl on CH) **c.** $CH_2{=}CHCH_2Cl$

8.12 HOW SOME SUBSTITUENTS ON A BENZENE RING CAN BE CHEMICALLY CHANGED

Benzene rings with substituents other than those listed in Section 8.6 can be prepared by first synthesizing one of those substituted benzenes and then chemically changing the substituent. For example, carbonyl groups and nitro groups can be reduced by catalytic hydrogenation (Section 5.12). When an organic compound is reduced, the number of C—H (or N—H) bonds increases.

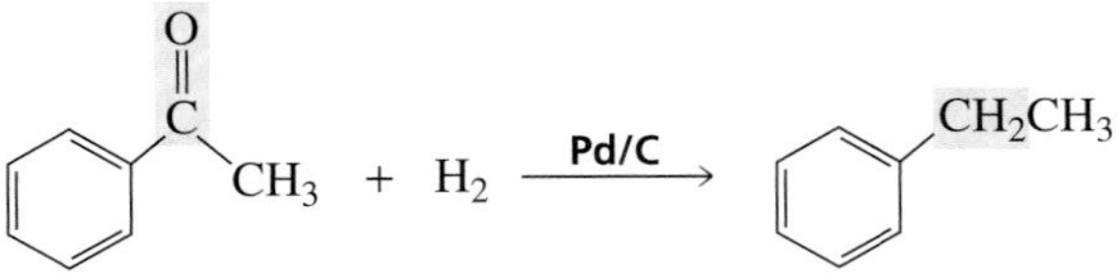

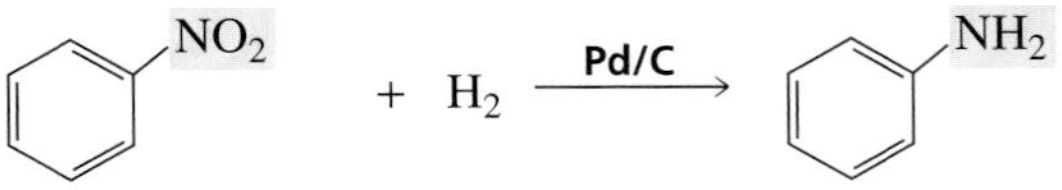

$$\text{C}_6\text{H}_5\text{–}\overset{\text{O}}{\overset{\|}{\text{C}}}\text{–}CH_3 + H_2 \xrightarrow{\text{Pd/C}} \text{C}_6\text{H}_5\text{–}CH_2CH_3$$

$$\text{C}_6\text{H}_5\text{–}NO_2 + H_2 \xrightarrow{\text{Pd/C}} \text{C}_6\text{H}_5\text{–}NH_2$$

Because benzene is an unusually stable compound, its double bonds do not react with H_2 under these conditions.

An alkyl group bonded to a benzene ring can be oxidized to a COOH group. When an organic compound is *oxidized,* either the number of C—O bonds increases or the number of C—H bonds decreases. Chromic acid (H_2CrO_4) is a commonly used oxidizing agent; the reaction requires heat. (Because the benzene ring is so stable, it will not be oxidized—only the alkyl group is oxidized.)

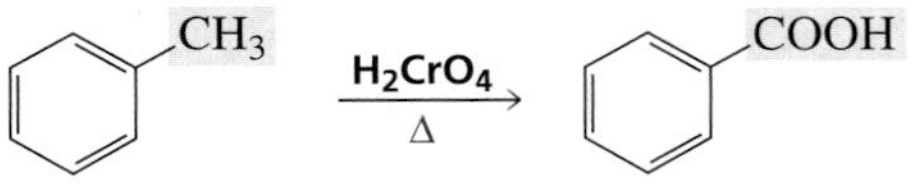

$$\underset{\textbf{toluene}}{\text{C}_6\text{H}_5\text{–}CH_3} \xrightarrow[\Delta]{H_2CrO_4} \underset{\textbf{benzoic acid}}{\text{C}_6\text{H}_5\text{–}COOH}$$

Regardless of the length of the alkyl substituent, it will be oxidized to a COOH group, provided that a hydrogen is bonded to the benzylic carbon.

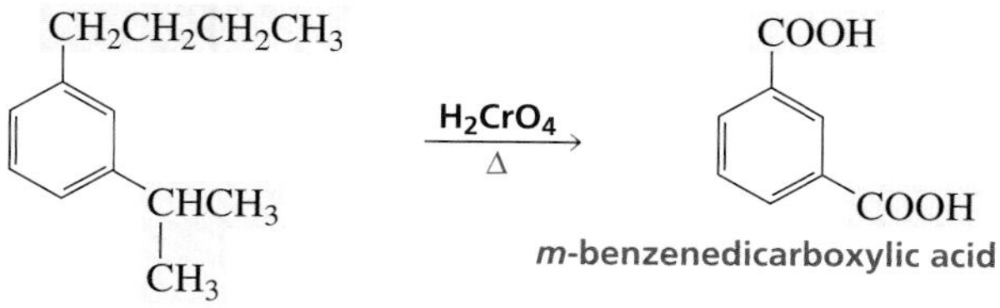

$CH_2CH_2CH_2CH_3$, $CHCH_3$, CH_3 $\xrightarrow[\Delta]{H_2CrO_4}$ COOH, COOH

m-butylisopropylbenzene → *m*-benzenedicarboxylic acid

If the alkyl group lacks a benzylic hydrogen, the oxidation reaction will not occur because the first step in the oxidation reaction is removal of a hydrogen from the benzylic carbon.

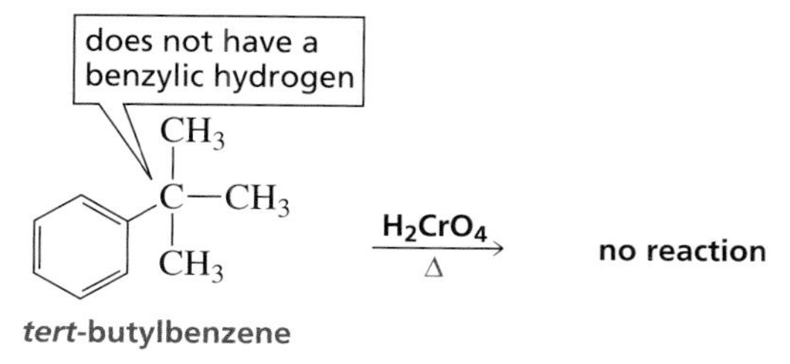

PROBLEM 12 ♦

Give the product of the following reactions:

a. 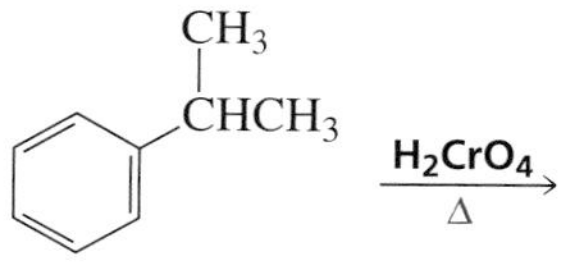 (benzene with $CH(CH_3)CH_3$: CH_3, $CHCH_3$) $\xrightarrow[\Delta]{H_2CrO_4}$

b. 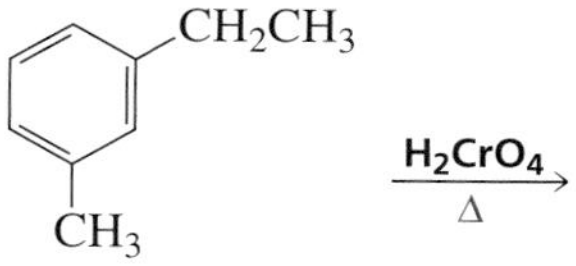 (benzene with CH_2CH_3 and CH_3) $\xrightarrow[\Delta]{H_2CrO_4}$

PROBLEM 13

Show how the following compounds could be prepared from benzene:

a. 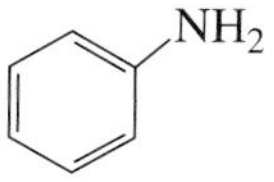 NH_2 **b.** 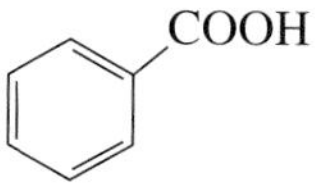 COOH **c.** 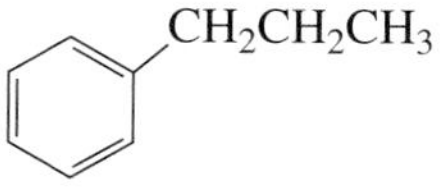 $CH_2CH_2CH_3$

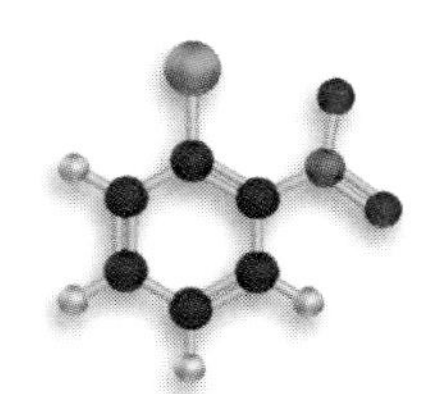

ortho-Chloronitrobenzene

8.13 THE NOMENCLATURE OF DISUBSTITUTED BENZENES

The relative positions of two substituents on a benzene ring can be indicated either by numbers or by the prefixes *ortho, meta*, and *para*. Adjacent substituents are called *ortho*, substituents separated by one carbon are called *meta*, and substituents located opposite one another are designated *para*. Often, only the abbreviations for these prefixes (*o*, *m*, *p*) are used in compounds' names.

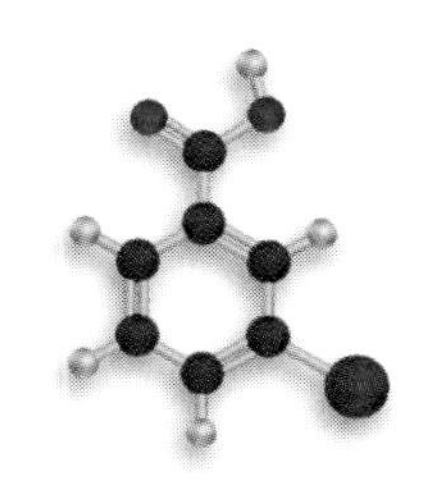

meta-Bromobenzoic acid

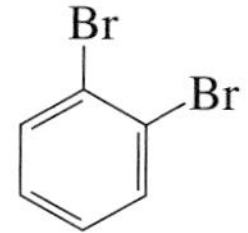

1,2-dibromobenzene
ortho-dibromobenzene
o-dibromobenzene

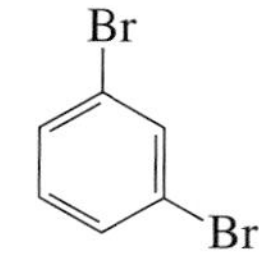

1,3-dibromobenzene
meta-dibromobenzene
m-dibromobenzene

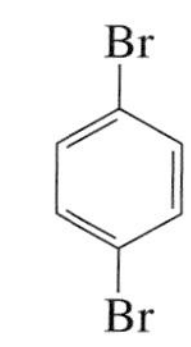

1,4-dibromobenzene
para-dibromobenzene
p-dibromobenzene

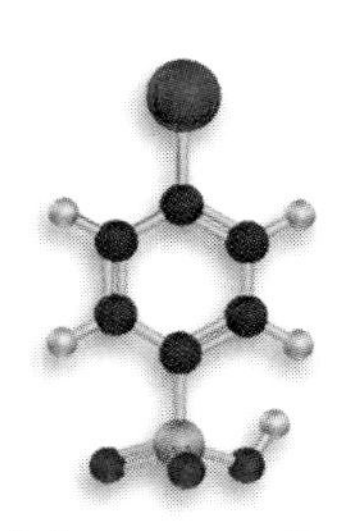

para-Iodobenzenesulfonic acid

If the two substituents are different, they are listed in alphabetical order. The first mentioned substituent is given the 1-position, and the ring is numbered in the direction that gives the second substituent the lowest possible number.

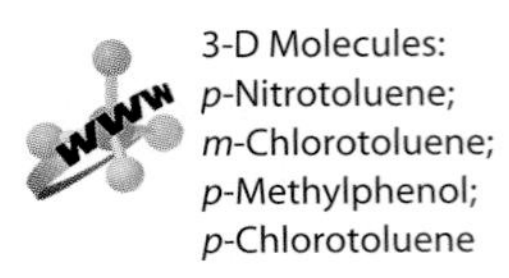

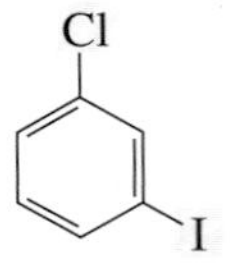

3-D Molecules:
p-Nitrotoluene;
m-Chlorotoluene;
p-Methylphenol;
p-Chlorotoluene

1-chloro-3-iodobenzene
meta-chloroiodobenzene

1-bromo-3-nitrobenzene
meta-bromonitrobenzene

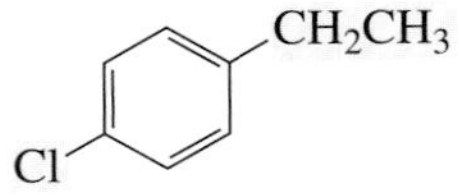

1-chloro-4-ethylbenzene
para-chloroethylbenzene

If one of the substituents can be incorporated into a name (Section 8.4), that name is used and the incorporated substituent is given the 1-position.

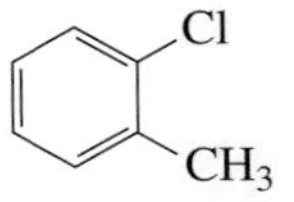

2-chlorotoluene
ortho-chlorotoluene
not
ortho-chloromethylbenzene

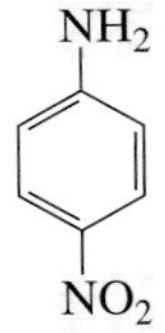

4-nitroaniline
para-nitroaniline
not
para-aminonitrobenzene

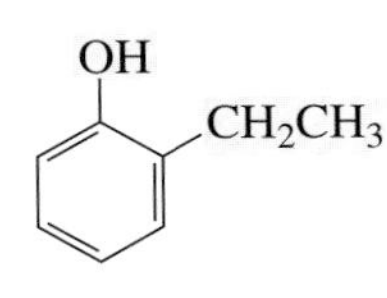

2-ethylphenol
ortho-ethylphenol
not
ortho-ethylhydroxybenzene

PROBLEM 14 ◆

Draw the structure of each of the following:

a. *p*-bromophenol
b. *o*-nitroaniline
c. 2-bromo-4-iodophenol
d. 2,5-dinitrobenzaldehyde

PROBLEM 15 ◆

Name the following:

a. 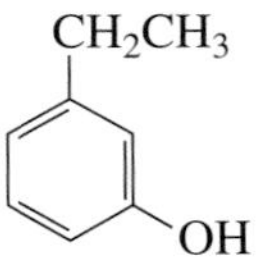 CH_2CH_3, OH
b. 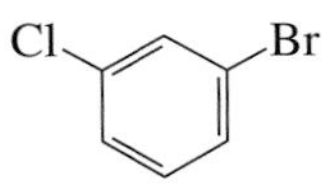 Cl, Br
c. 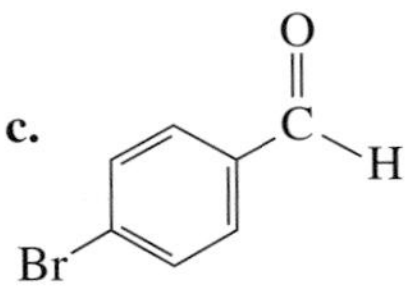 Br, C(=O)H
d. 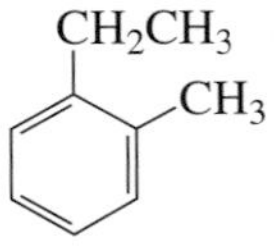 CH_2CH_3, CH_3

PROBLEM 16 ◆

Correct the following incorrect names:

a. 2,4,6-tribromobenzene
b. 3-hydroxynitrobenzene
c. *para*-methylbromobenzene
d. 1,6-dichlorobenzene

8.14 THE EFFECT OF SUBSTITUENTS ON REACTIVITY

Like benzene, substituted benzenes undergo the five electrophilic aromatic substitution reactions listed in Section 8.6: halogenation, nitration, sulfonation, acylation, and alkylation. Now we need to find out whether a substituted benzene is more reactive or less reactive than benzene itself. The answer depends on the substituent. Some substituents make the ring more reactive than benzene toward electrophilic aromatic substitution, and some make it less reactive than benzene.

The slow step of an electrophilic aromatic substitution reaction is the addition of an electrophile to the nucleophilic aromatic ring to form a carbocation intermediate (Section 8.6). *Substituents that donate electrons into the benzene ring stabilize both the carbocation intermediate and the transition state leading toward its formation* (Section 5.2), *thereby increasing the rate of electrophilic aromatic substitution;* these are called **activating substituents**. *In contrast, substituents that withdraw electrons from the benzene ring destabilize the carbocation intermediate and the transition state leading toward its formation, thereby decreasing the rate of electrophilic aromatic substitution;* these are called **deactivating substituents**.

Electron-donating substituents increase the reactivity of the benzene ring toward electrophilic aromatic substitution.

Electron-withdrawing substituents decrease the reactivity of the benzene ring toward electrophilic aromatic substitution.

relative rates of electrophilic substitution

Z donates electrons into the benzene ring — Z-substituted benzene > benzene > Y-substituted benzene — Y withdraws electrons from the benzene ring

There are two ways substituents can donate electrons—*inductively* or by *resonance.* Substituents can also withdraw electrons *inductively* or by *resonance.*

Donating and Withdrawing Electrons Inductively

If a substituent that is bonded to a benzene ring is *less electron withdrawing than a hydrogen*, the electrons in the σ bond that attaches the substituent to the benzene ring will move toward the ring more readily than will those in the σ bond that attaches a hydrogen to the ring. Such a substituent donates electrons inductively compared with a hydrogen. Donation of electrons through a σ bond is called **inductive electron donation**. We have seen that alkyl substituents (such as CH_3) donate electrons inductively compared with a hydrogen (Section 5.2).

If a substituent that is bonded to a benzene ring is *more electron withdrawing than a hydrogen*, it will draw the σ electrons away from the benzene ring more strongly than a hydrogen will. Withdrawal of electrons through a σ bond is called **inductive electron withdrawal**. The $^+NH_3$ group is an example of a substituent that withdraws electrons inductively because it is more electronegative than a hydrogen.

Donating and Withdrawing Electrons by Resonance

If a substituent has a lone pair on the atom directly attached to the benzene ring, the lone pair can be delocalized into the ring; these substituents are said to **donate electrons by resonance**. Substituents such as NH_2, OH, OR, and Cl donate electrons by resonance. These substituents also withdraw electrons inductively because the atom attached to the benzene ring is more electronegative than a hydrogen.

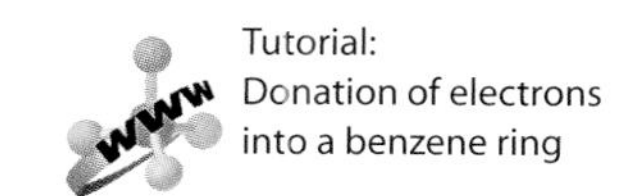

donating electrons by resonance into a benzene ring

$:\ddot{O}CH_3$ ⟷ $^+\ddot{O}CH_3$ ⟷ $^+\ddot{O}CH_3$ ⟷ $^+\ddot{O}CH_3$ ⟷ $:\ddot{O}CH_3$

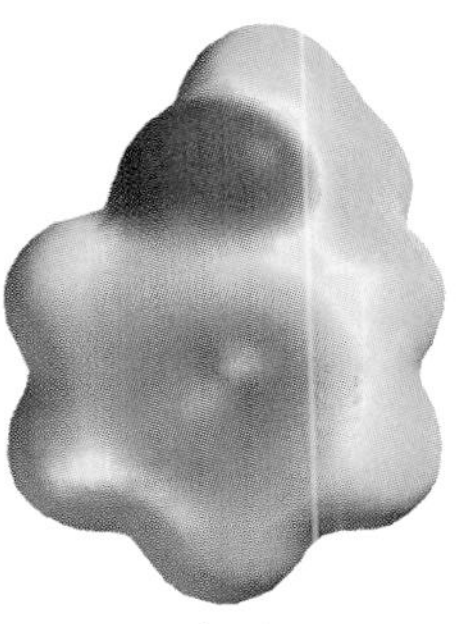

anisole

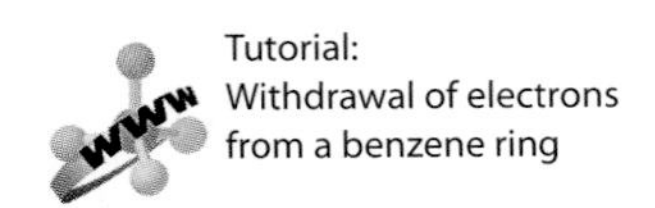

If a substituent is attached to the benzene ring by an atom that is doubly or triply bonded to a more electronegative atom, the π electrons of the ring can be delocalized onto the substituent; these substituents are said to **withdraw electrons by resonance**. Substituents such as C=O, C≡N, SO_3H, and NO_2 withdraw electrons by resonance (see the structures on page 217). These substituents also withdraw electrons inductively because the atom attached to the benzene ring has a full or partial positive charge and, therefore, is more electronegative than a hydrogen.

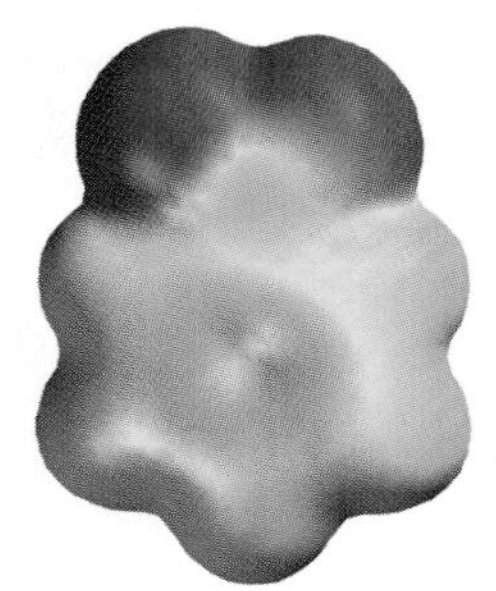

nitrobenzene

withdrawing electrons by resonance from a benzene ring

PROBLEM 17 ◆

For each of the following substituents, indicate whether it donates electrons inductively, withdraws electrons inductively, donates electrons by resonance, or withdraws electrons by resonance. (Effects should be compared with that of a hydrogen; remember that many substituents can be characterized in more than one way.)

a. Br **b.** CH_2CH_3 **c.** $\overset{O}{\overset{\|}{C}}CH_3$ **d.** $NHCH_3$ **e.** OCH_3 **f.** $\overset{+}{N}(CH_3)_3$

Relative Reactivity of Substituted Benzenes

The substituents shown in Table 8.1 are listed according to how they affect the reactivity of the benzene ring toward electrophilic aromatic substitution compared with benzene—in which the substituent is a hydrogen. *The* **activating substituents** *make the benzene ring more reactive toward electrophilic substitution; the* **deactivating substituents** *make the benzene ring less reactive.* Remember that activating substituents donate electrons into the ring and deactivating substituents withdraw electrons from the ring.

All the *activating substituents* (except for alkyl substituents) donate electrons into the ring by resonance and withdraw electrons from the ring inductively. The fact that these substituents have been found experimentally to make the benzene ring more reactive indicates that their electron donation into the ring by resonance is more significant than their inductive electron withdrawal from the ring.

We have seen that an alkyl substituent, compared with a hydrogen, donates electrons inductively.

The halogens are *weakly deactivating substituents*; they also donate electrons into the ring by resonance and withdraw electrons from the ring inductively. Because the halogens have been found experimentally to make the benzene ring less reactive, we can conclude that they withdraw electrons inductively more strongly than they donate electrons by resonance.

All the *substituents that are more strongly deactivating than the halogens* withdraw electrons both inductively and by resonance except for the ammonium ions

($^{+}NH_3$, $^{+}NH_2R$, $^{+}NHR_2$, and $^{+}NR_3$). The ammonium ions have no resonance effect, but the positive charge on the nitrogen atom causes them to strongly withdraw electrons inductively.

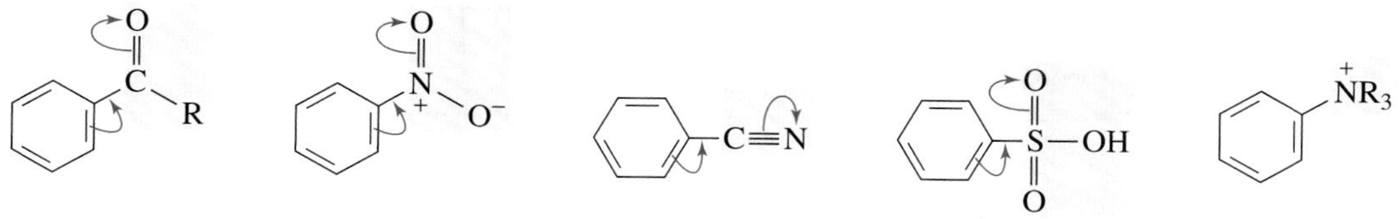

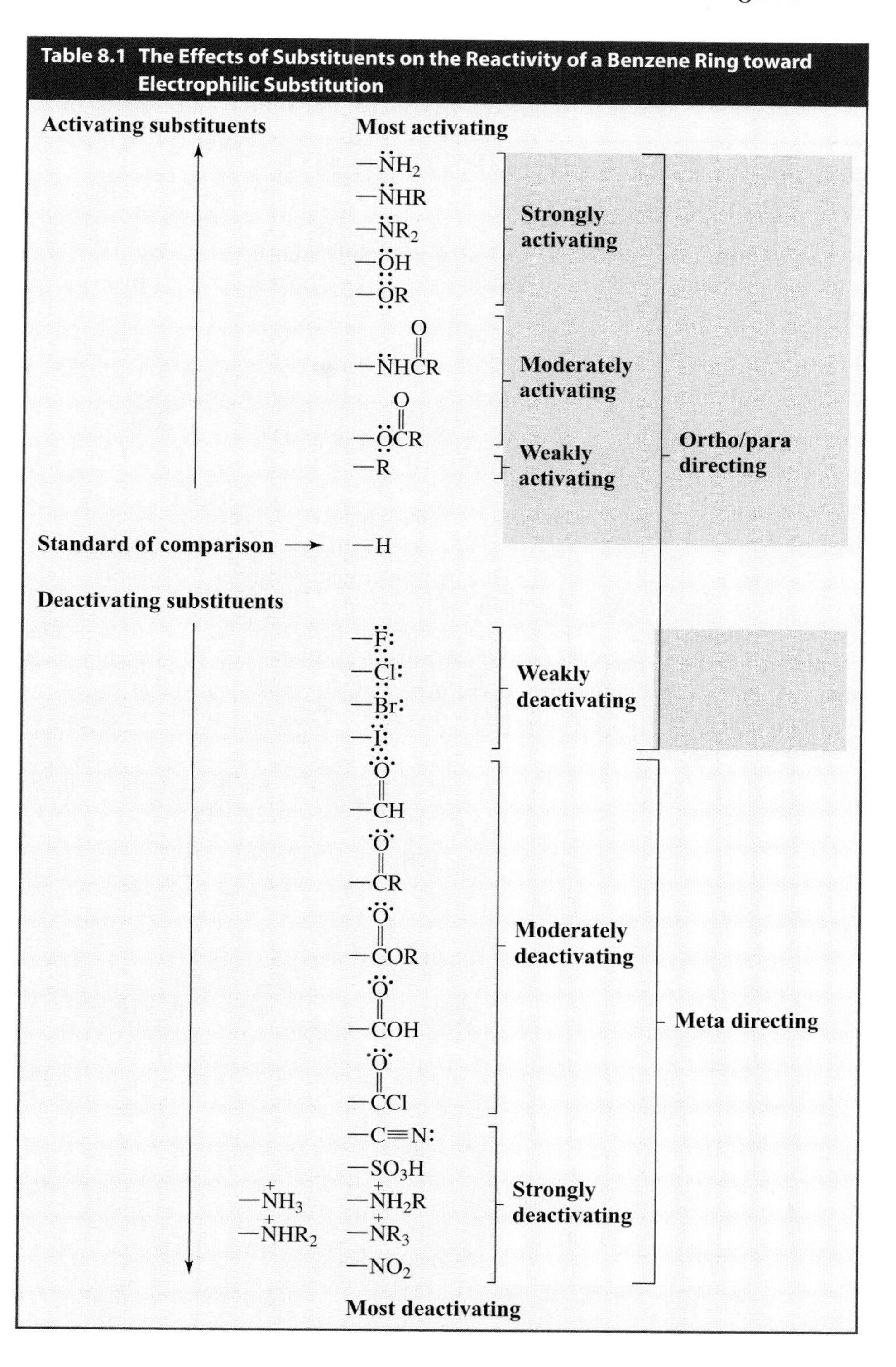

Table 8.1 The Effects of Substituents on the Reactivity of a Benzene Ring toward Electrophilic Substitution

	Substituent	Effect on reactivity	Directing effect
Activating substituents (↑)	**Most activating**		
	$-\ddot{N}H_2$	Strongly activating	Ortho/para directing
	$-\ddot{N}HR$	Strongly activating	Ortho/para directing
	$-\ddot{N}R_2$	Strongly activating	Ortho/para directing
	$-\ddot{O}H$	Strongly activating	Ortho/para directing
	$-\ddot{O}R$	Strongly activating	Ortho/para directing
	$-\ddot{N}HC(=O)R$	Moderately activating	Ortho/para directing
	$-\ddot{O}C(=O)R$	Moderately activating	Ortho/para directing
	$-R$	Weakly activating	Ortho/para directing
Standard of comparison →	$-H$		
Deactivating substituents (↓)	$-\ddot{F}:$	Weakly deactivating	Ortho/para directing
	$-\ddot{Cl}:$	Weakly deactivating	Ortho/para directing
	$-\ddot{Br}:$	Weakly deactivating	Ortho/para directing
	$-\ddot{I}:$	Weakly deactivating	Ortho/para directing
	$-CH(=O)$	Moderately deactivating	Meta directing
	$-CR(=O)$	Moderately deactivating	Meta directing
	$-COR(=O)$	Moderately deactivating	Meta directing
	$-COH(=O)$	Moderately deactivating	Meta directing
	$-CCl(=O)$	Moderately deactivating	Meta directing
	$-C{\equiv}N:$	Strongly deactivating	Meta directing
	$-SO_3H$	Strongly deactivating	Meta directing
$-\overset{+}{N}H_3$	$-\overset{+}{N}H_2R$	Strongly deactivating	Meta directing
$-\overset{+}{N}HR_2$	$-\overset{+}{N}R_3$	Strongly deactivating	Meta directing
	$-NO_2$	Strongly deactivating	Meta directing
	Most deactivating		

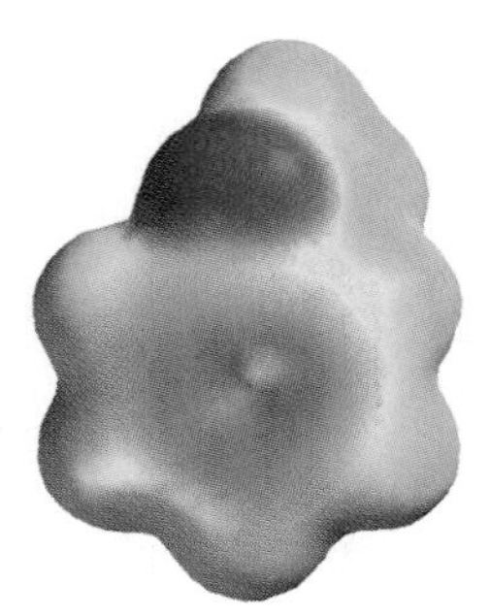
anisole

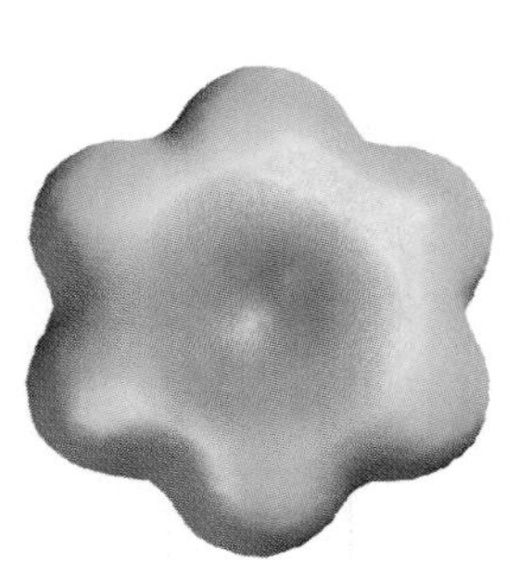
benzene

nitrobenzene

Take a minute to compare the electrostatic potential maps for anisole, benzene, and nitrobenzene. Notice that the electron-donating substituent of anisole (OCH_3) makes the ring more red (more negative), whereas the electron-withdrawing substituent of nitrobenzene (NO_2) makes the ring less red (less negative).

PROBLEM 18 ♦

List the following compounds in order of decreasing reactivity toward electrophilic aromatic substitution:

a. benzene, phenol, toluene, nitrobenzene, bromobenzene

b. dichloromethylbenzene, difluoromethylbenzene, toluene, chloromethylbenzene

8.15 THE EFFECT OF SUBSTITUENTS ON ORIENTATION

When a substituted benzene undergoes an electrophilic substitution reaction, where does the new substituent attach itself? Is the product of the reaction the ortho isomer, the meta isomer, or the para isomer?

X-benzene + Y^+ ⟶ ortho isomer or meta isomer or para isomer

The substituent already attached to the benzene ring determines the location of the new substituent. The attached substituent will have one of two effects: it will direct an incoming substituent either to the ortho *and* para positions, or it will direct an incoming substituent to the meta position. All activating substituents and the weakly deactivating halogens are **ortho–para directors**, and all substituents that are more deactivating than the halogens are **meta directors**. Thus, the substituents can be divided into three groups:

All activating substituents are ortho–para directors.

1. All *activating substituents* direct an incoming electrophile to the ortho and para positions.

toluene + Br_2 $\xrightarrow{FeBr_3}$ *o*-bromotoluene + *p*-bromotoluene

The weakly deactivating halogens are ortho–para directors.

2. The *weakly deactivating* halogens also direct an incoming electrophile to the ortho and para positions.

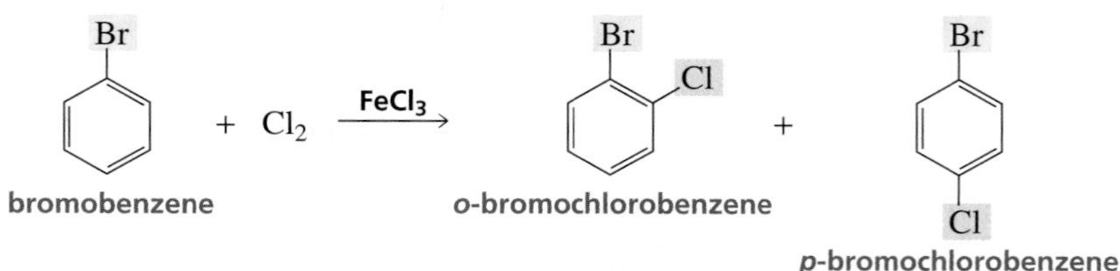

All deactivating substituents (except the halogens) are meta directors.

3. All *moderately deactivating and strongly deactivating* substituents direct an incoming electrophile to the meta position.

acetophenone + HNO_3 $\xrightarrow{H_2SO_4}$ *m*-nitroacetophenone

nitrobenzene + Br_2 $\xrightarrow{FeBr_3}$ *m*-bromonitrobenzene

Tutorial: Terms for the reactions of substituted benzenes

To understand why a substituent directs an incoming electrophile to a particular position, we must look at the stability of the carbocation intermediate, because as Figure 8.2 on page 207 shows, formation of the carbocation is the rate-determining step (Section 4.8).

When a substituted benzene undergoes an electrophilic substitution reaction, three different carbocation intermediates can be formed: an *ortho*-substituted carbocation, a *meta*-substituted carbocation, and a *para*-substituted carbocation (Figure 8.3). The relative stabilities of the three carbocations enable us to determine the preferred pathway of the reaction because the more stable the carbocation, the more stable the transition state for its formation, and the more rapidly it will be formed (Section 5.2).

When the substituent is one that can donate electrons by *resonance*, the carbocations formed by putting the incoming electrophile on the ortho and para positions have a fourth resonance contributor (highlighted in Figure 8.3). This is an especially stable resonance contributor because it is the only one whose atoms (except for hydrogen) all have complete octets (that is, all have outer shells that contain eight electrons; Section 1.3); it is obtained only by directing an incoming substituent to the ortho and para positions. Therefore, *all substituents that donate electrons by resonance are ortho–para directors.*

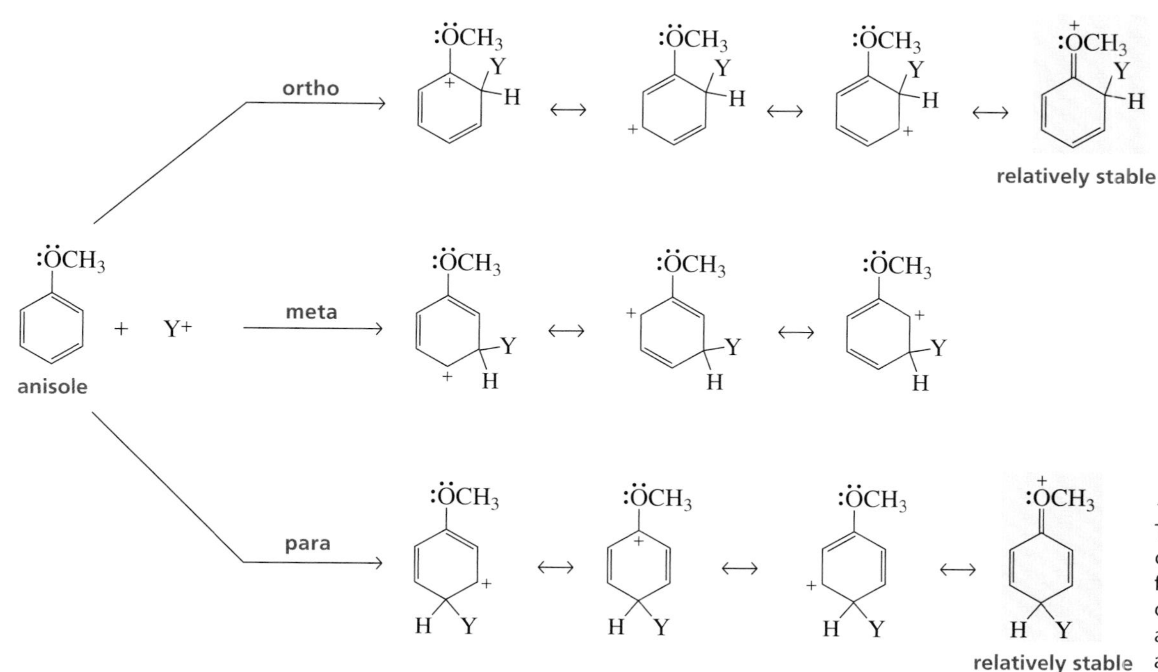

◀ **Figure 8.3**
The structures of the carbocation intermediates formed from the reaction of an electrophile with anisole at the ortho, meta, and para positions.

When the substituent is an alkyl group, the resonance contributors that are highlighted in Figure 8.4 are the most stable. In those contributors, the alkyl group is attached directly to the positively charged carbon and can stabilize it by inductive electron donation. A relatively stable resonance contributor is obtained only when the incoming group is directed to an ortho or para position. Therefore, *alkyl substituents are ortho–para directors.*

ortho, meta, para; most stable; toluene + Y^+

▶ **Figure 8.4**
The structures of the carbocation intermediates formed from the reaction of an electrophile with toluene at the ortho, meta, and para positions.

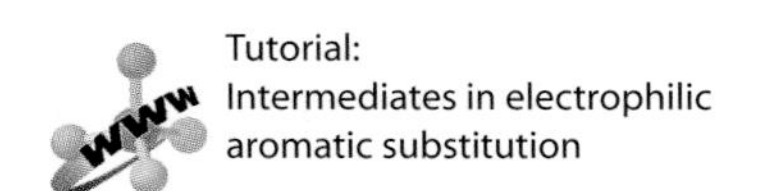

Tutorial:
Intermediates in electrophilic aromatic substitution

Substituents with a positive charge or a partial positive charge on the atom attached to the benzene ring will withdraw electrons inductively from the benzene ring, and most will withdraw electrons by resonance as well. For all such substituents, the resonance contributors highlighted in Figure 8.5 are the least stable because they have a positive charge on each of two adjacent atoms, so the most stable carbocation is

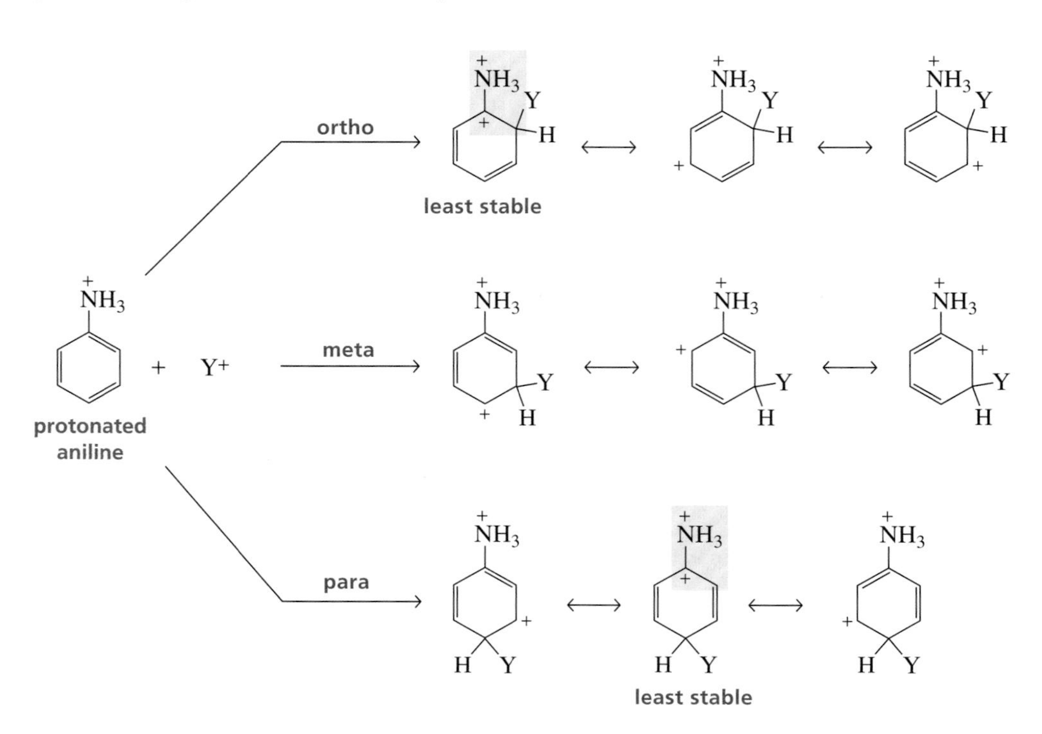

▶ **Figure 8.5**
The structures of the carbocation intermediates formed from the reaction of an electrophile with protonated aniline at the ortho, meta, and para positions.

formed when the incoming electrophile is directed to the meta position. Thus, *all substituents that withdraw electrons (except for the halogens, which are ortho–para directors because they donate electrons by resonance) are meta directors.*

Notice that the three possible carbocation intermediates in Figures 8.4 and 8.5 are the same, except for the substituent. The nature of that substituent determines whether the resonance contributors with the substituent directly attached to the positively charged carbon are the most stable (as with electron-donating substituents) or the least stable (as with electron-withdrawing substituents).

In summary, as shown in Table 8.1:

- all the activating substituents and the weakly deactivating halogens are ortho–para directors.
- all substituents more deactivating than the halogens are meta directors.

In other words, all substituents that donate electrons either by resonance or inductively are ortho–para directors; all substituents that cannot donate electrons are meta directors.

All substituents that donate electrons into the ring either inductively or by resonance are ortho–para directors.

All substituents that cannot donate electrons into the ring either inductively or by resonance are meta directors.

You do not need to resort to memorization to be able to identify a substituent as an ortho–para director or a meta director. It is easy to tell them apart: all ortho–para directors, except for alkyl groups, have at least one lone pair on the atom directly attached to the ring; all meta directors have a positive charge or a partial positive charge on the atom attached to the ring. Take a few minutes to examine the substituents listed in Table 8.1 to see that this is true.

PROBLEM 19

What product(s) would result from nitration of each of the following?

a. propylbenzene
b. bromobenzene
c. benzaldehyde
d. benzenesulfonic acid
e. cyclohexylbenzene
f. benzoic acid

8.16 SYNTHESIZING DISUBSTITUTED BENZENES

Designing the synthesis of a disubstituted benzene requires careful consideration of the order in which the substituents are to be placed on the ring. For example, if you want to synthesize *meta*-bromobenzenesulfonic acid, the sulfonic acid group has to be placed on the ring first, because that group will direct the bromo substituent to the desired meta position.

$$\text{benzene} \xrightarrow[\Delta]{H_2SO_4} \text{C}_6\text{H}_5SO_3H \xrightarrow[FeBr_3]{Br_2} m\text{-BrC}_6\text{H}_4SO_3H$$

m-bromobenzenesulfonic acid

However, if the desired product is *para*-bromobenzenesulfonic acid, the order of the two reactions must be reversed because only the bromo substituent is an ortho–para director.

$$\text{benzene} \xrightarrow[FeBr_3]{Br_2} \text{C}_6\text{H}_5Br \xrightarrow[\Delta]{H_2SO_4} p\text{-BrC}_6\text{H}_4SO_3H + o\text{-BrC}_6\text{H}_4SO_3H$$

p-bromobenzenesulfonic acid

o-bromobenzenesulfonic acid

Another question to consider is, at what point in a reaction sequence should a substituent be chemically modified? In the synthesis of *para*-chlorobenzoic acid from toluene, the methyl group is oxidized after it directs the chloro substituent to the para position. (*ortho*-Chlorobenzoic acid is also formed in this reaction.)

$$\text{toluene} \xrightarrow[\text{FeCl}_3]{\text{Cl}_2} p\text{-ClC}_6\text{H}_4\text{CH}_3 \xrightarrow[\Delta]{\text{H}_2\text{CrO}_4} \textit{para}\text{-chlorobenzoic acid}$$

In the synthesis of *meta*-chlorobenzoic acid, the methyl group is oxidized before chlorination because a meta director is needed to obtain the desired product.

$$\text{C}_6\text{H}_5\text{CH}_3 \xrightarrow[\Delta]{\text{H}_2\text{CrO}_4} \text{C}_6\text{H}_5\text{COOH} \xrightarrow[\text{FeCl}_3]{\text{Cl}_2} \textit{meta}\text{-chlorobenzoic acid}$$

Both substituents of *meta*-nitroacetophenone are meta directors. However, Friedel–Crafts reactions are the slowest of the electrophilic aromatic substitution reactions. Therefore, if the ring has a meta director, it will be too unreactive to undergo a Friedel–Crafts reaction. Thus the Friedel–Crafts acylation reaction must be carried out first because if the ring were nitrated first, the benzene ring of nitrobenzene would be too deactivated to undergo a Friedel–Crafts reaction.

$$\text{C}_6\text{H}_6 + \text{CH}_3\overset{\text{O}}{\overset{\|}{\text{C}}}\text{Cl} \xrightarrow{\text{AlCl}_3} \text{C}_6\text{H}_5\overset{\text{O}}{\overset{\|}{\text{C}}}\text{CH}_3 \xrightarrow[\text{H}_2\text{SO}_4]{\text{HNO}_3} m\text{-nitroacetophenone}$$

PROBLEM-SOLVING STRATEGY

Show how *meta*-ethylbenzenesulfonic acid could be synthesized from benzene.

The sulfonic acid group cannot be put on the ring first because a ring with a meta-directing sulfonic acid group cannot undergo a Friedel–Crafts reaction. The ethyl group is an ortho–para director; therefore, the two-carbon substituent must be put on the ring using a Friedel–Crafts acylation reaction rather than a Friedel–Crafts alkylation reaction so that the sulfonic acid group can be directed to the meta position. Reducing the carbonyl group forms the desired product.

$$\text{C}_6\text{H}_6 \xrightarrow[\text{AlCl}_3]{\text{CH}_3\overset{\text{O}}{\overset{\|}{\text{C}}}\text{Cl}} \text{C}_6\text{H}_5\overset{\text{O}}{\overset{\|}{\text{C}}}\text{CH}_3 \xrightarrow[\Delta]{\text{H}_2\text{SO}_4} m\text{-HO}_3\text{SC}_6\text{H}_4\overset{\text{O}}{\overset{\|}{\text{C}}}\text{CH}_3 \xrightarrow[\text{Pd/C}]{\text{H}_2} \textit{meta}\text{-ethylbenzenesulfonic acid}$$

Now continue on to Problem 20.

PROBLEM 20 ♦

Show how the following compounds could be synthesized from benzene:
a. *m*-chloronitrobenzene
b. *p*-chloronitrobenzene
c. *p*-ethylbenzene sulfonic acid

PROBLEM 21 *SOLVED*

Give the product(s) obtained from the reaction of each of the following compounds with one equivalent of Br_2 and a $FeBr_3$ catalyst:

a.

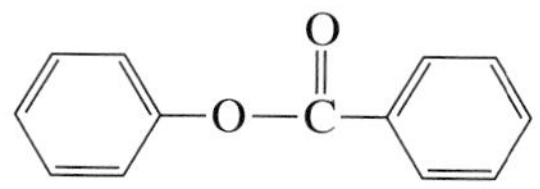

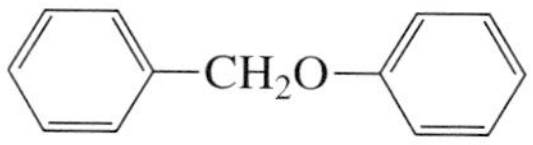

b.

c.

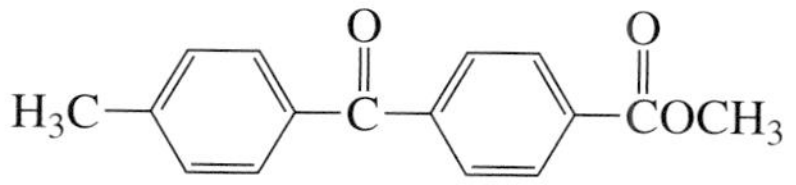

Solution to 21a The left-hand ring is attached to a substituent that activates the ring by donating electrons into the ring by resonance. The right-hand ring is attached to a substituent that deactivates the ring by withdrawing electrons from the ring by resonance.

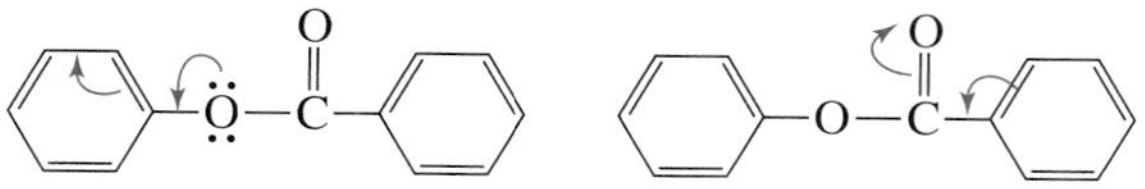

Therefore, the left-hand ring is more reactive toward electrophilic aromatic substitution. The substituent on the left-hand ring will direct the bromine to the ortho and para positions, giving the following products.

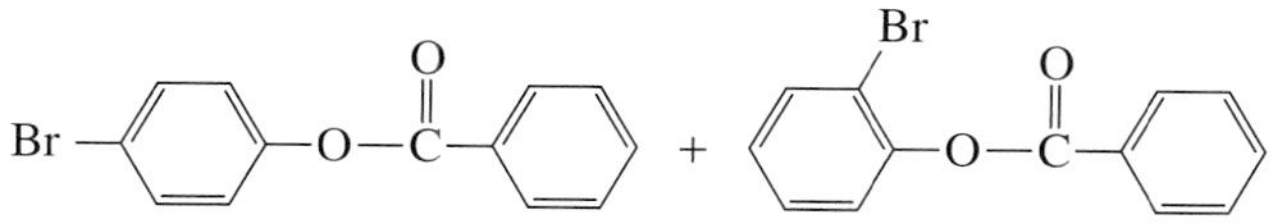

8.17 THE EFFECT OF SUBSTITUENTS ON pK_a

When a substituent either withdraws electrons from or donates electrons into a benzene ring, the pK_a values of substituted phenols, benzoic acids, and protonated anilines will reflect this withdrawal or donation.

Electron-withdrawing groups stabilize a base and, therefore, increase the strength of its conjugate acid. Electron-donating groups destabilize a base and this decreases the strength of its conjugate acid. (Remember, the stronger the acid, the more stable (weaker) its conjugate base.)

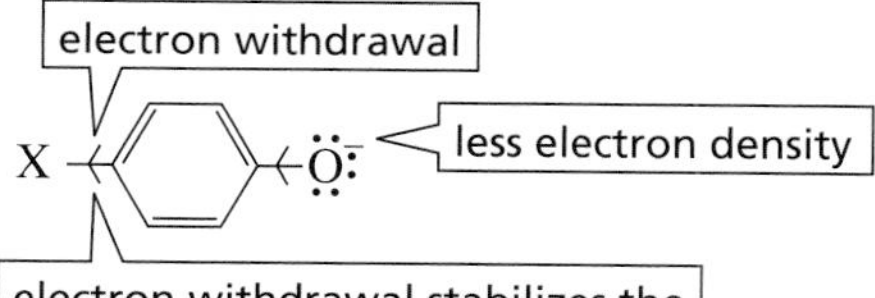
electron withdrawal
X
less electron density
electron withdrawal stabilizes the base by decreasing the electron density on the oxygen

electron donation

greater electron density

electron donation destabilizes the base by increasing the electron density on the oxygen

For example, the pK_a of phenol in H_2O at 25° C is 9.95. The pK_a of *para*-nitrophenol is lower (7.14) because the nitro substituent withdraws electrons from the ring, whereas the pK_a of *para*-methyl phenol is higher (10.19) because the methyl substituent donates electrons into the ring.

p-OCH_3 phenol	p-CH_3 phenol	phenol	p-Cl phenol	p-HC=O phenol	p-NO_2 phenol
OH / OCH_3	OH / CH_3	OH	OH / Cl	OH / HC=O	OH / NO_2
$pK_a = 10.20$	$pK_a = 10.19$	$pK_a = 9.95$ phenol	$pK_a = 9.38$	$pK_a = 7.66$	$pK_a = 7.14$

Take a minute to compare the effect a substituent has on the reactivity of a benzene ring toward electrophilic aromatic substitution with the effect the substituent has on the pK_a of phenol. Notice that the more strongly deactivating the substituent, the lower the pK_a of the phenol; and the more strongly activating the substituent, the higher the pK_a of the phenol. In other words, *electron withdrawal decreases reactivity toward electrophilic aromatic substitution and increases acidity, whereas electron donation increases reactivity toward electrophilic aromatic substitution and decreases acidity.*

A similar substituent effect on pK_a is observed for substituted benzoic acids and substituted protonated anilines: electron-withdrawing substituents increase acidity; electron-donating substituents decrease acidity.

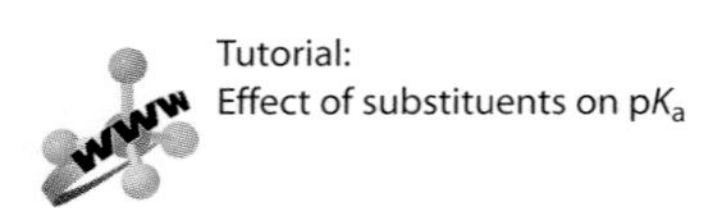

Tutorial:
Effect of substituents on pK_a

COOH / OCH_3	COOH / CH_3	COOH	COOH / Br	COOH / $CH_3C=O$	COOH / NO_2
$pK_a = 4.47$	$pK_a = 4.34$	$pK_a = 4.20$	$pK_a = 4.00$	$pK_a = 3.70$	$pK_a = 3.44$

$\overset{+}{N}H_3$ / OCH_3	$\overset{+}{N}H_3$ / CH_3	$\overset{+}{N}H_3$	$\overset{+}{N}H_3$ / Br	$\overset{+}{N}H_3$ / HC=O	$\overset{+}{N}H_3$ / NO_2
$pK_a = 5.29$	$pK_a = 5.07$	$pK_a = 4.58$	$pK_a = 3.91$	$pK_a = 1.76$	$pK_a = 0.98$

The more deactivating (electron withdrawing) the substituent, the more it increases the acidity of a COOH, an OH, or an $^{+}NH_3$ group attached to a benzene ring.

The more activating (electron donating) the substituent, the more it decreases the acidity of a COOH, an OH, or an $^{+}NH_3$ group attached to a benzene ring.

PROBLEM 22 ◆

Which of the compounds in each of the following pairs is more acidic?

a. $CH_3\overset{O}{\overset{\|}{C}}OH$ or $ClCH_2\overset{O}{\overset{\|}{C}}OH$

b. $CH_3CH_2\overset{O}{\overset{\|}{C}}OH$ or $H_3\overset{+}{N}CH_2\overset{O}{\overset{\|}{C}}OH$

c. $FCH_2\overset{O}{\overset{\|}{C}}OH$ or $ClCH_2\overset{O}{\overset{\|}{C}}OH$

d. $H\overset{O}{\overset{\|}{C}}OH$ or $CH_3\overset{O}{\overset{\|}{C}}OH$

SUMMARY

To be classified as **aromatic**, a compound must have an uninterrupted cyclic cloud of π electrons that contains an *odd number of pairs* of π electrons.

A **heterocyclic compound** is a cyclic compound in which one or more of the ring atoms is a **heteroatom**—an atom other than carbon. Pyridine, pyrrole, furan, and thiophene are aromatic heterocyclic compounds.

Benzene's aromaticity causes it to undergo **electrophilic aromatic substitution reactions**. Electrophilic addition reactions that are characteristic of alkenes and dienes would lead to much less stable nonaromatic products. The most common electrophilic aromatic substitution reactions are halogenation, nitration, sulfonation, and Friedel–Crafts acylation and alkylation. Once the electrophile is generated, all electrophilic aromatic substitution reactions take place by the same two-step mechanism: (1) the aromatic compound reacts with an electrophile, forming a carbocation intermediate; and (2) a base pulls off a proton from the carbon that formed the bond with the electrophile.

Some monosubstituted benzenes are named as substituted benzenes (for example, bromobenzene, nitrobenzene); others have names that incorporate the substituent (for example, toluene, phenol, aniline, anisole). The relative positions of two substituents on a benzene ring are indicated in the compound's name either by numbers or by the prefixes *ortho*, *meta*, and *para*.

Bromination or **chlorination** requires a Lewis acid catalyst. **Sulfonation** with sulfuric acid places an SO_3H group on the ring. **Nitration** with nitric acid requires sulfuric acid as a catalyst. An acyl chloride is used for a **Friedel–Crafts acylation**, a reaction that places an acyl group on a benzene ring; an alkyl halide is used for a **Friedel–Crafts** alkylation, a reaction that places an alkyl group on a benzene ring.

Benzene rings with substituents other than halo, nitro, sulfonic acid, alkyl, and acyl can be prepared by synthesizing one of these substituted benzenes and then chemically changing the substituent.

The nature of the substituent affects both the reactivity of the benzene ring and the placement of an incoming substituent: the rate of electrophilic aromatic substitution is increased by electron-donating substituents and decreased by electron-withdrawing substituents. Substituents can donate or withdraw electrons **inductively** or by **resonance**.

The stability of the carbocation intermediate determines the position to which the substituent directs an incoming electrophile. All activating substituents and the weakly deactivating halogens are **ortho–para directors**; all substituents more deactivating than the halogens are **meta directors**.

In planning the synthesis of disubstituted benzenes, the order in which the substituents are placed on the ring and the point in a reaction sequence at which a substituent is chemically modified are important considerations.

The acidity of substituted benzoic acids, phenols, and anilinium ions is increased by electron-withdrawing substituents and decreased by electron-donating substituents.

SUMMARY OF REACTIONS

1. Electrophilic aromatic substitution reactions:

a. Halogenation (Section 8.7)

$$C_6H_6 + Br_2 \xrightarrow{FeBr_3} C_6H_5Br + HBr$$

$$C_6H_6 + Cl_2 \xrightarrow{FeCl_3} C_6H_5Cl + HCl$$

b. Nitration and sulfonation (Sections 8.8 and 8.9)

$$C_6H_6 + HNO_3 \xrightarrow{H_2SO_4} C_6H_5NO_2 + H_2O$$

$$C_6H_6 + H_2SO_4 \overset{\Delta}{\rightleftharpoons} C_6H_5SO_3H + H_2O$$

c. Friedel–Crafts acylation and alkylation (Sections 8.10 and 8.11)

$$C_6H_6 + RCOCl \xrightarrow{AlCl_3} C_6H_5COR + HCl$$

$$C_6H_6 + RCl \xrightarrow{AlCl_3} C_6H_5R + HCl$$

2. Reactions of a substituent on a benzene ring (Section 8.12)

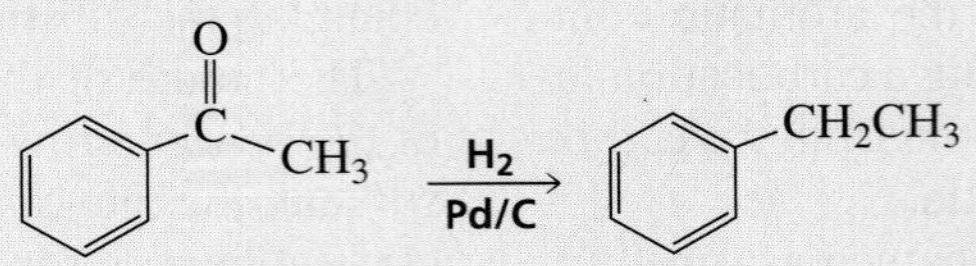

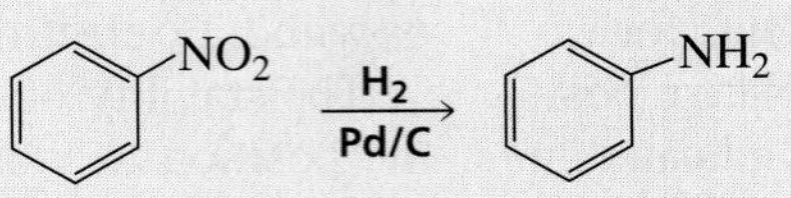

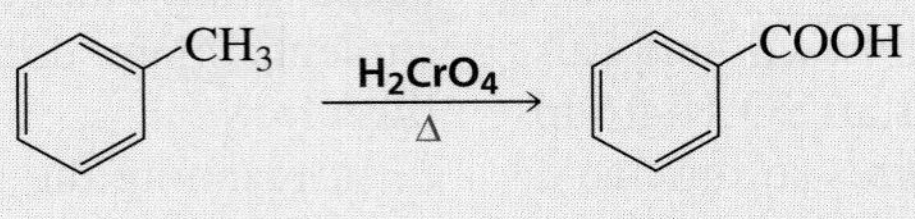

$C_6H_5COCH_3 \xrightarrow[\text{Pd/C}]{H_2} C_6H_5CH_2CH_3$

$C_6H_5NO_2 \xrightarrow[\text{Pd/C}]{H_2} C_6H_5NH_2$

$C_6H_5CH_3 \xrightarrow[\Delta]{H_2CrO_4} C_6H_5COOH$

PROBLEMS

23. Which of the following are aromatic?

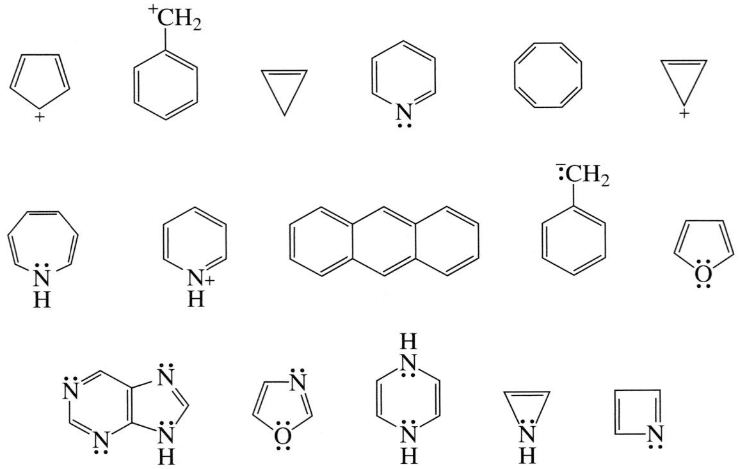

$^+CH_2$

$:\bar{C}H_2$

N N+ H H

H N N N N O N O N N H H H N

24. Give the product of the reaction of benzene with the following reagents:

a. $CH_3CHCH_3 + AlCl_3$ (with Cl on the central carbon)

b. $CH_3CH{=}CH_2 + HF$

c. $CH_3\overset{O}{\overset{\|}{C}}Cl + AlCl_3$

25. Which ion in each of the following pairs is more stable? Explain your choice.

a. 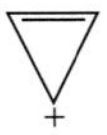 or  **b.**  or 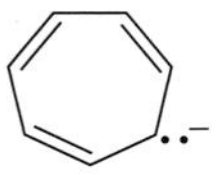 **c.** 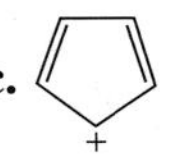 or

26. Draw the structure of each of the following:

a. *m*-ethylphenol
b. *p*-nitrobenzenesulfonic acid
c. *o*-bromoaniline
d. *p*-bromobenzoic acid
e. (*E*)-2-phenyl-2-pentene
f. 2,4-dichlorotoluene

27. Name the following:

a. COOH Br

b. CH_3 Br

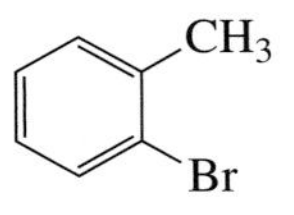

c. CH_3

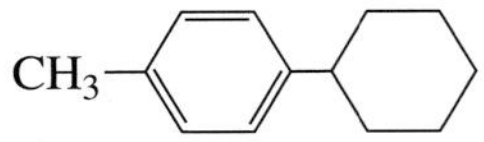

28. Give the product(s) of the following reactions:

a. benzoic acid + HNO_3/H_2SO_4
b. isopropylbenzene + cyclohexene + HF
c. cyclohexylbenzene + $Br_2/FeBr_3$
d. phenol + H_2SO_4 + Δ
e. ethylbenzene + $Br_2/FeBr_3$

29. Rank the following anions in order of decreasing basicity:

CH_3–C₆H₄–O^- $\quad$ CH_3O–C₆H₄–O^- $\quad$ $CH_3C(=O)$–C₆H₄–O^- $\quad$ Br–C₆H₄–O^-

30. Show how the following compounds could be synthesized from benzene:

a. *p*-nitrotoluene
b. *m*-chlorobenzenesulfonic acid
c. 3-phenylpentane

31. Draw the structures of the following groups of compounds and arrange them in order of decreasing reactivity toward electrophilic aromatic substitution:

a. benzene, ethylbenzene, chlorobenzene, nitrobenzene
b. 1-chloro-2,4-dinitrobenzene, 2,4-dinitrophenol, 2,4-dinitrotoluene
c. benzene, benzoic acid, phenol, propylbenzene
d. *p*-nitrotoluene, 2-chloro-4-nitrotoluene, 2,4-dinitrotoluene, *p*-chlorotoluene

32. Give the products of the following reactions:

a.
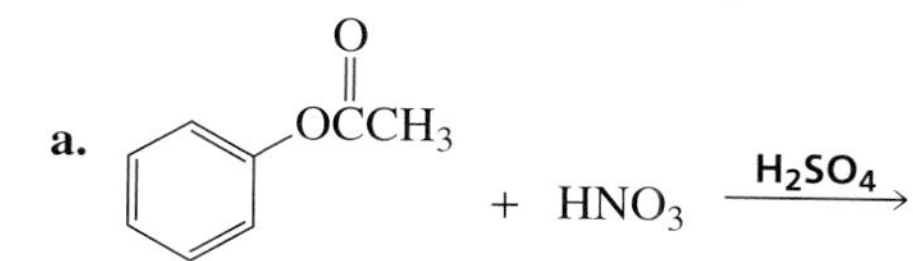

b.
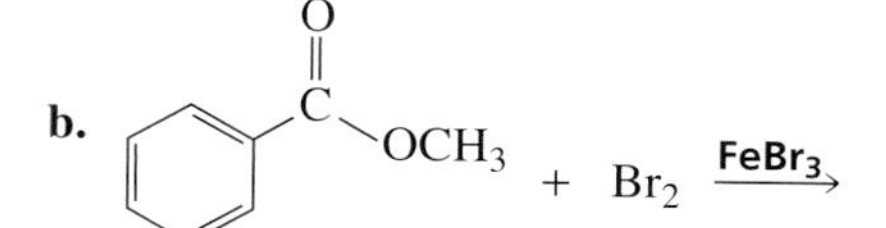

c. CF_3-benzene + Cl_2 $\xrightarrow{FeCl_3}$

33. For each of the statements in Column I, choose a substituent from Column II that fits the description for the following compound:

Z–C₆H₅

Column I

a. Z donates electrons inductively, but does not donate or withdraw electrons by resonance.
b. Z withdraws electrons inductively and withdraws electrons by resonance.
c. Z deactivates the ring and directs ortho–para.
d. Z withdraws electrons inductively, donates electrons by resonance, and activates the ring.
e. Z withdraws electrons inductively, but does not donate or withdraw electrons by resonance.

Column II

OH
Br
$^+NH_3$
CH_2CH_3
NO_2

34. Show how the following compounds could be synthesized from benzene:

a. *p*-nitrobenzoic acid
b. *m*-nitrobenzoic acid
c. *m*-chloroaniline
d. *p*-chlorotoluene
e. *m*-bromoethylbenzene
f. *o*-bromoethylbenzene

35. List the following compounds in order of decreasing reactivity toward electrophilic aromatic substitution: dichloromethylbenzene, difluoromethylbenzene, toluene, chloromethylbenzene

36. Which of the compounds in each of the following pairs is more acidic?

a.
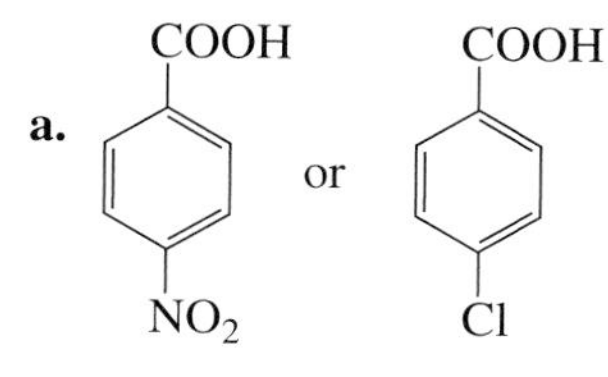

b.
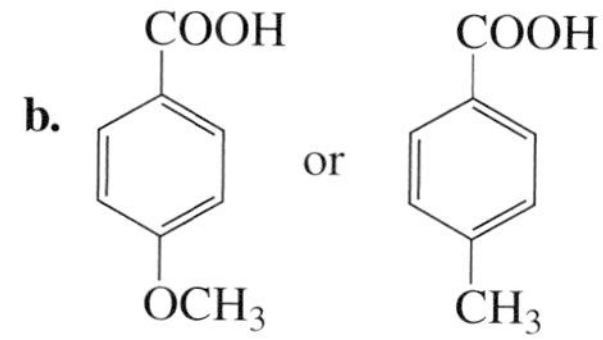

37. For each horizontal row of substituted benzenes shown below, indicate

a. the one that would be the most reactive in an electrophilic aromatic substitution reaction.
b. the one that would be the least reactive in an electrophilic aromatic substitution reaction.
c. the one that would yield the highest percentage of meta product.

1.
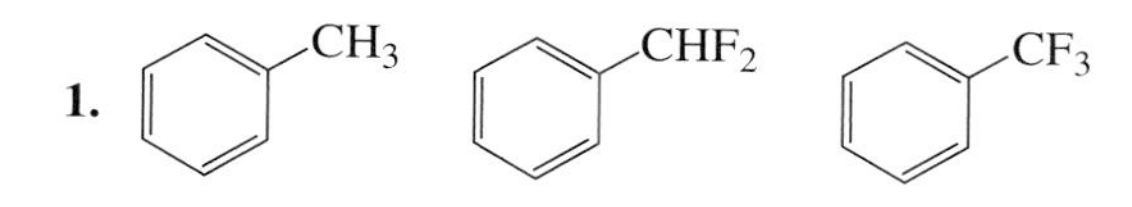

2. 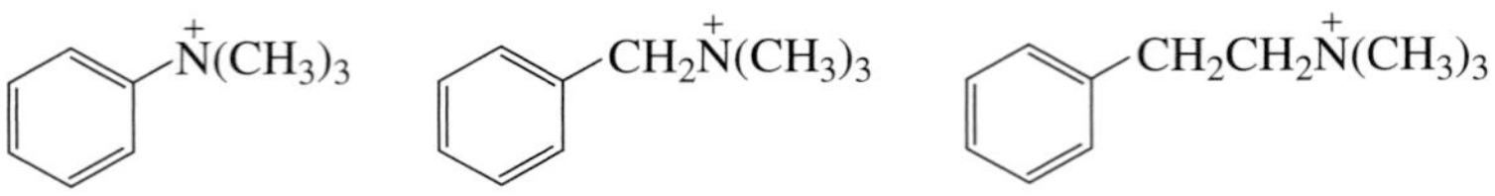 $C_6H_5\overset{+}{N}(CH_3)_3$ $C_6H_5CH_2\overset{+}{N}(CH_3)_3$ $C_6H_5CH_2CH_2\overset{+}{N}(CH_3)_3$

3. $C_6H_5OCH_2CH_3$ $C_6H_5CH_2OCH_3$ $C_6H_5\overset{O}{\overset{\|}{C}}OCH_3$

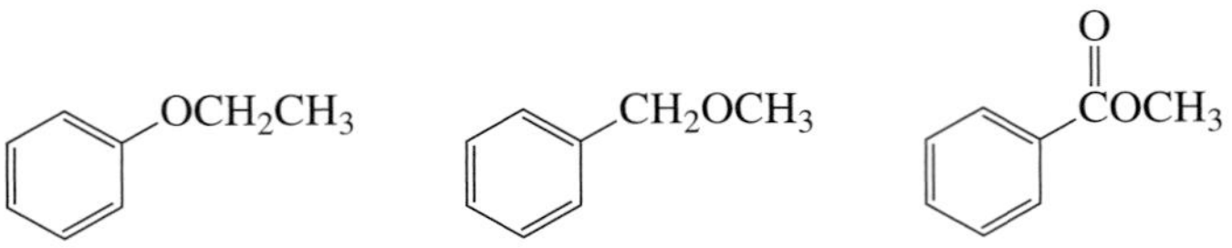

38. Which of the following compounds will react with HBr more rapidly?

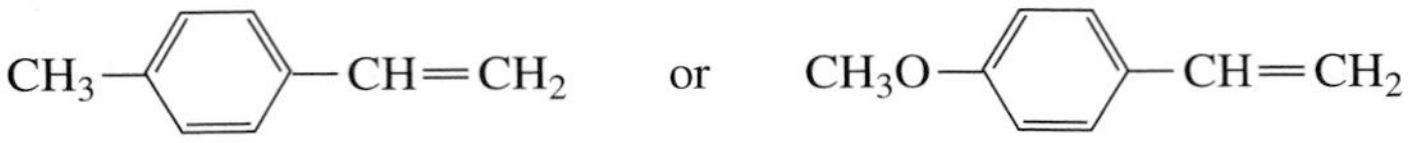

$CH_3-C_6H_4-CH{=}CH_2$ or $CH_3O-C_6H_4-CH{=}CH_2$

39. How could the following compound be synthesized when one of the reactants is:
a. toluene **b.** benzene

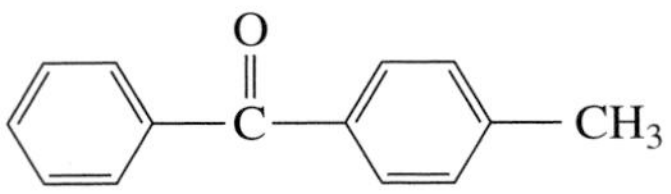

$C_6H_5-\overset{O}{\overset{\|}{C}}-C_6H_4-CH_3$

40. Give the product of each of the following reactions:

a. 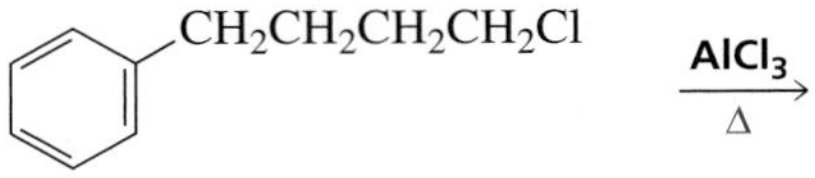 $C_6H_5CH_2CH_2CH_2CH_2Cl \xrightarrow[\Delta]{AlCl_3}$

b. 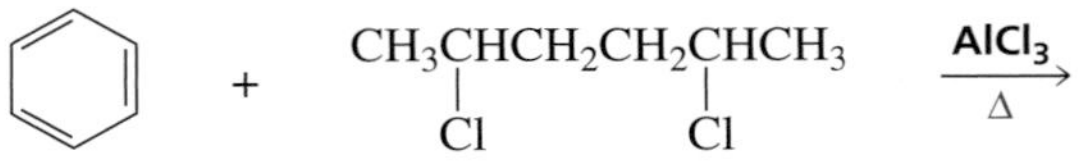 $C_6H_6 + CH_3\underset{Cl}{\underset{|}{C}}HCH_2CH_2\underset{Cl}{\underset{|}{C}}HCH_3 \xrightarrow[\Delta]{AlCl_3}$

41. Which compound in each of the following pairs is a stronger base? Why?

a. 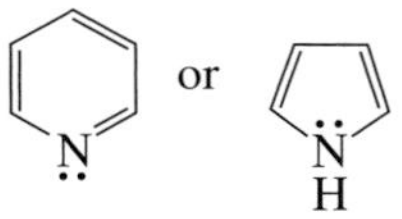 pyridine or pyrrole

b. 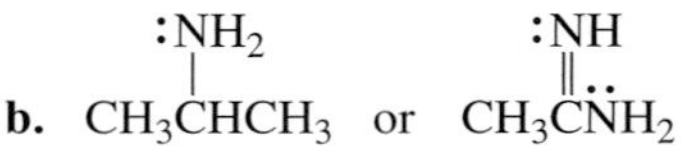 $CH_3\overset{:NH_2}{\overset{|}{C}}HCH_3$ or $CH_3\overset{:NH}{\overset{\|}{C}}\ddot{N}H_2$

42. Propose a mechanism for the following reaction:

a. 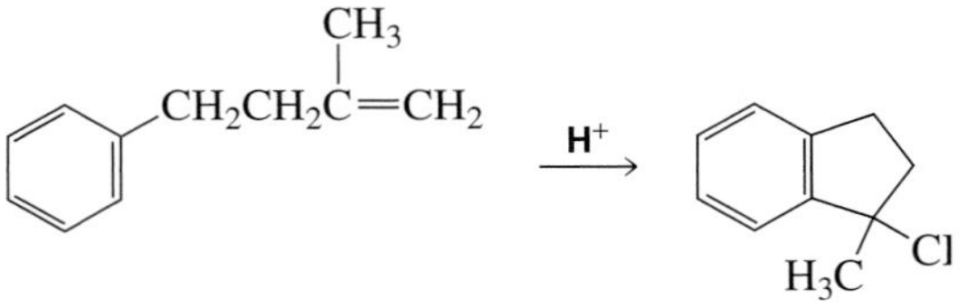 $C_6H_5CH_2CH_2\overset{CH_3}{\overset{|}{C}}{=}CH_2 \xrightarrow{H^+}$ 1-chloro-1-methylindane (H_3C, Cl)

43. Give the products of the following reactions:

a. 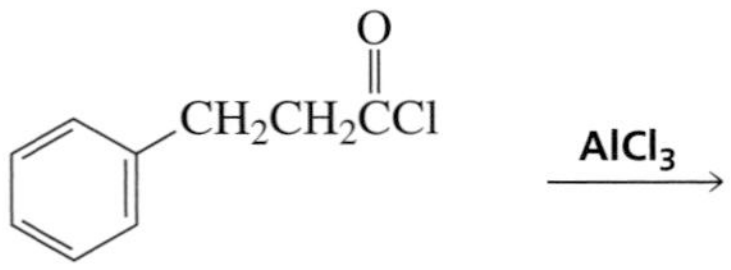 $C_6H_5CH_2CH_2\overset{O}{\overset{\|}{C}}Cl \xrightarrow{AlCl_3}$

b. 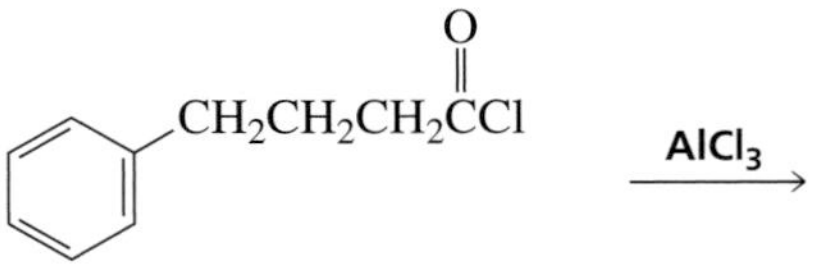 $C_6H_5CH_2CH_2CH_2\overset{O}{\overset{\|}{C}}Cl \xrightarrow{AlCl_3}$

44. Propose a mechanism for the following reaction:

$C_6H_6 \xrightarrow{DCl} C_6D_6$

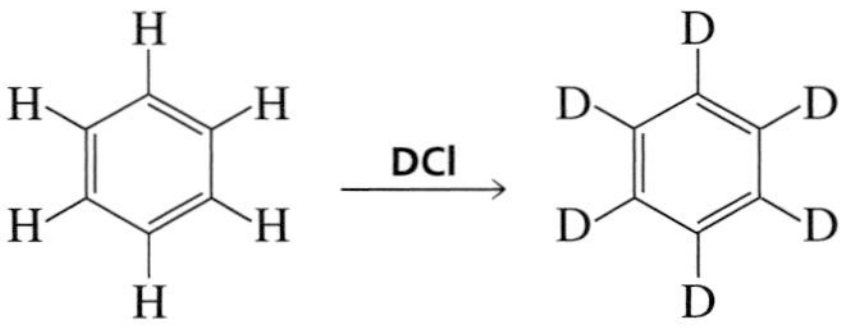

Substitution and Elimination Reactions of Alkyl Halides

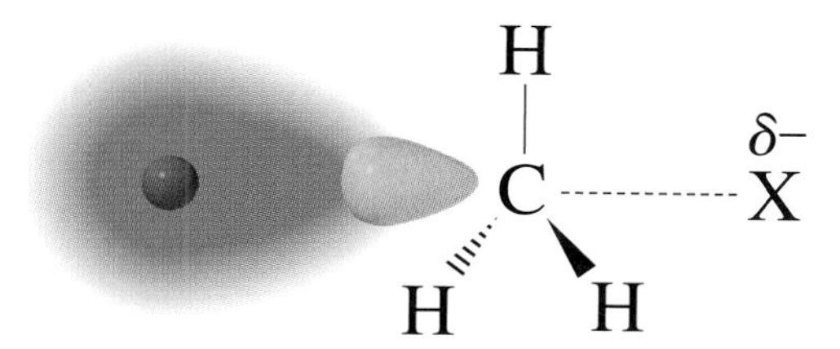

Organic compounds that have an electronegative atom or an electron-withdrawing group bonded to an sp^3 hybridized carbon undergo *substitution reactions* and/or *elimination reactions*.

In a **substitution reaction**, the electronegative atom or electron-withdrawing group is replaced by another atom or group. In an **elimination reaction**, the electronegative atom or electron-withdrawing group is eliminated, along with a hydrogen from an adjacent carbon. The atom or group that is *substituted* or *eliminated* in these reactions is called a **leaving group**. The substitution reaction is more precisely called a **nucleophilic substitution reaction** because the atom or group that replaces the leaving group is a nucleophile.

$$RCH_2CH_2X + Y^- \xrightarrow{\text{a substitution reaction}} RCH_2CH_2Y + X^-$$

$$RCH_2CH_2X + Y^- \xrightarrow{\text{an elimination reaction}} RCH{=}CH_2 + HY + X^-$$

(X = the leaving group)

Alkyl halides are a good family of compounds with which to start the study of substitution and elimination reactions because they have relatively good leaving groups; that is, the halide ions are easily displaced. After studying the reactions of alkyl halides, you will be prepared to look, in Chapter 10, at the substitution and elimination reactions of compounds with poor leaving groups—those that are more difficult to displace.

alkyl halides

R—F	R—Cl	R—Br	R—I
an alkyl fluoride	**an alkyl chloride**	**an alkyl bromide**	**an alkyl iodide**

Substitution reactions are important in organic chemistry because they make it possible to convert readily available alkyl halides into a wide variety of other compounds. Substitution reactions are also important in the cells of plants and animals. We will see, however, that because alkyl halides are insoluble in water and cells exist in predominantly aqueous environments, biological systems use compounds in which the group that is replaced is more polar than a halogen and, therefore, more soluble in water.

SURVIVAL COMPOUNDS

Several marine organisms, including sponges, corals, and algae, synthesize organohalides (halogen containing organic compounds) that they use to deter predators. For example, red algae synthesize a toxic, foul-tasting organohalide that keeps predators from eating them. One predator that is not deterred, however, is a mollusk called a sea hare. After consuming red algae, a sea hare converts the algae's organohalide into a structurally similar compound it uses for its own defense. Unlike other mollusks, a sea hare does not have a shell. Its method of defense is to surround itself with a slimy material that contains the organohalide, thereby protecting itself from carnivorous fish.

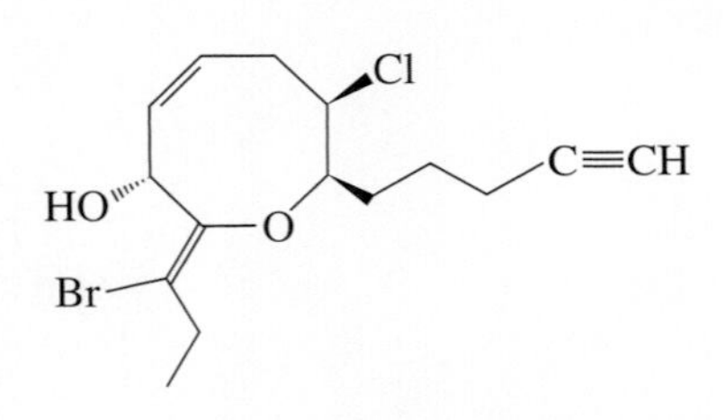

synthesized by red algae

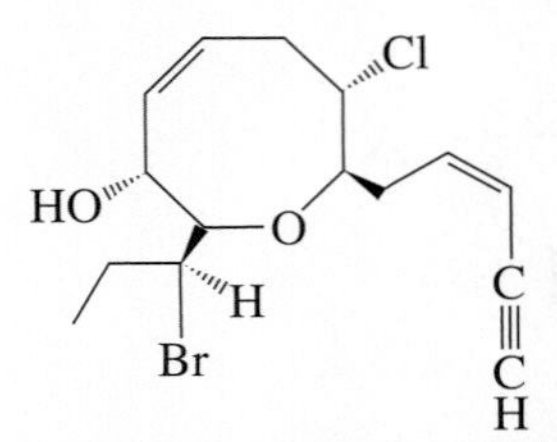

synthesized by the sea hare

a sea hare

ARTIFICIAL BLOOD

Clinical trials are currently underway to test the use of perfluorocarbons—alkanes in which all the hydrogens have been replaced by fluorines—as a substitute for blood. One compound under study has been found to be more effective than hemoglobin in carrying oxygen to cells and in transporting carbon dioxide to the lungs. Artificial blood has several advantages: it is safe from disease, it can be administered to any blood type, its availability is not dependent on blood donors, and it can be stored longer than whole blood, which is good for only about 40 days.

9.1 HOW ALKYL HALIDES REACT

Fluorine, chlorine, and bromine are all more electronegative than carbon. Consequently, when carbon is bonded to any of these elements, the two atoms do not share their bonding electrons equally. Because the more electronegative halogen has a larger share of the electrons, it has a partial negative charge (δ^-) and the carbon to which it is bonded has a partial positive charge (δ^+).*

$$\overset{\delta+}{RCH_2}-\overset{\delta-}{X} \qquad X = F, Cl, Br, I$$

a polar bond

It is this polar carbon–halogen bond that causes alkyl halides to undergo substitution and elimination reactions. We will see that there are two mechanisms for the substitution reaction and two mechanisms for the elimination reaction. First, we will look at the substitution reaction.

9.2 THE MECHANISM OF AN S_N2 REACTION

Perhaps you have been wondering how the mechanism of a reaction is determined. We can learn a great deal about the mechanism of a reaction by studying the factors that affect the rate of the reaction.

* The large electron cloud of the iodine atom is easily distorted causing the C—I bond to react as if it were polar, even though carbon and iodine have the same electronegativity (see Table 1.3 on page 8).

The rate of a nucleophilic substitution reaction, such as the reaction of methyl bromide with hydroxide ion, depends on the concentrations of *both* reagents. If the concentration of methyl bromide in the reaction mixture is doubled, the rate of the nucleophilic substitution reaction doubles. Similarly, if the concentration of the nucleophile (hydroxide ion) is doubled, the rate of the reaction also doubles. If the concentrations of both reactants are doubled, the rate of the reaction quadruples.

$$\underset{\text{methyl bromide}}{CH_3Br} + HO^- \longrightarrow \underset{\text{methyl alcohol}}{CH_3OH} + Br^-$$

When you know the relationship between the rate of a reaction and the concentration of the reactants, you can write a **rate law** for the reaction. Because the rate of the reaction of methyl bromide with hydroxide ion is dependent on the concentration of both reactants, the rate law for the reaction is

$$\textbf{rate} \propto \textbf{[alkyl halide][nucleophile]}$$

The proportionality sign ($\propto$) can be replaced by an equals sign and a proportionality constant. The proportionality constant (k) is called the **rate constant**. The magnitude of the rate constant for a particular reaction indicates how difficult it is for the reactants to overcome the energy barrier of the reaction—how hard it is to reach the transition state. The larger the rate constant, the lower is the energy barrier and, therefore, the easier it is to reach the transition state (see Figure 9.2 on page 233).

$$\textbf{rate} = k\ \textbf{[alkyl halide][nucleophile]}$$

The rate law tells us which molecules are involved in the rate-determining step of the reaction. From the rate law for the reaction of methyl bromide with hydroxide ion, for example, we know that both methyl bromide and hydroxide ion are involved in the rate-determining step.

BIOGRAPHY

Edward Davies Hughes (1906–1963), *born in northern Wales, earned a Ph.D. from the University of Wales and a D.Sc. from the University of London, where he worked with Sir Christopher Ingold. He was a professor of chemistry at University College, London.*

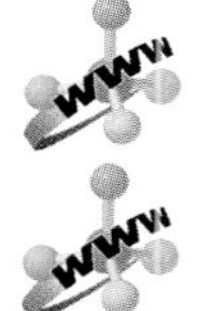

Tutorial: S_N2

Movie: Bimolecular reaction

PROBLEM 1 ♦

a. How is the rate of the reaction affected if the concentration of hydroxide ion is tripled?
b. How is the rate of the reaction affected if the concentration of methyl bromide is changed from 1.00 M to 0.50 M?

The reaction of methyl bromide with hydroxide ion is an example of an **S_N2 reaction**, where S stands for substitution, N for nucleophilic, and 2 for bimolecular. **Bimolecular** means that two molecules are involved in the rate-determining step. In 1937, Edward Hughes and Christopher Ingold proposed a mechanism for an S_N2 reaction. Remember that a mechanism describes the step-by-step process by which reactants are converted into products. It is a theory that fits the experimental evidence pertaining to the reaction. Hughes and Ingold based their mechanism for an S_N2 reaction on the following three pieces of experimental evidence:

1. The rate of the reaction depends on the concentration of the alkyl halide *and* on the concentration of the nucleophile. This means that both reactants are involved in the rate-determining step.
2. As the hydrogens of methyl bromide are successively replaced with methyl groups, the rate of the reaction with a given nucleophile becomes progressively slower (Table 9.1).
3. The reaction of an alkyl halide in which the halogen is bonded to an asymmetric center leads to the formation of only one stereoisomer, and the configuration of the asymmetric center in the product is inverted relative to its configuration in the reacting alkyl halide.

BIOGRAPHY

Sir Christopher Ingold (1893–1970) *was born in Ilford, England. In addition to determining the mechanism of the S_N2 reaction, he was a member of the group that developed the R,S system of nomenclature for enantiomers (see page 147). He also participated in developing resonance theory.*

Table 9.1 Relative Rates of S_N2 Reactions for Several Alkyl Halides

$$R{-}Br + Cl^- \xrightarrow{S_N2} R{-}Cl + Br^-$$

Alkyl halide	Class of alkyl halide	Relative rate
$CH_3{-}Br$	methyl	1200
$CH_3CH_2{-}Br$	primary	40
$CH_3CH_2CH_2{-}Br$	primary	16
$CH_3CH(CH_3){-}Br$	secondary	1
$(CH_3)_3C{-}Br$	tertiary	too slow to measure

Hughes and Ingold proposed the following mechanism for an S_N2 reaction.

mechanism for the S_N2 reaction of an alkyl halide

The nucleophile attacks the back side of the carbon that is bonded to the leaving group and displaces the leaving group.

$$H\ddot{O}:^- + CH_3{-}\ddot{Br}: \longrightarrow CH_3{-}OH + :\ddot{Br}:^-$$

leaving group

How does Hughes and Ingold's mechanism account for the three observed pieces of experimental evidence? The mechanism shows the alkyl halide and the nucleophile coming together in the transition state of the one-step reaction. Therefore, increasing the concentration of either of them makes their collision more probable. Thus, the reaction is dependent on the concentration of both reactants, exactly as observed.

$$HO^- + {>}C{-}Br \longrightarrow \left[\overset{\delta-}{HO}{-}{-}{-}C{-}{-}{-}\overset{\delta-}{Br}\right]^{\ddagger} \longrightarrow HO{-}C{<} + Br^-$$

transition state

3-D Molecules: Methyl chloride; *t*-Butyl chloride

Because the nucleophile attacks the back side of the carbon that is bonded to the halogen, bulky substituents attached to this carbon will decrease the nucleophile's access to the back side of the carbon and will therefore decrease the rate of the reaction (Figure 9.1). This explains why substituting methyl groups for the hydrogens in methyl bromide progressively slows the rate of the substitution reaction (Table 9.1).

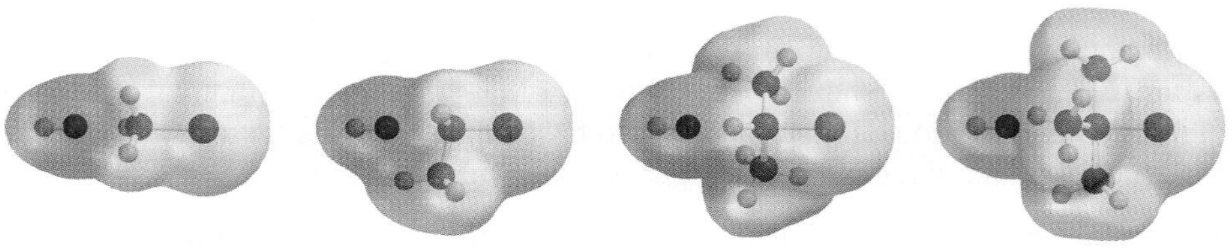

▶ **Figure 9.1**
The approach of HO^- to a methyl halide, a primary alkyl halide, a secondary alkyl halide, and a tertiary alkyl halide. Increasing the bulk of the substituents bonded to the carbon that is undergoing nucleophilic attack decreases access to the back side of the carbon, thereby decreasing the rate of the S_N2 reaction.

Steric effects are caused by the fact that groups occupy a certain volume of space (Section 3.8). A steric effect that decreases reactivity is called **steric hindrance**. Steric hindrance occurs when groups are in the way at the reaction site. Steric hindrance causes alkyl halides to have the following relative reactivities in an S_N2 reaction because, *generally*, primary alkyl halides are less sterically hindered than secondary alkyl halides, which, in turn, are less hindered than tertiary alkyl halides:

Steric hindrance causes methyl halides and primary alkyl halides to be the most reactive in S_N2 reactions.

relative reactivities of alkyl halides in an S_N2 reaction

most reactive > methyl halide > 1° alkyl halide > 2° alkyl halide > 3° alkyl halide < least reactive

The three alkyl groups of a tertiary alkyl halide make it impossible for the nucleophile to come within bonding distance of the tertiary carbon, so tertiary alkyl halides are unable to undergo S_N2 reactions.

Tertiary alkyl halides cannot undergo S_N2 reactions.

The reaction coordinate diagrams for the S_N2 reaction of *unhindered* methyl bromide (Figure 9.2a) and for that of a *sterically hindered* secondary alkyl bromide (Figure 9.2b) show that steric hindrance raises the energy of the transition state, slowing the reaction.

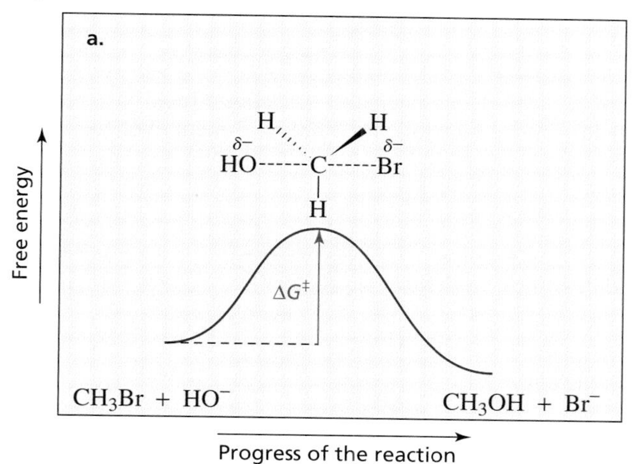

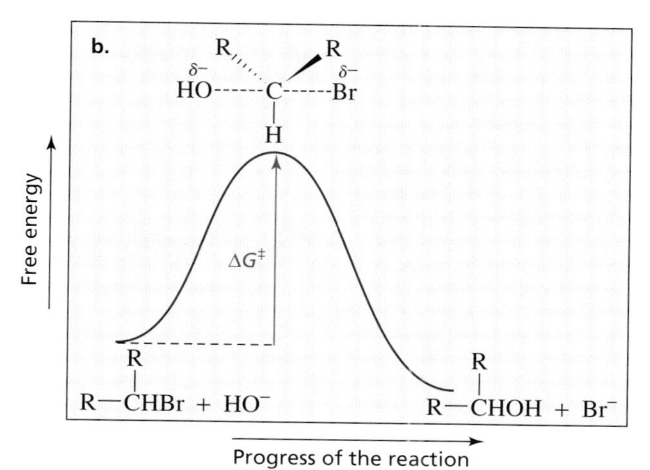

▲ **Figure 9.2**
Reaction coordinate diagrams for (a) the S_N2 reaction of methyl bromide with hydroxide ion; (b) the S_N2 reaction of a sterically hindered secondary alkyl bromide with hydroxide ion.

BIOGRAPHY

Viktor Meyer (1848–1897) *was born in Germany. To prevent him from becoming an actor, his parents persuaded him to enter the University of Heidelberg, where he earned a Ph.D. in 1867 at the age of 18. He was a professor of chemistry at the Universities of Stuttgart and Heidelberg. He coined the term "stereochemistry" for the study of molecular shapes and was the first to describe the effect of steric hindrance on a reaction.*

Figure 9.3 shows that as the nucleophile approaches the back side of the carbon of methyl bromide, the C—H bonds begin to move away from the nucleophile and its attacking electrons. By the time the transition state is reached, the C—H bonds are all in the same plane. As the nucleophile gets closer to the carbon and the bromine moves farther away from it, the C—H bonds continue to move in the same direction. Eventually, the bond between carbon and the nucleophile is fully formed, and the bond between carbon and bromine is completely broken. The carbon at which substitution occurs has *inverted its configuration* during the course of the reaction, just like an umbrella tends to invert in a windstorm.

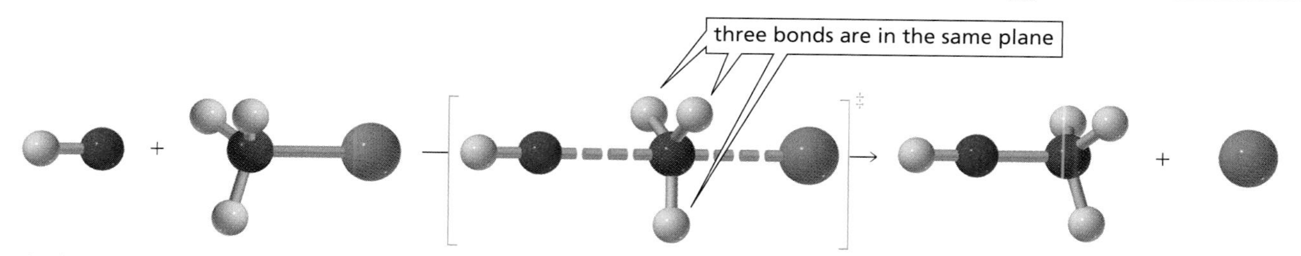

▲ **Figure 9.3**
An S_N2 reaction between hydroxide ion and methyl bromide.

Because an S_N2 reaction takes place with **inversion of configuration**, only one substitution product is formed when an alkyl halide whose halogen atom is bonded to an asymmetric center undergoes an S_N2 reaction. The configuration of that product is inverted relative to the configuration of the alkyl halide. For example, the substitution product obtained from the reaction of hydroxide ion with (*R*)-2-bromopentane is (*S*)-2-pentanol. (The "ol" suffix indicates that the compound is an alcohol, and the "2" tells us that the OH group is bonded to the #2 carbon.) Thus, the proposed mechanism also accounts for the observed configuration of the product.

To draw the inverted product, draw the mirror image of the *reactant* and replace the halogen with the nucleophile.

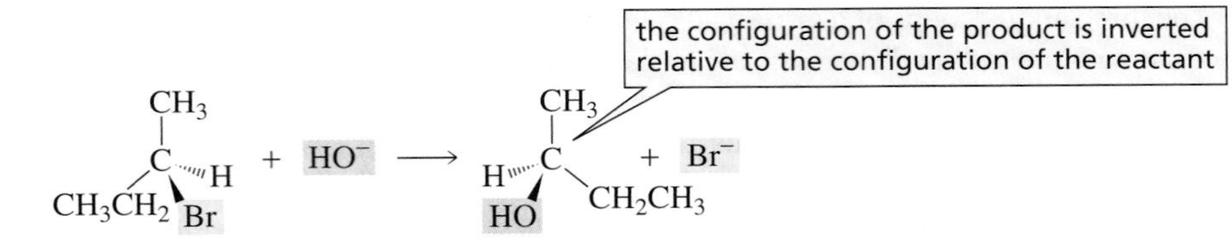

the configuration of the product is inverted relative to the configuration of the reactant

(*R*)-2-bromobutane + HO^- ⟶ (*S*)-2-butanol + Br^-

PROBLEM 2 ◆

Arrange the following alkyl bromides in order of decreasing reactivity in an S_N2 reaction: 1-bromo-2-methylbutane, 1-bromo-3-methylbutane, 2-bromo-2-methylbutane, and 1-bromopentane.

PROBLEM 3 ◆ SOLVED

Determine the product that would be formed from the S_N2 reaction of

a. 2-bromobutane and hydroxide ion.
b. (*R*)-2-bromobutane and hydroxide ion.
c. (*S*)-3-chlorohexane and hydroxide ion.
d. 3-iodopentane and hydroxide ion.

Solution to 3a The product is 2-butanol. Because the reaction is an S_N2 reaction, we know that the configuration of the product is inverted relative to the configuration of the reactant. The configuration of the reactant is not specified, however, so we cannot specify the configuration of the product.

the configuration is not specified

$$CH_3CH(Br)CH_2CH_3 + HO^- \longrightarrow CH_3CH(OH)CH_2CH_3 + Br^-$$

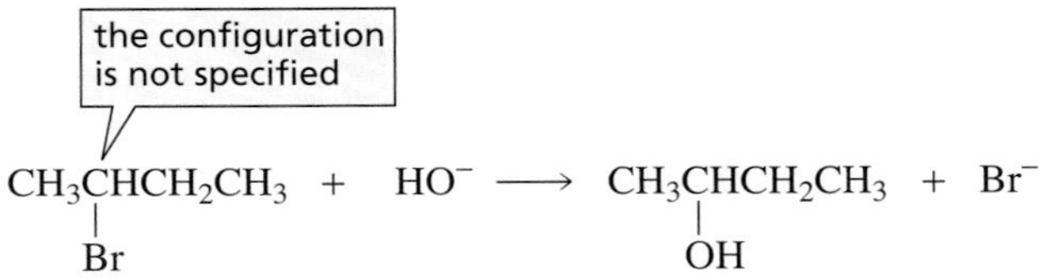

9.3 FACTORS THAT AFFECT S_N2 REACTIONS

The Leaving Group

If an alkyl iodide, an alkyl bromide, an alkyl chloride, and an alkyl fluoride, all having the same alkyl group, were allowed to react with the same nucleophile under the same conditions, we would find that the alkyl iodide is the most reactive and the alkyl fluoride is the least reactive.

Reaction	relative rates of reaction
$HO^- + RCH_2I \longrightarrow RCH_2OH + I^-$	30,000
$HO^- + RCH_2Br \longrightarrow RCH_2OH + Br^-$	10,000
$HO^- + RCH_2Cl \longrightarrow RCH_2OH + Cl^-$	200
$HO^- + RCH_2F \longrightarrow RCH_2OH + F^-$	1

The weaker the base, the better it is as a leaving group.

Stable bases are weak bases.

The only difference among these four reactions is the nature of the leaving group. From the relative reaction rates, we can see that the iodide ion is the best leaving group and the fluoride ion is the worst. This brings us to an important rule in organic chemistry—one that you will encounter frequently: *the weaker the basicity of a group, the better is its leaving ability.* The reason leaving ability depends on basicity is because *weak bases are stable bases*; they readily bear the electrons they formerly shared with a proton. Because weak bases do not share their electrons well, a weak base is not bonded as strongly to the carbon as a strong base would be, and a weaker bond is more easily broken (Section 2.6).

We have seen that the iodide ion is the weakest base of the halide ions and the fluoride ion is the strongest (Section 2.6). Therefore, alkyl iodides are the most reactive of the alkyl halides, and alkyl fluorides are the least reactive.

relative reactivities of alkyl halides in an S_N2 reaction

most reactive → RI > RBr > RCl > RF ← least reactive

The Nucleophile

When we talk about atoms or molecules that have lone-pair electrons, sometimes we call them bases and sometimes we call them nucleophiles. What is the difference between a base and a nucleophile?

Basicity is a measure of how well a species (a **base**) shares its lone pair with a proton. The stronger the base, the better it shares its electrons. **Nucleophilicity** is a measure of how readily a species (a **nucleophile**) is able to attack an electron-deficient atom. In the case of an S_N2 reaction, nucleophilicity is a measure of how readily the nucleophile attacks an sp^3 carbon bonded to a leaving group. Because the nucleophile attacks the sp^3 carbon in the rate-determining step of an S_N2 reaction, the rate of the reaction will depend on the strength of the nucleophile: *the better the nucleophile, the faster will be the S_N2 reaction.*

In general, *stronger bases are better nucleophiles*. For example, a species with a negative charge is a stronger base *and* a better nucleophile than a species that has the same attacking atom but that is neutral. Thus, HO^- is a stronger base and a better nucleophile than H_2O.

stronger base, better nucleophile		weaker base, poorer nucleophile
HO^-	>	H_2O
CH_3O^-	>	CH_3OH
$^-NH_2$	>	NH_3
$CH_3CH_2NH^-$	>	$CH_3CH_2NH_2$

If hydrogens are attached to the elements in the second row of the periodic table, the resulting compounds have the following relative acidities (Section 2.6):

relative acid strengths

weakest acid → NH_3 < H_2O < HF

Because the weakest acid has the strongest conjugate base (Section 2.6), the conjugate bases have the following relative *base strengths* and relative *nucleophilicities*:

relative base strengths and relative nucleophilicities

strongest base → $^-NH_2$ > HO^- > F^-

best nucleophile → $^-NH_2$

Note that the amide anion is the strongest base, as well as the best nucleophile.

PROBLEM 4 *SOLVED*

List the following species in order of decreasing nucleophilicity:

$$C_6H_5O^- \quad CH_3OH \quad HO^- \quad CH_3\overset{\overset{O}{\|}}{C}O^-$$

Solution Let's first divide the nucleophiles into groups. There are three nucleophiles with negatively charged oxygens and one with a neutral oxygen. We know that the poorest nucleophile is the one with the neutral oxygen. So we need to rank the three nucleophiles with negatively charged oxygens, which we can do by considering the pK_a values of their conjugate acids. A carboxylic acid is a stronger acid than phenol, which is a stronger acid than water (Section 7.9). Because water is the weakest acid, its conjugate base is the strongest base and the best nucleophile. Thus, the relative nucleophilicities are

$$HO^- > C_6H_5O^- > CH_3\overset{\overset{O}{\|}}{C}O^- > CH_3OH$$

PROBLEM 5 ◆

For each of the following pairs of S_N2 reactions, indicate which reaction occurs faster:

a. $CH_3CH_2Br + H_2O$ or $CH_3CH_2Br + HO^-$

b. $CH_3\underset{\underset{CH_3}{|}}{C}HCH_2Br + HO^-$ or $CH_3CH_2\underset{\underset{CH_3}{|}}{C}HBr + HO^-$

c. $CH_3CH_2Cl + I^-$ or $CH_3CH_2Br + I^-$

d. $CH_3CH_2CH_2I + HO^-$ or $CH_3CH_2CH_2Br + HO^-$

Many different kinds of nucleophiles can react with alkyl halides. Therefore, a wide variety of organic compounds can be synthesized by means of S_N2 reactions.

$$CH_3CH_2Cl + HO^- \longrightarrow CH_3CH_2OH + Cl^-$$
an alcohol

$$CH_3CH_2Br + HS^- \longrightarrow CH_3CH_2SH + Br^-$$
a thiol

$$CH_3CH_2I + RO^- \longrightarrow CH_3CH_2OR + I^-$$
an ether

$$CH_3CH_2Br + RS^- \longrightarrow CH_3CH_2SR + Br^-$$
a thioether

$$CH_3CH_2Cl + {}^-NH_2 \longrightarrow CH_3CH_2NH_2 + Cl^-$$
a primary amine

$$CH_3CH_2Br + {}^-C{\equiv}CR \longrightarrow CH_3CH_2C{\equiv}CR + Br^-$$
an alkyne

$$CH_3CH_2I + {}^-C{\equiv}N \longrightarrow CH_3CH_2C{\equiv}N + I^-$$
a nitrile

PROBLEM 6 ◆

What is the product of the reaction of ethyl bromide with each of the following nucleophiles?

a. $CH_3CH_2CH_2O^-$ **b.** $CH_3C{\equiv}C^-$ **c.** $(CH_3)_3N$ **d.** $CH_3CH_2S^-$

ENVIRONMENTAL ADAPTATION

The microorganism *Xanthobacter* has learned to use the alkyl halides that reach the ground as industrial pollutants as a source of carbon. The microorganism synthesizes an enzyme that uses the alkyl halide as a starting material to produce other carbon-containing compounds that it needs via S_N2 reactions.

WHY CARBON INSTEAD OF SILICON?

There are two reasons living organisms are composed primarily of carbon, oxygen, hydrogen, and nitrogen: the *fitness* of these elements for specific roles in life processes and their *availability* in the environment. Of the two reasons, we can see that fitness was more important than availability because carbon rather than silicon became the fundamental building block of living organisms, even though silicon, which is just below carbon in the periodic table, is more than 140 times more abundant than carbon in the Earth's crust.

Abundance (atoms/100 atoms)

Element	In living organisms	In Earth's crust
H	49	0.22
C	25	0.19
O	25	47
N	0.3	0.1
Si	0.03	28

Why are hydrogen, carbon, oxygen, and nitrogen so well suited for the roles they play in living organisms? First and foremost, they are among the smallest atoms that form covalent bonds, and carbon, oxygen, and nitrogen can also form multiple bonds. Because the atoms are small and can form multiple bonds, they form strong bonds that give rise to stable molecules. The compounds that make up living organisms must be stable and, therefore, slow to react if the organisms are to survive.

Silicon has almost twice the diameter of carbon, so silicon forms longer and weaker bonds. Consequently, an S_N2 reaction at silicon would occur much more rapidly than an S_N2 reaction at carbon. Moreover, silicon has another problem. The end product of carbon metabolism is CO_2. The analogous product of silicon metabolism would be SiO_2. But unlike carbon that is double bonded to oxygen in CO_2, silicon is only singly bonded to oxygen in SiO_2. Therefore, silicon dioxide molecules polymerize to form quartz (sea sand). It is hard to imagine that life could exist, much less proliferate, if animals exhaled sand instead of CO_2!

9.4 THE MECHANISM OF AN S_N1 REACTION

Given our understanding of S_N2 reactions, we would expect the rate of the reaction of *tert*-butyl bromide with water to be very slow because water is a poor nucleophile and *tert*-butyl bromide is sterically hindered to attack by a nucleophile. It turns out, however, that the reaction is surprisingly fast. In fact, it is over one million times faster than the reaction of methyl bromide (a compound with no steric hindrance) with water (Table 9.2). Clearly, the reaction must be taking place by a mechanism different from that of an S_N2 reaction.

$$(CH_3)_3C{-}Br + H_2O \longrightarrow (CH_3)_3C{-}OH + HBr$$

tert-butyl bromide → *tert*-butyl alcohol

We have seen that in order to determine the mechanism of a reaction, we need to find out what factors affect the rate of the reaction, and we also need to know the configuration of the products of the reaction. Finding that doubling the concentration of the alkyl halide doubles the rate of the reaction but changing the concentration of the nucleophile has no effect on the rate of the reaction, allows us to write the rate law for the reaction:

$$\textbf{rate} = k[\textbf{alkyl halide}]$$

Because the rate law for the reaction of *tert*-butyl bromide with water differs from the rate law for the reaction of methyl bromide with hydroxide ion (Section 9.2), the two reactions must have different mechanisms. We have seen that the reaction between methyl bromide and hydroxide ion is an S_N2 reaction. The reaction between *tert*-butyl bromide and water is an $\mathbf{S_N1}$ reaction, where "S" stands for substitution, "N" stands for nucleophilic, and "1" stands for unimolecular. **Unimolecular** means that only one

Tutorial: S_N1

Table 9.2 Relative Rates of S_N1 Reactions for Several Alkyl Bromides (solvent is H_2O, nucleophile is H_2O)

Alkyl bromide	Class of alkyl bromide	Relative rate
$CH_3\overset{CH_3}{\underset{CH_3}{C}}-Br$	tertiary	1,200,000
$CH_3\underset{CH_3}{CH}-Br$	secondary	11.6
CH_3CH_2-Br	primary	1.00*
CH_3-Br	methyl	1.05*

*Although the rate of the S_N1 reaction of this compound with water is 0, a small rate is observed as a result of an S_N2 reaction.

molecule is involved in the rate-determining step. The mechanism of an S_N1 reaction is based on the following experimental evidence:

1. The rate law shows that the rate of the reaction depends only on the concentration of the alkyl halide. This means that the rate-determining step of the reaction involves only the alkyl halide.
2. When the methyl groups of *tert*-butyl bromide are successively replaced by hydrogens, the rate of the S_N1 reaction decreases progressively (Table 9.2). This is opposite of the pattern of reactivity exhibited by alkyl halides in S_N2 reactions (Table 9.1).
3. The reaction of an alkyl halide in which the halogen is bonded to an asymmetric center forms two stereoisomers: one with the same relative configuration at the asymmetric center as the reacting alkyl halide, the other with the inverted configuration.

Unlike an S_N2 reaction, where the leaving group departs and the nucleophile approaches *at the same time*, the leaving group in an S_N1 reaction departs *before* the nucleophile approaches.

mechanism for the S_N1 reaction of an alkyl halide

$$CH_3-\overset{CH_3}{\underset{CH_3}{C}}-Br \overset{\text{slow}}{\rightleftharpoons} CH_3-\overset{CH_3}{\underset{CH_3}{C^+}} + H_2\ddot{O}: \xrightarrow{\text{fast}} CH_3-\overset{CH_3}{\underset{CH_3}{C}}-\overset{+}{\ddot{O}}H \overset{\text{fast}}{\rightleftharpoons} CH_3-\overset{CH_3}{\underset{CH_3}{C}}-\ddot{O}H + H^+$$

$+ Br^-$

nucleophile attacks the carbocation

C—Br bond breaks

proton dissociation

- In the first step of an S_N1 reaction of an alkyl halide, the carbon–halogen bond breaks such that both of its electrons stay with the halogen. As a result, a carbocation intermediate is formed.
- In the second step, the nucleophile reacts rapidly with the carbocation to form a protonated alcohol.
- Because a protonated alcohol is a strong acid, it will lose a proton. Thus, the final product is an alcohol.

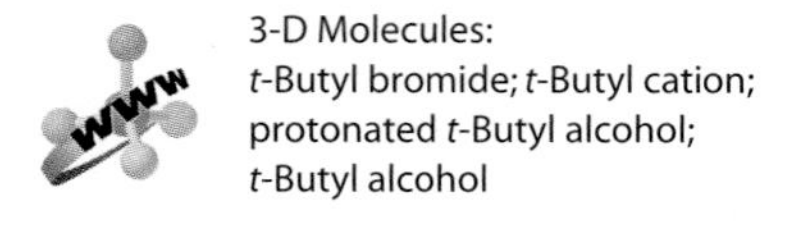

3-D Molecules: *t*-Butyl bromide; *t*-Butyl cation; protonated *t*-Butyl alcohol; *t*-Butyl alcohol

Because the rate of an S_N1 reaction depends only on the concentration of the alkyl halide, the first step must be the slow (rate-determining) step. The nucleophile, therefore, is not involved in the rate-determining step, so its concentration has no effect on the rate of the reaction. If you look at the reaction coordinate diagram in Figure 9.4, you will see why increasing the rate of the second step will not make an S_N1 reaction go any faster.

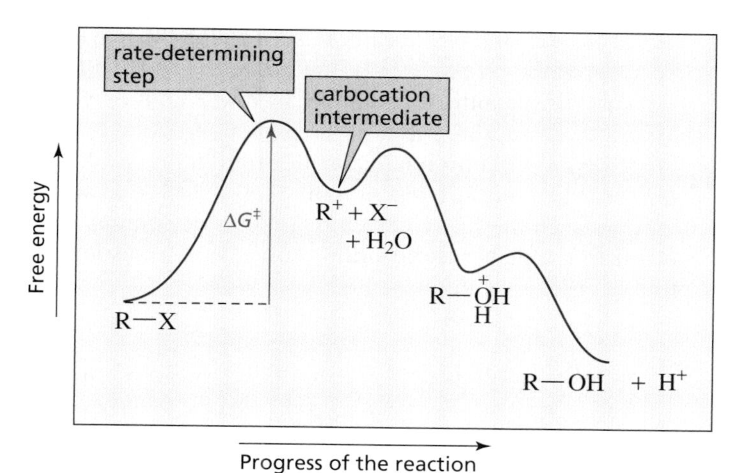

◀ **Figure 9.4** A reaction coordinate diagram for an S_N1 reaction.

How does the mechanism for an S_N1 reaction account for the three pieces of experimental evidence? First, because the alkyl halide is the only species that participates in the rate-determining step, the mechanism agrees with the observation that the rate of the reaction depends only on the concentration of the alkyl halide; it does not depend on the concentration of the nucleophile.

Second, the mechanism shows that a carbocation is formed in the rate-determining step. We know that a tertiary carbocation is more stable and is therefore formed more easily than a secondary carbocation, which in turn is more stable and formed more easily than a primary carbocation (Section 5.2). Tertiary alkyl halides, therefore, are more reactive than secondary alkyl halides, which are more reactive than primary alkyl halides. This relative order of reactivity agrees with the observation that the rate of an S_N1 reaction decreases as the methyl groups of *tert*-butyl bromide are successively replaced by hydrogens (Table 9.2).

relative reactivities of alkyl halides in an S_N1 reaction

most reactive > 3° alkyl halide > 2° alkyl halide > 1° alkyl halide < least reactive

Carbocation stability: 3° > 2° > 1°

In reality, primary carbocations and methyl cations are so unstable that primary alkyl halides and methyl halides do not undergo S_N1 reactions. (The very slow reactions reported for ethyl bromide and methyl bromide in Table 9.2 are S_N2 reactions.)

Primary alkyl halides cannot undergo S_N1 reactions

Third, the positively charged carbon of the carbocation intermediate is sp^2 hybridized, which means the three bonds connected to it are in the same plane. In the second step of the S_N1 reaction, the nucleophile can approach the carbocation from either side of the plane.

C—Br ← alkyl halide

carbocation intermediate

$H_2\ddot{O}$: (a) → C^+ ← (b) $:\ddot{O}H_2$ + Br^-

inverted configuration relative to the configuration of the alkyl halide

same configuration as the alkyl halide

H^+ + HO—C ⇌ $H\overset{+}{O}$—C (H) (a) (b) C—$\overset{+}{O}H$ (H) ⇌ C—OH + H^+

If the nucleophile attacks the side of the carbon from which the leaving group departed (labeled b in the depiction on page 239), the product will have the same relative configuration as that of the reacting alkyl halide. If, however, the nucleophile attacks the opposite side of the carbon (labeled a in the depiction on page 239), the product will have the inverted configuration relative to the configuration of the alkyl halide. We can now understand why an S_N1 reaction of an alkyl halide in which the leaving group is attached to an asymmetric center forms two stereoisomers: attack of the nucleophile on one side of the planar carbocation forms one stereoisomer, and attack on the other side produces the other stereoisomer.

an asymmetric center

R′–C(H)(R)(Br) + H_2O ⟶ R′–C(H)(HO)(R) + R′–C(H)(R)(OH) + HBr

if the leaving group in an S_N1 reaction is attached to an asymmetric center, a pair of enantiomers will be formed as products

PROBLEM 7 ◆

Arrange the following alkyl bromides in order of decreasing reactivity in an S_N1 reaction: isopropyl bromide, propyl bromide, *tert*-butyl bromide, methyl bromide.

PROBLEM 8 ◆

a. How many stereoisomers would be obtained from the S_N1 reaction of 2-bromopentane with CH_3OH?

b. How many stereoisomers would be obtained from the S_N1 reaction of 3-bromopentane with CH_3OH?

9.5 FACTORS THAT AFFECT S_N1 REACTIONS

The Leaving Group

Because the rate-determining step of an S_N1 reaction is the dissociation of the alkyl halide to form a carbocation, two factors affect the rate of the reaction: (1) the ease with which the leaving group dissociates from the carbon and (2) the stability of the carbocation that is formed. In the preceding section, we saw how carbocation stability affects the rate of the reaction. But how do we rank the relative reactivity of a series of alkyl halides with different leaving groups that dissociate to form the same carbocation?

As in the case of the S_N2 reaction, there is a direct relationship between basicity and leaving ability in an S_N1 reaction: the weaker the base, the less tightly it is bonded to the carbon and the more easily the carbon–halogen bond can be broken. As a result, an alkyl iodide is the most reactive, and an alkyl fluoride is the least reactive of the alkyl halides in both S_N1 and S_N2 reactions.

relative reactivities of alkyl halides in an S_N1 reaction

most reactive ⟶ RI > RBr > RCl > RF ⟵ least reactive

The Nucleophile

We have seen that the rate-determining step of an S_N1 reaction is formation of the carbocation. Because the nucleophile comes into play *after* the rate-determining step, the strength of the nucleophile has no effect on the rate of an S_N1 reaction (Figure 9.4).

PROBLEM 9 ◆

Arrange the following alkyl halides in order of decreasing reactivity in an S_N1 reaction: 2-bromopentane, 2-chloropentane, 1-chloropentane, 3-bromo-3-methylpentane.

9.6 COMPARING THE S_N2 AND S_N1 REACTIONS OF ALKYL HALIDES

The characteristics of S_N2 and S_N1 reactions are compared in Table 9.3. Remember that the "2" in "S_N2" and the "1" in "S_N1" refer to the number of molecules in the rate-determining step and not to the number of steps in the mechanism. In fact, the opposite is true: An S_N2 reaction proceeds by a *one*-step mechanism, and an S_N1 reaction proceeds by a *two*-step mechanism with a carbocation intermediate.

Because the nucleophile appears in the rate law of an S_N2 reaction, the better the nucleophile, the greater will be the rate constant for the reaction and, therefore, the faster the reaction will be. Because the nucleophile does *not* appear in the rate law of an S_N1 reaction, the strength of the nucleophile has no effect on the rate of an S_N1 reaction.

We have seen that if the halogen is attached to an asymmetric center, the product of an S_N2 reaction will have a configuration that is inverted relative to that of the reactant.

the configuration is inverted relative to that of the reactant

CH_2CH_3–C(H)(Br)–CH_3 + HO^- $\xrightarrow{S_N2 \text{ conditions}}$ CH_2CH_3–C(H)(HO)–CH_3 + Br^-

(*S*)-2-bromobutane (*R*)-2-butanol

Movies: S_N1 inversion; S_N1 retention

If the same reaction is carried out under S_N1 conditions, two substitution products are obtained: one with the same relative configuration as the reactant and one with the inverted configuration.

If the leaving group is attached to an asymmetric center, an S_N2 reaction forms the isomer with the inverted configuration.

If the leaving group is attached to an asymmetric center, an S_N1 reaction forms a pair of enantiomers.

product with the inverted configuration — product with the same configuration

CH_2CH_3–C(H)(Br)–CH_3 + H_2O $\xrightarrow{S_N1 \text{ conditions}}$ CH_2CH_3–C(H)(HO)–CH_3 + CH_2CH_3–C(H)(OH)–CH_3 + HBr

(*S*)-2-bromobutane (*R*)-2-butanol (*S*)-2-butanol

Table 9.3 Comparison of S_N2 and S_N1 Reactions

S_N2	S_N1
A one-step mechanism	A two-step mechanism
A bimolecular rate-determining step	A unimolecular rate-determining step
The better the nucleophile, the faster the rate of the reaction	The strength of the nucleophile does not affect the rate of the reaction
Reactivity order: methyl > 1° > 2°	Reactivity order: 3° > 2°
(No reaction with 3°)	(No reaction with 1° or methyl)
The product has inverted configuration relative to that of the reactant	The products have both the same and inverted configurations relative to that of the reactant
Leaving group: $I^- > Br^- > Cl^- > F^-$	Leaving group: $I^- > Br^- > Cl^- > F^-$

The difference between the products obtained from an S_N1 reaction and from an S_N2 reaction is a little easier to visualize in the case of cyclic compounds. For example, when *cis*-1-bromo-4-methylcyclohexane undergoes an S_N2 reaction, only the trans product is obtained because the carbon bonded to the leaving group is attacked by the nucleophile only on its back side.

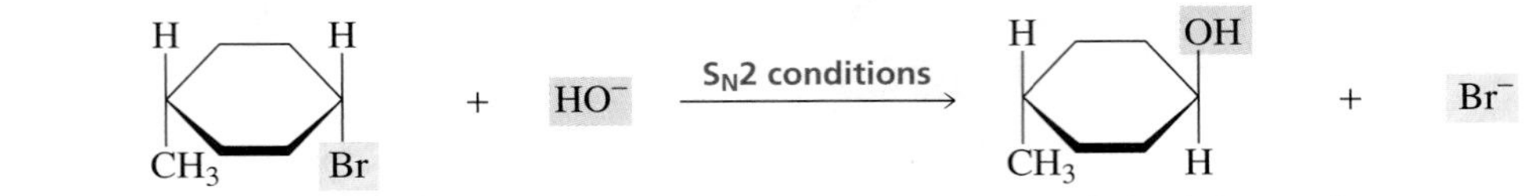

cis-1-bromo-4-methylcyclohexane + HO^- —(S_N2 conditions)→ *trans*-4-methylcyclohexanol + Br^-

However, when *cis*-1-bromo-4-methylcyclohexane undergoes an S_N1 reaction, both the cis and the trans products are formed because the nucleophile can approach the carbocation intermediate from either side.

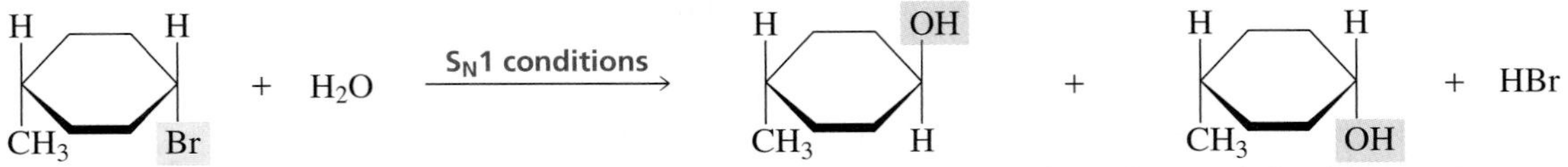

cis-1-bromo-4-methylcyclohexane + H_2O —(S_N1 conditions)→ *trans*-4-methylcyclohexanol + *cis*-4-methylcyclohexanol + HBr

PROBLEM 10 ◆

Which alkyl halide in each of the following pairs reacts faster in an S_N2 reaction?

a. $CH_3CH_2CH_2Br$ or $CH_3CH_2CHCH_3$ (with Br on CH)

b. $CH_3CHCH_2CHCH_3$ (Br on C2, CH_3 on C4) or $CH_3CH_2CH_2CCH_3$ (CH_3 and Br on C4)

c. chlorocyclohexane or bromocyclohexane

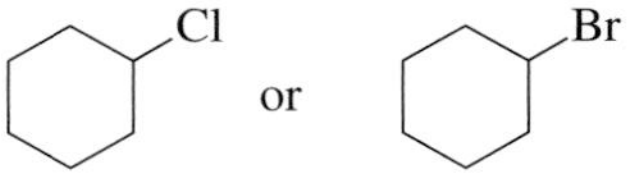

d. CH_3CCH_2Cl (two CH_3 on central C) or $CH_3CCH_2CH_2Cl$ (two CH_3 on C)

PROBLEM 11 ◆

Which alkyl halide in each of the pairs in Problem 10 reacts faster in an S_N1 reaction?

PROBLEM 12

Give the substitution products that will be obtained from the following reactions if:

a. the reaction is carried out under conditions that favor an S_N2 reaction.

b. the reaction is carried out under conditions that favor an S_N1 reaction.

1. *trans*-1-iodo-4-methylcyclohexane + sodium methoxide/methanol
2. *cis*-1-chloro-3-methylcyclobutane + sodium hydroxide/water

9.7 ELIMINATION REACTIONS OF ALKYL HALIDES

In addition to undergoing nucleophilic substitution reactions, alkyl halides also undergo elimination reactions. In an elimination reaction, the halogen (X) is removed from one carbon and a hydrogen is removed from an *adjacent* carbon. A double bond is formed between the two carbons from which the atoms are eliminated. Therefore, when an alkyl halide undergoes an elimination reaction, *the product is an alkene.*

$$CH_3CH_2CH_2X + Y^- \xrightarrow{\text{substitution}} CH_3CH_2CH_2Y + X^-$$

$$CH_3CH_2CH_2X + Y^- \xrightarrow{\text{elimination}} CH_3CH{=}CH_2 + HY + X^-$$

new double bond

The E2 Reaction

Movie: An E2 reaction

Just as there are two nucleophilic substitution reactions, S_N1 and S_N2, there are two elimination reactions, E1 and E2. The reaction of *tert*-butyl bromide with hydroxide ion is an example of an **E2 reaction**; "E" stands for *elimination* and "2" stands for *bimolecular*.

$$(CH_3)_3CBr + HO^- \longrightarrow CH_2{=}C(CH_3)CH_3 + H_2O + Br^-$$

tert-butyl bromide → 2-methylpropene

The rate of the E2 reaction depends on the concentrations of both *tert*-butyl bromide and hydroxide ion.

rate = *k*[alkyl halide][base]

The following mechanism agrees with the observation that both *tert*-butyl bromide and hydroxide ion are involved in the rate-determining step of the reaction:

mechanism for the E2 reaction of an alkyl halide

$$HO^- + H\text{–}CH_2\text{–}C(CH_3)_2\text{–}Br \longrightarrow CH_2{=}C(CH_3)CH_3 + H_2O + Br^-$$

a proton is removed

a double bond is formed

Br^- is eliminated

3-D Molecule: 2-methylpropene

- The base removes a proton from a carbon that is adjacent to the carbon bonded to the halogen. As the proton is removed, the electrons that it shared with carbon move toward the adjacent carbon that is bonded to the halogen. As these electrons move toward the carbon, the halogen leaves, taking its bonding electrons with it.

When the reaction is over, the electrons that were originally bonded to the hydrogen in the reactant have formed a π bond in the product.

The E1 Reaction

Movie: An E1 reaction

The second kind of elimination reaction that alkyl halides can undergo is an E1 reaction. The reaction of *tert*-butyl bromide with water to form 2-methylpropene is an example of an **E1 reaction**: "E" stands for *elimination* and "1" stands for *unimolecular*.

$$CH_3-\underset{Br}{\overset{CH_3}{C}}-CH_3 + H_2O \longrightarrow CH_2{=}\overset{CH_3}{C}-CH_3 + H_3O^+ + Br^-$$

tert-butyl bromide → 2-methylpropene

The rate of the reaction depends only on the concentration of the alkyl halide.

rate = k[alkyl halide]

We therefore know that only the alkyl halide takes part in the rate-determining step of the reaction. The following mechanism agrees with this observation: because the first step is the rate-determining step, an increase in the concentration of the base—which participates only in the second step of the reaction—has no effect on the rate of the reaction.

mechanism for the E1 reaction of an alkyl halide

$$CH_3-\underset{Br}{\overset{CH_3}{C}}-CH_3 \overset{slow}{\rightleftharpoons} \underset{H}{CH_2}-\overset{CH_3}{\overset{}{C^+}}-CH_3 + Br^- \xrightarrow{fast} CH_2{=}\overset{CH_3}{C}-CH_3 + H_3O^+$$

$H_2\ddot{O}:$

the alkyl halide dissociates, forming a carbocation

the base removes a proton

- The alkyl halide dissociates, forming a carbocation.
- The base forms the elimination product by removing a proton from a carbon that is adjacent to the positively charged carbon.

The weaker the base, the better it is as a leaving group.

Alkyl iodides are the most reactive and alkyl fluorides are the least reactive in both E2 and E1 reactions, because weaker bases are better leaving groups (Section 9.3).

relative reactivities of alkyl halides in E2 and E1 reactions

most reactive → RI > RBr > RCl > RF ← least reactive

increasing reactivity (←)

PROBLEM 13 ♦

a. Which alkyl halide would you expect to be more reactive in an E2 reaction?

$$CH_3\underset{CH_3}{CHCl} \quad \text{or} \quad CH_3\underset{CH_3}{CHBr}$$

b. Which would be more reactive in an E1 reaction?

INVESTIGATING NATURALLY OCCURRING HALOGEN-CONTAINING COMPOUNDS

Like many other natural products—that is, compounds produced in nature—certain organohalides found in marine organisms have interesting and potent biological activity. Cyclocinamide A, one such organohalide, comes from a creature called an orange-encrusting sponge. This compound, as well as a host of analogs, has impressive antitumor properties that are currently being exploited in the development of new anticancer drugs.

Jasplankinolide, another organohalide found in a sponge, modulates the formation and depolymerization of actin microtubules. Microtubules are found in all cells and are used for motile events, such as transportation of vesicles, migration, and cell division. Jasplankinolide is therefore being used to further our understanding of these processes. Notice that cyclocinamide A has four asymmetric centers and jasplankinolide has six. In view of the great diversity of marine life, the ocean probably contains many compounds with useful medicinal properties waiting to be discovered by scientists.

cyclocinamide A

jasplankinolide

9.8 PRODUCTS OF ELIMINATION REACTIONS

In an elimination reaction, the hydrogen is removed from a β-carbon. (The α-carbon is the carbon bonded to the halogen; the β-carbon is the carbon adjacent to the α-carbon.) An alkyl halide such as 2-bromopropane has two β-carbons from which a hydrogen can be removed in an E2 reaction. Because the two β-carbons are identical, the proton can be removed with equal ease from either one. The product of this elimination reaction is propene.

β-carbons

$$CH_3\underset{\underset{Br}{|}}{C}HCH_3 + CH_3O^- \longrightarrow CH_3CH{=}CH_2 + CH_3OH + Br^-$$

α-carbon

2-bromopropane — propene

In contrast, 2-bromobutane has two structurally different β-carbons from which a hydrogen can be removed. Therefore, when 2-bromobutane reacts with a base, two elimination products are formed: 2-butene and 1-butene. This E2 reaction is *regioselective* because more of one constitutional isomer is formed than the other (Section 5.3).

β-carbons

$$CH_3\underset{\underset{Br}{|}}{C}HCH_2CH_3 + CH_3O^- \xrightarrow{CH_3OH} CH_3CH{=}CHCH_3 + CH_2{=}CHCH_2CH_3 + CH_3OH + Br^-$$

2-bromobutane — 2-butene 80% — 1-butene 20%

Which of the alkenes is obtained in greater yield? The answer is just what we would predict: the more stable alkene is formed in greater yield, because it has the more stable transition state leading to its formation and therefore is formed faster (Figure 9.5).

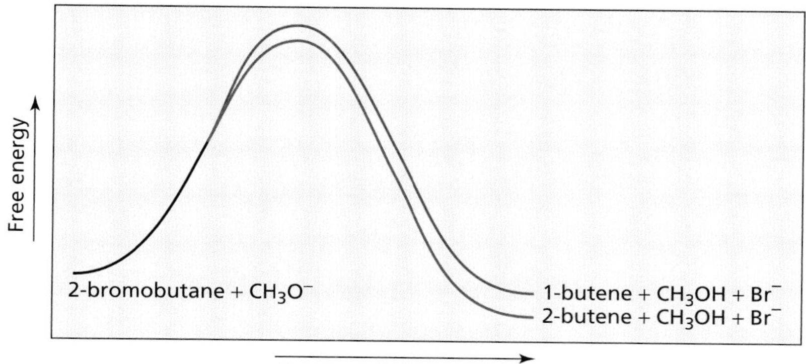

◀ **Figure 9.5**
A reaction coordinate diagram for the E2 reaction of 2-bromobutane and methoxide ion.

The most stable alkene is generally the most substituted alkene.

The most stable alkene is obtained by removing a hydrogen from the β-carbon that is bonded to the fewest hydrogens.

Tutorial:
E2 Elimination regiochemistry

Alexander M. Zaitsev (1841–1910) *was born in Kazan, Russia. (The German transliteration, Saytzeff, is sometimes used for his family name.) He received a Ph.D. from the university of Leipzig in 1866. Zaitsev was a professor of chemistry first at the University of Kazan and later at the University of Kiev.*

We have seen that the stability of an alkene depends on the number of alkyl substituents bonded to its sp^2 carbons: the greater the number of alkyl substituents, the more stable is the alkene (Section 4.6). Therefore, 2-butene, with a total of two methyl substituents bonded to its sp^2 carbons, is more stable than 1-butene, with one ethyl substituent. Thus, 2-butene is formed faster than 1-butene.

Alexander M. Zaitsev, a nineteenth-century Russian chemist, devised a shortcut to predict the most substituted alkene product. He pointed out *that the more substituted alkene is obtained when a hydrogen is removed from the β-carbon that is bonded to the fewest hydrogens*. This is called Zaitsev's rule.

Because elimination from a tertiary alkyl halide typically leads to a more substituted alkene than does elimination from a secondary alkyl halide, and elimination from a secondary alkyl halide generally leads to a more substituted alkene than does elimination from a primary alkyl halide, the relative reactivities of alkyl halides in an E2 reaction are as follows:

relative reactivities of alkyl halides in an E2 reaction

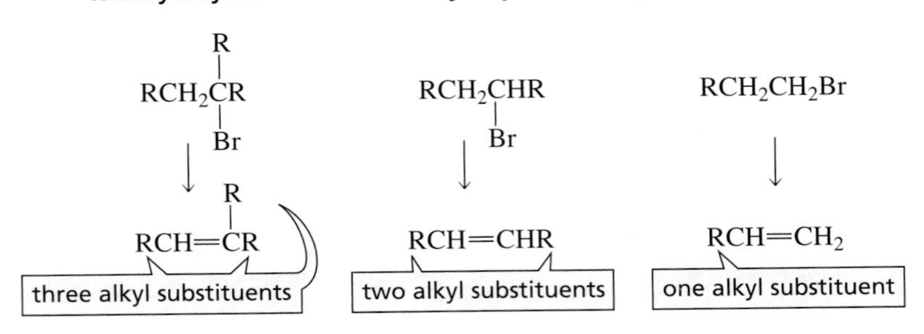

The major product of an E1 reaction, such as the reaction of 2-chloro-2-methylbutane with water, is also the more stable alkene. 2-Methyl-2-butene (with a total of three alkyl substituents) is more stable than 2-methyl-1-butene (with only two alkyl substituents).

$$CH_3CH_2C(CH_3)(Cl)CH_3 + H_2O \longrightarrow CH_3CH{=}C(CH_3)CH_3 + CH_3CH_2C(CH_3){=}CH_2 + H_3O^+ + Cl^-$$

2-chloro-2-methylbutane → 2-methyl-2-butene (major product) + 2-methyl-1-butene (minor product)

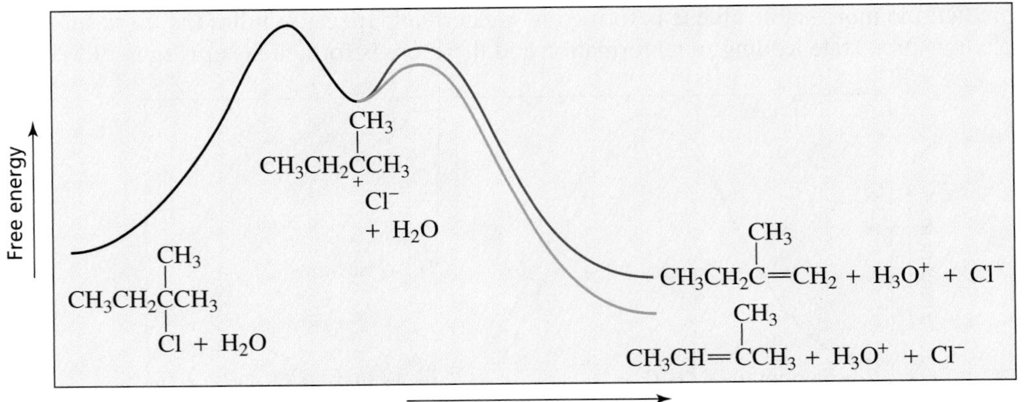

We see again that the more stable alkene is formed in greater yield, because the more stable alkene has the more stable transition state leading to its formation (Figure 9.6).

3-D Molecule:
2-Chloro-2-methylbutane

3-D Molecule:
2-Methyl-2-butene

▶ Figure 9.6
A reaction coordinate diagram for the E1 reaction of 2-chloro-2-methylbutane. The major product is the more substituted alkene because its greater stability causes the transition state leading to its formation to be more stable.

Because the first step of an E1 reaction is the rate-limiting step, the more stable the carbocation formed in the first step, the faster will be the rate of the E1 reaction. Therefore, alkyl halides have the following relative reactivities in an E1 reaction since a tertiary carbocation is more stable, and is therefore easier to form, than a secondary carbocation, which in turn is more stable and easier to form than a primary carbocation (Section 5.2).

relative reactivities of alkyl halides in an E1 reaction

tertiary alkyl halide > **secondary alkyl halide** > **primary alkyl halide**

Thus, for *both* E2 and E1 reactions, tertiary alkyl halides are the most reactive and primary alkyl halides are the least reactive, and the major product is the most stable alkene.

The major product of an E2 or E1 reaction is the most stable alkene.

PROBLEM 14 ◆

Which of the following compounds would react faster in an
a. E1 reaction? **b.** E2 reaction? **c.** S_N1 reaction? **d.** S_N2 reaction?

1.

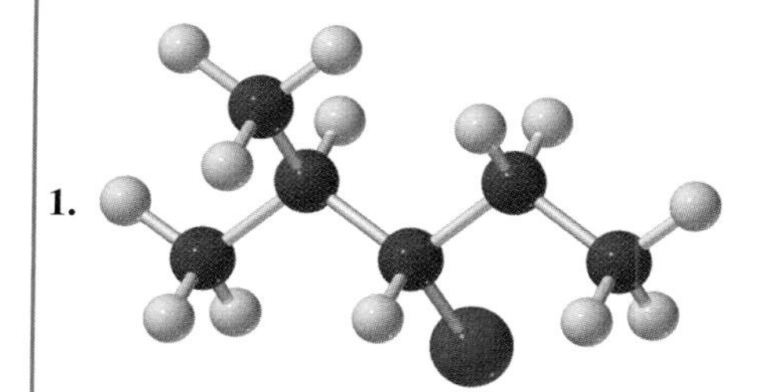

2.

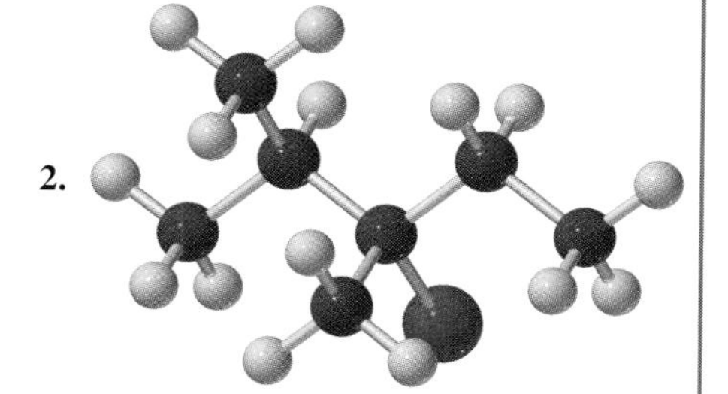

PROBLEM 15 ◆ SOLVED

What would be the major elimination product obtained from the reaction of each of the following alkyl halides with hydroxide ion?

a. $CH_3CH_2CH_2CH_2\underset{\displaystyle Br}{\underset{|}{C}}HCH_3$

b. $CH_3CH_2CH_2\overset{\displaystyle CH_3}{\overset{|}{\underset{\displaystyle Cl}{\underset{|}{C}}}}CH_3$

c. $CH_3\underset{\displaystyle Br}{\underset{|}{C}}HCH_2\underset{\displaystyle CH_3}{\underset{|}{C}}HCH_3$

d. $CH_3\overset{\displaystyle CH_3}{\overset{|}{\underset{\displaystyle CH_3}{\underset{|}{C}}}}-\underset{\displaystyle Br}{\underset{|}{C}}HCH_3$

Solution to 15a More 2-hexene will be formed than 1-hexene, because 2-hexene is more stable since it has more alkyl substituents bonded to its sp^2 carbons.

$$CH_3CH_2CH_2CH_2\underset{\displaystyle Br}{\underset{|}{C}}HCH_3 \xrightarrow{HO^-} \underset{\text{2-hexene, major product}}{CH_3CH_2CH_2CH{=}CHCH_3} + \underset{\text{1-hexene}}{CH_3CH_2CH_2CH_2CH{=}CH_2}$$

PROBLEM 16 ◆

Three alkenes are formed from the E1 reaction of 3-bromo-2,3-dimethylpentane. Give the structures of the alkenes, and rank them according to the amount that would be formed. (Ignore stereoisomers.)

The major stereoisomer formed in E2 or E1 reaction is the alkene with the bulkiest substituents on opposite sides of the double bond.

We saw that 2-butene is the major elimination product when 2-bromobutane reacts with methoxide ion (page 245). 2-Butene, however, has two stereoisomers, (*E*)-2-butene and (*Z*)-2-butene. Which of the stereoisomers is obtained in greater yield? Again, we find that the more stable product is the one formed in greater yield. Recall that the more stable alkene is the one with the *bulkiest groups on opposite sides of the double bond* (Section 4.6). Therefore, more (*E*)-2-butene is formed than (*Z*)-2-butene.

Tutorial:
E2 Stereochemistry

$$\underset{H_3C}{\overset{H}{}}\!\!>\!C\!=\!C\!<\!\!\underset{H}{\overset{CH_3}{}}$$

(*E*)-2-butene
more stable

$$\underset{H_3C}{\overset{H}{}}\!\!>\!C\!=\!C\!<\!\!\underset{CH_3}{\overset{H}{}}$$

(*Z*)-2-butene
less stable

PROBLEM 17 ***SOLVED***

For each of the major elimination products determined in Problem 15 that can exist as stereoisomers, which stereoisomer is obtained in greater yield?

Solution to 17a 2-Hexene has two stereoisomers. More (*E*)-2-hexene will be formed than (*Z*)-2-hexene because (*E*)-2-hexene is more stable, since the bulkiest substituents are on opposite sides of the double bond.

$$\underset{H}{\overset{CH_3CH_2CH_2}{}}\!\!>\!C\!=\!C\!<\!\!\underset{CH_3}{\overset{H}{}}$$

(*E*)-2-hexene
major product

$$\underset{H}{\overset{CH_3CH_2CH_2}{}}\!\!>\!C\!=\!C\!<\!\!\underset{H}{\overset{CH_3}{}}$$

(*Z*)-2-hexene

9.9 COMPARING THE E2 AND E1 REACTIONS OF ALKYL HALIDES

The characteristics of E2 and E1 reactions are compared in Table 9.4. As with S_N2 and S_N1 reactions, the "2" in E2 and the "1" in E1 refer to the number of molecules in the rate-determining step, not to the number of steps in the mechanism: an E2 reaction

Table 9.4 Comparison of E2 and E1 Reactions

E2	E1
A one-step mechanism	A two-step mechanism
A bimolecular rate-determining step	A unimolecular rate-determining step
The stronger the base, the faster the rate of the reaction	The strength of the base does not affect the rate of the reaction
Reactivity order: 3° > 2°	Reactivity order: 3° > 2°
(No reaction with 1°)	(No reaction with 1°)
The product is the more stable alkene, formed by removing a hydrogen from the β-carbon bonded to the fewest hydrogens	The product is the more stable alkene, formed by removing a hydrogen from the β-carbon bonded to the fewest hydrogens
If the alkene can exist as *E* and *Z* isomers, the stereoisomer with the bulkiest groups on opposite sides of the double bond will be formed in greater yield	If the alkene can exist as *E* and *Z* isomers, the stereoisomer with the bulkiest groups on opposite sides of the double bond will be formed in greater yield
Leaving group: $I^- > Br^- > Cl^- > F^-$	Leaving group: $I^- > Br^- > Cl^- > F^-$

proceeds by a *one*-step mechanism, and an E1 reaction proceeds by a *two*-step mechanism with a carbocation intermediate.

Because the base appears in the rate law of an E2 reaction, the stronger the base, the greater will be the rate constant for the reaction and, therefore, the faster the reaction will be. Because the base does *not* appear in the rate law of an E1 reaction, the strength of the base has no effect on the rate of an E1 reaction.

PROBLEM 18 ◆

Which alkyl halide in each of the pairs in Problem 10 reacts faster in an E2 reaction?

PROBLEM 19 ◆

Which alkyl halide in the pairs in Problem 10 parts a, b, and c reacts faster in an E1 reaction?

Tutorial:
Common terms for E1 and E2 reactions

9.10 DOES AN ALKYL HALIDE UNDERGO S_N2/E2 REACTIONS OR S_N1/E1 REACTIONS?

We have seen that alkyl halides can undergo four types of reactions: S_N2, S_N1, E2, and E1. As a result, you may feel a bit overwhelmed when you are asked to predict the products of the reaction of a given alkyl halide and a nucleophile/base. Let's therefore pause to organize what we know about the reactions of alkyl halides to make it a little easier for you to predict their products. Notice, in the following discussion, that HO^- is called a nucleophile in a substitution reaction (because it attacks a carbon) and a base in an elimination reaction (because it removes a proton).

To predict the products of the reaction of an alkyl halide, the first thing you must decide is whether the reaction conditions favor S_N2/E2 or S_N1/E1 reactions. We will see that the conditions that favor an S_N2 reaction also favor an E2 reaction and that the conditions that favor an S_N1 reaction also favor an E1 reaction. Therefore, S_N2/E2 occur together and S_N1/E1 occur together.

Two factors determine whether S_N2/E2 or S_N1/E1 reactions predominate: (1) the *concentration* of the nucleophile/base and (2) the *reactivity* of the nucleophile/base. To understand how these two factors determine the set of reactions that predominates, we must look at the overall rate law for the reaction. The overall rate law is the sum of the individual rate laws for the S_N1, S_N2, E1, and E2 reactions. (Subscripts have been added to the rate constants to indicate that they have different values.)

$$\textbf{rate} = k_1[\text{alkyl halide}] + k_2[\text{alkyl halide}][\text{nucleophile}] + k_3[\text{alkyl halide}] + k_4[\text{alkyl halide}][\text{base}]$$

contribution to the rate by an S_N1 reaction | contribution to the rate by an S_N2 reaction | contribution to the rate by an E1 reaction | contribution to the rate by an E2 reaction

From the overall rate law, you can see that increasing the *concentration* of the nucleophile/base has no effect on the rate of the S_N1 and E1 reactions, because the concentration of the nucleophile/base is not in their rate laws. In contrast, increasing the *concentration* of the nucleophile/base increases the rate of the S_N2 and E2 reactions, because the concentration of the nucleophile/base is in their rate laws. Similarly, increasing the *reactivity* of the nucleophile/base has no effect on the rate of S_N1 and E1 reactions, because the slow step in these reactions does not involve the nucleophile/base. It does, however, increase the rate of the S_N2 and E2 reactions by increasing the value of the rate constants (k_2 and k_4), because a more reactive nucleophile/base is better able to displace the leaving group. In summary:

Tutorial:
S_N2 promoting factors

- S_N2 and E2 reactions are favored by a high concentration of a good nucleophile/strong base.
- S_N1 and E1 reactions are favored by a poor nucleophile/weak base because a poor nucleophile/weak base disfavors S_N2 and E2 reactions.

Tutorial:
E2 Promoting factors

S_N2 and E2 reactions of alkyl halides are favored by a high concentration of a good nucleophile/strong base.

S_N1 and E1 reactions of alkyl halides are favored by a poor nucleophile/weak base.

Look back at the S_N1 and E1 reactions in previous sections and notice that they all have poor nucleophiles/weak bases (H_2O, CH_3OH), whereas the S_N2 and E2 reactions have good nucleophiles/strong bases (HO^-, CH_3O^-). In other words, a good nucleophile/strong base is used to encourage an S_N2/E2 reaction, and a poor nucleophile/weak base is used to encourage an S_N1/E1 reaction by discouraging the competing S_N2/E2 reaction.

PROBLEM-SOLVING STRATEGY

This problem will give you practice in determining whether a substitution reaction will take place by an S_N1 or an S_N2 pathway.

Give the configuration(s) of the substitution product(s) that will be formed from the reactions of the following alkyl halides with the indicated nucleophile:

a. $H_3C-C(CH_2CH_3)(H)(Br)$ + CH_3O^- (high concentration) ⟶ $H-C(CH_2CH_3)(CH_3)(OCH_3)$

Because a high concentration of a good nucleophile is used, we can predict that the reaction is an S_N2 reaction. Therefore, the product will have the inverted configuration relative to the configuration of the reactant. (An easy way to draw the inverted product is to draw the mirror image of the reacting alkyl halide and then put the nucleophile in the same location as the leaving group.)

b. $H_3C-C(CH_2CH_3)(H)(Br)$ + CH_3OH ⟶ $H_3C-C(CH_2CH_3)(H)(OCH_3)$ + $H-C(CH_2CH_3)(CH_3)(OCH_3)$

Because a poor nucleophile is used, we can predict that the reaction is an S_N1 reaction. Therefore, we will obtain two substitution products—one with the same configuration and one with the inverted configuration—relative to the configuration of the reactant.

c. $CH_3CH_2CHCH_2CH_3$ (with I on C3) + CH_3OH ⟶ $CH_3CH_2CHCH_2CH_3$ (with OCH_3 on C3)

The poor nucleophile allows us to predict that the reaction is an S_N1 reaction. However, the product does not have an asymmetric center, so it does not have stereoisomers. Therefore, only one substitution product will be formed. (The same substitution product would have been obtained if the reaction had been an S_N2 reaction.)

Now continue on to Problem 20.

PROBLEM 20

Give the configuration of the substitution product(s) that will be obtained from the reaction of the following alkyl halides with the indicated nucleophile:

a. $H_3C-C(CH_2CH_2CH_3)(H)(Br)$ + $CH_3CH_2CH_2O^-$ (high concentration) ⟶

b. $H-C(CH_2CH_3)(CH_3)(Br)$ + CH_3O^- (high concentration) ⟶

c. $H-C(CH_2CH_3)(CH_3)(Br)$ + CH_3OH ⟶

PROBLEM 21 *SOLVED*

What is the major elimination product that would be obtained from the reactions shown in Problem 20?

Solution to 21a In determining the major elimination product, we do not need to be concerned whether it is an E1 or an E2 reaction, because both form the same major product. The major elimination product is 2-pentene. (Remember, the major elimination product is obtained by removing a hydrogen from the β-carbon bonded to the fewest hydrogens.) 2-Pentene has two stereoisomers. More (*E*)-2-pentene will be formed than (*Z*)-2-pentene because (*E*)-2-pentene is more stable, since the largest substituents are on opposite sides of the double bond.

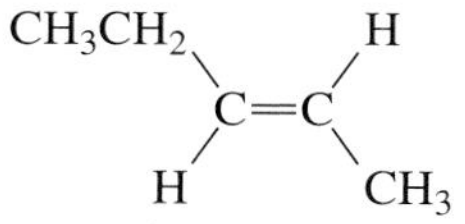

$$(CH_3CH_2)(H)C{=}C(H)(CH_3) \qquad (CH_3CH_2)(H)C{=}C(CH_3)(H)$$

(*E*)-2-pentene major product (*Z*)-2-pentene

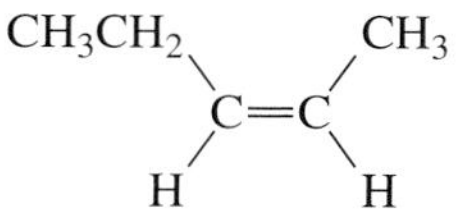

9.11 DOES AN ALKYL HALIDE UNDERGO A SUBSTITUTION REACTION, AN ELIMINATION REACTION, OR BOTH SUBSTITUTION AND ELIMINATION REACTIONS?

Having decided whether the reaction conditions favor S_N2/E2 reactions or S_N1/E1 reactions (Section 9.10), you must next decide whether the alkyl halide will form the substitution product, the elimination product, or both substitution and elimination products. *The relative amounts of substitution and elimination products depend on whether the alkyl halide is primary, secondary, or tertiary.*

S_N2/E2 Conditions

Let's first consider conditions that lead to S_N2/E2 reactions: a high concentration of a good nucleophile/strong base. A negatively charged species can act as a nucleophile and attack the back side of the α-carbon to form the substitution product, or it can act as a base and remove a proton from a β-carbon to form the elimination product. Thus, the two reactions compete with each other. In fact, they both occur for the same reason: the electron-withdrawing halogen gives the carbon to which it is bonded a partial positive charge.

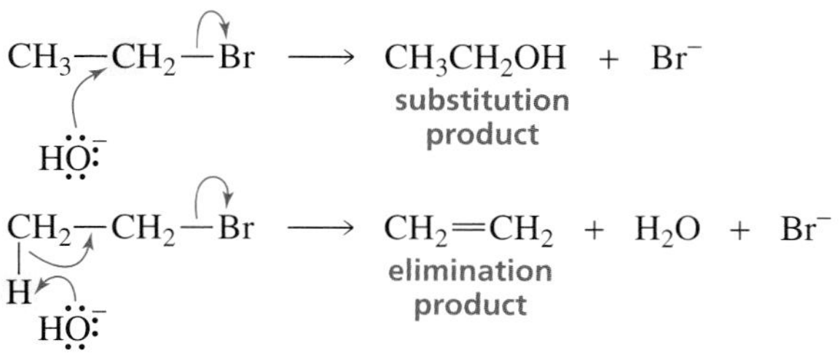

$$CH_3{-}CH_2{-}Br + HO^- \longrightarrow CH_3CH_2OH + Br^-$$

substitution product

$$H{-}CH_2{-}CH_2{-}Br + HO^- \longrightarrow CH_2{=}CH_2 + H_2O + Br^-$$

elimination product

The relative reactivities of alkyl halides in S_N2 and E2 reactions are shown in Table 9.5. Because a *primary* alkyl halide is the most reactive in an S_N2 reaction and the least reactive in an E2 reaction, a primary alkyl halide forms mainly the substitution product in a reaction carried out under conditions that favor S_N2/E2 reactions. In other words, substitution wins the competition.

Primary alkyl halides undergo primarily substitution under S_N2/E2 conditions.

a primary alkyl halide

$$CH_3CH_2CH_2Br + CH_3O^- \xrightarrow{CH_3OH} CH_3CH_2CH_2OCH_3 + CH_3CH{=}CH_2 + CH_3OH + Br^-$$

propyl bromide; methyl propyl ether 90%; propene 10%

Table 9.5 Relative Reactivities of Alkyl Halides

In an S_N2 reaction:	1° > 2° > 3°	In an S_N1 reaction:	3° > 2° > 1°
In an E2 reaction:	3° > 2° > 1°	In an E1 reaction:	3° > 2° > 1°

A *secondary* alkyl halide forms both substitution and elimination products under S_N2/E2 conditions.

a secondary alkyl halide

$$\underset{\text{2-chloropropane}}{CH_3\underset{|}{\overset{Cl}{C}}HCH_3} + CH_3O^- \xrightarrow[CH_3OH]{} \underset{\substack{\text{ethyl isopropyl ether}\\ 25\%}}{CH_3\overset{OCH_2CH_3}{C}HCH_3} + \underset{\substack{\text{propene}\\ 75\%}}{CH_3CH{=}CH_2} + CH_3OH + Cl^-$$

Tertiary alkyl halides undergo only elimination under S_N2/E2 conditions.

A *tertiary* alkyl halide is the least reactive of the alkyl halides in an S_N2 reaction and the most reactive in an E2 reaction (Table 9.5). Consequently, *only* the elimination product is formed when a tertiary alkyl halide reacts with a nucleophile/base under S_N2/E2 conditions.

a tertiary alkyl halide

$$\underset{\text{2-bromo-2-methylpropane}}{CH_3\underset{CH_3}{\overset{CH_3}{C}}Br} + CH_3CH_2O^- \xrightarrow[CH_3CH_2OH]{} \underset{\substack{\text{2-methylpropene}\\ 100\%}}{CH_3\overset{CH_3}{C}{=}CH_2} + CH_3CH_2OH + Br^-$$

PROBLEM 22 ◆

Indicate whether the alkyl halides listed will give both substitution and elimination products, primarily substitution products, only elimination products, or no products when they react with methanol under S_N2/E2 conditions.

a. 1-bromobutane
b. 1-bromo-2-methylpropane
c. 2-bromobutane
d. 2-bromo-2-methylpropane

PROBLEM 23

Draw the stereoisomers that would be obtained in greatest yield from the reaction of the following alkyl chlorides with hydroxide ion:

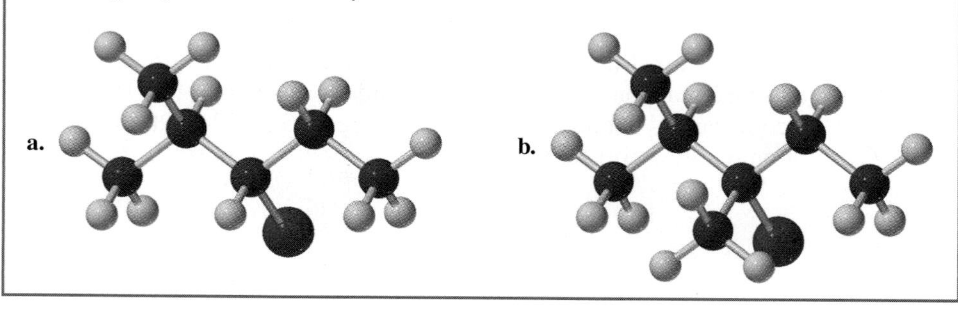

S_N1/E1 Conditions

Now let's look at what happens when conditions favor S_N1/E1 reactions: a poor nucleophile/weak base. In S_N1/E1 reactions, the alkyl halide dissociates to form a carbocation, which can then either combine with the nucleophile to form the substitution product or lose a proton to form the elimination product.

$$-\underset{H}{\overset{|}{C}}-\overset{|}{\underset{|}{C}}-Br \longrightarrow -\underset{H}{\overset{|}{C}}-\overset{|}{\underset{|}{C^+}} + Br^- \begin{cases} \xrightarrow[\text{substitution}]{H_2O} -\underset{H}{\overset{|}{C}}-\overset{|}{\underset{|}{C}}-OH + H_3O^+ \\ \xrightarrow[\text{elimination}]{H_2O} >C=C< + H_3O^+ \end{cases}$$

Alkyl halides have the same order of reactivity in S_N1 reactions as they do in E1 reactions (Table 9.5), because both reactions have the same rate-determining step—dissociation of the alkyl halide to form a carbocation. This means that all alkyl halides that react under S_N1/E1 conditions will give both substitution and elimination products. Remember that primary alkyl halides do not undergo S_N1/E1 reactions because primary carbocations are too unstable to be formed.

Primary alkyl halides do not form carbocations; therefore they cannot undergo S_N1 and E1 reactions.

Table 9.6 summarizes the products obtained when alkyl halides react with nucleophiles/bases under S_N2/E2 and S_N1/E1 conditions.

Table 9.6 Summary of the Products Expected in Substitution and Elimination Reactions

Class of alkyl halide	Products under S_N2/E2 Conditions	Products under S_N1/E1 Conditions
Primary alkyl halide	Primarily substitution	Cannot undergo S_N1/E1 reactions
Secondary alkyl halide	Both substitution and elimination	Both substitution and elimination
Tertiary alkyl halide	Only elimination	Both substitution and elimination

The stereoisomers obtained from substitution and elimination reactions are summarized in Table 9.7.

Table 9.7 Stereochemistry of Substitution and Elimination Reactions

Reaction	Products
S_N1	Both stereoisomers (*R* and *S*) are formed.
E1	Both *E* and *Z* stereoisomers are formed, but more of the stereoisomer with the bulkiest groups on opposite sides of the double bond is formed.
S_N2	Only the inverted product is formed.
E2	Both *E* and *Z* stereoisomers are formed, but more of the stereoisomer with the bulkiest groups on opposite sides of the double bond is formed.

PROBLEM 24 ♦

Indicate whether the alkyl halides listed will give both substitution and elimination products, primarily substitution products, only elimination products, or no products when they react with sodium methoxide under S_N1/E1 conditions.

1. 1-bromobutane
2. 1-bromo-2-methylpropane
3. 2-bromobutane
4. 2-bromo-2-methylpropane

PROBLEM 25

Which of the following reactions will go faster if the concentration of the nucleophile is increased?

a. (H, Br-substituted alkyl cyclohexane) + CH_3O^- ⟶ (H, OCH_3-substituted alkyl cyclohexane) + Br^-

b. (Br-substituted alkane) + CH_3S^- ⟶ (SCH_3-substituted alkane) + Br^-

c. (1-bromo-1-methylcyclohexane, Br) + $CH_3\overset{O}{\overset{\|}{C}}O^-$ ⟶ ($O\overset{O}{\overset{\|}{C}}CH_3$-substituted methylcyclohexane) + Br^-

9.12 SOLVENT EFFECTS

S_N1 and E1 reactions are faster in protic polar solvents. A polar solvent has a partial positive charge and a partial negative charge; a *protic* polar solvent is a polar solvent with a hydrogen bonded to an oxygen or a nitrogen. Water (H_2O) and alcohols (ROH) are examples of protic polar solvents. Protic polar solvents cluster around the carbocation intermediate, with the negative poles of the solvent surrounding the positive charge of the carbocation (Figure 9.7a).

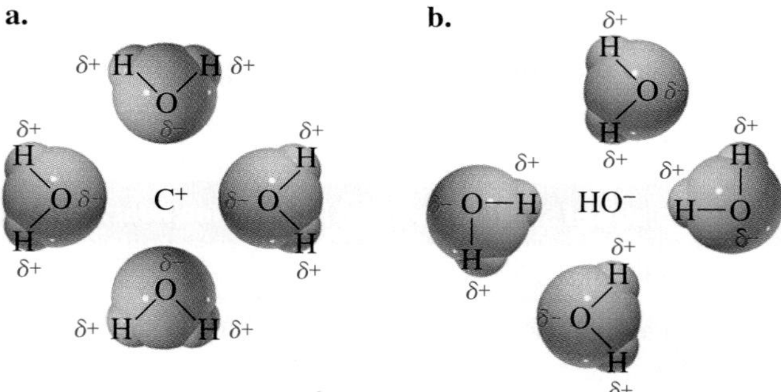

▶ **Figure 9.7**
a. The interactions of a protic polar solvent with a positively charged species.
b. The interactions of a protic polar solvent with a negatively charged species.

When a carbocation interacts with a protic polar solvent, the positive charge on the carbocation is spread out to the surrounding solvent molecules. Spreading out the charge stabilizes the charged species. Stabilizing the carbocation intermediate lowers the height of the energy barrier (the energy "hill") for the rate-determining step, causing the reaction to go faster.

In contrast, S_N2 and E2 reactions are slowed down by protic polar solvents. S_N2 and E2 reactions require a strong nucleophile/base. Most such reagents are negatively charged. The positive poles of a protic polar solvent surround the negatively charged nucleophile/base, thereby spreading out its charge and stabilizing it (Figure 9.7b). Stabilizing a reactant increases the height of the energy barrier for the reaction, causing the reaction to be slower. We would like, therefore, to carry out S_N2 and E2 reactions in a nonpolar solvent. However, negatively charged species generally will not dissolve in a nonpolar solvent. Therefore, an aprotic polar solvent is used. Because an aprotic polar solvent does not have a hydrogen bonded to an oxygen or a nitrogen, it interacts less effectively with an ion than does a polar protic solvent. In fact, aprotic polar solvents such as DMSO and DMF interact very poorly with nucleophiles/bases because the partial positive charge on the solvent is located on the inside of the molecule, and is therefore less accessible to the negatively charged nucleophile/base.

Tutorial: Common terms

the δ– is on the surface of the molecule

the δ+ is not very accessible

N,N-dimethylformamide DMF

dimethyl sulfoxide DMSO

We have now seen that when a reaction can undergo both S_N1/E1 and S_N2/E2 reactions, the S_N1/E1 reactions will be favored by a poor (neutral) nucleophile/weak base in a protic polar solvent, whereas the S_N2/E2 reactions will be favored by a high concentration of a good nucleophile/strong base in an aprotic polar solvent.

S_N1 and E1 reactions of alkyl halides are favored by a poor nucleophile/weak base in a protic polar solvent.

S_N2 and E2 reactions of alkyl halides are favored by a high concentration of a good nucleophile/strong base in an aprotic polar solvent.

PROBLEM 26

Which reaction in each of the following pairs will take place more rapidly?

a. $CH_3Br + HO^- \xrightarrow{\text{DMSO}} CH_3OH + Br^-$

$CH_3Br + HO^- \xrightarrow{\text{EtOH}} CH_3OH + Br^-$

b. $CH_3Br + NH_3 \longrightarrow CH_3\overset{+}{N}H_3 + Br^-$

$CH_3Br + H_2O \longrightarrow CH_3OH + HBr$

c. $CH_3Br + NH_3 \xrightarrow{\text{Et}_2\text{O}} CH_3\overset{+}{N}H_3 + Br^-$

$CH_3Br + NH_3 \xrightarrow{\text{EtOH}} CH_3\overset{+}{N}H_3 + Br^-$

9.13 USING SUBSTITUTION REACTIONS TO SYNTHESIZE ORGANIC COMPOUNDS

In Section 9.4, you saw that nucleophilic substitution reactions of alkyl halides can lead to a wide variety of organic compounds. For example, ethers are synthesized when an *alkyl halide* reacts with an *alkoxide ion*. This reaction, discovered by Alexander Williamson in 1850, is still considered one of the best ways to synthesize an ether.

Williamson ether synthesis

$$\underset{\text{alkyl halide}}{R{-}Br} + \underset{\text{alkoxide ion}}{R{-}O^-} \longrightarrow \underset{\text{ether}}{R{-}O{-}R} + Br^-$$

The alkoxide ion (RO^-) for the **Williamson ether synthesis** can be prepared using sodium metal.

$$2\,ROH + 2\,Na \longrightarrow 2\,RO^- + 2\,Na^+ + H_2$$

The Williamson ether synthesis is a nucleophilic substitution reaction. It requires a high concentration of a good nucleophile, which indicates that it is an S_N2 reaction. If you want to synthesize an ether such as butyl propyl ether, you have a choice of starting

BIOGRAPHY

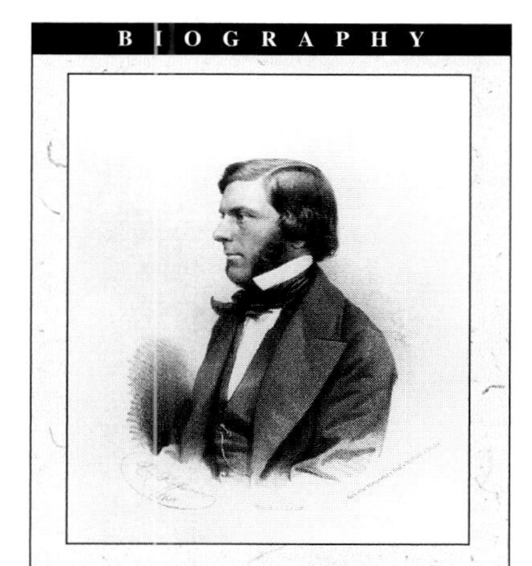

Alexander William Williamson (1824–1904) *was born in London to Scottish parents. As a child, he lost an arm and the use of an eye. Williamson embarked on the study of medicine, but midway through he changed his mind and decided to study chemistry. He received a Ph.D. from the University of Giessen in 1846. In 1849, he became a professor of chemistry at University College, London.*

materials: you can use either a propyl halide and butoxide ion or a butyl halide and propoxide ion.

$$\underset{\text{propyl bromide}}{CH_3CH_2CH_2Br} + \underset{\text{butoxide ion}}{CH_3CH_2CH_2CH_2O^-} \longrightarrow \underset{\text{butyl propyl ether}}{CH_3CH_2CH_2OCH_2CH_2CH_2CH_3} + Br^-$$

$$\underset{\text{butyl bromide}}{CH_3CH_2CH_2CH_2Br} + \underset{\text{propoxide ion}}{CH_3CH_2CH_2O^-} \longrightarrow \underset{\text{butyl propyl ether}}{CH_3CH_2CH_2OCH_2CH_2CH_2CH_3} + Br^-$$

However, if you want to synthesize *tert*-butyl ethyl ether, the starting materials must be an ethyl halide and *tert*-butoxide ion. If you tried to use a *tert*-butyl halide and ethoxide ion as reactants, you would not obtain any ether, because the reaction of a tertiary alkyl halide under S_N2/E2 conditions forms only the elimination product. Consequently, a Williamson ether synthesis should be designed in such a way that *the less hindered alkyl group is provided by the alkyl halide and the more hindered alkyl group comes from the alkoxide ion.*

In ether synthesis, the less hindered group should be provided by the alkyl halide.

$$\underset{\text{ethyl bromide}}{CH_3CH_2Br} + \underset{\textit{tert}\text{-butoxide ion}}{(CH_3)_3CO^-} \longrightarrow \underset{\textit{tert}\text{-butyl ethyl ether}}{CH_3CH_2OC(CH_3)_3} + \underset{\text{ethene}}{CH_2{=}CH_2} + (CH_3)_3COH + Br^-$$

$$\underset{\text{ethoxide ion}}{CH_3CH_2O^-} + \underset{\textit{tert}\text{-butyl bromide}}{(CH_3)_3CBr} \longrightarrow \underset{\text{2-methylpropene}}{CH_2{=}C(CH_3)CH_3} + CH_3CH_2OH + Br^-$$

We saw in Section 5.14 that alkynes can be synthesized by the reaction of an acetylide anion with an alkyl halide.

$$CH_3CH_2C{\equiv}C^- + CH_3CH_2CH_2Br \longrightarrow CH_3CH_2C{\equiv}CCH_2CH_2CH_3 + Br^-$$

Now that you know that this is an S_N2 reaction (the alkyl halide reacts with a high concentration of a good nucleophile), you can understand why you were told that it is best to use primary alkyl halides and methyl halides in the reaction. These alkyl halides are the only ones that form primarily the desired substitution product.

PROBLEM 27

How could the following ethers be prepared using an alkyl halide and an alcohol?

a. $CH_3OCH(CH_3)CH_3$

b. $CH_3CH_2CH(CH_3)OCH_2CH_2CH_3$

c. cyclohexyl–OCH_3

d. $CH_3CH_2OCH_2CH(CH_3)CH_2CH_2CH_3$

PROBLEM 28

Rank the following alkyl halides in order of decreasing reactivity with methoxide ion in an S_N2 reaction. (Label the most reactive alkyl halide #1.)

$CH_3CH(CH_3)CH_2Cl$ $\quad CH_3CH_2CH_2CH_2Br$ $\quad CH_3CH(Cl)CH_2CH_3$ $\quad CH_3C(CH_3)(Cl)CH_3$ $\quad CH_3CH_2CH_2CH_2Cl$

9.14 BIOLOGICAL METHYLATING REAGENTS

If an organic chemist wanted to put a methyl group on a nucleophile (Nu^-), methyl iodide would most likely be the methylating agent used. Of the methyl halides, methyl iodide has the most easily displaced leaving group because I^- is the weakest base of the halide ions. The reaction would be a simple S_N2 reaction.

$$\ddot{\bar{N}}u + CH_3{-}I \longrightarrow CH_3{-}Nu + I^-$$

In a living cell, however, methyl iodide is not available. Methyl halides are only slightly soluble in water, so they are not found in the predominantly aqueous environments of biological systems. Instead, biological systems use *S*-adenosylmethionine (SAM), a water-soluble compound, as a methylating agent. (A less common biological methylating agent is discussed in Section 17.11.) Although it looks much more complicated than methyl iodide, it performs the same function: it transfers a methyl group to a nucleophile. Notice that the methyl group of SAM is attached to a positively charged sulfur, which can readily accept the electrons left behind when the methyl group is transferred. In other words, the methyl group is attached to a very good leaving group, allowing biological methylation to take place at a reasonable rate.

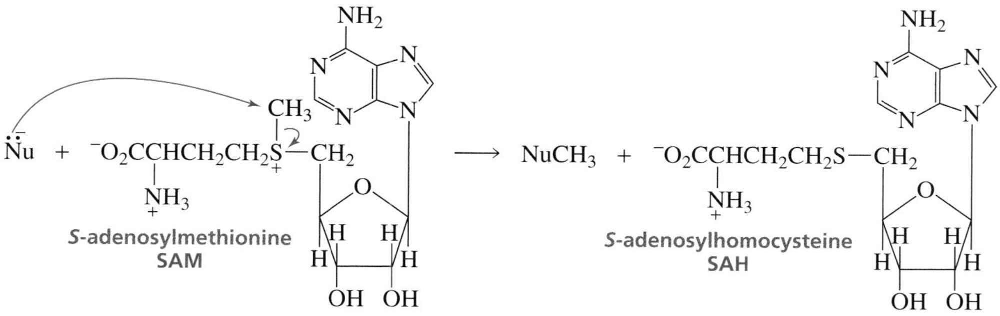

$$\ddot{\bar{N}}u + {}^-O_2CCHCH_2CH_2\overset{+}{S}(CH_3){-}CH_2\text{-(adenosine)} \longrightarrow NuCH_3 + {}^-O_2CCHCH_2CH_2S{-}CH_2\text{-(adenosine)}$$

S-adenosylmethionine
SAM

S-adenosylhomocysteine
SAH

A specific example of a methylation reaction that takes place in biological systems is the conversion of noradrenaline (norepinephrine) to adrenaline (epinephrine), using SAM to provide the methyl group. Noradrenaline and adrenaline are hormones that control glycogen metabolism; they are released into the bloodstream in response to stress. Adrenaline is more potent than noradrenaline.

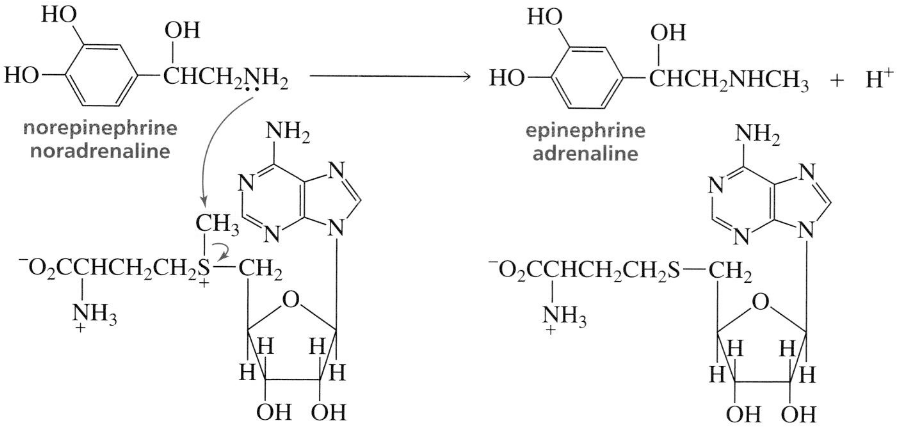

$$(HO)_2C_6H_3{-}CH(OH)CH_2\ddot{N}H_2 + \text{SAM} \longrightarrow (HO)_2C_6H_3{-}CH(OH)CH_2NHCH_3 + H^+ + \text{SAH}$$

norepinephrine
noradrenaline

epinephrine
adrenaline

ERADICATING TERMITES

Alkyl halides can be very toxic to biological organisms. For example, methyl bromide is used to kill termites and other pests. Methyl bromide works by methylating the NH_2 and SH groups of enzymes, thereby destroying the enzyme's ability to catalyze necessary biological reactions. Unfortunately, methyl bromide has been found to deplete the ozone layer (Section 5.9), so its production has recently been banned in developed countries, and developing countries will have until 2015 to phase out its use.

S-ADENOSYLMETHIONINE: A NATURAL ANTIDEPRESSANT

Marketed under the name SAMe (pronounced Sammy), *S*-adenosylmethionine is sold in many health food and drug stores as a treatment for depression and arthritis. Although SAMe has been used clinically in Europe for more than two decades, it has not been rigorously evaluated in the United States and thus has not been approved by the FDA. It can be sold, however, because the FDA does not prohibit the sale of most naturally occurring substances as long as the marketer does not make therapeutic claims. SAMe has also been found to be effective in the treatment of liver diseases, such as diseases caused by alcohol and the hepatitis C virus. The attenuation of liver injuries is accompanied by increased levels of glutathione in the liver. Glutathione is an important biological antioxidant. SAM is required for the synthesis of cysteine, an amino acid that, in turn, is required for the synthesis of glutathione (Section 16.6).

SUMMARY

Alkyl halides undergo two kinds of **nucleophilic substitution reactions**: S_N2 and S_N1. In both reactions, a nucleophile substitutes for a halogen. An S_N2 reaction is bimolecular: two molecules are involved in the rate-limiting step; an S_N1 reaction is unimolecular: one molecule is involved in the rate-limiting step.

The rate of an **S_N2 reaction** depends on the concentration of both the alkyl halide and the nucleophile. An S_N2 reaction has a one-step mechanism: the nucleophile attacks the back side of the carbon that is attached to the halogen. The rate of an S_N2 reaction is affected by steric hindrance: the bulkier the groups at the back side of the carbon undergoing attack, the slower is the reaction. Tertiary alkyl halides, therefore, cannot undergo S_N2 reactions. An S_N2 reaction takes place with **inversion of configuration**.

The rate of an **S_N1 reaction** depends only on the concentration of the alkyl halide. The halogen departs in the first step, forming a carbocation that is attacked by a nucleophile in the second step. The rate of an S_N1 reaction depends on the ease of carbocation formation. Tertiary alkyl halides, therefore, are more reactive than secondary alkyl halides because tertiary carbocations are more stable than secondary carbocations. Primary carbocations are so unstable that primary alkyl halides cannot undergo S_N1 reactions. An S_N1 reaction forms both inverted and noninverted products.

The rates of both S_N2 and S_N1 reactions are influenced by the nature of the **leaving group**. Weak bases are the best leaving groups because weak bases form the weakest bonds. Thus, the weaker the basicity of the leaving group, the faster the reaction will occur. Therefore, the relative reactivities of alkyl halides that differ only in the halogen atom are $RI > RBr > RCl > RF$ in both S_N2 and S_N1 reactions.

Basicity is a measure of how well a species shares its lone pair with a proton. **Nucleophilicity** is a measure of how readily a species is able to attack an electron-deficient atom. In general, the stronger base is the better nucleophile.

In addition to undergoing nucleophilic substitution reactions, alkyl halides undergo elimination reactions, in which the halogen is removed from one carbon, a hydrogen is removed from an adjacent carbon, and a double bond is formed between the two carbons from which the atoms were eliminated. The product of an elimination reaction is therefore an alkene. There are two important elimination reactions, E1 and E2.

An **E2 reaction** is a one-step reaction; the proton and the halide ion are removed in the same step, so no intermediate is formed. In an **E1 reaction**, the alkyl halide dissociates, forming a carbocation intermediate. In a second step, a base removes a proton from a carbon that is adjacent to the positively charged carbon.

The major product of an elimination reaction is the more stable alkene—the one formed when a proton is removed from the β-carbon that is bonded to the fewest hydrogens. If both *E* and *Z* isomers are possible for the product, the one with the bulkiest groups on opposite sides of the double bond is more stable and, therefore, will be formed in greater yield.

Predicting which products are formed when an alkyl halide undergoes a reaction begins with determining whether the conditions favor S_N2/E2 or S_N1/E1 reactions. S_N2/E2 reactions are favored by a high concentration of a good nucleophile/strong base in an **aprotic polar solvent**, whereas S_N1/E1 reactions are favored by a poor nucleophile/weak base in a **protic polar solvent**.

When S_N2/E2 reactions are favored, primary alkyl halides form mainly substitution products, secondary alkyl halides form both substitution and elimination products, and tertiary alkyl halides form only elimination products. When S_N1/E1 conditions are favored, secondary and tertiary alkyl halides form both substitution and elimination products; primary alkyl halides do not undergo S_N1/E1 reactions.

SUMMARY OF REACTIONS

1. S_N2 reaction: a one-step mechanism

$$\ddot{\text{Nu}}^- + -\text{C}-\text{X} \longrightarrow -\text{C}-\text{Nu} + \text{X}^-$$

Relative reactivities of alkyl halides: $CH_3X > 1° > 2° > 3°$

Only the inverted product is formed.

2. S_N1 reaction: a two-step mechanism with a carbocation intermediate

$$-\text{C}-\text{X} \longrightarrow -\text{C}^+ + \text{X}^- \xrightarrow{\ddot{\text{Nu}}^-} -\text{C}-\text{Nu}$$

Relative reactivities of alkyl halides: $3° > 2° > 1° > CH_3X$

Both the inverted and noninverted products are formed.

3. E2 reaction: a one-step mechanism

$$\ddot{\text{B}}^- + -\underset{|}{\overset{\text{H}}{\text{C}}}-\text{C}-\text{X} \longrightarrow \text{C}=\text{C} + \text{BH} + \text{X}^-$$

Relative reactivities of alkyl halides: $3° > 2° > 1°$

Both *E* and *Z* stereoisomers are formed; the isomer with the bulkiest groups on opposite sides of the double bond will be formed in greater yield.

4. E1 reaction: a two-step mechanism with a carbocation intermediate

$$-\underset{\text{H}}{\text{C}}-\text{C}-\text{X} \longrightarrow -\underset{\text{H}}{\text{C}}-\text{C}^+ + \text{X}^- \xrightarrow{\ddot{\text{B}}} \text{C}=\text{C} + {}^+\text{BH}$$

Relative reactivities of alkyl halides: $3° > 2° > 1°$

Both *E* and *Z* stereoisomers are formed; the isomer with the bulkiest groups on opposite sides of the double bond will be formed in greater yield.

Competing S_N2 and E2 Reactions

Primary alkyl halides: mainly substitution
Secondary alkyl halides: substitution and elimination
Tertiary alkyl halides: only elimination

Competing S_N1 and E1 Reactions

Primary alkyl halides: cannot undergo S_N1 or E1 reactions
Secondary alkyl halides: substitution and elimination
Tertiary alkyl halides: substitution and elimination

PROBLEMS

29. Which reaction in each of the following pairs will take place more rapidly?

a. $CH_3Br + CH_3O^- \longrightarrow CH_3OCH_3 + Br^-$

$CH_3Br + CH_3OH \longrightarrow CH_3OCH_3 + HBr$

b. $CH_3I + NH_3 \longrightarrow CH_3\overset{+}{N}H_3 + I^-$

$CH_3Cl + NH_3 \longrightarrow CH_3\overset{+}{N}H_3 + Cl^-$

c. $CH_3Br + CH_3NH_2 \longrightarrow CH_3\overset{+}{N}H_2CH_3 + Br^-$

$CH_3Br + CH_3OH \longrightarrow CH_3OCH_3 + HBr$

30. Give the product of the reaction of methyl bromide with each of the following nucleophiles:

a. HO^-

b. $^-NH_2$

c. H_2S

d. HS^-

e. $CH_3CH_2O^-$

f. CH_3NH_2

31. Which is a better nucleophile?

a. H_2O or HO^-

b. NH_3 or $^-NH_2$

c. $CH_3\overset{O}{\overset{\|}{C}}O^-$ or $CH_3CH_2O^-$

d. 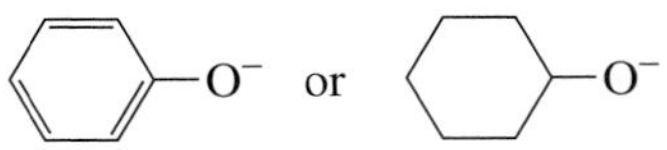 (benzene ring)$-O^-$ or (cyclohexane ring)$-O^-$

32. For each of the pairs in Problem 31, indicate which is a better leaving group.

33. What nucleophiles could be used to react with butyl bromide to prepare the following compounds?

a. $CH_3CH_2CH_2CH_2OH$

b. $CH_3CH_2CH_2CH_2OCH_3$

c. $CH_3CH_2CH_2CH_2SCH_2CH_3$

d. $CH_3CH_2CH_2CH_2C{\equiv}N$

e. $CH_3CH_2CH_2CH_2O\overset{O}{\overset{\|}{C}}CH_3$

f. $CH_3CH_2CH_2CH_2C{\equiv}CCH_3$

34. Which alkyl halide in each pair would you expect to be more reactive in an S_N2 reaction with a given nucleophile?

a. $CH_3CH_2\underset{I}{\underset{|}{C}}HCH_3$ or $CH_3CH_2\underset{Br}{\underset{|}{C}}HCH_3$

b. $CH_3CH_2\overset{CH_3}{\overset{|}{C}}HBr$ or $CH_3CH_2\overset{CH_2CH_3}{\overset{|}{C}}HBr$

c. $CH_3CH_2CH_2\overset{CH_3}{\overset{|}{C}}HBr$ or $CH_3CH_2\overset{CH_3}{\overset{|}{C}}HCH_2Br$

d. 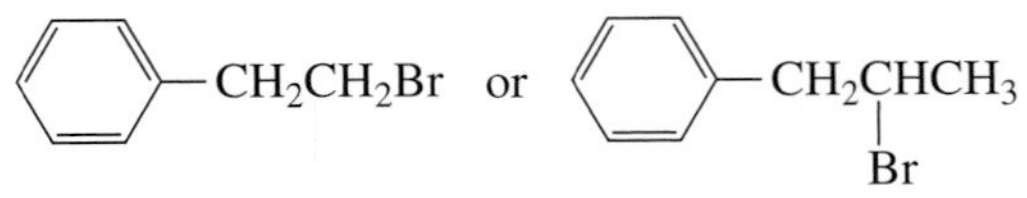 (benzene ring)$-CH_2CH_2Br$ or (benzene ring)$-CH_2\underset{Br}{\underset{|}{C}}HCH_3$

35. For each of the pairs in Problem 34, which compound would be more reactive in an S_N1 reaction?

36. For each of the following reactions, give the substitution products; if the products can exist as stereoisomers, show what stereoisomers are obtained:

a. (*R*)-2-bromopentane + high concentration of CH_3O^-
b. (*R*)-2-bromopentane + CH_3OH
c. *trans*-1-bromo-4-methylcyclohexane + high concentration of CH_3O^-
d. *trans*-1-bromo-4-methylcyclohexane + CH_3OH
e. 3-bromo-3-methylpentane + CH_3OH

37. Give the major product obtained when each of the following alkyl halides undergoes an E2 reaction:

a. 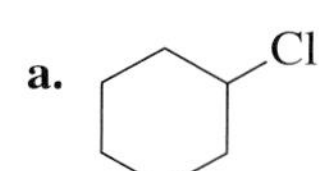

b. (cyclohexyl)CH_2Cl **c.** (1-chloro-1-methylcyclohexane) CH_3, Cl

38. Give the stereoisomer that would be obtained in greater yield when each of the following alkyl halides undergoes an E2 reaction:

a. $CH_3CH(Br)CH_2CH_3$ **b.** $CH_3CH(Cl)CH_2CH_3$ **c.** $CH_3CH(Cl)CH_2CH_2CH_3$

39. Which reactant in each of the following pairs will undergo an elimination reaction more rapidly?

a. $(CH_3)_3CCl \xrightarrow[H_2O]{HO^-}$ or $(CH_3)_3CI \xrightarrow[H_2O]{HO^-}$

b. $(CH_3)_3CBr \xrightarrow[H_2O]{HO^-}$ or $(CH_3)_2CHBr \xrightarrow[H_2O]{HO^-}$

40. a. Identify the three products that are formed when 2-bromo-2-methylpropane is dissolved in a mixture of 80% ethanol and 20% water.

b. Explain why the same products are obtained when 2-chloro-2-methylpropane is dissolved in a mixture of 80% ethanol and 20% water.

41. Starting with bromocyclohexane, how could the following compounds be prepared?

a. (cyclohexyl)$C{\equiv}CCH_3$ **b.** (cyclohexyl)OH **c.** (cyclohexyl)OCH_3 **d.** (cyclohexyl)$C{\equiv}N$ **e.** (cyclohexene)

42. For each of the following reactions, give the major elimination product; if the product can exist as stereoisomers, indicate which stereoisomer is obtained in greater yield:

a. (*R*)-2-bromohexane + high concentration of HO^-
b. (*R*)-2-bromohexane + H_2O
c. 3-bromo-3-methylpentane + high concentration of HO^-
d. 3-bromo-3-methylpentane + H_2O

43. The rate of reaction of methyl iodide with quinuclidine was measured in nitrobenzene, and then the rate of reaction of methyl iodide with triethylamine was measured in the same solvent. The concentration of the reagents was the same in both experiments.

a. Which reaction was faster?
b. Which reaction had the larger rate constant?

quinuclidine

$CH_3CH_2N(CH_2CH_3)CH_2CH_3$
triethylamine

44. Which substitution reaction in each of the following pairs will occur more rapidly?

a. $(CH_3)_3CCH_2Cl \xrightarrow{CH_3S^-} (CH_3)_3CCH_2SCH_3 + Cl^-$
or
$(CH_3)_3CCH_2Cl \xrightarrow{(CH_3)_2CHS^-} (CH_3)_3CCH_2SCH(CH_3)_2 + Cl^-$

b. $CH_3CH_2CH_2CH_2Cl \xrightarrow{HO^-} CH_3CH_2CH_2CH_2OH + Cl^-$
or
$CH_3CH_2OCH_2Cl \xrightarrow{HO^-} CH_3CH_2OCH_2OH + Cl^-$

45. Which of the following is more reactive in an E2 reaction?

a. $C_6H_5CH_2CH(Br)CH_3$ or $C_6H_5CH_2CH_2CH_2Br$

b. $CH_3CH_2CH(Br)CH_3$ or $CH_2{=}CHCH_2CH(Br)CH_3$

46. Would you expect methoxide ion to be a better nucleophile if it were dissolved in CH_3OH or in DMSO? Why?

47. a. Explain why 1-bromo-2,2-dimethylpropane has difficulty undergoing either S_N2 or S_N1 reactions.

b. Can it undergo E2 and E1 reactions?

48. Which stereoisomer would be obtained in greater yield from an E2 reaction of each of the following alkyl halides?

a. 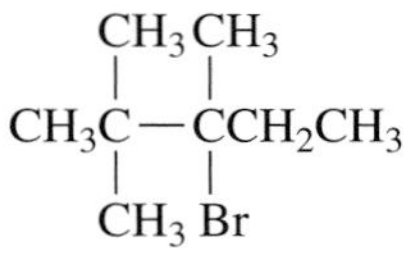 $CH_3C(CH_3)_2-C(CH_3)(Br)CH_2CH_3$

b. 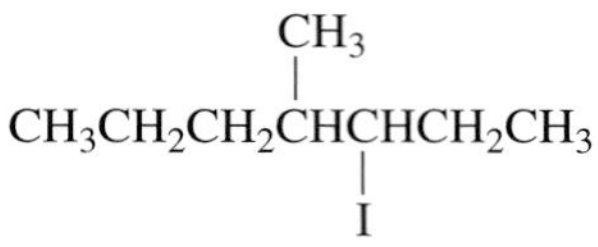 $CH_3CH_2CH_2CH(CH_3)CH(I)CH_2CH_3$

49. An ether can be prepared by an S_N2 reaction of an alkyl halide with an alkoxide ion (RO^-). Which set of alkyl halide and alkoxide ion would give you a better yield of cyclopentyl methyl ether?

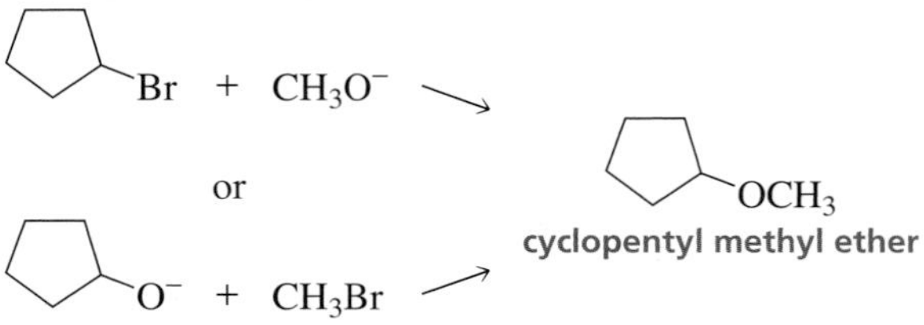

cyclopentyl–Br + CH_3O^- →

or

cyclopentyl–O^- + CH_3Br →

cyclopentyl–OCH_3
cyclopentyl methyl ether

50. Dr. Don T. Doit wanted to synthesize the anesthetic 2-ethoxy-2-methylpropane. He used ethoxide ion and 2-chloro-2-methylpropane for his synthesis and ended up with no ether. What was the product of his synthesis? What reagents should he have used?

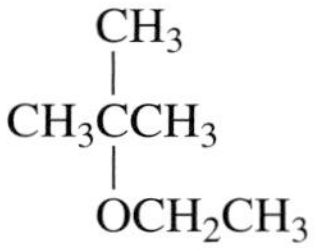

$CH_3C(CH_3)(OCH_2CH_3)CH_3$
2-ethoxy-2-methylpropane

51. Which alkyl halide undergoes an E1 reaction more rapidly?

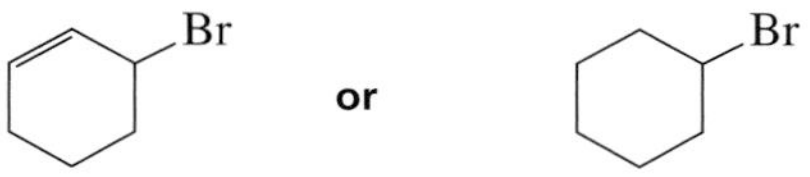

3-bromocyclohexene (Br) **or** bromocyclohexane (Br)

52. In Section 9.14, we saw that *S*-adenosylmethionine (SAM) methylates the nitrogen atom of noradrenaline to form adrenaline, a more potent hormone. If SAM methylates an OH group on the benzene ring instead, it completely destroys noradrenaline's activity. Give the mechanism for the methylation of the OH group by SAM.

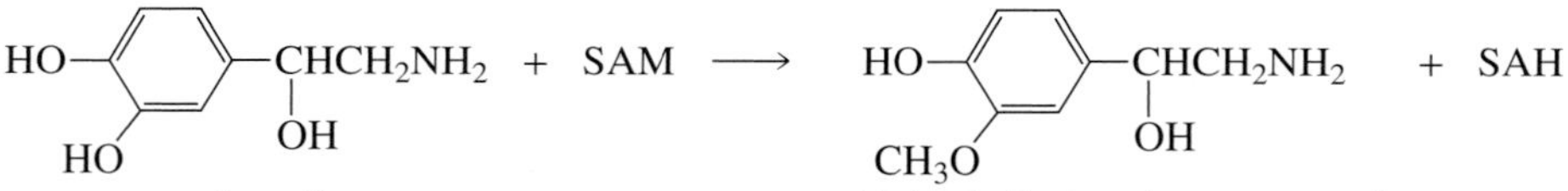

$(HO)_2C_6H_3-CH(OH)CH_2NH_2$ + SAM ⟶ $HO(CH_3O)C_6H_3-CH(OH)CH_2NH_2$ + SAH

noradrenaline — **a biologically inactive compound**

53. Give the substitution products obtained from the reaction of each of the following alkyl halides with ethanol:

a. 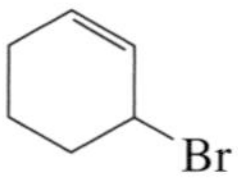 3-bromocyclohexene (Br) **b.** 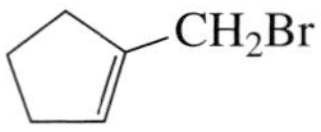 cyclopentenyl–CH_2Br **c.** 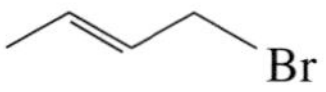 allylic Br chain (Br)

54. Show how the following compounds could be synthesized using the given starting materials:

a. $CH_3CH_2CH_2CH_2Br$ ⟶ $CH_3CH_2CH_2CH_2NH_2$

b. 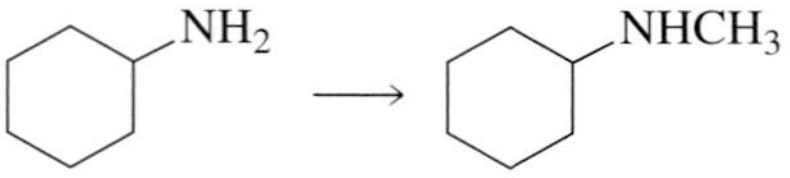 cyclohexyl–NH_2 ⟶ cyclohexyl–$NHCH_3$

c. 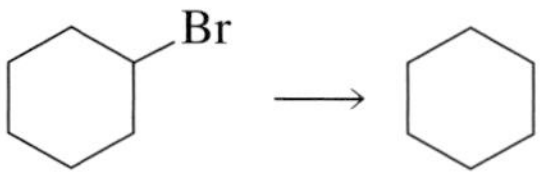 cyclohexyl–Br ⟶ cyclohexene

55. A cyclic compound can be formed by an intramolecular reaction. An intramolecular reaction is one in which the two reacting groups are in the same molecule. Give the structure of the ether that would be formed from each of the following intramolecular reactions.

a. $BrCH_2CH_2CH_2CH_2O^-$ ⟶ ether **b.** $ClCH_2CH_2CH_2CH_2CH_2O^-$ ⟶ ether

Reactions of Alcohols, Amines, Ethers, and Epoxides

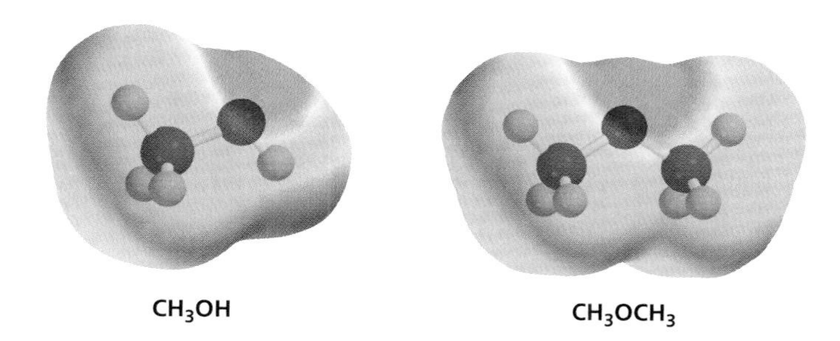

CH_3OH CH_3OCH_3

We have seen that alkyl halides undergo substitution and elimination reactions because of their electron-withdrawing halogen atoms (Chapter 9). Compounds with other electron-withdrawing groups also undergo substitution and elimination reactions. The relative reactivity of these compounds depends on the electron-withdrawing group. For example, an alcohol (ROH) has an electron-withdrawing OH group. An OH group, however, is much more basic than a halogen, so we will see that it is much harder to displace.

10.1 THE NOMENCLATURE OF ALCOHOLS

Before we look at the reactions of alcohols, we need to learn how to name them. An **alcohol** is a compound in which a hydrogen of an alkane has been replaced by an OH group. We have seen that alcohols are classified as **primary**, **secondary**, or **tertiary**, depending on whether the OH group is bonded to a primary, secondary, or tertiary carbon—the same way alkyl halides are classified (Section 3.5).

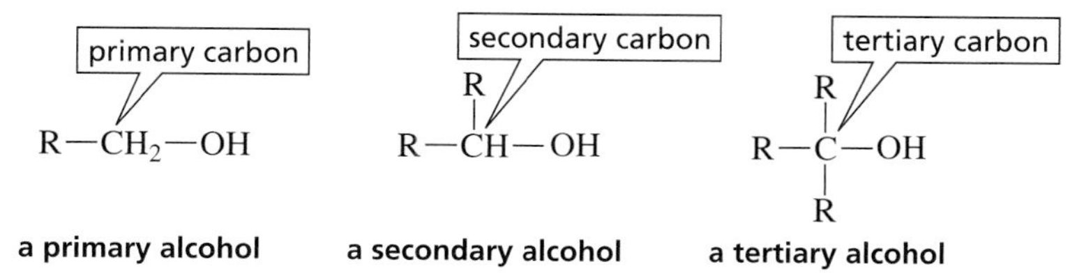

The common name of an alcohol consists of the name of the alkyl group to which the OH group is attached, followed by the word "alcohol."

CH_3CH_2OH — ethyl alcohol

$CH_3CH_2CH_2OH$ — propyl alcohol

$CH_3CH(CH_3)OH$ — isopropyl alcohol

The **functional group** is the center of reactivity in an organic compound. In an alcohol, the OH is the functional group. The IUPAC system uses the suffix "ol" to denote the OH group. Thus, the systematic name of an alcohol is obtained by replacing

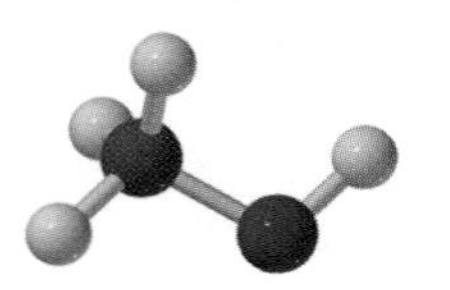
methyl alcohol

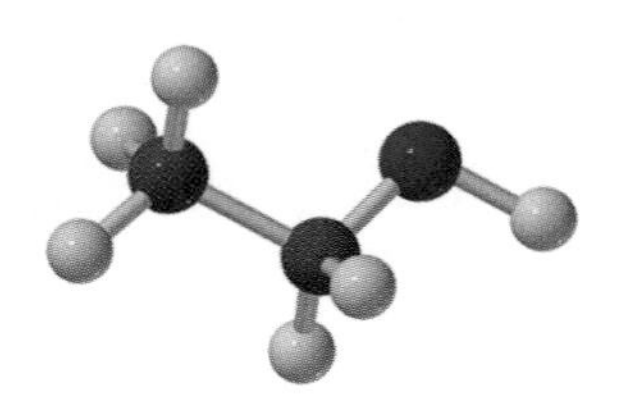
ethyl alcohol

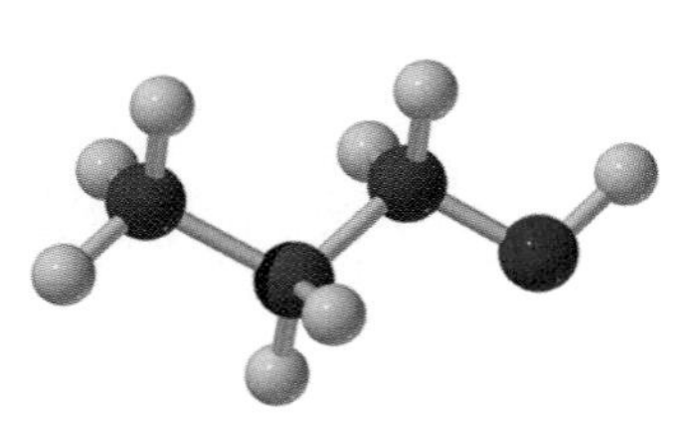
propyl alcohol

the "e" at the end of the name of the parent hydrocarbon with the suffix "ol." This should remind you of the use of the suffix "ene" to denote the functional group of an alkene (Section 4.2).

CH_3OH — **methanol**

CH_3CH_2OH — **ethanol**

When necessary, the position of the functional group is indicated by a number.

$CH_3CH_2CHCH_2CH_3$ (OH on C-3)
3-pentanol

Let's review the rules used to name a compound that has a functional group suffix:

1. The parent hydrocarbon is the longest chain containing the functional group. The parent chain is numbered in the direction that gives the *functional group suffix the lowest possible number.*

$\overset{1}{C}H_3\overset{2}{C}H\overset{3}{C}H_2\overset{4}{C}H_3$ (OH on C-2)
2-butanol

$\overset{5}{C}H_3\overset{4}{C}H_2\overset{3}{C}H_2\overset{2}{C}H\overset{1}{C}H_2OH$ (CH_2CH_3 on C-2)
2-ethyl-1-pentanol

The longest continuous chain has six carbons, but the longest continuous chain containing the OH functional group has five carbons so the compound is named as a pentanol.

2. If there is a functional group suffix and a substituent, the functional group suffix gets the lowest possible number.

$HO\overset{1}{C}H_2\overset{2}{C}H_2\overset{3}{C}H_2Br$
3-bromo-1-propanol

$Cl\overset{4}{C}H_2\overset{3}{C}H_2\overset{2}{C}H\overset{1}{C}H_3$ (OH on C-2)
4-chloro-2-butanol

$\overset{5}{C}H_3\overset{4}{C}(CH_3)_2\overset{3}{C}H_2\overset{2}{C}H\overset{1}{C}H_3$ (OH on C-2)
4,4-dimethyl-2-pentanol

3. If counting in either direction gives the same number for the functional group suffix, the chain is numbered in the direction that gives a substituent the lowest possible number. Notice that a number is not needed to designate the position of a functional group suffix in a cyclic compound, because it is assumed to be at the 1-position.

$CH_3CHCHCH_2CH_3$ (Cl, OH)
2-chloro-3-pentanol
not
4-chloro-3-pentanol

$CH_3CH_2CH_2CHCH_2CHCH_3$ (OH, CH_3)
2-methyl-4-heptanol
not
6-methyl-4-heptanol

(cyclohexane ring with CH_3 and OH)
3-methylcyclohexanol
not
5-methylcyclohexanol

4. If there is more than one substituent, the substituents are stated in alphabetical order.

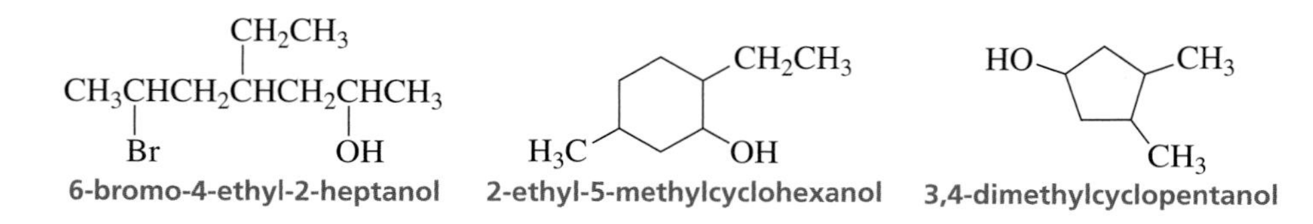

6-bromo-4-ethyl-2-heptanol **2-ethyl-5-methylcyclohexanol** **3,4-dimethylcyclopentanol**

PROBLEM 1

Draw the structures of straight-chain alcohols that have from one to six carbons, and then give each of them a common name and a systematic name.

PROBLEM 2 ♦

Give each of the following a systematic name, and indicate whether each is a primary, secondary, or tertiary alcohol:

a. $CH_3CH_2CH_2CH_2CH_2OH$

b. 4-methylcyclohexanol structure: cyclohexane ring with HO and CH_3 substituents

c. $CH_3C(CH_3)(OH)CH_2CH_2CH_2Cl$

d. $CH_3CH_2CH_2CH(CH_2OH)CH_2CH_3$

e. $CH_3CH(CH_3)CH_2CH(OH)CH_2CH_3$

f. $CH_3CH(CH_3)CH_2CH(OH)CH_2CH(CH_3)CH_2CH_3$

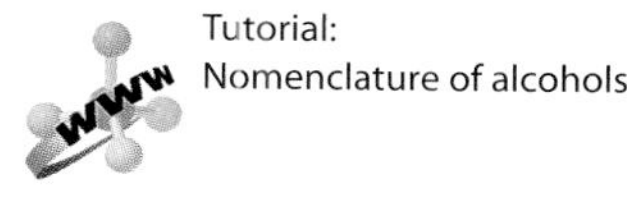

Tutorial: Nomenclature of alcohols

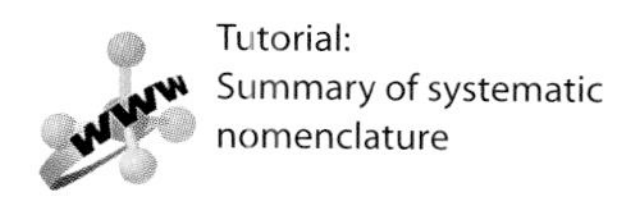

Tutorial: Summary of systematic nomenclature

PROBLEM 3 ♦

Write the structures of all the tertiary alcohols with molecular formula $C_6H_{14}O$, and give each a systematic name.

10.2 SUBSTITUTION REACTIONS OF ALCOHOLS

An **alcohol** has a strongly basic group (HO^-) that cannot be displaced by a nucleophile. Therefore, an alcohol cannot undergo a nucleophilic substitution reaction.

$$CH_3{-}\ddot{O}H + Br^- \not\rightarrow CH_3{-}Br + HO^-$$

(a strongly basic leaving group: OH; HO^-: strong base)

However, if the alcohol's OH group is converted into a group that is a weaker base (and therefore a better leaving group), a nucleophilic substitution reaction can occur. One way to convert an OH group into a weaker base is to protonate it by adding acid to the solution. Protonation changes the leaving group from HO^- to H_2O, which is a weak enough base to be displaced by a nucleophile. The substitution reaction is slow and requires heat (except in the case of tertiary alcohols) if it is to take place in a reasonable period of time.

The weaker the base, the more easily it can be displaced.

$$CH_3{-}\ddot{O}H + HBr \rightleftharpoons CH_3{-}\overset{+}{O}H_2 \;\; Br^- \xrightarrow{\Delta} CH_3{-}Br + H_2O$$

(OH of CH_3OH: poor leaving group; protonated OH: good leaving group, a weakly basic leaving group; H_2O: weak base)

Because the OH group of the alcohol has to be protonated before it can be displaced by a nucleophile, only weakly basic nucleophiles (I^-, Br^-, Cl^-) can be used in the substitution reaction. Moderately and strongly basic nucleophiles (NH_3, RNH_2, CH_3O^-) cannot be used because they would also be protonated in the acidic solution and, once protonated, would no longer be nucleophiles ($^+NH_4$, RNH_3^+) or would be poor nucleophiles (CH_3OH).

PROBLEM 4 ◆

Why are NH_3 and CH_3NH_2 no longer nucleophiles when they are protonated?

Primary, secondary, and tertiary alcohols all undergo nucleophilic substitution reactions with HI, HBr, and HCl to form alkyl halides.

$$CH_3CH_2CH_2OH + HI \xrightarrow{\Delta} CH_3CH_2CH_2I + H_2O$$

1-propanol, a primary alcohol → 1-iodopropane

$$\text{cyclohexanol} + HBr \xrightarrow{\Delta} \text{bromocyclohexane} + H_2O$$

cyclohexanol, a secondary alcohol → bromocyclohexane

$$CH_3CH_2C(CH_3)_2OH + HCl \longrightarrow CH_3CH_2C(CH_3)_2Cl + H_2O$$

2-methyl-2-butanol, a tertiary alcohol → 2-chloro-2-methylbutane

The mechanism of the substitution reaction depends on the structure of the alcohol. Secondary and tertiary alcohols undergo S_N1 reactions—an S_N1 reaction of a protonated alcohol.

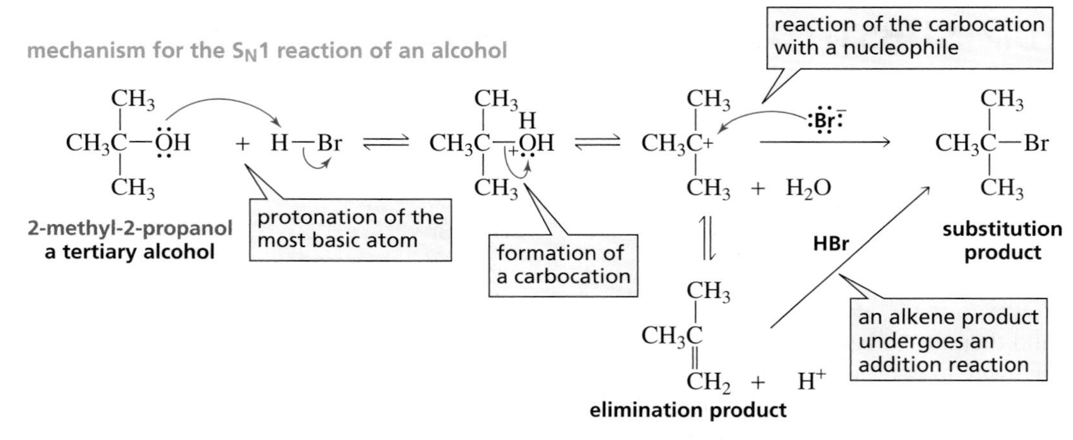

- An acid always reacts with an organic molecule in the same way: it protonates the most basic atom in the reactant.
- Weakly basic water is the leaving group that is expelled, forming a carbocation.
- The carbocation has two possible fates: it can combine with a nucleophile and form a substitution product, or it can lose a proton and form an elimination product.

Secondary and tertiary alcohols undergo S_N1 reactions with hydrogen halides.

Although the reaction can form both a substitution product and an elimination product, only the substitution product is actually obtained because any alkene formed in an elimination reaction will undergo a subsequent addition reaction with HBr to form more of the substitution product (Section 5.1).

Tertiary alcohols undergo substitution reactions with hydrogen halides faster than secondary alcohols do, because tertiary carbocations are easier to form than secondary carbocations (Section 9.4). Thus, the reaction of a tertiary alcohol with a hydrogen halide proceeds readily at room temperature, whereas the reaction of a secondary alcohol with a hydrogen halide has to be heated to have the reaction occur at the same rate.

Carbocation stability: 3° > 2° > 1°

$$\mathrm{CH_3\underset{\underset{\displaystyle OH}{|}}{\overset{\overset{\displaystyle CH_3}{|}}{C}}CH_2CH_3 + HBr \longrightarrow CH_3\underset{\underset{\displaystyle Br}{|}}{\overset{\overset{\displaystyle CH_3}{|}}{C}}CH_2CH_3 + H_2O}$$

$$\mathrm{CH_3\underset{\underset{\displaystyle OH}{|}}{C}HCH_2CH_3 + HBr \xrightarrow{\Delta} CH_3\underset{\underset{\displaystyle Br}{|}}{C}HCH_2CH_3 + H_2O}$$

Primary alcohols cannot undergo S_N1 reactions because primary carbocations are too unstable to be formed (Section 9.4). Therefore, when a primary alcohol reacts with a hydrogen halide, it must do so in an S_N2 reaction.

Primary alcohols undergo S_N2 reactions with hydrogen halides.

mechanism for the S_N2 reaction of an alcohol

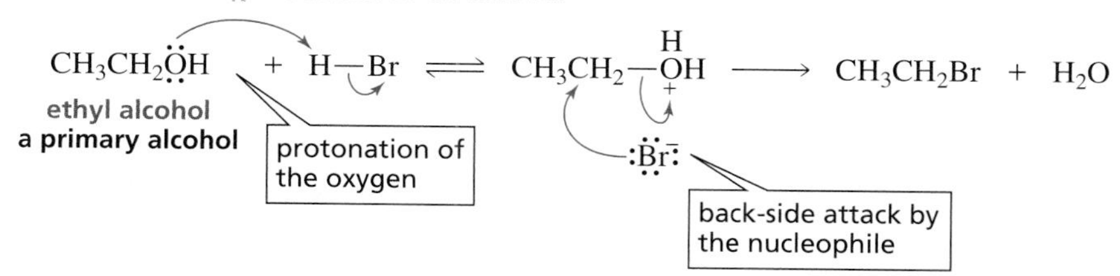

- The acid protonates the most basic atom in the reactant.
- The nucleophile hits the back side of the carbon and displaces the leaving group.

PROBLEM 5 *SOLVED*

Using the pK_a values of the conjugate acids of the leaving groups (the pK_a of HBr is -9; the pK_a of H_2O is 15.7; the pK_a of H_3O^+ is -1.7), explain the difference in reactivity in substitution reactions between:

a. CH_3Br and CH_3OH

b. $CH_3\overset{+}{O}H_2$ and CH_3OH

Solution to 5a The conjugate acid of the leaving group of CH_3Br is HBr; its pK_a is -9; the conjugate acid of the leaving group of CH_3OH is H_2O; its pK_a is 15.7. Because HBr is a much stronger acid than H_2O, Br^- is a much weaker base than HO^-. (Recall the stronger the acid, the weaker is its conjugate base.) Therefore, Br^- is a much better leaving group than HO^-, causing CH_3Br to be much more reactive in a substitution reaction than CH_3OH.

PROBLEM 6 ◆

Give the major product of each of the following reactions:

a. $\mathrm{CH_3CH_2\underset{\underset{\displaystyle OH}{|}}{C}HCH_3 + HBr \xrightarrow{\Delta}}$

b. 1-methylcyclopentanol (cyclopentane ring bearing CH_3 and OH on the same carbon) $+ \mathrm{HCl} \longrightarrow$

PROBLEM 7 *SOLVED*

Show how 1-butanol can be converted into the following compounds:

a. $CH_3CH_2CH_2CH_2OCH_3$

b. $\mathrm{CH_3CH_2CH_2CH_2O\overset{\overset{\displaystyle O}{\|}}{C}CH_2CH_3}$

c. $CH_3CH_2CH_2CH_2NHCH_2CH_3$

d. $CH_3CH_2CH_2CH_2C{\equiv}N$

Solution to 7a Because the OH group of 1-butanol is too basic to be substituted, the alcohol must first be converted into an alkyl halide. The alkyl halide has a leaving group that can be substituted by CH_3O^-, the nucleophile required to obtain the desired product.

$$\mathrm{CH_3CH_2CH_2CH_2OH \xrightarrow[\Delta]{HBr} CH_3CH_2CH_2CH_2Br \xrightarrow{CH_3O^-} CH_3CH_2CH_2CH_2OCH_3}$$

10.3 ELIMINATION REACTIONS OF ALCOHOLS: DEHYDRATION

An alcohol can undergo an elimination reaction by losing an OH from one carbon and an H from an adjacent carbon. The product of the reaction is an alkene. Overall, this amounts to the elimination of a molecule of water. Loss of water from a molecule is called **dehydration**. Dehydration of an alcohol requires an acid catalyst and heat. Sulfuric acid (H_2SO_4) is a commonly used acid catalyst.

$$\underset{\displaystyle \text{OH}}{CH_3CH_2\underset{|}{C}HCH_3} \xrightleftharpoons[\Delta]{\mathbf{H_2SO_4}} CH_3CH{=}CHCH_3 + H_2O$$

The dehydration of a secondary or a tertiary alcohol is an E1 reaction.

mechanism for the E1 dehydration of an alcohol

$$\underset{:\ddot{O}H}{CH_3\underset{|}{C}HCH_3} + H{-}OSO_3H \rightleftharpoons \underset{+:OH \atop H}{CH_3\underset{|}{C}HCH_3} \underset{H_2\ddot{O}:}{\rightleftharpoons} H{-}CH_2{-}\overset{}{\underset{+}{C}}HCH_3 \rightleftharpoons CH_2{=}CHCH_3$$

HSO_4^- $H_2O + H_2SO_4$

formation of a carbocation

a carbocation

protonation of the most basic atom

a base removes a proton from a β-carbon

- The acid protonates the most basic atom in the reactant. As we saw earlier, protonation converts the very poor leaving group (HO^-) into a good leaving group (H_2O).
- Water departs, leaving behind a carbocation.
- A base in the reaction mixture (water is the base present in the highest concentration) removes a proton from a β-carbon (a carbon adjacent to the positively charged carbon), forming an alkene and regenerating the acid catalyst.

An acid protonates the most basic atom in a molecule.

Secondary and tertiary alcohols undergo dehydration by an E1 pathway.

Notice that the dehydration reaction is an E1 reaction of a protonated alcohol.

When more than one elimination product can be formed, the major product is the more stable alkene—the one obtained by removing a proton from the β-carbon bonded to the fewest hydrogens (Figure 10.1).

$$\underset{\displaystyle \text{OH}}{\overset{\displaystyle CH_3}{CH_3\overset{|}{\underset{|}{C}}CH_2CH_3}} \xrightleftharpoons[\Delta]{\mathbf{H_3PO_4}} \underset{84\%}{\overset{\displaystyle CH_3}{CH_3\overset{|}{C}{=}CHCH_3}} + \underset{16\%}{\overset{\displaystyle CH_3}{CH_2{=}\overset{|}{C}CH_2CH_3}} + H_2O$$

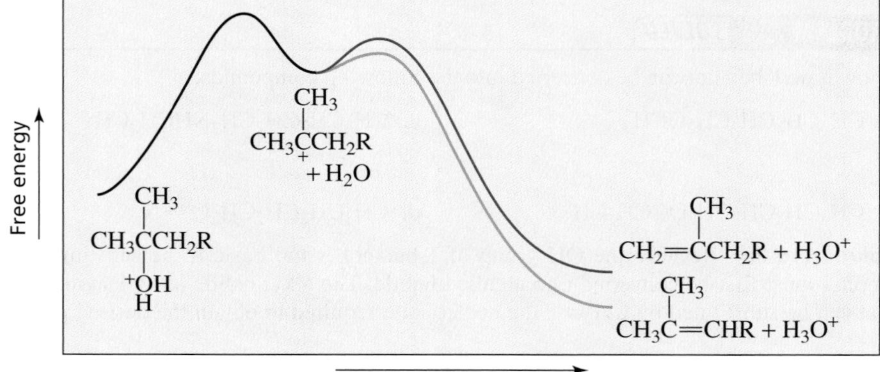

▶ **Figure 10.1**
The reaction coordinate diagram for the dehydration of a protonated alcohol. The major product is the more substituted alkene because the transition state leading to its formation is more stable, allowing that alkene to be formed more rapidly.

Because the rate-determining step in the dehydration of a secondary or a tertiary alcohol is formation of a carbocation intermediate, the rate of dehydration parallels the ease with which the carbocation is formed. Tertiary alcohols are the easiest to dehydrate because tertiary carbocations are more stable and are, therefore, easier to form than secondary and primary carbocations (Section 9.4).

relative ease of dehydration

R_3COH (a tertiary alcohol) > R_2CHOH (a secondary alcohol) > RCH_2OH (a primary alcohol)

← increasing ease of dehydration

While the dehydration of a tertiary or a secondary alcohol is an E1 reaction, the dehydration of a primary alcohol is an E2 reaction because primary carbocations are too unstable to be formed.

Primary alcohols undergo dehydration by an E2 pathway.

mechanism for the E2 dehydration and the competing substitution (S_N2) reaction

$$CH_3CH_2\ddot{O}H + H{-}OSO_3H \rightleftharpoons CH_2{-}CH_2{-}\overset{+}{\ddot{O}}H_2 \xrightarrow{E2} CH_2{=}CH_2 + H_2O + H_2SO_4$$

elimination product

protonation of the most basic atom

base removes a proton from a β-carbon (by $^-{:}\ddot{O}SO_3H$)

$$CH_3CH_2\ddot{O}H + CH_3CH_2{-}\overset{+}{O}H_2 \xrightarrow{S_N2} CH_3CH_2\overset{+}{\ddot{O}}(H)CH_2CH_3 \longrightarrow CH_3CH_2OCH_2CH_3 + H_3O^+$$

substitution product

back-side attack by the nucleophile

$H_2\ddot{O}{:}$

proton dissociation

- The acid protonates the most basic atom in the reactant.
- A base removes a proton in the elimination reaction.
- An ether is also obtained; it is the product of a competing S_N2 reaction, since primary alkyl halides are the ones most likely to form substitution products in S_N2/E2 reactions (Section 9.11).

PROBLEM 8 ♦

List the following alcohols in order of decreasing rate of dehydration in the presence of acid:

cyclohexyl–CH_2OH 1-methylcyclohexanol (CH_3, OH)   2-methylcyclohexanol (CH_3, OH)

Movie: Dehydration

GRAIN ALCOHOL AND WOOD ALCOHOL

When ethanol is ingested, it acts on the central nervous system. Moderate amounts affect judgment and lower inhibitions. Higher amounts interfere with motor coordination and cause slurred speech and amnesia. Still higher amounts cause nausea and loss of consciousness. Ingesting very large amounts of ethanol interferes with spontaneous respiration and can be fatal.

The ethanol in alcoholic beverages is produced by the fermentation of glucose, generally obtained from grapes or from grains such as corn, rye, and wheat (which is why ethanol is also known as grain alcohol). Grains are cooked in the presence of malt (sprouted barley) to convert much of their starch into glucose. Yeast is added to convert the glucose into ethanol and carbon dioxide (Section 18.5).

$$\underset{\text{glucose}}{C_6H_{12}O_6} \xrightarrow{\text{yeast enzymes}} 2\ \underset{\text{ethanol}}{CH_3CH_2OH} + 2\ CO_2$$

The kind of beverage produced (white or red wine, beer, scotch, bourbon, champagne) depends on the plant species providing the glucose, whether the CO_2 formed in the fermentation is allowed to escape, whether other substances are added, and how the beverage is purified (by sedimentation, for wines; by distillation, for scotch and bourbon).

The tax imposed on liquor would make ethanol a prohibitively expensive laboratory reagent. Laboratory alcohol, therefore, is not taxed, because ethanol is needed in a wide variety of commercial processes. Although this alcohol is not taxed, it is carefully regulated by the federal government to make certain that it is not used for the preparation of alcoholic beverages. Denatured alcohol—ethanol that has been made undrinkable by the addition of a denaturant such as benzene or methanol—is not taxed, but the added impurities make it unfit for many laboratory uses.

Methanol, also known as wood alcohol because at one time it was obtained by heating wood in the absence of oxygen, is highly toxic. Ingesting even very small amounts can cause blindness, and ingesting as little as an ounce can be fatal.

PROBLEM 9

Heating an alcohol with H_2SO_4 is a good way to prepare a symmetrical ether such as diethyl ether.

a. Explain why it is not a good way to prepare an unsymmetrical ether such as ethyl propyl ether.

b. How would you synthesize ethyl propyl ether?

The products obtained from the acid-catalyzed elimination (dehydration) of an alcohol are identical to those obtained from the elimination reaction of an alkyl halide. That is, both the E and Z stereoisomers are obtained as products. The reaction produces more of the stereoisomer in which the bulkier group on each of the sp^2 carbons are on opposite sides of the double bond; that stereoisomer, being more stable, is formed more rapidly, since the transition state leading to its formation is more stable.

$$\underset{\text{2-butanol}}{CH_3CH_2CH(OH)CH_3} \underset{\Delta}{\overset{H_2SO_4}{\rightleftharpoons}} CH_3CH_2\overset{+}{C}HCH_3 + H_2O \longrightarrow \underset{\substack{(E)\text{-2-butene} \\ 74\%}}{(H_3C)(H)C{=}C(H)(CH_3)} + \underset{\substack{(Z)\text{-2-butene} \\ 23\%}}{(H_3C)(H)C{=}C(CH_3)(H)} + \underset{\substack{\text{1-butene} \\ 3\%}}{CH_3CH_2CH{=}CH_2} + H^+$$

Alcohols and ethers undergo S_N1 reactions unless they would have to form a methyl or primary carbocation, in which case they undergo S_N2 reactions.

We can summarize what we have learned about the mechanisms by which alcohols undergo substitution and elimination reactions: they react by S_N1 and E1 pathways, unless they cannot do so. In other words, 3° and 2° alcohols undergo S_N1 and E1 reactions; 1° alcohols, because they cannot form primary carbocations, have to undergo S_N2 and E2 reactions.

BIOLOGICAL DEHYDRATIONS

Dehydration reactions occur in many important biological processes. Instead of being catalyzed by strong acids, which would not be available to a cell, they are catalyzed by enzymes. Fumarase, for example, is the enzyme that catalyzes the dehydration of malate in the citric acid cycle. The citric acid cycle is a series of reactions that oxidize compounds derived from carbohydrates, fatty acids, and amino acids (Section 18.7).

$$\underset{\text{malate}}{{}^-OOC{-}CH_2{-}CH(OH){-}COO^-} \xrightleftharpoons{\text{fumarase}} \underset{\text{fumarate}}{{}^-OOC{-}CH{=}CH{-}COO^-} + H_2O$$

Enolase, another enzyme, catalyzes the dehydration of α-phosphoglycerate in glycolysis (Section 18.4). Glycolysis is a series of reactions that prepare glucose for entry into the citric acid cycle.

$$\underset{\alpha\text{-phosphoglycerate}}{CH_2(OH){-}CH(OPO_3^{2-}){-}COO^-} \xrightleftharpoons{\text{enolase}} \underset{\text{phosphoenolpyruvate}}{CH_2{=}C(OPO_3^{2-}){-}COO^-} + H_2O$$

PROBLEM 10 ◆

Give the major product formed when each of the following alcohols is heated in the presence of H_2SO_4:

a. $CH_3CH_2C(CH_3)(OH){-}CH(CH_3)CH_3$

b. (cyclohexene ring bearing an OH group)

PROBLEM 11 ◆

The following compound is heated in the presence of H_2SO_4:

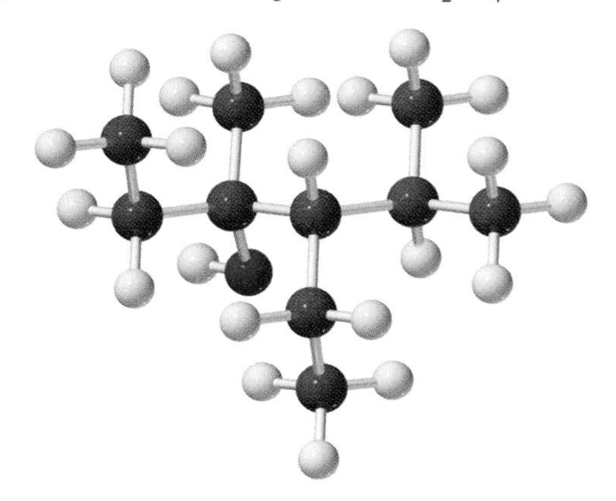

a. What constitutional isomer is produced in greatest yield?
b. What stereoisomer is produced in greatest yield?

10.4 OXIDATION OF ALCOHOLS

We have seen that a **reduction reaction** *increases* the number of C—H bonds in a compound (Section 5.12). Oxidation is the reverse of reduction. Therefore, an **oxidation reaction** *decreases* the number of C—H bonds (or increases the number of C—O bonds).

Secondary alcohols are oxidized to *ketones*. Chromic acid (H_2CrO_4) is the reagent commonly used to oxidize alcohols.

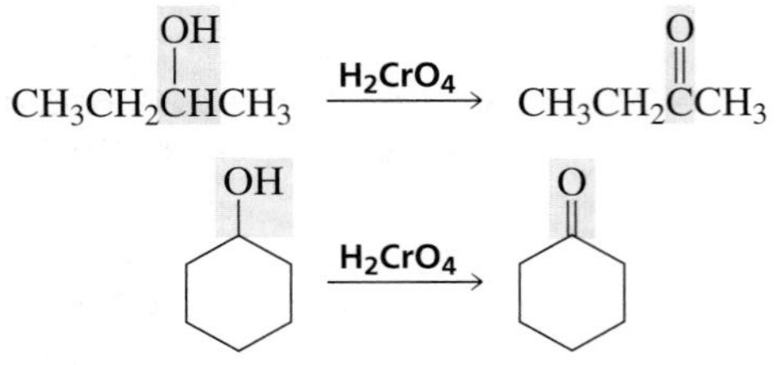

$$CH_3CH_2CH(OH)CH_3 \xrightarrow{H_2CrO_4} CH_3CH_2C(=O)CH_3$$

$$\text{cyclohexanol} \xrightarrow{H_2CrO_4} \text{cyclohexanone}$$

secondary alcohols — ketones

Secondary alcohols are oxidized to ketones.

Primary alcohols are initially oxidized to *aldehydes* by this reagent. The reaction, however, does not stop at the aldehyde. Instead, the aldehyde is further oxidized to a *carboxylic acid.*

$$\underset{\text{a primary alcohol}}{CH_3CH_2CH_2CH_2OH} \xrightarrow{H_2CrO_4} \underset{\text{an aldehyde}}{\left[CH_3CH_2CH_2CH(=O)\right]} \xrightarrow[\text{oxidation}]{\text{further}} \underset{\text{a carboxylic acid}}{CH_3CH_2CH_2C(=O)OH}$$

The oxidation of a primary alcohol will stop at the aldehyde if pyridinium chlorochromate (PCC) is used as the oxidizing agent in a solvent such as dichloromethane (CH_2Cl_2).

Primary alcohols are oxidized to aldehydes and carboxylic acids.

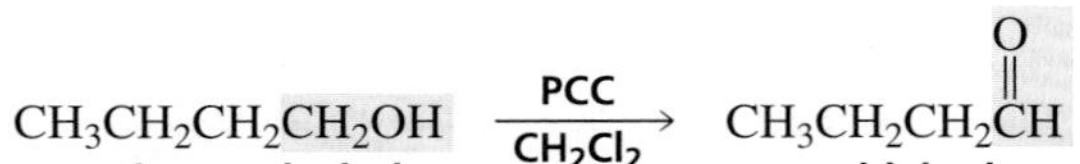

$$\underset{\text{a primary alcohol}}{CH_3CH_2CH_2CH_2OH} \xrightarrow[CH_2Cl_2]{PCC} \underset{\text{an aldehyde}}{CH_3CH_2CH_2CH(=O)}$$

Notice that in the oxidation of either a primary or a secondary alcohol, a hydrogen is removed from the carbon to which the OH is attached. The carbon bearing the OH group in a tertiary alcohol is not bonded to a hydrogen, so its OH group cannot be oxidized to a carbonyl group.

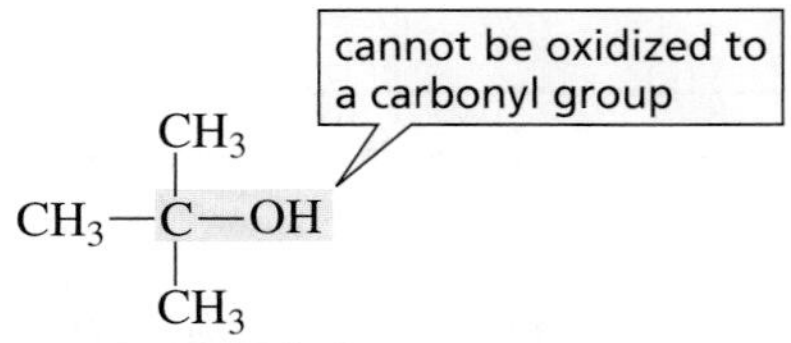

$$\underset{\text{a tertiary alcohol}}{(CH_3)_3C{-}OH}$$

cannot be oxidized to a carbonyl group

PROBLEM 12 ♦

Give the product formed from the reaction of each of the following compounds with chromic acid:

a. 3-pentanol **b.** 1-pentanol **c.** cyclohexanol **d.** benzyl alcohol

PROBLEM 13 ♦

What alcohol would be required to synthesize each of the following compounds?

a. $CH_3CH_2C(=O)CH_3$ **b.** cyclohexanone **c.** $CH_3CH_2CH_2C(=O)OH$

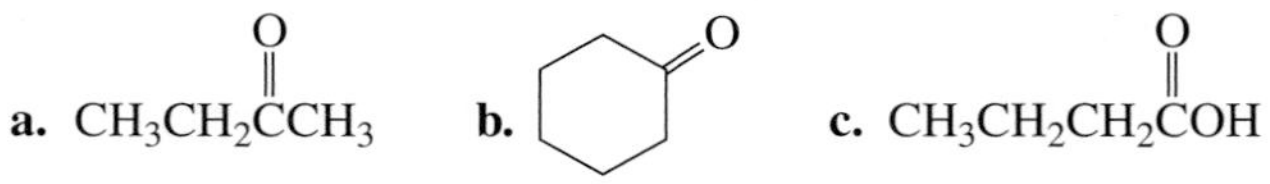

BLOOD ALCOHOL CONTENT

As blood passes through the arteries in our lungs, an equilibrium is established between alcohol in the blood and alcohol in the breath. Therefore, if the concentration of one is known, the concentration of the other can be estimated.

The test that law enforcement agencies use to approximate a person's blood alcohol level is based on the oxidation of breath ethanol. An oxidizing agent impregnated onto an inert material is enclosed within a sealed glass tube. When the test is to be administered, the ends of the tube are broken off and replaced with a mouthpiece at one end and a balloon-type bag at the other. The person being tested blows into the mouthpiece until the bag is filled with air.

Any ethanol in the breath is oxidized as it passes through the column. When ethanol is oxidized, the red-orange dichromate ion is reduced to green chromic ion. The greater the concentration of alcohol in the breath, the farther the green color spreads through the tube.

$$CH_3CH_2OH + \underset{\text{red orange}}{Cr_2O_7^{2-}} \xrightarrow{H^+} CH_3\overset{\overset{\displaystyle O}{\|}}{C}OH + \underset{\text{green}}{Cr^{3+}}$$

If the person fails this test—determined by the extent to which the green color spreads through the tube—a more accurate Breathalyzer test is administered. The Breathalyzer test also depends on the oxidation of breath ethanol, but it provides more accurate results because it is quantitative. In the test, a known volume of breath is bubbled through an acidic solution of sodium dichromate, and the concentration of the green chromic ion is measured precisely with a UV/Vis spectrophotometer (Section 7.12).

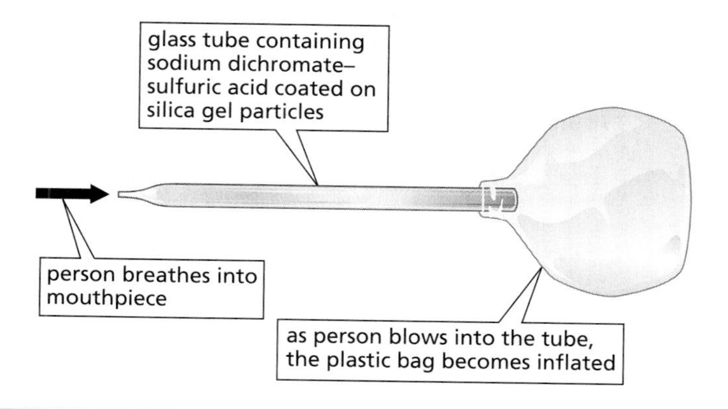
glass tube containing sodium dichromate–sulfuric acid coated on silica gel particles
person breathes into mouthpiece
as person blows into the tube, the plastic bag becomes inflated

10.5 AMINES DO NOT UNDERGO SUBSTITUTION OR ELIMINATION REACTIONS

We have just seen that alcohols are much less reactive than alkyl halides in substitution and elimination reactions. Amines are even *less reactive* than alcohols. The relative reactivities of an alkyl fluoride (the least reactive of the alkyl halides because it has the poorest leaving group), an alcohol, and an amine can be appreciated by comparing the pK_a values of the conjugate acids of their leaving groups, recalling that the weaker the acid, the stronger is its conjugate base and the poorer it is as a leaving group. The leaving group of an amine ($^-NH_2$) is such a strong base that amines cannot undergo substitution or elimination reactions.

relative reactivities

most reactive RCH_2F > RCH_2OH > RCH_2NH_2 least reactive

RCH_2F	RCH_2OH	RCH_2NH_2
HF	H_2O	NH_3
pK_a = 3.2	pK_a = 15.7	pK_a = 36

The stronger the base, the poorer it is as a leaving group.

Protonation of the amino group makes it a better leaving group, but not nearly as good a leaving group as a protonated alcohol, which is ~13 pK_a units more acidic than a protonated amine. Therefore, unlike the leaving group of a protonated alcohol, the leaving group of a protonated amine cannot be replaced by a halide ion or dissociate to form a carbocation.

$$\underset{pK_a = -2.4}{CH_3CH_2\overset{+}{O}H_2} \quad > \quad \underset{pK_a = 11.2}{CH_3CH_2\overset{+}{N}H_3}$$

Although amines cannot undergo substitution or elimination reactions, they are extremely important organic compounds. (The lone pair on the nitrogen atom allows it to act as both a base and a nucleophile.)

Amines are the most common organic bases. We have seen that protonated amines have pK_a values of about 11 (Section 2.3) and that protonated anilines have pK_a values of about 5 (Section 7.9). Neutral amines have very high pK_a values. For example, the pK_a of methylamine is 40.

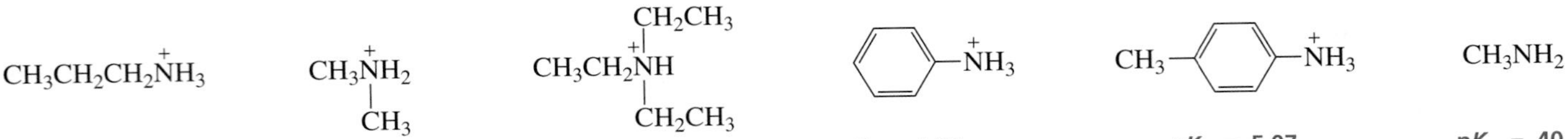

$CH_3CH_2CH_2\overset{+}{N}H_3$	$CH_3\overset{+}{N}H_2CH_3$	$CH_3CH_2\overset{+}{N}H(CH_2CH_3)_2$	$C_6H_5\overset{+}{N}H_3$	$CH_3C_6H_4\overset{+}{N}H_3$	CH_3NH_2
$pK_a = 10.8$	$pK_a = 10.9$	$pK_a = 11.1$	$pK_a = 4.58$	$pK_a = 5.07$	$pK_a = 40$

Amines react as nucleophiles in a wide variety of reactions. For example, they react as nucleophiles with alkyl halides in S_N2 reactions.

an S_N2 reaction

$$CH_3CH_2Br + CH_3NH_2 \longrightarrow CH_3CH_2\overset{+}{N}H_2CH_3 + Br^-$$

We will see that they also react as nucleophiles with a wide variety of carbonyl compounds (Sections 11.7, 11.8, and 12.7).

PROBLEM 14

Why can protonated amino groups not be displaced by strongly basic nucleophiles such as HO^-?

ALKALOIDS

Alkaloids are amines found in the leaves, bark, roots, or seeds of many plants. Examples include caffeine (found in tea leaves, coffee beans, and cola nuts), nicotine (found in tobacco leaves), and cocaine (obtained from the coca bush in the rainforest areas of Colombia, Peru, and Bolivia). Ephedrine, a bronchodilator, is an alkaloid obtained from *Ephedra sinica*, a plant found in China. Morphine is an alkaloid obtained from opium, the juice derived from a species of poppy (Section 21.3).

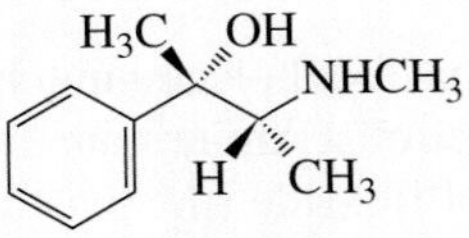

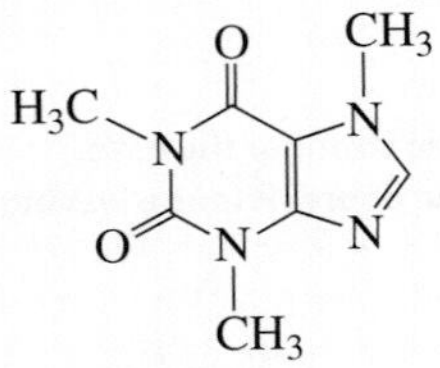

caffeine

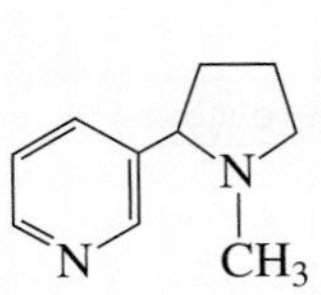

nicotine

ephedrine

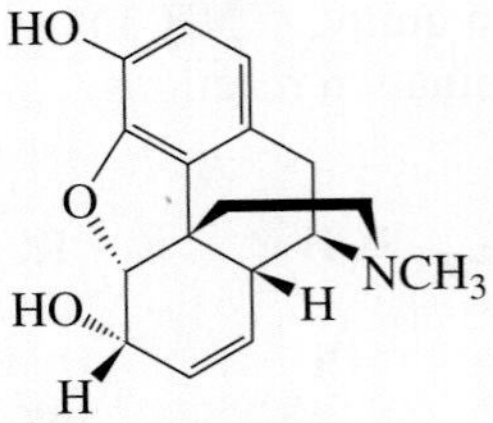

morphine

10.6 NOMENCLATURE OF ETHERS

An **ether** is a compound in which an oxygen is bonded to two alkyl substituents. The common name of an ether consists of the names of the two alkyl substituents (in alphabetical order), followed by the word "ether." The smallest ethers are almost always named by their common names.

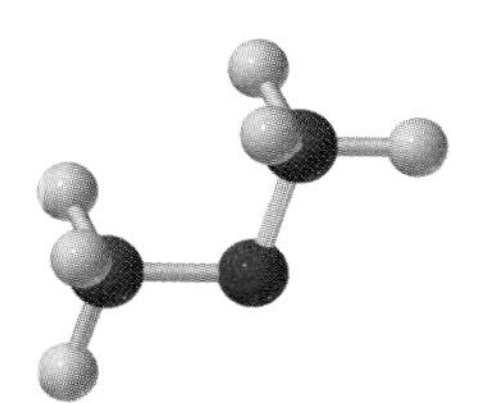
dimethyl ether

$CH_3OCH_2CH_3$ ethyl methyl ether

$CH_3CH_2OCH_2CH_3$ diethyl ether

$CH_3CH(CH_3)CH_2OC(CH_3)_2CH_3$ *tert*-butyl isobutyl ether

The IUPAC system names an ether as an alkane with an RO substituent. The substituents are named by replacing the "yl" ending in the name of the alkyl substituent with "oxy."

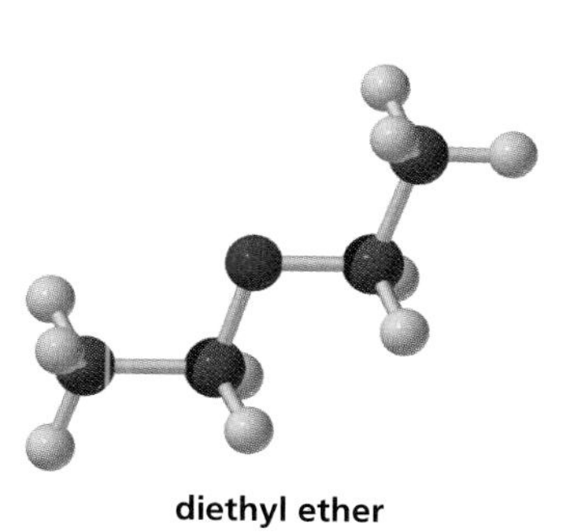
diethyl ether

$CH_3O—$ methoxy

$CH_3CH_2O—$ ethoxy

$CH_3CH(CH_3)O—$ isopropoxy

$CH_3CH_2CH(CH_3)O—$ *sec*-butoxy

$CH_3C(CH_3)_2O—$ *tert*-butoxy

$CH_3CH(OCH_3)CH_2CH_3$ 2-methoxybutane

$CH_3CH_2CH(CH_3)CH_2CH_2OCH_2CH_3$ 1-ethoxy-3-methylpentane

PROBLEM 15 ♦

a. Give the systematic name for the following ethers:

1. $CH_3OCH_2CH_3$

2. $CH_3CH_2OCH_2CH_3$

3. $CH_3CH_2CH_2CH_2CH(OCH_3)CH_2CH_2CH_3$

4. $CH_3CH_2CH_2OCH_2CH_2CH_2CH_3$

b. Do all of these ethers have common names?

c. What are their common names?

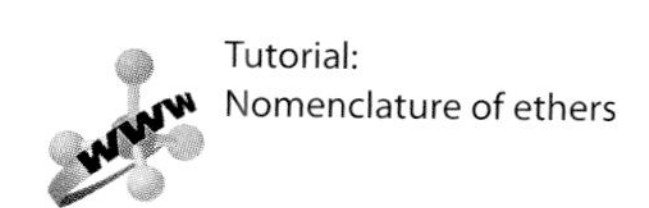
Tutorial:
Nomenclature of ethers

ANESTHETICS

Because diethyl ether (commonly known simply as ether) is a short-lived muscle relaxant, it was at one time widely used as an inhalation anesthetic. However, it takes effect slowly and has a slow and unpleasant recovery period, so other compounds, such as enflurane, isoflurane, and halothane, have replaced it as an anesthetic. Even so, diethyl ether is still used where trained anesthesiologists are scarce, because it is the safest anesthetic for an untrained person to administer. Anesthetics interact with the nonpolar molecules of cell membranes, causing the membranes to swell, which interferes with their permeability.

$CH_3CH_2OCH_2CH_3$ "ether"

$CF_3CHClOCHF_2$ isoflurane

$CHClFCF_2OCHF_2$ enflurane

$CF_3CHClBr$ halothane

Sodium pentothal (also called thiopental sodium) is commonly used as an intravenous anesthetic. The onset of anesthesia and the loss of consciousness occur within seconds of its administration. Care must be taken when administering sodium pentothal because the dose for effective anesthesia is 75% of the lethal dose. Because of its toxicity, it cannot be used as the sole anesthetic. It is generally used to induce anesthesia before an inhalation anesthetic is administered. Propofol is an anesthetic that has all the properties of the "perfect anesthetic": it can be used as the sole anesthetic by intravenous drip, it has a rapid and pleasant induction period and a wide margin of safety, and recovery from the drug also is rapid and pleasant.

(continued...)

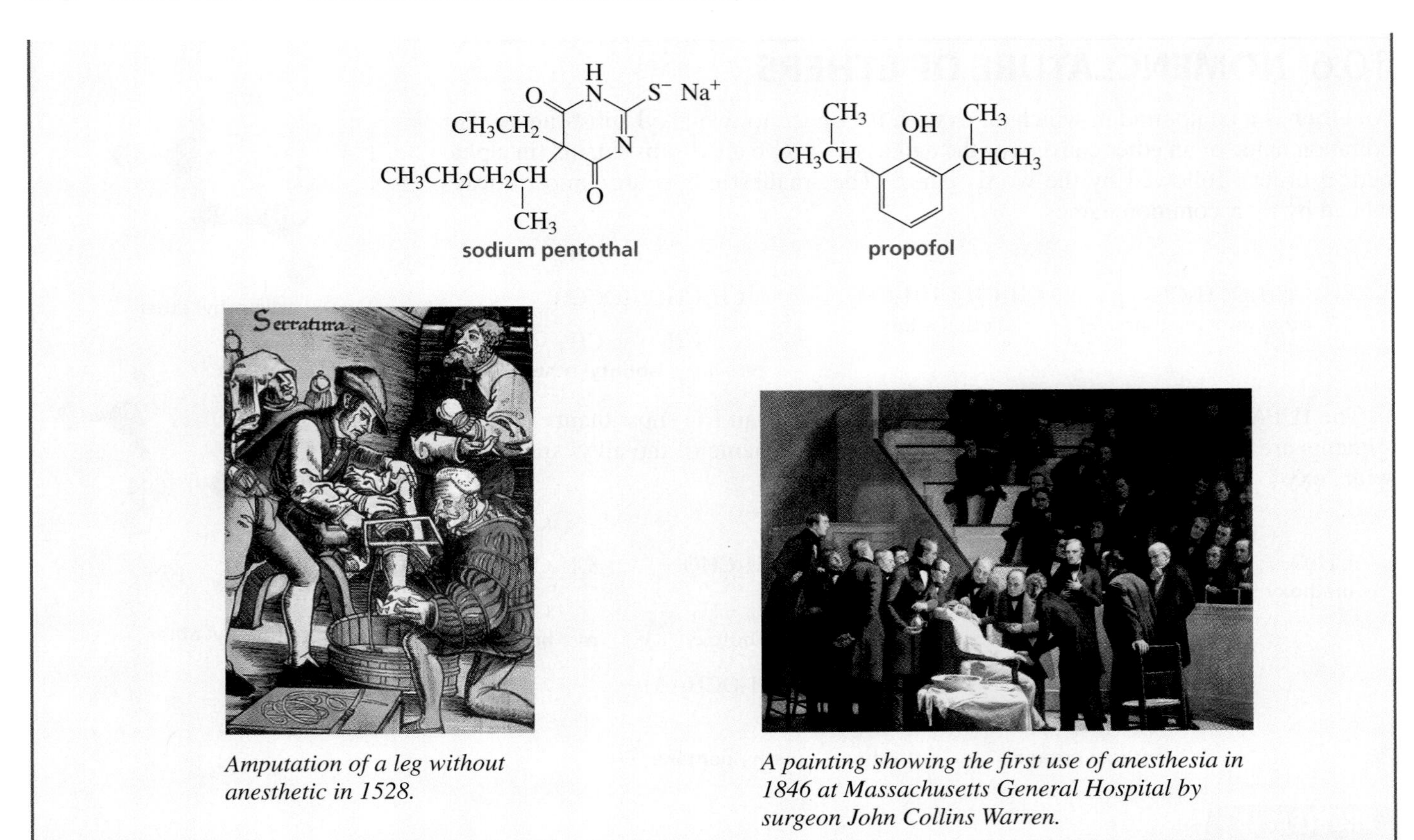

Amputation of a leg without anesthetic in 1528.

A painting showing the first use of anesthesia in 1846 at Massachusetts General Hospital by surgeon John Collins Warren.

10.7 NUCLEOPHILIC SUBSTITUTION REACTIONS OF ETHERS

The OR group of an **ether** and the OH group of an alcohol have nearly the same basicity because the conjugate acids of these two groups have similar pK_a values. (The pK_a of CH_3OH is 15.5, and the pK_a of H_2O is 15.7.) Both groups are strong bases, so both are very poor leaving groups. Consequently, ethers, like alcohols, need to be activated before they can undergo nucleophilic substitution.

$$R-\ddot{\underset{..}{O}}-H \qquad R-\ddot{\underset{..}{O}}-R$$

an alcohol   an ether

Ethers, like alcohols, can be activated by protonation. Ethers, therefore, can undergo nucleophilic substitution reactions with HBr or HI. As with alcohols, the reaction of ethers with hydrogen halides is slow. The reaction mixture must be heated to cause the reaction to occur at a reasonable rate.

$$R-O-R' + HI \rightleftharpoons R-\overset{H}{\underset{+}{O}}-R' \;\; I^- \xrightarrow{\Delta} R-I + R'-OH$$

poor leaving group   good leaving group

What happens *after* the ether is protonated depends on the structure of the ether. If departure of ROH creates a relatively stable carbocation (such as a tertiary carbocation), an S_N1 reaction occurs.

ether cleavage: an S_N1 reaction

$$(CH_3)_3C\text{—}\ddot{O}CH_3 + H\text{—}I \rightleftharpoons (CH_3)_3C\text{—}\overset{+}{\underset{H}{O}}CH_3 \xrightarrow{S_N1} (CH_3)_3C^{+} + CH_3\ddot{O}H \xrightarrow{:\ddot{I}:^-} (CH_3)_3C\text{—}\ddot{I}:$$

protonation

carbocation formation

attack by a nucleophile

- Protonation converts the very basic RO^- leaving group into the less basic ROH leaving group.
- The leaving group departs.
- The halide ion combines with the carbocation.

However, if departure of the leaving group would create an unstable carbocation (such as a methyl or primary carbocation), the leaving group cannot depart. It has to be displaced by the halide ion. In other words, an S_N2 reaction occurs.

ether cleavage: an S_N2 reaction

$$CH_3\ddot{O}CH_2CH_2CH_3 + H\text{—}I \rightleftharpoons CH_3\text{—}\overset{+}{\underset{H}{O}}CH_2CH_2CH_3 \xrightarrow{S_N2} CH_3\ddot{I}: + CH_3CH_2CH_2OH$$

protonation

$:\ddot{I}:^-$ nucleophile attacks the less sterically hindered carbon

- Protonation converts the very basic RO^- leaving group into the less basic ROH leaving group.
- The halide ion preferentially attacks the less sterically hindered of the two alkyl groups.

In summary, ethers are cleaved by an S_N1 pathway, unless the instability of the carbocation causes the reaction to follow an S_N2 pathway.

Because the only reagents that react with ethers are hydrogen halides, ethers are frequently used as solvents. Some common ether solvents are shown in Table 10.1.

Table 10.1 Some Ethers That Are Used as Solvents

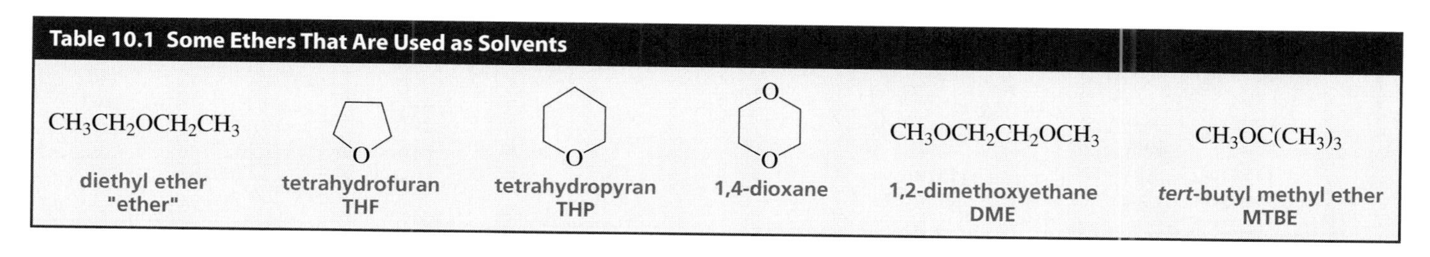

$CH_3CH_2OCH_2CH_3$	(structure)	(structure)	(structure)	$CH_3OCH_2CH_2OCH_3$	$CH_3OC(CH_3)_3$
diethyl ether "ether"	tetrahydrofuran THF	tetrahydropyran THP	1,4-dioxane	1,2-dimethoxyethane DME	*tert*-butyl methyl ether MTBE

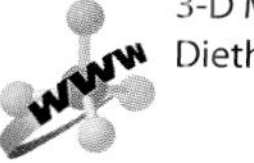

3-D Molecules: Diethyl ether; Tetrahydrofuran

PROBLEM 16 ***SOLVED***

Explain why methyl propyl ether forms both methyl iodide and propyl iodide when it is heated with excess HI.

Solution We have just seen that the S_N2 reaction of methyl propyl ether with an equivalent of HI forms methyl iodide and propyl alcohol because the methyl group is less sterically hindered to attack by the iodide ion. When there is excess HI, the alcohol product of this first reaction can react with HI in another S_N2 reaction (Section 10.1). Thus, the major products are methyl iodide and propyl iodide.

$$CH_3CH_2CH_2OCH_3 \xrightarrow{HI} CH_3CH_2CH_2OH + CH_3I \xrightarrow{HI} CH_3CH_2CH_2I + H_2O$$

3-D Molecule: [15]-Crown-5; [12]-Crown-4

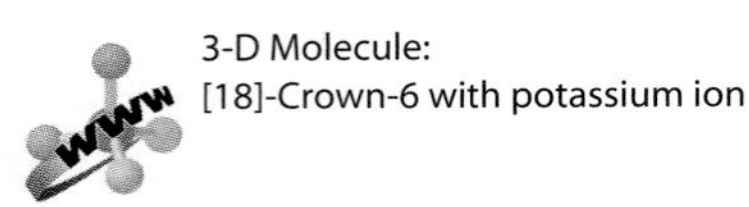

PROBLEM 17 ◆

Give the major products obtained from heating each of the following ethers with one equivalent of HI:

a. $CH_3CH(CH_3)CH_2OCH_2CH_3$ **b.** (tetrahydropyran, O in six-membered ring) **c.** $CH_3C(CH_3)_2OCH_2CH_3$ **d.** (2,2-dimethyltetrahydrofuran: five-membered ring with O, bearing CH_3 and CH_3)

AN UNUSUAL ANTIBIOTIC

Crown ethers are cyclic compounds containing several ether linkages around a central cavity. A crown ether specifically binds certain metal ions or organic molecules, depending on the cavity's size. The crown ether is called the "host," and the species it binds is called the "guest." Because the ether linkages are chemically inert, the crown ether can bind the guest without reacting with it.

Na^+ **guest** + **host** [15]-crown-5 cavity diameter = 1.7–2.2 Å ⟶ Na^+ complex

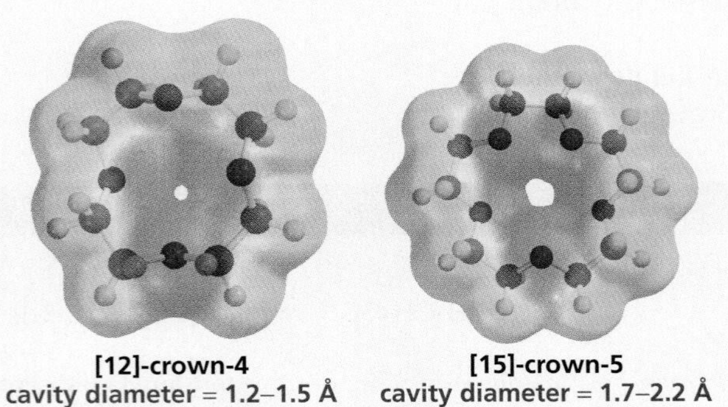

[12]-crown-4
cavity diameter = 1.2–1.5 Å

[15]-crown-5
cavity diameter = 1.7–2.2 Å

An antibiotic is a compound that interferes with the growth of microorganisms. Nonactin is a naturally occurring antibiotic that owes its biological activity to its ability to disrupt the carefully maintained levels of electrolytes on the inside and outside of a cell. Normal cell function requires gradients between the concentrations of potassium and sodium ions inside the cell and those outside the cell. To achieve these gradients, potassium ions are pumped in and sodium ions are pumped out.

Nonactin disrupts one of these gradients by acting like a crown ether. Nonactin's diameter is such that it specifically binds potassium ions. The eight oxygen atoms that point into the cavity and interact with K^+ are highlighted in the structure shown below. The outside of nonactin is nonpolar, so it can easily transport K^+ ions out of the cell through the nonpolar cell membrane. The consequent decrease in the concentration of K^+ within the cell causes the bacterium to die.

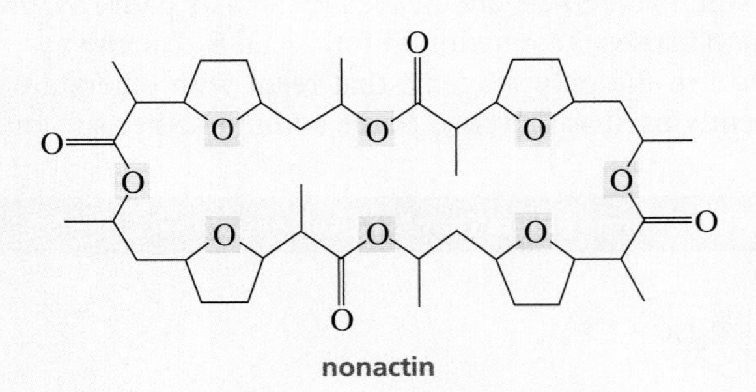

nonactin

10.8 NUCLEOPHILIC SUBSTITUTION REACTIONS OF EPOXIDES

An **epoxide** is an ether in which the oxygen atom is incorporated into a three-membered ring. The common name of an epoxide is obtained by adding "oxide" to the common name of the alkene, assuming that the oxygen atom is where the π bond of an alkene would be. The simplest epoxide is ethylene oxide.

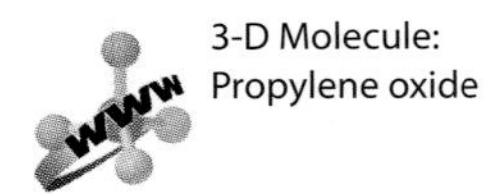

$H_2C{=}CH_2$ **ethylene** | $H_2C{-}CH_2$ (with bridging O) **ethylene oxide** | $H_2C{=}CHCH_3$ **propylene** | $H_2C{-}CHCH_3$ (with bridging O) **propylene oxide**

Alternatively, an epoxide can be named as an alkane, with an "epoxy" prefix that identifies the carbons to which the oxygen is attached.

H_2C—$CHCH_2CH_3$ (epoxide O bridging C1–C2)
1,2-epoxybutane

CH_3CH—$CHCH_3$ (epoxide O bridging C2–C3)
2,3-epoxybutane

H_2C—$C(CH_3)_2$ (epoxide O bridging C1–C2)
1,2-epoxy-2-methylpropane

PROBLEM 18 ◆

Draw the structure of the following:
a. cyclohexene oxide
b. 2,3-epoxy-2-methylpentane

An epoxide is formed from the reaction of an alkene with a *peroxyacid*. A **peroxyacid** is a carboxylic acid with an extra oxygen atom. It is this oxygen that is transferred to the alkene in order to from the epoxide. The reaction increases the number of C—O bonds in the reactant. It is, therefore, an oxidation reaction (Section 10.4).

$$RCH{=}CH_2 + RC({=}O)OOH \longrightarrow RCH{-}CH_2\ (\text{epoxide}) + RC({=}O)OH$$

an alkene, a peroxyacid, an epoxide, a carboxylic acid

Although an epoxide and an ether have the same leaving group, epoxides are much more reactive than ethers in nucleophilic substitution reactions because the strain in the three-membered ring is relieved when the ring opens (Figure 10.2). Epoxides, therefore, readily undergo nucleophilic substitution reactions with a wide variety of nucleophiles.

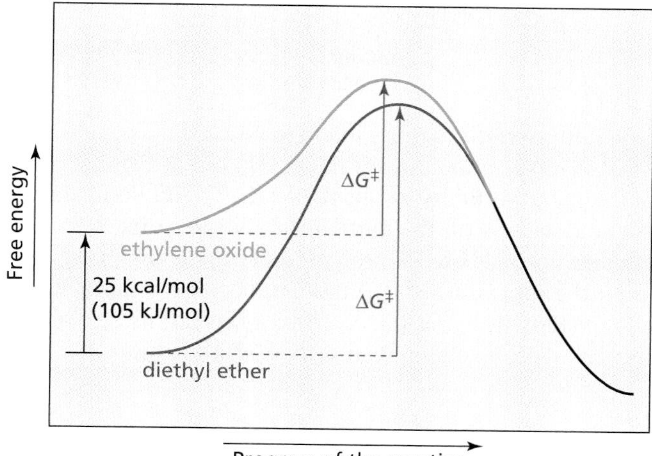

◀ **Figure 10.2**
The reaction coordinate diagrams for nucleophilic attack of hydroxide ion on ethylene oxide and on diethyl ether. The greater reactivity of the epoxide is a result of the strain in the three-membered ring, which increases the epoxide's free energy.

Epoxides, like other ethers, react with hydrogen halides.

epoxide ring opening: acidic conditions

$$H_2C{-}CH_2\ (\text{epoxide}) + H{-}\ddot{Br}{:} \rightleftharpoons H_2C{-}CH_2\ (\text{protonated epoxide}, {:}\overset{+}{O}H) + {:}\ddot{Br}{:}^- \longrightarrow H\ddot{O}CH_2CH_2\ddot{Br}{:}$$

protonation of the epoxide oxygen atom

back-side attack by the nucleophile

- Again we see that the first step in a mechanism, in which an acid is one of the reagents, is protonation of the most basic atom.
- The protonated epoxide undergoes back-side attack by the halide ion.

Because epoxides are so much more reactive than ethers, the reaction takes place readily at room temperature, unlike the reaction of an ether with a hydrogen halide, which requires heat.

Protonated epoxides are so reactive that they can be opened by poor nucleophiles, such as H_2O and alcohols.

$$H_2C\overset{O}{—}CH_2 \overset{H^+}{\rightleftharpoons} H_2C\overset{\overset{H}{O^+}}{—}CH_2 \xrightarrow{H_2\ddot{O}:} CH_2CH_2\overset{+}{O}H_2 \rightleftharpoons HOCH_2CH_2OH + H^+$$

1,2-ethanediol
ethylene glycol

the strongly acidic species loses a proton

$$CH_3CH\overset{O}{—}CHCH_3 \overset{H^+}{\rightleftharpoons} CH_3CH\overset{\overset{H}{O^+}}{—}CHCH_3 \xrightarrow{CH_3\ddot{O}H} CH_3CH(OH)CH(\overset{+}{O}HCH_3)CH_3 \rightleftharpoons CH_3CH(OH)CH(OCH_3)CH_3 + H^+$$

3-methoxy-2-butanol

If different substituents are attached to the two carbons of the protonated epoxide and the nucleophile is something other than H_2O, the product obtained from nucleophilic attack on the 2-position of the ring will be different from that obtained from attack on the 1-position. The major product is the one resulting from nucleophilic attack on the *more substituted* carbon.

$$\underset{3\quad 2\qquad 1}{CH_3CH\overset{O}{—}CH_2} \overset{H^+}{\rightleftharpoons} CH_3CH\overset{\overset{H}{O^+}}{—}CH_2 \xrightarrow{CH_3OH} CH_3CH(OCH_3)CH_2OH + CH_3CH(OH)CH_2OCH_3 + H^+$$

2-methoxy-1-propanol (major product); 1-methoxy-2-propanol (minor product)

The more substituted carbon is more likely to be attacked because, after the epoxide is protonated, it is so reactive that one of the C—O bonds begins to break even before the nucleophile has an opportunity to attack. As the C—O bond starts to break, a partial positive charge develops on the carbon that is losing its share of oxygen's electrons. The protonated epoxide breaks preferentially in the direction that puts the partial positive charge on the more substituted carbon, because a more substituted carbocation is more stable. (Recall that tertiary carbocations are more stable than secondary carbocations, which are more stable than primary carbocations.)

$$CH_3CH\overset{\overset{H}{O^+}}{—}CH_2 \longrightarrow CH_3\overset{\delta+}{C}H\overset{\overset{H}{\delta+O}}{—}CH_2 \xrightarrow{CH_3\ddot{O}H} CH_3CH(\overset{+}{O}HCH_3)CH_2OH \rightleftharpoons CH_3CH(OCH_3)CH_2OH + H^+$$

developing secondary carbocation; major product

$$CH_3CH\overset{\overset{H}{O^+}}{—}CH_2 \longrightarrow CH_3CH\overset{\overset{H}{O\delta+}}{—}\overset{\delta+}{C}H_2 \xrightarrow{CH_3\ddot{O}H} CH_3CH(OH)CH_2\overset{+}{O}HCH_3 \rightleftharpoons CH_3CH(OH)CH_2OCH_3 + H^+$$

developing primary carbocation; minor product

The best way to describe the reaction is to say that it occurs by a pathway that is partially S_N1 and partially S_N2. It is not a pure S_N1 reaction because a carbocation intermediate is not fully formed; it is not a pure S_N2 reaction because the leaving group begins to depart before the compound is attacked by the nucleophile.

Although an ether must be protonated before it can undergo a nucleophilic substitution reaction (Section 10.7), epoxides can undergo nucleophilic substitution reactions without first being protonated because of the strain in the three-membered ring. When a nucleophile attacks an unprotonated epoxide, the reaction is a pure S_N2 reaction.

epoxide ring opening: basic conditions

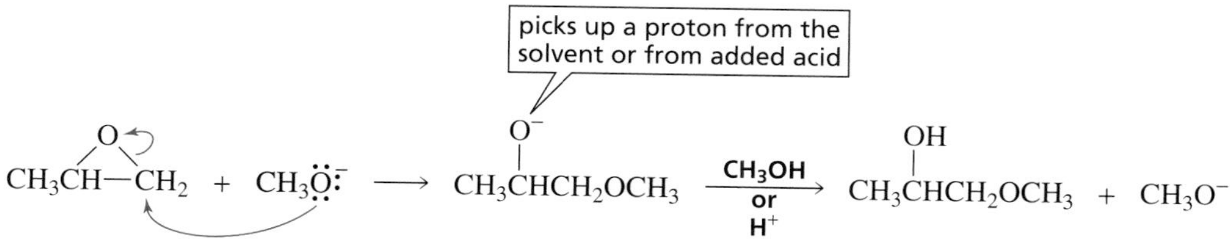

picks up a proton from the solvent or from added acid

$$CH_3CH(O)CH_2 + CH_3\ddot{O}^- \longrightarrow CH_3CH(O^-)CH_2OCH_3 \xrightarrow[\text{or } H^+]{CH_3OH} CH_3CH(OH)CH_2OCH_3 + CH_3O^-$$

- The C—O bond does not begin to break until the carbon is attacked by the nucleophile. The nucleophile is more likely to attack the *less substituted* carbon because the less substituted carbon is more accessible to attack, because it is less sterically hindered.
- The alkoxide ion picks up a proton from the solvent or from an acid added after the reaction is over.

Thus, the site of nucleophilic attack on an unsymmetrical epoxide under basic conditions (when the epoxide *is not* protonated) is different from the site of nucleophilic attack under acidic conditions (when the epoxide *is* protonated).

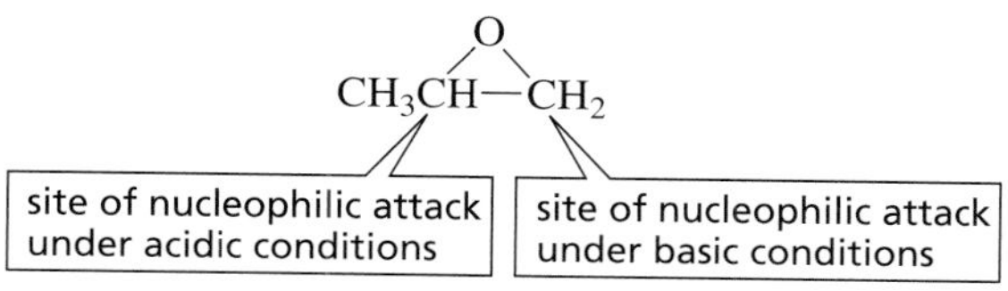

$$CH_3CH(O)CH_2$$

site of nucleophilic attack under acidic conditions

site of nucleophilic attack under basic conditions

Epoxides are useful reagents because they can react with a wide variety of nucleophiles, leading to the formation of a wide variety of products.

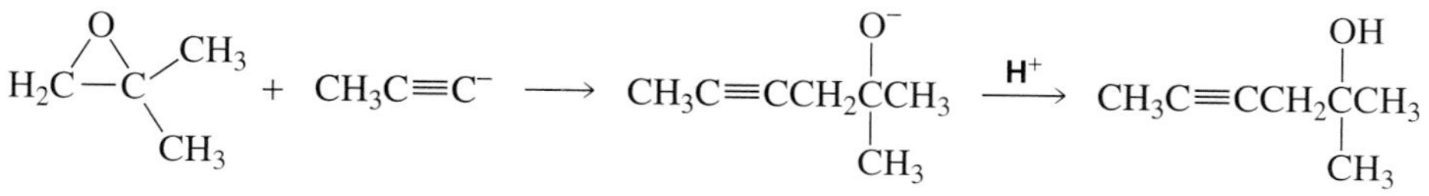

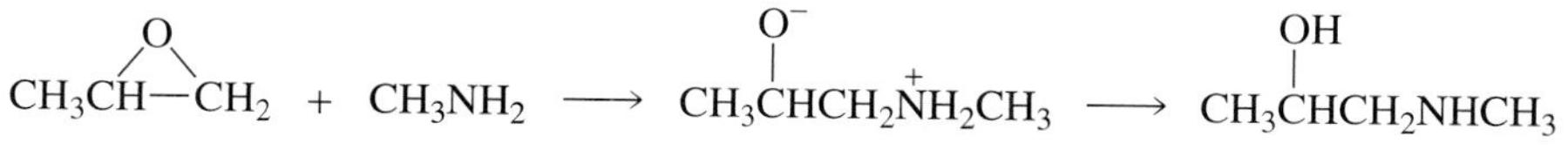

$$H_2C(O)C(CH_3)_2 + CH_3C{\equiv}C^- \longrightarrow CH_3C{\equiv}CCH_2C(O^-)(CH_3)CH_3 \xrightarrow{H^+} CH_3C{\equiv}CCH_2C(OH)(CH_3)CH_3$$

$$CH_3CH(O)CH_2 + CH_3NH_2 \longrightarrow CH_3CH(O^-)CH_2\overset{+}{N}H_2CH_3 \longrightarrow CH_3CH(OH)CH_2NHCH_3$$

Epoxides also are important in biological processes because they are reactive enough to be attacked by nucleophiles under the conditions found in living systems (Section 10.8).

PROBLEM 19 ◆

Give the major product of each of the following reactions:

a. 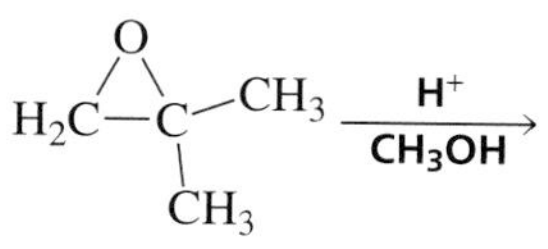 $H_2C(O)C(CH_3)CH_3 \xrightarrow[CH_3OH]{H^+}$

b. $H_2C(O)C(CH_3)CH_3 \xrightarrow[CH_3OH]{CH_3O^-}$

c. $H(H_3C)C(O)C(CH_3)CH_3 \xrightarrow[CH_3OH]{H^+}$

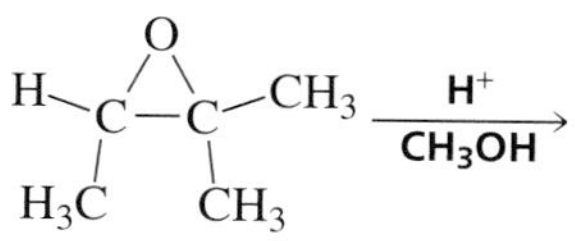

d. 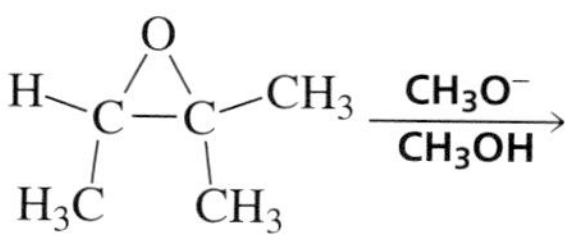 $H(H_3C)C(O)C(CH_3)CH_3 \xrightarrow[CH_3OH]{CH_3O^-}$

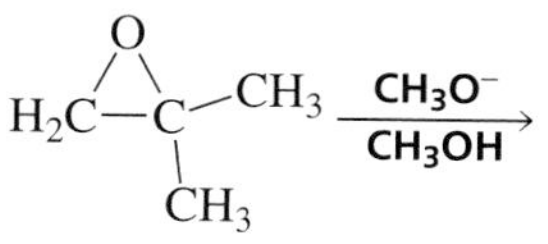

PROBLEM 20 ◆

Would you expect the reactivity of a five-membered ring ether such as tetrahydrofuran (Table 10.1) to be more similar to an epoxide or to a noncyclic ether?

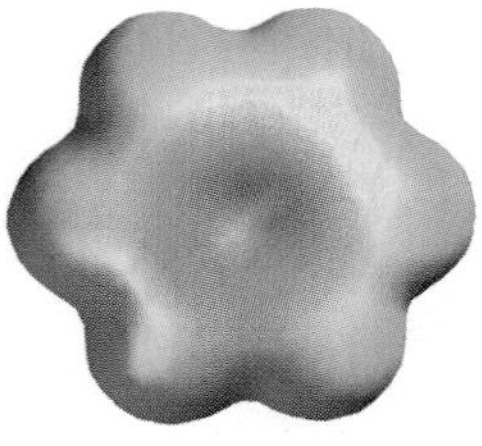

benzene

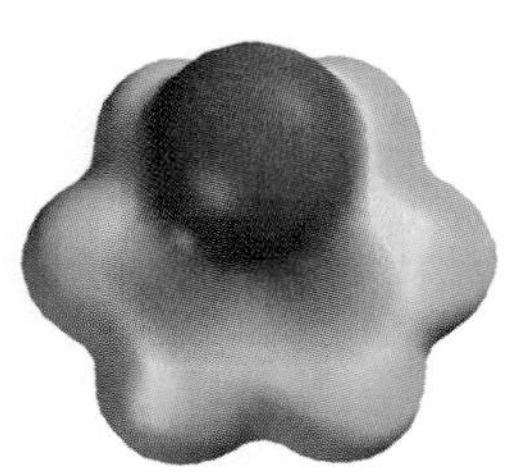

benzene oxide

10.9 USING CARBOCATION STABILITY TO DETERMINE THE CARCINOGENICITY OF AN ARENE OXIDE

After an aromatic hydrocarbon such as benzene is ingested or inhaled, it is converted into an *arene oxide* by an enzyme called cytochrome P_{450}. An **arene oxide** is a compound in which one of the "double bonds" of the aromatic ring has been converted into an epoxide. Formation of an arene oxide is the first step in changing an aromatic compound that enters the body as a foreign substance (for example, a drug, cigarette smoke, or automobile exhaust) into a more water-soluble compound that can eventually be eliminated.

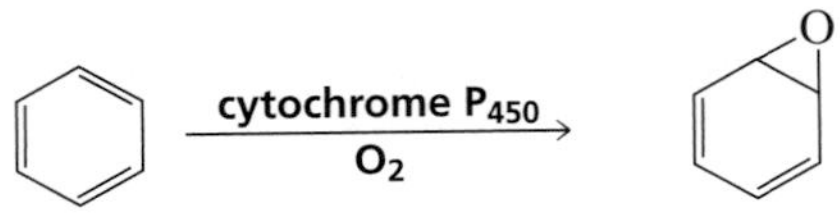

cytochrome P_{450}, O_2

benzene → benzene oxide, an arene oxide

An arene oxide can react in two different ways. It can react as a typical epoxide, undergoing attack by a nucleophile to form an addition product (Section 10.8). Alternatively, it can rearrange to form a phenol, which epoxides cannot do.

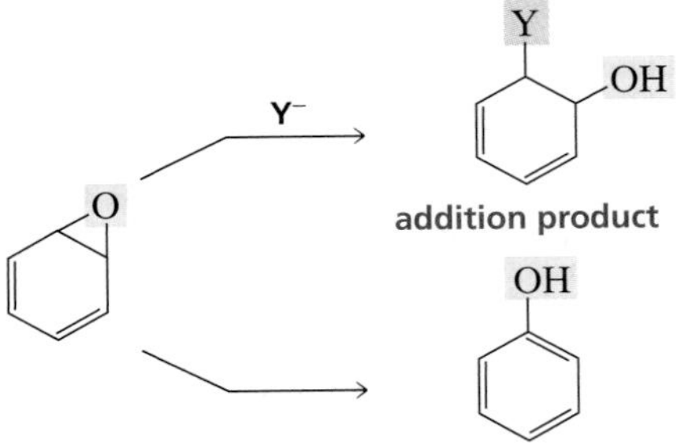

Y^- → addition product

→ rearranged product

When an arene oxide reacts with a nucleophile, the nucleophile attacks the three-member ring and forms an addition product.

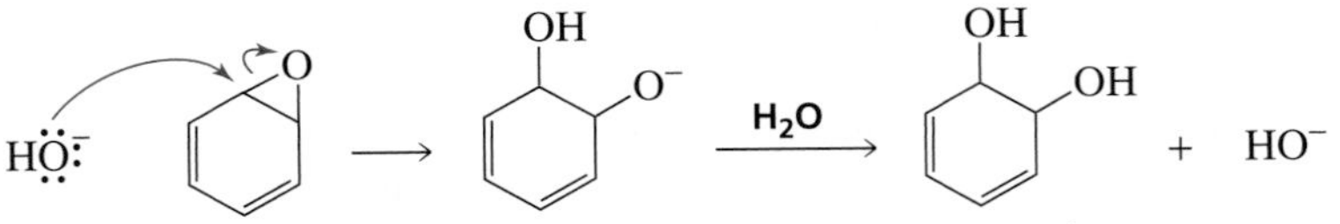

HO^- → ; H_2O → addition product + HO^-

When an arene oxide undergoes rearrangement, the three-membered epoxide ring opens, picking up a proton from a species in the solution (HB^+). A base in the solution (:B) removes a proton from the carbocation intermediate; the product is phenol.

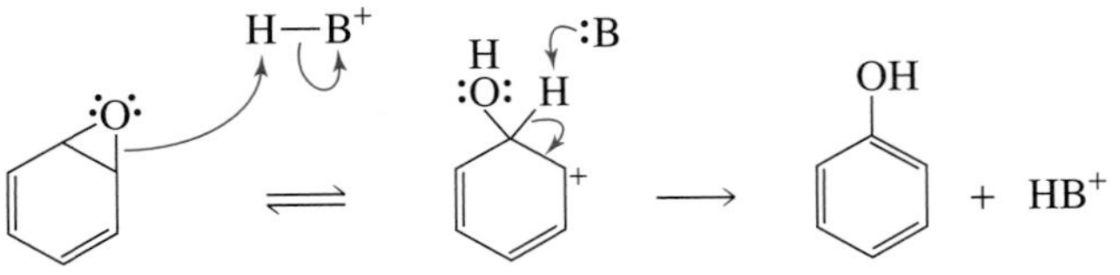

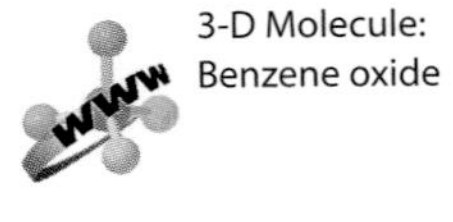

3-D Molecule: Benzene oxide

$H{-}B^+$, :B

benzene oxide ⇌ a carbocation → phenol + HB^+

Because formation of the carbocation is the rate-determining step, the rate of formation of phenol depends on the stability of the carbocation. The more stable the carbocation, the easier it is to open the epoxide ring and form phenol.

Some aromatic hydrocarbons are carcinogens—compounds that cause cancer. Investigation has revealed, however, that the hydrocarbons themselves are not carcinogenic; the actual carcinogens are the arene oxides into which the hydrocarbons are converted.

How do arene oxides cause cancer? We have seen that nucleophiles react with arene oxides to form addition products. 2′-Deoxyguanosine, a component of DNA, has a nucleophilic NH_2 group that is known to react with certain arene oxides. Once a molecule of 2′-deoxyguanosine becomes covalently attached to an arene oxide, the 2′-deoxyguanosine can no longer fit into the DNA double helix. As a result, the genetic code will not be properly transcribed (Section 20.6), which can lead to mutations that cause cancer. Cancer results when cells lose their ability to control their growth and reproduction.

a segment of DNA

an arene oxide

2′-deoxyguanosine

covalently attached to the arene oxide

Not all arene oxides are carcinogenic. Whether a particular arene oxide is carcinogenic depends on the relative rates of its two reaction pathways: rearrangement to a phenol and reaction with a nucleophile to form an addition product. Arene oxide rearrangement leads to phenols that are not carcinogenic, whereas formation of addition products from nucleophilic attack by DNA can lead to cancer-causing products. Thus, if the rate of arene oxide rearrangement is faster than the rate of nucleophilic attack by DNA, the arene oxide will be harmless. However, if the rate of nucleophilic attack is faster than the rate of rearrangement, the arene oxide will likely be a carcinogen.

Because the rate of arene oxide rearrangement depends on the stability of the carbocation formed in the first step of the rearrangement, *an arene oxide's cancer-causing potential depends on the stability of this carbocation*. If the carbocation is relatively stable, rearrangement will be fast and the arene oxide will most likely not be carcinogenic. On the other hand, if the carbocation is relatively unstable, rearrangement will be slow and the arene oxide will more likely exist long enough to be attacked by nucleophiles and, therefore, be carcinogenic. This means that the more reactive the arene oxide (the more easily it opens to form a carbocation), the less likely it is to be carcinogenic.

The more stable the carbocation formed when the epoxide ring of an arene oxide opens, the less likely it is that the arene oxide is carcinogenic.

PROBLEM-SOLVING STRATEGY

Which compound is more likely to be carcinogenic?

OCH_3 or NO_2

To determine the compound most likely to be carcinogenic, we must compare the stabilities of the carbocations formed when the epoxide rings open. The compound with the least stable carbocation is the one most apt to be carcinogenic.

The methoxy group stabilizes the carbocation by donating electrons into the ring by resonance. In contrast, the nitro group destabilizes the carbocation by withdrawing electrons from the ring by resonance. Carbocation formation leads to the harmless product. Thus, the nitro-substituted compound, with a less stable (harder-to-form) carbocation, is more likely to be carcinogenic.

(continued...)

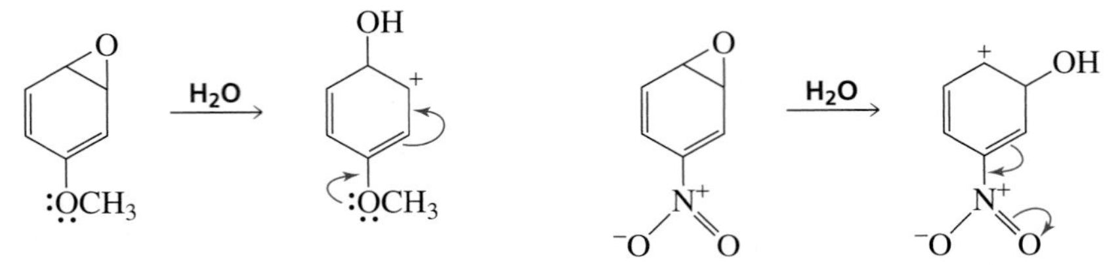

H_2O →

:ÖCH₃ / :ÖCH₃

H_2O →

OH

Now continue on to Problem 21.

PROBLEM 21 ♦

Which compound is more likely to be carcinogenic? (*Hint:* Read the box on benzo[*a*]pyrene to see why the 4,5-epoxide is harmful.)

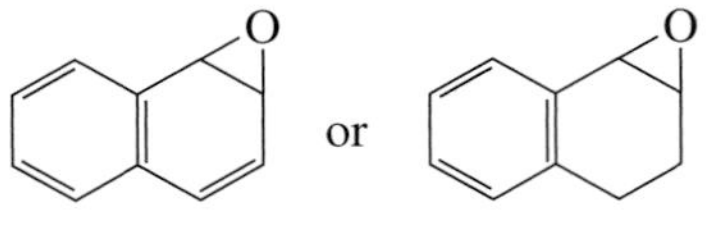

or

BENZO[*a*]PYRENE AND CANCER

Benzo[*a*]pyrene is one of the most carcinogenic of the aromatic hydrocarbons. It is formed whenever an organic compound is not completely burned. For example, benzo[*a*]pyrene is found in cigarette smoke, automobile exhaust, and charcoal-broiled meat. Several arene oxides can be formed from benzo[*a*]pyrene. The two most harmful are the 4,5-oxide and the 7,8-oxide. It has been suggested that people who develop lung cancer as a result of smoking may have a higher than normal concentration of cytochrome P_{450} in their lung tissue.

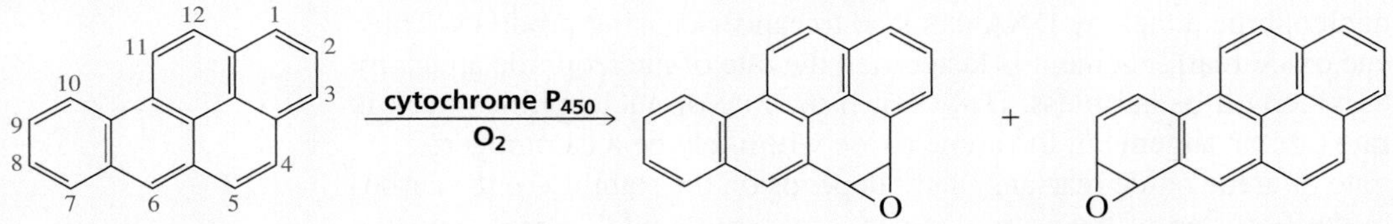

cytochrome P_{450}, O_2 →

benzo[*a*]pyrene → 4,5-benzo[*a*]pyrene oxide + 7,8-benzo[*a*]pyrene oxide

The 4,5-oxide is harmful because it forms a carbocation that cannot be stabilized by electron delocalization without destroying the aromaticity of an adjacent benzene ring. Thus, the carbocation is relatively unstable, so the epoxide tends not to open until it is attacked by a nucleophile (the carcinogenic pathway).

The 7,8-oxide is harmful because it reacts with water to form a diol, which then forms a diol epoxide. The diol epoxide does not readily undergo rearrangement (the harmless pathway), because it opens to a carbocation that is destabilized by the electron-withdrawing OH groups. Therefore, the diol epoxide can exist long enough to be attacked by nucleophiles (the carcinogenic pathway).

H_2O, epoxide hydrolase →

a diol

O_2 ↓ cytochrome P_{450}

a diol epoxide

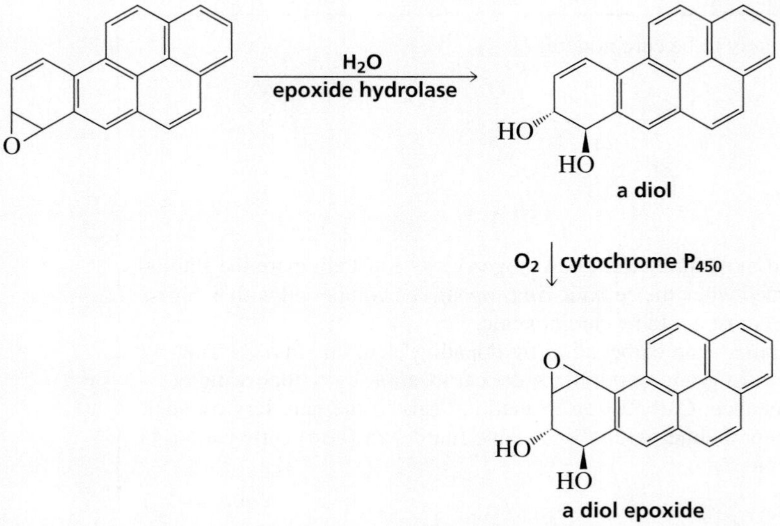

CHIMNEY SWEEPS AND CANCER

In 1775, British physician Percival Potts became the first to recognize that environmental factors can cause cancer when he observed that chimney sweeps had a higher incidence of scrotum cancer than the male population as a whole. He theorized that something in the chimney soot was causing cancer. We now know that it was benzo[*a*]pyrene.

Titch Cox, the chimney sweep responsible for cleaning the 800 chimneys at Buckingham Palace.

SUMMARY

The leaving groups of **alcohols** and **ethers** are stronger bases than halide ions are, so alcohols and ethers are less reactive than alkyl halides and have to be protonated before they can undergo a substitution or an elimination reaction. **Epoxides** do not have to be activated, because ring strain increases their reactivity. Amines cannot undergo substitution or elimination reactions because their leaving groups ($^-NH_2$, ^-NHR, $^-NR_2$) are very strong bases.

Primary, secondary, and tertiary alcohols undergo nucleophilic substitution reactions with HI, HBr, and HCl to form alkyl halides. These are S_N1 reactions in the case of tertiary and secondary alcohols and S_N2 reactions in the case of primary alcohols.

An alcohol undergoes an elimination reaction if heated with an acid. **Dehydration** (elimination of a water molecule) is an E1 reaction in the case of tertiary and secondary alcohols and an E2 reaction in the case of primary alcohols. Tertiary alcohols are the easiest to dehydrate, and primary alcohols are the hardest. The major product is the more substituted alkene. If the alkene has stereoisomers, both the *E* and *Z* stereoisomers are formed, but the one with the bulkiest groups on opposite sides of the double bond predominates.

Chromic acid oxidizes secondary alcohols to ketones and primary alcohols to carboxylic acids. PCC oxidizes primary alcohols to aldehydes.

Ethers can undergo nucleophilic substitution reactions with HBr or HI. If departure of the leaving group creates a relatively stable carbocation, an S_N1 reaction occurs; otherwise, an S_N2 reaction occurs.

Epoxides undergo nucleophilic substitution reactions. Under basic conditions, the least sterically hindered ring-carbon is attacked; under acidic conditions, the most substituted ring-carbon is attacked. **Arene oxides** undergo rearrangement to form phenols and nucleophilic attack to form addition products. An arene oxide's cancer-causing potential depends on the stability of the carbocation formed during rearrangement.

SUMMARY OF REACTIONS

1. Nucleophilic substitution reactions of *alcohols* (Section 10.2).

$$\mathrm{ROH + HBr \xrightarrow{\Delta} RBr}$$

$$\mathrm{ROH + HI \xrightarrow{\Delta} RI}$$

$$\mathrm{ROH + HCl \xrightarrow{\Delta} RCl}$$

relative rate: tertiary > secondary > primary

2. Elimination reactions of *alcohols*: dehydration (Section 10.3).

$$\mathrm{-\underset{H}{\overset{|}{\underset{|}{C}}}-\underset{OH}{\overset{|}{\underset{|}{C}}}- \underset{\Delta}{\overset{H_2SO_4}{\rightleftharpoons}} \;\rangle C{=}C\langle\; + H_2O}$$

relative rate: tertiary > secondary > primary

3. Oxidation of *alcohols* (Section 10.4).

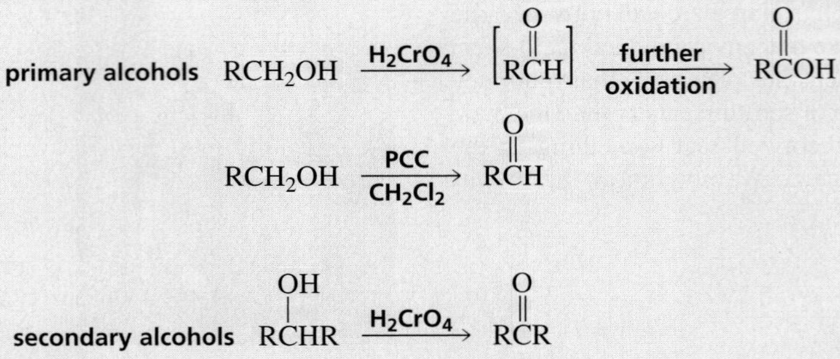

$$\text{primary alcohols}\quad RCH_2OH \xrightarrow{H_2CrO_4} \left[\overset{O}{\overset{\|}{RCH}}\right] \xrightarrow[\text{oxidation}]{\text{further}} \overset{O}{\overset{\|}{RCOH}}$$

$$RCH_2OH \xrightarrow[CH_2Cl_2]{PCC} \overset{O}{\overset{\|}{RCH}}$$

$$\text{secondary alcohols}\quad \overset{OH}{\overset{|}{RCHR}} \xrightarrow{H_2CrO_4} \overset{O}{\overset{\|}{RCR}}$$

4. Nucleophilic substitution reactions of *ethers* (Section 10.7).

$$ROR' + HX \xrightarrow{\Delta} ROH + R'X$$

$$HX = HBr \text{ or } HI$$

5. Nucleophilic substitution reactions of *epoxides* (Section 10.8)

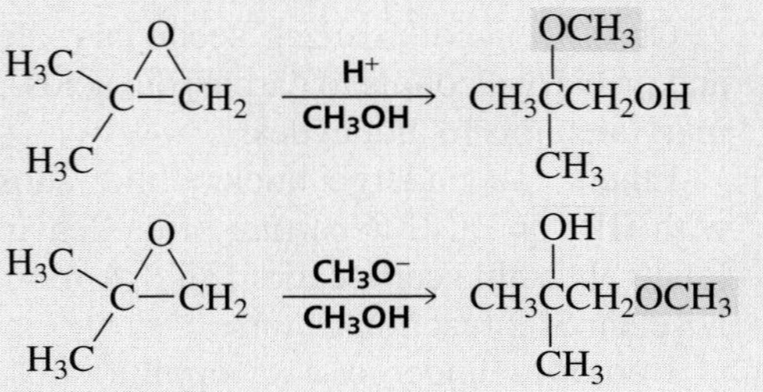

$$(H_3C)_2C\overset{O}{—}CH_2 \xrightarrow[CH_3OH]{H^+} CH_3\overset{OCH_3}{\overset{|}{C}}(CH_3)CH_2OH$$

under acidic conditions, the nucleophile attacks the more substituted ring-carbon

$$(H_3C)_2C\overset{O}{—}CH_2 \xrightarrow[CH_3OH]{CH_3O^-} CH_3\overset{OH}{\overset{|}{C}}(CH_3)CH_2OCH_3$$

under basic conditions, the nucleophile attacks the less sterically hindered ring-carbon

6. Reactions of *arene oxides*: ring opening and rearrangement (Section 10.9).

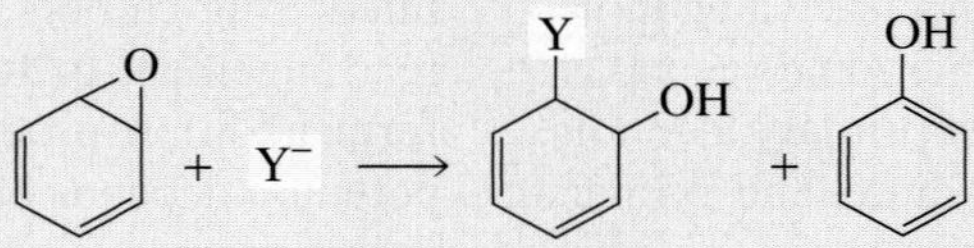

arene oxide + Y^- ⟶ (Y, OH-substituted cyclohexadiene) + (phenol, OH)

PROBLEMS

22. Give the product of each of the following reactions:

a. 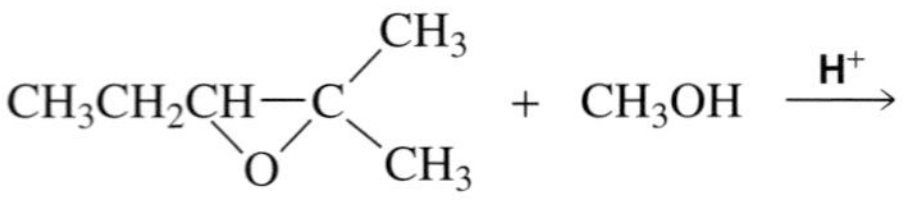 $CH_3CH_2CH\overset{O}{—}C(CH_3)_2 + CH_3OH \xrightarrow{H^+}$

b. $CH_3\underset{CH_3}{\underset{|}{C}}HCH_2OCH_3 + HI \xrightarrow{\Delta}$

c. cyclohexyl–$CH_2CH_2OH \xrightarrow{H_2CrO_4}$

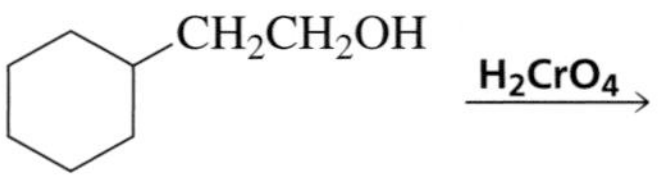

d. 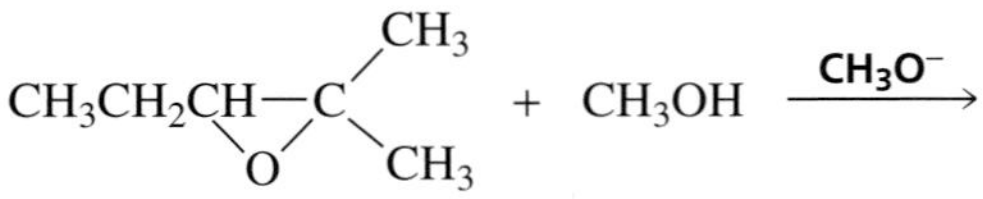 $CH_3CH_2CH\overset{O}{—}C(CH_3)_2 + CH_3OH \xrightarrow{CH_3O^-}$

e. 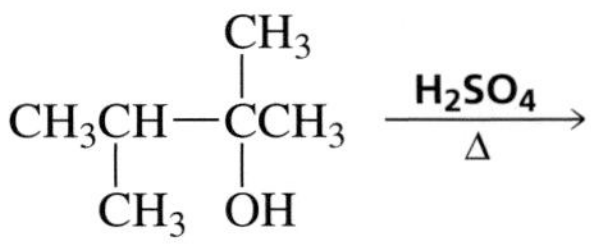 $CH_3\underset{CH_3}{\underset{|}{C}}H—\overset{CH_3}{\overset{|}{\underset{OH}{\underset{|}{C}}}}CH_3 \xrightarrow[\Delta]{H_2SO_4}$

f. 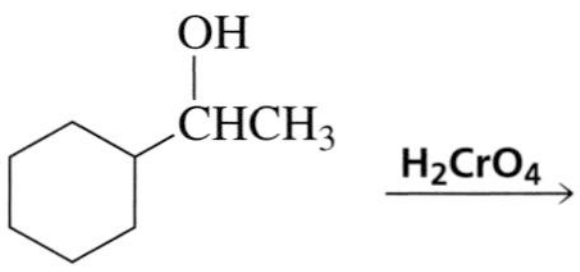 cyclohexyl–$\overset{OH}{\overset{|}{C}}HCH_3 \xrightarrow{H_2CrO_4}$

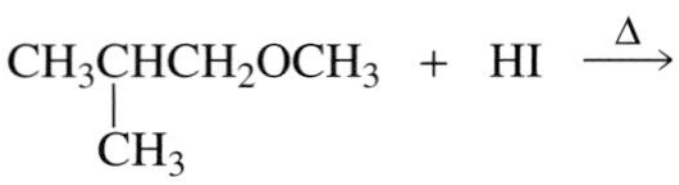

23. Give common and systematic names for each of the following:

a. $CH_3CHOCH_2CH_2CH_3$
$\quad\quad | $
$\quad CH_3$

b. $CH_3CH_2CH_2CH_2OCH_2CH_3$

c. $CH_3CH_2CHOCH_3$
$\quad\quad\quad | $
$\quad\quad CH_3$

d. $CH_3CHOCHCH_3$
$\quad | \quad | $
$\quad CH_3\ CH_3$

24. Which alcohol in each pair will undergo dehydration more rapidly when heated with H_2SO_4?

a. 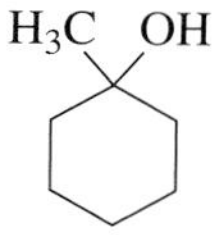 1-methylcyclohexanol (ring carbon bearing H_3C and OH) or 2-methylcyclohexanol (ring with CH_3 and adjacent OH)

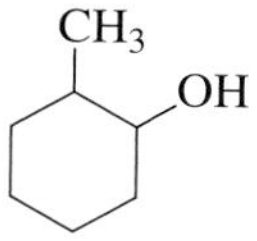

b. 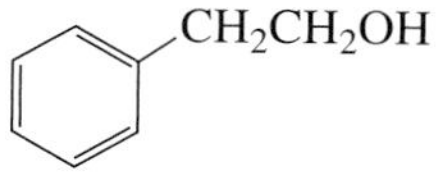 phenyl–CH_2CH_2OH or phenyl–$CHCH_3$ (with OH on the CH)

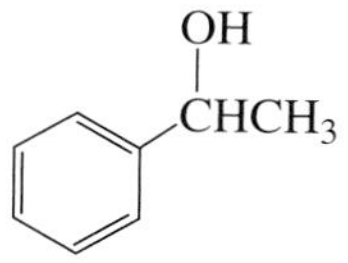

c. $CH_3CH_2CHCH_3$ (OH on C-2) or $CH_3CCH_2CH_3$ (with CH_3 and OH on C-2)

d. 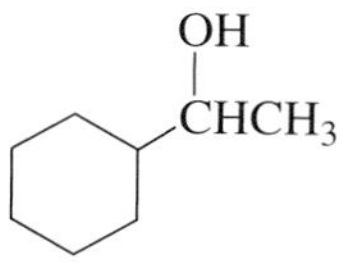 cyclohexyl–$CHCH_3$ (with OH on the CH) or phenyl–$CHCH_3$ (with OH on the CH)

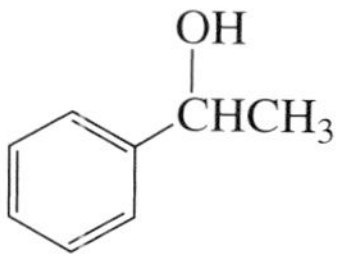

25. Name each of the following:

a. $CH_3CH_2CHOCH_2CH_3$ (with $CH_2CH_2CH_2CH_3$ on the CH)

b. 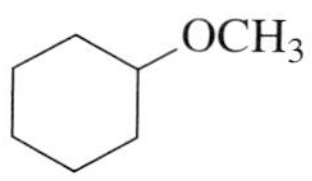 cyclohexyl–OCH_3

c. $CH_3CHCH_2CH_2CH_2OH$ (with CH_3 on C-2)

d. $CH_3CHOCH_2CH_2CHCH_3$ (with CH_3 on each CH)

e. 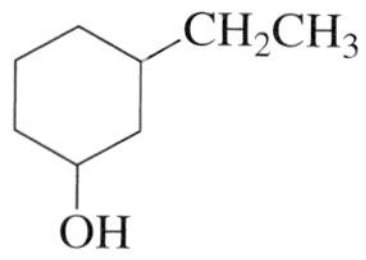 cyclohexane ring bearing CH_2CH_3 and OH (1,3-positions)

f. $CH_3CHOCHCH_2CH_2CH_3$ (with CH_3 on each CH)

26. Using the given starting material, any necessary inorganic reagents, and any carbon-containing compounds with no more than two carbon atoms, indicate how the following syntheses could be carried out:

a. 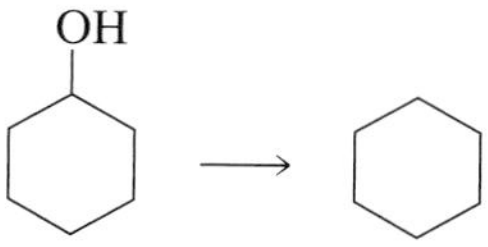 cyclohexanol $\longrightarrow$ cyclohexane

b. $CH_3CH_2C{\equiv}CH \longrightarrow CH_3CH_2C{\equiv}CCH_2CH_3$

c. $CH_3CH_2C{\equiv}CH \longrightarrow CH_3CH_2C{\equiv}CCH_2CH_2OH$

27. Draw structures for the following:

a. diisopropyl ether

b. allyl vinyl ether

c. *sec*-butyl isobutyl ether

d. benzyl phenyl ether

28. If any of the ethers in Problem 27 can exist as stereoisomers, draw the stereoisomers.

29. Give the product of each of the following reactions:

a. 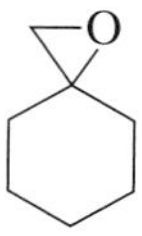 cyclohexane spiro-epoxide $\xrightarrow[CH_3OH]{CH_3O^-}$

b. $CH_3COCH_2CH_3$ (with two CH_3 on the C) $+$ HBr $\xrightarrow{\Delta}$

c. $CH_3CH_2CHCCH_3$ (OH on the CH; two CH_3 on the C) $\xrightarrow[\Delta]{H_2SO_4}$

d. $CH_3CHCH_2OCH_3$ (with CH_3 on the CH) $+$ HI $\xrightarrow{\Delta}$

e. 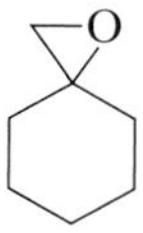 cyclohexane spiro-epoxide $\xrightarrow[CH_3OH]{H^+}$

f. $CH_3CH_2CHOCCH_3$ (CH_3 on the CH; two CH_3 on the C) $+$ HI $\xrightarrow{\Delta}$

g. 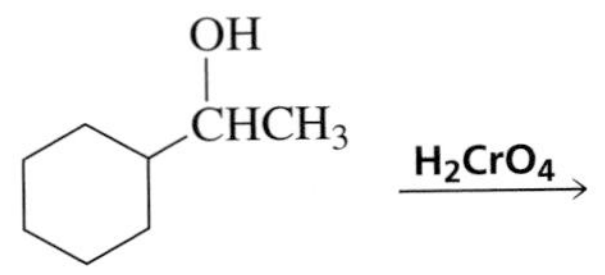 (cyclohexyl)$CH(OH)CH_3$ $\xrightarrow{H_2CrO_4}$

h. (cyclohexyl)CH_2CH_2OH $\xrightarrow{H_2CrO_4}$

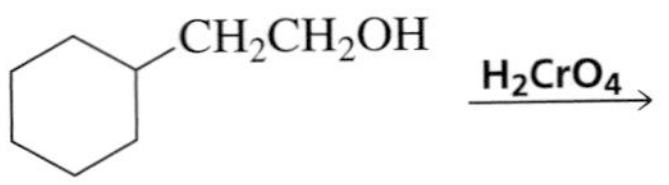

30. Draw structures for the following:
- **a.** *trans*-4-methylcyclohexanol
- **b.** 3-ethoxy-1-propanol

31. Give the product formed from the reaction of each of the following compounds with chromic acid:
- **a.** 3-methyl-2-pentanol
- **b.** butanol
- **c.** 2-methylcyclohexanol

32. Which alcohol in each pair will undergo dehydration more rapidly when heated with H_2SO_4?

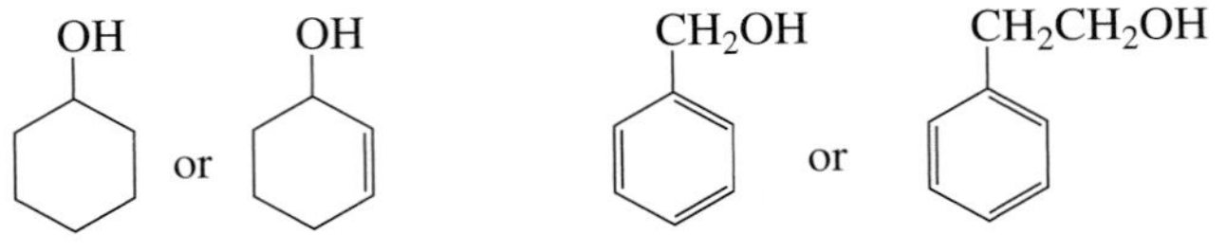

OH (cyclohexanol) or OH (cyclohexenol) $\qquad$ CH_2OH (benzyl) or CH_2CH_2OH (phenyl)

33. Propose a mechanism for the following reaction:

$$CH_3CH(Cl)CH\overset{O}{—}CH_2 + CH_3O^- \xrightarrow{CH_3OH} CH_3CH\overset{O}{—}CHCH_2OCH_3 + Cl^-$$

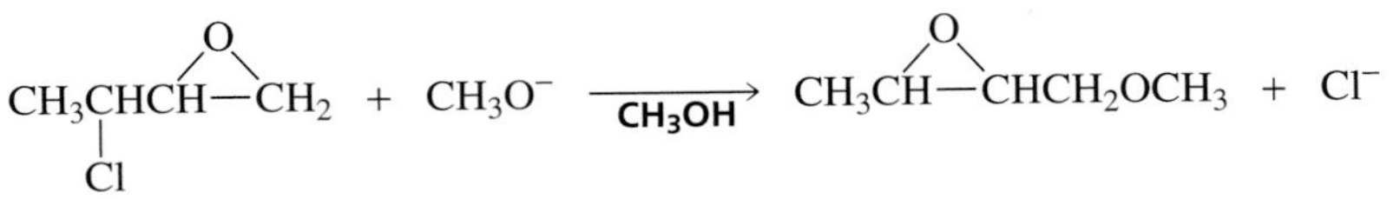

34. The observed relative reactivities of primary, secondary, and tertiary alcohols with a hydrogen halide are $3° > 2° > 1°$. If a secondary alcohol underwent an S_N2 reaction rather than an S_N1 reaction with a hydrogen halide, what would be the relative reactivities of the three classes of alcohols?

35. Give the major product expected from the reaction of 1,2-epoxybutane with each of the following reagents:
- **a.** 0.1 M HCl
- **b.** CH_3OH/H^+
- **c.** CH_3OH/CH_3O^-
- **d.** 0.1 M NaOH

36. Name each of the following:

a. 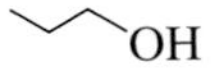 OH $\qquad$ b. 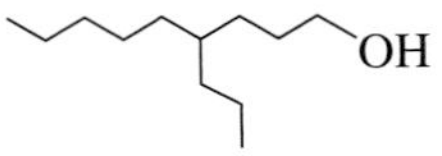 OH $\qquad$ c. 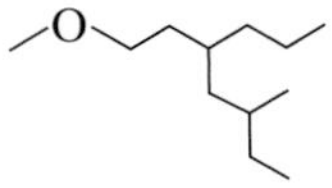 O

37. When ethyl ether is heated with excess HI for several hours, the only organic product obtained is ethyl iodide. Explain why ethyl alcohol is not obtained as a product.

38. Ethylene oxide reacts readily with HO^- because of the strain in the three-membered ring. Explain why cyclopropane, with approximately the same amount of strain, does not react with HO^-.

39. Propose a mechanism for each of the following reactions:

a. 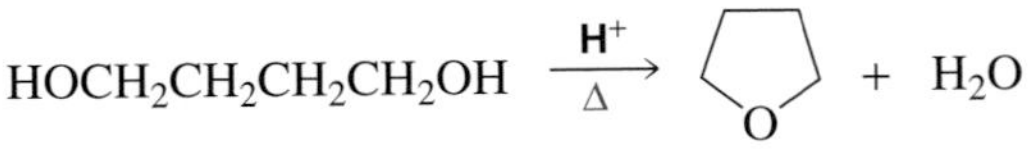 $HOCH_2CH_2CH_2CH_2OH \xrightarrow[\Delta]{H^+}$ (tetrahydrofuran, O) $+ H_2O$

b. 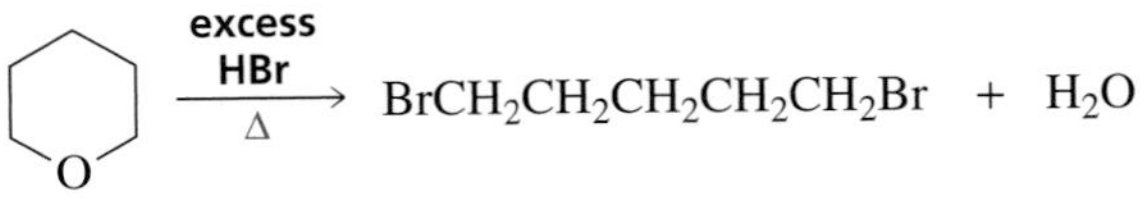 (tetrahydropyran, O) $\xrightarrow[\Delta]{\text{excess HBr}} BrCH_2CH_2CH_2CH_2CH_2Br + H_2O$

40. Triethylene glycol is one of the products obtained from the reaction of ethylene oxide and hydroxide ion. Propose a mechanism for its formation.

$$H_2C\overset{O}{—}CH_2 + HO^- \longrightarrow \underset{\text{triethylene glycol}}{HOCH_2CH_2OCH_2CH_2OCH_2CH_2OH}$$

41. Explain why the major product obtained from the acid-catalyzed dehydration of 1-butanol is 2-butene.

42. What alkenes would you expect to be obtained from the acid-catalyzed dehydration of 1-hexanol?

43. Propose a mechanism for the following reaction:

$\xrightarrow[H_2O]{H^+}$

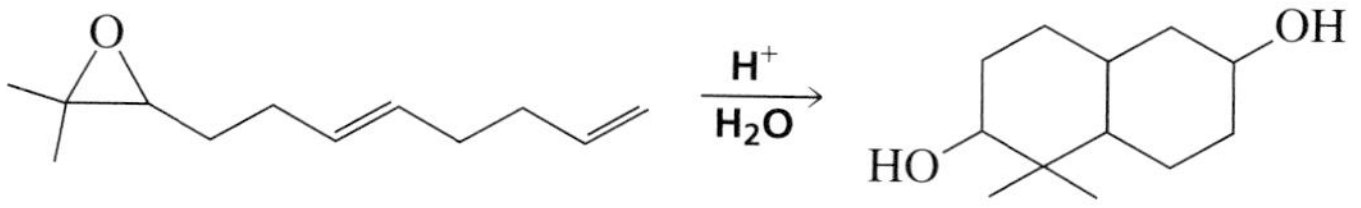

44. Which of the following ethers would be obtained in greatest yield directly from alcohols?

$CH_3OCH_2CH_2CH_3$ $\quad CH_3CH_2OCH_2CH_2CH_3$ $\quad CH_3CH_2OCH_2CH_3$ $\quad CH_3OC(CH_3)_2CH_3$

45. Explain why (*S*)-2-butanol forms a racemic mixture when it is heated in sulfuric acid.

46. Two stereoisomers are obtained from the reaction of cyclopentene oxide and dimethylamine. The *R,R*-isomer is used in the manufacture of eclanamine, an antidepressant. What other stereoisomer is obtained?

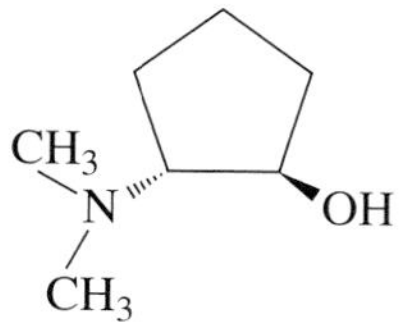

R,R-isomer

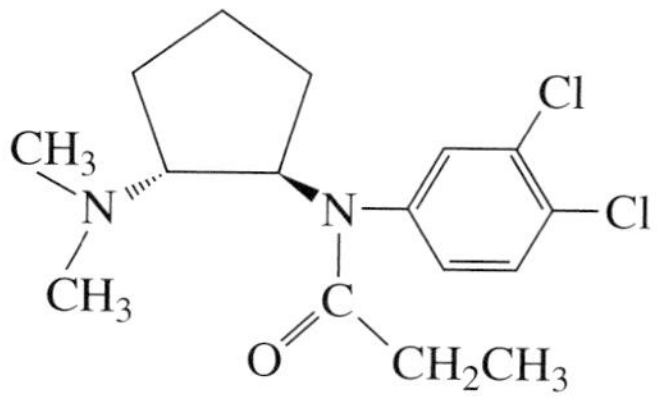

eclanamine

47. Explain why more 1-naphthol than 2-naphthol is obtained from the rearrangement of naphthalene oxide.

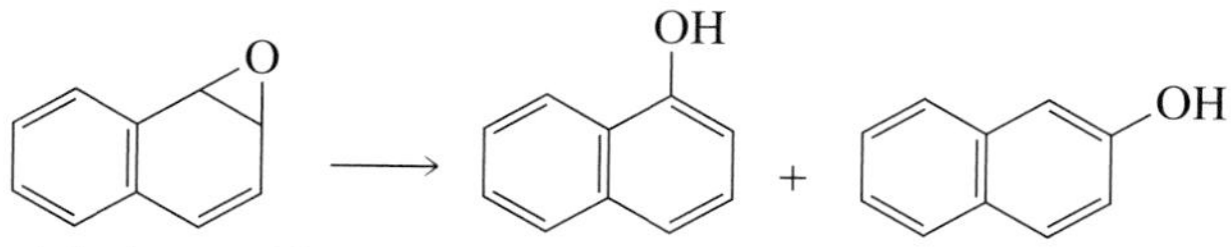

naphthalene oxide → 1-naphthol 90% + 2-naphthol 10%

48. Three arene oxides can be obtained from phenanthrene.

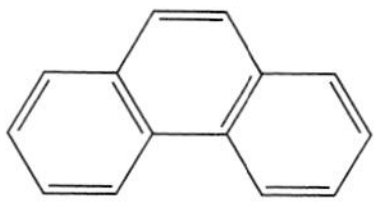

phenanthrene

a. Give the structures of the three phenanthrene oxides.
b. What phenols can be obtained from each phenanthrene oxide?
c. If a phenanthrene oxide can lead to the formation of more than one phenol, which phenol is obtained in greater yield?
d. Which of the three phenanthrene oxides is most likely to be carcinogenic?

49. a. Propose a mechanism for the following reaction:

$CH_2CH_2CH_2Br$ $\xrightarrow{CH_3O^-}$ CH_2OCH_3 $+ \ Br^-$

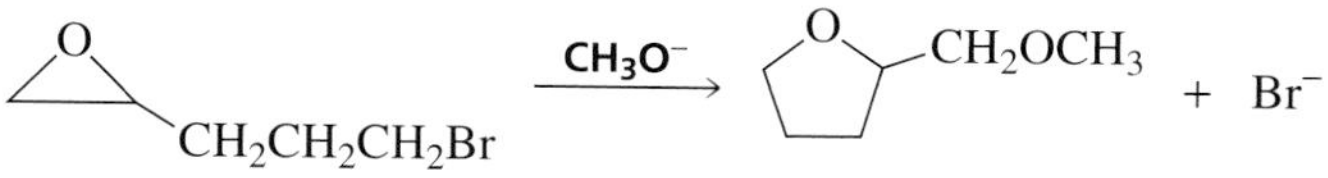

b. A small amount of a product containing a six-membered ring is also formed. Give the structure of that product.
c. Why is so little six-membered ring product formed?

50. Show how each of the following compounds could be prepared from bromocyclohexane.

a. 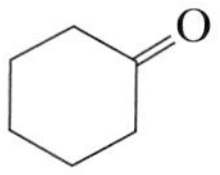 O **b.** 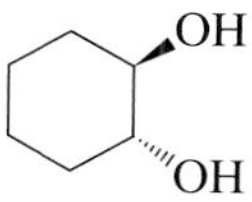 OH, OH

CHAPTER 11

Carbonyl Compounds I

Nucleophilic Acyl Substitution

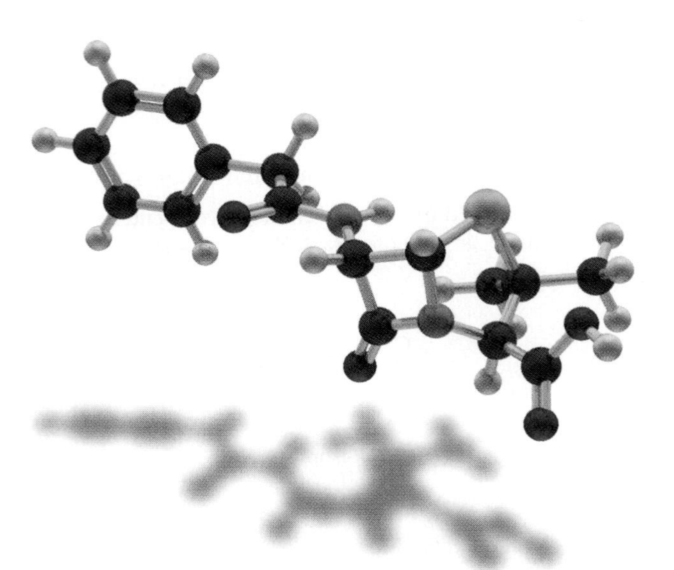

Penicillin G

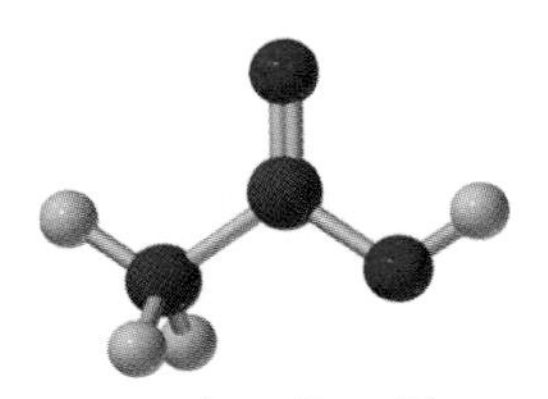

a carboxylic acid

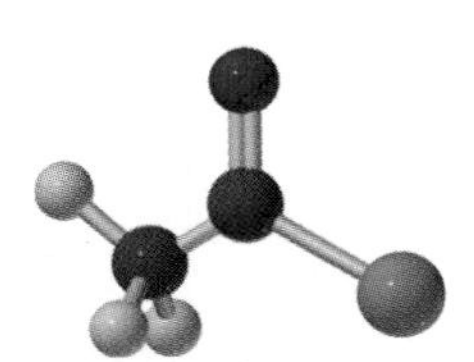

an acyl chloride

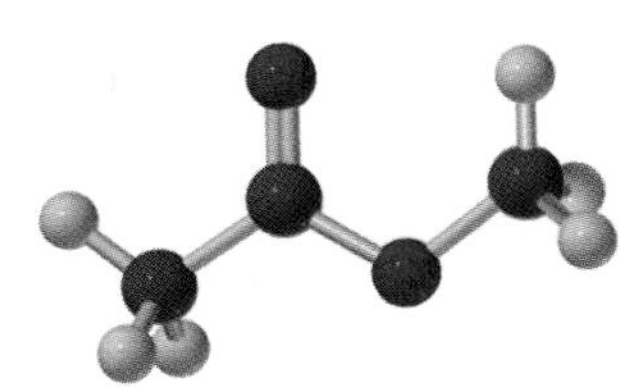

an ester

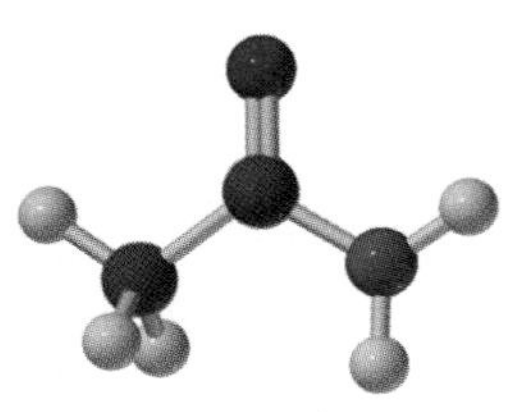

an amide

The **carbonyl group**—a carbon double bonded to an oxygen—is probably the most important functional group. Compounds containing carbonyl groups—called **carbonyl compounds**—are abundant in nature. Many play important roles in biological processes. Hormones, vitamins, amino acids, proteins, drugs, and flavorings are just a few of the carbonyl compounds that affect us daily.

An **acyl group** consists of a carbonyl group attached to an alkyl group (R) or an aryl group (Ar).

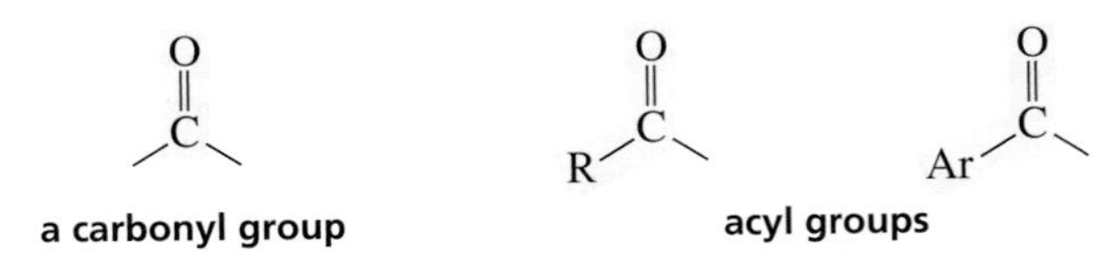

a carbonyl group

acyl groups

The group (or atom) attached to the acyl group strongly affects the reactivity of the carbonyl compound. In fact, carbonyl compounds can be divided into two classes determined by that group: Class I carbonyl compounds are those in which the acyl group is attached to a group (or atom) that *can* be replaced by another group. Carboxylic acids, acyl chlorides, esters, and amides belong to this class. All of these compounds contain a group (OH, Cl, OR, NH_2, NHR, NR_2) that can be replaced by a nucleophile.

carbonyl compounds with groups that can be replaced by a nucleophile

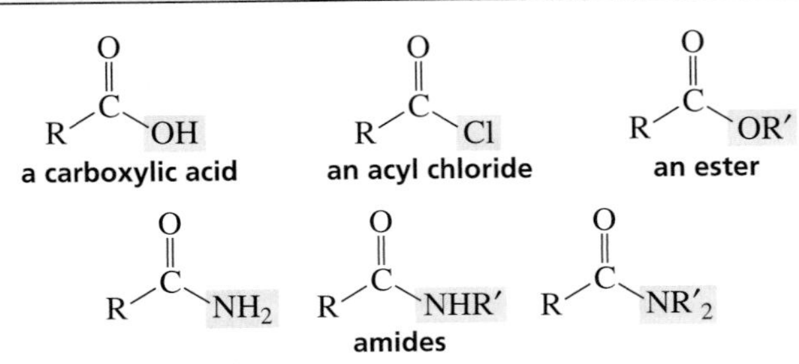

a carboxylic acid

an acyl chloride

an ester

amides

Acyl chlorides, esters, and amides are all called **carboxylic acid derivatives** because they differ from a carboxylic acid only in the nature of the group that has replaced the OH group of the carboxylic acid.

Class II carbonyl compounds are those in which the acyl group is attached to a group that *cannot* be readily replaced by another group. Aldehydes and ketones belong to this class. The H bonded to the acyl group of an aldehyde and the R bonded to the acyl group of a ketone cannot be readily replaced by a nucleophile.

cannot be replaced by a nucleophile

$$\underset{\text{an aldehyde}}{R-\overset{O}{\overset{\|}{C}}-H} \qquad \underset{\text{a ketone}}{R'-\overset{O}{\overset{\|}{C}}-R}$$

This chapter discusses the reactions of Class I carbonyl compounds. We will see that these compounds undergo substitution reactions, because they have an acyl group attached to a group that can be replaced (substituted) by a nucleophile. The reactions of Class II carbonyl compounds—aldehydes and ketones—will be considered in Chapter 12, where we will see that these compounds *do not* undergo substitution reactions because their acyl group is attached to a group that *cannot* be replaced by a nucleophile.

11.1 THE NOMENCLATURE OF CARBOXYLIC ACIDS AND CARBOXYLIC ACID DERIVATIVES

We will look at the names of carboxylic acids first because their names form the basis of the names of other carbonyl compounds.

Naming Carboxylic Acids

The functional group of a carboxylic acid is called a **carboxyl group**.

$$\underset{\text{a carboxyl group}}{-\overset{O}{\overset{\|}{C}}-OH} \qquad -COOH \qquad -CO_2H$$

carboxyl groups are frequently shown in abbreviated forms

In systematic (IUPAC) nomenclature, a **carboxylic acid** is named by replacing the terminal "e" of the alkane name with "oic acid." For example, the one-carbon alkane is methan*e*, so the one-carbon carboxylic acid is methan*oic acid*.

	$H-\overset{O}{\overset{\|}{C}}-OH$	$CH_3-\overset{O}{\overset{\|}{C}}-OH$	$CH_3CH_2-\overset{O}{\overset{\|}{C}}-OH$	$CH_3CH_2CH_2-\overset{O}{\overset{\|}{C}}-OH$
systematic name:	methanoic acid	ethanoic acid	propanoic acid	butanoic acid
common name:	formic acid	acetic acid	propionic acid	butyric acid

$CH_3CH_2CH_2CH_2-\overset{O}{\overset{\|}{C}}-OH$	$CH_3CH_2CH_2CH_2CH_2-\overset{O}{\overset{\|}{C}}-OH$
pentanoic acid	hexanoic acid
valeric acid	caproic acid

Carboxylic acids containing six or fewer carbons are frequently called by their common names. These names were chosen by early chemists to describe some feature of the compound, usually its origin. For example, formic acid is found in ants, bees, and other stinging insects; its name comes from *formica*, which is Latin for "ant." Acetic acid—contained in vinegar—got its name from *acetum*, the Latin word for "vinegar." Propionic acid is the smallest acid that shows some of the characteristics of the larger fatty acids (Section 19.1); its name comes from the Greek words *pro* ("the first") and *pion* ("fat"). Butyric acid is found in rancid butter; the Latin word for "butter" is *butyrum*. Caproic acid is found in goat's milk; if you have the occasion to smell both a goat and caproic acid, you will find that they have similar odors. *Caper* is the Latin word for "goat."

In systematic (IUPAC) nomenclature, the position of a substituent is designated by a number. The carbonyl carbon is always the C-1 carbon. In common nomenclature, the position of a substituent is designated by a lowercase Greek letter, and the carbonyl carbon is not given a designation. The carbon adjacent to the carbonyl carbon is the **α-carbon**, the carbon adjacent to the α-carbon is the β-carbon, and so on.

α = alpha
β = beta
γ = gamma
δ = delta
ε = epsilon

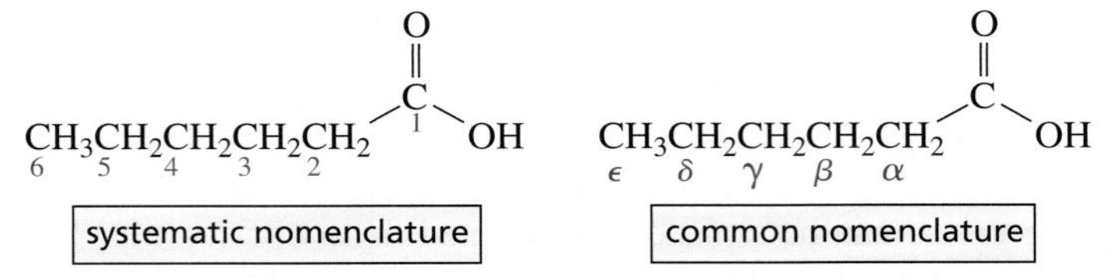

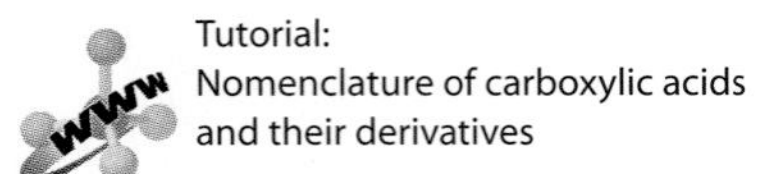

Tutorial:
Nomenclature of carboxylic acids and their derivatives

Take a careful look at the following examples to make sure that you understand the difference between systematic and common nomenclature:

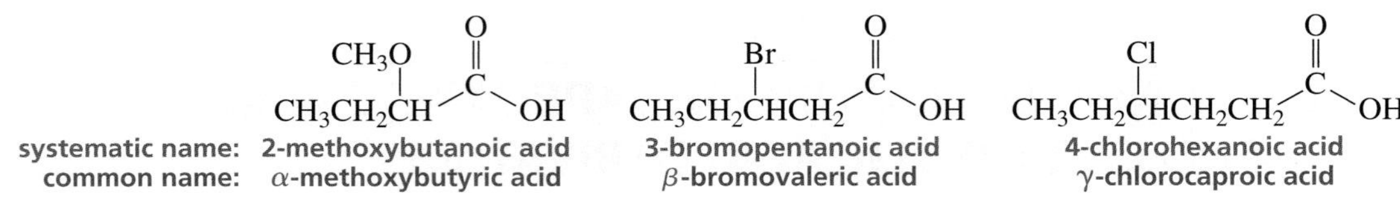

Naming Acyl Chlorides

Acyl chlorides have a Cl in place of the OH group of a carboxylic acid. They are named by taking the acid name and replacing "ic acid" with "yl chloride."

CH_3COCl — systematic name: ethanoyl chloride; common name: acetyl chloride

$CH_3CH_2CH(CH_3)CH_2COCl$ — systematic name: 3-methylpentanoyl chloride; common name: β-methylvaleryl chloride

Naming Esters

An **ester** has an OR group in place of the OH group of a carboxylic acid. In naming an ester, the name of the group (R) attached to the **carboxyl oxygen** is stated first, followed by the name of the acid with "ic acid" replaced by "ate."

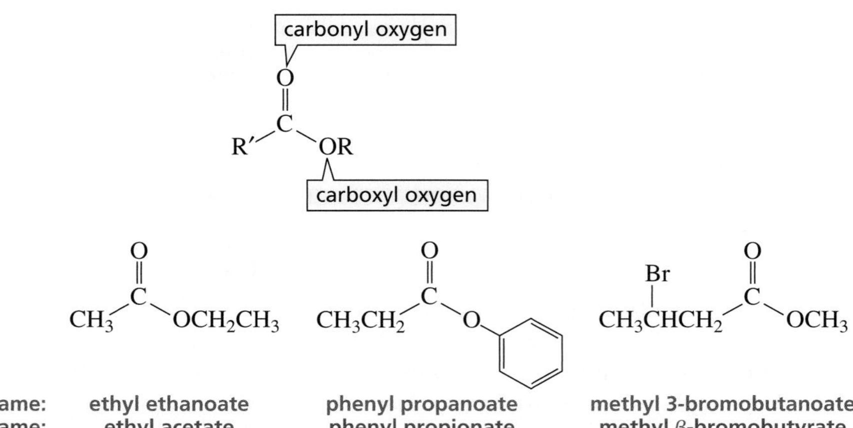

Salts of carboxylic acids are named in the same way. The cation is named first, followed by the name of the acid with "ic acid" replaced by "ate."

$HC(=O)O^- Na^+$ $\qquad$ $CH_3C(=O)O^- K^+$

systematic name:	sodium methanoate	potassium ethanoate
common name:	sodium formate	potassium acetate

Naming Amides

An **amide** has an NH_2, NHR, or NR_2 group in place of the OH group of a carboxylic acid. Amides are named by taking the acid name and replacing "oic acid" or "ic acid" with "amide."

systematic name:	ethanamide	4-chlorobutanamide
common name:	acetamide	γ-chlorobutyramide

3-D Molecules: Acetyl chloride; Methyl acetate; Acetic acid; Acetamide

If a substituent is bonded to the nitrogen, the name of the substituent is stated first (if there is more than one substituent bonded to the nitrogen, they are stated alphabetically), followed by the name of the amide. The name of each substituent is preceded by a capital *N* to indicate that the substituent is bonded to a nitrogen.

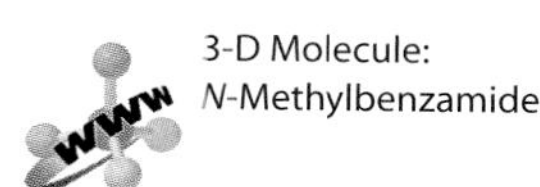

3-D Molecule: *N*-Methylbenzamide

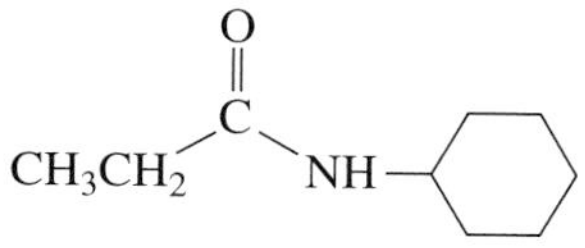

$CH_3CH_2C(=O)NH$–cyclohexyl $\qquad$ $CH_3CH_2CH_2CH_2C(=O)N(CH_3)CH_2CH_3$

N-cyclohexylpropanamide 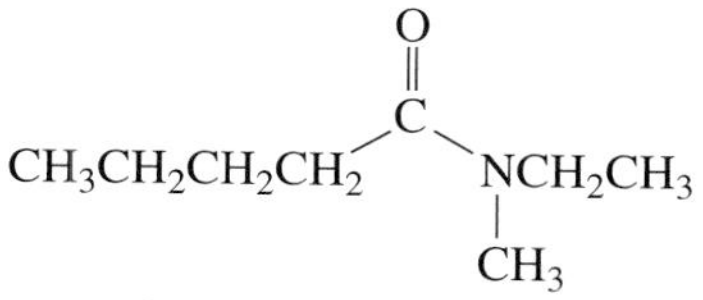 $\qquad$ *N*-ethyl-*N*-methylpentanamide

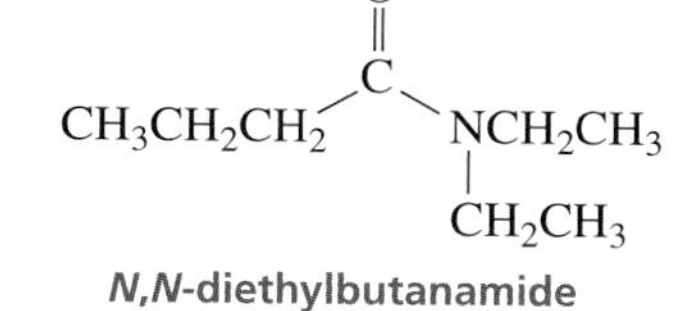

N,N-diethylbutanamide

PROBLEM 1 ◆

Name the following:

a. $CH_3CH_2\overset{O}{\overset{\|}{C}}NH_2$

b. $CH_3CH_2CH_2\overset{O}{\overset{\|}{C}}OCH_2\underset{}{\overset{CH_3}{\overset{|}{C}}}HCH_3$

c. 

d. $CH_3CH_2CH_2CH_2\overset{O}{\overset{\|}{C}}Cl$

e. $CH_3CH_2CH_2CH_2CH_2\overset{O}{\overset{\|}{C}}N(CH_3)_2$

f. $CH_3CH_2\underset{CH_3}{\underset{|}{C}}H\overset{O}{\overset{\|}{C}}NHCH_2CH_3$

PROBLEM 2 ◆

Write the structure of each of the following:

a. phenyl acetate
b. sodium acetate
c. *N*-benzylethanamide
d. ethyl 2-chloropentanoate
e. β-bromobutyramide
f. propanoyl chloride

11.2 THE STRUCTURES OF CARBOXYLIC ACIDS AND CARBOXYLIC ACID DERIVATIVES

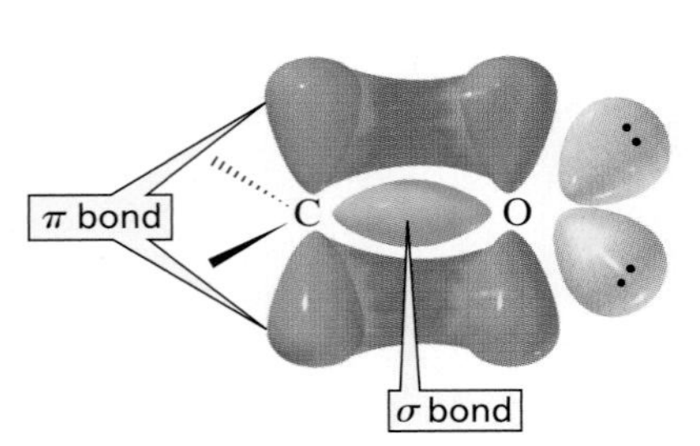

▲ **Figure 11.1**
Bonding in a carbonyl group. The π bond is formed by side-to-side overlap of a *p* orbital of carbon with a *p* orbital of oxygen.

The **carbonyl carbon** in carboxylic acids and carboxylic acid derivatives is sp^2 hybridized. It uses its three sp^2 orbitals to form σ bonds to the carbonyl oxygen, the α-carbon, and a substituent (Y). The three atoms attached to the carbonyl carbon are in the same plane, and the bond angles are each approximately 120°.

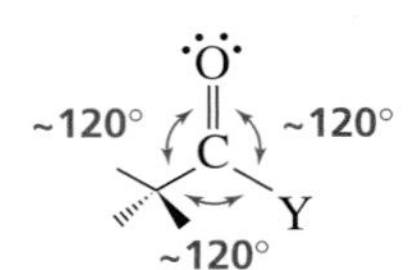

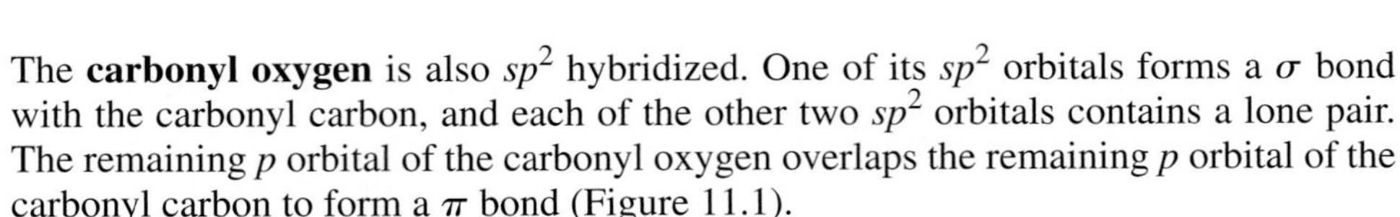

The **carbonyl oxygen** is also sp^2 hybridized. One of its sp^2 orbitals forms a σ bond with the carbonyl carbon, and each of the other two sp^2 orbitals contains a lone pair. The remaining *p* orbital of the carbonyl oxygen overlaps the remaining *p* orbital of the carbonyl carbon to form a π bond (Figure 11.1).

Esters, carboxylic acids, and amides each have two major resonance contributors.

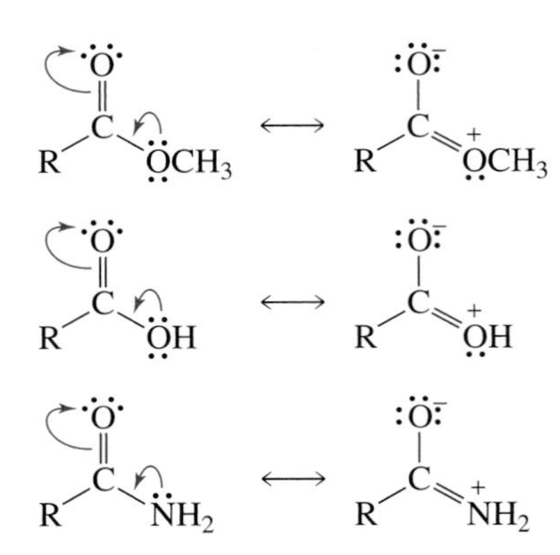

The resonance contributor on the right makes a greater contribution to the hybrid in the amide than in the ester or the carboxylic acid because the less electronegative nitrogen atom can better accommodate a positive charge.

PROBLEM 3 ♦

Which is longer, the carbon–oxygen single bond in a carboxylic acid or the carbon–oxygen bond in an alcohol? Why?

CH_3COOH (O=C(CH$_3$)OH) $\qquad$ $CH_3{-}OH$

PROBLEM 4 ♦

What are the relative lengths of the three carbon–oxygen bonds in methyl acetate?

$CH_3C(=O){-}O{-}CH_3$

11.3 THE PHYSICAL PROPERTIES OF CARBONYL COMPOUNDS

The acid properties of carboxylic acids were discussed in Sections 2.3 and 7.9. Recall that carboxylic acids have pK_a values of approximately 3–5 (Appendix II). The boiling points and other physical properties of carbonyl compounds are listed in Appendix I. Carbonyl compounds have the following relative boiling points:

relative boiling points

amide > carboxylic acid >> ester ~ acyl chloride ~ aldehyde ~ ketone

The boiling points of an ester, acyl chloride, ketone, and aldehyde are similar and are lower than the boiling point of an alcohol with a comparable molecular weight because only the alcohol molecules can form hydrogen bonds with each other. The boiling points of these four carbonyl compounds are higher than the boiling point of the same-size ether because of the polar carbonyl group.

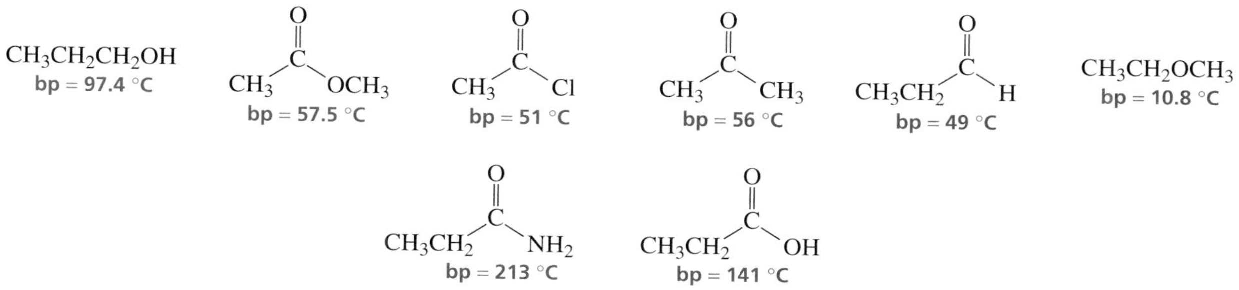

$CH_3CH_2CH_2OH$ bp = 97.4 °C

CH_3COOCH_3 bp = 57.5 °C

CH_3COCl bp = 51 °C

CH_3COCH_3 bp = 56 °C

CH_3CH_2CHO bp = 49 °C

$CH_3CH_2OCH_3$ bp = 10.8 °C

$CH_3CH_2CONH_2$ bp = 213 °C

CH_3CH_2COOH bp = 141 °C

Carboxylic acids have relatively high boiling points because each molecule can form two hydrogen bonds.

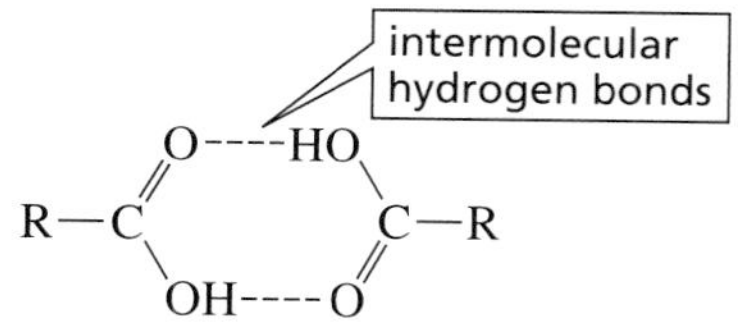

intermolecular hydrogen bonds

Amides have the highest boiling points, because they have strong dipole–dipole interactions, since the resonance contributor with separated charges contributes significantly to the overall structure of the compound (Section 11.2). In addition, if the nitrogen of an amide is bonded to a hydrogen, hydrogen bonds will form between the molecules.

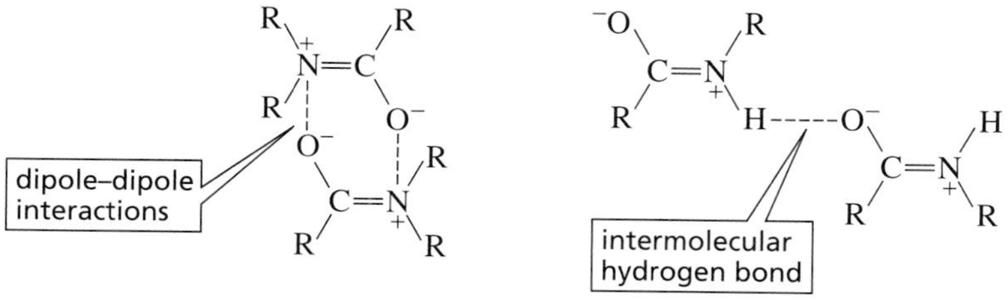

dipole–dipole interactions

intermolecular hydrogen bond

Carboxylic acid derivatives are soluble in solvents such as ethers, chlorinated alkanes, and aromatic hydrocarbons. Like alcohols and ethers, carbonyl compounds with fewer than about four carbons per oxygen atom are soluble in water (Section 3.7).

11.4 CARBOXYLIC ACIDS AND CARBOXYLIC ACID DERIVATIVES FOUND IN NATURE

Acyl halides are much more reactive than carboxylic acids and esters, which, in turn, are more reactive than amides. We will see the reason for these differences in reactivity in Section 11.6.

Because of their high reactivity, acyl halides are not found in nature. Carboxylic acids, on the other hand, being less reactive *are* found widely in nature. For example, glucose is metabolized to pyruvic acid (Section 18.4). (*S*)-(+)-Lactic acid is the compound responsible for the burning sensation felt in muscles during anaerobic exercise, and it is also found in sour milk. Spinach and other leafy green vegetables are rich in oxalic acid. Succinic acid and citric acid are important intermediates in the citric acid cycle, a series of reactions in biological systems that oxidize carbohydrates, fatty acids, and amino acids to CO_2 (Section 18.7). Citrus fruits are rich in citric acid; the concentration is greatest in lemons, less in grapefruit, and still less in oranges.

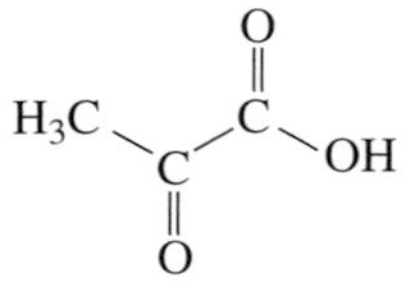

pyruvic acid

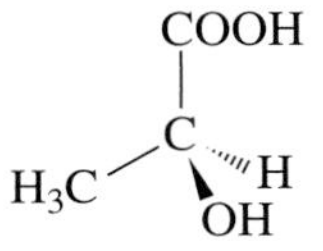

(*S*)-(+)-lactic acid

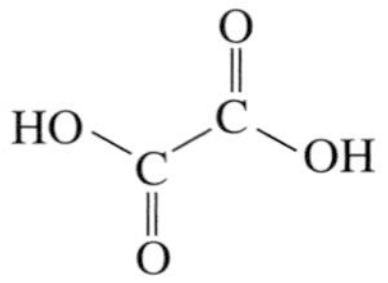

oxalic acid

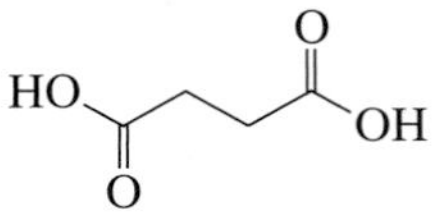

succinic acid

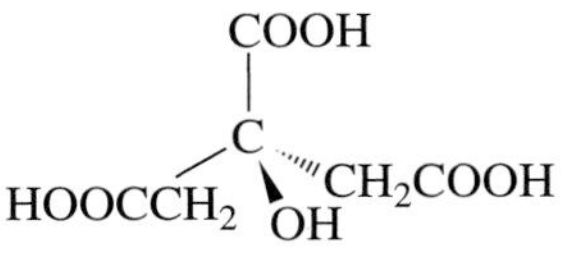

citric acid

(*S*)-(-)-Malic acid is responsible for the sharp taste of unripe apples and pears. As these fruits ripen, the amount of malic acid decreases and the amount of sugar increases. The inverse relationship between the levels of malic acid and sugar is important for the propagation of the plant: malic acid prevents animals from eating the fruit until it becomes ripe, at which time its seeds are mature enough to germinate when they are scattered about. Prostaglandins are locally acting hormones that have several different physiological functions (Section 11.9), such as regulating inflammation, blood pressure, pain, and fever.

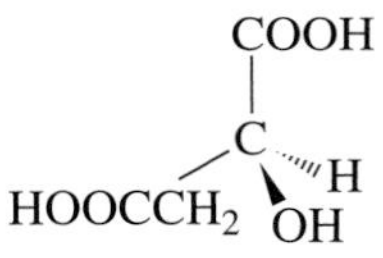

(*S*)-(–)-malic acid

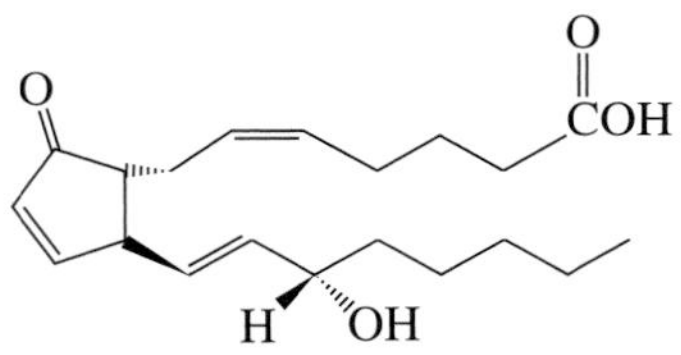

prostaglandin A_2

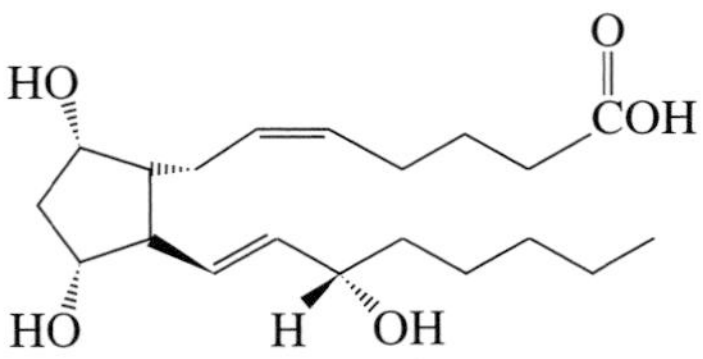

prostaglandin $F_{2\alpha}$

Esters are also commonly found in nature. The aromas of many flowers and fruits are due to esters. (See Problems 19 and 30.)

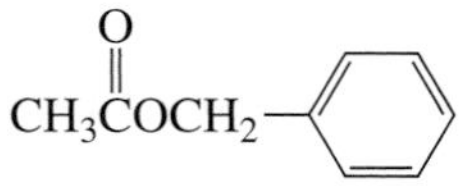

benzyl acetate
jasmine

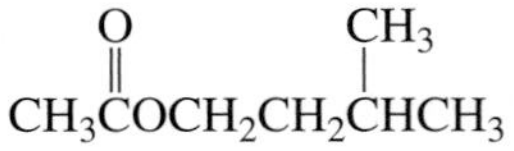

isopentyl acetate
banana

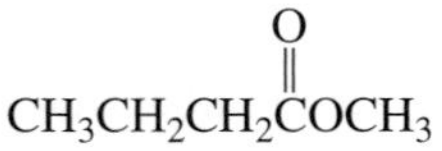

methyl butyrate
apple

Carboxylic acids with an amino group on the α-carbon are commonly called **amino acids**. Amino acids are linked together by amide bonds to form peptides and proteins (Section 16.0).

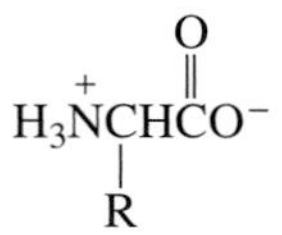

amide bonds

an amino acid

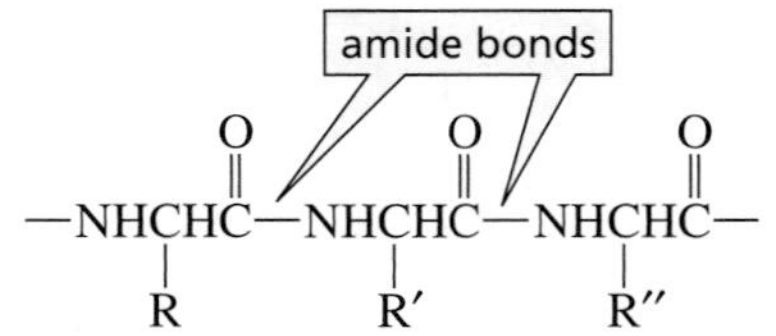

general structure for a peptide or a protein

Caffeine, another naturally occurring amide, is found in cocoa and in coffee beans. Penicillin G has two amide bonds; the four-membered ring amide is the reactive part of the molecule (Section 11.13).

caffeine

piperine
the major component of black pepper

penicillin G

THE DISCOVERY OF PENICILLIN

Sir Alexander Fleming (1881–1955), born in Scotland, was a professor of bacteriology at University College, London. The story is told that one day Fleming was about to throw away a culture of staphylococcal bacteria that had been contaminated by a rare strain of the mold *Penicillium notatum*. He noticed that the bacteria had disappeared wherever there was a particle of mold. This suggested to him that the mold must have produced an antibacterial substance. Ten years later, Howard Florey and Ernest Chain isolated the active substance—penicillin G (Section 11.13)—but the delay allowed sulfa drugs to be the first antibiotics (Section 21.4).

After penicillin G was found to cure bacterial infections in mice, it was used successfully in 1941 on nine cases of human bacterial infections. By 1943, penicillin G was being produced for the military and was first used for war casualties in Sicily and Tunisia. The drug became available to the civilian population in 1944. The pressure of the war made determination of penicillin G's structure a priority because once its structure was determined, large quantities of the drug could conceivably be synthesized.

Fleming, Florey, and Chain shared the 1945 Nobel Prize in physiology or medicine. Chain also discovered penicillinase, the enzyme that destroys penicillin (Section 11.13). Although Fleming is generally given credit for the discovery of penicillin, there is clear evidence that the germicidal activity of the mold was recognized in the nineteenth century by Lord Joseph Lister (1827–1912), the English physician renowned for the introduction of aseptic surgery.

Sir Alexander Fleming (1881–1955) *was born in Scotland, the seventh of eight children of a farmer. In 1902, he received a legacy from an uncle that, together with a scholarship, allowed him to study medicine at the University of London. He subsequently became a professor there in 1928. He was knighted in 1944.*

Sir Howard W. Florey (1898–1968) *was born in Australia and received a medical degree from the University of Adelaide. He went to England as a Rhodes Scholar and studied at both Oxford and Cambridge Universities. He became a professor of pathology at the University of Sheffield in 1931 and then at Oxford in 1935. Knighted in 1944, he was given a peerage in 1965 that made him Baron Florey of Adelaide.*

Ernest B. Chain (1906–1979) *was born in Germany and received a Ph.D. from Friedrich-Wilhelm University in Berlin. In 1933, he left Germany for England because Hitler had come to power. He studied at Cambridge, and in 1935 Florey invited him to Oxford. In 1948, he became the director of an institute in Rome, but he returned to England in 1961 to become a professor at the University of London.*

DALMATIANS: DON'T TRY TO FOOL MOTHER NATURE

When amino acids are metabolized, the excess nitrogen is concentrated into uric acid, a compound with five amide bonds. A series of enzyme-catalyzed reactions degrades uric acid all the way to ammonium ion. The extent to which uric acid is degraded in animals depends on the species. Birds, reptiles, and insects excrete excess nitrogen as uric acid. Mammals excrete excess nitrogen as allantoin. Excess nitrogen in aquatic animals is excreted as allantoic acid, urea, or ammonium salts.

uric acid
excreted by:
birds, reptiles, insects

enzyme →

allantoin

mammals

enzyme →

allantoic acid

marine vertebrates

enzyme →

urea

cartilaginous fish, amphibia

↓ enzyme

$^{+}NH_4X^{-}$
ammonium salt
marine invertebrates

(*continued...*)

Dalmatians, unlike other mammals, excrete high levels of uric acid. This is because breeders of Dalmatians have selected dogs that have no white hairs in their black spots, and the gene that causes the white hairs is linked to the gene that causes uric acid to be converted to allantoin. Dalmatians, therefore, are susceptible to gout, a painful buildup of uric acid in joints.

11.5 HOW CARBOXYLIC ACIDS AND CARBOXYLIC ACIDS COMPOUNDS REACT

The reactivity of carbonyl compounds is due to the polarity of the carbonyl group that results from oxygen being more electronegative than carbon. The carbonyl carbon is, therefore, electron deficient (an electrophile), so we can safely predict that it will be attacked by nucleophiles.

$$\underset{\text{R}\quad\delta+\quad\text{Y}}{\overset{\overset{\delta-}{\text{O}}}{\overset{\|}{\text{C}}}}$$

When a nucleophile attacks the carbonyl carbon of a carboxylic acid derivative, the weakest bond in the molecule—the carbon–oxygen π bond—breaks, and an intermediate is formed. The intermediate is called a **tetrahedral intermediate** because the sp^2 carbon in the reactant has become an sp^3 carbon in the intermediate. Generally, *a compound that has an sp^3 carbon bonded to an oxygen atom will be unstable if the sp^3 carbon is bonded to another electronegative atom.* The tetrahedral intermediate, therefore, is unstable because Y and Z are both electronegative atoms. A lone pair on the oxygen reforms the π bond, and either Y^- or Z^- is expelled with its bonding electrons.

A compound that has an sp^3 carbon bonded to an oxygen atom generally will be unstable if the sp^3 carbon is bonded to a second electronegative atom.

$$\text{R}-\overset{\overset{\text{O}}{\|}}{\text{C}}-\text{Y} + \text{Z}^- \rightleftharpoons \text{R}-\overset{\overset{\text{O}^-}{|}}{\underset{\underset{\text{Z}}{|}}{\text{C}}}-\text{Y} \rightleftharpoons \text{R}-\overset{\overset{\text{O}}{\|}}{\text{C}}-\text{Z} + \text{Y}^-$$

sp2

nucleophile attacks the carbonyl carbon

sp3

a group is expelled

a tetrahedral intermediate

sp2

Whether Y^- or Z^- is expelled from the tetrahedral intermediate depends on their relative basicities. The weaker base is expelled preferentially, making this another example of the principle we first saw in Section 9.3: *the weaker the base, the better it is as a leaving group.* Because a weak base does not share its electrons as well as a strong base does, a weaker base forms a weaker bond—one that is easier to break. If Z^- is a much weaker base than Y^-, Z^- will be expelled and the reaction can be written as follows:

$$\text{R}-\overset{\overset{\text{O}}{\|}}{\text{C}}-\text{Y} + \text{Z}^- \rightleftharpoons \text{R}-\overset{\overset{\text{O}^-}{|}}{\underset{\underset{\text{Z}}{|}}{\text{C}}}-\text{Y}$$

Z^- is a weaker base than Y^-, so Z^- is expelled

In this case, no new product is formed. The nucleophile attacks the carbonyl carbon, but the tetrahedral intermediate expels the attacking nucleophile and reforms the reactants.

On the other hand, if Y^- is a much weaker base than Z^-, Y^- will be expelled and a new product will be formed.

Y^- is a weaker base than Z^-, so Y^- is expelled

$$RC(=O)Y + Z^- \rightleftharpoons R-C(O^-)(Y)-Z \rightleftharpoons RC(=O)Z + Y^-$$

This reaction is called a **nucleophilic acyl substitution reaction** because a nucleophile (Z^-) has replaced the substituent (Y^-) that was attached to the acyl group in the reactant.

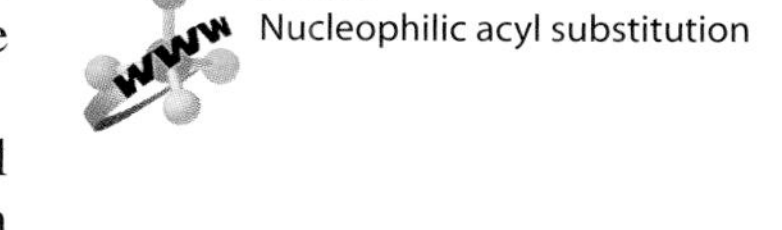

If the basicities of Y^- and Z^- are similar, some molecules of the tetrahedral intermediate will expel Y^- and others will expel Z^-. When the reaction is over, both reactant and product will be present.

the basicities of Y^- and Z^- are similar

$$RC(=O)Y + Z^- \rightleftharpoons R-C(O^-)(Y)-Z \rightleftharpoons RC(=O)Z + Y^-$$

We can, therefore, make the following general statement about the reactions of carboxylic acid derivatives: *a carboxylic acid derivative will undergo a nucleophilic acyl substitution reaction, provided that the newly added group in the tetrahedral intermediate is not a weaker base than the group that is attached to the acyl group in the reactant.*

Tutorial:
Free energy diagrams for nucleophilic acyl substitution reactions

PROBLEM-SOLVING STRATEGY

The pK_a of HCl is −7; the pK_a of CH_3OH is 15.5. What is the product of the reaction of acetyl chloride with CH_3O^-?

In order to determine what the product of the reaction will be, we need to compare the basicities of the two groups that will be in the tetrahedral intermediate in order to see which one will be eliminated. Because HCl is a stronger acid than CH_3OH, Cl^- is a weaker base than CH_3O^-. Cl^-, therefore, will be eliminated from the tetrahedral intermediate, so the product of the reaction will be methyl acetate.

$$CH_3C(=O)Cl + CH_3O^- \longrightarrow CH_3-C(O^-)(Cl)-OCH_3 \longrightarrow CH_3C(=O)OCH_3 + Cl^-$$

acetyl chloride → methyl acetate

Now continue on to Problem 5.

PROBLEM 5

a. The pK_a of HCl is −7; the pK_a of H_2O is 15.7. What is the product of the reaction of acetyl chloride with HO^-?

b. The pK_a of NH_3 is 36; the pK_a of H_2O is 15.7. What is the product of the reaction of acetamide with HO^-?

11.6 RELATIVE REACTIVITIES OF CARBOXYLIC ACIDS AND CARBOXYLIC ACID DERIVATIVES

We have just seen that there are two steps in a nucleophilic acyl substitution reaction: formation of a tetrahedral intermediate and collapse of the tetrahedral intermediate. The weaker the base attached to the acyl group, the easier it is for *both steps* of the reaction to take place. In other words, the reactivity of a carboxylic acid derivative depends on the basicity of the substituent attached to the acyl group: the less basic the substituent, the more reactive is the carboxylic acid derivative. (The pK_a values of the conjugate acids of the leaving groups of Class I carbonyl compounds are shown in Table 11.1.)

The weaker the base, the better it is as a leaving group.

relative basicities of the leaving groups

weakest base $Cl^- < {}^-OR \sim {}^-OH < {}^-NH_2$ strongest base

relative reactivities of carboxylic acid derivatives

most reactive — RCOCl (acyl chloride) > RCOOR′ (ester) ~ RCOOH (carboxylic acid) > RCONH$_2$ (amide) — least reactive

relative reactivity: acyl chloride > ester ~ carboxylic acid > amide

Table 11.1 The pK_a Values of the Conjugate Acids of the Leaving Groups of Carbonyl Compounds

Carbonyl compound	Leaving group	Conjugate acid of the leaving group	pK_a
Class I			
RCOCl	Cl^-	HCl	−7
RCOOR′	$^-OR'$	R′OH	~15–16
RCOOH	^-OH	H_2O	15.7
RCONH$_2$	$^-NH_2$	NH_3	36
Class II			
RCOH	H^-	H_2	~40
RCOR	R^-	RH	~60

How does having a weak base attached to the acyl group make the *first* step of the nucleophilic acyl substitution reaction easier? A weaker base is a more electronegative base (Section 2.6). Therefore, it is better at withdrawing electrons inductively from the carbonyl carbon, and electron withdrawal increases the carbonyl carbon's susceptibility to nucleophilic attack.

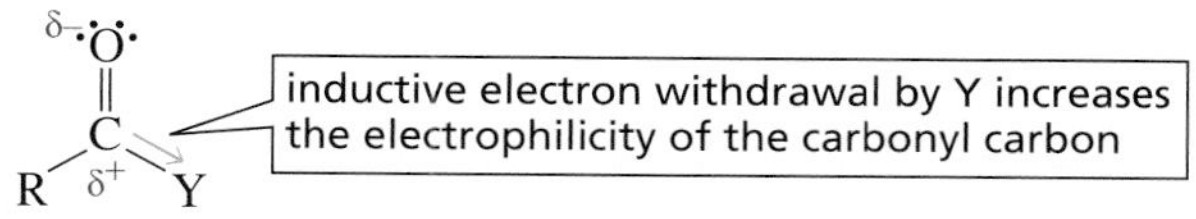

inductive electron withdrawal by Y increases the electrophilicity of the carbonyl carbon

A weak base attached to the acyl group also makes the *second* step of the nucleophilic acyl substitution reaction easier because weak bases—which form weak bonds—are easier to expel when the tetrahedral intermediate collapses.

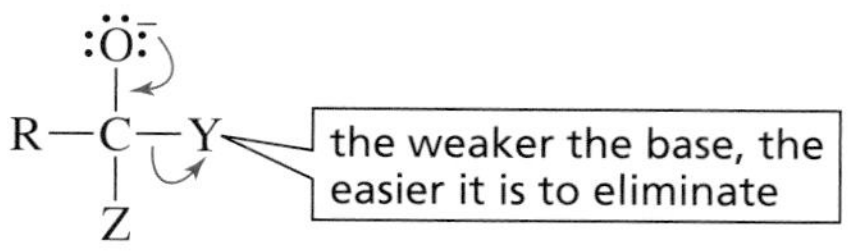

the weaker the base, the easier it is to eliminate

In Section 11.5, we saw that in a nucleophilic acyl substitution reaction, the nucleophile that forms the tetrahedral intermediate must not be a weaker base than the base that is already there. This means that *a carboxylic acid derivative can be converted into a less reactive carboxylic acid derivative, but not into one that is more reactive*. For example, an acyl chloride can be converted into an ester because an alkoxide ion, such as methoxide ion, is a stronger base than a chloride ion.

For a carboxylic acid derivative to undergo a nucleophilic acyl substitution reaction, the incoming nucleophile must not be a much weaker base than the group that is to be replaced.

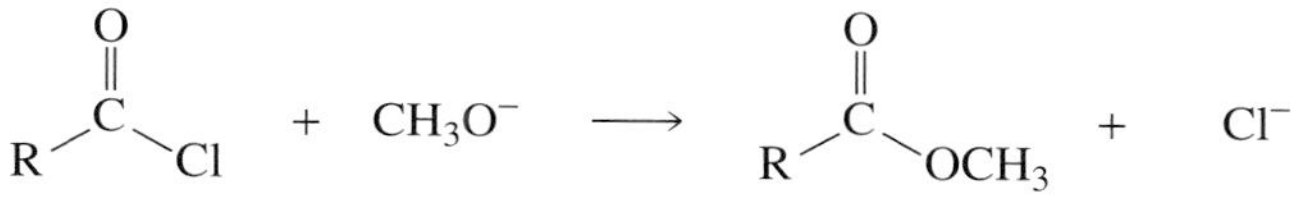

$$RCOCl + CH_3O^- \longrightarrow RCOOCH_3 + Cl^-$$

An ester, however, cannot be converted into an acyl chloride because a chloride ion is a weaker base than an alkoxide ion.

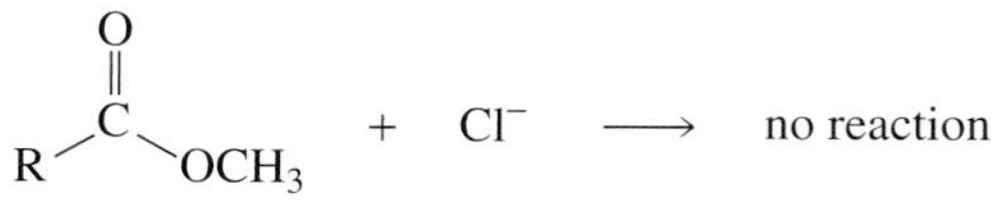

$$RCOOCH_3 + Cl^- \longrightarrow \text{no reaction}$$

PROBLEM 6 ♦

What will be the product of a nucleophilic acyl substitution reaction—a new carboxylic acid derivative, a mixture of two carboxylic acid derivatives, or no reaction—if the new group in the tetrahedral intermediate is the following?

a. a stronger base than the group that is already there
b. a weaker base than the group that is already there
c. similar in basicity to the group that is already there

PROBLEM 7 ♦

Using the pK_a values in Table 11.1, predict the products of the following reactions:

a. 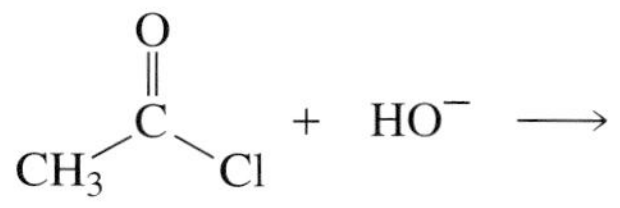 $CH_3COCl + HO^- \longrightarrow$

b.  $CH_3CONHCH_3 + Cl^- \longrightarrow$

11.7 THE REACTIONS OF ACYL CHLORIDES

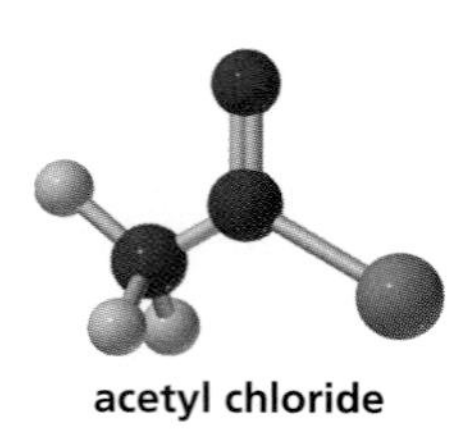

Acyl chlorides react with alcohols to form esters, with water to form carboxylic acids, and with amines to form amides because in each case the incoming nucleophile is a stronger base than the departing chloride ion.

$$CH_3C(=O)Cl + CH_3OH \longrightarrow CH_3C(=O)OCH_3 + HCl$$

acetyl chloride → methyl acetate

$$CH_3CH_2CH_2C(=O)Cl + H_2O \longrightarrow CH_3CH_2CH_2C(=O)OH + HCl$$

butyryl chloride → butyric acid

3-D Molecule: Benzoyl chloride

$$CH_3CH_2C(=O)Cl + 2\,CH_3NH_2 \longrightarrow CH_3CH_2C(=O)NHCH_3 + CH_3\overset{+}{N}H_3\,Cl^-$$

propionyl chloride → *N*-methylpropionamide

All carboxylic acid derivatives undergo nucleophilic acyl substitution reactions by one of the two following mechanisms. The only difference in the two mechanism is whether the nucleophile is neutral or charged.

mechanism for the conversion of an acyl chloride into an ester (with a neutral nucleophile)

$$CH_3C(=O)Cl + CH_3\ddot{O}H \rightleftharpoons CH_3-C(O^-)(Cl)-\overset{+}{O}(H)CH_3 \rightleftharpoons CH_3-C(O^-)(Cl)-\ddot{O}CH_3 + H^+ \longrightarrow CH_3C(=O)\ddot{O}CH_3 + Cl^-$$

formation of a tetrahedral intermediate

proton dissociation

the weaker base is expelled

weaker base

- The nucleophile attacks the carbonyl carbon, forming a tetrahedral intermediate.
- The proton dissociates before the tetrahedral intermediate collapses.
- The tetrahedral intermediate collapses, expelling the weaker base.

If the nucleophile is negatively charged, there is no need for the proton dissociation step.

mechanism for the conversion of an acyl chloride into an ester (with a negatively charged nucleophile)

$$CH_3C(=O)Cl + CH_3\ddot{O}^- \rightleftharpoons CH_3-C(O^-)(Cl)-\ddot{O}CH_3 \longrightarrow CH_3C(=O)\ddot{O}CH_3 + Cl^-$$

The tetrahedral intermediate eliminates the weakest base.

- The nucleophile attacks the carbonyl carbon, forming a tetrahedral intermediate.
- The tetrahedral intermediate collapses, expelling the weaker base.

Notice that the reaction of an acyl chloride with an amine to form an amide is carried out with twice as much amine as acyl chloride, because the HCl formed as a product of the reaction will protonate any amine that has yet to react. Protonated amines are not nucleophiles, so they cannot react with the acyl chloride. Using twice as much amine guarantees that there is enough unprotonated amine to react with all the acyl chloride.

PROBLEM 8 SOLVED

Two amides are obtained from the reaction of acetyl chloride with a mixture of ethylamine and propylamine. Identify the amides.

Solution Either of the amines can react with acetyl chloride, so both *N*-ethylacetamide and *N*-propylacetamide are formed.

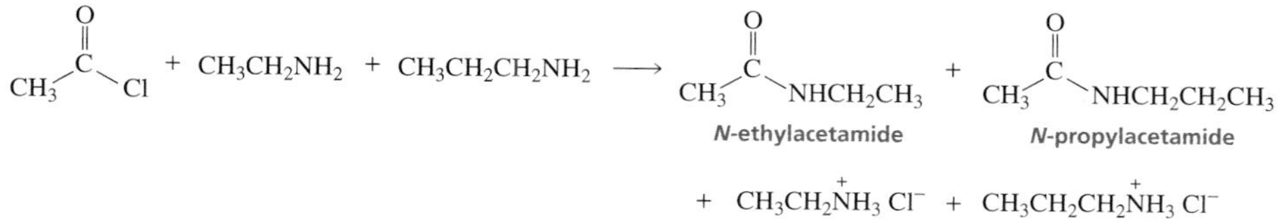

$$CH_3COCl + CH_3CH_2NH_2 + CH_3CH_2CH_2NH_2 \longrightarrow CH_3CONHCH_2CH_3 + CH_3CONHCH_2CH_2CH_3$$

N-ethylacetamide, N-propylacetamide

$$+ CH_3CH_2\overset{+}{N}H_3\ Cl^- + CH_3CH_2CH_2\overset{+}{N}H_3\ Cl^-$$

PROBLEM 9

Write the mechanism for each of the following reactions:
a. the reaction of acetyl chloride with water to form acetic acid
b. the reaction of acetyl chloride with methylamine to form *N*-methylacetamide

PROBLEM 10 ♦

Starting with acetyl chloride, what nucleophile would you use to make each of the following compounds?

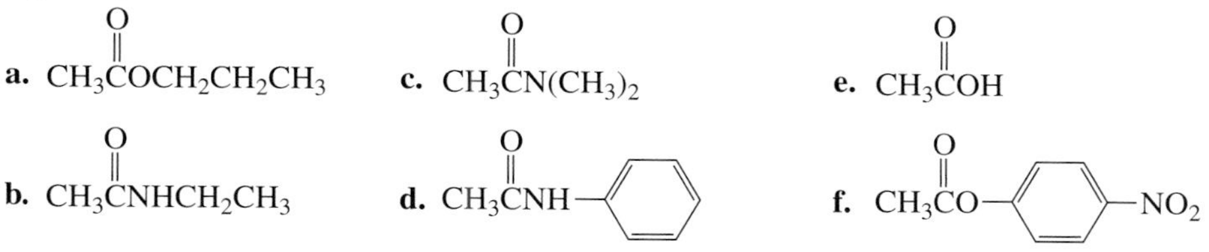

a. $CH_3\overset{O}{\overset{\|}{C}}OCH_2CH_2CH_3$
b. $CH_3\overset{O}{\overset{\|}{C}}NHCH_2CH_3$
c. $CH_3\overset{O}{\overset{\|}{C}}N(CH_3)_2$
d. $CH_3\overset{O}{\overset{\|}{C}}NH-C_6H_5$
e. $CH_3\overset{O}{\overset{\|}{C}}OH$
f. $CH_3\overset{O}{\overset{\|}{C}}O-C_6H_4-NO_2$

11.8 THE REACTIONS OF ESTERS

Esters do not react with Cl^- because it is a much weaker base than the RO^- leaving group of the ester (Table 11.1).

An ester reacts with water to form a carboxylic acid and an alcohol. This is an example of a *hydrolysis* reaction. A **hydrolysis reaction** is a reaction with water that converts one compound into two compounds (*lysis* is Greek for "breaking down").

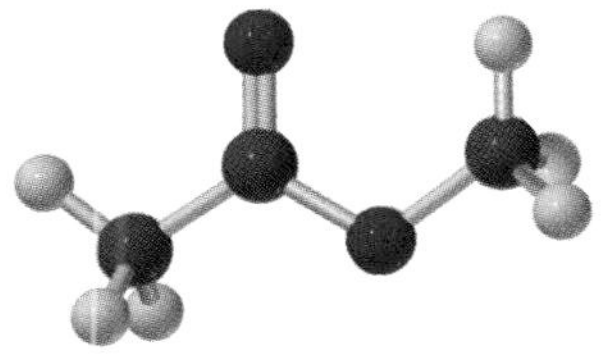

methyl acetate

a hydrolysis reaction

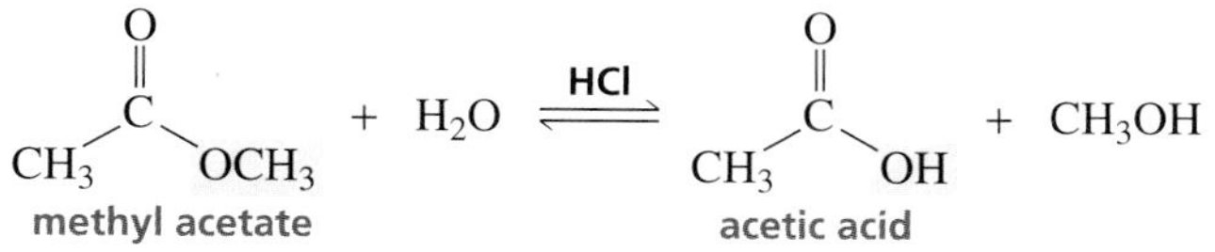

$$CH_3COOCH_3 + H_2O \underset{}{\overset{HCl}{\rightleftharpoons}} CH_3COOH + CH_3OH$$

methyl acetate, acetic acid

An ester reacts with an alcohol to form a new ester and a new alcohol. This is an example of an **alcoholysis** reaction—a reaction with an alcohol that converts one compound into two compounds. This particular alcoholysis reaction is also called a **transesterification reaction** because one ester is converted to another ester.

a transesterification reaction

$$C_6H_5COOCH_3 + CH_3CH_2OH \overset{HCl}{\rightleftharpoons} C_6H_5COOCH_2CH_3 + CH_3OH$$

methyl benzoate, ethyl benzoate

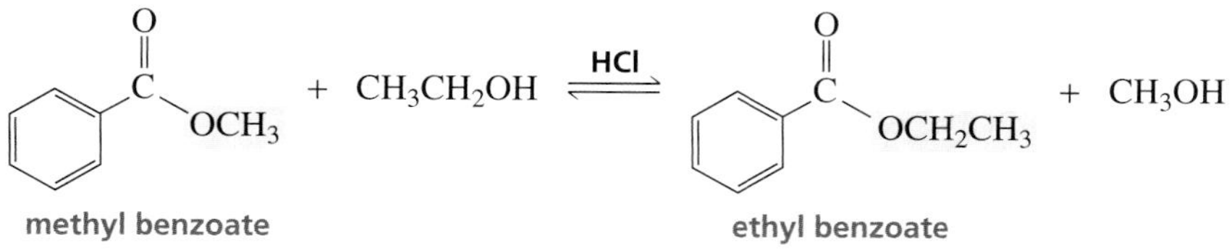

Both the hydrolysis and the alcoholysis of an ester are very slow reactions because water and alcohols are poor nucleophiles and esters have very basic (poor) leaving groups. These reactions are therefore always catalyzed when carried out in the laboratory. Both hydrolysis and alcoholysis of an ester can be catalyzed by an acid (Section 11.9).

Esters also react with amines to form amides. A reaction with an amine that converts one compound into two compounds is called **aminolysis**.

an aminolysis reaction

$$CH_3CH_2C(=O)OCH_2CH_3 + CH_3NH_2 \longrightarrow CH_3CH_2C(=O)NHCH_3 + CH_3CH_2OH$$

ethyl propionate → *N*-methylpropionamide

The reaction of an ester with an amine is not as slow as the reaction of an ester with water or an alcohol, because an amine is a better nucleophile. This is fortunate, because the reaction of an ester with an amine cannot be catalyzed by an acid. The acid will protonate the amine and a protonated amine is not a nucleophile, so it cannot react with the ester.

NERVE IMPULSES, PARALYSIS, AND INSECTICIDES

After an impulse is transmitted between two nerve cells, an ester called acetylcholine must be rapidly hydrolyzed to enable the recipient cell to receive another impulse.

$$CH_3C(=O)OCH_2CH_2\overset{+}{N}(CH_3)_3 + H_2O \xrightarrow{\text{acetylcholinesterase}} CH_3C(=O)O^- + HOCH_2CH_2\overset{+}{N}(CH_3)_3$$

acetylcholine

Acetylcholinesterase, the enzyme that catalyzes this hydrolysis, has a CH_2OH group that is necessary for its catalytic activity. Diisopropyl fluorophosphate (DFP), a military nerve gas used during World War II, inactivates acetylcholinesterase by reacting with the CH_2OH group. When the enzyme is inactivated, the nerve impulses cannot be transmitted properly, and paralysis occurs. DFP is extremely toxic. Its LD_{50} (the lethal dose for 50% of the test animals) is only 0.5 mg/kg of body weight.

$$\text{enzyme}{-}CH_2OH + F{-}P(=O)(OCH(CH_3)_2)_2 \longrightarrow \text{enzyme}{-}CH_2O{-}P(=O)(OCH(CH_3)_2)_2 + HF$$

active enzyme; DFP; inactive enzyme

Malathion and parathion, widely used as insecticides, are compounds related to DFP. The LD_{50} of malathion is 2800 mg/kg. Parathion is more toxic, with an LD_{50} of 2 mg/kg.

$$CH_3CH_2OC(=O)CH(CH_2C(=O)OCH_2CH_3){-}S{-}P(=S)(OCH_3)_2$$

malathion

$$O_2N{-}C_6H_4{-}O{-}P(=S)(OCH_2CH_3)_2$$

parathion

BIODEGRADABLE POLYMERS

Biodegradable polymers are polymers that can be broken into small segments by enzyme-catalyzed reactions. The enzymes are produced by microorganisms. The carbon–carbon bonds of chain-growth polymers are inert to enzyme-catalyzed reactions, so the polymers are nonbiodegradable unless bonds that can be broken by enzymes are inserted into the polymer. Then when the polymer is buried as waste, microorganisms in the ground can degrade the polymer. One method used to make a polymer biodegradable inserts ester groups into it. For example, if the acetal (Section 12.9) shown below is added to an alkene that is undergoing radical polymerization (Section 5.16), ester groups will be inserted into the polymer, forming "weak links" that are susceptible to enzyme-catalyzed ester hydrolysis.

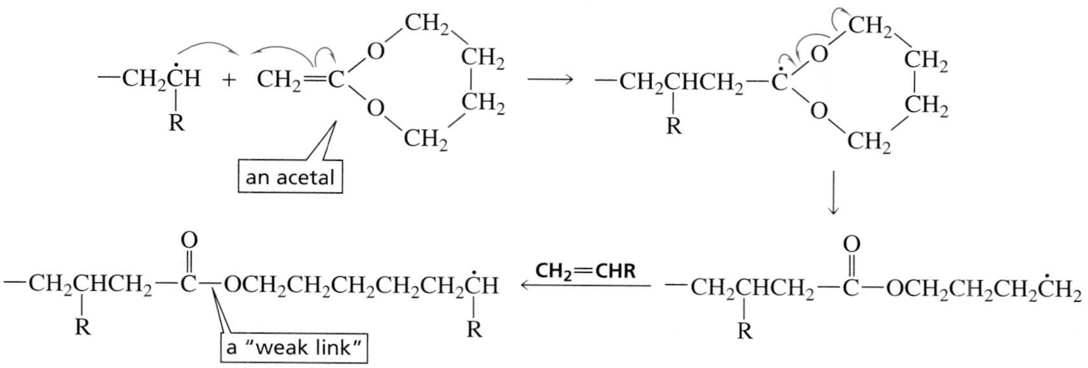

an acetal

$CH_2=CHR$

a "weak link"

PROBLEM 11

Write the mechanism for each of the following reactions:

a. the noncatalyzed hydrolysis of methyl propionate

b. the aminolysis of phenyl formate, using methylamine

PROBLEM 12 *SOLVED*

a. List the following esters in order of decreasing reactivity toward hydrolysis:

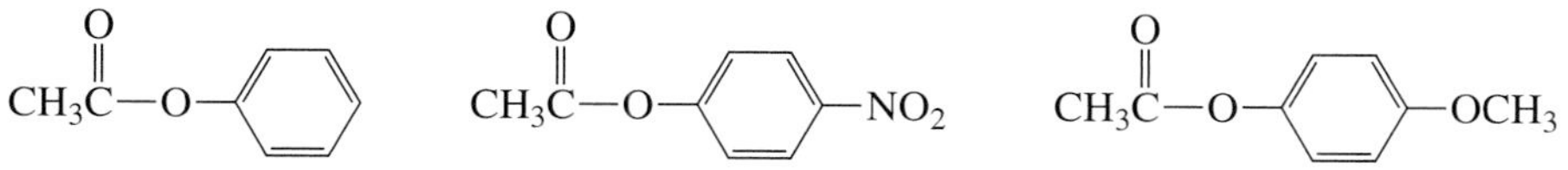

b. How would the rate of hydrolysis of the *para*-methylphenyl ester compare with the rates of hydrolysis of these three esters?

Solution to 12a Both *formation of the tetrahedral intermediate* and *collapse of the tetrahedral intermediate* are fastest for the ester with the electron-withdrawing nitro substituent and slowest for the ester with the electron-donating methoxy substituent.

Formation of the tetrahedral intermediate: An electron-withdrawing substituent increases the ester's susceptibility to nucleophilic attack, and an electron-donating substituent decreases its susceptibility.

Collapse of the tetrahedral intermediate: Electron withdrawal increases acidity and electron donation decreases acidity (Section 8.17). Therefore, *para*-nitrophenol with a strong electron-withdrawing group is a stronger acid than phenol, which in turn is a stronger acid than *para*-methoxyphenol with a strong electron-donating group. Therefore, the *para*-nitrophenoxide ion is the weakest base and the best leaving group of the three, whereas the *para*-methoxyphenoxide ion is the strongest base and the worst leaving group. Thus,

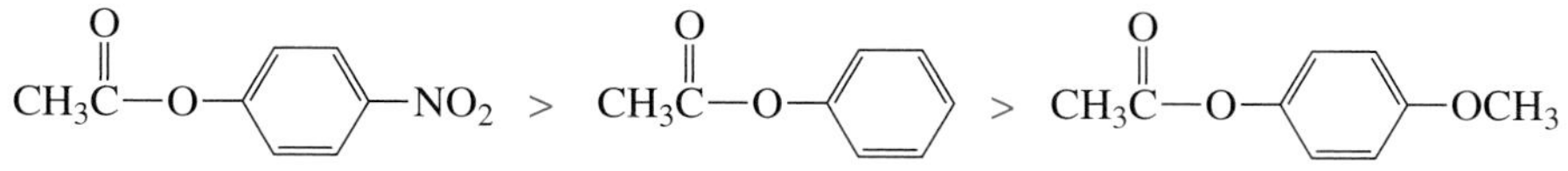

Solution to 12b The methyl substituent donates electrons inductively to the benzene ring, but donates electrons to a lesser extent than does the methoxy substituent (Section 8.14).

Therefore, the rate of hydrolysis of the methyl-substituted ester is slower than the rate of hydrolysis of the unsubstituted ester, but faster than the rate of hydrolysis of the methoxy-substituted ester.

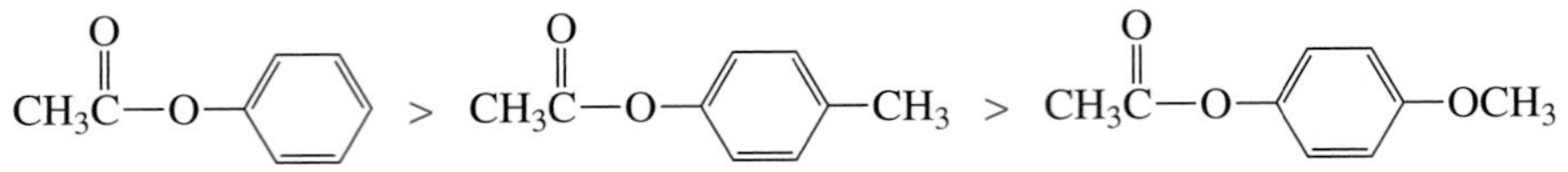

$CH_3C(=O)-O-C_6H_5 > CH_3C(=O)-O-C_6H_4-CH_3 > CH_3C(=O)-O-C_6H_4-OCH_3$

PROBLEM 13 ◆

Which ester is more reactive toward hydrolysis?

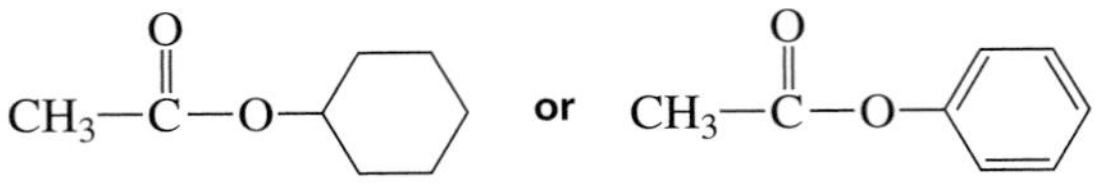

$CH_3-C(=O)-O-C_6H_{11}$ or $CH_3-C(=O)-O-C_6H_5$

11.9 ACID-CATALYZED ESTER HYDROLYSIS

We have seen that esters hydrolyze slowly because water is a poor nucleophile and esters have very basic leaving groups. The rate of hydrolysis can be increased by acid.

When an acid is added to a reaction, the first thing that happens is the acid protonates the atom in the reactant that has the greatest electron density, that is, the most basic atom (Section 10.2). The resonance contributors of the ester show that the atom with the greatest electron density is the carbonyl oxygen.

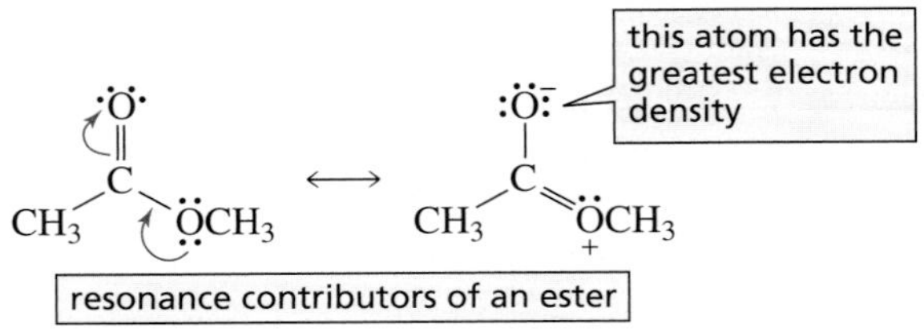

this atom has the greatest electron density

resonance contributors of an ester

The mechanism for acid-catalyzed ester hydrolysis is shown below.

mechanism for acid-catalyzed ester hydrolysis

the acid protonates the carbonyl oxygen

the nucleophile attacks the carbonyl carbon

proton dissociation

tetrahedral intermediate I

either OH or OCH_3 can be protonated

tetrahedral intermediate II

the weaker base is expelled

tetrahedral intermediate III

proton dissociation

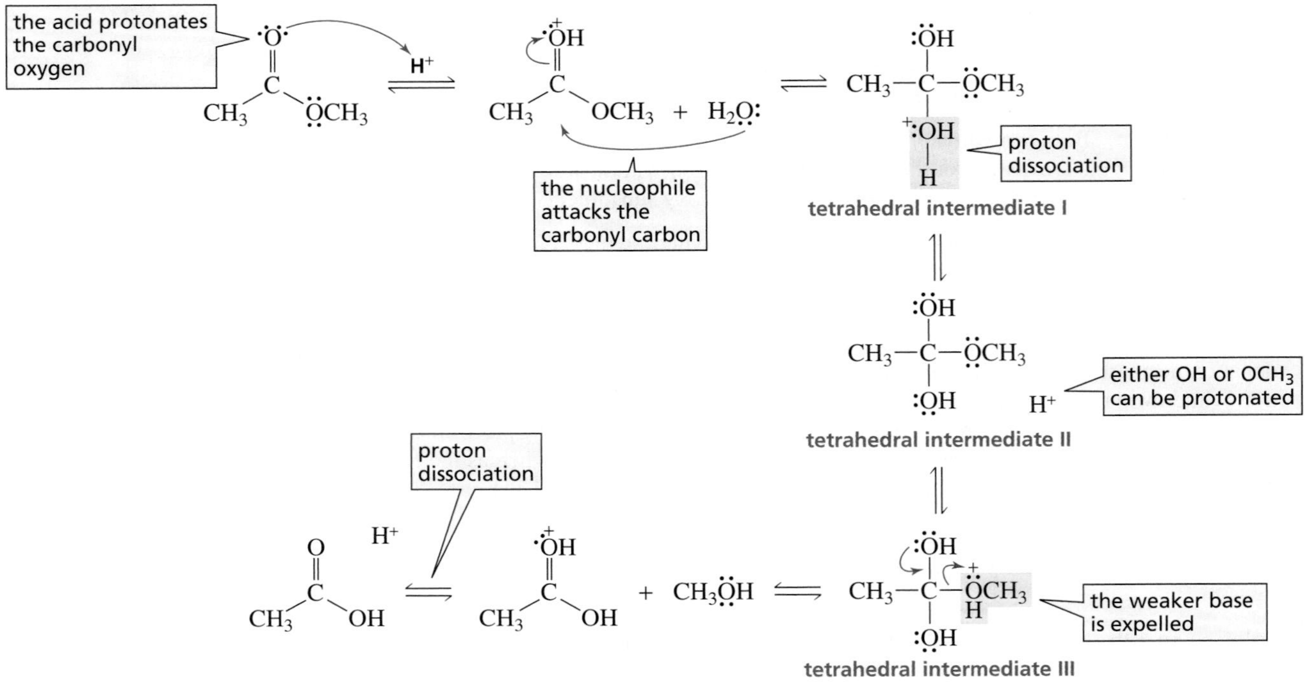

- The acid protonates the carbonyl oxygen.
- The nucleophile (H_2O) attacks the carbonyl carbon of the protonated carbonyl group, forming a protonated tetrahedral intermediate (tetrahedral intermediate I).
- Tetrahedral intermediate I is in equilibrium with its nonprotonated form (tetrahedral intermediate II).
- Once tetrahedral intermediate II has been formed, either the OH or the OCH_3 group of this intermediate I can be protonated, in one case reforming tetrahedral intermediate I (OH is protonated), and in the other case forming tetrahedral intermediate III (OCH_3 is protonated).
- When tetrahedral intermediate I collapses, it expels H_2O in preference to CH_3O^- (because H_2O is a weaker base) and reforms the ester. When tetrahedral intermediate III collapses, it expels CH_3OH rather than HO^- (because CH_3OH is a weaker base) and forms the carboxylic acid.

Because H_2O and CH_3OH have approximately the same basicity, it will be as likely for tetrahedral intermediate I to form and then collapse to reform the ester as it will for tetrahedral intermediate III to form and then collapse to form the carboxylic acid. Consequently, when the reaction has reached equilibrium, both ester and carboxylic acid will be present in approximately equal amounts.

$$CH_3C(=O)OCH_3 + H_2O \overset{HCl}{\rightleftharpoons} CH_3C(=O)OH + CH_3OH$$

both ester and carboxylic acid will be present in approximately equal amounts when the reaction has reached equilibrium

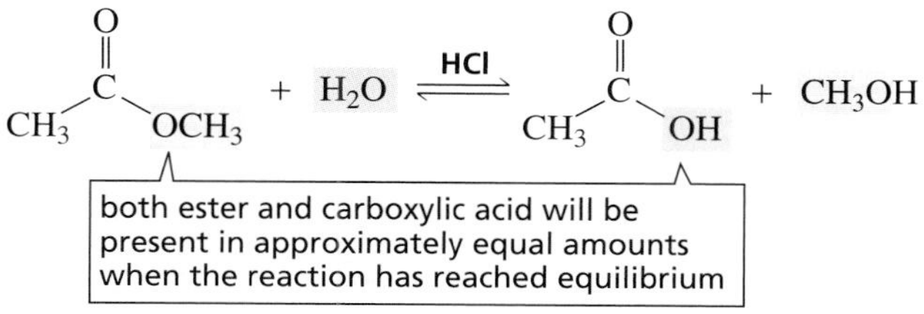

Now let's see how the acid increases the rate of ester hydrolysis. The acid is a catalyst. Recall that **catalyst** is a substance that increases the rate of a reaction without being consumed or changed in the overall reaction (Section 4.8). For a catalyst to increase the rate of a reaction, it must increase the rate of the slow step of the reaction, because changing the rate of a fast step will not affect the rate of the overall reaction. There are two relatively slow steps in the mechanism: formation of the tetrahedral intermediate and collapse of the tetrahedral intermediate. The acid increases the rates of both slow steps.

The acid increases *the rate of formation of the tetrahedral intermediate* by protonating the carbonyl oxygen. Protonated carbonyl groups are more susceptible than nonprotonated carbonyl groups to nucleophilic attack because a positively charged oxygen is more electron withdrawing than a neutral oxygen. Increased electron withdrawal by the oxygen makes the carbonyl carbon more electron deficient, which increases its attractiveness to nucleophiles.

protonation of the carbonyl oxygen increases the susceptibility of the carbonyl carbon to nucleophilic attack

$$CH_3C(=\overset{+}{O}H)OCH_3 \qquad CH_3C(=O)OCH_3$$

more susceptible to attack by a nucleophile

less susceptible to attack by a nucleophile

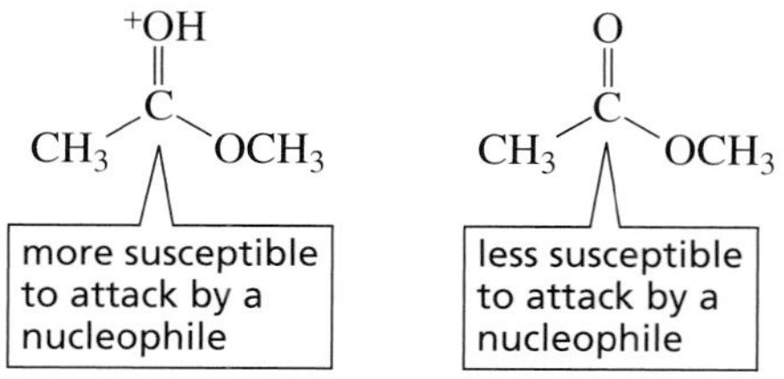

An acid catalyst increases the reactivity of a carbonyl group.

An acid catalyst can make a group a better leaving group.

The acid increases *the rate of collapse of the tetrahedral intermediate* by decreasing the basicity of the leaving group, so the group is more easily eliminated. In the acid-catalyzed hydrolysis of an ester, the leaving group is CH_3OH, a weaker base than the leaving group (CH_3O^-) in the uncatalyzed reaction.

tetrahedral intermediate in acid-catalyzed ester hydrolysis

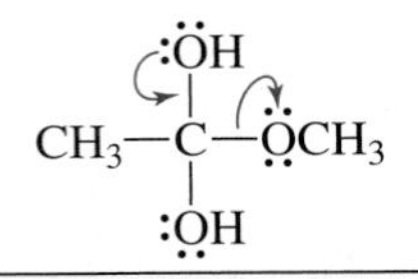

tetrahedral intermediate in uncatalyzed ester hydrolysis

ASPIRIN

Aspirin, found naturally in willow bark and myrtle leaves, is one of the oldest and most commonly used drugs. As early as the fifth century BC, Hippocrates wrote about the curative powers of willow bark. Even so, aspirin's mode of action was not discovered until 1971, when it was found that the anti-inflammatory and fever-reducing activity of aspirin and related compounds called NSAIDs (nonsteroidal anti-inflammatory agents) were due to a transesterification reaction that blocks prostaglandin synthesis. Prostaglandins have several different biological functions, one being to stimulate inflammation and another to induce fever. The enzyme prostaglandin synthase catalyzes the conversion of arachidonic acid into PGH_2, the precursor of all prostaglandins and the related thromboxanes.

arachidonic acid —prostaglandin synthase→ PGH_2 → prostaglandins, thromboxanes

Prostaglandin synthase is composed of two enzymes. One of the enzymes—cyclooxygenase—has a CH_2OH group that is necessary for enzymatic activity. The CH_2OH group reacts with aspirin in a transesterification reaction that inactivates the enzyme. When the enzyme is inactivated, prostaglandins cannot be synthesized, and inflammation and fever are suppressed. Because aspirin inhibits the formation of PGH_2, it inhibits the synthesis of thromboxanes, compounds involved in blood clotting. Presumably, this is why low levels of aspirin have been reported to reduce the incidence of strokes and heart attacks that result from blood clot formation. Aspirin's activity as an anticoagulant is why doctors caution patients not to take aspirin for several days before surgery.

acetylsalicylate (aspirin) + $HOCH_2$–enzyme (active) —transesterification→ $CH_3C(=O)OCH_2$–enzyme (acetylated enzyme, inactive) + salicylate

NSAIDs, such as ibuprofen (the active ingredient in Advil, Motrin, and Nuprin) and naproxen (the active ingredient in Aleve), also inhibit the synthesis of prostaglandins.

ibuprofen

naproxen

(continued...)

Both aspirin and NSAIDs inhibit the synthesis of both the prostaglandins produced under normal physiological conditions and the prostaglandins produced in response to stress. One prostaglandin regulates the production of acid in the stomach, so when prostaglandin synthesis stops, the acidity of the stomach can rise above normal levels. Celebrex, a relatively new drug, inhibits only the enzyme (cyclooxygenase-2) that produces prostaglandin in response to stress. Thus, inflammatory conditions now can be treated without some of the harmful side effects. This drug is known as a COX-2 inhibitor.

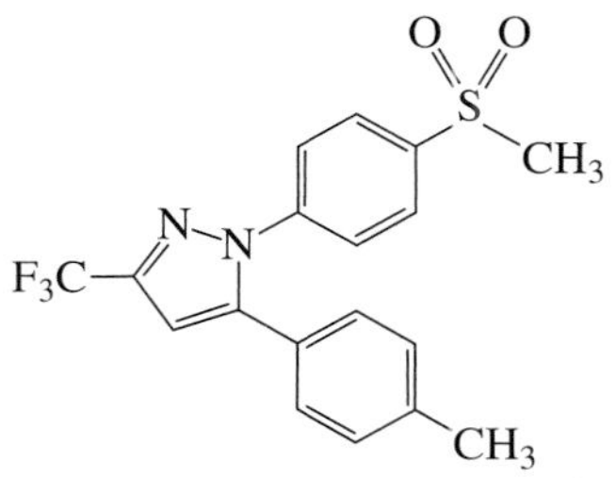

F_3C, N, N, O, S, O, CH_3, CH_3

Celebrex®

PROBLEM 14 ◆

What products would be formed from the acid-catalyzed hydrolysis of the following esters?

a. 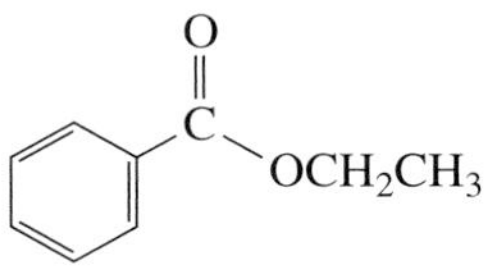 $C_6H_5C(=O)OCH_2CH_3$

b. 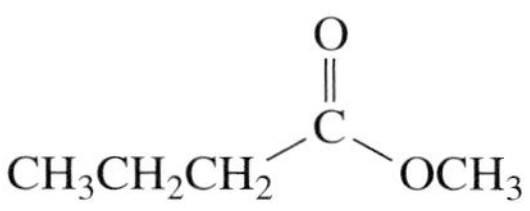 $CH_3CH_2CH_2C(=O)OCH_3$

PROBLEM 15 ◆

What product would be formed from the acid-catalyzed hydrolysis of the following cyclic ester?

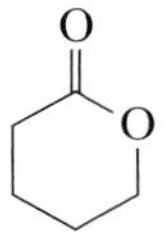

(six-membered cyclic ester, δ-valerolactone: ring O adjacent to C=O)

Transesterification

Transesterification—the reaction of an ester with an alcohol—is also catalyzed by acid. The mechanism for transesterification is identical to the mechanism for ester hydrolysis, except that the nucleophile is ROH rather than H_2O. As in the hydrolysis of an ester, the two leaving groups in the tetrahedral intermediate formed in transesterification have approximately the same basicity. Consequently, when the reaction reaches equilibrium, both esters will be present in approximately equal amounts.

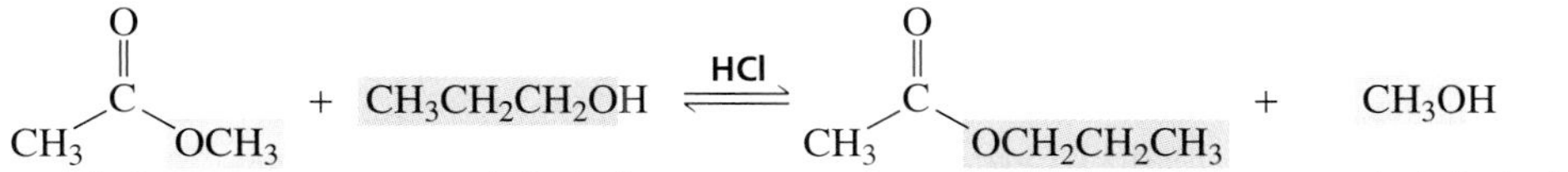

$$\underset{\text{methyl acetate}}{CH_3C(=O)OCH_3} + \underset{\text{propyl alcohol}}{CH_3CH_2CH_2OH} \overset{HCl}{\rightleftharpoons} \underset{\text{propyl acetate}}{CH_3C(=O)OCH_2CH_2CH_3} + \underset{\text{methyl alcohol}}{CH_3OH}$$

PROBLEM 16 ◆

Give the products of the following reaction:

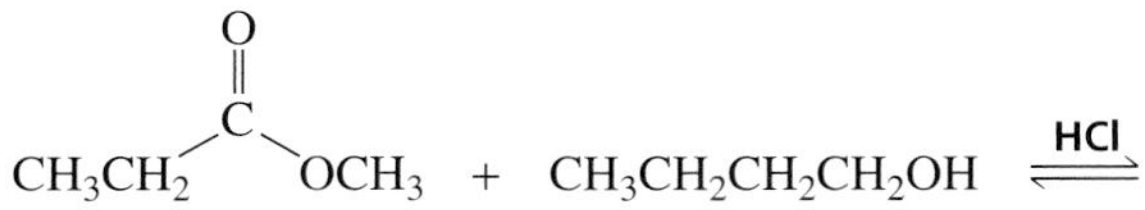

$$CH_3CH_2C(=O)OCH_3 + CH_3CH_2CH_2CH_2OH \overset{HCl}{\rightleftharpoons}$$

PROBLEM 17

Write the mechanism for the acid-catalyzed transesterification reaction of methyl acetate with ethyl alcohol.

11.10 SOAPS, DETERGENTS, AND MICELLES

Fats and **oils** are triesters of glycerol. Glycerol contains three alcohol groups and therefore can form three ester groups. When the ester groups of a fat or an oil are hydrolyzed in a basic solution, glycerol and carboxylate ions are formed The carboxylic acids that are bonded to glycerol in fats and oils have long, unbranched R groups. Because they are obtained from fats, unbranched long-chain carboxylic acids are called **fatty acids**. In Section 19.1, we will see that the difference between a fat and an oil resides in the structure of the fatty acid.

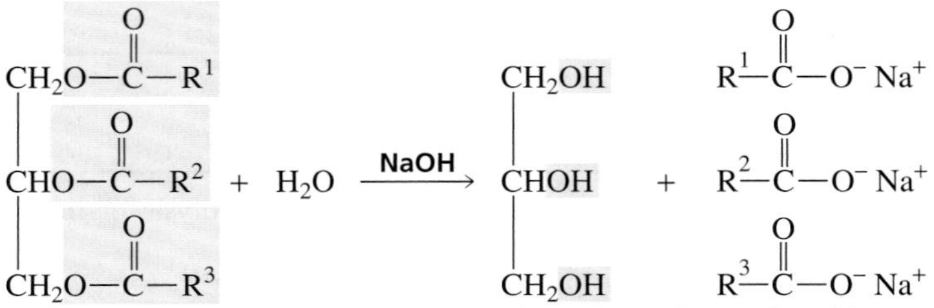

$$\begin{array}{l} CH_2O-\overset{O}{\overset{\|}{C}}-R^1 \\ CHO-\overset{O}{\overset{\|}{C}}-R^2 \\ CH_2O-\overset{O}{\overset{\|}{C}}-R^3 \end{array} + H_2O \xrightarrow{NaOH} \begin{array}{l} CH_2OH \\ CHOH \\ CH_2OH \end{array} + \begin{array}{l} R^1-\overset{O}{\overset{\|}{C}}-O^- Na^+ \\ R^2-\overset{O}{\overset{\|}{C}}-O^- Na^+ \\ R^3-\overset{O}{\overset{\|}{C}}-O^- Na^+ \end{array}$$

a fat or an oil — **glycerol** — **sodium salts of fatty acids** / **soap**

Soaps are sodium or potassium salts of fatty acids. Thus, soaps are obtained when fats or oils are hydrolyzed under basic conditions. The hydrolysis of an ester in a basic solution is called **saponification** (the Latin word for "soap" is *sapo*). Three of the most common soaps are:

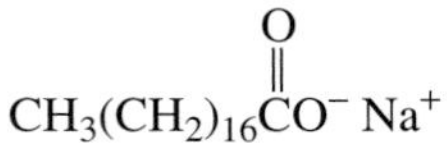

$$CH_3(CH_2)_{16}\overset{O}{\overset{\|}{C}}O^-\ Na^+$$
sodium stearate

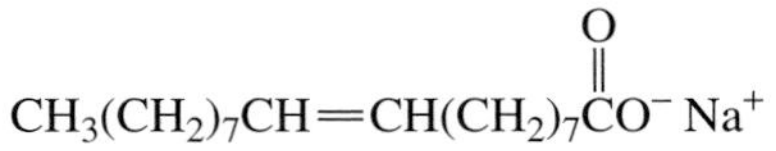

$$CH_3(CH_2)_7CH{=}CH(CH_2)_7\overset{O}{\overset{\|}{C}}O^-\ Na^+$$
sodium oleate

$$CH_3(CH_2)_4CH{=}CHCH_2CH{=}CH(CH_2)_7\overset{O}{\overset{\|}{C}}O^-\ Na^+$$
sodium linoleate

Long-chain carboxylate ions do not exist as individual ions in aqueous solution. Instead, they arrange themselves in spherical clusters called **micelles**, as shown in Figure 11.2. Each micelle contains 50 to 100 long-chain carboxylate ions and resembles a large ball: the polar heads of the carboxylate ions, each accompanied by a counterion, are on the outside of the ball because of their attraction for water, whereas the nonpolar tails are buried in the interior of the ball to minimize their contact with water. Soap has cleansing ability because nonpolar oil molecules that carry dirt dissolve in the nonpolar interior of the micelle and are washed away with the micelle during rinsing.

Because the surface of the micelle is negatively charged, the individual micelles repel each other instead of clustering to form larger aggregates. However, in "hard" water—water containing high concentrations of calcium and magnesium ions—micelles do form aggregates. In hard water, therefore, soaps form a precipitate that we know as "bathtub ring" or "soap scum."

The formation of soap scum in hard water led to a search for synthetic materials that would have the cleansing properties of soap, but would not form scum when they encountered calcium and magnesium ions. The synthetic "soaps" that were developed, known as **detergents** (from the Latin *detergere*, which means "to wipe off"), are salts of benzenesulfonic acids. Calcium and magnesium salts of benzenesulfonic acids do not form aggregates.

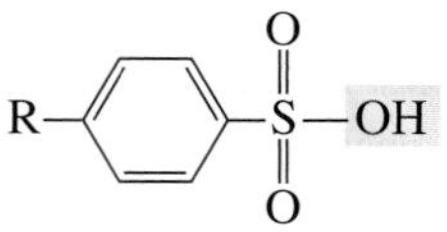

$$R-C_6H_4-\overset{O}{\underset{O}{\overset{\|}{\underset{\|}{S}}}}-OH$$
a benzenesulfonic acid

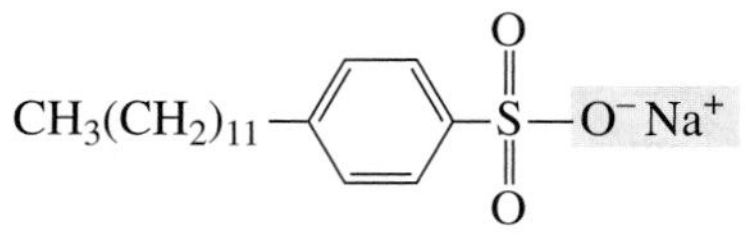

$$CH_3(CH_2)_{11}-C_6H_4-\overset{O}{\underset{O}{\overset{\|}{\underset{\|}{S}}}}-O^-\ Na^+$$
a detergent

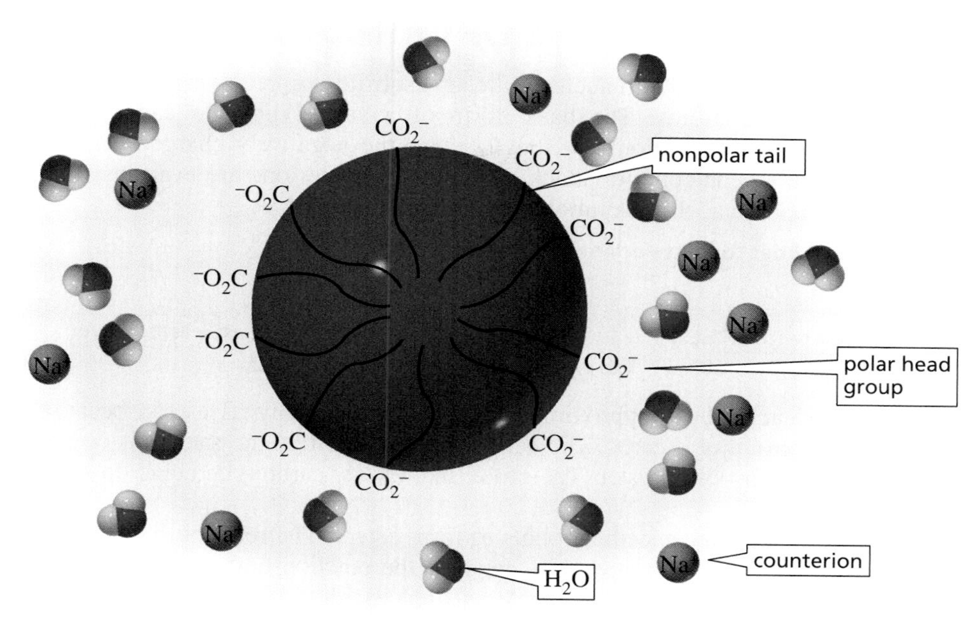

◀ **Figure 11.2**
In aqueous solution, soap molecules form micelles. The polar heads (carboxylate groups) of the soap molecules form the micelle's outer surface; the nonpolar tails (fatty acid R groups) extend into the micelle's interior.

After the initial introduction of detergents into the marketplace, it was discovered that those with straight-chain alkyl groups are biodegradable, whereas those with branched-chain alkyl groups are not. Therefore, to prevent detergents from polluting rivers and lakes, detergents should be made only with straight-chain alkyl groups.

MAKING SOAP

For thousands of years, soap was prepared by heating animal fat with wood ashes. Because the ashes contain potassium carbonate, the solution that results is basic. In the modern commercial method of making soap, fats or oils are boiled in aqueous sodium hydroxide. Sodium chloride is then added to precipitate the soap, which is dried and pressed into bars. Perfume can be added for scented soaps, dyes can be added for colored soaps, sand can be added for scouring soaps, and air can be blown into the soap to make it float in water.

Making soap.

PROBLEM 18 SOLVED

An oil obtained from coconuts is unusual in that all three fatty acid components are identical. The molecular formula of the oil is $C_{45}H_{86}O_6$. What is the molecular formula of the carboxylate ion obtained when the oil is saponified?

Solution When the oil is saponified, it forms glycerol and three equivalents of carboxylate ion. In losing glycerol, the fat loses three carbons and five hydrogens. Thus, the three equivalents of carboxylate ion have a combined molecular formula of $C_{42}H_{81}O_6$. Dividing by three gives a molecular formula of $C_{14}H_{27}O_2$ for the carboxylate ion.

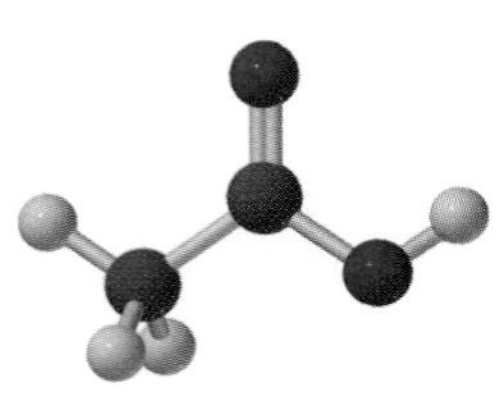
acetic acid

11.11 THE REACTIONS OF CARBOXYLIC ACIDS

Carboxylic acids can undergo nucleophilic acyl substitution reactions only when they are in their acidic forms. The basic form of a carboxylic acid cannot undergo nucleophilic acyl substitution reactions because the negatively charged carboxylate ion is resistant to nucleophilic attack. Thus, carboxylate ions are even less reactive toward nucleophilic acyl substitution reactions than are amides.

relative reactivities toward nucleophilic acyl substitution

$$\text{most reactive} \quad RC(=O)OH > RC(=O)NH_2 > RC(=O)O^- \quad \text{least reactive}$$

Carboxylic acids have approximately the same reactivity as esters, because the HO^- leaving group of a carboxylic acid has approximately the same basicity as the RO^- leaving group of an ester. Therefore, like esters, carboxylic acids do not react with chloride ions.

Carboxylic acids react with alcohols to form esters. The reaction must be carried out in an acidic solution, not only to catalyze the reaction but also to keep the carboxylic acid in its acidic form so that it will react with the nucleophile. The mechanism for the acid-catalyzed reaction of a carboxylic acid and an alcohol to form an ester and water is the exact reverse of the mechanism for the acid-catalyzed hydrolysis of an ester to form a carboxylic acid and an alcohol (Section 11.9).

Because the tetrahedral intermediate formed in this reaction has two potential leaving groups of approximately the same basicity, both carboxylic acid and ester will be present in approximately equal amounts when the reaction has reached equilibrium. Emil Fischer (Section 15.8) was the first to discover that an ester could be prepared by reacting a carboxylic acid with an alcohol in the presence of an acid catalyst. The reaction, therefore, is called a **Fischer esterification**.

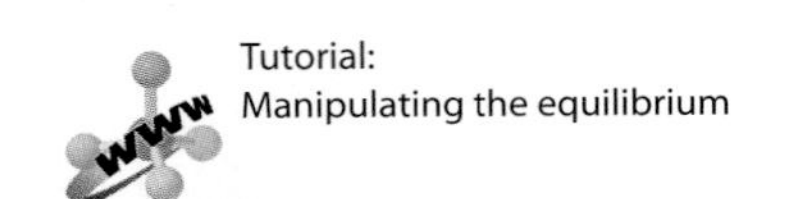

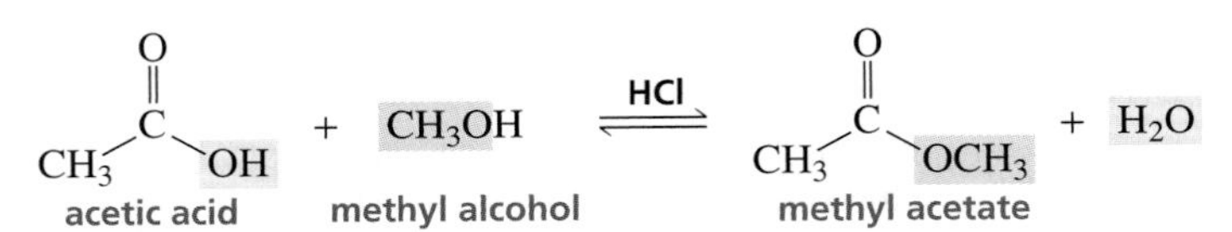

Carboxylic acids do not undergo nucleophilic acyl substitution reactions with amines. Because a carboxylic acid is an acid and an amine is a base, the carboxylic acid immediately donates a proton to the amine when the two compounds are mixed. The resulting ammonium carboxylate salt is the final product of the reaction; the carboxylate ion is unreactive, and the protonated amine is not a nucleophile.

$$CH_3C(=O)OH + CH_3CH_2NH_2 \longrightarrow CH_3C(=O)O^-\ H_3\overset{+}{N}CH_2CH_3$$

an ammonium carboxylate salt

PROBLEM 19 ♦

Using an acyl chloride and an alcohol, show how the following esters could be synthesized:

a. methyl butyrate (odor of apples) **b.** octyl acetate (odor of oranges)

PROBLEM 20

Referring to the mechanism for the acid-catalyzed hydrolysis of methyl acetate, write the mechanism—showing all the curved arrows—for the acid-catalyzed reaction of acetic acid and methyl alcohol to form methyl acetate.

PROBLEM 21 *SOLVED*

Loss of water from two molecules of a carboxylic acid forms an **acid anhydride**. Acid anhydrides are also carboxylic acid derivatives—the OH group of the carboxylic acid has been replaced with a carboxylate group. Thus, the carboxylate group is the leaving group of an acid anhydride.

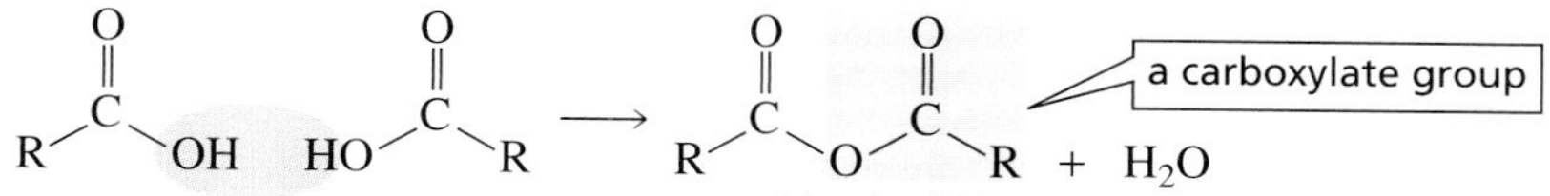

$$RC(=O)OH + HOC(=O)R \longrightarrow RC(=O)OC(=O)R + H_2O$$

an acid anhydride (O–C(=O)R: a carboxylate group)

What is the product of the reaction of acid anhydride with
a. an alcohol?
b. an amine?

Solution to 21a When an acid anhydride reacts with an alcohol, the two potential leaving groups in the tetrahedral intermediate will be a carboxylate ion and an alkoxide ion. The carboxylate ion is the weaker base, so the product of the reaction will be an ester and a carboxylic acid.

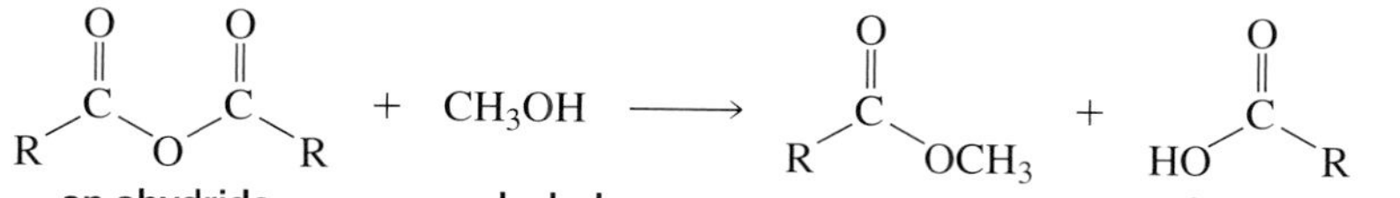

$$RC(=O)OC(=O)R + CH_3OH \longrightarrow RC(=O)OCH_3 + HOC(=O)R$$

an ahydride; an alcohol; an ester; a carboxylic acid

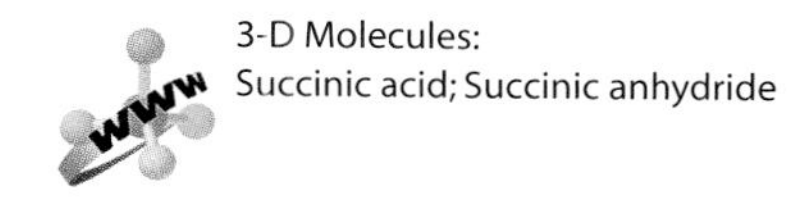

11.12 THE REACTIONS OF AMIDES

Amides are very unreactive compounds, which is comforting since proteins, which impart strength to biological structures, are composed of amino acids linked together by amide bonds (Section 16.0). Amides do not react with chloride ions, alcohols, or water because, in each case, the incoming nucleophile is a weaker base than the leaving group of the amide (Table 11.1).

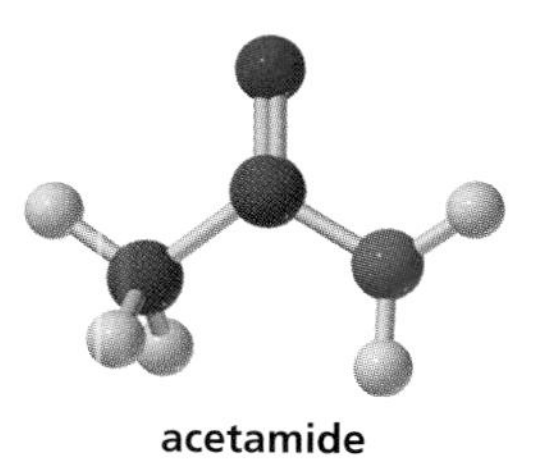
acetamide

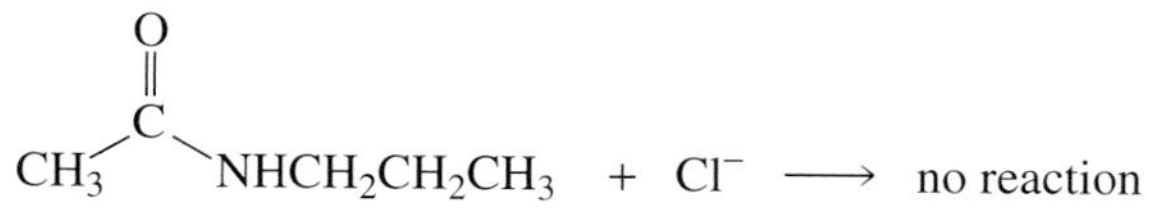

$$CH_3C(=O)NHCH_2CH_2CH_3 + Cl^- \longrightarrow \text{no reaction}$$

N-propylacetamide

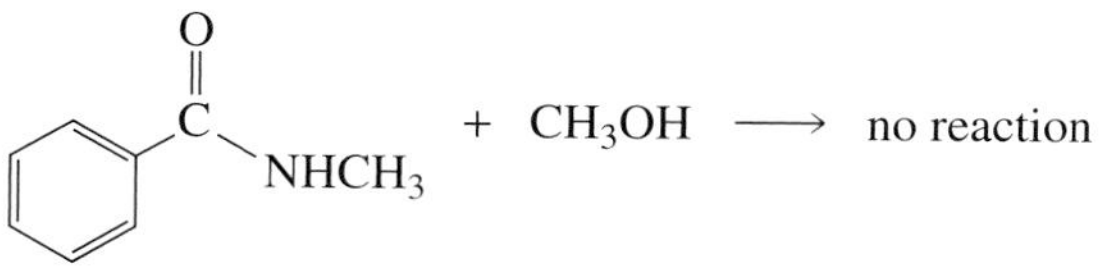

$$C_6H_5C(=O)NHCH_3 + CH_3OH \longrightarrow \text{no reaction}$$

N-methylbenzamide

$$CH_3CH_2C(=O)NHCH_2CH_3 + H_2O \longrightarrow \text{no reaction}$$

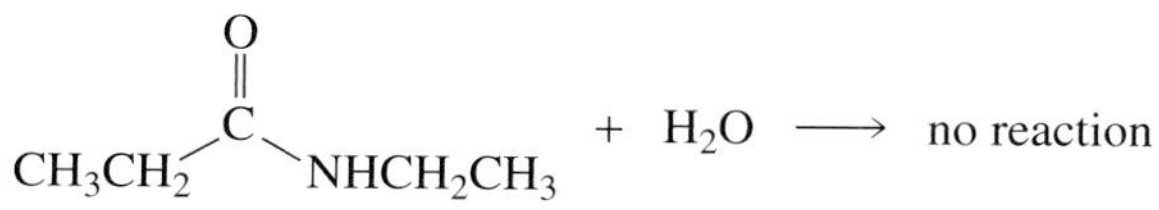

N-ethylpropanamide

Amides do, however, react with water and alcohols if the reaction mixture is heated in the presence of an acid. The reason for this will be explained in the next section.

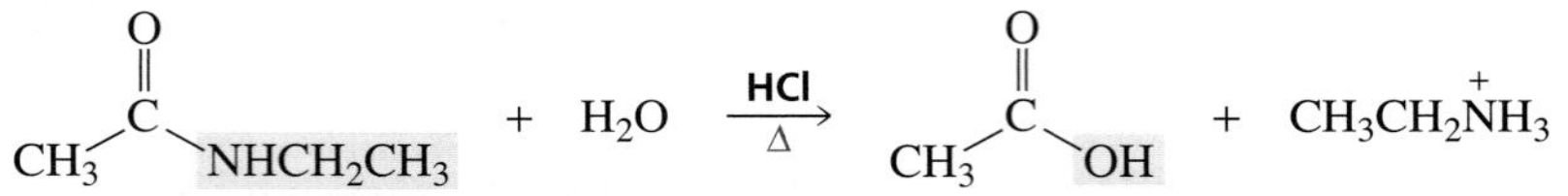

$$CH_3C(=O)NHCH_2CH_3 + H_2O \xrightarrow[\Delta]{HCl} CH_3C(=O)OH + CH_3CH_2\overset{+}{N}H_3$$

N-ethylacetamide

$$C_6H_5C(=O)NHCH_3 + CH_3CH_2OH \xrightarrow[\Delta]{HCl} C_6H_5C(=O)OCH_2CH_3 + CH_3\overset{+}{N}H_3$$

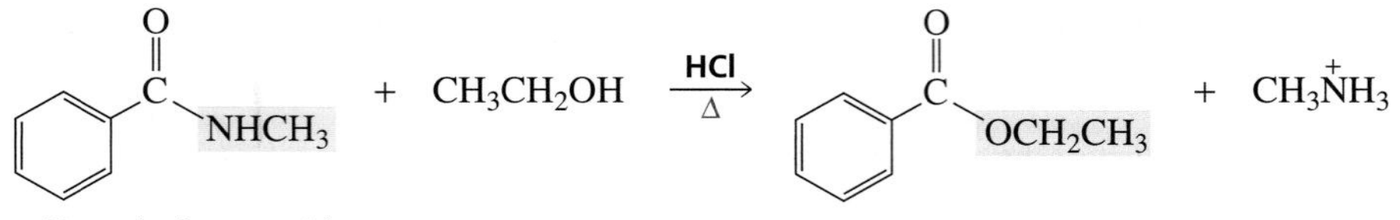

N-methylbenzamide

PROBLEM 22 ♦

What acyl chloride and what amine would be required to synthesize the following amides?

a. *N*-ethylbutanamide

b. *N,N*-dimethylbenzamide

PROBLEM 23 ♦

Which of the following reactions would lead to the formation of an amide?

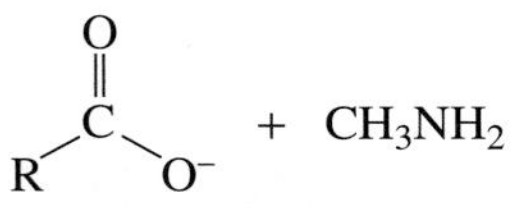

1. 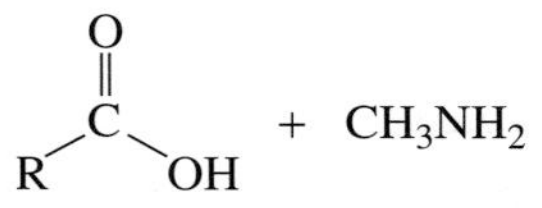 $RC(=O)OH + CH_3NH_2$

2. $RC(=O)OCH_3 + CH_3NH_2$

3. $RC(=O)O^- + CH_3NH_2$

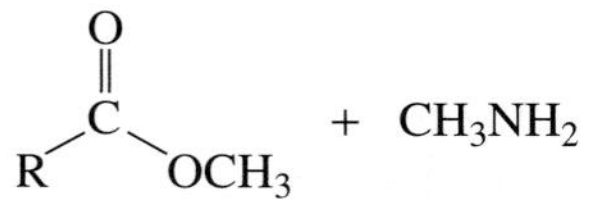

4. 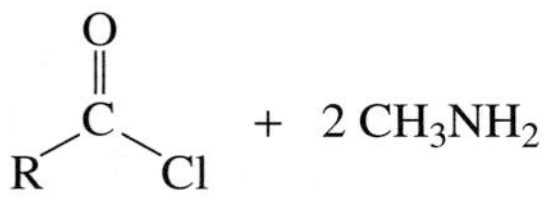 $RC(=O)Cl + 2\ CH_3NH_2$

NATURE'S SLEEPING PILL

Melatonin, a naturally occurring amide, is a hormone synthesized by the pineal gland from the amino acid tryptophan. Melatonin regulates the dark–light clock in our brains that governs such things as the sleep–wake cycle, body temperature, and hormone production.

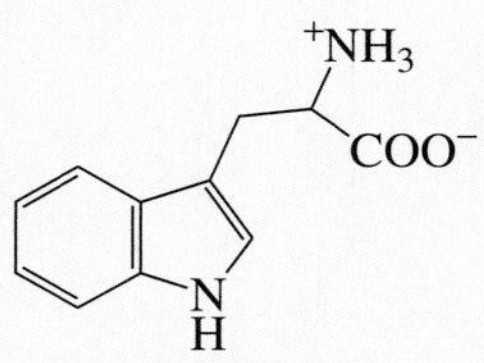

tryptophan
an amino acid

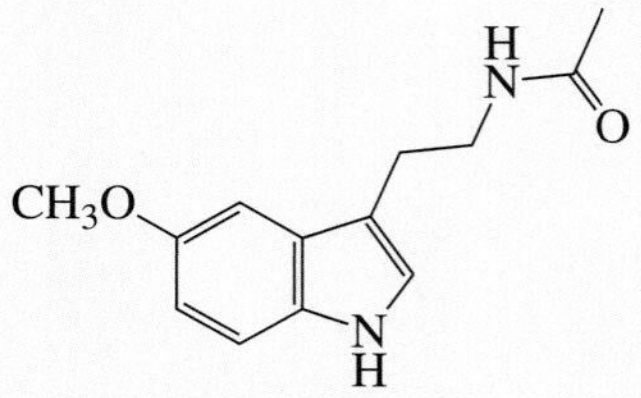

melatonin

Melatonin levels increase from evening to night and then decrease as morning approaches. People with high levels of melatonin sleep longer and more soundly than those with low levels. The concentration of the hormone in the blood varies with age—6-year-olds have more than five times the concentration that 80-year-olds have—which is one of the reasons young people have less trouble sleeping than older people. Melatonin supplements are used to treat insomnia, jet lag, and seasonal affective disorder.

11.13 ACID-CATALYZED AMIDE HYDROLYSIS

The mechanism for the acid-catalyzed hydrolysis of an amide is similar to the mechanism for the acid-catalyzed hydrolysis of an ester (Section 11.9).

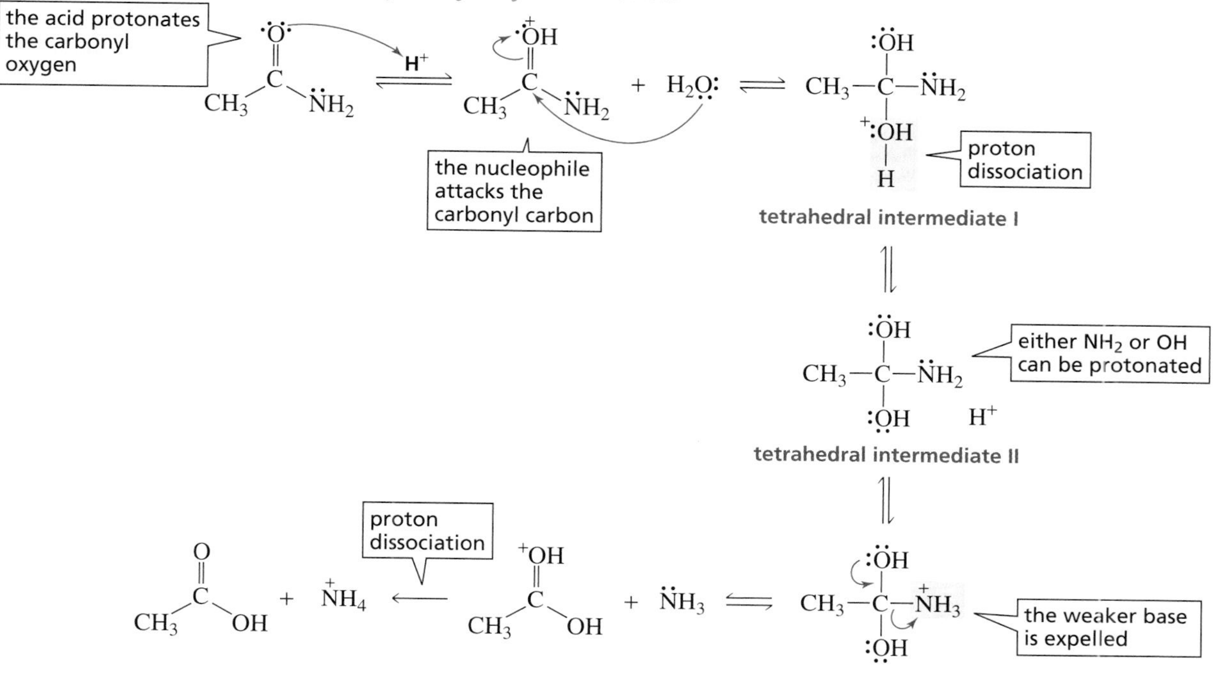

- The acid protonates the carbonyl oxygen, increasing the susceptibility of the carbonyl carbon to nucleophilic attack.
- Nucleophilic attack by water on the carbonyl carbon leads to tetrahedral intermediate I, which is in equilibrium with its nonprotonated form, tetrahedral intermediate II.
- Reprotonation of tetrahedral intermediate II can occur either on oxygen to reform tetrahedral intermediate I or on nitrogen to form tetrahedral intermediate III. Protonation on nitrogen is favored because the NH_2 group is a stronger base than the OH group.
- Of the two possible leaving groups in tetrahedral intermediate III (HO^- and NH_3), NH_3 is the weaker base, so it is expelled, forming the carboxylic acid as the final product.
- Since the reaction is carried out in an acidic solution, NH_3 will be protonated after it is expelled from the tetrahedral intermediate. This prevents the reverse reaction from occurring, because $^+NH_4$ is not a nucleophile.

Let's take a minute to see why an amide cannot be hydrolyzed without a catalyst. In the uncatalyzed reaction, the amide would not be protonated. Therefore, water, a very poor nucleophile, would have to attack a neutral amide that is much less susceptible to nucleophilic attack than a protonated amide would be. In addition, the NH_2 group of the tetrahedral intermediate would not be protonated in the uncatalyzed reaction. Therefore, HO^- is the group that would be expelled from the tetrahedral intermediate, because HO^- is a weaker base than $^-NH_2$. This would reform the amide.

$CH_3-C(OH)(OH)-\overset{+}{N}H_3$

tetrahedral intermediate in acid-catalyzed amide hydrolysis

$CH_3-C(OH)(OH)-NH_2$

tetrahedral intermediate in uncatalyzed amide hydrolysis

An amide reacts with an alcohol in the presence of acid (page 314) for the same reason that it reacts with water in the presence of acid. The only difference in the mechanisms of the two reactions is the nucleophile, which is water in one case and an alcohol in the other.

PENICILLIN AND DRUG RESISTANCE

Penicillin contains an amide in a four-membered ring. The strain in this ring increases the amide's reactivity. It is thought that the antibiotic activity of penicillin results from its ability to put an acyl group on a CH_2OH group of an enzyme that has a role in the synthesis of bacterial cell walls. This inactivates the enzyme, and actively growing bacteria die because they are unable to synthesize functional cell walls. Penicillin has no effect on mammalian cells because mammalian cells are not enclosed by cell walls. Penicillins are stored at cold temperatures to minimize hydrolysis of the four-membered ring.

enzyme active —$CH_2\ddot{O}H$ + penicillin (RCNH, H, H, S, CH_3, CH_3, N, O, COO^- K^+) ⟶ enzyme inactive —CH_2O—C(=O)— (RCNH, H, S, CH_3, CH_3, HN, COO^- K^+)

Bacteria that are resistant to penicillin secrete penicillinase, an enzyme that catalyzes the hydrolysis of penicillin's four-membered ring. Penicillinoic acid, the ring-opened product, has no antibacterial activity.

penicillin + H_2O $\xrightarrow{\text{penicillinase}}$ penicillinoic acid (^-O, O, H_2N^+)

PENICILLINS IN CLINICAL USE

More than 10 different penicillins are currently in clinical use. They differ only in the group (R) attached to the carbonyl group. Some of these penicillins are shown here. In addition to their structural differences, the penicillins differ in the organisms against which they are most effective. They also differ in their susceptibility to penicillinase. For example, ampicillin, a *synthetic* penicillin, is clinically effective against bacteria that are resistant to penicillin G, the *naturally occurring* penicillin discussed in Section 11.4. Almost 19% of humans are allergic to penicillin G.

R		R		R	
C_6H_5—CH_2—	penicillin G	C_6H_5—CH(NH_2)—	ampicillin	HO—C_6H_4—CH(NH_2)—	amoxicillin
(phenyl, N, O, CH_3 isoxazole)	oxacillin	(Cl-phenyl, N, O, CH_3 isoxazole)	cloxacillin	CH_2=$CHCH_2SCH_2$—	penicillin O
				C_6H_5—OCH_2—	penicillin V

Penicillin V is a *semisynthetic* penicillin in clinical use. It is not a naturally occurring penicillin, but it is also not a true synthetic penicillin because chemists do not synthesize it. The *Penicillium* mold synthesizes it after the mold is fed 2-phenoxyethanol, the compound needed for the side chain.

C_6H_5—OCH_2CH_2OH (2-phenoxyethanol) $\xrightarrow{\textit{Penicillium}\text{ mold}}$ C_6H_5—OCH_2C(=O)NH— (H, H, S, CH_3, CH_3, N, O, COO^- K^+) (penicillin V)

11.14 THE SYNTHESIS OF CARBOXYLIC ACID DERIVATIVES

Of the various classes of carbonyl compounds discussed in this chapter—acyl chlorides, esters, carboxylic acids, and amides—carboxylic acids are the most commonly available. However, we have seen that carboxylic acids are relatively unreactive toward nucleophilic acyl substitution reactions because the OH group of a carboxylic acid is a strong base and thus a poor leaving group. Therefore, chemists need a way to activate carboxylic acids so that they can readily undergo nucleophilic acyl substitution reactions.

Because acyl chlorides are the most reactive of the carboxylic acid derivatives, the easiest way to synthesize any other carboxylic acid derivative is to add the appropriate nucleophile to an acyl chloride. Consequently, organic chemists activate carboxylic acids by converting them into acyl chlorides.

A carboxylic acid can be converted into an acyl chloride by heating it with thionyl chloride ($SOCl_2$). This reagent replaces the OH group with a Cl.

$$CH_3COOH + SOCl_2 \xrightarrow{\Delta} CH_3COCl + SO_2 + HCl$$

acetic acid, thionyl chloride, acetyl chloride

Once the acyl chloride has been prepared, esters and amides can be synthesized simply by adding the appropriate nucleophile (Section 11.7).

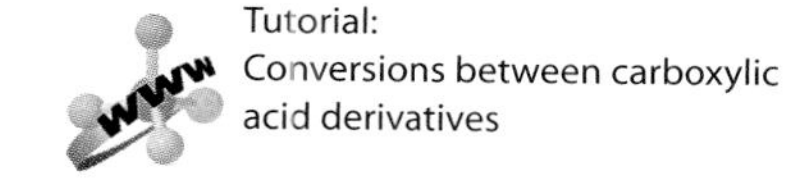

$$RCOCl + ROH \longrightarrow RCOOR + HCl$$

an ester

$$RCOCl + 2\,RNH_2 \longrightarrow RCONHR + R\overset{+}{N}H_3\ Cl^-$$

an amide

PROBLEM 24◆

How would you synthesize the following compounds starting with a carboxylic acid?

a. $CH_3CH_2COOC_6H_5$ **b.** $C_6H_5CONHCH_2CH_3$

SYNTHETIC POLYMERS

Synthetic polymers play important roles in our daily lives. Polymers are compounds that are made by linking together many small molecules called monomers. We have seen that **chain-growth polymers** are made by adding monomers to the end of a growing chain (Section 5.16).

A second kind of synthetic polymer, called a **step-growth polymer**, is made with monomers that have reactive functional groups at each end. The functional groups form ester or amide bonds between the monomers. Nylon and Dacron are examles of step-growth polymers. Nylon is a polyamide. Dacron is a polyester.

$$H_3\overset{+}{N}(CH_2)_5CO^- \xrightarrow[-H_2O]{\Delta} -NH(CH_2)_5C(=O)-[NH(CH_2)_5C(=O)]_n-NH(CH_2)_5C(=O)-$$

6-aminohexanoic acid; nylon 6, a polyamide

(continued...)

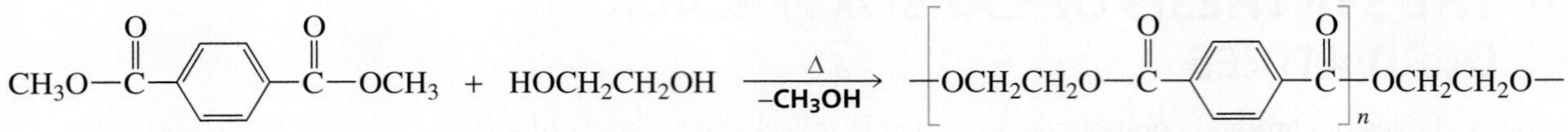

$$CH_3O-\overset{O}{\overset{\|}{C}}-C_6H_4-\overset{O}{\overset{\|}{C}}-OCH_3 + HOCH_2CH_2OH \xrightarrow[-CH_3OH]{\Delta} \left[-OCH_2CH_2O-\overset{O}{\overset{\|}{C}}-C_6H_4-\overset{O}{\overset{\|}{C}}- \right]_n -OCH_2CH_2O-$$

dimethyl terephthalate; 1,2-ethanediol ethylene glycol; poly(ethylene terephthalate) Dacron® a polyester

Synthetic polymers have taken the place of metals, fabrics, glass, ceramics, wood, and paper, allowing us to have a greater variety and larger quantities of materials than nature could have provided. New polymers are continually being designed to fit human needs. For example, Kevlar and Lexan are relatively new step-growth polymers. Kevlar has a tensile strength greater than steel. It is used for high-performance skis and bulletproof vests.

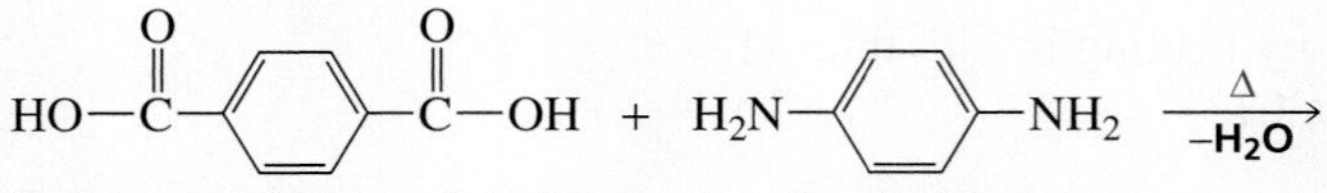

$$HO-\overset{O}{\overset{\|}{C}}-C_6H_4-\overset{O}{\overset{\|}{C}}-OH + H_2N-C_6H_4-NH_2 \xrightarrow[-H_2O]{\Delta}$$

1,4-benzenedicarboxylic acid; 1,4-diaminobenzene

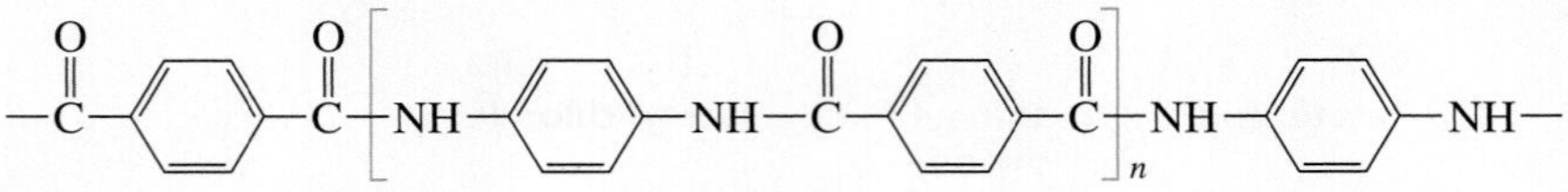

$$-\overset{O}{\overset{\|}{C}}-C_6H_4-\overset{O}{\overset{\|}{C}}-\left[NH-C_6H_4-NH-\overset{O}{\overset{\|}{C}}-C_6H_4-\overset{O}{\overset{\|}{C}}- \right]_n NH-C_6H_4-NH-$$

Kevlar®

Lexan is a strong and transparent polymer used for such things as traffic light lenses and compact disks.

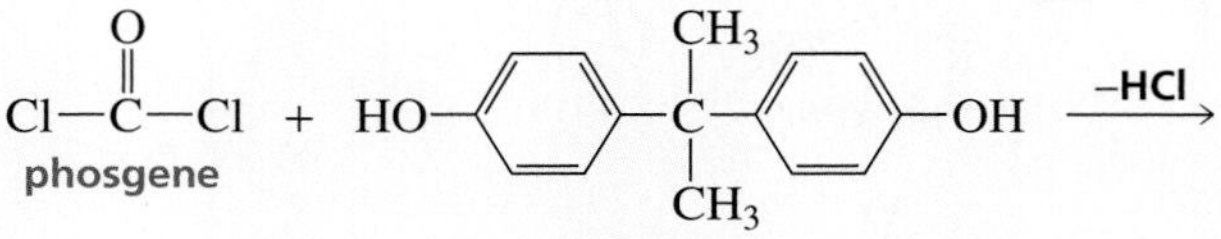

$$Cl-\overset{O}{\overset{\|}{C}}-Cl + HO-C_6H_4-C(CH_3)_2-C_6H_4-OH \xrightarrow{-HCl}$$

phosgene; bisphenol A

$$-C_6H_4-C(CH_3)_2-C_6H_4-O-\left[\overset{O}{\overset{\|}{C}}-O-C_6H_4-C(CH_3)_2-C_6H_4-O- \right]_n \overset{O}{\overset{\|}{C}}-O-C_6H_4-C(CH_3)_2-C_6H_4-O-\overset{O}{\overset{\|}{C}}-$$

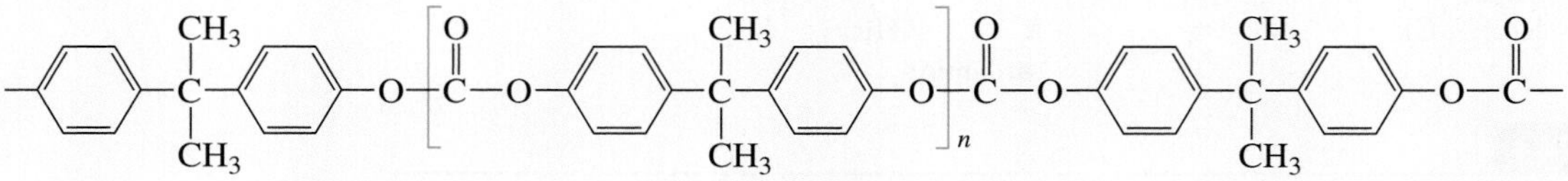

Lexan®

DISSOLVING SUTURES

Dissolving sutures, such as dexon and poly(dioxanone) (PDS), are synthetic polymers that now are routinely used in surgery. The many ester groups they contain are slowly hydrolyzed to small molecules that subsequently are metabolized to compounds easily excreted by the body. Patients no longer have to undergo a second medical procedure that was required to remove the sutures when traditional suture materials were used.

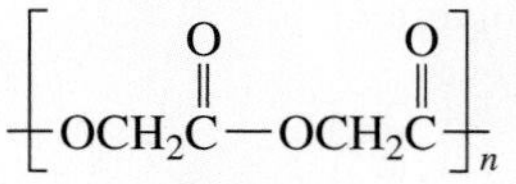

$$\left[OCH_2\overset{O}{\overset{\|}{C}}-OCH_2\overset{O}{\overset{\|}{C}} \right]_n$$

Dexon

$$\left[OCH_2CH_2OCH_2\overset{O}{\overset{\|}{C}}-OCH_2CH_2OCH_2\overset{O}{\overset{\|}{C}} \right]_n$$

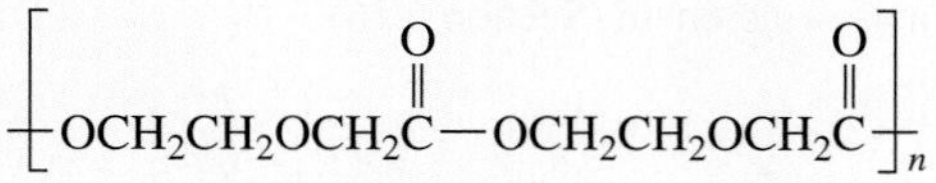

PDS

Depending on their structures, these synthetic sutures lose 50% of their strength after two to three weeks and are completely absorbed within three to six months.

11.15 NITRILES

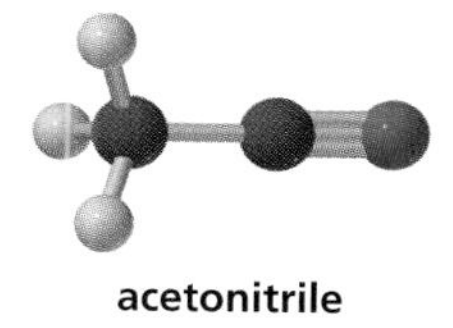

Nitriles are compounds that contain a C≡N functional group. Nitriles are considered carboxylic acid derivatives because, like all the other carboxylic acid derivatives, they react with water to form carboxylic acids. They are even less reactive than amides, but hydrolyze when heated with water and an acid.

$$CH_3CH_2C{\equiv}N + H_2O \xrightarrow[\Delta]{HCl} CH_3CH_2C(=O)OH + \overset{+}{N}H_4$$

Nitriles are named by adding "nitrile" to the parent alkane name. Notice that the triple-bonded carbon of the nitrile group is counted in the number of carbons in the longest continuous chain.

	$CH_3C{\equiv}N$	$CH_3CH(CH_3)CH_2CH_2CH_2C{\equiv}N$	$CH_2{=}CHC{\equiv}N$
systematic name:	ethanenitrile	5-methylhexanenitrile	propenenitrile
common name:	acetonitrile		acrylonitrile

Nitriles can be prepared from an S_N2 reaction of alkyl halide with cyanide ion. Because a nitrile can be hydrolyzed to carboxylic acid, you now know how to convert an alkyl halide into a carboxylic acid. Notice that the carboxylic acid has one more carbon than the alkyl halide.

$$\underset{\text{ethyl bromide}}{CH_3CH_2Br} + \underset{\text{cyanide ion}}{^{-}C{\equiv}N} \xrightarrow{S_N2\text{ reaction}} \underset{\text{propanenitrile}}{CH_3CH_2C{\equiv}N} \xrightarrow[\Delta]{HCl,\ H_2O} \underset{\text{propanoic acid}}{CH_3CH_2C(=O)OH}$$

A nitrile can be reduced to a primary amine by the same reagents that reduce an alkyne to an alkane (Section 5.12).

$$\underset{\text{pentanenitrile}}{CH_3CH_2CH_2CH_2C{\equiv}N} \xrightarrow[Pt/C]{H_2} \underset{\text{pentylamine}}{CH_3CH_2CH_2CH_2CH_2NH_2}$$

PROBLEM 25 ◆

Which alkyl halides form the carboxylic acids listed below after reaction with sodium cyanide followed by heating the product in an acidic aqueous solution?

a. butyric acid **b.** 4-methylpentanoic acid

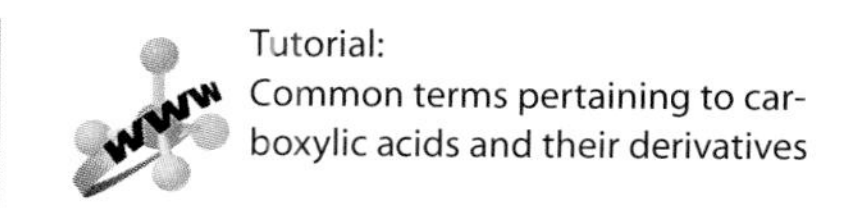

Tutorial:
Common terms pertaining to carboxylic acids and their derivatives

SUMMARY

A **carbonyl group** is a carbon double bonded to an oxygen; an **acyl group** is a carbonyl group attached to an alkyl or aryl group. **Acyl chlorides**, **esters**, and **amides** are called **carboxylic acid derivatives** because they differ from a carboxylic acid only in the nature of the group that has replaced the OH group of the carboxylic acid.

Carbonyl compounds can be placed in one of two classes. Class I carbonyl compounds contain a group that can be replaced by another group; carboxylic acids and carboxylic acid derivatives belong to this class. Class II carbonyl compounds do not contain a group that can be replaced by another group; aldehydes and ketones belong to this class.

The reactivity of carbonyl compounds resides in the polarity of the carbonyl group; the carbonyl carbon has a partial positive charge that is attractive to nucleophiles. Class I carbonyl compounds undergo nucleophilic acyl substitution reactions, in which a nucleophile replaces the substituent that was attached to the acyl group in the reactant. All Class I carbonyl compounds react with nucleophiles in the same way: the nucleophile attacks the carbonyl carbon, forming an unstable tetrahedral intermediate, which reforms a carbonyl compound by eliminating the weakest base.

A carboxylic acid derivative will undergo a nucleophilic acyl substitution reaction provided that the newly added group in the tetrahedral intermediate is not a much weaker base than the group that is attached to the acyl group in the reactant. The weaker the base attached to the acyl group, the more easily both steps of the nucleophilic acyl substitution reaction can take place. The relative reactivities toward nucleophilic acyl substitution: acyl chlorides > esters and carboxylic acids > amides > carboxylate ions.

Hydrolysis, **alcoholysis**, and **aminolysis** are reactions in which water, alcohols, and amines, respectively, convert one compound into two compounds. A **transesterification reaction** converts one ester to another ester. Reacting a carboxylic acid with an alcohol and an acid catalyst is called a **Fischer esterification**.

The rate of hydrolysis or alcoholysis can be increased by an acid. An acid increases the rate of formation of the tetrahedral intermediate by protonating the carbonyl oxygen, which increases the electrophilicity of the carbonyl group, and by decreasing the basicity of the leaving group, which makes it easier to eliminate.

Amides are unreactive compounds but do react with water and alcohols if the reaction mixture is heated in the presence of an acid. Nitriles are harder to hydrolyze than amides. Carboxylic acids are activated by being converted to acyl chlorides.

SUMMARY OF REACTIONS

1. Reactions of acyl halides (Section 11.7)

$$RC(=O)Cl + CH_3OH \longrightarrow RC(=O)OCH_3 + HCl$$

$$RC(=O)Cl + H_2O \longrightarrow RC(=O)OH + HCl$$

$$RC(=O)Cl + 2\ CH_3NH_2 \longrightarrow RC(=O)NHCH_3 + CH_3\overset{+}{N}H_3\ Cl^-$$

2. Reactions of esters (Sections 11.8–11.9)

$$RC(=O)OR + CH_3OH \underset{}{\overset{HCl}{\rightleftharpoons}} RC(=O)OCH_3 + ROH$$

$$RC(=O)OR + H_2O \overset{HCl}{\rightleftharpoons} RC(=O)OH + ROH$$

$$RC(=O)OR + CH_3NH_2 \longrightarrow RC(=O)NHCH_3 + ROH$$

3. Reactions of carboxylic acids (Section 11.11)

$$RC(=O)OH + CH_3OH \overset{HCl}{\rightleftharpoons} RC(=O)OCH_3 + H_2O$$

$$RC(=O)OH + CH_3NH_2 \longrightarrow RC(=O)O^-\ H_3\overset{+}{N}CH_3$$

4. Reactions of amides (Sections 11.12 and 11.13)

$$RC(=O)NH_2 + H_2O \xrightarrow[\Delta]{HCl} RC(=O)OH + \overset{+}{N}H_4\,Cl^-$$

$$RC(=O)NH_2 + CH_3OH \xrightarrow[\Delta]{HCl} RC(=O)OCH_3 + \overset{+}{N}H_4\,Cl^-$$

5. Activation of carboxylic acids (Section 11.14)

$$RC(=O)OH + SOCl_2 \xrightarrow[\Delta]{} RC(=O)Cl + SO_2 + HCl$$

6. Hydrolysis of nitriles (Section 11.15)

$$RC{\equiv}N + H_2O \xrightarrow[\Delta]{HCl} RC(=O)OH + \overset{+}{N}H_4$$

PROBLEMS

26. Write a structure for each of the following:

a. *N,N*-dimethylhexanamide
b. 3,3-dimethylhexanamide
c. 3-methylpentanoyl chloride
d. 2-bromohexanoic acid
e. sodium acetate
f. propionyl chloride

27. Name the following:

a. $CH_3CH_2CH(CH_2CH_3)CH_2CH_2CH_2C(=O)OH$

b. $CH_3CH_2C(=O)OCH_2CH_2CH_3$

c. 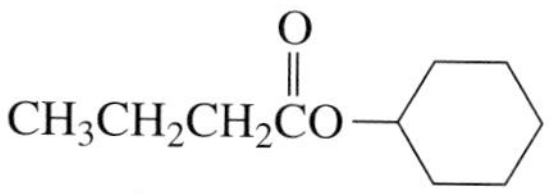 $CH_3CH_2CH_2C(=O)O$–cyclohexyl

d. $CH_3CH_2CH_2C(=O)N(CH_3)_2$

e. $CH_3CH_2CH_2CH_2C(=O)Cl$

f. $CH_3CH_2CH(CH_3)CH_2C(=O)OCH_3$

g. $CH_2{=}CHCH_2C(=O)NHCH_3$

h. 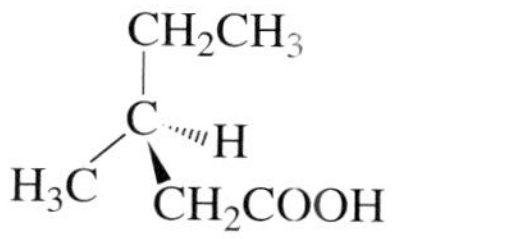 H_3C–C(H)(CH_2CH_3)–CH_2COOH (stereocenter with wedge/hash bonds)

i. $CH_3CH_2CH_2CH_2C{\equiv}N$

28. What products would be formed from the reaction of acetyl chloride with the following reagents?

a. water
b. excess dimethylamine
c. excess aniline
d. cyclohexanol
e. 4-chlorophenol
f. isopropyl alcohol

29. **a.** List the following esters in order of decreasing reactivity in the first step of a nucleophilic acyl substitution reaction (formation of the tetrahedral intermediate):

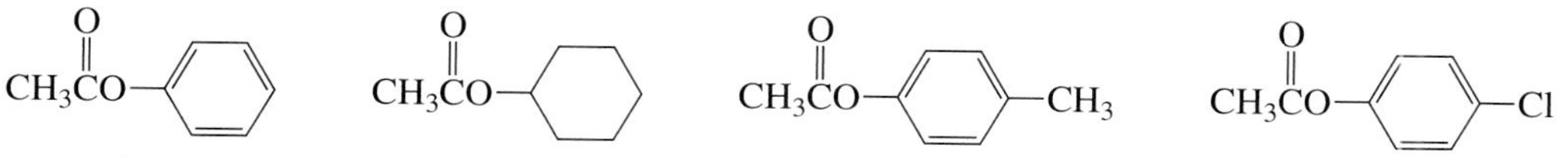

A: $CH_3C(=O)O$–phenyl
B: $CH_3C(=O)O$–cyclohexyl
C: $CH_3C(=O)O$–C_6H_4–CH_3 (para)
D: $CH_3C(=O)O$–C_6H_4–Cl (para)

b. List the same esters in order of decreasing reactivity in the last step of a nucleophilic acyl substitution reaction (collapse of the tetrahedral intermediate).

30. Using an alcohol for one method and an alkyl halide for the other, show two ways to make each of the following esters:

a. propyl acetate (odor of pears)
b. ethyl butyrate (odor of pineapple)
c. isopentyl acetate (odor of bananas)
d. methyl phenylethanoate (odor of honey)

31. Which compound would you expect to have a higher boiling point, the ester or the carboxylic acid?

$$CH_3\overset{O}{\overset{\|}{C}}OCH_3 \qquad CH_3\overset{O}{\overset{\|}{C}}OH$$

32. If propionyl chloride is added to one equivalent of methylamine, only a 50% yield of *N*-methylpropanamide is obtained. If, however, the acyl chloride is added to two equivalents of methylamine, the yield of *N*-methylpropanamide is almost 100%. Explain these observations.

33. What reagents would you use to convert methyl propanoate into the following compounds?

a. isopropyl propanoate
b. sodium propanoate
c. *N*-ethylpropanamide
d. propanoic acid

34. Aspartame, the sweetener used in the commercial products NutraSweet and Equal, is 160 times sweeter than sucrose. What products would be obtained if aspartame were hydrolyzed completely in an aqueous solution of HCl?

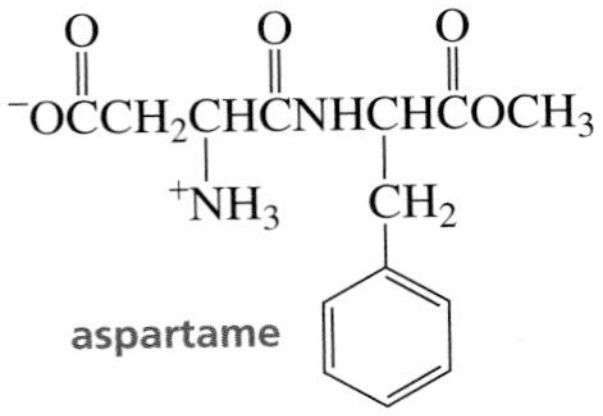

$$^{-}O\overset{O}{\overset{\|}{C}}CH_2\underset{^{+}NH_3}{\underset{|}{C}H}\overset{O}{\overset{\|}{C}}NH\underset{CH_2C_6H_5}{\underset{|}{C}H}\overset{O}{\overset{\|}{C}}OCH_3$$

aspartame

35. a. Which of the following reactions will not give the carbonyl product shown?

1. $CH_3\overset{O}{\overset{\|}{C}}NH_2 + Cl^- \longrightarrow CH_3\overset{O}{\overset{\|}{C}}Cl$

2. $CH_3\overset{O}{\overset{\|}{C}}OH + CH_3NH_2 \longrightarrow CH_3\overset{O}{\overset{\|}{C}}NHCH_3$

3. $CH_3\overset{O}{\overset{\|}{C}}OCH_3 + CH_3NH_2 \longrightarrow CH_3\overset{O}{\overset{\|}{C}}NHCH_3$

4. $CH_3\overset{O}{\overset{\|}{C}}OCH_3 + Cl^- \longrightarrow CH_3\overset{O}{\overset{\|}{C}}Cl$

5. $CH_3\overset{O}{\overset{\|}{C}}Cl + H_2O \longrightarrow CH_3\overset{O}{\overset{\|}{C}}OH$

6. $CH_3\overset{O}{\overset{\|}{C}}NHCH_3 + H_2O \longrightarrow CH_3\overset{O}{\overset{\|}{C}}OH$

b. Which of the reactions that do not occur can be made to occur if an acid catalyst is added to the reaction mixture?

36. Identify the major and minor products of the following reaction:

$$\text{2-piperidinyl ring bearing } \underset{CH_3}{CHOH} \text{ and } CH_2CH_3 + CH_3\overset{O}{\overset{\|}{C}}Cl \longrightarrow$$

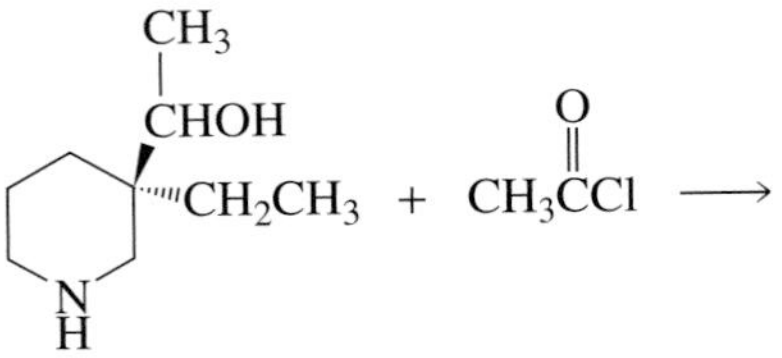

37. D. N. Kursanov, a Russian chemist, studied the hydrolysis of the following ester in a 1.0 M solution of sodium hydroxide, and he was able to prove that the bond that is broken in the reaction is the acyl C—O bond rather than the alkyl C—O bond:

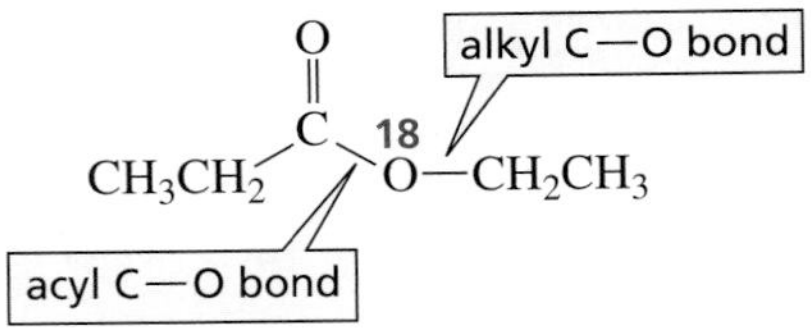

$$CH_3CH_2\overset{O}{\overset{\|}{C}}{-}^{18}O{-}CH_2CH_3$$

alkyl C—O bond

acyl C—O bond

a. Which of the products contained the ^{18}O label?
b. What product would have contained the ^{18}O label if the alkyl C—O bond had broken?

38. Give the products of the following reactions:

a. 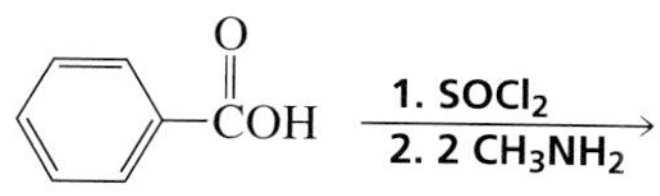 $C_6H_5COOH \xrightarrow[2.\ 2\ CH_3NH_2]{1.\ SOCl_2}$

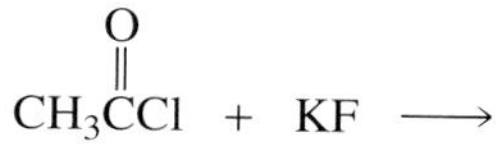

b. CH_3CCl (C=O) + KF ⟶

c. (γ-butyrolactam, cyclic amide with NH) + $H_2O \xrightarrow[\Delta]{HCl}$

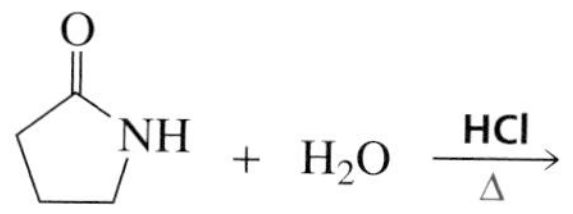

d. 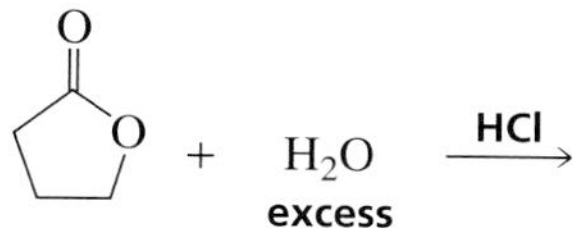 (γ-butyrolactone) + H_2O (excess) $\xrightarrow{HCl}$

39. Which of the reaction coordinate diagrams represents the reaction of an ester with chloride ion?

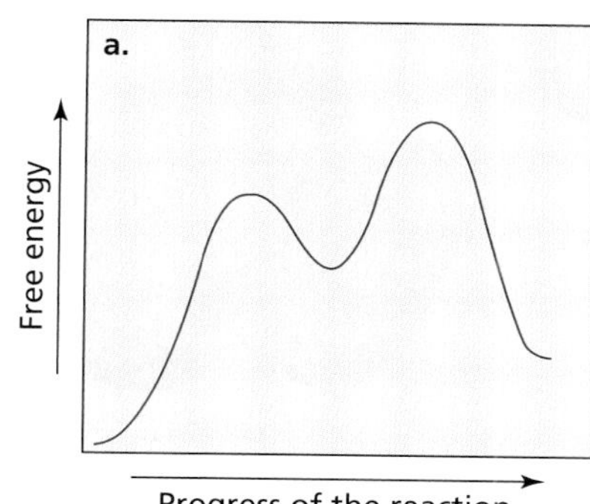

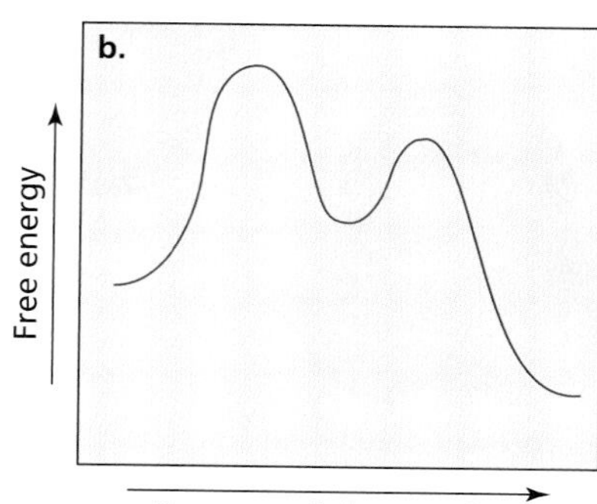

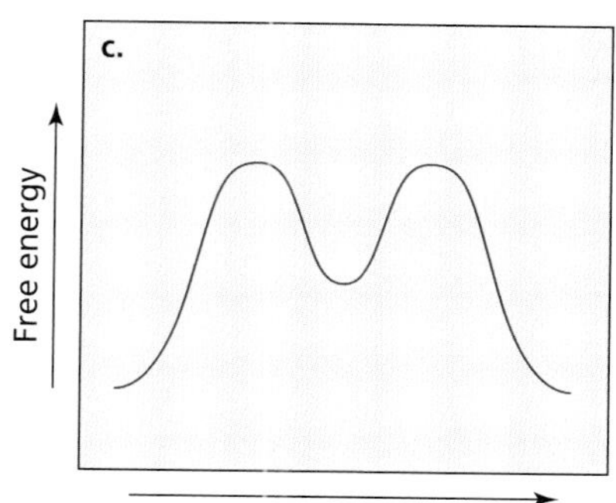

40. Which ester is more reactive, methyl acetate or phenyl acetate?

41. List the following amides in order of decreasing reactivity toward acid-catalyzed hydrolysis:

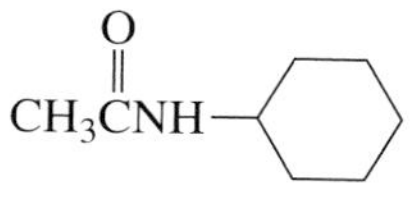

A: CH_3CNH–cyclohexyl (C=O)

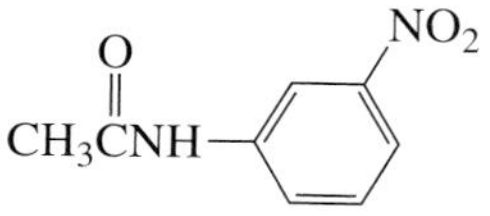

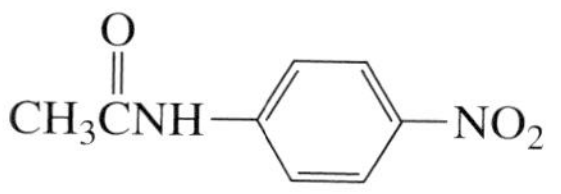

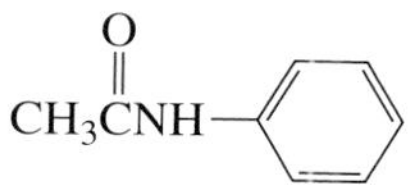

B: CH_3CNH–phenyl with meta-NO_2

C: CH_3CNH–phenyl with para-NO_2

D: CH_3CNH–phenyl

42. Give the products of the following reactions:

a. 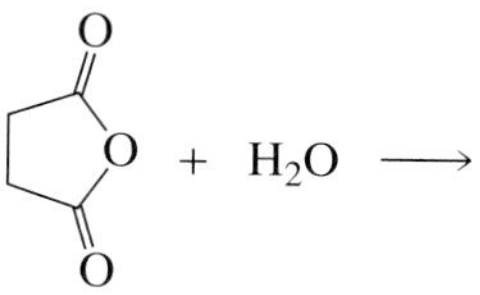 (succinic anhydride) + H_2O ⟶

b. 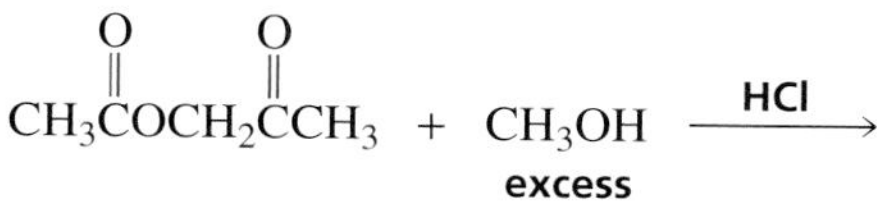 $CH_3COCH_2CCH_3$ (two C=O) + CH_3OH (excess) $\xrightarrow{HCl}$

43. An aqueous solution of a primary or secondary amine reacts with an acyl chloride to form an amide as the major product. However, if the amine is tertiary, an amide is not formed. What product *is* formed? Explain.

44. Is the acid-catalyzed hydrolysis of acetamide a reversible or an irreversible reaction? Explain.

45. What product would you expect to obtain from each of the following reactions?

a.  $CH_3CH_2CH(OH)CH_2CH_2CH_2COOH \xrightarrow{HCl}$

b. 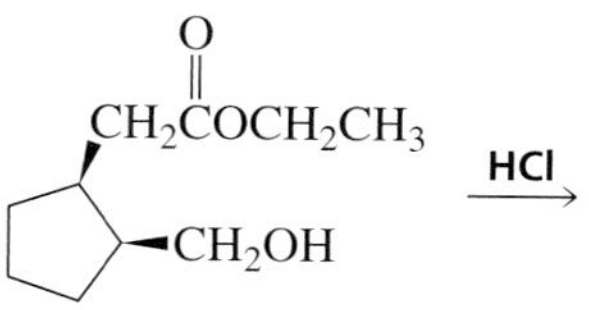 (cyclopentane bearing cis $CH_2COCH_2CH_3$ (C=O) and CH_2OH substituents) $\xrightarrow{HCl}$

46. a. When a carboxylic acid is dissolved in isotopically labeled water (H_2O^{18}) in the presence of an acid catalyst, the label is incorporated into both oxygens of the acid. Propose a mechanism to account for this.

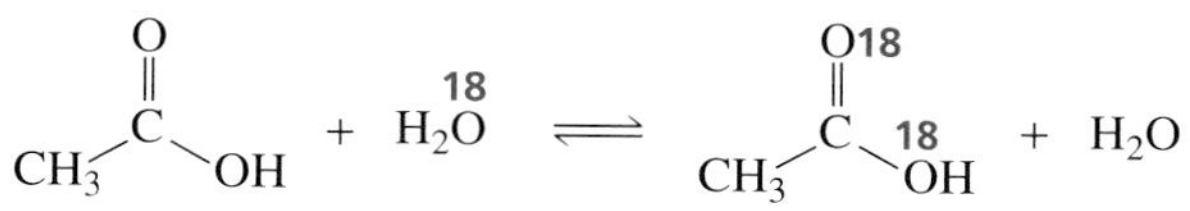

$$CH_3C(=O)OH + H_2{}^{18}O \rightleftharpoons CH_3C(={}^{18}O){}^{18}OH + H_2O$$

b. If a carboxylic acid is dissolved in isotopically labeled methanol ($CH_3{}^{18}OH$) and an acid catalyst is added, where will the label reside in the product?

CHAPTER 12

Carbonyl Compounds II

Reactions of Aldehydes and Ketones • More Reactions of Carboxylic Acid Derivatives

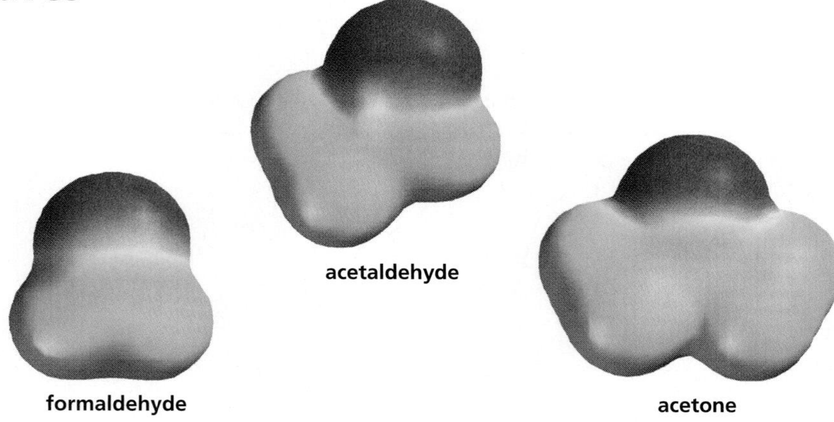

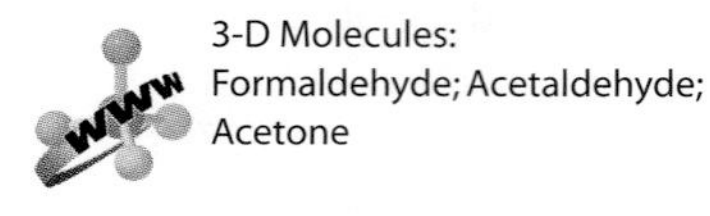

At the beginning of Chapter 11, we saw that carbonyl compounds—compounds that possess a carbonyl group (C=O)—can be divided into two classes: Class I carbonyl compounds, which have a group that can be replaced by a nucleophile, and Class II carbonyl compounds, which do not have a group that can be replaced by a nucleophile. Class II carbonyl compounds comprise aldehydes and ketones.

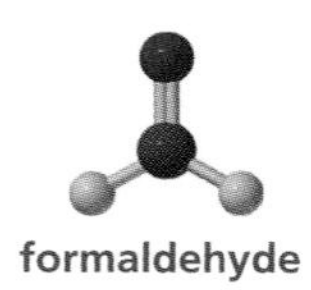
formaldehyde

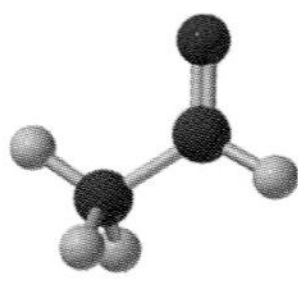
acetaldehyde

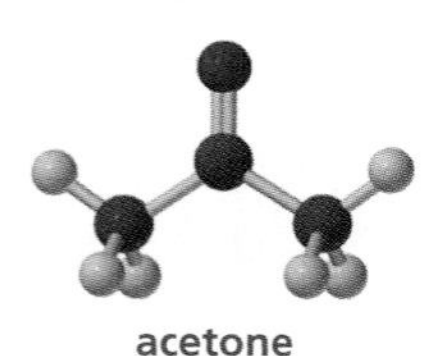
acetone

The carbonyl carbon of the simplest aldehyde, formaldehyde, is bonded to two hydrogens. The carbonyl carbon in all other **aldehydes** is bonded to a hydrogen and to an alkyl (or an aryl) group. The carbonyl carbon of a **ketone** is bonded to two alkyl (or aryl) groups. Aldehydes and ketones *do not have* a group that can be replaced by another group because hydride ions (H^-) and carbanions (R^-) are too basic to be displaced by nucleophiles under normal conditions.

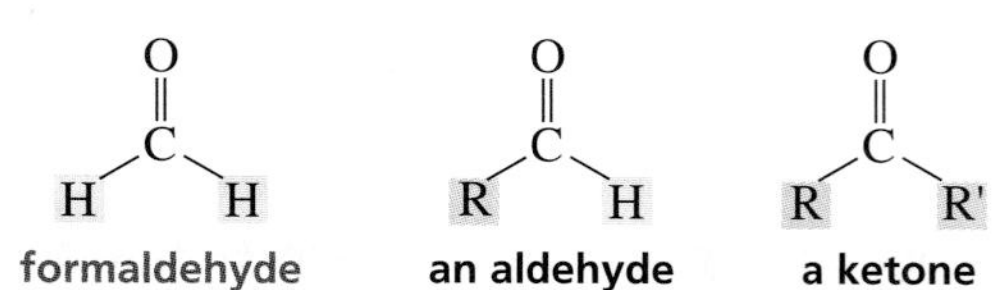
formaldehyde — an aldehyde — a ketone

The physical properties of aldehydes and ketones are discussed in Section 11.3 (see also Appendix I).

Many compounds found in nature have aldehyde or ketone functional groups. Aldehydes have pungent odors, whereas ketones tend to smell sweet. Vanillin and cinnamaldehyde are examples of naturally occurring aldehydes. A whiff of vanilla extract will allow you to appreciate the pungent odor of vanilla. The ketones carvone and camphor are responsible for the characteristic sweet odors of spearmint leaves, caraway seeds, and the leaves of the camphor tree.

vanillin
vanilla flavoring

cinnamaldehyde
cinnamon flavoring

camphor

(*R*)-(–)-carvone
spearmint oil

(*S*)-(+)-carvone
caraway seed oil

In ketosis, a pathological condition that can occur in people with diabetes, the body produces more acetoacetate than can be metabolized. The excess acetoacetate breaks down to acetone (a ketone) and CO_2. Ketosis can be recognized by the smell of acetone on a person's breath.

$$CH_3COCH_2COO^- + H_2O \longrightarrow CH_3COCH_3 + CO_2 + HO^-$$

acetoacetate acetone

Two ketones that are of biological importance illustrate how a small difference in structure can be responsible for a large difference in biological activity: progesterone is a sex hormone synthesized primarily in the ovaries, whereas testosterone is a sex hormone synthesized primarily in the testes.

progesterone
a female sex hormone

testosterone
a male sex hormone

12.1 THE NOMENCLATURE OF ALDEHYDES AND KETONES

Naming Aldehydes

The systematic name of an aldehyde is obtained by replacing the final "e" on the name of the parent hydrocarbon with "al." For example, a one-carbon aldehyde is methan*al*; a two-carbon aldehyde is ethan*al*. The position of the carbonyl carbon does not have to be designated because it is always at the end of the parent hydrocarbon and therefore always has the 1-position.

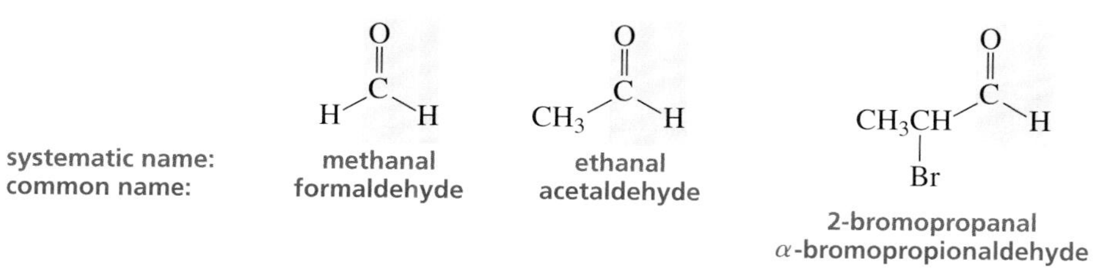

systematic name:	methanal	ethanal	2-bromopropanal
common name:	formaldehyde	acetaldehyde	α-bromopropionaldehyde

The common name of an aldehyde is the same as the common name of the corresponding carboxylic acid (Section 11.1), except that "aldehyde' is substituted for "ic acid" (or "oic acid"). When common names are used, the position of a substituent is designated by a lowercase Greek letter. The carbonyl carbon is not given a letter; the carbon adjacent to the carbonyl carbon is the α-carbon.

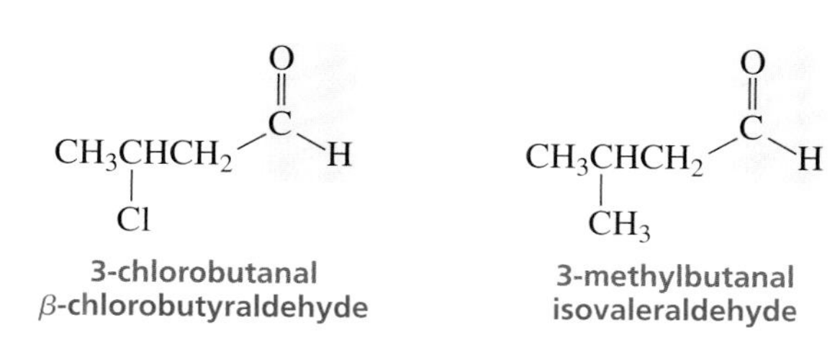

systematic name:	3-chlorobutanal	3-methylbutanal
common name:	β-chlorobutyraldehyde	isovaleraldehyde

Naming Ketones

The systematic name of a ketone is obtained by replacing the final "e" on the end of the name of the parent hydrocarbon with "one." The chain is numbered in the direction that gives the carbonyl carbon the smaller number. In the case of cyclic ketones, however, a number is not necessary because the carbonyl carbon is assumed to be at the 1-position. Derived names are often used for ketones: the substituents attached to the carbonyl group are stated in alphabetical order, followed by "ketone."

3-D Molecules: 3-Methyl-2-butanone; 2,4-Dimethyl-3-pentanone

CH_3COCH_3 | $CH_3CH_2COCH_2CH_2CH_3$ | $CH_3CH(CH_3)CH_2CH_2CH_2COCH_3$

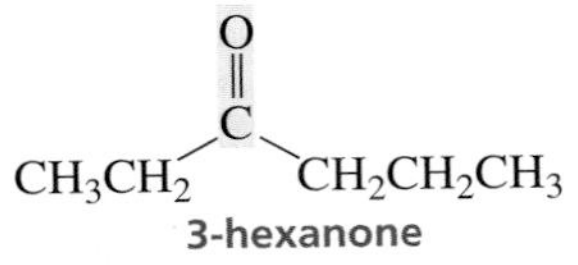

systematic name:	propanone	3-hexanone	6-methyl-2-heptanone
common name:	acetone		
derived name:	dimethyl ketone	ethyl propyl ketone	isohexyl methyl ketone

Tutorial: Nomenclature of aldehydes and ketones

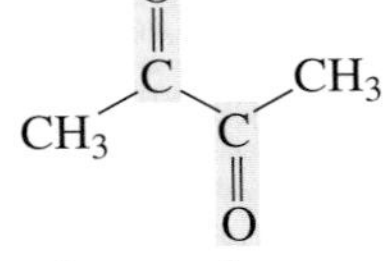

$CH_3COCOCH_3$ | $CH_3COCH_2COCH_3$

systematic name:	cyclohexanone	butanedione	2,4-pentanedione
common name:			acetylacetone

Only a few ketones have common names. The smallest ketone, propanone, is usually referred to by its common name, acetone. Acetone is a widely used laboratory solvent.

BUTANEDIONE: AN UNPLEASANT COMPOUND

Fresh perspiration is odorless. The smells we associate with perspiration result from a chain of events initiated by bacteria that are always present on our skin. These bacteria produce lactic acid, which creates an acidic environment that allows other bacteria to break down the components of perspiration, producing compounds with the unappealing odors we associate with armpits and sweaty feet. One such compound is butanedione.

PROBLEM 1 ♦

Why are numbers not used to designate the positions of the functional groups in propanone and butanedione?

PROBLEM 2 ♦

Name the following:

a. $CH_3CH_2CH_2\overset{O}{\overset{\|}{C}}CH_2CH_2CH_3$

b. $C_6H_5{-}CH_2CH_2CH_2\overset{O}{\overset{\|}{C}}H$

PROBLEM 3 ♦

Give two names for each of the following:

a. $CH_3CH_2\underset{CH_3}{\underset{|}{C}}HCH_2\overset{O}{\overset{\|}{C}}H$

b. $CH_3\underset{CH_3}{\underset{|}{C}}HCH_2\overset{O}{\overset{\|}{C}}CH_2CH_2CH_3$

c. $CH_3CH_2\overset{CH_2CH_3}{\overset{|}{C}}HCH_2CH_2\overset{O}{\overset{\|}{C}}H$

d. $CH_3CH_2\overset{O}{\overset{\|}{C}}CH_2CH_2\underset{CH_3}{\underset{|}{C}}HCH_3$

12.2 RELATIVE REACTIVITIES OF CARBONYL COMPOUNDS

We have seen that the carbonyl group is polar because oxygen, being more electronegative than carbon, has a greater share of the double bond's electrons (Section 11.5). The partial positive charge on the carbonyl carbon causes it to be attacked by nucleophiles. The electron deficiency of the carbonyl carbon is indicated by the blue regions in the electrostatic potential maps on page 324.

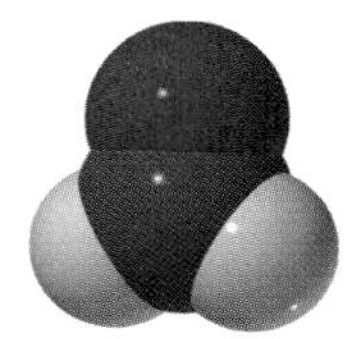

formaldehyde

$$\mathrm{R(C{=}\ddot{O}^{\delta-})R(H)},\ \delta+\ \text{on C} \quad \longleftarrow \ \ddot{:}\mathrm{Nu}$$

acetaldehyde

An aldehyde has a greater partial positive charge on its carbonyl carbon than does a ketone because a hydrogen is electron withdrawing compared with an alkyl group (Section 5.2). An aldehyde, therefore, is more reactive than a ketone toward nucleophilic attack.

relative reactivities

most reactive → $\mathrm{HC(=O)H}$ (formaldehyde) > $\mathrm{RC(=O)H}$ (an aldehyde) > $\mathrm{RC(=O)R'}$ (a ketone) ← least reactive

acetone

Steric factors also contribute to the greater reactivity of an aldehyde. The carbonyl carbon of an aldehyde is more accessible to the nucleophile than is the carbonyl carbon of a ketone because the hydrogen attached to the carbonyl carbon of an aldehyde is smaller than the second alkyl group attached to the carbonyl carbon of a ketone.

For the same reason, ketones with small alkyl groups bonded to the carbonyl carbon are more reactive than ketones with large alkyl groups.

relative reactivities

most reactive → $\mathrm{H_3CC(=O)CH_3}$ > $\mathrm{H_3CC(=O)CH(CH_3)CH_3}$ > $\mathrm{CH_3CH(CH_3)C(=O)CH(CH_3)CH_3}$ ← least reactive

How does the reactivity of an aldehyde or a ketone toward nucleophiles compare with the reactivity of the carbonyl compounds whose reactions we looked at in Chapter 11? Aldehydes and ketones are in the middle; they are less reactive than acyl chlorides, but more reactive than esters, carboxylic acids, and amides.

relative reactivities of carbonyl compounds

acyl chloride > aldehyde > ketone > ester ~ carboxylic acid > amide > carboxylate ion

(acyl chloride: most reactive; carboxylate ion: least reactive)

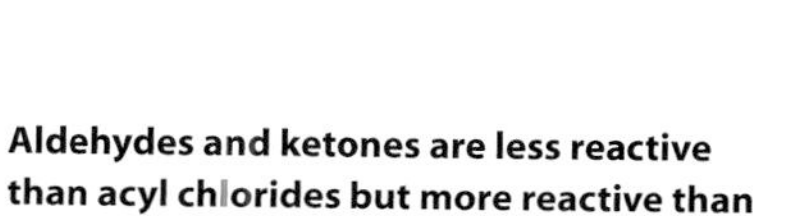

Aldehydes and ketones are less reactive than acyl chlorides but more reactive than esters, carboxylic acids, and amides.

PROBLEM 4 ♦

Which ketone in each pair is more reactive

a. 2-heptanone or 4-heptanone?

b. 4-methyl-3-hexanone or 5-methyl-3-hexanone?

12.3 HOW ALDEHYDES AND KETONES REACT

In Section 11.5, we saw that the carbonyl group of a carboxylic acid or a carboxylic acid derivative is attached to a group that can be replaced by another group. Therefore, these compounds react with nucleophiles to form substitution products.

$$\mathrm{RC(=O)Y} + \mathrm{Z^-} \rightleftharpoons \mathrm{R{-}C(O^-)(Z){-}Y} \rightleftharpoons \mathrm{RC(=O)Z} + \mathrm{Y^-}$$

product of nucleophilic acyl substitution

In contrast, the carbonyl group of an aldehyde or a ketone is attached to a group that is too strong a base (H^- or R^-) to be eliminated under normal conditions, so it cannot be replaced by another group. Consequently, aldehydes and ketones react with nucleophiles to form addition products, not substitution products. Aldehydes and ketones therefore undergo **nucleophilic addition** reactions, whereas carboxylic acid derivatives undergo **nucleophilic acyl substitution** reactions.

Aldehydes and ketones undergo nucleophilic addition reactions.

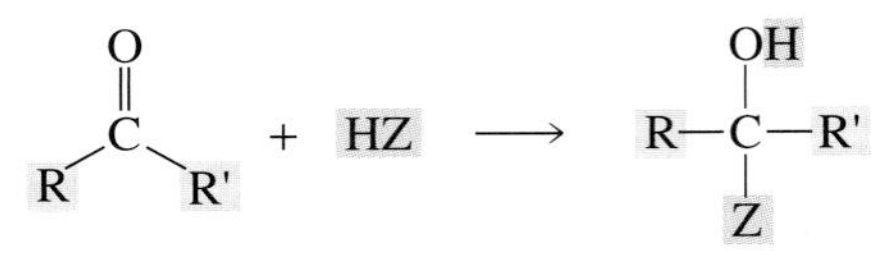

product of nucleophilic addition

Francis August Victor Grignard (1871–1935) *was born in France, the son of a sail maker. He received a Ph.D. from the University of Lyons in 1901. His synthesis of the first Grignard reagent was announced in 1900. During the next five years, some 200 papers were published about Grignard reagents. He was a professor of chemistry at the University of Nancy and later at the University of Lyons. He received the Nobel Prize in chemistry in 1912. During World War I, Grignard was drafted into the French army, where he developed a method to detect war gases.*

12.4 GRIGNARD REAGENTS

Few reactions in organic chemistry result in the formation of new carbon–carbon bonds. Consequently, those reactions that do are very important to synthetic organic chemists when they need to synthesize large organic molecules from smaller molecules. The addition of a *carbon nucleophile* to a carbonyl compound is an example of a reaction that forms a new C—C bond and therefore forms a product with more carbon atoms than the starting material.

Grignard reagents are the most widely used carbon nucleophiles. They are prepared by adding an alkyl halide to magnesium shavings being stirred in diethyl ether. This reaction inserts a magnesium between the carbon and the halogen. Grignard reagents react as if they were carbanions. Recall that a carbanion is a species containing a negatively charged carbon (Section 1.4).

$$CH_3CH_2Br \xrightarrow[Et_2O]{Mg} CH_3CH_2MgBr$$

CH_3CH_2MgBr **reacts as if it were** $CH_3\ddot{C}H_2^{-}$ $\overset{+}{M}gBr$

Grignard reagents have a nucleophilic carbon because the carbon is bonded to an atom that is *less* electronegative than carbon. (See Table 1.3 on page 8.) The nucleophilic carbon of a Grignard reagent reacts with electrophiles.

$$\overset{\delta-}{CH_3CH_2}{-}\overset{\delta+}{M} + E^+ \longrightarrow CH_3CH_2{-}E + M^+$$

less electronegative than carbon

nucleophile

electrophile

Compare the carbon in Grignard reagent that is nucleophilic because it is attached to an atom that is *less* electronegative that carbon with the carbon in an alkyl halide that is electrophilic because it is attached to an atom that is *more* electronegative than carbon. We have seen that the electrophilic carbon of an alkyl halide reacts with nucleophiles (Section 9.1).

$$\overset{\delta+}{CH_3CH_2}-\overset{\delta-}{Z} + Y^- \longrightarrow CH_3CH_2-Y + Z^-$$

more electronegative than carbon

electrophile

nucleophile

Grignard reagents are such strong bases that they will react immediately with any acid that is present in the reaction mixture—even with trace amounts of very weak acids such as water, alcohols, and amines. When this happens, the Grignard reagent is converted into an alkane.

$$CH_3CH_2\underset{Br}{\underset{|}{C}}HCH_3 \xrightarrow[Et_2O]{Mg} CH_3CH_2\underset{MgBr}{\underset{|}{C}}HCH_3 \xrightarrow{H_2O} CH_3CH_2CH_2CH_3$$

PROBLEM 5 ◆

Give the products of the following reactions:

a. $CH_3CH_2MgBr + H_2O \longrightarrow$

b. $CH_3CH_2MgBr + CH_3OH \longrightarrow$

c. $CH_3CH_2MgBr + CH_3NH_2 \longrightarrow$

12.5 THE REACTIONS OF CARBONYL COMPOUNDS WITH GRIGNARD REAGENTS

The Reactions of Aldehydes and Ketones with Grignard Reagents

The mechanism for the reaction of a Grignard reagent with an aldehyde or a ketone is shown below.

mechanism for the reaction of an aldehyde or a ketone with a Grignard reagent

$$\underset{R\quad R}{\overset{:O:}{\overset{\|}{C}}} + R'-MgBr \longrightarrow R-\underset{R'}{\overset{:\ddot{O}:^-\ \overset{+}{M}gBr}{C}}-R \xrightarrow{H_3O^+} R-\underset{R'}{\overset{:\ddot{O}H}{C}}-R$$

- Nucleophilic attack of a Grignard reagent on the carbonyl carbon forms an alkoxide ion that is complexed with magnesium ion.
- Protonation of the alkoxide ion forms an alcohol.

The reaction is a nucleophilic addition reaction: the nucleophile has added to the carbonyl carbon. Notice that the tetrahedral alkoxide ion is stable because it does not have a group that can be eliminated. (In Section 11.5, we saw that a tetrahedral compound is unstable only if the newly formed sp^3 carbon is attached to an oxygen *and* to another electronegative atom.)

A tetrahedral compound with its sp^3 carbon bonded to an oxygen and to another electronegative atom is unstable.

When a Grignard reagent reacts with *formaldehyde*, the product of the addition reaction is a *primary alcohol.*

$$\underset{\text{formaldehyde}}{HCHO} + \underset{\text{butylmagnesium bromide}}{CH_3CH_2CH_2CH_2{-}MgBr} \longrightarrow \underset{\text{an alkoxide ion}}{CH_3CH_2CH_2CH_2CH_2\ddot{O}{:}^- \overset{+}{Mg}Br} \xrightarrow{H_3O^+} \underset{\text{1-pentanol, a primary alcohol}}{CH_3CH_2CH_2CH_2CH_2\ddot{O}H}$$

When a Grignard reagent reacts with an *aldehyde other than formaldehyde*, the product of the addition reaction is a *secondary alcohol*.

$$\underset{\text{propanal}}{CH_3CH_2CHO} + \underset{\text{propylmagnesium bromide}}{CH_3CH_2CH_2{-}MgBr} \longrightarrow CH_3CH_2CH(\ddot{O}{:}^- \overset{+}{Mg}Br)CH_2CH_2CH_3 \xrightarrow{H_3O^+} \underset{\text{3-hexanol, a secondary alcohol}}{CH_3CH_2CH(\ddot{O}H)CH_2CH_2CH_3}$$

When a Grignard reagent reacts with a *ketone*, the product of the addition reaction is a *tertiary alcohol.*

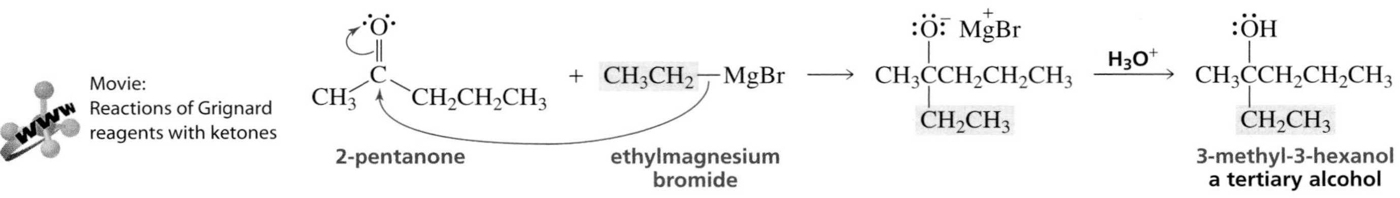

In the following reactions, the reagents are numbered in order of use, indicating that the acid is not added until after the Grignard reagent has reacted with the carbonyl compound:

$$\underset{\text{3-pentanone}}{CH_3CH_2COCH_2CH_3} \xrightarrow[\text{2. } H_3O^+]{\text{1. } CH_3MgBr} \underset{\text{3-methyl-3-pentanol}}{CH_3CH_2C(OH)(CH_3)CH_2CH_3}$$

$$\underset{\text{butanal}}{CH_3CH_2CH_2CHO} \xrightarrow[\text{2. } H_3O^+]{\text{1. } C_6H_5{-}MgBr} \underset{\text{1-phenyl-1-butanol}}{CH_3CH_2CH_2CH(OH){-}C_6H_5}$$

The reaction of a Grignard reagent with a carbonyl compound can produce compounds with a variety of structures because both the structure of the carbonyl compound and the structure of the Grignard reagent can be varied.

PROBLEM 6 ♦

What products would be formed when the following compounds react with CH_3MgBr followed by the addition of dilute acid?

a. $CH_3CH_2CH_2CH_2CHO$ **b.** $CH_3CH_2CH_2COCH_3$ **c.** cyclohexanone

PROBLEM 7 ♦

We saw that 3-methyl-3-hexanol can be synthesized from the reaction of 2-pentanone with ethylmagnesium bromide. What two other combinations of ketone and Grignard reagent could be used to prepare the same tertiary alcohol?

The Reactions of Esters and Acyl Chlorides with Grignard Reagents

In addition to reacting with aldehydes and ketones (Class II carbonyl compounds), Grignard reagents react with esters and acyl chlorides (Class I carbonyl compounds that have a group that can be replaced by another group).

Esters and acyl chlorides undergo two successive reactions with a Grignard reagent. For example, when an ester reacts with a Grignard reagent, the first reaction is a *nucleophilic acyl substitution reaction* because an ester, unlike an aldehyde or a ketone, has a group that can be replaced by the Grignard reagent; the second reaction is a *nucleophilic addition reaction*. The mechanism for the overall reaction is shown below.

mechanism for the reaction of an ester with a Grignard reagent

$CH_3CH_2C(=O)OCH_3$ (an ester) + $CH_3{-}MgBr$ ⟶ $CH_3CH_2C(O^-\,{}^+MgBr)(CH_3)OCH_3$ ⟶ $CH_3CH_2C(=O)CH_3$ (a ketone) + CH_3O^-

a group is expelled from the tetrahedral intermediate

product of nucleophilic acyl substitution

$CH_3CH_2C(=O)CH_3$ + $CH_3{-}MgBr$ ⟶ $CH_3CH_2C(O^-\,{}^+MgBr)(CH_3)CH_3$ $\xrightarrow{H_3O^+}$ $CH_3CH_2C(OH)(CH_3)CH_3$ (a tertiary alcohol)

product of nucleophilic addition

Movie: Reaction of a Grignard reagent with an ester

- Nucleophilic attack by the Grignard reagent forms a tetrahedral intermediate, which is unstable because the sp^3 carbon is attached to an oxygen and to another electronegative atom (Section 11.5).
- The tetrahedral intermediate expels a methoxide ion, forming a ketone. The reaction does not stop at the ketone stage, however, because ketones are more reactive than esters toward nucleophilic attack.
- The ketone reacts with a second molecule of Grignard reagent.
- Protonation of the alkoxide ion forms a tertiary alcohol.

Because the tertiary alcohol is formed as a result of two successive reactions with a Grignard reagent, the alcohol has two identical groups bonded to the tertiary carbon.

Tertiary alcohols are also formed from the reaction of two equivalents of a Grignard reagent with an acyl chloride. The first equivalent replaces the Cl in a nucleophilic acyl substitution reaction; the second equivalent reacts in a nucleophilic addition reaction.

$CH_3CH_2CH_2C(=O)Cl$ (butyryl chloride) $\xrightarrow[2.\ H_3O^+]{1.\ 2\ CH_3CH_2MgBr}$ $CH_3CH_2CH_2C(OH)(CH_2CH_3)CH_2CH_3$ (3-ethyl-3-hexanol)

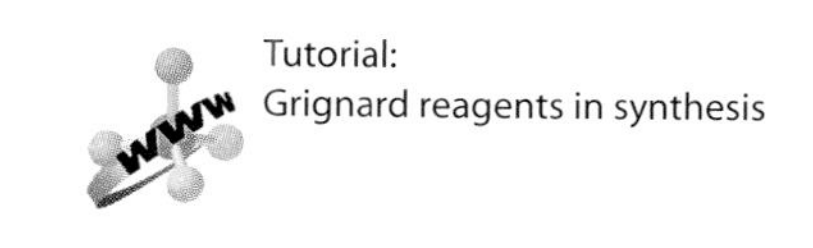

SYNTHESIZING ORGANIC COMPOUNDS

Organic chemists synthesize compounds for many reasons: to study their properties, to answer a variety of chemical questions, or because they have useful properties. One reason chemists synthesize natural products is to provide us with larger supplies of those compounds than nature can produce. For example, Taxol—a compound that has been successful in treating ovarian cancer, breast cancer, and certain forms of lung cancer—is extracted from the bark of *Taxus*, the yew tree found in the Pacific Northwest. The supply of natural Taxol is limited because yew trees are uncommon, grow very slowly, and stripping the bark kills the tree. Moreover, the bark of a 40-foot tree, which may have taken 200 years to grow, provides only 0.5 g of the drug. In addition, *Taxus* forests serve as habitats for the spotted owl, an endangered species, so harvesting the trees would accelerate the owl's demise. Once chemists were successful in determining the structure of Taxol, efforts were undertaken to synthesize it in order to make it more widely available. Several syntheses have been successful.

(continued...)

Taxol

Once a compound has been synthesized, chemists can study its properties to learn how it works; then they can design and synthesize safer or more potent analogs. For example, chemists have found that the anticancer activity of Taxol is substantially reduced if its four ester groups are hydrolyzed. This gives one small clue as to how the molecule functions.

PROBLEM 8 *SOLVED*

a. Which of the following tertiary alcohols cannot be prepared from the reaction of an ester with excess Grignard reagent?

1. $CH_3C(OH)(CH_3)CH_2CH_3$

2. $CH_3C(OH)(CH_3)CH_3$

3. $CH_3CH_2C(OH)(CH_3)CH_2CH_2CH_3$

4. $CH_3CH_2C(OH)(CH_3)CH_2CH_3$

5. $CH_3C(OH)(CH_2CH_3)CH_2CH_2CH_2CH_3$

6. $C_6H_5C(OH)(CH_3)C_6H_5$

b. For those alcohols that can be prepared by the reaction of an ester with excess Grignard reagent, what ester and what Grignard reagent should be used?

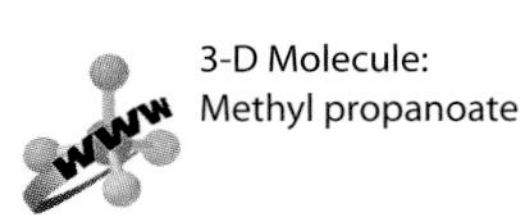

3-D Molecule: Methyl propanoate

Solution to 8a A tertiary alcohol is obtained from the reaction of an ester with two equivalents of a Grignard reagent. Therefore, tertiary alcohols prepared in this way must have two identical substituents on the carbon to which the OH is bonded because two substituents come from the Grignard reagent. Alcohols (3) and (5) cannot be prepared in this way because they do not have two identical substituents.

Solution to 8b(1) Methyl propanoate and excess methylmagnesium bromide.

PROBLEM 9 ◆

Which of the following secondary alcohols can be prepared from the reaction of methyl formate with excess Grignard reagent?

$CH_3CH_2CH(OH)CH_3$ $\quad CH_3CH(OH)CH_3$ $\quad CH_3CH(OH)CH_2CH_2CH_3$ $\quad CH_3CH_2CH(OH)CH_2CH_3$

PROBLEM-SOLVING STRATEGY

Why does a Grignard reagent not add to the carbonyl carbon of a carboxylic acid?

We know that Grignard reagents add to carbonyl carbons. If we find that a Grignard reagent does not add to the carbonyl carbon of a particular compound, we can conclude that the Grignard reagent must react more rapidly with another part of the molecule. A carboxylic acid has an acidic proton that reacts rapidly with the Grignard reagent, converting it to an alkane.

$$CH_3C(=O)O{-}H + CH_3CH_2{-}MgBr \longrightarrow CH_3C(=O)O^- \; {}^+MgBr + CH_3CH_3$$

Now continue on to Problem 10.

PROBLEM 10 ♦

Which of the following compounds will not undergo a nucleophilic addition reaction with a Grignard reagent?

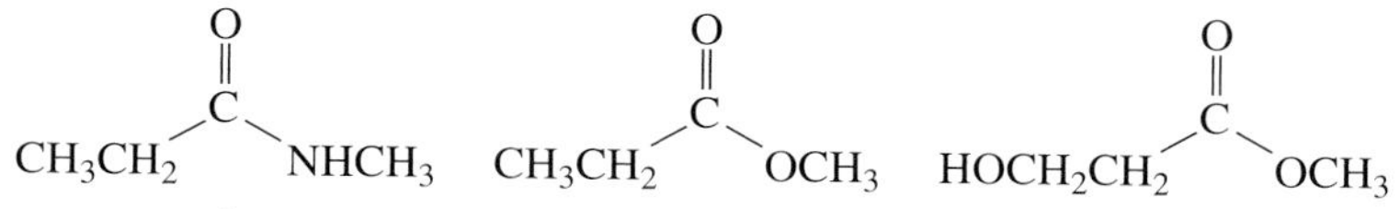

A: $CH_3CH_2C(=O)NHCH_3$ B: $CH_3CH_2C(=O)OCH_3$ C: $HOCH_2CH_2C(=O)OCH_3$

SEMISYNTHETIC DRUGS

Taxol is a difficult molecule to synthesize because of its complicated structure. Chemists have made the synthesis a lot easier by allowing the common English yew shrub to carry out the first part of the synthesis. A precursor of the drug is extracted from the shrub's needles, and the precursor is converted to Taxol in a four-step procedure in the laboratory. Thus, the precursor is isolated from a renewable resource, whereas the drug itself could be obtained only by killing a slow-growing tree. This is an example of how chemists have learned to synthesize compounds jointly with nature.

12.6 THE REACTIONS OF CARBONYL COMPOUNDS WITH HYDRIDE ION

Reactions of Aldehydes and Ketones with Hydride Ion

A hydride ion is another strongly basic nucleophile (Table 11.1 on page 300) that reacts with aldehydes and ketones to form nucleophilic addition products. The mechanism for the reaction is shown below.

mechanism for the reaction of an aldehyde or a ketone with hydride ion

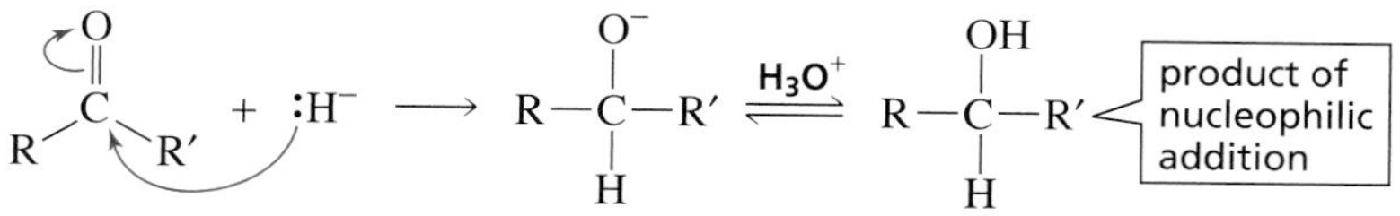

$$RC(=O)R' + :H^- \longrightarrow R-C(O^-)(H)-R' \xrightleftharpoons{H_3O^+} R-C(OH)(H)-R'$$

product of nucleophilic addition

- Nucleophilic attack of hydride ion on the carbonyl carbon forms an alkoxide ion.
- Subsequent protonation by an acid produces an alcohol. The overall reaction adds H_2 to the carbonyl group.

Recall that the addition of hydrogen to an organic compound is a **reduction reaction** (Section 5.12). Aldehydes and ketones are generally reduced using sodium borohydride ($NaBH_4$) as the source of hydride ion. Aldehydes are reduced to primary alcohols, and ketones are reduced to secondary alcohols. Notice that the acid is not added to the reaction mixture until after the hydride ion has reacted with the carbonyl compound.

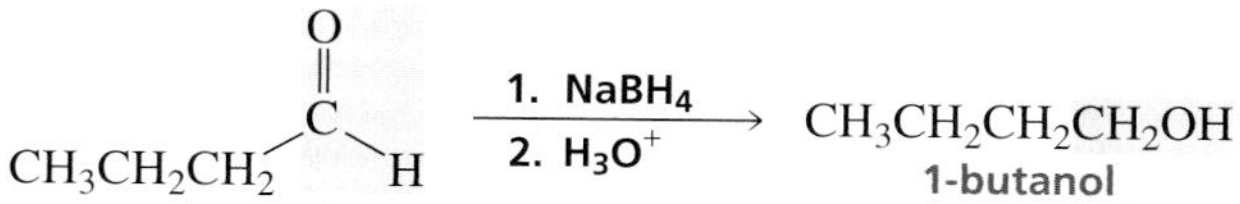

$$CH_3CH_2CH_2C(=O)H \xrightarrow[2.\ H_3O^+]{1.\ NaBH_4} CH_3CH_2CH_2CH_2OH$$

butanal, an aldehyde → 1-butanol, a primary alcohol

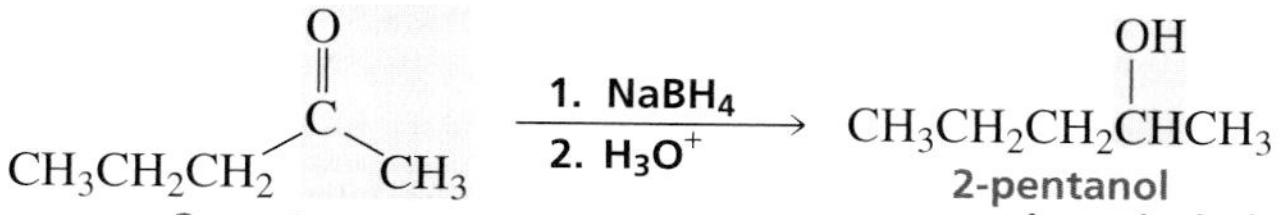

$$CH_3CH_2CH_2C(=O)CH_3 \xrightarrow[2.\ H_3O^+]{1.\ NaBH_4} CH_3CH_2CH_2CH(OH)CH_3$$

2-pentanone, a ketone → 2-pentanol, a secondary alcohol

PROBLEM 11 ♦

What alcohols are obtained from the reduction of the following compounds with sodium borohydride followed by addition of acid?

a. 2-methylpropanal
b. cyclohexanone
c. benzaldehyde
d. methyl phenyl ketone

Reactions of Carboxylic Acids and Carboxylic Acid Derivatives with Hydride Ion

Because Class I carbonyl compounds have a group that can be replaced by another group, they undergo two successive reactions with hydride ion, just like they undergo two successive reactions with a Grignard reagent (Section 12.4). For this reason, the reaction of an acyl chloride with sodium borohydride forms an alcohol.

$$CH_3CH_2CH_2C(=O)Cl \xrightarrow[2.\ H_3O^+]{1.\ NaBH_4} CH_3CH_2CH_2CH_2OH$$

butanoyl chloride → 1-butanol

The mechanism for the reaction is shown below.

mechanism for the reaction of an acyl chloride with hydride ion

$$CH_3CH_2C(=O)Cl + :H^- \longrightarrow CH_3CH_2C(O^-)(H)Cl \longrightarrow CH_3CH_2C(=O)H + Cl^-$$

an acyl chloride; a group is expelled from the tetrahedral intermediate; an aldehyde (product of nucleophilic acyl substitution)

$$CH_3CH_2C(=O)H \xrightarrow{:H^-} CH_3CH_2CH(O^-)H \xrightarrow{H_3O^+} CH_3CH_2CH_2OH$$

a primary alcohol (product of nucleophilic addition)

- The acyl chloride undergoes a nucleophilic acyl substitution reaction because it has a group (Cl^-) that can be replaced by hydride ion. The product of this reaction is an aldehyde.
- The aldehyde then undergoes a nucleophilic addition reaction with a second equivalent of hydride ion, forming an alkoxide ion.
- Protonation of the alkoxide ion gives a primary alcohol.

Sodium borohydride ($NaBH_4$) is not a sufficiently strong hydride donor to react with carbonyl compounds that are less reactive than aldehydes and ketones. Therefore, esters, carboxylic acids, and amides must be reduced with lithium aluminum hydride ($LiAlH_4$), a more reactive hydride donor.

The reaction of an ester with $LiAlH_4$ produces two alcohols, one corresponding to the acyl portion of the ester and one corresponding to the alkyl portion.

Esters and acyl chlorides undergo two successive reactions with hydride ion and with Grignard reagents.

$$CH_3CH_2C(=O)OCH_3 \xrightarrow[2.\ H_3O^+]{1.\ LiAlH_4} CH_3CH_2CH_2OH + CH_3OH$$

methyl propanoate (an ester) → 1-propanol + methanol

The reaction of a carboxylic acid with $LiAlH_4$ forms a single primary alcohol.

$$CH_3C(=O)OH \xrightarrow[2.\ H_3O^+]{1.\ LiAlH_4} CH_3CH_2OH$$

acetic acid → ethanol

Amides also undergo two successive additions of hydride ion when they react with $LiAlH_4$. Overall, the reaction converts a carbonyl group into a CH_2 group. Therefore, the product of the reaction is an amine. Primary, secondary, or tertiary amines can be formed, depending on the number of substituents bonded to the nitrogen of the amide.

$$C_6H_5C(=O)NH_2 \xrightarrow[2.\ H_2O]{1.\ LiAlH_4} C_6H_5CH_2NH_2$$

benzamide → benzylamine, a primary amine

$$CH_3C(=O)NHCH_3 \xrightarrow[2.\ H_2O]{1.\ LiAlH_4} CH_3CH_2NHCH_3$$

N-methylacetamide → ethylmethylamine, a secondary amine

Notice that H_2O rather than H_3O^+ is used in the second step of the reaction. If H_3O^+ is used, the product will be an ammonium ion rather than an amine. (Remember that the pK_a of a protonated amine is about 10; See Table 7.1 on page 188.)

$$CH_3C(=O)NHCH_3 \xrightarrow[2.\ H_3O^+]{1.\ LiAlH_4} CH_3CH_2\overset{+}{N}H_2CH_3$$

PROBLEM 12 ♦

What amides would you treat with $LiAlH_4$ in order to prepare the following amines?

a. benzylmethylamine
b. ethylamine
c. diethylamine
d. triethylamine

12.7 THE REACTIONS OF ALDEHYDES AND KETONES WITH AMINES

An aldehyde or a ketone reacts with a *primary amine* to form an imine. An **imine** is a compound with a carbon–nitrogen double bond.

$$\text{H–C–C=O} + R\text{–}NH_2 \underset{}{\overset{\text{trace } H^+}{\rightleftharpoons}} \text{H–C–C=N–R} + H_2O$$

an aldehyde or a ketone + a primary amine ⇌ an imine

A C=N group is similar to a C=O group (compare Figure 12.1 with Figure 11.1 on page 294). The imine nitrogen is sp^2 hybridized. One of its sp^2 orbitals forms a σ bond with the imine carbon, one forms a σ bond with a substituent, and the third contains a lone pair. The p orbital of nitrogen and the p orbital of carbon overlap to form a π bond.

An aldehyde or a ketone reacts with a *secondary amine* to form an enamine (pronounced "ENE-amine"). An **enamine** is a tertiary amine with a double bond in the α,β-position relative to the nitrogen atom. Notice that the double bond is in the part of the molecule that comes from the aldehyde or ketone, not from the part that is provided by the secondary amine. The name "enamine" comes from "ene" + "amine," with the "e" omitted in order to avoid two successive vowels.

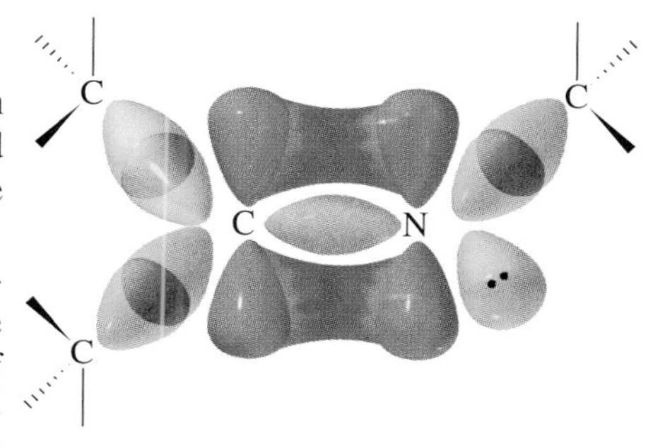

▲ **Figure 12.1**
Bonding in an imine.

$$\text{(H)C–C=O} + R_2NH \xrightleftharpoons{\text{trace } H^+} \text{C=C–NR}_2 + H_2O$$

an aldehyde or a ketone + a secondary amine ⇌ an enamine (α,β-position)

When you first look at the products of imine and enamine formation, they appear to be quite different. However, when you look at the mechanisms for the reactions, you will see that the mechanisms are exactly the same except for the site from which a proton is lost in the last step.

Primary Amines Form Imines

Aldehydes and ketones react with primary amines to form imines.

Aldehyde and ketones react with primary amines to form imines. The reaction requires a trace amount of acid.

$$C_6H_5CH{=}O + CH_3CH_2NH_2 \xrightleftharpoons{\text{trace } H^+} C_6H_5CH{=}NCH_2CH_3 + H_2O$$

benzaldehyde (an aldehyde) + ethylamine (a primary amine) ⇌ an imine

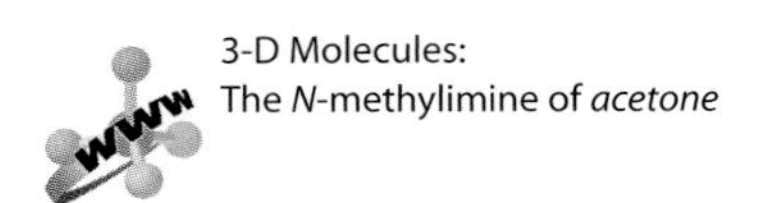

$$(CH_3CH_2)_2C{=}O + H_2NCH_2C_6H_5 \xrightleftharpoons{\text{trace } H^+} (CH_3CH_2)_2C{=}NCH_2C_6H_5 + H_2O$$

3-pentanone (a ketone) + benzylamine (a primary amine) ⇌ an imine

The mechanism for imine formation is shown below.

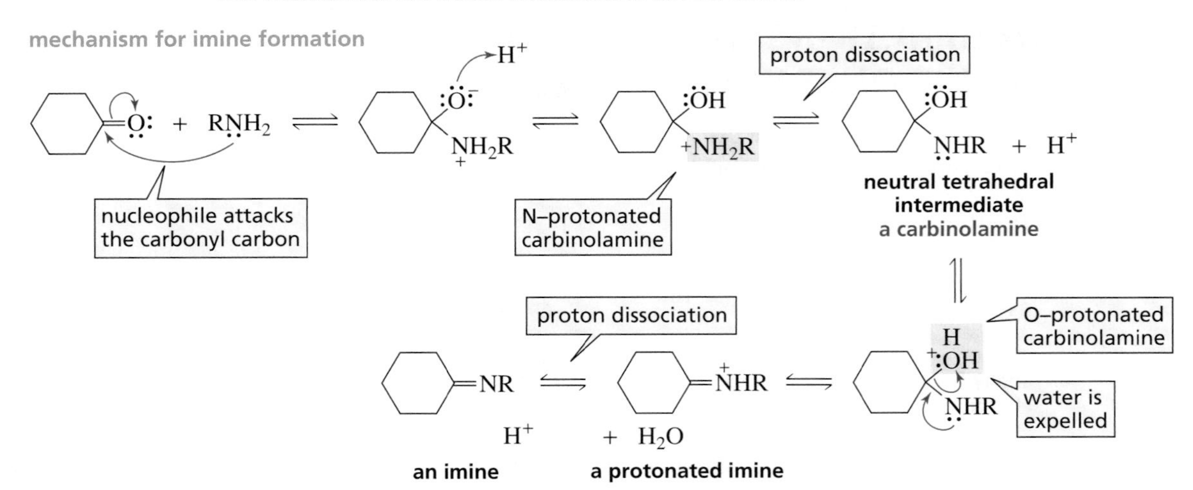

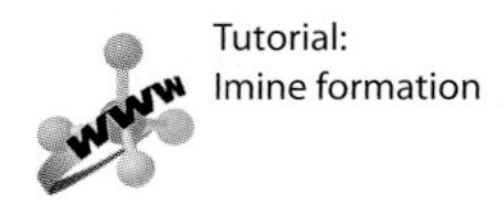

- The amine attacks the carbonyl carbon.
- Gain of a proton by the alkoxide ion and loss of a proton by the ammonium ion form a neutral tetrahedral intermediate.
- The neutral tetrahedral intermediate, called a *carbinolamine*, is in equilibrium with two protonated forms because either its nitrogen atom or its oxygen atom can be protonated.
- Elimination of water from the oxygen-protonated intermediate forms a protonated imine.
- The protonated imine loses a proton to yield the imine.

Unlike the stable tetrahedral compounds that are formed when a Grignard reagent or a hydride ion adds to an aldehyde or a ketone, the tetrahedral compound formed when an amine adds to an aldehyde or a ketone is unstable because the newly formed sp^3 carbon is bonded to an oxygen and to another electronegative atom (a nitrogen).

stable tetrahedral compounds

unstable tetrahedral compounds

In an acidic aqueous solution, an imine is hydrolyzed back to the carbonyl compound and amine. Notice that the amine is protonated because the solution is acidic.

An imine undergoes acid-catalyzed hydrolysis to form a carbonyl compound and a primary amine.

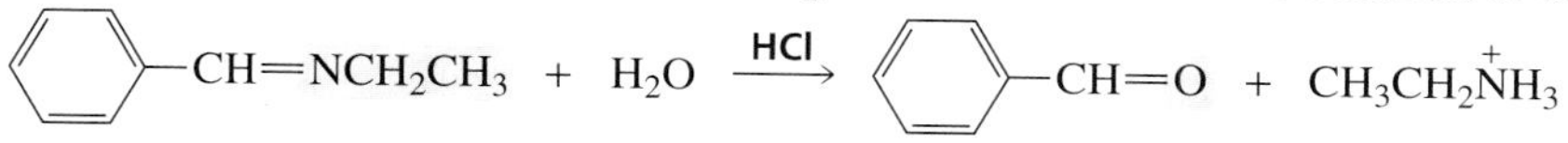

$$C_6H_5CH{=}NCH_2CH_3 + H_2O \xrightarrow{HCl} C_6H_5CH{=}O + CH_3CH_2\overset{+}{N}H_3$$

Imine formation and hydrolysis are important reactions in biological systems. For example, we will see that imine hydrolysis is why DNA contains A, G, C, and T nucleotides, whereas RNA contains A, G, C, and U nucleotides (Section 20.9).

Secondary Amines Form Enamines

Aldehydes and ketones react with secondary amines to form enamines. Like imine formation, the reaction requires a trace amount of an acid catalyst.

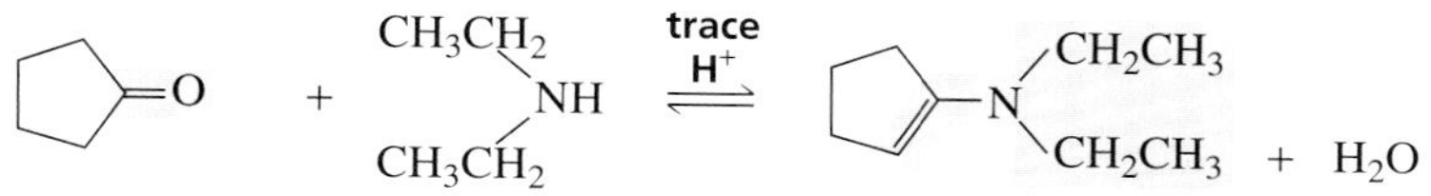

cyclopentanone + diethylamine (a secondary amine) $\underset{}{\overset{\text{trace } H^+}{\rightleftharpoons}}$ an enamine + H_2O

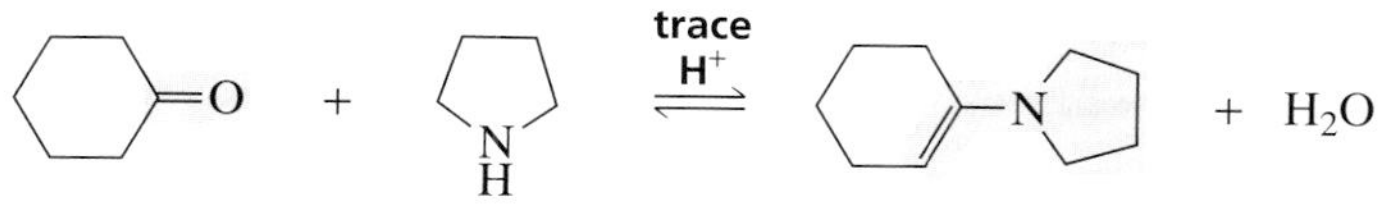

cyclohexanone + pyrrolidine (a secondary amine) $\overset{\text{trace } H^+}{\rightleftharpoons}$ an enamine + H_2O

Aldehydes and ketones react with secondary amines to form enamines.

The mechanism for enamine formation is exactly the same as that for imine formation, except for the last step of the mechanism.

mechanism for enamine formation

nucleophile attacks the carbonyl carbon

proton dissociation

N-protonated carbinolamine

neutral tetrahedral intermediate

a carbinolamine

O-protonated carbinolamine

water is expelled

an enamine

this intermediate cannot lose a proton from N, so it loses a proton from an α-carbon

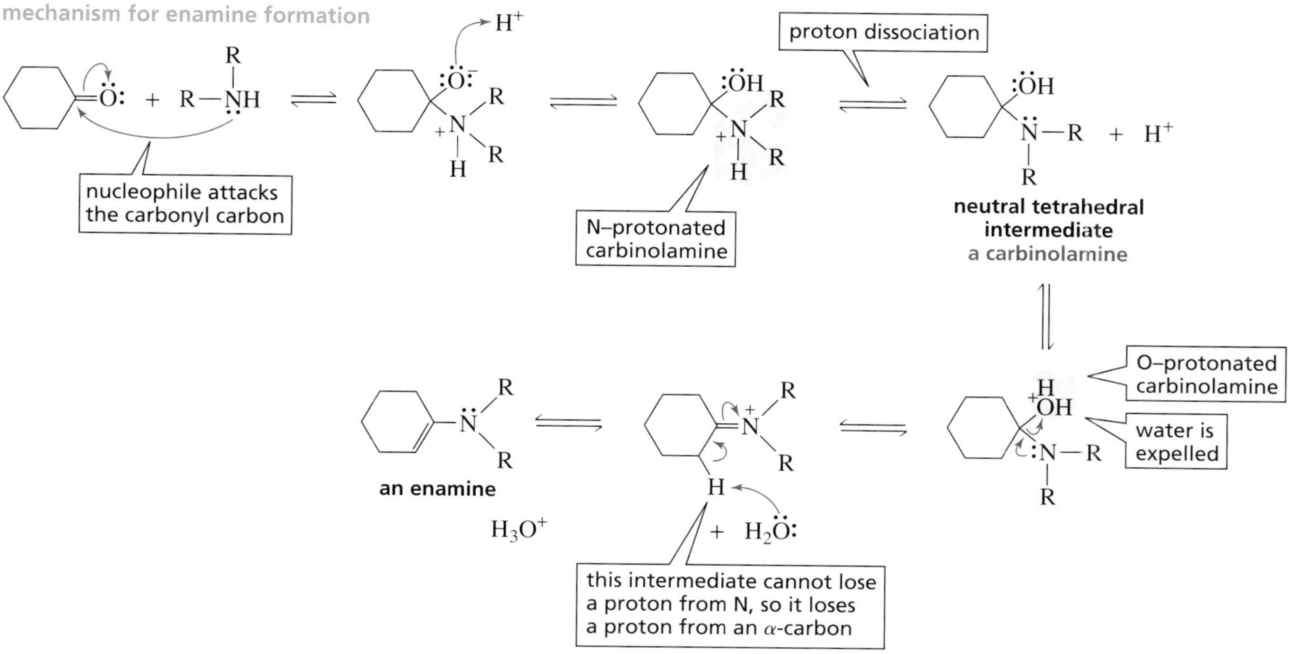

- The amine attacks the carbonyl carbon.
- Gain of a proton by the alkoxide ion and loss of a proton by the ammonium ion forms a neutral tetrahedral intermediate.
- The neutral tetrahedral intermediate is in equilibrium with two protonated forms because either its nitrogen atom or its oxygen atom can be protonated.
- Elimination of water from the oxygen-protonated intermediate forms a compound with positively charged nitrogen.
- When a primary amine reacts with an aldehyde or a ketone, the compound with a positively charged nitrogen loses a proton from nitrogen in the last step of the mechanism, forming a neutral imine. However, when the amine is secondary, the positively charged nitrogen is not bonded to a hydrogen. In this case, a stable neutral molecule is obtained by removing a proton from the α-carbon of the compound with the positively charged nitrogen. An enamine is the result.

An enamine undergoes acid-catalyzed hydrolysis to form a carbonyl compound and a secondary amine.

In an aqueous acidic solution, an enamine is hydrolyzed back to the carbonyl compound and secondary amine, a reaction that is similar to the acid-catalyzed hydrolysis of an imine back to the carbonyl compound and primary amine. Again, the amine is protonated because the solution is acidic.

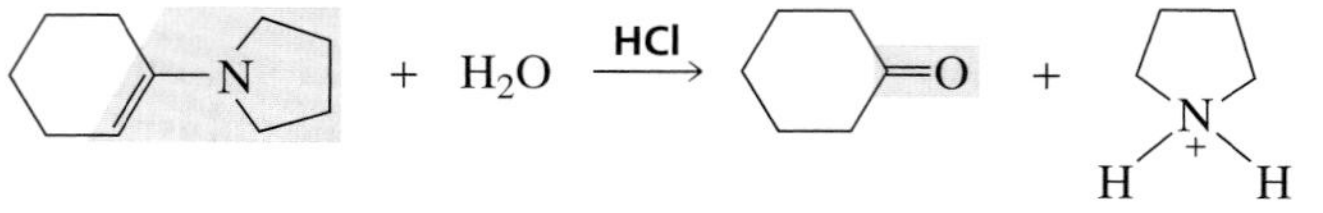

$$\text{(1-cyclohexenyl)pyrrolidine} + H_2O \xrightarrow{\text{HCl}} \text{cyclohexanone} + \text{pyrrolidinium } (H_2N^+)$$

PROBLEM 13 ♦

Give the products of the following reactions. (A trace amount of acid is present in each case.)

a. cyclopentanone + ethylamine
b. cyclopentanone + diethylamine
c. 3-pentanone + hexylamine
d. 3-pentanone + cyclohexylamine

IMINES IN BIOLOGICAL SYSTEMS

Almost all biological reactions are catalyzed by enzymes. Some enzymes need the help of a coenzyme, which is an organic molecule derived from a vitamin, to carry out the catalysis (Sections 17.1 and 17.4). For example, pyridoxal phosphate, a coenzyme derived from vitamin B_6, is required by enzymes that metabolize amino acids (Section 17.9). Pyridoxal phosphate is attached to the enzyme by an imine linkage.

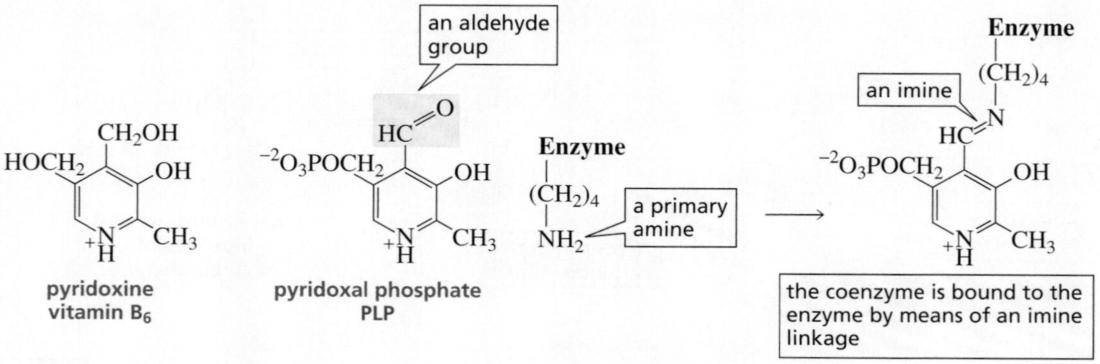

pyridoxine
vitamin B_6

pyridoxal phosphate
PLP

an aldehyde group

a primary amine

an imine

the coenzyme is bound to the enzyme by means of an imine linkage

Glucose is metabolized to pyruvate, which can be converted to the amino acid known as alanine in a two-step process. First, pyruvate reacts with NH_3 to form an imine, which is then reduced by an enzyme to the amino acid.

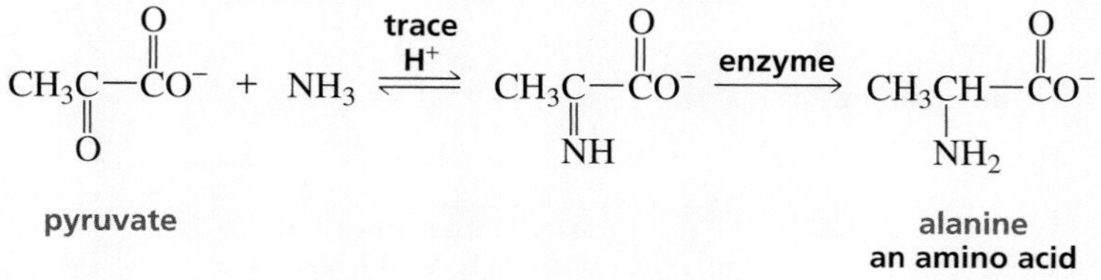

$$CH_3C(=O)CO^- + NH_3 \underset{}{\overset{\text{trace } H^+}{\rightleftharpoons}} CH_3C(=NH)CO^- \xrightarrow{\text{enzyme}} CH_3CH(NH_2)CO^-$$

pyruvate

alanine
an amino acid

12.8 THE REACTIONS OF ALDEHYDES AND KETONES WITH WATER

Most hydrates are too unstable to be isolated.

The addition of water to an aldehyde or a ketone forms a *hydrate*. A **hydrate** is a molecule with two OH groups on the same carbon. Hydrates of aldehydes or ketones are generally too unstable to be isolated because the tetrahedral carbon is attached to two electron-withdrawing (oxygen) atoms.

$$\underset{\text{an aldehyde or a ketone}}{R-\overset{\overset{O}{\|}}{C}-R\,(H)} + H_2O \underset{}{\overset{HCl}{\rightleftharpoons}} \underset{\text{a hydrate}}{R-\underset{OH}{\overset{OH}{|}}{C}-R\,(H)}$$

Water is a poor nucleophile and therefore adds relatively slowly to a carbonyl group. The rate of the reaction can be increased by an acid catalyst. Keep in mind that the catalyst affects the *rate* at which an aldehyde or a ketone is converted to a hydrate; it has no effect on the *amount* of aldehyde or ketone converted to hydrate (Section 4.8). The mechanism for the reaction is shown below.

mechanism for acid-catalyzed hydrate formation

$$CH_3-\overset{\overset{\ddot{O}:}{\|}}{C}-H + H^+ \rightleftharpoons CH_3-\overset{\overset{+\ddot{O}H}{\|}}{C}-H + H_2\ddot{O}: \rightleftharpoons CH_3-\underset{\overset{+}{:}OH_2}{\overset{OH}{C}}-H \rightleftharpoons CH_3-\underset{OH}{\overset{OH}{C}}-H + H^+$$

the acid protonates the carbonyl oxygen

the nucleophile attacks the carbonyl carbon

proton dissociation

- The acid protonates the carbonyl oxygen, which makes the carbonyl carbon more susceptible to nucleophilic attack (Figure 12.2).
- Water attacks the carbonyl carbon.
- Loss of a proton from the protonated tetrahedral intermediate gives the hydrate.

The extent to which an aldehyde or a ketone is hydrated in an aqueous solution depends on the substituents attached to the carbonyl compound. For example, only 0.2% of acetone is hydrated at equilibrium, but 99.9% of formaldehyde is hydrated. Bulky substituents and electron-donating substituents (for example, the methyl groups of acetone) *decrease* the percentage of hydrate present at equilibrium, whereas small substituents and electron-withdrawing substituents (the hydrogens of formaldehyde) *increase* it.

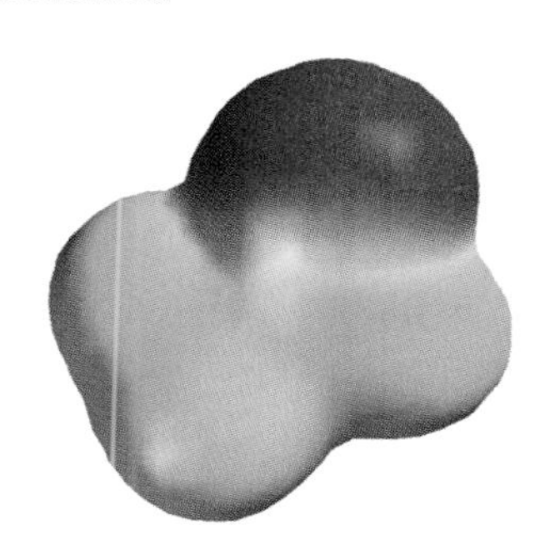

$$CH_3-\overset{\overset{O}{\|}}{C}-H$$

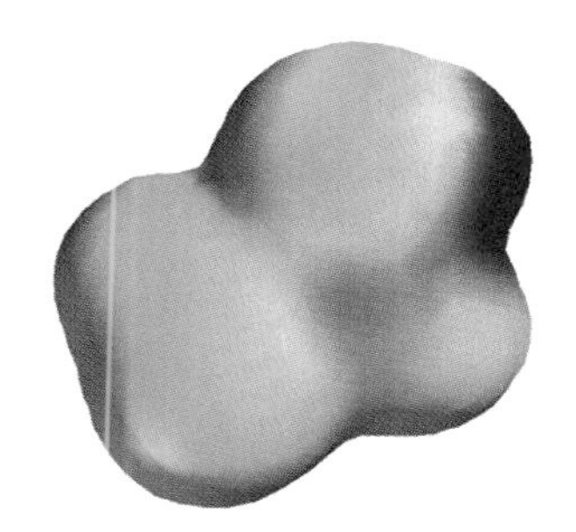

$$CH_3-\overset{\overset{+OH}{\|}}{C}-H$$

▲ **Figure 12.2**
The electrostatic potential maps show that the carbonyl carbon of the protonated aldehyde is more susceptible to nucleophilic attack (the blue is more intense) than the carbonyl carbon of the unprotonated aldehyde.

PRESERVING BIOLOGICAL SPECIMENS

A 37% solution of formaldehyde in water, known as *formalin*, was commonly used in the past to preserve biological specimens. Formaldehyde is an eye and a skin irritant, however, so formalin has been replaced in many biology laboratories by other preservatives. One preservative frequently used is a solution of 2 to 5% phenol in ethanol with added antimicrobial agents.

PROBLEM 14 ♦

When trichloroacetaldehyde is dissolved in water, almost all of it is converted to the hydrate. Chloral hydrate, the product of the reaction, is a sedative that can be lethal. A cocktail laced with it is commonly known—in detective novels, at least—as a "Mickey Finn." Explain why an aqueous solution of trichloroacetaldehyde is almost all hydrate.

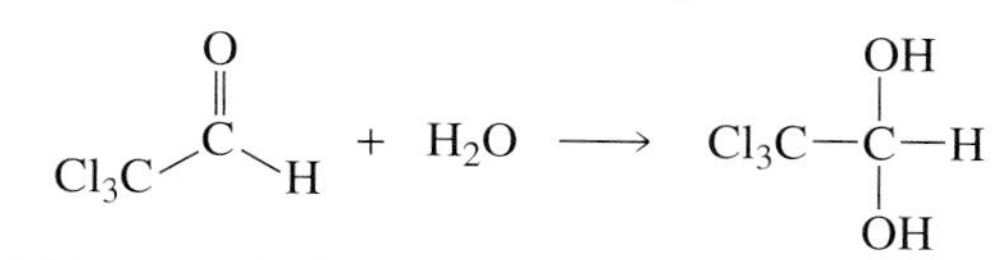

3-D Molecules:
Acetone; Acetone hydrate

PROBLEM 15 ♦

Which of the following ketones forms the most hydrate in an aqueous solution?

$CH_3O-C_6H_4-C(=O)-C_6H_4-OCH_3$ $\quad$ $C_6H_5-C(=O)-C_6H_5$ $\quad$ $O_2N-C_6H_4-C(=O)-C_6H_4-NO_2$

12.9 THE REACTIONS OF ALDEHYDES AND KETONES WITH ALCOHOLS

Aldehydes react with alcohols to form hemiacetals and acetals.

The product formed when one equivalent of an alcohol adds to an *aldehyde* is called a **hemiacetal**. The product formed when a second equivalent of alcohol is added is called an **acetal** (ass-ett-AL). Like water, an alcohol is a poor nucleophile, so an acid catalyst is required for the reaction to take place at a reasonable rate.

$$CH_3CHO + CH_3OH \underset{}{\overset{HCl}{\rightleftharpoons}} CH_3-C(OH)(OCH_3)-H \underset{}{\overset{CH_3OH,\ HCl}{\rightleftharpoons}} CH_3-C(OCH_3)_2-H + H_2O$$

an aldehyde — a hemiacetal — an acetal

Ketones react with alcohols to form hemiketals and ketals.

When the carbonyl compound is a *ketone* instead of an aldehyde, the addition products are called a **hemiketal** and a **ketal**, respectively.

$$CH_3COCH_3 + CH_3OH \overset{HCl}{\rightleftharpoons} CH_3-C(OH)(OCH_3)-CH_3 \overset{CH_3OH,\ HCl}{\rightleftharpoons} CH_3-C(OCH_3)_2-CH_3 + H_2O$$

a ketone — a hemiketal — a ketal

Hemi is the Greek word for "half." When one equivalent of alcohol has added to an aldehyde or a ketone, the compound is halfway to the final acetal or ketal, which contains groups from two equivalents of alcohol.

The mechanism for ketal (or acetal) formation is shown below.

mechanism for acid-catalyzed acetal or ketal formation

$$CH_3COCH_3 \overset{H^+}{\rightleftharpoons} CH_3C(=\overset{+}{O}H)CH_3 + CH_3\ddot{O}H \rightleftharpoons CH_3-C(OH)(\overset{+}{O}HCH_3)-CH_3 \rightleftharpoons CH_3-C(OH)(OCH_3)-CH_3 + H^+$$

the acid protonates the carbonyl oxygen; the nucleophile attacks the carbonyl carbon; proton dissociation; a hemiketal; either the OH or the OCH_3 can be protonated

$$CH_3-C(OCH_3)(\overset{+}{O}H_2)-CH_3 \rightleftharpoons CH_3C(=\overset{+}{O}CH_3)CH_3 + H_2O \overset{CH_3\ddot{O}H}{\rightleftharpoons} CH_3-C(OCH_3)(\overset{+}{O}HCH_3)-CH_3 \rightleftharpoons CH_3-C(OCH_3)_2-CH_3 + H^+$$

water is expelled; an O-alkylated intermediate; the nucleophile attacks the carbonyl carbon; proton dissociation; a ketal

- The acid protonates the carbonyl oxygen to make the carbonyl carbon more susceptible to nucleophilic attack (Figure 12.2).
- The alcohol attacks the carbonyl carbon.
- Loss of a proton from the protonated tetrahedral intermediate gives the hemiacetal (or hemiketal).

- Because the reaction is carried out in an acidic solution, the hemiacetal (or hemiketal) is in equilibrium with its protonated form. The two oxygen atoms of the hemiacetal (or hemiketal) are equally basic, so either one can be protonated.
- Loss of water from the tetrahedral intermediate with a protonated OH group forms an O-alkylated intermediate that is very reactive because of its positively charged oxygen atom. Nucleophilic attack on this compound by a second molecule of alcohol, followed by loss of a proton, forms the acetal (or ketal).

Although the tetrahedral carbon of an acetal (or ketal) is bonded to two oxygen atoms, causing us to predict that it is not stable, the acetal (or ketal) can be isolated if the water that is eliminated from the hemiacetal (or hemiketal) is removed from the reaction mixture. This is because, if water is not available, the only compound the acetal (or ketal) can form is the O-alkylated species, which is less stable than the acetal (or ketal).

The acetal (or ketal) can be hydrolyzed back to the aldehyde (or ketone) in an acidic aqueous solution.

$$CH_3CH_2C(OCH_2CH_3)_2CH_3 + \underset{\text{excess}}{H_2O} \underset{}{\overset{\text{HCl}}{\rightleftharpoons}} CH_3CH_2C(=O)CH_3 + 2\,CH_3CH_2OH$$

CARBOHYDRATES

When you study carbohydrates in Chapter 15, you will see that the individual sugar units in a carbohydrate are held together by acetal or ketal linkages. For example, the reaction of an alcohol group and an aldehyde group of D-glucose forms a cyclic D-glucose molecule with a hemiacetal linkage. Molecules of cyclic D-glucose are then hooked up as a result of the reaction of the OH group of one molecule with the hemiacetal group of another, resulting in the formation of an acetal linkage.

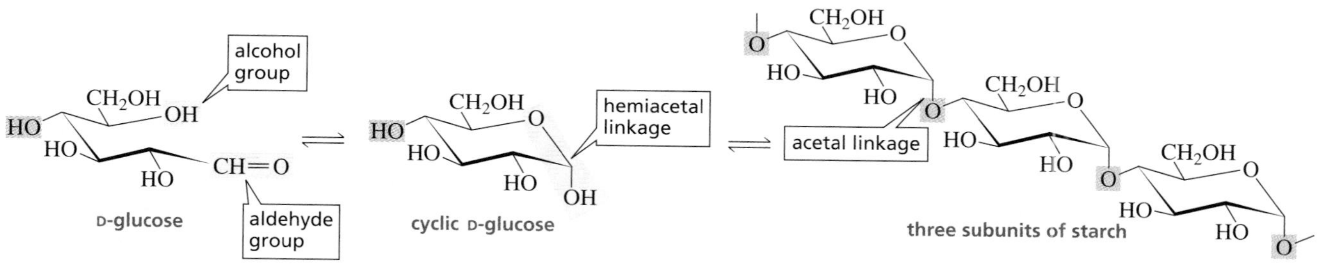

D-glucose (alcohol group; aldehyde group) ⇌ cyclic D-glucose (hemiacetal linkage) ⇌ three subunits of starch (acetal linkage)

PROBLEM 16

Show the mechanism for the acid-catalyzed hydrolysis of an acetal.

PROBLEM 17 ♦

Which of the following are

a. hemiacetals? **b.** acetals? **c.** hemiketals? **d.** ketals? **e.** hydrates?

1. $CH_3-C(OH)(OCH_3)-CH_3$

2. $CH_3-C(OCH_2CH_3)(OCH_2CH_3)-H$

3. $CH_3-C(OCH_3)(OCH_3)-H$

4. $CH_3-C(OH)(OH)-CH_3$

5. $CH_3-C(OCH_3)(OCH_3)-CH_3$

6. $CH_3-C(OH)(OH)-H$

7. $CH_3-C(OH)(OCH_3)-H$

8. $CH_3-C(OH)(OCH_3)-CH_2CH_3$

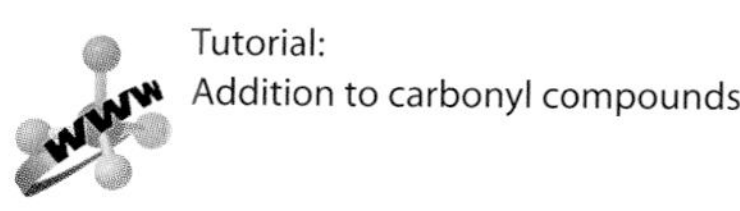

Tutorial:
Addition to carbonyl compounds

12.10 NUCLEOPHILIC ADDITION TO α,β-UNSATURATED CARBONYL COMPOUNDS

The resonance contributors for an α,β-unsaturated carbonyl compound show that the molecule has two electrophilic sites: the carbonyl carbon and the β-carbon.

$$RCH{=}CH{-}C({=}O){-}R \longleftrightarrow RCH{=}CH{-}\overset{+}{C}(O^-){-}R \longleftrightarrow R\overset{+}{C}H{-}CH{=}C(O^-){-}R$$

an α,β-unsaturated carbonyl compound

electrophilic site

electrophilic site

This means that if an aldehyde or a ketone has a double bond in the α,β-position, a nucleophile can add either to the carbonyl carbon or to the β-carbon.

Nucleophilic addition to the carbonyl carbon is called **direct addition** or **1,2-addition**.

direct addition

$$Y\!:^- + RCH{=}CH{-}C({=}O){-}R \longrightarrow RCH{=}CH{-}C(O^-)(Y){-}R \xrightarrow{H_3O^+} RCH{=}CH{-}C(OH)(Y){-}R$$

Nucleophilic addition to the β-carbon is called **conjugate addition** or **1,4-addition**, because addition occurs at the 1- and 4-positions (that is, across the conjugated system). After 1,4-addition has occurred, the product—an enol—tautomerizes to a ketone (or to an aldehyde) because the keto tautomer is more stable than the enol tautomer (Section 5.11).

conjugate addition

$$Y\!:^- + RCH{=}CH{-}C({=}O){-}R \longrightarrow \left[RCH(Y){-}\ddot{C}H^-{-}C({=}O){-}R \longleftrightarrow RCH(Y){-}CH{=}C(O^-){-}R \right]$$

resonance contributors

$$\downarrow H_3O^+$$

$$\underset{\textbf{keto tautomer}}{RCH(Y){-}CH_2{-}C({=}O){-}R} \rightleftharpoons \underset{\textbf{enol tautomer}}{RCH(Y){-}CH{=}C(OH){-}R}$$

Nucleophiles that are weak bases form conjugate addition products.

Nucleophiles that are weak bases form conjugate addition products. The overall reaction amounts to addition to the carbon–carbon double bond, with the nucleophile adding to the β-carbon of the double bond and a proton from the reaction mixture adding to the α-carbon.

$$CH_2{=}CH{-}C({=}O){-}CH_3 + HBr \longrightarrow BrCH_2CH_2{-}C({=}O){-}CH_3$$

$$\text{2-cyclopentenone} + CH_3SH \longrightarrow \text{3-(methylthio)cyclopentanone } (SCH_3)$$

In general, nucleophiles that are strong bases form direct addition products. Ethyl alcohol is used in the second step to protonate the alkoxide ion.

$$\text{cyclohexenone} \xrightarrow[\text{2. EtOH}]{\text{1. } CH_3MgBr} \text{1-methylcyclohex-2-en-1-ol (HO, } CH_3\text{)}$$

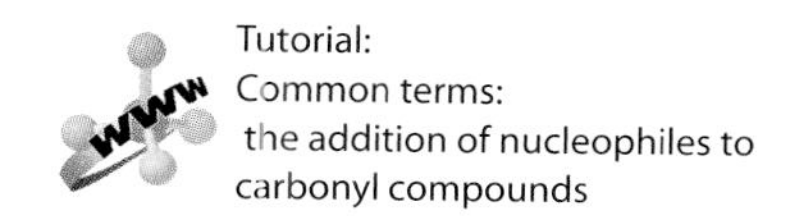

Tutorial:
Common terms:
the addition of nucleophiles to carbonyl compounds

ANTICANCER DRUGS

Two compounds—vernolepin and helenalin—owe their effectiveness as anticancer drugs to conjugate addition reactions.

Cancer cells are cells that have lost their ability to control their growth; therefore, they proliferate rapidly. DNA polymerase is an enzyme that a cell needs in order to make a copy of its DNA for a new cell. DNA polymerase has an SH group at its active site (Section 16.6) and each of these drugs has two α,β-unsaturated carbonyl groups. When an SH group of DNA polymerase reacts with one of the α,β-unsaturated carbonyl groups of vernolepin or helenalin, the enzyme is inactivated, and cells cannot proliferate because theey cannot make copies of their DNA.

vernolepin

helenalin

active enzyme–HS + α-methylene lactone $\xrightarrow{\text{conjugate addition}}$ inactive enzyme–S–CH_2–lactone

PROBLEM 18 ◆

Give the major product of each of the following reactions.

a. bicyclic α,β-unsaturated ketone $\xrightarrow{\text{HBr}}$

b. bicyclic α,β-unsaturated ketone $\xrightarrow[\text{2. EtOH}]{\text{1. } CH_3MgBr}$

12.11 CONJUGATE ADDITION REACTIONS IN BIOLOGICAL SYSTEMS

Several reactions in biological systems involve conjugate addition to α,β-unsaturated carbonyl compounds. Below are examples of two of them. The first occurs in gluconeogenesis—the synthesis of glucose from pyruvate (Section 18.4). The second occurs in the oxidation of fatty acids (Section 18.3).

$$H_2C{=}C(OPO_3^{2-})CO^- (\text{C=O}) + H_2O \overset{\text{enolase}}{\rightleftharpoons} CH_2(OH)CH(OPO_3^{2-})\overset{O}{\overset{\|}{C}}O^-$$

$$CH_3(CH_2)_nCH{=}CH\overset{O}{\overset{\|}{C}}SCoA + H_2O \overset{\text{enoyl-CoA hydratase}}{\rightleftharpoons} CH_3(CH_2)_nCH(OH)CH_2\overset{O}{\overset{\|}{C}}SCoA$$

SUMMARY

Aldehydes and **ketones** have an acyl group attached to a group (H or R) that cannot be readily replaced by another group. Steric and electronic factors cause aldehydes to be more reactive than ketones toward nucleophilic attack. Aldehydes and ketones are less reactive than acyl chlorides and are more reactive than esters, carboxylic acids, and amides.

Aldehydes and ketones undergo **nucleophilic addition reactions** with Grignard reagents and with hydride ion. In contrast, esters and acyl chlorides undergo **nucleophilic acyl substitution** reactions with these nucleophiles, forming aldehydes or ketones that undergo a **nucleophilic addition** reaction with a second equivalent of the nucleophile.

The tetrahedral intermediate formed by attack of a nucleophile on a carbonyl compound is stable if the newly formed tetrahedral carbon is not bonded to a second electronegative atom or group and is generally unstable if it is.

Grignard reagents react with aldehydes to form secondary alcohols and with ketones, esters, and acyl chlorides to form tertiary alcohols. Hydride ion reduces aldehydes, acyl chlorides, and carboxylic acids to primary alcohols, ketones to secondary alcohols, and amides to amines.

Aldehydes and ketones react with primary amines to form **imines** and with secondary amines to form **enamines**. The mechanisms are the same, except for the site from which a proton is lost in the last step of the reaction. Imines and enamines are hydrolyzed under acidic conditions back to the carbonyl compound and amine.

Aldehydes and ketones undergo acid-catalyzed addition of water to form hydrates. Most hydrates are too unstable to be isolated. Acid-catalyzed addition of an alcohol to an aldehyde forms **hemiacetals** and **acetals**, and to a ketone forms **hemiketals** and **ketals**.

In general, nucleophiles that are strong bases form **direct addition** products when they react with α,β-unsaturated carbonyl compound, and nucleophiles that are weak bases form **conjugate addition** products.

SUMMARY OF REACTIONS

1. Reactions of *carbonyl compounds* with Grignard reagents (Section 12.5)

a. Reaction of *formaldehyde* with a Grignard reagent forms a *primary alcohol*:

$$H-\overset{O}{\overset{\|}{C}}-H \xrightarrow[2.\ H_3O^+]{1.\ CH_3MgBr} CH_3CH_2OH$$

b. Reaction of an *aldehyde* (other than formaldehyde) with a Grignard reagent forms a *secondary alcohol:*

$$R-\overset{O}{\overset{\|}{C}}-H \xrightarrow[2.\ H_3O^+]{1.\ CH_3MgBr} R-\underset{CH_3}{\underset{|}{\overset{OH}{\overset{|}{C}}}}-H$$

c. Reaction of a *ketone* with a Grignard reagent forms a *tertiary alcohol*:

$$R-\overset{O}{\overset{\|}{C}}-R' \xrightarrow[2.\ H_3O^+]{1.\ CH_3MgBr} R-\underset{CH_3}{\underset{|}{\overset{OH}{\overset{|}{C}}}}-R'$$

d. Reaction of an *ester* with a Grignard reagent forms a *tertiary alcohol* with two identical substituents:

$$R-\overset{O}{\overset{\|}{C}}-OR' \xrightarrow[2.\ H_3O^+]{1.\ 2\ CH_3MgBr} R-\underset{CH_3}{\underset{|}{\overset{OH}{\overset{|}{C}}}}-CH_3$$

e. Reaction of an *acyl chloride* with a Grignard reagent forms a *tertiary alcohol* with two identical substituents:

$$R-\overset{O}{\overset{\|}{C}}-Cl \xrightarrow[2.\ H_3O^+]{1.\ 2\ CH_3MgBr} R-\underset{CH_3}{\underset{|}{\overset{OH}{\overset{|}{C}}}}-CH_3$$

2. Reactions of *carbonyl compounds* with hydride ion donors (Section 12.6)

a. Reaction of an *aldehyde* with sodium borohydride forms a *primary alcohol*:

$$\mathrm{RC(=O)H} \xrightarrow[\text{2. } H_3O^+]{\text{1. } NaBH_4} \mathrm{RCH_2OH}$$

b. Reaction of a *ketone* with sodium borohydride forms a *secondary alcohol*:

$$\mathrm{RC(=O)R} \xrightarrow[\text{2. } H_3O^+]{\text{1. } NaBH_4} \mathrm{R{-}CH(OH){-}R}$$

c. Reaction of an *acyl chloride* with sodium borohydride forms a *primary alcohol*:

$$\mathrm{RC(=O)Cl} \xrightarrow[\text{2. } H_3O^+]{\text{1. } NaBH_4} \mathrm{R{-}CH_2{-}OH}$$

d. Reaction of an *ester* with lithium aluminum hydride forms *two alcohols*:

$$\mathrm{RC(=O)OR'} \xrightarrow[\text{2. } H_3O^+]{\text{1. } LiAlH_4} \mathrm{RCH_2OH} + \mathrm{R'OH}$$

e. Reaction of a *carboxylic acid* with lithium aluminum hydride forms a *primary alcohol*:

$$\mathrm{RC(=O)OH} \xrightarrow[\text{2. } H_3O^+]{\text{1. } LiAlH_4} \mathrm{R{-}CH_2{-}OH}$$

f. Reaction of an *amide* with lithium aluminum hydride forms an *amine*:

$$\mathrm{RC(=O)NH_2} \xrightarrow[\text{2. } H_2O]{\text{1. } LiAlH_4} \mathrm{R{-}CH_2{-}NH_2}$$

$$\mathrm{RC(=O)NHR'} \xrightarrow[\text{2. } H_2O]{\text{1. } LiAlH_4} \mathrm{R{-}CH_2{-}NHR'}$$

$$\mathrm{RC(=O)NR'R''} \xrightarrow[\text{2. } H_2O]{\text{1. } LiAlH_4} \mathrm{R{-}CH_2{-}N(R''){-}R'}$$

3. Reactions of *aldehydes* and *ketones* with amines (Section 12.7)

a. Reaction with a *primary amine* forms an *imine*:

$$\mathrm{R_2C{=}O} + \mathrm{H_2NR} \underset{H^+}{\overset{\text{trace}}{\rightleftharpoons}} \mathrm{R_2C{=}NR} + \mathrm{H_2O}$$

b. Reaction with a *secondary amine* forms an *enamine*:

$$\mathrm{R(-CH<)C{=}O} + \mathrm{RNHR} \underset{H^+}{\overset{\text{trace}}{\rightleftharpoons}} \mathrm{R(-C{=})C{-}NR_2} + \mathrm{H_2O}$$

4. Reaction of an *aldehyde* or a *ketone* with water forms a *hydrate* (Section 12.8).

$$\mathrm{RC(=O)R'} + \mathrm{H_2O} \overset{\text{HCl}}{\rightleftharpoons} \mathrm{R{-}C(OH)_2{-}R'}$$

5. Reaction of an *aldehyde* or a *ketone* with excess alcohol forms an *acetal* or a *ketal* (Section 12.9).

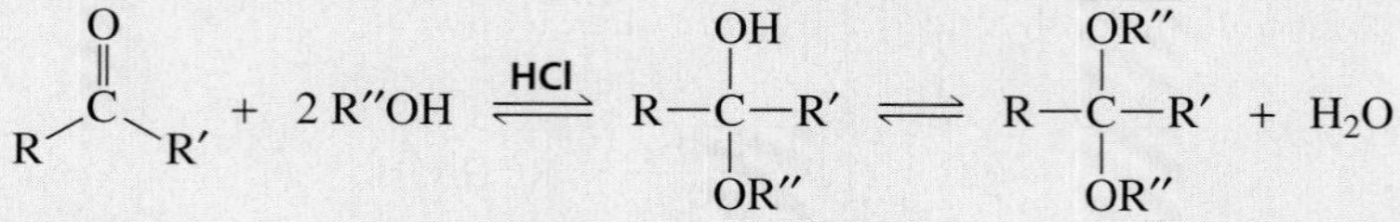

$$RC(=O)R' + 2\,R''OH \underset{}{\overset{HCl}{\rightleftharpoons}} R-C(OH)(OR'')-R' \rightleftharpoons R-C(OR'')_2-R' + H_2O$$

6. Reaction of α,β-unsaturated carbonyl compounds with a nucleophile (Section 12.10):

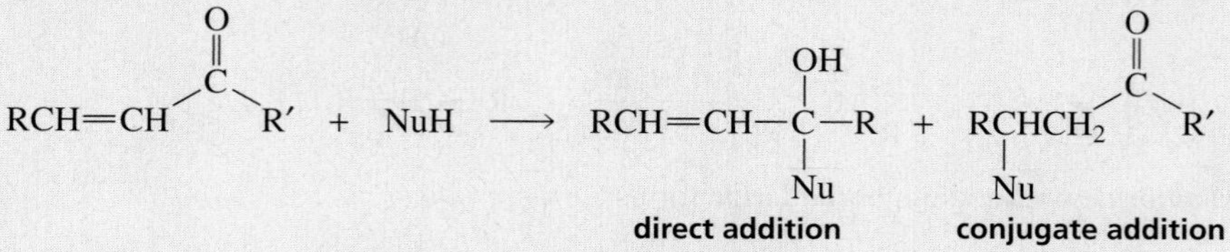

$$RCH{=}CH-C(=O)-R' + NuH \longrightarrow \underset{\text{direct addition}}{RCH{=}CH-C(OH)(Nu)-R} + \underset{\text{conjugate addition}}{RCH(Nu)CH_2-C(=O)-R'}$$

Nucleophiles that are strong bases (H^-, RMgBr) form direct addition products; nucleophiles that are weak bases (RSH, RNH_2, Br^-) form conjugate addition products.

PROBLEMS

19. Draw the structure for each of the following:

- **a.** isobutyraldehyde
- **b.** 4-octanone
- **c.** 4-bromohexanal
- **d.** 4-bromo-3-heptanone
- **e.** 3-methylcyclohexanone
- **f.** 2,4-pentanedione

20. Give the products of each of the following reactions:

a. $CH_3CH_2CH(=O) + \underset{\text{excess}}{CH_3CH_2OH} \xrightarrow{HCl}$

b. $CH_3CH_2C(=O)CH_3 \xrightarrow[\text{2. } H_3O^+]{\text{1. } NaBH_4}$

c. $CH_3CH_2CH_2C(=O)OCH_2CH_3 \xrightarrow[\text{2. } H_3O^+]{\text{1. } LiAlH_4}$

d. 2-cyclohexenone + HBr $\longrightarrow$

21. Give an example of each of the following:

- **a.** a hemiacetal
- **b.** an imine
- **c.** a ketal
- **d.** an enamine

22. List the following in order of decreasing reactivity toward nucleophilic attack:

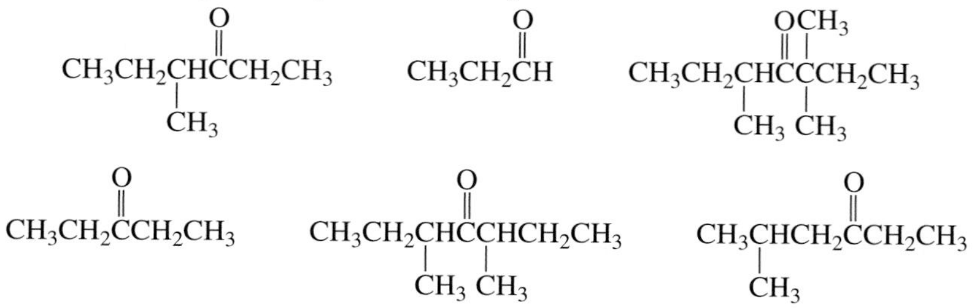

$CH_3CH_2CH(CH_3)C(=O)CH_2CH_3$ $\quad CH_3CH_2CH(=O)$ $\quad CH_3CH_2CH(CH_3)C(=O)CH(CH_3)CH_2CH_3$

$CH_3CH_2C(=O)CH_2CH_3$ $\quad CH_3CH_2CH(CH_3)C(=O)CH(CH_3)CH_2CH_3$ $\quad CH_3CH(CH_3)CH_2C(=O)CH_2CH_3$

23. Show the reagents required to form the primary alcohol.

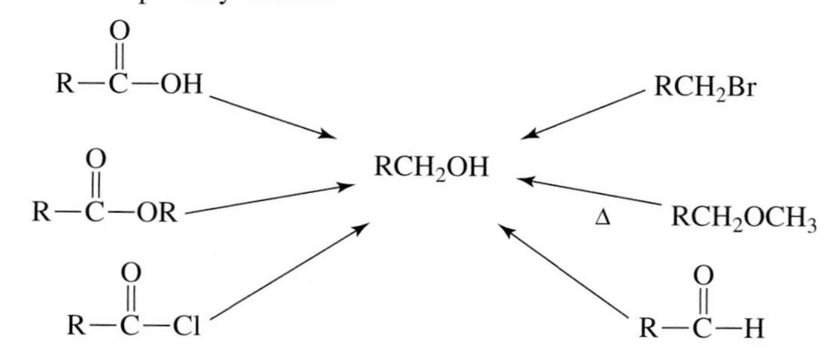

24. Fill in the boxes:

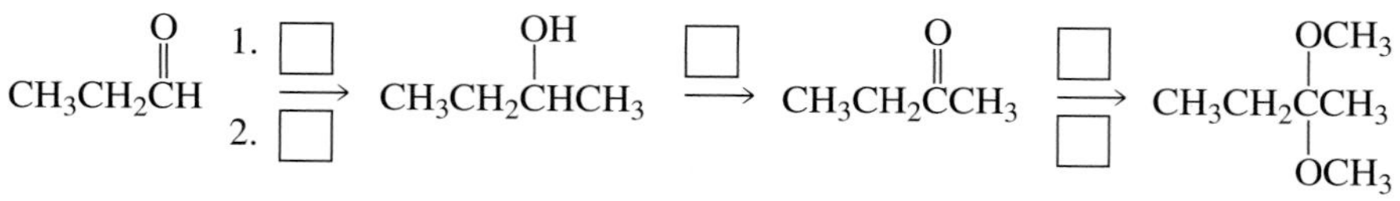

$$CH_3CH_2CH(=O) \xrightarrow[\text{2. } \square]{\text{1. } \square} CH_3CH_2CH(OH)CH_3 \xrightarrow{\square} CH_3CH_2C(=O)CH_3 \xrightarrow[\square]{\square} CH_3CH_2C(OCH_3)_2CH_3$$

25. Using cyclohexanone as the starting material, describe how each of the following could be synthesized:

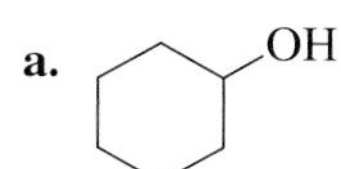
a.

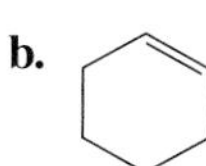
b.

c.

d.

e.

f. (Show 2 methods)

26. a. How many isomers are obtained from the reaction of 2-pentanone with ethylmagnesium bromide followed by addition of dilute acid?

b. How many isomers are obtained from the reaction of 2-pentanone with methylmagnesium bromide followed by addition of dilute aqueous acid?

27. How would you convert *N*-methylbenzamide into the following compounds?

a. *N*-methylbenzylamine **b.** benzoic acid **c.** methyl benzoate **d.** benzyl alcohol

28. Propose a mechanism for the following reaction:

$$HOCH_2CH_2CH_2CH_2CHO \xrightarrow[CH_3OH]{HCl}$$

29. List the following in order of decreasing amount of hydrate formed in an acidic aqueous solution:

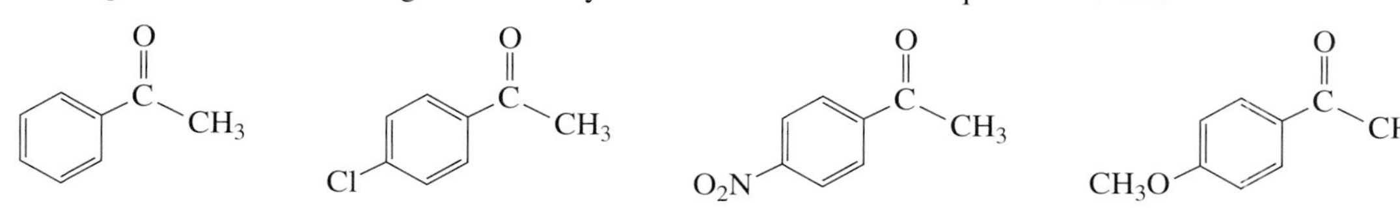

30. Fill in the boxes:

$$CH_3OH \xrightarrow{\square} CH_3Br \xrightarrow[\square]{\square} \square \xrightarrow[2.\ \square]{1.\ \square} CH_3CH_2CH_2OH$$

31. Give the products of the following reactions:

a. $C_6H_5C(CH_2CH_3){=}NCH_2CH_3 + H_2O \xrightarrow{HCl}$

b. $CH_3CH_2COCH_3 \xrightarrow[2.\ H_3O^+]{1.\ CH_3CH_2MgBr}$

c. $CH_3CH_2COOCH_3 \xrightarrow[2.\ H_3O^+]{1.\ CH_3CH_2MgBr \text{ excess}}$

d. $C_6H_5COCH_2CH_3 + CH_3NH_2 \xrightarrow{\text{trace } H^+}$

32. List three different sets of reagents (each set consisting of a carbonyl compound and a Grignard reagent) that could be used to prepare each of the following tertiary alcohols:

a. $CH_3CH_2C(OH)(C_6H_5)CH_2CH_2CH_2CH_3$

b. $CH_3CH_2C(OH)(CH_2CH_3)CH_2CH_2CH_3$

33. Give the product of the reaction of 3-methyl-2-cyclohexenone with each of the following reagents:

a. CH_3CH_2SH **b.** HBr **c.** H_2, Pd/C

34. Give the product of each of the following reactions:

a. 2-pyrrolidinone $\xrightarrow[2.\ H_2O]{1.\ LiAlH_4}$

b. cyclohexanone $+ CH_3CH_2NH_2 \xrightarrow{\text{trace } H^+}$

c. cyclohexanone $+ (CH_3CH_2)_2NH \xrightarrow{\text{trace } H^+}$

d. $CH_3C(CH_3){=}CHCOCH_3 + HBr \longrightarrow$

35. Indicate how the following compounds could be prepared from the given starting materials:

a. 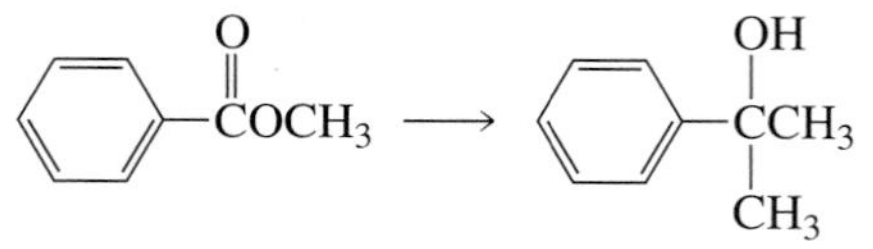 $C_6H_5COCH_3$ (with C=O) ⟶ $C_6H_5C(OH)(CH_3)CH_3$

b. 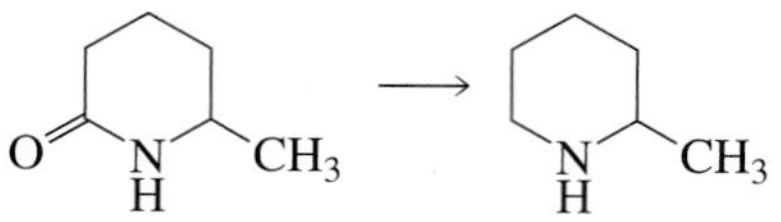 6-methyl-2-piperidinone ⟶ 2-methylpiperidine

36. Give the products of the following reactions. Show all stereoisomers that are formed.

a. $CH_3CH_2\overset{O}{\overset{\|}{C}}CH_2CH_2CH_2CH_3$ $\xrightarrow[2.\ H_3O^+]{1.\ NaBH_4}$

b. 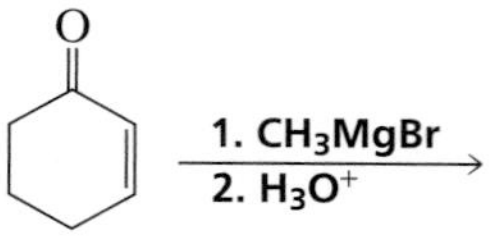 cyclohexenone $\xrightarrow[2.\ H_3O^+]{1.\ CH_3MgBr}$

37. Propose a mechanism for the following reaction:

dihydropyran + CH_3CH_2OH $\xrightarrow{HCl}$ 2-ethoxytetrahydropyran (OCH_2CH_3)

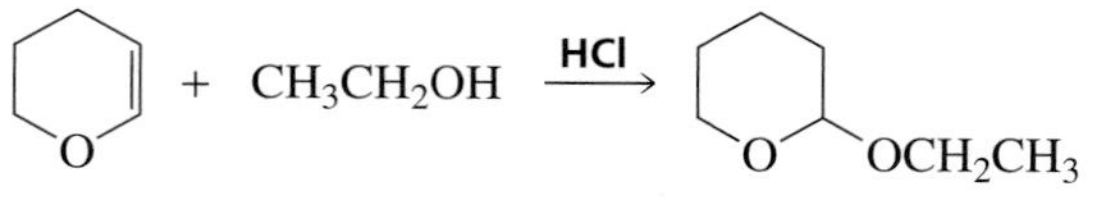

38. Indicate how the following compounds could be prepared from the given starting materials:

a. $CH_3CH_2CH_2CH_2Br$ ⟶ $CH_3CH_2CH_2CH_2\overset{O}{\overset{\|}{C}}OH$

b. $CH_3CH_2CH_2CH_2Br$ ⟶ $CH_3CH_2CH_2CH_2CH_2NH_2$

39. What class of alcohol (primary, secondary, or tertiary) is formed from the reaction of methyl formate with excess Grignard reagent followed by addition of dilute acid?

40. Put the appropriate compound in each box:

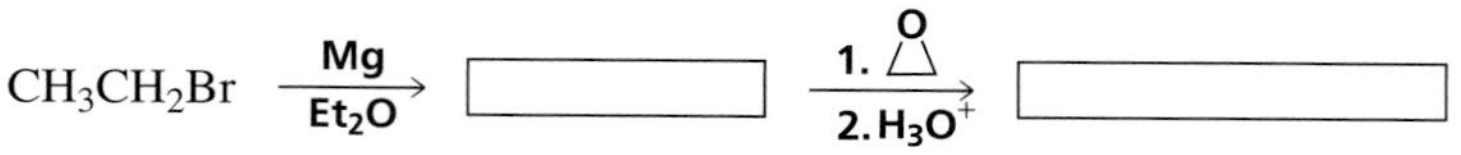

CH_3CH_2Br $\xrightarrow[Et_2O]{Mg}$ [ ] $\xrightarrow[2.\ H_3O^+]{1.\ \text{ethylene oxide}}$ [ ]

41. What alcohol would be formed from the reaction of the following Grignard reagent with ethylene oxide followed by the addition of acid?

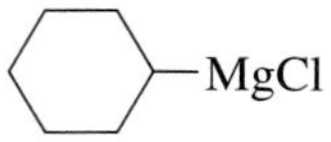

cyclohexyl–MgCl

42. a. Write the mechanism for the following reactions:
1. The acid-catalyzed hydrolysis of an imine to a carbonyl compound and a protonated primary amine.
2. The acid-catalyzed hydrolysis of an enamine to a carbonyl compound and a protonated secondary amine.

b. How do these mechanisms differ?

43. Which of the following alkyl halides could be successfully used to prepare a Grignard reagent?

a. $HOCH_2CH_2CH_2CH_2Br$

b. $BrCH_2CH_2CH_2\overset{O}{\overset{\|}{C}}OH$

c. $CH_3\underset{CH_3}{\underset{|}{N}}CH_2CH_2CH_2Br$

44. The pK_a values of oxaloacetic acid are 2.22 and 3.98.

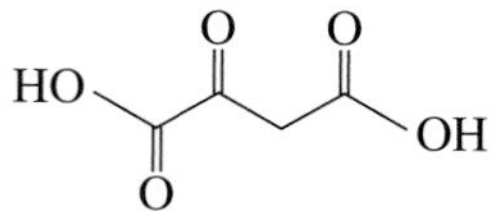

$HOOC{-}CO{-}CH_2{-}COOH$

oxaloacetic acid

a. Which carboxyl group is more acidic?

b. The amount of hydrate present in an aqueous solution of oxaloacetic acid depends on the pH of the solution: 95% at pH = 0, 81% at pH = 1.3, 35% at pH = 3.1, 13% at pH = 4.7, 6% at pH = 6.7, and 6% at pH = 12.7. Explain this pH dependence.

45. Propose a reasonable mechanism for the following reaction:

$CH_3\overset{O}{\overset{\|}{C}}CH_2CH_2\overset{O}{\overset{\|}{C}}OCH_2CH_3$ $\xrightarrow[2.\ H_3O^+]{1.\ CH_3MgBr}$ 5,5-dimethyl-γ-butyrolactone (H_3C, H_3C) + CH_3CH_2OH

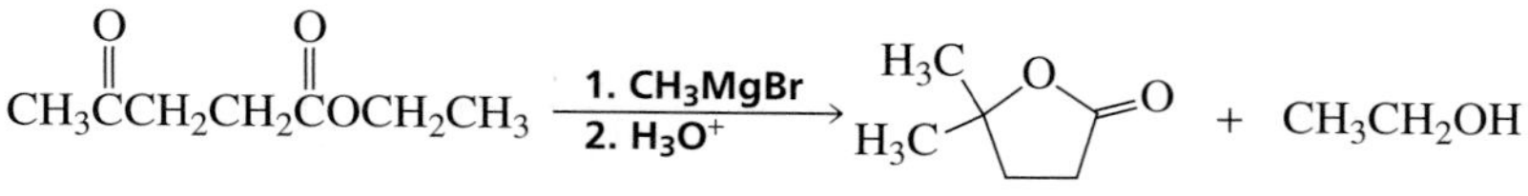

Carbonyl Compounds III

Reactions at the α-Carbon

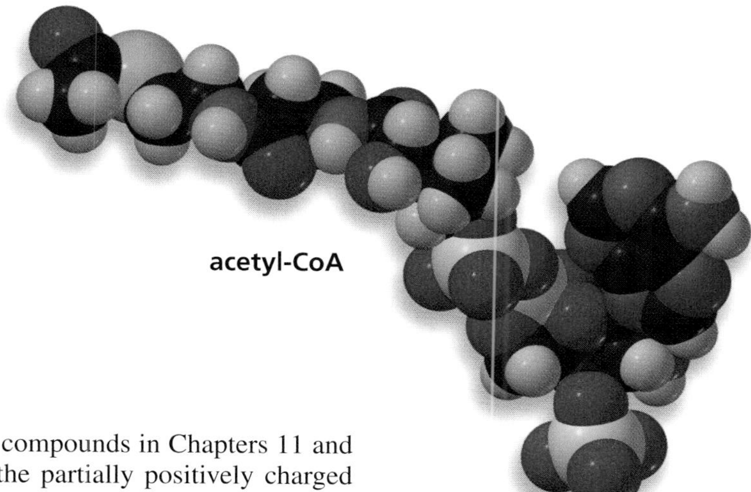

When we looked at the reactions of carbonyl compounds in Chapters 11 and 12, we saw that their site of reactivity is the partially positively charged carbonyl carbon, which is attacked by nucleophiles.

$$RCH_2\overset{\delta+}{C}(=\overset{\delta-}{O})R + Nu:^- \rightleftharpoons RCH_2-C(O^-)(Nu)-R$$

Aldehydes, ketones, and esters have a second site of reactivity. A hydrogen bonded to a carbon *adjacent* to a carbonyl carbon is sufficiently acidic to be removed by a strong base. The carbon adjacent to a carbonyl carbon is called an **α-carbon**. A hydrogen bonded to an α-carbon is called an **α-hydrogen**.

$$R-CH(H)-C(=O)-R + :\bar{base} \rightleftharpoons R-\ddot{C}H^{-}-C(=O)-R + baseH$$

an α-carbon

an α-hydrogen

13.1 THE ACIDITY OF AN α-HYDROGEN

Hydrogen and carbon have similar electronegativities, which means that the electrons binding them together are shared almost equally by the two atoms. Consequently, a hydrogen bonded to a carbon is usually not acidic. This is particularly true for hydrogens bonded to sp^3 carbons because these carbons are the most similar to hydrogen in electronegativity (Section 5.13). The high pK_a of ethane provides evidence for the low acidity of a hydrogen bonded to an sp^3 carbon.

$$CH_3CH_3 \quad pK_a > 60$$

A hydrogen bonded to an sp^3 carbon adjacent to a carbonyl carbon, however, is much more acidic than hydrogens bonded to other sp^3 carbons. For example, the pK_a value for dissociation of a proton from an α-carbon of an aldehyde or a ketone ranges from 16 to

20, and the pK_a value for dissociation of a proton from the α-carbon of an ester is about 25 (Table 13.1). Notice that although an α-hydrogen is more acidic than most other carbon-bound hydrogens, it is less acidic than a hydrogen of water (pK_a = 15.7).

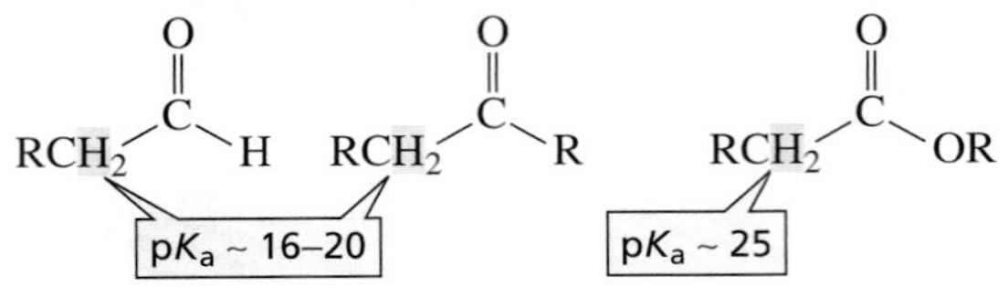

$$RCH_2C(=O)H \quad RCH_2C(=O)R \qquad RCH_2C(=O)OR$$

pK_a ~ 16–20 (aldehyde, ketone) pK_a ~ 25 (ester)

Table 13.1 The pK_a Values of Some Carbon Acids

	pK_a		pK_a
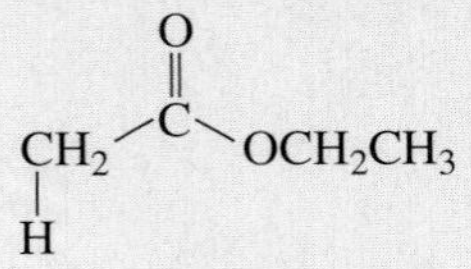 $CH_2(H)C(=O)OCH_2CH_3$	25	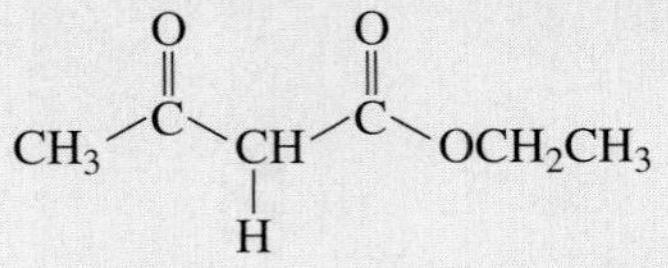 $CH_3C(=O)CH(H)C(=O)OCH_2CH_3$	10.7
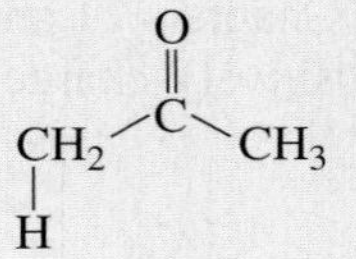 $CH_2(H)C(=O)CH_3$	20	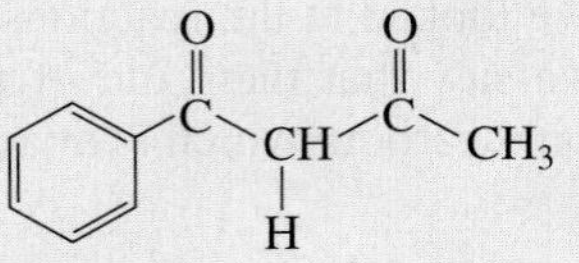 $C_6H_5C(=O)CH(H)C(=O)CH_3$	9.4
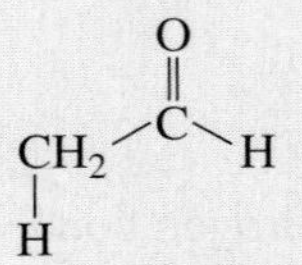 $CH_2(H)C(=O)H$	17	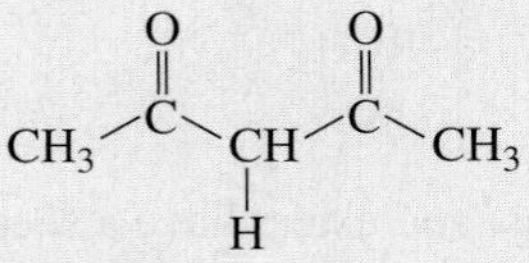 $CH_3C(=O)CH(H)C(=O)CH_3$	8.9
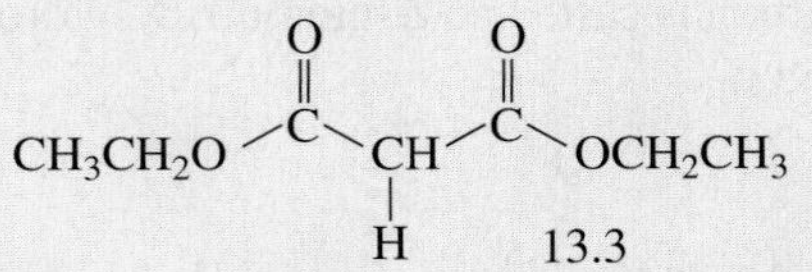 $CH_3CH_2OC(=O)CH(H)C(=O)OCH_2CH_3$	13.3	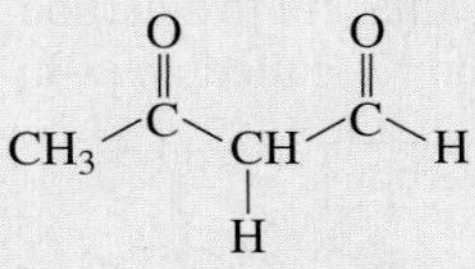 $CH_3C(=O)CH(H)C(=O)H$	5.9

Why is a hydrogen bonded to an sp^3 carbon adjacent to a carbonyl carbon so much more acidic than hydrogens bonded to other sp^3 carbons? An α-hydrogen is more acidic because the base formed when a proton is removed from an α-carbon is more stable than the base formed when a proton is removed from other sp^3 carbons. As we have seen, the more stable the base, the stronger is its conjugate acid (Section 2.6).

Why is the base formed when a proton is removed from α-carbon more stable? When a proton is removed from ethane, the electrons left behind reside solely on a carbon atom. Because carbon is not very electronegative, a carbanion is unstable. As a result, the pK_a of its conjugate acid is very high.

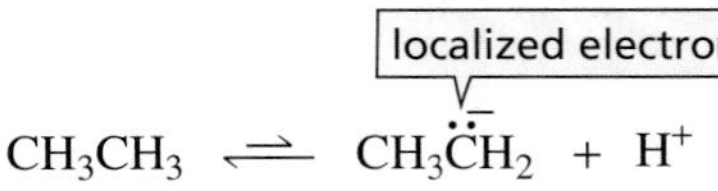

localized electrons

$$CH_3CH_3 \rightleftharpoons CH_3\ddot{C}H_2^- + H^+$$

In contrast, when a proton is removed from a carbon adjacent to a carbonyl carbon, two factors combine to increase the stability of the base that is formed. First, the electrons left behind, when the proton is removed, are delocalized, and we have seen that electron delocalization increases stability (Section 7.6). More importantly, the electrons are delocalized onto an oxygen, an atom that is better able to accommodate them because it is more electronegative than carbon.

electrons are better accommodated on O than on C

delocalized electrons

resonance contributors

Now we can understand why aldehydes and ketones ($pK_a = 16 - 20$) are more acidic than esters ($pK_a = 25$). The electrons left behind when a proton is removed from the α-carbon of an ester are not as readily delocalized onto the carbonyl oxygen as they would be in an aldehyde or a ketone. This is because the oxygen of the OR group of the ester also has a lone pair that can be delocalized onto the carbonyl oxygen. Thus, the two pairs of electrons compete for delocalization onto the same oxygen.

The α-hydrogen of a ketone or an aldehyde is more acidic than the α-hydrogen of an ester.

delocalization of a lone pair on oxygen

delocalization of the α-carbon's negative charge

resonance contributors

If the α-carbon is *between* two carbonyl groups, the acidity of an α-hydrogen is even greater (Table 13.1). For example, the pK_a value for dissociation of a proton from the α-carbon of 2,4-pentanedione, a compound with an α-carbon between two ketone carbonyl groups, is 8.9; the pK_a value for dissociation of a proton from the α-carbon of ethyl 3-oxobutyrate, a compound with an α-carbon between a ketone carbonyl group and an ester carbonyl group, is 10.7.

$pK_a = 8.9$

2,4-pentanedione
acetylacetone
a β-diketone

$pK_a = 10.7$

ethyl 3-oxobutyrate
ethyl acetoacetate
a β-keto ester

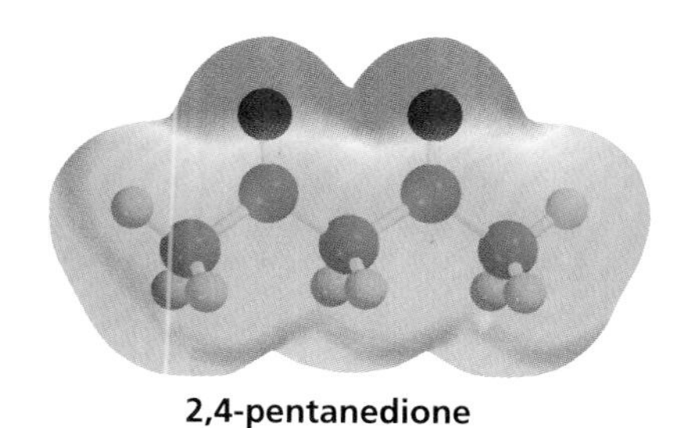

2,4-pentanedione

The acidity of α-hydrogens bonded to carbons flanked by two carbonyl groups increases because the electrons left behind when the proton is removed can be delocalized onto *two* oxygen atoms.

resonance contributors for the 2,4-pentanedione anion

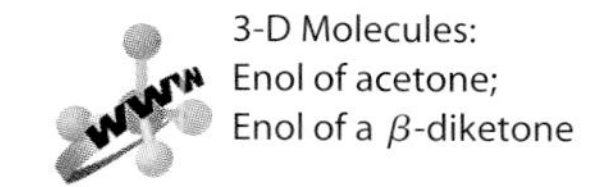

3-D Molecules:
Enol of acetone;
Enol of a β-diketone

PROBLEM 1

Identify the most acidic hydrogen in each compound.

PROBLEM 2 ◆

a. Which compound is a stronger acid?
b. Which compound has a greater pK_a value?

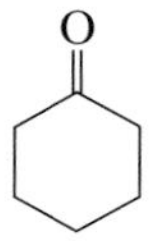

O ... or ... O, O

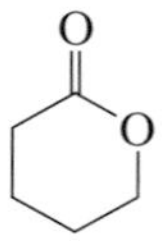

PROBLEM 3 ◆

Why is 2,4-pentanedione a stronger acid than ethyl 3-oxobutyrate?

PROBLEM-SOLVING STRATEGY

Explain why HO^- cannot remove a proton from the α-carbon of a carboxylic acid.

Finding that HO^- cannot remove a proton from the α-carbon of a carboxylic acid suggests that HO^- reacts with another portion of the molecule more rapidly. Because the proton on the carboxyl group is more acidic ($pK_a \sim 5$) than the proton on the α-carbon, we can conclude that HO^- removes a proton from the carboxyl group rather than from the α-carbon.

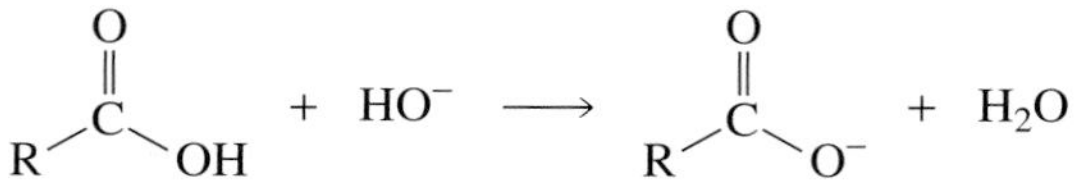

$$RC(=O)OH + HO^- \longrightarrow RC(=O)O^- + H_2O$$

Now continue on to Problem 4.

PROBLEM 4 ◆

Explain why a proton can be removed from the α-carbon of *N,N*-dimethylethanamide but not from the α-carbon of either *N*-methylethanamide or ethanamide.

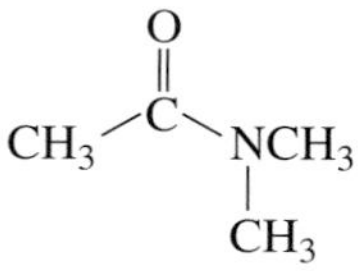

$CH_3C(=O)N(CH_3)CH_3$ — ***N,N*-dimethylethanamide**

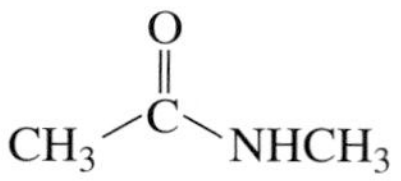

$CH_3C(=O)NHCH_3$ — ***N*-methylethanamide**

$CH_3C(=O)NH_2$ — **ethanamide**

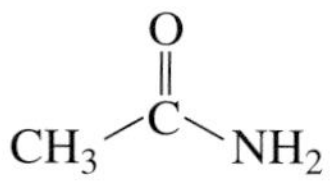

PROBLEM 5 ◆

Explain why HO^- cannot remove a proton from the α-carbon of an acyl chloride.

13.2 KETO–ENOL TAUTOMERS

A ketone exists in equilibrium with its enol tautomer. Recall that **tautomers** are isomers that are in rapid equilibrium (Section 5.11). Keto–enol tautomers differ in the location of a double bond and a hydrogen.

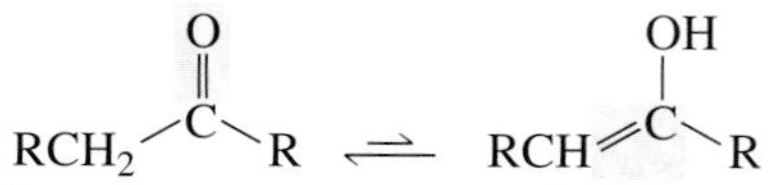

$$RCH_2C(=O)R \rightleftharpoons RCH{=}C(OH)R$$

keto tautomer — **enol tautomer**

For most ketones, the enol tautomer is much less stable than the keto tautomer. For example, an aqueous solution of acetone exists as an equilibrium mixture of more than 99.9% keto tautomer and less than 0.1% enol tautomer.

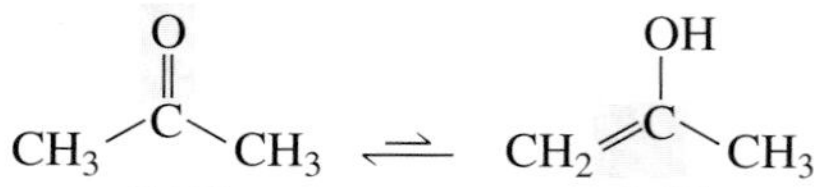

$$CH_3C(=O)CH_3 \rightleftharpoons CH_2{=}C(OH)CH_3$$

$>99.9\%$ **keto tautomer** — $<0.1\%$ **enol tautomer**

Phenol is unusual in that its enol tautomer is *more* stable than its keto tautomer because the enol tautomer is aromatic, but the keto tautomer is not.

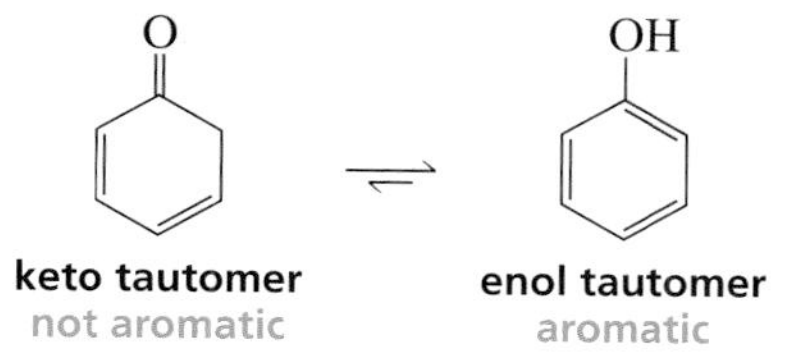

keto tautomer
not aromatic

enol tautomer
aromatic

13.3 KETO–ENOL INTERCONVERSION

Now that we know that a hydrogen bonded to a carbon adjacent to a carbonyl carbon is somewhat acidic, we can understand why keto and enol tautomers interconvert as we first saw in Chapter 5. The interconversion of keto and enol tautomers is called **keto–enol interconversion** or **tautomerization**. The interconversion can be catalyzed by either a base or an acid. The mechanism for base-catalyzed keto–enol interconversion is shown below.

mechanism for base-catalyzed keto–enol interconversion

removal of a proton from the α-carbon

protonation of oxygen

keto tautomer

enolate ion

enol tautomer

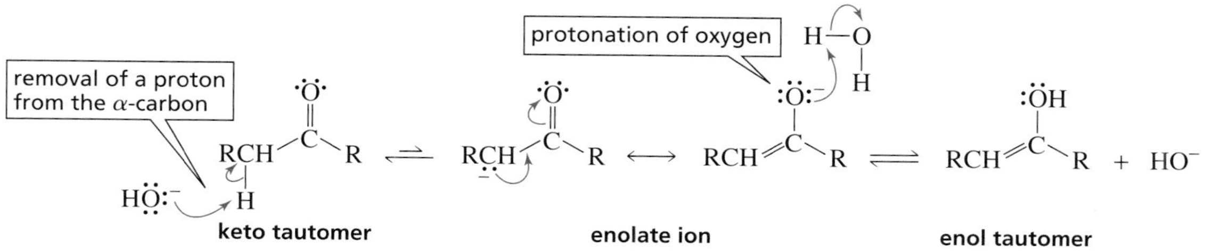

- Hydroxide ion removes a proton from the α-carbon of the keto tautomer, forming an anion called an **enolate ion**. The enolate ion has two resonance contributors.
- Protonation on oxygen forms the enol tautomer, whereas protonation on the α-carbon reforms the keto tautomer.

The mechanism for acid-catalyzed keto–enol interconversion follows.

mechanism for acid-catalyzed keto–enol interconversion

protonation of oxygen

keto tautomer

enol tautomer

removal of a proton from the α-carbon

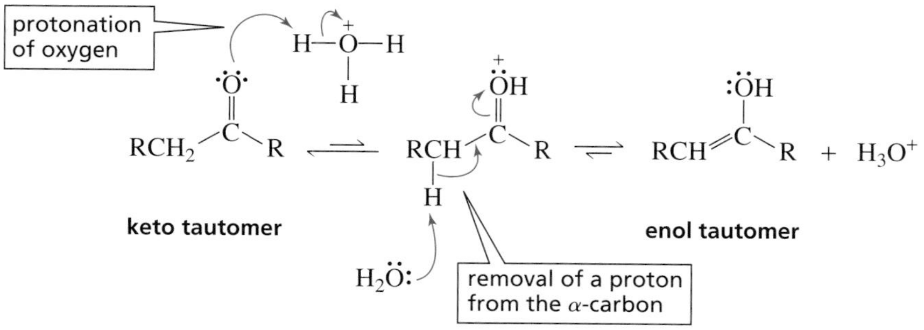

- The acid protonates the carbonyl oxygen of the keto tautomer.
- Water removes a proton from the α-carbon, forming the enol.

Notice that the steps are reversed in the base- and acid-catalyzed reactions. In the base-catalyzed reaction, the base removes a proton from the α-carbon in the first step and the oxygen is protonated in the second step. In the acid-catalyzed reaction, the oxygen is protonated in the first step and the proton is removed from the α-carbon in the second step.

PROBLEM 6 ♦

Draw the enol tautomer for each of the following:

a. $CH_3CH_2\overset{O}{\overset{\|}{C}}CH_2CH_3$ **b.** $C_6H_5\overset{O}{\overset{\|}{C}}CH_3$ **c.**

PROBLEM 7 ♦

Draw the two enol tautomers for the following compound. Which one is more stable?

13.4 ALKYLATION OF ENOLATE IONS

The resonance contributors of an enolate ion show that it has two electron-rich sites: the α-carbon and the oxygen.

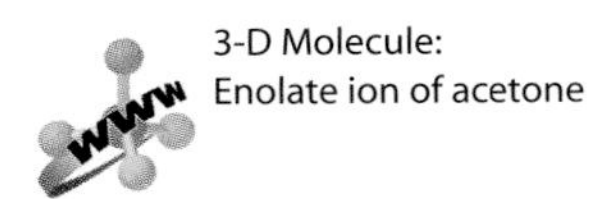

3-D Molecule: Enolate ion of acetone

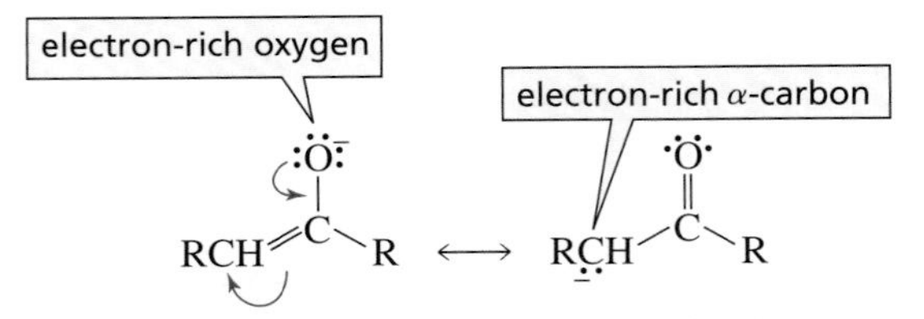

resonance contributors of an enolate ion

Which nucleophilic site (C or O) reacts with an electrophile depends on the electrophile. Protonation occurs preferentially on oxygen because of the greater concentration of negative charge on the more electronegative oxygen atom. However, when the electrophile is something other than a proton, carbon is more likely to be the nucleophile because carbon is a better nucleophile than oxygen.

Alkylation of the α-carbon of a carbonyl compound is an important reaction because it gives us another way to form a carbon–carbon bond. Alkylation is carried out by first removing a proton from the α-carbon with a strong base and then adding the appropriate alkyl halide. Because the alkylation is an S_N2 reaction, it works best with primary alkyl halides and methyl halides (Section 9.2).

Enolate ions can be alkylated on the α-carbon.

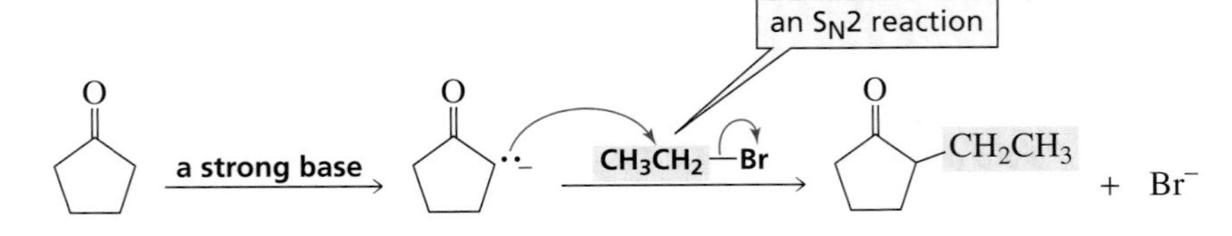

THE SYNTHESIS OF ASPIRIN

In the first step in the industrial synthesis of aspirin, a phenolate ion reacts with carbon dioxide under pressure to form *o*-hydroxybenzoic acid, also known as salicylic acid. Salicylic acid reacts with acetic anhydride to form acetylsalicylic acid (aspirin).

During World War I, the Bayer Company bought as much phenol as it could on the international market, knowing that eventually all the phenol could be used to manufacture aspirin. This left little phenol available for other countries to purchase for the synthesis of 2,4,6-trinitrophenol, a common explosive at that time.

(continued...)

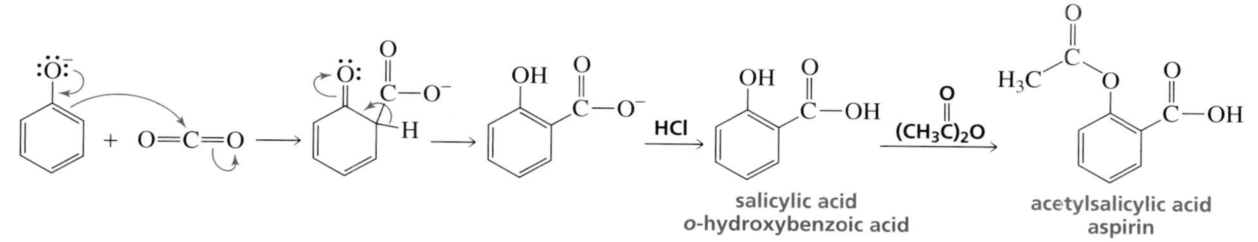

$$\text{phenolate} + O{=}C{=}O \longrightarrow \longrightarrow \xrightarrow{HCl} \text{salicylic acid} \xrightarrow{(CH_3\overset{O}{\overset{\|}{C}})_2O} \text{acetylsalicylic acid}$$

salicylic acid
o-hydroxybenzoic acid

acetylsalicylic acid
aspirin

PROBLEM 8

Draw the contributing resonance structures for the enolate ion of
a. 3-pentanone **b.** cyclohexanone

PROBLEM 9 ◆

Give the product that would be formed if the enolate ion of each compound in Problem 8 were treated with ethyl bromide.

13.5 AN ALDOL ADDITION FORMS β-HYDROXYALDEHYDES AND β-HYDROXYKETONES

We saw in Chapter 12 that the carbonyl carbon of aldehydes and ketones is an electrophile. We have just seen that a proton can be removed from the α-carbon of an aldehyde or a ketone, converting the α-carbon into a nucleophile. An **aldol addition** is a reaction in which *both* of these activities are observed: one molecule of a carbonyl compound—after a proton is removed from an α-carbon—reacts as a *nucleophile* and attacks the *electrophilic* carbonyl carbon of a second molecule of the carbonyl compound.

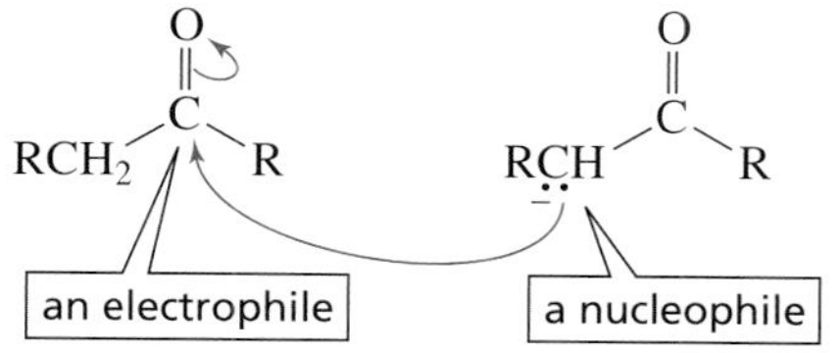

$$RCH_2\overset{O}{\overset{\|}{C}}R \qquad R\overset{-}{\ddot{C}}H\overset{O}{\overset{\|}{C}}R$$

an electrophile

a nucleophile

Thus, an aldol addition is a reaction between two molecules of an *aldehyde* or two molecules of a *ketone*. When the reactant is an aldehyde, the addition product is a β-hydroxyaldehyde, which is why the reaction is called an aldol addition ("ald" for aldehyde, "ol" for alcohol). When the reactant is a ketone, the addition product is a β-hydroxyketone. Notice that the reaction forms a new C—C bond between the α-carbon of one molecule and the carbon that formerly was the carbonyl carbon of the other molecule.

aldol additions

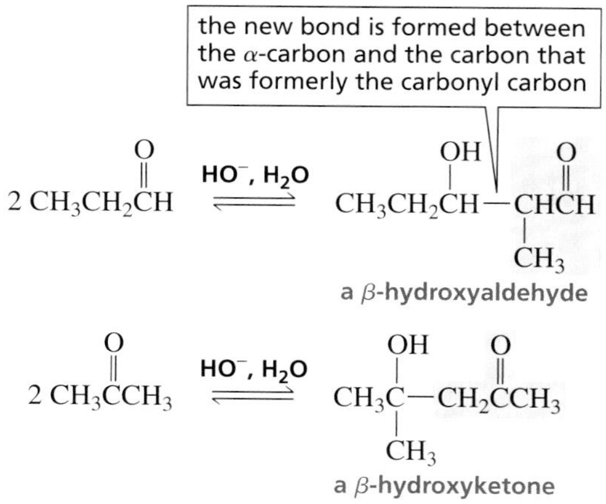

the new bond is formed between the α-carbon and the carbon that was formerly the carbonyl carbon

$$2\ CH_3CH_2\overset{O}{\overset{\|}{C}}H \xrightleftharpoons{HO^-,\ H_2O} CH_3CH_2\overset{OH}{\overset{|}{C}}H-\underset{CH_3}{\underset{|}{C}}H\overset{O}{\overset{\|}{C}}H$$

a β-hydroxyaldehyde

$$2\ CH_3\overset{O}{\overset{\|}{C}}CH_3 \xrightleftharpoons{HO^-,\ H_2O} CH_3\underset{CH_3}{\underset{|}{\overset{OH}{\overset{|}{C}}}}-CH_2\overset{O}{\overset{\|}{C}}CH_3$$

a β-hydroxyketone

The mechanism of the reaction is shown below.

mechanism for the aldol addition (aldehyde)

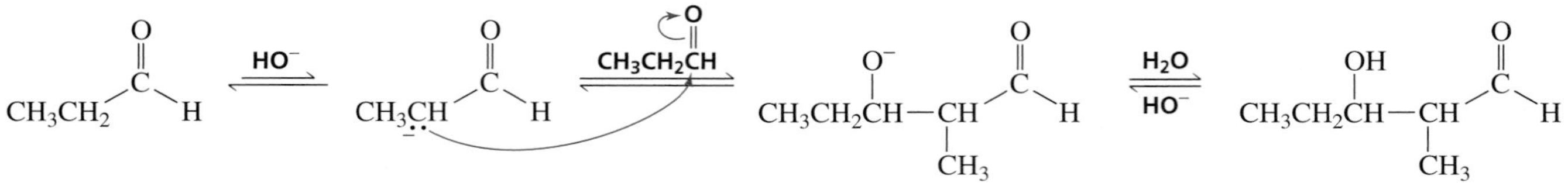

$$CH_3CH_2C(=O)H \xrightleftharpoons{HO^-} CH_3\ddot{C}^-HC(=O)H \xrightleftharpoons{CH_3CH_2CH=O} CH_3CH_2CH(O^-)-CH(CH_3)-C(=O)H \xrightleftharpoons[HO^-]{H_2O} CH_3CH_2CH(OH)-CH(CH_3)-C(=O)H$$

a β-hydroxyaldehyde

The new C—C bond formed in an aldol addition connects the α-carbon of one molecule and the carbon that formerly was the carbonyl carbon of the other molecule.

- A base removes a proton from an α-carbon, creating an enolate ion (a nucleophile).
- The enolate ion adds to the carbonyl carbon of a second molecule of the carbonyl compound.
- The negatively charged oxygen is protonated by the solvent.

Because an aldol addition occurs between two molecules of the same carbonyl compound, the product has twice as many carbons as the reacting aldehyde or ketone.

mechanism for the aldol addition (ketone)

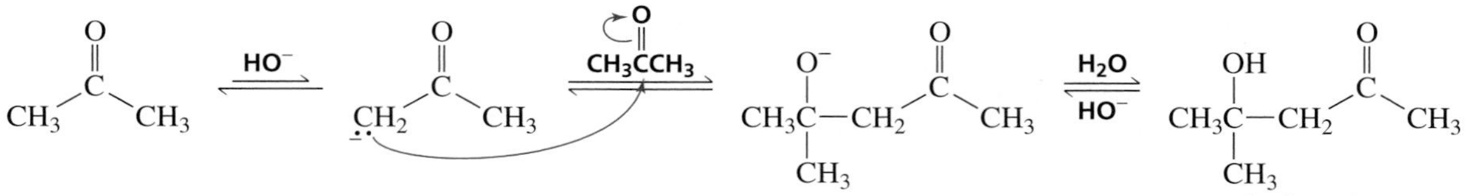

$$CH_3C(=O)CH_3 \xrightleftharpoons{HO^-} \ddot{C}^-H_2C(=O)CH_3 \xrightleftharpoons{CH_3CCH_3 (=O)} CH_3C(O^-)(CH_3)-CH_2-C(=O)CH_3 \xrightleftharpoons[HO^-]{H_2O} CH_3C(OH)(CH_3)-CH_2-C(=O)CH_3$$

a β-hydroxyketone

3-D Molecules:
β-Hydroxyaldehyde;
β-Hydroxyketone

PROBLEM 10

Show the aldol addition product that would be formed from each of the following compounds:

a. $CH_3CH_2CH_2CH_2C(=O)H$

b. $CH_3CH(CH_3)CH_2CH_2C(=O)H$

c. $CH_3CH_2C(=O)CH_2CH_3$

d. cyclohexanone

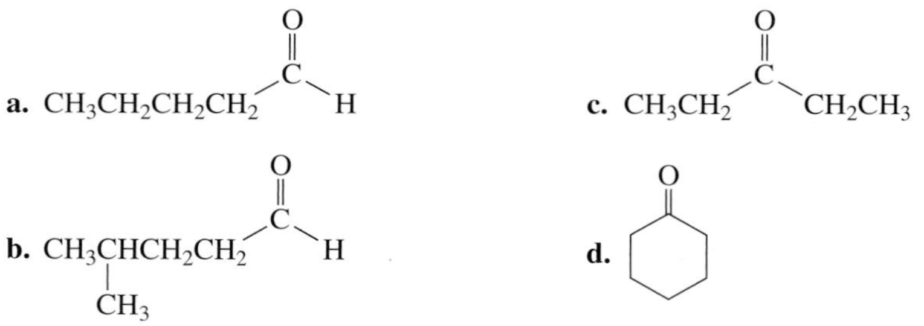

PROBLEM 11 ♦

For each of the following compounds, indicate the aldehyde or ketone from which it would be formed by an aldol addition:

a. 2-ethyl-3-hydroxyhexanal
b. 4-hydroxy-4-methyl-2-pentanone
c. 2,4-dicyclohexyl-3-hydroxybutanal
d. 5-ethyl-5-hydroxy-4-methyl-3-heptanone

13.6 DEHYDRATION OF THE PRODUCT OF AN ALDOL ADDITION

We have seen that alcohols are dehydrated when they are heated with acid (Section 10.3). The β-hydroxyaldehyde and β-hydroxyketone products of aldol additions are easier to dehydrate than many other alcohols because the double bond formed as the result of dehydration is conjugated with a carbonyl group. Conjugation increases the stability of the product (Section 6.7) and therefore makes it easier to form. If the product of an aldol addition is dehydrated, the overall reaction is called an

aldol condensation. A **condensation reaction** is a reaction that combines two molecules by forming a new C—C bond while removing a small molecule (usually water or an alcohol). Notice that an aldol condensation forms an α,β-unsaturated aldehyde or an α,β-unsaturated ketone.

An aldol addition product loses water to form an aldol condensation product.

$$2\ CH_3CH_2CHO \underset{}{\overset{HO^-,\ H_2O}{\rightleftharpoons}} CH_3CH_2CH(OH)-CH(CH_3)-CHO \xrightarrow[\Delta]{H_3O^+} CH_3CH_2CH=C(CH_3)-CHO + H_2O$$

a β-hydroxyaldehyde

an α,β-unsaturated aldehyde

conjugated double bonds

PROBLEM 12 ♦

Give the product obtained from the aldol condensation of cyclohexanone.

PROBLEM 13 SOLVED

How could you prepare the following compounds using a starting material containing no more than three carbons?

a. $CH_3CH_2CH=C(CH_3)-CHO$

b. $CH_3C(=O)CH_2C(=O)OH$

Tutorial: Aldol reactions—synthesis

Solution to 13a A compound with the correct six-carbon skeleton can be obtained if a three-carbon aldehyde undergoes an aldol addition. Dehydration of the addition product forms the desired α,β-unsaturated aldehyde.

$$CH_3CH_2CHO \xrightarrow{HO^-,\ H_2O} CH_3CH_2CH(OH)CH(CH_3)CHO \xrightarrow[\Delta]{H_3O^+} CH_3CH_2CH=C(CH_3)CHO$$

13.7 A CLAISEN CONDENSATION FORMS A β-KETO ESTER

When two molecules of an *ester* undergo a condensation reaction, the reaction is called a **Claisen condensation**. The product of a Claisen condensation is a β-keto ester.

$$2\ CH_3CH_2C(=O)OCH_2CH_3 \xrightarrow[2.\ HCl]{1.\ CH_3CH_2O^-} CH_3CH_2C(=O)CH(CH_3)C(=O)OCH_2CH_3 + CH_3CH_2OH$$

a β-keto ester

the new bond is formed between the α-carbon and the carbon that formerly was the carbonyl carbon

In a Claisen condensation, as in an aldol addition, one molecule of carbonyl compound is converted into an enolate ion that reacts with a second molecule of the carbonyl compound.

The new C—C bond formed in a Claisen condensation connects the α-carbon of one molecule and the carbon that formerly was the carbonyl carbon of the other molecule.

mechanism for the Claisen condensation

$$\mathrm{CH_3CH(H)C(=O)OCH_3} \underset{}{\overset{\mathrm{CH_3\ddot{O}:^-}}{\rightleftharpoons}} \mathrm{CH_3\bar{C}HC(=O)OCH_3} + \mathrm{CH_3OH} \overset{\mathrm{CH_3CH_2COCH_3\ (C=O)}}{\rightleftharpoons} \mathrm{CH_3CH_2C(O^-)(OCH_3)-CH(CH_3)-C(=O)OCH_3}$$

$$\rightleftharpoons \mathrm{CH_3CH_2C(=O)-CH(CH_3)-C(=O)OCH_3} + \mathrm{CH_3O^-}$$

BIOGRAPHY

Ludwig Claisen (1851–1930) *was born in Germany and received a Ph.D. from the University of Bonn, studying under Kekulé (Section 7.1). Claisen was a professor of chemistry at the University of Bonn, Owens College (Manchester, England), the University of Munich, the University of Aachen, the University of Kiel, and the University of Berlin.*

- A strong base removes a proton from an α-carbon, forming an enolate ion.
- The enolate ion attacks the carbonyl carbon of a second molecule of ester. The new C—C bond is formed between the α-carbon of one molecule and the carbon that formerly was the carbonyl carbon of the other molecule.
- The negatively charged oxygen reforms the carbon–oxygen π bond and expels the $^-$OR group.

When the reaction is over, HCl is added to protonate the $^-$OR group. The base employed in a Claisen condensation corresponds to the leaving group of the ester so that the reactant will not change if the base were to act as a nucleophile and attack the carbonyl group.

$$\mathrm{CH_3C(=\ddot{O}:)OCH_3} + \mathrm{CH_3\ddot{O}:^-} \rightleftharpoons \mathrm{CH_3-C(:\ddot{O}:^-)(\ddot{O}CH_3)-:\ddot{O}CH_3}$$

Notice that after nucleophilic attack, the Claisen condensation and the aldol addition differ. In the Claisen condensation, *the negatively charged oxygen* reforms the carbon–oxygen π bond. In the aldol addition, *the negatively charged oxygen* obtains a proton from the solvent.

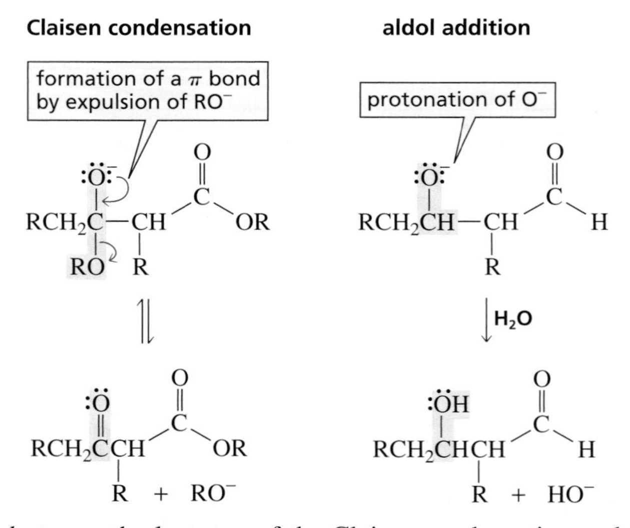

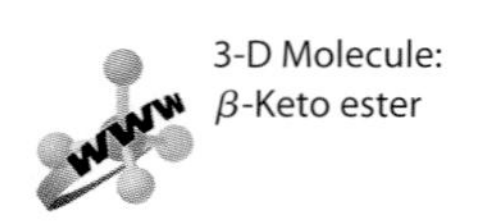

3-D Molecule: β-Keto ester

The difference between the last step of the Claisen condensation and the last step of the aldol addition arises from the difference between esters and aldehydes or ketones. With esters, the carbon bonded to the negatively charged oxygen is also bonded to a group that can be expelled. With aldehydes or ketones, the carbon bonded to the negatively charged oxygen is not bonded to a group that can be expelled. Thus, the Claisen condensation is a *nucleophilic substitution reaction*, whereas the aldol addition is a *nucleophilic addition reaction*.

PROBLEM 14 ◆

Give the products of the following reactions:

a. 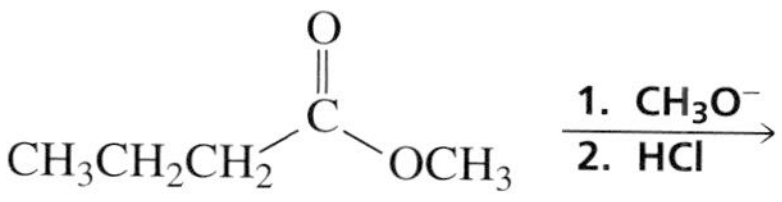 $CH_3CH_2CH_2C(=O)OCH_3$ $\xrightarrow[\text{2. HCl}]{\text{1. } CH_3O^-}$

b. 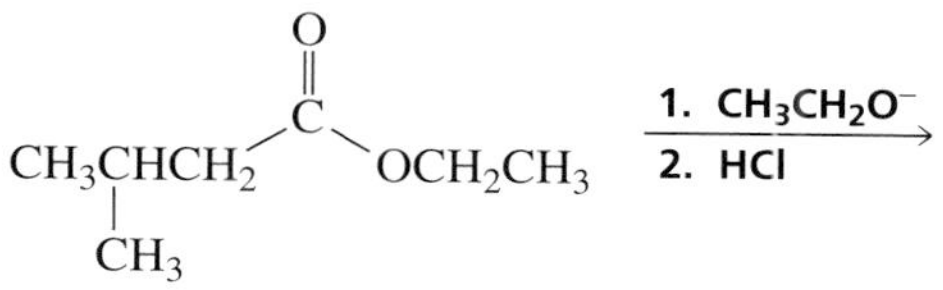 $CH_3CH(CH_3)CH_2C(=O)OCH_2CH_3$ $\xrightarrow[\text{2. HCl}]{\text{1. } CH_3CH_2O^-}$

PROBLEM 15 ◆

Which of the following esters cannot undergo a Claisen condensation?

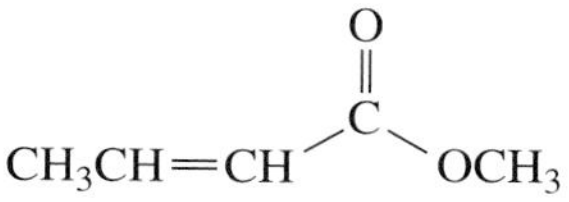

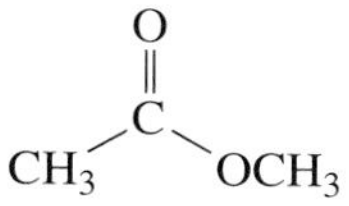

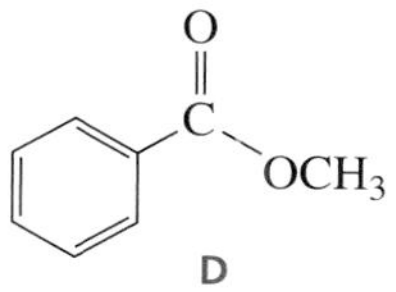

A: $CH_3CH{=}CHC(=O)OCH_3$

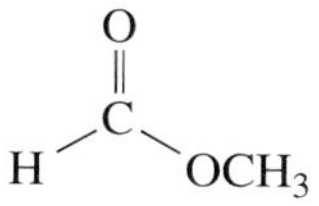

B: $HC(=O)OCH_3$

C: $CH_3C(=O)OCH_3$

D: $C_6H_5C(=O)OCH_3$

PROBLEM 16 ◆

What starting materials were used to make the following β-keto ester?

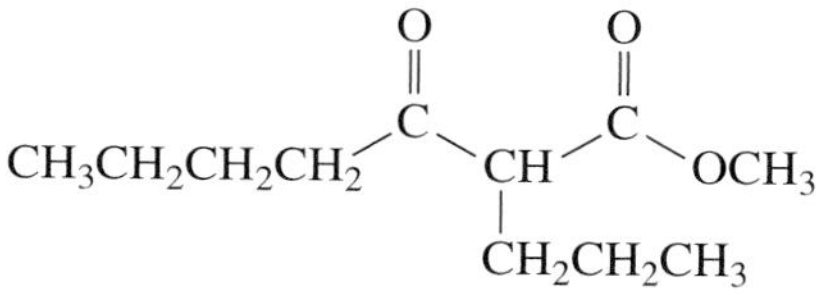

$CH_3CH_2CH_2CH_2C(=O)CH(CH_2CH_2CH_3)C(=O)OCH_3$

Tutorial:
Claisen reactions—synthesis

13.8 CARBOXYLIC ACIDS WITH A CARBONYL GROUP AT THE 3-POSITION CAN BE DECARBOXYLATED

Carboxylate ions do not lose CO_2, for the same reason that alkanes such as ethane do not lose a proton—because the leaving group would be a carbanion. Carbanions are very strong bases and therefore are very poor leaving groups.

$CH_3CH_2{-}H$ $\qquad$ $CH_3CH_2C(=O)\ddot{O}:^-$

If, however, the CO_2 group is bonded to a carbon adjacent to a carbonyl carbon, the CO_2 group can be removed, because the electrons left behind can be delocalized onto the carbonyl oxygen. Consequently, carboxylate ions with a carbonyl group at the 3-position lose CO_2 when they are heated. Loss of CO_2 from a molecule is called **decarboxylation**.

removing CO_2 from an α-carbon

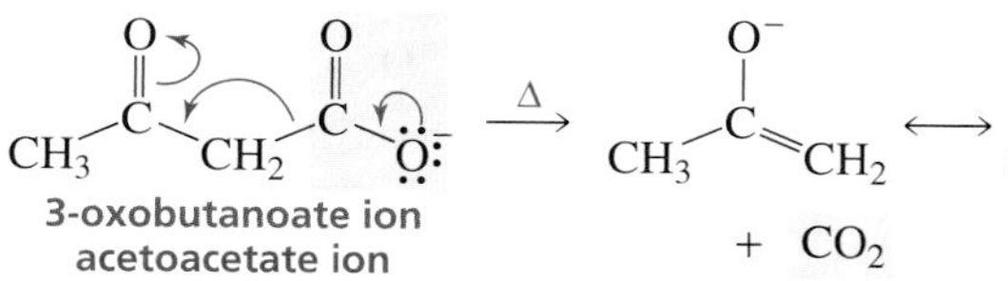

$$CH_3C(=O)CH_2C(=O)O^- \xrightarrow{\Delta} CH_3C(O^-){=}CH_2 \longleftrightarrow CH_3C(=O)\bar{C}H_2 \; + \; CO_2$$

3-oxobutanoate ion
acetoacetate ion

A carboxylic acid with a carbonyl group in the 3-position loses CO_2 when heated.

Notice the similarity between removal of CO_2 from an α-carbon adjacent to a carbonyl carbon and removal of a proton from an α-carbon. In both reactions, a substituent—CO_2 in one case, H^+ in the other—is removed from an α-carbon and its bonding electrons are delocalized onto an oxygen.

removing a proton from an α-carbon

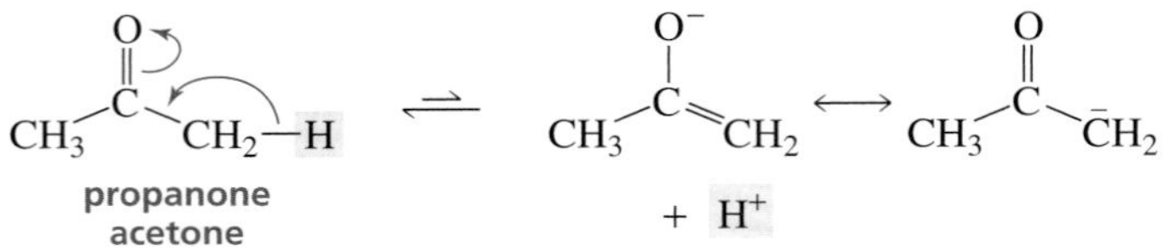

$$CH_3COCH_2{-}H \rightleftharpoons CH_3C(O^-){=}CH_2 \longleftrightarrow CH_3C(=O)\bar{C}H_2 \quad + H^+$$

propanone
acetone

Decarboxylation is even easier if the reaction is carried out under acidic conditions, because the reaction is catalyzed by an intramolecular transfer of a proton from the carboxyl group to the carbonyl oxygen. The enol that is formed immediately tautomerizes to a ketone.

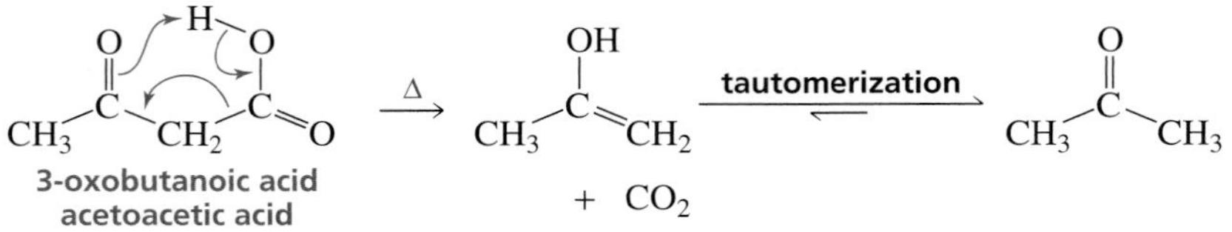

$$CH_3COCH_2COOH \xrightarrow{\Delta} CH_3C(OH){=}CH_2 + CO_2 \xrightleftharpoons{\text{tautomerization}} CH_3COCH_3$$

3-oxobutanoic acid
acetoacetic acid
a β-keto acid

In summary, carboxylic acids and carboxylate ions with a carbonyl group at the 3-position lose CO_2 when they are heated.

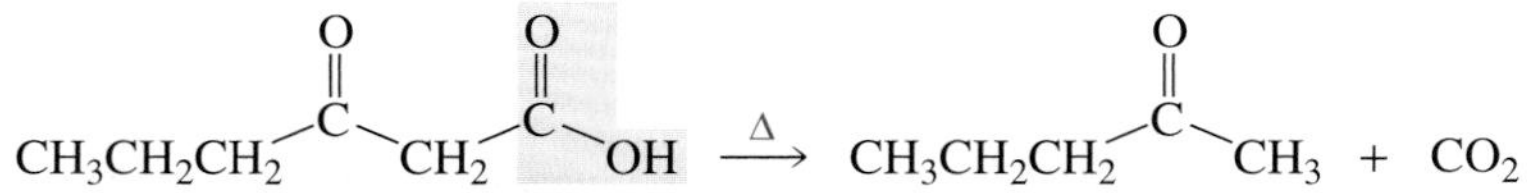

$$CH_3CH_2CH_2COCH_2COOH \xrightarrow{\Delta} CH_3CH_2CH_2COCH_3 + CO_2$$

3-oxohexanoic acid → 2-pentanone

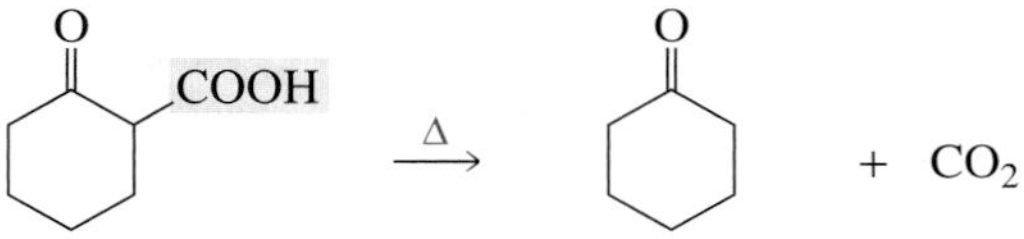

2-oxocyclohexanecarboxylic acid $\xrightarrow{\Delta}$ cyclohexanone $+ CO_2$

$$HOOC{-}CH(CH_3){-}COOH \xrightarrow{\Delta} CH_3CH_2COOH + CO_2$$

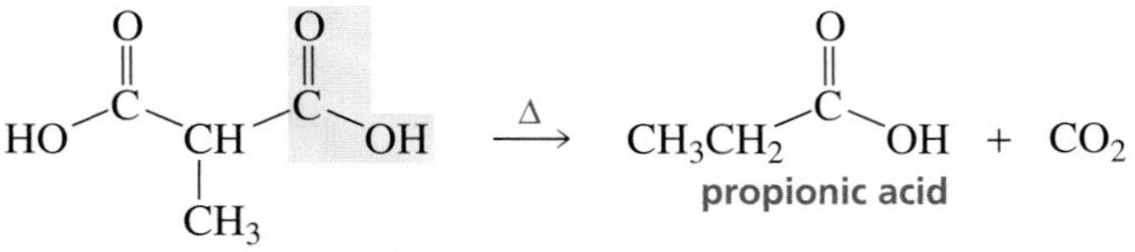

α-methylmalonic acid → propionic acid

PROBLEM 17 ♦

Which of the following compounds would be expected to lose CO_2 when heated?

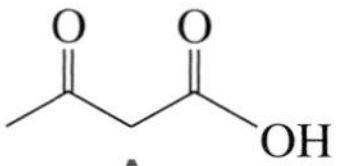

A

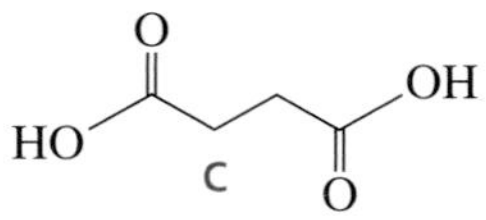

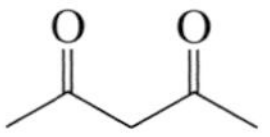

B

C

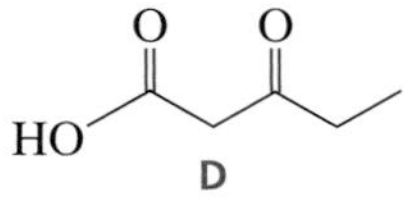

D

13.9 THE MALONIC ESTER SYNTHESIS: A METHOD TO SYNTHESIZE A CARBOXYLIC ACID

A combination of two of the reactions discussed in this chapter—alkylation of an α-carbon and decarboxylation of a carboxylic acid that has a carbonyl group at the 3-position—can be used to prepare carboxylic acids of any desired chain length. The procedure is called the **malonic ester synthesis** because the starting material for the synthesis is the diethyl ester of malonic acid.

The first two carbons of the carboxylic acid being synthesized come from malonic ester, and the rest of the carboxylic acid comes from the alkyl halide used in the second step of the synthesis.

malonic ester synthesis

$$C_2H_5O\text{-}C(=O)\text{-}CH_2\text{-}C(=O)\text{-}OC_2H_5 \xrightarrow[\text{3. HCl, } H_2O, \Delta]{\text{1. } CH_3CH_2O^-\text{ 2. RBr}} R\text{-}CH_2\text{-}C(=O)\text{-}OH$$

diethyl malonate
malonic ester

from malonic ester

from the alkyl halide

A malonic ester synthesis forms a carboxylic acid with two more carbons than the alkyl halide.

The steps in the malonic ester synthesis are shown below.

$$C_2H_5O\text{-}C(=O)\text{-}CH_2\text{-}C(=O)\text{-}OC_2H_5 \xrightarrow{CH_3CH_2O^-} C_2H_5O\text{-}C(=O)\text{-}\ddot{C}H^-\text{-}C(=O)\text{-}OC_2H_5 \xrightarrow{R-Br} C_2H_5O\text{-}C(=O)\text{-}CH(R)\text{-}C(=O)\text{-}OC_2H_5 + Br^-$$

removal of a proton from the α-carbon

alkylation of the α-carbon

an α-substituted malonic ester

$$\xrightarrow{\text{HCl, } H_2O, \Delta} HO\text{-}C(=O)\text{-}CH(R)\text{-}C(=O)\text{-}OH + 2\ CH_3CH_2OH$$

hydrolysis

an α-substituted malonic acid

$$\xrightarrow{\Delta} R\text{-}CH_2\text{-}C(=O)\text{-}OH + CO_2$$

decarboxylation

- A base easily removes a proton from the α-carbon because it is flanked by two ester groups ($pK_a = 13$).
- The resulting α-carbanion reacts with an alkyl halide, forming an α-substituted malonic ester. Because alkylation is an S_N2 reaction, it works best with primary alkyl halides and methyl halides (Section 9.2).
- Heating the α-substituted malonic ester in an acidic aqueous solution hydrolyzes both ester groups.
- Further heating decarboxylates the α-substituted malonic acid.

Tutorial:
Malonic ester synthesis

PROBLEM 18 ◆

What alkyl bromide should be used in the malonic ester synthesis of each of the following carboxylic acids?

a. propanoic acid **b.** 3-phenylpropanoic acid **c.** 4-methylpentanoic acid

13.10 THE ACETOACETIC ESTER SYNTHESIS: A METHOD TO SYNTHESIZE A METHYL KETONE

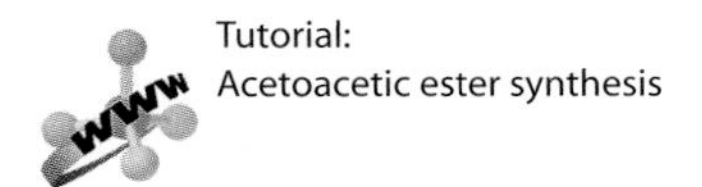

Tutorial: Acetoacetic ester synthesis

The only difference between the acetoacetic ester synthesis and the malonic ester synthesis is the use of acetoacetic ester rather than malonic ester as the starting material. The difference in starting material causes the product of the **acetoacetic ester synthesis** to be a *methyl ketone* rather than a *carboxylic acid*. The carbonyl group of the methyl ketone and the carbons on either side of it come from acetoacetic ester; the rest of the ketone comes from the alkyl halide used in the second step of the synthesis.

An acetoacetic ester synthesis forms a methyl ketone with two more carbons than the alkyl halide.

acetoacetic ester synthesis

$$CH_3COCH_2COOC_2H_5 \xrightarrow[\text{2. RBr} \quad \text{3. HCl, } H_2O,\ \Delta]{\text{1. } CH_3CH_2O^-} R{-}CH_2COCH_3$$

ethyl 3-oxobutanoate
ethyl acetoacetate
"acetoacetic ester"

from the alkyl halide (R); from acetoacetic ester (CH_2COCH_3)

The steps in the acetoacetic ester synthesis and the malonic ester synthesis are similar. The last step in the acetoacetic ester synthesis is decarboxylation of a substituted acetoacetic acid rather than a substituted malonic acid.

$$CH_3COCH_2COOC_2H_5 \xrightarrow{CH_3CH_2O^-} CH_3CO\ddot{C}^-HCOOC_2H_5 \xrightarrow{R{-}Br} CH_3COCH(R)COOC_2H_5 + Br^-$$

removal of a proton from the α-carbon

alkylation of the α-carbon

$$\xrightarrow{\text{HCl, } H_2O,\ \Delta} CH_3COCH(R)COOH + CH_3CH_2OH$$

hydrolysis

$$CH_3COCH(R)COOH \xrightarrow{\Delta} CH_3COCH_2{-}R + CO_2$$

decarboxylation

PROBLEM 19 ***SOLVED***

Starting with methyl propanoate, how could you prepare 4-methyl-3-heptanone?

$$CH_3CH_2COOCH_3 \xrightarrow{?} CH_3CH_2COCH(CH_3)CH_2CH_2CH_3$$

methyl propanoate → 4-methyl-3-heptanone

Solution Because the starting material is an ester and the target molecule has more carbon atoms than the starting material, a Claisen condensation appears to be a good way to start this synthesis. The Claisen condensation forms a β-keto ester that can be easily alkylated at the desired carbon because it is flanked by two carbonyl groups. Acid-catalyzed hydrolysis will form a carboxylic acid that will decarboxylate when it is heated.

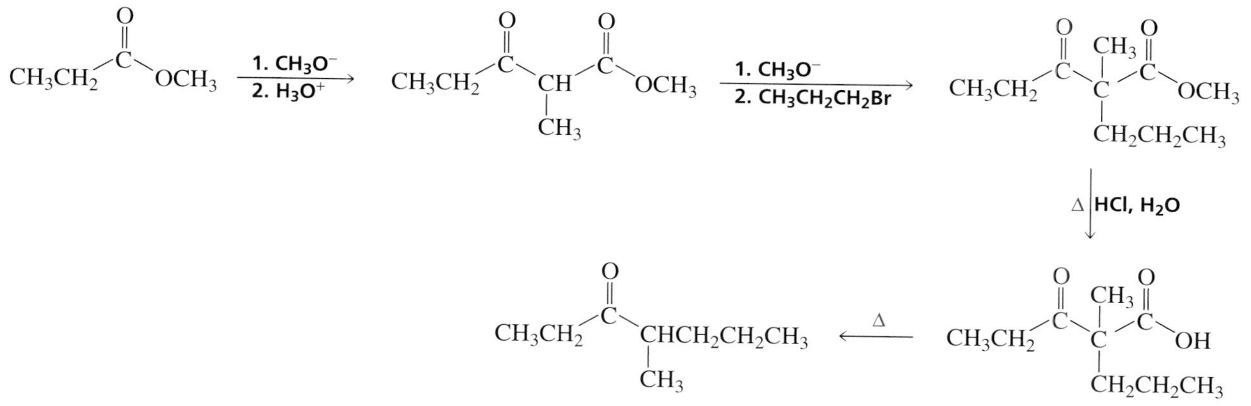

$$CH_3CH_2COCH_3 \xrightarrow[\text{2. } H_3O^+]{\text{1. } CH_3O^-} CH_3CH_2COCH(CH_3)COCH_3 \xrightarrow[\text{2. } CH_3CH_2CH_2Br]{\text{1. } CH_3O^-} CH_3CH_2COC(CH_3)(CH_2CH_2CH_3)COCH_3$$

$$\xrightarrow[\Delta]{HCl,\ H_2O} CH_3CH_2COC(CH_3)(CH_2CH_2CH_3)COOH \xrightarrow{\Delta} CH_3CH_2COCH(CH_3)CH_2CH_2CH_3$$

PROBLEM 20 ♦

What alkyl bromide should be used in the acetoacetic ester synthesis of each of the following methyl ketones?

a. 2-pentanone **b.** 2-octanone **c.** 4-phenyl-2-butanone

13.11 REACTIONS AT THE α-CARBON IN BIOLOGICAL SYSTEMS

Many reactions that occur in biological systems involve reactions at the α-carbon—the kinds of reactions you have studied in this chapter. We will now look at a few examples.

A Biological Aldol Addition

Glucose, the most abundant sugar found in nature, is synthesized in biological systems from two molecules of pyruvate. The series of reactions that convert two molecules of pyruvate into glucose is called **gluconeogenesis**. The reverse process—the breakdown of glucose into two molecules of pyruvate—is called **glycolysis** (Section 18.4).

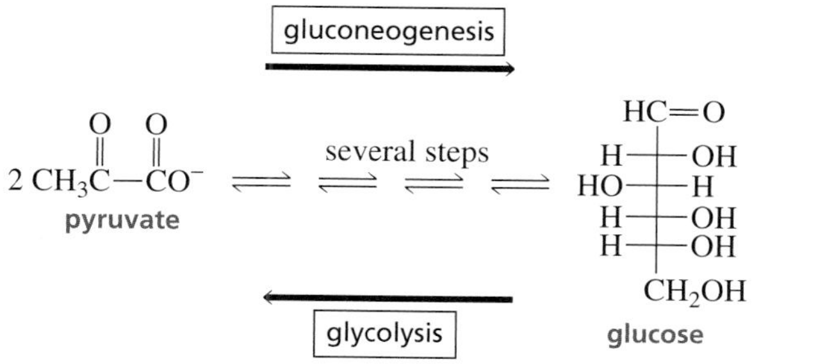

$$2\ CH_3COCOO^- \underset{\text{glycolysis}}{\overset{\text{gluconeogenesis}}{\rightleftharpoons}} \text{(several steps)} \text{ glucose}$$

pyruvate — glucose

Because glucose has twice as many carbons as pyruvate, you should not be surprised to learn that one of the steps in the biosynthesis of glucose is an aldol addition. An enzyme called aldolase catalyzes an aldol addition between dihydroxyacetone phosphate and glyceraldehyde-3-phosphate (Section 17.3). The product is fructose-1,6-diphosphate, which is subsequently converted to glucose.

3-D Molecule: Aldolase

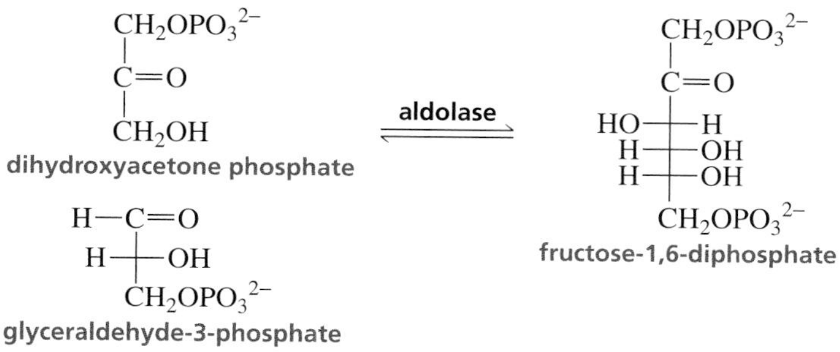

$$\text{dihydroxyacetone phosphate} + \text{glyceraldehyde-3-phosphate} \overset{\text{aldolase}}{\rightleftharpoons} \text{fructose-1,6-diphosphate}$$

dihydroxyacetone phosphate ($CH_2OPO_3^{2-}$–C=O–CH_2OH); glyceraldehyde-3-phosphate (H–C=O, H–C–OH, $CH_2OPO_3^{2-}$); fructose-1,6-diphosphate

PROBLEM 21

Propose a mechanism for the formation of fructose-1,6-diphosphate from dihydroxyacetone phosphate and glyceraldehyde-3-phosphate, using HO^- as the catalyst.

A Biological Aldol Condensation

Collagen is the most abundant protein in mammals, amounting to about one-fourth of the total protein. It is the major fibrous component of bone, teeth, skin, cartilage, and tendons. Individual collagen molecules can be isolated only from tissues of young animals. As animals age, the individual molecules become cross-linked, which is why meat from older animals is tougher than meat from younger ones. Collagen cross-linking is an example of an aldol condensation.

Before collagen molecules can cross-link, their ammonium groups must be converted to aldehyde groups. The enzyme that catalyzes this reaction is called lysyl oxidase. An aldol condensation between two aldehyde groups results in a cross-linked protein.

collagen — $CH(CH_2)_2CH_2CH_2\overset{+}{N}H_3$ (ammonium groups) $H_3\overset{+}{N}CH_2CH_2(CH_2)_2CH$ $\xrightarrow{\text{lysyl oxidase}}$ $CH(CH_2)_2CH_2CH{=}O$ $O{=}CHCH_2(CH_2)_2CH$ $\xrightarrow{\text{aldol condensation}}$ $CH(CH_2)_2CH_2CH{=}C(CH_2)_2CH$ with $HC{=}O$ branch

cross-linked collagen

A Biological Claisen Condensation

Fatty acids are long-chain, unbranched carboxylic acids (Sections 11.10 and 19.1). Most naturally occurring fatty acids contain an even number of carbons because they are synthesized from acetic acid, which has two carbons.

Because the biological reaction occurs at physiological pH (7.3), the reactant is acetate—acetic acid without its proton. We have seen that carboxylate ions are very unreactive toward nucleophilic attack (Section 11.11). Biological systems can activate carboxylate ions by converting them into *thioesters*. A **thioester** is an ester with sulfur in place of the carboxylate oxygen. This reaction requires ATP; ATP puts a leaving group on the carboxylate ion that can be replaced by the thiol (Section 18.2). A **thiol** is an alcohol with sulfur in place of the oxygen. The particular *thiol* used to make the thioester shown below is called coenzyme A.

$$CH_3C(=O)O^- + CoASH \xrightarrow{ATP} CH_3C(=O)SCoA$$

acetate + coenzyme A → acetyl-CoA

One of the necessary reactants for fatty acid synthesis is malonyl-CoA, which is obtained by carboxylation of acetyl-CoA (Section 17.8).

$$CH_3C(=O)SCoA + HCO_3^- \longrightarrow {}^-OC(=O)CH_2C(=O)SCoA$$

acetyl-CoA → malonyl-CoA

Before fatty acid synthesis can occur, the acyl groups of acetyl-CoA and malonyl-CoA are transferred to other thiols by means of a transesterification reaction (Section 11.8).

transesterification reactions

$$CH_3C(=O)SCoA + RSH \longrightarrow CH_3C(=O)SR + CoASH$$

$$^{-}OC(=O)CH_2C(=O)SCoA + RSH \longrightarrow {}^{-}OC(=O)CH_2C(=O)SR + CoASH$$

A molecule of acetyl thioester and a molecule of malonyl thioester are the reactants for the first round in the biosynthesis of a fatty acid.

$$\underset{\text{a two-carbon thioester}}{CH_3C(=O)SR} + {}^{-}OC(=O)CH_2C(=O){-}SR \longrightarrow CH_3C(O^{-})(SR){-}CH_2C(=O){-}SR + CO_2 \longrightarrow CH_3C(=O)CH_2C(=O)SR$$

$$\xrightarrow{\text{reduction}} CH_3CH(OH){-}CH_2C(=O)SR \xrightarrow{\text{dehydration}} CH_3CH{=}CHC(=O)SR \xrightarrow{\text{reduction}} \underset{\text{a four-carbon thioester}}{CH_3CH_2CH_2C(=O)SR}$$

- The first step is a Claisen condensation. We have seen that the nucleophile needed for a Claisen condensation is obtained by using a strong base to remove an α-hydrogen. Strong bases are not available in living cells, however, because biological reactions take place at neutral pH. Thus, the required nucleophile is generated by removing CO_2—rather than a proton—from the α-carbon of malonyl thioester. (Recall that a carboxylic acid with a carbonyl group in the 3-position is easily decarboxylated; see Section 13.8.)
- The product of the condensation reaction undergoes a reduction, a dehydration, and a second reduction to give a four-carbon thioester.
- The four-carbon thioester and a molecule of malonyl thioester are the reactants for the second round. Again, the product of the condensation reaction undergoes a reduction, a dehydration, and a second reduction, this time to form a six-carbon thioester.

$$CH_3CH_2CH_2C(=O)SR + {}^{-}OC(=O)CH_2C(=O)SR \xrightarrow{\text{Claisen condensation}} CH_3CH_2CH_2C(=O)CH_2C(=O)SR + CO_2$$

$$\xrightarrow{\text{1. reduction; 2. dehydration; 3. reduction}} CH_3CH_2CH_2CH_2CH_2C(=O)SR$$

The sequence of reactions is repeated, and each time two more carbons are added to the chain. Now we can understand why naturally occurring fatty acids are unbranched and generally contain an even number of carbons.

Once a thioester with the appropriate number of carbons is obtained, it undergoes a transesterification reaction with glycerol in order to form fats, oils, and phospholipids (see Sections 11.10, 19.3, and 19.5).

PROBLEM 22 ♦

Palmitic acid is a straight-chain saturated 16-carbon fatty acid. How many moles of malonyl-CoA are required for the synthesis of one mole of palmitic acid?

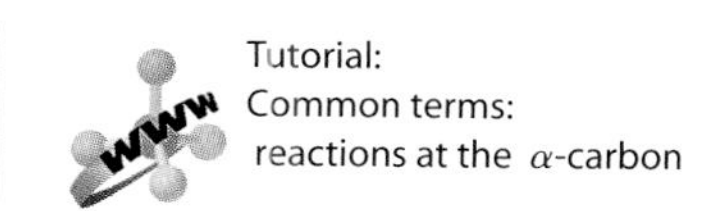

www
Tutorial:
Common terms:
reactions at the α-carbon

PROBLEM 23 ♦

a. If the biosynthesis of palmitic acid were carried out with CD_3COSR and nondeuterated malonyl thioester, how many deuteriums would be incorporated into palmitic acid?

b. If the biosynthesis of palmitic acid were carried out with $^{-}OOCCD_2COSR$ and nondeuterated acetyl thioester, how many deuteriums would be incorporated into palmitic acid?

SUMMARY

A hydrogen bonded to an **α-carbon** of an aldehyde, ketone, or ester is sufficiently acidic to be removed by a strong base, because the base that is formed when the proton is removed is stabilized by delocalization of its negative charge onto an oxygen. Aldehydes and ketones ($pK_a = 16 - 20$) are more acidic than esters ($pK_a \sim 25$). **β-Diketones** ($pK_a \sim 9$) and **β-keto esters** ($pK_a \sim 11$) are even more acidic.

Keto-enol interconversion, also called **tautomerization**, can be catalyzed by acids or by bases. Generally, the **keto tautomer** is more stable. **Enolate ions** can be alkylated on the α-carbon.

In an **aldol addition**, the enolate ion of an aldehyde or a ketone attacks the carbonyl carbon of a second molecule of aldehyde or ketone, forming a β-hydroxyaldehyde or a β-hydroxyketone. The new C—C bond is formed between the α-carbon of one molecule and the carbon that formerly was the carbonyl carbon of the other molecule. The product of an aldol addition can be dehydrated to give an **aldol condensation** product. In a **Claisen condensation**, the enolate ion of an ester attacks the carbonyl carbon of a second molecule of ester, eliminating an ^{-}OR group to form a β-keto ester.

Carboxylic acids with a carbonyl group at the 3-position **decarboxylate** when they are heated. Carboxylic acids can be prepared by a **malonic ester synthesis**; the α-carbon of the diester is alkylated and the α-substituted malonic ester undergoes acid-catalyzed hydrolysis and decarboxylation; the resulting carboxylic acid has two more carbons than the alkyl halide. Similarly, methyl ketones can be prepared by an **acetoacetic ester synthesis**; the carbonyl group and the carbons on either side of it come from acetoacetic ester, and the rest of the methyl ketone comes from the alkyl halide.

SUMMARY OF REACTIONS

1. Base-catalyzed keto-enolization interconversion (Section 13.3)

$$RCH_2C(=O)R \underset{}{\overset{HO^-}{\rightleftharpoons}} RCH{=}C(OH)R$$

2. Acid-catalyzed keto-enolization interconversion (Section 13.3)

$$RCH_2C(=O)R \underset{}{\overset{H_3O^+}{\rightleftharpoons}} RCH{=}C(OH)R$$

3. Alkylation of an enolate ion (Section 13.4)

$$RCH_2C(=O)R \xrightarrow[\text{2. } RCH_2X]{\text{1. strong base}} RCH(CH_2R)C(=O)R \quad X = \text{halogen}$$

4. Aldol addition (Section 13.5)

$$2\ RCH_2C(=O)H \overset{HO^-}{\rightleftharpoons} RCH_2CH(OH)CH(R)C(=O)H$$

5. Dehydration of the product of an aldol addition (Section 13.6)

$$RCH_2CH(OH)CH(R)C(=O)H \underset{\Delta}{\overset{H_3O^+ \text{ or } HO^-}{\rightleftharpoons}} RCH_2CH{=}C(R)C(=O)H + H_2O$$

6. Claisen condensation (Section 13.7)

$$2\ RCH_2C(=O)OCH_3 \xrightarrow[\text{2. HCl}]{\text{1. } CH_3O^-} RCH_2C(=O)CH(R)C(=O)OCH_3 + CH_3OH$$

7. Decarboxylation of a carboxylic acid with a carbonyl group at the 3-position (Section 13.8)

$$RC(=O)CH_2C(=O)OH \xrightarrow{\Delta} RC(=O)CH_3 + CO_2$$

8. Malonic ester synthesis: preparation of carboxylic acids (Section 13.9)

$$C_2H_5OC(=O)CH_2C(=O)OC_2H_5 \xrightarrow[\text{3. HCl, } H_2O, \Delta]{\text{1. } CH_3CH_2O^- \text{ 2. RBr}} RCH_2C(=O)OH + CO_2$$

9. Acetoacetic ester synthesis: preparation of methyl ketones (Section 13.10)

$$CH_3C(=O)CH_2C(=O)OC_2H_5 \xrightarrow[\text{3. HCl, } H_2O, \Delta]{\text{1. } CH_3CH_2O^- \text{ 2. RBr}} RCH_2C(=O)CH_3 + CO_2$$

PROBLEMS

24. Write a structure for each of the following:
    a. β-keto ester
    b. the enol tautomer of cyclopentanone
    c. the carboxylic acid obtained from the malonic ester synthesis when the alkyl halide is propyl bromide

25. Draw the enol tautomers for each of the following compounds. If the compound has more than one enol tautomer, indicate which one is more stable.

    a. $CH_3CH_2C(=O)CH_2C(=O)CH_2CH_3$
    b. $C_6H_5CH_2C(=O)CH_3$
    c. 2-methylcyclohexanone (ring with C=O and CH_3)

26. List the compounds in each of the following groups in order of decreasing acidity:

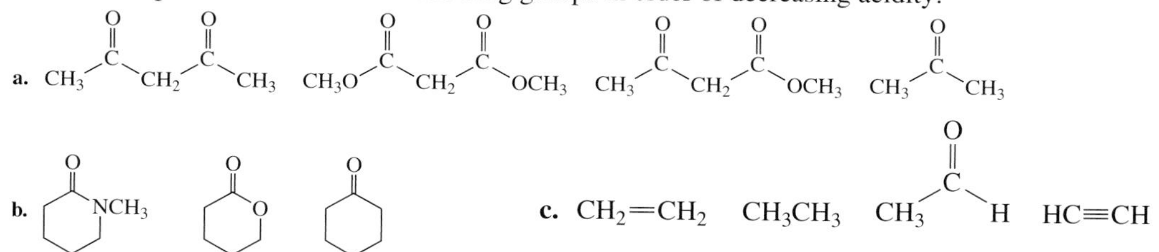

27. Show how hexanoic acid can be prepared from a malonic ester synthesis.

28. Arachidic acid is a saturated 20-carbon fatty acid. How many moles of malonyl-CoA are required for the synthesis of one mole of arachidic acid?

29. a. If the biosynthesis of arachidic acid were carried out with CD_3COSR and nondeuterated malonyl thioester, how many deuteriums would be incorporated into arachidic acid?
    b. If the biosynthesis of arachidic acid were carried out with $^-OOCCD_2COSR$ and nondeuterated acetyl thioester, how many deuteriums would be incorporated into arachidic acid?

30. The pK_a of a hydrogen bonded to the sp^3 carbon of propene is 42, which is greater than that of any of the compounds listed in Table 13.1, but less than the pK_a of an alkane. Explain.

31. Give the structures of the four β-keto esters that would be obtained from a mixture of methyl acetate and methyl propanoate in a solution of $NaOCH_3$ in methanol.

32. A β,γ-unsaturated carbonyl compound rearranges to a more stable conjugated α,β-unsaturated compound in the presence of either an acid or a base.
    a. Propose a mechanism for the base-catalyzed rearrangement.
    b. Propose a mechanism for the acid-catalyzed rearrangement.

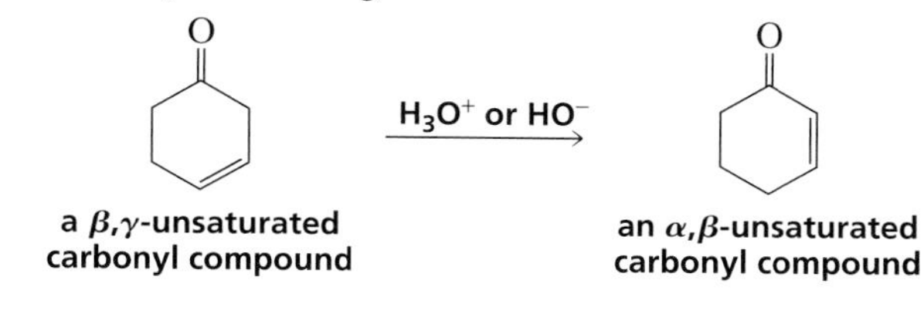

33. Both 2,6-heptanedione and 2,8-nonanedione form a product with a six-membered ring when treated with sodium hydroxide. Give the structure of the six-membered ring products.

34. a. Explain why a racemic mixture of 2-methyl-1-phenyl-1-butanone is formed when (*R*)-2-methyl-1-phenyl-1-butanone is dissolved in a basic aqueous solution.

b. Give an example of another ketone that would form a racemic mixture in a basic aqueous solution.

35. An intramolecular Claisen condensation is called a Dieckmann condensation. Give the mechanism for the following Dieckmann condensation.

$CH_2CH_2COCH_3$ (C=O) / CH_2 / CH_2 / CH_2COCH_3 (C=O) $\xrightarrow[\text{2. HCl}]{\text{1. } CH_3O^-}$ (cyclohexanone ring with $COCH_3$) + CH_3OH

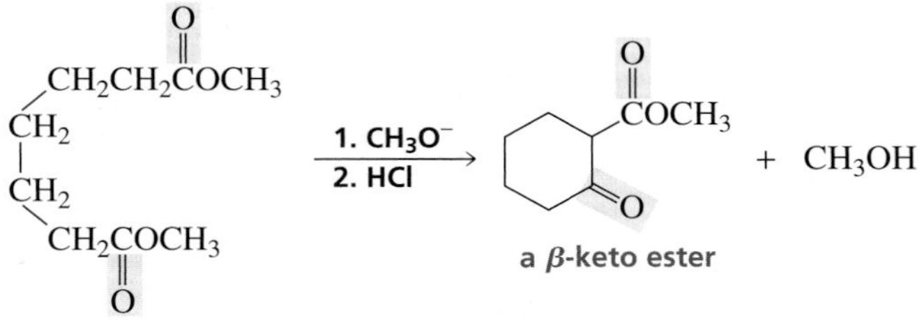

a 1,7-diester — a β-keto ester

36. What product is formed when a 1,6-diester undergoes a Dieckmann condensation?

37. Show how the following compounds could be synthesized from the given starting materials:

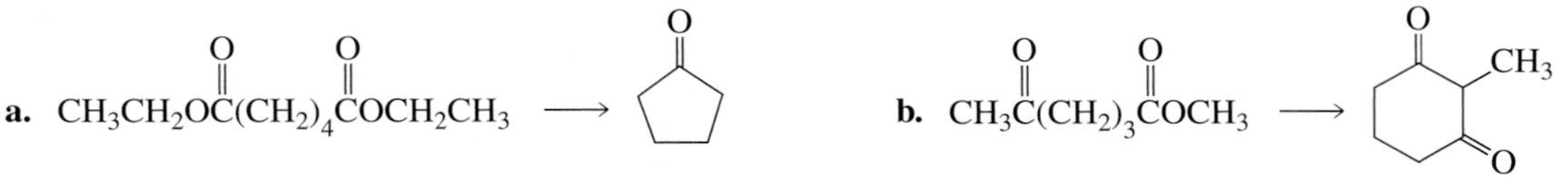

a. $CH_3CH_2O\overset{O}{\overset{\|}{C}}(CH_2)_4\overset{O}{\overset{\|}{C}}OCH_2CH_3 \longrightarrow$ cyclopentanone

b. $CH_3\overset{O}{\overset{\|}{C}}(CH_2)_3\overset{O}{\overset{\|}{C}}OCH_3 \longrightarrow$ 2-acetylcyclohexanone (CH_3)

38. Give the products of the following reactions:

a. diethyl heptanedioate: (1) sodium ethoxide; (2) HCl

b. diethyl 2-ethylhexanedioate: (1) sodium ethoxide; (2) HCl

c. diethyl malonate: (1) sodium ethoxide; (2) isobutyl bromide; (3) HCl, $H_2O + \Delta$

d. 2,7-octanedione + aqueous sodium hydroxide

39. Which would require a higher temperature, decarboxylation of a β-dicarboxylic acid or decarboxylation of a β-keto acid?

40. What compound is formed when a dilute solution of cyclohexanone is shaken with NaOD in D_2O for several hours?

41. Explain why the following carboxylic acid cannot be prepared by the malonic ester synthesis:

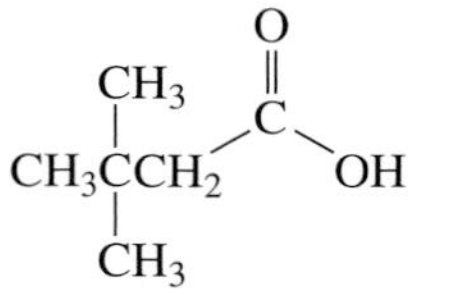

$(CH_3)_3CCH_2\overset{O}{\overset{\|}{C}}OH$

42. Give the structures of the four β-hydroxyaldehydes that would be obtained from a mixture of butanal and pentanal in a basic aqueous solution.

43. a. Draw the enol tautomer of 2,4-pentanedione.

b. Most ketones form less than 1% enol in an aqueous solution. Explain why the enol tautomer of 2,4-pentanedione is much more prevalent (15%).

44. Give the major product of the following reaction:

$H\overset{O}{\overset{\|}{C}}H$ (excess) $+ CH_3CH_2\overset{O}{\overset{\|}{C}}H \xrightarrow[H_2O]{HO^-}$

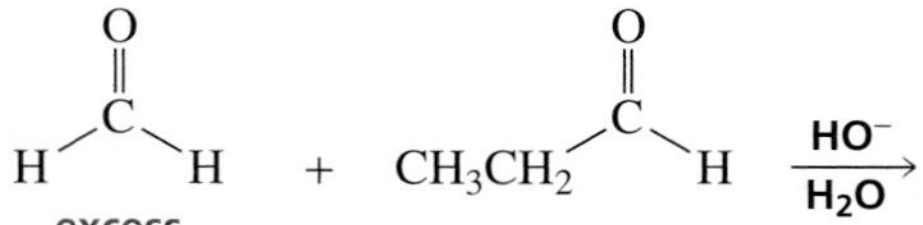

45. Ninhydrin reacts with an amino acid to form a purple-colored compound. Propose a mechanism to account for the formation of the colored compound.

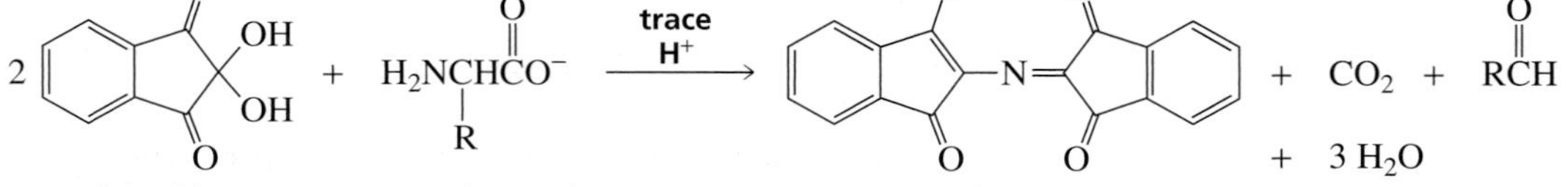

2 ninhydrin + H_2NCHCO^- (R; an amino acid) $\xrightarrow{\text{trace } H^+}$ purple-colored compound + CO_2 + $R\overset{O}{\overset{\|}{C}}H$ + $3\,H_2O$

Determining the Structures of Organic Compounds

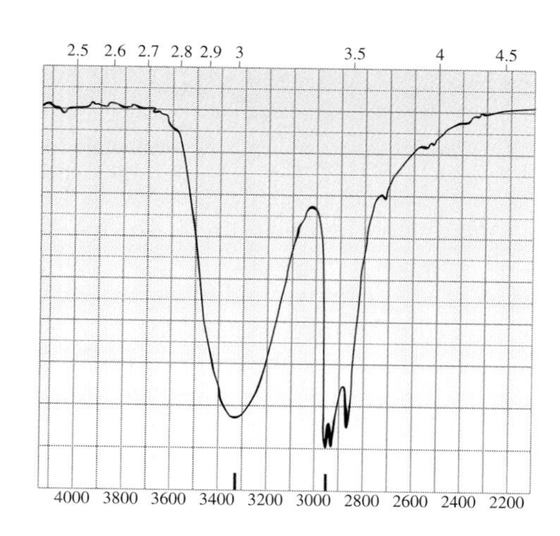

Determining the structures of organic compounds is an important part of organic chemistry. Whenever a chemist synthesizes a compound, its structure must be confirmed. For example, you were told that a ketone is formed when an alkyne undergoes the acid-catalyzed addition of water (Section 5.11). But how was it determined that the product of that reaction is actually a ketone?

Scientists search the world for new compounds with physiological activity. If a promising compound is found, its structure needs to be determined. Without knowing its structure, chemists cannot design ways to synthesize the compound, nor can they undertake studies to provide insights into its biological behavior.

At one time, determining the structure of an organic compound was a daunting task and required a relatively large amount of the compound—a real problem for the analysis of compounds that were difficult to obtain. Today, a number of different instrumental techniques are used to determine the structures of organic compounds. These techniques can be performed quickly on small amounts of a compound. We have already discussed one such technique: ultraviolet/visible (UV/Vis) spectroscopy, which provides information about organic compounds with conjugated double bonds (Section 7.10). In this chapter, we will look at three more instrumental techniques.

- **mass spectrometry**, which allows us to determine the *molecular mass* and the *molecular formula* of a compound, as well as some of its *structural features*,
- **infrared (IR) spectroscopy**, which allows us to determine the *kinds of functional groups* in a compound, and
- **nuclear magnetic resonance (NMR) spectroscopy**, which provides information about the carbon–hydrogen framework of a compound.

We will be referring to different classes of organic compounds as we discuss various instrumental techniques; these classes are listed inside the back cover of the book for easy reference.

14.1 MASS SPECTROMETRY

One of the most valuable uses of mass spectrometry is to tell us the molecular weight and molecular formula of a compound. In addition, as we will see, it can tell us some things about the compound's structure.

In mass spectrometry, a small amount of a compound is introduced into an instrument called a mass spectrometer, where it is vaporized and bombarded by a stream of high-energy electrons. When the electron beam hits a molecule, it knocks out an electron, producing a **molecular ion**. A molecular ion is a **radical cation**, a species with an unpaired electron and a positive charge.

$$\underset{\textbf{molecule}}{\mathrm{M}} \xrightarrow{\textbf{electron beam}} \underset{\substack{\textbf{molecular ion} \\ \text{a radical cation}}}{\mathrm{M}^{\dot{+}}} + \underset{\textbf{electron}}{e^-}$$

Electron bombardment injects so much kinetic energy into the molecular ions that most of them break apart into smaller cations, radicals, neutral molecules, and other radical cations. Not surprisingly, the bonds most likely to break are the weakest ones and those that result in the formation of the most stable products. All the *positively charged fragments* of the molecule are drawn between two negatively charged plates, which accelerate the fragments into an analyzer tube and out the ion exit slit to the collector (Figure 14.1). The more stable the fragment, the more likely it is to arrive at the collector without breaking down further. Neutral fragments are not attracted to the negatively charged plates and therefore are not accelerated. They are eventually pumped out of the spectrometer.

The mass spectrometer records a **mass spectrum**—a graph of the *relative abundance* of each fragment plotted against its *mass-to-charge ratio* (m/z). Because only positively charged fragments are accelerated, the charge (z) on essentially all the fragments that reach the collector is +1. Therefore, the m/z value is the molecular mass (m) of the fragment.

A mass spectrum records only positively charged fragments.

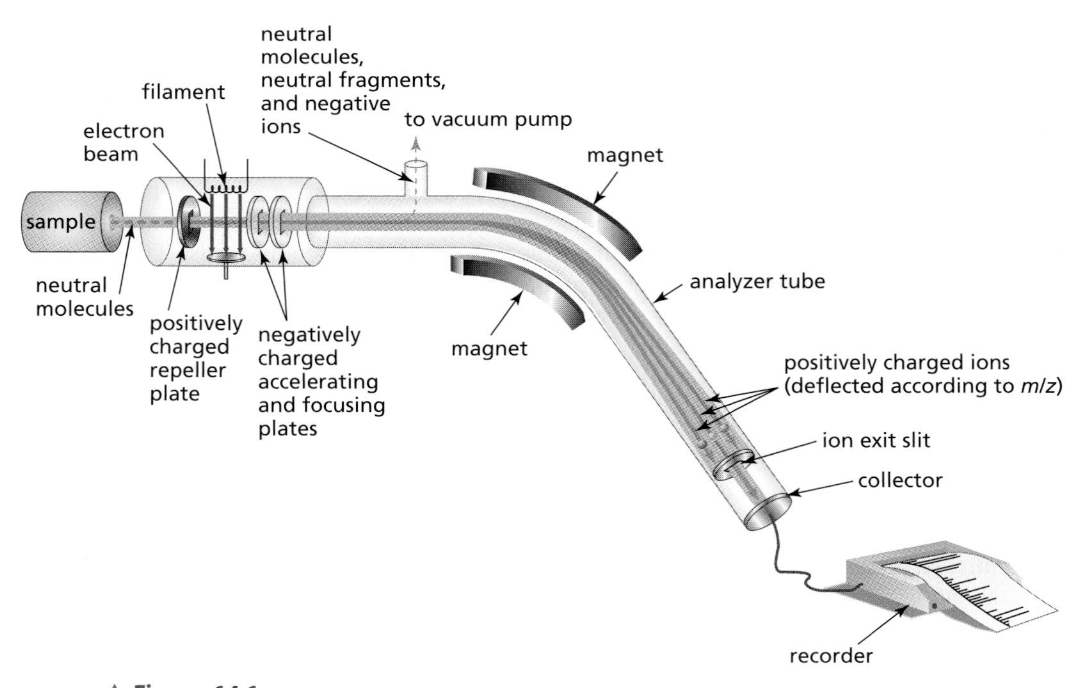

▲ **Figure 14.1**
Schematic diagram of a mass spectrometer. A beam of high-energy electrons causes molecules to ionize and fragment. Positively charged fragments pass through the analyzer tube. Changing the magnetic field strength allows the separation of fragments of varying mass-to-charge ratio.

PROBLEM 1 ◆

Which of the following fragments produced in a mass spectrometer would be accelerated through the analyzer tube?

$CH_3\dot{C}H_2$ $\quad CH_3CH_2\overset{+}{C}H_2$ $\quad [CH_3CH_2CH_3]^{\cdot+}$ $\quad \dot{C}H_2CH{=}CH_2$ $\quad \overset{+}{C}H_2CH{=}CH_2$

14.2 THE MASS SPECTRUM • FRAGMENTATION

The mass spectrum of pentane is shown in Figure 14.2. Each *m/z* value in the spectrum is the **molecular mass** of one of the fragments.

The peak with the highest *m/z* value—in this case, at $m/z = 72$—is the molecular ion (M), the fragment that results when an electron is knocked out of the molecule. (The extremely tiny peak at $m/z = 73$ will be explained later.) *The m/z value of the molecular ion gives the molecular mass of the compound.* Since it is not known what bond loses the electron, the molecular ion is written in brackets and the positive charge and unpaired electron are assigned to the entire structure.

The *m/z* value of the molecular ion gives the molecular mass of the compound.

$$CH_3CH_2CH_2CH_2CH_3 \xrightarrow{\text{electron beam}} [CH_3CH_2CH_2CH_2CH_3]^{\cdot+} + e^-$$

molecular ion
$m/z = 72$

Peaks with smaller *m/z* values—called **fragment ion peaks**—represent positively charged fragments of the molecular ion. The **base peak** is the tallest peak, due to its having the greatest relative abundance.

A mass spectrum gives us structural information about the compound because the *m/z* values and relative abundances of the fragments depend on the strength of the molecular ion's bonds and the stability of the fragments. *Weak bonds break in preference to strong bonds, and bonds that break to form more stable fragments break in preference to those that form less stable fragments.*

For example, all the C—C bonds in the molecular ion formed from pentane have about the same strength. However, the C-2—C-3 bond is more likely to break than the C-1—C-2 bond because C-2—C-3 fragmentation leads to a *primary* carbocation and a *primary* radical, which together are more stable than the *primary* carbocation and *methyl* radical (or *primary* radical and *methyl* cation) obtained from C-1—C-2 fragmentation. (Like carbocations, the relative stabilities of radicals are in the order: 3° > 2° > 1° > methyl.) Ions formed by C-2—C-3 fragmentation have $m/z = 43$ or 29,

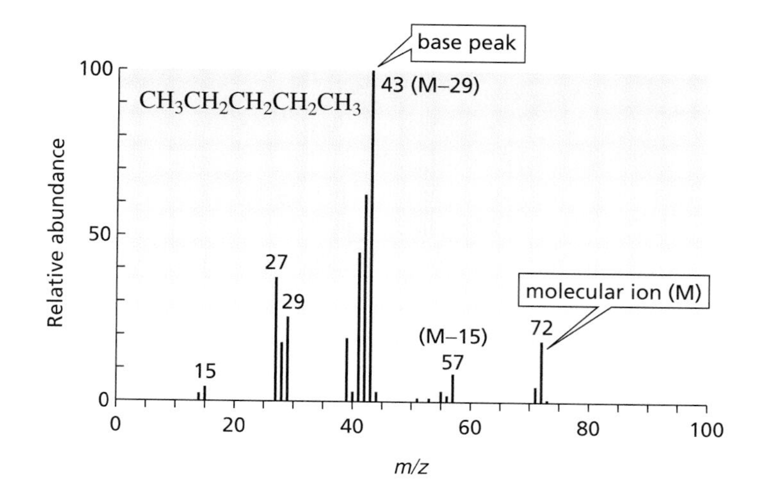

◀ **Figure 14.2**
The mass spectrum of pentane. The base peak represents the fragment that appears in greatest abundance. The *m/z* value of the molecular ion gives the molecular mass of the compound.

whereas ions formed by C-1—C-2 fragmentation have $m/z = 57$ or 15. The base peak of 43 in the mass spectrum of pentane indicates the greater likelihood of C-2—C-3 fragmentation.

The way a molecular ion fragments depends on the strength of its bonds and the stability of the fragments.

$$[\overset{1}{C}H_3\overset{2}{C}H_2\overset{3}{C}H_2\overset{4}{C}H_2\overset{5}{C}H_3]^{+\bullet}$$
molecular ion
$m/z = 72$

$$\longrightarrow CH_3\dot{C}H_2 + \overset{+}{C}H_2CH_2CH_3 \quad (m/z = 43)$$
$$\longrightarrow CH_3\overset{+}{C}H_2\ (m/z = 29) + \dot{C}H_2CH_2CH_3$$
$$\longrightarrow \dot{C}H_3 + \overset{+}{C}H_2CH_2CH_2CH_3 \quad (m/z = 57)$$
$$\longrightarrow \overset{+}{C}H_3\ (m/z = 15) + \dot{C}H_2CH_2CH_2CH_3$$

A method for identifying fragment ions makes use of the difference between the *m/z* value of a given fragment ion and that of the molecular ion. For example, the fragment ion with $m/z = 43$ in the mass spectrum of pentane is 29 units smaller than the molecular ion ($72 - 43 = 29$). An ethyl radical ($CH_3\dot{C}H_2$) has a molecular mass of 29 (because the mass numbers of C and H are 12 and 1, respectively). Thus, the peak at 43 can be attributed to the molecular ion minus an ethyl radical. Similarly, the fragment ion with $m/z = 57$ can be attributed to the molecular ion minus a methyl radical ($72 - 57 = 15$). Peaks at $m/z = 15$ and $m/z = 29$ are readily recognizable as being due to methyl and ethyl cations, respectively.

Because 2-methylbutane has the same molecular formula as pentane, it, too, has a molecular ion with $m/z = 72$ (Figure 14.3). Its mass spectrum is similar to that of pentane, with one notable exception: the peak at $m/z = 57$, indicating loss of a methyl radical, is much more intense.

2-Methylbutane is more likely than pentane to lose a methyl radical because, when it does, a *secondary* carbocation is formed. In contrast, when pentane loses a methyl radical, a less stable *primary* carbocation is formed.

$$[CH_3\underset{|}{\overset{CH_3}{C}}HCH_2CH_3]^{+\bullet} \longrightarrow CH_3\overset{+}{C}HCH_2CH_3 + \dot{C}H_3$$
molecular ion $m/z = 72$ → $m/z = 57$

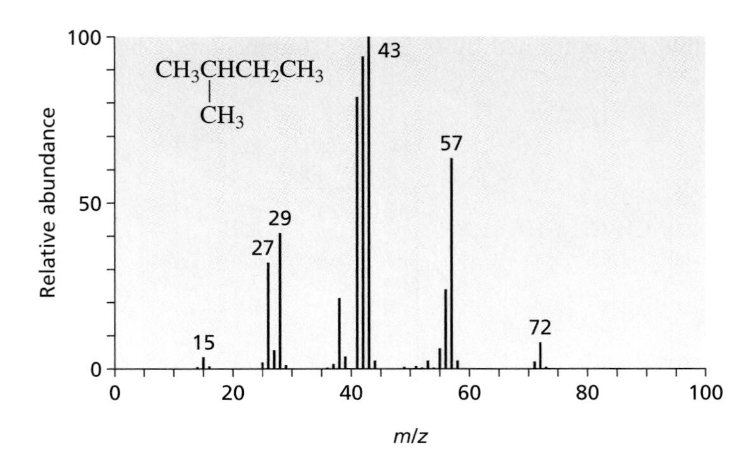

▶ **Figure 14.3**
The mass spectrum of 2-methylbutane.

PROBLEM 2 ♦

What is the likeliest *m/z* value for the base peak in the mass spectrum of 3-methylpentane?

PROBLEM 3 *SOLVED*

The mass spectra of two very stable cycloalkanes both show a molecular ion peak at $m/z = 98$. One spectrum shows a base peak at $m/z = 69$; the other shows a base peak at $m/z = 83$. Identify the cycloalkanes.

Solution The molecular formula for a cycloalkane is C_nH_{2n}. Because the molecular mass of both cycloalkanes is 98, their molecular formulas must be C_7H_{14} ($7 \times 12 = 84$; $84 + 14 = 98$). A base peak of 69 means the loss of an ethyl substituent ($98 - 69 = 29$), whereas a base peak of 83 means the loss of a methyl substituent ($98 - 83 = 15$). Because the two cycloalkanes are known to be very stable, we assume they do not have three- or four-membered rings. A seven-carbon cycloalkane with a base peak signifying the loss of an ethyl substituent must be ethylcyclopentane. A seven-carbon cycloalkane with a base peak signifying the loss of a methyl substituent must be methylcyclohexane.

CH_2CH_3 (ethylcyclopentane) $\xrightarrow{\text{electron beam}}$ cyclopentyl cation ($m/z = 69$) + $CH_3\dot{C}H_2$

CH_3 (methylcyclohexane) $\xrightarrow{\text{electron beam}}$ cyclohexyl cation ($m/z = 83$) + $\dot{C}H_3$

14.3 ISOTOPES IN MASS SPECTROMETRY

Although the molecular ions of pentane and 2-methylbutane both have m/z values of 72, each spectrum shows a very small peak at $m/z = 73$ (Figures 14.2 and 14.3). This peak is called an M + 1 peak because the ion responsible for it is one unit heavier than the molecular ion. The M + 1 peak owes its presence to the fact that one out of every 100 carbons is ^{13}C instead of ^{12}C. Because mass spectrometry records individual molecules, any molecule containing ^{13}C will appear at M + 1.

The isotopic distributions of several elements commonly found in organic compounds are shown in Table 14.1. The presence of a large M + 2 peak is evidence of a compound containing either chlorine or bromine, because each of these elements has a high percentage of a naturally occurring isotope that is two units heavier than the most abundant isotope. From the natural abundance of the isotopes of chlorine and bromine in Table 14.1, one can conclude that if the M + 2 peak is one-third the height of the

Table 14.1 The Natural Abundance of Isotopes Commonly Found in Organic Compounds

Element	Natural abundance		Element	Natural abundance	
Carbon	^{12}C 98.89%	^{13}C 1.11%	Fluorine	^{19}F 100%	
Hydrogen	^{1}H 99.99%	^{2}H 0.01%	Chlorine	^{35}Cl 75.77%	^{37}Cl 24.23%
Nitrogen	^{14}N 99.64%	^{15}N 0.36%	Bromine	^{79}Br 50.69%	^{81}Br 49.31%
Oxygen	^{16}O 99.76%	^{17}O 0.04%	Iodine	^{127}I 100%	
Sulfur	^{32}S 95.0%	^{33}S 0.76%			

molecular ion peak, then the compound contains one chlorine atom because the natural abundance of ^{37}Cl is one-third that of ^{35}Cl. If the M and M + 2 peaks are about the same height, then the compound contains one bromine atom because the natural abundances of ^{79}Br and ^{81}Br are about the same.

PROBLEM 4 ◆

Identify the primary alkyl halide responsible for the mass spectrum in Figure 14.4.

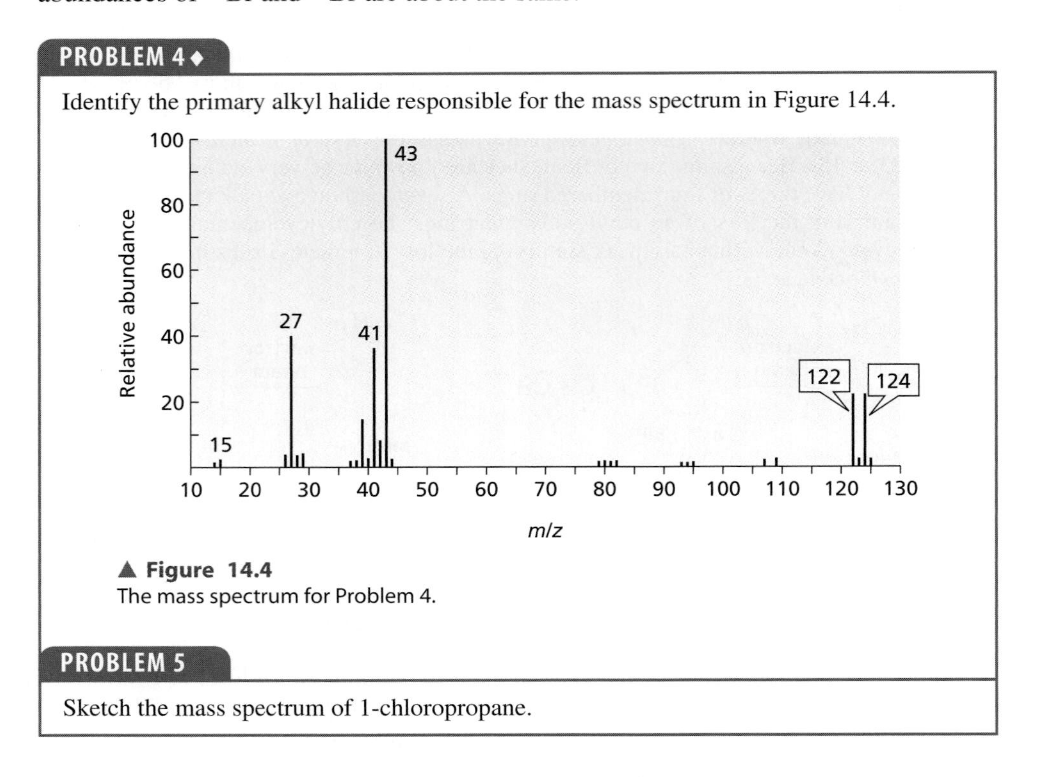

▲ **Figure 14.4**
The mass spectrum for Problem 4.

PROBLEM 5

Sketch the mass spectrum of 1-chloropropane.

14.4 FRAGMENTATION PATTERNS

Each functional group has characteristic fragmentation patterns that can help identify a compound. The patterns began to be recognized after the mass spectra of many compounds containing a particular functional group had been studied. We will look at the fragmentation patterns of ketones as an example.

Electron bombardment is most likely to dislodge a lone-pair electron, if a molecule has any, because a molecule does not hold onto its lone-pair electrons as tightly as it holds onto its bonding electrons. Therefore, when a ketone is bombarded by electrons, the molecular ion is formed when one of oxygen's lone-pair electrons is knocked out of the molecule.

$$\underset{\text{2-pentanone}}{CH_3CH_2CH_2-\overset{\ddot{O}:}{\overset{\|}{C}}-CH_3} \xrightarrow{-e^-} \underset{m/z = 86}{CH_3CH_2CH_2-\overset{\dot{O}:^{+}}{\overset{\|}{C}}-CH_3} \longrightarrow \begin{cases} CH_3CH_2\dot{C}H_2 + \underset{m/z = 43}{CH_3-C\equiv\overset{+}{O}:} \\ \underset{m/z = 71}{CH_3CH_2CH_2-C\equiv\overset{+}{O}:} + \dot{C}H_3 \end{cases}$$

The molecular ion fragments at the C—C bond adjacent to the C=O bond, with each of the carbons retaining one of the electrons. The C—C bond is the bond most easily broken because the species that is formed is a relatively stable cation, since its positive charge is shared by two atoms.

$$CH_3C\equiv\overset{+}{O}: \longleftrightarrow CH_3\overset{+}{C}=\ddot{O}:$$

If one of the alkyl groups attached to a carbonyl carbon has a γ-hydrogen, a cleavage, which goes through a favorable six-membered-ring transition state, may occur. In this rearrangement, the bond between the α-carbon and the β-carbon breaks, with each of the carbons retaining one of the electrons, and a hydrogen atom from the γ-carbon migrates to the oxygen atom. Again, fragmentation has occurred in a way that produces a cation with a positive charge shared by two atoms.

Recall that an arrowhead with one barb represents the movement of one electron.

$$H_2C(\gamma)(H)-H_2C(\beta)-CH_2(\alpha)-\overset{\overset{+\cdot}{O:}}{\overset{\|}{C}}-CH_3 \xrightarrow{\text{rearrangement}} H_2C{=}CH_2 + \cdot CH_2-\overset{\overset{H-\overset{+}{O:}}{\|}}{C}-CH_3$$

$m/z = 86$ $\qquad$ $m/z = 58$

PROBLEM 6 ♦

Identify the ketone responsible for the mass spectrum in Figure 14.5.

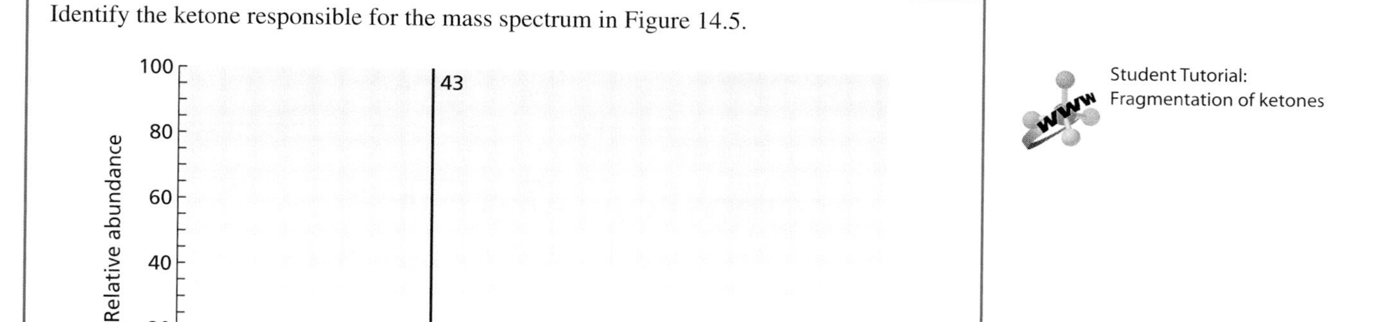

▲ **Figure 14.5**
The mass spectrum for Problem 6.

Student Tutorial: Fragmentation of ketones

PROBLEM 7

How could their mass spectra distinguish the following compounds?

$CH_3CH_2\overset{O}{\overset{\|}{C}}CH_2CH_3$ $\qquad$ $CH_3\overset{O}{\overset{\|}{C}}CH_2CH_2CH_3$ $\qquad$ $CH_3\overset{O}{\overset{\|}{C}}\underset{CH_3}{\underset{|}{C}H}CH_3$

14.5 HIGH-RESOLUTION MASS SPECTROMETRY CAN REVEAL MOLECULAR FORMULAS

High-resolution mass spectrometers can determine the *exact molecular mass* of a fragment to a precision of 0.0001 atomic mass units (amu), making it possible to distinguish between compounds with the same nominal mass (mass to the nearest whole number). For example, the following listing shows six compounds that have a nominal molecular mass of 122 amu, but each of them has a different exact molecular mass. There are computer programs that can determine the molecular formula of a compound from its exact molecular mass.

Some Compounds with a Nominal Molecular Mass of 122 Amu and Their Exact Molecular Masses and Molecular Formulas

Exact molecular mass (amu)	122.1096	122.0845	122.0732	122.0368	122.0579	122.0225
Molecular formula	C_9H_{14}	$C_7H_{10}N_2$	$C_8H_{10}O$	$C_7H_6O_2$	$C_4H_{10}O_4$	$C_4H_{10}S_2$

MASS SPECTROMETRY IN FORENSICS

Forensic science is the application of science for the purpose of justice. Mass spectrometry is an important tool of the forensic scientist. It is used to analyze body fluids for the presence and levels of drugs and other toxic substances. It can also identify the presence of drugs in hair, which increases the window of detection from hours and days (after which body fluids are no longer useful) to months and even years. It was employed for the first time in 1955 to detect drugs in athletes at a cycling competition in France. (Twenty percent of those tests were positive.) Mass spectrometry is also used to identify residues of arson fires and explosives from post-explosion residues, and to analyze paints, adhesives, and fibers.

PROBLEM 8 ♦

Which molecular formula has an exact molecular mass of 86.1096 amu: C_6H_{14}, $C_4H_{10}N_2$, or $C_4H_6O_2$? Exact masses (in amu): $^{1}H = 1.007825$, $^{12}C = 12.00000$; $^{14}N = 14.0031$, $^{16}O = 15.9949$.

14.6 SPECTROSCOPY AND THE ELECTROMAGNETIC SPECTRUM

Spectroscopy is the study of the interaction of matter and *electromagnetic radiation*. A continuum of different types of electromagnetic radiation—each type associated with a particular energy range—constitutes the electromagnetic spectrum (Figure 14.6).

The various kinds of electromagnetic radiation can be characterized briefly as follows:

- *Cosmic rays* are discharged by the sun; they have the highest energy.
- *γ-Rays* (gamma rays) are emitted from the nuclei of certain radioactive elements; because of their high energy, they can severely damage biological organisms.
- *X-Rays,* somewhat lower in energy than γ-rays, are less harmful, except in high doses. Low-dose X-rays are used to examine the internal structure of organisms. The denser the tissue, the more it blocks X-rays.
- *Ultraviolet (UV) light,* a component of sunlight, is responsible for sunburns; repeated exposure to it can cause skin cancer by damaging DNA molecules in skin cells.

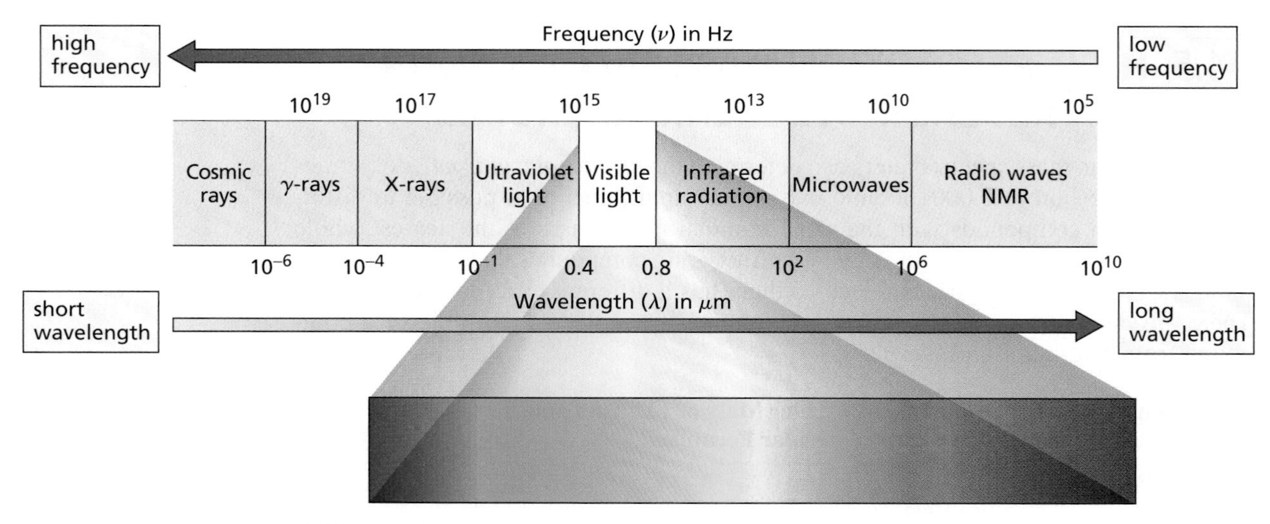

▲ Figure 14.6
The electromagnetic spectrum.

- *Visible light* is the electromagnetic radiation we see.
- We feel *infrared radiation* as heat.
- We cook with *microwaves* and use them in radar.
- *Radio waves* have the lowest energy of the different kinds of electromagnetic radiation. We use them for radio and television communication, digital imaging, remote control devices, and wireless linkages for laptop computers. Radio waves are also used in NMR spectroscopy and in magnetic resonance imaging (MRI).

Electromagnetic radiation can be characterized by either its frequency (ν) or its wavelength (λ). **Frequency** is defined as the number of wave crests that pass by a given point in one second. **Wavelength** is the distance from any point on one wave to the corresponding point on the next wave.

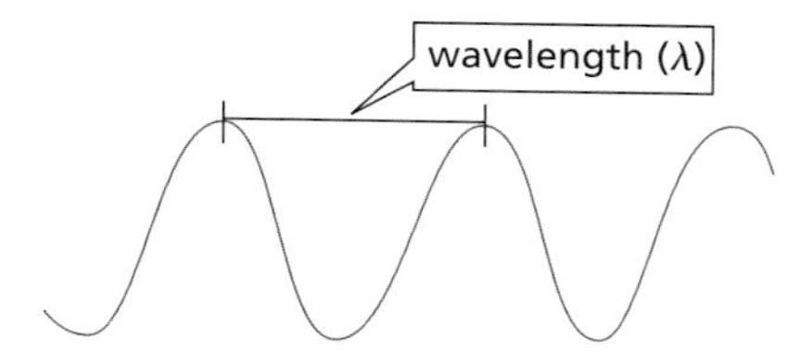

The relationship between the energy (E) and the frequency (ν) or wavelength (λ) of the electromagnetic radiation is described by the equation

$$E = h\nu = \frac{hc}{\lambda}$$

where h is the proportionality constant called *Planck's constant*, named after the German physicist who discovered the relationship, and c is the speed of light. The frequency of electromagnetic radiation, therefore, is equal to the speed of light (c) divided by the radiation's wavelength:

$$\nu = \frac{c}{\lambda} \qquad c = 3 \times 10^{10}\ \text{cm/s}$$

Another way to describe the *frequency* of electromagnetic radiation—and the one most often used in infrared spectroscopy—is **wavenumber** ($\tilde{\nu}$), the number of waves in 1 centimeter. Therefore, it has units of reciprocal centimeters (cm^{-1}). The relationship between wavenumber (in cm^{-1}) and wavelength (in μm) is given by the equation

$$\tilde{\nu}(\text{cm}^{-1}) = \frac{10^4}{\lambda(\mu\text{m})} \qquad (\text{because } 1\ \mu\text{m} = 10^{-4}\ \text{cm})$$

So *high frequencies*, *large wavenumbers*, and *short wavelengths* are associated with *high energy*.

High frequencies, large wavenumbers, and short wavelengths are associated with high energy.

PROBLEM 9 ♦

One of the following depicts the waves associated with infrared radiation, and one depicts the waves associated with ultraviolet light (Section 7.10). Which is which?

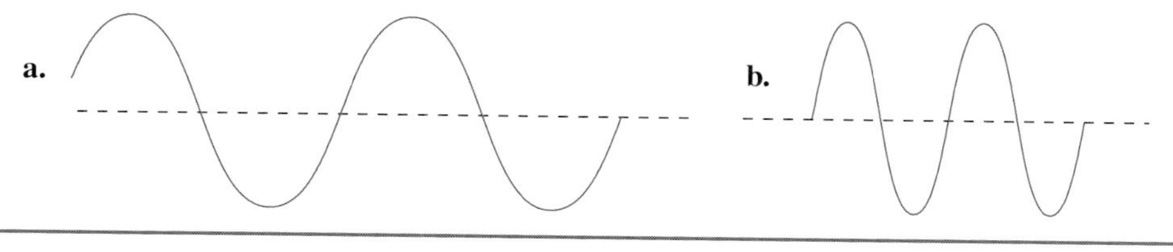

PROBLEM 10 ♦

a. Which is higher in energy, electromagnetic radiation with wavenumber 100 cm^{-1} or with wavenumber 2000 cm^{-1}?
b. Which is higher in energy, electromagnetic radiation with wavelength 9 μm or with wavelength 8 μm?
c. Which is higher in energy, electromagnetic radiation with wavenumber 3000 cm^{-1} or with wavelength 2 μm?

14.7 INFRARED SPECTROSCOPY

The length of a bond between two atoms is only an average length because in reality a bond behaves as if it were a vibrating spring. A bond vibrates with both stretching and bending motions. A *stretch* is a vibration occurring along the line of the bond that changes the bond length. A *bend* is a vibration that does *not* occur along the line of the bond.

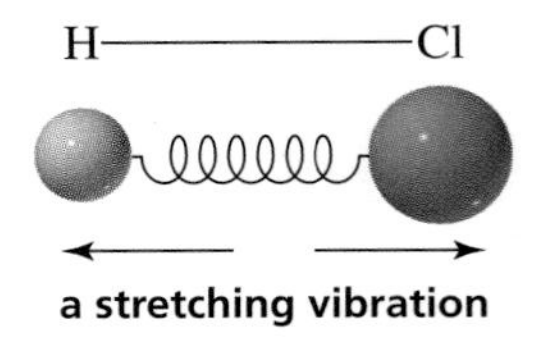

a stretching vibration

Each stretching and bending vibration of a bond in a molecule occurs with a characteristic frequency. **Infrared radiation** has just the right range of frequencies to correspond to the frequencies of the stretching and bending vibrations of organic molecules. When a compound is bombarded with radiation of a frequency that exactly matches the frequency of one of its vibrations, the molecule absorbs energy. By experimentally determining the wavenumbers of the energy absorbed by a particular compound, we can ascertain what kinds of bonds it has. For example, the stretching vibration of a C═O bond absorbs energy with wavenumber ~1700 cm^{-1}, whereas the stretching vibration of an O—H bond absorbs energy with wavenumber ~3450 cm^{-1} (Figure 14.7).

An infrared (IR) spectrum can be divided into two areas. The left-hand two-thirds of the spectrum (4000–1400 cm^{-1}) is where most of the functional groups show absorption bands. This is called the **functional group region**. The right-hand third (1400–600 cm^{-1}) of the spectrum is called the **fingerprint region** because it is characteristic of the compound as a whole, just as a fingerprint is characteristic of an individual.

Because more energy is required to stretch a bond than to bend it, **absorption bands** for stretching vibrations are found in the functional group region, whereas

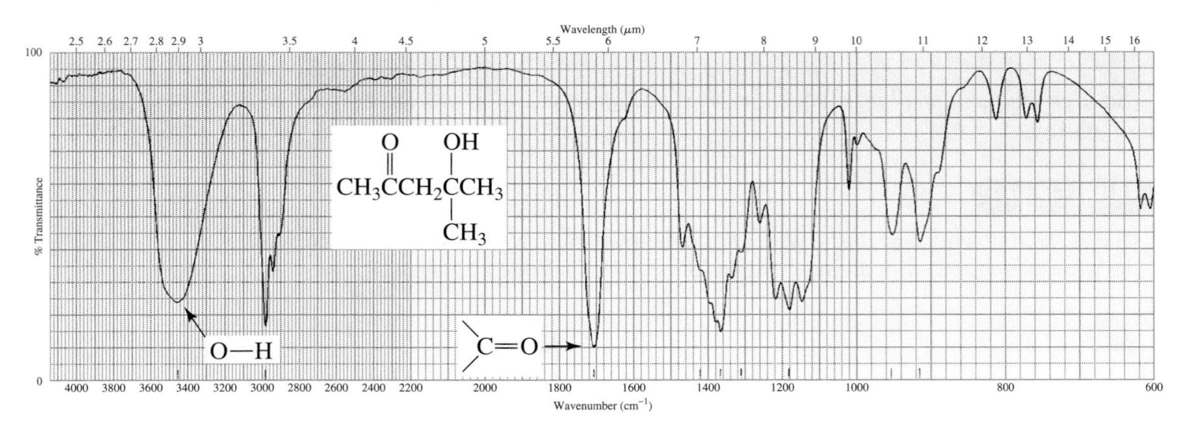

▲ **Figure 14.7**
The infrared spectrum of 4-hydroxy-4-methyl-2-pentanone.

absorption bands for bending vibrations are typically found in the fingerprint region. Therefore, stretching vibrations are the ones most often used to determine what kinds of bonds a molecule has. The frequencies of the **stretching frequencies** associated with different types of bonds are shown in Table 14.2.

It takes more energy to stretch a bond than to bend it.

Table 14.2 Important IR Stretching Frequencies

Type of bond	Wavenumber (cm^{-1})	Intensity
C≡N	2260–2220	medium
C≡C	2260–2100	medium to weak
C=C	1680–1600	medium
C=N	1650–1550	medium
	~1600 and ~1500–1430	strong to weak
C=O	1780–1650	strong
C—O	1250–1050	strong
C—N	1230–1020	medium
O—H (alcohol)	3650–3200	strong, broad
O—H (carboxylic acid)	3300–2500	strong, very broad
N—H	3500–3300	medium, broad
C—H	3300–2700	medium

14.8 CHARACTERISTIC INFRARED ABSORPTION BANDS

IR spectra can be quite complex because the stretching and bending vibrations of each bond in a molecule can produce an absorption band. Obviously, there is a lot more to infrared spectroscopy than we can cover here. In this chapter, we will look at some characteristic absorption bands so you will be able to tell something about the structure of a compound that gives a particular IR spectrum. You can, however, find an extensive table of characteristic functional group frequencies in Appendix III.

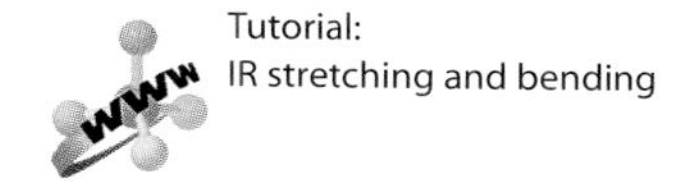

Tutorial:
IR stretching and bending

Effect of Bond Order

Bond order affects bond strength; therefore bond order affects the position of an absorption band. A C≡C bond is stronger than a C=C bond, so a C≡C bond stretches at a higher frequency ($\sim 2100\ cm^{-1}$) than does a C=C bond ($\sim 1650\ cm^{-1}$); C—C bonds show stretching vibrations in the region from 1200 to 800 cm^{-1}, but being weak and very common, these vibrations are of little value in identifying compounds. Similarly, a C=O bond stretches at a higher frequency ($\sim 1700\ cm^{-1}$) than does a C—O bond ($\sim 1100\ cm^{-1}$), and a C≡N bond stretches at a higher frequency ($\sim 2200\ cm^{-1}$) than does a C=N bond ($\sim 1600\ cm^{-1}$), which in turn stretches at a higher frequency than does a C—N bond ($\sim 1100\ cm^{-1}$) (Table 14.2).

Stronger bonds show absorption bands at larger wavenumbers.
C≡N ~2200 cm^{-1}
C=N ~1600 cm^{-1}
C—N ~1100 cm^{-1}

PROBLEM 11 ♦

Which will occur at a larger wavenumber?

a. a C≡C stretch or a C=C stretch
b. a C—H stretch or a C—H bend
c. a C=N stretch or a C≡N stretch
d. a C=O stretch or a C—O stretch

Resonance Effects

Table 14.2 shows a range of wavenumbers for the frequency of the stretching vibration for each functional group because the exact position of a group's absorption band depends on other structural features of the molecule. For example, the IR spectrum in Figure 14.8 shows that the carbonyl group (C=O) of 2-pentanone absorbs at 1720 cm^{-1}, whereas the IR spectrum in Figure 14.9 shows that the carbonyl group of 2-cyclohexenone absorbs at a lower frequency (1680 cm^{-1}).

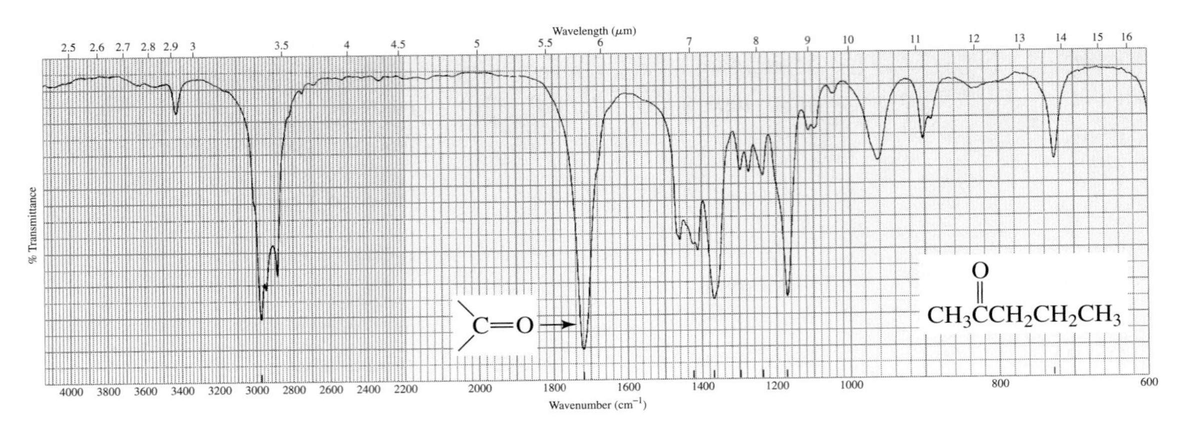

▲ **Figure 14.8**
The IR spectrum of 2-pentanone. The intense absorption band at ~1720 cm^{-1} indicates a C=O bond.

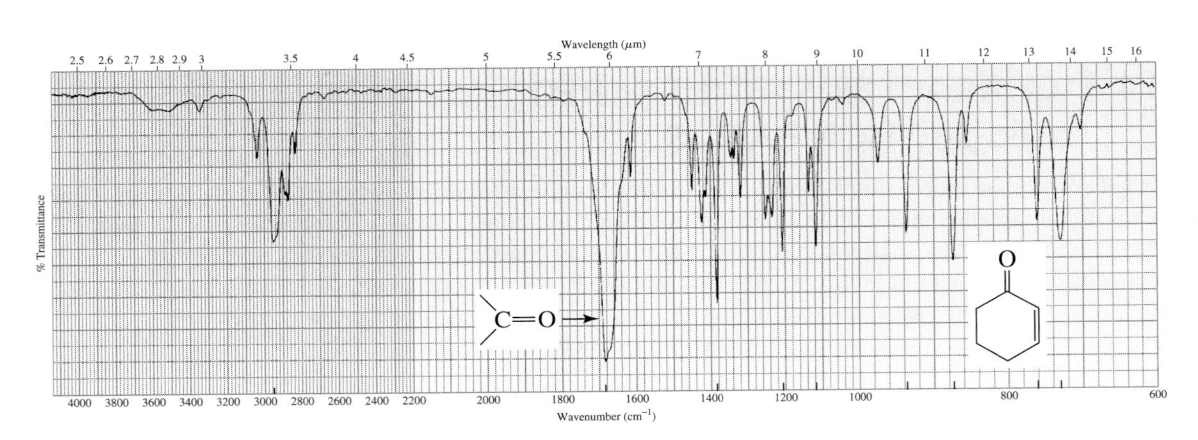

▲ **Figure 14.9**
The IR spectrum of 2-cyclohexenone. Electron delocalization gives its carbonyl group less double-bond character, so it absorbs at a lower frequency (~1680 cm^{-1}) than does a carbonyl group with localized electrons (~1720 cm^{-1}).

2-Cyclohexenone's carbonyl group absorbs at a lower frequency because it has less double-bond character due to electron delocalization. A single bond is weaker than a double bond, so a carbonyl group with significant single-bond character will stretch at a lower frequency than will one with little or no single-bond character.

$CH_3\overset{O}{\overset{\|}{C}}CH_2CH_2CH_3$

2-pentanone

C=O at 1720 cm^{-1}

2-cyclohexenone

C=O at 1680 cm^{-1}

PROBLEM-SOLVING STRATEGY

Which will occur at a larger wavenumber, the C—N stretch of an amine or the C—N stretch of an amide?

To answer this kind of question, we need to see if electron delocalization causes one of the bonds to be other than a pure single bond. When we do that, we see that electron delocalization causes the C—N bond of the amide to have partial double-bond character, whereas there is no electron delocalization in the amine. The C—N stretch of an amide, therefore, will occur at a larger wavenumber.

$R{-}\ddot{N}H_2$

no electron delocalization

$R{-}C({=}\ddot{O}{:}){-}\ddot{N}H_2 \longleftrightarrow R{-}C({-}{:}\ddot{O}{:}^-){=}\overset{+}{N}H_2$

electron delocalization causes the C—N bond to have partial double-bond character

Now continue on to Problem 12.

PROBLEM 12

Why is the C—O absorption band of 1-hexanol at a smaller wavenumber ($1060\ cm^{-1}$) than the C—O absorption band of pentanoic acid ($1220\ cm^{-1}$)?

PROBLEM 13 ♦

Which will occur at a larger wavenumber:

a. the C—O stretch of phenol or the C—O stretch of cyclohexanol?
b. the C=O stretch of a ketone or the C=O stretch of an amide?
c. the stretch or the bend of the C—O bond in ethanol?

PROBLEM 14 ♦

Which would show an absorption band at a larger wavenumber: a carbonyl group bonded to an sp^3 carbon or a carbonyl group bonded to an sp^2 carbon?

14.9 THE INTENSITY OF ABSORPTION BANDS

The **intensity of an absorption band** depends on the polarity of the bond: the more polar the bond, the greater is the intensity of the absorption. An O—H bond will show a more intense absorption band than an N—H bond because the O—H bond is more polar. Similarly, an N—H bond will show a more intense absorption than a C—H bond because the N—H bond is more polar.

The more polar the bond, the more intense is the absorption.

relative bond polarities
relative intensities of IR absorption

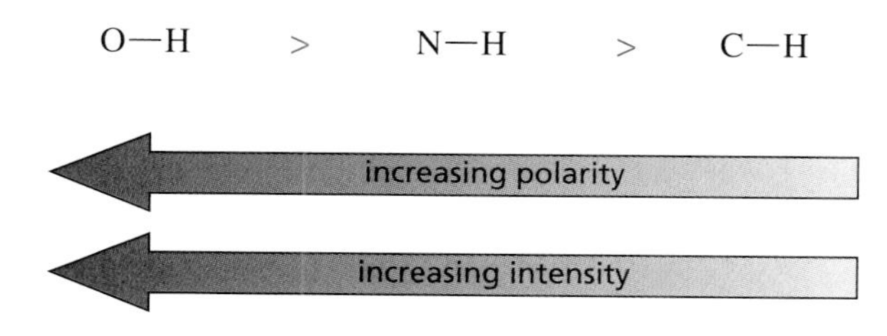

PROBLEM 15 ♦

Which would be expected to be more intense, the stretching vibration of a C=O bond or the stretching vibration of a C=C bond?

14.10 C—H ABSORPTION BANDS

The strength of a C—H bond depends on the hybridization of the carbon: a C—H bond is stronger when the carbon is *sp* hybridized than when it is sp^2 hybridized, which in turn is stronger than when the carbon is sp^3 hybridized. More energy is needed to stretch a stronger bond, and this is reflected in the C—H stretch absorption bands, which occur at ~ 3300 cm^{-1} when the carbon is *sp* hybridized, at ~3100 cm^{-1} when the carbon is sp^2 hybridized, and at ~2900 cm^{-1} when the carbon is sp^3 hybridized (Table 14.3).

Table 14.3 IR Absorptions of Carbon–Hydrogen Bonds

Carbon–Hydrogen Stretching Vibrations	Wavenumber (cm^{-1})
C≡C—H	~3300
C=C—H	3100–3020
C—C—H	2960–2850
R—C(=O)—H	~2820 and ~2720

A useful step in the analysis of a spectrum is to look at the absorption bands in the vicinity of 3000 cm^{-1}. Figures 14.10 and 14.11 show the IR spectra for methylcyclohexane and cyclohexene, respectively. The only absorption band in the vicinity of 3000 cm^{-1} in Figure 14.10 is slightly to the right of that value. This tells us that the compound has hydrogens bonded to sp^3 carbons, but none bonded to sp^2 or to *sp* carbons. Figure 14.11 shows absorption bands slightly to the left and slightly to the right of 3000 cm^{-1}, indicating that the compound that produced that spectrum contains hydrogens bonded to sp^2 and sp^3 carbons.

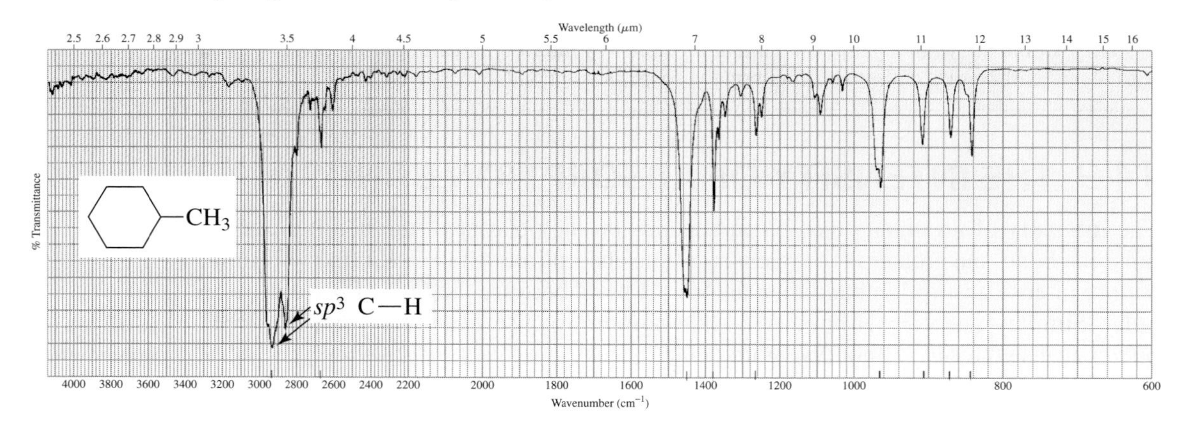

▶ **Figure 14.10**
The IR spectrum of methylcyclohexane.

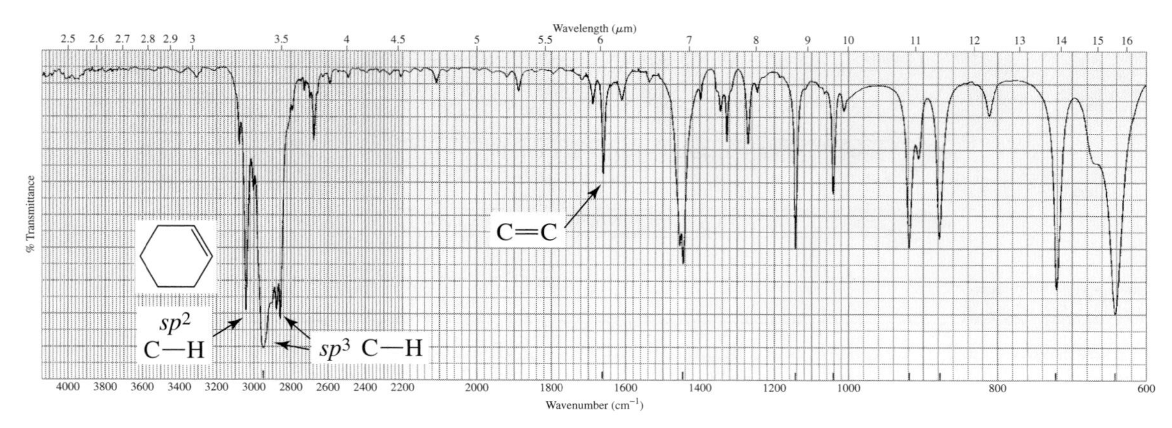

▶ **Figure 14.11**
The IR spectrum of cyclohexene.

The stretch of the C—H bond in an aldehyde group shows two absorption bands, one at ~2820 cm^{-1} and the other at ~2720 cm^{-1} (Figure 14.12). This makes aldehydes relatively easy to identify because essentially no other absorption occurs at these wavenumbers.

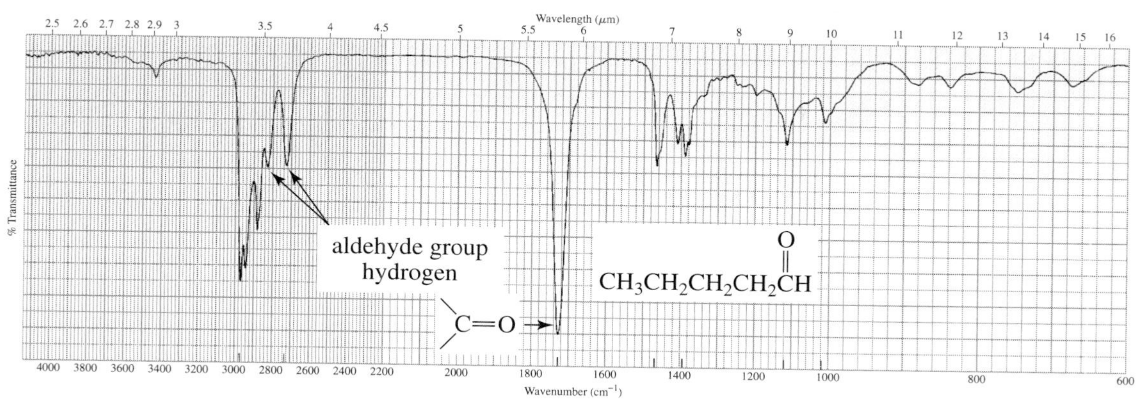

▲ **Figure 14.12**
The IR spectrum of pentanal. The absorptions at ~2820 and ~2720 cm^{-1} readily identify an aldehyde group. Note also the intense absorption band at ~1730 cm^{-1} indicating a C=O bond.

14.11 THE SHAPE OF ABSORPTION BANDS

The **shape of an absorption band** can be helpful in identifying the compound responsible for an IR spectrum. For example, both O—H and N—H bonds stretch at wavenumbers above 3100 cm^{-1}, but the shapes of their stretching absorption bands are distinctive. Notice the difference in the shape of these absorption bands in the IR spectra of 1-hexanol (Figure 14.13), pentanoic acid (Figure 14.14), and isopentylamine (Figure 14.15). An N—H absorption band (~3300 cm^{-1}) is narrower and less intense than an O—H absorption band (~3300 cm^{-1}), and the O—H absorption band of a carboxylic acid (~3300–2500 cm^{-1}) is broader than the O—H absorption band of an alcohol. Notice that two absorption bands are detectable in Figure 14.15 for the N—H stretch because there are two N—H bonds in the compound.

The position, intensity, and shape of an absorption band are helpful in identifying functional groups.

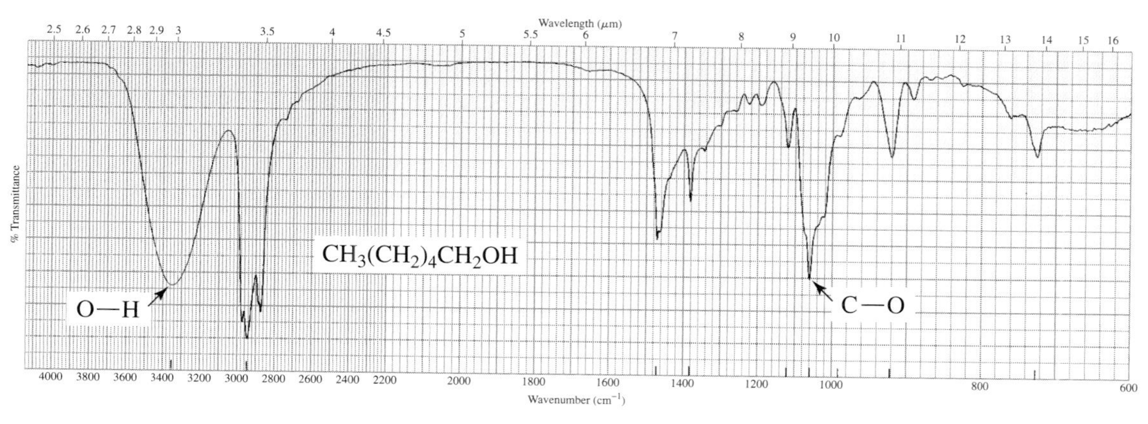

▲ **Figure 14.13**
The IR spectrum of 1-hexanol.

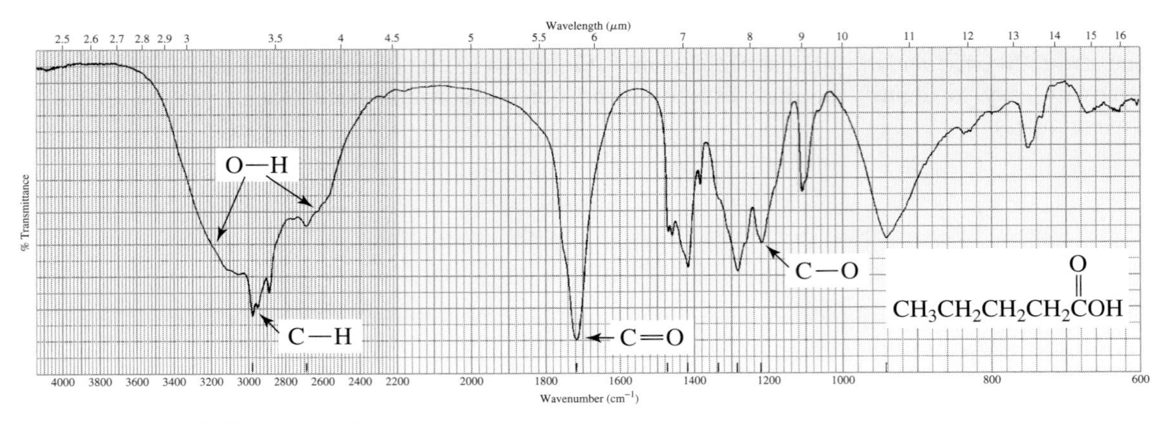

▲ **Figure 14.14**
The IR spectrum of pentanoic acid.

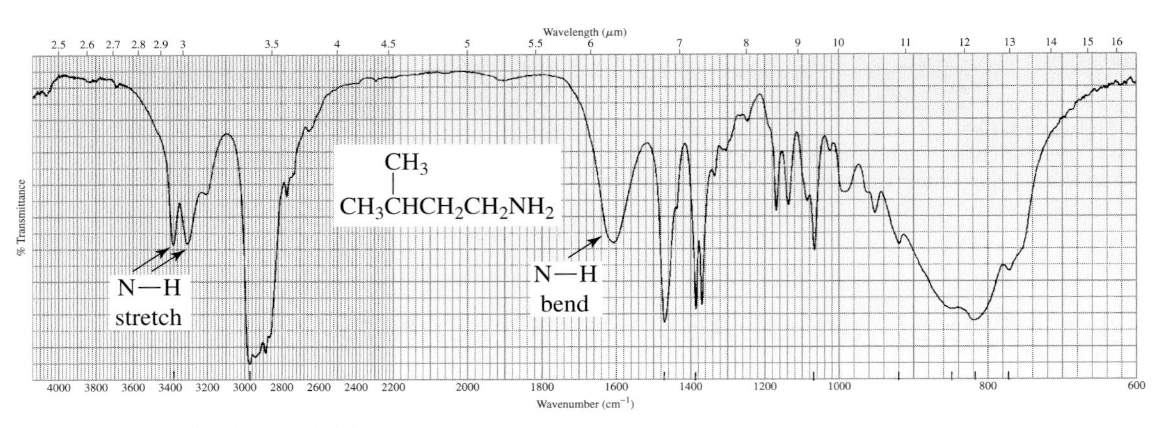

▲ **Figure 14.15**
The IR spectrum of isopentylamine. The double peak at ~3300 cm^{-1} indicates the presence of two N—H bonds.

PROBLEM 16

Why is the O—H stretch of a carboxylic acid broader than the O—H stretch of an alcohol?

14.12 THE ABSENCE OF ABSORPTION BANDS

The absence of an absorption band can be as useful as the presence of an absorption band in identifying a compound by IR spectroscopy. For example, the IR spectrum in Figure 14.16 shows a strong absorption at ~1100 cm^{-1}, indicating the presence of a C—O bond (Table 14.3). Clearly, the compound is not an alcohol because there is no absorption above 3100 cm^{-1}. Nor is it an ester or any other kind of carbonyl compound because there is no absorption at ~1700 cm^{-1}. The compound has no C≡C, C=C, C≡N, C=N, or C—N bonds. We may deduce, then, that the compound is an ether. Its C—H absorption bands show that it has hydrogens only on sp^3 carbons. The compound is in fact diethyl ether.

PROBLEM 17 ♦

a. An oxygen-containing compound shows an absorption band at ~1700 cm^{-1} and no absorption bands at ~3300 cm^{-1}, ~2700 cm^{-1}, or ~1100 cm^{-1}. What class of compound is it?
b. A nitrogen-containing compound shows no absorption band at ~3400 cm^{-1} and no absorption bands between 1700 cm^{-1} and 1600 cm^{-1}. What class of compound is it?

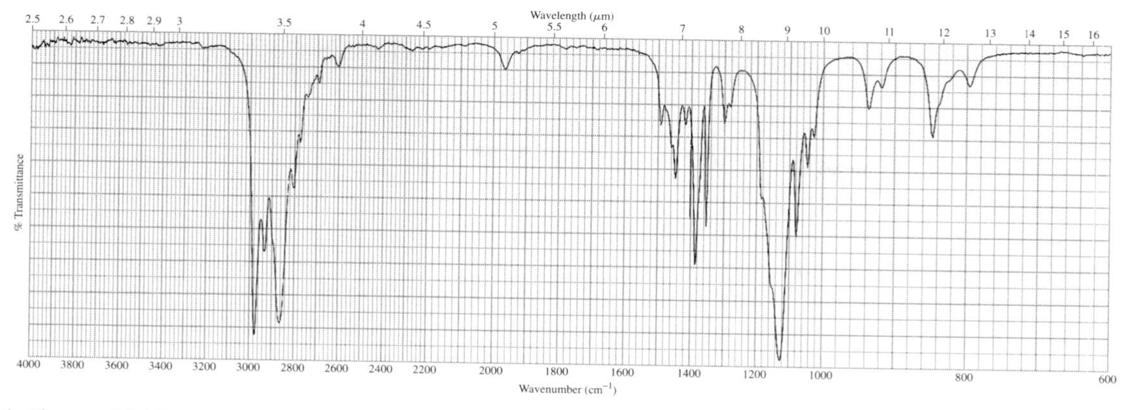

▲ **Figure 14.16**
The IR spectrum of diethyl ether.

14.13 IDENTIFYING INFRARED SPECTRA

We will now look at a few IR spectra and see what we can deduce about the structure of the compounds that give rise to the spectra. We might not be able to identify the compound precisely, but when we are told what it is, its structure should fit our observations.

Tutorial: IR spectra

Compound 1. The absorption in the 3000 cm^{-1} region in Figure 14.17 indicates that hydrogens are attached to sp^2 carbons (3050 cm^{-1}) but not to sp^3 carbons. The absorptions at 1600 cm^{-1} and 1460 cm^{-1} indicate that the compound has a benzene ring. The absorptions at 2810 cm^{-1} and 2730 cm^{-1} show that the compound is an aldehyde. The absorption band for the carbonyl group (C═O) is lower (1700 cm^{-1}) than normal (1720 cm^{-1}), so the carbonyl group has partial single-bond character. Thus, it must be attached directly to the benzene ring. The compound is benzaldehyde.

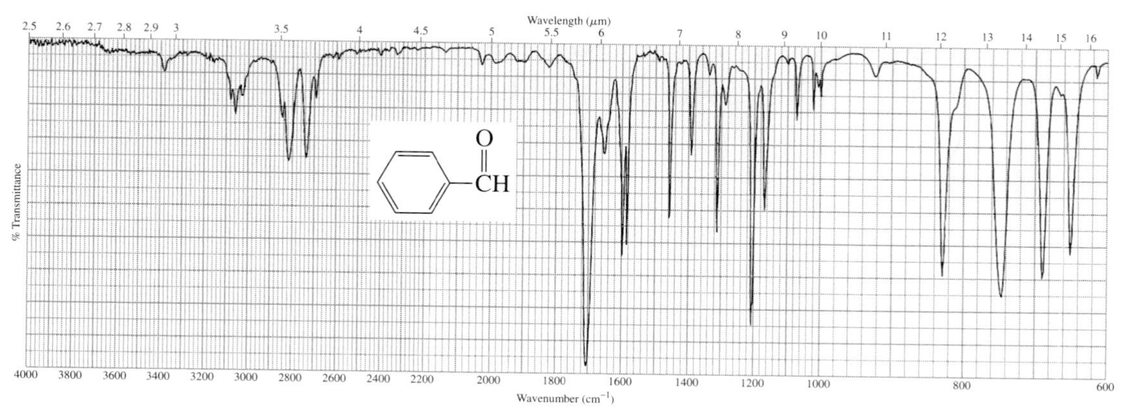

▲ **Figure 14.17**
The IR spectrum of Compound 1.

Compound 2. The absorptions in the 3000 cm^{-1} region in Figure 14.18 indicate that hydrogens are attached to sp^3 carbons (2950 cm^{-1}) but not to sp^2 carbons. The shape of the strong absorption band at 3300 cm^{-1} is characteristic of an O—H group of an alcohol. The absorption at ~ 2100 cm^{-1} indicates that the compound has a triple bond. The sharp absorption band at 3300 cm^{-1} indicates that the compound has a hydrogen attached to an *sp* carbon, so we know it is a terminal alkyne. The compound is 2-propyn-1-ol.

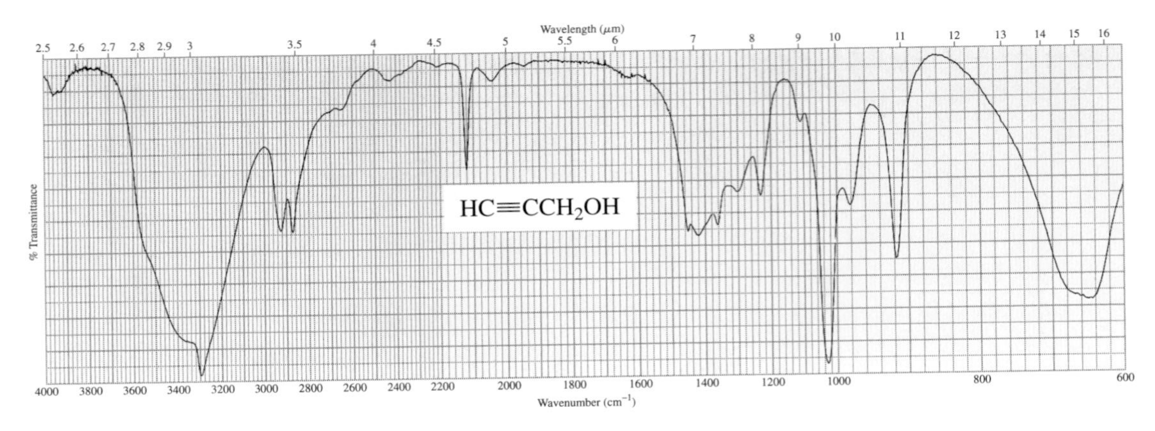

▲ **Figure 14.18**
The IR spectrum of Compound 2.

Tutorial:
Spectra match

PROBLEM 18

How could IR spectroscopy distinguish between the following compounds?

a. a ketone and an aldehyde

b. benzene and toluene

c. cyclohexene and cyclohexane

d. a primary amine and a tertiary amine

PROBLEM 19

For each of the following pairs of compounds, give one absorption band that could be used to distinguish between them:

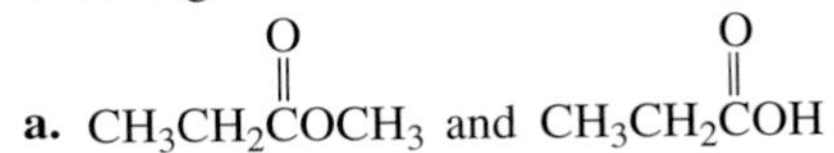

a. $CH_3CH_2\overset{O}{\overset{\|}{C}}OCH_3$ and $CH_3CH_2\overset{O}{\overset{\|}{C}}OH$

c. $CH_3CH_2C{\equiv}CCH_3$ and $CH_3CH_2C{\equiv}CH$

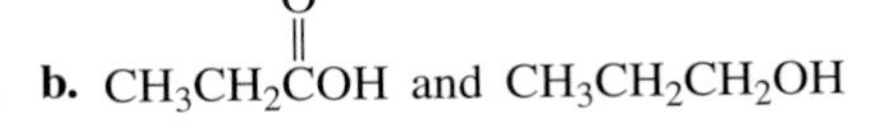

b. $CH_3CH_2\overset{O}{\overset{\|}{C}}OH$ and $CH_3CH_2CH_2OH$

d. 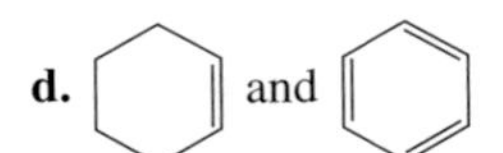

14.14 AN INTRODUCTION TO NMR SPECTROSCOPY

A spinning charged ^1H or ^{13}C nucleus generates a magnetic field similar to the magnetic field of a small bar magnet. In the absence of an applied magnetic field, the magnetic moments associated with the nuclear spins are randomly oriented. However, when placed between the poles of a magnet (Figure 14.19), the nuclear magnetic moments align either *with* or *against* the applied magnetic field. Those that align with the field are in the lower-energy **α-spin state**; those that align against the field are in the higher-energy **β-spin state** because more energy is needed to align against the field than with it. More nuclei are in the α-spin state than in the β-spin state. The difference in the populations is very small (about 20 out of 1 million nuclei) but is sufficient to form the basis of NMR spectroscopy.

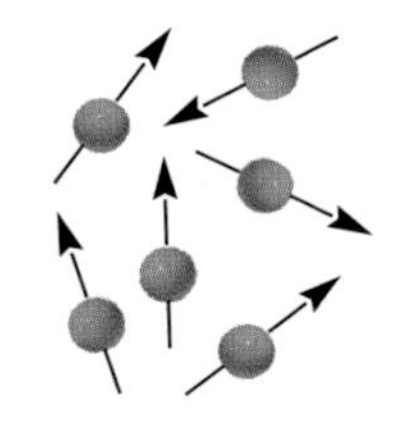

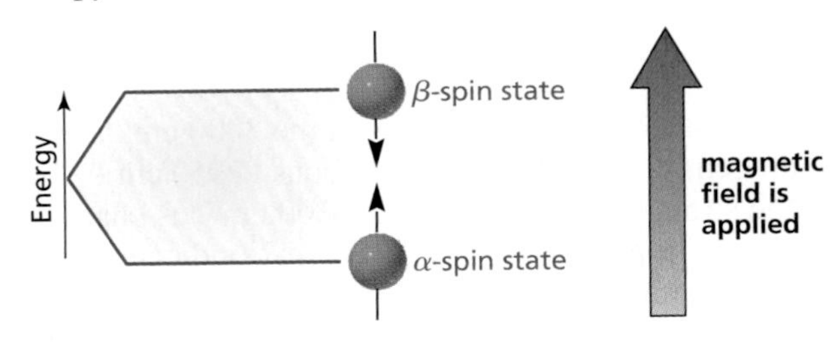

▶ **Figure 14.19**
In the absence of an applied magnetic field, the magnetic moments of the nuclei are randomly oriented.
In the presence of an applied magnetic field, the magnetic moments of the nuclei line up with or against the field.

The energy difference (ΔE) between the α- and β-spin states depends on the strength of the **applied magnetic field (B_0)**, measured in tesla (T). The greater the strength of the magnetic field, the greater is the difference in energy between the α- and β-spin states (Figure 14.20).

Earth's magnetic field is 5×10^{-5} T, measured at the equator. Its maximum surface magnetic field is 7×10^{-5} T, measured at the south magnetic pole.

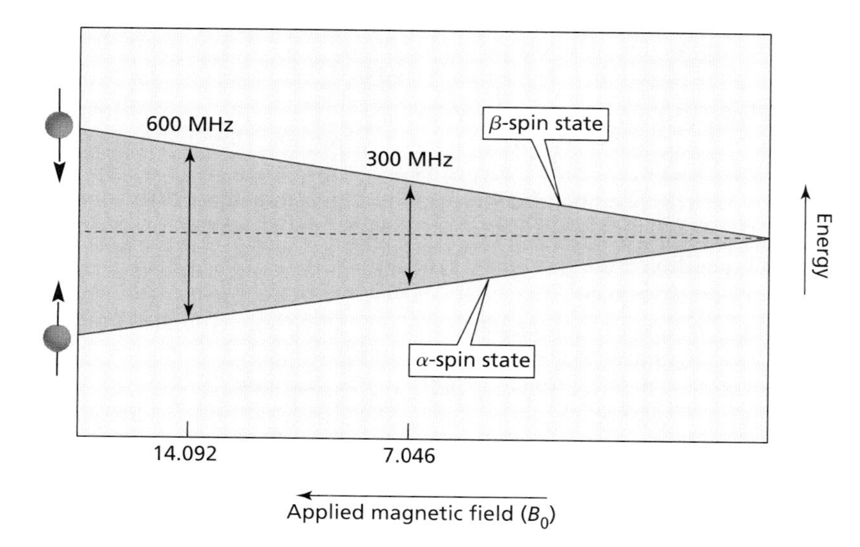

◀ **Figure 14.20**
The greater the strength of the applied magnetic field, the greater is the difference in energy between the α- and β-spin states.

When a sample is subjected to a pulse of radiation whose energy corresponds to the difference in energy (ΔE) between the α- and β-spin states, nuclei in the α-spin state are promoted to the β-spin state. When the nuclei absorb the radiation, they flip their spin, causing them to generate signals whose frequency depends on the difference in energy (ΔE) between the α- and β-spin states. The NMR spectrometer detects these signals and displays them as a plot of signal frequency versus intensity; this plot is the NMR spectrum. The nuclei are said to be *in resonance* with the radiation, hence the term **nuclear magnetic resonance (NMR)**. In this context, "resonance" refers to the flipping back and forth of nuclei between the α- and β-spin states in response to the radiation.

Today's NMR spectrometers operate at frequencies between 60 and 950 MHz. The greater the operating frequency of the instrument—and the stronger the magnet—the better is the resolution of the NMR spectrum.

NIKOLA TESLA (1856–1943)

Nikola Tesla was born in Croatia, the son of a clergyman. He immigrated to the United States in 1884 and became a citizen in 1891. He was a proponent of alternating current to distribute electricity and bitterly fought Thomas Edison, who promoted direct current. Tesla was granted a patent for developing the radio in 1900, but Guglielmo Marconi was also given a patent for its development in 1904. Not until shortly after his death was Tesla's patent upheld by the U.S. Supreme Court. Tesla also is given credit for developing neon and fluorescent lighting, the electron microscope, the refrigerator motor, and the Tesla coil, a type of transformer for changing the voltage of alternating current.

Nikola Tesla in his laboratory

14.15 SHIELDING CAUSES DIFFERENT HYDROGENS TO SHOW SIGNALS AT DIFFERENT FREQUENCIES

The frequency of an NMR signal depends on the strength of the applied magnetic field experienced by the nucleus (Figure 14.20). Thus, if all the hydrogens in an organic compound were to experience the applied magnetic field to the same degree, they would all give signals with the same frequency. If this were the case, all NMR spectra would consist of one signal, which would tell us nothing about the structure of the compound, except that it contains hydrogens.

A nucleus, however, is embedded in a cloud of electrons that partly *shields* it from the applied magnetic field. Fortunately for chemists, the **shielding** varies for different hydrogens within a molecule. In other words, all the hydrogens do not experience the same applied magnetic field.

What causes shielding? In a magnetic field, the electrons circulate about the nuclei and induce a local magnetic field that acts in opposition to the applied magnetic field and subtracts from it. The **effective magnetic field**—the amount of magnetic field that the nuclei actually "sense" through the surrounding electronic environment—is, therefore, somewhat smaller than the applied field:

$$B_{\text{effective}} = B_{\text{applied}} - B_{\text{local}}$$

This means that the greater the electron density of the environment in which the proton* is located, the greater B_{local} is and the more the proton is shielded from the applied magnetic field. Thus, protons in electron-rich environments sense a *smaller effective magnetic field.* They, therefore, will require a *lower frequency* to come into resonance—that is, flip their spins—because ΔE is smaller (Figure 14.20). Protons in electron-poor environments sense a *larger effective magnetic field* and, therefore, will require a *higher frequency* to come into resonance because ΔE is larger.

The larger the magnetic field sensed by the proton, the higher is the frequency of the signal.

An NMR spectrum gives a signal for each proton in a different environment. Protons in electron-rich environments are more shielded from the applied magnetic field and appear at lower frequencies (on the right-hand side of the spectrum) (Figure 14.21). Protons in electron-poor environments are less shielded from the applied magnetic field and appear at higher frequencies (on the left-hand side of the spectrum). For example, the signal for the methyl protons of CH_3F occurs at a higher frequency than the signal for the methyl protons of CH_3Br because fluorine is more electron withdrawing than bromine; thus, the protons in CH_3F are in a less electron-rich environment.

deshielded = less shielded

these protons sense a larger effective magnetic field, so come into resonance at a higher frequency
these protons sense a smaller effective magnetic field, so come into resonance at a lower frequency
Intensity
deshielded nuclei
shielded nuclei
"downfield"
"upfield"
Frequency

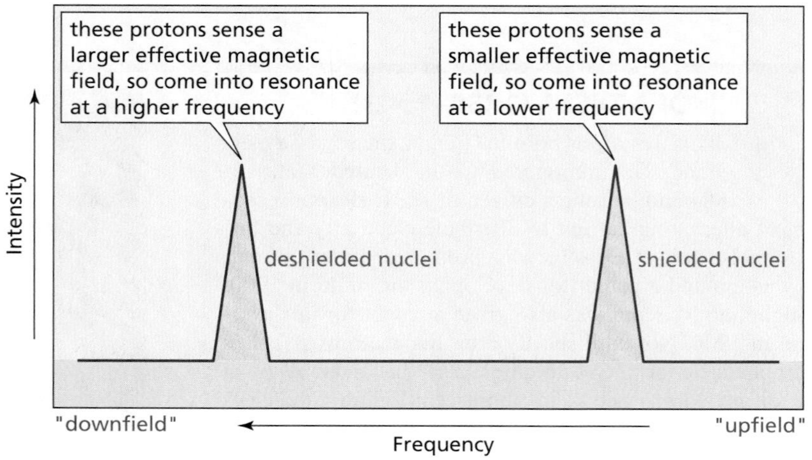

▶ **Figure 14.21**
Shielded nuclei come into resonance at lower frequencies than deshielded nuclei.

* The terms *proton* and *hydrogen* are both used to describe covalently bonded hydrogens in discussions of NMR spectroscopy.

The terms *upfield* and *downfield* are used to describe the position of a signal: **upfield** means farther to the right-hand side of the spectrum, and **downfield** means farther to the left-hand side of the spectrum.

14.16 THE NUMBER OF SIGNALS IN AN ^{1}H NMR SPECTRUM

Protons in the same environment are called **chemically equivalent protons**. For example, 1-bromopropane has three different sets of chemically equivalent protons: (1) the three methyl protons are chemically equivalent because of rotation about the C—C bonds; (2) the two methylene (CH_2) protons on the middle carbon are chemically equivalent; and (3) the two methylene protons on the carbon bonded to the bromine atom make up the third set of chemically equivalent protons.

chemically equivalent protons

$CH_3CH_2CH_2Br$

chemically equivalent protons — chemically equivalent protons

Each set of chemically equivalent protons in a compound gives rise to a signal in the ^1H NMR spectrum of that compound. Because 1-bromopropane has three sets of chemically equivalent protons, it has three signals in its ^1H NMR spectrum.

Each set of chemically equivalent protons gives rise to a signal.

2-Bromopropane has two sets of chemically equivalent protons and, therefore, it has two signals in its ^1H NMR spectrum; the six methyl protons in 2-bromopropane are equivalent, so they give rise to only one signal; the hydrogen bonded to the middle carbon gives the second signal. Ethyl methyl ether has three sets of chemically equivalent protons: the methyl protons on the carbon adjacent to the oxygen, the methylene (CH_2) protons on the carbon adjacent to the oxygen, and the methyl protons on the carbon that is one carbon removed from the oxygen. The chemically equivalent protons in the following compounds are designated by the same letter:

$\overset{a}{CH_3}\overset{b}{CH_2}\overset{c}{CH_2}Br$ — three signals

$\overset{a}{CH_3}\overset{b}{CH}(Br)\overset{a}{CH_3}$ — two signals

$\overset{a}{CH_3}\overset{c}{CH_2}O\overset{b}{CH_3}$ — three signals

$\overset{a}{CH_3}O\overset{a}{CH_3}$ — one signal

$(\overset{a}{CH_3})_3CO\overset{b}{CH_3}$ — two signals

You can tell how many sets of chemically equivalent protons a compound has from the number of signals in its ^1H NMR spectrum.

PROBLEM-SOLVING STRATEGY

How many signals would you expect to see in the ^1H NMR spectrum of ethylbenzene?

$CH_3CH_2-C_6H_5$

To determine the number of signals you would expect to see in the spectrum, replace each hydrogen in turn by another atom (here we use Br) and name the resulting compound. The number of different names corresponds to the number of signals in the ^1H NMR spectrum. We get five different names for the bromosubstituted benzenes, so we expect to see five signals in the ^1H NMR spectrum of ethylbenzene.

$BrCH_2CH_2$–C_6H_5 **1-bromo-2-phenylethane**

$CH_3CH(Br)$–C_6H_5 **1-bromo-1-phenylethane**

CH_3CH_2–C_6H_4–Br (ortho) **1-bromo-2-ethylbenzene**

CH_3CH_2–C_6H_4–Br (meta) **1-bromo-3-ethylbenzene**

CH_3CH_2–C_6H_4–Br (para) **1-bromo-4-ethylbenzene**

CH_3CH_2–C_6H_4–Br (meta) **1-bromo-3-ethylbenzene**

CH_3CH_2–C_6H_4–Br (ortho) **1-bromo-2-ethylbenzene**

Now continue on to Problem 20.

PROBLEM 20 ♦

How many signals would you expect to see in the ^{1}H NMR spectrum of each of the following compounds?

a. $CH_3CH_2CH_2CH_3$

b. $BrCH_2CH_2Br$

c. $CH_3CH_2CH_2\overset{O}{\overset{\|}{C}}CH_3$

d. $CH_3CH_2CH(Cl)CH_2CH_3$

e. $CH_3CH(CH_3)CH_2CH(CH_3)CH_3$

f. C_6H_5–NO_2

PROBLEM 21 ♦

How could you distinguish the ^{1}H NMR spectra of the following compounds?

A: $CH_3OCH_2OCH_3$

B: CH_3OCH_3

C: $CH_3OCH_2C(CH_3)_2CH_2OCH_3$

14.17 THE CHEMICAL SHIFT TELLS US HOW FAR THE SIGNAL IS FROM THE REFERENCE SIGNAL

A small amount of an inert **reference compound** is added to the sample tube containing the compound whose NMR spectrum is to be taken. The positions of the signals in an NMR spectrum are defined according to how far they are from the signal of the reference compound. The most commonly used reference compound is tetramethylsilane (TMS).

$(CH_3)_4Si$

tetramethylsilane (TMS)

The methyl protons of TMS are in a more electron-rich environment than are most protons in organic molecules because silicon is less electronegative than carbon (electronegativities of 1.8 and 2.5, respectively). Consequently, the signal for the methyl protons of TMS is at a lower frequency than most other signals (that is, it appears to the right of the other signals).

The position at which a signal occurs in an NMR spectrum is called the *chemical shift*. The **chemical shift** is a measure of how far the signal is from the reference TMS signal. The most common scale for chemical shifts is the δ (delta) scale. The chemical shift is determined by measuring the distance from the TMS peak in hertz and dividing by the operating frequency of the instrument in megahertz. Because the units are Hz/MHz, a chemical shift has units of parts per million (ppm) of the operating frequency:

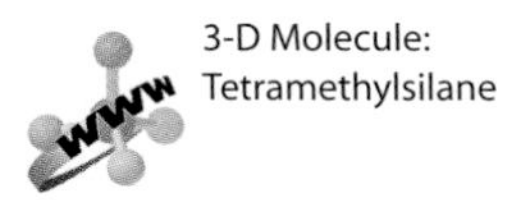

3-D Molecule: Tetramethylsilane

$$\delta = \text{chemical shift (ppm)} = \frac{\text{distance downfield from TMS (Hz)}}{\text{operating frequency of the spectrometer (MHz)}}$$

Most proton chemical shifts fall in the range from 0 to 10 ppm.

The ^1H NMR spectrum for 1-bromo-2,2-dimethylpropane in Figure 14.22 shows that the chemical shift of the methyl protons is at 1.05 ppm and that the chemical shift of the methylene protons is at 3.28 ppm. *Notice that low-frequency (upfield, shielded) signals have small δ (ppm) values, whereas high-frequency (downfield, deshielded) signals have large δ values.*

The greater the chemical shift (δ), the higher the frequency.

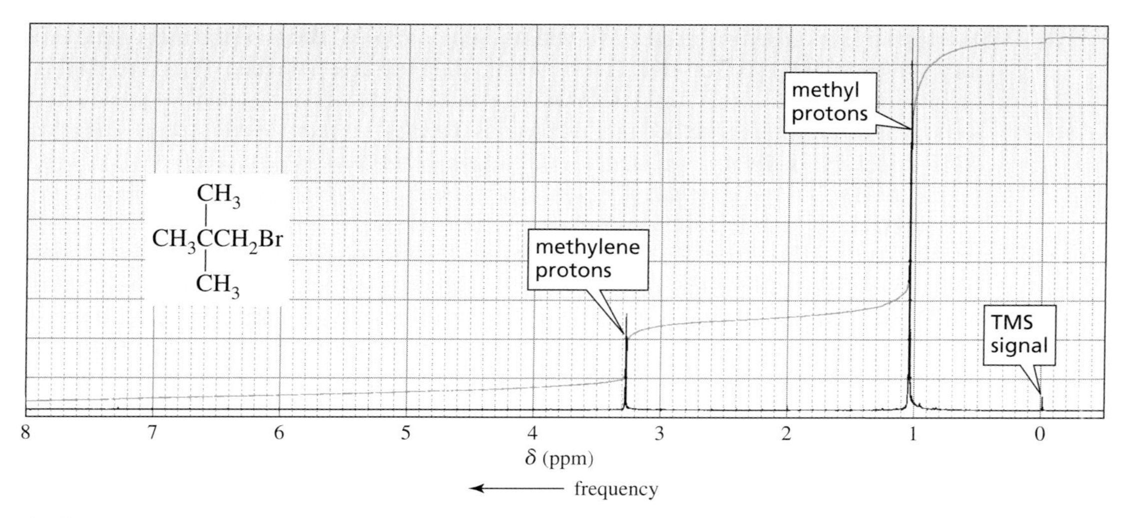

▲ Figure 14.22
^1H NMR spectrum of 1-bromo-2,2-dimethylpropane. The TMS signal is a reference signal from which chemical shifts are measured; it defines the zero position on the scale.

The advantage of the δ scale is that the chemical shift of a given nucleus is *independent of the operating frequency of the NMR spectrometer*. Thus, the chemical shift of the methyl protons of 1-bromo-2,2-dimethylpropane is at 1.05 ppm in both a 60-MHz and a 360-MHz instrument. The following diagram will help you keep track of the terms associated with NMR spectroscopy:

The chemical shift (δ) is independent of the operating frequency of the spectrometer.

protons in electron-poor environments	protons in electron-rich environments
deshielded protons	shielded protons
downfield	upfield
high frequency	low frequency
large δ values	small δ values

← δ
← frequency

PROBLEM 22 ♦

A signal has been reported to occur at 600 Hz downfield from the TMS signal in an NMR spectrometer with a 300-MHz operating frequency.

a. What is the chemical shift of the signal?
b. What would its chemical shift be in an instrument operating at 100 MHz?

PROBLEM 23 ♦

If two signals differ by 1.5 ppm in a 300-MHz spectrometer, by how much do they differ in a 100-MHz spectrometer?

PROBLEM 24 ♦

Where would you expect to find the ^1H NMR signal of $(CH_3)_2Mg$ relative to the TMS signal? (*Hint:* Magnesium is even less electronegative than silicon.)

14.18 THE RELATIVE POSITIONS OF ^{1}H NMR SIGNALS

The ^{1}H NMR spectrum of 1-bromo-2,2-dimethylpropane in Figure 14.22 has two signals because the compound has two different kinds of protons. The methylene protons are in a less electron-rich environment than are the methyl protons because the methylene protons are closer to the electron-withdrawing bromine. Because the methylene protons are in a less electron-rich environment, they are less shielded from the applied magnetic field. The signal for these protons, therefore, occurs at a higher frequency than the signal for the more shielded methyl protons. *Remember that the right-hand side of an NMR spectrum is the low-frequency side, where protons in electron-rich environments (more shielded) show a signal. The left-hand side is the high-frequency side, where protons in electron-poor environments (less shielded) show a signal* (Figure 14.21).

Protons in electron-poor environments show signals at high frequencies.

We would expect the ^{1}H NMR spectrum of 1-nitropropane to have three signals because the compound has three different kinds of protons. The closer the protons are to the electron-withdrawing nitro group, the less they are shielded from the applied magnetic field, so the higher the frequency at which their signal will appear. Thus, the protons closest to the nitro group show a signal at the highest frequency (4.37 ppm), and the ones farthest from the nitro group show a signal at the lowest frequency (1.04 ppm).

Electron withdrawal causes NMR signals to appear at higher frequencies (at larger δ values).

1.04 ppm 2.07 ppm 4.37 ppm

$CH_3CH_2CH_2NO_2$

Compare the chemical shifts of the methylene protons immediately adjacent to the halogen in each of the following alkyl halides. The position of the signal depends on the electronegativity of the halogen—the more electronegative the halogen, the higher is the frequency of the signal. Thus, the signal for the methylene protons adjacent to fluorine (the most electronegative of the halogens) occurs at the highest frequency, whereas the signal for the methylene protons adjacent to iodine (the least electronegative of the halogens) occurs at the lowest frequency.

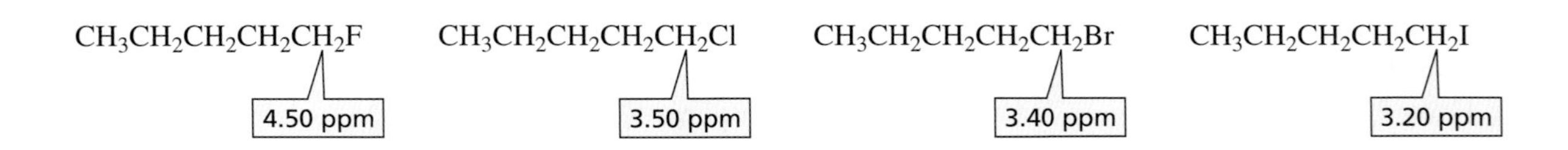

PROBLEM 25

a. Which set of protons in each of the following compounds is the least shielded?

1. $CH_3CH_2CH_2Cl$ **2.** $CH_3CH_2C(=O)OCH_3$ **3.** $CH_3CH(Br)CH(Br)Br$

b. Which set of protons in each compound is the most shielded?

PROBLEM 26 ♦

One of the spectra in Figure 14.23 is due to 1-chloropropane, and the other to 1-iodopropane. Which is which?

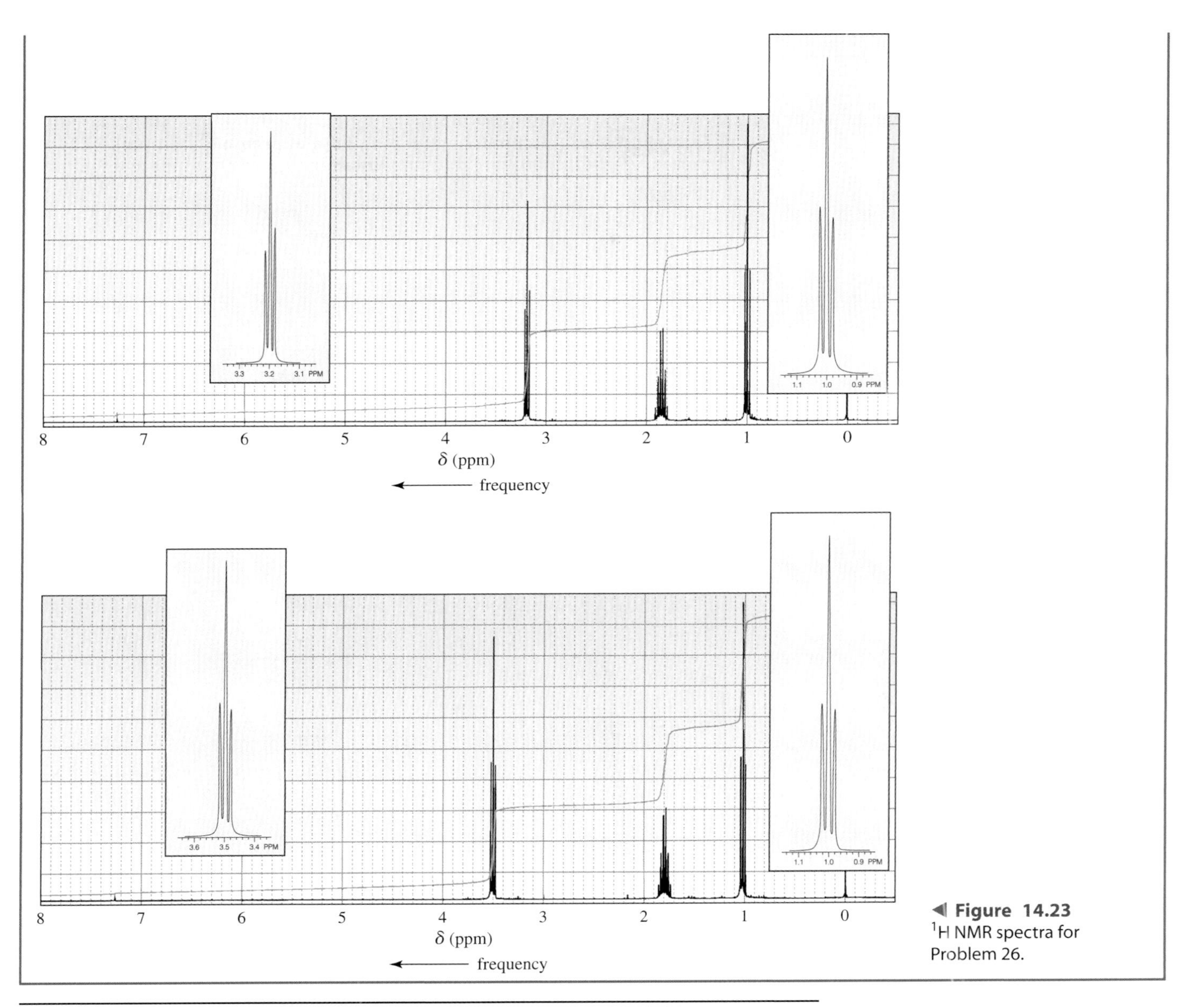

◀ **Figure 14.23**
^{1}H NMR spectra for Problem 26.

14.19 THE CHARACTERISTIC VALUES OF CHEMICAL SHIFTS

Approximate values of chemical shifts for different kinds of protons are shown in Table 14.4 on page 394. (A more extensive compilation is given in Appendix III.) An ^{1}H NMR spectrum can be divided into seven regions, one of which is empty. If you can remember the kinds of protons that appear in each region, you will be able to tell what kinds of protons a molecule has from a quick look at its NMR spectrum.

12–9.0	9.0–8.5	8.5–6.5	6.5–4.5	4.5–2.5	2.5–1.5	1.5–0
–C(=O)–H –C(=O)–OH		benzene–H	C=C–H **vinylic**	–C(Z)–H **Z = O, N, halogen**	C(=O)–C–H C=C–C–H **allylic**	–C–C–H **saturated**

δ (ppm)

Table 14.4 Approximate Values of Chemical Shifts for ^{1}H NMR[a]

Type of proton	Approximate chemical shift (ppm)	Type of proton	Approximate chemical shift (ppm)
$(CH_3)_4Si$	0	C_6H_5—H (benzene ring—H)	6.5–8.5
—CH_3	0.9	—C(=O)—H	9.0–10
—CH_2—	1.3	I—C—H	2.5–4
—CH— (with one more bond to C)	1.4	Br—C—H	2.5–4
—C=C—CH_3	1.7	Cl—C—H	3–4
—C(=O)—CH_3	2.1	F—C—H	4–4.5
C_6H_5—CH_3 (benzene ring—CH_3)	2.3	RNH_2	Variable, 1.5–4
—C≡C—H	2.4	ROH	Variable, 2–5
R—O—CH_3	3.3	ArOH	Variable, 4–7
R—C(R)=CH_2	4.7	—C(=O)—OH	Variable, 10–12
R—C(R)=C(R)—H	5.3	—C(=O)—NH_2	Variable, 5–8

[a]The values are approximate because they are affected by neighboring substituents.

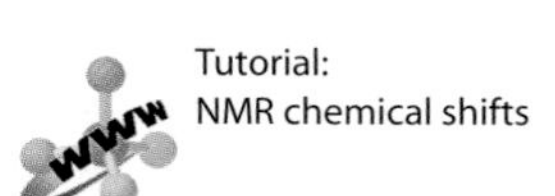
Tutorial:
NMR chemical shifts

In a similar environment, the signal for a methine proton occurs at a higher frequency than the signal for methylene protons, which in turn occurs at a higher frequency than the signal for methyl protons.

—CH— (methine) —CH_2— (methylene) —CH_3 (methyl)

Table 14.4 shows that the chemical shift of a methine proton (a hydrogen attached to an sp^3 carbon that is attached to three carbons) is more deshielded and, therefore, shows a chemical shift at a higher frequency than the chemical shift of methylene protons (hydrogens attached to an sp^3 carbon that is attached to two carbons) in a similar environment. Similarly, the chemical shift of methylene protons is at a higher frequency than is the chemical shift of methyl protons (hydrogens attached to an sp^3 carbon that is attached to one carbon) in a similar environment.

methine proton: C—C(C)(C)—H, 1.55 ppm

methylene proton: H—C(C)(C)—H, 1.20 ppm

methyl proton: H—C(H)(C)—H, 0.85 ppm

For example, the ^{1}H NMR spectrum of butanone shows three signals. The signal at the lowest frequency is the signal for the ***a*** protons of butanone; these protons are farthest from the electron-withdrawing carbonyl group. (In correlating an NMR spectrum with a structure, the set of protons responsible for the signal at the lowest frequency will be labeled ***a***, the next set will be labeled ***b***, the next set ***c***, and so on.) The ***b*** and ***c*** protons are the same distance from the carbonyl group, but the signal for the ***b*** protons is at a lower frequency because methyl protons appear at a lower frequency than do methylene protons in a similar environment.

$$\underset{a}{CH_3}\underset{c}{CH_2}\overset{O}{\overset{\|}{C}}\underset{b}{CH_3}$$

butanone

$$\overset{b}{CH_3}O\overset{c}{C}H\overset{a}{CH_3} \quad (\text{with } \underset{a}{CH_3} \text{ on the } CH)$$

2-methoxypropane

The signal for the ***a*** protons of 2-methoxypropane is the signal at the lowest frequency in the ^{1}H NMR spectrum of this compound because these protons are farthest from the electron-withdrawing oxygen. The ***b*** and ***c*** protons are the same distance from the oxygen, but the signal for the ***b*** protons appears at a lower frequency because, in a similar environment, methyl protons appear at a lower frequency than does a methine proton. (Take a minute to study the *a*, *b*, and *c* designations on the compounds on page 389.)

PROBLEM 27 ♦

In each of the following compounds, which of the underlined protons has the greater chemical shift (that is, the higher frequency signal)?

a. $CH_3C\underline{H}C\underline{H}Br$ (Br on each CH)

b. $C\underline{H_3}CHOC\underline{H_3}$ (CH_3 on the CH)

c. $CH_3C\underline{H_2}CHC\underline{H_3}$ (Cl on the CH)

PROBLEM 28 ♦

In each of the following pairs of compounds, which of the underlined protons has the greater chemical shift (that is, the higher frequency signal)?

a. $CH_3CH_2C\underline{H_2}Cl$ or $CH_3CH_2C\underline{H}CH_3$ (Cl on the CH)

b. $CH_3CH_2C\underline{H_2}Cl$ or $CH_3CH_2C\underline{H_2}Br$

PROBLEM 29

For each compound, without referring to Table 14.4, label the proton that gives the signal at the lowest frequency *a*, the next lowest *b*, and so on.

a. $CH_3CH_2CH_2\overset{O}{\overset{\|}{C}}CH_3$

b. $CH_3CH_2CHCH_2CH_3$ (OCH_3 on the CH)

c. $CH_3CH_2CH_2\overset{O}{\overset{\|}{C}}OCH_3$

d. $CH_3CHCHCH_3$ (CH_3 on the first CH, Cl on the second CH)

e. $CH_3CHCH_2OCH_3$ (CH_3 on the CH)

f. $CH_3CH_2CH_2OCHCH_3$ (CH_3 on the CH)

14.20 THE INTEGRATION OF NMR SIGNALS REVEALS THE RELATIVE NUMBER OF PROTONS CAUSING THE SIGNAL

The two signals in the ^{1}H NMR spectrum of 1-bromo-2,2-dimethylpropane in Figure 14.22 are not the same size because *the area under each signal is proportional to the number of protons giving rise to the signal.* The spectrum is shown again in Figure 14.24. The area under the signal occurring at the lower frequency is larger because the signal is caused by *nine* methyl protons, whereas the smaller, higher-frequency signal results from *two* methylene protons.

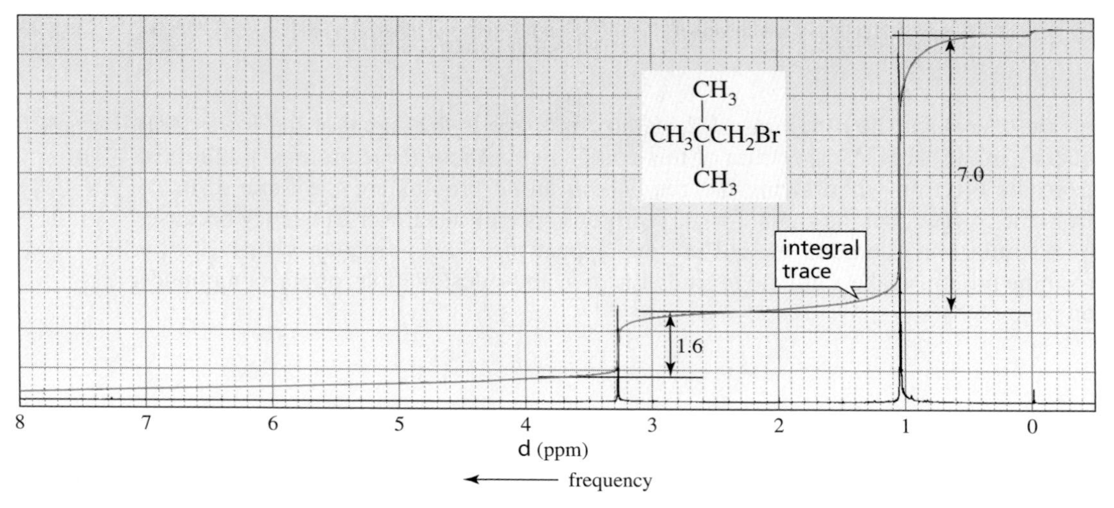

▲ **Figure 14.24**
Analysis of the integral trace in the ^{1}H NMR spectrum of 1-bromo-2,2-dimethylpropane.

The area under each signal can be determined by integration. An ^{1}H NMR spectrometer is equipped with a computer that calculates the integrals electronically and then displays them as an integral trace superimposed on the original spectrum (Figure 14.24). The height of each step in the integral trace is proportional to the number of protons giving rise to the signal. The heights of the integration steps in Figure 14.24, for example, tell us that the ratio of the integrals is approximately 1.6:7.0 = 1:4.4. We multiply the ratio by a number that will cause all the numbers making up the ratio to be close to whole numbers—in this case, we multiply by 2. That means that the ratio of protons in the compound is 2:8.8, which is rounded to 2:9, as there can be only whole numbers of protons. (The measured integrals are approximate because of experimental error.) Modern spectrometers print the integrals as numbers on the spectrum; see Figure 14.26 on page 398.

The **integration** tells us the *relative* number of protons that give rise to each signal, not the *absolute* number. In other words, integration cannot distinguish between 1,1-dichloroethane and 1,2-dichloro-2-methylpropane because both compounds will show an integral ratio of 1:3.

$$CH_3{-}\underset{\displaystyle Cl}{\underset{|}{CH}}{-}Cl \qquad\qquad CH_3{-}\overset{\displaystyle CH_3}{\overset{|}{\underset{\displaystyle Cl}{\underset{|}{C}}}}{-}CH_2Cl$$

1,1-dichloroethane
ratio of protons = 1:3

1,2-dichloro-2-methylpropane
ratio of protons 2:6 = 1:3

PROBLEM 30

How can integration distinguish the ^{1}H NMR spectra of the following compounds?

$$CH_3-\underset{CH_3}{\overset{CH_3}{\underset{|}{\overset{|}{C}}}}-CH_2Br \qquad CH_3-\underset{Br}{\overset{CH_3}{\underset{|}{\overset{|}{C}}}}-CH_2Br \qquad CH_3-\underset{CH_2Br}{\overset{CH_2Br}{\underset{|}{\overset{|}{C}}}}-CH_2Br$$

PROBLEM 31 ♦

The ^{1}H NMR spectrum shown in Figure 14.25 corresponds to one of the following compounds. Which compound is responsible for this spectrum?

$HC{\equiv}C-C_6H_4-C{\equiv}CH$ (A) $\quad CH_3-C_6H_4-CH_3$ (B) $\quad ClCH_2-C_6H_4-CH_2Cl$ (C) $\quad Br_2CH-C_6H_4-CHBr_2$ (D)

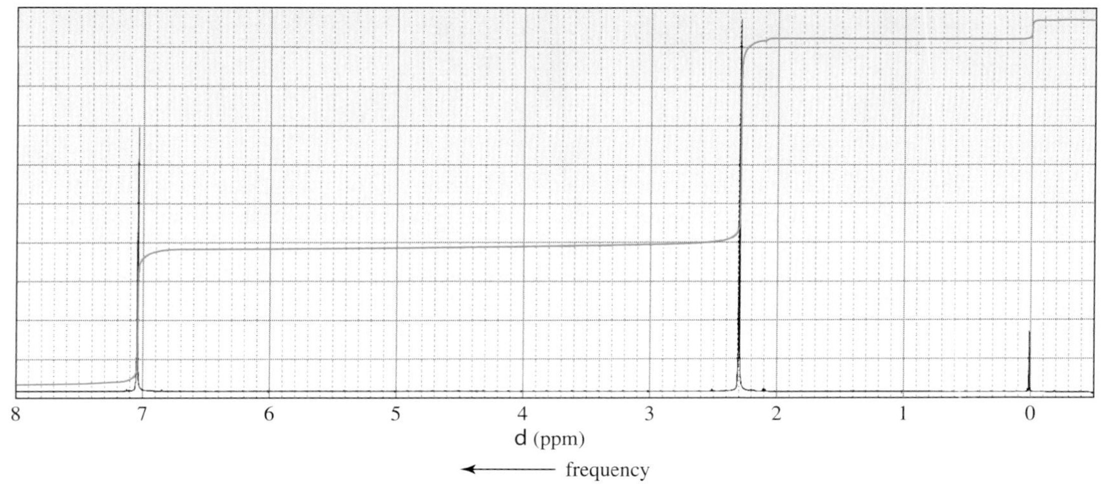

▲ **Figure 14.25**
^{1}H NMR spectrum for Problem 31.

14.21 THE SPLITTING OF THE SIGNALS IS DESCRIBED BY THE N + 1 RULE

Notice that the shapes of the signals in the ^{1}H NMR spectrum of 1,1-dichloroethane (Figure 14.26) are different from the shapes of the signals in the ^{1}H NMR spectrum of 1-bromo-2,2-dimethylpropane (Figure 14.24). Both signals in Figure 14.24 are **singlets** (meaning each is composed of a single peak). In contrast, the signal for the methyl protons of 1,1-dichloroethane (the lower-frequency signal) is split into two peaks (a **doublet**), and the signal for the methine proton is split into four peaks (a **quartet**). (Magnifications of the doublet and quartet are shown as insets in Figure 14.26.)

Splitting is caused by protons bonded to adjacent carbons. The splitting of a signal is described by the ***N* + 1 rule**, where N is the number of *equivalent* protons bonded to *adjacent* carbons. By "equivalent protons," we mean that the protons bonded to an adjacent carbon are equivalent to each other, but not equivalent to the proton giving rise to the signal. Both signals in Figure 14.24 are singlets; the three methyl groups of 1-bromo-2,2-dimethylpropane give an unsplit signal because they are attached to a

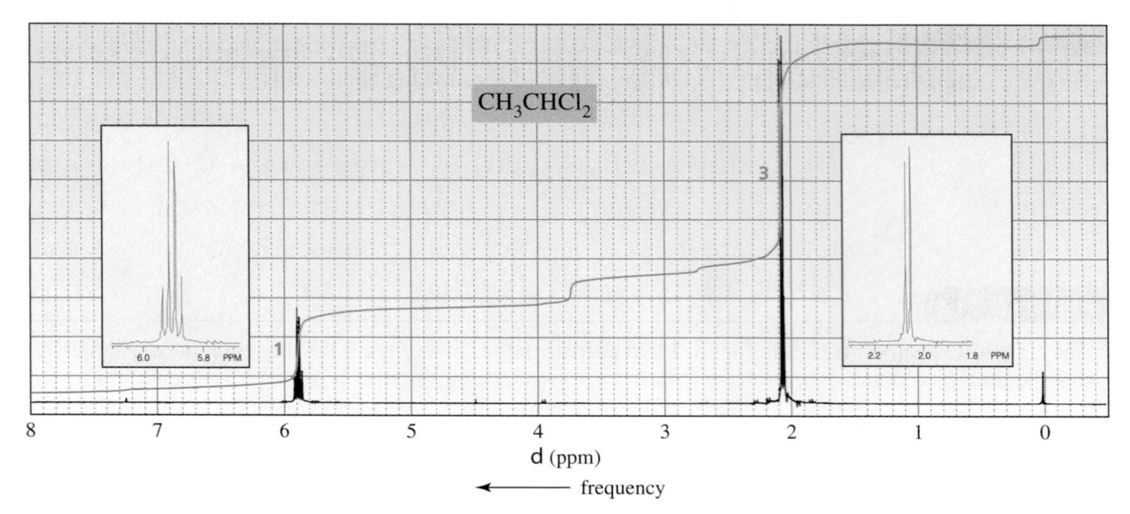

▲ **Figure 14.26**
^{1}H NMR spectrum of 1,1-dichloroethane. The higher-frequency signal is an example of a quartet; the lower-frequency signal is a doublet.

carbon that is not bonded to a hydrogen; the methylene group also gives an unsplit signal because it too is attached to a carbon that is not bonded to a hydrogen ($N = 0$, so $N + 1 = 1$). In contrast, the carbon adjacent to the methyl group in 1,1-dichloroethane (Figure 14.26) is bonded to one proton, so the signal for the methyl protons is split into a doublet ($N = 1$, so $N + 1 = 2$). The carbon adjacent to the carbon bonded to the methine proton is bonded to three equivalent protons, so the signal for the methine proton is split into a quartet ($N = 3$, so $N + 1 = 4$). The number of peaks in a signal is called the **multiplicity** of the signal. Splitting is always mutual: if the ***a*** protons split the ***b*** protons, then the ***b*** protons must split the ***a*** protons. The ***a*** and ***b*** protons, in this case, are coupled protons. **Coupled protons** split each other's signal. Coupled protons are on adjacent carbons.

An ^{1}H NMR signal is split into $N + 1$ peaks, where N is the number of equivalent protons bonded to adjacent carbons.

Coupled protons split each other's signal.

Coupled protons are bonded to adjacent carbons.

Keep in mind that it is not the number of protons giving rise to a signal that determines the multiplicity of the signal; rather, it is the number of protons bonded to the immediately adjacent carbons that determines the multiplicity. For example, the signal for the ***a*** protons in the following compound will be split into three peaks (a **triplet**) because the adjacent carbon is bonded to two hydrogens. The signal for the ***b*** protons will appear as a quartet because the adjacent carbon is bonded to three hydrogens, and the signal for the ***c*** protons will be a singlet.

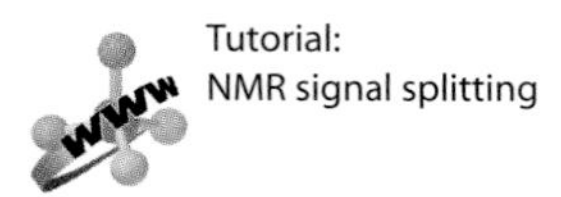

$$\underset{a}{CH_3}\underset{b}{CH_2}\overset{O}{\overset{\|}{C}}O\underset{c}{CH_3}$$

A signal for a proton is never split by *equivalent* protons. For example, the ^{1}H NMR spectrum of bromomethane shows one singlet. The three methyl protons are chemically equivalent, and chemically equivalent protons do not split each other's signal. The four protons in 1,2-dichloroethane are also chemically equivalent, so its ^{1}H NMR spectrum also shows one singlet.

Equivalent protons do not split each other's signal.

CH_3Br	$ClCH_2CH_2Cl$
bromomethane	**1,2-dichloroethane**

each compound has an NMR spectrum that shows one singlet because equivalent protons do not split each other's signals

The ^{1}H NMR spectra of two carboxylic acids with molecular formula $C_3H_5O_2Cl$ are shown in Figure 14.27. Identify the carboxylic acids. (The "offset" notation means that the signal has been moved to the right by the indicated amount.)

a.

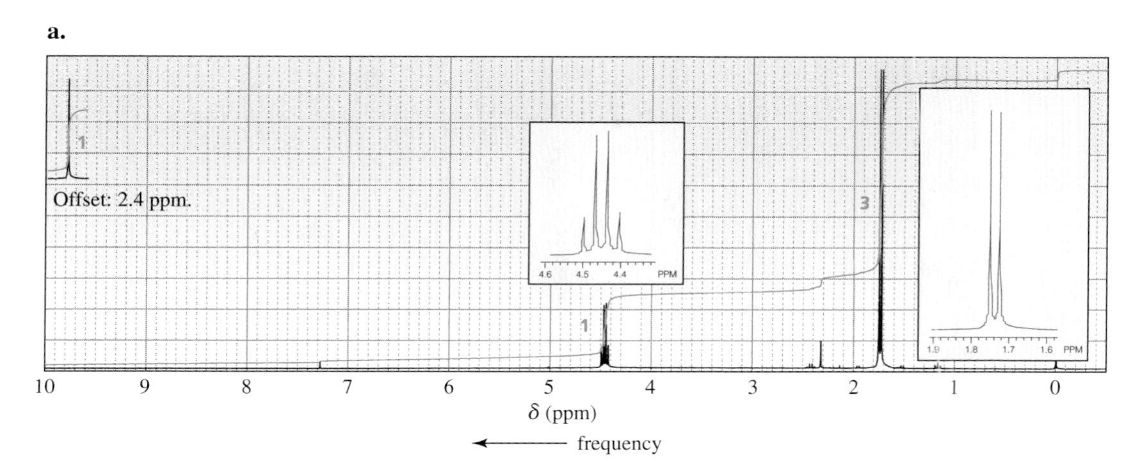

b.

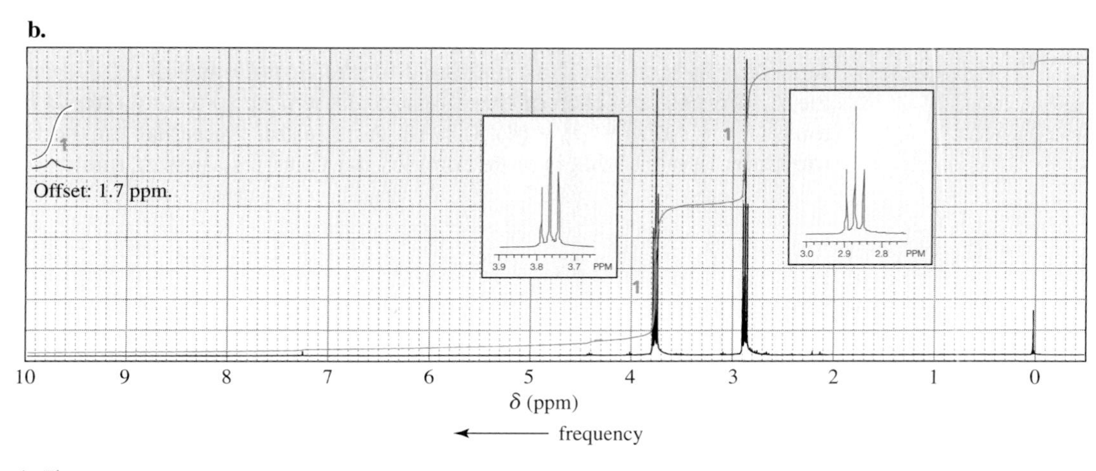

▲ **Figure 14.27**
^{1}H NMR spectra for Problem 32.

14.22 MORE EXAMPLES OF ^{1}H NMR SPECTRA

We will now look at a few more spectra to give you additional practice in analyzing ^{1}H NMR spectra.

There are two signals in the ^{1}H NMR spectrum of 1,3-dibromopropane (Figure 14.28). The signal for the H_b protons is split into a triplet by the two hydrogens on the adjacent carbon. The carbons adjacent to the carbon bonded to the H_a protons are both bonded to protons. The protons on one of these carbons are equivalent to the protons on the other. Because the two sets of protons are equivalent, the $N + 1$ rule is applied to both sets at the same time. In other words, N is equal to the sum of the equivalent protons on both carbons. So the signal for the H_a protons is split into a quintet ($4 + 1 = 5$).

The ^{1}H NMR spectrum of isopropyl butanoate shows five signals (Figure 14.29). The signal for the H_a protons is split into a triplet by the H_c protons. The signal for the H_b protons is split into a doublet by the H_e proton. The signal for the H_d protons is split into a triplet by the H_c protons, and the signal for the H_e proton is split into a septet by the H_b protons. The signal for the H_c protons is split by both the H_a and H_d protons. Because the H_a and H_d protons are not equivalent, the $N + 1$ rule has to be applied

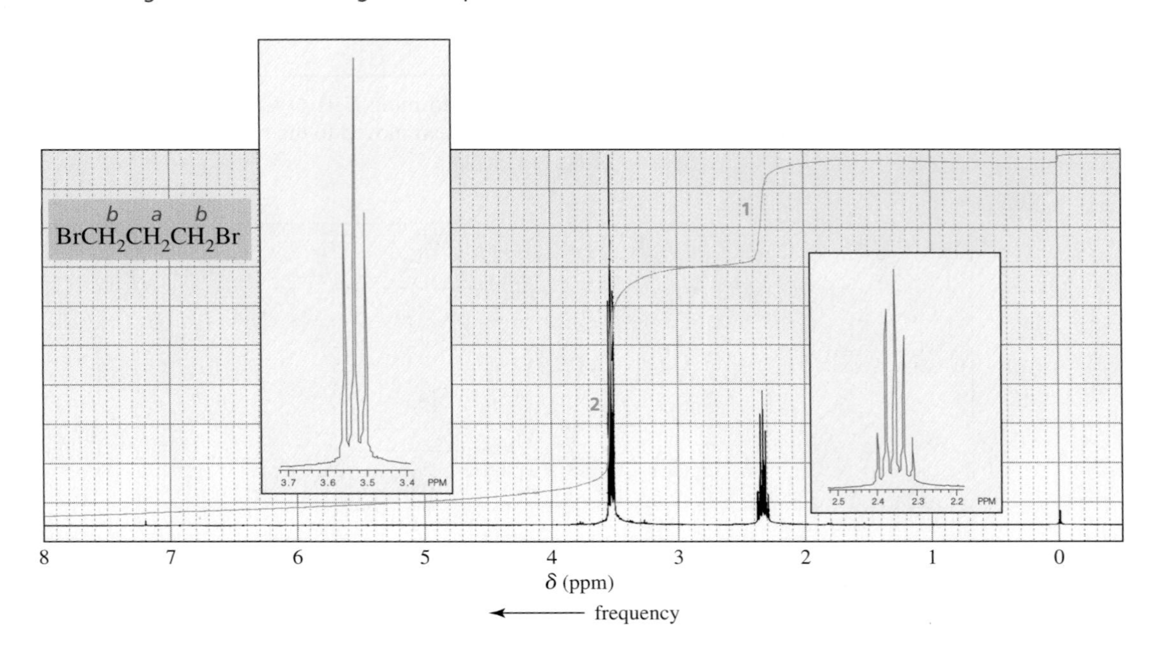

▶ **Figure 14.28**
^{1}H NMR spectrum of 1, 3-dibromopropane.

separately to each set. Thus, the signal for the H_c protons will be split into a quartet by the H_a protons, and each of these four peaks will be split into a triplet by the H_d protons: $(N_a + 1)(N_d + 1) = (4)(3) = 12$. As a result, the signal for the H_c protons is a **multiplet** (a signal that is more complex than a triplet, quartet, quintet, or such).

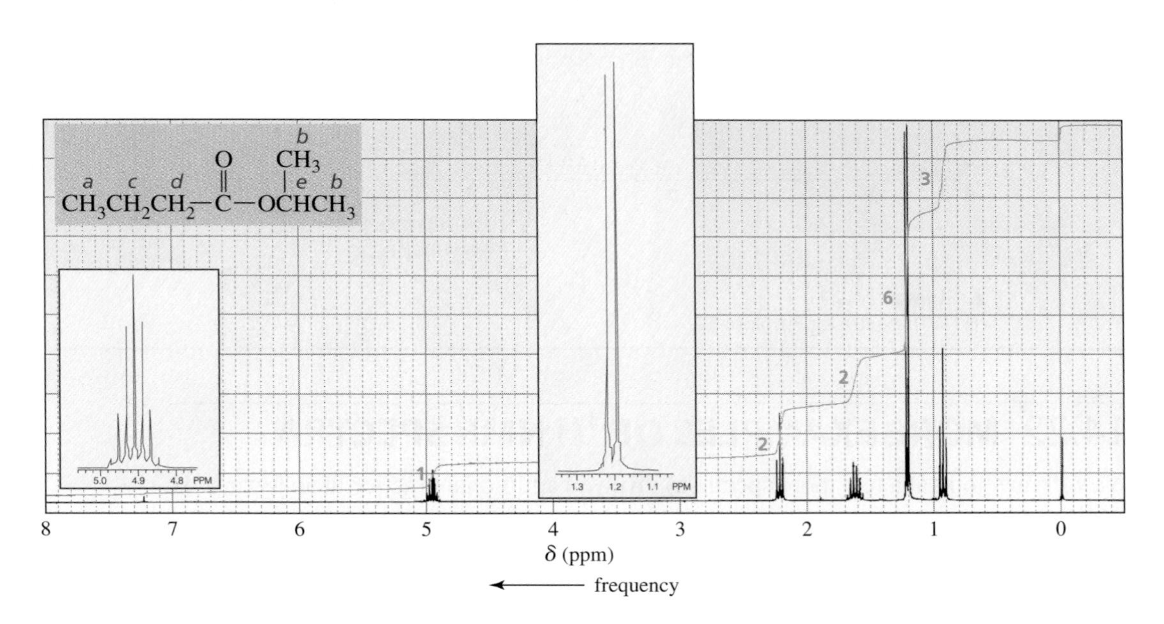

▶ **Figure 14.29**
^{1}H NMR spectrum of isopropyl butanoate.

Ethylbenzene has five sets of chemically equivalent protons (Figure 14.30). We see the expected triplet for the H_a protons and the quartet for the H_b protons. (This is a characteristic pattern for an ethyl group.) The five protons on the benzene ring are not all in the same environment, so we expect to see three signals for them: one for the H_c protons, one for the H_d protons, and one for the H_e protons. However, we do not see three distinct signals because their environments are not sufficiently different to allow them to appear as separate signals.

Notice that the signals for the benzene ring protons in Figure 14.30 occur in the 6.5–8.5 ppm region. Other kinds of protons usually do not resonate in this region, so signals in this region of an ^{1}H NMR spectrum indicate that the compound probably contains an aromatic ring.

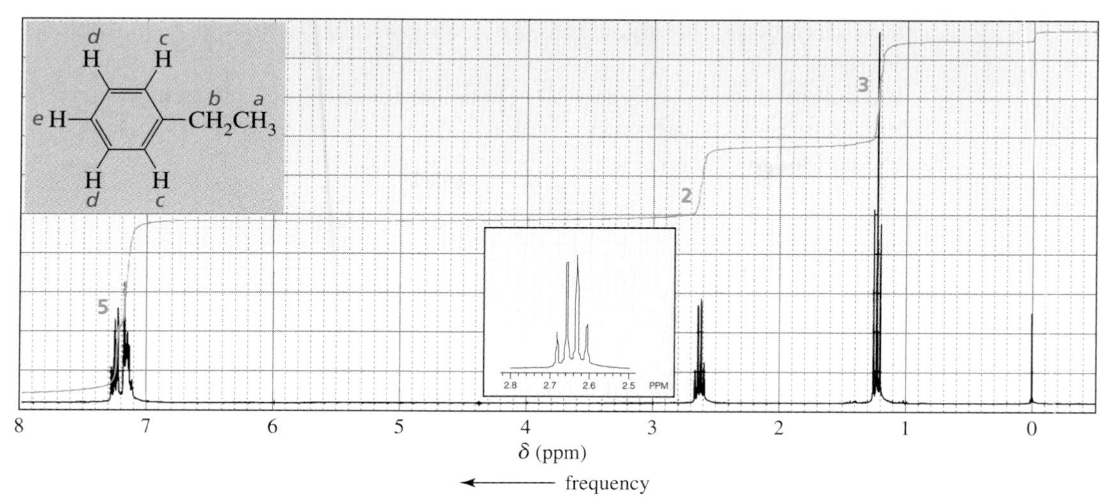

◀ **Figure 14.30** ^{1}H NMR spectrum of ethylbenzene. The signals for the H_c, H_d, and H_e protons overlap.

PROBLEM 33

Indicate the number of signals and the multiplicity of each signal in the ^{1}H NMR spectrum of each of the following compounds:

a. $ICH_2CH_2CH_2Br$ **b.** $ClCH_2CH_2CH_2Cl$ **c.** $ICH_2CH_2CHBr_2$

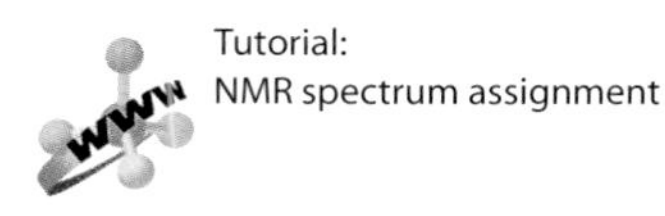

Tutorial:
NMR spectrum assignment

PROBLEM 34

Predict the splitting patterns for the signals given by each of the compounds in Problem 20.

PROBLEM 35

How could ^{1}H NMR spectra distinguish between the following compounds?

NO_2 Br   A NO_2 Br   B NO_2 Br   C

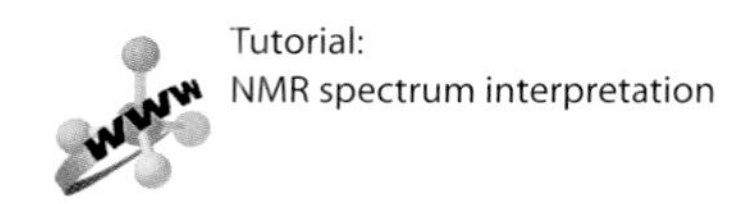

Tutorial:
NMR spectrum interpretation

PROBLEM 36

How would the ^{1}H NMR spectra for the four compounds with molecular formula $C_3H_6Br_2$ differ?

14.23 COUPLING CONSTANTS IDENTIFY COUPLED PROTONS

The distance, in hertz, between two adjacent peaks of a split NMR signal is called the **coupling constant** (denoted by J). The coupling constant for H_a being split by H_b is denoted by J_{ab}. The signals of coupled protons (protons that split each other's signal) have the same coupling constant; in other words, $J_{ab} = J_{ba}$ (Figure 14.31). Coupling constants are useful in analyzing complex NMR spectra because protons on adjacent carbons can be identified by their identical coupling constants. Coupling constants range from 0 to 15 Hz.

Let's now summarize the kind of information that can be obtained from an ^{1}H NMR spectrum:

1. The number of signals indicates the number of different kinds of protons in the compound.
2. The position of a signal indicates the kind of proton(s) responsible for the signal (methyl, methylene, methine, allylic, vinylic, aromatic, and so on) and the kinds of neighboring substituents.

3. The integration of the signal tells the relative number of protons responsible for the signal.
4. The multiplicity of the signal ($N + 1$) tells the number of equivalent protons (N) bonded to adjacent carbons.
5. The coupling constants identify coupled protons.

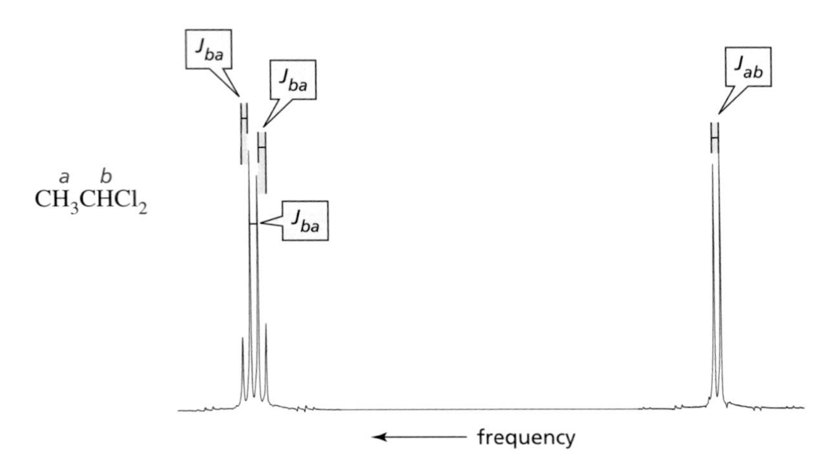

▶ **Figure 14.31**
The H_a and H_b protons of 1,1-dichloroethane are coupled protons, so their signals have the same coupling constant, $J_{ab} = J_{ba}$.

PROBLEM-SOLVING STRATEGY

Identify the compound with molecular formula $C_9H_{10}O$ that gives the IR and ^1H NMR spectra in Figure 14.32.

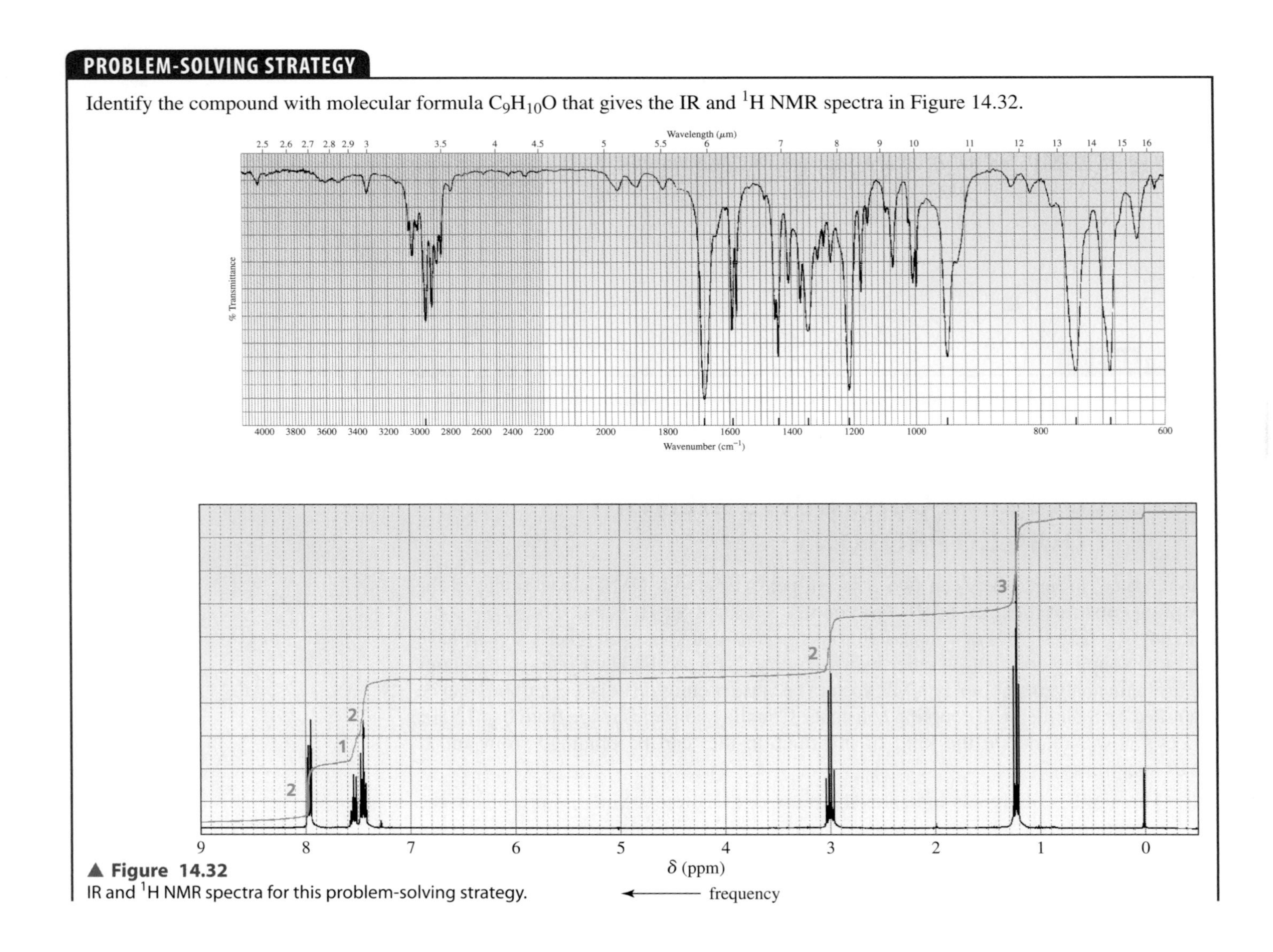

▲ **Figure 14.32**
IR and ^1H NMR spectra for this problem-solving strategy.

The best way to approach this kind of problem is to identify whatever structural features you can from the molecular formula and the IR spectrum and then use the information from the ^{1}H NMR spectrum to expand on that knowledge. From the molecular formula and IR spectrum, we learn that the compound is a ketone: it has a carbonyl group at ~1680 cm^{-1}, only one oxygen, and no absorption bands at ~2820 and ~2720 cm^{-1} that would indicate an aldehyde. The carbonyl group absorption is at a lower frequency than normal, which suggests that it has partial single-bond character as a result of electron delocalization, indicating that it is attached to an sp^2 carbon. The compound contains a benzene ring (C—H absorption >3000 cm^{-1}, absorption at ~1600 cm^{-1} and 1440 cm^{-1}), and it has hydrogens bonded to sp^3 carbons (C—H absorption < 3000 cm^{-1}). In the NMR spectrum, the triplet at ~1.2 ppm and the quartet at ~3.0 ppm indicate the presence of an ethyl group that is attached to an electron-withdrawing group. The signals in the 7.4 to 8.0 ppm region confirm the presence of a benzene ring. From this information, we can conclude that the compound is the ketone shown here. The integration ratio (5:2:3) confirms this answer.

$C_6H_5C(=O)CH_2CH_3$

Now continue on to Problem 37.

PROBLEM 37 ◆

Identify the compound with molecular formula $C_8H_{10}O$ that gives the IR and ^{1}H NMR spectra shown in Figure 14.33.

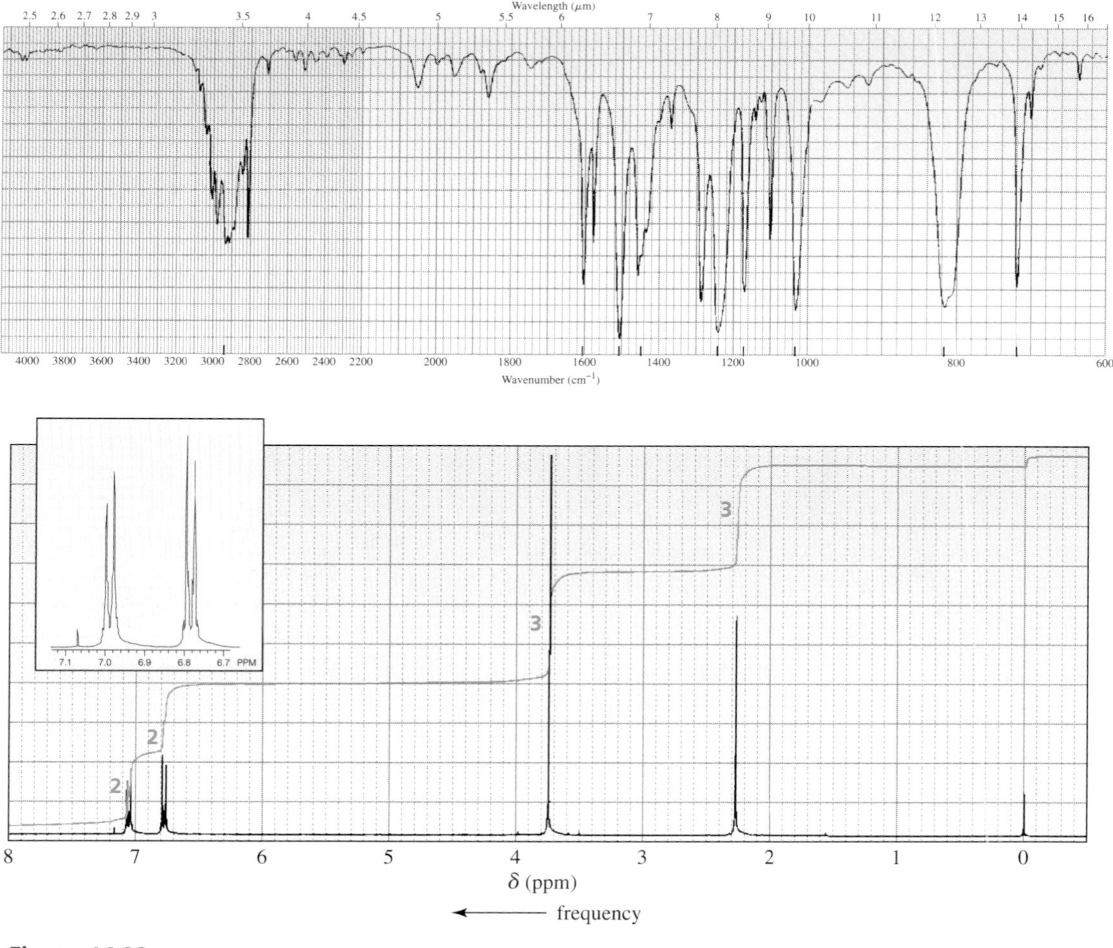

▲ **Figure 14.33**
IR and ^{1}H NMR spectra for Problem 37.

14.24 ^{13}C NMR SPECTROSCOPY

The number of signals in a ^{13}C NMR spectrum tells you how many different kinds of carbons a compound has—just as the number of signals in an ^{1}H NMR spectrum tells you how many different kinds of hydrogens a compound has. The principles behind ^{1}H NMR and ^{13}C NMR spectroscopy are essentially the same. There are, however, some differences that make ^{13}C NMR easier to interpret.

One advantage to ^{13}C NMR spectroscopy is that the chemical shifts of carbon atoms range over about 220 ppm (Table 14.5), compared with over about 12 ppm for ^{1}H NMR (Table 14.4). This means that signals are less likely to overlap. Notice that aldehyde (190–200 ppm) and ketone (205–220 ppm) carbonyl groups can be easily distinguished from other carbonyl groups.

Table 14.5 Approximate Values of Chemical Shifts for ^{13}C NMR

Type of carbon	Approximate chemical shift (ppm)	Type of carbon	Approximate chemical shift (ppm)
$(CH_3)_4Si$	0	C—I	0–40
R—CH_3	8–35	C—Br	25–65
R—CH_2—R	15–50	C—Cl C—N C—O	35–80 40–60 50–80
R_2CH—R (R—CH(R)—R)	20–60	R(N<)C=O	165–175
R_3C—R (R—C(R)(R)—R)	30–40	R(RO)C=O	165–175
C≡C	65–85	R(HO)C=O	175–185
C=C	100–150	R(H)C=O	190–200
aromatic C	110–170	R(R)C=O	205–220

The reference compound used in ^{13}C NMR is TMS, the same reference compound used in ^{1}H NMR. You will find it helpful when analyzing a ^{13}C NMR spectrum to divide it into five regions and remember the kind of carbons that show signals in each region.

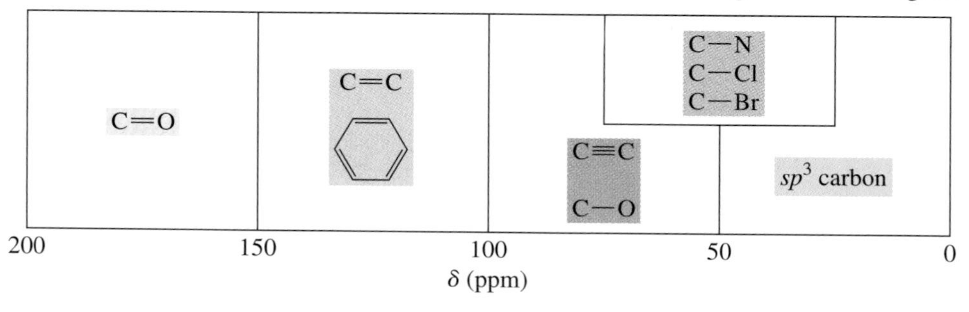

A disadvantage of ^{13}C NMR spectroscopy is that, unless special techniques are used, the area under a ^{13}C NMR signal is *not* proportional to the number of carbons giving rise to the signal. Therefore, the number of carbons giving rise to a ^{13}C NMR signal cannot routinely be determined by integration.

The ^{13}C NMR spectrum of 2-butanol is shown in Figure 14.34. 2-Butanol has carbons in four different environments, so there are four signals in the spectrum. The relative positions of the signals depend on the same factors that determine the relative positions of the proton signals in an ^{1}H NMR spectrum. Carbons in electron-rich environments produce low-frequency signals, and carbons close to electron-withdrawing groups produce high-frequency signals. This means that the signals for the carbons of 2-butanol are in the same relative order as the signals for the protons bonded to those carbons in the ^{1}H NMR spectrum. Thus, the carbon of the methyl group farthest away from the electron-withdrawing OH group gives the lowest-frequency signal. The other methyl carbon comes next in order of increasing frequency, followed by the methylene carbon; the carbon attached to the OH group gives the highest-frequency signal.

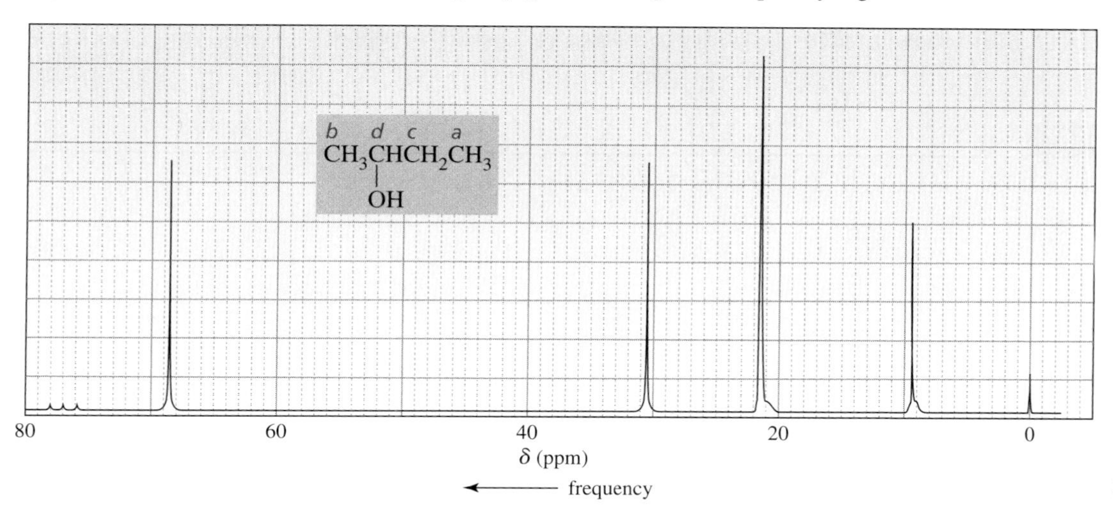

◀ Figure 14.34
^{13}C NMR spectrum of 2-butanol.

The signals in a ^{13}C NMR spectrum are not split by neighboring carbons. If the spectrometer is run in a *proton-coupled* mode, however, each signal will be split by the *hydrogens* bonded to the carbon that produces the signal. The multiplicity of the signal is determined by the $N + 1$ rule. The **proton-coupled ^{13}C NMR spectrum** of 2-butanol is shown in Figure 14.35. (The triplet at 78 ppm is produced by the solvent.) The signals for the methyl carbons are each split into a quartet because each methyl carbon is bonded to three hydrogens ($3 + 1 = 4$). The signal for the methylene carbon is split into a triplet ($2 + 1 = 3$), and the signal for the carbon bonded to the OH group is split into a doublet ($1 + 1 = 2$).

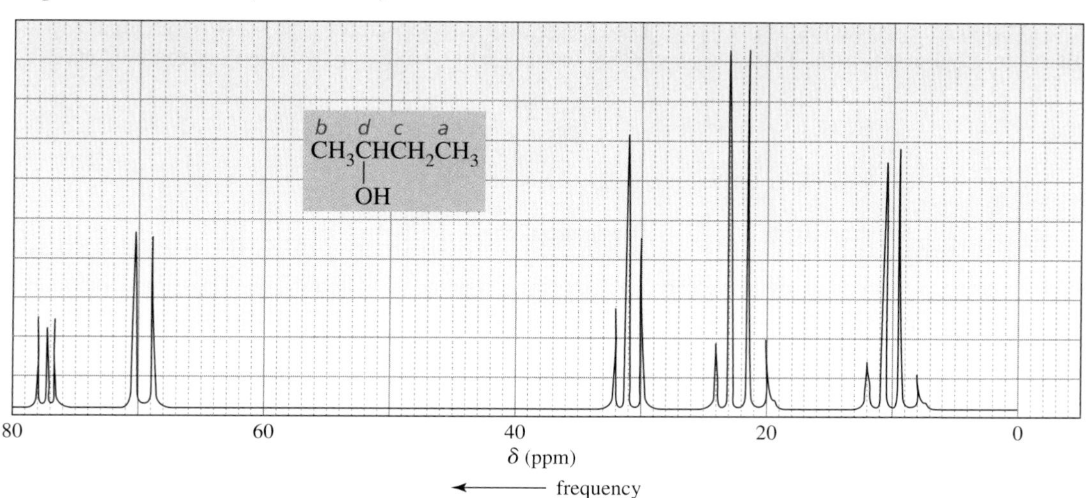

◀ Figure 14.35
Proton-coupled ^{13}C NMR spectrum of 2-butanol: when the spectrometer is run in a proton-coupled mode, splitting is observed in the ^{13}C NMR spectrum.

PROBLEM 38 ◆

Answer the two questions for each compound:

a. How many signals are in its ^{13}C NMR spectrum?

b. Which signal is at the lowest frequency?

1. $CH_3CH_2CH_2Br$
2. $CH_3CH_2OCH_3$
3. $CH_3CH_2COCH_3$ (with C=O)
4. $CH_3C(CH_3)_2OCH_3$
5. $CH_3CHBrCH_3$
6. $CH_3COCH_2CH_2COCH_3$

PROBLEM 39

Describe the proton-coupled ^{13}C NMR spectrum for compounds 1, 2, and 3 in Problem 38, showing relative values (not absolute values) of chemical shifts.

PROBLEM-SOLVING STRATEGY

Identify the compound with molecular formula $C_9H_{10}O_2$ that gives the following ^{13}C NMR spectrum:

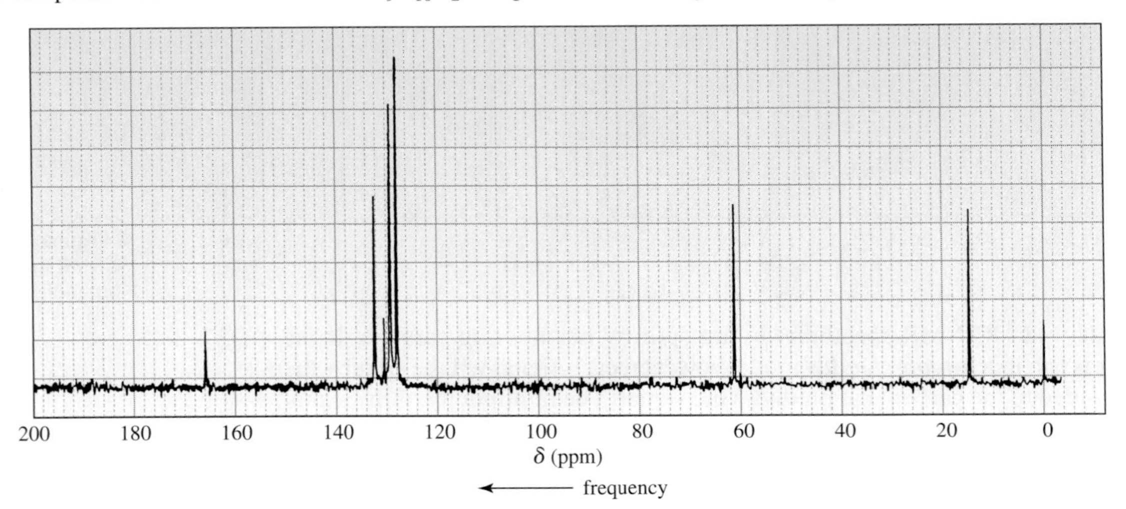

First, pick out the signals that can be identified with certainty. For example, the two oxygen atoms in the molecular formula and the carbonyl carbon signal at 166 ppm indicate that the compound is an ester. The four signals at about 130 ppm suggest that the compound has a benzene ring with a single substituent. (One signal is for the carbon to which the substituent is attached, one is for the two adjacent carbons, and so on.) Subtracting those fragments (C_6H_5 and CO_2) from the molecular formula of the compound leaves C_2H_5, the molecular formula of an ethyl substituent. Therefore, we know that the compound is either ethyl benzoate or phenyl propanoate.

$C_6H_5COCH_2CH_3$ (with C=O) — **ethyl benzoate**

$CH_3CH_2COC_6H_5$ (with C=O) — **phenyl propanoate**

Since the signal for the methylene group is at about 60 ppm, we can conclude that it is adjacent to an oxygen. Thus, the compound must be ethyl benzoate.

Now continue on to Problem 40.

PROBLEM 40 ♦

Identify the compound in Figure 14.36 from its molecular formula and its ^{13}C NMR spectrum.

$C_{11}H_{22}O$

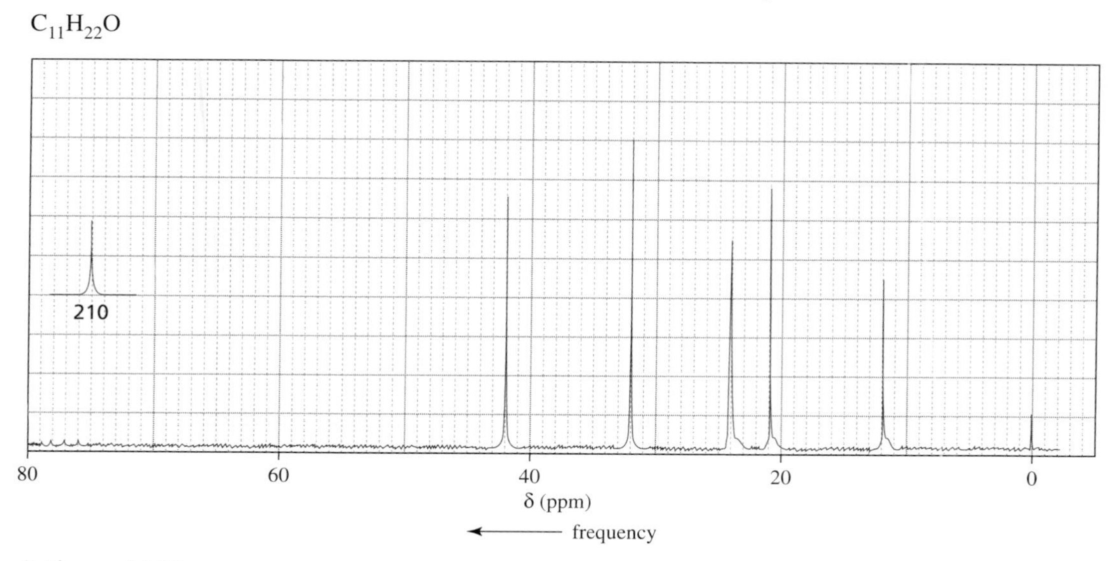

▲ **Figure 14.36**
The ^{13}C NMR spectra for Problem 40.

MAGNETIC RESONANCE IMAGING

NMR has become an important tool in medical diagnosis because it allows physicians to examine internal organs and structures without resorting to invasive surgery or the harmful ionizing radiation of X-rays. When NMR was first introduced into clinical practice in 1981, the selection of an appropriate name was a matter of some debate. Because many nonscientists associate the word "nuclear" with harmful radiation (radioactivity), the "N" was dropped from the medical application of NMR, which is known as **magnetic resonance imaging (MRI)**. The spectrometer is called an **MRI scanner**.

An MRI scanner consists of a magnet sufficiently large to surround a person entirely, along with an apparatus for exciting the nuclei, modifying the magnetic field, and receiving signals. (By comparison, the NMR spectrometer used by chemists is only large enough to accommodate a 5-mm glass tube.) Different tissues yield different signals, and the signals can be attributed to a specific site of origin in the patient, allowing a cross-sectional image of the patient's body to be constructed. MRI can produce an image showing any cross section of the body, regardless of the patient's position in the machine, allowing optimum visualization of the anatomical feature of interest.

Most of the signals in an MRI scan originate from the hydrogens of water molecules because tissues contain far more of these hydrogens than they do hydrogens of organic compounds. The difference in the way water is bound in different tissues is what produces much of the variation in signal between different organs, as well as the variation in signal between healthy and diseased tissue (Figure 14.37). MRI scans, therefore, can sometimes provide much more information than images obtained by other means. For example, MRI can provide detailed images of blood vessels. Flowing fluids, such as blood, respond differently to excitation in an MRI scanner than do stationary tissues, and proper processing will result in the display only of the moving fluids. Images of other fluid-filled structures, such as the biliary tree, can be obtained, which allow the visualization of stones that can obstruct passage of bile into the bowel, and eliminate the need for more invasive diagnostic techniques.

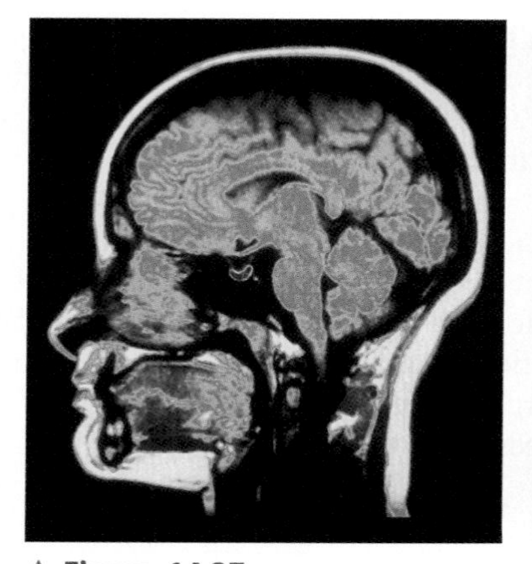

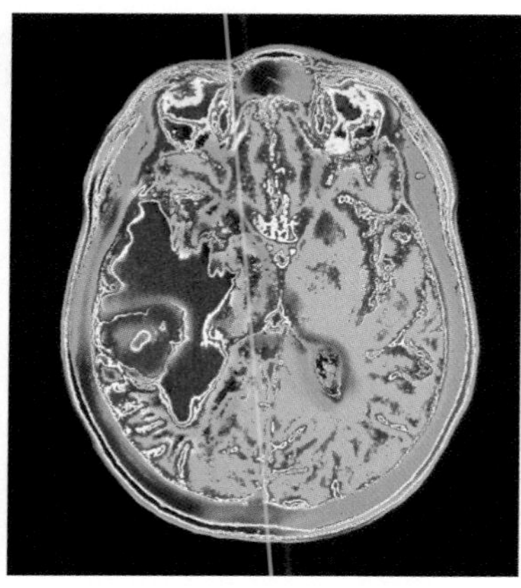

▲ **Figure 14.37**
(a) An MRI of a normal brain. The pituitary is highlighted (pink).
(b) An MRI of an axial section through the brain showing a tumor (purple) surrounded by damaged, fluid-filled tissue (red).

SUMMARY

Mass spectrometry allows us to determine the *molecular mass* and the *molecular formula* of a compound, as well as some if its structural features. In this analytical method, a small sample of the compound is vaporized and then ionized as a result of an electron being removed from each molecule, producing a **molecular ion**, which is a radical cation. Many of the molecular ions break apart into cations, radicals, neutral molecules, and other radical cations. The bonds most likely to break are the weakest ones and those that result in the formation of the most stable products.

The mass spectrometer records a **mass spectrum**, which is a graph of the relative abundance of each positively charged fragment plotted against its m/z value. The molecular ion peak (M) represents the molecular ion, and its m/z value gives the molecular mass of the compound. Peaks with smaller m/z values—**fragment ion peaks**—represent positively charged fragments of the molecule. The **base peak** is the peak with the greatest intensity. High-resolution mass spectrometers determine the exact molecular mass, which allows a compound's molecular formula to be determined.

The $M + 1$ peak occurs because of the naturally occurring ^{13}C isotope of carbon. A large $M + 2$ peak is evidence of a compound containing either chlorine or bromine; if it is one-third the height of the M peak, the compound contains one chlorine atom; if the M and $M + 2$ peaks are about the same height, the compound contains one bromine atom. Each functional group has characteristic fragmentation patterns. Electron bombardment is most likely to dislodge a lone-pair electron.

Spectroscopy is the study of the interaction of matter and **electromagnetic radiation**. A continuum of different types of electromagnetic radiation constitutes the electromagnetic spectrum. High-energy radiation is associated with *high frequencies, large wavenumbers*, and *short wavelengths*.

Infrared spectroscopy identifies the kinds of functional groups in a compound. Each stretching and bending vibration of a bond occurs with a characteristic frequency. It takes more energy to stretch a bond than to bend it. When a compound is bombarded with radiation of a frequency that exactly matches the frequency of one of its vibrations, the molecule absorbs energy and exhibits an **absorption band** in its IR spectrum. The position, intensity, and shape of an absorption band help identify functional groups. Stronger bonds show absorption bands at larger wavenumbers. The intensity of an absorption band depends on the polarity of the bond.

NMR spectroscopy is used to identify the carbon–hydrogen framework of an organic compound. When a sample is placed in a magnetic field, nuclei aligning with the field are in the lower-energy α-spin state; those aligning against the field are in the higher-energy β-spin state. The energy difference between the spin states depends on the strength of the **applied magnetic field**. When subjected to radiation with energy corresponding to the energy difference between the spin states, nuclei in the α-spin state are promoted to the β-spin state and emit signals whose frequency depends on the difference in energy between the spin states. An **NMR spectrometer** detects and displays these signals as a plot of their frequency versus their intensity—an **NMR spectrum**.

Each set of chemically equivalent protons gives rise to a signal, so the number of signals in an ^{1}H NMR spectrum indicates the number of different kinds of protons in a compound. The **chemical shift** is a measure of how far the signal is from the reference TMS signal.

The larger the magnetic field sensed by the proton, the higher is the frequency of the signal. The electron density of the environment in which the proton is located **shields** the proton from the applied magnetic field. Therefore, a proton in an electron-rich environment shows a signal at a lower frequency than a proton near electron-withdrawing groups. Low-frequency (upfield) signals have small δ (ppm) values; high-frequency (downfield) signals have large δ values. Thus, the position of a signal indicates the kind of proton(s) responsible for the signal and the kinds of neighboring substituents. **Integration** tells us the relative number of protons that give rise to each signal.

The **multiplicity** of a signal (the number of peaks in the signal) indicates the number of protons bonded to adjacent carbons. Multiplicity is described by the $\boldsymbol{N} + \mathbf{1}$ **rule**, where N is the number of equivalent protons bonded to adjacent carbons. The **coupling constant (*J*)** is the distance between two adjacent peaks of a split NMR signal. Coupled protons have the same coupling constant.

The number of signals in a ^{13}C NMR spectrum tells how many kinds of carbons a compound has. Carbons in electron-rich environments produce low-frequency signals; carbons close to electron-withdrawing groups produce high-frequency signals. ^{13}C NMR signals are not normally split by directly attached protons, unless the spectrometer is run in a proton-coupled mode.

PROBLEMS

41. For each of the following pairs of compounds, identify one IR absorption band that could be used to distinguish between them:

a. $CH_3CH_2\overset{O}{\overset{\|}{C}}H$ and $CH_3CH_2\overset{O}{\overset{\|}{C}}CH_3$

b. cyclohexyl$-\overset{O}{\overset{\|}{C}}-H$ and phenyl$-\overset{O}{\overset{\|}{C}}-H$

c. $CH_3CH_2\overset{O}{\overset{\|}{C}}NH_2$ and $CH_3CH_2\overset{O}{\overset{\|}{C}}OCH_3$

d. $CH_3CH_2CH_2OH$ and $CH_3CH_2OCH_3$

42. In the mass spectrum of each of the following compounds, which would be more intense, the peak at $m/z = 57$ or the peak at $m/z = 71$?

a. 3-methylpentane **b.** 2-methylpentane

43. Each of the IR spectra presented in Figures 14.38 and 14.39 is accompanied by a set of four compounds. In each case, indicate which of the four compounds is responsible for the spectrum.

a. $CH_3CH_2CH_2C{\equiv}CCH_3$ $CH_3CH_2CH_2CH_2OH$ $CH_3CH_2CH_2CH_2C{\equiv}CH$ $CH_3CH_2CH_2\overset{O}{\overset{\|}{C}}OH$

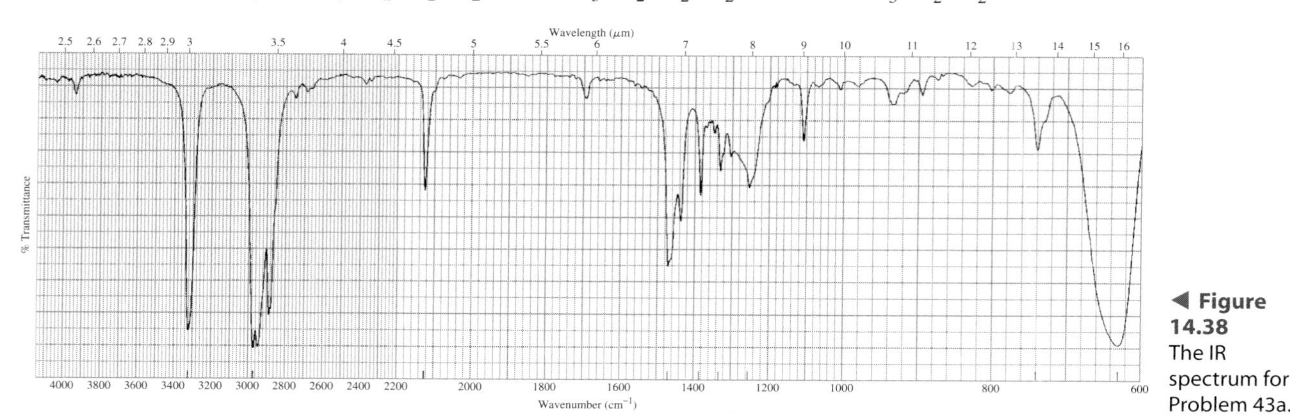

◀ **Figure 14.38** The IR spectrum for Problem 43a.

b. $CH_3CH_2\overset{O}{\overset{\|}{C}}OH$ $CH_3CH_2\overset{O}{\overset{\|}{C}}OCH_2CH_3$ $CH_3CH_2\overset{O}{\overset{\|}{C}}H$ $CH_3CH_2\overset{O}{\overset{\|}{C}}CH_3$

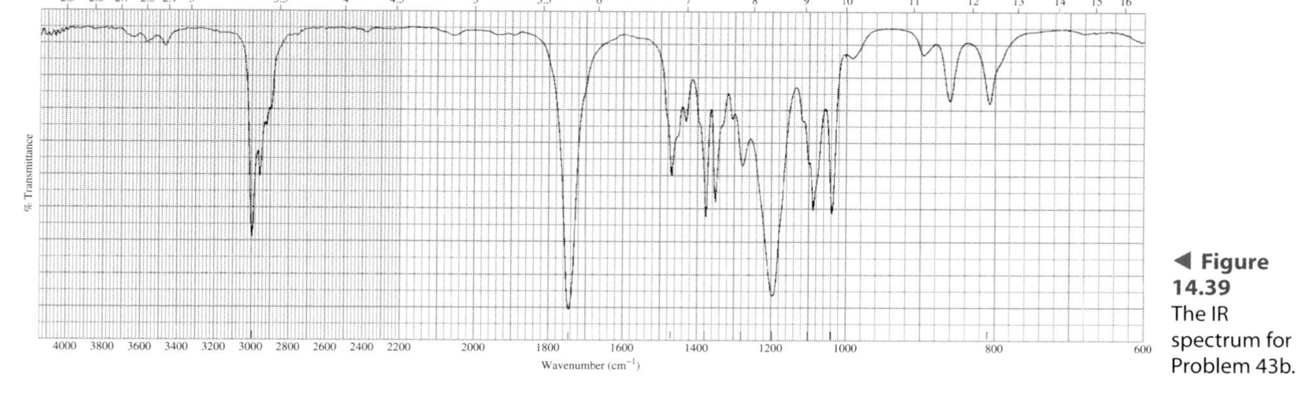

◀ **Figure 14.39** The IR spectrum for Problem 43b.

44. Norlutin and Enovid are ketones that suppress ovulation. Consequently, they have been used clinically as contraceptives. For which of these compounds would you expect the infrared carbonyl absorption (C=O stretch) to be at a higher frequency? Explain.

OH C≡CH O Norlutin®

OH C≡CH O Enovid®

45. For each of the following pairs of compounds, identify one IR absorption band that could be used to distinguish between them:

a. $CH_3CH_2\overset{O}{\overset{\|}{C}}OCH_3$ and $CH_3CH_2\overset{O}{\overset{\|}{C}}CH_3$

b. and

c. $CH_3CH_2CH{=}CHCH_3$ and $CH_3CH_2C{\equiv}CCH_3$

d. $CH_3CH_2CH{=}CH_2$ and $CH_3CH_2CH_2CH_3$

46. a. How could you use IR spectroscopy to determine whether the following reaction had occurred?

$$\overset{O}{\overset{\|}{C}}H \xrightarrow[\text{HO}^-,\ \Delta]{\text{NH}_2\text{NH}_2} CH_3$$

b. After purifying the product, how could you determine whether all the NH_2NH_2 had been removed?

47. What would distinguish the mass spectrum of 2,2-dimethylpropane from those of pentane and 2-methylbutane?

48. Five compounds are shown for each of the IR spectra in Figures 14.40, 14.41, and 14.42. Indicate which of the five compounds is responsible for each spectrum:

a. $CH_3CH_2CH{=}CH_2$ $CH_3CH_2CH_2CH_2OH$ $CH_2{=}CHCH_2CH_2OH$ $CH_3CH_2CH_2OCH_3$ $CH_3CH_2CH_2\overset{O}{\overset{\|}{C}}OH$

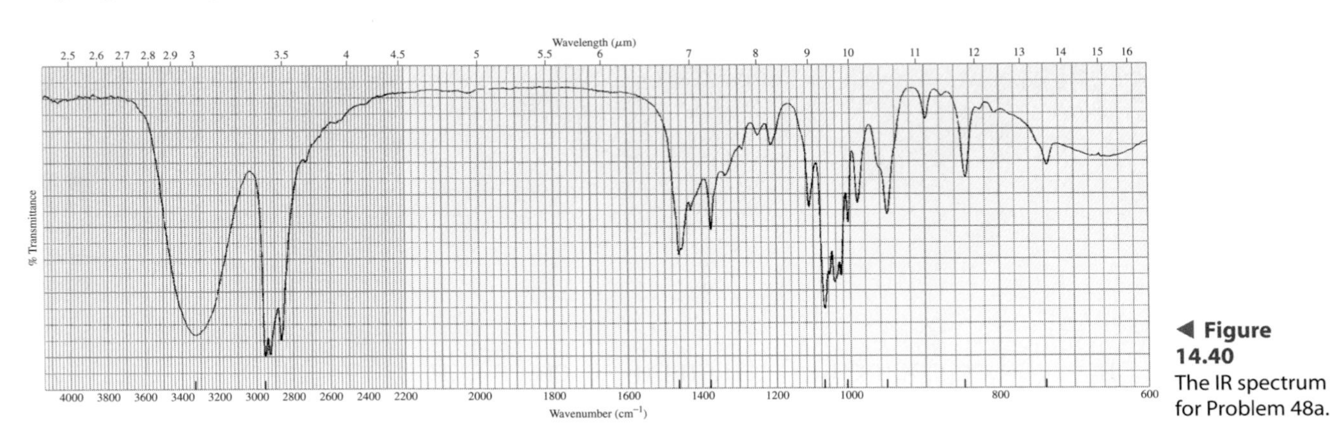

◀ **Figure 14.40**
The IR spectrum for Problem 48a.

b.

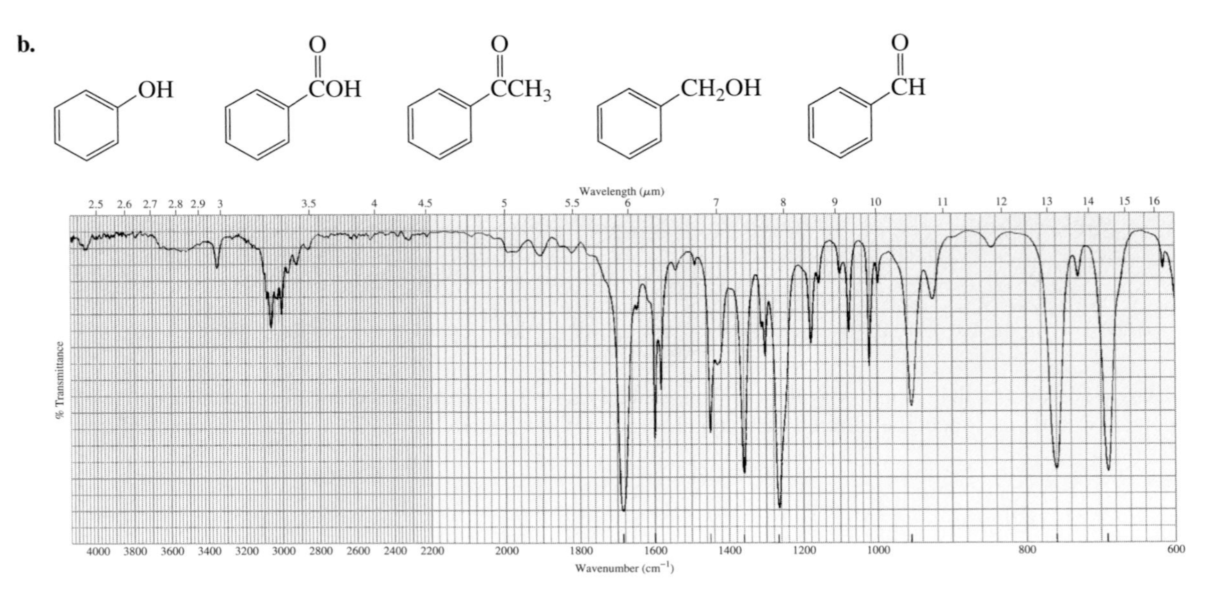

▲ **Figure 14.41**
The IR spectrum for Problem 48b.

c.

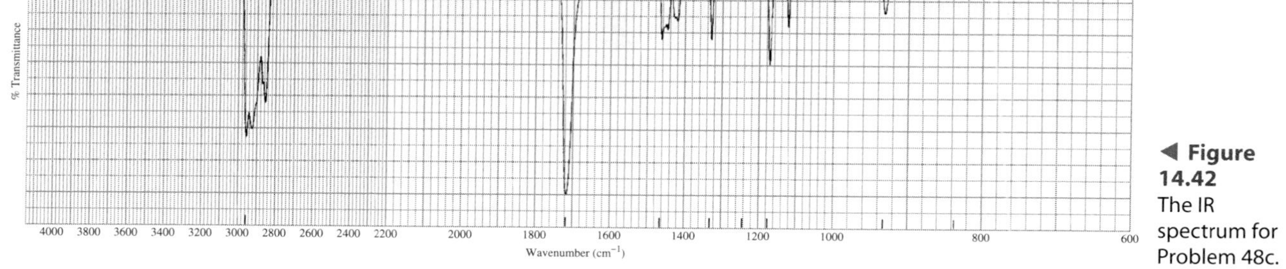

◀ **Figure 14.42** The IR spectrum for Problem 48c.

49. Each of the IR spectra shown in Figure 14.43 is the spectrum of one of the following compounds. Identify the compound that produced each spectrum.

CH_2CH (C=O) $\quad$ CNH_2 (C=O) $\quad$ CH_3CH_2CHCH (C=O; CH_3) $\quad$ $CH_3CH_2CNH_2$ (C=O)

a.

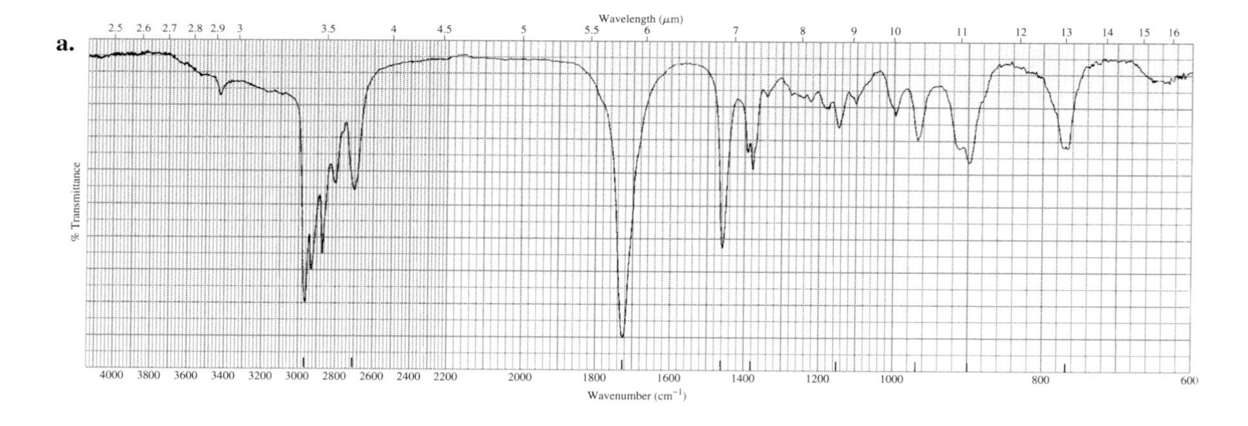

b.

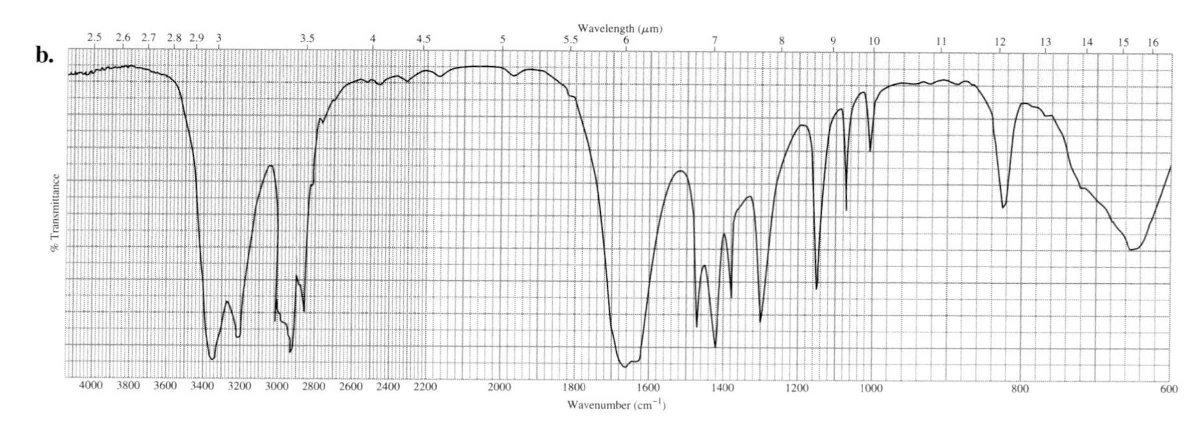

◀ **Figure 14.43** The IR spectra for Problem 49.

c.

Wavelength (μm)
% Transmittance
Wavenumber (cm^{-1})

▲ **Figure 14.43** *(cont.)*
The IR spectra for Problem 49.

50. What identifying characteristics would be present in the mass spectrum of a compound containing two bromine atoms?

51. Which one of the following compounds produced the IR spectrum shown in Figure 14.44?

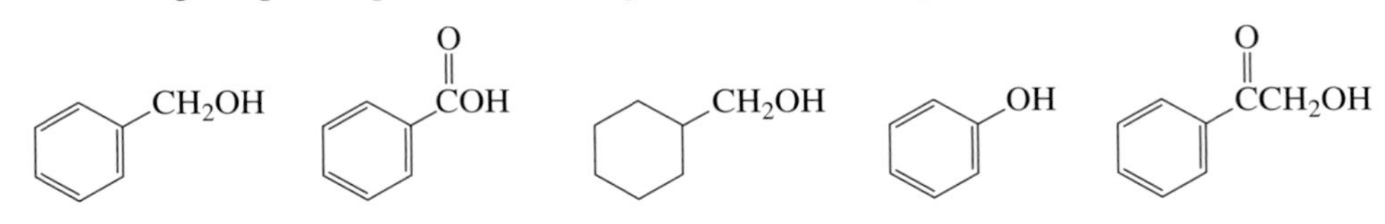

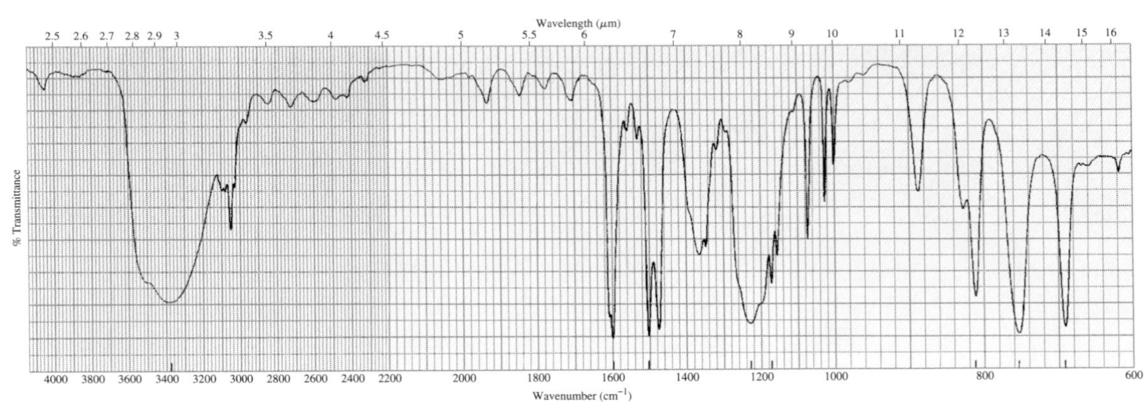

▲ **Figure 14.44**
The IR spectrum for Problem 51.

52. How could IR spectroscopy distinguish between 1,5-hexadiene and 2,4-hexadiene?

53. How does the frequency of the electromagnetic radiation used in NMR spectroscopy compare with that used in IR and UV/Vis spectroscopy?

54. How many signals are produced by each of the following compounds in its
a. ^{1}H NMR spectrum ? **b.** ^{13}C NMR spectrum?

1. CH_3–(benzene ring)–$OCHCH_3$ with CH_3 on the CH

2. (benzene ring)–C(=O)–OCH_2CH_3

3. (six-membered ring with O and =O)

4. (four-membered ring with O)

55. Label each set of chemically equivalent protons, using *a* for the set that will be at the lowest frequency (farthest upfield) in the ^1H NMR spectrum, *b* for the next lowest, and so on. Indicate the multiplicity of each signal.

a. CH_3CHNO_2 (with CH_3 on the CH)

b. $CH_3CH_2CH_2OCH_3$

c. $CH_3CHCCH_2CH_2CH_3$ (C=O at the third carbon; CH_3 on the CH)

d. $CH_3CH_2CH_2CCH_2Cl$ (C=O at the fourth carbon)

e. $ClCH_2CCHCl_2$ (two CH_3 groups on the central C)

f. $ClCH_2CH_2CH_2CH_2CH_2Cl$

56. Identify the following compounds. (Relative integrals are given from left to right across the spectrum.)

a. The ^1H NMR spectrum of a compound with molecular formula $C_4H_{10}O_2$ has two singlets with an area ratio of 2:3.

b. The ^1H NMR spectrum of a compound with molecular formula $C_8H_6O_2$ has two singlets with an area ratio of 1:2.

57. Match each of the ^1H NMR spectra with one of the following compounds:

$CH_3CH_2CCH_3$ (C=O at the third carbon) $\quad$ CH_3CNO_2 (two CH_3 groups on the C) $\quad$ $CH_3CH_2CCH_2CH_3$ (C=O at the middle carbon) $\quad$ $CH_3CH_2CH_2NO_2$ $\quad$ $CH_3CH_2NO_2$ $\quad$ CH_3CHBr (with CH_3 on the CH)

a.

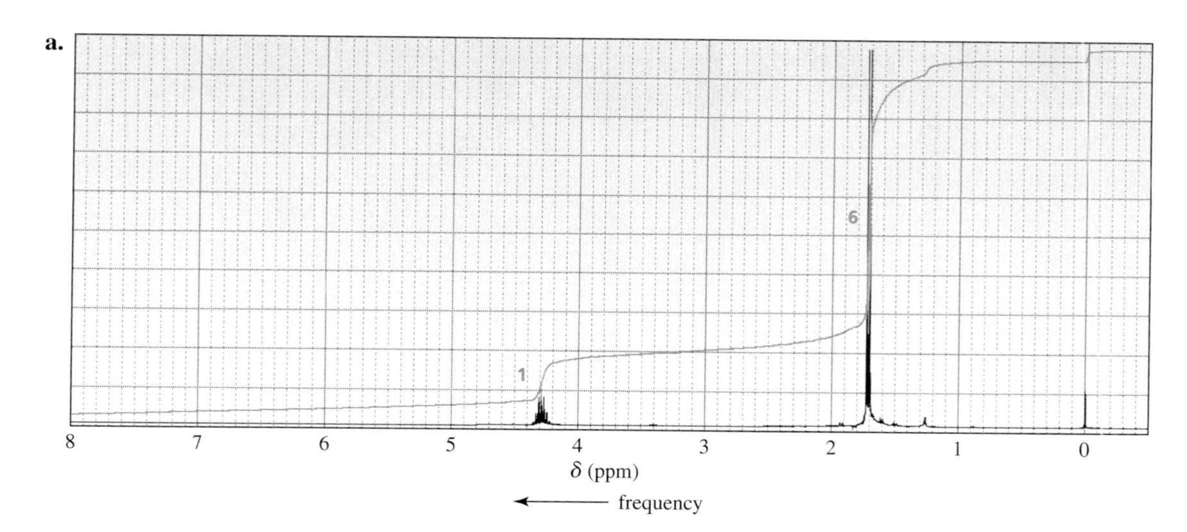

b.

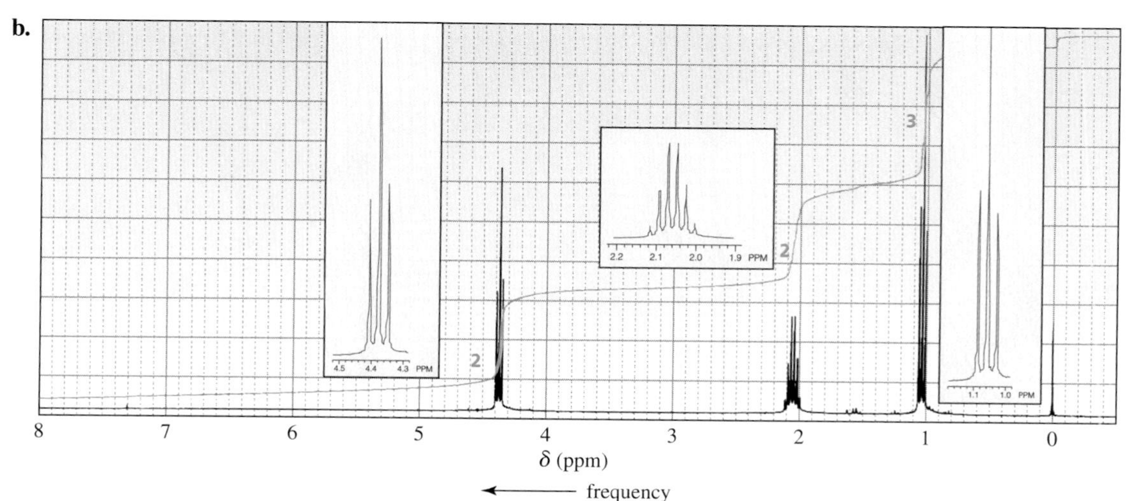

c.

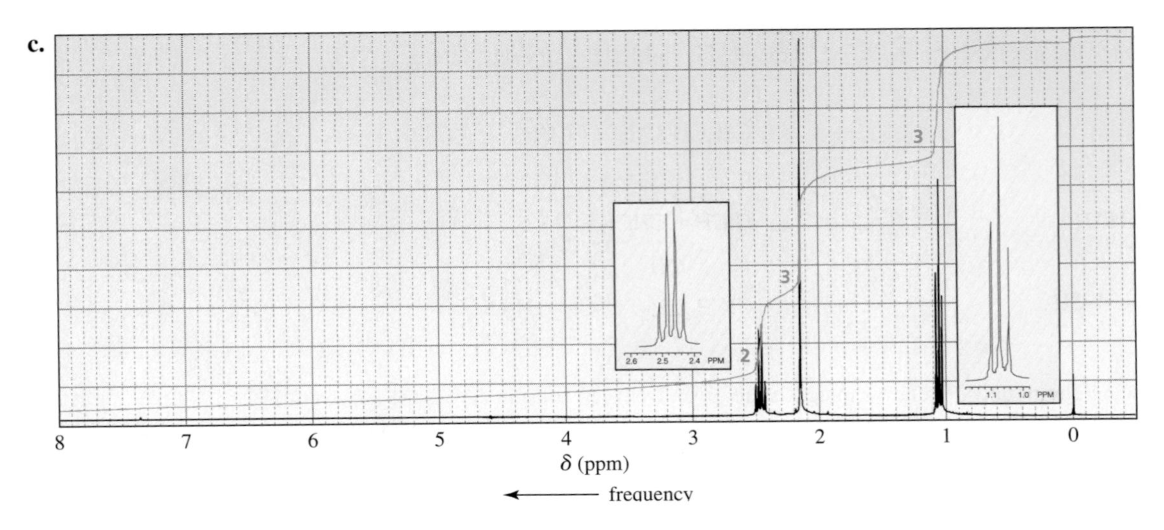

58. How can 1,2-, 1,3-, and 1,4-dinitrobenzene be distinguished by

a. ^{1}H NMR spectroscopy? **b.** ^{13}C NMR spectroscopy?

59. The ^{1}H NMR spectra of three isomers with molecular formula C_4H_9Br are shown here. Which isomer produces which spectrum?

a.

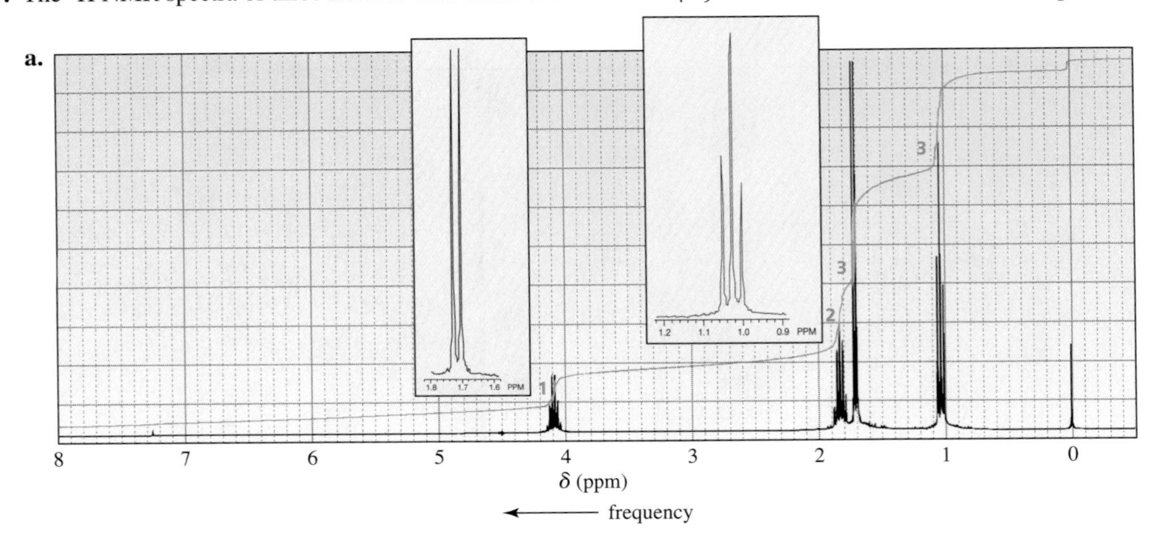

b.

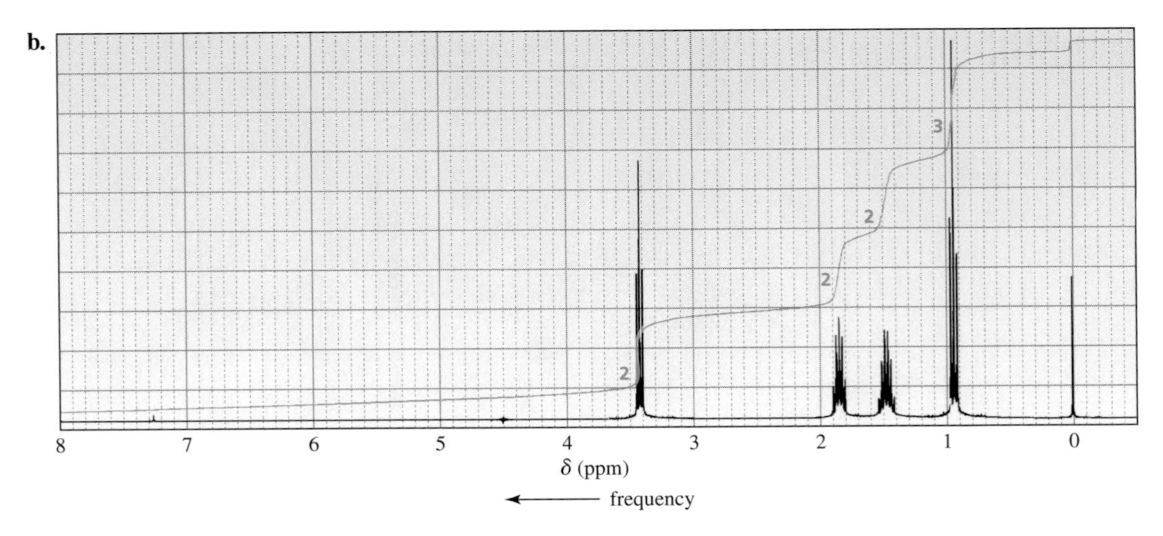

c.

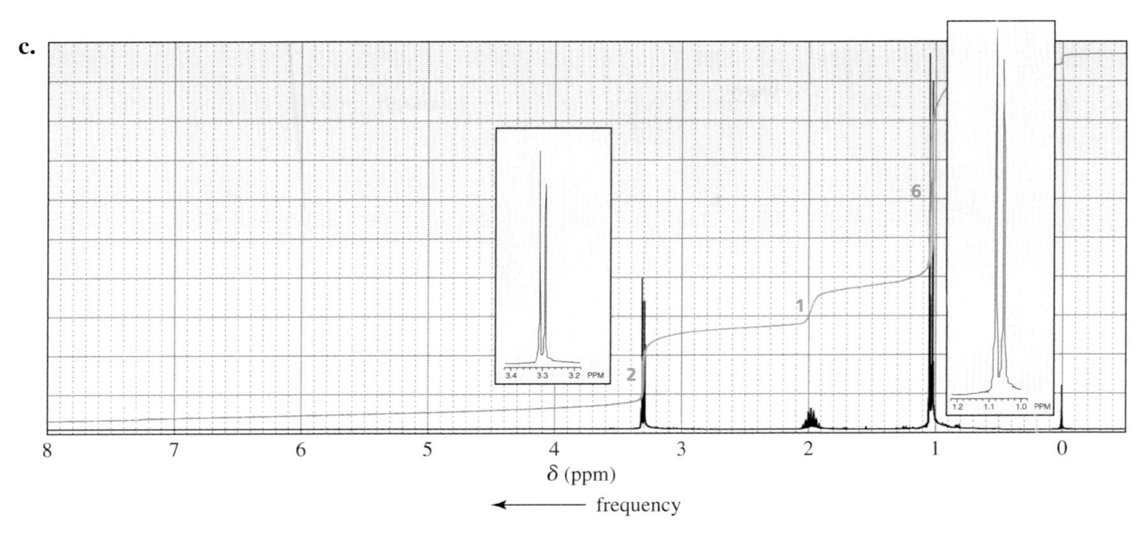

60. How could ^1H NMR distinguish between the compounds in each of the following pairs?

a. $CH_3CH_2CH_2OCH_3$ and $CH_3CH_2OCH_2CH_3$

b. $BrCH_2CH_2CH_2Br$ and $BrCH_2CH_2CH_2NO_2$

c. $CH_3CH(CH_3)-CH(CH_3)CH_3$ and $CH_3C(CH_3)_2CH_2CH_3$

d. $CH_3-C(CH_3)_2-C(=O)-OCH_3$ and $CH_3-C(OCH_3)_2-CH_3$

61. Identify each of the following compounds from the ^1H NMR data and molecular formula. The number of hydrogens responsible for each signal is shown in parentheses.

a. $C_4H_8Br_2$ 1.97 ppm (6) singlet
3.89 ppm (2) singlet

b. C_8H_9Br 2.01 ppm (3) doublet
5.14 ppm (1) quartet
7.35 ppm (5) broad singlet

c. $C_5H_{10}O_2$ 1.15 ppm (3) triplet
1.25 ppm (3) triplet
2.33 ppm (2) quartet
4.13 ppm (2) quartet

62. Compound A, with molecular formula C_4H_9Cl, shows two signals in its ^{13}C NMR spectrum. Compound B, an isomer of compound A, shows four signals, and in the proton-coupled mode, the signal at the highest frequency is a doublet. Identify compounds A and B.

63. How could ^1H NMR spectroscopy distinguish between the compounds in each of the following pairs?

a.

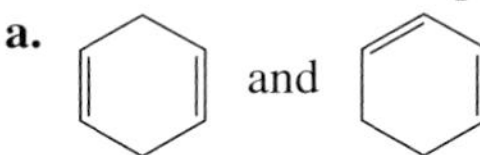

b. $CH_3-C_6H_4-CH_3$ and $CH_3-C_6H_4-OCH_3$

64. The ^1H NMR spectra of three isomers with molecular formula $C_7H_{14}O$ are shown here. Which isomer produces which spectrum?

a.

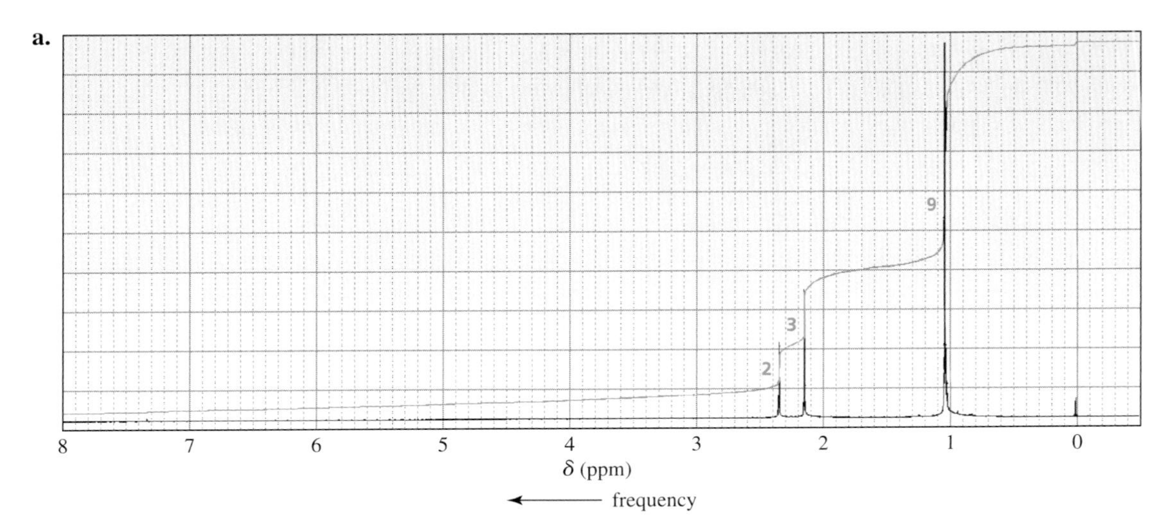

b.

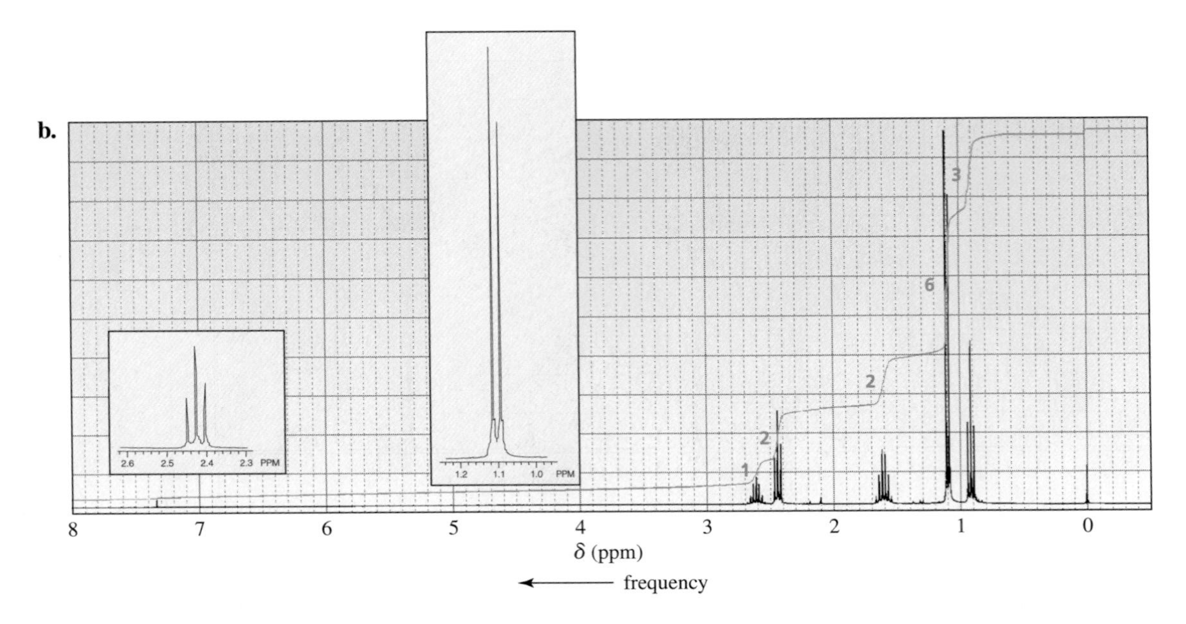

c.

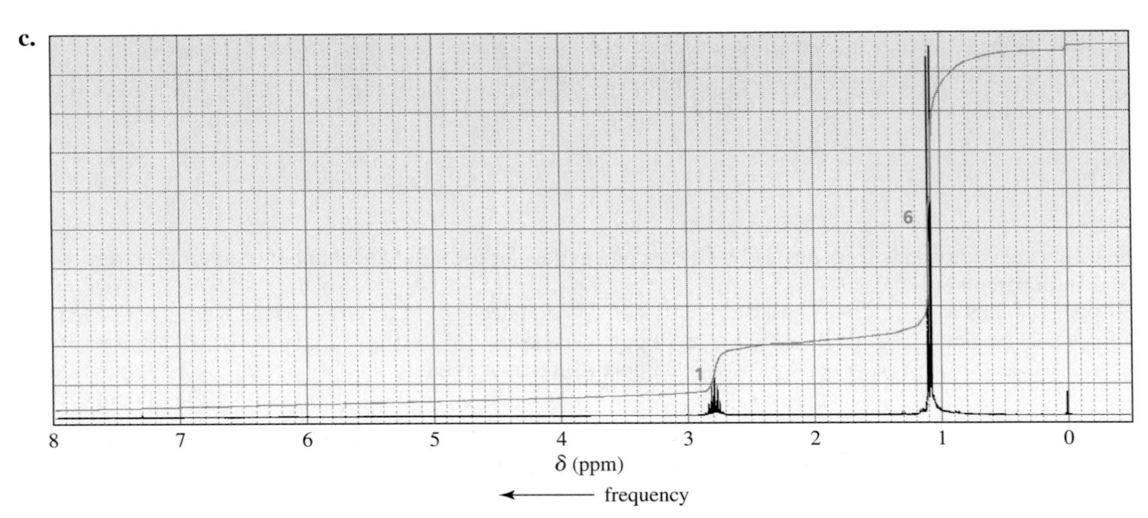

65. How could you use ^1H NMR spectroscopy to distinguish the following esters?

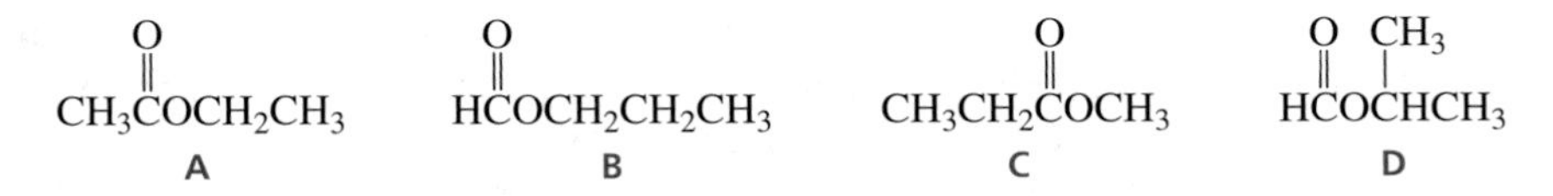

66. Identify the compound with molecular formula $C_7H_{14}O$ that gives the following proton-coupled ^{13}C NMR spectrum.

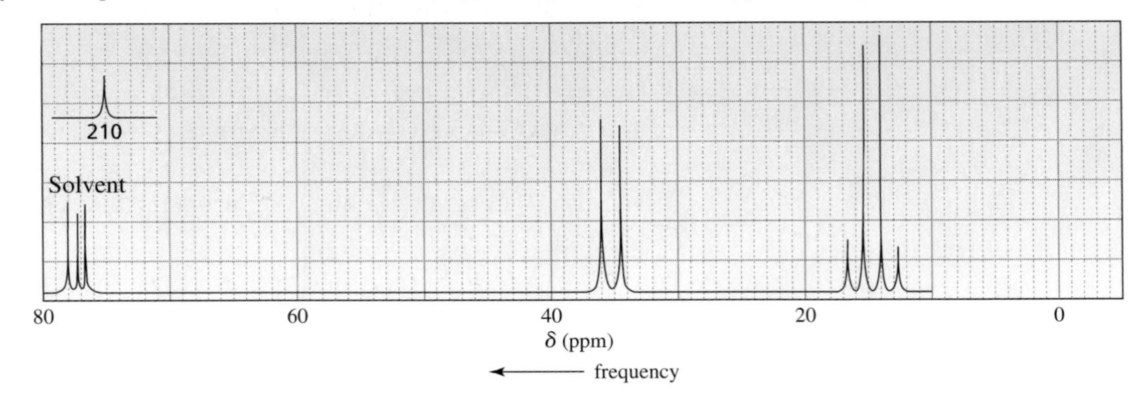

67. An alkyl halide reacts with an alkoxide ion to form a compound whose ^{1}H NMR spectrum is shown here. Identify the alkyl halide and the alkoxide ion.

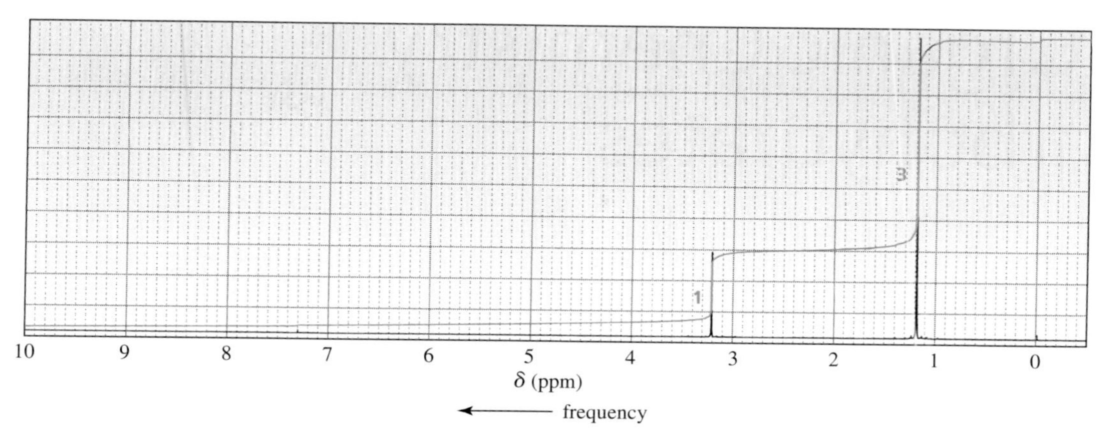

68. Determine the structure of the following unknown compound based on its molecular formula and its IR and ^{1}H NMR spectra.

$C_6H_{12}O_2$

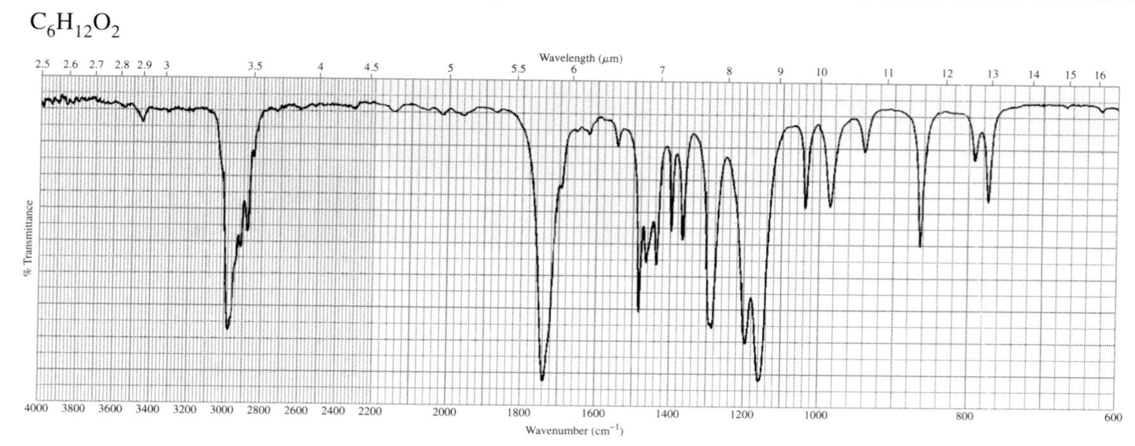

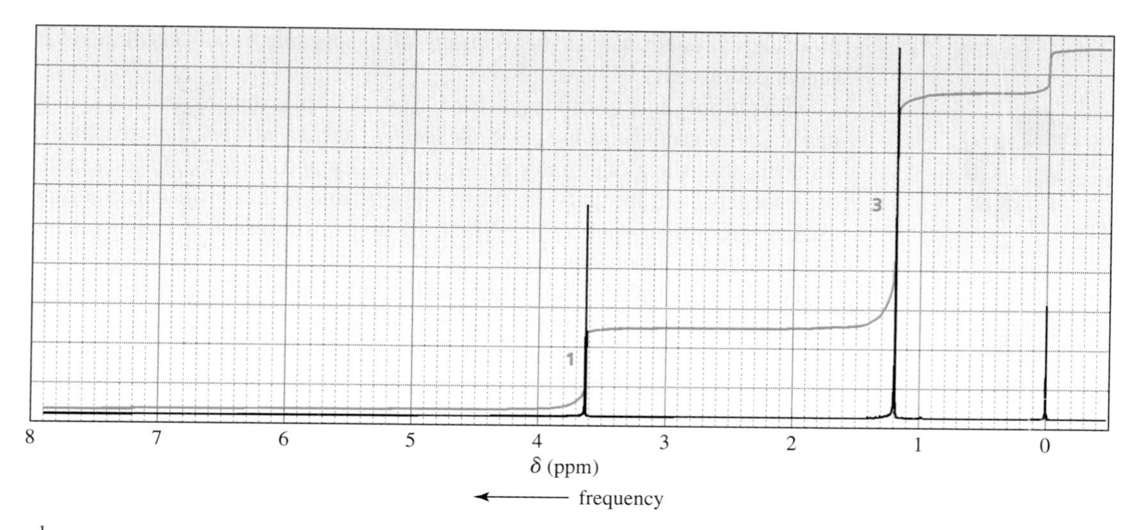

69. How could ^{1}H NMR be used to prove that the addition of HBr to propene follows the rule that says that the electrophile adds to the sp^2 carbon bonded to the greater number of hydrogens (Section 5.3)?

CHAPTER 15

The Organic Chemistry of Carbohydrates

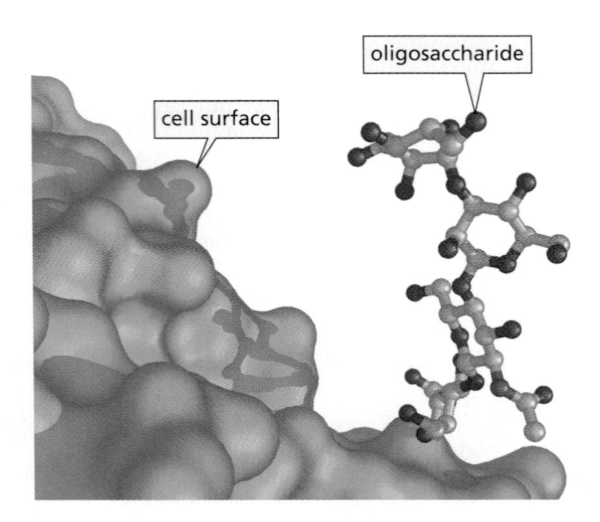

Bioorganic compounds are organic compounds found in biological systems. Their structures can be quite complex, and yet their reactivity is governed by the same principles as the comparatively simple organic molecules we have discussed so far. The organic reactions that chemists carry out in the laboratory are in many ways just like those performed by nature inside a living cell. In other words, bioorganic reactions can be thought of as organic reactions that take place in tiny flasks called cells.

D-glucose

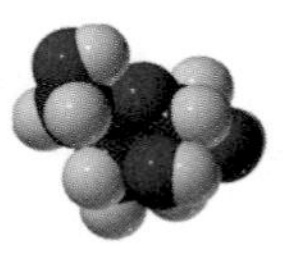
D-fructose

Most bioorganic compounds have more complicated structures than those of the organic compounds you are used to seeing, but do not let the structures fool you into thinking that their chemistry must be equally complicated. One reason the structures of bioorganic compounds are more complicated is that bioorganic compounds must be able to recognize each other. Much of their structure is for that purpose, a function called **molecular recognition**.

The first group of bioorganic compounds we will look at are *carbohydrates*—the most abundant class of compounds in the biological world, making up more than 50% of the dry weight of the Earth's biomass. Carbohydrates are important constituents of all living organisms and have a variety of different functions. Some are important structural components of cells; others act as recognition sites on cell surfaces. For example, the first event in any of our lives was a sperm recognizing a carbohydrate on the outer surface of an egg. Other carbohydrates serve as a major source of metabolic energy. The leaves, fruits, seeds, stems, and roots of plants, for instance, contain carbohydrates that plants use for their own metabolic needs but that can also serve the metabolic needs of the animals that eat the plants.

Carbohydrates are polyhydroxy aldehydes such as D-glucose, polyhydroxy ketones such as D-fructose, and compounds such as sucrose formed by linking polyhydroxy aldehydes or polyhydroxy ketones together (Section 15.13). The chemical structures of carbohydrates are commonly represented by *wedge-and-dash structures* or by *Fischer projections*. A **Fischer projection** represents an asymmetric center as the point of intersection of two perpendicular lines. Horizontal lines represent the bonds that project out of the plane of the paper toward the viewer, and vertical lines represent the bonds that extend back from the plane of the paper away from the viewer.

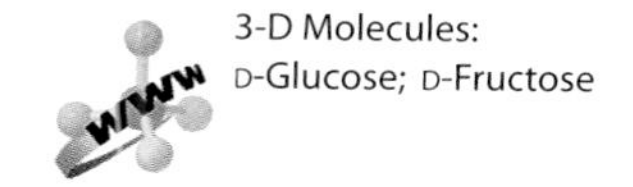


```
   HC=O              HC=O                 CH2OH               CH2OH
    |                 |                     |                   |
 H—C—OH            H—+—OH                  C=O                 C=O
    |                 |                     |                   |
HO—C—H            HO—+—H               HO—C—H              HO—+—H
    |                 |                     |                   |
 H—C—OH            H—+—OH                H—C—OH              H—+—OH
    |                 |                     |                   |
 H—C—OH            H—+—OH                H—C—OH              H—+—OH
    |                 |                     |                   |
  CH2OH             CH2OH                 CH2OH               CH2OH
```

wedge-and-dash structure | Fischer projection | wedge-and-dash structure | Fischer projection

D-glucose
a polyhydroxy aldehyde

D-fructose
a polyhydroxy ketone

The most abundant carbohydrate in nature is D-glucose. Animals obtain D-glucose from food that contains D-glucose, such as plants. Plants produce D-glucose by *photosynthesis*. During photosynthesis, plants take up water through their roots and use carbon dioxide from the air to synthesize D-glucose and oxygen. Because photosynthesis is the reverse of the process used by organisms to obtain energy—specifically, the oxidation of D-glucose to carbon dioxide and water—plants require energy to carry out photosynthesis. Plants obtain the energy they need for photosynthesis from sunlight, captured by chlorophyll molecules in green plants. Photosynthesis uses the CO_2 that animals exhale as waste and generates the O_2 that animals inhale to sustain life. Nearly all the oxygen in the atmosphere has been released by photosynthetic processes.

$$\underset{\text{D-glucose}}{C_6H_{12}O_6} + 6\,O_2 \underset{\text{photosynthesis}}{\overset{\text{oxidation}}{\rightleftharpoons}} 6\,CO_2 + 6\,H_2O + \text{energy}$$

15.1 THE CLASSIFICATION OF CARBOHYDRATES

The terms *carbohydrate, saccharide*, and *sugar* are used interchangeably. "Saccharide" comes from the word for "sugar" in several early languages (*sarkara* in Sanskrit, *sakcharon* in Greek, and *saccharum* in Latin).

There are two classes of carbohydrates: *simple carbohydrates* and *complex carbohydrates*. **Simple carbohydrates** are **monosaccharides** (single sugars), whereas **complex carbohydrates** contain 2 or more monosaccharides linked together. **Disaccharides** have 2 monosaccharides linked together, **oligosaccharides** have 3 to 10 (*oligos* is Greek for "few"), and **polysaccharides** have more than 10. Disaccharides, oligosaccharides, and polysaccharides can be broken down to monosaccharides by hydrolysis.

a monosaccharide subunit

$$\underset{\text{polysaccharide}}{—M—M—M—M—M—M—M—M—M—} \xrightarrow{\text{hydrolysis}} \underset{\text{monosaccharide}}{x\,M}$$

A *monosaccharide* can be a polyhydroxy aldehyde such as D-glucose or a polyhydroxy ketone such as D-fructose. Polyhydroxy aldehydes are called **aldoses** ("ald" is for aldehyde; "ose" is the suffix for a sugar), whereas polyhydroxy ketones are called **ketoses**. Monosaccharides are also classified according to the number of carbons they contain: those with three carbons are **trioses**, those with four carbons are **tetroses**, those with five carbons are **pentoses**, and those with six and seven carbons are **hexoses** and **heptoses**, respectively. Therefore, a six-carbon polyhydroxy aldehyde such as D-glucose is an aldohexose, whereas a six-carbon polyhydroxy ketone such as D-fructose is a ketohexose.

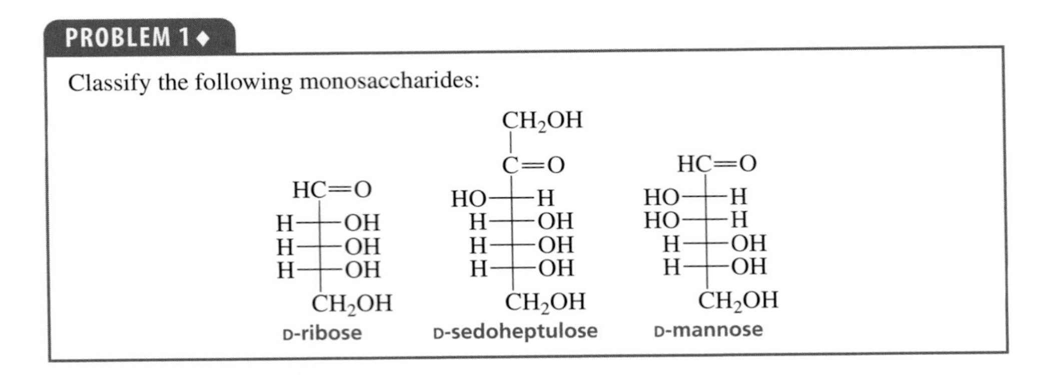

15.2 THE D and L Notation

The smallest aldose, and the only one whose name does not end in "ose," is glyceraldehyde, an aldotriose.

asymmetric center

$HOCH_2CH(OH)CH=O$

glyceraldehyde

A carbon to which four different groups are attached is an asymmetric center.

Because glyceraldehyde has an asymmetric center, it can exist as a pair of enantiomers. We know that the isomer on the left below has the *R* configuration because the arrow drawn from the highest priority substituent (OH) to the next highest priority substituent (CH=O) is clockwise (Section 6.6). To determine the configuration of a Fischer projection, after drawing an arrow from the highest priority to the next highest priority substituent, we need to look at the lowest priority substituent. Only if the lowest priority substituent is on a *vertical bond* does a clockwise arrow specify *R* and a counterclockwise arrow specify *S*. If the smallest priority substituent is on a *horizontal bond*, as it is here, then the answer you get from the direction of the arrow is opposite of the usual answer—clockwise is *S* and counterclockwise is *R*.

clockwise is ***R***

H is on a horizontal bond so counter-clockwise is ***R***

(*R*)-(+)-glyceraldehyde (*S*)-(−)-glyceraldehyde

perspective formulas

(*R*)-(+)-glyceraldehyde (*S*)-(−)-glyceraldehyde

Fischer projections

The notations D and L are used to describe the configurations of carbohydrates. In a Fischer projection of a monosaccharide, the carbonyl group is always placed on top (in the case of aldoses) or as close to the top as possible (in the case of ketoses). Examine the Fischer projection of galactose shown below and note that the compound has four asymmetric centers (C-2, C-3, C-4, and C-5). *If the OH group attached to the bottom-most asymmetric center (the carbon second from the bottom) is on the right, the compound is a D-sugar. If that OH group is on the left, the compound is an* L-*sugar*. Almost all sugars found in nature are D-sugars. Notice that the mirror image of a D-sugar is an L-sugar.

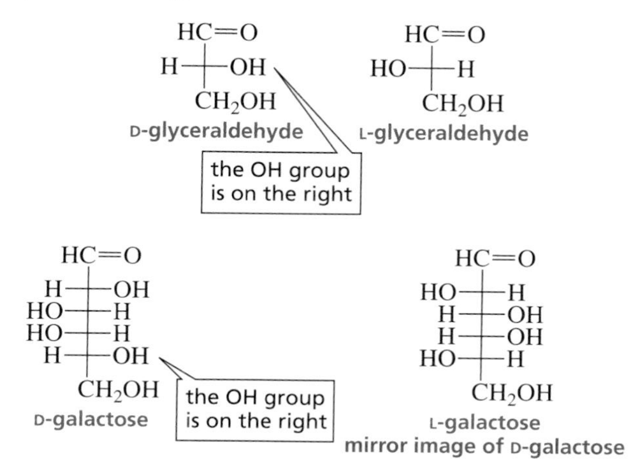

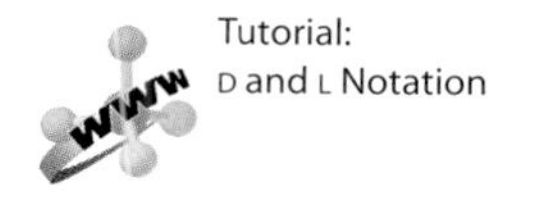

Tutorial:
D and L Notation

Emil Fischer and his colleagues studied carbohydrates in the late nineteenth century, when techniques for determining the configurations of compounds were not available. Fischer arbitrarily assigned the *R*-configuration to the dextrorotatory isomer of glyceraldehyde that we call D-glyceraldehyde. He turned out to be correct: D-glyceraldehyde is (R)-$(+)$-glyceraldehyde, and L-glyceraldehyde is (S)-$(-)$-glyceraldehyde.

Like *R* and *S*, the symbols D and L indicate the configuration of an asymmetric center, but they do not indicate whether the compound rotates the plane of polarization of polarized light to the right $(+)$ or to the left $(+)$ (Section 6.7). For example, D-glyceraldehyde is dextrorotatory, whereas D-lactic acid is levorotatory. In other words, optical rotation, like melting or boiling points, is a physical property of a compound, whereas *R*, *S*, D, and L are conventions humans use to indicate the configuration about an asymmetric center.

HC=O
H—OH
CH_2OH
D-(+)-glyceraldehyde

COOH
H—OH
CH_3
D-(−)-lactic acid

The common name of the monosaccharide, together with the D or L designation, completely defines its structure, because the configurations of all the asymmetric centers are implicit in the common name.

PROBLEM 2

Draw Fischer projections of L-glucose and L-fructose.

15.3 THE CONFIGURATIONS OF ALDOSES

Aldotetroses have two asymmetric centers and therefore four stereoisomers. Two of the stereoisomers are D-sugars and two are L-sugars. The aldotetroses are called erythrose and threose.

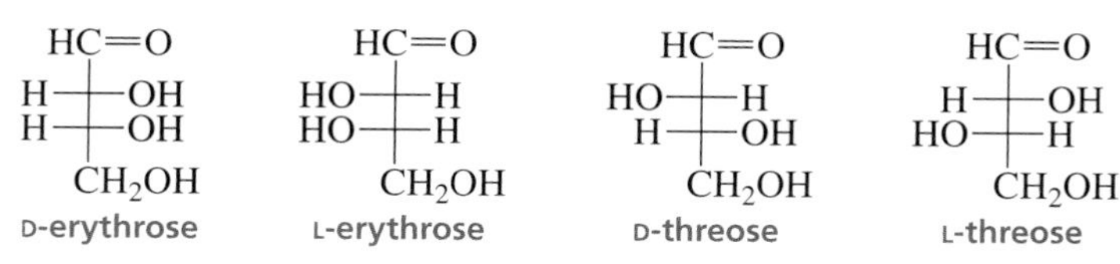

Aldopentoses have three asymmetric centers and therefore eight stereoisomers (four pairs of enantiomers); and aldohexoses have four asymmetric centers and 16 stereoisomers (eight pairs of enantiomers). The four D-aldopentoses and the eight D-aldohexoses are shown in Table 15.1.

Table 15.1 Configurations of the D-Aldoses

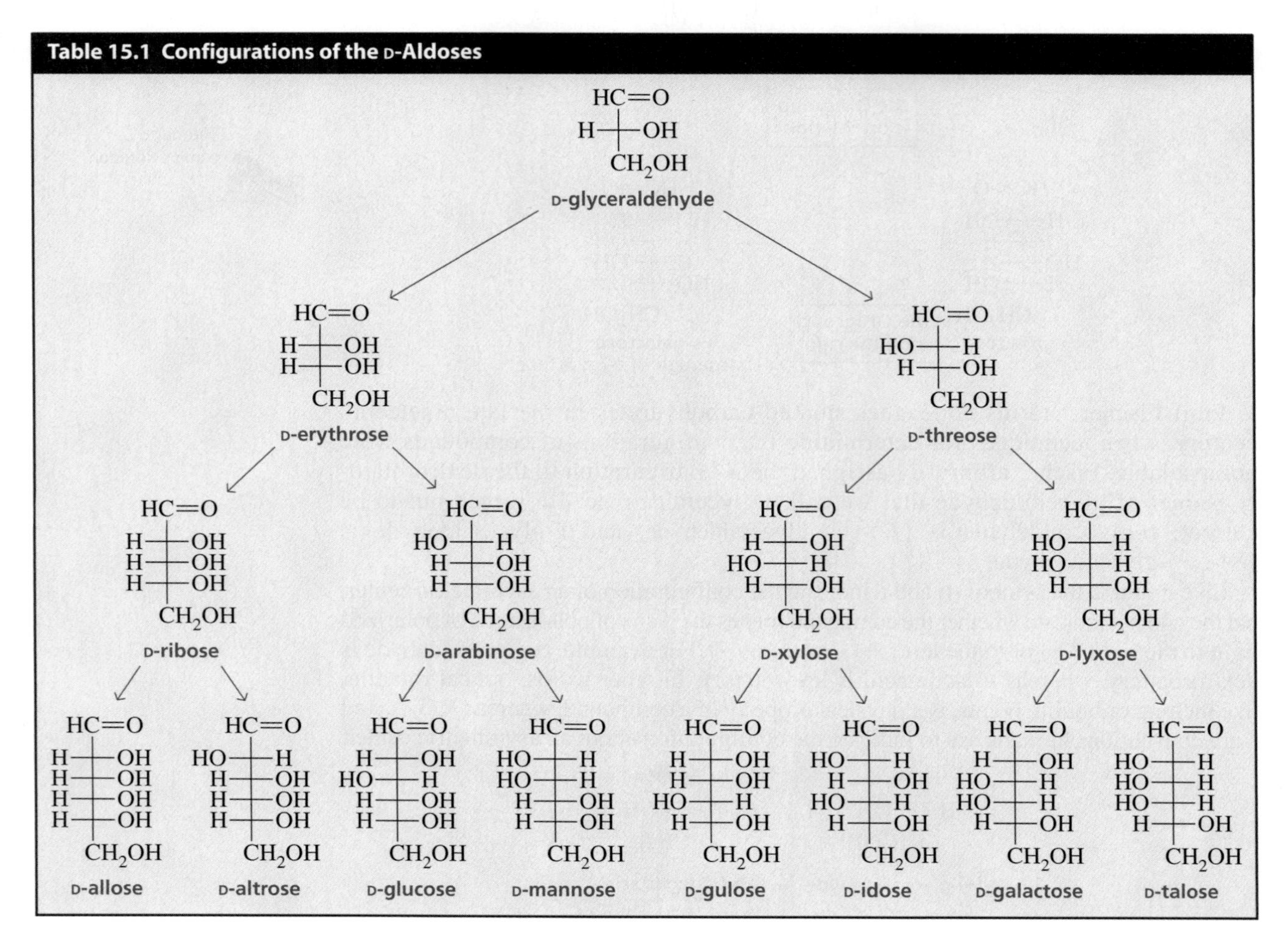

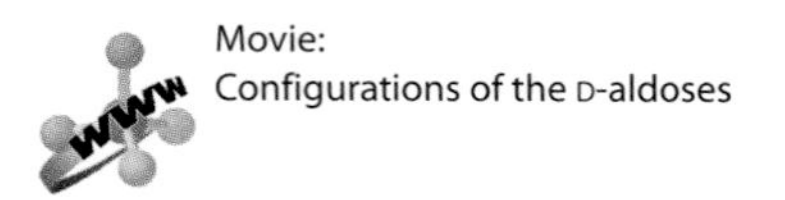
Movie: Configurations of the D-aldoses

Diastereomers that differ in configuration at only one asymmetric center are called **epimers**. For example, D-ribose and D-arabinose are C-2 epimers because they differ in configuration only at C-2; D-idose and D-talose are C-3 epimers.

D-ribose D-arabinose
C-2 epimers

D-idose D-talose
C-3 epimers

D-Mannose is the C-2 epimer of D-glucose.

D-Galactose is the C-4 epimer of D-glucose.

D-Glucose, D-mannose, and D-galactose are the most common aldohexoses in biological systems. An easy way to learn their structures is to memorize the structure of D-glucose and then remember that D-mannose is the C-2 epimer of D-glucose and D-galactose is the C-4 epimer of D-glucose.

PROBLEM 3 ◆

a. What sugar is the C-3 epimer of D-xylose?
b. What sugar is the C-5 epimer of D-allose?

15.4 THE CONFIGURATIONS OF KETOSES

Naturally occurring ketoses have the ketone group in the 2-position. The configurations of the naturally occurring ketoses are shown in Table 15.2. A ketose has one fewer asymmetric center than does an aldose with the same number of carbons. Therefore, a ketose has only half as many stereoisomers as an aldose with the same number of carbons.

Table 15.2 Configurations of the D-Ketoses

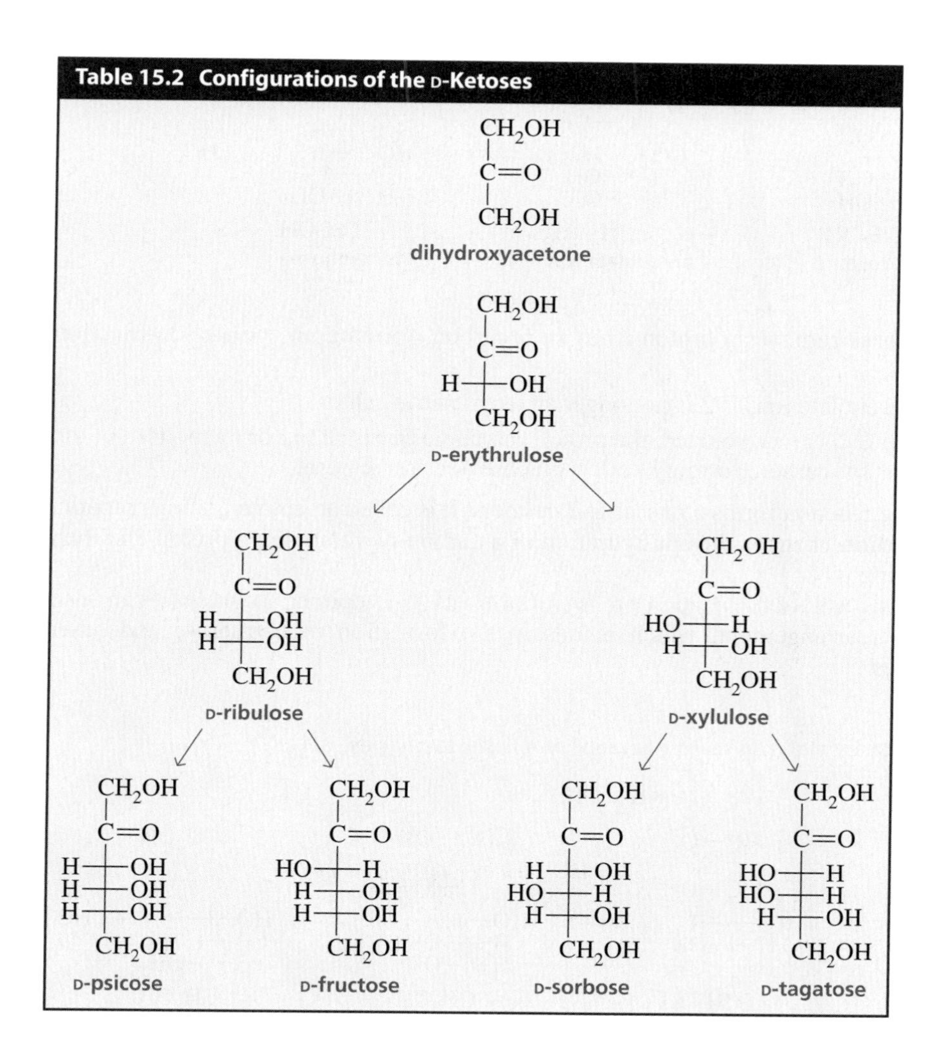

PROBLEM 4 ♦

What sugar is the C-3 epimer of D-fructose?

PROBLEM 5 ♦

How many stereoisomers are possible for

a. a 2-ketoheptose? **b.** an aldoheptose? **c.** a ketotriose?

15.5 THE REACTIONS OF MONOSACCHARIDES IN BASIC SOLUTIONS

Monosaccharides cannot undergo reactions with basic reagents because in a basic solution, a monosaccharide is converted to a complex mixture of polyhydroxy aldehydes and polyhydroxy ketones. Let's first look at what happens to D-glucose in a basic solution, beginning with its conversion to its C-2 epimer.

mechanism for the base-catalyzed epimerization of a monosaccharide

D-glucose ⇌ an enolate ion ⇌ D-mannose + HO^-

- The base removes a proton from an α-carbon, forming an enolate ion (Section 13.3).
- In the enolate ion, C-2 is no longer an asymmetric center.
- When C-2 is reprotonated, the proton can come from the top or the bottom of the planar sp^2 carbon, forming both D-glucose and D-mannose.

Because the reaction forms a pair of C-2 epimers, it is called an epimerization reaction. **Epimerization** changes the configuration of a carbon by removing a proton and then reprotonating it.

In a basic solution, in addition to forming its C-2 epimer, D-glucose can also undergo a rearrangement, which results in the formation of D-fructose and other ketohexoses.

mechanism for the base-catalyzed rearrangement of a monosaccharide

D-glucose ⇌ an enolate ion ⇌ an enediol ⇌ D-fructose + HO^-

- The base removes a proton from an α-carbon, forming an enolate ion (Section 13.3).
- The enolate ion can be protonated on the oxygen (instead of on the carbon as in the mechanism shown above) to form an enediol.
- The enediol has two OH groups that can form a carbonyl group. Tautomerization of the OH at C-1 reforms D-glucose (Section 5.5); tautomerization of the OH group at C-2 forms D-fructose.

Another rearrangement, initiated by a base removing a proton from C-3 of D-fructose, forms an enediol that can tautomerize to give a ketose with the carbonyl group at C-2 or C-3. Thus, the carbonyl group can be moved up and down the chain.

PROBLEM 6 ♦

When D-tagatose is added to a basic aqueous solution, an equilibrium mixture of three monosaccharides is obtained. What are these monosaccharides?

PROBLEM 7

Write the mechanism for the base-catalyzed conversion of D-fructose into D-glucose and D-mannose.

15.6 THE OXIDATION–REDUCTION REACTIONS OF MONOSACCHARIDES

Because they contain *alcohol* functional groups and *aldehyde* or *ketone* functional groups, the reactions of monosaccharides are an extension of what you have already learned about the reactions of alcohols, aldehydes, and ketones. For example, an aldehyde group in a monosaccharide can be oxidized or reduced and can react with nucleophiles to form imines, hemiacetals, and acetals. As you read this section and the ones that follow, dealing with the reactions of monosaccharides, you will find cross-references to earlier discussions of simpler organic compounds undergoing the same reactions. Go back and look at these earlier discussions when they are mentioned; they will make learning about carbohydrates a lot easier.

Reduction Reactions

The carbonyl group in aldoses and ketoses can be reduced by sodium borohydride ($NaBH_4$; Section 12.6). The product of the reduction is a polyalcohol, known as an **alditol**. Reduction of an aldose forms one alditol. For example, the reduction of D-mannose forms D-mannitol, the alditol found in mushrooms, olives, and onions. Reduction of a ketose forms two alditols because the reaction creates a new asymmetric center in the product. For example, the reduction of D-fructose forms D-mannitol and D-glucitol, the C-2 epimer of D-mannitol. D-Glucitol—also called sorbitol—is about 60% as sweet as sucrose. It is found in plums, pears, cherries, and berries and is used as a sugar substitute in the manufacture of candy.

HC=O
HO—H
HO—H
H—OH
H—OH
CH_2OH
D-mannose

→ 1. $NaBH_4$ 2. H_3O^+

CH_2OH
HO—H
HO—H
H—OH
H—OH
CH_2OH
D-mannitol
an alditol

← 1. $NaBH_4$ 2. H_3O^+

CH_2OH
C=O
HO—H
H—OH
H—OH
CH_2OH
D-fructose

→ 1. $NaBH_4$ 2. H_3O^+

CH_2OH
H—OH
HO—H
H—OH
H—OH
CH_2OH
D-glucitol
an alditol

PROBLEM 8 ♦

What products are obtained from the reduction of
a. D-idose? **b.** D-sorbose?

PROBLEM 9 ♦

What other monosaccharide is reduced only to the alditol obtained from the reduction of D-galactose?

PROBLEM 10 ♦

What monosaccharide is reduced to two alditols, one of which is the alditol obtained from the reduction of D-talose?

Oxidation Reactions

Aldoses can be distinguished from ketoses by observing what happens to the color of an aqueous solution of bromine when it is added to the sugar. Br_2 is a mild oxidizing agent and easily oxidizes an aldehyde group, but it cannot oxidize ketones or alcohols. Consequently, if a small amount of an aqueous solution of Br_2 is added to an unknown monosaccharide, the reddish-brown color of Br_2 will disappear if the monosaccharide is an aldose because Br_2 will be reduced to Br^-, which is colorless. If the red color persists, indicating no reaction with Br_2, the monosaccharide is a ketose.

HC=O
H—OH
HO—H
H—OH
H—OH
CH_2OH
D-glucose

\+ Br_2 (red) $\xrightarrow{H_2O}$

COOH
H—OH
HO—H
H—OH
H—OH
CH_2OH
D-gluconic acid
an aldonic acid

\+ 2 Br^- (colorless)

If an oxidizing agent stronger than Br_2 is used (such as HNO_3), the primary alcohol will also be oxidized.

HC=O
H—OH
HO—H
H—OH
H—OH
CH_2OH
D-glucose

$\xrightarrow[\Delta]{HNO_3}$

COOH
H—OH
HO—H
H—OH
H—OH
COOH
D-glucaric acid
an aldaric acid

PROBLEM 11 ◆

a. Name an aldohexose other than D-glucose that is oxidized to D-glucaric acid by nitric acid.
b. What is another name for D-glucaric acid?
c. Name another pair of aldohexoses that are oxidized to identical aldaric acids.

MEASURING THE BLOOD GLUCOSE LEVELS OF DIABETICS

Glucose in the bloodstream reacts with an NH_2 group of hemoglobin to form an imine (Section 12.7) that subsequently undergoes an irreversible rearrangement to a more stable α-aminoketone known as hemoglobin-A_{IC}.

HC=O
H—OH
HO—H
H—OH
H—OH
CH_2OH
D-glucose

$\xrightarrow[\text{trace } H^+]{NH_2\text{–hemoglobin}}$

HC=N–hemoglobin
H—OH
HO—H
H—OH
H—OH
CH_2OH

$\xrightarrow{\text{rearrangement}}$

CH_2NH–hemoglobin
C=O
HO—H
H—OH
H—OH
CH_2OH
hemoglobin-A_{Ic}

Insulin is the hormone that regulates the level of glucose—and thus the amount of hemoglobin-A_{IC}—in the blood. Diabetes is a condition in which the body does not produce sufficient insulin or in which the insulin it produces does not function properly. Because people with untreated diabetes have increased blood glucose levels, they also have a higher concentration of hemoglobin-A_{IC} than people without diabetes. Thus, measuring the hemoglobin-A_{IC} level is a way to determine whether the blood glucose level of a diabetic patient is being controlled.

Cataracts, a common complication in diabetics, are caused by the reaction of glucose with the NH_2 group of proteins in the lens of the eye. Some think the arterial rigidity common in old age may be attributable to a similar reaction of glucose with the NH_2 group of proteins.

15.7 LENGTHENING THE CHAIN: THE KILIANI–FISCHER SYNTHESIS

The carbon chain of an aldose can be increased by one carbon in a **Kiliani–Fischer synthesis**. In other words, tetroses can be converted into pentoses, and pentoses can be converted into hexoses.

the modified Kiliani–Fischer synthesis

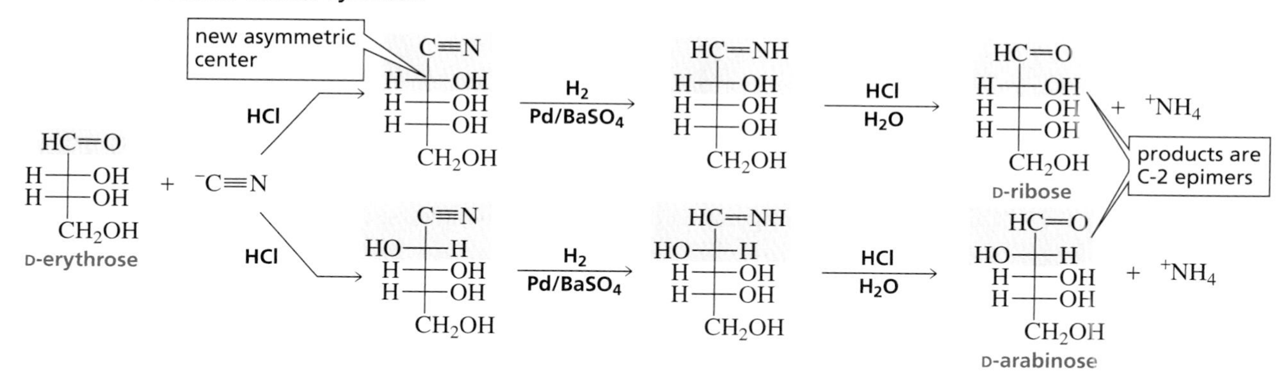

- In the first step of the synthesis, cyanide ion adds to the carbonyl group. This reaction converts the carbonyl carbon in the starting material to an asymmetric center. (The OH bonded to C-2 in the product can be on the right or on the left in the Fischer projection.) Consequently, two products are formed that differ only in their configuration at C-2. The configurations of the other asymmetric centers do not change because no bond to any of the asymmetric centers is broken during the course of the reaction.
- The C≡N bond is reduced to an imine, using a partially deactivated palladium catalyst so that the imine is not further reduced to an amine (Section 5.12). The two imines are hydrolyzed to aldoses (Section 12.7).

Notice that the synthesis leads to a pair of C-2 epimers because the first step of the synthesis converts the carbonyl carbon in the starting material to an asymmetric center.

BIOGRAPHY

Heinrich Kiliani (1855–1945) *was born in Germany. He received a Ph.D. from the University of Munich, studying under Professor Emil Erlenmeyer. Kiliani became a professor of chemistry at the University of Freiburg.*

The Kiliani–Fischer synthesis leads to a pair of C-2 epimers.

PROBLEM 12 ♦

What monosaccharides would be formed in a Kiliani–Fischer synthesis starting with each of these compounds?

a. D-xylose **b.** L-threose

15.8 STEREOCHEMISTRY OF GLUCOSE: THE FISCHER PROOF

In 1891, Emil Fischer (Sections 15.2 and 15.7) determined the stereochemistry of glucose using one of the most brilliant examples of reasoning in the history of chemistry. He chose (+)-glucose for his study because it is the most common monosaccharide found in nature.

Fischer knew that (+)-glucose is an aldohexose, but 16 different structures can be written for an aldohexose. Which of them represents the structure of (+)-glucose? The 16 stereoisomers of an aldose are actually eight pairs of enantiomers, so if you know the structures of one set of eight, you automatically know the structures of the other set of eight. Therefore, Fischer needed to consider only one set of eight. He considered the eight stereoisomers that have their C-5 OH group on the right in the Fischer projection (the stereoisomers shown below that we now call the D-sugars).

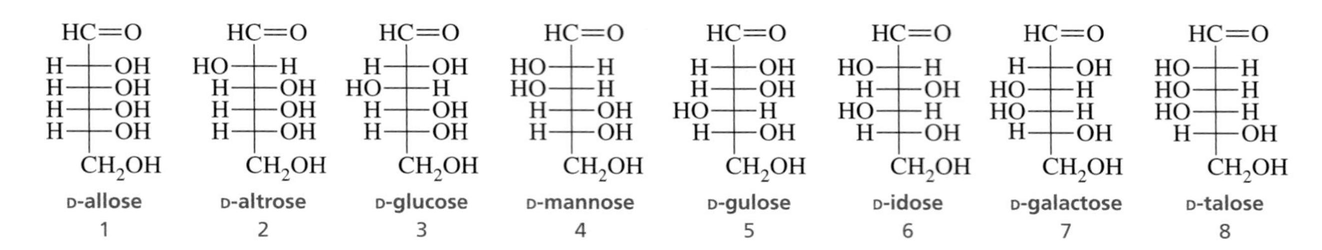

BIOGRAPHY

Emil Fischer (1852–1919) *was born in a village near Cologne, Germany. He became a chemist against the wishes of his father, a successful merchant, who wanted him to enter the family business. Fischer was a professor of chemistry at the Universities of Erlangen, Würzburg, and Berlin and in 1902 received the Nobel Prize in chemistry for his work on sugars. During World War I, he organized German chemical production. Two of his three sons died in that war.*

One of these is (+)-glucose, and its mirror image is (−)-glucose. It would not be possible to determine whether (+)-glucose was D-glucose or L-glucose until 1951. Fischer used the following information to determine glucose's stereochemistry—that is, to determine the configuration of each of its asymmetric centers.

1. When the Kiliani–Fischer synthesis is performed on the sugar known as (−)-arabinose, the two sugars known as (+)-glucose and (+)-mannose are obtained. This means that (+)-glucose and (+)-mannose are C-2 epimers; in other words, they have the same configuration at C-3, C-4, and C-5. Consequently, (+)-glucose and (+)-mannose have to be one of the following pairs: sugars 1 and 2, 3 and 4, 5 and 6, or 7 and 8.
2. (+)-Glucose and (+)-mannose are both oxidized by nitric acid to optically active aldaric acids. The aldaric acids of sugars 1 and 7 would not be optically active because each has a plane of symmetry. (A compound containing a plane of symmetry is achiral, meaning it has a superimposable mirror image; see Section 6.10.) Excluding sugars 1 and 7 means that (+)-glucose and (+)-mannose must be sugars 3 and 4 or 5 and 6.
3. Since (+)-glucose and (+)-mannose are the products obtained when the Kiliani–Fischer synthesis is carried out on (−)-arabinose, Fischer knew that if (−)-arabinose has the structure shown below on the left, (+)-glucose and (+)-mannose are sugars 3 and 4. On the other hand, if (−)-arabinose has the structure shown on the right, (+)-glucose and (+)-mannose are sugars 5 and 6.

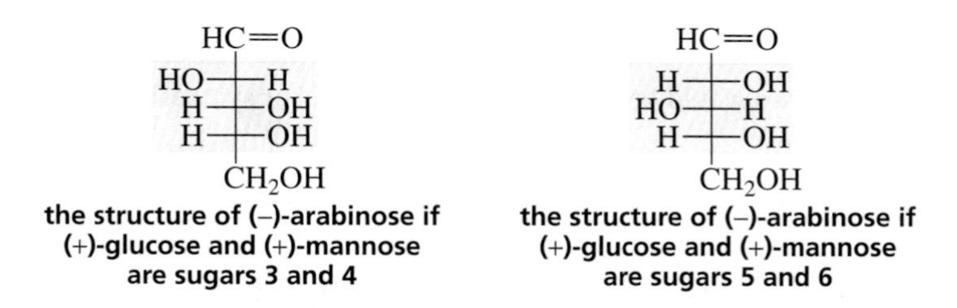

the structure of (−)-arabinose if (+)-glucose and (+)-mannose are sugars 3 and 4   **the structure of (−)-arabinose if (+)-glucose and (+)-mannose are sugars 5 and 6**

When (−)-arabinose is oxidized with nitric acid, the aldaric acid that results is optically active. This means that the aldaric acid does *not* have a plane of symmetry. Therefore, (−)-arabinose must have the structure shown on the left because the aldaric acid of the sugar on the right has a plane of symmetry. Thus, (+)-glucose and (+)-mannose are represented by sugars 3 and 4.

4. Now the only question remaining was whether (+)-glucose was sugar 3 or sugar 4. To answer this question, Fischer had to develop a chemical method for interchanging the aldehyde and primary alcohol groups of an aldohexose. When he chemically interchanged those groups on the sugar known as (+)-glucose, he obtained an aldohexose that was different from (+)-glucose. When he chemically interchanged the

aldehyde and primary alcohol groups of (+)-mannose, he still had (+)-mannose. Therefore, he concluded that (+)-glucose is sugar 3 because interchanging its aldehyde and primary alcohol groups leads to a different sugar (L-gulose).

HC=O
H—OH
HO—H
H—OH
H—OH
CH_2OH
D-glucose

reverse the aldehyde and primary alcohol groups →

CH_2OH
H—OH
HO—H
H—OH
H—OH
HC=O
L-gulose
drawn upside down

=

HC=O
HO—H
HO—H
H—OH
HO—H
CH_2OH
L-gulose

If (+)-glucose is sugar 3, (+)-mannose must be sugar 4. As predicted, when the aldehyde and primary alcohol groups of sugar 4 are interchanged, the same sugar is obtained.

HC=O
HO—H
HO—H
H—OH
H—OH
CH_2OH
D-mannose

reverse the aldehyde and primary alcohol groups →

CH_2OH
HO—H
HO—H
H—OH
H—OH
HC=O
D-mannose
drawn upside down

=

HC=O
HO—H
HO—H
H—OH
H—OH
CH_2OH
D-mannose

Using similar reasoning, Fischer went on to determine the stereochemistry of the other aldohexoses. He received the Nobel Prize in chemistry in 1902 for this achievement. His original guess that (+)-glucose is a D-sugar was later shown to be correct, so all of his structures are correct. If he had been wrong and (+)-glucose had been an L-sugar, his contribution to the stereochemistry of aldoses would still have had the same importance, but all his stereochemical assignments would have had to be reversed.

GLUCOSE/DEXTROSE

André Dumas first used the term *glucose* in 1838 to refer to the sweet compound that comes from honey and grapes. Later, Kekulé (Section 7.1) decided that it should be called dextrose because it was dextrorotatory. When Fischer studied the sugar, he called it glucose, and chemists have called it glucose ever since, although dextrose is often found on food labels.

Jean-Baptiste-André Dumas (1800–1884), *born in France, was apprenticed to an apothecary but left that position to study chemistry in Switzerland. He became a professor of chemistry at the University of Paris and at the Collège de France and was the first French chemist to teach laboratory courses. In 1848, Dumas left science for a political career. He became a senator, master of the French mint, and mayor of Paris.*

15.9 MONOSACCHARIDES FORM CYCLIC HEMIACETALS

D-Glucose exists in three different forms: the open-chain form that we have been discussing (Table 15.1) and two cyclic forms—α-D-glucose and β-D-glucose. We know that the two cyclic forms are different because they have different melting points and different specific rotations (Section 6.8).

How can D-glucose exist in a cyclic form? In Section 12.9, we saw that an aldehyde reacts with an equivalent of an alcohol to form a hemiacetal. A monosaccharide such as D-glucose has an aldehyde group and several alcohol groups. The alcohol group bonded to C-5 of D-glucose reacts with the aldehyde group. The reaction forms two cyclic (six-membered ring) hemiacetals.

Groups on the *right* in a Fischer projection are *down* in a planar cyclic structure.

Groups on the *left* in a Fischer projection are *up* in a planar cyclic structure.

To see that the OH group on C-5 is in the proper position to attack the aldehyde group, we need to convert the Fischer projection of D-glucose to a flat ring structure. To do this, draw the primary alcohol group *up* from the back left-hand corner. Groups on the *right* in a Fischer projection are *down* in the cyclic structure, whereas groups on the *left* in a Fischer projection are *up* in the cyclic structure.

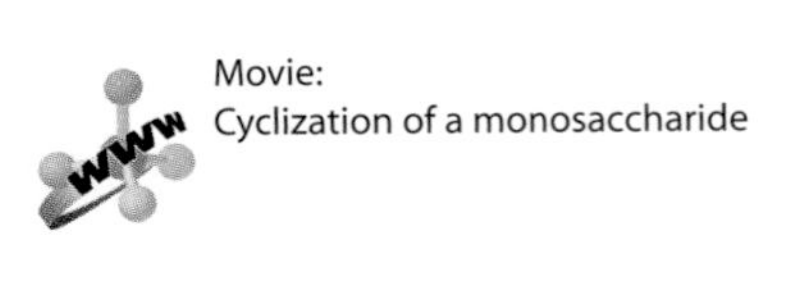

D-glucose

D-glucose
0.02%

anomeric carbon; new asymmetric center

α-D-glucose
36%
a Haworth projection

anomeric carbon; new asymmetric center

β-D-glucose
64%
a Haworth projection

Why are there two different cyclic forms? Two different hemiacetals are formed because the carbonyl carbon of the open-chain aldehyde becomes a new asymmetric center in the cyclic hemiacetal. If the OH group bonded to the new asymmetric center (C-1) is down, the hemiacetal is α-D-glucose; if the OH group is up, the hemiacetal is β-D-glucose. The mechanism for cyclic hemiacetal formation is the same as the mechanism for hemiacetal formation between individual aldehyde and alcohol molecules (Section 12.9).

α-D-Glucose and β-D-glucose are anomers. **Anomers** are two sugars that differ in configuration only at the carbon that was the carbonyl carbon in the open-chain form. This carbon is called the **anomeric carbon**. The prefixes α- and β- denote the configuration about the anomeric carbon. Anomers, like epimers, differ in configuration at only one carbon. Notice that the anomeric carbon is the only carbon in the molecule that is bonded to two oxygens.

In an aqueous solution, the open-chain aldehyde is in equilibrium with the two cyclic hemiacetals. However, because formation of the cyclic hemiacetals proceeds nearly to completion, very little glucose exists in the open-chain form (about 0.02%). Even so, the sugar still undergoes the reactions discussed in the previous sections because the reagents react with the small amount of open-chain aldehyde that is present. As the open-chain aldehyde reacts, the equilibrium shifts to produce more open-chain aldehyde, which can then undergo reaction. Eventually, all the glucose molecules react by way of the open-chain aldehyde.

When crystals of pure α-D-glucose are dissolved in water, the specific rotation gradually changes from +112.2 to +52.7. When crystals of pure β-D-glucose are dissolved in water, the specific rotation gradually changes from +18.7 to +52.7. This change in rotation occurs because, in water the hemiacetal opens to form the aldehyde and, when the aldehyde recyclizes, both α-D-glucose and β-D-glucose can be formed. Eventually, the three forms of glucose reach equilibrium concentrations. The specific rotation of the equilibrium mixture is +52.7. This is why the same specific rotation results, whether the crystals originally dissolved in water are α-D-glucose or β-D-glucose. A slow change in optical rotation to an equilibrium value is called **mutarotation**.

If an aldose can form a five- or a six-membered ring, it will exist predominantly as a cyclic hemiacetal in solution. D-Ribose is an example of an aldose that forms five-membered ring hemiacetals: α-D-ribose and β-D-ribose.

D-ribose = **D-ribose** ⇌ **α-D-ribose** + **β-D-ribose**

Six-membered-ring sugars are called **pyranoses**, and five-membered-ring sugars are called **furanoses**. These names come from *pyran* and *furan*, respectively, the names of the five- and six-membered-ring cyclic ethers shown in the margin. Consequently, α-D-glucose is also called α-D-glucopyranose, and α-D-ribose is also called α-D-ribofuranose. The prefix α- indicates the configuration about the anomeric carbon, and "pyranose" or "furanose" indicates the size of the ring.

pyran

furan

α-D-glucose
α-D-glucopyranose

α-D-ribose
α-D-ribofuranose

Ketoses also exist in solution predominantly in cyclic forms. For example, D-fructose forms a five-membered-ring hemiketal as a result of reaction of its OH group at C-5 with its ketone carbonyl group (Section 12.9). If the OH group bonded to the new asymmetric center (C-2) is down, the hemiketal is α-D-glucose; if the OH group is up, the hemiketal is β-D-glucose.

α-D-fructose
α-D-fructofuranose

β-D-fructose
β-D-fructofuranose

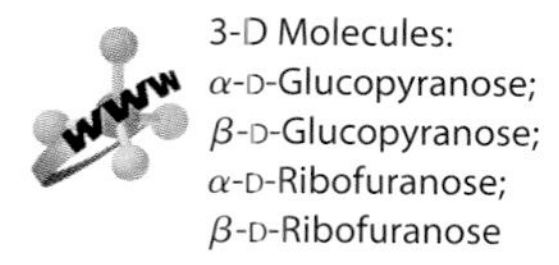

WWW
3-D Molecules:
α-D-Glucopyranose;
β-D-Glucopyranose;
α-D-Ribofuranose;
β-D-Ribofuranose

The planar representations are useful because they show clearly whether the OH groups on the ring are cis or trans to each other. Five-membered rings are nearly planar, so furanoses are represented fairly accurately by the planar structures. The planar representations, however, are structurally misleading for pyranoses because a six-membered ring is not flat but exists preferentially in a chair conformation (Section 3.10).

PROBLEM 13 *SOLVED*

4-Hydroxyaldehydes and 5-hydroxyaldehydes exist primarily as cyclic hemiacetals. Draw the structure of the cyclic hemiacetal formed by each of the following:

a. 4-hydroxybutanal **b.** 5-hydroxypentanal

Solution to 13a Draw the reactant with the alcohol and carbonyl groups on the same side of the molecule. Then look to see what size ring will form. Two cyclic products are obtained because the carbonyl carbon of the reactant has been converted into a new asymmetric center in the product.

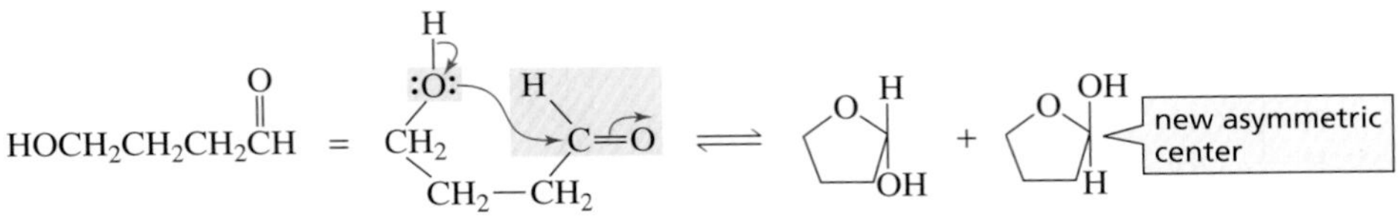

$HOCH_2CH_2CH_2CH{=}O$ = [cyclization scheme] ⇌ [two cyclic products]

new asymmetric center

15.10 GLUCOSE IS THE MOST STABLE ALDOHEXOSE

Drawing D-glucose in its chair conformation shows why it is the most common aldohexose in nature. To convert the planar representation of D-glucose into a chair conformation, start by drawing the chair so that the backrest is on the left and the footrest is on the right. Then place the ring oxygen at the back right-hand corner and the primary alcohol group in the equatorial position. The primary alcohol group is the largest of all the substituents, and large substituents are more stable in the equatorial position because there is less steric strain in that position (Section 3.11). Because the OH group bonded to C-4 is trans to the primary alcohol group (this is easily seen in the flat six-membered ring representation), the C-4 OH group is also in the equatorial position. (Recall from Section 3.12 that 1,2-diequatorial substituents are trans to one another.) The C-3 OH group is trans to the C-4 OH group, so the C-3 OH group is also in the equatorial position. As you move around the ring, you will find that all the OH substituents in β-D-glucose are in equatorial positions. The axial positions are all occupied by hydrogens, which require little space and therefore experience little steric strain. No other aldohexose exists in such a strain-free conformation. This means that β-D-glucose is the most stable of all the D-aldohexoses, so we should not be surprised that it is the most prevalent aldohexose in nature.

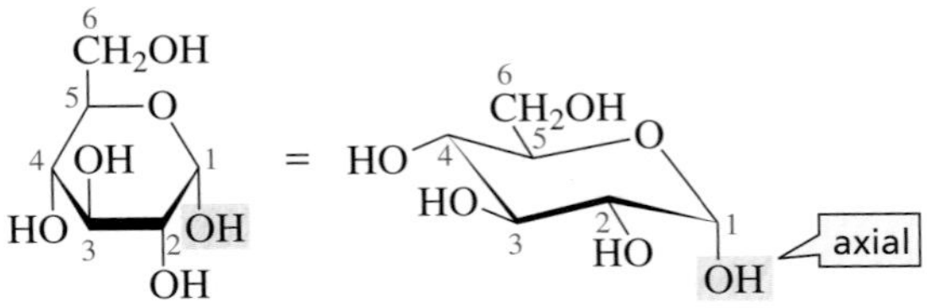

axial

α-D-glucose
chair conformation

The α-position is down in a planar representation and axial in a chair conformation.

The β-position is up in a planar representation and equatorial in a chair conformation.

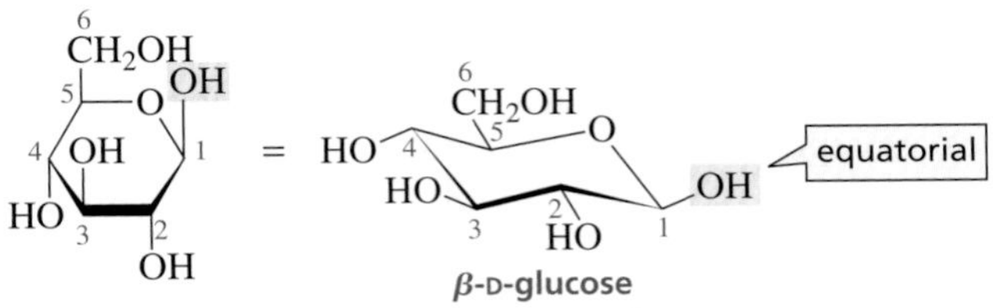

equatorial

β-D-glucose
chair conformation

The chair conformations show why there is more β-D-glucose than α-D-glucose in an aqueous solution at equilibrium. The OH group bonded to the anomeric carbon is in the equatorial position in β-D-glucose, whereas it is in the axial position in α-D-glucose. Therefore, β-D-glucose is more stable than α-D-glucose, so β-D-glucose predominates at equilibrium in an aqueous solution.

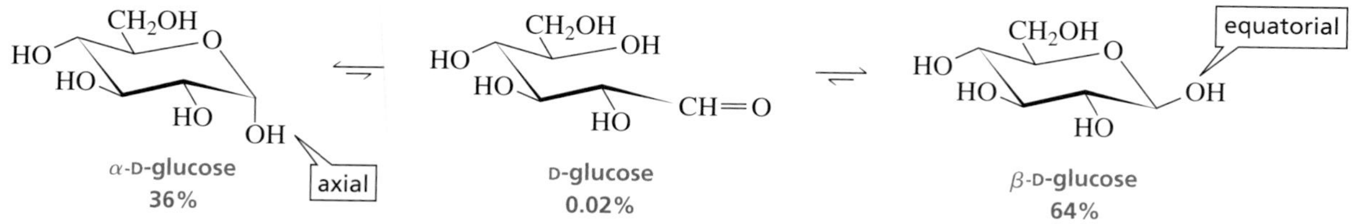

α-D-glucose 36% ⇌ D-glucose 0.02% ⇌ β-D-glucose 64%

If you remember that all the OH groups in β-D-glucose are in equatorial positions, you will find it easy to draw the chair conformation of any other pyranose. For example, if you want to draw α-D-galactose, you would put all the OH groups in equatorial positions, except the OH groups at C-4 (because galactose is a C-4 epimer of glucose) and at C-1 (because it is the α-anomer). You would put these two OH groups in axial positions.

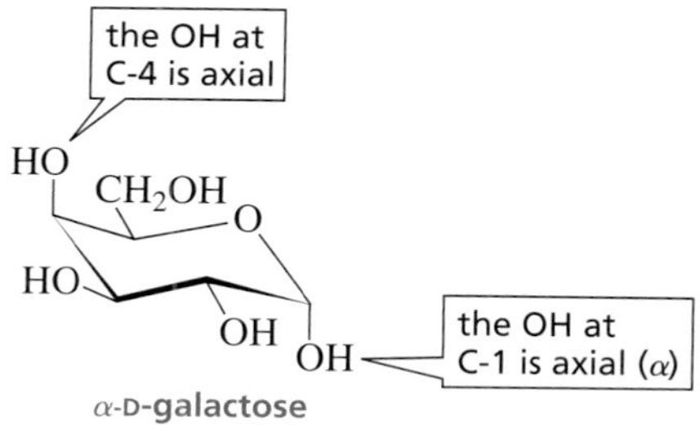

α-D-galactose

PROBLEM 14◆ SOLVED

Which OH groups are in the axial position in
a. β-D-mannopyranose?
b. β-D-idopyranose?
c. α-D-allopyranose?

Solution To 14a All the OH groups in β-D-glucose are in equatorial positions. Because β-D-mannose is a C-2 epimer of β-D-glucose, only the C-2 OH group of β-D-mannose is in the axial position.

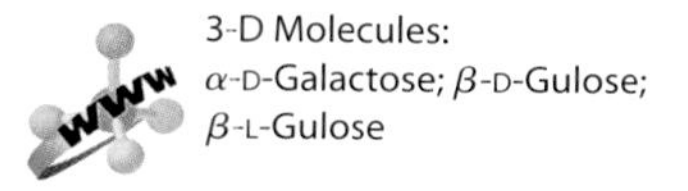

3-D Molecules: α-D-Galactose; β-D-Gulose; β-L-Gulose

15.11 THE FORMATION OF GLYCOSIDES

In the same way that a hemiacetal (or hemiketal) reacts with an alcohol to form an acetal (or ketal) (Section 12.9), the cyclic hemiacetal (or hemiketal) formed by a monosaccharide can react with an alcohol to form an acetal (or ketal). The acetal or ketal of a sugar is called a **glycoside**, and the bond between the anomeric carbon and the alkoxy oxygen is called a **glycosidic bond**. Glycosides are named by replacing the "e" ending of the sugar's name with "ide." Thus, a glycoside of glucose is a glucoside, a glycoside of galactose is a galactoside, and so on. If the pyranose or furanose name is used, the acetal (or ketal) is called a **pyranoside** or a **furanoside**.

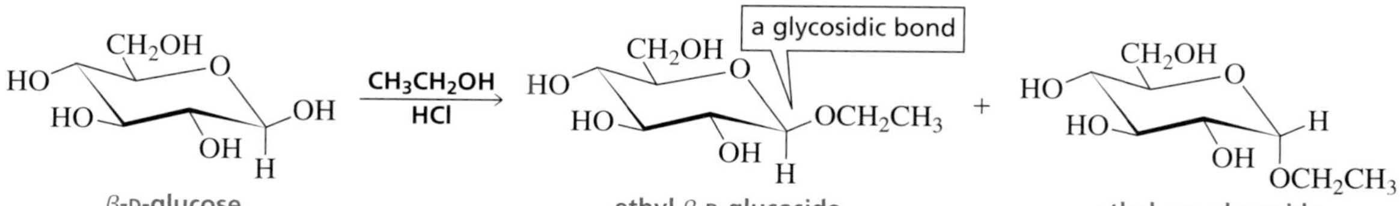

β-D-glucose
β-D-glucopyranose
$\xrightarrow[HCl]{CH_3CH_2OH}$
ethyl β-D-glucoside
ethyl β-D-glucopyranoside
+
ethyl α-D-glucoside
ethyl α-D-glucopyranoside

Notice that the reaction of a single anomer with an alcohol leads to the formation of both the α- and β-glycosides. The mechanism of the reaction shows why both glycosides are formed.

mechanism for glycoside formation

$CH_3CH_2\ddot{O}H$ comes in from the top

$CH_3CH_2\ddot{O}H$ comes in from the bottom

a β-glycoside

an α-glycoside

- The OH group bonded to the anomeric carbon becomes protonated in the acidic solution.
- A lone pair on the ring oxygen helps expel a molecule of water. The anomeric carbon is now sp^2 hybridized, so that part of the molecule is planar (Section 6.13).
- When the alcohol approaches from the top of the plane, the β-glycoside is formed; when it approaches from the bottom of the plane, the α-glycoside is formed.

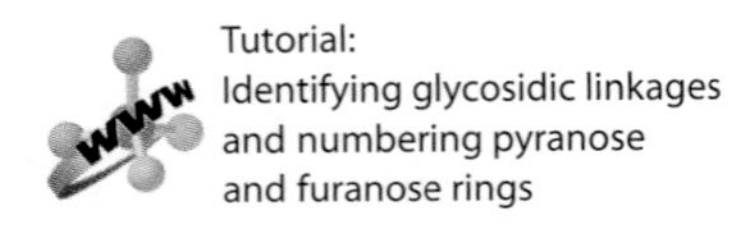

Tutorial:
Identifying glycosidic linkages and numbering pyranose and furanose rings

Notice that the mechanism is the same as that shown for acetal formation in Section 12.9.

15.12 REDUCING AND NONREDUCING SUGARS

Because glycosides are acetals (or ketals), they are not in equilibrium with the open-chain aldehyde (or ketone) in aqueous solution. Without being in equilibrium with a compound with a carbonyl group, they cannot be oxidized by Br_2. Glycosides, therefore, are nonreducing sugars—they cannot reduce Br_2.

In contrast, hemiacetals (or hemiketals) are in equilibrium with the open-chain sugars in aqueous solution, so they can reduce Br_2. In summary, as long as a sugar has an aldehyde, a ketone, a hemiacetal, or a hemiketal group, it is able to reduce an oxidizing agent and therefore is classified as a **reducing sugar**. Without one of these groups, it is a **nonreducing sugar**.

A sugar with an aldehyde, a ketone, a hemiacetal, or a hemiketal group is a reducing sugar. A sugar without one of these groups is a nonreducing sugar.

PROBLEM 15 ◆ SOLVED

Name the following compounds, and indicate whether each is a reducing sugar or a nonreducing sugar:

a.

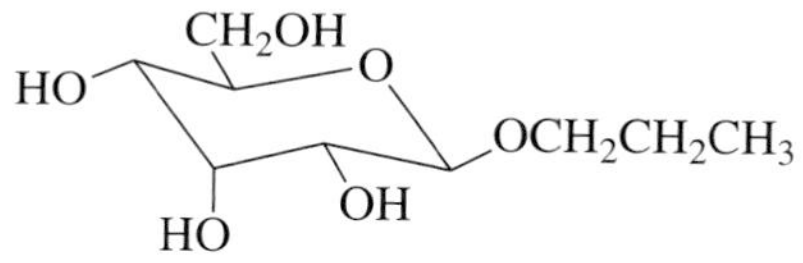

b.

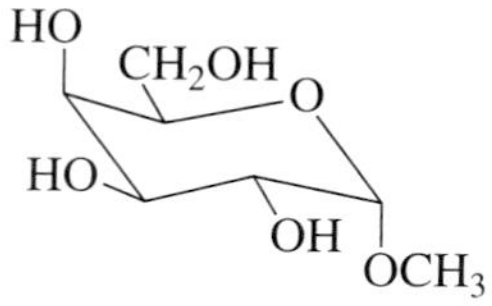

Solution To 15a The only OH group in an axial position in part a is the one at C-3. Therefore, this sugar is the C-3 epimer of D-glucose, which is D-allose. The substituent at the anomeric carbon is in the β-position. Thus, the sugar's name is propyl β-D-alloside or propyl β-D-allopyranoside. Because the sugar is an acetal, it is a nonreducing sugar.

15.13 DISACCHARIDES

If the hemiacetal group of a monosaccharide forms an acetal by reacting with an alcohol group of another monosaccharide, the glycoside that is formed is a disaccharide. **Disaccharides** are compounds consisting of two monosaccharide subunits hooked together by a glycosidic linkage. For example, maltose is a disaccharide obtained from the hydrolysis of starch. It contains two D-glucose subunits connected by a glycosidic bond. This particular glycosidic bond is called an **α-1,4′-glycosidic linkage** because the linkage is between C-1 of one sugar subunit and C-4 of the other, and the oxygen bonded to the anomeric carbon is in the α-position. The "prime" superscript indicates that C-4 is not in the same ring as C-1. *Remember that the α-position is axial and the β-position is equatorial when a sugar is shown in a chair conformation.*

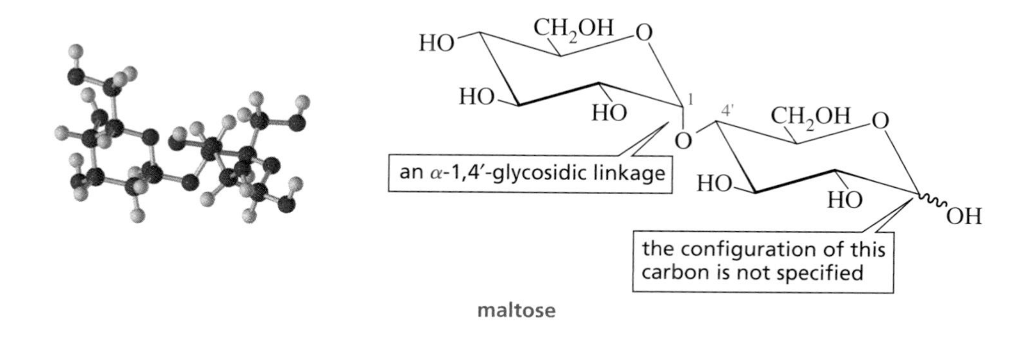

maltose

Note that the structure of maltose does not specify the configuration of the anomeric carbon that is not an acetal (the anomeric carbon of the subunit on the right that is marked with a wavy line), because maltose can exist in both the α and β forms. In α-maltose, the OH group bonded to this anomeric carbon is in the axial position. In β-maltose, the OH group is in the equatorial position.

Cellobiose, a disaccharide obtained from the hydrolysis of cellulose, also contains two D-glucose subunits. Cellobiose differs from maltose in that the two glucose subunits are hooked together by a **β-1,4′-glycosidic linkage**. Thus, the only difference in the structures of maltose and cellobiose is the configuration of the glycosidic linkage. Like maltose, cellobiose exists in both α and β forms because the OH group bonded to the anomeric carbon not involved in acetal formation can be in either the axial position (in α-cellobiose) or the equatorial position (in β-cellobiose).

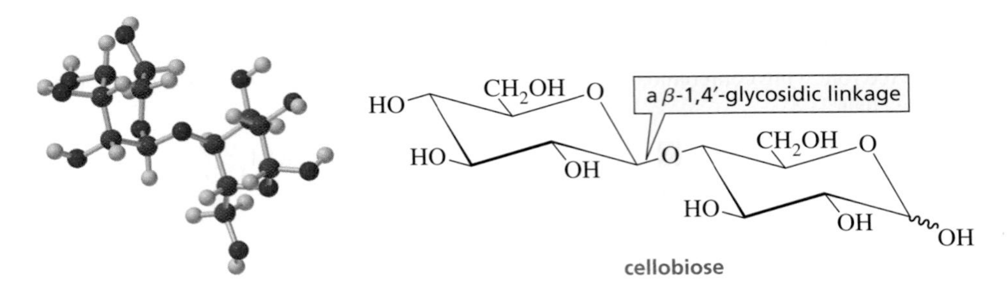

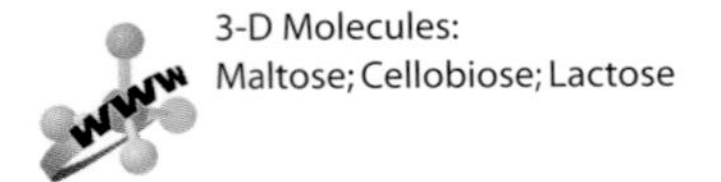

3-D Molecules: Maltose; Cellobiose; Lactose

Lactose is a disaccharide found in milk. It constitutes 4.5% of cow's milk by weight and 6.5% of human milk. One of the subunits of lactose is D-galactose, and the other is D-glucose. The D-galactose subunit is an acetal, and the D-glucose subunit is a hemiacetal. The subunits are joined by a β-1,4′-glycosidic linkage.

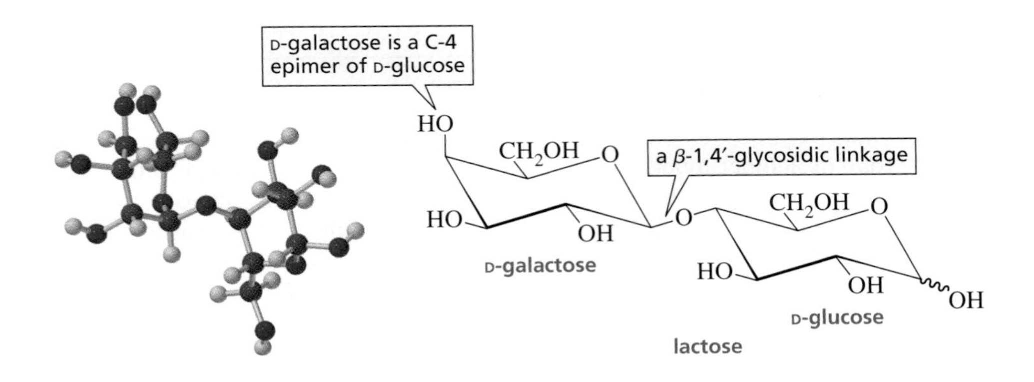

LACTOSE INTOLERANCE

Lactase is an enzyme that specifically breaks the β-1,4′-glycosidic linkage of lactose. Cats and dogs lose their intestinal lactase when they become adults; they are then no longer able to digest lactose. Consequently, when they are fed milk or milk products, the undegraded lactose causes digestive problems such as bloating, abdominal pain, and diarrhea. These problems occur because only monosaccharides can pass into the bloodstream, so lactose has to pass undigested into the large intestine. When humans have stomach flu or other intestinal disturbances, they can temporarily lose their lactase, thereby becoming lactose intolerant. Some humans lose their lactase permanently as they mature. Lactose intolerance is much more common in people whose ancestors came from nondairy-producing countries. For example, only 3% of Danes are lactose intolerant, compared with 90% of all Chinese and Japanese and 97% of Thais. This is why you are not likely to find any dairy items on Chinese menus.

GALACTOSEMIA

After lactose is degraded into glucose and galactose, the galactose must be converted into glucose before it can be used by cells. Individuals who do not have the enzyme that converts galactose into glucose have the genetic disease known as galactosemia. Without this enzyme, galactose accumulates in the bloodstream. This condition can cause mental retardation and even death in infants. Galactosemia is treated by excluding galactose from the diet.

The most common disaccharide, sucrose, is the substance we know as table sugar. Obtained from sugar beets and sugar cane, sucrose consists of a D-glucose subunit and a D-fructose subunit linked by a glycosidic bond between C-1 of glucose (in the α-position) and C-2 of fructose (in the β-position). About 90 million tons of sucrose are produced in the world each year.

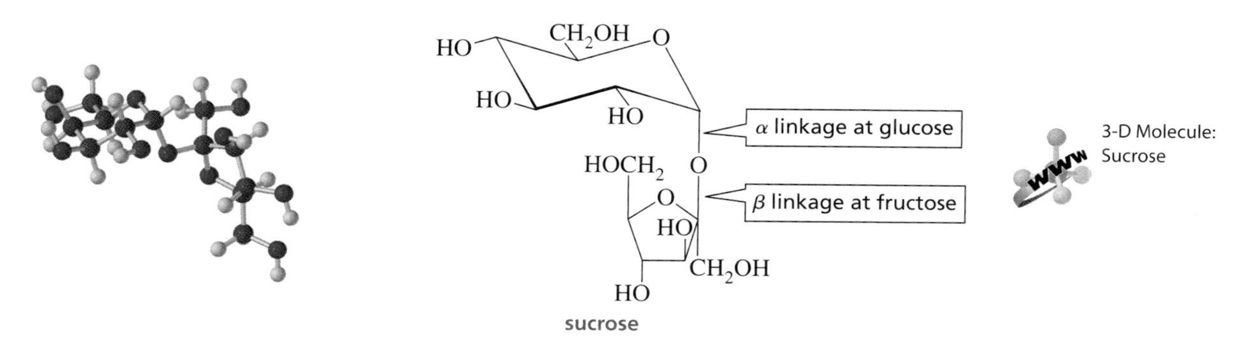

sucrose

Sucrose has a specific rotation of +66.5. When it is hydrolyzed, the resulting 1:1 mixture of glucose and fructose has a specific rotation of −22.0. Because the sign of the rotation changes when sucrose is hydrolyzed, a 1:1 mixture of glucose and fructose is called *invert sugar*. The enzyme that catalyzes the hydrolysis of sucrose is called invertase. Honeybees have invertase, so the honey they produce is a mixture of sucrose, glucose, and fructose. Because fructose is sweeter than sucrose, invert sugar is sweeter than sucrose. Some foods are advertised as containing fructose instead of sucrose, which means that they achieve the same level of sweetness with a lower sugar (lower calorie) content.

15.14 POLYSACCHARIDES

Polysaccharides contain as few as 10 or as many as several thousand monosaccharide units joined together by glycosidic linkages. The most common polysaccharides are starch and cellulose.

Starch is the major component of flour, potatoes, rice, beans, corn, and peas. It is a mixture of two different polysaccharides: amylose (about 20%) and amylopectin (about 80%). Amylose is composed of unbranched chains of D-glucose units joined by α-1,4′-glycosidic linkages.

an α-1,4′-glycosidic linkage

three subunits of amylose

Amylopectin is a branched polysaccharide. Like amylose, it is composed of chains of D-glucose units joined by α-1,4′-glycosidic linkages. Unlike amylose, amylopectin also contains **α-1,6′-glycosidic linkages**. These linkages create the branches in the polysaccharide (Figure 15.1).

an α-1,6′-glycosidic linkage

five subunits of amylopectin

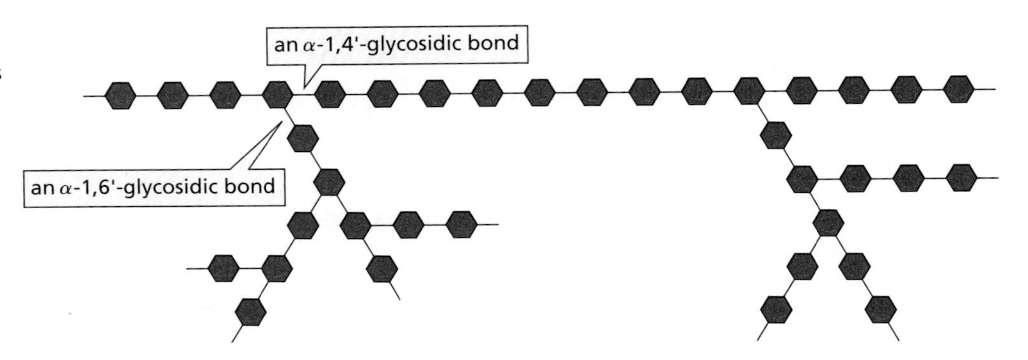

▶ **Figure 15.1**
Branching in amylopectin. The hexagons represent glucose units. They are joined by α-1,4′- and α-1,6′-glyclosidic bonds.

Living cells oxidize D-glucose in the first of a series of processes that provide them with energy (Section 18.4). When animals have more D-glucose than they need for energy, they convert the excess D-glucose into a polymer called glycogen. Glycogen has a structure similar to that of amylopectin, but glycogen has more branches (Figure 15.2). The high degree of branching in glycogen has important physiological consequences. When an animal needs energy, many individual glucose units can be simultaneously removed from the ends of many branches. Plants convert excess D-glucose into starch.

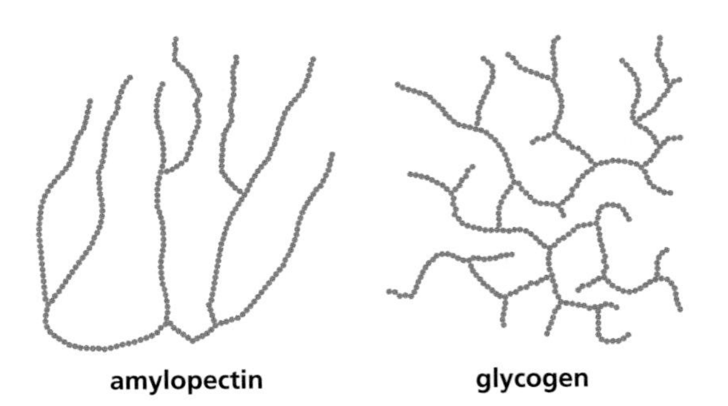

▶ **Figure 15.2**
A comparison of the branching in amylopectin and glycogen.

Cellulose is the structural material of higher plants. Cotton, for example, is composed of about 90% cellulose, and wood is about 50% cellulose. Like amylose, cellulose is composed of unbranched chains of D-glucose units. Unlike amylose, however, the glucose units in cellulose are joined by β-1,4′-glycosidic linkages rather than by α-1,4′-glycosidic linkages.

a β-1,4′-glycosidic linkage

three subunits of cellulose

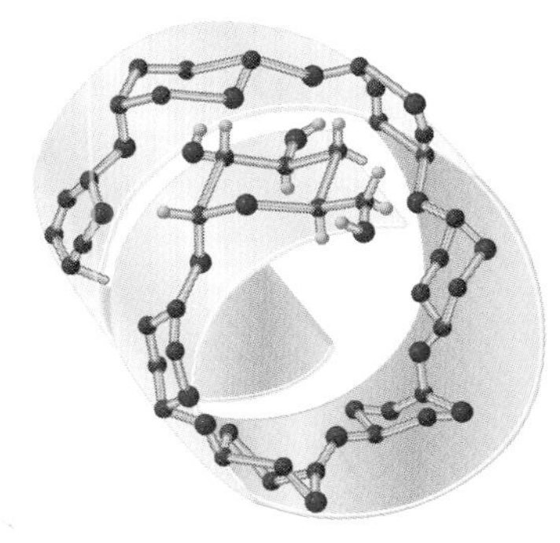

▲ **Figure 15.3**
The α-1,4′-glycosidic linkages in amylose cause it to form a left-handed helix. Many of its OH groups form hydrogen bonds with water molecules.

All mammals have the enzyme (α-glucosidase) that hydrolyzes the α-1,4′-glycosidic linkages that join glucose units in amylose, amylopectin, and glycogen, but they do not have the enzyme (β-glucosidase) that hydrolyzes β-1,4′-glycosidic linkages. As a result, mammals *cannot* obtain the glucose they need by eating cellulose. However, bacteria that possess β-glucosidase inhabit the digestive tracts of grazing animals, so cows can eat grass and horses can eat hay to meet their nutritional requirements for glucose. Termites also harbor bacteria that break down the cellulose in the wood they eat.

The different glycosidic linkages in starch and cellulose give these compounds very different physical properties. The α-linkages in starch cause amylose to form a helix that promotes hydrogen bonding of its OH groups to water molecules (Figure 15.3). As a result, starch is soluble in water.

On the other hand, the β-linkages in cellulose promote the formation of intramolecular hydrogen bonds. Consequently, these molecules line up in linear arrays (Figure 15.4), held together by hydrogen bonds between adjacent chains. These large aggregates cause cellulose to be insoluble in water. The strength of these bundles of polymer chains makes cellulose an effective structural material. Processed cellulose is also used for the production of paper and cellophane.

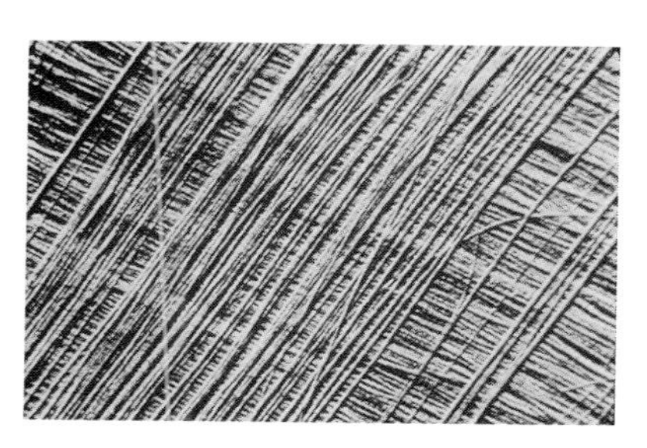

Strands of cellulose in a plant fiber.

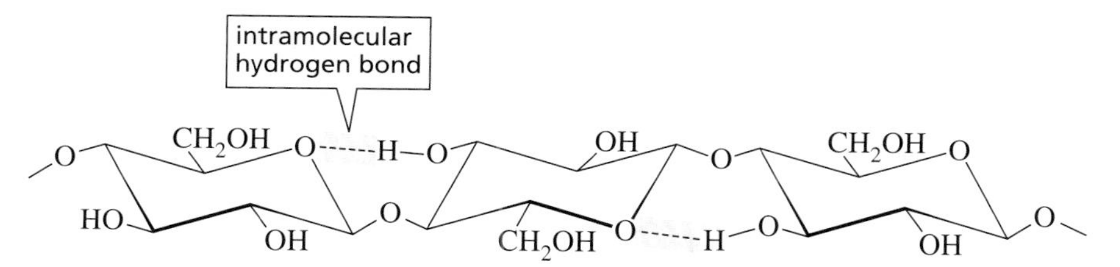

◀ **Figure 15.4**
The β-1,4′-glycosidic linkages in cellulose form intramolecular hydrogen bonds, which cause the molecules to line up in linear arrays.

Chitin is a polysaccharide that is structurally similar to cellulose. It is the major structural component of the shells of crustaceans (such as lobsters, crabs, and shrimp) and the exoskeletons of insects and other arthropods. Like cellulose, chitin has β-1,4′-glycosidic linkages. It differs from cellulose in that it has an *N*-acetylamino group instead of an OH group at the C-2 position. The β-1,4′-glycosidic linkages give chitin its structural rigidity.

three subunits of chitin

The shell of this bright orange crab from Australia is composed largely of chitin.

WHY THE DENTIST IS RIGHT

Bacteria found in the mouth have an enzyme that converts sucrose into a polysaccharide called dextran. Dextran is made up of glucose units joined mainly through α-1,3′- and α-1,6′-glycosidic linkages. About 10% of dental plaque is composed of dextran. This is the chemical basis of why your dentist cautions you not to eat candy.

CONTROLLING FLEAS

Several different drugs have been developed to help pet owners control fleas. One of these drugs is lufenuron, the active ingredient in Program. Lufenuron interferes with the flea's production of chitin. The consequence is fatal for the flea because its exoskeleton is composed primarily of chitin.

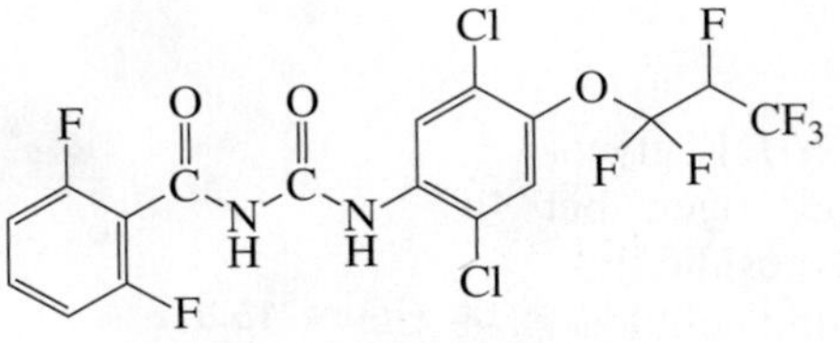

Cl F O O F O CF$_3$ C N C N F F H H Cl F

lufenuron

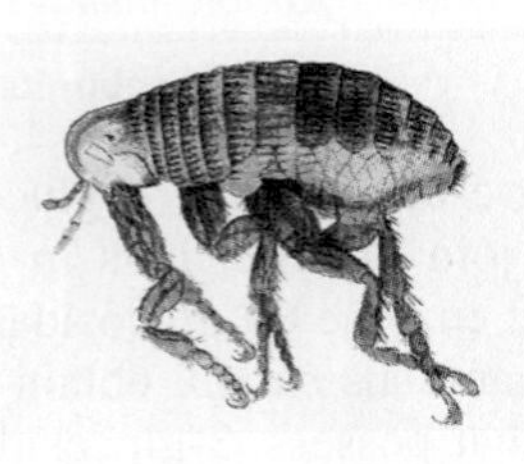

PROBLEM 16

What is the main structural difference between

a. amylose and cellulose?
b. amylose and amylopectin?
c. amylopectin and glycogen?
d. cellulose and chitin?

15.15 SOME NATURALLY OCCURRING PRODUCTS DERIVED FROM CARBOHYDRATES

Deoxy sugars are sugars in which one of the OH groups is replaced by a hydrogen (*deoxy* means "without oxygen"). 2-Deoxyribose is an important example of a deoxy sugar; it is missing the oxygen at the C-2 position. D-Ribose is the sugar component of ribonucleic acid (RNA), whereas 2-deoxyribose is the sugar component of deoxyribonucleic acid (DNA) (Section 20.1).

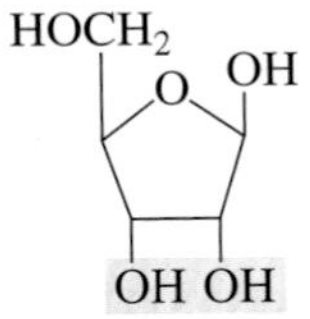

$HOCH_2$ O OH

OH OH

β-D-ribose

$HOCH_2$ O OH

OH

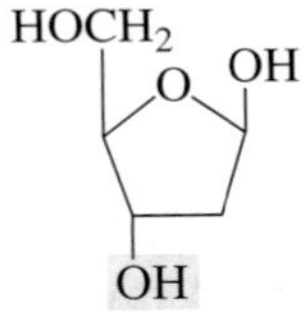

β-D-2-deoxyribose

HEPARIN

Heparin is a polysaccharide found principally in cells that line arterial walls. It is released when an injury occurs in order to prevent excessive blood clot formation. Heparin is widely used clinically as an anticoagulant.

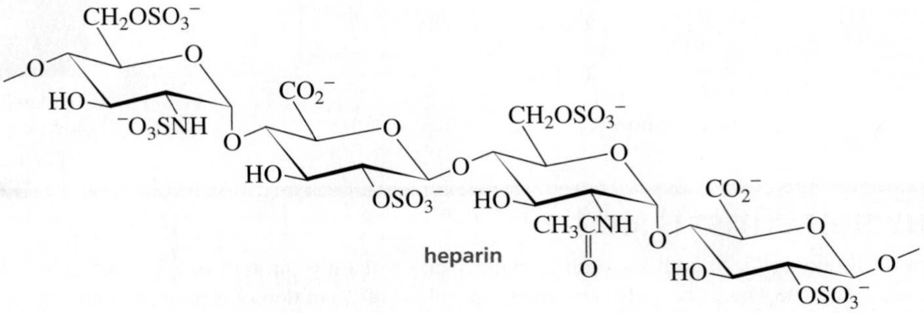

$CH_2OSO_3^-$ O O HO ^-O_3SNH CO_2^- O O HO OSO_3^- O $CH_2OSO_3^-$ O HO CH_3CNH O CO_2^- O O HO OSO_3^- O

heparin

In **amino sugars**, one of the OH groups is replaced by an amino group. *N*-Acetylglucosamine—the subunit of chitin—is an example of an amino sugar (Section 15.14). Some important antibiotics contain amino sugars. For example, the three subunits of the antibiotic gentamicin are deoxyamino sugars. Notice that the middle subunit is missing the ring oxygen, so it is not really a sugar.

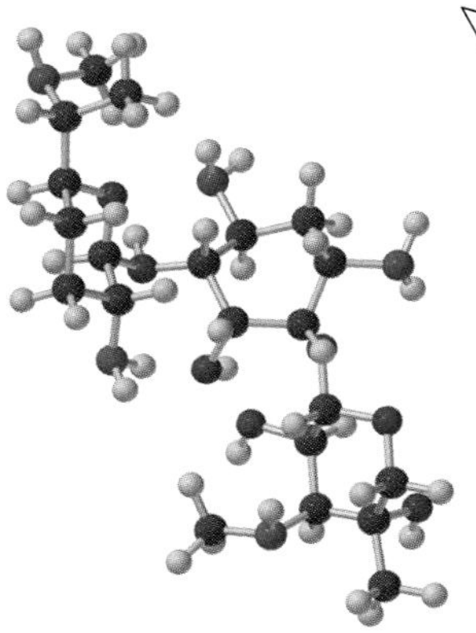

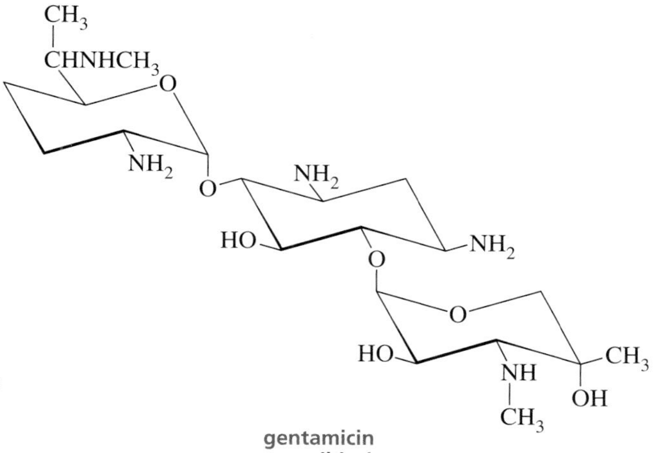

gentamicin
an antibiotic

3-D Molecule: Gentamicin

L-Ascorbic acid (vitamin C) is synthesized from D-glucose in plants and in the livers of most vertebrates. Humans, monkeys, and guinea pigs do not have the enzymes necessary for the biosynthesis of vitamin C, so they must obtain the vitamin in their diets. The L-configuration of ascorbic acid refers to the configuration at C-5, which was C-2 in D-glucose.

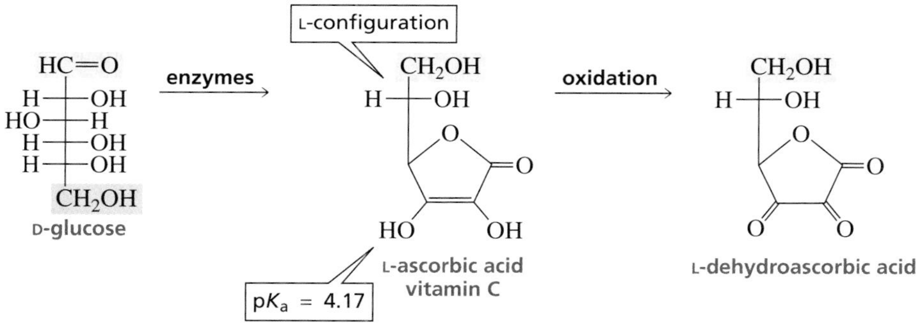

D-glucose → (enzymes) → L-ascorbic acid, vitamin C → (oxidation) → L-dehydroascorbic acid

L-configuration

pK_a = 4.17

Although L-ascorbic acid does not have a carboxylic acid group, it is an acidic compound because the pK_a of the C-3 OH group is 4.17. L-Ascorbic acid is readily oxidized—notice that when it is oxidized, it loses hydrogens—to L-dehydroascorbic acid, which is also physiologically active. If the five-membered ring is opened by hydrolysis, all vitamin C activity is lost. Therefore, not much intact vitamin C survives in food that has been thoroughly cooked. Worse, if the food is cooked in water and then drained, the water-soluble vitamin is thrown out with the water!

VITAMIN C

Vitamin C is an antioxidant because it prevents oxidation reactions by radicals. It traps radicals formed in aqueous environments (Section 5.17), preventing harmful oxidation reactions the radicals would cause. Not all the physiological functions of vitamin C are known; however, we do know that vitamin C is required for the synthesis of collagen, which is the structural protein of skin, tendons, connective tissue, and bone. Vitamin C is abundant in citrus fruits and tomatoes, but when it is not present in the diet, lesions appear on the skin, bleeding occurs about the gums, in the joints, and under the skin, and any wounds heal slowly. The condition, known as *scurvy*, was the first disease to be treated by adjusting the diet. British sailors who shipped out to sea after the late 1700s were required to eat limes to prevent it (which is how they came to be called "limeys"). *Scorbutus* is Latin for "scurvy"; *ascorbic*, therefore, means "no scurvy."

15.16 CARBOHYDRATES ON CELL SURFACES

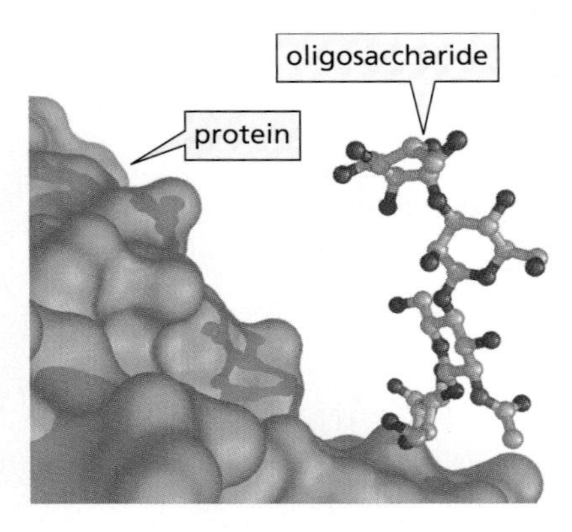

Many cells have short oligosaccharide chains on their surface that enable the cells to recognize and interact with other cells and with invading viruses and bacteria. These oligosaccharides are linked to the surface of the cell by the reaction of an OH or an NH_2 group of a cell-membrane protein with the anomeric carbon of a cyclic sugar. Proteins bonded to oligosaccharides are called **glycoproteins**.

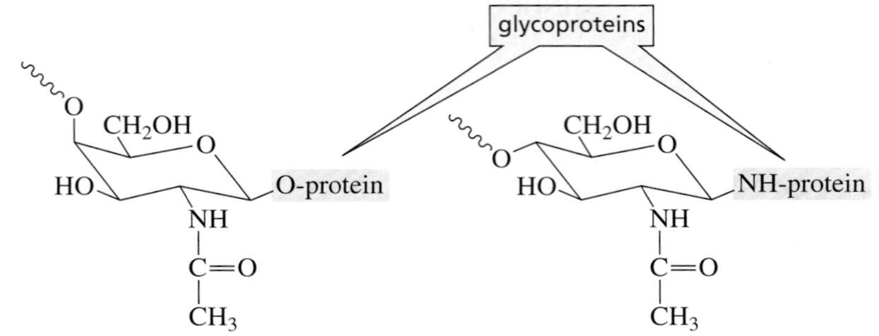

Carbohydrates on the surfaces of cells provide a way for cells to recognize one another. The interaction between surface carbohydrates has been found to play a role in activities as diverse as infection, prevention of infection, fertilization, inflammatory diseases such as rheumatoid arthritis and septic shock, and blood clotting. The goal of the HIV protease inhibitor drugs, for example, is to prevent HIV from recognizing cells by means of their surface oligosaccharide and then penetrating them. The fact that several known antibiotics contain amino sugars (Section 15.15) suggests that they function by recognizing target cells. Carbohydrate interactions also are involved in the regulation of cell growth, so changes in membrane glycoproteins are thought to be correlated with malignant transformations.

Differences in blood type (A, B, or O) are actually differences in the sugars bound to the surfaces of red blood cells. Each type of blood is associated with a different carbohydrate structure (Figure 15.5). Type AB blood is a mixture of type A and type B.

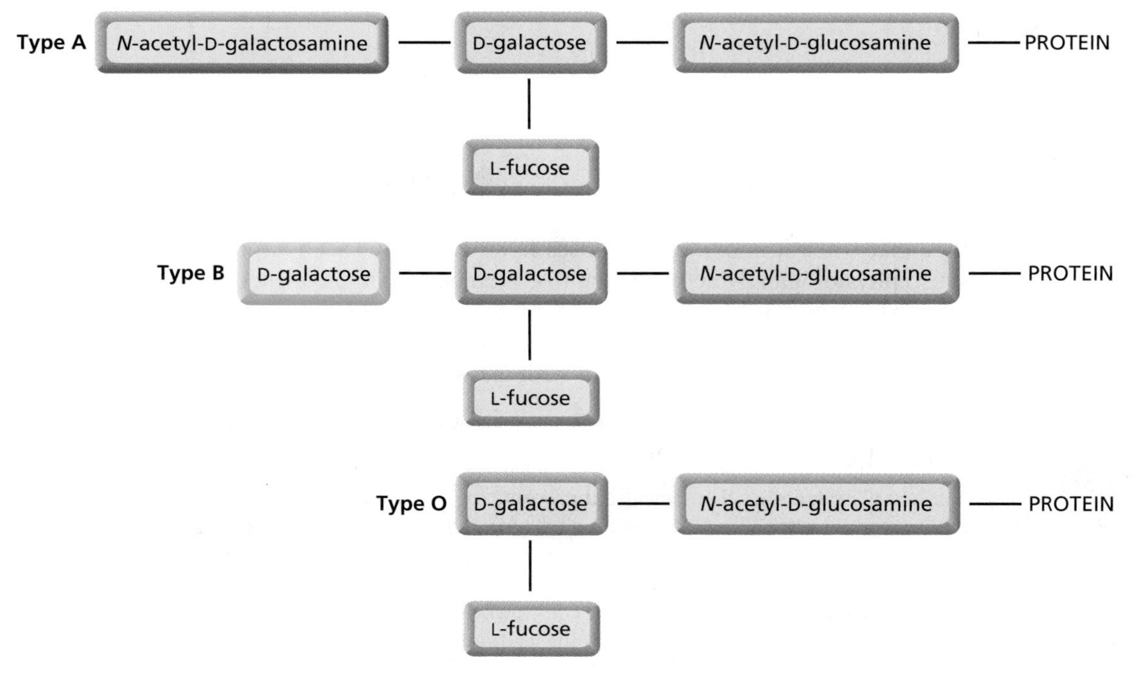

▲ **Figure 15.5**
Blood type is determined by the nature of the sugar on the surfaces of red blood cells. Fucose is 6-deoxygalactose.

Antibodies are proteins that are synthesized by the body in response to foreign substances, called *antigens*. Interaction with the antibody causes the antigen to either precipitate it or flag it for destruction by immune system cells. This is why, for example, blood cannot be transferred from one person to another unless the blood types of the donor and acceptor are compatible. Otherwise the donated blood will be considered a foreign substance and provoke an immune response.

Looking at Figure 15.5, we can see why the immune system of type A people recognizes type B blood as foreign and vice versa. The immune system of people with type A, B, or AB blood does not, however, recognize type O blood as foreign, because the carbohydrate in type O blood is also a component of types A, B, and AB blood. Thus, anyone can accept type O blood, so people with type O blood are called universal donors. People with type AB blood can accept types AB, A, B, and O blood, so they are referred to as universal acceptors.

PROBLEM 17

Refer to Figure 15.5 to answer the following questions:

a. People with type O blood can donate blood to anyone, but they cannot receive blood from everyone. From whom can they *not* receive blood?

b. People with type AB blood can receive blood from anyone, but they cannot give blood to everyone. To whom can they *not* give blood?

15.17 SYNTHETIC SWEETENERS

For a molecule to taste sweet, it must bind to a receptor on a taste bud cell of the tongue. The binding of this molecule causes a nerve impulse to pass from the taste bud to the brain, where the molecule is interpreted as being sweet. Sugars differ in their degree of "sweetness." Compared with the sweetness of glucose, which is assigned a relative value of 1.00, the sweetness of sucrose is 1.45, and that of fructose, the sweetest of all sugars, is 1.65. Lactose, the sugar found in milk, is only one-sixth as sweet as glucose.

Developers of synthetic sweeteners must evaluate potential products in terms of several factors—such as toxicity, stability, and cost—in addition to taste. Saccharin (Sweet'N Low), the first synthetic sweetener, was discovered accidentally by Ira Remsen and his student Constantine Fahlberg at Johns Hopkins University in 1878. Fahlberg was studying the oxidation of ortho-substituted toluenes in Remsen's laboratory when he found that one of his newly synthesized compounds had an extremely sweet taste. (As strange as it may seem today, at one time it was common for chemists to taste compounds in order to characterize them.) He called this compound saccharin; it was eventually found to be about 300 times sweeter than glucose. Notice that, in spite of its name, saccharin is *not* a saccharide.

BIOGRAPHY

Ira Remsen (1846–1927) *was born in New York. After receiving an M.D. from Columbia University, he decided to become a chemist. He earned a Ph.D. in Germany and then returned to the United States in 1872 to accept a faculty position at Williams College. In 1876, he became a professor of chemistry at the newly established Johns Hopkins University, where he initiated the first center for chemical research in the United States. He later became the second president of Johns Hopkins.*

ACCEPTABLE DAILY INTAKE

The FDA has established an acceptable daily intake (ADI) value for many of the food ingredients it clears for use. The ADI is the amount of the substance that people can consume safely, each day of his or her life. For example, the ADI for acesulfame potassium is 15 mg/kg/day. This means a 132-lb person can consume the amount of acesulfame-potassium every day that would be found in two gallons of an artificially sweetened beverage. The ADI for sucralose is also 15 mg/kg/day.

saccharin

CH_3CH_2O—⟨benzene ring⟩—$NHCNH_2$ (C=O)

dulcin

acesulfame potassium

$^-OCCH_2CHCNHCHCOCH_3$ ($^+NH_3$, CH_2–phenyl)

aspartame

⟨cyclohexane ring⟩—$NHSO_3^-\ Na^+$

sodium cyclamate

sucralose

Because it has little caloric value, saccharin became an important substitute for sucrose after becoming commercially available in 1885. The chief nutritional problem in the West was—and still is—the overconsumption of sugar and its consequences: obesity, heart disease, and dental decay. Saccharin is also a boon to people with diabetes, who must limit their consumption of sucrose and glucose. Although the toxicity of saccharin had not been studied carefully when the compound was first marketed (our current concern with toxicity is a fairly recent development), extensive studies since then have shown saccharin to be harmless. In 1912, saccharin was temporarily banned in the United States, not because of any concern about its toxicity, but because of a concern that people would miss out on the nutritional benefits of sugar.

Dulcin was the second synthetic sweetener to be discovered (in 1884). Even though it did not have the bitter, metallic aftertaste associated with saccharin, it never achieved much popularity. Dulcin was taken off the market in 1951 in response to concerns about its toxicity.

Sodium cyclamate became a widely used nonnutritive sweetener in the 1950s, but was banned in the United States some 20 years later in response to two studies that appeared to show that large amounts of sodium cyclamate cause liver cancer in mice.

Aspartame (NutraSweet, Equal), about 200 times sweeter than sucrose, was approved by the U.S. Food and Drug Administration (FDA) in 1981. Because aspartame contains a phenylalanine subunit, it should not be used by people with the genetic disease known as PKU (Section 18.6).

Acesulfame potassium (Sweet and Safe, Sunette, Sweet One) was approved in 1988. Also called ascesulfame-K, it, too, is about 200 times sweeter than glucose. It has less aftertaste than saccharin and is more stable than aspartame at high temperatures.

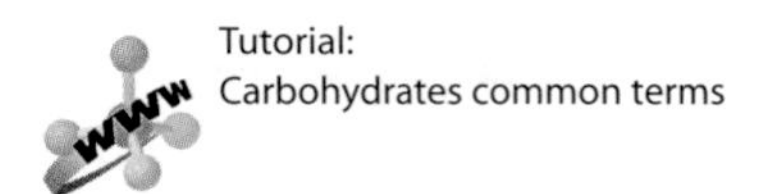

Tutorial:
Carbohydrates common terms

Sucralose (Splenda) is 600 times sweeter than glucose; it is the most recently approved (1991) synthetic sweetener. It maintains its sweetness in foods stored for long periods and at temperatures used in baking. Sucralose is made from sucrose by selectively replacing three of sucrose's OH groups with chlorines. During the chlorination, the 4-position of the glucose ring becomes inverted, so sucralose is a galactoside, not a glucoside. The body does not recognize sucralose as a carbohydrate so instead of being metabolized it is eliminated from the body unchanged.

The fact that these synthetic sweeteners have such different structures shows that the sensation of sweetness is not induced by a single molecular shape.

SUMMARY

Carbohydrates are the most abundant class of compounds in the biological world. They are polyhydroxy aldehydes (**aldoses**) and polyhydroxy ketones (**ketoses**), or compounds formed by linking up aldoses and ketoses. The notations D and L describe the configuration of the bottom-most asymmetric center of a **monosaccharide** in a Fischer projection; the configurations of the other carbons are implicit in the common name. Most naturally occurring sugars are D-sugars. Naturally occurring ketoses have the ketone group in the 2-position. **Epimers** differ in configuration at only one asymmetric center.

In a basic solution, a monosaccharide is converted to a complex mixture of polyhydroxy aldehydes and polyhydroxy ketones. Reduction of an aldose forms one **alditol**; reduction of a ketose forms two alditols. Br_2 oxidizes aldoses but not ketoses. Br_2 oxidizes the aldehyde group of aldoses; nitric acid oxidizes the aldehyde and primary alcohol groups. The **Kiliani–Fischer synthesis** increases the carbon chain of an aldose by one carbon—it forms C-2 epimers.

The aldehyde or ketone group of a **monosaccharide** reacts with one of its OH groups to form cyclic hemiacetals or hemiketals: glucose forms α-D-glucose and β-D-glucose. The α-position is axial when a sugar is shown in a chair conformation and down when the sugar is shown in a planar representation; the β-position is equatorial when a sugar is shown in a chair conformation and up when the sugar is shown in a planar representation. α-D-Glucose and β-D-glucose are **anomers**—they differ in configuration only at the **anomeric carbon**, which is the carbon that was the carbonyl carbon in the open-chain form. Six-membered-ring sugars are **pyranoses**; five-membered-ring sugars are **furanoses**. The most abundant monosaccharide in nature is D-glucose. All the OH groups in β-D-glucose are in equatorial positions. A slow change in optical rotation to an equilibrium value is called **mutarotation**.

The cyclic hemiacetal (or hemiketal) can react with an alcohol to form an acetal (or ketal), called a **glycoside**. If the name "pyranose" or "furanose" is used, the acetal (or ketal) is called a **pyranoside** or a **furanoside**. The bond between the anomeric carbon and the alkoxy oxygen is called a **glycosidic bond**.

Disaccharides consist of two monosaccharides hooked together by a glycosidic linkage. Maltose has an **α-1,4′-glycosidic linkage** between two glucose subunits; cellobiose has a **β-1,4′-glycosidic linkage** between two glucose subunits. The most common disaccharide is sucrose; it has a D-glucose subunit and a D-fructose subunit linked by their anomeric carbons.

Polysaccharides contain as few as 10 or as many as several thousand monosaccharides joined together by glycosidic linkages. Starch is composed of amylose and amylopectin. Amylose has unbranched chains of D-glucose units joined by α-1,4′-glycosidic linkages. Amylopectin, too, has chains of D-glucose units joined by α-1,4′-glycosidic linkages, but it also has **α-1,6′-glycosidic linkages** that create branches. Glycogen is similar to amylopectin but has more branches. Cellulose has unbranched chains of D-glucose units joined by β-1,4′-glycosidic linkages. The α-linkages cause amylose to form a helix; the β-linkages allow the molecules of cellulose to form intramolecular hydrogen bonds.

The surfaces of many cells contain short oligosaccharide chains that allow the cells to interact with each other. These oligosaccharides are linked to the cell surface by protein groups. Proteins bonded to oligosaccharides are called **glycoproteins**.

SUMMARY OF REACTIONS

1. Base-catalyzed epimerization (Section 15.5)

$$\begin{array}{c}HC{=}O\\ H{-}C{-}OH\\ (CHOH)_n\\ CH_2OH\end{array} \underset{H_2O}{\overset{HO^-}{\rightleftharpoons}} \begin{array}{c}HC{=}O\\ HO{-}C{-}H\\ (CHOH)_n\\ CH_2OH\end{array}$$

2. Base-catalyzed rearrangement (Section 15.5)

$$\begin{array}{c}HC{=}O\\ CHOH\\ (CHOH)_n\\ CH_2OH\end{array} \underset{H_2O}{\overset{HO^-}{\rightleftharpoons}} \begin{array}{c}HC{-}OH\\ \Vert\\ C{-}OH\\ (CHOH)_n\\ CH_2OH\end{array} \underset{HO^-}{\overset{H_2O}{\rightleftharpoons}} \begin{array}{c}CH_2OH\\ C{=}O\\ (CHOH)_n\\ CH_2OH\end{array}$$

3. Reduction (Section 15.6)

$$\begin{array}{c} CH_2OH \\ | \\ C{=}O \\ | \\ (CHOH)_n \\ | \\ CH_2OH \end{array} \xrightarrow[\textbf{2. } H_3O^+]{\textbf{1. } NaBH_4} \begin{array}{c} CH_2OH \\ | \\ CHOH \\ | \\ (CHOH)_n \\ | \\ CH_2OH \end{array}$$

4. Oxidation (Section 15.6)

a.

$$\begin{array}{c} HC{=}O \\ | \\ (CHOH)_n \\ | \\ CH_2OH \end{array} \xrightarrow[H_2O]{Br_2} \begin{array}{c} COOH \\ | \\ (CHOH)_n \\ | \\ CH_2OH \end{array} + 2\ Br^-$$

b.

$$\begin{array}{c} HC{=}O \\ | \\ (CHOH)_n \\ | \\ CH_2OH \end{array} \xrightarrow[\Delta]{HNO_3} \begin{array}{c} COOH \\ | \\ (CHOH)_n \\ | \\ COOH \end{array}$$

5. Chain elongation: the Kiliani–Fischer synthesis (Section 15.7)

$$\begin{array}{c} HC{=}O \\ | \\ (CHOH)_n \\ | \\ CH_2OH \end{array} \xrightarrow[\textbf{3. } H_3O^+]{\textbf{1. } NaC{\equiv}N/HCl \quad \textbf{2. } H_2,\ Pd/BaSO_4} \begin{array}{c} HC{=}O \\ | \\ (CHOH)_{n+1} \\ | \\ CH_2OH \end{array}$$

6. Acetal (and ketal) formation (Section 15.11)

$$\text{pyranose (CH}_2\text{OH, HO, HO, OH, OH)} \xrightarrow[ROH]{HCl} \text{pyranoside (CH}_2\text{OH, HO, HO, OR, OH)} + \text{pyranoside (CH}_2\text{OH, HO, HO, OH, OR)}$$

PROBLEMS

18. Give the product or products that are obtained when D-galactose reacts with the following substances:

a. nitric acid **b.** $NaBH_4$ followed by H_3O^+ **c.** Br_2 in water **d.** ethanol + HCl

19. What sugar is the C-4 epimer of L-gulose?

20. Identify the sugars in each description:

a. an aldopentose that is not D-arabinose forms D-arabinitol when it is reduced with $NaBH_4$.

b. a sugar that is not D-altrose forms D-altraric acid when it reacts with nitric acid.

c. a ketose, when reduced with $NaBH_4$, forms D-altritol and D-allitol.

21. Answer the following questions about the eight aldopentoses:

a. Which are enantiomers? **b.** Which form an optically active compound when oxidized with nitric acid?

22. What other monosaccharide is reduced only to the alditol obtained from the reduction of D-talose?

23. What monosaccharide is reduced to two alditols, one of which is the alditol obtained from the reduction of D-mannose?

24. What monosaccharides would be formed in a Kiliani–Fischer synthesis starting with

a. D-xylose? **b.** L-threose?

25. Name the following compounds, and indicate whether each is a reducing sugar or a nonreducing sugar:

a. HO, CH_2OH, O, OH, HO, OH

b. $HOCH_2$, O, OCH_2CH_3, CH_2OH, OH OH

26. The reaction of D-ribose with one equivalent of methanol plus HCl forms four products. Give the structures of the products.

27. Dr. Isent T. Sweet isolated a monosaccharide and determined that it had a molecular weight of 150. Much to his surprise, he found that it was not optically active. What is the structure of the monosaccharide?

28. Indicate whether each of the following is D-glyceraldehyde or L-glyceraldehyde, assuming that the horizontal bonds point toward you and the vertical bonds point away from you:

a. HC=O; $HOCH_2$—C—OH; H

b. H; HO—C—CH_2OH; HC=O

c. CH_2OH; HO—C—H; HC=O

29. D-Glucuronic acid is found widely in plants and animals. One of its functions is to detoxify poisonous HO-containing compounds by reacting with them in the liver to form glucuronides. Glucuronides are water soluble and therefore readily excreted. After one ingests a poison such as turpentine, morphine, or phenol, the glucuronides of these compounds are found in the urine. Give the structure of the glucuronide formed by the reaction of β-D-glucuronic acid and phenol.

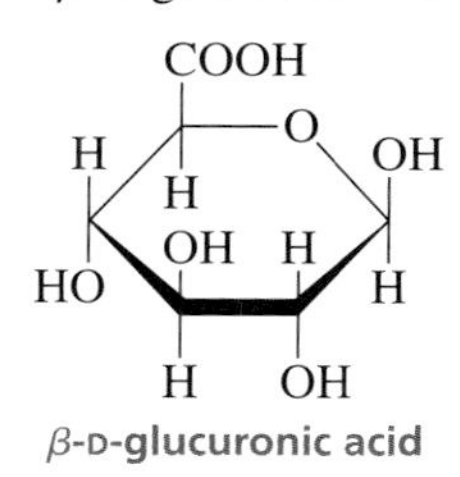

β-D-glucuronic acid

30. In order to synthesize D-galactose, Professor Amy Losse went to the stockroom to get some D-lyxose to use as a starting material. She found that the labels had fallen off the bottles containing D-lyxose and D-xylose. How could she determine which bottle contained D-lyxose?

31. Hyaluronic acid, a component of connective tissue, is the fluid that lubricates the joints. It is an alternating polymer of *N*-acetyl-D-glucosamine and D-glucuronic acid joined by β-1,3′-glycosidic linkages. Draw a short segment of hyaluronic acid.

32. How many aldaric acids are obtained from the 16 aldohexoses?

33. Give the structure of the cyclic hemiacetal formed by each of the following:

a. 4-hydroxypentanal **b.** 4-hydroxyheptanal

34. Explain why the C-3 OH group of vitamin C is more acidic than the C-2 OH group.

35. Calculate the percentages of α-D-glucose and β-D-glucose present at equilibrium from the specific rotations of α-D-glucose, β-D-glucose, and the equilibrium mixture. Compare your values with those given in Section 15.9. (*Hint:* The specific rotation of the mixture equals the specific rotation of α-D-glucose times the fraction of glucose present in the α-form plus the specific rotation of β-D-glucose times the fraction of glucose present in the β-form.)

36. Predict whether D-altrose exists preferentially as a pyranose or a furanose. (*Hint:* The most stable arrangement for a five-membered ring is for all the adjacent substituents to be trans.)

CHAPTER 16

The Organic Chemistry of Amino Acids, Peptides, and Proteins

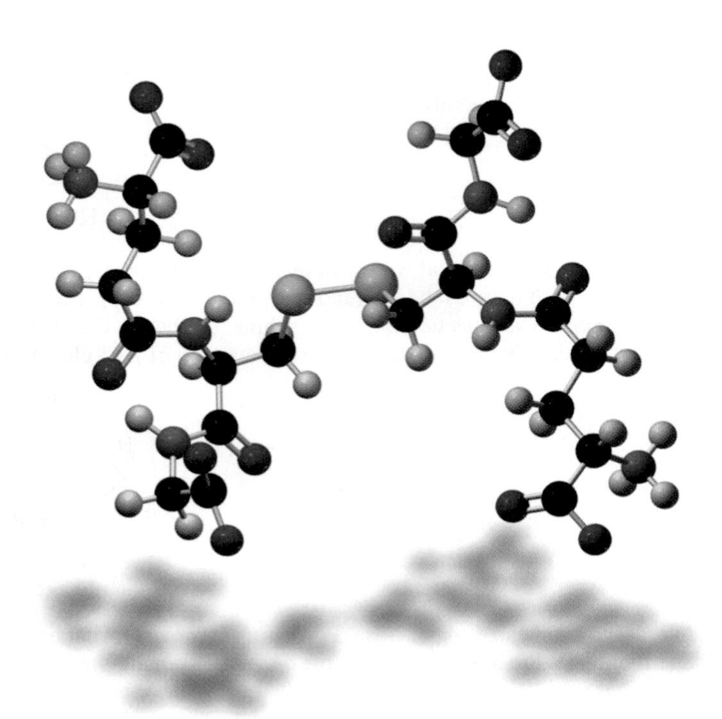

oxidized glutathione

The three kinds of polymers prevalent in nature are polysaccharides, proteins, and nucleic acids. You have already learned about polysaccharides, which are naturally occurring polymers of monosaccharide subunits (Section 15.1). We will now look at proteins and the structurally similar, but shorter, peptides.

Peptides and **proteins** are polymers of **amino acids**. The amino acids are linked together by amide bonds. An **amino acid** is a carboxylic acid with an amino group on the α-carbon. The repeating units are called **amino acid residues**.

$$\mathrm{R{-}\underset{\underset{{}^{+}NH_3}{|}}{CH}{-}\overset{\overset{O}{\|}}{C}{-}OH}$$

a protonated
α-aminocarboxylic acid
an amino acid

amide bonds

$$\mathrm{{-}\underset{\underset{R}{|}}{NHCH}{-}\overset{\overset{O}{\|}}{C}{-}\underset{\underset{R'}{|}}{NHCH}{-}\overset{\overset{O}{\|}}{C}{-}\underset{\underset{R''}{|}}{NHCH}{-}\overset{\overset{O}{\|}}{C}{-}O^{-}}$$

amino acids are linked together by amide bonds

Amino acid polymers can be composed of any number of amino acid residues. A **dipeptide** contains 2 amino acid residues, a **tripeptide** contains 3, an **oligopeptide** contains 3 to 10, and a **polypeptide** contains many amino acid residues. Proteins are naturally occurring polypeptides that are made up of 40 to 4000 amino acid residues. Proteins and peptides serve many functions in biological systems (Table 16.1).

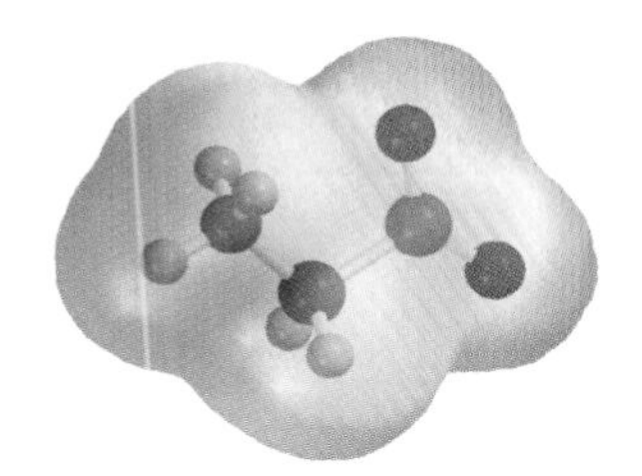

glycine

Proteins can be divided roughly into two classes. **Fibrous proteins** contain long chains of polypeptides arranged in bundles; these proteins are insoluble in water. All structural proteins are fibrous proteins. **Globular proteins** tend to have roughly spherical shapes and are soluble in water. Essentially all enzymes are globular proteins.

Table 16.1 Examples of the Many Functions of Proteins in Biological Systems

Structural proteins	These proteins impart strength to biological structures or protect organisms from their environment. For example, collagen is the major component of bones, muscles, and tendons; keratin is the major component of hair, hooves, feathers, fur, and the outer layer of skin.
Protective proteins	Snake venoms and plant toxins protect their owners from predators. Blood-clotting proteins protect the vascular system when it is injured. Antibodies and peptide antibiotics protect us from disease.
Enzymes	Enzymes are proteins that catalyze the reactions that occur in living systems.
Hormones	Some of the hormones, such as insulin, that regulate the reactions that occur in living systems are proteins.
Proteins with physiological functions	These proteins are responsible for physiological functions such as the transport and storage of oxygen in the body, the storage of oxygen in the muscles, and the contraction of muscles.

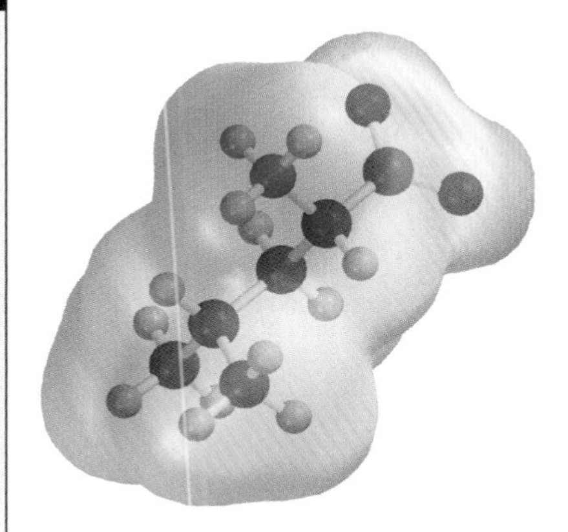

leucine

16.1 THE CLASSIFICATION AND NOMENCLATURE OF AMINO ACIDS

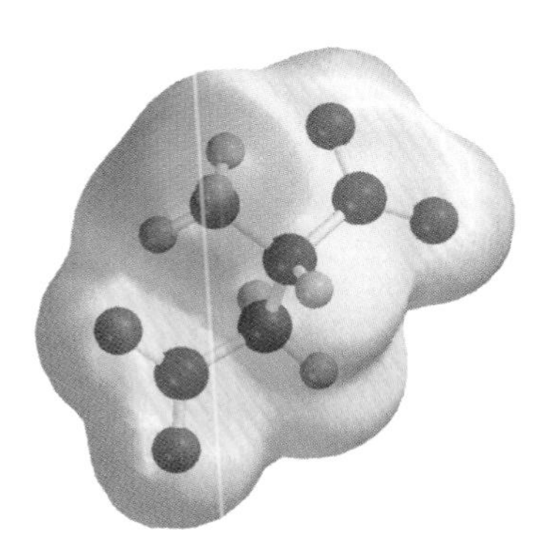

aspartate

The structures of the 20 most common naturally occurring amino acids and the frequency with which each occurs in proteins are shown in Table 16.2. Other amino acids occur in nature, but only infrequently. Note that the amino acids differ only in the substituent (R) attached to the α-carbon. The wide variation in these substituents (called **side chains**) is what gives proteins their great structural diversity and, as a consequence, their great functional diversity (Table 16.1).

The amino acids are almost always called by their common names. Often, the name tells you something about the amino acid. For example, glycine got its name as a result of its sweet taste (*glykos* is Greek for "sweet"). Asparagine was first found in asparagus, and tyrosine was isolated from cheese (*tyros* is Greek for "cheese"). Table 16.2 shows that each of the amino acids has both a three-letter abbreviation (the first three letters of the name, in most cases) and a single-letter abbreviation.

Notice that, in spite of its name, isoleucine does *not* have an isobutyl substituent; it has a *sec*-butyl substituent. Leucine is the amino acid that has an isobutyl substituent. Proline is the only amino acid that is a secondary amine.

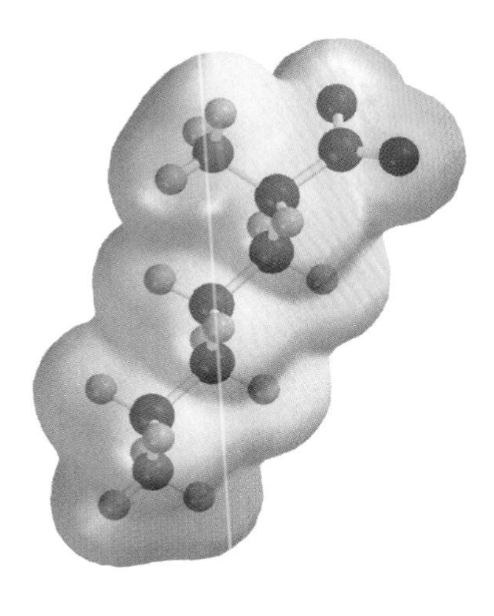

lysine

Ten amino acids are *essential amino acids*; these are denoted by an asterisk (*) in Table 16.2. We humans must obtain these 10 **essential amino acids** from our diets because we either cannot synthesize them at all or cannot synthesize them in adequate amounts. For example, we must have a dietary source of phenylalanine because we cannot synthesize benzene rings. However, we do not need tyrosine in our diets because we can synthesize the necessary amounts from phenylalanine. Although humans can synthesize arginine, it is needed for growth in greater amounts than can be synthesized. So arginine is an essential amino acid for children but a nonessential amino acid for adults.

Table 16.2 The Most Common Naturally Occurring Amino Acids
The amino acids are shown in the form that predominates at physiological pH (7.3).

	Formula	Name	Abbreviations		Average relative abundance in proteins
Aliphatic side chain amino acids	$H{-}CH({}^{+}NH_3){-}C(=O)O^-$	Glycine	Gly	G	7.5%
	$CH_3{-}CH({}^{+}NH_3){-}C(=O)O^-$	Alanine	Ala	A	9.0%
	$CH_3CH(CH_3){-}CH({}^{+}NH_3){-}C(=O)O^-$	Valine*	Val	V	6.9%
	$CH_3CH(CH_3)CH_2{-}CH({}^{+}NH_3){-}C(=O)O^-$	Leucine*	Leu	L	7.5%
	$CH_3CH_2CH(CH_3){-}CH({}^{+}NH_3){-}C(=O)O^-$	Isoleucine*	Ile	I	4.6%
Hydroxy-containing amino acids	$HOCH_2{-}CH({}^{+}NH_3){-}C(=O)O^-$	Serine	Ser	S	7.1%
	$CH_3CH(OH){-}CH({}^{+}NH_3){-}C(=O)O^-$	Threonine*	Thr	T	6.0%
Sulfur-containing amino acids	$HSCH_2{-}CH({}^{+}NH_3){-}C(=O)O^-$	Cysteine	Cys	C	2.8%
	$CH_3SCH_2CH_2{-}CH({}^{+}NH_3){-}C(=O)O^-$	Methionine*	Met	M	1.7%

(Continued)

Table 16.2 Continued

	Formula	Name	Abbreviations		Average relative abundance in proteins
Acidic amino acids	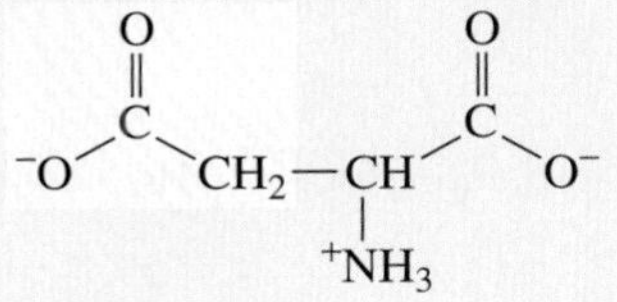 $^{-}OOC{-}CH_2{-}CH(^{+}NH_3){-}COO^{-}$	Aspartate (aspartic acid)	Asp	D	5.5%
	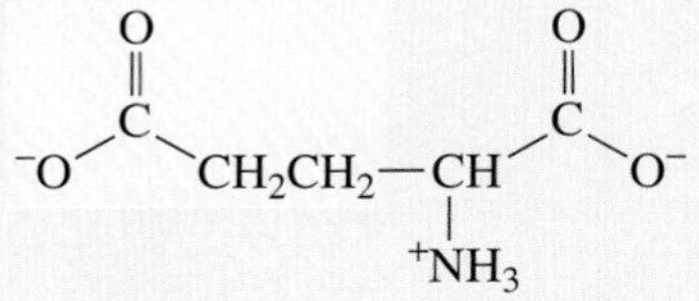 $^{-}OOC{-}CH_2CH_2{-}CH(^{+}NH_3){-}COO^{-}$	Glutamate (glutamic acid)	Glu	E	6.2%
Amides of acidic amino acids	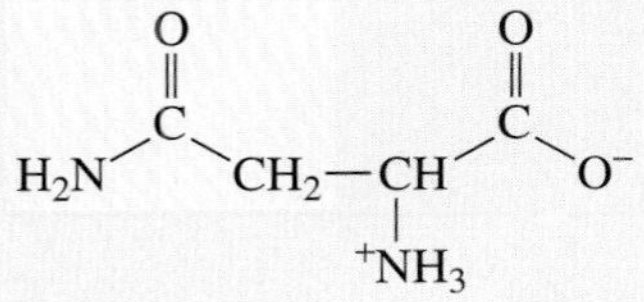 $H_2N{-}CO{-}CH_2{-}CH(^{+}NH_3){-}COO^{-}$	Asparagine	Asn	N	4.4%
	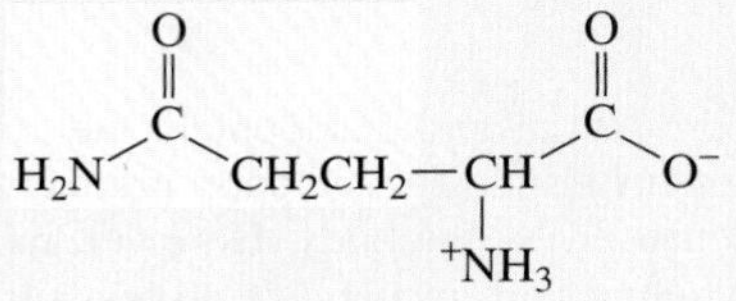 $H_2N{-}CO{-}CH_2CH_2{-}CH(^{+}NH_3){-}COO^{-}$	Glutamine	Gln	Q	3.9%
Basic amino acids	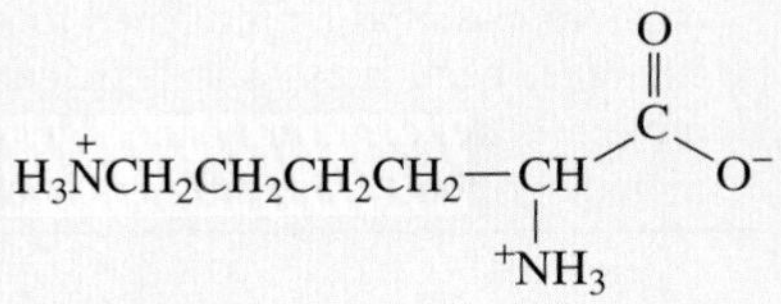 $H_3\overset{+}{N}CH_2CH_2CH_2CH_2{-}CH(^{+}NH_3){-}COO^{-}$	Lysine*	Lys	K	7.0%
	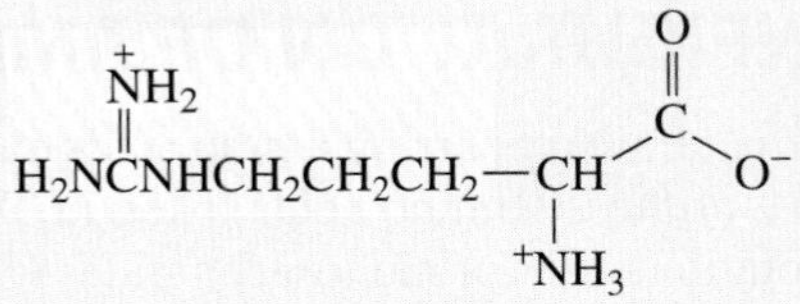 $H_2N\overset{\overset{+}{N}H_2}{\overset{\Vert}{C}}NHCH_2CH_2CH_2{-}CH(^{+}NH_3){-}COO^{-}$	Arginine*	Arg	R	4.7%
Benzene-containing amino acids	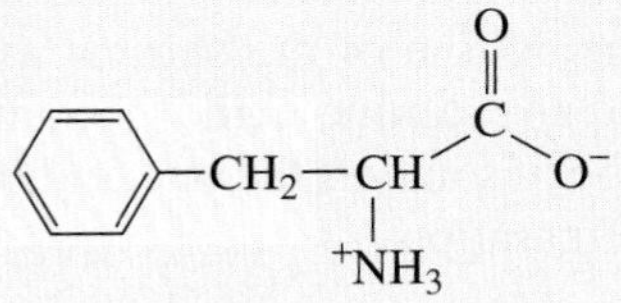 $C_6H_5{-}CH_2{-}CH(^{+}NH_3){-}COO^{-}$	Phenylalanine*	Phe	F	3.5%
	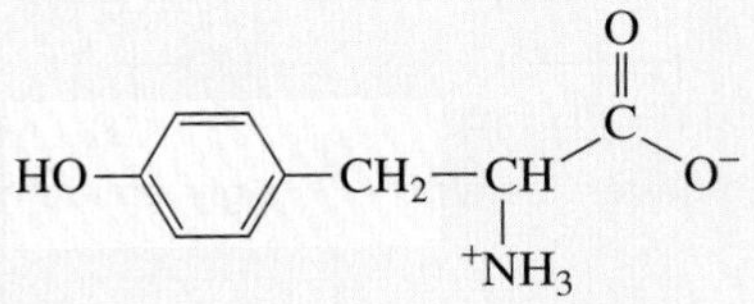 $HO{-}C_6H_4{-}CH_2{-}CH(^{+}NH_3){-}COO^{-}$	Tyrosine	Tyr	Y	3.5%
Heterocylic amino acids	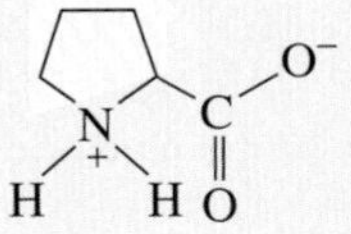 Pyrrolidine ring with ring $\overset{+}{N}H_2$; the ring carbon bonded to N carries H and COO^{-}	Proline	Pro	P	4.6%

(*Continued*)

Table 16.2 Continued

Formula	Name	Abbreviations		Average relative abundance in proteins
O ‖ C CH_2-CH O^- $^+NH_3$ N NH	Histidine*	His	H	2.1%
O ‖ C CH_2-CH O^- $^+NH_3$ N H	Tryptophan*	Trp	W	1.1%

* An essential amino acid.

PROTEINS AND NUTRITION

Proteins are an important component of our diets. Dietary protein is hydrolyzed in the body to individual amino acids. Some of these amino acids are used to synthesize proteins needed by the body, some are broken down further to supply energy to the body, and some are used as starting materials for the synthesis of nonprotein compounds that the body needs, such as thyroxine (Section 8.7), adrenaline, and melanin (Section 18.6).

Not all proteins contain the same amino acids. Most proteins from meat and dairy products contain all the amino acids needed by the body. However, most proteins from vegetable sources are *incomplete* proteins; they contain too little of one or more essential amino acids to support human growth. For example, bean protein is deficient in methionine, and wheat protein is deficient in lysine. Therefore, a balanced diet must include proteins from different sources.

16.2 THE CONFIGURATION OF AMINO ACIDS

Naturally occurring monosaccharides have the D-configuration.

Naturally occurring amino acids have the L-configuration.

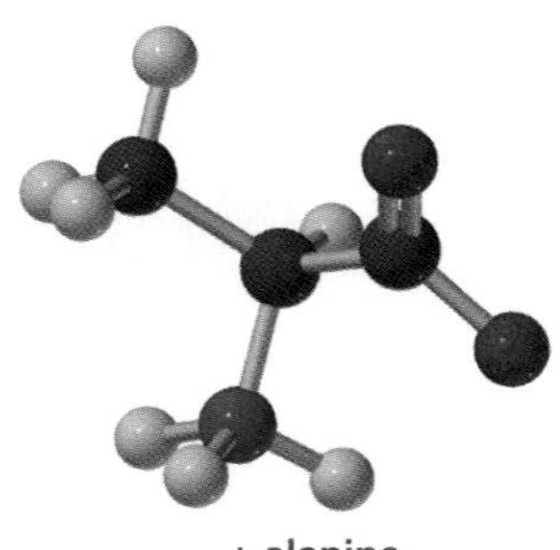
L-alanine
an amino acid

The *α-carbon* of all the naturally occurring amino acids except glycine is an asymmetric center. Therefore, 19 of the 20 amino acids listed in Table 16.2 can exist as enantiomers. The D and L notation used for monosaccharides (Section 15.2) is also used for amino acids. An amino acid drawn in a Fischer projection with the carboxyl group on the top and the R group on the bottom of the vertical axis is a **D-amino acid** if the amino group is on the right, and an **L-amino acid** if the amino group is on the left. Unlike monosaccharides, where the D isomer is the one found in nature, most amino acids found in nature have the L configuration.

D-glyceraldehyde	L-glyceraldehyde
HC=O H—┼—OH CH_2OH	HC=O HO—┼—H CH_2OH

D-amino acid	L-amino acid
COO^- H—┼—$\overset{+}{N}H_3$ R	COO^- $H_3\overset{+}{N}$—┼—H R

Why D-sugars and L-amino acids? Although it makes no difference which isomer nature "selected" to be synthesized, it is important that the same isomer is synthesized by all organisms. For example, since mammals ended up having L-amino acids, L-amino acids must be the isomers synthesized by the organisms that mammals depend on for food.

AMINO ACIDS AND DISEASE

The Chamorro people of Guam have a high incidence of a syndrome that resembles amyotrophic lateral sclerosis (ALS or Lou Gehrig's disease) with elements of Parkinson's disease and dementia. This syndrome developed during World War II when, as a result of food shortages, the tribe ate large quantities of *Cycas circinalis* seeds. These seeds contain β-methylamino-L-alanine, an amino acid that binds to glutamate receptors. When monkeys are given β-methylamino-L-alanine, they develop some of the features of this syndrome. There is hope that, by studying this unusual amino acid, we may gain an understanding of how ALS and Parkinson's disease arise.

COO^-, CH_3, C, H, $^+NH_3$
L-alanine

COO^-, $CH_3\overset{+}{N}H_2CH_2$, C, H, $^+NH_3$
β-methylamino-L-alanine

A PEPTIDE ANTIBIOTIC

Gramicidin S, an antibiotic produced by a strain of bacteria, is a cyclic decapeptide. Notice that one of its residues is ornithine, an amino acid not listed in Table 16.2 because it occurs rarely in nature. Ornithine resembles lysine but has one less methylene group in its side chain. Notice also that the antibiotic contains two D-amino acids.

L-Val, L-Orn, L-Leu, D-Phe, L-Pro, L-Val, D-Orn, L-Leu, L-Phe, L-Pro
gramicidin S

$H_3\overset{+}{N}CH_2CH_2CH_2CH(^+NH_3)C(=O)O^-$
ornithine

PROBLEM 1 ◆

Which isomer—(*R*)-alanine or (*S*)-alanine—is L-alanine?

PROBLEM 2 SOLVED

Threonine has two asymmetric centers and therefore has four stereoisomers.

	1	2	3	4
top	COO^-	COO^-	COO^-	COO^-
C-2	H—C—$\overset{+}{N}H_3$	$H_3\overset{+}{N}$—C—H	H—C—$\overset{+}{N}H_3$	$H_3\overset{+}{N}$—C—H
C-3	H—C—OH	HO—C—H	HO—C—H	H—C—OH
bottom	CH_3	CH_3	CH_3	CH_3

Naturally occurring L-threonine is (2*S*,3*R*)-threonine. Which of the stereoisomers is L-threonine?

Solution Stereoisomer number 1 has the *R* configuration at both C-2 and C-3 because in both cases the arrow drawn from the highest to the next highest priority substituent is counterclockwise. In both cases, counterclockwise signifies *R* because the lowest-priority substituent (H) is on a horizontal bond (Section 15.2). Therefore, the configuration of (2*S*,3*R*)-threonine is the opposite of that in stereoisomer number 1 at C-2 and the same as that in stereoisomer number 1 at C-3. Thus, L-threonine is stereoisomer number 4. Notice that the $^+NH_3$ group is on the left, just as we would expect for the Fischer projection of an L-amino acid.

3-D Molecules: Common naturally occurring amino acids.

PROBLEM 3 ◆

Do any other amino acids in Table 16.2 have more than one asymmetric center?

16.3 THE ACID–BASE PROPERTIES OF AMINO ACIDS

Every amino acid has a carboxyl group and an amino group, and each group can exist in an acidic form or a basic form, depending on the pH of the solution in which the amino acid is dissolved. Some amino acids, such as aspartate and glutamate, also have an ionizable side chain (Table 16.3).

Table 16.3 The pK_a Values of Amino Acids

Amino acid	pK_a α-COOH	pK_a α-$\overset{+}{N}H_3$	pK_a side chain
Alanine	2.34	9.69	—
Arginine	2.17	9.04	12.48
Asparagine	2.02	8.84	—
Aspartic acid	2.09	9.82	3.86
Cysteine	1.92	10.46	8.35
Glutamic acid	2.19	9.67	4.25
Glutamine	2.17	9.13	—
Glycine	2.34	9.60	—
Histidine	1.82	9.17	6.04
Isoleucine	2.36	9.68	—
Leucine	2.36	9.60	—
Lysine	2.18	8.95	10.79
Methionine	2.28	9.21	—
Phenylalanine	2.16	9.18	—
Proline	1.99	10.60	—
Serine	2.21	9.15	—
Threonine	2.63	9.10	—
Tryptophan	2.38	9.39	—
Tyrosine	2.20	9.11	10.07
Valine	2.32	9.62	—

The acidic form (with the proton) predominates if the pH of the solution is less than the pK_a of the compound, and the basic form (without the proton) predominates if the pH of the solution is greater than the pK_a of the compound.

We have seen that compounds exist primarily in their acidic forms (that is, with their protons attached) in solutions that are more acidic than their pK_a values and primarily in their basic forms (that is, without their protons) in solutions that are more basic than their pK_a values (Section 2.7). The carboxyl groups of the amino acids have pK_a values of approximately 2, and the protonated amino groups have pK_a values near 9. Both groups, therefore, will be in their acidic forms in a very acidic solution (pH ~ 0). At pH = 7, the pH of the solution is greater than the pK_a of the carboxyl group, but less than the pK_a of the protonated amino group. The carboxyl group, therefore, will be in its basic form, and the amino group will be in its acidic form. In a strongly basic solution (pH ~ 11), both groups will be in their basic forms.

$$\underset{\text{pH = 0}}{\mathrm{R{-}CH(\overset{+}{N}H_3){-}C({=}O){-}OH}} \rightleftharpoons \underset{\substack{\text{a zwitterion} \\ \text{pH = 7}}}{\mathrm{R{-}CH(\overset{+}{N}H_3){-}C({=}O){-}O^-}} + H^+ \rightleftharpoons \underset{\text{pH = 11}}{\mathrm{R{-}CH(NH_2){-}C({=}O){-}O^-}} + H^+$$

Notice that an amino acid can never exist as an uncharged compound, regardless of the pH of the solution. To be uncharged, an amino acid would have to lose a proton from an $^{+}NH_3$ group with a pK_a of about 9 before it would lose a proton from a COOH group with a pK_a of about 2. This clearly is impossible: a weak acid (pK_a = 9) cannot be more acidic than a strong acid (pK_a = 2). Therefore, at physiological pH (7.3) an amino acid such as alanine exists as a dipolar ion, called a *zwitterion*. A **zwitterion** is a compound that has a negative charge on one atom and a positive charge on a nonadjacent atom. (The name comes from *zwitter*, German for "hermaphrodite" or "hybrid.")

PROBLEM 4

Why are the carboxylic acid groups of the amino acids so much more acidic (pK_a ~ 2) than a carboxylic acid such as acetic acid (pK_a = 4.76)?

PROBLEM 5

Explain why, unlike most amines and carboxylic acids, amino acids are insoluble in diethyl ether.

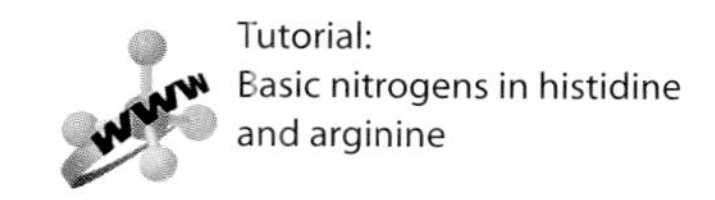

PROBLEM 6 ◆ *SOLVED*

Draw the predominant form for each of the following amino acids at physiological pH (7.3):

a. aspartic acid **b.** glutamine **c.** arginine

Solution To 6a Both carboxyl groups are in their basic forms because the pH of the solution is greater than their pK_a values. The protonated amino group is in its acidic form because the pH of the solution is less than its pK_a value.

$$^{-}O-\overset{O}{\overset{\|}{C}}-CH_2\underset{^{+}NH_3}{\underset{|}{CH}}-\overset{O}{\overset{\|}{C}}-O^{-}$$

PROBLEM 7

Draw the form of glutamic acid that predominates in a solution with the following pH:

a. pH = 0 **b.** pH = 3 **c.** pH = 6 **d.** pH = 11

16.4 THE ISOELECTRIC POINT

The **isoelectric point** (pI) of an amino acid is the pH at which it has no net charge. In other words, it is the pH at which the amount of positive charge on an amino acid exactly balances the amount of negative charge:

pI (isoelectric point) = pH at which there is no net charge

The pI of an amino acid that does *not* have an ionizable side chain—such as alanine—is midway between its two pK_a values.

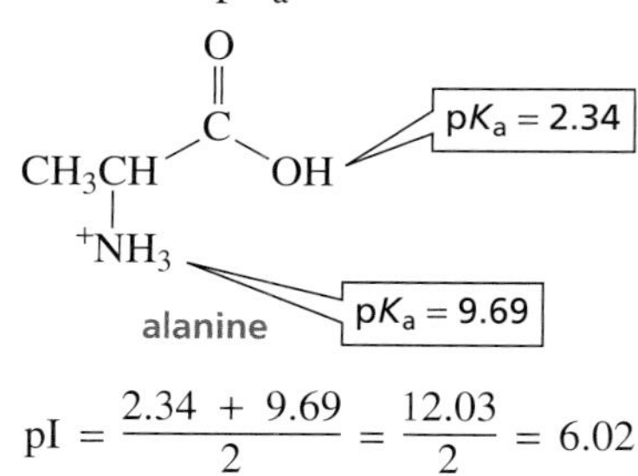

$$pI = \frac{2.34 + 9.69}{2} = \frac{12.03}{2} = 6.02$$

The pI of an amino acid that *has* an ionizable side chain is the average of the pK_a values of the similarly ionizing groups (either positively charged groups becoming uncharged groups or uncharged groups becoming negatively charged groups). For example, the pI of lysine is the average of the pK_a values of the two groups that are positively charged in their acidic form and uncharged in their basic form. The pI of glutamic acid, on the other hand, is the average of the pK_a values of the two groups that are uncharged in their acidic form and negatively charged in their basic form.

$$\overset{+}{H_3N}CH_2CH_2CH_2CH_2CH(\overset{+}{N}H_3)COOH$$

lysine: $pK_a = 10.79$ (side-chain $\overset{+}{N}H_3$); $pK_a = 2.18$ (COOH); $pK_a = 8.95$ (α-$\overset{+}{N}H_3$)

$$pI = \frac{8.95 + 10.79}{2} = \frac{19.74}{2} = 9.87$$

$$HOOCCH_2CH_2CH(\overset{+}{N}H_3)COOH$$

glutamic acid: $pK_a = 4.25$ (side-chain COOH); $pK_a = 2.19$ (COOH); $pK_a = 9.67$ ($\overset{+}{N}H_3$)

$$pI = \frac{2.19 + 4.25}{2} = \frac{6.44}{2} = 3.22$$

PROBLEM 8 ♦

Calculate the pI of the following amino acids:

a. asparagine
b. arginine
c. serine
d. aspartic acid

PROBLEM 9 ♦

a. Which amino acid has the lowest pI value?
b. Which amino acid has the highest pI value?

16.5 SEPARATION OF AMINO ACIDS

A mixture of amino acids can be separated by several different techniques. Two are discussed below.

Electrophoresis

Electrophoresis separates amino acids on the basis of their pI values. A few drops of a solution of an amino acid mixture are applied to the middle of a piece of filter paper or to a gel. When the paper or the gel is placed in a buffered solution between two electrodes and an electric field is applied (Figure 16.1), an amino acid with a pI greater

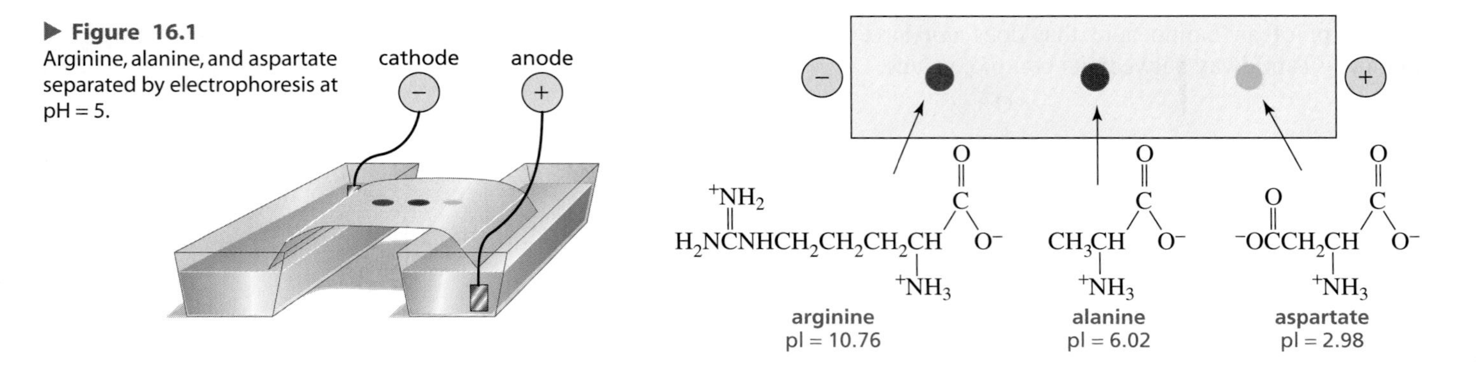

▶ **Figure 16.1**
Arginine, alanine, and aspartate separated by electrophoresis at pH = 5.

than the pH of the solution will have an overall positive charge and will migrate toward the cathode (the negative electrode). The farther the amino acid's pI is from the pH of the buffered solution, the more positive the amino acid will be and the farther it will migrate toward the cathode. An amino acid with a pI less than the pH of the buffered solution will have an overall negative charge and will migrate toward the anode (the positive electrode). If two molecules have the same charge, the larger one will move more slowly during electrophoresis because the same charge has to move a greater mass.

An amino acid will be positively charged if the pH of the solution is less than the pI of the amino acid and will be negatively charged if the pH of the solution is greater than the pI of the amino acid.

Considering that amino acids are colorless, how can we detect them after they have been separated? After the amino acids have been separated by electrophoresis, the filter paper is sprayed with ninhydrin and dried in a warm oven. Most amino acids form a purple product when they are heated with ninhydrin. The number of different kinds of amino acids in the mixture is determined by the number of colored spots on the filter paper (Figure 16.1). The individual amino acids are identified by their location on the paper compared with a standard.

the reaction of an amino acid with ninhydrin to form a colored product

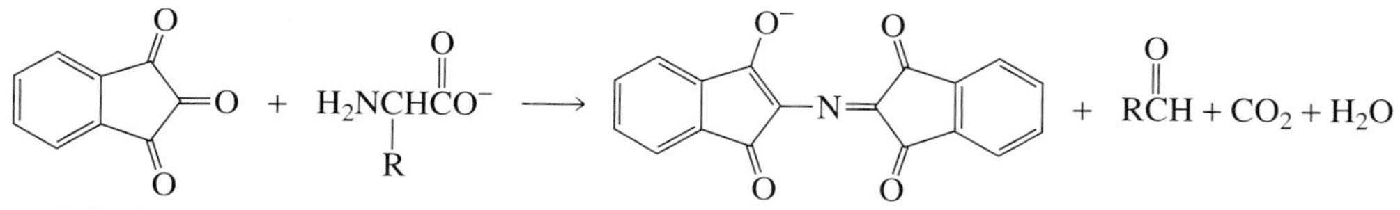

ninhydrin + H_2NCHCO^- (R) ⟶ purple-colored product + $RCHO + CO_2 + H_2O$

ninhydrin   an amino acid   purple-colored product

PROBLEM 10 ♦

What aldehyde is formed when valine is treated with ninhydrin?

Ion-Exchange Chromatography

A technique called **ion-exchange chromatography** can not only identify amino acids, but it can also determine the relative amounts of the amino acids present in the mixture. This technique employs a column packed with an insoluble resin. A solution of a mixture of amino acids is loaded onto the top of the column, and a series of buffer solutions of increasing pH are poured through the column. The amino acids separate because they flow through the column at different rates, as explained below.

The resin is a chemically inert material with charged side chains. The structure of a commonly used resin is shown in Figure 16.2. If a mixture of lysine and glutamate in a solution with a pH of 6 were to be loaded onto the column, glutamate would travel down the column rapidly because its negatively charged side chain would be repelled by the negatively charged sulfonic acid groups of the resin. The positively charged side chain of lysine, on the other hand, would cause that amino acid to be retained on

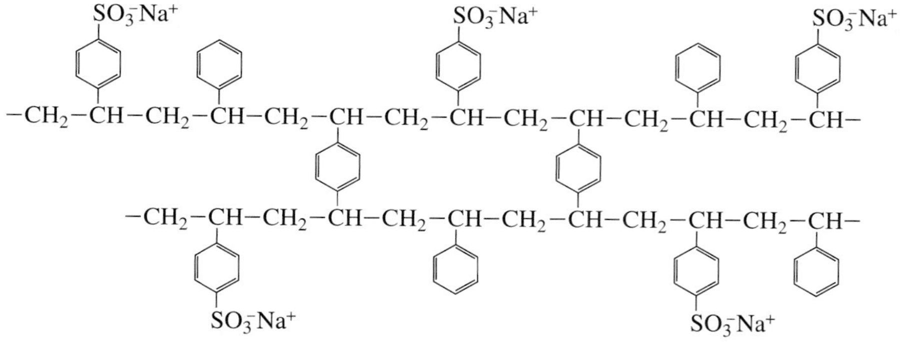

$SO_3^-Na^+$ $SO_3^-Na^+$ $SO_3^-Na^+$

$-CH_2-CH-CH_2-CH-CH_2-CH-CH_2-CH-CH_2-CH-CH_2-CH-CH_2-CH-$

$-CH_2-CH-CH_2-CH-CH_2-CH-CH_2-CH-CH_2-CH-CH_2-CH-$

$SO_3^-Na^+$ $SO_3^-Na^+$

◀ **Figure 16.2**
A section of a cation-exchange resin.

the column. This kind of resin is called a **cation-exchange resin** because it exchanges the Na^+ counterions of the SO_3^- groups for the positively charged species traveling through the column. In addition, the relatively nonpolar nature of the column causes it to retain nonpolar amino acids longer than polar amino acids.

An **amino acid analyzer** is an instrument that automates ion-exchange chromatography. When a solution of an amino acid mixture passes through the column of an amino acid analyzer containing a cation-exchange resin, the amino acids move through the column at different rates, depending on their overall charge. The solution that flows out of the column is collected in fractions, and these are collected often enough that a different amino acid ends up in each one (Figure 16.3).

Cations bind most strongly to cation-exchange resins.

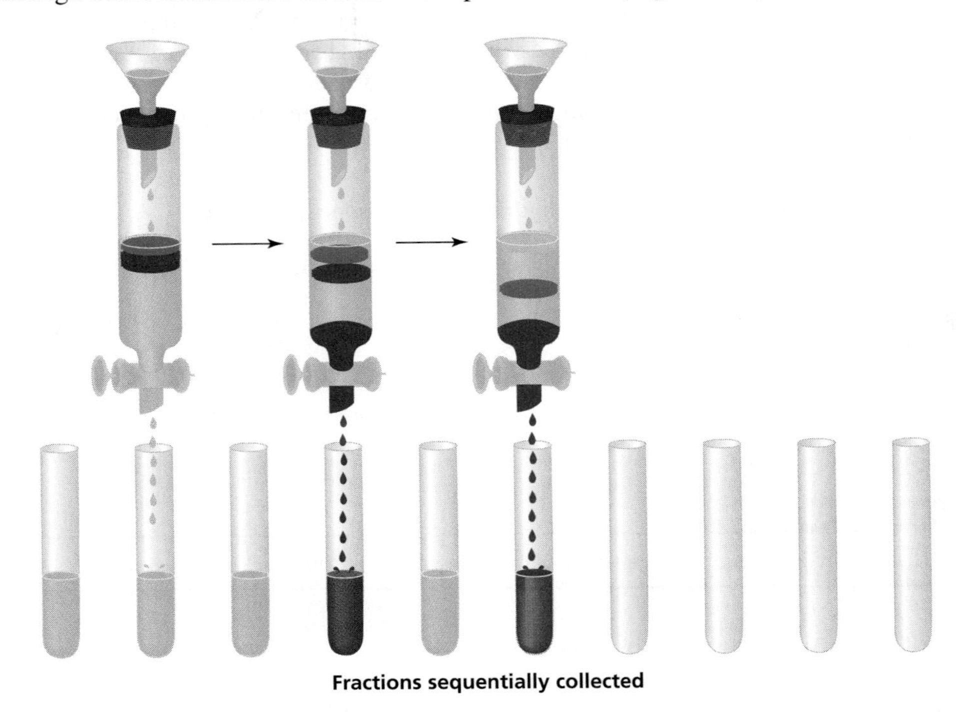

▶ **Figure 16.3**
Separation of amino acids by ion-exchange chromatography.

If ninhydrin is added to each of the fractions, the relative amount of each amino acid in each fraction can be determined using visible spectroscopy (Section 7.10) to measure the intensity of the absorbance of the colored compound formed by the reaction of the amino acid with ninhydrin (Figure 16.4).

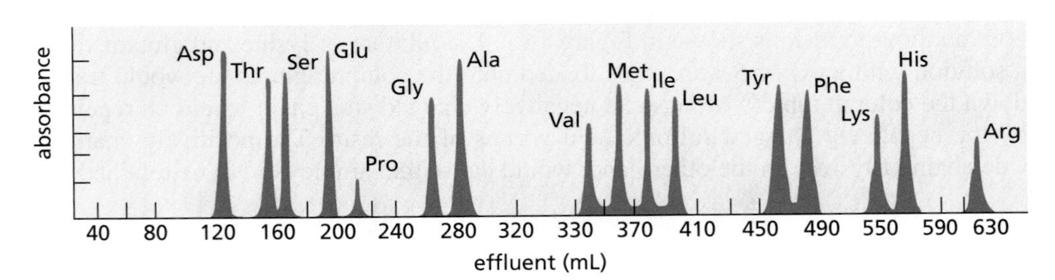

▲ **Figure 16.4**
A typical chromatogram obtained from the separation of a mixture of amino acids using an automated amino acid analyzer.

Tutorial:
Electrophoresis and pI

PROBLEM 11

Explain the order of elution (with a solution of pH = 4) of each of the following pairs of amino acids on a column packed with the resin shown in Figure 16.2:

a. glutamate before threonine **b.** alanine before leucine

WATER SOFTENERS: EXAMPLES OF CATION-EXCHANGE CHROMATOGRAPHY

Water-softening systems contain a column with a cation-exchange resin that has been flushed with concentrated sodium chloride. When "hard water" (water with high levels of calcium and magnesium ions; see Section 11.10) passes through the column, the resin binds magnesium and calcium ions more tightly than it binds sodium ions. The water softener thus removes magnesium and calcium ions from the water and replaces them with sodium ions. The resin must be recharged from time to time by being flushed with concentrated sodium chloride, thereby replacing the bound magnesium and calcium ions with sodium ions.

16.6 PEPTIDE BONDS AND DISULFIDE BONDS

Peptide bonds and disulfide bonds are the only covalent bonds that join amino acid residues together in a peptide or a protein.

Peptide Bonds

The amide bonds that link amino acid residues are called **peptide bonds**. By convention, peptides and proteins are written with the free amino group (of the **N-terminal amino acid**) on the left and the free carboxyl group (of the **C-terminal amino acid**) on the right.

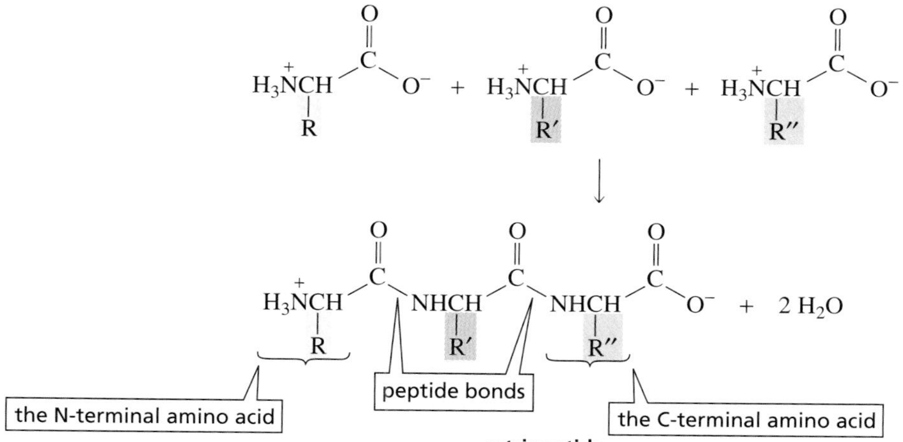

$$H_3\overset{+}{N}CH(R)C(=O)O^- + H_3\overset{+}{N}CH(R')C(=O)O^- + H_3\overset{+}{N}CH(R'')C(=O)O^-$$

$$\downarrow$$

$$H_3\overset{+}{N}CH(R)C(=O)NHCH(R')C(=O)NHCH(R'')C(=O)O^- + 2\,H_2O$$

the N-terminal amino acid

peptide bonds

the C-terminal amino acid

a tripeptide

When the identities of the amino acids in a peptide are known but their sequence is not known, the amino acids are written separated by commas. When the sequence of amino acids is known, the amino acids are written connected by hyphens. In the pentapeptide represented on the right below, valine is the N-terminal amino acid and histidine is the C-terminal amino acid. The amino acids are numbered starting with the N-terminal end. The glutamate residue is therefore referred to as Glu 4 because it is the fourth amino acid from the N-terminal end.

Glu, Cys, His, Val, Ala

the pentapeptide contains the indicated amino acids, but their sequence is not known

Val-Cys-Ala-Glu-His

the amino acids in the pentapeptide have the indicated sequence

A peptide bond has about 40% double-bond character because of electron delocalization (Section 7.4).

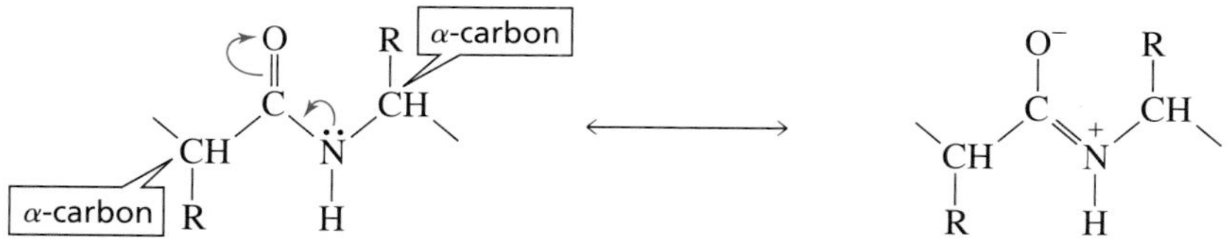

$$-CH(R)-C(=O)-\ddot{N}(H)-CH(R)- \longleftrightarrow -CH(R)-C(O^-)=\overset{+}{N}(H)-CH(R)-$$

α-carbon

α-carbon

resonance contributors

The partial double-bond character prevents free rotation about the peptide bond; therefore, the carbon and nitrogen atoms of the peptide bond and the two atoms to which each is attached are held rigidly in a plane (Figure 16.5).

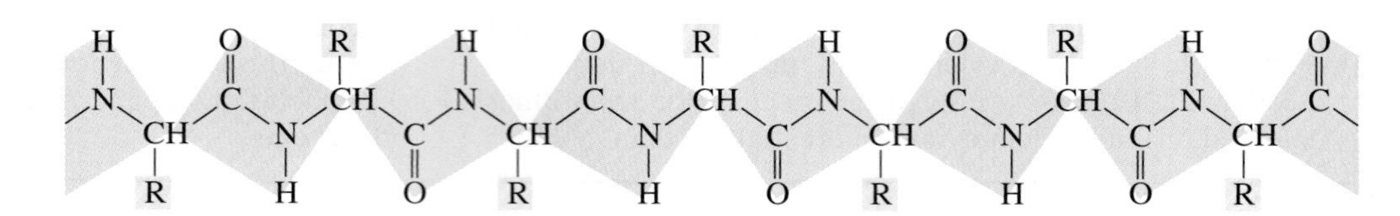

▶ **Figure 16.5**
A segment of a polypeptide chain. Colored squares indicate the plane defined by each peptide bond. Notice that the R groups bonded to the α-carbons are on alternate sides of the peptide backbone.

PROBLEM 12

Draw Gly-Val and Val-Gly.

PROBLEM 13

Draw the tetrapeptide Ala-Thr-Asp-Asn and indicate the peptide bonds.

PROBLEM 14◆

Using the three-letter abbreviations, write the six tripeptides consisting of Ala, Gly, and Met.

PROBLEM 15

Which bonds in the backbone of a peptide can rotate freely?

ENKEPHALINS

Enkephalins are pentapeptides that are synthesized by the body to control pain. They decrease the body's sensitivity to pain by binding to receptors in certain brain cells. Part of the three-dimensional structures of enkephalins must be similar to those of morphine and related painkillers such as Demerol because they bind to the same receptors (Section 21.6).

Tyr-Gly-Gly-Phe-Leu	Tyr-Gly-Gly-Phe-Met
leucine enkephalin	**methionine enkephalin**

Disulfide Bonds

When thiols are oxidized under mild conditions, they form disulfides. A **disulfide** is a compound with an S—S bond.

$$2\,R{-}SH \xrightarrow{\text{mild oxidation}} RS{-}SR$$

a thiol → **a disulfide**

Disulfides are reduced to thiols. Thiols are oxidized to disulfides.

Because thiols can be oxidized to disulfides, disulfides can be reduced to thiols.

$$RS{-}SR \xrightarrow{\text{reduction}} 2\,R{-}SH$$

a disulfide → **a thiol**

Cysteine is an amino acid that contains a thiol group. Two cysteine molecules therefore can be oxidized to a disulfide. This disulfide is called cystine.

$$2\,HSCH_2CH(\overset{+}{N}H_3)C(=O)O^- \xrightarrow{\text{mild oxidation}} {}^-OC(=O)CH(\overset{+}{N}H_3)CH_2S{-}SCH_2CH(\overset{+}{N}H_3)C(=O)O^-$$

cysteine a thiol → **cystine a disulfide**

Two cysteine residues in a protein can be oxidized to a disulfide, creating a bond known as a **disulfide bridge**. Disulfide bridges contribute to the overall shape of a protein by linking cysteine residues found in different parts of the peptide backbone, as shown in Figure 16.6.

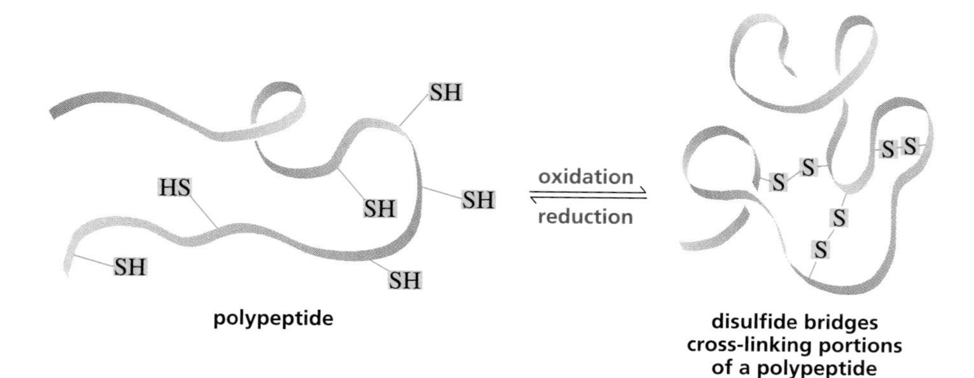

◀ **Figure 16.6**
Disulfide bridges cross-linking portions of a peptide.

The hormone insulin, secreted by the pancreas, controls the level of glucose in the blood by regulating glucose metabolism. Insulin is a polypeptide with two peptide chains. It has three disulfide bridges, two of which hold the two chains together.

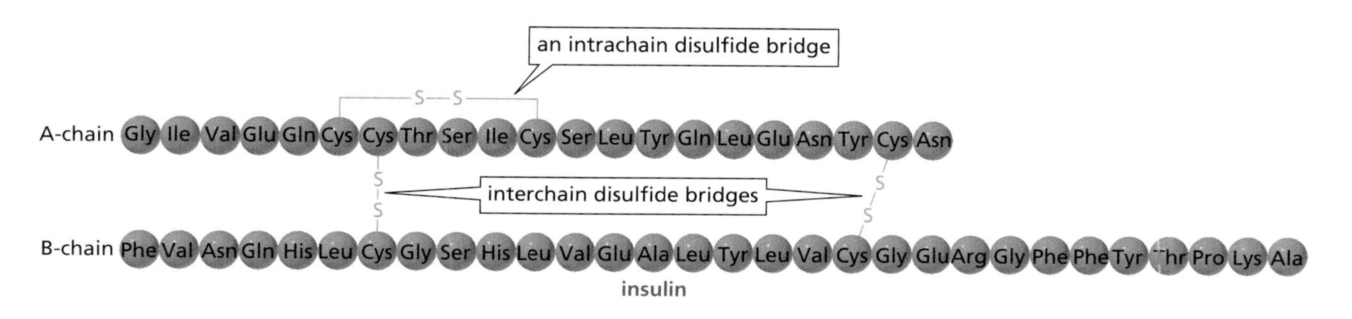

HAIR: STRAIGHT OR CURLY?

Hair is made up of a protein called keratin that contains an unusually large number of cysteine residues. These furnish keratin with many disulfide bridges that preserve its three-dimensional structure. People can alter the structure of their hair (if they think it is either too straight or too curly) by changing the location of these disulfide bridges. This change is accomplished by first applying a reducing agent to the hair to reduce all the disulfide bridges in the protein strands. Then, after rearranging the hair into the desired shape (using curlers to curl it or combing it straight to uncurl it), an oxidizing agent is applied to form new disulfide bridges. The new disulfide bridges maintain the hair in its new shape. When this treatment is applied to straight hair, it is called a "permanent." When it is applied to curly hair, it is called "hair straightening."

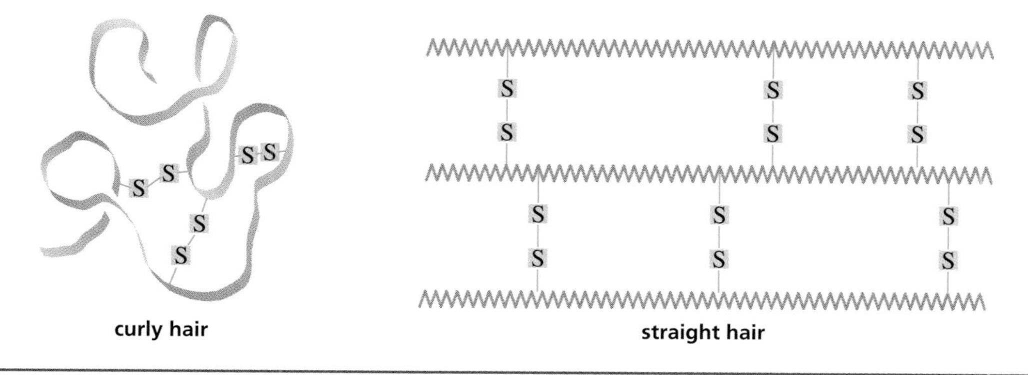

PEPTIDE HORMONES

Bradykinin, vasopressin, and oxytocin are peptide hormones. They are all nonapeptides. Bradykinin inhibits the inflammation of tissues. Vasopressin controls blood pressure by regulating the contraction of smooth muscle; it is also an antidiuretic. Oxytocin induces labor in pregnant women by stimulating the uterine muscle to contract, and it also stimulates milk production in nursing mothers. Vasopressin and oxytocin both have a disulfide bridge, and their C-terminal amino acids contain amide rather than carboxyl groups. Notice that the C-terminal amide group is indicated by writing "NH_2" after the name of the C-terminal amino acid. In spite of their very different physiological effects, vasopressin and oxytocin differ only by two amino acids.

bradykinin Arg-Pro-Pro-Gly-Phe-Ser-Pro-Phe-Arg

vasopressin Cys-Tyr-Phe-Gln-Asn-Cys-Pro-Arg-Gly-NH_2 (S—S bridge between the two Cys)

oxytocin Cys-Tyr-Ile-Gln-Asn-Cys-Pro-Leu-Gly-NH_2 (S—S bridge between the two Cys)

Oxytocin was the first small peptide to be synthesized. This was accomplished in 1953 by **Vincent du Vigneaud (1901–1978)**, *who later synthesized vasopressin. Du Vigneaud was born in Chicago and was a professor at George Washington University Medical School and later at Cornell University Medical College. For synthesizing these nonapeptides, he received the Nobel Prize in chemistry in 1955.*

PROBLEM 16 ♦

Glutathione is a tripeptide whose function is to destroy harmful oxidizing agents in the body. Oxidizing agents are thought to be responsible for some of the effects of aging and to play a causative role in cancer. Glutathione removes oxidizing agents by reducing them and as a result glutathione is oxidized, resulting in the formation of a disulfide bond between two glutathione molecules (see pages 448 and 460).

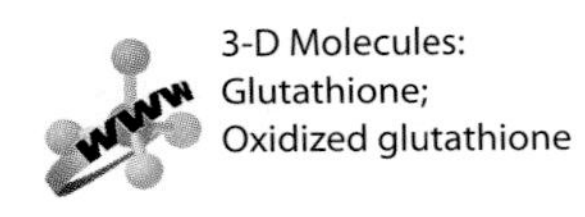

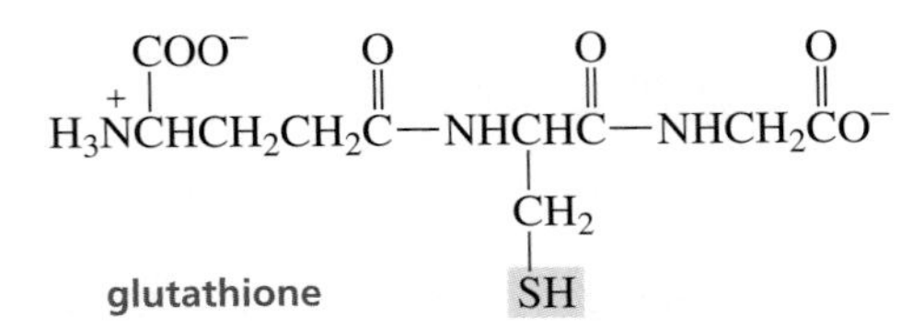

a. What amino acids make up glutathione?
b. What is unusual about glutathione's structure? (If you cannot answer this question, draw the structure you would expect for the tripeptide, and compare your structure with the actual structure of glutathione.)

16.7 THE STRATEGY OF PEPTIDE BOND SYNTHESIS: N-PROTECTION AND C-ACTIVATION

One difficulty in synthesizing a polypeptide once its structure is known is that the amino acids have two functional groups that enable them to combine in various ways. For example, suppose you wanted to make the dipeptide Gly-Ala. That dipeptide is only one of four possible dipeptides that could be formed from a mixture of alanine and glycine.

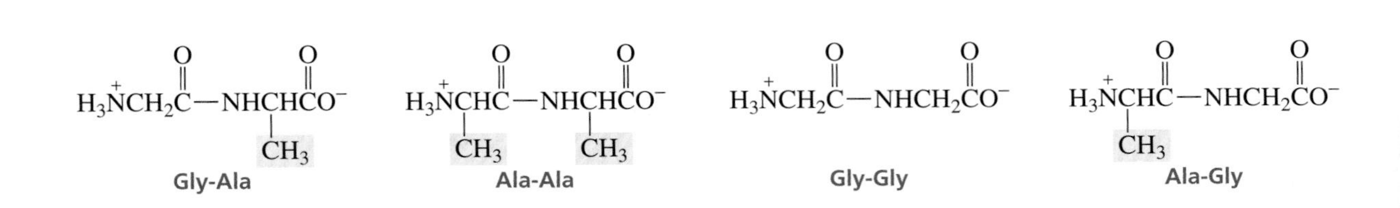

If the amino group of the amino acid that is to be on the N-terminal end (in this case, Gly) is protected, it will not be available to form a peptide bond. If the carboxyl group of this same amino acid is activated before the second amino acid is added, the amino group of the added amino acid (in this case, Ala) will react with the activated carboxyl group of glycine in preference to reacting with a nonactivated carboxyl group of another alanine molecule.

glycine: $H_2NCH_2CO^-$ (C=O) alanine: $H_2NCH(CH_3)CO^-$ (C=O)

protect

activate

peptide bond is formed between these groups

The N-terminal amino acid must have its amino group protected and its carboxyl group activated.

The reagent most often used to protect the amino group of an amino acid is di-*tert*-butyl dicarbonate. Its popularity is due to the ease with which the protecting group can be removed when the need for protection is over.

$$(CH_3)_3C{-}O{-}\overset{O}{\overset{\|}{C}}{-}O{-}\overset{O}{\overset{\|}{C}}{-}O{-}C(CH_3)_3 + H_2NCH_2\overset{O}{\overset{\|}{C}}O^- \longrightarrow (CH_3)_3C{-}O{-}\overset{O}{\overset{\|}{C}}{-}NHCH_2\overset{O}{\overset{\|}{C}}O^- + CO_2 + HO{-}C(CH_3)_3$$

di-*tert*-butyl dicarbonate glycine N-protected glycine

protecting group

The reagent most often used to activate the carboxyl group is dicyclohexylcarbodiimide (DCC). DCC activates a carboxyl group by putting a good leaving group on the carbonyl carbon.

$$(CH_3)_3C{-}O{-}\overset{O}{\overset{\|}{C}}{-}NHCH_2\overset{O}{\overset{\|}{C}}OH + C_6H_{11}N{=}C{=}NC_6H_{11} \xrightleftharpoons{\text{proton transfer}} (CH_3)_3C{-}O{-}\overset{O}{\overset{\|}{C}}{-}NHCH_2\overset{O}{\overset{\|}{C}}O^- + C_6H_{11}N{=}C{=}\overset{+}{N}HC_6H_{11}$$

N-protected amino acid

dicyclohexylcarbodiimide DCC

$$\downarrow$$

$$(CH_3)_3C{-}O{-}\overset{O}{\overset{\|}{C}}{-}NHCH_2\overset{O}{\overset{\|}{C}}O{-}C({=}NC_6H_{11}){-}NHC_6H_{11}$$

protected

activated

activating group

After the amino acid has its N-terminal group protected and its C-terminal group activated, the second amino acid is added. The unprotected amino group of the second amino acid attacks the activated carboxyl group and forms a new peptide bond.

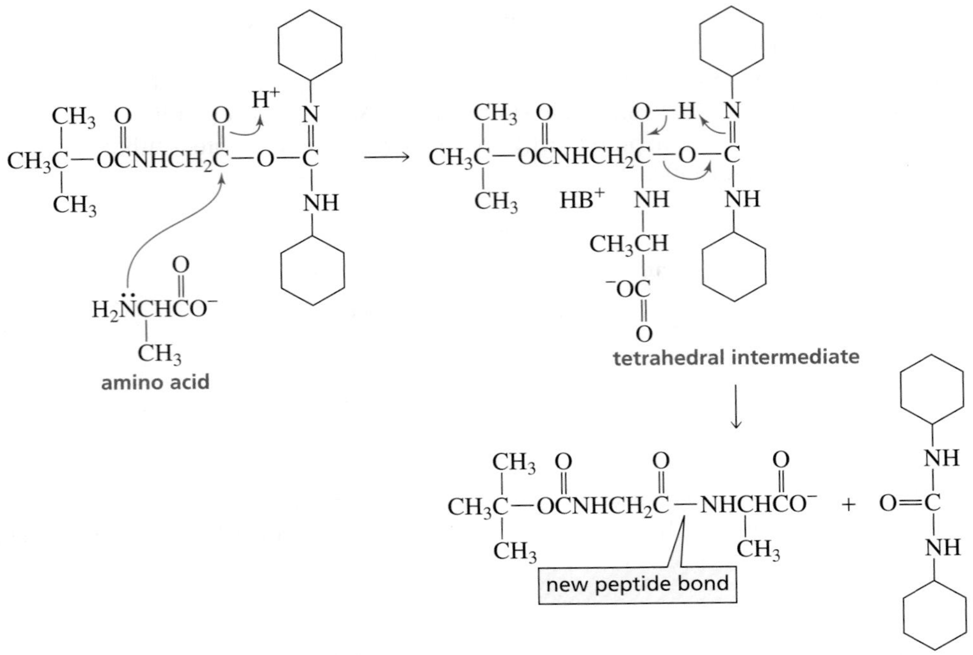

amino acid

tetrahedral intermediate

new peptide bond

Amino acids can be added to the growing C-terminal end by repeating these two steps: activating the carboxyl group of the C-terminal amino acid of the peptide by treating it with DCC and then adding a new amino acid.

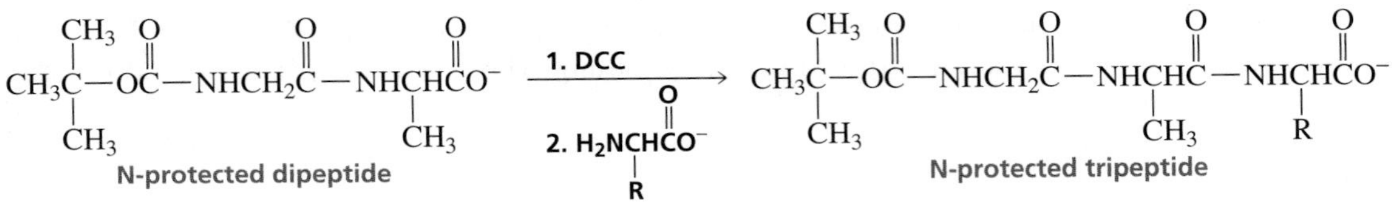

1. DCC

2. H_2NCHCO^- (R)

N-protected dipeptide

N-protected tripeptide

When the desired number of amino acids has been added to the chain, the protecting group on the N-terminal amino acid is removed.

Theoretically, one should be able to make as long a peptide as desired with this technique. Reactions do not produce 100% yields, however, and the yields are further decreased during the purification process. After each step of the synthesis, the peptide must be purified to prevent subsequent unwanted reactions with leftover reagents. Assuming that each amino acid can be added to the growing end of the peptide chain in an 80% yield, the overall yield of a nonapeptide will be only 17%. It is clear that large polypeptides could never be synthesized in this way.

Number of amino acids	**2**	**3**	**4**	**5**	**6**	**7**	**8**	**9**
Overall yield	80%	64%	51%	41%	33%	26%	21%	17%

Since the early 1980s, it has been possible to synthesize proteins by genetic engineering techniques. Strands of DNA, introduced into bacterial cells, will cause the cells to produce large amounts of a desired protein (Section 20.12). For example, mass quantities of human insulin are produced from genetically modified *E. coli*.

NUTRASWEET

The synthetic sweetener aspartame, or NutraSweet (Section 15.17), is the methyl ester of a dipeptide of L-aspartate and L-phenylalanine. Aspartame is about 200 times sweeter than glucose. The ethyl ester of the same dipeptide is not sweet. If a D-amino acid is substituted for either of the L-amino acids of aspartame, the resulting dipeptide is bitter rather than sweet.

$H_3\overset{+}{N}CH(CH_2COO^-)C(=O)-NHCH(CH_2C_6H_5)C(=O)OCH_3$

aspartame
NutraSweet

PROBLEM 17 ◆

What dipeptides would be formed by heating a mixture of valine and N-protected leucine?

PROBLEM 18 ◆

Calculate the overall yield of a nonapeptide if the yield for the addition of each amino acid to the chain is 70%.

16.8 AN INTRODUCTION TO PROTEIN STRUCTURE

Proteins are described by four levels of structure. The **primary structure** of a protein is the sequence of amino acids in the chain and the location of all the disulfide bridges. **Secondary structure** describes the regular conformations assumed by segments of the protein's backbone when it folds. The **tertiary structure** is the three-dimensional structure of the entire polypeptide. If a protein has more than one polypeptide chain, it also has quaternary structure. The **quaternary structure** of a protein is the way the individual protein chains are arranged with respect to each other.

PRIMARY STRUCTURE AND EVOLUTION

When scientists examine the primary structures of proteins that carry out the same function in different organisms, they can correlate the number of amino acid differences in the proteins to the closeness of the taxonomic relationship between the species. For example, cytochrome *c*, a protein that transfers electrons in biological oxidations, has about 100 amino acid residues. Yeast cytochrome *c* differs by 48 amino acids from horse cytochrome *c*, whereas duck cytochrome *c* differs by only two amino acids from chicken cytochrome *c*. Thus, ducks and chickens have a much closer taxonomic relationship than horses and yeast. Likewise, the cytochrome *c* in chickens and that in turkeys have identical primary structures. Humans and chimpanzees also have identical cytochrome *c*'s, differing by one amino acid from the cytochrome *c* of the rhesus monkey.

16.9 DETERMINING THE PRIMARY STRUCTURE OF A PEPTIDE OR PROTEIN

One of the most widely used methods to identify the N-terminal amino acid of a peptide or protein is to treat the protein with phenyl isothiocyanate (PITC), more commonly known as **Edman's reagent**. This reagent reacts with the N-terminal amino group, and the N-terminal amino acid is cleaved off as a PTH–amino acid, leaving behind a peptide with one fewer amino acid.

$$C_6H_5-N{=}C{=}S + H_2\ddot{N}CH(R)C(=O)-NHCH(R')C(=O)-NHCH(R'')C(=O)- \longrightarrow C_6H_5-NH-C(=S)-\ddot{N}HCH(R)C(=O)-NHCH(R')C(=O)-NHCH(R'')C(=O)-$$

phenyl isothiocyanate
PITC
Edman's reagent

$\downarrow$ HF

BIOGRAPHY

Insulin was the first protein to have its primary sequence determined. This was done in 1953 by **Frederick Sanger**, *who was awarded the 1958 Nobel Prize in chemistry for his work. Sanger was born in England in 1918 and received a Ph.D. from Cambridge University, where he has worked for his entire career. He also received a share of the 1980 Nobel Prize in chemistry (Section 20.10) for being the first to sequence a DNA molecule (with 5375 nucleotide pairs).*

PTH–amino acid + $H_3\overset{+}{N}CH(R')C(=O)—NHCH(R'')C(=O)—$

peptide without the original N-terminal amino acid

Because each amino acid has a different substituent (R), each amino acid forms a different PTH–amino acid. The particular PTH–amino acid can be identified by chromatography using known standards. Several successive Edman degradations can be carried out on a protein. An automated instrument known as a *sequenator* allows about 50 successive Edman degradations to be performed. The entire primary sequence cannot be determined in this way, however, because side products accumulate that interfere with the results.

PROBLEM 19

In determining the primary structure of insulin, what would lead you to conclude that it had more than one polypeptide chain?

Once the N-terminal amino acid has been identified, a sample of the protein is hydrolyzed with dilute acid. This treatment, called **partial hydrolysis**, hydrolyzes only some of the peptide bonds. The resulting fragments are separated, and the amino acid composition of each is determined by using electrophoresis or an amino acid analyzer (Section 16.2). The sequence of the original protein can then be deduced by lining up the peptides and looking for points of overlap.

PROBLEM-SOLVING STRATEGY

A nonapeptide undergoes partial hydrolysis to give peptides whose amino acid compositions are shown below. Reaction of the intact nonapeptide with Edman's reagent releases PTH–Leu. What is the sequence of the nonapeptide?

1. Pro, Ser **3.** Met, Ala, Leu **5.** Glu, Ser, Val, Pro **7.** Met, Leu
2. Gly, Glu **4.** Gly, Ala **6.** Glu, Pro, Gly **8.** His, Val

- Because we know that the N-terminal amino acid is Leu, we need to look for a fragment that contains Leu. Fragment (7) tells us that Met is next to Leu and fragment (3) tells us that Ala is next to Met.
- Now we look for a fragment that contains Ala. Fragment (4) contains Ala and tells us that Gly is next to Ala.
- From fragment (2), we know that Glu comes next. Glu is in both fragments (5) and (6).
- Fragment (5) has three amino acids we have yet to place in the growing peptide (Ser, Val, Pro), but fragment (6) has only one, so from fragment (6), we know that Pro is the next amino acid.
- Fragment (1) tells us that the next amino acid is Ser; now we can use fragment (5).
- Fragment (5) tells us that the next amino acid is Val, and fragment (8) tells us that His is the last (C-terminal) amino acid.
- Thus, the amino acid sequence of the nonapeptide is

Leu-Met-Ala-Gly-Glu-Pro-Ser-Val-His

Now continue on to Problem 20.

PROBLEM 20 ♦

A decapeptide undergoes partial hydrolysis to give peptides whose amino acid compositions are shown. Reaction of the intact decapeptide with Edman's reagent releases PTH-Gly. What is the sequence of the decapeptide?

a. Ala, Trp **c.** Pro, Val **e.** Trp, Ala, Arg **g.** Glu, Ala, Leu
b. Val, Pro, Asp **d.** Ala, Glu **f.** Arg, Gly **h.** Met, Pro, Leu, Glu

The C-terminal amino acid of a peptide or protein can be identified by treating the protein with carboxypeptidase A. A **peptidase** is an enzyme that catalyzes the hydrolysis of a peptide bond. Carboxypeptidase A catalyzes the hydrolysis of the C-terminal peptide bond, cleaving off the C-terminal amino acid, as long as it is *not* arginine or lysine.

site where carboxypeptidase cleaves

$$-\mathrm{NHCH(R)C(=O)}-\mathrm{NHCH(R')C(=O)}-\mathrm{NHCH(R'')CO^-}$$

Trypsin and chymotrypsin are peptidases that catalyze the hydrolysis of specific peptide bonds (Table 16.4). Trypsin, for example, catalyzes the hydrolysis of the peptide bond on the C-side (right-hand side) of only arginine or lysine residues.

C-side of lysine

C-side of arginine

$$-\mathrm{NHCH(R)C(=O)}-\mathrm{NHCH(CH_2CH_2CH_2CH_2\overset{+}{N}H_3)C(=O)}-\mathrm{NHCH(R')C(=O)}-\mathrm{NHCH(R'')C(=O)}-\mathrm{NHCH(CH_2CH_2CH_2NHC(=\overset{+}{N}H_2)NH_2)C(=O)}-\mathrm{NHCH(R''')C(=O)}-$$

Table 16.4 Specificity of Peptide or Protein Cleavage

Reagent	Specificity
Chemical reagents	
Edman's reagent	removes the N-terminal amino acid
Cyanogen bromide	hydrolyzes on the C-side of Met
Peptidases*	
Carboxypeptidase A	removes the C-terminal amino acid (not Arg or Lys)
Trypsin	hydrolyzes on the C-side of Arg and Lys
Chymotrypsin	hydrolyzes on the C-side of amino acids that contain aromatic six-membered rings (Phe, Tyr, Trp)

* Cleavage will not occur if Pro is at the hydrolysis site.

Thus, trypsin will catalyze the hydrolysis of three peptide bonds in the following peptide, creating a hexapeptide, a dipeptide, and two tripeptides.

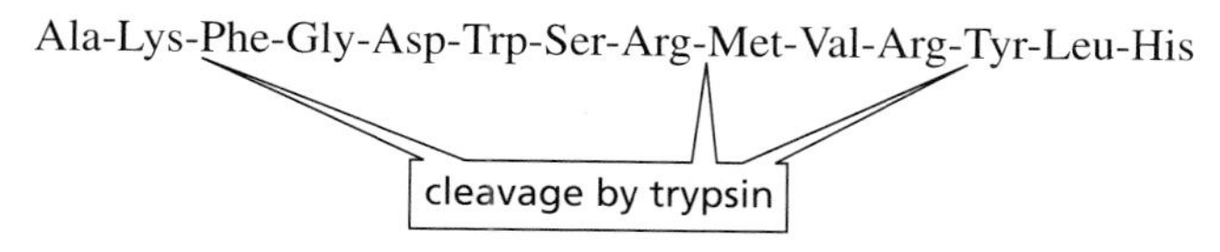

Chymotrypsin catalyzes the hydrolysis of the peptide bond on the C-side of amino acids that contain aromatic six-membered rings (Phe, Tyr, Trp).

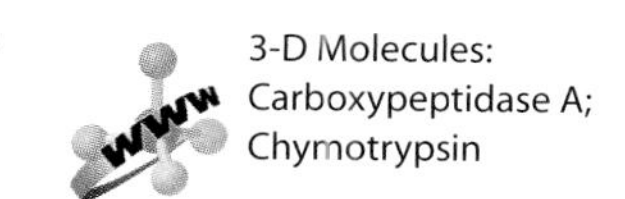

3-D Molecules: Carboxypeptidase A; Chymotrypsin

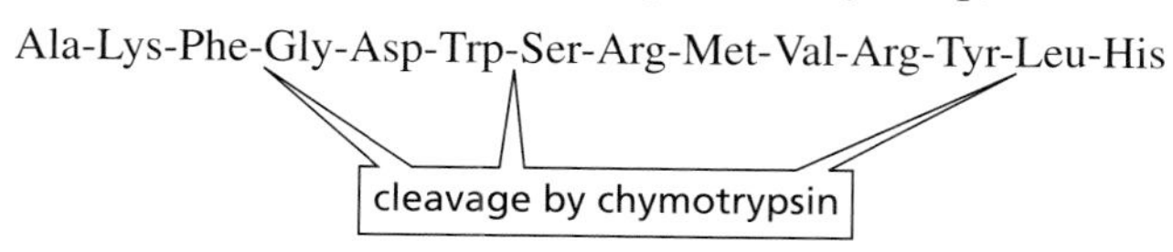

Peptidases will not catalyze the hydrolysis of a peptide bond if proline is at the hydrolysis site. These enzymes recognize the appropriate hydrolysis site by its shape and charge, and the cyclic structure of proline causes the hydrolysis site to have an unrecognizable three-dimensional shape (Section 17.1).

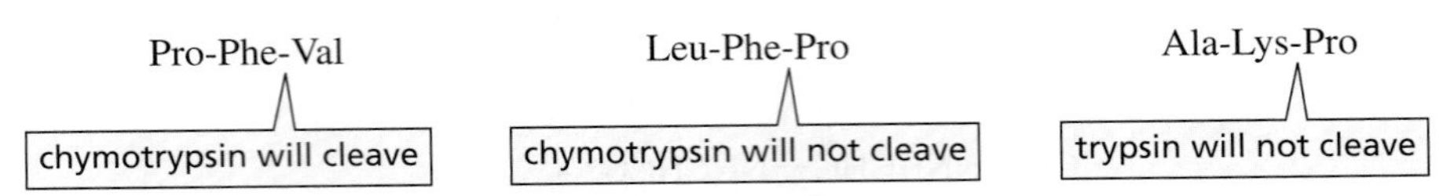

Cyanogen bromide (BrC≡N) causes the hydrolysis of the peptide bond on the C-side of a methionine residue. Cyanogen bromide is not an enzyme and therefore does not recognize the substrate by its shape, so it will still cleave the peptide bond if proline is at the cleavage site.

Ala-Lys-Phe-Gly-Asp-Trp-Ser-Arg-Met-Val-Arg-Tyr-Leu-His

cleavage by cyanogen bromide

The mechanism for cleavage of a peptide bond by cyanogen bromide is shown below.

mechanism for the cleavage of a peptide bond by cyanogen bromide

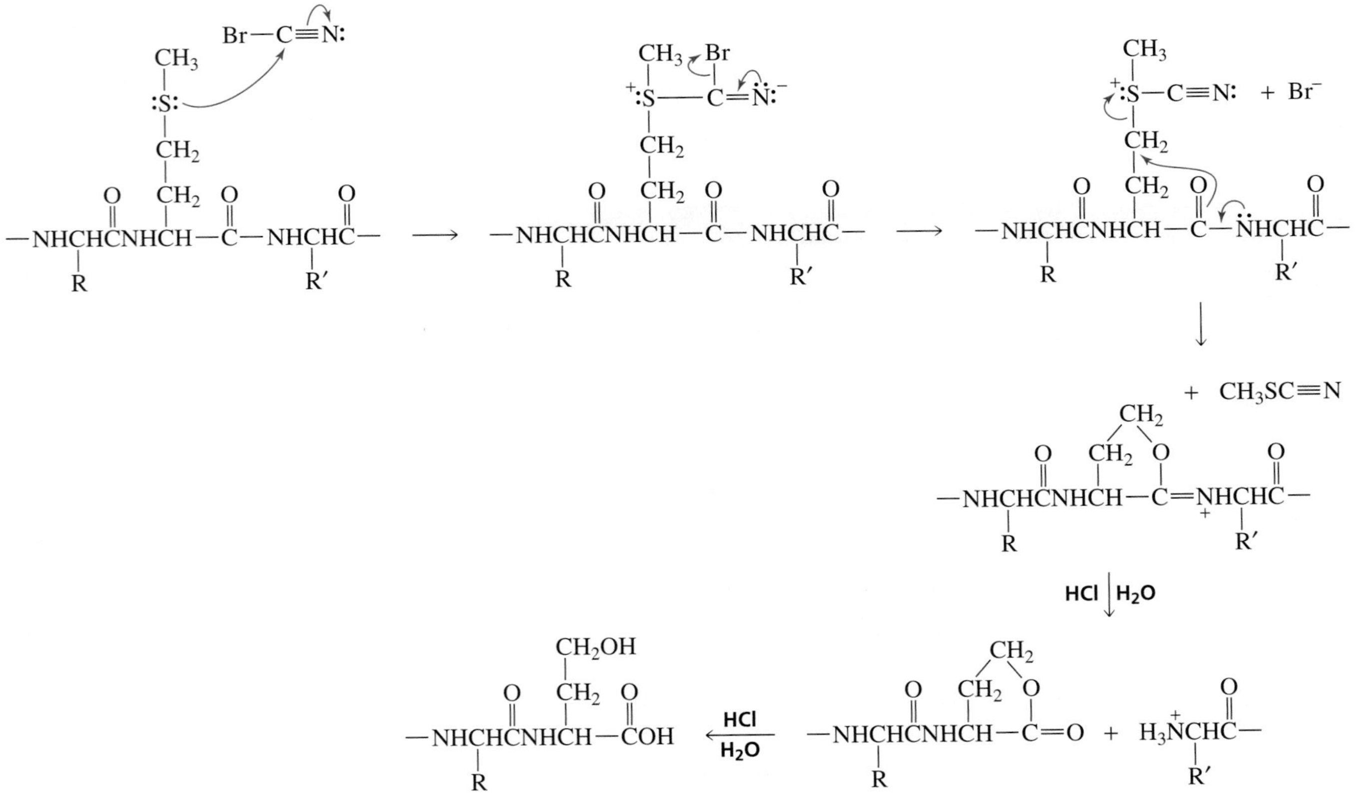

- The highly nucleophilic sulfur of methionine attacks the carbon of cyanogen bromide.
- Nucleophilic attack by the oxygen on the methylene group, resulting in departure of the weakly basic leaving group, forms a five-membered ring.
- Acid-catalyzed hydrolysis of the imine cleaves the protein (Section 12.7).
- Further hydrolysis causes the cyclic ester to open to a carboxyl group and an alcohol group (Section 11.9).

PROBLEM 21 ♦

Indicate the peptides that would result from cleavage by the indicated reagent:

a. His-Lys-Leu-Val-Glu-Pro-Arg-Ala-Gly-Ala by trypsin
b. Leu-Gly-Ser-Met-Phe-Pro-Tyr-Gly-Val by chymotrypsin

PROBLEM 22 *SOLVED*

Determine the amino acid sequence of a polypeptide from the following data:

Acid hydrolysis gives Ala, Arg, His, 2 Lys, Leu, 2 Met, Ser, Thr, Val.
Carboxypeptidase A releases Val.
Edman's reagent releases PTH-Leu.

Cleavage with cyanogen bromide gives three peptides with the following amino acid compositions:

1. His, Lys, Met
2. Thr, Val
3. Ala, Arg, Leu, Lys, Met, Ser

Trypsin-catalyzed hydrolysis gives three peptides and a single amino acid:

4. Arg, Leu, Ser
5. Met, Thr, Val
6. Lys
7. Ala, His, Lys, Met

Solution Acid-catalyzed hydrolysis shows that the polypeptide has 11 amino acids. The N-terminal amino acid is Leu (revealed by Edman's reagent), and the C-terminal amino acid is Val (revealed by carboxypeptidase A).

Leu ___ ___ ___ ___ ___ ___ ___ ___ ___ Val

Because cyanogen bromide cleaves on the C-side of Met, any peptide containing Met must have Met as its C-terminal amino acid, and the peptide that does not contain Met must be the C-terminal peptide. We know that peptide 3 is the N-terminal peptide because it contains Leu. Since it is a hexapeptide, we know that the sixth amino acid in the 11-amino acid polypeptide is Met. We also know that the ninth amino acid is Met because cyanogen bromide cleavage gave the dipeptide Thr, Val. The cyanogen bromide data also tell us that Thr is the tenth amino acid.

Ala, Arg, Lys, Ser (positions 2–5); His, Lys (positions 7–8)

Leu ___ ___ ___ ___ Met ___ ___ Met Thr Val

Because trypsin cleaves on the C-side of Arg and Lys, any peptide containing Arg or Lys must have that amino acid as its C-terminal amino acid. Therefore, Arg is the C-terminal amino acid of peptide 4, so we now know that the first three amino acids are Leu-Ser-Arg. We also know that the next two are Lys-Ala because if they were Ala-Lys, trypsin cleavage would give an Ala, Lys dipeptide. The trypsin data also identify the positions of His and Lys.

Leu Ser Arg Lys Ala Met His Lys Met Thr Val

PROBLEM 23 ♦

Determine the primary structure of an octapeptide from the following data:

Acid hydrolysis gives 2 Arg, Leu, Lys, Met, Phe, Ser, Tyr.
Carboxypeptidase A releases Ser.
Edman's reagent releases Leu.

Cyanogen bromide forms two peptides with the following amino acid compositions:

1. Arg, Phe, Ser
2. Arg, Leu, Lys, Met, Tyr

Trypsin forms the following two amino acids and two peptides:

3. Arg
4. Ser
5. Arg, Met, Phe
6. Leu, Lys, Tyr

16.10 THE SECONDARY STRUCTURE OF PROTEINS

hydrogen bonding between peptide groups

Secondary structure describes the repetitive conformations assumed by segments of the backbone chain of a peptide or protein. In other words, the secondary structure describes how segments of the backbone fold. The conformations are stabilized by hydrogen bonding between peptide groups, that is, between the hydrogen attached to a nitrogen of one amino acid residue and the carbonyl oxygen of another (Section 3.7).

α-Helix

One type of secondary structure is the ***α*-helix**. In an α-helix, the backbone of the polypeptide coils around the long axis of the protein molecule. The substituents on the α-carbons of the amino acids protrude outward from the helix, thereby minimizing steric hindrance (Figure 16.7a). The helix is stabilized by hydrogen bonds; each hydrogen attached to an amide nitrogen is hydrogen bonded to a carbonyl oxygen of an amino acid four amino acids away (Figure 16.7b).

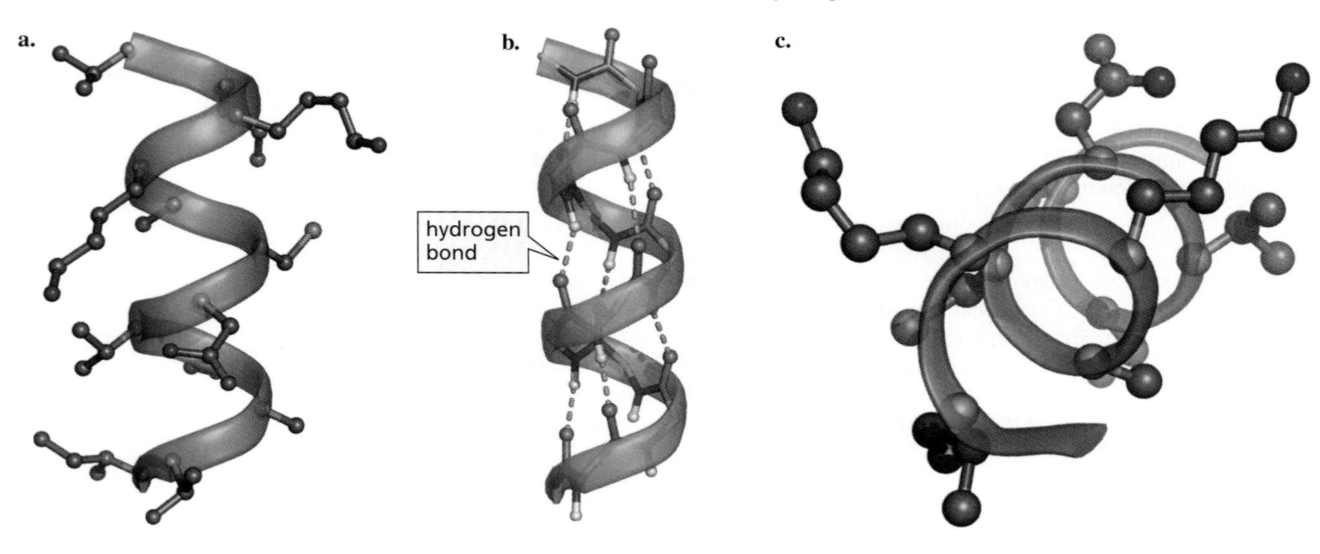

▲ **Figure 16.7**
(a) A segment of a protein in an α-helix. (b) The helix is stabilized by hydrogen bonding between peptide groups. (c) Looking at the longitudinal axis of an α-helix.

β-Pleated Sheet

The second type of secondary structure is the ***β*-pleated sheet**. In a β-pleated sheet, the polypeptide backbone is extended in a zigzag structure resembling a series of pleats. The hydrogen bonding in a β-pleated sheet occurs between neighboring peptide chains (Figure 16.8).

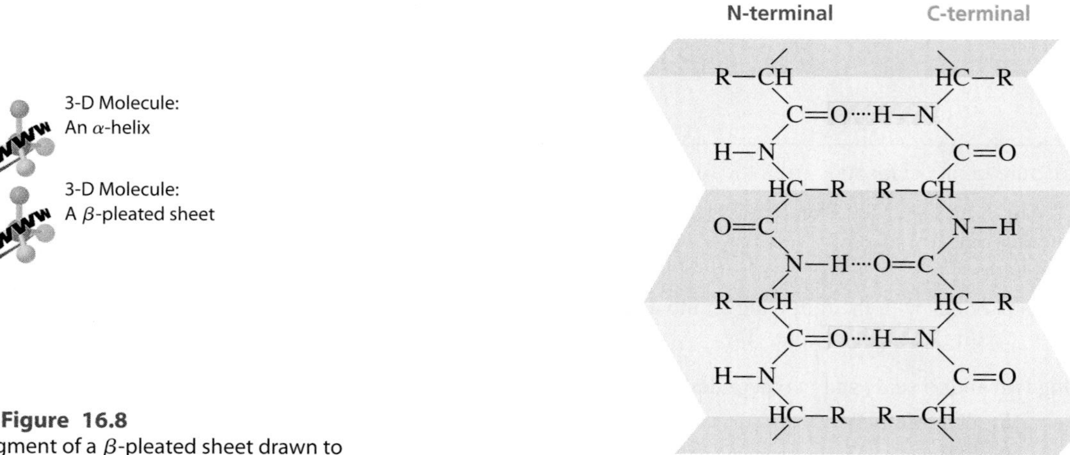

▶ **Figure 16.8**
Segment of a β-pleated sheet drawn to illustrate its pleated character.

Because the substituents (R) on the α-carbons of the amino acids on adjacent chains of a β-pleated sheet are close to each other, the substituents must be small if the chains are to nestle closely enough together to maximize hydrogen-bonding interactions. Silk, for example, contains a large proportion of relatively small amino acids (glycine and alanine) and therefore has large segments of β-pleated sheet.

Wool and the fibrous protein of muscle are examples of proteins with secondary structures that are almost all α-helices. Consequently, these proteins can be stretched. In contrast, proteins with secondary structures that are predominantly β-pleated sheets, such as silk and spider webs, cannot be stretched because a β-pleated sheet is almost fully extended.

Generally, less than half of the protein's backbone is arranged in a defined secondary structure, whether an α-helix or a β-pleated sheet. Most of the rest of the protein, however, though highly ordered, is difficult to describe (Figure 16.9).

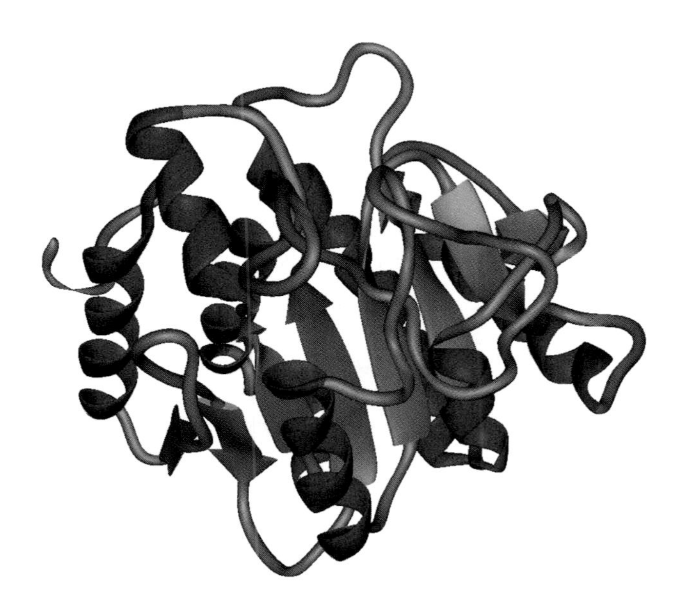

◀ **Figure 16.9**
The backbone structure of carboxypeptidase A: α-helical segments are purple; β-pleated sheets are indicated by flat green arrows pointing in the N⟶C direction.

16.11 THE TERTIARY STRUCTURE OF PROTEINS

The **tertiary structure** of a protein is the three-dimensional arrangement of all the atoms in the protein (Figure 16.10). Proteins fold spontaneously in solution to maximize their stability. Every time there is a stabilizing interaction between two atoms, free energy is released. The more free energy released, the more stable is the protein. Consequently, a protein tends to fold in a way that maximizes the number of stabilizing interactions.

The stabilizing interactions in a protein include disulfide bonds, hydrogen bonds, electrostatic attractions (attractions between opposite charges), and hydrophobic interactions (attractions between nonpolar groups). Stabilizing interactions can occur between peptide groups (atoms in the backbone of the protein), between side-chain groups (α-substituents), and between peptide groups and side-chain groups (Figure 16.11). Because the side-chain groups help determine how a protein folds, the tertiary structure of a protein is determined by its primary structure.

Most proteins exist in aqueous environments. Therefore, they tend to fold in a way that exposes the maximum number of polar groups to the surrounding water and buries the nonpolar groups in the interior of the protein, away from water.

BIOGRAPHY

Max Perutz (1914–2002) *was born in Austria. In 1936, because of the rise of Nazism, he moved to England where he received a Ph.D. from Cambridge University and became a professor there. He and John Kendrew shared the 1962 Nobel Prize in chemistry. They had used X-ray diffraction to determine, for the first time, the tertiary structure of a protein.*

BIOGRAPHY

John Kendrew (1917–1997) *was born in England and was educated at Cambridge University, where Max Perutz was working on the structure of hemoglobin (completed in 1959). Perutz assigned the work on myoglobin, a smaller protein, to Kendrew, who completed it in 1957.*

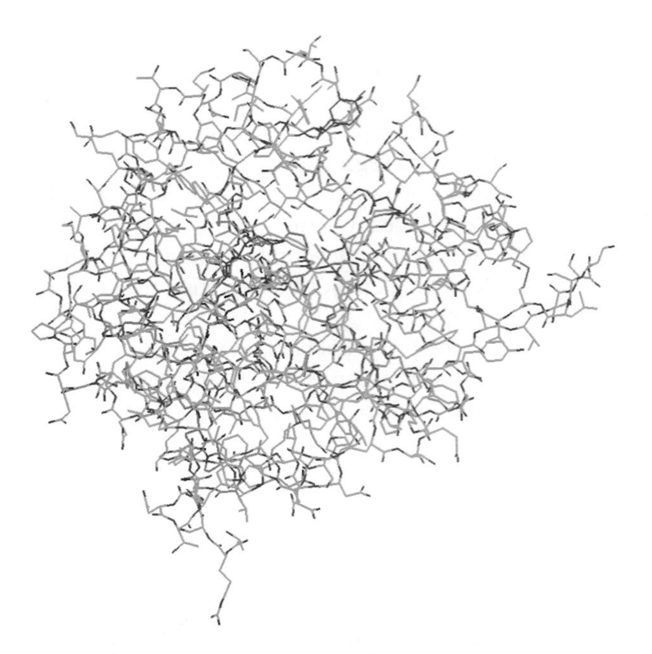

▲ **Figure 16.10**
The three-dimensional structure of carboxypeptidase A.

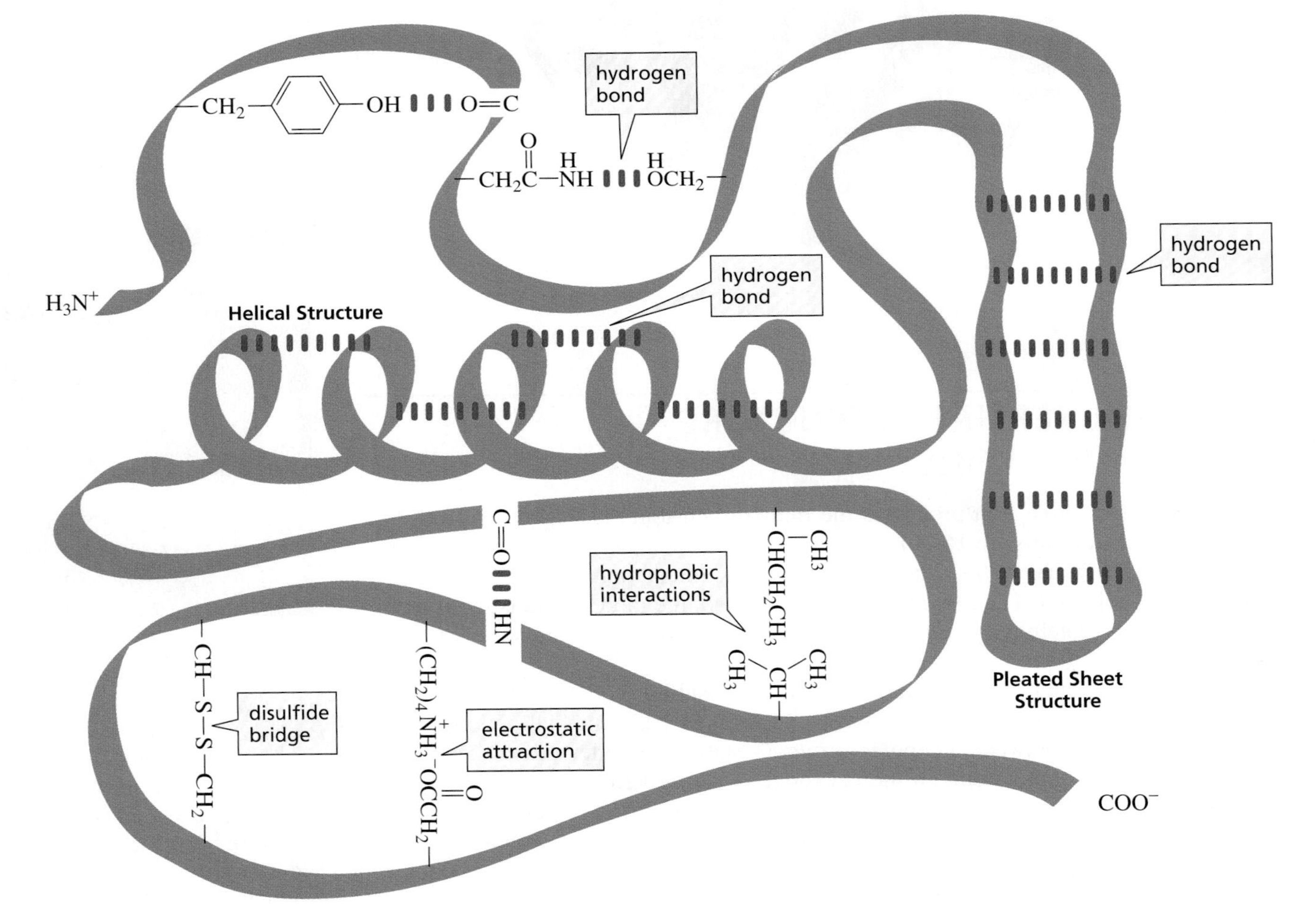

▲ **Figure 16.11**
Stabilizing interactions responsible for the tertiary structure of a protein.

PROBLEM 24

How would a protein that resides in the interior of a membrane fold, compared with the water-soluble protein just discussed? (*Hint:* Membranes are nonpolar; see Section 19.4.)

16.12 THE QUATERNARY STRUCTURE OF PROTEINS

Some proteins have more than one peptide chain. The individual chains are called **subunits**. The subunits are held together by the same kinds of interactions that hold the individual protein chains in a particular three-dimensional conformation: hydrophobic interactions, hydrogen bonding, and electrostatic attractions. The **quaternary structure** of a protein describes the way the subunits are arranged in space. Some of the possible arrangements of the six subunits of a hexamer, for example, are shown here:

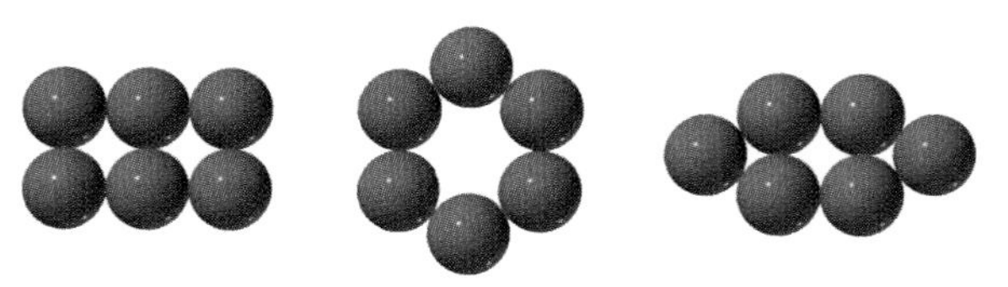

possible quaternary structures for a hexamer

Hemoglobin has four subunits; its quaternary structure is shown in Figure 16.12.

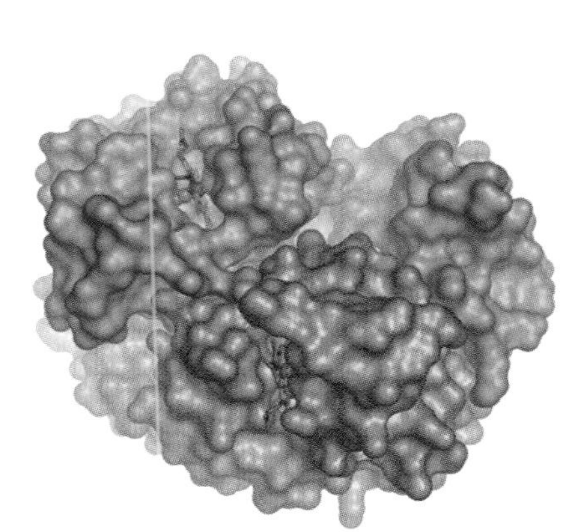

▲ **Figure 16.12**
A representation of the quaternary structure of hemoglobin. The orange and green represent the polypeptide chains; there are two identical orange subunits and two identical green subunits. Two of the porphyrin rings (gray beads; Section 8.3) are visible, ligated to iron (pink) and bonded to oxygen (red).

PROBLEM 25 ♦

a. Which water-soluble protein or subunit would have the greatest percentage of polar amino acids—a spherical protein, a cigar-shaped protein, or a subunit of a hexamer?
b. Which would have the smallest percentage of polar amino acids?

16.13 PROTEIN DENATURATION

Destroying the highly organized tertiary structure of a protein is called **denaturation**. Anything that breaks the bonds maintaining the three-dimensional shape of the protein will cause the protein to denature (unfold). Because these bonds are weak, proteins are easily denatured. The following are some of the ways that proteins can be denatured:

- Changing the pH denatures proteins because it changes the charges on many of the side chains. This disrupts electrostatic attractions and hydrogen bonds.
- Certain reagents such as urea denature proteins by forming hydrogen bonds to the protein groups that are stronger than the hydrogen bonds formed between the groups.
- Organic solvents denature proteins by disrupting hydrophobic interactions.
- Proteins can also be denatured by heat or by agitation; both disrupt the attractive forces. A well-known example is the change that occurs to the white of an egg when it is heated or whipped.

SUMMARY

Peptides and **proteins** are polymers of **amino acids** linked together by **peptide** (amide) **bonds**. A **dipeptide** contains 2 amino acid residues, and a **polypeptide** contains many amino acid residues. Proteins have 40 to 4000 amino acid residues. The **amino acids** differ only in the substituent attached to the α-carbon. Almost all amino acids found in nature have the L configuration.

The carboxyl groups of the amino acids have pK_a values of ~2, and the protonated amino groups have pK_a values of ~9. At physiological pH, an amino acid exists as a

zwitterion. A few amino acids have side chains with ionizable hydrogens. The **isoelectric point** (pI) of an amino acid is the pH at which the amino acid has no net charge. A mixture of amino acids can be separated based on their pI's by **electrophoresis**. Separation can also be achieved by **ion-exchange chromatography**. An **amino acid analyzer** is an instrument that automates ion-exchange chromatography.

The amide bonds that link amino acid residues are called **peptide bonds**. Two cysteine residues can be oxidized to cystine that contains a **disulfide bridge**. By convention, peptides and proteins are written with the free amino group (the **N-terminal amino acid**) on the left and the free carboxyl group (the **C-terminal amino acid**) on the right.

To synthesize a peptide bond, the amino group of the N-terminal amino acid must be protected (by *t*-BOC) and its carboxyl group activated (with DCC). A second amino acid is added to form a dipeptide. Amino acids can be added to the growing C-terminal end by repeating the same two steps: activating the carboxyl group of the C-terminal amino acid with DCC and adding a new amino acid.

The **primary structure** of a protein is the sequence of its amino acids and the location of all its disulfide bridges. The N-terminal amino acid of a peptide or protein can be identified with **Edman's reagent**; the C-terminal amino acid can be identified with carboxypeptidase A. **Partial hydrolysis** hydrolyzes only some of the peptide bonds. A **peptidase** catalyzes the hydrolysis of a peptide bond.

The **secondary structure** of a protein describes how local segments of the protein's backbone fold. An α-helix and a β-pleated sheet are types of secondary structures. A protein folds so as to maximize the number of stabilizing interactions: **disulfide bonds, hydrogen bonds, electrostatic attractions**, and **hydrophobic interactions**. The **tertiary structure** of a protein is the three-dimensional arrangement of all the atoms in the protein. The **quaternary structure** of a protein describes the way the **subunits** are arranged with respect to each other.

PROBLEMS

26. Draw the predominant form of the following amino acids at physiological pH (7.3):
a. lysine **b.** arginine **c.** tyrosine

27. What is the pI of serine?

28. Indicate the peptides that would result from cleavage by the indicated reagent:
a. Val-Arg-Gly-Met-Arg-Ala-Ser by carboxypeptidase A
b. Ser-Phe-Lys-Met-Pro-Ser-Ala-Asp by cyanogen bromide
c. Arg-Ser-Pro-Lys-Lys-Ser-Glu-Gly by trypsin

29. a. Which isomer—(*R*)-aspartate or (*S*)-aspartate—is L-aspartate?
b. Can a general statement be made relating *R* and *S* to D and L?

30. Aspartame has a pI of 5.9. Draw its most prevalent form at physiological pH.

31. Draw the form of aspartate that predominates at
a. pH = 1.0 **c.** pH = 6.0
b. pH = 2.6 **d.** pH = 11.0

32. a. Why is the pK_a of the glutamate side chain greater than the pK_a of the aspartate side chain?
b. Why is the pK_a of the arginine side chain greater than the pK_a of the lysine side chain?

33. Which would be a more effective buffer at physiological pH, a solution of 0.1 M glycylglycylglycylglycine or a solution of 0.2 M glycine?

34. a. At pH = 11, only one of arginine's nitrogen atoms is protonated. Which one is it?
b. At pH = 5, one of the nitrogens in histidine's five-membered ring is protonated. Which one is it? (*Hint:* Localized electrons are more apt to be protonated than are delocalized electrons.)

$H_2\ddot{N}C(=\ddot{N}H)\ddot{N}HCH_2CH_2CH_2CH(NH_2)CO_2^-$

arginine

histidine: imidazole ring (:N, :NH) bearing CH_2CHCOO^- with $^+NH_3$

histidine

35. Identify the location and type of charge on the hexapeptide Lys-Ser-Asp-Cys-His-Tyr at
a. pH = 7 **b.** pH = 5 **c.** pH = 9

36. a. Which amino acid has the greatest amount of negative charge at pH = 6.20?
b. Which amino acid—glycine or methionine—has a greater negative charge at pH = 6.20?

37. Explain the order of elution (with a buffer of pH 4) of each of the following pairs of amino acids on a column packed with the cation-exchange resin shown in Figure 16.2:

a. aspartate before serine
b. glycine before alanine
c. valine before leucine
d. tyrosine before phenylalanine

38. Reaction of a polypeptide with carboxypeptidase A releases Met. The polypeptide undergoes partial hydrolysis to give the following peptides. What is the sequence of the polypeptide?

a. Ser, Lys, Trp
b. Gly, His, Ala
c. Glu, Val, Ser
d. Leu, Glu, Ser
e. Met, Ala, Gly
f. Ser, Lys, Val
g. Glu, His
h. Leu, Lys, Trp
i. Lys, Ser
j. Glu, His, Val
k. Trp, Leu, Glu
l. Ala, Met

39. Glycine has pK_a values of 2.3 and 9.6. Would you expect the pK_a values of glycylglycine to be higher or lower than these values?

40. The following polypeptide was hydrolyzed by trypsin:

Gly-Ser-Asp-Ala-Leu-Pro-Gly-Ile-Thr-Ser-Arg-Asp-Val-Ser-Lys-Val-Glu-
Tyr-Phe-Glu-Ala-Gly-Arg-Ser-Glu-Phe-Lys-GluPro-Arg-Leu-Tyr-
Met-Lys-Val-Glu-Gly-Arg-Pro-Val-Ser-Ala-Gly-Leu-Trp

a. How many fragments are obtained from the peptide?
b. In what order would the fragments be eluted from an anion-exchange column using a buffer of pH = 5?

41. After breaking the disufide bridges in a polypeptide, it is found to have two polypeptide chains with the following primary sequences:

Val-Met-Tyr-Ala-Cys-Ser-Phe-Ala-Glu-Ser
Ser-Cys-Phe-Lys-Cys-Trp-Lys-Tyr-Cys-Phe-Arg-Cys-Ser

Treatment of the original intact polypeptide with chymotrypsin yields the following peptides:

a. Ala, Glu, Ser
b. 2 Phe, 2 Cys, Ser
c. Tyr, Val, Met
d. Arg, Ser, Cys
e. Ser, Phe, 2 Cys, Lys, Ala, Trp
f. Tyr, Lys

Determine the positions of the disulfide bridges in the original polypeptide.

42. Determine the amino acid sequence of a polypeptide from the following data:

a. Complete hydrolysis of the peptide yields the following amino acids: Ala, Arg, Gly, 2 Lys, Met, Phe, Pro, 2 Ser, Tyr, Val.
b. Treatment with Edman's reagent gives PTH-Val.
c. Carboxypeptidase A releases Ala.
d. Treatment with cyanogen bromide yields the following two peptides:
  1. Ala, 2 Lys, Phe, Pro, Ser, Tyr
  2. Arg, Gly, Met, Ser, Val
e. Treatment with trypsin yields the following three peptides:
  1. Gly, Lys, Met, Tyr
  2. Ala, Lys, Phe, Pro, Ser
  3. Arg, Ser, Val
f. Treatment with chymotrypsin yields the following three peptides:
  1. 2 Lys, Phe, Pro
  2. Arg, Gly, Met, Ser, Tyr, Val
  3. Ala, Ser

43. What products are obtained when aspartame (page 464) is hydrolyzed in the presence of an acid catalyst?

44. Dr. Kim S. Tree was preparing a manuscript for publication in which she reported that the pI of the tripeptide Lys-Lys-Lys was 10.6. One of her students pointed out that there must be an error in her calculations because the pK_a of the ε-amino group of lysine is 10.8 and the pI of the tripeptide has to be greater than any of its individual pK_a values. Was the student correct?

45. Show the steps in the synthesis of the tetrapeptide Leu-Phe-Lys-Val.

46. a. Calculate the overall yield of a nonapeptide if the yield for the addition of each amino acid to the chain is 70%.
b. What would be the overall yield of a peptide containing 15 amino acid residues if the yield for the incorporation of each is 80%?

47. Why is proline never found in an α-helix?

48. Explain why the following are not found in an α-helix: two adjacent glutamates, two adjacent aspartates, or a glutamate adjacent to an aspartate.

49. Explain the difference in the pK_a values of the carboxyl groups of alanine, serine, and cysteine.

50. The C-terminal end of a protein extends into the aqueous environment surrounding the protein. The C-terminal amino acids are Gln, Asp, 2 Ser, and three nonpolar amino acids. Assuming that the $\Delta G°$ for the formation of a hydrogen bond is −3 kcal/mol and the $\Delta G°$ for removal of a hydrophobic group from water is −4 kcal/mol, calculate the $\Delta G°$ for folding the C-terminal end of the protein into the interior of the protein under the following conditions:

a. Each of the polar groups forms one intramolecular hydrogen bond.
b. All but two of the polar groups form intramolecular hydrogen bonds.

CHAPTER 17

How Enzymes Catalyze Reactions • The Organic Chemistry of the Vitamins

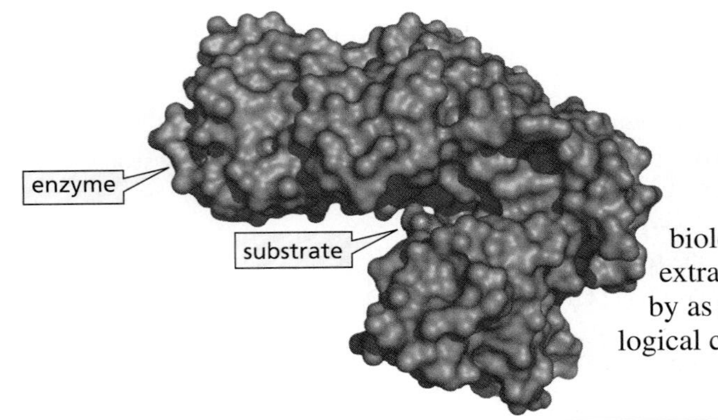

Essentially all organic reactions that occur in biological systems require a catalyst. Most biological catalysts are **enzymes**, which are globular proteins (Section 16.0). Each biological reaction is catalyzed by a different enzyme. Enzymes are extraordinarily good catalysts—they can increase the rate of a reaction by as much as 10^{16}! In contrast, rate enhancements achieved by nonbiological catalysts are seldom greater than 10^4 fold.

17.1 ENZYME-CATALYZED REACTIONS

The reactant of an enzyme-catalyzed reaction is called a **substrate**.

$$\text{substrate} \xrightarrow{\text{enzyme}} \text{product}$$

The enzyme binds the substrate in a pocket called the **active site**. (See the above picture.) All the bond-making and bond-breaking steps of the reaction occur while the substrate is bound to the active site. Enzymes differ from nonbiological catalysts in that they are specific for the substrate whose reaction they catalyze (Section 6.14). Not all enzymes, however, have the same degree of specificity. Some are specific for a single compound and will not tolerate even the slightest variation in structure, whereas other enzymes catalyze the reaction of a family of compounds with related structures. The specificity of an enzyme for its substrate is an example of the phenomenon known as **molecular recognition**—the ability of one molecule to recognize another.

The specificity of an enzyme results from its conformation and the particular **amino acid side chains** (α-substituents, Section 16.1) that are at the active site. For example, an amino acid with a negatively charged side chain can associate with a positively charged group on the substrate, an amino acid side chain with a hydrogen-bond donor can associate with a hydrogen-bond acceptor on the substrate, and a hydrophobic amino acid side chain can associate with a hydrophobic group on the substrate.

Various factors contribute to the remarkable catalytic ability of enzymes. Some of the most important are:

- Reacting groups are brought together at the active site in the proper orientation for reaction.
- Some of the amino acid side chains of the enzyme serve as catalysts. These are positioned relative to the substrate precisely where they are needed for catalysis.
- Amino acid side chains can stabilize transition states and intermediates—by van der Waals interactions, electrostatic interactions, and hydrogen bonding—which makes them easier to form (Section 16.11).

We will now look at the mechanisms for two enzyme-catalyzed reactions in order to understand how the amino acid side chains at the active site act as catalytic groups. As you examine these reactions, notice that they are similar to reactions that you have seen organic compounds undergo. *If you refer back to sections referenced throughout this chapter, you will see that much of the organic chemistry you have learned applies to the reactions of compounds found in the biological world.*

17.2 THE MECHANISM FOR GLUCOSE-6-PHOSPHATE ISOMERASE

The names of most enzymes end in "ase," and the enzyme's name tells you something about the reaction it catalyzes. For example, the enzyme glucose-6-phosphate isomerase catalyzes an isomerization reaction that converts glucose-6-phosphate to fructose-6-phosphate. Recall that the open-chain form of glucose is an aldohexose, whereas the open-chain form of fructose is a ketohexose (Sections 15.3 and 15.4). Therefore, the enzyme that catalyzes this reaction—glucose-6-phosphate isomerase—converts an aldose to a ketose. In solution, the sugars exist predominantly in their cyclic forms, so the enzyme must open the six-membered-ring sugar and convert it to the five-membered-ring sugar. Glucose-6-phosphate isomerase is known to have at least three catalytic groups at its active site, one functioning as an acid catalyst and two acting as base catalysts (Figure 17.1). The reaction takes place as follows:

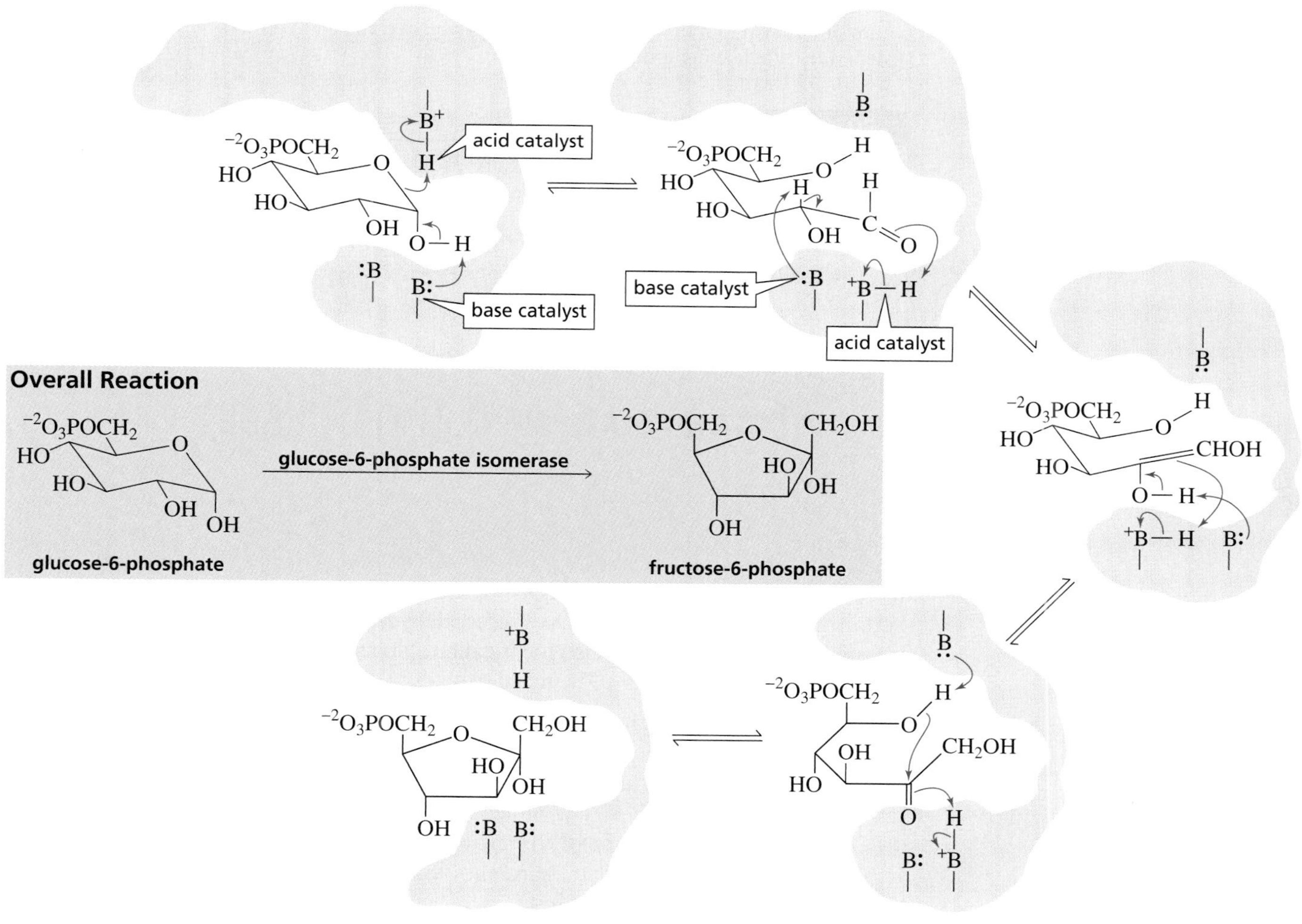

▲ **Figure 17.1**
Proposed mechanism for the isomerization of glucose-6-phosphate to fructose-6-diphosphate.

A proton is donated to the reactant in an acid-catalyzed reaction.

A proton is removed from the reactant in a base-catalyzed reaction.

- The first step is a ring-opening reaction. A base catalyst (a histidine side chain) removes a proton from the OH group on C-1, and an **acid catalyst** (a protonated lysine side chain) aids the departure of the leaving group by protonating it, thereby making it a weaker base and therefore a better leaving group (Section 10.2).
- In the second step of the reaction, a **base catalyst** (a glutamate residue) removes a proton from the α-carbon of the aldehyde. Recall that α-hydrogens are relatively acidic (Section 13.1).
- In the next step, the enol is converted to a ketone (Section 13.3).
- In the final step of the reaction, the conjugate acid of the base catalyst employed in the first step and the conjugate base of the acid employed in the first step catalyze ring closure.

PROBLEM 1 *SOLVED*

Which of the following amino acid side chains can aid the departure of a leaving group by protonating it?

$-CH_2CH_2SCH_3$ **1** $\quad -CH(CH_3)_2$ **2** $\quad -CH_2-$(imidazolium, $\overset{+}{N}H$, NH) **3** $\quad -CH_2\overset{O}{\overset{\|}{C}}OH$ **4**

Solution Side chains **1** and **2** do not have an acidic proton, so they cannot aid the departure of a leaving group. Side chains **3** and **4** do have an acidic proton, so they can aid the departure of a leaving group.

PROBLEM 2 ◆

Which of the following amino acid side chains can help remove a proton from the α-carbon of an aldehyde?

$-CH_2\overset{O}{\overset{\|}{C}}NH_2$ **1** $\quad$ $-C_6H_4-O^-$ **2** $\quad -CH_2-$(imidazole, N, NH) **3** $\quad -\overset{CH_3}{\overset{|}{C}}HCH_3$ **4**

17.3 THE MECHANISM FOR ALDOLASE

The substrate for the first in a series of enzyme-catalyzed reactions known as glycolysis (Section 18.4) is a six-carbon compound (D-glucose). The final product of glycolysis is two molecules of a three-carbon compound (pyruvate). Therefore, at some point in the series of enzyme-catalyzed reactions, a six-carbon compound must be cleaved into two three-carbon compounds. The enzyme *aldolase* catalyzes this cleavage. Aldolase converts fructose-1,6-diphosphate into glyceraldehyde-3-phosphate and dihydroxyacetone phosphate (Figure 17.2). The enzyme is called aldolase because the reverse reaction is an aldol addition (Section 13.5). The reaction takes place as follows:

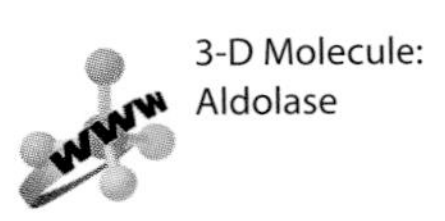

3-D Molecule: Aldolase

- In the first step of the reaction, fructose-1,6-diphosphate forms an imine with a lysine residue at the active site of the enzyme (Section 12.7).
- A tyrosine residue functions as a base catalyst in the step that cleaves the bond between C-3 and C-4. The molecule of glyceraldehyde-3-phosphate formed in this step dissociates from the enzyme.
- The enamine intermediate rearranges to an imine, with the tyrosine residue now functioning as an acid catalyst.
- Hydrolysis of the imine (Section 12.7) releases dihydroxyacetone phosphate, the other three-carbon product.

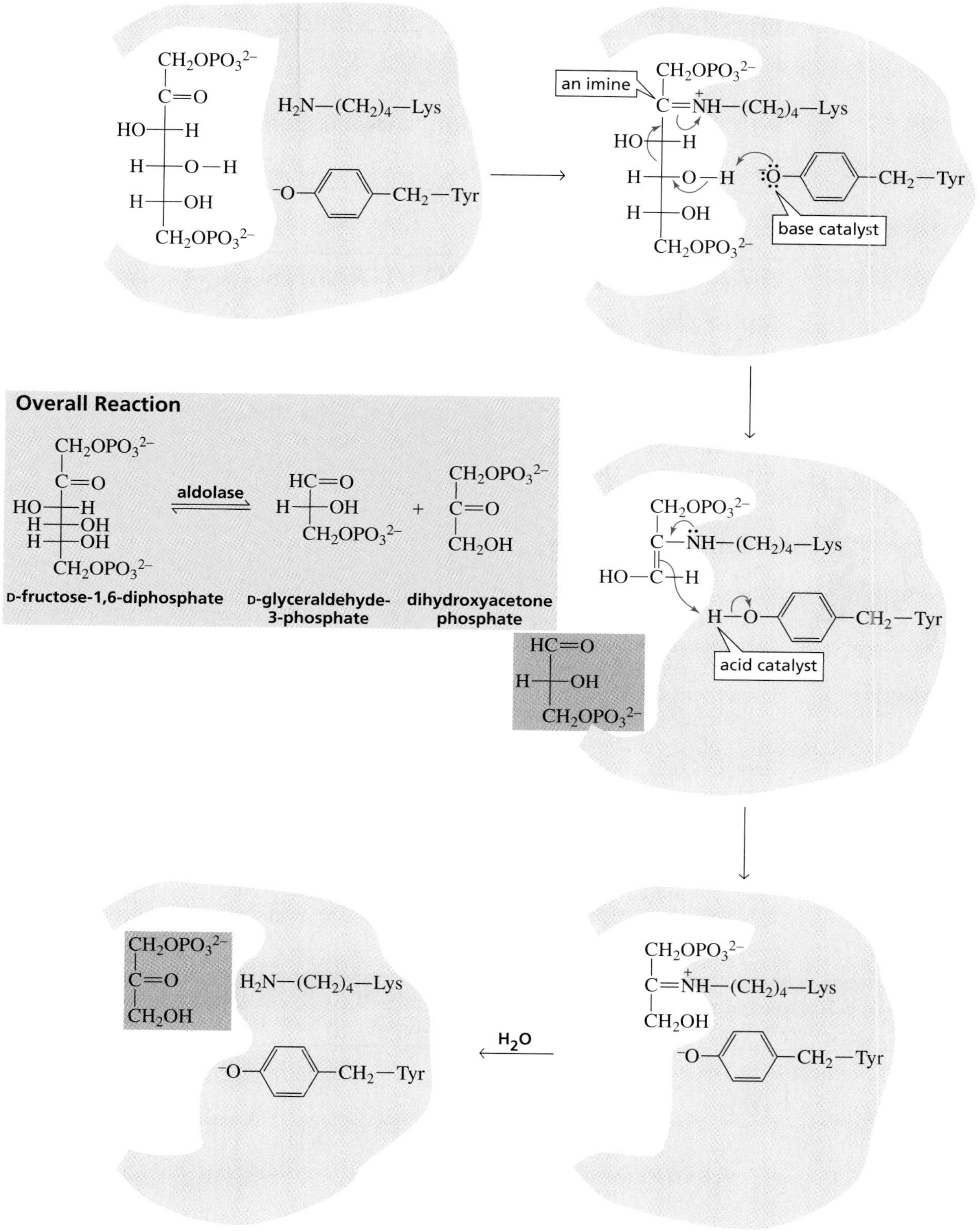

an imine

base catalyst

acid catalyst

H_2O

Overall Reaction

D-fructose-1,6-diphosphate $\xrightleftharpoons{\text{aldolase}}$ D-glyceraldehyde-3-phosphate + dihydroxyacetone phosphate

▲ **Figure 17.2**
Proposed mechanism for the aldolase-catalyzed cleavage of fructose-1,6-diphosphate to form glyceraldehyde-3-phosphate and dihydroxyacetone phosphate.

PROBLEM 3 ◆

Which of the following amino acid side chains can form an imine with the substrate?

$$\underset{\mathbf{1}}{-CH_2\overset{O}{\overset{\|}{C}}NH_2} \qquad \underset{\mathbf{2}}{-(CH_2)_4NH_2} \qquad \underset{\mathbf{3}}{-(CH_2)_3NH\overset{NH}{\overset{\|}{C}}NH_2} \qquad \underset{\mathbf{4}}{-CH_2OH}$$

17.4 COENZYMES AND VITAMINS

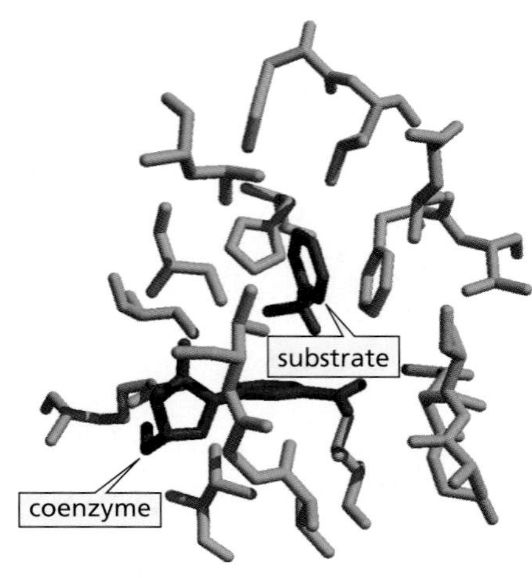

▲ **Figure 17.3** A substrate and a coenzyme bound at the active site of an enzyme.

Many enzymes cannot catalyze a reaction without the help of a *coenzyme*. **Coenzymes** are organic molecules that assist enzymes in catalyzing certain reactions that cannot be catalyzed by the amino acid side chains of the enzyme alone. Enzymes that require a coenzyme to catalyze a reaction bind both the substrate and the coenzyme at the active site (Figure 17.3).

Coenzymes are derived from organic compounds commonly known as *vitamins* (Table 17.1). A **vitamin** is a substance needed in small amounts for normal body function that the body cannot synthesize. The body synthesizes the coenzyme from the vitamin.

Early nutritional studies divided vitamins into two classes: water-soluble vitamins and water-insoluble vitamins. Vitamins A, D, E, and K are water insoluble. Vitamin K is the only water-insoluble vitamin currently known to be a precursor for a coenzyme. Vitamin A is required for proper vision, vitamin D regulates calcium and phosphate metabolism, and vitamin E is an antioxidant. Because they do not function as precursors for coenzymes, vitamins A, D, and E are not discussed in this chapter. (See, however, Section 5.17.)

All the water-soluble vitamins except vitamin C are precursors for coenzymes. In spite of its name, vitamin C is not actually a vitamin because it is required in fairly high amounts and most mammals are able to synthesize it (Section 15.15). Humans and guinea pigs cannot synthesize it, however, so it must be included in their diets. We have seen that vitamin C is a radical inhibitor (Section 5.17).

It is hard to overdose on water-soluble vitamins because the body can eliminate any reasonable excess. One can, however, overdose on water-insoluble vitamins because they are *not* easily eliminated by the body and can accumulate in cell membranes and other nonpolar components of the body. For example, excess vitamin D causes

Table 17.1 The Vitamins, Their Coenzymes, and the Reactions They Catalyze

Vitamin	Coenzyme	Reaction catalyzed	Human deficiency disease
Niacin (vitamin B_3)	NAD^+	Oxidation	Pellagra
	NADH	Reduction	
Riboflavin (vitamin B_2)	FAD	Oxidation	Skin inflammation
	$FADH_2$	Reduction	
Thiamine (vitamin B_1)	Thiamine pyrophosphate (TPP)	Two-carbon transfer	Beriberi
Pantothenic acid	Coenzyme A (CoASH)	Activates carboxylic acids for acyl transfer	—
Biotin (vitamin H)	Biotin	Carboxylation	—
Pyridoxine (vitamin B_6)	Pyridoxal phosphate (PLP)	Transamination and other reactions of amino acids	Anemia
Vitamin B_{12}	Coenzyme B_{12}	Isomerization	Pernicious anemia
Folic acid	Tetrahydrofolate (THF)	One-carbon transfer	Megaloblastic anemia
Vitamin K	Vitamin KH_2	Carboxylation	Internal bleeding

calcification of soft tissues. The kidneys are particularly susceptible to calcification, which eventually leads to kidney failure. Vitamin D is formed in the skin as a result of a photochemical reaction caused by the ultraviolet rays from the sun.

VITAMIN B_1

Christiaan Eijkman (1858–1930) was a member of a medical team that was sent to the East Indies to study beriberi in 1886. At that time, all diseases were thought to be caused by microorganisms. When the microorganism that caused beriberi could not be found, the team left the East Indies, but Eijkman stayed behind to become the director of a new bacteriological laboratory there.

In 1896, Eijkman accidentally discovered the cause of beriberi when he noticed that chickens used in the laboratory had developed symptoms characteristic of the disease. He found that the symptoms had developed when a cook had started feeding the chickens rice meant for hospital patients. The symptoms disappeared when a new cook resumed giving chicken feed to the chickens. Later it was recognized that thiamine (vitamin B_1) is present in rice husks but not in polished rice. For these advances, Eijkman shared the 1929 Nobel Prize in physiology or medicine with Frederick Hopkins.

Christiaan Eijkman

"VITAMINE"—AN AMINE REQUIRED FOR LIFE

Sir Frederick G. Hopkins (1861–1947), born in England, was the first to suggest that diseases such as rickets and scurvy might result from the absence of substances in the diet that are needed only in very small quantities. The first such compound recognized to be essential in the diet was an amine (thiamine, page 485), which led to the incorrect assumption that all such compounds were amines. They therefore, were called vitamines ("amines required for life"). The *e* was later dropped from the name. Hopkins's hypothesis later became known as the "vitamin concept," for which he received a share of the 1929 Nobel Prize in physiology or medicine. He also originated the concept of essential amino acids.

Sir Frederick Hopkins

17.5 NIACIN: THE VITAMIN NEEDED FOR MANY OXIDATION-REDUCTION REACTIONS

Any enzyme that catalyzes an oxidation or a reduction reaction requires a coenzyme because none of the amino acid side chains are oxidizing or reducing agents. The coenzyme serves as the oxidizing or reducing agent. The enzyme's role is to hold the substrate and coenzyme together so that the oxidation or reduction reaction can take place (Figure 17.3).

The coenzyme most commonly used by enzymes *to catalyze oxidation reactions* is **nicotinamide adenine dinucleotide (NAD^+)**. NAD^+ is composed of two nucleotides linked together through their phosphate groups. A **nucleotide** consists of a heterocyclic compound attached to C-1 of a phosphorylated ribose (Section 20.1). In a **heterocyclic compound**, one or more of the ring atoms is an atom other than carbon.

NAD^+ is an oxidizing agent.

pyridine

nicotinamide

responsible for the "+" in NAD^+

adenine

NAD^+

NADH

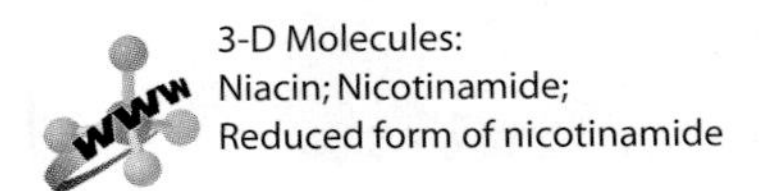
3-D Molecules: Niacin; Nicotinamide; Reduced form of nicotinamide

The heterocyclic component of one of the nucleotides of NAD^+ is nicotinamide, and the heterocyclic component of the other is adenine. This accounts for the coenzyme's name (**n**icotinamide **a**denine **d**inucleotide). The positive charge in the NAD^+ abbreviation indicates the positively charged nitrogen of the substituted pyridine ring.

a nucleotide   **adenine**   **niacinamide nicotinamide**   **niacin nicotinic acid**

The adenine nucleotide for the coenzyme is provided by ATP. Niacin (vitamin B_3) is the portion of the coenzyme that the body cannot synthesize and must acquire through the diet. Niacinamide is a nutritionally equivalent form of niacin.

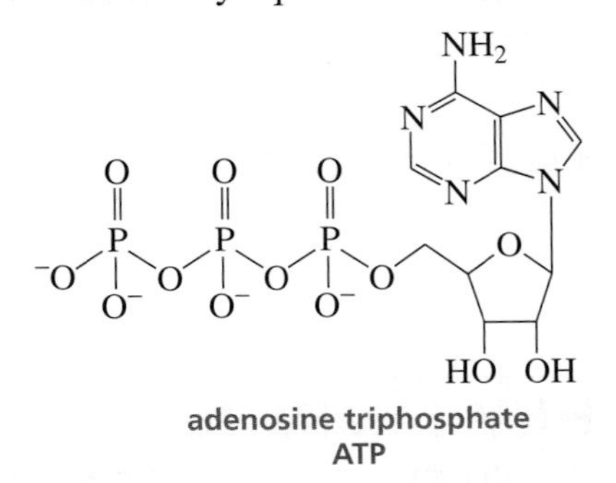
adenosine triphosphate
ATP

NICOTINIC ACID BECOMES NIACIN

When bread companies started adding nicotinic acid to their bread, they insisted that its name be changed to niacin because nicotinic acid sounded too much like nicotine and they did not want their vitamin-enriched bread to be associated with a harmful substance.

NADH is a reducing agent.

When NAD^+ oxidizes a substrate, the coenzyme is reduced to NADH. **NADH** is a reducing agent; it is used as a coenzyme by certain enzymes that catalyze reduction reactions.

$$\text{substrate}_{\text{reduced}} + NAD^+ \xrightleftharpoons{\text{enzyme}} \text{substrate}_{\text{oxidized}} + NADH + H^+$$

Malate dehydrogenase is an example of an enzyme that catalyzes an oxidation reaction; it catalyzes the oxidation of the *secondary alcohol group* of malate to a *ketone group* (Section 10.4). The oxidizing agent in this reaction is NAD^+. Many enzymes that catalyze oxidation reactions are called **dehydrogenases**. Recall that the number of C—H bonds decreases in an oxidation reaction (Section 10.4). In other words, dehydrogenases remove hydrogen.

$$^{-}OOCCH_2CH(OH){-}COO^{-} + NAD^+ \xrightleftharpoons{\text{malate dehydrogenase}} {}^{-}OOCCH_2C(=O){-}COO^{-} + NADH + H^+$$

malate → oxaloacetate

How does this oxidation reaction take place? When a substrate is being *oxidized*, it donates a hydride ion (H^-) to the 4-position of the pyridine ring of NAD^+. The pyridine ring, therefore, is reduced. The rest of the NAD^+ molecule has the job of binding the coenzyme to the proper site on the enzyme. A basic amino acid side chain of the enzyme can help the oxidation reaction by removing a proton from the oxygen atom of the substrate.

a basic group of an amino acid side chain

oxidation of substrate
reduction of coenzyme

PROBLEM 4 ◆

What is the product of the following reaction?

$$^{-}OOCCH(OH)CH(COO^{-})CH_2COO^{-} + NAD^+ \xrightarrow{\text{isocitrate dehydrogenase}}$$

isocitrate

The mechanism for reduction by NADH is the reverse of the mechanism for oxidation by NAD^+. When a substrate is being *reduced*, the dihydropyridine ring of NADH donates a hydride ion from its 4-position to the substrate. The dihydropyridine ring, therefore, is oxidized. An acidic amino acid side chain of the enzyme aids the reduction reaction by donating a proton to the substrate.

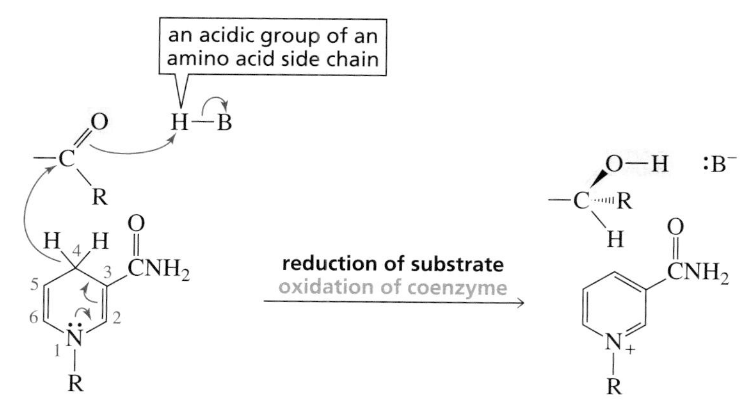

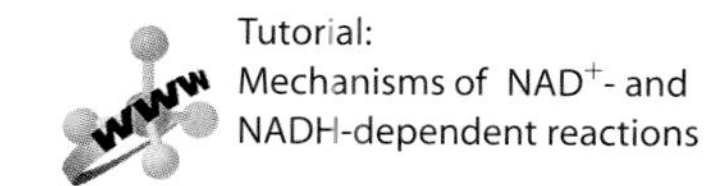

Tutorial:
Mechanisms of NAD+- and NADH-dependent reactions

Because NADH reduces a compound by donating a hydride ion, it can be considered the biological equivalent of $NaBH_4$ or $LiAlH_4$, the hydride donors we have seen used as reducing reagents in nonbiological reactions (Section 12.6).

NIACIN DEFICIENCY

A deficiency in niacin causes pellagra, a disease that begins with dermatitis and ultimately causes insanity and death. More than 120,000 cases of pellagra were reported in the United States in 1927, mainly among poor people with unvaried diets. A factor known to be present in preparations of vitamin B prevented pellagra, but it was not until 1937 that the factor was identified as nicotinic acid. Mild deficiencies slow down metabolism, which is a potential contributing factor in obesity.

PROBLEM 5 ♦

What is the product of the following reaction?

$$CH_3\overset{O}{\overset{\|}{C}}-\overset{O}{\overset{\|}{C}}O^- \xrightarrow[\text{NADH + H}^+]{\text{enzyme}}$$

17.6 VITAMIN B_2

FAD is an oxidizing agent.

Flavin adenine dinucleotide (FAD) is another coenzyme used *to oxidize substrates.* For example, FAD is the coenzyme used by succinate dehydrogenase to oxidize succinate to fumarate.

$$^-O\overset{O}{\overset{\|}{C}}CH_2CH_2\overset{O}{\overset{\|}{C}}O^- + FAD \xrightarrow{\text{succinate dehydrogenase}} \;^-OOC(H)C{=}C(H)COO^- + FADH_2$$

succinate → fumarate

As its name indicates, FAD is a dinucleotide in which one of the heterocyclic compounds is flavin and the other is adenine. Notice that instead of ribose, the flavin nucleotide has a reduced ribose (a ribitol group). Flavin plus ribitol is the vitamin known as *riboflavin* or vitamin B_2. A vitamin B_2 deficiency causes inflammation of the skin.

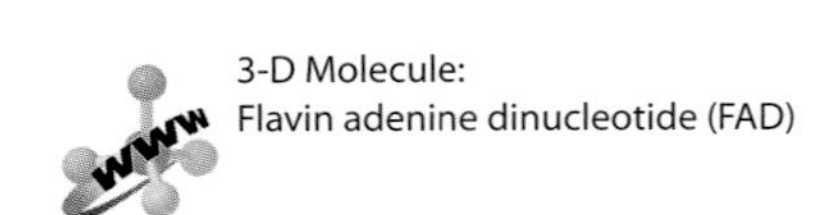

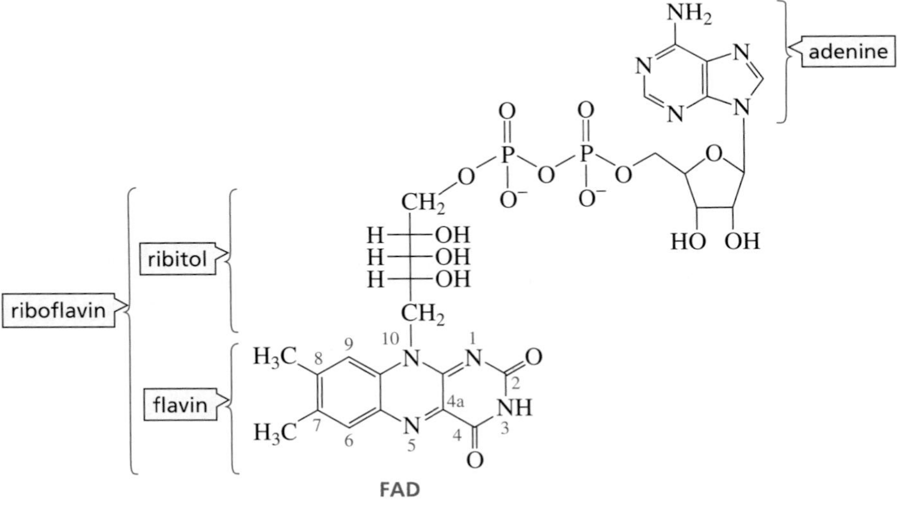

FAD

When FAD oxidizes a compound (S), FAD is reduced to $FADH_2$.

$$\text{FAD} + S_{red} \longrightarrow \text{FADH}_2 + S_{ox}$$

FAD FADH$_2$

How can we tell which enzymes use FAD and which use NAD^+ as the oxidizing coenzyme? A rough guideline is that NAD^+ is the coenzyme used in enzyme-catalyzed oxidation reactions involving carbonyl compounds (alcohols being oxidized to ketones, aldehydes, or carboxylic acids), while FAD is the coenzyme used in other types of oxidations.

PROBLEM 6 ◆

How many conjugated double bonds are there in
a. FAD? **b.** $FADH_2$?

PROBLEM 7 ◆

What is the product of the following reaction? (*Hint:* See Section 16.6.)

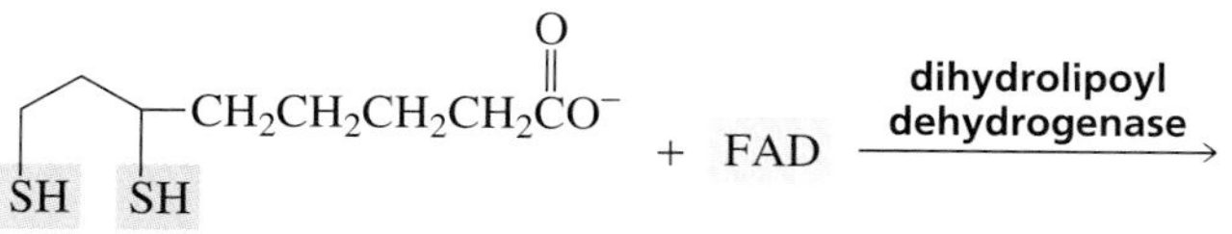

$$\text{dihydrolipoate (ring-opened dithiol with two SH groups)}-CH_2CH_2CH_2CH_2CO^- \;+\; FAD \xrightarrow{\text{dihydrolipoyl dehydrogenase}}$$

dihydrolipoate

17.7 VITAMIN B_1

Thiamine was the first of the B vitamins to be identified, so it became known as vitamin B_1. The absence of vitamin B_1 in the diet causes a disease called beriberi (Section 17.4), which damages the heart, impairs nerve reflexes, and in extreme cases causes paralysis.

Vitamin B_1 is used to form the coenzyme **thiamine pyrophosphate (TPP)**. TPP is the coenzyme required by enzymes that catalyze *the transfer of a two-carbon fragment from one species to another.*

Thiamine pyrophosphate (TPP) is required by enzymes that catalyze the transfer of a two-carbon fragment from one species to another.

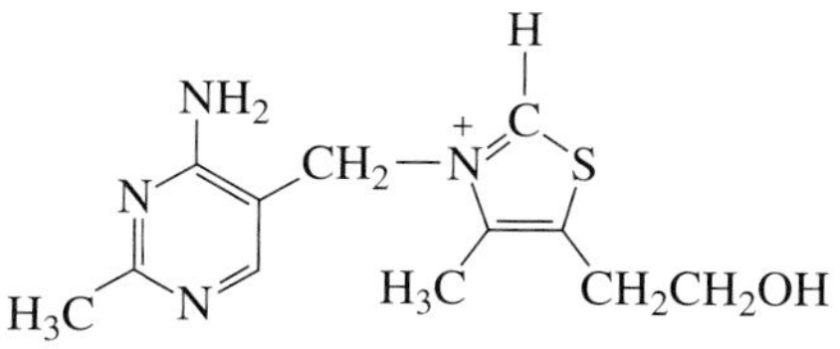

vitamin B_1 (thiazolium ring bearing CH_2CH_2OH)

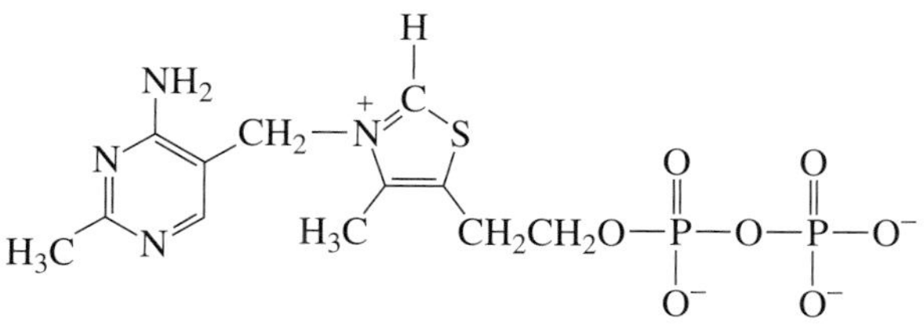

thiamine pyrophosphate
TPP (thiazolium ring bearing $CH_2CH_2O-P(=O)(O^-)-O-P(=O)(O^-)-O^-$)

Pyruvate decarboxylase is an example of an enzyme that requires thiamine pyrophosphate. This enzyme catalyzes the decarboxylation of pyruvate and transfers the remaining two-carbon fragment to a proton, resulting in the formation of acetaldehyde.

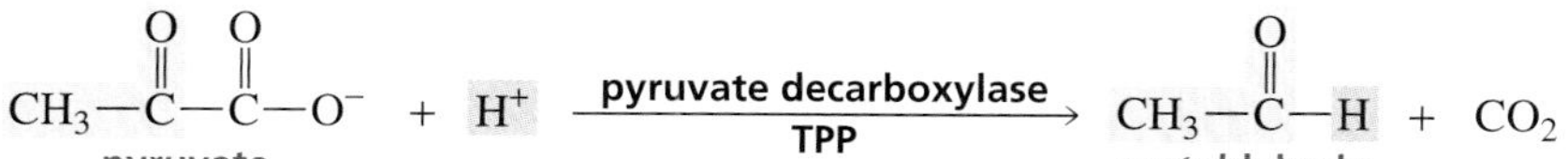

$$\underset{\text{pyruvate}}{CH_3-\overset{O}{\overset{\|}{C}}-\overset{O}{\overset{\|}{C}}-O^-} + H^+ \xrightarrow[\text{TPP}]{\text{pyruvate decarboxylase}} \underset{\text{acetaldehyde}}{CH_3-\overset{O}{\overset{\|}{C}}-H} + CO_2$$

You may wonder why an α-keto acid such as pyruvate can be decarboxylated, since the electrons left behind when CO_2 is removed cannot be delocalized onto the carbonyl oxygen (Section 13.8). The mechanism for the reaction, shown on the next page, shows that the coenzyme provides a site for the delocalization of electrons.

mechanism for pyruvate decarboxylase

resonance contributor

resonance contributor
resonance-stabilized carbanion

- A base removes a proton from the coenzyme, forming a nucleophile.
- The nucleophile attacks the substrate's electrophilic ketone group. An acidic side chain of the enzyme helps the reaction.
- The intermediate thus formed can easily undergo decarboxylation because the electrons left behind when CO_2 is removed can be delocalized onto the positively charged nitrogen of the coenzyme.
- The decarboxylated product is stabilized by electron delocalization.
- Protonation of the resonance-stabilized carbanion and a subsequent elimination reaction form acetaldehyde and regenerate the coenzyme.

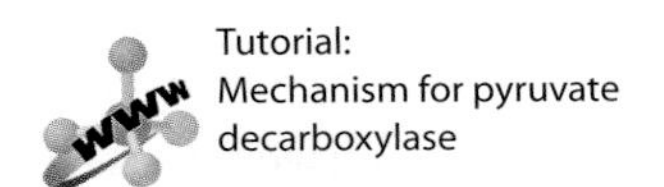

The *pyruvate dehydrogenase system* is a group of three enzymes and five coenzymes. One of the coenzymes is TPP. The overall reaction catalyzes the decarboxylation of pyruvate and transfers the remaining two-carbon fragment to coenzyme A, resulting in the formation of acetyl-CoA.

$$CH_3-\overset{O}{\overset{\|}{C}}-\overset{O}{\overset{\|}{C}}-O^- + CoASH \xrightarrow{\text{pyruvate dehydrogenase system}} CH_3-\overset{O}{\overset{\|}{C}}-SCoA + CO_2$$

pyruvate — acetyl-CoA

Coenzyme A (CoASH) activates a carboxylic acid by converting it to a thioester.

Coenzyme A (CoASH) is the coenzyme that activates carboxylic acids by converting them to thioesters, which are much more reactive in nucleophilic acyl substitution reactions than are carboxylic acids, since the pK_a of the conjugate acid of the thiol leaving group of a thioester is ~ 10 (Section 11.6). The vitamin needed to make CoASH is pantothenate.

ADP — pantothenate — decarboxylated cysteine — adenine — phosphate

coenzyme A
CoASH

PROBLEM 8 ◆

a. What two-carbon fragment does pyruvate decarboxylase transfer to a proton?
b. What two-carbon fragment does the pyruvate dehydrogenase system transfer to coenzyme A?

17.8 VITAMIN H

Vitamin H (**biotin**) is an unusual vitamin because it can be synthesized by bacteria that live in the intestine. Consequently, biotin does not have to be included in our diet and deficiencies are rare. Biotin deficiencies, however, can be found in people who maintain a diet high in raw eggs. Egg whites contain a protein that binds biotin tightly and thereby prevents it from acting as a coenzyme. When eggs are cooked, the protein is denatured (Section 16.13), and the denatured protein does not bind biotin.

[Structure: biotin — bicyclic ring with HN–C(=O)–NH urea ring fused to a tetrahydrothiophene (S) ring, side chain $CH_2CH_2CH_2CH_2COH$ (C=O)]

biotin

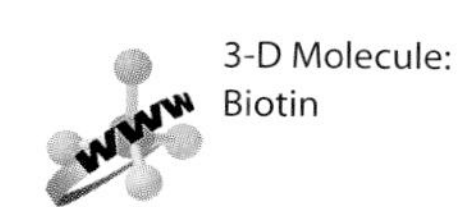

3-D Molecule: Biotin

Biotin is the coenzyme required by enzymes that catalyze *carboxylation of an α-carbon (a carbon adjacent to a carbonyl group)*. Therefore, the enzymes that require biotin as a coenzyme are called carboxylases. For example, acetyl-CoA carboxylase converts acetyl-CoA into malonyl-CoA. Biotin-requiring enzymes use bicarbonate (HCO_3^-) for the source of the carboxyl group that becomes attached to the substrate. The enzymes also require ATP and Mg^{2+}.

Biotin is required by enzymes that catalyze the carboxylation of a carbon adjacent to a carbonyl group.

$$CH_3-\overset{O}{\overset{\|}{C}}-SCoA + HCO_3^- + ATP \xrightarrow[\text{Mg}^{2+},\ \text{biotin}]{\text{acetyl-CoA carboxylase}} {}^-O-\overset{O}{\overset{\|}{C}}-CH_2-\overset{O}{\overset{\|}{C}}-SCoA + ADP + {}^-O-\overset{O}{\overset{\|}{P}}(-O^-)-OH$$

acetyl-CoA → malonyl-CoA

a carboxyl group has been added to the α-carbon

PROBLEM 9 ◆

Pyruvate carboxylase is a biotin-requiring enzyme. The enzyme's substrate is pyruvate. What is the product of the enzyme-catalyzed reaction?

$$CH_3\overset{O}{\overset{\|}{C}}-\overset{O}{\overset{\|}{C}}O^-$$

pyruvate

PROBLEM-SOLVING STRATEGY

How many moles of acetyl-CoA must be converted to malonyl-CoA in order to synthesize 1 mole of palmitic acid, a 16-carbon saturated fatty acid?

To answer this question, we need to recall how fatty acids are biosynthesized (Section 13.11). The biosynthesis starts with the reaction of a molecule of acetyl thioester and a molecule of malonyl thioester to form a four-carbon fatty acid. Each subsequent two-carbon unit is added by a molecule of malonyl thioester. Twelve more carbons are needed to form palmitic acid, so another six molecules of malonyl-CoA are required. Therefore, the synthesis of palmitic acid requires 7 moles of acetyl-CoA to be converted to malonyl-CoA.

Now continue on to Problem 10.

PROBLEM 10 ◆

How many moles of acetyl-CoA must be converted to malonyl-CoA in order to synthesize 1 mole of arachidic acid, a 20-carbon saturated fatty acid?

PROBLEM 11 ***SOLVED***

How many moles of ATP are needed to make 1 mole of palmitic acid?

Solution One mole of a 16-carbon fatty acid is synthesized from 1 mole of acetyl-CoA and 7 moles of malonyl-CoA. Each mole of malonyl-CoA synthesized from acetyl-CoA requires 1 mole of ATP for the carboxylation reaction. Therefore, 7 moles of ATP are needed to make 1 mole of palmitic acid.

17.9 VITAMIN B_6

3-D Molecule: Pyridoxal phosphate (PLP)

The coenzyme **pyridoxal phosphate (PLP)** is derived from the vitamin known as pyridoxine or vitamin B_6. Pyridoxal's "al" suffix indicates that the coenzyme is an aldehyde. A deficiency in vitamin B_6 causes anemia; severe deficiencies can cause seizures and death.

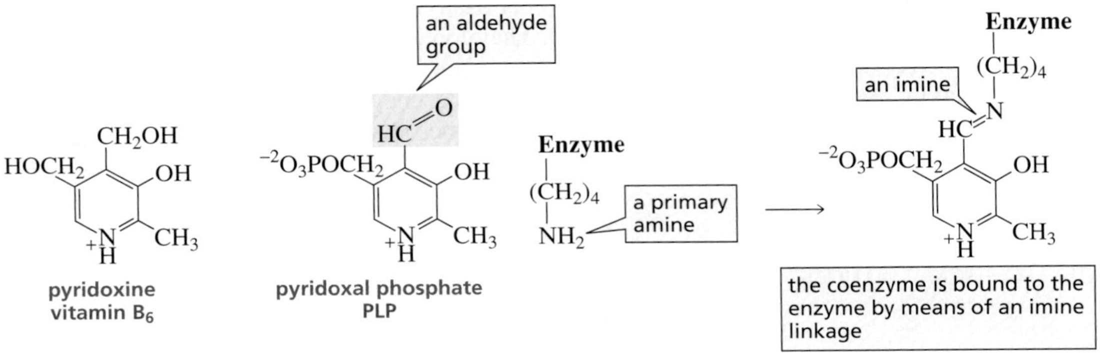

an aldehyde group

a primary amine

an imine

pyridoxine vitamin B_6

pyridoxal phosphate PLP

the coenzyme is bound to the enzyme by means of an imine linkage

Pyridoxal phosphate (PLP) is required by enzymes that catalyze certain reactions of amino acids.

PLP is required by enzymes that catalyze certain reactions of amino acids. One such reaction is decarboxylation.

decarboxylation

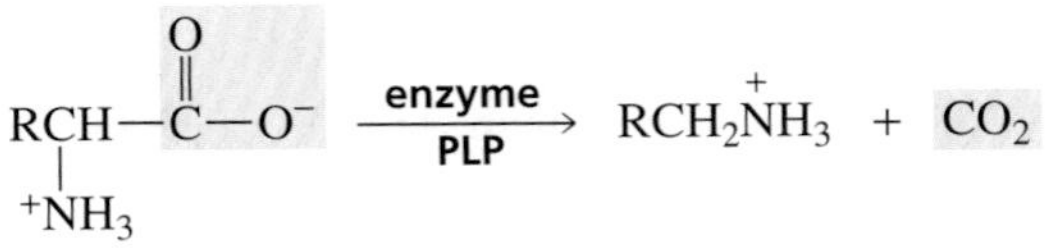

$$RCH(^{+}NH_3)-C(=O)-O^- \xrightarrow[\text{PLP}]{\text{enzyme}} RCH_2\overset{+}{N}H_3 + CO_2$$

The mechanism of the reaction shows that the coenzyme is needed to accept the electrons left behind when CO_2 is removed.

HEART ATTACKS: ASSESSING THE DAMAGE

After a heart attack, aminotransferases (see page 489) and other enzymes leak from the damaged cells of the heart into the bloodstream. The severity of the damage done to the heart can be determined from the concentrations of alanine aminotransferase and aspartate aminotransferase in the bloodstream.

mechanism for decarboxylation

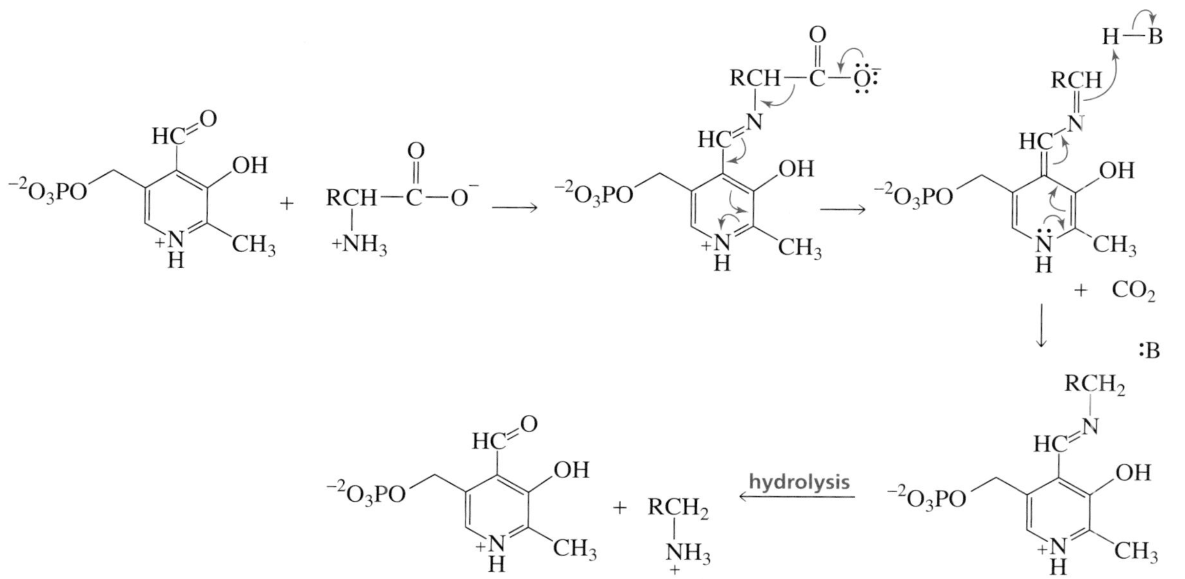

$$^{-2}O_3PO\text{-pyridoxal phosphate (HC=O)} + RCH(^{+}NH_3)-COO^- \longrightarrow \text{imine} \longrightarrow \text{decarboxylated intermediate} + CO_2 \xrightarrow{H-B} \text{imine of } RCH_2 \xrightarrow{\text{hydrolysis}} \text{PLP} + RCH_2NH_3^+$$

- The aldehyde group of the coenzyme forms an imine with the primary amino group of the substrate (Section 12.7).
- The electrons left behind when CO_2 is removed are delocalized onto the positively charged nitrogen of the coenzyme.
- Electron rearrangement and protonation of the α-carbon of the decarboxylated intermediate by an acidic side chain of the enzyme reestablishes the aromaticity of the six-membered ring.
- Hydrolysis of the imine releases the decarboxylated amino acid and reforms the coenzyme.

The first reaction in the metabolism of most amino acids is the replacement of the amino group of the amino acid by a ketone group. This is called **transamination** because the amino group removed from the amino acid is not lost, but is *transferred* to the ketone group of α-ketoglutarate, thereby forming glutamate. Transamination is one of the most common reactions that requires PLP.

transamination

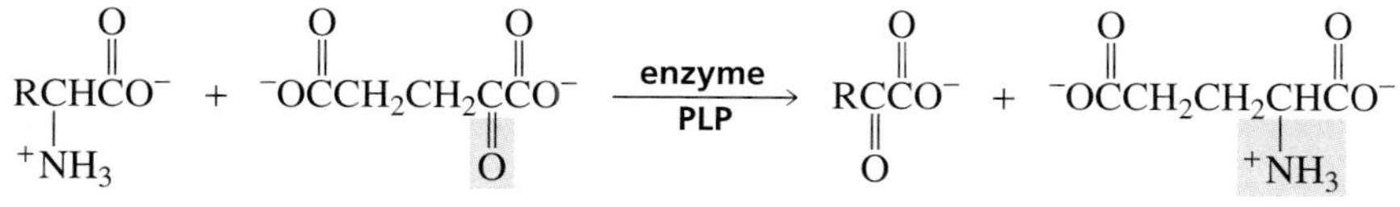

$$RCH(^{+}NH_3)CO_2^- + {}^-O_2CCH_2CH_2C(=O)CO_2^- \xrightarrow[\text{PLP}]{\text{enzyme}} RC(=O)CO_2^- + {}^-O_2CCH_2CH_2CH(^{+}NH_3)CO_2^-$$

α-ketoglutarate | glutamate

The enzymes that catalyze transaminations are called *aminotransferases*. Transamination allows the amino groups of the various amino acids to be collected into a single amino acid (glutamate) so that excess nitrogen can be easily excreted.

PROBLEM 12 ♦

α-Keto acids other than α-ketoglutarate can be used to accept the amino group of amino acids in enzyme-catalyzed transaminations. What amino acids are formed from the following α-keto acids?

a. $CH_3C(=O)-CO_2^-$ pyruvate

b. ${}^-O_2CCH_2C(=O)-CO_2^-$ oxaloacetate

Dorothy Crowfoot Hodgkin (1910–1994) *was born in Egypt to English parents. She received an undergraduate degree from Somerville College at Oxford University and earned a Ph.D. from Cambridge University. She performed the first three-dimensional calculations in crystallography, and she was the first to use computers to determine the structures of compounds, successfully determining the structures of penicillin, insulin, and vitamin* B_{12}*. For her work on vitamin* B_{12}*, she received the 1964 Nobel Prize in chemistry. Hodgkin was a professor of chemistry at Somerville, where one of her graduate students was former Prime Minister of England Margaret Thatcher. She was also a founding member of Pugwash, an organization whose purpose was to further communication between scientists on both sides of the Iron Curtain.*

17.10 VITAMIN B_{12}

Enzymes that catalyze certain rearrangement reactions require **coenzyme B_{12}**, a coenzyme derived from vitamin B_{12}. The vitamin has a $^-C{\equiv}N$ or HO^- group coordinated with cobalt. In coenzyme B_{12}, this group is replaced by a 5′-deoxyadenosyl group.

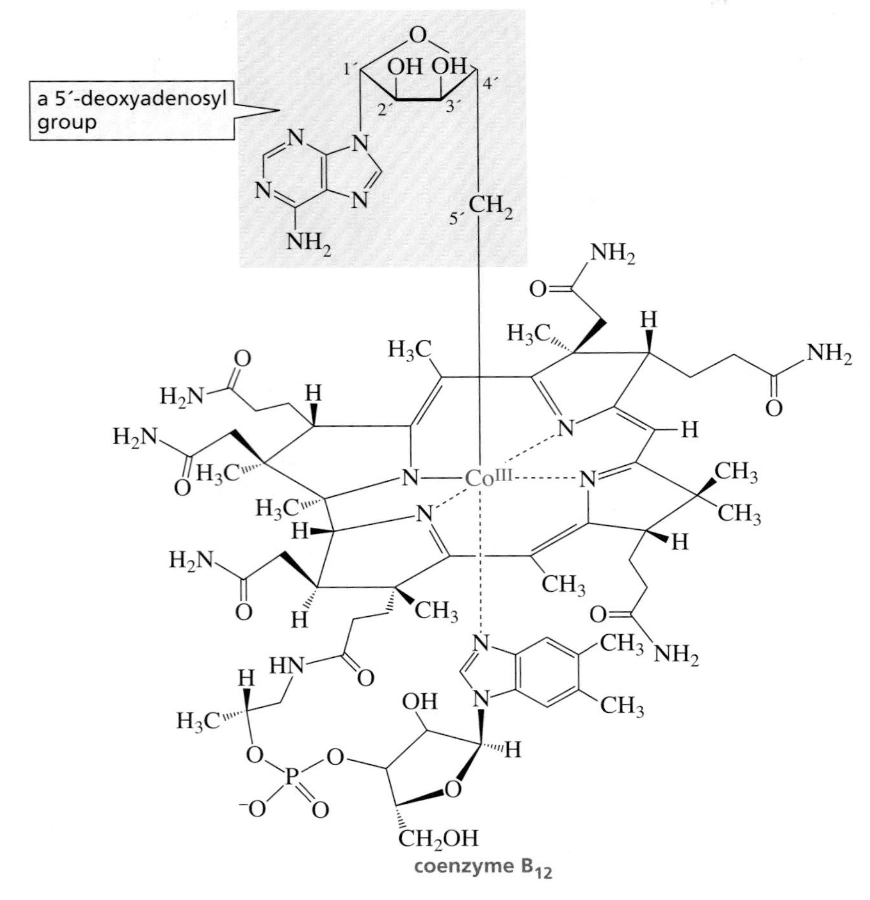

coenzyme B_{12}

Animals and plants cannot synthesize vitamin B_{12}; in fact, only a few microorganisms can synthesize it. Humans must obtain all their vitamin B_{12} from their diet, particularly from meat. A deficiency causes pernicious anemia. Because vitamin B_{12} is needed in only very small amounts, deficiencies caused by consumption of insufficient amounts of the vitamin are rare but have been found in vegetarians who eat no animal products. Most deficiencies are caused by the intestines' inability to absorb the vitamin.

Coenzyme B_{12} is required by enzymes that catalyze *reactions in which a group (Y) bonded to one carbon changes places with a hydrogen bonded to an adjacent carbon.*

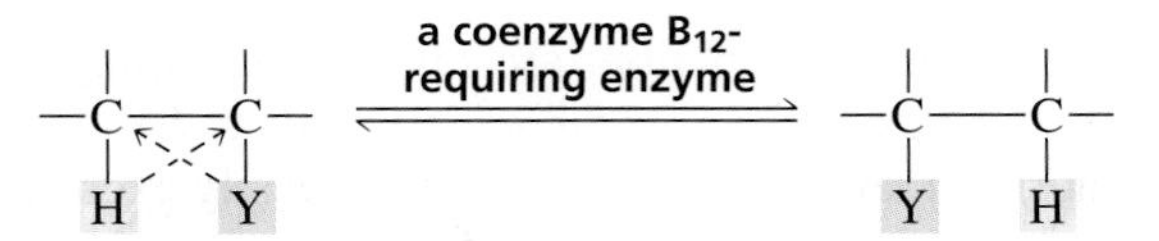

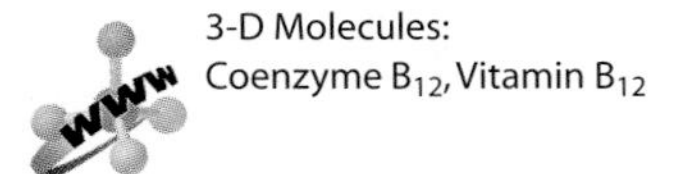

For example, in each of the following reactions, a COO^- group bonded to one carbon changes places with an H of an adjacent methyl group.

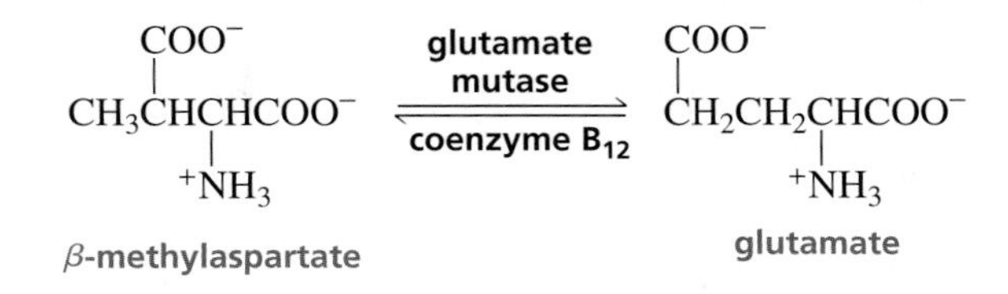

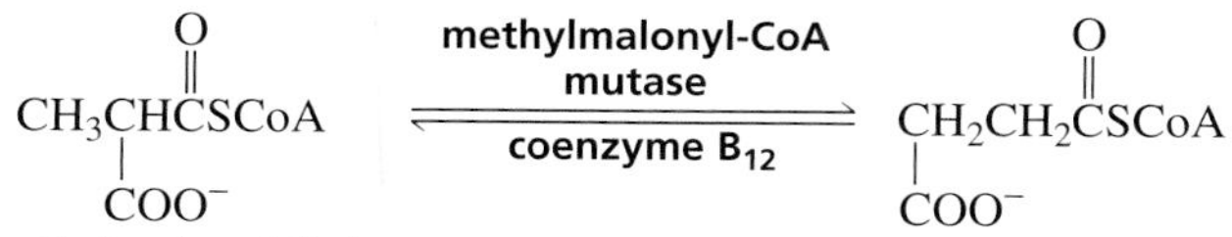

$$CH_3CH(COO^-)C(=O)SCoA \xrightleftharpoons[\text{coenzyme } B_{12}]{\text{methylmalonyl-CoA mutase}} CH_2(COO^-)CH_2C(=O)SCoA$$

methylmalonyl-CoA — succinyl-CoA

Coenzyme B_{12} is required by enzymes that catalyze the exchange of a hydrogen bonded to one carbon with a group bonded to an adjacent carbon.

PROBLEM 13 ♦

What groups are interchanged in the following enzyme-catalyzed reaction that requires coenzyme B_{12}?

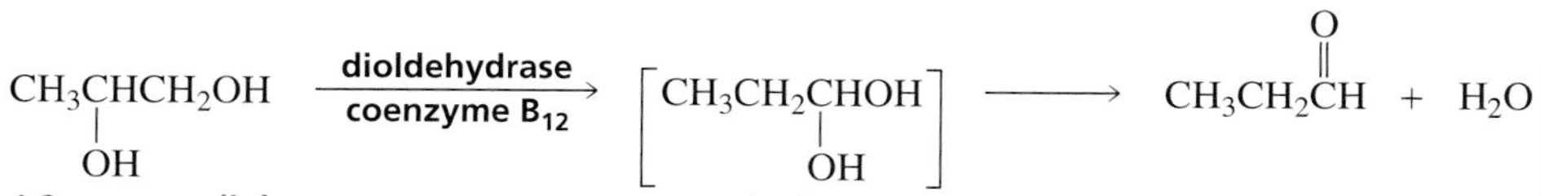

$$CH_3CH(OH)CH_2OH \xrightarrow[\text{coenzyme } B_{12}]{\text{dioldehydrase}} [CH_3CH_2CH(OH)OH] \longrightarrow CH_3CH_2CH(=O) + H_2O$$

1,2-propanediol — a hydrate — propanal

17.11 FOLIC ACID

Tetrahydrofolate (THF) is the coenzyme used by enzymes that catalyze reactions that *transfer a group containing a single carbon to their substrates*. The one-carbon group can be a methyl group (CH_3), a methylene group (CH_2), or a formyl group ($HC{=}O$). Tetrahydrofolate is produced by the reduction of two double bonds of folic acid (folate), its precursor vitamin. Bacteria can synthesize folate, but mammals cannot (Section 21.4).

Tetrahydrofolate (THF) is the coenzyme required by enzymes that catalyze the transfer of a group containing one carbon to their substrates.

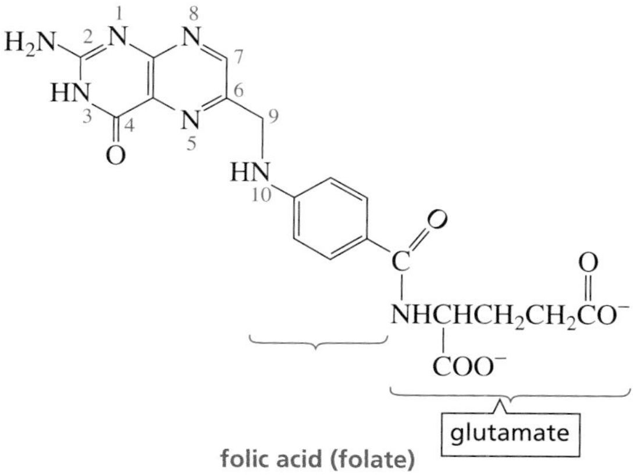

folic acid (folate)

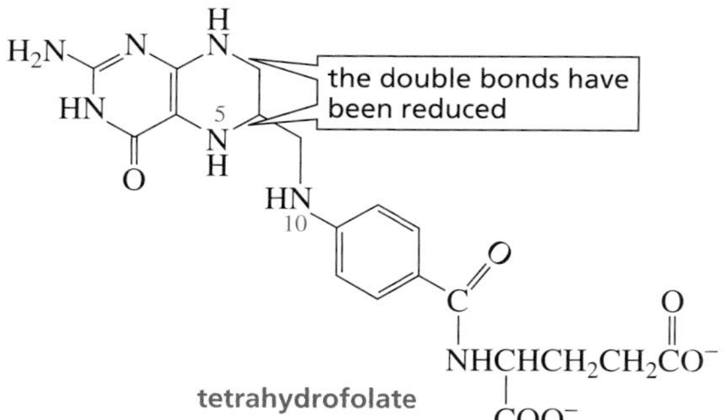

glutamate

tetrahydrofolate THF

the double bonds have been reduced

3-D Molecule: Tetrahydrofolate (THF)

Three THF-coenzymes are shown below: N^5-methyl-THF transfers a methyl group; N^5,N^{10}-methylene-THF transfers a methylene group; and N^5,N^{10}-methenyl-THF transfers a formyl group.

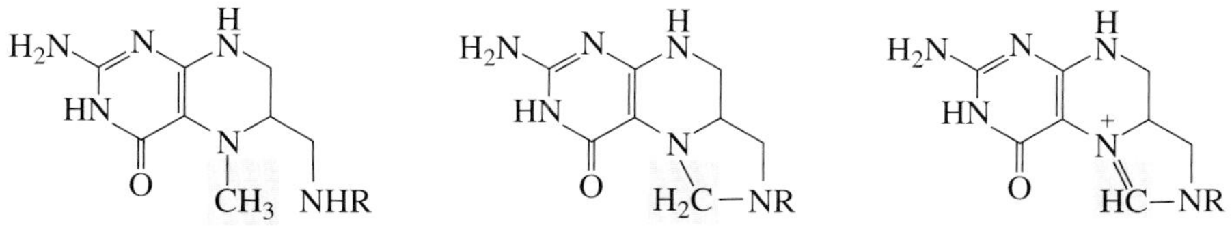

N^5-methyl-THF — N^5,N^{10}-methylene-THF — N^5,N^{10}-methenyl-THF

The coenzyme is required for the synthesis of the bases found in DNA and RNA and for the synthesis of aromatic amino acids (Section 16.1).

PROBLEM 14 ♦

Why is the coenzyme called tetrahydrofolate?

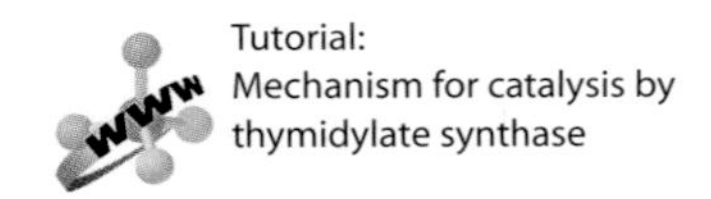

The heterocyclic bases in RNA are adenine, guanine, cytosine, and uracil (A, G, C, and U); the heterocyclic bases in DNA are adenine, guanine, cytosine, and thymine (A, G, C, and T). In other words, the heterocyclic bases in RNA and DNA are the same, except RNA contains U's, whereas DNA contains T's. Why DNA contains T's instead of U's is explained in Section 20.9.

The T's used for the biosynthesis of DNA are synthesized from U's by thymidylate synthase, an enzyme that requires N^5,N^{10}-methylene-THF as a coenzyme.

dUMP + N^5,N^{10}-methylene-THF —(thymidylate synthase)→ dTMP + dihydrofolate (DHF)

R′ = 2′-deoxyribose-5′-phosphate

THF has been oxidized to DHF

Even though the only structural difference between a T and a U is a *methyl* group, a T is synthesized by first transferring a *methylene* group to a U. The methylene group is then reduced to a methyl group. The coenzyme is the reducing agent, so it is oxidized (it loses two hydrogens). The oxidized coenzyme is dihydrofolate.

CANCER CHEMOTHERAPY

Cancer is the abnormal growth and proliferation of cells. Because cells cannot multiply if they cannot synthesize DNA, scientists searched for compounds that would inhibit thymidylate synthase. If a cell cannot make thymidine (T), it cannot synthesize DNA.

A common anticancer drug that inhibits thymidylate synthase is 5-fluorouracil. Thymidylate synthase reacts with uracil and 5-fluorouracil in the same way. However, the fluorine of 5-fluoruracil causes it to become permanently attached to the enzyme because the base cannot remove a F^+ in an elimination reaction. As a consequence, the reaction stops, leaving the enzyme permanently attached to 5-fluorouracil. The active site of the enzyme is now blocked with 5-fluorouracil and cannot bind dUMP. Therefore, dTMP can no longer be synthesized and, without dTMP, DNA cannot be synthesized.

Unfortunately, most anticancer drugs cannot discriminate between diseased and normal cells, so cancer chemotherapy is accompanied by debilitating side effects. However, because cancer cells, with their uncontrolled cell division are dividing more rapidly than normal cells, they are harder hit by cancer-fighting chemotherapeutic agents than normal cells are.

5-fluorouracil
5-FU

the enzyme has become irreversibly attached to the substrate

THE FIRST ANTIBIOTICS

Sulfonamides—commonly known as sulfa drugs—were introduced clinically in 1936 as the first effective antibiotics (Section 21.4). Notice that sulfanilamide, the first sulfonamide, is structurally similar to *p*-aminobenzoic acid.

a sulfonamide　　sulfanilamide　　*p*-aminobenzoic acid

Sulfanilamide acts by inhibiting the enzyme that incorporates *p*-aminobenzoic acid into folate. Because the enzyme cannot tell the difference between sulfanilamide and *p*-aminobenzoic acid, both compounds compete for the active site of the enzyme. Humans are not adversely affected by the drug because they do not synthesize folate; they get all their folate from their diets.

PROBLEM 15 ◆

What amino acid is formed by the following reaction?

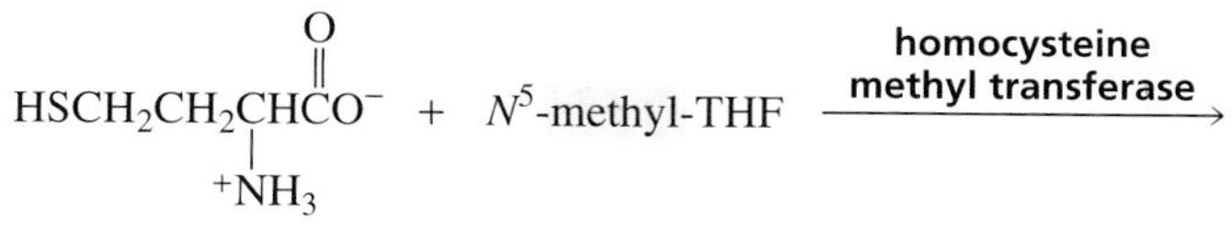

$$\mathrm{HSCH_2CH_2\underset{\underset{+NH_3}{|}}{C}H\overset{\overset{O}{\|}}{C}O^-} + N^5\text{-methyl-THF} \xrightarrow[\text{}]{\text{homocysteine methyl transferase}}$$

homocysteine

PROBLEM 16 ◆

What is the source of the methyl group in dTMP?

17.12 VITAMIN K

Vitamin K is required for proper clotting of blood. The letter K comes from *koagulation*, which is German for "clotting." A series of reactions utilizing six proteins causes blood to clot. This process requires these proteins to bind Ca^{2+}. Vitamin K is required for proper Ca^{2+} binding. Vitamin K is found in the leaves of green plants. Deficiencies in the vitamin are rare because it is synthesized by intestinal bacteria. **Vitamin KH_2** is the coenzyme form of the vitamin.

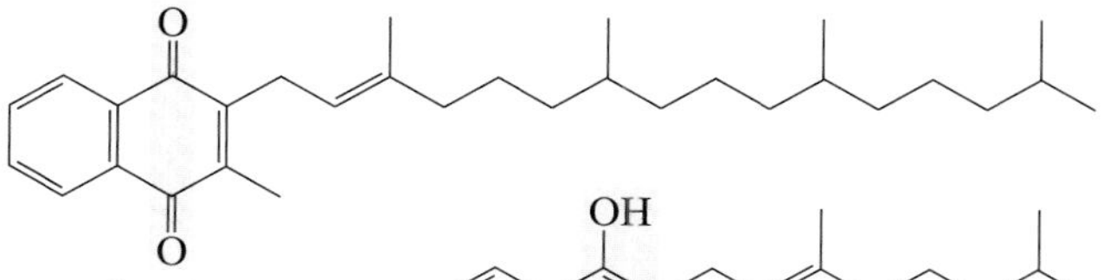

vitamin K

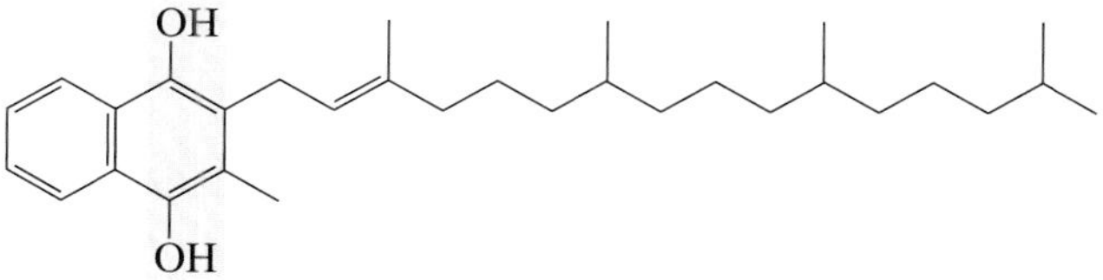

vitamin KH_2

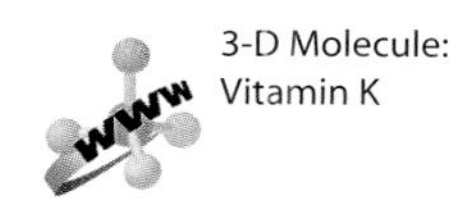

Vitamin KH_2 is the coenzyme for the enzyme that catalyzes *the carboxylation of the γ-carbon of glutamate side chains* in proteins, forming γ-carboxyglutamates. γ-Carboxyglutamates complex Ca^{2+} much more effectively than glutamates do. The enzyme uses CO_2 for the carboxyl group it puts on glutamate side chains. All the proteins responsible for blood clotting have several glutamates near their N-terminal ends. For example, prothrombin, a blood-clotting protein, has glutamates at positions 7, 8, 15, 17, 20, 21, 26, 27, 30, and 33.

Vitamin KH_2 is required by the enzyme that catalyzes the carboxylation of the γ-carbon of a glutamate side chain in a protein.

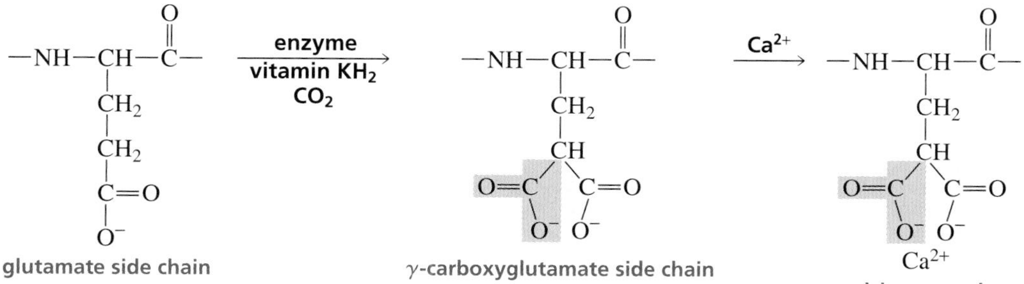

glutamate side chain $\xrightarrow[\text{vitamin } KH_2,\ CO_2]{\text{enzyme}}$ γ-carboxyglutamate side chain $\xrightarrow{Ca^{2+}}$ calcium complex

Warfarin (coumadin) and dicoumerol are used clinically as anticoagulants. They prevent clotting by inhibiting the enzyme that synthesizes vitamin KH_2 from vitamin K epoxide. Because the enzyme cannot tell the difference between vitamin K epoxide and warfarin (or dicoumerol), the two compounds compete for binding at the enzyme's active site. Warfarin is also a common rat poison, causing death by internal bleeding.

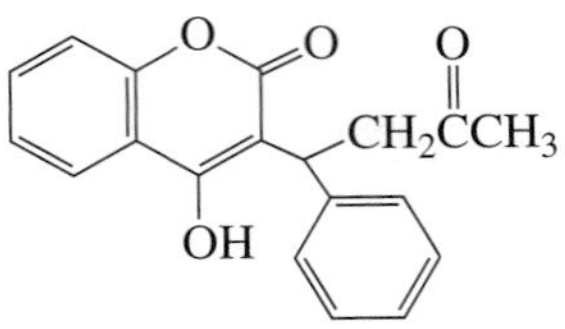

warfarin

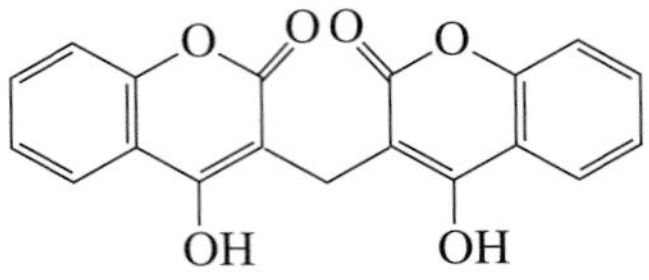

dicoumarol

TOO MUCH BROCCOLI

An article describing two women with diseases characterized by abnormal blood clotting reported that they did not improve when they were given warfarin. When questioned about their diets, one woman said that she ate at least a pound (0.45 kg) of broccoli every day, and the other ate broccoli soup and a broccoli salad every day. When broccoli was removed from their diets, warfarin became effective in preventing the abnormal clotting of their blood. Because broccoli is high in vitamin K, these patients had been getting enough dietary vitamin K to compete with the drug for the enzyme's active site, thereby making the drug ineffective.

SUMMARY

Essentially all organic reactions that occur in biological systems require a catalyst. Most biological catalysts are **enzymes**. The reactant of an enzyme-catalyzed reaction is called a **substrate**. The substrate specifically binds to the **active site** of the enzyme, and all the bond-making and bond-breaking steps of the reaction occur while it is at that site. Two important factors contribute to the remarkable catalytic ability of enzymes: (1) the reacting groups are brought together at the active site in the proper orientation for reaction and (2) the catalytic groups are in the proper position relative to the substrate, precisely where they are needed for catalysis.

Coenzymes assist enzymes in catalyzing a variety of reactions that cannot be catalyzed solely by the enzymes' amino acid side chains. Coenzymes are derived from **vitamins**, which are substances needed in small amounts for normal body function that the body cannot synthesize. All the water-soluble vitamins except vitamin C are precursors for coenzymes. Vitamin K is the only water-insoluble vitamin that is a precursor for a coenzyme.

The coenzymes used by enzymes to catalyze oxidation reactions are **NAD^+** and **FAD**; **NADH** and **$FADH_2$** catalyze reduction reactions. Many enzymes that catalyze oxidation reactions are called **dehydrogenases**. **Thiamine pyrophosphate (TPP)** is the coenzyme required by enzymes that catalyze the transfer of a two-carbon fragment. **Biotin** is the coenzyme required by enzymes that catalyze carboxylation of a carbon adjacent to a carbonyl group. **Pyridoxal phosphate (PLP)** is the coenzyme required by enzymes that catalyze certain reactions of amino acids, such as decarboxylation or transamination. In a **transamination reaction**, the amino group is removed from an amino acid and transferred to another molecule, leaving a ketone group in its place.

In a **coenzyme B_{12}**-requiring reaction, a group bonded to one carbon changes places with a hydrogen bonded to an adjacent carbon. **Tetrahydrofolate (THF)** is the coenzyme used by enzymes that catalyze the transfer of a group containing a single carbon—methyl, methylene, or formyl—to their substrates. **Vitamin KH_2** is the coenzyme for the enzyme that catalyzes the carboxylation of the γ-carbon of glutamate side chains, a reaction required for blood clotting.

PROBLEMS

17. From what vitamins are the following coenzymes derived?
a. NAD^+ **b.** FAD **c.** pyridoxal phosphate **d.** N^5-methylenetetrahydrofolate

18. Name two coenzymes that act as oxidizing agents.

19. Name two coenzymes that allow electrons to be delocalized.

20. Name the three one-carbon groups that various tetrahydrofolates are capable of donating to substrates.

21. What reaction, catalyzed by vitamin KH_2, is necessary for proper blood clotting?

22. What two coenzymes are used for carboxylation reactions?

23. For each of the following enzyme-catalyzed reactions, name the required coenzyme:

a. CH_3CSCoA (C=O) $\xrightarrow[\text{ATP, Mg}^{2+}\text{, HCO}_3^-]{\text{enzyme}}$ $^-OCCH_2CSCoA$ (both C=O)

b. $^-OCCHCSCoA$ (both C=O; CH_3 on CH) $\xrightarrow{\text{enzyme}}$ $^-OCCH_2CH_2CSCoA$ (both C=O)

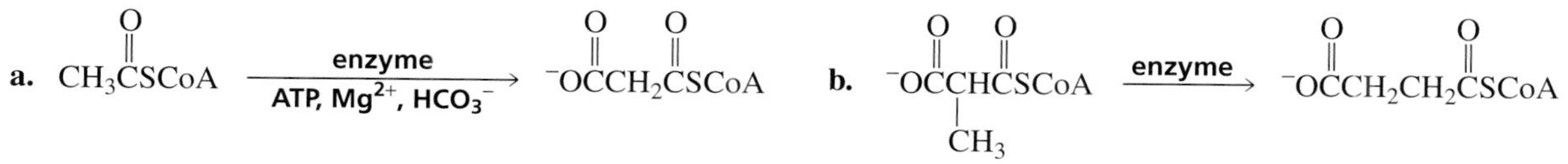

24. What is the product of the following reaction?

$$RCH{=}CHCSR \text{ (C=O)} + FADH_2 \xrightarrow{\text{enzyme}}$$

25. *S*-adenosylmethionine (SAM) is formed from the reaction between ATP and methionine (Section 9.14). The other product of the reaction is triphosphate. Propose a mechanism for this reaction. (*Hint:* It is an S_N2 reaction.)

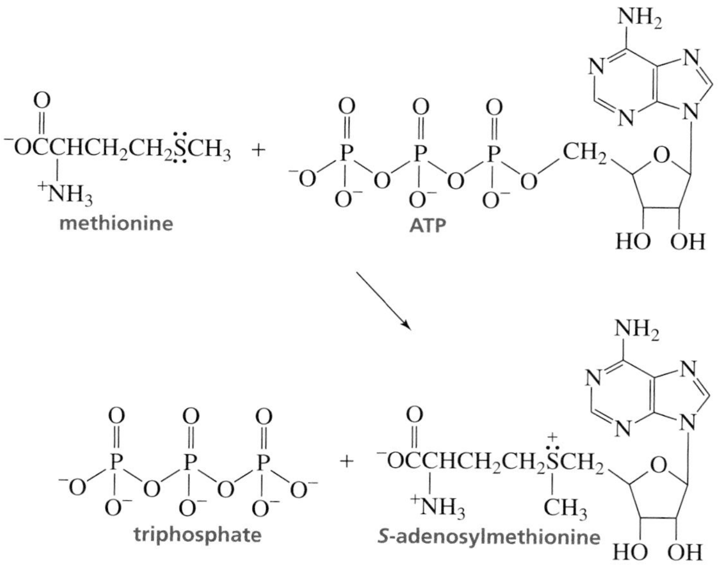

$^-OCCHCH_2CH_2\ddot{S}CH_3$ (C=O; $^+NH_3$ on CH) + ATP

methionine

ATP

$\longrightarrow$ triphosphate + $^-OCCHCH_2CH_2\overset{+}{S}CH_2$–ribose–adenine (C=O; $^+NH_3$ on CH; CH_3 on S; HO OH on ribose)

triphosphate

S-adenosylmethionine

26. For each of the following enzyme-catalyzed reactions, name the required coenzyme:

a. $^-OCCH_2CHCO^-$ ($^+NH_3$) + $^-OCCH_2CH_2CCO^-$ (C=O) $\xrightarrow{\text{enzyme}}$ $^-OCCH_2CCO^-$ (C=O) + $^-OCCH_2CH_2CHCO^-$ ($^+NH_3$)

b. CH_3CH_2CSCoA (C=O) $\xrightarrow{\text{enzyme}}$ $^-OCCHCSCoA$ (both C=O; CH_3 on CH)

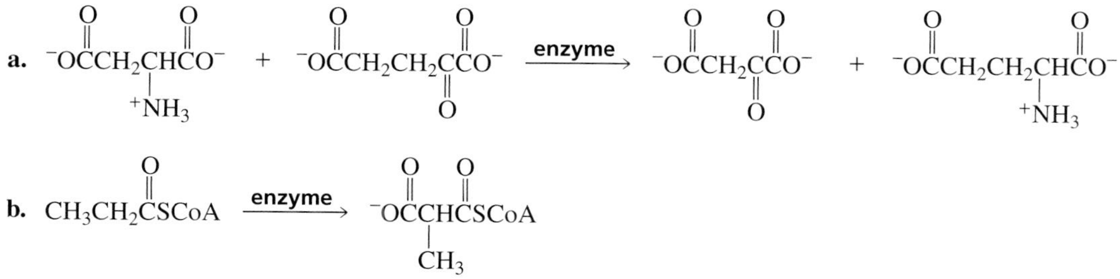

27. Which of the following compounds is more likely to lose CO_2?

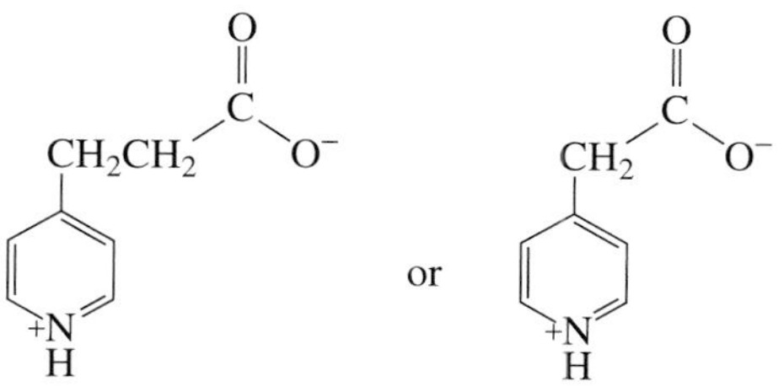

4-pyridinium (^+NH) with $CH_2CH_2C(=O)O^-$ or 4-pyridinium (^+NH) with $CH_2C(=O)O^-$

28. Propose a mechanism for the aldolase-catalyzed cleavage of fructose-1,6 diphosphate if it did not form an imine with the substrate. What is the advantage gained by imine formation?

29. In glycolysis, why must glucose-6-phosphate isomerize to fructose-6-phosphate before the cleavage reaction with aldolase occurs?

CHAPTER 18

The Organic Chemistry of Metabolic Pathways

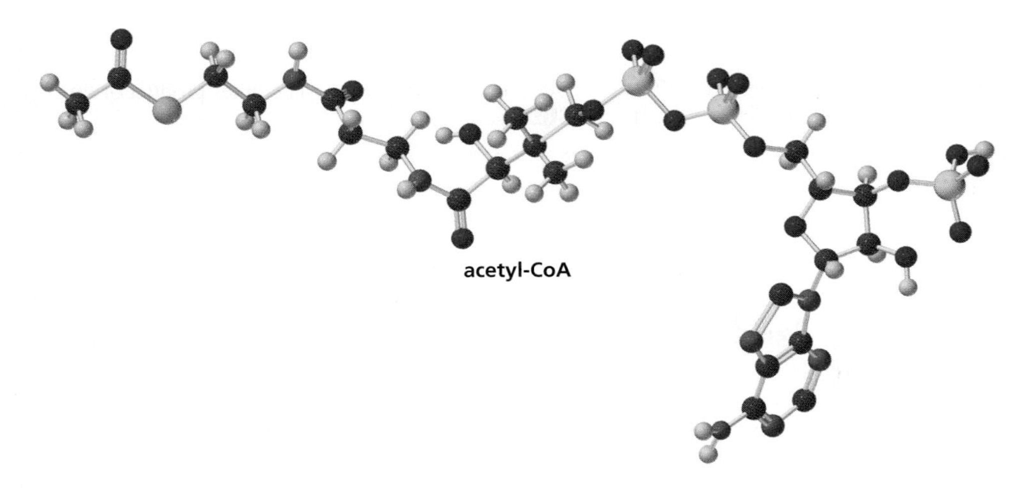

The reactions that living organisms carry out to obtain the energy they need and to synthesize the compounds they require are collectively known as **metabolism**. Metabolism can be divided into two parts: *catabolism* and *anabolism*. **Catabolic reactions** break down complex nutrient molecules to produce energy and simple precursor molecules for synthesis. **Anabolic reactions** synthesize complex biomolecules from simpler precursor molecules; these reactions require energy.

catabolism: complex molecules $\longrightarrow$ simple molecules + energy

anabolism: simple molecules + energy $\longrightarrow$ complex molecules

It is important to remember that almost every reaction that occurs in a living system is catalyzed by an enzyme. The enzyme holds the reactants and any necessary coenzymes in place, orienting the reacting functional groups and the catalyzing amino acid side chains in such a way that the enzyme-catalyzed reaction can take place (Section 17.1).

Most of the reactions that you will see in this chapter are reactions that you have studied in previous chapters. If you take the time to go back to the sections that are referred to in this chapter and review these reactions, you will see that many of the organic reactions done by cells are the same as the organic reactions done by chemists.

DIFFERENCES IN METABOLISM

Humans do not necessarily metabolize compounds in the same way as other species do. This becomes a significant problem when drugs are tested on animals (Section 21.4). For example, chocolate is metabolized to different compounds in humans and in dogs; the metabolites produced in humans are nontoxic, whereas those produced in dogs can be highly toxic. Differences in metabolism have been found even within the same species. For example, isoniazid—an antituberculosis drug—is metabolized by Eskimos much faster than by Egyptians. Current research is showing that men and women metabolize certain drugs differently. For example, kappa opioids—a class of painkillers—have been found to be about twice as effective in women as they are in men.

18.1 DIGESTION

The reactants required for all life processes ultimately come from our diet. In that sense, we really are what we eat. Catabolism can be divided into four stages (Figure 18.1). The *first stage* is called digestion. In this first stage, fats, carbohydrates, and proteins are hydrolyzed to fatty acids, monosaccharides, and amino acids, respectively. These reactions occur in the mouth, stomach, and small intestine.

In the first stage of catabolism, fats, carbohydrates, and proteins are hydrolyzed to fatty acids, monosaccharides, and amino acids.

Stage 1: fats → fatty acids; carbohydrates → monosaccharides; proteins → amino acids

Stage 2: fatty acids → acetyl-CoA + CO_2; monosaccharides → pyruvate → acetyl-CoA + CO_2; amino acids → pyruvate, acetyl-CoA + CO_2, citric acid cycle

Stage 3: citric acid cycle → CO_2, CO_2

Stage 4: NADH ⟶ 3 ATP + NAD^+

You are what you eat.

◀ **Figure 18.1**
The four stages of catabolism:
1. digestion
2. conversion of fatty acids, monosaccharides, and amino acids to compounds that can enter the citric acid cycle
3. the citric acid cycle
4. oxidative phosphorylation

In the *second stage of catabolism*, the products obtained from the first stage—fatty acids, monosaccharides, and amino acids—are converted to compounds that can enter the citric acid cycle. In order to enter the citric acid cycle, a compound must be either one of the compounds in the cycle itself—these are called citric acid cycle intermediates—or it must be acetyl-CoA or pyruvate.

In the second stage of catabolism, the products obtained from the first stage are converted to compounds that can enter the citric acid cycle.

Compounds that can enter the citric acid cycle are citric acid cycle intermediates, acetyl-CoA, and pyruvate.

Acetyl CoA is the only non–citric acid cycle intermediate that can enter the cycle; it enters by being converted to citrate, a citric acid cycle intermediate (Section 18.7). Pyruvate enters the citric acid cycle by being converted to acetyl-CoA; this is the reaction catalyzed by the pyruvate dehydrogenase system discussed in Section 17.7.

The *third stage of catabolism* is the citric acid cycle. In the citric acid cycle, the acetyl group of each molecule of acetyl-CoA is converted to two molecules of CO_2.

$$CH_3C(=O)SCoA \longrightarrow 2\ CO_2 + CoASH$$

acetyl-CoA

The citric acid cycle is the third stage of catabolism.

Metabolic energy is measured in terms of adenosine 5′-triphosphate (ATP). Cells get the energy (ATP) they need by using nutrient molecules to make ATP. Only a small amount of ATP is formed in the first three stages of catabolism. Most ATP is formed in the fourth stage of catabolism.

Metabolic energy is measured in terms of adenosine triphosphate (ATP).

We will see that many catabolic reactions are oxidation reactions. In the *fourth stage of catabolism*, every molecule of NADH formed in one of the earlier stages of catabolism (from NAD^+ being used to carry out an oxidation reaction) is converted into three molecules of ATP in a process called *oxidative phosphorylation*. Oxidative phosphorylation also converts every molecule of $FADH_2$ formed earlier (as a result of FAD carrying out an oxidation reaction) into two molecules of ATP. Thus, most of the energy (ATP) provided by fats, carbohydrates, and proteins is obtained in the fourth stage of catabolism.

Oxidative phosphorylation is the fourth stage of catabolism.

18.2 ATP • PHOSPHORYL TRANSFER REACTIONS

Without ATP, many important biological reactions could not occur. For example, the reaction of glucose with hydrogen phosphate to form glucose-6-phosphate does not occur because the 6-OH group of glucose would have to displace a very basic HO^- group from hydrogen phosphate.

too basic to be displaced

glucose + hydrogen phosphate ✗→

If ATP is present, the 6-OH group of glucose attacks the terminal phosphate of ATP, breaking a **phosphoanhydride bond**. The phosphoanhydride bond is a weaker bond than the P=O π bond, so the reaction is essentially an S_N2 reaction (Section 9.2). Because the phosphoanhydride bond is weak, ADP is a very good leaving group.

a phosphoanhydride bond

glucose + ATP (a good leaving group)

→ glucose-6-phosphate + ADP

3-D Molecule: Adenosine triphosphate (ATP)

Tutorial: Phosphoryl transfer reactions

ATP provides a reaction pathway involving a good leaving group for a reaction that cannot occur because of a poor leaving group.

The transfer of a phosphoryl group from ATP to glucose is an example of a **phosphoryl transfer reaction**. There are many phosphoryl transfer reactions in biological systems. The above example of a phosphoryl transfer reaction demonstrates the actual chemical function of ATP: *it provides a reaction pathway involving a good leaving group for a reaction that cannot occur (or would occur very slowly) because of a poor leaving group.*

PROBLEM 1 *SOLVED*

Why is ADP a weak base and, therefore, a good leaving group?

Solution When the bond attached to the leaving group breaks, the electrons that were the bonding electrons are not localized on the departing oxygen atom. Instead, they are delocalized—they are shared by two oxygen atoms. Electron delocalization stabilizes the leaving group, and since stable bases are weak bases, they are good leaving groups.

$$ROH + {}^-O-\overset{O}{\overset{\|}{P}}(O^-)-O-\overset{O}{\overset{\|}{P}}(O^-)-O-\overset{O}{\overset{\|}{P}}(O^-)-O-\text{adenine} \longrightarrow RO-\overset{O}{\overset{\|}{P}}(O^-)-O^- + {}^-O-\overset{O}{\overset{\|}{P}}(O^-)-O-\overset{O}{\overset{\|}{P}}(O^-)-O-\text{adenine}$$

electrons are delocalized

THE NOBEL PRIZE

The Nobel Prize may well be the most coveted award a scientist can receive. The award was established by **Alfred Bernhard Nobel (1833–1896)**, and the first prizes were awarded in 1901. Nobel was born in Stockholm, Sweden. When he was 9, he moved with his parents to St. Petersburg, where his father manufactured the torpedoes and submarine mines he had invented for the Russian government. As a young man, Alfred did research on explosives in a factory his father owned near Stockholm. In 1864, an explosion in the factory killed his younger brother, causing Alfred to look for ways to make explosives safer to handle and transport. When the Swedish government would not allow the factory to be rebuilt because of the many accidents that had occurred there, Nobel established an explosives factory in Germany. There, in 1867, he discovered that if nitroglycerin is mixed with diatomaceous earth, the mixture can be molded into sticks that cannot be set off without a detonating cap. Thus, Nobel invented dynamite. He also invented blasting gelatin and smokeless powder. Although he was the inventor of explosives used by the military, he was a strong supporter of peace movements.

The 355 patents Nobel held made him a wealthy man. He never married, and when he died, his will stipulated that the bulk of his estate ($9.2 million) be used to establish prizes to be awarded to those who "have conferred the greatest benefit on mankind." He instructed that the money be invested and the interest earned each year be divided into five equal portions "to be awarded to the persons having made the most important contributions in the fields of chemistry, physics, physiology or medicine, literature, and to the one who had done the most toward fostering fraternity among nations, the abolition of standing armies, and the holding and promotion of peace congresses." Nobel also stipulated that in awarding the prizes, no consideration be given to the nationality of the candidate, that each prize be shared by no more than three persons, and that no prize be awarded posthumously.

Nobel gave instructions that the prizes for chemistry and physics were to be awarded by the Royal Swedish Academy of Sciences, for physiology or medicine by the Karolinska Institute in Stockholm, for literature by the Swedish Academy, and for peace by a five-person committee appointed by the Norwegian Parliament. The deliberations are secret, and the decisions cannot be appealed. In 1969, the Swedish Central Bank established a prize in economics in Nobel's honor. The recipient of this prize is selected by the Royal Swedish Academy of Sciences. On December 10—the anniversary of Nobel's death—the prizes are awarded in Stockholm, except for the peace prize, which is awarded in Oslo.

Alfred Bernhard Nobel

18.3 THE CATABOLISM OF FATS

In the first stage of fat catabolism, a fat's three ester groups are hydrolyzed by enzymes to glycerol and three fatty acid molecules.

$$\underset{\text{a fat}}{\begin{array}{l} CH_2O{-}\overset{O}{\overset{\|}{C}}{-}R^1 \\ CHO{-}\overset{O}{\overset{\|}{C}}{-}R^2 \\ CH_2O{-}\overset{O}{\overset{\|}{C}}{-}R^3 \end{array}} + 3\,H_2O \longrightarrow \underset{\text{glycerol}}{\begin{array}{l} CH_2OH \\ CHOH \\ CH_2OH \end{array}} + \underset{\text{fatty acids}}{\begin{array}{l} R^1{-}\overset{O}{\overset{\|}{C}}{-}OH \\ R^2{-}\overset{O}{\overset{\|}{C}}{-}OH \\ R^3{-}\overset{O}{\overset{\|}{C}}{-}OH \end{array}}$$

Glycerol reacts with ATP to form glycerol-3-phosphate in the same way that glucose reacts with ATP to from glucose-3-phosphate (Section 18.2). The enzyme that catalyzes this reaction is called glycerol kinase. A **kinase** is an enzyme that puts a phosphoryl group on its substrate; thus, glycerol kinase puts a phosphate group on glycerol.

$$\underset{\text{glycerol}}{\begin{array}{l} CH_2OH \\ CHOH \\ CH_2OH \end{array}} \xrightarrow[\text{glycerol kinase}]{\text{ATP} \quad \text{ADP}} \underset{\text{glycerol-3-phosphate}}{\begin{array}{l} CH_2OH \\ CHOH \\ CH_2OPO_3^{2-} \end{array}} \xrightarrow[\text{glycerol phosphate dehydrogenase}]{NAD^+ \quad \text{NADH, } H^+} \underset{\text{dihydroxyacetone phosphate}}{\begin{array}{l} CH_2OH \\ C{=}O \\ CH_2OPO_3^{2-} \end{array}}$$

a phosphoryl group

The secondary alcohol group of glycerol-3-phosphate is then oxidized to a ketone; NAD^+ is the oxidizing agent. The enzyme that catalyzes this reaction is called glycerol phosphate dehydrogenase. Recall that a **dehydrogenase** is an enzyme that oxidizes its substrate (Section 17.5). The product of the reaction, dihydroxyacetone phosphate, is one of the compounds in the glycolytic pathway, so it can enter that pathway and be broken down further (Section 18.4).

Notice that when biochemical reactions are written, the only structures shown are those of the primary reactant and the primary product. The names of other reactants and products are abbreviated and placed on a curved arrow that intersects the reaction arrow.

PROBLEM 2

Show the mechanism for the reaction of glycerol with ATP to form glycerol-3-phosphate.

PROBLEM 3

The asymmetric center of glycerol-3-phosphate has the *R* configuration. Draw the structure of (*R*)-glycerol-3-phosphate.

Before a fatty acid can be metabolized, it must be activated by being converted into a thioester (Section 17.7).

$$\underset{\text{a fatty acid}}{RCH_2CH_2CH_2CH_2{-}\overset{O}{\overset{\|}{C}}{-}O^-} \xrightarrow[\text{acyl-CoA synthetase}]{\text{CoASH, ATP} \quad \text{ADP, } HPO_4^{2-}} \underset{\text{a thioester, a fatty acyl-CoA}}{RCH_2CH_2CH_2CH_2{-}\overset{O}{\overset{\|}{C}}{-}SCoA}$$

The mechanism for activation of a fatty acid is shown on page 501.

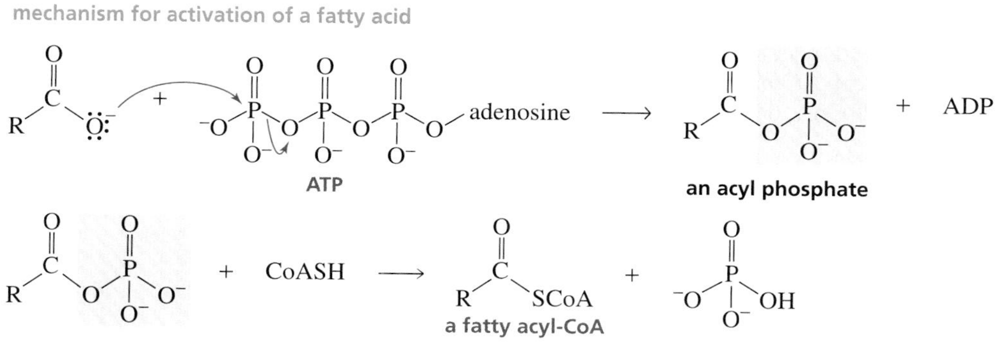

mechanism for activation of a fatty acid

ATP

an acyl phosphate

a fatty acyl-CoA

- The fatty acid reacts with ATP, in a phosphoryl transfer reaction (Section 18.2), in order to put a good leaving group on the fatty acid; the product is an acyl phosphate.
- The acyl phosphate reacts with CoASH in a nucleophilic acyl substitution reaction.

The fatty acyl-CoA is converted to acetyl-CoA in a repeating pathway called ***β*-oxidation**; *β*-oxidation is a series of four reactions. Each passage through the four reactions removes two carbons from the fatty acyl-CoA by converting them into acetyl-CoA (Figure 18.2). Each of the four reactions is catalyzed by a different enzyme.

Fatty acids are converted to molecules of acetyl-CoA.

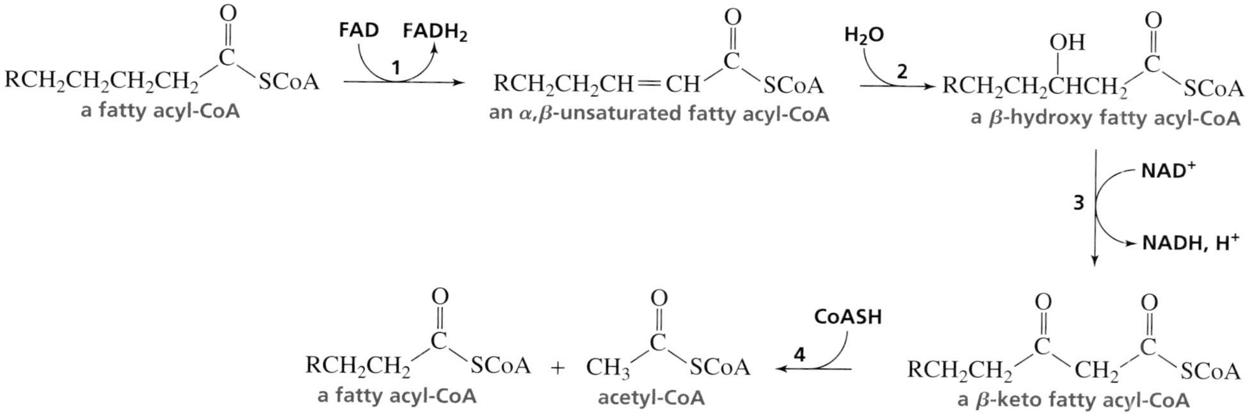

a fatty acyl-CoA

an *α*,*β*-unsaturated fatty acyl-CoA

a *β*-hydroxy fatty acyl-CoA

a fatty acyl-CoA

acetyl-CoA

a *β*-keto fatty acyl-CoA

▲ **Figure 18.2**
In ***β*-oxidation** a series of four enzyme-catalyzed reactions is repeated until the entire fatty acyl-CoA molecule has been converted to acetyl-CoA molecules. The enzymes that catalyze the reactions are:
1. acyl-CoA dehydrogenase
2. enoyl-CoA hydratase
3. 3-L-hydroxyacyl-CoA dehydrogenase
4. *β*-ketoacyl-CoA thiolase

1. The first reaction is an oxidation reaction that removes hydrogen from the *α*- and *β*-carbons, forming an *α*,*β*-unsaturated fatty acyl-CoA. The oxidizing agent is FAD, the oxidizing agent used primarily for non-carbonyl oxidations (Section 17.6.). The enzyme that catalyzes this reaction has been found to be deficient in 10% of babies that experience sudden infant death syndrome (SIDS).
2. The second reaction is the conjugate addition of water to the *α*,*β*-unsaturated fatty acyl-CoA (Section 12.10).
3. The third reaction is another oxidation reaction: NAD^+ oxidizes the secondary alcohol to a ketone.

4. The fourth reaction is the reverse of a Claisen condensation (Section 13.7), followed by conversion of the enol tautomer to the keto tautomer. The final product is acetyl-CoA and a fatty acyl-CoA with *two fewer* carbons than the starting fatty acyl-CoA. The mechanism for this reaction is shown below.

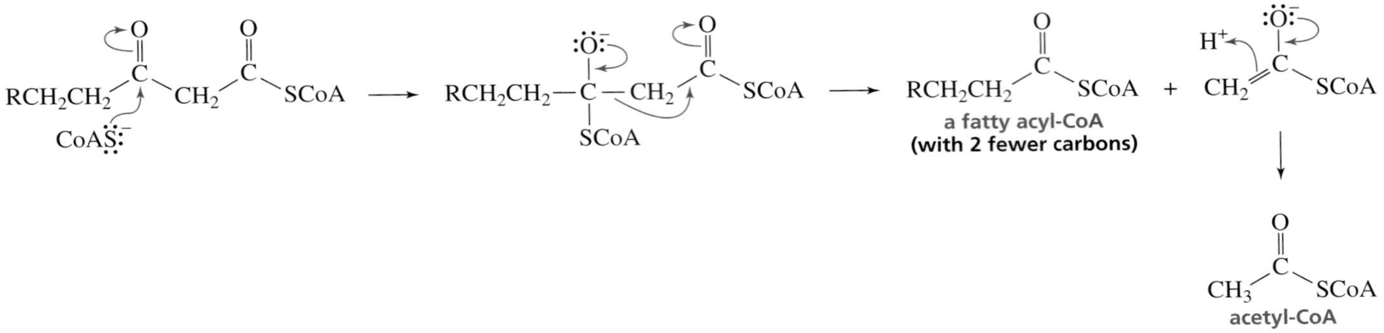

$RCH_2CH_2C(=O)CH_2C(=O)SCoA$ + $CoAS^-$ ⟶ $RCH_2CH_2-C(O^-)(SCoA)-CH_2C(=O)SCoA$ ⟶ $RCH_2CH_2C(=O)SCoA$ + $CH_2{=}C(O^-)SCoA$ (H^+) ⟶ $CH_3C(=O)SCoA$

a fatty acyl-CoA (with 2 fewer carbons)

acetyl-CoA

The four reactions are repeated, forming another molecule of acetyl-CoA and a fatty acyl-CoA that is now four carbons shorter than it was originally. Each time the series of four reactions is repeated, two more carbons are removed (as acetyl-CoA) from the fatty acyl-CoA. The series of reactions is repeated until the entire fatty acid has been converted into acetyl-CoA molecules.

We will see that acetyl-CoA enters the citric acid cycle by reacting with oxaloacetate (a citric acid cycle intermediate) to form citrate, another citric acid cycle intermediate (Section 18.7).

PROBLEM 4 ◆

Why does the OH group add to the β-carbon rather than to the α-carbon in the second reaction in the catabolism of fats? (*Hint*: See Section 12.10.)

PROBLEM 5 ◆

Palmitic acid is a 16-carbon saturated fatty acid. How many moles of acetyl-CoA are formed from the catabolism of 1 mole of palmitic acid?

PROBLEM 6 ◆

How many moles of NADH are formed from the β-oxidation of 1 mole of palmitic acid?

18.4 THE CATABOLISM OF CARBOHYDRATES

In the first stage of carbohydrate catabolism, the acetal groups that hold glucose subunits together are hydrolyzed, forming individual glucose molecules (Section 15.14).

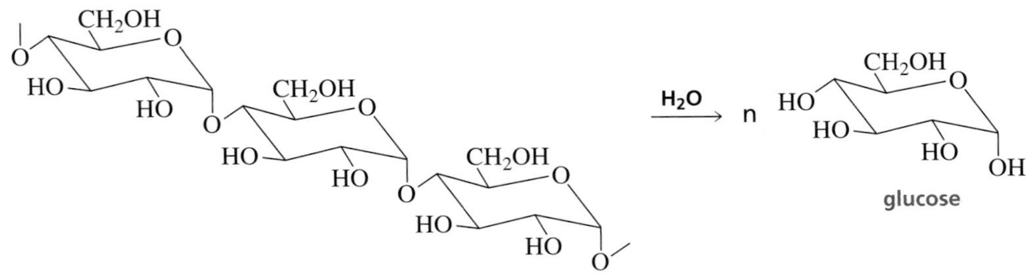

[polysaccharide chain of glucose units] $\xrightarrow{H_2O}$ n glucose

glucose

Each glucose molecule is converted to two molecules of pyruvate in a series of 10 reactions known as **glycolysis** or the **glycolytic pathway** (Figure 18.3).

Glucose is converted to two molecules of pyruvate.

glucose

ATP ADP 1

glucose-6-P

2

fructose-6-P

ATP 3 ADP

fructose-1,6-diP

4

dihydroxyacetone phosphate

5

glyceraldehyde-3-P

NADH, H^+ NAD^+, HPO_4^{2-} 6

1,3-diphosphoglycerate

ADP 7 ATP

3-phosphoglycerate

8

2-phosphoglycerate

9

phosphoenolpyruvate

ADP ATP 10

pyruvate

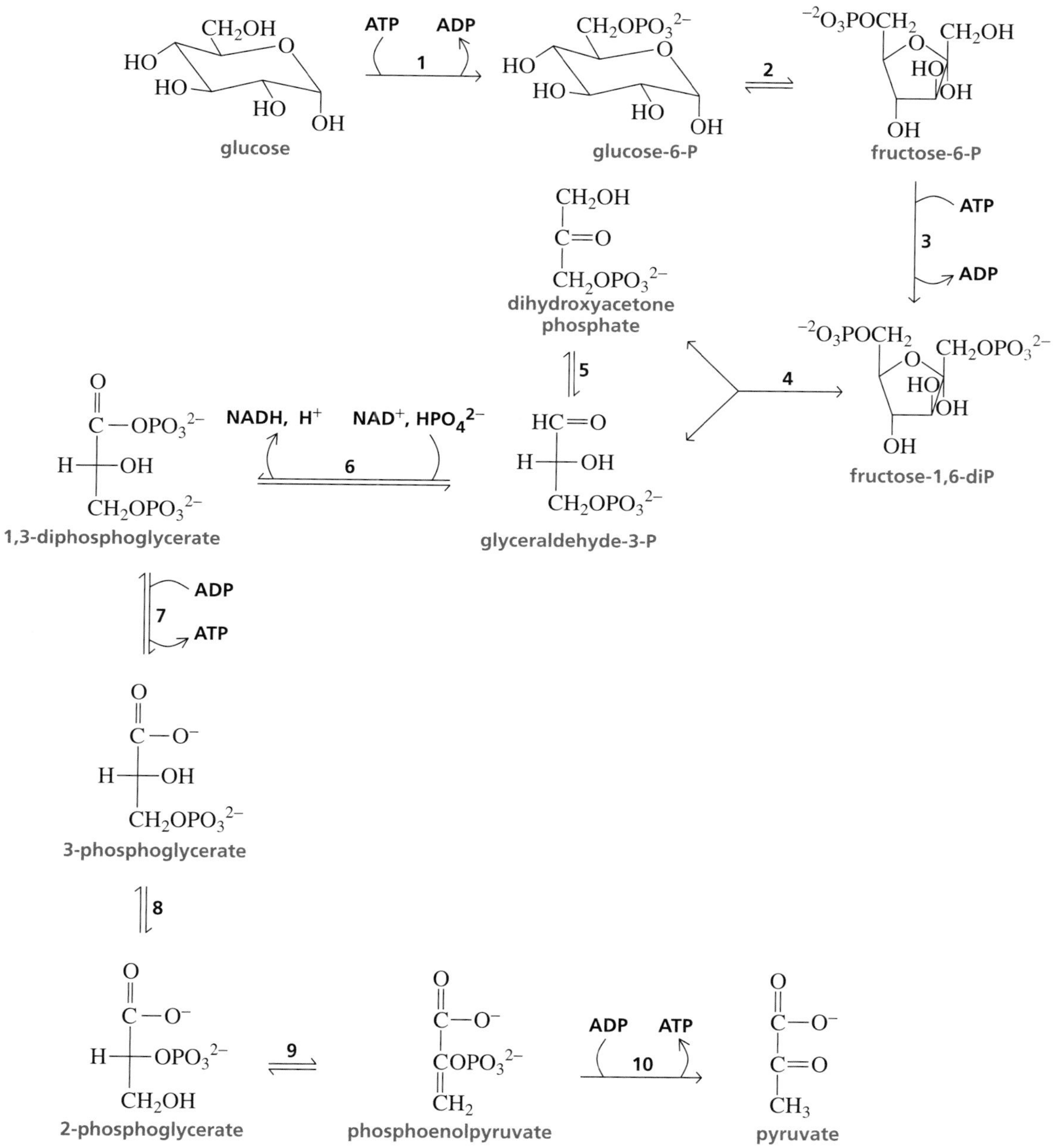

▲ **Figure 18.3**
Glycolysis, the series of enzyme-catalyzed reactions responsible for the conversion of 1 mole of glucose to 2 moles of pyruvate. The enzymes that catalyze the reactions are:

1. hexokinase
2. phosphoglucose isomerase
3. phosphofructokinase
4. aldolase
5. triosephosphate isomerase
6. glyceraldehyde-3-phosphate dehydrogenase
7. phosphoglycerate kinase
8. phosphoglycerate mutase
9. enolase
10. pyruvate kinase

1. In the first reaction, glucose is converted to glucose-6-phosphate, a reaction we just looked at in Section 18.2.
2. Glucose-6-phosphate then isomerizes to fructose-6-phosphate, a reaction whose mechanism we examined in Section 17.2.
3. In the third reaction, ATP puts a second phosphoryl group on fructose-6-phosphate. The product of the reaction is fructose-1,6-diphosphate.
4. The fourth reaction is the reverse of an aldol addition reaction. We looked at the mechanism of this reaction in Section 17.3.
5. Dihydroxyacetone phosphate, produced in the fourth reaction, is converted into glyceraldehyde-3-phosphate by forming an enol that then forms glyceraldehyde-3-phosphate (if the OH group at C-1 tautomerizes), or reforms dihydroxyacetone phosphate (if the OH group at C-2 tautomerizes). Compare this reaction with the base-catalyzed rearrangement shown in Section 15.5.

CH_2OH, $C{=}O$, $CH_2OPO_3^{2-}$ $\rightleftharpoons$ $HC{-}OH$, $C{-}OH$, $CH_2OPO_3^{2-}$ $\rightleftharpoons$ $HC{=}O$, $H{-}C{-}OH$, $CH_2OPO_3^{2-}$

dihydroxyacetone phosphate | an enol | glyceraldehyde-3-phosphate

Thus, overall, each molecule of D-glucose is converted to two molecules of glyceraldehyde-3-phosphate.

6. The aldehyde group of glyceraldehyde-3-phosphate is oxidized by NAD^+, forming 1,3-diphosphoglycerate. In this reaction, the aldehyde is oxidized to a carboxylic acid, which then forms an ester with phosphoric acid.

$O{=}C{-}H$, $H{-}C{-}OH$, $CH_2OPO_3^{2-}$ $\xrightleftharpoons[]{NAD^+,\ HPO_4^{2-} \quad NADH,\ H^+}$ $O{=}C{-}OPO_3^{2-}$, $H{-}C{-}OH$, $CH_2OPO_3^{2-}$

D-glyceraldehyde-3-phosphate | 1,3-diphosphoglycerate

7. In the seventh reaction, 1,3-diphosphoglycerate transfers a phosphoryl group to ADP, thereby forming ATP and 3-phosphoglycerate.

$O{=}C{-}O{-}PO_3^{2-}$, $H{-}C{-}OH$, $CH_2OPO_3^{2-}$ + $^-O{-}P(O)(O^-){-}O{-}P(O)(O^-){-}O{-}$adenine $\longrightarrow$ $O{=}C{-}O^-$, $H{-}C{-}OH$, $CH_2OPO_3^{2-}$ + $^-O{-}P(O)(O^-){-}O{-}P(O)(O^-){-}O{-}P(O)(O^-){-}O{-}$adenine

1,3-diphosphoglycerate | ADP | 3-phosphoglycerate | ATP

8. The eighth reaction is an isomerization: 3-phosphoglycerate is converted to 2-phosphoglycerate. The enzyme that catalyzes this reaction has a phosphoryl group on one of its amino acid side chains that it transfers to the 2-position of 3-phosphoglycerate to form an intermediate with two phosphoryl groups. The intermediate transfers the phosphoryl group on its 3-position back to the enzyme.

$O{=}C{-}O^-$, $H{-}C{-}OH$, $CH_2OPO_3^{2-}$ $\rightleftharpoons$ $O{=}C{-}O^-$, $H{-}C{-}OPO_3^{2-}$, $CH_2OPO_3^{2-}$ $\rightleftharpoons$ $O{=}C{-}O^-$, $H{-}C{-}OPO_3^{2-}$, CH_2OH

3-phosphoglycerate | an intermediate | 2-phosphoglycerate

9. The ninth reaction is a dehydration reaction that forms phosphoenolpyruvate. The HO^- group is protonated by an acid at the active site of the enzyme, which makes it a better leaving group (Section 10.2).

:B H—C(CO_2^-)(OPO_3^{2-})—CH_2—OH H—B^+ ⇌ HB^+ CH_2=C(CO_2^-)—OPO_3^{2-} + H_2O :B

2-phosphoglycerate **2-phosphoenolpyruvate**

10. In the last reaction of the glycolytic pathway, phosphoenolpyruvate transfers its phosphoryl group to ADP, forming ATP and pyruvate.

2-phosphoenolpyruvate + ADP (H^+) ⟶ pyruvate (CH_3—CO—CO_2^-) + ATP

2-phosphoenolpyruvate **ADP** **pyruvate** **ATP**

PROBLEM 7

Propose a mechanism for the third reaction in glycolysis: the formation of fructose-1,6-diphosphate from the reaction of fructose-6-diphosphate with ATP.

PROBLEM 8

a. Which steps in glycolysis require ATP?
b. Which steps in glycolysis produce ATP?

PROBLEM-SOLVING STRATEGY

How many molecules of ATP are obtained from each molecule of glucose that is metabolized to pyruvate?

We first need to count the number of ATPs that are *used* to convert glucose to pyruvate. We see that two are used: one to form glucose-1-phosphate and the other to form fructose-1,6-diphosphate. Next, we need to know how many ATPs are *formed*. Each glyceraldehyde-3-phosphate that is metabolized to pyruvate forms two ATPs. Because each glucose molecule forms two molecules of glyceraldehyde-3-phosphate, 4 ATPs are formed from each molecule of glucose. Subtracting the molecules used, we find that each molecule of glucose metabolized to pyruvate forms two molecules of ATP.

Now continue on to Problem 9.

PROBLEM 9 ♦

How many moles of NAD^+ are required to convert 1 mole of glucose to pyruvate?

18.5 THE FATES OF PYRUVATE

We have just seen that NAD^+ is used as an oxidizing agent in glycolysis. If glycolysis is to continue, the NADH produced as a result of the oxidation reaction has to be oxidized back to NAD^+ in order for NAD^+ to continue to be available as an oxidizing agent.

If oxygen is present, it is the oxidizing agent used to oxidize NADH back to NAD^+; this happens *in the fourth stage of catabolism.* If oxygen is not present—as occurs, for example in muscle cells when intense muscle activity causes all the oxygen to be depleted—pyruvate (the product of glycolysis) is used to oxidize NADH back to NAD^+. In the process, pyruvate is reduced to lactate (lactic acid). The acidic conditions caused by a buildup of lactic acid in muscles are responsible for the burning sensation you may have experienced when exercising.

$$\underset{\text{pyruvate}}{CH_3-\overset{O}{\overset{\|}{C}}-\overset{O}{\overset{\|}{C}}-O^-} \xrightarrow{\text{NADH, H}^+ \quad \text{NAD}^+} \underset{\text{lactate}}{CH_3-\overset{OH}{\overset{|}{C}H}-\overset{O}{\overset{\|}{C}}-O^-}$$

Under normal (aerobic) conditions, when oxygen rather than pyruvate is used to oxidize NADH to NAD^+, pyruvate is converted to acetyl-CoA, which then enters the citric acid cycle. This reaction is catalyzed by the pyruvate dehydrogenase system, a series of reactions that requires three enzymes and five coenzymes. We have seen that one of the coenzymes is thiamine pyrophosphate, the coenzyme required by enzymes that catalyze the transfer of a two-carbon fragment from one species to another (Section 17.7); in this reaction, the two carbons of the acetyl group of pyruvate are transferred to coenzyme A.

$$\underset{\text{pyruvate}}{CH_3-\overset{O}{\overset{\|}{C}}-\overset{O}{\overset{\|}{C}}-O^-} + \text{CoASH} \xrightarrow{\text{pyruvate dehydrogenase system}} \underset{\text{acetyl-CoA}}{CH_3-\overset{O}{\overset{\|}{C}}-\text{SCoA}} + CO_2$$

We have just seen that under anaerobic (no oxygen) conditions, pyruvate is reduced to lactate. Under these same conditions in yeast, however, pyruvate has a different fate: it is decarboxylated to acetaldehyde by pyruvate decarboxylase, an enzyme that requires thiamine pyrophosphate whose mechanism we looked at in Section 17.7. In this case, acetaldehyde is the compound that oxidizes NADH back to NAD^+ and, in the process, is reduced to ethanol—a reaction that has been used by humankind for thousands of years to produce wine, beer, and other fermented drinks.

$$\underset{\text{pyruvate}}{CH_3-\overset{O}{\overset{\|}{C}}-\overset{O}{\overset{\|}{C}}-O^-} \xrightarrow{H^+ \quad CO_2} \underset{\text{acetaldehyde}}{CH_3-\overset{O}{\overset{\|}{C}}-H} \xrightarrow{\text{NADH, H}^+ \quad \text{NAD}^+} \underset{\text{ethanol}}{CH_3CH_2OH}$$

PROBLEM 10 ♦

What functional group of pyruvate is reduced when pyruvate is converted to lactate?

PROBLEM 11 ♦

What coenzyme is required to convert pyruvate to acetaldehyde?

PROBLEM 12 ♦

What functional group of acetaldehyde is reduced when acetaldehyde is converted to ethanol?

PROBLEM 13

Propose a mechanism for the reduction of acetaldehyde by NADH to ethanol. (*Hint:* See Section 17.5.)

18.6 THE CATABOLISM OF PROTEINS

In the first stage of protein catabolism, proteins are hydrolyzed to amino acids.

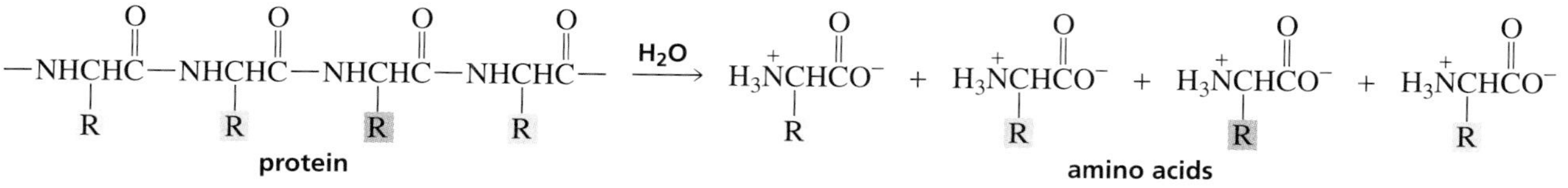

$$-NHCHC(=O)-NHCHC(=O)-NHCHC(=O)-NHCHC(=O)- \xrightarrow{H_2O} H_3\overset{+}{N}CHCO^- + H_3\overset{+}{N}CHCO^- + H_3\overset{+}{N}CHCO^- + H_3\overset{+}{N}CHCO^-$$

protein (each residue with side chain R) → amino acids (each with side chain R)

In the second stage of catabolism, the amino acids are converted to acetyl-CoA, pyruvate, or citric acid cycle intermediates, depending on the amino acid. These products of the second stage of catabolism then enter the citric acid cycle—the third stage of catabolism— and are further metabolized.

Amino acids are converted to acetyl-CoA, pyruvate, or citric acid cycle intermediates.

We will look at the catabolism of phenylalanine as an example of how an amino acid is metabolized (Figure 18.4). Phenylalanine is one of the essential amino acids and thus must be included in our diet (Section 16.1). The enzyme phenylalanine hydroxylase converts phenylalanine into tyrosine. Thus, tyrosine is not an essential amino acid unless our diet lacks phenylalanine.

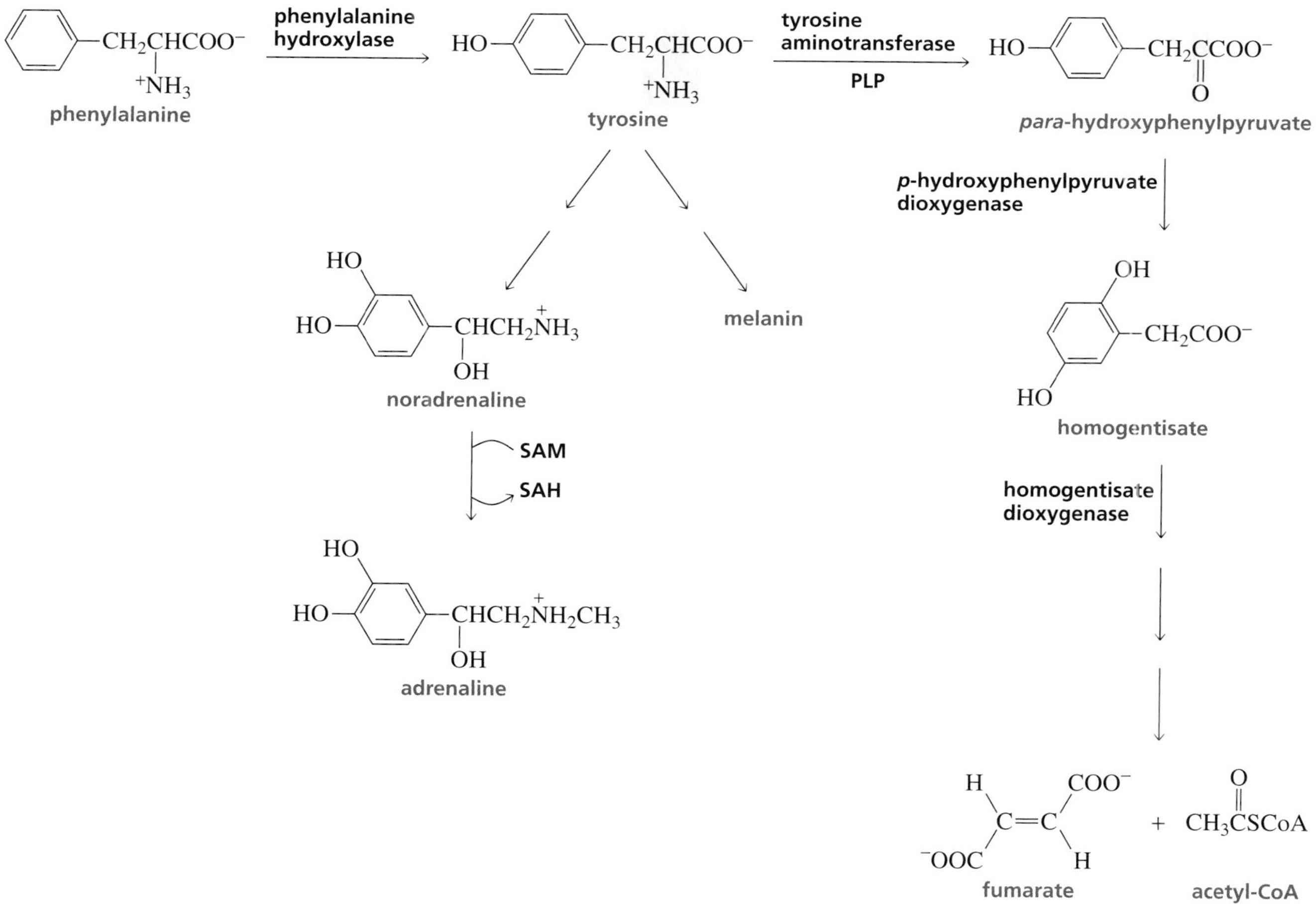

phenylalanine ($C_6H_5-CH_2CH(\overset{+}{N}H_3)COO^-$) —phenylalanine hydroxylase→ tyrosine ($HO-C_6H_4-CH_2CH(\overset{+}{N}H_3)COO^-$) —tyrosine aminotransferase, PLP→ *para*-hydroxyphenylpyruvate ($HO-C_6H_4-CH_2C(=O)COO^-$)

tyrosine → noradrenaline ($(HO)_2C_6H_3-CH(OH)CH_2\overset{+}{N}H_3$) —SAM → SAH→ adrenaline ($(HO)_2C_6H_3-CH(OH)CH_2\overset{+}{N}H_2CH_3$)

tyrosine → melanin

para-hydroxyphenylpyruvate —*p*-hydroxyphenylpyruvate dioxygenase→ homogentisate ($(HO)_2C_6H_3-CH_2COO^-$) —homogentisate dioxygenase→ → → fumarate ($^-OOC-CH=CH-COO^-$) + acetyl-CoA (CH_3CSCoA, with C=O)

▲ **Figure 18.4**
The catabolism of phenylalanine.

The first reaction in the catabolism of most amino acids is transamination, a reaction that replaces the amino group of the amino acid with a ketone group (Section 17.9). *para*-Hydroxyphenylpyruvate, the product of the transamination of tyrosine, is converted by a series of reactions to fumarate and acetyl-CoA. Fumarate is a citric acid cycle intermediate, so it can enter the citric acid cycle directly, and acetyl-CoA gets into the cycle by reacting with oxaloacetate to form citrate (Section 18.7). Each of the reactions in this catabolic pathway is catalyzed by a different enzyme.

In addition to being used for energy, the amino acids we ingest are also used for the synthesis of proteins (Section 20.8) and other compounds that the body needs. For example, tyrosine is used to synthesize neurotransmitters (dopamine and adrenaline) and melanin, the compound responsible for skin pigmentation. Recall that SAM (*S*-adenosylmethionine) is the reagent used to methylate noradrenaline in order to form adrenaline (Section 9.14).

PHENYLKETONURIA (PKU): AN INBORN ERROR OF METABOLISM

About one in every 20,000 babies is born without phenylalanine hydroxylase, the enzyme that converts phenylalanine into tyrosine. This genetic disease is called phenylketonuria (PKU). Without phenylalanine hydroxylase, the level of phenylalanine builds up; when it reaches a high concentration, it is transaminated to phenylpyruvate. The high level of phenylpyruvate found in urine gives the disease its name.

$$C_6H_5-CH_2CH(^{+}NH_3)COO^- \xrightarrow{\text{transamination}} C_6H_5-CH_2C(=O)COO^-$$

phenylalanine → phenylpyruvate

Within 24 hours after birth, all babies born in the United States are tested for high serum phenylalanine levels, which indicate a buildup of phenylalanine caused by an absence of phenylalanine hydroxylase. Babies with high levels are immediately put on a diet low in phenylalanine and high in tyrosine. As long as the phenylalanine level is kept under careful control for the first 5 to 10 years of life, the child will experience no adverse effects. You may have noticed the warning on containers of food that contain NutraSweet, announcing that it contains phenylalanine. (Recall that this sweetener is a methyl ester of a dipeptide of L-aspartate and L-phenylalanine; see Section 15.17.)

If the diet is not controlled, however, the baby will be severely mentally retarded by the time he or she is a few months old. Untreated children have paler skin and fairer hair than other members of their family because, without tyrosine, they cannot synthesize melanin, a black skin pigment. Half of untreated phenylketonurics are dead by age 20. When a woman with PKU becomes pregnant, she must return to the low phenylalanine diet she had as a child, because a high level of phenylalanine can cause abnormal development of the fetus.

ALCAPTONURIA

Another genetic disease that results from a deficiency of an enzyme in the pathway for phenylalanine degradation is alcaptonuria, which is caused by lack of homogentisate dioxygenase. The only ill effect of this enzyme deficiency is black urine. The urine of those afflicted with alcaptonuria turns black because the homogentisate they excrete immediately oxidizes in the air, forming a black compound.

PROBLEM 14 ◆

What coenzyme is required for transamination? (*Hint:* See Section 17.9.)

PROBLEM 15 ◆

When the amino acid known as alanine undergoes transamination, what compound is formed?

18.7 THE CITRIC ACID CYCLE

The **citric acid cycle**, as already noted, is the third stage of catabolism. In this series of eight reactions, the acetyl group of each molecule of acetyl-CoA—formed by the catabolism of fats, carbohydrates, and amino acids—is converted to two molecules of CO_2.

The acetyl group of each molecule of acetyl-CoA that enters the citric acid cycle is converted to two molecules of CO_2.

$$CH_3C(=O)SCoA \longrightarrow 2\,CO_2 + CoASH$$

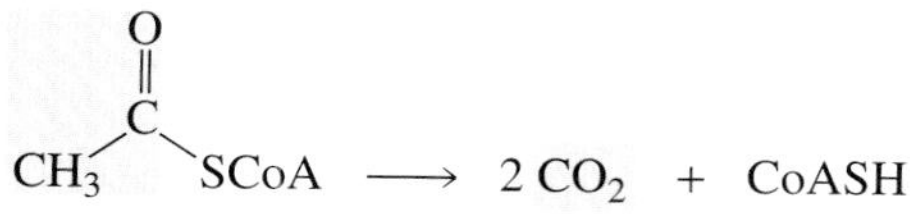

The series of eight reactions is called a *cycle* because the reactions comprise a closed loop in which the product of the eighth reaction (oxaloacetate) is a reactant for the first reaction (Figure 18.5).

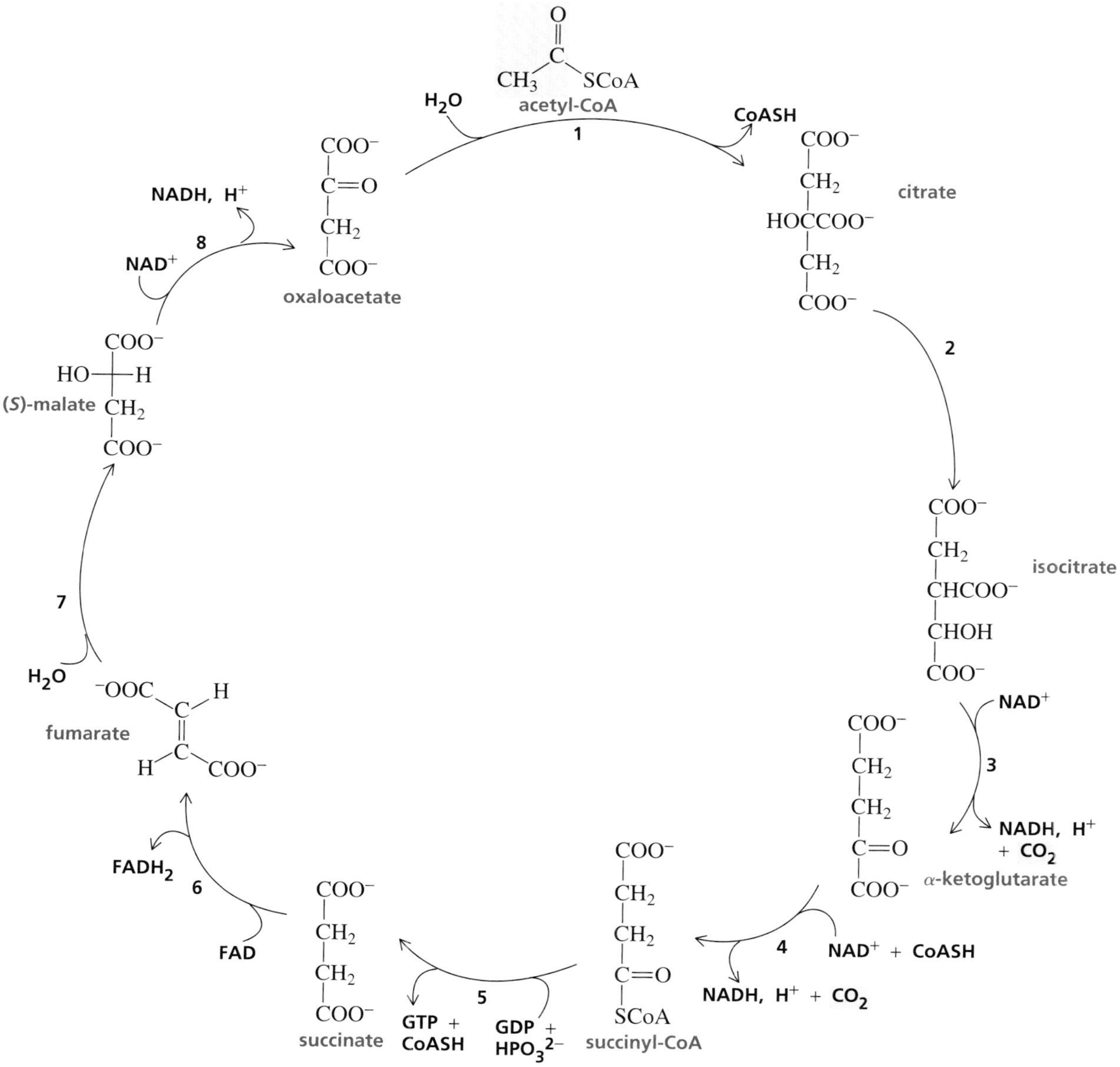

▲ **Figure 18.5 The citric acid cycle** is the series of enzyme-catalyzed reactions responsible for the oxidation of the acetyl group of acetyl-CoA to 2 molecules of CO_2. The enzymes that catalyze the reactions are:

1. citrate synthase
2. aconitase
3. isocitrate dehydrogenase
4. α-ketoglutarate dehydrogenase
5. succinyl-CoA synthetase
6. succinate dehydrogenase
7. fumarase
8. malate dehydrogenase

1. In the first reaction, acetyl-CoA reacts with oxaloacetate to form citrate. This rection is similar to an aldol addition (Section 13.5). The mechanism of the reaction shows that an aspartate side chain of the enzyme removes a proton from the α-carbon of acetyl-CoA, creating a nucleophile that attacks the carbonyl carbon of oxaloacetate. The carbonyl oxygen picks up a proton from a histidine side chain. The intermediate that results is hydrolyzed to citrate.

Asp—CO⁻ | H—His⁺ | $H—CH_2CSCoA$ acetyl-CoA | $O{=}C—COO^-$, CH_2, COO^- oxaloacetate → $HO—C(CH_2CSCoA)—COO^-$, CH_2, COO^- an intermediate —(H_2O, CoASH)→ $HO—C(CH_2COO^-)—COO^-$, CH_2, COO^- citrate

2. In the second reaction, citrate is converted to isocitrate, its isomer. The reaction takes place in two steps. The first step is a dehydration reaction (Section 10.3); a serine side chain of the enzyme removes a proton, and the OH leaving group is protonated by a histidine side chain to make it a weaker base (H_2O) and, therefore, a better leaving group. In the second step, conjugate addition of water to the intermediate forms isocitrate (Section 12.10).

H—His⁺ | Ser—O⁻ | citrate (COO^-, CH_2, $HO—C—COO^-$, $H—CH$, COO^-) → an intermediate (COO^-, CH_2, $C—COO^-$, ‖ CH, COO^-) —H_2O→ isocitrate (COO^-, CH_2, $H—C—COO^-$, $HO—CH$, COO^-)

3. The third reaction also has two steps. In the first step, the secondary alcohol group of isocitrate is oxidized to a ketone by NAD^+. In the second step, the ketone loses CO_2. We have seen that a COO^- group bonded to a carbon adjacent to a carbonyl carbon can be removed because the electrons left behind can be delocalized onto the carbonyl oxygen (Section 13.8). The resulting enolate ion tautomerizes to a ketone (Section 13.3).

isocitrate (COO^-, CH_2, $H—C—COO^-$, $HO—CH$, COO^-) —(NAD^+, NADH, H^+)→ (COO^-, CH_2, $H—C—C(=O)—O^-$, $O{=}C$, COO^-) —CO_2→ (COO^-, CH_2, $H—C$, ‖ $^-O—C$, COO^-) —H^+→ α-ketoglutarate (COO^-, CH_2, CH_2, $O{=}C$, COO^-)

4. In the fourth reaction, NAD^+ is again the oxidizing agent. This is the reaction that releases the second molecule of CO_2. It requires a group of enzymes and the same five coenzymes required by the pyruvate dehydrogenase system that forms acetyl-CoA (Section 18.5). The product of the reaction is succinyl-CoA.

$$\underset{\alpha\text{-ketoglutarate}}{{}^{-}OOC-C(=O)-CH_2-CH_2-COO^-} \xrightarrow[\text{NAD}^+ \;\to\; \text{NADH, H}^+]{\text{CoASH} \;\;\; CO_2} \underset{\text{succinyl-CoA}}{CoAS-C(=O)-CH_2-CH_2-COO^-}$$

5. The fifth reaction takes place in two steps. Hydrogen phosphate reacts with succinyl-CoA in a nucleophilic acyl substitution reaction (Section 11.5) to form succinyl phosphate. Succinyl phosphate transfers its phosphoryl group to the enzyme, which then transfers it to guanosine 5′-diphosphate (GDP) to form guanosine 5′-triphosphate (GTP).

$$\underset{\text{succinyl-CoA}}{{}^{-}OOC-CH_2-CH_2-C(=O)-SCoA} + {}^{-2}OPO_3H \xrightarrow{\text{CoASH}} {}^{-}OOC-CH_2-CH_2-C(=O)-OPO_3^{2-} \xrightarrow{\text{GDP} \;\to\; \text{GTP}} \underset{\text{succinate}}{{}^{-}OOC-CH_2-CH_2-C(=O)-O^-}$$

GTP transfers a phosphoryl group to ADP to form ATP.

$$\text{GTP} + \text{ADP} \rightleftharpoons \text{GDP} + \text{ATP}$$

6. In the sixth reaction, FAD oxidizes succinate to fumarate, a reaction we looked at in Section 17.6.
7. Conjugate addition of water to the double bond of fumarate forms (*S*)-malate (Section 12.10). We saw why only one enantiomer is formed in Section 6.14.
8. Oxidation of the secondary alcohol group of (*S*)-malate by NAD^+ forms oxaloacetate, returning the cycle to its starting point. Oxaloacetate now begins the cycle again, reacting with another molecule of acetyl-CoA to initiate the conversion of acetyl-CoA's acetyl group to another two molecules of CO_2.

Notice that reactions 6, 7, and 8 in the citric acid cycle are similar to reactions 1, 2, and 3 in the β-oxidation of fatty acids (Section 18.3).

PROBLEM 16 ◆

What functional group of isocitrate is oxidized in the third reaction of the citric acid cycle?

PROBLEM 17 ◆

How many molecules of CO_2 are formed from each molecule of acetyl-CoA that enters the citric acid cycle?

PROBLEM 18 ◆

The citric acid cycle is also called the tricarboxylic acid cycle. Which of the citric acid cycle intermediates are tricarboxylic acids?

18.8 OXIDATIVE PHOSPHORYLATION

In oxidative phosphorylation, each molecule of NADH is converted to three molecules of ATP and each molecule of $FADH_2$ is converted to two molecules of ATP.

Each round of the citric acid cycle forms 3 molecules of NADH, 1 molecule of $FADH_2$ and 1 molecule of ATP. The NADH and $FADH_2$ molecules undergo **oxidative phosphorylation**—the fourth stage of catabolism—which oxidizes them back to NAD^+ and FAD. For each NADH that undergoes oxidative phosphorylation, 3 molecules of ATP are formed, and for each $FADH_2$ that undergoes oxidative phosphorylation, 2 molecules of ATP are formed.

$$\text{NADH} \longrightarrow \text{NAD}^+ + 3\text{ ATP}$$

$$\text{FADH}_2 \longrightarrow \text{FAD} + 2\text{ ATP}$$

Thus, for every acetyl-CoA molecule that enters the citric acid cycle, 11 molecules of ATP are formed from NADH and $FADH_2$ and 1 molecule of ATP is formed in the cycle.

$$3\text{ NADH} + \text{FADH}_2 + \text{ATP} \longrightarrow 3\text{ NAD}^+ + \text{FAD} + 12\text{ ATP}$$

PROBLEM 19 ♦

How many molecules of ATP are obtained when the NADH and $FADH_2$ formed during the metabolism of one molecule of acetyl-CoA to CO_2 undergo oxidative phosphorylation?

18.9 ANABOLISM

Anabolism can be thought of as the reverse of catabolism. In anabolism, acetyl-CoA, pyruvate, and citric acid cycle intermediates are the starting materials for the synthesis of fatty acids, monosaccharides, and amino acids. These compounds are then used to form fats, carbohydrates, and proteins. The mechanisms utilized by biological systems to synthesize fats and proteins are discussed in Sections 13.11 and 20.8.

BASAL METABOLIC RATE

Your basal metabolic rate (BMR) is the number of calories you would burn if you stayed in bed all day. A person's BMR is affected by gender, age, and genetics; it is greater for men than for women, it is greater for young people than for old people, and some people are born with a faster metabolic rate than others. The BMR is also affected by the percentage of body fat: the higher the percentage, the lower the BMR. For humans, the average BMR is about 1600 kcal/day.

In addition to consuming sufficient calories to sustain your basal metabolism, you must also obtain energy for physical activities. The more active you are, the more calories you must consume in order to maintain your current weight. People who consume more calories than required by their BMR plus their level of physical activity, will gain weight. If they consume fewer calories, they will lose weight.

SUMMARY

Metabolism is the set of reactions that living organisms carry out to obtain energy and to synthesize the compounds they require. Metabolism can be divided into **catabolism** and **anabolism**. **Catabolic reactions** break down complex molecules to provide energy and simple molecules. **Anabolic reactions** require energy and lead to the synthesis of complex biomolecules from simple molecules.

ATP is a cell's most important source of chemical energy; ATP provides a reaction pathway involving a good leaving group for a reaction that would not otherwise occur because of a poor leaving group. This occurs by way of a **phosphoryl transfer reaction** in which a phosphoryl group of ATP is transferred to a nucleophile as a result of breaking a **phosphoanhydride bond**.

Catabolism can be divided into four stages. In the *first stage*, fats, carbohydrates, and proteins are hydrolyzed to fatty acids, monosaccharides, and amino acids, respectively. In the *second stage*, the products obtained from the first stage are converted to compounds that can enter the citric acid cycle (the *third stage* of catabolism). In order to enter the citric acid cycle, a compound must be either a citric acid cycle intermediate, acetyl-CoA, or pyruvate.

In the second stage of catabolism, a fatty acyl-CoA is converted to acetyl-CoA in a pathway called **β-oxidation**. The series of four reactions is repeated until the entire fatty acid has been converted to acetyl-CoA molecules. Glucose, in the second stage of catabolism, is converted to two molecules of pyruvate in a series of 10 reactions known as **glycolysis**.

Under normal (aerobic) conditions, pyruvate is converted to acetyl-CoA, which then enters the citric acid cycle.

Amino acids are metabolized to acetyl-CoA, pyruvate, or citric acid cycle intermediates, in the second stage of catabolism, depending on the amino acid. The amino acids we ingest are used for energy and for the synthesis of proteins and other compounds needed by the body.

The **citric acid cycle** is the third stage of catabolism. It is a series of eight reactions that converts the acetyl group of each molecule of acetyl-CoA that enters the cycle to two molecules of CO_2.

Metabolic energy is measured in terms of ATP. In **oxidative phosphorylation**, each NADH and $FADH_2$ formed as a result of oxidation reactions in the second and third stages of catabolism is converted into three molecules of ATP and two molecules of ATP, respectively.

PROBLEMS

20. Indicate whether an anabolic pathway or a catabolic pathway does the following:

a. produces energy in the form of ATP **b.** involves primarily oxidation reactions

21. Galactose can enter the glycolytic cycle, but it must first react with ATP to form galactose-1-phosphate. Propose a mechanism for this reaction.

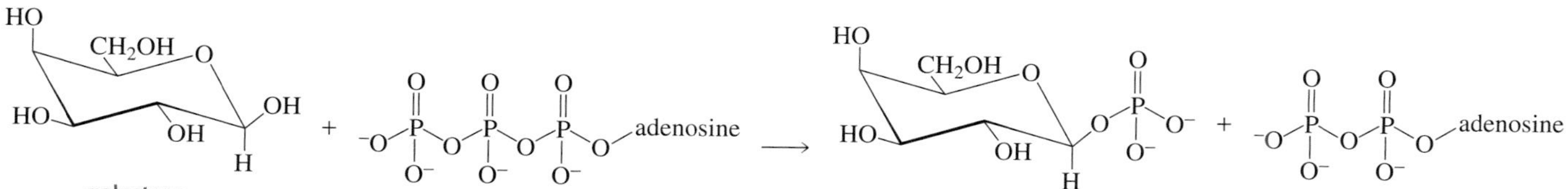

galactose + ATP ⟶ galactose-1-phosphate + ADP

22. Pyruvate is the leaving group in the tenth reaction of glycolysis. Why is it a good leaving group?

23. When pyruvate is reduced by NADH to lactate, which hydrogen in lactate comes from NADH?

24. What reactions in the citric acid cycle form a product with a new asymmetric center?

25. If the phosphorus atom in 3-phosphoglycerate is radioactively labeled, where will the radioactive label be when the reaction that forms 2-phosphoglycerate is over?

26. What carbon atoms of glucose end up as a carboxyl group in pyruvate?

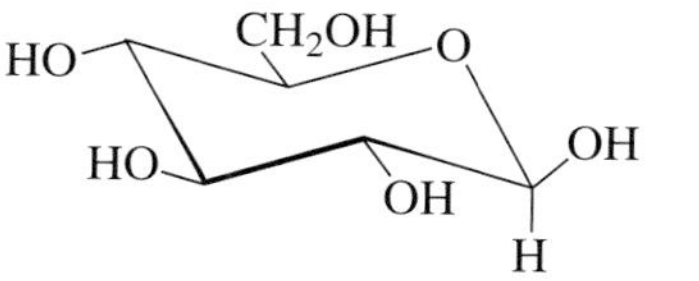

glucose

27. What carbon atoms of glucose end up in ethanol under anaerobic conditions in yeast?

28. How many molecules of acetyl-CoA are obtained from β-oxidation of one molecule of a 16-carbon saturated fatty acyl-CoA?

29. How many molecules of CO_2 are obtained from the complete metabolism of one molecule of a 16-carbon saturated fatty acyl-CoA?

30. How many molecules of ATP are obtained from β-oxidation of one molecule of a 16-carbon saturated fatty acyl-CoA?

31. How many molecules of NADH and $FADH_2$ are obtained from β-oxidation of one molecule of a 16-carbon saturated fatty acyl-CoA?

32. How many molecules of ATP are obtained from the NADH and $FADH_2$ formed in β-oxidation of one molecule of a 16-carbon saturated fatty acyl-CoA?

33. How many molecules of ATP are obtained from complete (including the fourth stage of catabolism) metabolism of one molecule of a 16-carbon saturated fatty acyl-CoA?

34. How many molecules of ATP are obtained from the complete (including the fourth stage of catabolism) metabolism of a molecule of glucose? (Note: one molecule of NADH is formed from each molecule of pyruvate that is converted to acetyl-CoA.)

35. Most fatty acids have an even number of carbon atoms and therefore are completely metabolized to acetyl-CoA. A fatty acid with an odd number of carbon atoms is metabolized to acetyl-CoA and one equivalent of propionyl-CoA. The following two reactions convert propionyl-CoA into succinyl-CoA, a citric acid cycle intermediate, so it can be further metabolized. Each of the enzyme-catalyzed reactions requires a coenzyme. Identify the coenzyme for each step. From what vitamins are the coenzymes derived?

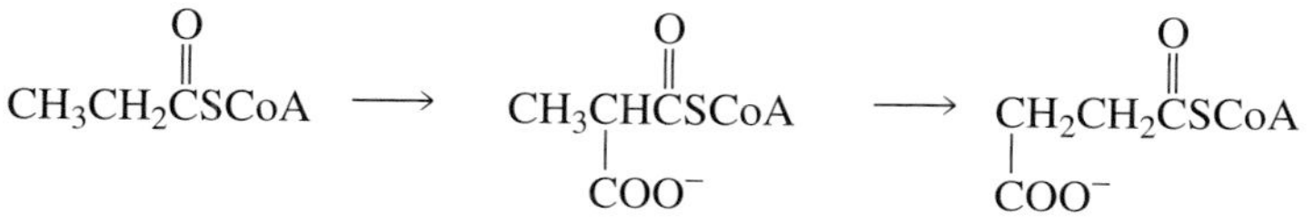

$$CH_3CH_2\overset{O}{\overset{\|}{C}}SCoA \longrightarrow CH_3\underset{COO^-}{\underset{|}{C}H}\overset{O}{\overset{\|}{C}}SCoA \longrightarrow \underset{COO^-}{\underset{|}{C}H_2}CH_2\overset{O}{\overset{\|}{C}}SCoA$$

propionyl-CoA ⟶ methylmalonyl-CoA ⟶ succinyl-CoA

CHAPTER 19

The Organic Chemistry of Lipids

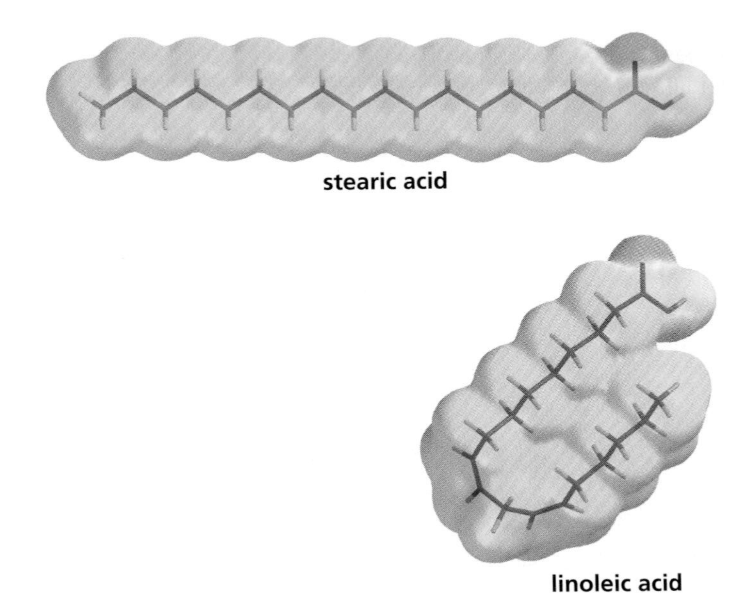

stearic acid

linoleic acid

Lipids are organic compounds, found in living organisms, that are soluble in nonpolar solvents. Because compounds are classified as lipids on the basis of a physical property—their solubility—rather than on the basis of their structures, lipids have a variety of structures and functions, as the following examples illustrate:

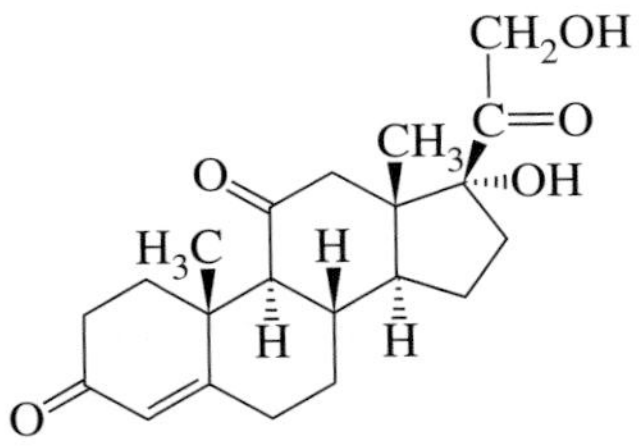

cortisone
a hormone

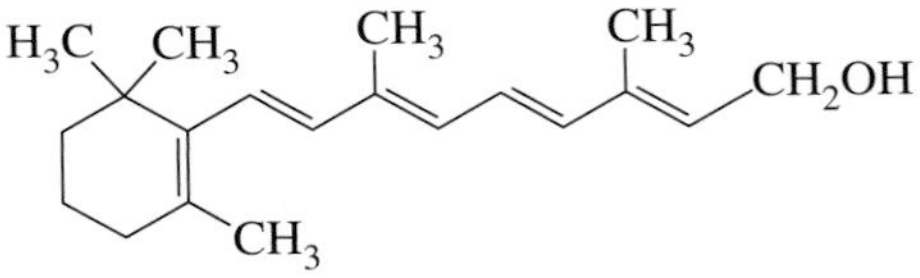

vitamin A
a vitamin

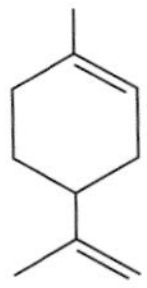

limonene
in orange and
lemon oils

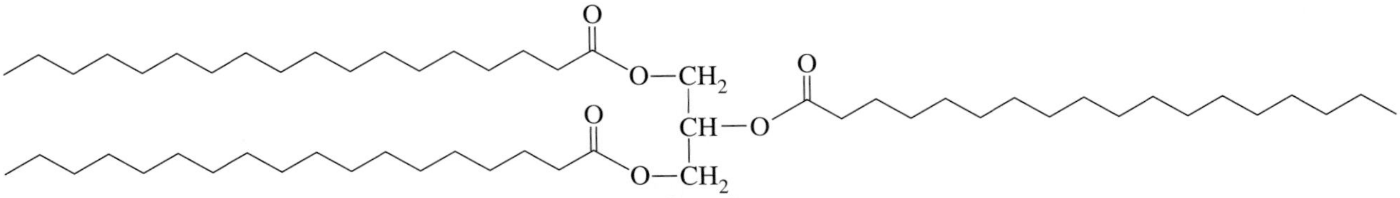

tristearin
a fat

The ability of lipids to dissolve in nonpolar solvents results from their significant hydrocarbon component—the part of the molecule responsible for its "oiliness" or "fattiness." The word *lipid* comes from the Greek *lipos*, which means "fat."

19.1 FATTY ACIDS

Fatty acids, one major group of lipids, are carboxylic acids with long hydrocarbon chains. The fatty acids most frequently found in nature are shown in Table 19.1. Most naturally occurring fatty acids are unbranched and contain an even number of carbon atoms because they are synthesized from acetate, a compound with two carbons (Section 13.11).

Fatty acids can be saturated with hydrogen (and therefore have no carbon–carbon double bonds) or unsaturated (and have carbon–carbon double bonds). Fatty acids with more than one double bond are called **polyunsaturated fatty acids**.

The melting points of saturated fatty acids increase with increasing molecular weight because of increased van der Waals interactions between the molecules (Section 3.7). The melting points of unsaturated fatty acids also increase with increasing molecular weight (Table 19.1).

Table 19.1 Common Naturally Occurring Fatty Acids

Number of carbons	Common name	Systematic name	Structure	Melting point °C
Saturated				
12	lauric acid	dodecanoic acid	COOH	44
14	myristic acid	tetradecanoic acid	COOH	58
16	palmitic acid	hexadecanoic acid	COOH	63
18	stearic acid	octadecanoic acid	COOH	69
20	arachidic acid	eicosanoic acid	COOH	77
Unsaturated				
16	palmitoleic acid	(9Z)-hexadecenoic acid	COOH	0
18	oleic acid	(9Z)-octadecenoic acid	COOH	13
18	linoleic acid	(9Z,12Z)-octadecadienoic acid	COOH	−5
18	linolenic acid	(9Z,12Z,15Z)-octadecatrienoic acid	COOH	−11
20	arachidonic acid	(5Z,8Z,11Z,14Z)-eicosatetraenoic acid	COOH	−50
20	EPA	(5Z,8Z,11Z,14Z,17Z)-eicosapentaenoic acid	COOH	−50

The double bonds in naturally occurring unsaturated fatty acids have the cis configuration and are always separated by one CH_2 group. The cis double bond produces a bend in the molecules, which prevents them from packing together as tightly as saturated fatty acids. As a result, unsaturated fatty acids have fewer intermolecular interactions and, therefore, have lower melting points than saturated fatty acids with comparable molecular weight (Table 19.1).

The melting points of unsaturated fatty acids decrease as the number of double bonds increases. For example, an 18-carbon fatty acid melts at 69 °C if it is saturated, at 13 °C if it has one double bond, at −5 °C if it has two double bonds, and at −11 °C if it has three double bonds.

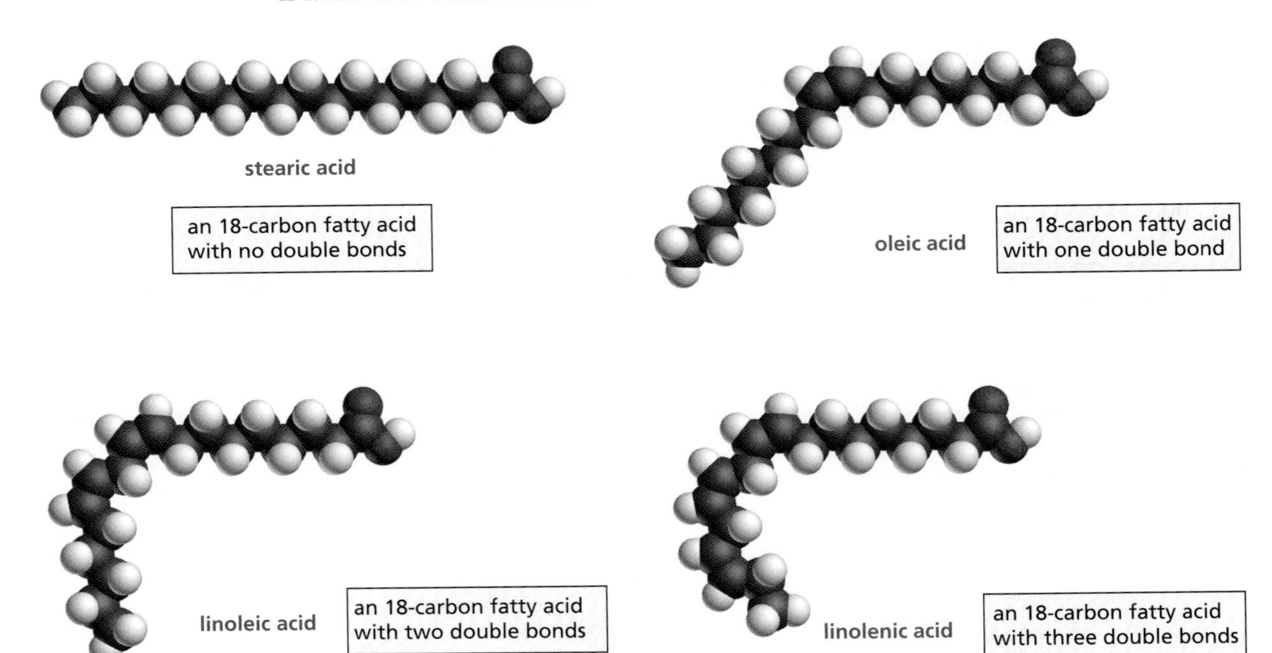

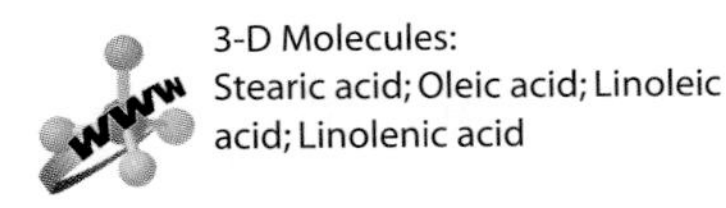

3-D Molecules: Stearic acid; Oleic acid; Linoleic acid; Linolenic acid

PROBLEM 1

Explain the difference in the melting points of the following fatty acids:

a. palmitic acid and stearic acid
b. palmitic acid and palmitoleic acid
c. oleic acid and linoleic acid

OMEGA FATTY ACIDS

The term *omega* indicates the position of the first double bond, counting from the methyl end, in an unsaturated fatty acid. For example, linoleic acid is called omega-6 fatty acid because its first double bond is after the sixth carbon, and linolenic acid is called omega-3 fatty acid because its first double bond is after the third carbon. Mammals lack the enzyme that introduces a double bond beyond C-9 (the carboxyl carbon is C-1). Linoleic acid and linolenic acids, therefore, are essential fatty acids for mammals: they cannot synthesize them, yet they require them for normal body function, so they must be included in their diets.

They must have linoleic and linolenic acids in their diets.

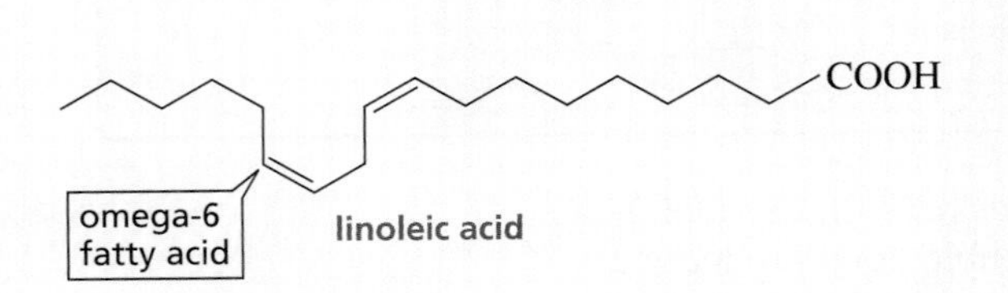

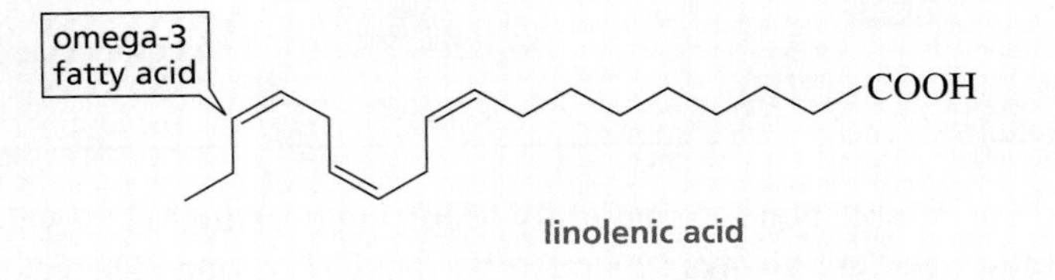

19.2 WAXES ARE HIGH-MOLECULAR WEIGHT ESTERS

Waxes are esters formed from long-chain carboxylic acids and long-chain alcohols. For example, beeswax, the structural material of beehives, has a 26-carbon carboxylic acid component and a 30-carbon alcohol component. Carnauba wax is a particularly hard wax because of its relatively high molecular weight; it has a 32-carbon carboxylic acid component and a 34-carbon alcohol component. Carnauba wax is widely used as a car wax and in floor polishes.

Layers of honeycomb in a beehive.

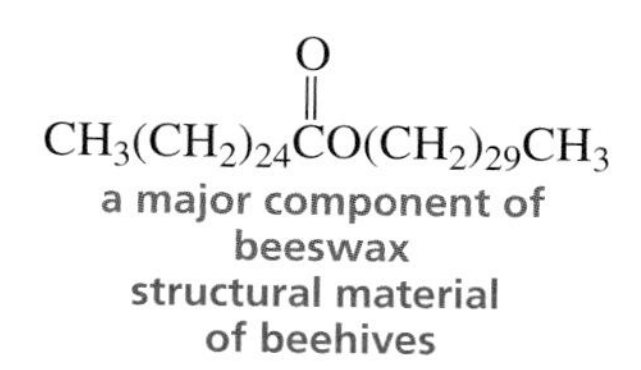

a major component of beeswax
structural material of beehives

$CH_3(CH_2)_{30}\overset{O}{\overset{\|}{C}}O(CH_2)_{33}CH_3$
a major component of carnauba wax
coating on the leaves of a Brazilian palm

$CH_3(CH_2)_{14}\overset{O}{\overset{\|}{C}}O(CH_2)_{15}CH_3$
a major component of spermaceti wax
from the heads of sperm whales

Waxes are common in living organisms. The feathers of birds are coated with wax to make them water repellent. Some vertebrates secrete wax in order to keep their fur lubricated and water repellent. Insects secrete a waterproof, waxy layer on the outside of their exoskeletons. Wax is also found on the surfaces of certain leaves and fruits, where it serves as a protectant against parasites and minimizes the evaporation of water.

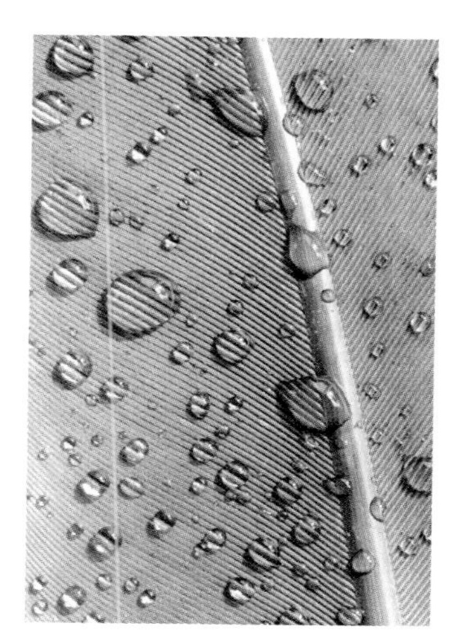

Raindrops on a feather.

19.3 FATS AND OILS

Triglycerides, also called **triacylglycerols**, are compounds in which each of the three OH groups of glycerol has formed an ester with a fatty acid (Section 19.1).

$$\begin{array}{l} CH_2-OH \\ | \\ CH-OH \\ | \\ CH_2-OH \end{array} \qquad \begin{array}{l} R^1-\overset{O}{\overset{\|}{C}}-OH \\ R^2-\overset{O}{\overset{\|}{C}}-OH \\ R^3-\overset{O}{\overset{\|}{C}}-OH \end{array} \qquad \begin{array}{l} CH_2-O-\overset{O}{\overset{\|}{C}}-R^1 \\ | \\ CH-O-\overset{O}{\overset{\|}{C}}-R^2 \\ | \\ CH_2-O-\overset{O}{\overset{\|}{C}}-R^3 \end{array}$$

glycerol — fatty acids — a triacylglycerol a fat or an oil

If the three fatty acid components of a triglyceride are the same, the compound is called a **simple triglyceride**. **Mixed triglycerides** contain two or three different fatty acid components and are more common than simple triglycerides. Not all triglyceride molecules from a single source are necessarily identical; substances such as lard and olive oil, for example, are mixtures of several different triglycerides (Table 19.2).

Triglycerides that are solids or semisolids at room temperature are called **fats**. Most fats are obtained from animals and are composed largely of triglycerides with fatty acid components that are either saturated or have only one double bond. The saturated fatty acid tails pack closely together, giving these triglycerides relatively high melting points, causing them to be solids at room temperature.

Table 19.2 Approximate Percentage of Fatty Acids in Some Common Fats and Oils

		Saturated fatty acids				Unsaturated fatty acids		
	mp (°C)	lauric C_{12}	myristic C_{14}	palmitic C_{16}	stearic C_{18}	oleic C_{18}	linoleic C_{18}	linolenic C_{18}
Animal fats								
Butter	32	2	11	29	9	27	4	—
Lard	30	—	1	28	12	48	6	—
Human fat	15	1	3	25	8	46	10	—
Whale blubber	24	—	8	12	3	35	10	—
Plant oils								
Corn	20	—	1	10	3	50	34	—
Cottonseed	−1	—	1	23	1	23	48	—
Linseed	−24	—	—	6	3	19	24	47
Olive	−6	—	—	7	2	84	5	—
Peanut	3	—	—	8	3	56	26	—
Safflower	−15	—	—	3	3	19	70	3
Sesame	−6	—	—	10	4	45	40	—
Soybean	−16	—	—	10	2	29	51	7

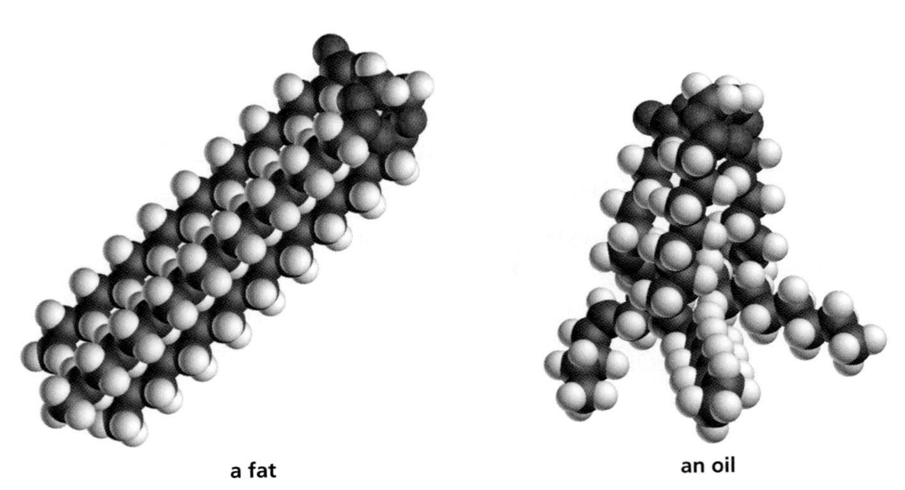

Liquid triglycerides are called **oils**. Oils typically come from plant products such as corn, soybeans, olives, and peanuts. They are composed primarily of triglycerides with fatty acid components that are unsaturated and therefore cannot pack tightly together. Consequently, they have relatively low melting points, causing them to be liquids at room temperature. Compare the approximate fatty acid compositions of the common fats and oils shown in Table 19.2.

Unsaturated fatty acids have lower melting points than saturated fatty acids.

Some or all of the double bonds of polyunsaturated oils can be reduced by catalytic hydrogenation. Margarine and shortening are prepared by hydrogenating vegetable oils, such as safflower oil or soybean oil, until they have the desired consistency. The hydrogenation reaction must be carefully controlled, however, because reducing all the carbon–carbon double bonds would produce a hard fat with the consistency of beef tallow. Some trans double bonds may be created during this process (Section 5.12).

$$\underset{\text{polyunsaturated oil}}{RCH{=}CHCH_2CH{=}CHCH_2CH{=}CH{-}} \xrightarrow[\text{Pd/C}]{H_2} \underset{\text{hydrogenated oil}}{RCH_2CH_2CH_2CH{=}CHCH_2CH_2CH_2{-}}$$

Vegetable oils have become popular in food preparation because some studies have linked the consumption of saturated fats with heart disease. Recent studies have shown that unsaturated fats may also be implicated in heart disease. However, one unsaturated fatty acid—a 20-carbon fatty acid with five double bonds, known as EPA and found in high concentrations in fish oils—is thought to lower the chance of developing certain forms of heart disease. Once consumed, dietary fat is hydrolyzed, regenerating glycerol and fatty acids (Section 18.3). We have seen that the hydrolysis of fats under basic conditions forms glycerol and salts of fatty acids that are commonly known as soap (Section 11.10).

This puffin's diet is high in fish oil.

PROBLEM 2 ♦

Which has a higher melting point, glyceryl tripalmitoleate or glyceryl tripalmitate?

PROBLEM 3 ♦

Draw the structure of an optically inactive fat that, when hydrolyzed, gives glycerol, one equivalent of lauric acid, and two equivalents of stearic acid.

PROBLEM 4 ♦

Draw the structure of an optically active fat that, when hydrolyzed, gives the same products as the fat in Problem 3.

Animals have a subcutaneous layer of fat cells that serves as both an energy source and an insulator. The fat content of the average man is about 21%, whereas the fat content of the average woman is about 25%. A fat provides about six times as much metabolic energy as an equal weight of hydrated glycogen because fats are less oxidized than carbohydrates, and because fats are nonpolar, they do not bind water. In contrast, two-thirds of the weight of stored glycogen is water (Section 15.14).

Humans can store sufficient fat to provide for the body's metabolic needs for two to three months, but can store only enough carbohydrate to provide for its metabolic needs for less than 24 hours. Carbohydrates, therefore, are used primarily as a quick, short-term energy source.

Polyunsaturated fats and oils are easily oxidized by O_2 by means of a radical chain reaction. In the initiation step, a radical removes a hydrogen from a CH_2 group that is flanked by two double bonds. This hydrogen is the one most easily removed because the unpaired electron is shared by three carbons. (Follow the green arrows to arrive at one of the resonance contributors and the red arrows to arrive at the other.)

$$RCH{=}CH{-}\underset{H}{CH}{-}CH{=}CH{-} + X\cdot \xrightarrow{\text{initiation}} RCH{=}CH{-}\dot{C}H{-}CH{=}CH{-} + HX$$

$$R\dot{C}H{-}CH{=}CH{-}CH{=}CH{-} \longleftrightarrow RCH{=}CH{-}CH{=}CH{-}\dot{C}H{-}$$

$$\downarrow O_2$$

$$CH_3CH_2CH_2\overset{O}{\overset{\|}{C}}OH$$ **and other short-chain carboxylic acids**

The reaction of fatty acids with O_2 causes them to become rancid. The unpleasant taste and smell associated with sour milk and rancid butter are due to the products of the oxidation reaction, which have strong odors.

PROBLEM 5

Draw the resonance contributors for the radical formed when a hydrogen atom is removed from C-10 of arachidonic acid.

3-D Molecule: Olestra

OLESTRA: NONFAT WITH FLAVOR

Chemists have been searching for ways to reduce the caloric content of foods without decreasing their flavor. Many people who believe that "no fat" is synonymous with "no flavor" think this is a worthy endeavor. Procter and Gamble spent 30 years and more than $2 billion to develop a fat substitute they named Olestra. After reviewing the results of more than 150 studies, the Food and Drug Administration (Section 21.10) approved the limited use of Olestra in snack foods in 1996.

Olestra is a semisynthetic compound. That is, Olestra itself does not exist in nature, but its components do. Developing a compound that can be made from units that are a normal part of our diet decreases the likelihood that the new compound will be toxic. Olestra is made by esterifying all the OH groups of sucrose with fatty acids obtained from cottonseed oil and soybean oil. Therefore, its component parts are table sugar and vegetable oil. Because its ester linkages are too sterically hindered to be hydrolyzed by digestive enzymes, Olestra tastes like fat but it cannot be digested, so it has no caloric value.

ester groups are too hindered to be hydrolyzed

glucose

fructose

Olestra

WHALES AND ECHOLOCATION

Whales have enormous heads, accounting for 33% of their total weight. They have large deposits of fat in their heads and lower jaws. This fat is very different from both the whale's normal body fat and its dietary fat. Because major anatomical modifications were necessary to accommodate this fat, it must have some important function for the animal. It is now believed that the fat is used for echolocation—the emitting of sounds in pulses in order to gain information by analyzing the returning echoes. The fat in the whale's head focuses the emitted sound waves in a directional beam, and the echoes are received by the fat organ in the lower jaw. This organ transmits the sound to the brain for processing and interpretation, providing the whale with information about the depth of the water, changes in the sea floor, and the location of the coastline. The fat deposits in the whale's head and jaw therefore give the animal a unique acoustic sensory system and allow it to compete successfully for survival with the shark, which also has a well-developed sense of sound direction.

Humpback whale breaching in S. Africa.

19.4 PHOSPHOLIPIDS ARE COMPONENTS OF MEMBRANES

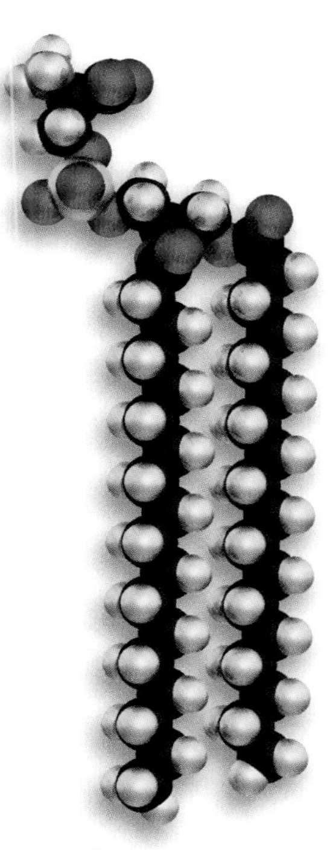

phosphatidylserine

For organisms to operate properly, some of their parts must be separated from other parts. On a cellular level, for example, the outside of the cell must be separated from the inside. "Greasy" lipid membranes serve as the barrier. In addition to isolating the cell's contents, membranes allow the selective transport of ions and organic molecules into and out of the cell.

Phosphoglycerides (also called phosphoacylglycerols), the major components of cell membranes, belong to a class of compounds called **phospholipids**—lipids that contain a phosphate group. Phosphoglycerides are similar to triglycerides except that a terminal OH group of glycerol is esterified with phosphoric acid rather than with a fatty acid, forming a phosphatidic acid. The C-2 carbon in phosphoglycerides has the *R* configuration.

$$\begin{array}{l} CH_2-O-\overset{\overset{\displaystyle O}{\|}}{C}-R^1 \\ | \\ CH-O-\overset{\overset{\displaystyle O}{\|}}{C}-R^2 \\ | \\ CH_2-O-\overset{\overset{\displaystyle O}{\|}}{\underset{\underset{\displaystyle O^-}{|}}{P}}-OH \end{array} \qquad \qquad HO-\overset{\overset{\displaystyle O}{\|}}{\underset{\underset{\displaystyle O^-}{|}}{P}}-OH$$

R configuration

an ester of phosphoric acid

a phosphatidic acid — phosphoric acid

Phosphatidic acids are the simplest phosphoglycerides and are present only in small amounts in membranes. The most common phosphoglycerides in membranes have a second phosphate ester linkage—they are **phosphodiesters**.

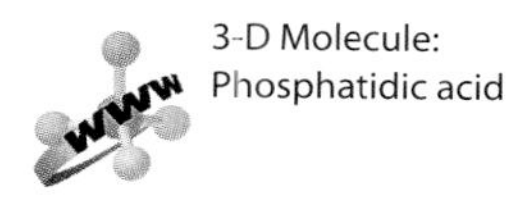

3-D Molecule:
Phosphatidic acid

phosphoglycerides

$$\begin{array}{l} CH_2-O-\overset{\overset{\displaystyle O}{\|}}{C}-R^1 \\ | \\ CH-O-\overset{\overset{\displaystyle O}{\|}}{C}-R^2 \\ | \\ CH_2-O-\overset{\overset{\displaystyle O}{\|}}{\underset{\underset{\displaystyle O^-}{|}}{P}}-OCH_2CH_2\overset{+}{N}H_3 \end{array} \quad \begin{array}{l} CH_2-O-\overset{\overset{\displaystyle O}{\|}}{C}-R^1 \\ | \\ CH-O-\overset{\overset{\displaystyle O}{\|}}{C}-R^2 \\ | \\ CH_2-O-\overset{\overset{\displaystyle O}{\|}}{\underset{\underset{\displaystyle O^-}{|}}{P}}-OCH_2CH_2\overset{\overset{\displaystyle CH_3}{|+}}{\underset{\underset{\displaystyle CH_3}{|}}{N}}CH_3 \end{array} \quad \begin{array}{l} CH_2-O-\overset{\overset{\displaystyle O}{\|}}{C}-R^1 \\ | \\ CH-O-\overset{\overset{\displaystyle O}{\|}}{C}-R^2 \\ | \\ CH_2-O-\overset{\overset{\displaystyle O}{\|}}{\underset{\underset{\displaystyle O^-}{|}}{P}}-OCH_2\underset{\underset{\displaystyle \overset{+}{N}H_3}{|}}{C}HCOO^- \end{array}$$

phosphate ester linkages

a phosphatidylethanolamine
a cephalin

a phosphatidylcholine
a lecithin

a phosphatidylserine

The alcohols most commonly used to form the second ester group are ethanolamine, choline, and serine. Phosphatidylethanolamines are also called *cephalins*, and phosphatidylcholines are called *lecithins*. Lecithins are added to foods such as mayonnaise to prevent the aqueous and fat components from separating.

The most common phosphoglycerides in membranes are phosphodiesters.

Phosphoglycerides form membranes by arranging themselves in a **lipid bilayer**. The polar heads of the phosphoglycerides are on both surfaces of the bilayer, and the fatty acid chains form the interior of the bilayer. Cholesterol—a lipid discussed in Section 19.7—is also found in the interior of the bilayer (Figure 19.1).

The fluidity of a membrane is controlled by the fatty acid components of the phosphoglycerides. Saturated fatty acids decrease membrane fluidity because their hydrocarbon chains pack closely together. Unsaturated fatty acids increase fluidity because they pack less closely together. Cholesterol also decreases fluidity (Section 3.13). Only animal membranes contain cholesterol, so they are more rigid than plant membranes.

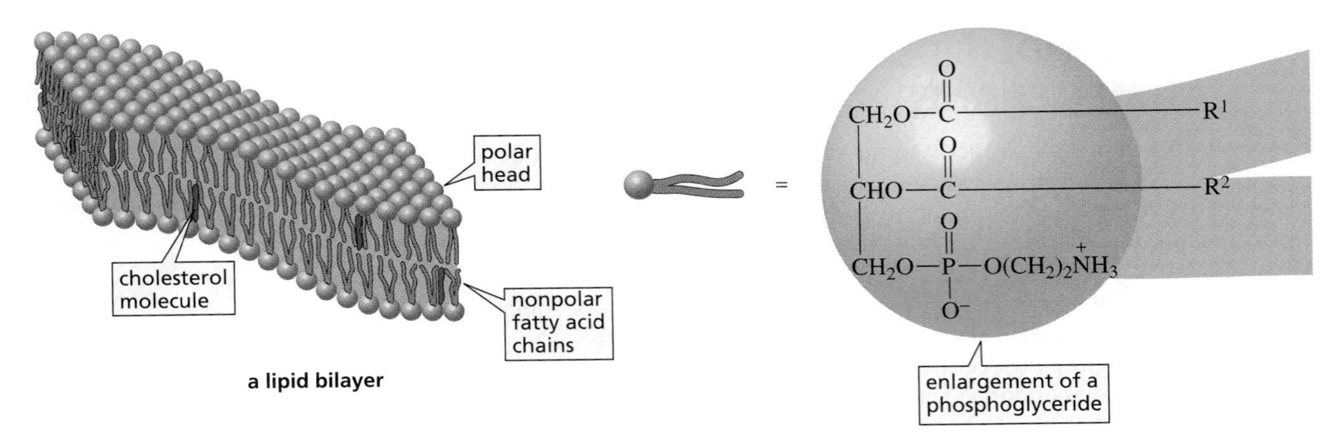

▲ **Figure 19.1**
A lipid bilayer.

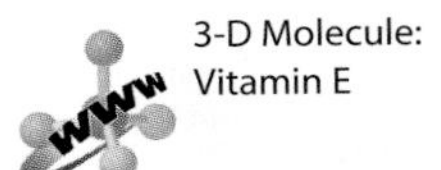

3-D Molecule: Vitamin E

The unsaturated fatty acid chains of phosphoglycerides are susceptible to reaction with O_2, similar to the reaction described on page 519 for fats and oils. This oxidation reaction leads to the degradation of membranes. Vitamin E is an important antioxidant that protects fatty acid chains from degradation through oxidation. Vitamin E, also called α-tocopherol, is classified as a lipid because it is soluble in nonpolar solvents (Section 5.17). It is able therefore to enter biological membranes; once there, it reacts more rapidly with oxygen than the bilayer phosphoglycerides do, thus preventing them from reacting with oxygen. The ability of vitamin E to react with oxygen more rapidly than fats do is the reason it is added as a preservative to many fat-containing foods.

SNAKE VENOM

The venom of some poisonous snakes contains a phospholipase, an enzyme that hydrolyzes an ester group of a phosphoglyceride. For example, both the eastern diamondback rattlesnake and the Indian cobra contain a phospholipase that hydrolyzes an ester bond of cephalins, causing the membranes of red blood cells to rupture.

An eastern diamondback rattlesnake.

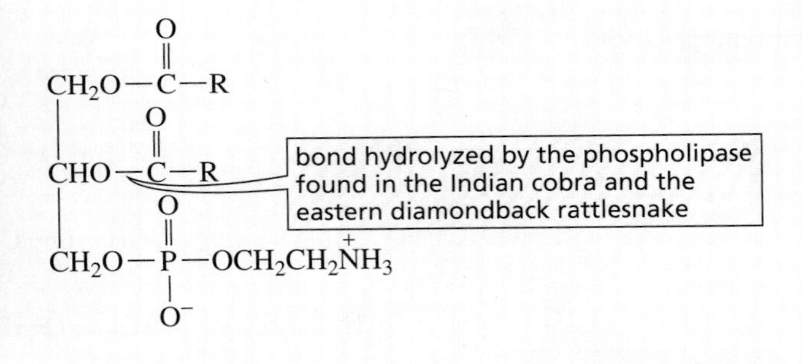

IS CHOCOLATE A HEALTH FOOD?

We have long been told that our diets should include lots of fruits and vegetables because they are good sources of antioxidants. Antioxidants protect against cardiovascular disease, cancer, and cataracts, and they are thought to slow the effects of aging. Recent studies show that chocolate also has high levels of antioxidants. On a weight basis, the concentration of antioxidants in chocolate is higher than the concentration in red wine or green tea and 20 times higher than the concentration in tomatoes. Another piece of good news for chocolate lovers is that stearic acid, the main fatty acid in chocolate, does not appear to raise blood cholesterol levels the way other saturated fatty acids do. Dark chocolate contains more than twice the level of antioxidants as milk chocolate. Unfortunately, white chocolate contains no antioxidants.

PROBLEM 6◆

Membranes contain proteins. Integral membrane proteins extend partly or completely through the membrane, whereas peripheral membrane proteins are found on the inner or outer surfaces of the membrane. What is the likely difference in the amino acid composition of integral and peripheral membrane proteins?

PROBLEM 7◆

A colony of bacteria accustomed to an environment at 25 °C was moved to an identical environment except that its temperature was 35 °C. The higher temperature increased the fluidity of the bacterial membranes. What could the bacteria do to regain their original membrane fluidity?

PROBLEM 8

The membrane phospholipids in deer and elk have a higher degree of unsaturation in cells closer to the hoof than in cells closer to the body. Explain how this trait can be important for survival.

19.5 TERPENES CONTAIN CARBON ATOMS IN MULTIPLES OF FIVE

Terpenes are a diverse class of lipids. More than 20,000 terpenes, many extracted from the oils of fragrant plants, are known. They can be hydrocarbons, or they can contain oxygen and be alcohols, ketones, or aldehydes. Certain terpenes have been used as spices, perfumes, and medicines for thousands of years.

HO — menthol, peppermint oil

OH — geraniol, geranium oil

zingiberene, oil of ginger

β-selinene, oil of celery

After analyzing a large number of terpenes, organic chemists realized that they contained carbon atoms in multiples of five. These naturally occurring compounds contain 10, 15, 20, 25, 30, and 40 carbon atoms, which suggests that there is a compound with five carbon atoms that serves as their building block. Further investigation showed that their structures are consistent with their being made by joining together isoprene units, usually in a head-to-tail fashion (The branched end of isoprene is called the head, and the unbranched end is called the tail.)

That isoprene units are linked in a head-to-tail fashion to form terpenes is known as the **isoprene rule**.

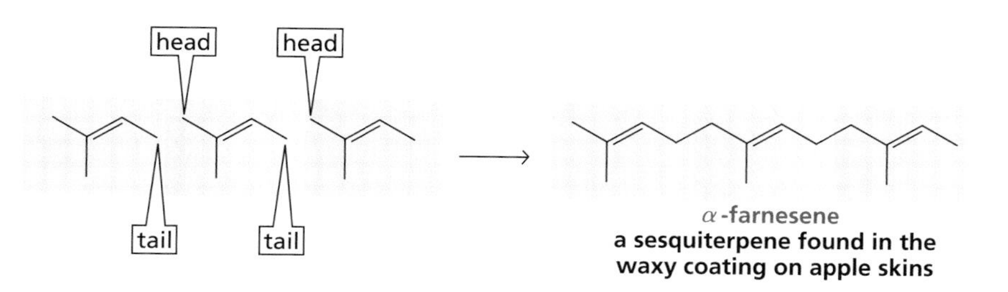

α-farnesene
a sesquiterpene found in the waxy coating on apple skins

BIOGRAPHY

Leopold Stephen Ružička (1887–1976) *was the first to recognize that many organic compounds contain multiples of five carbons. A Croatian, Ružička attended college in Switzerland and became a Swiss citizen in 1917. He was a professor of chemistry at the University of Utrecht in the Netherlands and later at the Federal Institute of Technology in Zürich. For his work on terpenes, he shared the 1939 Nobel Prize in chemistry with Adolph Butenandt (page 528).*

In the case of cyclic compounds, the linkage of the head of one isoprene unit to the tail of another is followed by an additional linkage to form the ring. The second linkage is not necessarily head-to-tail but is whatever linkage is necessary to form a stable five- or six-membered ring.

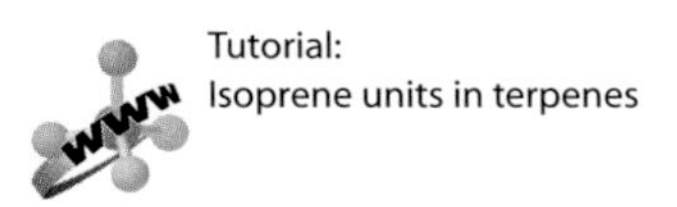

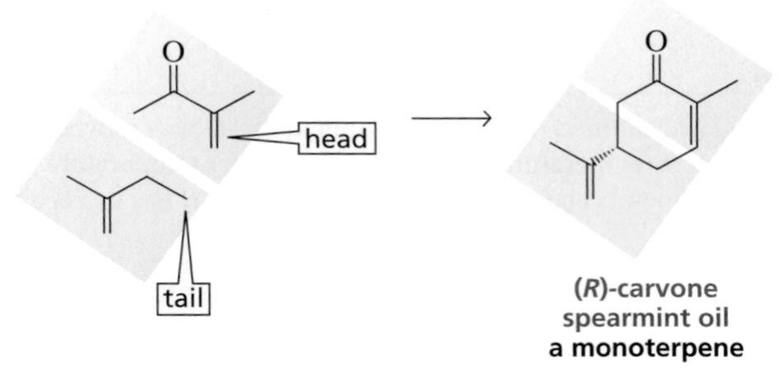

(*R*)-carvone
spearmint oil
a monoterpene

Terpenes are classified according to the number of carbons they contain. **Monoterpenes** are composed of two isoprene units, so they have 10 carbons. **Sesquiterpenes**, with 15 carbons, are composed of three isoprene units. Many fragrances and flavorings found in plants are monoterpenes and sesquiterpenes. These compounds are known as *essential oils*.

A monoterpene has 10 carbon atoms.

Triterpenes (with 30 carbons) and **tetraterpenes** (with 40 carbons) have important biological roles. For example, **squalene**, a triterpene, is the precursor of cholesterol, which is the precursor of all the other steroid molecules (Section 19.7).

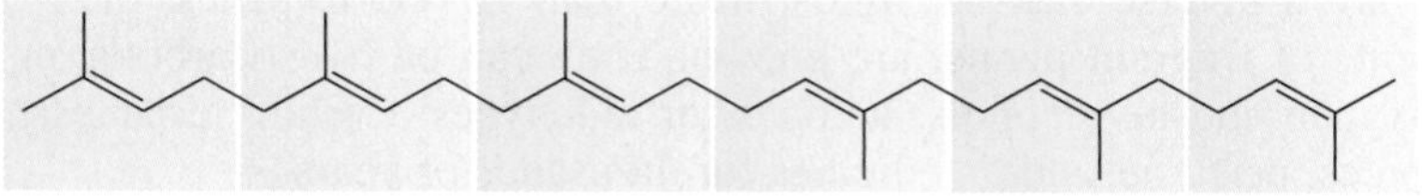

squalene

Carotenoids are tetraterpenes. Lycopene, the compound responsible for the red coloring of tomatoes and watermelon, and β-carotene, the compound that causes carrots and apricots to be orange, are examples of carotenoids. Cleaving β-carotene forms two molecules of vitamin A, the vitamin that plays an important role in vision.

The many conjugated double bonds in lycopene and β-carotene cause the compounds to be colored. β-carotene and other colored compounds are found in the leaves of trees, but their characteristic colors are usually obscured by the green color of chlorophyll. In the fall, when chlorophyll degrades, the colors become apparent. β-carotene is also the coloring agent used in margarine (Section 7.12).

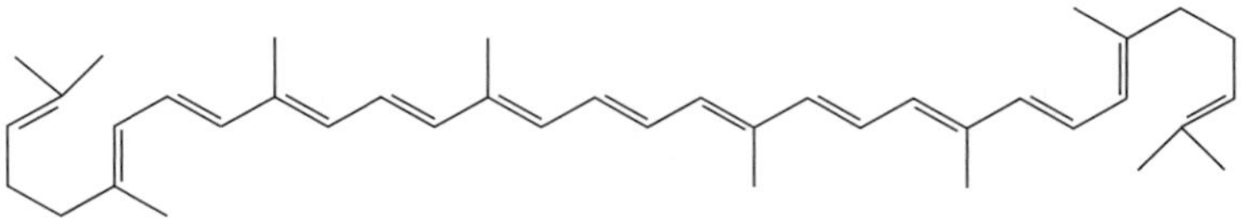

lycopene

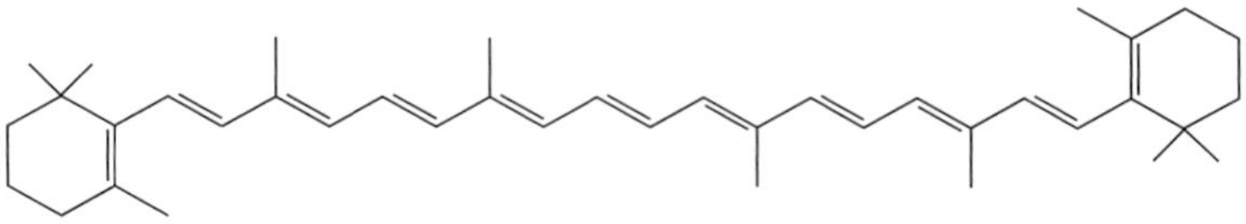

β-carotene

PROBLEM 9 ♦ SOLVED

a. What class of lipid is α-farnesene?
b. What class of lipid is spearmint oil?

Solution To 9a α-Farnesene has 15 carbons, so it is a sesquiterpene.

PROBLEM 10 SOLVED

Mark off the isoprene units in zingiberene and menthol. Their structures are on page 523.

Solution For zingiberene, we find

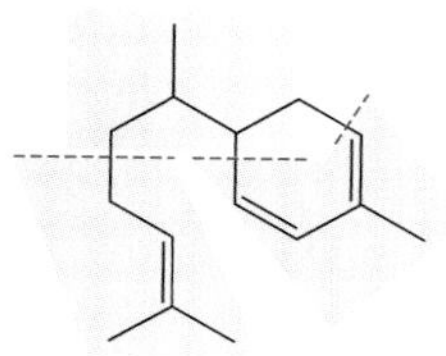

19.6 HOW TERPENES ARE BIOSYNTHESIZED

The five-carbon compound used for the biosynthesis of terpenes is actually not isoprene; it isopentenyl pyrophosphate, which can be thought of as isoprene to which a good leaving group had been added.

Terpenes are biosynthesized from isopentenyl pyrophosphate.

The starting material for the biosynthesis of terpenes is **dimethylallyl pyrophosphate**, so some **isopentenyl pyrophosphate** must be converted to dimethylallyl pyrophosphate. This conversion takes place through an enzyme-catalyzed isomerization reaction that involves the addition of a proton to the sp^2 carbon of isopentenyl pyrophosphate that is bonded to the greater number of hydrogens (Section 5.3) and the elimination of a proton from the carbocation intermediate in accordance with Zaitsev's rule (Section 9.8).

Addition of a proton and loss of a proton convert isopentenyl pyrophosphate into dimethylallyl pyrophosphate.

a good leaving group

isopentenyl pyrophosphate $\xrightarrow{H^+}$ ... ⟶ dimethylallyl pyrophosphate + H^+

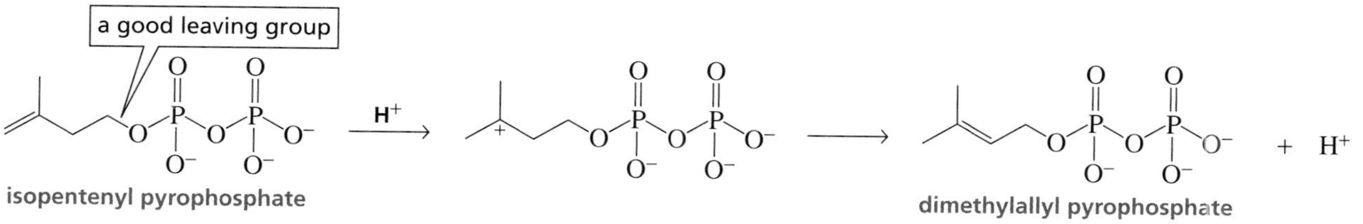

The reaction of dimethylallyl pyrophosphate with isopentenyl pyrophosphate forms geranyl pyrophosphate, a 10-carbon compound, in a two-step process.

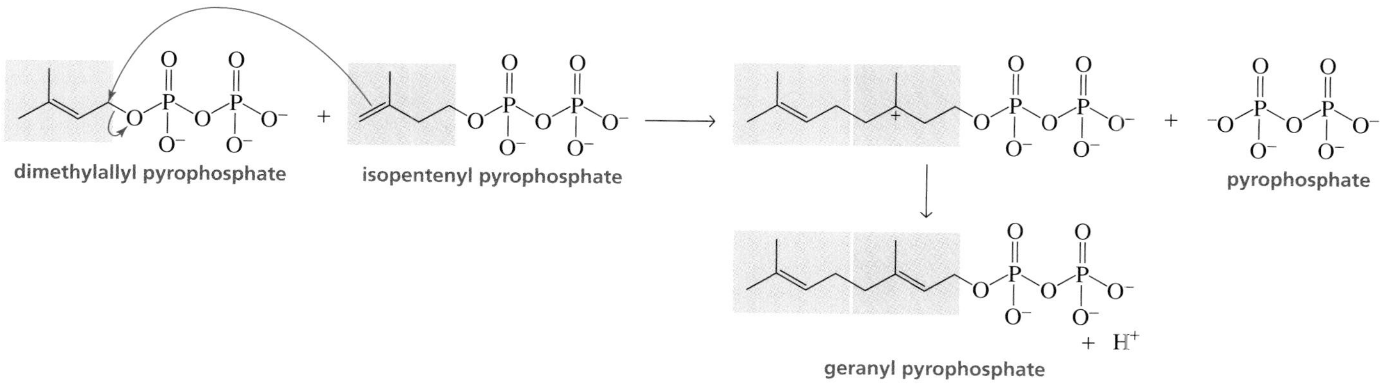

dimethylallyl pyrophosphate + isopentenyl pyrophosphate ⟶ ... + pyrophosphate

⟶ geranyl pyrophosphate + H^+

- Isopentenyl pyrophosphate acts as a nucleophile and displaces a pyrophosphate group from dimethylallyl pyrophosphate. Pyrophosphate is a stable base and therefore an excellent leaving group.
- A proton is removed in the next step, resulting in the formation of geranyl pyrophosphate.

In the same two-step process, geranyl pyrophosphate reacts with another molecule of isopentenyl pyrophosphate to form farnesyl pyrophosphate, a 15-carbon compound.

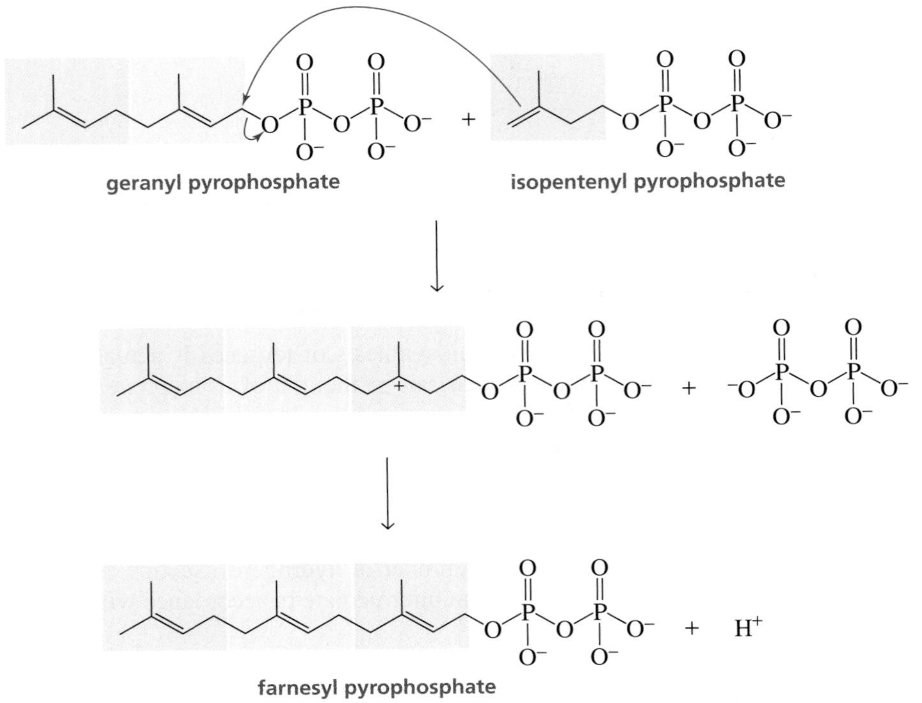

geranyl pyrophosphate + isopentenyl pyrophosphate

farnesyl pyrophosphate + H^+

Cholesterol is the precursor of all other steroids.

Two molecules of farnesyl pyrophosphate form **squalene**, a 30-carbon compound. The reaction is catalyzed by an enzyme that joins the two molecules in a tail-to-tail linkage. As we noted earlier, squalene is the precursor of cholesterol, and cholesterol is the precursor of all other steroids.

farnesyl pyrophosphate + farnesyl pyrophosphate

squalene synthase

tail to tail

squalene

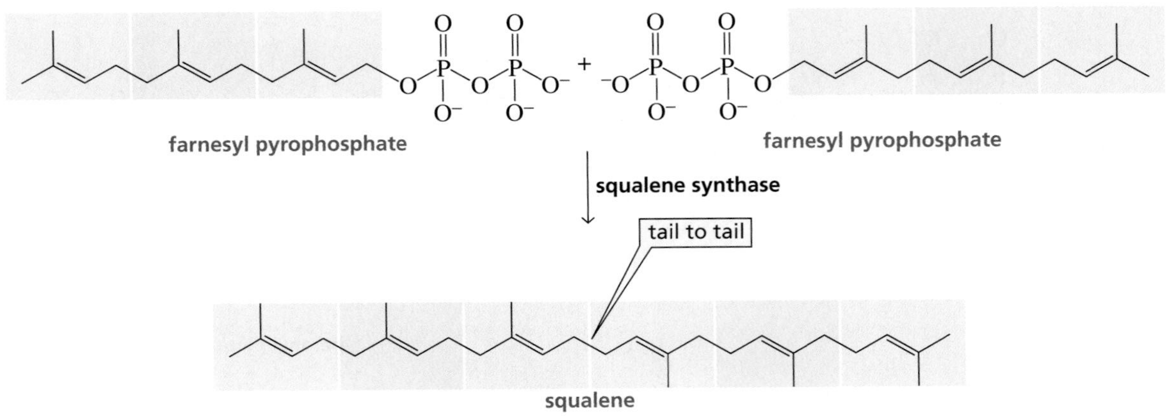

The following scheme shows how some of the many monoterpenes can be synthesized from geranyl pyrophosphate:

geranyl pyrophosphate $\xrightarrow{H_2O}$ geraniol (in rose and geranium oils) $\xrightarrow{\text{reduction}}$ citronellol (in rose and geranium oils) $\xrightarrow{\text{oxidation}}$ citronellal (in lemon oil)

geranyl pyrophosphate $\xrightarrow{H^+}$ carbocation $\xrightarrow{H_2O}$ α-terpineol (in juniper oil) $\xrightarrow[H_2O]{H^+}$ terpin hydrate (a common constituent of cough medicine)

carbocation $\longrightarrow$ limonene (in orange and lemon oils) + H^+ $\xrightarrow{\text{oxidation}}$ ketone $\xrightarrow{\text{reduction}}$ menthol (in peppermint oil)

19.7 STEROIDS ARE CHEMICAL MESSENGERS

Hormones are chemical messengers—organic compounds synthesized in glands and delivered by the bloodstream to target tissues in order to stimulate or inhibit some process. Many hormones are **steroids**. Because steroids are nonpolar compounds, they too are lipids. Their nonpolar character allows them to cross cell membranes, so they can leave the cells in which they are synthesized and enter their target cells.

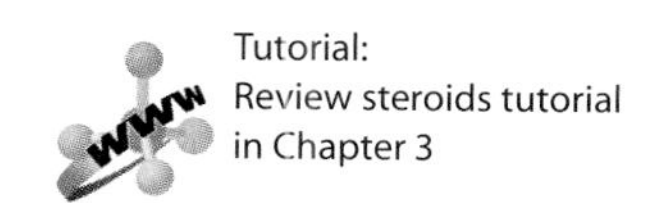

Tutorial:
Review steroids tutorial in Chapter 3

All steroids contain a tetracyclic ring system composed of three six-membered rings and one five-membered ring. In steroids, the B, C, and D rings are all trans fused. In most naturally occurring steroids, the A and B rings are also trans fused. We have seen that trans fused rings are more stable than cis fused rings (Section 3.13). The rigid conformations of the cyclohexane rings prevents them from undergoing ring-flip (Section 3.10).

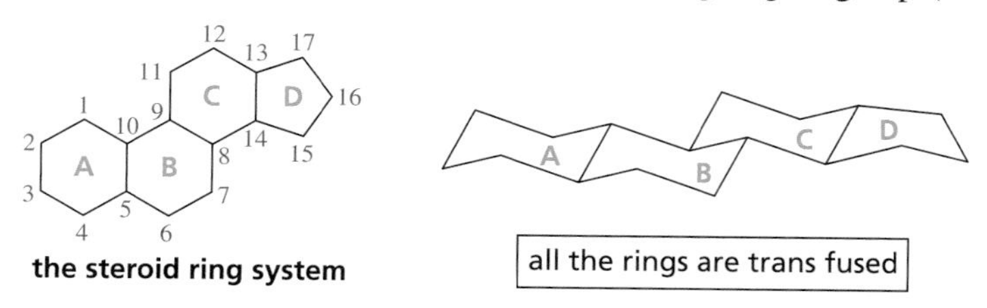

the steroid ring system

all the rings are trans fused

The most abundant member of the steroid family in animals is **cholesterol**, the precursor of all other steroids. Cholesterol is biosynthesized from squalene, a triterpene (Section 19.6), and is an important component of cell membranes (Figure 19.1); its ring structure causes cholesterol to be more rigid than other membrane lipids. Because

Cholesterol is synthesized from squalene.

cholesterol has eight asymmetric centers, 256 stereoisomers are possible, but only one exists in nature (Chapter 6, Problem 21). We have seen how cholesterol is related to heart disease and how high cholesterol is treated clinically (Section 3.13).

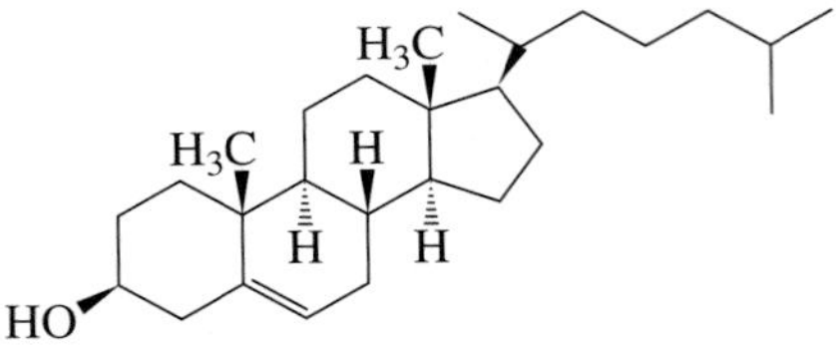

H_3C H_3C H H H HO

cholesterol

BIOGRAPHY

Two German chemists, **Heinrich Otto Wieland (1877–1957)** *and* **Adolf Windaus (1876–1959),** *each received a Nobel Prize in chemistry (Wieland in 1927 and Windaus in 1928) for work that led to the determination of the structure of cholesterol.* **Heinrich Wieland,** *the son of a chemist, was a professor at the University of Munich, where he showed that bile acids were steroids and determined their individual structures. During World War II, he remained in Germany but was openly anti-Nazi.* **Adolf Windaus** *originally intended to be a physician, but the experience of working with Emil Fischer for a year changed his mind. Windaus discovered that vitamin D was a steroid, and he was the first to recognize that vitamin B_1 contained sulfur.*

The steroid hormones can be divided into five classes: glucocorticoids, mineralocorticoids, androgens, estrogens, and progestins.

Glucocorticoids, as their name suggests, are involved in glucose metabolism; they also participate in the metabolism of proteins and fatty acids. Cortisone is an example of a glucocorticoid. Because of its anti-inflammatory effect, it is used clinically to treat arthritis and other inflammatory conditions.

mineralocorticoids

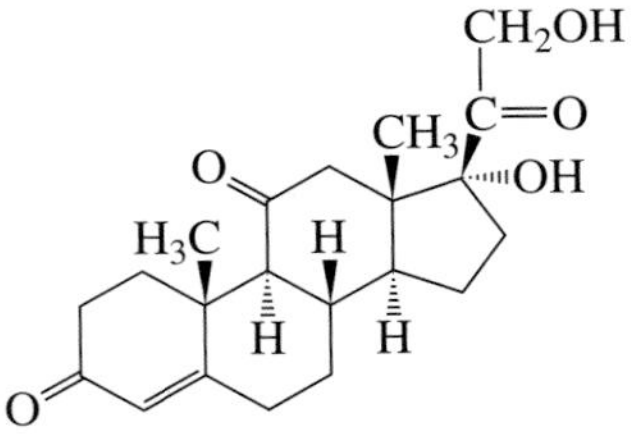

CH_2OH $C{=}O$ CH_3 OH O H_3C H H H O

cortisone

O CH_2OH HC $C{=}O$ HO H_3C H H H O

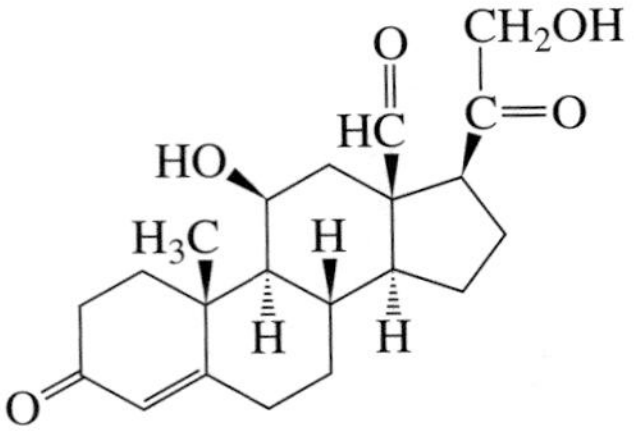

aldosterone

Mineralocorticoids cause increased reabsorption of Na^+, Cl^- and HCO_3^- by the kidneys, leading to an increase in blood pressure.

The sex hormones known as *androgens* are secreted primarily by the testes. They are responsible for the development of male secondary sex characteristics during puberty, including muscle growth.

androgens

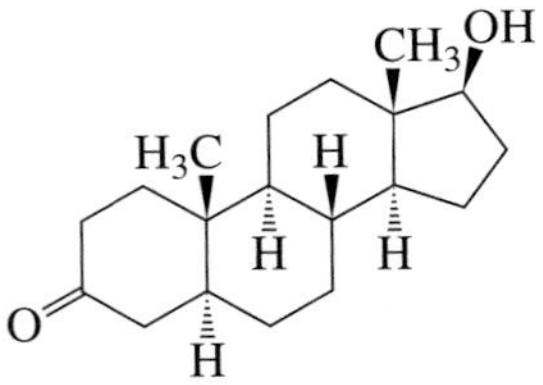

CH_3 OH H_3C H H H O H

5 α-dihydrotestosterone

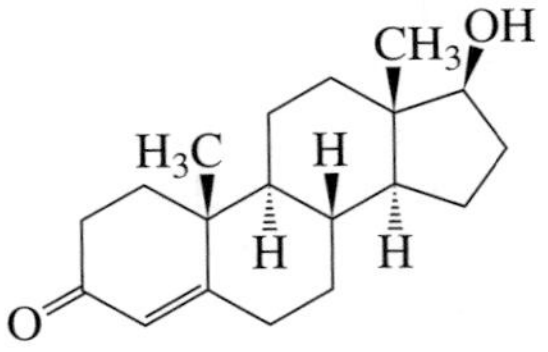

CH_3 OH H_3C H H H O

testosterone

BIOGRAPHY

Adolf Friedrich Johann Butenandt (1903–1995) *was born in Germany. He shared the 1939 Nobel Prize in chemistry with Leopold Ružička* *(see page 523)* *for isolating and determining the structures of estrone, androsterone, and progesterone. Forced by the Nazi government to refuse the prize, Butenandt accepted it after World War II. He was the director of the Kaiser Wilhelm Institute in Berlin and later was a professor at the Universities of Tübingen and Munich.*

The sex hormones known as *estrogens* are secreted primarily by the ovaries and are responsible for the development of female secondary sex characteristics. They also regulate the menstrual cycle. Progesterone prepares the lining of the uterus for implantation of an ovum and is essential for the maintenance of pregnancy. It also prevents ovulation during pregnancy.

estrogens

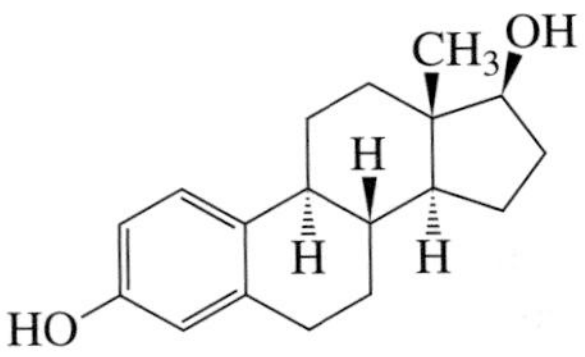

CH_3 OH H H H HO

estradiol

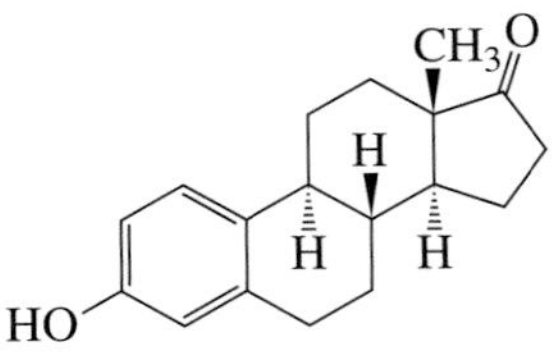

CH_3 O H H H HO

estrone

CH_3 CH_3 $C{=}O$ H_3C H H H O

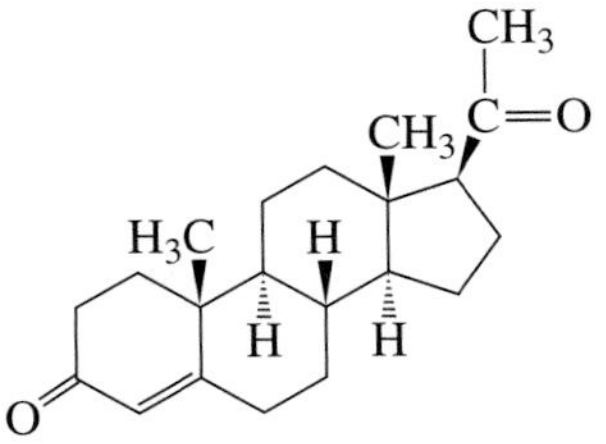

progesterone

Although the various steroid hormones have similar structures, they have remarkably different physiological effects. For example, the only difference between testosterone and progesterone is the substituent on the five-membered ring, and the only difference between 5α-dihydrotestosterone and estradiol is a methyl group and three hydrogens, but these compounds make the difference between being male and being female. These examples illustrate the extreme specificity of biochemical reactions.

In addition to being the precursor of all the steroid hormones, cholesterol is the precursor of the *bile acids*. In fact, the word *cholesterol* is derived from the Greek words *chole* meaning "bile" and *stereos* meaning "solid." The bile acids—cholic acid and chenodeoxycholic acid—are synthesized in the liver, stored in the gallbladder, and secreted into the small intestine, where they act as emulsifying agents so that fats and oils can be digested by water-soluble digestive enzymes. Cholesterol is also the precursor of vitamin D. A deficiency in vitamin D, which can be prevented by getting enough sun, causes a disease known as rickets.

bile acids

cholic acid

chenodeoxycholic acid

LIPOPROTEINS

Lipoproteins are composed of lipids (cholesterol, triglycerides, and phospholipids) and proteins. The density of a lipoprotein depends on the relative amounts of lipid and protein. Proteins are more dense than lipids, so the higher the protein component of a lipoprotein, the more dense it will be. In Section 3.13, we saw that low-density lipoprotein (LDL) is the so-called bad cholesterol, and high density lipoprotein (HDL) is the so-called good cholesterol. LDL is composed of about 75% lipid and 25% protein; HDL is composed of about 50% lipid and 50% protein.

PROBLEM 11 ◆

What functional groups are in aldosterone?

19.8 SYNTHETIC STEROIDS

The potent physiological effects of steroids led scientists, in their search for new drugs, to synthesize steroids that are not available in nature and to investigate their physiological effects. Two such drugs, stanozolol and Dianabol, have the same muscle-building effect as testosterone. Steroids that aid in the development of muscle are called *anabolic steroids*. These drugs are available by prescription and are used to treat people suffering from traumas accompanied by muscle deterioration. The same drugs have been administered to athletes and racehorses to increase their muscle mass. Stanozolol was the drug detected in several athletes in the 1988 Olympics. Anabolic steroids, when taken in relatively high dosages, have been found to cause liver tumors, personality disorders, and testicular atrophy.

stanozolol

Dianabol®

Many synthetic steroids have been found to be much more potent than natural steroids. Norethindrone, for example, is better than progesterone in arresting ovulation. Another synthetic steroid, mifepristone, terminates pregnancy within the first nine weeks of gestation. Mefipristone is also called RU 486; the name comes from Roussel-Uclaf, the French pharmaceutical company where it was first synthesized in 1980, and an arbitrary lab serial number. Notice that these compounds have structures similar to that of progesterone.

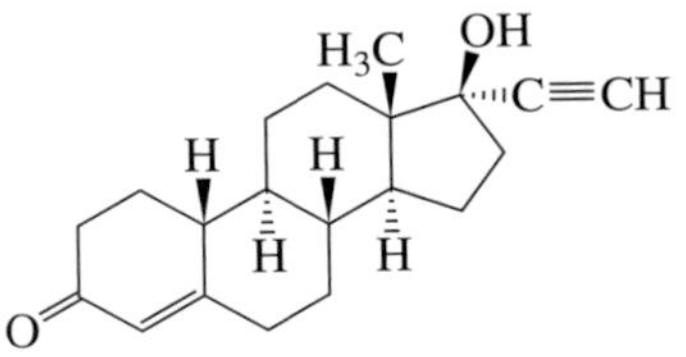

norethindrone

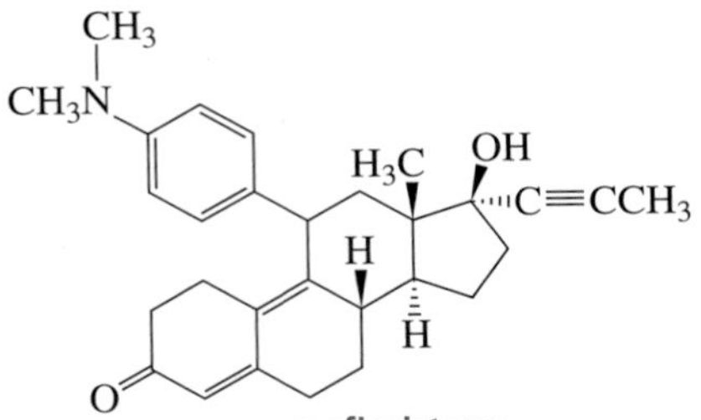

mefipristone
RU 486

PROBLEM 12

How do testosterone, a natural muscle-building steroid, and stanozolol, a synthetic muscle-building steroid, differ in structure?

SUMMARY

Lipids are organic compounds, found in living organisms, that are soluble in nonpolar solvents. One class of lipids, **fatty acids**, are carboxylic acids with long unbranched hydrocarbon chains. Double bonds in fatty acids have the cis configuration. Fatty acids with more than one double bond are called **polyunsaturated fatty acids**. Double bonds in naturally occurring unsaturated fatty acids are separated by one CH_2 group. **Waxes** are esters formed from long-chain carboxylic acids and long-chain alcohols.

Triglycerides are compounds in which the three OH groups of glycerol are esterified with fatty acids. Triglycerides that are solids or semisolids at room temperature are called **fats**; liquid triglycerides are called **oils.** Some or all of the double bonds of polyunsaturated oils can be reduced by catalytic hydrogenation. **Phosphoglycerides** differ from triglycerides in that the terminal **OH** group of glycerol is esterified with phosphoric acid instead of a fatty acid. Phosphoglycerides form membranes by arranging themselves in a **lipid bilayer**. **Phospholipids** are lipids that contain a phosphate group.

Terpenes contain carbon atoms in multiples of five. They are made by joining together five-carbon units, usually in a head-to-tail fashion; this is called the **isoprene rule**. **Monoterpenes**—terpenes with two isoprene units—have 10 carbons; **sesquiterpenes** have 15. **Squalene**, a **triterpene** (a terpene with 30 carbons), is a precursor of steroid molecules. Lycopene and β-carotene are **tetraterpenes** (40 carbons) called **carotenoids**.

The five-carbon compound used for the synthesis of terpenes is isopentenyl pyrophosphate. The reaction of **dimethylallyl pyrophosphate** (formed from isopentenyl pyrophosphate) with **isopentenyl pyrophosphate** forms geranyl pyrophosphate, a 10-carbon compound, which can react with another molecule of isopentenyl pyrophosphate to form farnesyl pyrophosphate, a 15-carbon compound. Two molecules of farnesyl pyrophosphate form **squalene**, a 30-carbon compound. Squalene is the precursor of **cholesterol**.

Hormones are chemical messengers. Many hormones are **steroids**. All steroids contain a tetracyclic ring system. The most abundant member of the steroid family in animals is **cholesterol**, the precursor to all other steroids.

PROBLEMS

13. Give the product that would be obtained from the reaction of cholesterol with each of the following reagents:

a. H_2, Pd/C **b.** acetyl chloride **c.** H_2SO_4, Δ **d.** H_2O, H^+ **e.** a peroxyacid

14. Do all triglycerides have the same number of asymmetric centers?

15. Cardiolipins are found in heart muscles. Give the products formed when a cardiolipin undergoes complete acid-catalyzed hydrolysis.

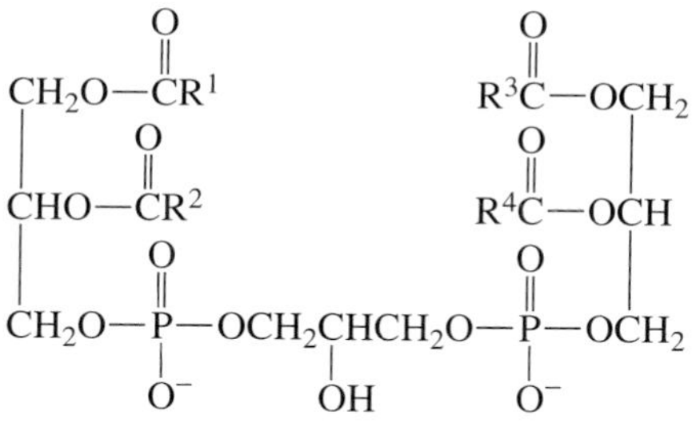

a cardiolipin

16. Nutmeg contains a simple, fully saturated triglyceride with a molecular weight of 722. Draw its structure.

17. The asymmetric center of phospholipids has the *R* configuration. Which of the following is a lecithin?

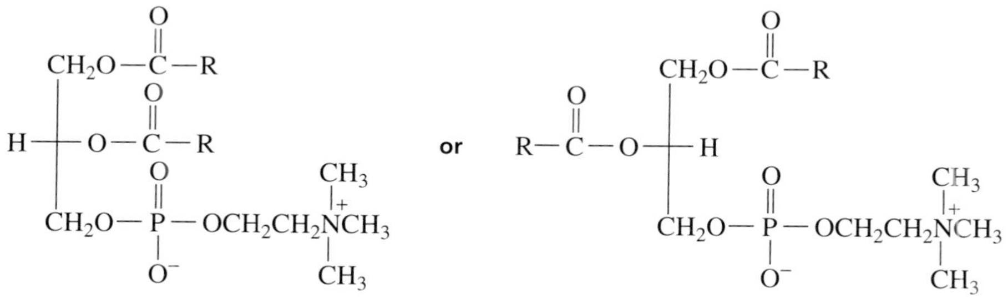

or

18. Mark off the isoprene units in geraniol (page 523).

19. One of the linkages in squalene is tail-to-tail, not head-to-tail. What does this suggest about how squalene is synthesized in nature? (*Hint:* Locate the position of the tail-to-tail linkage.)

20. Mark off the isoprene units in lycopene and β-carotene. Can you detect a similarity in the way in which squalene, lycopene, and β-carotene are biosynthesized?

21. 5-Androstene-3,17-dione is isomerized to 4-androstene-3,17-dione by hydroxide ion. Propose a mechanism for this reaction.

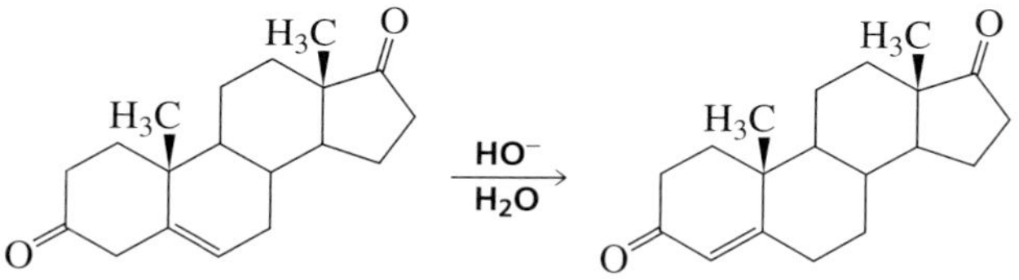

5-androstene-3,17-dione $\xrightarrow[H_2O]{HO^-}$ 4-androstene-3,17-dione

22. a. How many different triglycerides are there in which one of the fatty acid components is lauric acid and two are myristic acid?

b. How many different triglycerides are there in which one of the fatty acid components is lauric acid, one is myristic acid, and one is palmitic acid?

23. Sphingolipids, also found in membranes, contain sphingosine instead of glycerol. One class of sphingolipids, sphingomyelins, are a major constituent of the coating around nerve fibers. In a sphingomyelin, the amino group of sphingosine is bonded to the acyl group of a fatty acid and the primary alcohol group forms an ester with phosphoric acid, which also forms an ester with choline. Draw the structure of a sphingomyelin.

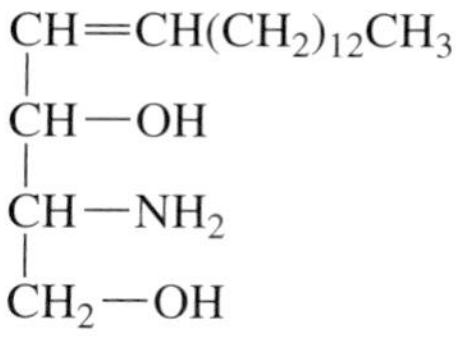

$CH{=}CH(CH_2)_{12}CH_3$
$CH{-}OH$
$CH{-}NH_2$
$CH_2{-}OH$

sphingosine

24. Diethylstilbestrol (DES) was given to pregnant women to prevent miscarriage, until it was found that the drug caused cancer in both the mothers and their female children. DES has estradiol activity even though it is not a steroid. Draw DES in a way that shows that it is structurally similar to estradiol.

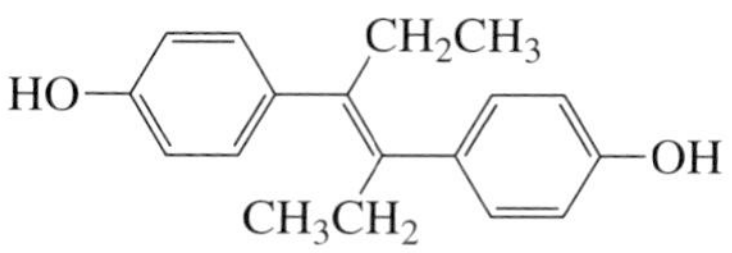

diethylstilbestrol
DES

CHAPTER 20

The Chemistry of the Nucleic Acids

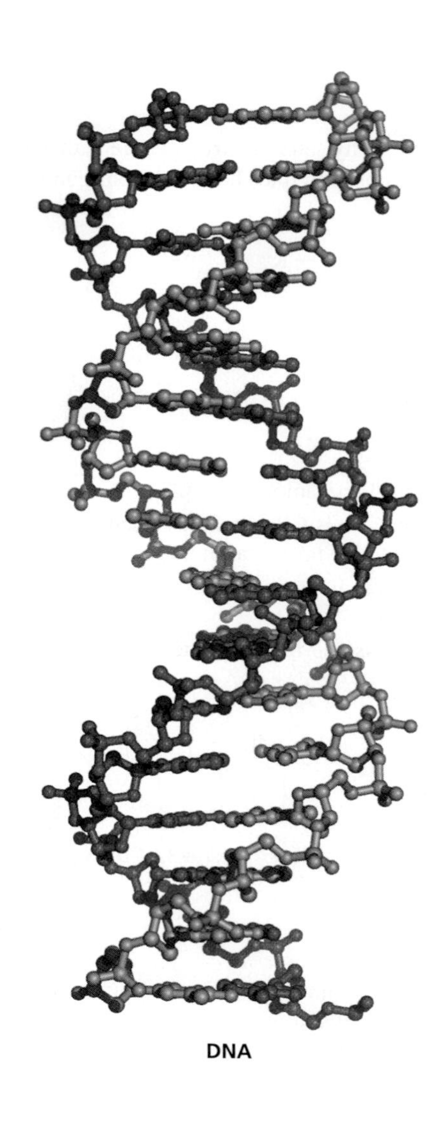

DNA

There are two types of nucleic acids: **deoxyribonucleic acid (DNA)** and **ribonucleic acid (RNA)**. DNA encodes an organism's entire hereditary information and controls the growth and division of cells. The genetic information stored in DNA is transcribed into RNA. This information can then be translated for the synthesis of all the proteins needed for cellular structure and function.

DNA was first isolated in 1869 from the nuclei of white blood cells. Because it was found in the nucleus and was acidic, it was called *nucleic acid*. Eventually, scientists found that the nuclei of all cells contain DNA, but it wasn't until 1944 that they realized that nucleic acids are the carriers of genetic information. In 1953, James Watson and Francis Crick described the three-dimensional structure of DNA—the famed double helix.

20.1 NUCLEOSIDES AND NUCLEOTIDES

Nucleic acids are chains of five-membered-ring sugars linked by forming a **phosphodiester** with phosphoric acid (Figure 20.1). The anomeric carbon of each sugar is bonded to a nitrogen of a heterocyclic compound in a β-glycosidic linkage (Section 15.11). Because the heterocyclic compounds are amines, they are commonly referred to as **bases**. In RNA, the five-membered-ring sugar is D-ribose. In DNA, it is 2′-deoxy-D-ribose (D-ribose without an OH group in the 2′-position).

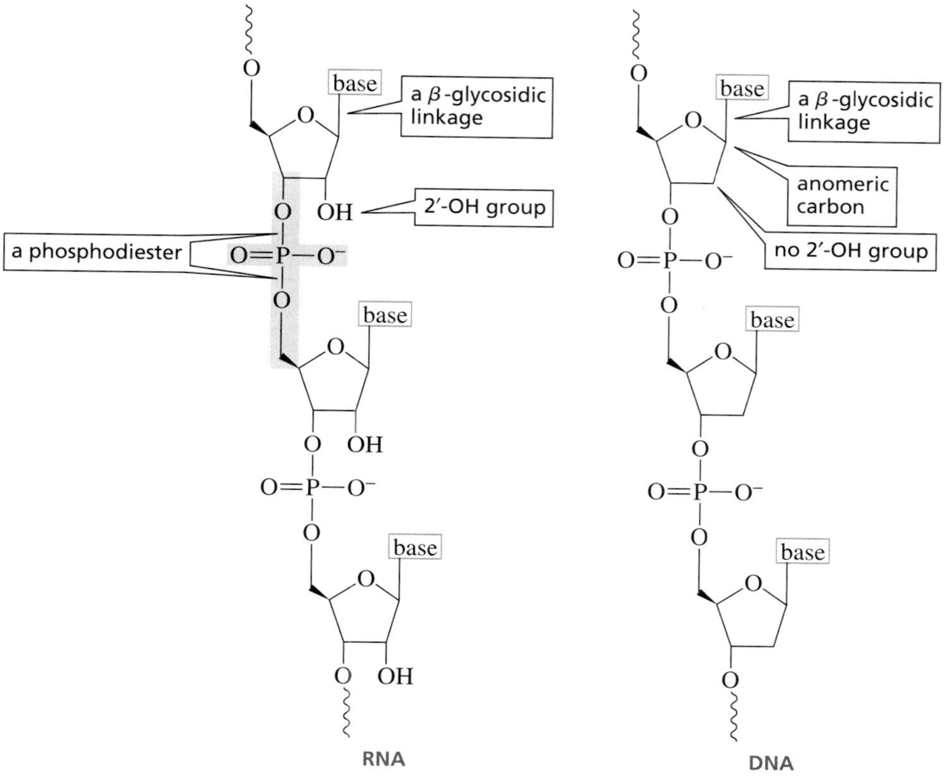

◀ **Figure 20.1**
Nucleic acids consist of a chain of five-membered-ring sugars linked by phosphate groups. Each sugar (D-ribose in RNA, 2′-deoxy-D-ribose in DNA) is bonded to a heterocyclic amine (a base) in a β-glycosidic linkage.

The vast differences in heredity between different species and between different members of the same species are determined by the sequence of the bases in DNA. Surprisingly, there are only four bases in DNA: two are substituted purines (adenine and guanine), and two are substituted pyrimidines (cytosine and thymine).

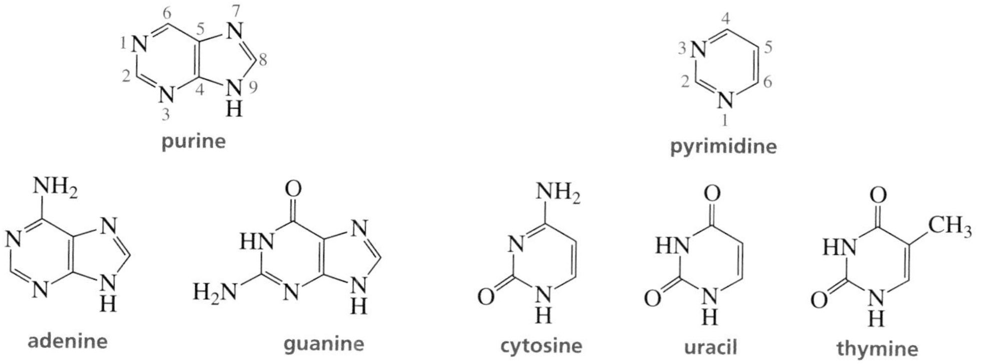

THE STRUCTURE OF DNA: WATSON, CRICK, FRANKLIN, AND WILKINS

James D. Watson was born in Chicago in 1928. He graduated from the University of Chicago at the age of 19 and received a Ph.D. three years later from Indiana University. In 1951, as a postdoctoral fellow at Cambridge University, Watson worked on determining the three-dimensional structure of DNA.

Francis H. C. Crick (1916–2004) was born in Northampton, England. Originally trained as a physicist, Crick did research on radar during World War II. After the war, deciding that the most interesting problem in science was the physical basis of life, he entered Cambridge University to study the structure of biological molecules by X-ray analysis. He was a graduate student when he carried out his portion of the work that led to the proposal of the double helical structure of DNA. He received a Ph.D. in chemistry in 1953.

Rosalind Franklin (1920–1958) was born in London. She graduated from Cambridge University and in 1942 accepted a position as a research officer in the British Coal Utilisation Research Association. After the war, she studied X-ray diffraction techniques in Paris. In 1951 she returned to England and accepted a position to develop an X-ray diffraction unit in the biophysics department at King's College. Her X-ray studies showed that DNA was a helix with phosphate groups on the outside of the molecule. Franklin died without knowing the role her work had played in determining the structure of DNA and without being recognized for her contributions.

Watson and Crick received the 1962 Nobel Prize in medicine or physiology with Maurice Wilkins for determining the double helical structure of DNA. Wilkins (1916–2004), who contributed X-ray studies that confirmed the double helical structure, was born in New Zealand and moved to England six years later with his parents. During World War II he joined other British scientists who were working with American scientists on the development of the atomic bomb.

James Watson (left) and Francis Crick (right)

Rosalind Franklin

RNA also contains only four bases. Three (adenine, guanine, and cytosine) are the same as those in DNA, but the fourth base in RNA is uracil instead of thymine. Notice that thymine and uracil differ only by a methyl group. (Thymine is 5-methyluracil.) The reason DNA contains thymine instead of uracil is explained in Section 20.9.

nucleoside = base + sugar

A compound containing a base bonded to D-ribose or to 2′-deoxy-D-ribose is called a **nucleoside**. The ring positions of the sugar component of a nucleoside are indicated by primed numbers to distinguish them from the ring positions of the base. This is why the sugar component of DNA is referred to as 2′-deoxy-D-ribose.

nucleosides

adenosine

guanosine

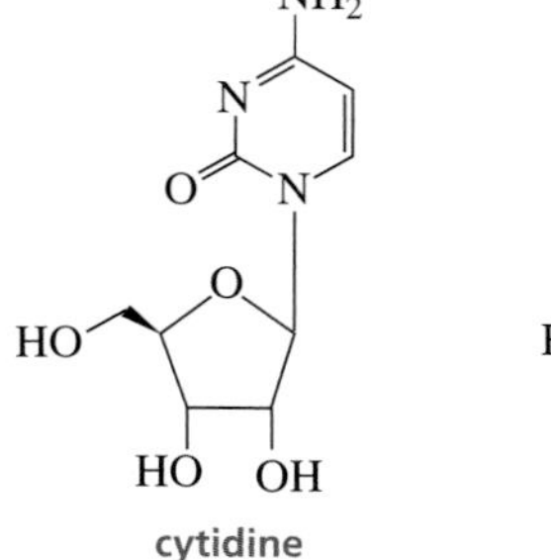

cytidine

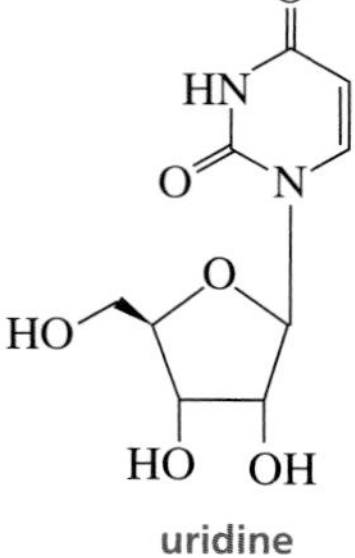

uridine

2′-deoxyadenosine 2′-deoxyguanosine 2′-deoxycytidine thymidine

A **nucleotide** is a nucleoside with either the 5′- or the 3′-OH group bonded in an ester linkage to phosphoric acid. The nucleotides of RNA—where the sugar is D-ribose—are more precisely called **ribonucleotides**, whereas the nucleotides of DNA—where the sugar is 2′-deoxy-D-ribose are called **deoxyribonucleotides**.

nucleotide = base + sugar + phosphate

nucleotides

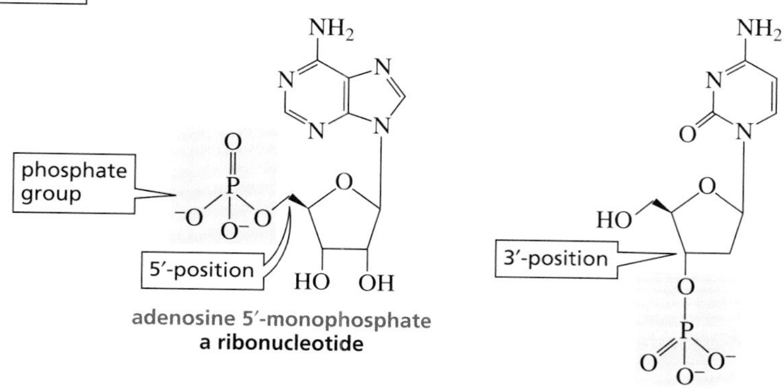

adenosine 5′-monophosphate
a ribonucleotide

2′-deoxycytidine 3′-monophosphate
a deoxyribonucleotide

3-D Molecules: Bases; Nucleosides; Nucleotides

Nucleotides can exist as monophosphates, diphosphates, and triphosphates. They are named by adding *monophosphate* or *diphosphate* or *triphosphate* to the name of the nucleoside.

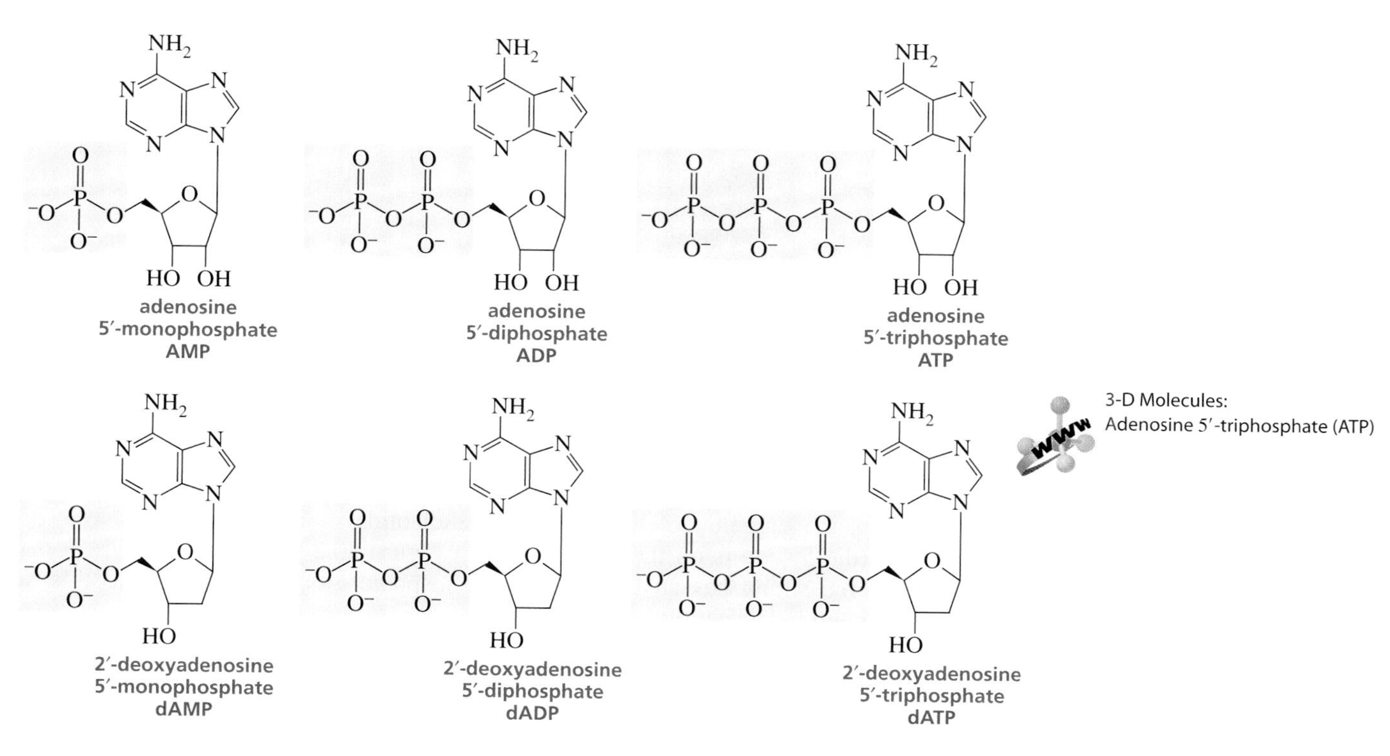

adenosine 5′-monophosphate AMP

adenosine 5′-diphosphate ADP

adenosine 5′-triphosphate ATP

2′-deoxyadenosine 5′-monophosphate dAMP

2′-deoxyadenosine 5′-diphosphate dADP

2′-deoxyadenosine 5′-triphosphate dATP

3-D Molecules: Adenosine 5′-triphosphate (ATP)

PROBLEM 1

Draw the structure for each of the following:

a. dCDP **c.** dUMP **e.** guanosine 5′-triphosphate

b. dTTP **d.** UDP **f.** adenosine 3′-monophosphate

20.2 NUCLEIC ACIDS ARE COMPOSED OF NUCLEOTIDE SUBUNITS

We have seen that **nucleic acids** are composed of long strands of nucleotide subunits linked by phosphodiester bonds (Figure 20.1). A **dinucleotide** contains two nucleotide subunits, an **oligonucleotide** contains three to ten subunits, and a **polynucleotide** contains many subunits. DNA and RNA are polynucleotides.

DNA is synthesized in the 5′ ⟶ 3′ direction.

Nucleotide triphosphates are the starting materials for the biosynthesis of nucleic acids. The nucleotides are linked as a result of nucleophilic attack by a 3′-OH group of one nucleotide triphosphate on the α-phosphorus (the phosphorus closest to ribose) of another nucleotide triphosphate, breaking a phosphoanhydride bond and eliminating pyrophosphate (Figure 20.2). This means that the growing polymer is synthesized in the 5′ ⟶ 3′ direction; in other words, new nucleotides are added to the 3′-end.

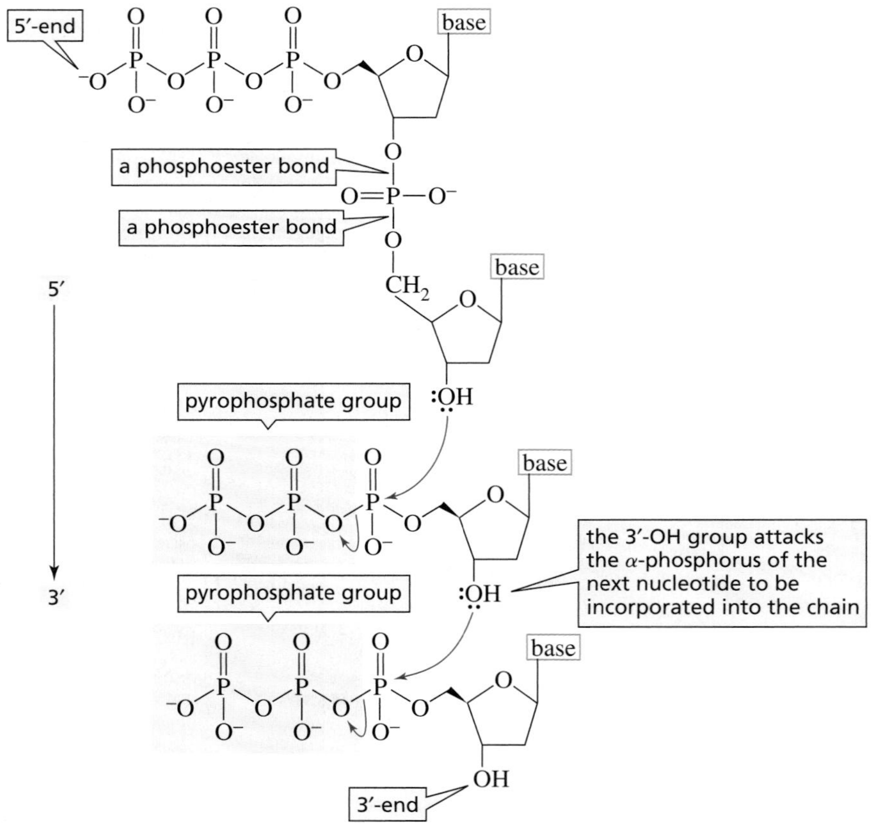

▶ **Figure 20.2**
Addition of nucleotides to a growing strand of DNA. Biosynthesis occurs in the 5′ ⟶ 3′ direction.

The **primary structure** of a nucleic acid is the sequence of bases in the strand. By convention, the sequence of bases in a polynucleotide is written in the 5′ ⟶ 3′ direction (the 5′-end is on the left). Notice in Figure 20.2 that the nucleotide at the 5′-end of the strand has an unlinked 5′-triphosphate group, and the nucleotide at the 3′-end will have an unlinked 3′-hydroxyl group.

ATGAGCCATGTAGCCTAATCGGC

5′-end 3′-end

DNA consists of two strands of nucleic acids, with the sugar–phosphate backbone on the outside and the bases on the inside. The strands are antiparallel (they run in opposite directions) and are held together by hydrogen bonds between the bases on one strand and the bases on the other strand (Figure 20.3). Adenine (A) always pairs with thymine (T), and guanine (G) always pairs with cytosine (C). This means the two strands are *complementary*: where there is an A in one strand, there is a T in the opposing strand, and where there is a G in one strand, there is a C in the other strand. Thus, if you know the sequence of bases in one strand, you can figure out the sequence of bases in the other strand.

Why does A pair with T? Why does G pair with C? First of all, the width of the double-stranded molecule is relatively constant, so a purine must pair with a pyrimidine. If the larger purines paired, the strand would bulge; if the smaller pyrimidines paired, the strands would have to pull in to bring the two pyrimidines close enough to form hydrogen bonds. But what causes A to pair with T rather than with C (the other pyrimidine)? The base pairing is dictated by hydrogen bonding. Adenine forms two hydrogen bonds with thymine but would form only one hydrogen bond with cytosine. Guanine forms three hydrogen bonds with cytosine but would form only one hydrogen bond with thymine (Figure 20.4).

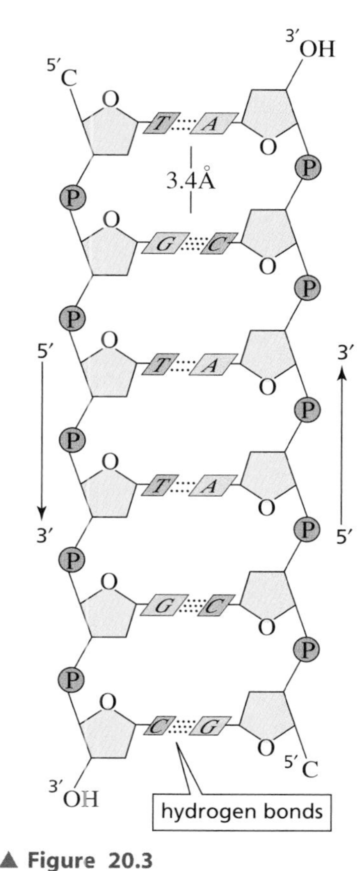

▲ **Figure 20.3**
The sugar–phosphate backbone of DNA is on the outside, and the bases are on the inside. A (a purine) always pairs with a T (a pyrimidine); G (a purine) always pairs with a C (a pyrimidine). The two strands are antiparallel—they run in opposite directions.

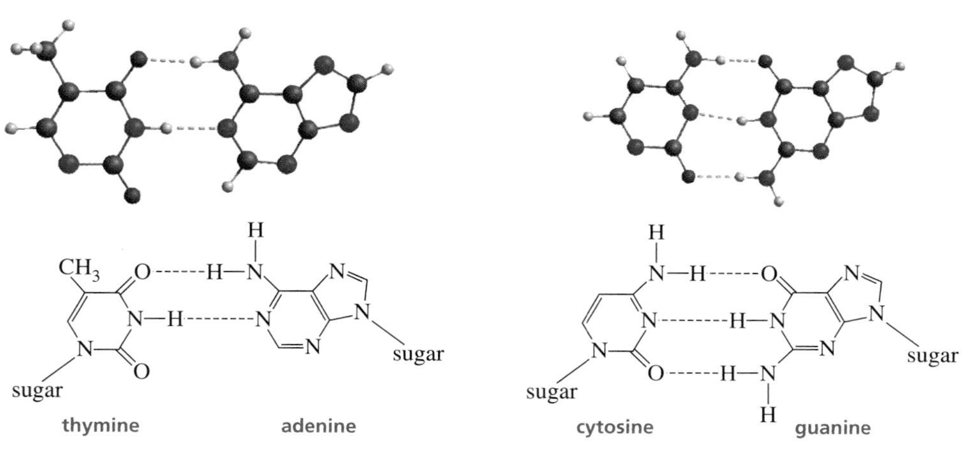

▲ **Figure 20.4**
Base pairing in DNA: adenine and thymine form two hydrogen bonds; cytosine and guanine form three hydrogen bonds.

The two antiparallel DNA strands are not linear but are twisted into a helix around a common axis (see Figure 20.5a). The base pairs are planar and parallel to each other on the inside of the helix (Figure 20.5c). The secondary structure is therefore known as a **double helix**. The double helix resembles a circular staircase: the base pairs are the rungs, and the sugar–phosphate backbones are the handrails.

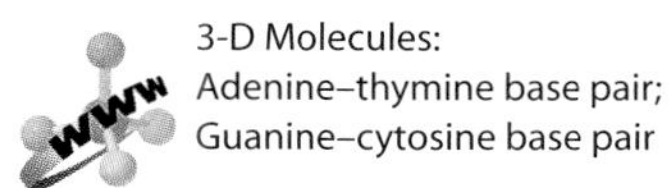

3-D Molecules: Adenine–thymine base pair; Guanine–cytosine base pair

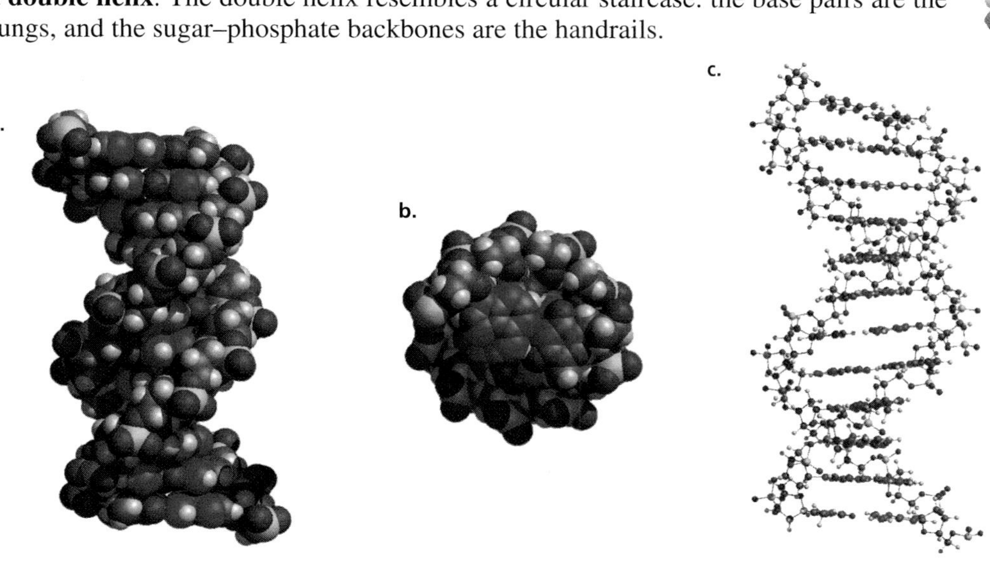

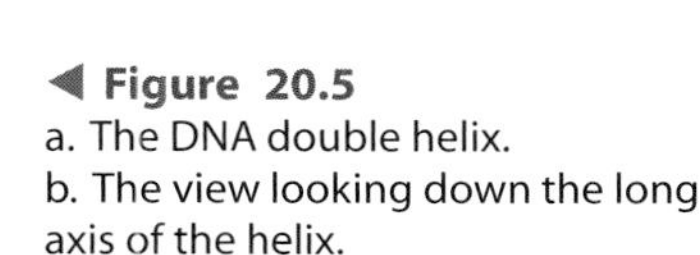

3-D Molecule: DNA double helix

◀ **Figure 20.5**
a. The DNA double helix.
b. The view looking down the long axis of the helix.
c. The bases are planar and parallel on the inside of the helix.

PROBLEM 2 ♦

If one of the strands of DNA has the following sequence of bases running in the 5′ ⟶ 3′ direction:

5′—G—G—A—C—A—A—T—C—T—G—C—3′

a. What is the sequence of bases in the complementary strand?
b. What base is closest to the 5′-end in the complementary strand?

20.3 WHY DNA DOES NOT HAVE A 2′-OH GROUP

DNA is stable, but RNA is not stable because the 2′-OH group of ribose acts as a nucleophile and causes RNA to be easily cleaved (Figure 20.6). This explains why the 2′-OH group is absent in DNA. DNA must remain intact throughout the life span of a cell to preserve the genetic information. Easy cleavage of DNA would have disastrous consequences for the cell and for life itself. RNA, in contrast, is synthesized as it is needed and is degraded once it has served its purpose.

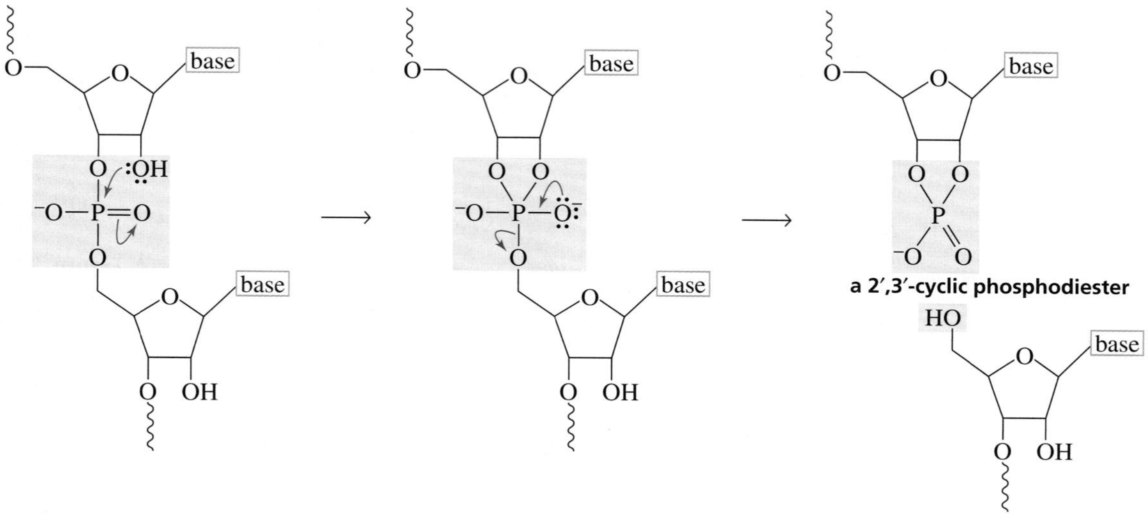

▲ **Figure 20.6**
Cleavage of RNA by the 2′-OH group. RNA is thought to undergo cleavage 3 billion times faster than DNA.

20.4 THE BIOSYNTHESIS OF DNA IS CALLED REPLICATION

The genetic information of a human cell is contained in 23 pairs of chromosomes. Each chromosome is composed of several thousand **genes** (segments of DNA). The total DNA of a human cell—the **human genome**—contains 3.1 billion base pairs.

Part of the excitement created by Watson and Crick's proposed structure for DNA was due to the fact that the structure immediately suggested how DNA is able to pass on genetic information to succeeding generations. Because the two strands are complementary, both carry the same genetic information. Both strands serve as templates for the synthesis of complementary new strands (Figure 20.7). The new (daughter) DNA molecules are identical to the original (parent) molecule—they contain all the original genetic information. The synthesis of identical copies of DNA is called **replication**.

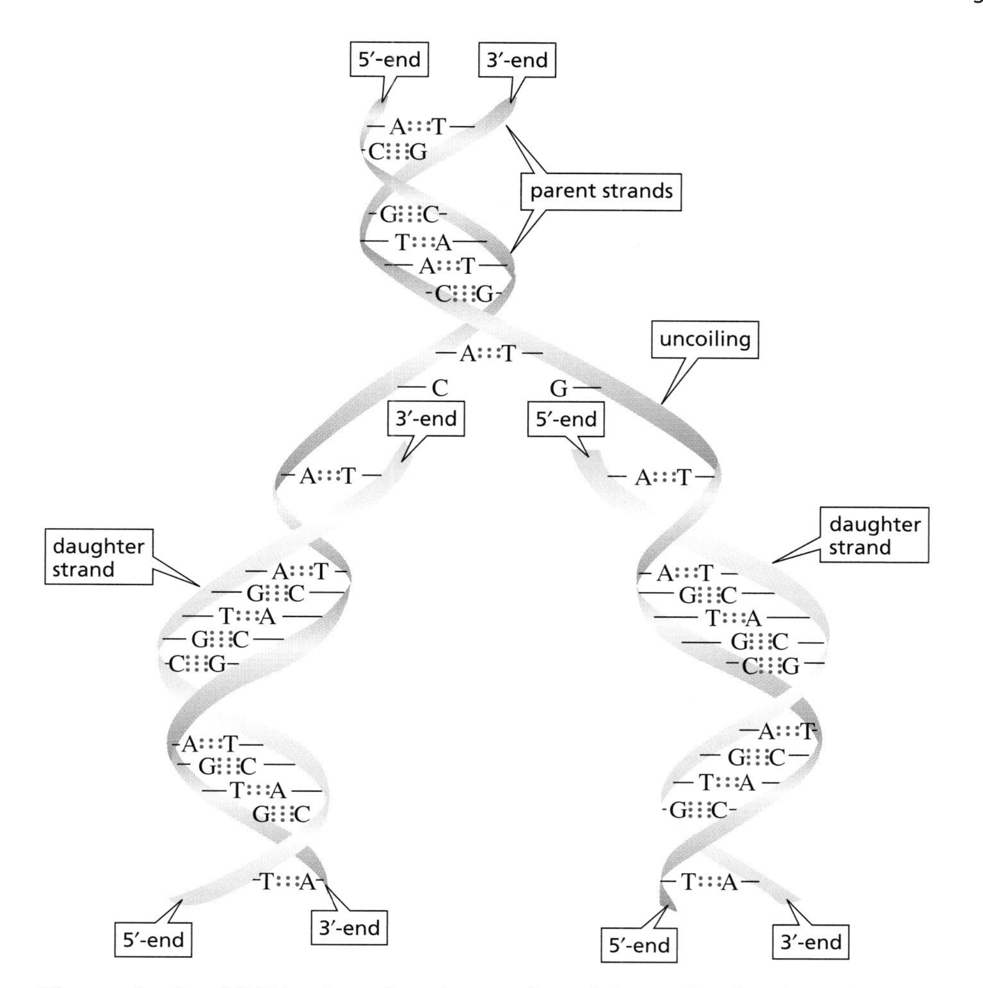

◀ **Figure 20.7**
Replication of DNA. The daughter strand on the left is synthesized continuously in the 5′ ⟶ 3′ direction; the daughter strand on the right is synthesized discontinuously in the 5′ ⟶ 3′ direction.

The synthesis of DNA takes place in a region of the molecule where the strands have started to separate. Because a nucleic acid can be synthesized only in the 5′ ⟶ 3′ direction (Figure 20.2), only the strand on the left in Figure 20.7 is synthesized continuously in a single piece (because it is synthesized in the 5′ ⟶ 3′ direction). The other strand needs to grow in a 3′ ⟶ 5′ direction, so it is synthesized discontinuously in small pieces. Each piece is synthesized in the 5′ ⟶ 3′ direction, and the fragments are joined together by an enzyme called DNA ligase. Each of the two resulting new molecules of DNA—called daughter molecules—contain one of the original parent strands (blue strand in Figure 20.7) plus a newly synthesized strand (green strand). This process is called **semiconservative replication**.

BIOGRAPHY

Erwin Chargaff (1905–2002) *carried out the work showing that the number of adenines in DNA equals the number of thymines and the number of guanines equals the number of cytosines. Chargaff was born in Austria and received a Ph.D. from the University of Vienna. To escape Hitler, he came to the United States in 1935, becoming a professor at Columbia University College of Physicians and Surgeons.*

PROBLEM 3

Using a dark line for the original parental DNA and a wavy line for DNA synthesized from parental DNA, show what the population of DNA molecules would look like in the fourth generation of replicated molecules.

20.5 DNA AND HEREDITY

If DNA contains hereditary information, there must be a method to decode that information. The decoding occurs in two steps.

1. The sequence of the bases in DNA provides a blueprint for the sequence of bases in RNA; the synthesis of RNA from a DNA blueprint is called **transcription**.
2. The sequence of bases in RNA determines the sequence of amino acids in a protein; the synthesis of proteins from an RNA blueprint is called **translation**.

Transcription: DNA ⟶ RNA

Translation: RNA ⟶ protein

Don't confuse transcription and translation: these words are used just as they are used in English. Transcription (DNA to RNA) is copying *within the same language*—in this case the language is nucleotides. Translation (RNA to protein) is *changing to another language*—the language of amino acids. First we will look at transcription.

20.6 THE BIOSYNTHESIS OF RNA IS CALLED TRANSCRIPTION

Transcription starts when DNA unwinds at a particular site to give two single strands. One of the strands is called the **sense strand**. The complementary strand is called the **template strand**. The template strand is read in the 3′ ⟶ 5′ direction, so that RNA can be synthesized in the 5′ ⟶ 3′ direction (Figure 20.8). The bases in the template strand specify the bases that need to be incorporated into RNA, following the same base-pairing principle that allows replication of DNA. For example, each guanine in the template strand specifies the incorporation of a cytosine into RNA, and each adenine in the template strand specifies the incorporation of a uracil into RNA. (Recall that, in RNA, uracil is used instead of thymine.) Because both RNA and the sense strand of DNA are complementary to the template strand, RNA and the sense strand of DNA have the same base sequence, except that RNA has a uracil wherever the sense strand has a thymine.

RNA is synthesized in the 5′ ⟶ 3′ direction.

▶ **Figure 20.8**
Transcription: using DNA as a blueprint for the synthesis of RNA.

sense strand
DNA
5′ ATCGGACCTAGAGGCC GA TCTCGATACACGCA TG CTAGA 3′
3′ TAGCCTGGATCTCCGG CT AGAGCTATGTGCGT AC GATCT 5′
template strand
UCUCGAUACACG OH 3′
5′ pppA CCUAGAGGCC GA
RNA
direction of transcription

BIOGRAPHY

Elizabeth Keller (1918–1997) *was the first to recognize that tRNA had a cloverleaf structure. She received a B.S. from the University of Chicago in 1940 and a Ph.D. from Cornell University Medical College in 1948. She worked at the Huntington Memorial Laboratory of Massachusetts General Hospital and at the United States Public Health Service. Later, she became a professor at MIT and then at Cornell University.*

Photo Source: Division of Rare and Manuscript Collections, Cornell University Library

PROBLEM 4

Why do both thymine and uracil specify the incorporation of adenine?

20.7 THERE ARE THREE KINDS OF RNA

RNA molecules are much shorter than DNA molecules and are generally single-stranded. There are three kinds of RNA:

- **messenger RNA (mRNA)**, whose sequence of bases determines the sequence of amino acids in a protein
- **ribosomal RNA (rRNA)**, a structural component of ribosomes, which are the particles on which the biosynthesis of proteins takes place
- **transfer RNA (tRNA)**, the carrier of amino acids for protein synthesis

tRNA molecules are much smaller than mRNA or rRNA molecules; a tRNA contains only 70 to 90 nucleotides. The single strand of tRNA is folded into a characteristic cloverleaf structure (Figure 20.9a). All tRNAs have a CCA sequence at the 3′-end. The three bases at the bottom of the loop directly opposite the 5′- and 3′-ends are called an **anticodon** (Figures 20.9a and 20.9b).

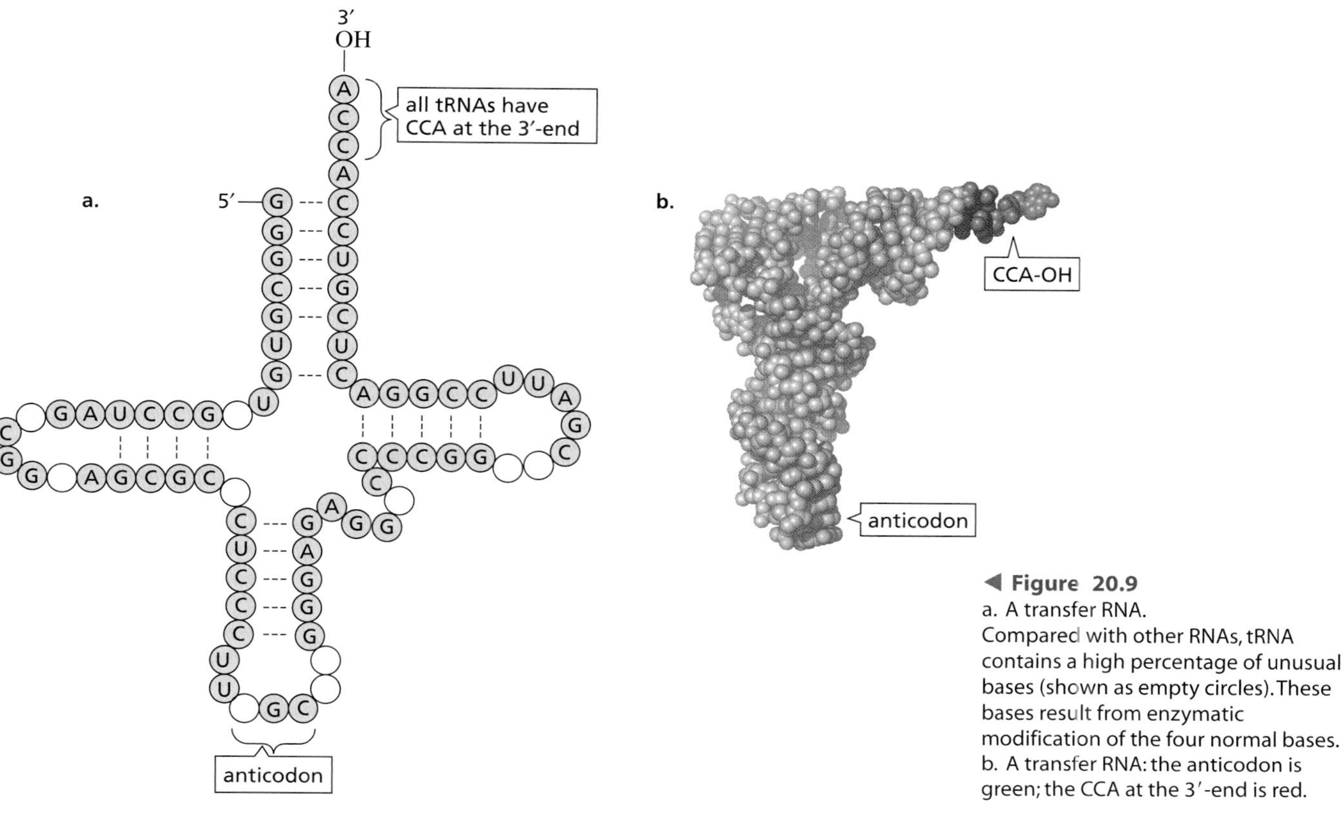

◀ Figure 20.9
a. A transfer RNA.
Compared with other RNAs, tRNA contains a high percentage of unusual bases (shown as empty circles). These bases result from enzymatic modification of the four normal bases.
b. A transfer RNA: the anticodon is green; the CCA at the 3′-end is red.

Each tRNA can carry an amino acid bound as an ester to its terminal 3′-OH group. The amino acid will be inserted into a protein during protein biosynthesis. Each tRNA can carry only one particular amino acid. A tRNA that carries alanine is designated as $tRNA^{Ala}$.

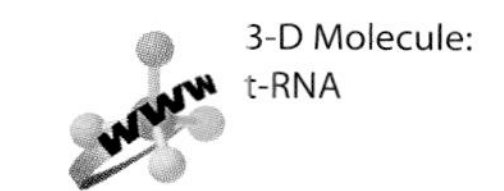

3-D Molecule: t-RNA

The attachment of an amino acid to a tRNA is catalyzed by an enzyme called aminoacyl-tRNA synthetase. The mechanism for the reaction is shown below.

the mechanism for aminoacyl-tRNA synthetase

an amino acid + ATP ⟶ an acyl adenylate + pyrophosphate

pyrophosphate —H_2O→ 2 $^-O-PO(O^-)-OH$

an acyl adenylate + HÖ—ACC—tRNA (3′-OH group of A; 5′; tRNA) ⟶ a tetrahedral intermediate

a tetrahedral intermediate ⟶ an aminoacyl-tRNA + AMP

BIOGRAPHY

The genetic code was worked out independently by **Marshall Nirenberg** *and* **Har Gobind Khorana**, *for which they shared the 1968 Nobel Prize in physiology or medicine.* **Robert W. Holley**, *who worked on the structure of tRNA molecules, also shared that year's prize.*

- The carboxylate group of the amino acid is activated by attacking the α-phosphorus of ATP; this puts a good leaving group on the amino acid.
- A nucleophilic acyl substitution reaction occurs in which the 3′-OH group of tRNA attacks the carbonyl carbon of the acyl adenylate, forming a tetrahedral intermediate (Section 11.5).
- The aminoacyl-tRNA is formed when AMP is expelled from the tetrahedral intermediate.

20.8 THE BIOSYNTHESIS OF PROTEINS IS CALLED TRANSLATION

BIOGRAPHY

Marshall Nirenberg *was born in New York in 1927. He received a bachelor's degree from the University of Florida and a Ph.D. from the University of Michigan. He is a scientist at the National Institutes of Health.*

A protein is synthesized from its N-terminal end to its C-terminal end by a process that reads the bases along the mRNA strand in the 5′ ⟶ 3′ direction. Each amino acid that is to be incorporated into a protein is specified by one or more three-base sequences called **codons**. The bases are read consecutively and are never skipped. The three-base sequences and the amino acids each sequence codes for are known as the **genetic code** (Table 20.1). A codon is written with the 5′-nucleotide on the left. For example, the codon UCA on mRNA codes for the amino acid serine, whereas CAG codes for glutamine. **Stop codons** tell the cell to "stop protein synthesis here."

Table 20.1 The Genetic Code

5′-Position	Middle position				3′-Position
	U	C	A	G	
U	Phe	Ser	Tyr	Cys	U
	Phe	Ser	Tyr	Cys	C
	Leu	Ser	Stop	Stop	A
	Leu	Ser	Stop	Trp	G
C	Leu	Pro	His	Arg	U
	Leu	Pro	His	Arg	C
	Leu	Pro	Gln	Arg	A
	Leu	Pro	Gln	Arg	G
A	Ile	Thr	Asn	Ser	U
	Ile	Thr	Asn	Ser	C
	Ile	Thr	Lys	Arg	A
	Met	Thr	Lys	Arg	G
G	Val	Ala	Asp	Gly	U
	Val	Ala	Asp	Gly	C
	Val	Ala	Glu	Gly	A
	Val	Ala	Glu	Gly	G

BIOGRAPHY

Har Gobind Khorana *was born in India in 1922. He received a bachelor's and a master's degree from Punjab University and a Ph.D. from the University of Liverpool. In 1960, he joined the faculty at the University of Wisconsin and later became a professor at MIT.*

How the information in mRNA is translated into a polypeptide is shown in Figure 20.10. In this figure, serine was the last amino acid incorporated into the growing polypeptide chain. Serine was specified by the AGC codon because the anticodon of the tRNA that carries serine is GCU (3′-UCG-5′). (Remember that a base sequence is read in the 5′ ⟶ 3′ direction, so the sequence of bases in an anticodon must be read from right to left.)

a protein is synthesized in the N ⟶ C direction.

Tutorial: Translation

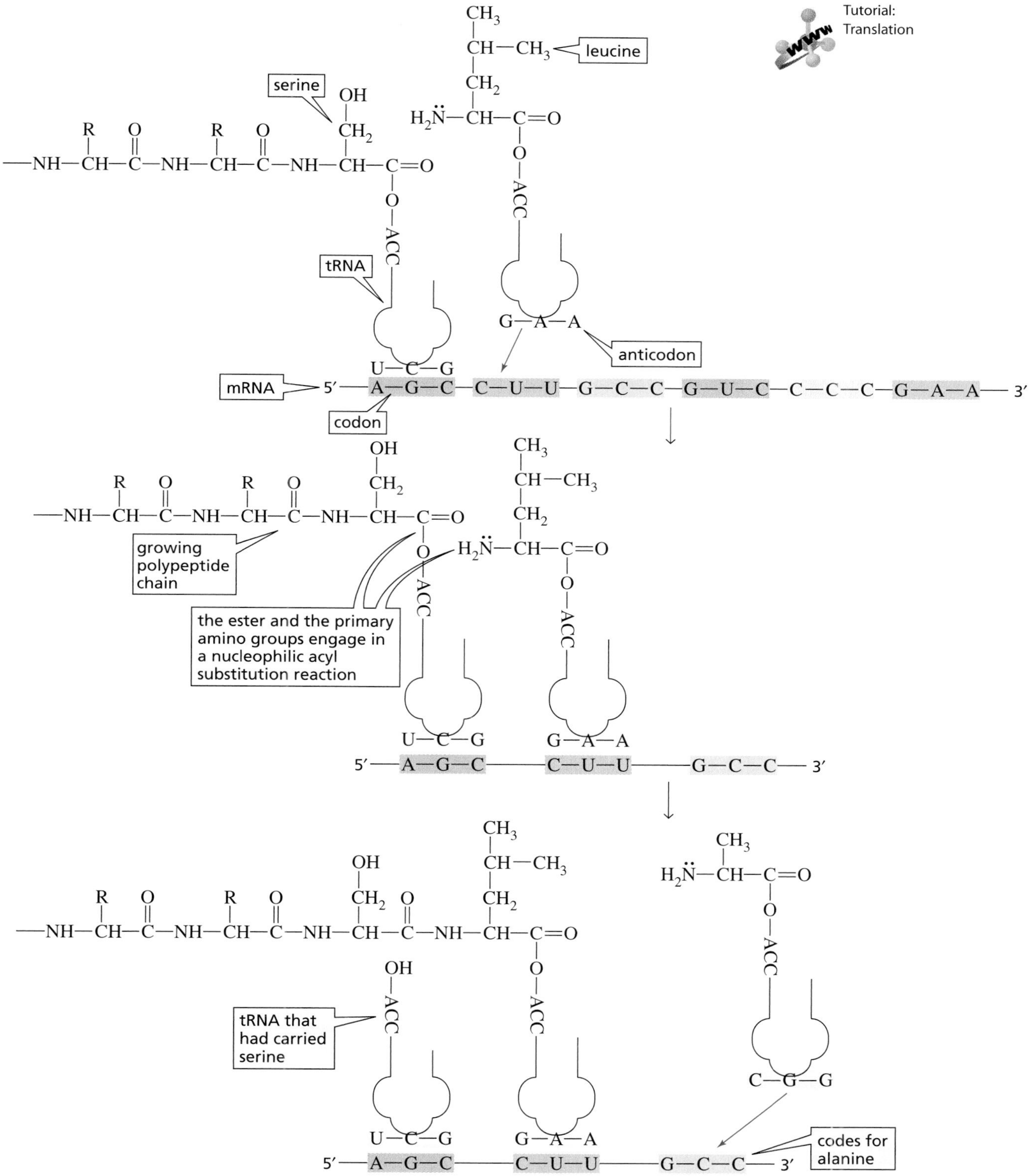

leucine
serine
tRNA
anticodon
mRNA
codon
growing polypeptide chain
the ester and the primary amino groups engage in a nucleophilic acyl substitution reaction
tRNA that had carried serine
codes for alanine

▲ **Figure 20.10**
Translation. The sequence of bases in mRNA determines the sequence of amino acids in a protein.

BIOGRAPHY

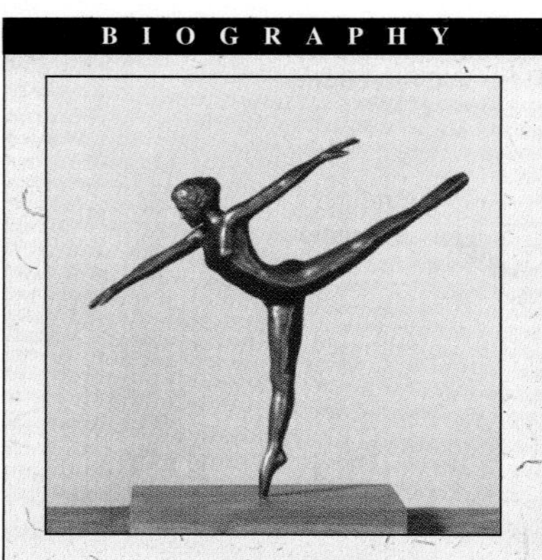

A sculpture done by Robert Holley

Robert W. Holley (1922–1993) *was born in Illinois and received a bachelor's degree from the University of Illinois and a Ph.D. from Cornell University. During World War II, he worked on the synthesis of penicillin at Cornell Medical School. He was a professor at Cornell and later at the University of California, San Diego. He was also a noted sculptor.*

- The next codon, CUU, signals for a tRNA with an anticodon of AAG (3′-GAA- 5′). That particular tRNA carries leucine. The amino group of leucine reacts in an enzyme-catalyzed nucleophilic acyl substitution reaction (Section 11.8) with the ester on the adjacent tRNA, displacing the tRNA.
- The next codon (GCC) brings in a tRNA carrying alanine. The amino group of alanine displaces the tRNA that brought in leucine.

Subsequent amino acids are brought in one at a time in the same way, with the codon in mRNA specifying the amino acid to be incorporated by complementary base pairing with the anticodon of the tRNA that carries that amino acid.

PROBLEM 5 ◆

If methionine is the first amino acid incorporated into an oligopeptide, what oligopeptide is encoded for by the following stretch of mRNA?

5′—G—C—A—U—G—G—A—C—C—C—C—G—U—U—A—U—
U—A—A—A—C—A—C—3′

PROBLEM 6 ◆

Four C's occur in a row in the segment of mRNA in Problem 5. What oligopeptide would be formed from the mRNA if one of the four C's were cut out of the strand?

PROBLEM 7

UAA is a stop codon. Why does the UAA sequence in mRNA in Problem 5 not cause protein synthesis to stop?

PROBLEM 8 ◆

Write the sequences of bases in the sense strand of DNA that resulted in the mRNA in Problem 5.

SICKLE CELL ANEMIA

Sickle cell anemia is an example of the damage that can be caused by a change in a single base of DNA. It is a hereditary disease caused when a GAG triplet becomes a GTG triplet in the sense strand of a section of DNA that codes for the protein component of hemoglobin. As a consequence, the mRNA codon becomes GUG—which signals for incorporation of valine—rather than GAG, which would have signaled for incorporation of glutamate. The change from a polar glutamate to a nonpolar valine is sufficient to change the shape of the deoxyhemoglobin molecule, which induces aggregation and subsequent precipitation of deoxyhemoglobin in red blood cells. This stiffens the cells, making it difficult for them to squeeze through a capillary. Blocked capillaries cause severe pain and can be fatal.

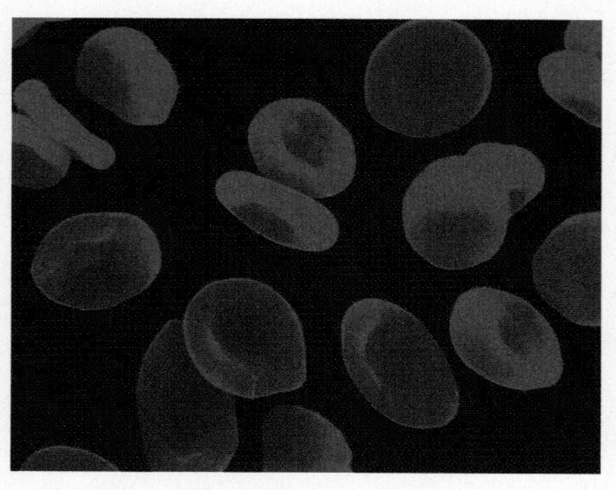

Normal red blood cells

Sickle red blood cells

ANTIBIOTICS THAT ACT BY INHIBITING TRANSLATION

Puromycin is a naturally occurring antibiotic, one of several that acts by inhibiting translation. It does so by mimicking the 3′-CCA-aminoacyl portion of a tRNA, fooling the enzyme into transferring the growing peptide chain to the amino group of puromycin rather than to the amino group of the incoming 3′-CCA-aminoacyl tRNA. As a result, protein synthesis stops. Because puromycin blocks protein synthesis in eukaryotes as well as in prokaryotes, it is poisonous to humans and therefore is not a clinically useful antibiotic. To be clinically useful, an antibiotic must affect protein synthesis only in prokaryotic cells.

puromycin

Clinically useful antibiotics	Mode of action
Tetracycline	Prevents the aminoacyl-tRNA from binding to the ribosome
Erythromycin	Prevents the incorporation of new amino acids into the protein
Streptomycin	Inhibits the initiation of protein synthesis
Chloramphenicol	Prevents the new peptide bond from being formed

20.9 WHY DNA CONTAINS THYMINE INSTEAD OF URACIL

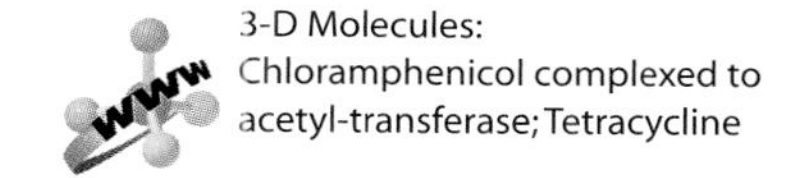

WWW
3-D Molecules:
Chloramphenicol complexed to acetyl-transferase; Tetracycline

Thymines are synthesized from uracils in a process that requires considerable energy (Section 17.11). Therefore, there must be a good reason for DNA to contain thymine instead of uracil.

The presence of thymine instead of uracil in DNA prevents potentially lethal mutations. Cytosine can tautomerize (Section 13.3) to form an imine, and hydrolysis of the imine forms uracil (Section 12.7). The overall reaction is called a **deamination** because it removes an amino group.

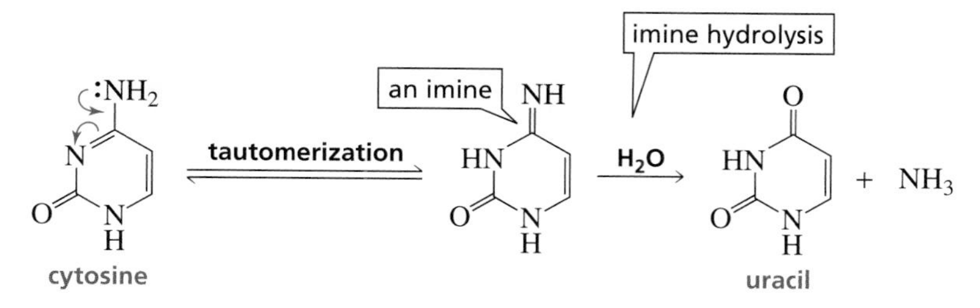

If a C in DNA is deaminated to a U, the U will specify the incorporation of an A into the daughter strand during replication instead of the G that would have been specified by C. Fortunately, there is an enzyme that recognizes a U in DNA as a "mistake" before an incorrect base can be inserted into the daughter strand. The enzyme cuts out the U and replaces it with a C. If U's were normally found in DNA, the enzyme could not distinguish between a normal U and a U formed by deamination of a C. Having T's in place of U's in DNA allows the U's that are found in DNA to be recognized as mistakes.

Unlike DNA, which replicates itself, any mistake in RNA does not survive for long because RNA is constantly being degraded and then resynthesized from the DNA template. Therefore, it is not worth spending the extra energy to incorporate T's into RNA.

BIOGRAPHY

Frederick Sanger *(see Section 15.8) and* **Walter Gilbert** *shared half of the 1980 Nobel Prize in chemistry for their work on DNA sequencing. The other half went to* **Paul Berg**, *who had developed a method of cutting nucleic acids at specific sites and recombining the fragments in new ways, a technique known as recombinant DNA technology.*

BIOGRAPHY

Walter Gilbert *was born in Boston in 1932. He received a master's degree in physics from Harvard and a Ph.D. in mathematics from Cambridge University. In 1958 he joined the faculty at Harvard, where he became interested in molecular biology.*

BIOGRAPHY

Paul Berg *was born in New York in 1926. He received a Ph.D. from Western Reserve University (now Case Western Reserve University). He joined the faculty at Washington University in St. Louis in 1955 and became a professor of biochemistry at Stanford in 1959.*

20.10 HOW THE BASE SEQUENCE OF DNA IS DETERMINED

If the sequence of one million base pairs could be determined each day, it would take more than 10 years to complete the sequence of the human genome. Clearly, DNA molecules are too large to sequence as a unit. Therefore, DNA is first cleaved at specific base sequences, and the resulting DNA fragments are then sequenced individually. The enzymes that cleave DNA at specific base sequences are called **restriction endonucleases**, and the DNA fragments they produce are called **restriction fragments**. Several hundred restriction endonucleases are now known; a few examples, the base sequence each recognizes and the point of cleavage in that base sequence, are shown here.

restriction enzyme	recognition sequence
*Alu*I	AG|CT TC|GA
*Fnu*DI	GG|CC CC|GG
*Pst*I	CTGCA|G G|ACGTC

The base sequences that most restriction endonucleases recognize are *palindromes*. A palindrome is a word or a group of words that reads the same forward and backward. "Toot" and "race car" are examples of palindromes.[1] A restriction endonuclease recognizes a piece of DNA in which *the template strand is a palindrome of the sense strand*. In other words, the sequence of bases in the template strand (reading from right to left) is identical to the sequence of bases in the sense strand (reading from left to right).

PROBLEM 9 ♦

Which of the following base sequences would most likely be recognized by a restriction endonuclease?

a. ACGCGT
b. ACGGGT
c. ACGGCA
d. ACACGT
e. ACATCGT
f. CCAACC

The restriction fragments can be sequenced using a procedure known as the **dideoxy method**. This method involves generating fragments whose length depends on the last base added to the fragment.

In the dideoxy method, a small piece of DNA called a primer, labeled at the 5′-end with ^{32}P, is added to the restriction fragment whose sequence is to be determined. Next, the four 2′-deoxyribonucleoside triphosphates are added, and then DNA polymerase—the enzyme that adds nucleotides to a strand of DNA—is added as well. A small amount of the 2′,3′-dideoxynucleoside triphosphate of one of the bases is also added to the reaction mixture. A 2′,3′-dideoxynucleoside triphosphate has no OH group at either the 2′- or 3′-position.

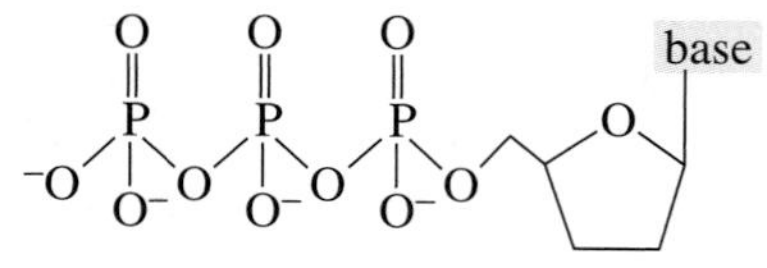

a 2′,3′-dideoxynucleoside triphosphate

[1]Some other palindromes are "Mom," "Dad," "Bob," "Lil," "radar," "noon," "wow," "poor Dan in a droop", "a man, a plan, a canal, Panama," "Sex at noon taxes," and "He lived as a devil, eh?"

Nucleotides will be added to the primer by base pairing with the restriction fragment. (In the procedure described here, 2′,3′-ddATP was added to the mixture.) Synthesis will stop if 2′,3′-ddATP is added to the growing chain instead of dATP, because the 2′,3′-dideoxy analog does not have a 3′-OH to which additional nucleotides can be added. Therefore, three different chain-terminated fragments will be obtained from the DNA restriction fragment shown here.

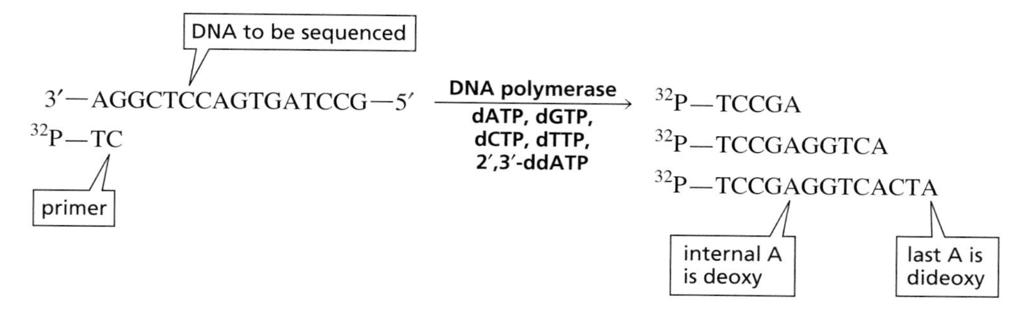

The procedure is repeated three more times using 2′,3′-ddGTP, next 2′,3′-ddCTP, and then 2′,3′-ddTTP.

PROBLEM 10

What labeled fragments would be obtained from the segment of DNA shown above if 2′,3′-ddGTP had been added to the reaction mixture instead of 2′,3′-ddATP?

The chain-terminated fragments obtained from each of the four experiments are loaded onto separate lanes of a gel: the fragments obtained using 2′,3′-ddATP are loaded onto one lane, the fragments obtained using 2′,3′-ddGTP onto another lane, and so on (Figure 20.11a). The smaller fragments fit through the spaces in the gel relatively easily and therefore travel through the gel faster, while the larger fragments pass through the gel more slowly.

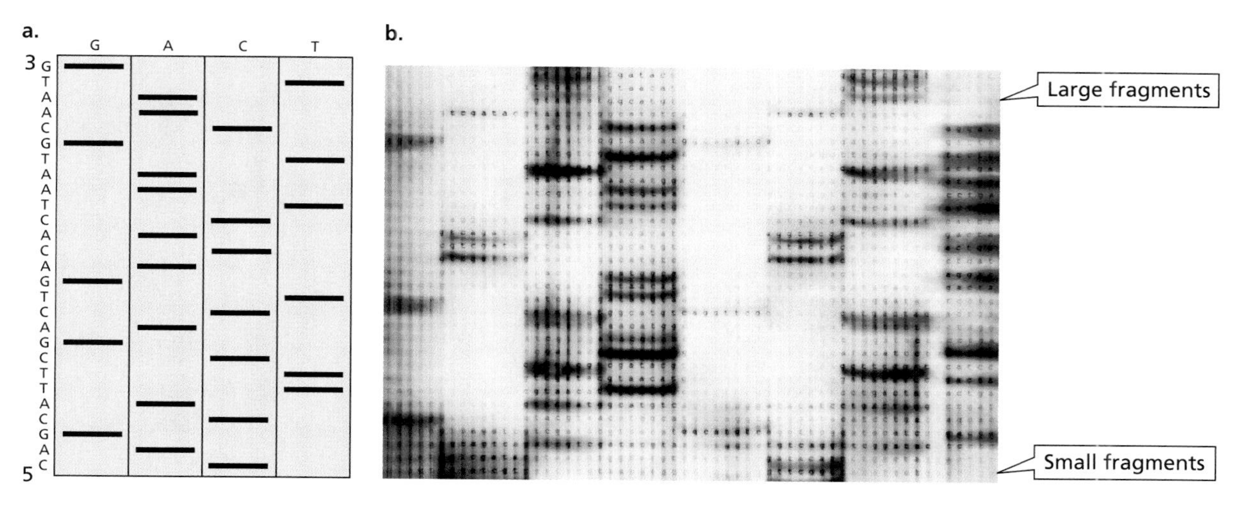

▲ **Figure 20.11**
a. A schematic drawing of an autoradiograph
b. An actual autoradiograph

After the fragments have been separated, the gel is placed in contact with a photographic plate. Radiation from ^{32}P causes a dark spot to appear on the plate opposite the location of each labeled fragment in the gel. This technique is called autoradiography, and the exposed photographic plate is known as an **autoradiograph** (Figure 20.11b).

The sequence of bases in the original restriction fragment can be read directly from the autoradiograph. The identity of each base is determined by noting the column where each successive dark spot (larger piece of labeled fragment) appears, starting at the bottom of the gel to identify the base added to the primer. The sequence of the fragment of DNA responsible for the autoradiograph in Figure 20.11a is shown on the left-hand side of the figure.

Once the sequence of bases in a restriction fragment is determined, the results can be checked by using the same process to obtain the base sequence of the fragment's complementary strand. The base sequence in the original piece of DNA can be determined by repeating the entire procedure with a different restriction endonuclease and noting overlapping fragments.

20.11 POLYMERASE CHAIN REACTION (PCR)

The PCR (polymerase chain reaction) technique, developed in 1983, allows scientists to amplify DNA (that is, to make many copies) in a very short time. With PCR, sufficient DNA for analysis can be obtained from a single hair or sperm.

PCR is done by adding the following to a solution containing the segment of DNA to be amplified (the target DNA):

- a large excess of primers (short pieces of DNA) that base pair with (anneal with) the short nucleotide sequences at the two ends of the piece of DNA to be amplified
- the four deoxyribonucleotide triphosphates (dATP, dGTP, dCTP, dTTP)
- a heat-stable DNA polymerase

The following three steps are then carried out (Figure 20.12):

BIOGRAPHY

Kary B. Mullis *received the Nobel Prize in chemistry in 1993 for his invention of PCR. He was born in North Carolina in 1944 and received a B.S. from the Georgia Institute of Technology in 1966 and a Ph.D. from the University of California, Berkeley. After holding several postdoctoral positions, he joined the Cetus Corporation in Emeryville, California. He conceived the idea of PCR while driving from Berkeley to Mendocino. Currently, he consults and lectures about biotechnology.*

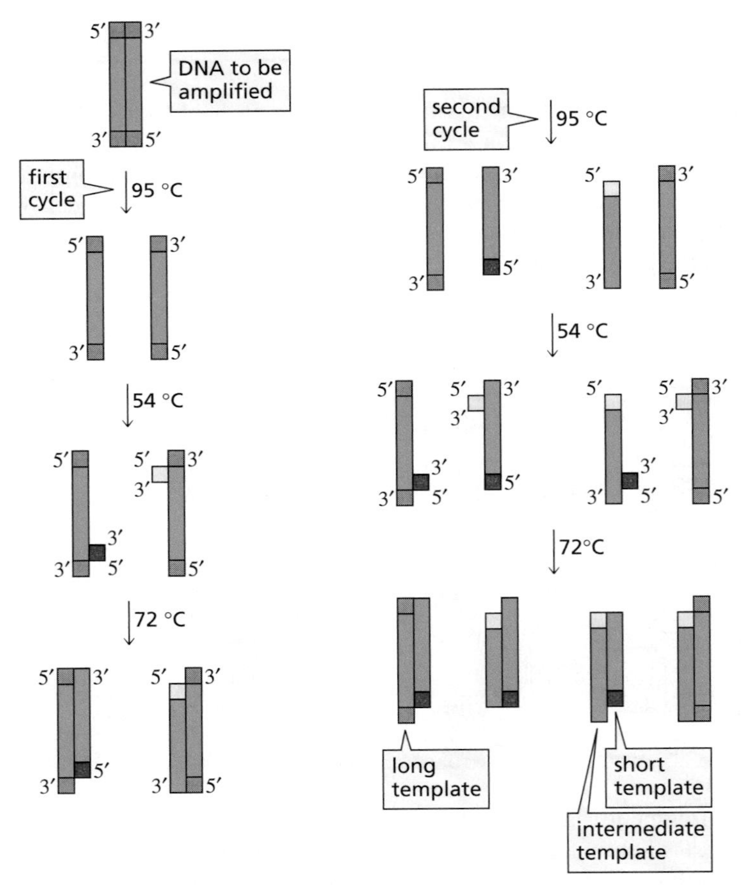

Figure 20.12
Two cycles of the polymerase chain reaction. The first cycle creates long and intermediate templates; the second cycle creates intermediate and short templates. (The short templates are copies of the segment of DNA selected to be amplified.) As the number of cycles increases, the number of short templates increases. The final product consists predominantly of short templates.

- *Strand Separation:* The solution is heated to 95 °C, causing double-stranded DNA to separate into two single strands.
- *Base Pairing of the Primers:* The solution is cooled to 54 °C, allowing the primers (red and yellow boxes in Figure 20.12) to pair with the bases at the 3′-end of the target (green) DNA.

- *DNA Synthesis:* The solution is heated to 72 °C, a temperature at which DNA polymerase catalyzes the addition of nucleotides to the primer. Notice that because the primers attach to the 3′-end of the target DNA, copies of the target DNA are synthesized in the required 5′ ⟶ 3′ direction.

The solution is then heated to 95 °C to begin a second cycle; the second cycle produces four copies of double-stranded DNA. The third cycle, therefore, starts with eight single strands of DNA and produces sixteen single strands. It takes about an hour to complete 30 cycles at which point DNA has been amplified a billionfold.

PCR has a wide range of clinical uses. It can be used to detect mutations that lead to cancer, to diagnose genetic diseases, to reveal the presence of HIV that would be missed by an antibody assay, to monitor cancer chemotherapy, and to rapidly identify an infectious disease.

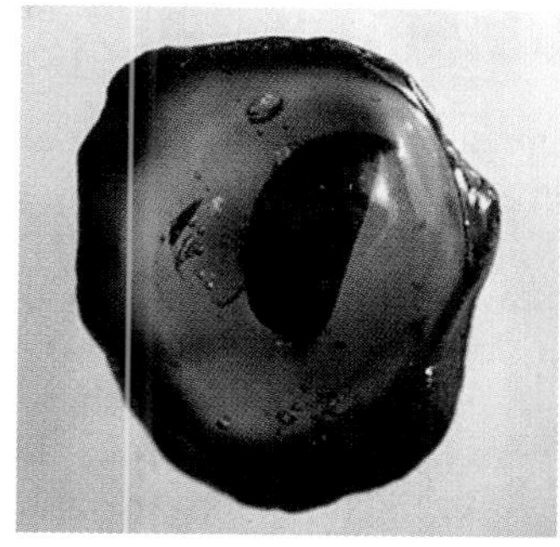

The DNA from a 40-million-year-old leaf preserved in amber was amplified by PCR and then sequenced.

DNA FINGERPRINTING

Forensic chemists use PCR to compare DNA samples collected at the scene of a crime with the DNA of the suspected perpetrator. The base sequence of segments of noncoding DNA varies from individual to individual. Forensic laboratories have identified 13 of these segments that are the most accurate for identification purposes. If the sequence of bases from two DNA samples are the same, the chance is about 80 billion to one that they are from the same individual. DNA fingerprinting is also used to establish paternity, accounting for about 100,000 DNA profiles a year.

20.12 GENETIC ENGINEERING

Recombinant DNA molecules are DNA molecules (natural or synthetic) that have been attached to a small piece of carrier DNA in a compatible host cell and allowed to replicate millions of times. Recombinant DNA technology—also known as **genetic engineering**—has many practical applications. For example, replicating the DNA for human insulin in this way allows the synthesis of large amounts of the human insulin.

Agriculture is benefiting from genetic engineering as crops are being produced with new genes that increase their resistance to drought and insects. For example, genetically engineered cotton crops are resistant to the cotton bollworm, and genetically engineered corn is resistant to the corn rootworm. Genetically modified organisms (GMOs) have been responsible for a nearly 50% reduction in agricultural chemical sales in the United States. On the other hand, many people worry about the potential consequences of changing the genes of any species.

RESISTING HERBICIDES

Glyphosate, the active ingredient in a well-known herbicide called Roundup, kills weeds by inhibiting an enzyme that plants need to synthesize phenylalanine and tryptophan, amino acids required for growth. Studies are being done with corn and cotton that have been genetically engineered to tolerate the herbicide. Then, when fields are sprayed with glyphosate, the weeds are killed but not the crops. These crops have been given a gene that produces an enzyme that makes glyphosate inactive by acetylating it with acetyl-CoA.

Corn genetically engineered to resist the herbicide glyphosate by acetylating it.

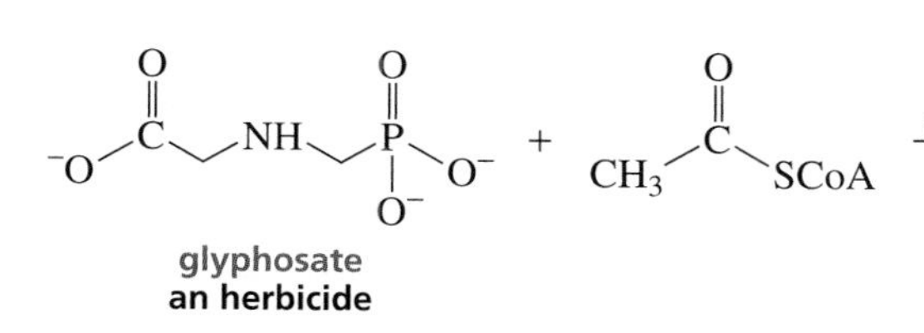

glyphosate
an herbicide

enzyme ⟶ N-acetylglyphosate + CoASH

N-acetylglyphosate
harmless to plants

SUMMARY

There are two types of nucleic acids: **deoxyribonucleic acid (DNA)** and **ribonucleic acid (RNA)**. The genetic information stored in DNA is **transcribed** into RNA. The information thus contained in the RNA strand can then be **translated** for the synthesis of proteins.

A **nucleoside** contains a base bonded to D-ribose or to 2′-deoxy-D-ribose. A **nucleotide** is a nucleoside with either the 5′- or the 3′-OH group bonded to phosphoric acid by an ester linkage. **Nucleic acids** are composed of long strands of nucleotide subunits linked by phosphodiester bonds. These linkages join the 3′-OH group of one nucleotide to the 5′-OH group of the next nucleotide. A **dinucleotide** contains two nucleotide subunits, an **oligonucleotide** contains three to ten subunits, and a **polynucleotide** contains many subunits. DNA contains 2′-deoxy-D-ribose, while RNA contains D-ribose. The difference in the sugars causes DNA to be stable and RNA to be easily cleaved.

The **primary structure** of a nucleic acid is the sequence of bases in its strand. DNA contains **A**, **G**, **C**, and **T**; RNA contains **A**, **G**, **C**, and **U**. The presence of T instead of U in DNA prevents mutations caused by imine hydrolysis of C to form U. DNA is double-stranded. The strands run in opposite directions and are twisted into a helix. The bases are confined to the inside of the helix, and the sugar and phosphate groups are on the outside. The strands are held together by hydrogen bonds between bases of opposing strands. The two strands—one is called a **sense strand** and the other a **template strand**—are complementary: **A** pairs with **T**, and **G** pairs with **C**. DNA is synthesized in the 5′ ⟶ 3′ direction by a process called **semiconservative replication**.

The sequence of bases in DNA provides the blueprint for the synthesis (**transcription**) of RNA. RNA is synthesized in the 5′ ⟶ 3′ direction by transcribing the DNA template strand in the 3′ ⟶ 5′ direction. There are three kinds of RNA: messenger RNA, ribosomal RNA, and transfer RNA. Protein synthesis (**translation**) proceeds from the N-terminal end to the C-terminal end by a process that reads the bases along the mRNA strand in the 5′ ⟶ 3′ direction. Each three-base sequence—**a codon**—specifies the particular amino acid to be incorporated into a protein. A tRNA carries the amino acid bound as an ester to its 3′-terminal position. The codons and the amino acids they specify are known as the **genetic code**.

Restriction endonucleases cleave DNA at specific palindromes forming **restriction fragments**. The sequence of bases in the restriction fragments can be determined by the **dideoxy method**. The **polymerase chain reaction (PCR)** amplifies segments of DNA, making billions of copies in a very short time.

PROBLEMS

11. Name the following:

a. 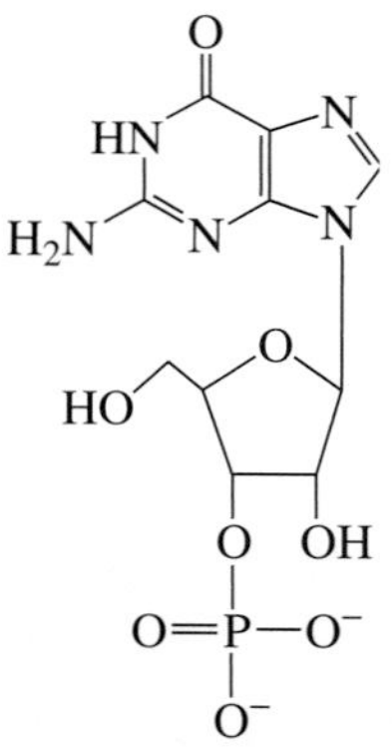 **b.** 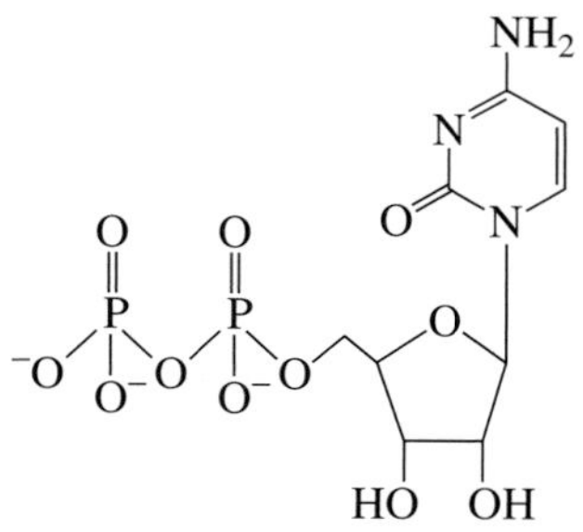 **c.** 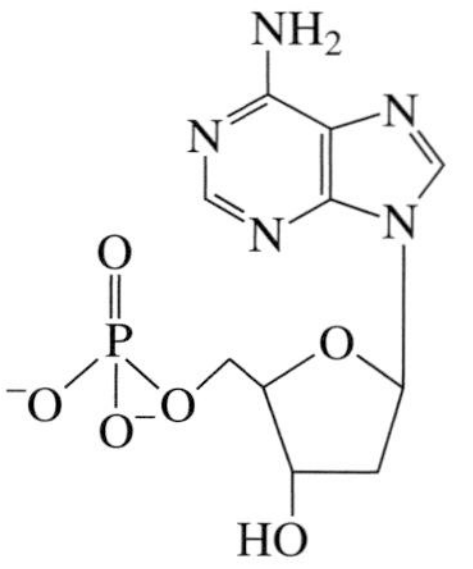 **d.**

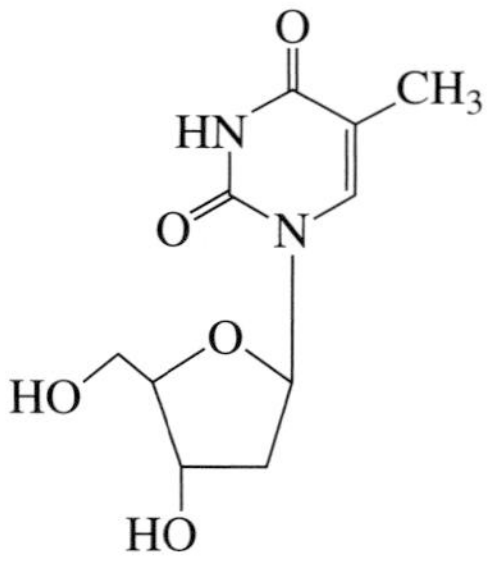

12. What nonapeptide is coded for by the following piece of mRNA?

5′—AAA—GUU—GGC—UAC—CCC—GGA—AUG—GUG—GUC—3′

13. What is the sequence of bases in the template strand of DNA that codes for the mRNA in Problem 12?

14. What is the sequence of bases in the sense strand of DNA that codes for the mRNA in Problem 12?

15. What would be the C-terminal amino acid if the codon at the 3′-end of the mRNA in Problem 12 underwent the following mutation:

a. if the first base was changed to A?
b. if the second base was changed to A?
c. if the third base was changed to A?
d. if the third base was changed to G?

16. What would be the base sequence of the segment of DNA that is responsible for the biosynthesis of the following hexapeptide?

Gly-Ser-Arg-Val-His-Glu

17. Match the codon with the anticodon:

Codon	Anticodon
AAA	ACC
GCA	CCU
CUU	UUU
AGG	AGG
CCU	UGA
GGU	AAG
UCA	GUC
GAC	UGC

18. Using the single-letter abbreviations for the amino acids in Table 16.2, write the sequence of amino acids in a tetrapeptide represented by the first four different letters in your first name. Do not use any letter twice. (Because not all letters are assigned to amino acids, you might have to use one or two letters in your last name.) Write the sequence of bases in mRNA that would result in the synthesis of that tetrapeptide. Write the sequence of bases in the sense strand of DNA that would result in formation of that fragment of mRNA.

19. Which of the following pairs of dinucleotides are present in equal amounts in DNA?

a. CC and GG **b.** CG and GT **c.** CA and TG **d.** CG and AT **e.** GT and CA **f.** TA and AT

20. The amino acid sequences of peptide fragments obtained from a normal protein and from the same protein synthesized by a defective gene were compared. They were found to differ in only one peptide fragment; their amino acid sequences are shown here:

Normal: Gln-Tyr-Gly-Thr-Arg-Tyr-Val

Mutant: Gln-Ser-Glu-Pro-Gly-Thr

a. What is the defect in DNA?

b. It was later determined that the normal peptide fragment is an octapeptide with a C-terminal Val-Leu. What is the C-terminal amino acid of the mutant peptide?

21. List the possible codons on mRNA that specify each amino acid in the oligopeptide in Problem 5 and the anticodon on the tRNA that carries that amino acid.

22. Indicate whether each functional group of the five heterocyclic bases in nucleic acids can function as a hydrogen bond acceptor (A), a hydrogen bond donor (D), or both (D/A).

23. Using the D, A, and D/A designations in Problem 22, explain how base pairing would be affected if the bases existed in the enol form.

24. The 2′,3′-cyclic phosphodiester, which is formed when RNA is hydrolyzed (Figure 20.6), reacts with water, forming a mixture of nucleotide 2′- and 3′-phosphates. Propose a mechanism for this reaction.

25. Adenine can be deaminated to hypoxanthine, and guanine can be deaminated to xanthine. Draw structures for hypoxanthine and xanthine.

26. Explain why thymine cannot be deaminated.

27. In acidic solutions, nucleosides are hydrolyzed to a sugar and a heterocyclic base. Propose a mechanism for this reaction.

28. Why is the codon a triplet rather than a doublet or a quartet?

29. 5-Bromouracil, a highly mutagenic compound, is used in cancer chemotherapy. When administered to a patient, it is converted to a nucleotide triphosphate and incorporated into DNA in place of thymine, which it resembles sterically. Why does it cause mutations? (*Hint:* The bromo substituent causes the enol tautomer to be more stable than the keto tautomer.)

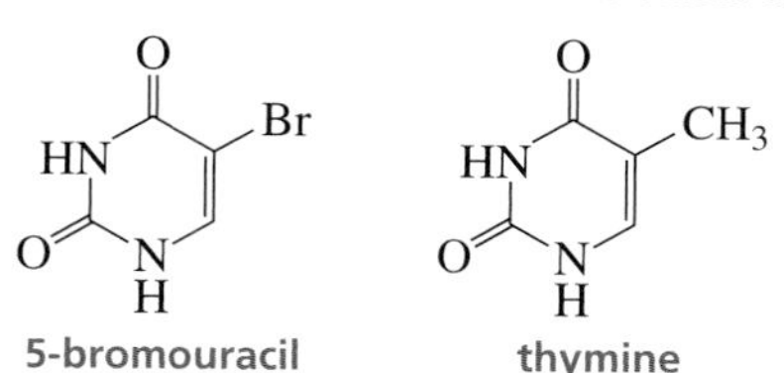

30. Which cytosine in the following sense strand of DNA could cause the most damage to the organism if it were deaminated?

5′—A—T—G—T—C—G—C—T—A—A—T—C—3′

31. The first amino acid incorporated into a polypeptide chain during its biosynthesis in prokaryotes is *N*-formylmethionine. Explain the purpose of the formyl group.

32. Why doesn't DNA unravel completely before replication begins?

CHAPTER 21

The Organic Chemistry of Drugs

Discovery and Design

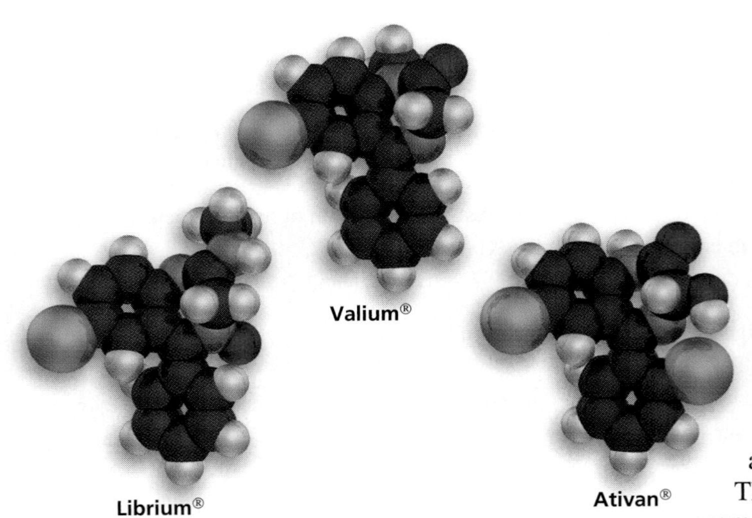

A **drug** is any substance that is absorbed by the body and then changes or enhances a physical or psychological function in the body. A drug can be a gas, a liquid, or a solid; it can have a simple structure or a complicated one. Humans have used drugs for thousands of years to alleviate pain and illness. By trial and error, people learned which herbs, berries, roots, and bark could be used for medicinal purposes. The knowledge about natural medicines was passed down from generation to generation without any understanding of how the drugs actually worked. Those who dispensed the drugs—medicine men and women, shamans, and witch doctors—were important members of every civilization. However, the drugs available to them were just a small fraction of the drugs available to us today.

Even at the beginning of the twentieth century, there were no drugs for the dozens of functional, degenerative, neurological, and psychiatric disorders from which people suffer; no hormone therapies; no vitamins; and—most significantly—no effective drug for the cure of any infectious disease. Local anesthetics had just been discovered, and there were only two analgesics to relieve major pain. One reason that families had many children was because some of the children were bound to succumb to childhood diseases. Life spans were generally short. In 1900, for example, the average life expectancy in the United States was 46 years for a man and 48 years for a woman. In 1920, about 80 of every 100,000 children died before their fifteenth birthday—most as a result of infections in their first year of life. Now there is a drug for almost every disease, and the effectiveness of this medicinal arsenal is reflected in current life expectancies: 74 years for a man and 79 years for a woman. Now only about four of every 100,000 children die before the age of 15, mainly from cancer, accidents, and inherited diseases.

The shelves of a typical modern pharmacy are stocked with almost 2000 preparations, most of which contain a single active ingredient, usually an organic compound. These medicines can be swallowed, injected, inhaled, or absorbed through the skin. More than 3 billion prescriptions are dispensed in the United States each year. The most widely *prescribed drugs* are listed in Table 21.1, in order of the number of prescriptions written. Worldwide, antibiotics are the most widely prescribed class of drugs. In the developed world, heart drugs and cholesterol-reducing drugs are the most widely prescribed, partly because they are generally taken for the remainder of the patient's life. In recent years, prescriptions for psychotropic drugs have decreased as doctors have become more aware of problems associated with addiction, and prescriptions for asthma have increased, reflecting a greater incidence (or awareness) of the disease. The U.S. market accounts for 50% of pharmaceutical sales. Now that this market is aging—nearly 30% of the U.S. population is over 50 years old—the demand is rising for drugs that treat conditions such as high cholesterol levels, hypertension, diabetes, and osteoarthritis.

Table 21.1 The Most Widely Dispensed Drugs in the United States in 2006 in Decreasing Order of the Number of Prescriptions Written

Brand name	Generic name	Structure	Use
Lipitor	atorvastatin		cholesterol-reducing drug
Vicodin	hydrocodone and acetaminophen		analgesic
Toprol-XL	metoprolol		β-adrenergic blocking agent (antiarrhythmic, antihypertensive)
Norvasc	amlodipine		calcium channel blocker (antihypertensive)
Amoxil	amoxicillin		antibiotic
Synthroid	levothyroxine		treatment of hypothyroidism
Nexium	esomeprazole (the *S*-enantiomer of omeprazole)		treatment of acid reflux

Table 21.1 Continued

Brand name	Generic name	Structure	Use
Lexapro	escitalopram	F; O; N; N≡C	antidepressant
Proventil	albuterol	HO; HO; OH; N H	bronchodilator
Singulair	montelukast	OH; O; S; H; Cl; N; HO	treatment of asthma and relief of symptoms of seasonal allergies
Prinivil	lisinopril	NH_2; HOOC; H COOH H; H; N; N H; O	antihypertensive
Ambien	zolpidem	N; N; O; N	hypnotic (treatment for insomnia)
Zyrtec	cetirizine	Cl; N; N; O; COOH	antihistamine

21.1 NAMING DRUGS

The most accurate names for drugs are the chemical names that define their structures. However, these names are too long and complicated to appeal to the general public, or even to their physicians. The pharmaceutical company therefore chooses a brand name for any drug it develops. A **brand name** identifies the drug as a commercial product and distinguishes it from other products. Only the company that holds the patent can use the brand name for marketing the product. It is in the company's best interest to choose a name that is easy to remember and pronounce so that when the patent expires, the public will continue to request the drug by that name.

Each drug is also given a **generic name** that any pharmaceutical company can use to identify the product. The pharmaceutical company that developed the drug is allowed to choose the generic name from a list of 10 names provided by an independent group. It is in the company's best interest to choose the generic name that is hardest to pronounce and least likely to be remembered, so that physicians and consumers will continue to use the familiar brand name. Brand names must always be capitalized, whereas generic names are not capitalized.

Drug manufacturers are permitted to patent and retain exclusive rights to the drugs they develop. A patent is valid for 20 years. Once the patent expires, other drug companies can market the drug under the generic name or under their own brand name, which is called a branded generic name and is a generic name that only they can use. For example, the antibiotic ampicillin is sold as Penbritin by the company that held the original patent. Now that the patent has expired, the drug is sold by other companies as Ampicin, Ampilar, Amplital, Binotal, Nuvapen, Pentrex, Ultrabion, Viccillin, and 30 other branded generic names.

The over-the-counter drugs that line the shelves of drug stores are available without a prescription. They are often mixtures containing one or more active ingredients (generic or brand-name drugs), plus sweeteners and inert fillers. For example, the preparations called Advil (Whitehall Laboratories), Motrin (Upjohn), and Nuprin (Bristol-Meyers Squibb) all contain ibuprofen, a mild analgesic and anti-inflammatory drug. Ibuprofen was patented in Britain in 1964 by Boots, Inc., and the U.S. Food and Drug Administration (FDA) approved its use as a nonprescription drug in 1984.

21.2 LEAD COMPOUNDS

The goal of the medicinal chemist is to find compounds that have potent effects on specific diseases while producing no or minimal side effects. In other words, a drug must react selectively with its target and have few, if any, negative effects. A drug must get to the right place in the body, at the right concentration, and at the right time. Therefore, a drug must have the appropriate solubility and other physical and chemical properties to allow it to be transported to the target cell. For example, if a drug is to be taken orally, it must be insensitive to the acid conditions of the stomach, and it also must resist enzymatic degradation by the liver before it reaches its target. Finally, it must eventually be excreted as is or be degraded to harmless compounds that can be excreted.

3-D Molecule: Foxglove

Medicinal agents used by humans since ancient times provided the starting point for the development of our current arsenal of drugs. The active ingredients were isolated from the herbs, berries, roots, and bark used in traditional medicine. Foxglove, for instance, furnished digitoxin, a cardiac stimulant. The bark of the cinchona tree yielded quinine for relief from malaria. Willow bark contains salicylates used to control fever and pain. A milky fluid obtained from the oriental opium poppy provided morphine for severe pain and codeine for the control of a cough. By 1882, more than 50 different herbs were commonly used to make medicines. Many were grown in the gardens of religious establishments that treated the sick.

Foxglove

Scientists still search the world for plants and berries and the oceans for flora and fauna that might yield new medicinal compounds. Taxol, a compound isolated from the bark of the Pacific yew tree, is a recently recognized anticancer agent (Section 12.5). Almost half the new drugs approved in recent years were natural products or derived from natural products.

Once a naturally occurring drug is isolated and its structure determined, it can serve as a prototype in a search for other biologically active compounds. The prototype is called a **lead compound** (that is, it plays a leading role in the search). Analogs of the lead compound are synthesized and tested to see if they are more effective than the lead compound or if they have fewer side effects. An analog may have a different substituent than the lead compound, a branched chain instead of a straight chain, a different ring system, or some other structural difference.

21.3 MOLECULAR MODIFICATION

Coca leaves

3-D Molecule: Coca leaves

Producing analogs by changing the structure of the lead compound is called **molecular modification**. In a classic example of this process, a number of synthetic local anesthetics were developed from the lead compound, cocaine. Cocaine comes from the leaves of a bush native to the highlands of the South American Andes. Cocaine is a highly effective local anesthetic, but it produces undesirable effects on the central nervous system (CNS), ranging from initial euphoria to severe depression. By dissecting the cocaine molecule step by step—removing the methoxycarbonyl group and cleaving the seven-membered-ring system—scientists identified the portion of the molecule that carries the local anesthetic activity but does not induce the damaging CNS effects. This knowledge provided an improved lead compound—a different ester of benzoic acid.

cocaine
lead compound

improved lead compound

Hundreds of related esters were then synthesized. Successful anesthetics obtained through this process were benzocaine, a topical anesthetic, and procaine, commonly known by the brand name Novocain.

anesthetics

Benzocaine®

procaine
Novocain®

lidocaine
Xylocaine®

Because the ester group of procaine is hydrolyzed relatively rapidly by enzymes that catalyze ester hydrolysis, procaine has a short half-life. Researchers therefore focused next on synthesizing compounds with less easily hydrolyzed amide groups (Section 11.6). In this way, lidocaine, one of the most widely used injectable anesthetics, was discovered. The rate at which lidocaine is hydrolyzed is further decreased by its two *ortho*-methyl substituents, which sterically hinder the reactive carbonyl group.

Later, physicians recognized that the action of an anesthetic administered in vivo (in a living organism) could be lengthened considerably if it were administered along with epinephrine. Because epinephrine is a vasoconstrictor, it reduces the circulation of blood, allowing the drug to remain at its targeted site for a longer period.

In screening the structurally modified cocaine analogs for biological activity, scientists were surprised to find that replacing the ester linkage of procaine with an amide linkage led to a compound—procainamide hydrochloride—that had activity as a cardiac depressant as well as activity as a local anesthetic. Procainamide hydrochloride is currently used clinically as an antiarrhythmic.

procainamide hydrochloride

Morphine, the most widely used analgesic for severe pain, is the standard by which other painkilling medications are measured. Although scientists have learned how to synthesize morphine, all commercial morphine is obtained from opium, a milky fluid exuded by a species of poppy (see page 2). Morphine occurs in opium at concentrations as high as 10%. Methylating one of the OH groups of morphine produces codeine, which has one-tenth the analgesic activity of morphine but profoundly inhibits the cough reflex. Although 3% of opium is codeine, most commercial codeine is obtained by methylating morphine. Putting an acetyl group on one of the OH groups of morphine produces a compound with a similar reduced potency. Putting an acetyl group on both OH groups forms heroin, which is much more potent than morphine. Heroin is less polar than morphine and therefore crosses the blood–brain barrier faster, resulting in a more rapid "high." Heroin has been banned in most countries because it is widely abused. It is synthesized by using acetic anhydride to put the acetyl groups on morphine (Section 11.11). Therefore, both heroin and acetic acid are formed as products. Drug enforcement agencies use dogs trained to recognize the pungent odor of acetic acid.

analgesics

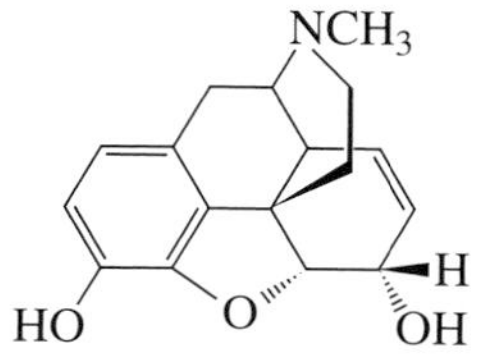

morphine

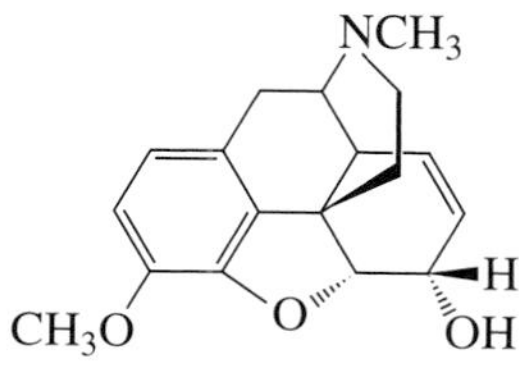

codeine

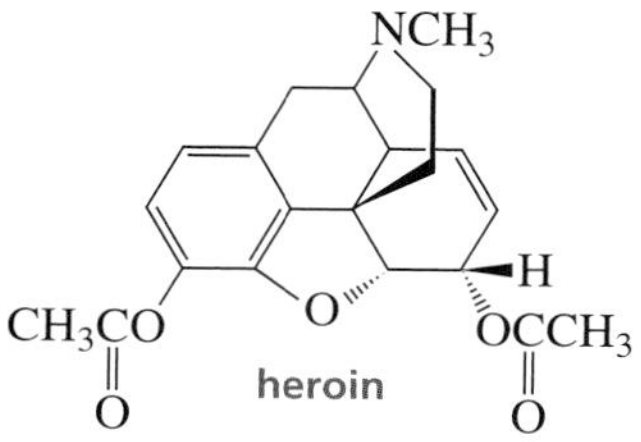

heroin

Molecular modification of codeine led to dextromethorphan, the active ingredient in most cough medicines. Etorphine was synthesized when scientists realized that analgesic potency was related to the ability of a nonpolar segment of the drug to bind to the nonpolar portion of the opiate receptor (Section 21.6). Etorphine is about 2000 times more potent than morphine, but it is not safe for use by humans. It has been used to tranquilize elephants and other large animals. Pentazocine is useful in obstetrics because it dulls the pain of labor but does not depress the respiration of the infant as morphine does.

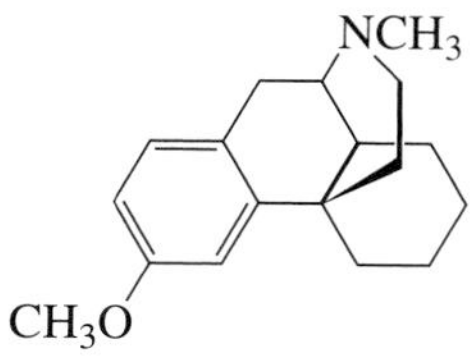

dextromethorphan

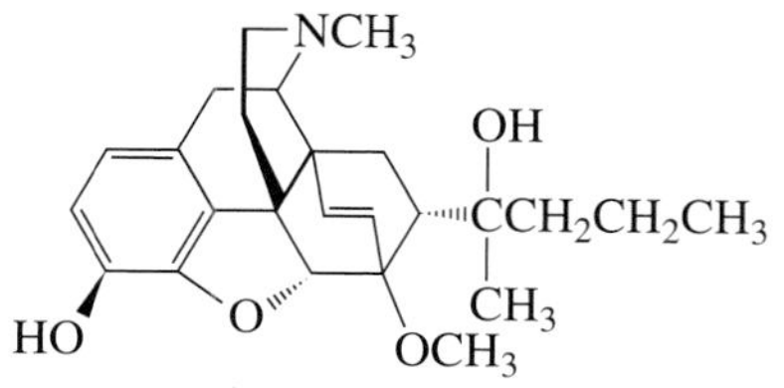

etorphine

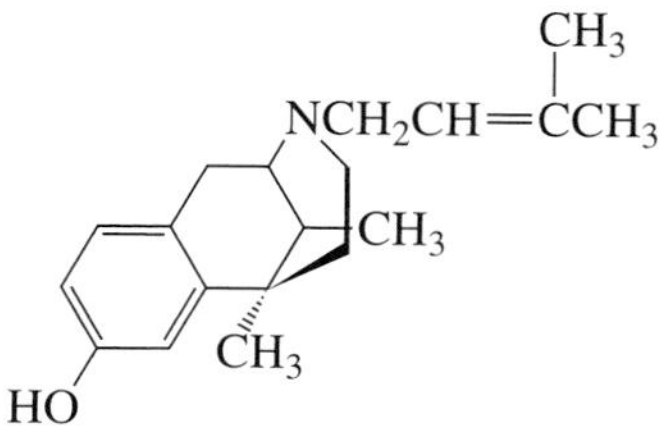

pentazocine

German scientists synthesized methadone in 1944 in an attempt to find a drug to treat muscle spasms. No one recognized until 10 years later—after building molecular models—that methadone and morphine have similar shapes. Unlike morphine, however, methadone can be administered orally. Methadone also has a considerably longer half-life (24 to 26 hours) than morphine (2 to 4 hours). Repeated doses of methadone have cumulative effects, so it can be used in lower doses and at longer intervals. Because of these properties, methadone is used to treat chronic pain and the withdrawal symptoms of heroin addiction. Reducing the carbonyl group of methadone and acetylating it forms α-acetylmethadol. The levo (-) isomer of this compound can suppress withdrawal symptoms for 72 hours (Section 6.12). When the methadone analog Darvon was introduced, it was initially thought to be the long-sought-after nonaddicting painkiller. However, it was later found to have no therapeutic advantage over less toxic and more effective analgesics.

methadone

α-acetylmethadol

isomethadone
Darvon®

Notice that morphine and all the compounds prepared by molecular modification of morphine have a structural feature in common: an aromatic ring attached to a quaternary carbon that is attached to a tertiary amine two carbons away.

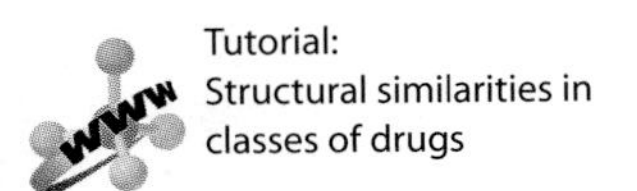

Tutorial: Structural similarities in classes of drugs

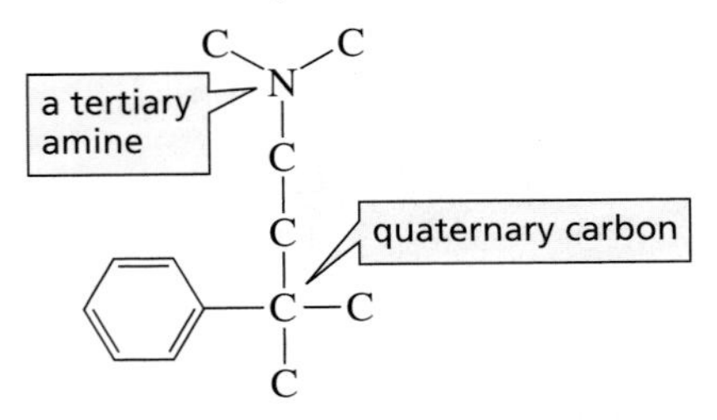

21.4 RANDOM SCREENING

Most lead compounds are found by screening thousands of compounds randomly. A **random screen**, also known as a **blind screen**, is a search for a pharmacologically active compound conducted without the benefit of any information about what chemical structures might show activity. The first blind screen was carried out by Paul Ehrlich, who was searching for a "magic bullet" that would kill trypanosomes—the microorganisms that cause African sleeping sickness—without harming their human host. After testing more than 900 compounds against trypanosomes, Ehrlich tested some of his compounds against other bacteria. Compound 606 (salvarsan) was found to be dramatically effective against the microorganisms that cause syphilis, an often deadly disease and incurable at the time, that was exacting a toll on human health, much in the way HIV is doing today.

An important part of random screening is recognizing an effective compound. This requires the development of an assay for the desired biological activity. Some assays can be done in vitro ("in glass," that is, in a test tube or flask)—for example, searching for a compound that will inhibit a particular enzyme. Others are done in vivo (in a living organism)—for instance, searching for a compound that will save a mouse from a lethal dose of a virus. One problem with in vivo assays is that drugs can be metabolized differently by different animals (Section 18.0). Thus, an effective drug in a mouse may be less effective or even useless in a human. Another problem is the guesswork involved in regulating the dosages of both the virus and the drug. If the dosage of the virus is too high, it might kill the mouse in spite of the presence of a biologically active compound that could save the animal. If the dosage of the potential drug is too high, the drug might kill the mouse even though a lower dosage would have saved it.

BIOGRAPHY

Paul Ehrlich (1854–1915) *was a German bacteriologist. He received a medical degree from the University of Leipzig and was a professor at the University of Berlin. In 1892, he developed an effective diphtheria antitoxin. For his work on immunity, he received the 1908 Nobel Prize in physiology or medicine.*

The observation that azo dyes effectively dyed wool fibers (animal protein) gave scientists the idea that such dyes might selectively bind to bacterial proteins, too, and in the process perhaps harm the bacteria. Well over 10,000 dyes were screened in vitro in antibacterial tests, but none showed any antibiotic activity. At that point, some scientists suggested screening the dyes in vivo, saying that what physicians really needed were antibacterial agents that would cure infections in humans and animals, not in test tubes.

In vivo studies were therefore done in mice that had been infected with a bacterial culture. Now the luck of the investigators improved. Several dyes turned out to counteract gram-positive infections. The least toxic of these, Prontosil (a bright red dye), became the first drug to treat bacterial infections.

NH_2 H_2N $N{=}N$ SO_2NH_2

Prontosil®

The fact that Prontosil was inactive in vitro but active in vivo should have been recognized as a sign that the dye was converted to an active compound by the mammalian organism, but this did not occur to the bacteriologists, who were content to have found a useful antibiotic. When scientists at the Pasteur Institute later investigated Prontosil, they noted that mice given the drug did not excrete a red compound. Urine analysis showed that the mice instead excreted *para*-acetamidobenzenesulfonamide, a colorless compound. Chemists knew that anilines are acetylated in vivo, so they prepared the nonacetylated compound (sulfanilamide). When sulfanilamide was tested in mice infected with streptococcus, all the mice were cured, whereas untreated control mice died. Sulfanilamide was the first of the sulfa drugs—the first class of antibiotic.

$CH_3C(=O)-NH$ SO_2NH_2

para-acetamidobenzenesulfonamide

H_2N SO_2NH_2

para-aminobenzenesulfonamide
sulfanilamide

Sulfanilamide acts by inhibiting the bacterial enzyme that synthesizes folic acid (Section 17.11). Thus, sulfanilamide is a **bacteriostatic drug**, a drug that inhibits the further growth of bacteria, rather than a **bactericidal drug**, a drug that kills the bacteria. Sulfanilamide inhibits the enzyme because it is similar in size to the carboxylic acid that should be incorporated into folic acid. Many successful drugs have been designed through this strategy of similar-size replacements.

BIOGRAPHY

Gerhard Domagk (1895–1964) *was a research scientist at I. G. Farbenindustrie, a German manufacturer of dyes and other chemicals. He carried out studies that showed Prontosil to be an effective antibacterial agent. His daughter, who was dying of a streptococcal infection as a result of cutting her finger, was the first patient to receive the drug and be cured by it. Prontosil received wider fame when it was used in 1936 to save the life of Franklin D. Roosevelt, Jr., son of the U.S. president. Domagk received the Nobel Prize in physiology or medicine in 1939, but Hitler did not allow Germans to accept Nobel Prizes because Carl von Ossietsky, a German who was in a concentration camp, had been awarded the Nobel Prize for peace in 1935. Domagk was eventually able to accept the prize in 1947 but, because of the time that had elapsed, was not given the monetary award.*

DRUG SAFETY

In October 1937 patients who had obtained sulfanilamide from a company in Tennessee were experiencing excruciating abdominal pains before slipping into fatal comas. The FDA asked Eugene Geiling, a pharmacologist at the University of Chicago, and his graduate student Frances Kelsey to investigate. They found that the drug company was dissolving sulfanilamide in diethylene glycol, a sweet-tasting liquid, to make the drug easy to swallow. However, the safety of diethylene glycol in humans had never been tested. It turned out to be a deadly poison. Interestingly, Frances Kelsey was the person who prevented thalidomide from being marketed in the United States (Section 6.12)

At the time of the sulfanilamide investigation, there was no legislation to prevent the sale of medicines with lethal effects, but in June 1938, the Federal Food, Drug, and Cosmetic Act was enacted. This legislation required all new drugs to be thoroughly tested for effectiveness and safety before being marketed.

21.5 SERENDIPITY IN DRUG DEVELOPMENT

BIOGRAPHY

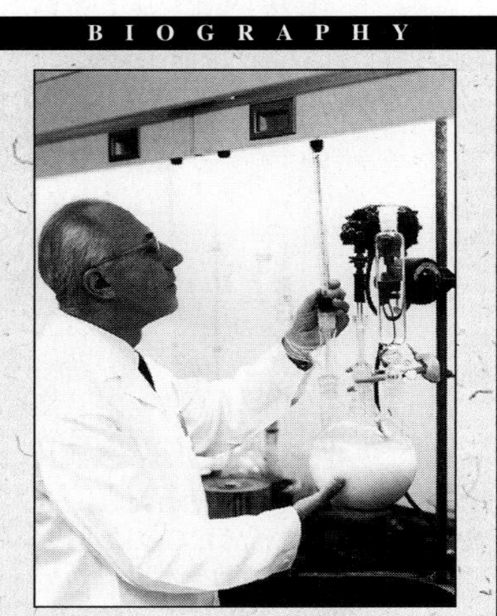

Leo H. Sternbach (1908–2005) *was born in Austria. In 1918, after World War I and the dissolution of the Austro–Hungarian Empire, Sternbach's father moved to Krakow in the re-created Poland and obtained a concession to open a pharmacy. As a pharmacist's son, Sternbach was accepted at the Jagiellonian University School of Pharmacy, where he received both a master's degree in pharmacy and a Ph.D. in chemistry. With discrimination against Jewish scientists growing in Eastern Europe in 1937, Sternbach moved to Switzerland to work with Ružička (p. 523) at the ETH (the Swiss Federal Institute of Technology). In 1941, Hoffmann–LaRoche brought Sternbach and several other scientists out of Europe. Sternbach became a research chemist at LaRoche's U.S. headquarters in Nutley, New Jersey, where he later became the director of medicinal chemistry.*

Many drugs have been discovered accidentally. Nitroglycerin, the drug used to relieve the symptoms of angina pectoris (heart pain), was discovered when workers handling nitroglycerin in the explosives industry experienced severe headaches. Investigation revealed that the headaches were caused by nitroglycerin's ability to produce a marked dilation of blood vessels. The pain associated with an angina attack results from the inability of the blood vessels to supply the heart adequately with blood. Nitroglycerin relieves the discomfort by dilating cardiac blood vessels.

CH_2-ONO_2
$CH-ONO_2$
CH_2-ONO_2
nitroglycerin

The tranquilizer Librium is another drug that was discovered accidentally. Leo Sternbach synthesized a series of quinazoline 3-oxides, but none of them showed any pharmacological activity. One of the compounds was not submitted for testing because it was not the quinazoline 3-oxide he had set out to synthesize. Two years after the project was abandoned, a laboratory worker came across this compound while cleaning up the lab, and Sternbach decided that he might as well submit it for testing before it was thrown away. The compound was shown to have tranquilizing properties and, when its structure was investigated, was found to be a benzodiazepine 4-oxide. Methylamine, instead of displacing the chloro substituent in a substitution reaction to form a quinazoline 3-oxide, had added to the imine group of the six-membered ring, causing the ring to open and reclose to a seven-membered ring. The compound was given the brand name Librium when it was put into clinical use in 1960.

a quinazoline 3-oxide $\xrightarrow{CH_3NH_2}$ (an addition reaction occurred) $\longrightarrow$ $\downarrow$ a benzodiazepine 4-oxide, chlordiazepoxide, Librium® (1960)

a quinazoline 3-oxide $\xrightarrow{CH_3NH_2}$ ✗ (a substitution reaction did not occur)

Librium was structurally modified in an attempt to find other tranquilizers. One successful modification produced Valium, a tranquilizer almost 10 times more potent than Librium. Currently, there are eight benzodiazepines in clinical use as tranquilizers in the United States and some 15 others abroad. Rohypnol is one of the so-called date-rape drugs.

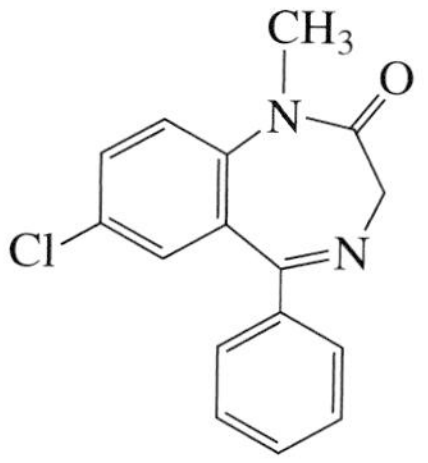

diazepam
Valium® (1963)

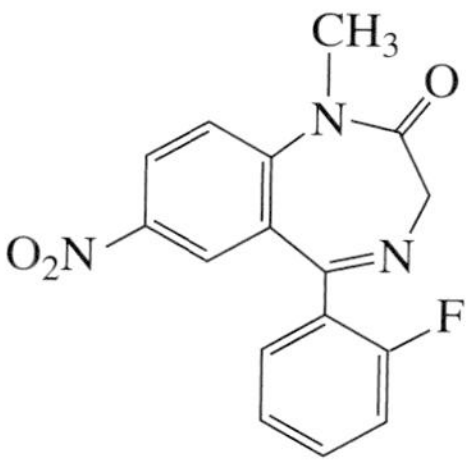

flunitrazepam
Rohypnol® (1963)

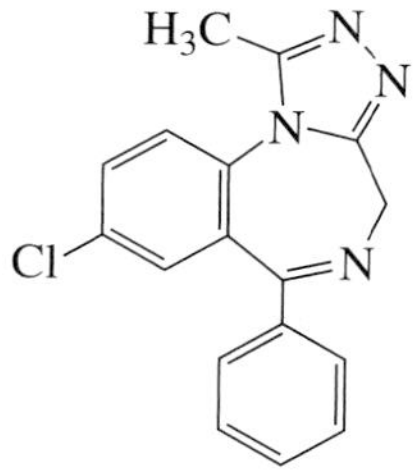

alprazolam
Xanax® (1970)

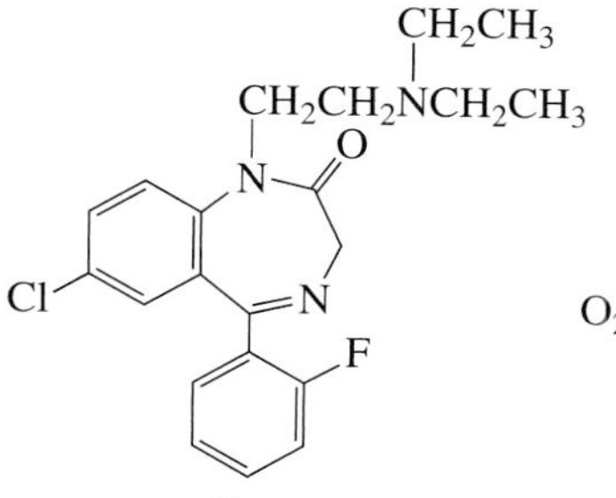

flurazepam
Dalmane® (1970)

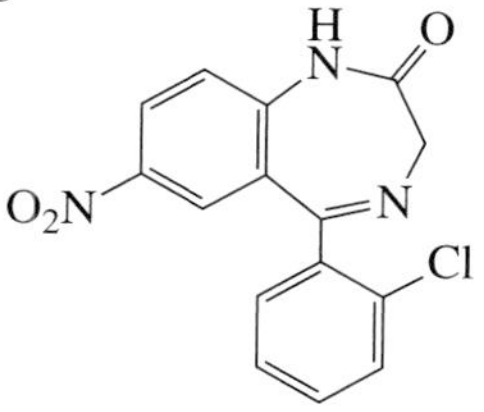

clonazepam
Klonopin® (1975)

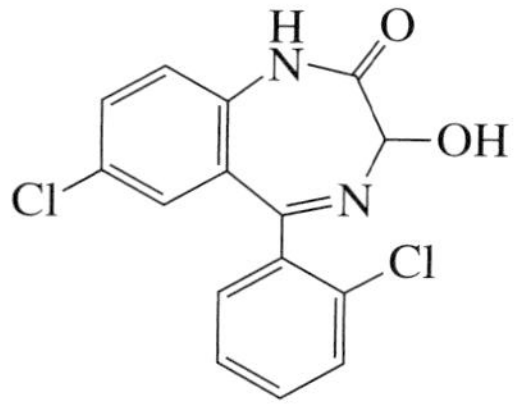

lorazepam
Ativan® (1977)

Viagra is a recent example of serendipity in drug development. Viagra was in clinical trials as a drug for heart ailments. When the clinical trials were canceled because Viagra turned out to be ineffective as a heart drug, those enrolled in the trials refused to return the remaining tablets. The pharmaceutical company then realized that the drug had other marketable effects. About 75% of new drugs are approved for something other than the original target.

21.6 RECEPTORS

Many drugs exert their physiological effects by binding to a specific cellular binding site called a **receptor**, whose role is to trigger a response in a cell. That is why a small amount of a drug can bring about a measurable physiological effect. Because most receptors are chiral, different enantiomers of a drug can have different effects (Section 6.12). Drug receptors are often lipoproteins or glycoproteins (Section 15.16). Some receptors are part of cell membranes, while others are found in the cytoplasm—the cell contents outside the nucleus. Nucleic acids—particularly DNA—also act as receptors for certain kinds of drugs. Because not all cells have the same receptors, drugs have considerable specificity. For example, epinephrine has intense effects on cardiac muscle, but almost no effect on muscle in other parts of the body.

A drug interacts with its receptor by means of the same kinds of bonding interactions—hydrogen bonding, electrostatic attractions, and van der Waals interactions—that we encountered in other examples of molecular recognition (Section 17.1). The most important factor in the interaction between a drug and a receptor is a snug fit: the greater the affinity of a drug for its binding site, the higher is the drug's potential biological activity. Two drugs for which DNA is a receptor are chloroquine (an antimalarial) and 3,6-diaminoacridine (an antibacterial). These flat cyclic compounds can slide into the DNA double helix between base pairs—like a card being inserted into a deck of playing cards—and interfere with the normal replication of DNA.

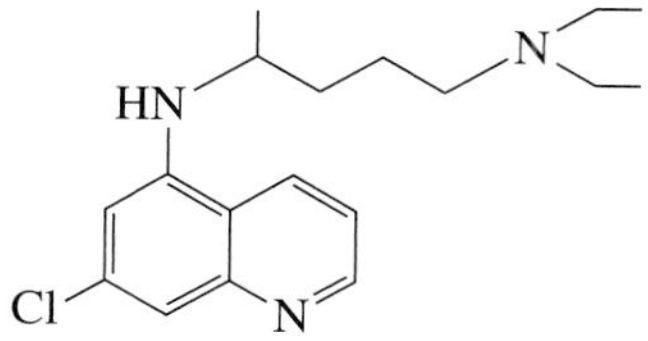

chloroquine

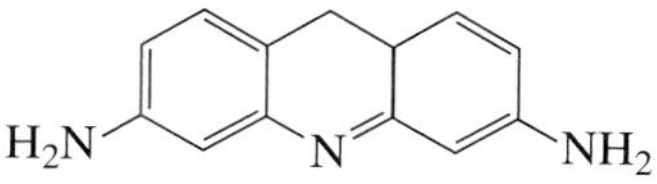

3,6-diaminoacridine

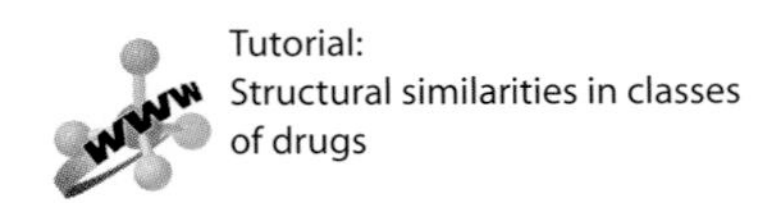

Tutorial:
Structural similarities in classes of drugs

When scientists know something about the molecular basis of drug action—such as how a particular drug interacts with a receptor—they can design and synthesize compounds that might have a desired biological activity. For example, when excess histamine is produced by the body, it causes the symptoms associated with allergic responses and with the common cold. This is thought to be the result of the protonated ethylamino group anchoring the histamine molecule to a negatively charged portion of the histamine receptor.

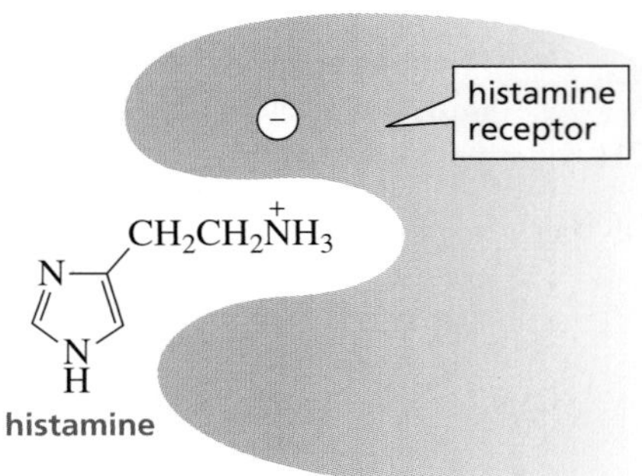

Drugs that interfere with the natural action of histamine—called antihistamines—bind to the histamine receptor but do not trigger the same response as histamine. Like histamine, these drugs have a protonated amino group that binds to the receptor. The drugs also have bulky groups that keep the histamine molecule from approaching the receptor.

antihistamines

Cl^-

diphenhydramine
Benadryl®

promethazine
Promine®

promazine
Talofen®

Excess histamine production by the body also causes the hypersecretion of stomach acid by the cells of the stomach lining, leading to the development of ulcers. The antihistamines that block the histamine receptors, thereby preventing the allergic responses associated with excess histamine production, have no effect on HCl production. This fact led scientists to conclude that a second kind of histamine receptor—a histamine H_2-receptor—triggers the release of acid into the stomach.

Because 4-methylhistamine was found to cause weak inhibition of HCl secretion, it was used as a lead compound. About 500 molecular modifications were performed over a 10-year period before four clinically useful antiulcer agents were found. Two of these are Tagamet and Zantac. Notice that steric blocking of the receptor site is not a factor in these compounds. Compared with the antihistamines, the effective antiulcer drugs have more polar rings and longer side chains. Tagamet has the same imidazole ring as 4-methylhistidine, but a different side chain. Zantac has a different heterocyclic ring and a side chain that is similar to that of Tagamet.

BIOGRAPHY

Sir James W. Black *was born in Great Britain in 1924 and is a professor of physiology at King's College Hospital Medical School of the University of London. He led the project that discovered the H_2-receptor antagonists. For this and the development of beta-blockers, he received the Nobel Prize in physiology or medicine in 1988.*

4-methylhistamine

cimetidine
Tagamet®

ranitidine
Zantac®

Research that implicated serotonin in the generation of migraine attacks led to the development of drugs that bind to serotonin receptors. Sumatriptan, introduced in 1991, relieved not only the pain associated with migraines, but also many of migraine's other symptoms, including nausea and sensitivity to light and sound.

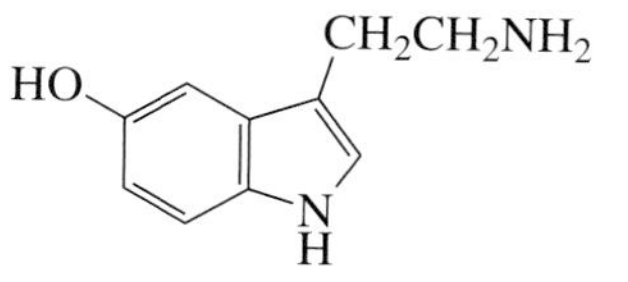

serotonin

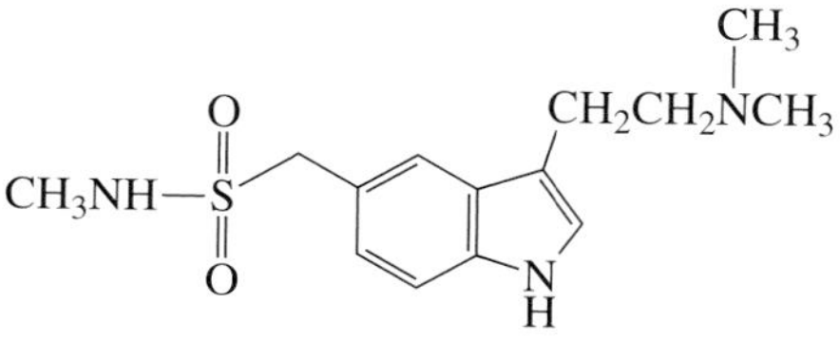

sumatriptan
Imitrex®

The success of sumatriptan spurred a search for other antimigraine agents by molecular modification, and three new triptans were introduced in 1997 and 1998. These second-generation triptans showed some improvements over sumatriptan—specifically, a longer half-life, reduced cardiac side effects, and improved CNS penetration.

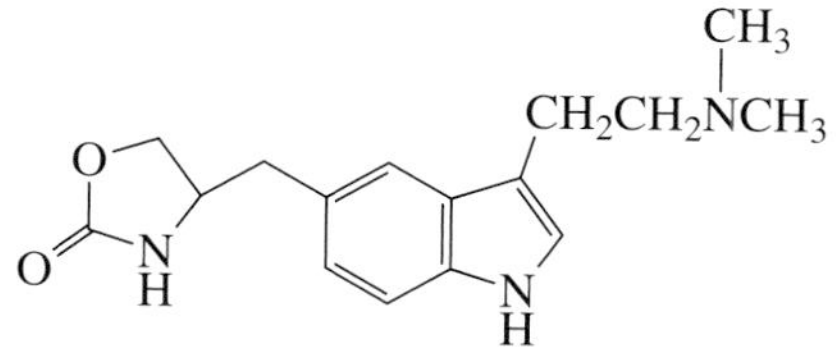

zolmitriptan
Zomig®

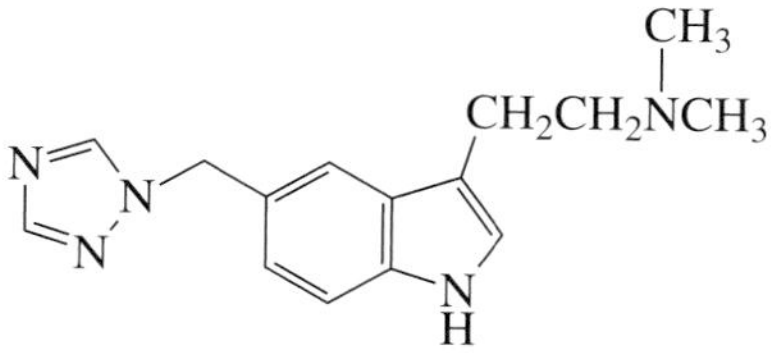

rizatriptan
Maxalt®

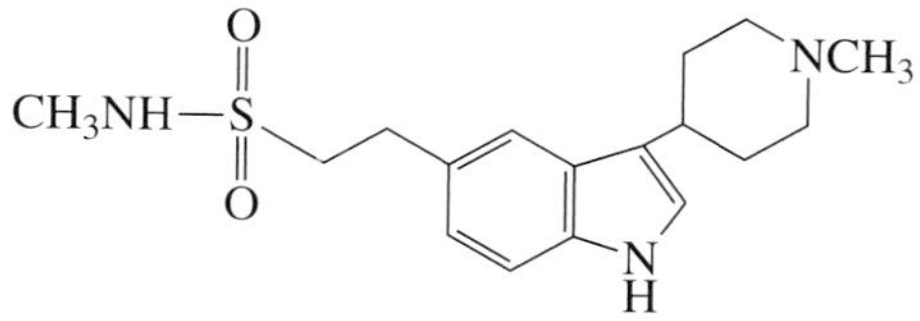

naratriptan
Amerge®

In screening modified compounds, it is not unusual to find a compound with a completely different pharmacological activity than the lead compound. For example, a molecular modification of a sulfonamide, an antibiotic (Section 21.4), led to tolbutamide, a drug with hypoglycemic activity.

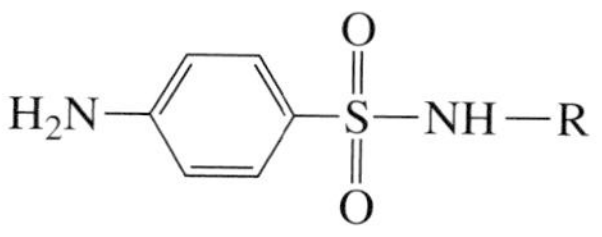

a sulfonamide

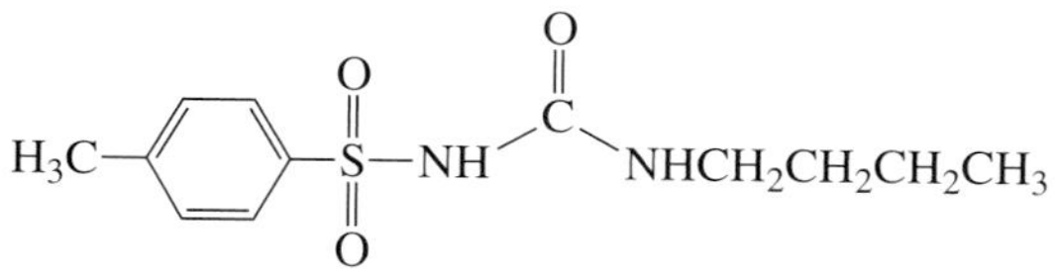

tolbutamide

Molecular modification of promethazine (page 562), an antihistamine, led to chlorpromazine, a drug that was found to lower body temperature. This drug found clinical use in chest surgery, where patients previously had to be cooled down by wrapping them in cold, wet sheets. Because wrapping in cold, wet sheets had been an old method of calming psychotic patients, a French psychiatrist tried the drug on some of his psychiatric patients. He found that chlorpromazine was able to suppress psychotic symptoms to the point that the patients assumed almost normal behavioral characteristics. Schizophrenic hallucinations and delusions went into remission, allowing many institutionalized patients to return to society. The excitement over chlorpromazine was short-lived, however, as patients on the drug soon developed uncoordinated involuntary movements. After thousands of molecular modifications, thioridazine was found to have the appropriate calming effect with less problematic side effects. It is now in clinical use as an antipsychotic.

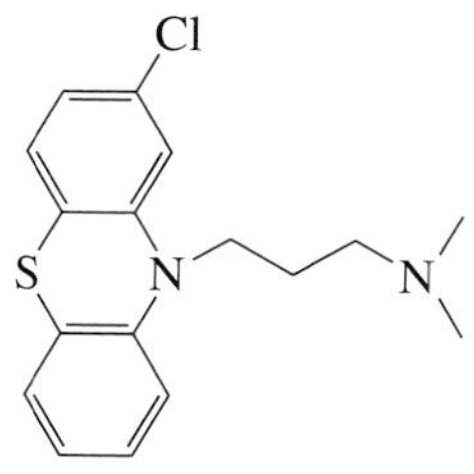

chlorpromazine
Thorazine®

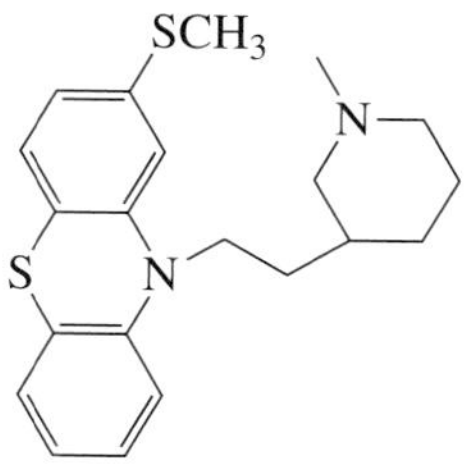

thioridazine
Mellaril®

Sometimes a drug initially developed for one purpose is later found to have properties that make it a better drug for a different purpose. Beta-blockers were originally intended to be used to alleviate the pain associated with angina by reducing the amount of work done by the heart. Later, they were found to have antihypertensive properties, so now they are used primarily to manage hypertension.

21.7 DRUG RESISTANCE

Typically, a bacterial strain takes 15 to 20 years to become resistant to an antibiotic. The fluoroquinolones, the last class of antibiotics to be discovered until recently, were discovered more than 30 years ago, so **drug resistance** has become an increasingly important problem in medicinal chemistry. More and more bacteria have become resistant to all antibiotics.

The antibiotic activity of the fluoroquinolones results from their ability to inhibit DNA gyrase, an enzyme required for transcription (Section 20.6). Humans are not harmed by the drug because the bacterial and mammalian forms of the enzyme are sufficiently different that the fluoroquinolones inhibit only the bacterial enzyme.

There are many different fluoroquinolones. All have fluorine substituents, which increase the lipophilicity of the drug to enable it to penetrate into tissues and cells. If either the carboxyl group or the double bond in the 4-pyridinone ring is removed, all activity is lost. Changing the substituents on the piperazine ring can cause the drug to be excreted by the kidney rather than by the liver, which is useful to patients with impaired liver function. The substituents on the piperazine ring also affect the half-life of the drug (the time it takes for half of the drug to lose its reactivity).

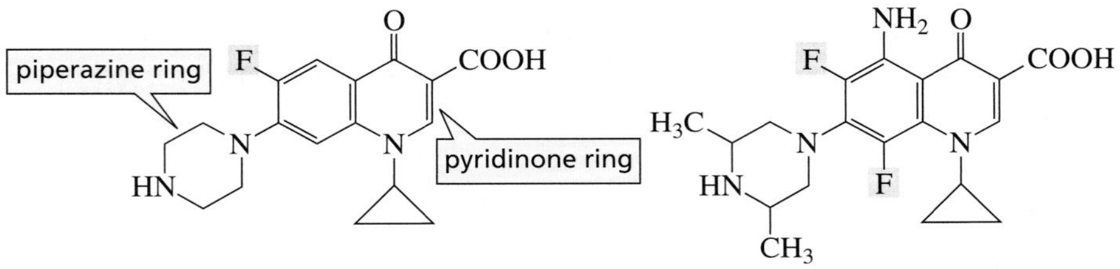

piperazine ring

pyridinone ring

ciprofloxacin
Cipro®
active against gram-negative bacteria

sparfloxacin
Zagam®
active against gram-negative bacteria and gram-positive bacteria

The approval of Zyvox by the FDA in 2000 was met with great relief by the medical community. Zyvox is the first in a new family of antibiotics: the oxazolidinones. In clinical trials, Zyvox was found to cure 75% of the patients infected with bacteria that had become resistant to all other antibiotics. Another new class of antibiotic became available in 2005 when the FDA approved Cubicin, the first of the lipopeptide antibiotics, with eleven amino acids (9 L and 2 D) and a ten-carbon acyl group attached to one of the α-amino groups.

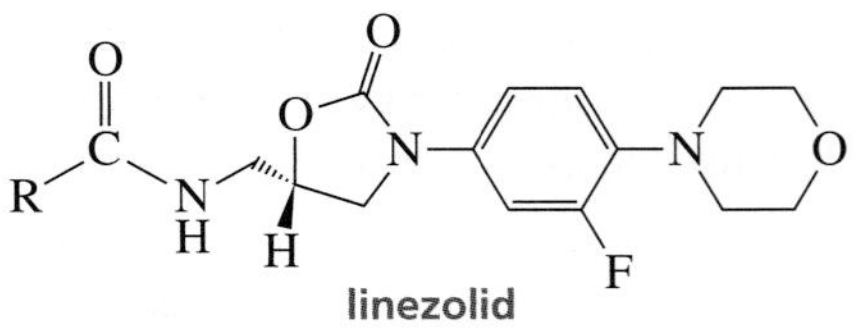

linezolid
Zyvox®

Zyvox is a synthetic compound designed by scientists to inhibit bacterial growth at a point different from that at which any other antibiotic exerts its effect. Zyvox inhibits the initiation of protein synthesis by preventing the formation of the complex between mRNA, the first amino-acid-bearing tRNA, and the ribosome where protein synthesis takes place (Sections 20.7 and 20.8). Because of the drug's new mode of activity, resistance is expected to be rare at first and, hopefully, slow to emerge.

21.8 MOLECULAR MODELING

Because the shape of a molecule determines whether it will be recognized by a receptor and, therefore, whether it will exhibit biological activity, compounds with similar biological activity often have similar structures. Because computers can draw molecular models of compounds on a video display and move them around to assume different conformations, computer **molecular modeling** allows more rational drug design. There are computer programs that allow chemists to scan existing collections of thousands of compounds to find those with the appropriate structural and conformational properties. For example, the binding of netropsin, an antibiotic with a wide range of antimicrobial activity, to DNA is shown in Figure 21.1. Retonavir, a drug used to treat the AIDS virus, inactivates HIV protease, an enzyme essential for the maturation of the virus, by binding to its active site (Figure 21.2).

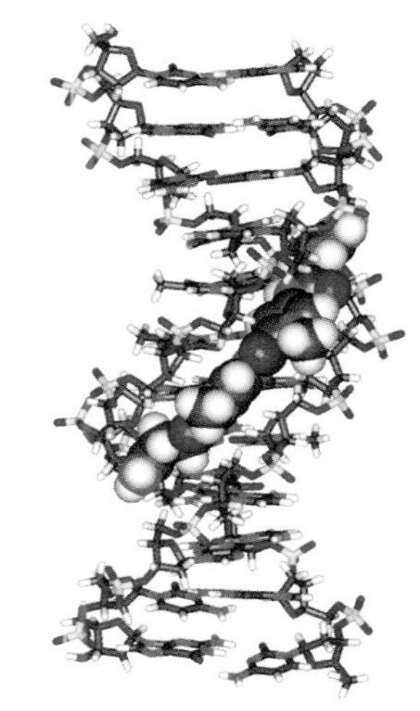

▲ **Figure 21.1**
The antibiotic netropsin bound to DNA.

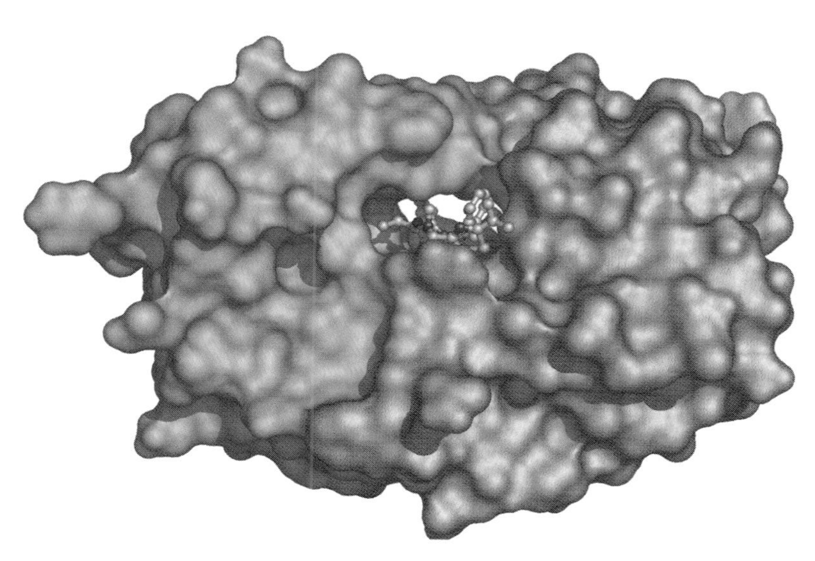

▲ **Figure 21.2**
Retonavir, a drug used to treat the AIDS virus, binds to the active site of HIV protease.

Being able to visualize the fit between the compound and the receptor may suggest modifications that can be made to the compound that might result in more favorable binding. In this way, the selection of compounds to be synthesized to screen for biological activity can be more rational, leading to faster discovery of pharmacologically active compounds. Molecular modeling will become more valuable as scientists learn more about receptor sites.

21.9 ANTIVIRAL DRUGS

Relatively few clinically useful drugs have been developed for viral infections. This slow progress is due to the nature of viruses and the way they replicate. Most **antiviral drugs** are analogs of nucleosides, interfering with the virus's DNA or RNA synthesis. In this way, they prevent the virus from replicating. For example, acyclovir, the drug used against herpes viruses, has a three-dimensional shape similar to guanine. Acyclovir can therefore fool the virus into incorporating it instead of guanine into the virus's DNA. Once this happens, the DNA strand can no longer grow, because acyclovir lacks a 3′-OH group (Section 20.2).

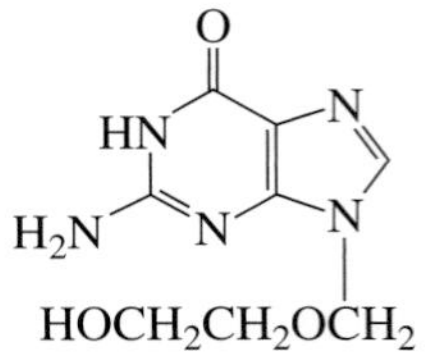

acyclovir
Aclovir®
used against herpes simplex infections

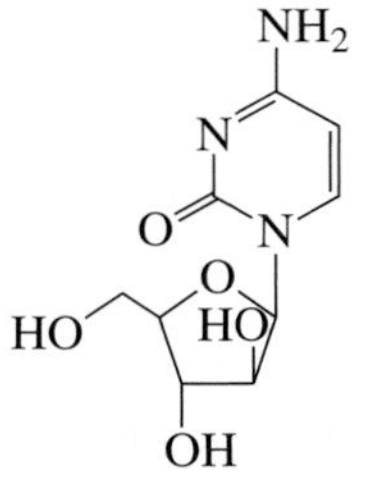

cytarabine
Cytosar®
used against acute myelocytic leukemia

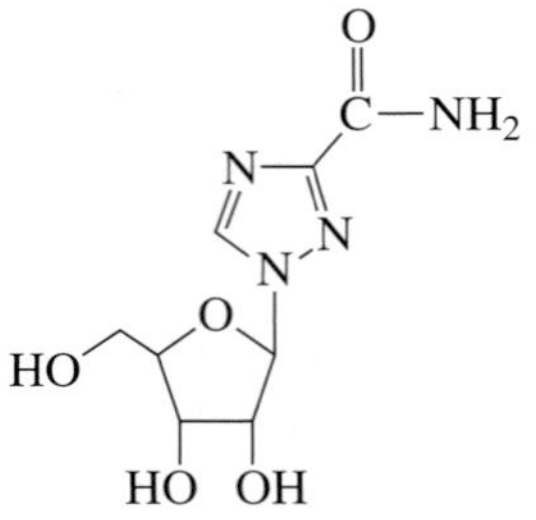

ribavirin
Viramid®
a broad-spectrum antiviral agent

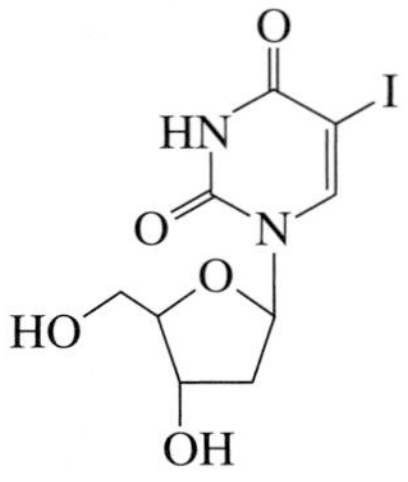

idoxuridine
Herplex®
approved for topical ophthalmic use

Cytarabine, used for acute myelocytic leukemia, competes with cytosine for incorporation into viral DNA. Cytarabine contains an arabinose rather than a ribose (Table 15.1). Because the 2′-OH group is in the β-position (recall that the 2′-OH group of a ribonucleoside is in the α-position), the bases in DNA are not able to stack properly.

Ribavirin is a broad-spectrum antiviral agent that interferes with the synthesis of GTP and, therefore, with the synthesis of all nucleic acids. It is used to treat children with chronic hepatitis C.

Idoxuridine is used for the topical treatment of ocular infections. Idoxuridine has an iodo group in place of the methyl group of thymine and is incorporated into DNA in place of thymine. Chain elongation can continue because idoxuridine has a 3′-OH group, but the resulting DNA is more easily broken and is also not transcribed properly.

21.10 THE ECONOMICS OF DRUGS • GOVERNMENT REGULATIONS

The average cost of launching a new drug is $100 million to $500 million. The manufacturer has to recover this cost quickly because the starting date of a patent is the date the drug is first discovered. A patent is good for 20 years from the date it is applied for, but because it takes an average of 12 years after the initial discovery of a drug before it reaches the marketplace, the patent protects the discoverer of the drug for an average of only 8 years. It is only during the 8 years of patent protection that drug sales can provide income to cover the initial costs and pay for research on new drugs. In addition, the average commercial lifetime of a drug is only 15 to 20 years. After that, the drug is generally replaced by a new and improved drug. Only about 1 in 3 drugs makes a profit for the company.

Why does it cost so much to develop a new drug? First of all, the Food and Drug Administration (FDA) has very high standards that must be met before a drug is approved for a particular use (Section 21.4). An important factor leading to the high price of many drugs is the low success rate in progressing from the initial concept to an approved product: only 1 or 2 of every 100 compounds tested become lead compounds; out of 100 structural modifications of a lead compound, only 1 is worthy of further study. For every 10,000 compounds evaluated in animal studies, only 10 will get to clinical trials.

Clinical trials consist of three phases: phase I evaluates the effectiveness, safety, side effects, and dosage levels in up to 100 healthy volunteers; phase II investigates the effectiveness, safety, and side effects in 100 to 500 volunteers, who have the condition the drug is meant to treat; and phase III establishes the effectiveness and appropriate dosage of the drug and monitors adverse reactions in several thousand volunteer patients. For every 10 compounds that enter clinical trials, only one satisfies the increasingly stringent requirements to become a marketable drug.

ORPHAN DRUGS

Because of the high cost of developing a drug, pharmaceutical companies are reluctant to carry out research on drugs for rare diseases. Even if a company were to find an effective drug for a disease, there would be no way to recoup the expenditure, because of the limited demand. Fortunately for persons suffering from such diseases, the U.S. Congress, in 1983, passed the Orphan Drugs Act. The act creates public subsidies to fund research and provides tax credits for the development and marketing of drugs—called **orphan drugs**—for diseases or conditions that affect fewer than 200,000 people. In addition, the act stipulates that the company that develops the drug has seven years of exclusive marketing rights. In the 10 years prior to the passage of this act, fewer than 10 orphan drugs were developed. Today, more than 250 are available, and more than 1000 others are in development.

Drugs originally developed as orphan drugs include AZT (to treat AIDS), Taxol (to treat ovarian cancer), Exosurf Neonatal (to treat respiratory distress syndrome in infants), Opticrom (to treat corneal swelling), and Epogen (to treat anemia).

SUMMARY

A **drug** is a compound that interacts with a biological molecule, to trigger a physiological response. Each drug has a **brand name** that can be used only by the holder of the patent, which is valid for 20 years. Once a patent expires, other drug companies can market the drug under a **generic name** that can be used by any pharmaceutical company. The Food and Drug Administration (FDA) sets high standards that must be met before it approves a drug for a particular use.

The prototype for a new drug is called a **lead compound**. Changing the structure of a lead compound is called **molecular modification**. A **random screen** is a search for a pharmacologically active lead compound without the benefit of any information about what structures might show activity.

Many drugs exert their physiological effects by binding to a specific binding site called a **receptor**. Most **antiviral drugs** are analogs of nucleosides, interfering with DNA or RNA synthesis and thereby preventing the virus from replicating.

A **bacteriostatic drug** inhibits the further growth of bacteria; a **bactericidal drug** kills the bacteria. In recent years, many bacteria have become resistant to all antibiotics, so **drug resistance** has become an increasingly important problem in medicinal chemistry.

PROBLEMS

1. What is the chemical name of each of the following drugs?
   a. benzocaine (Section 21.3) **b.** idoxuridine (Section 21.9)

2. Based on the lead compound for the development of procaine and lidocaine, propose structures for other compounds that you would like to see tested for use as anesthetics.

3. Which of the following compounds is more likely to exhibit activity as a tranquilizer?

   CH_3O–(benzodiazepine) or $CH_3C(=O)$–(benzodiazepine)

4. Which compound is more likely to be a general anesthetic?

   $CH_3CH_2CH_2OH$ or $CH_3OCH_2CH_3$

5. The therapeutic index of a drug is the ratio of the lethal dose to the therapeutic dose. Therefore, the higher the therapeutic index, the greater is the margin of safety of the drug. The lethal dose of tetrahydrocannabinol in mice is 2.0 g/kg, and the therapeutic dose is 20 mg/kg. The lethal dose of sodium pentathol in mice is 100 mg/kg, and the therapeutic dose is 30 mg/kg. Which is the safer drug?

6. Explain how each of the antiviral drugs shown in Section 21.9 differs from the naturally occurring nucleoside that it most closely resembles.

7. Show a mechanism for the formation of a benzodiazepine 4-oxide from the reaction of a quinazoline 3-oxide with methylamine.

Appendix I

Physical Properties of Organic Compounds

Physical Properties of Alkenes

Name	Structure	mp (°C)	bp (°C)	Density (g/mL)
Ethene	$CH_2{=}CH_2$	−169	−104	
Propene	$CH_2{=}CHCH_3$	−185	−47	
1-Butene	$CH_2{=}CHCH_2CH_3$	−185	−6.3	
1-Pentene	$CH_2{=}CH(CH_2)_2CH_3$	−138	30	0.641
1-Hexene	$CH_2{=}CH(CH_2)_3CH_3$	−140	64	0.673
1-Heptene	$CH_2{=}CH(CH_2)_4CH_3$	−119	94	0.697
1-Octene	$CH_2{=}CH(CH_2)_5CH_3$	−101	122	0.715
1-Nonene	$CH_2{=}CH(CH_2)_6CH_3$	−81	146	0.730
1-Decene	$CH_2{=}CH(CH_2)_7CH_3$	−66	171	0.741
cis-2-Butene	*cis*-$CH_3CH{=}CHCH_3$	−180	37	0.650
trans-2-Butene	*trans*-$CH_3CH{=}CHCH_3$	−140	37	0.649
Methylpropene	$CH_2{=}C(CH_3)_2$	−140	−6.9	0.594
cis-2-Pentene	*cis*-$H_3CH{=}CHCH_2CH_3$	−180	37	0.650
trans-2-Pentene	*trans*-$CH_3CH{=}CHCH_2CH_3$	−140	37	0.649
Cyclohexene		−104	83	0.811

Physical Properties of Alkynes

Name	Structure	mp (°C)	bp (°C)	Density (g/mL)
Ethyne	$HC{\equiv}CH$	−82	−84.0	
Propyne	$HC{\equiv}CCH_3$	−101.5	−23.2	
1-Butyne	$HC{\equiv}CCH_2CH_3$	−122	8.1	
2-Butyne	$CH_3C{\equiv}CCH_3$	−32	27	0.694
1-Pentyne	$HC{\equiv}C(CH_2)_2CH_3$	−98	39.3	0.695
2-Pentyne	$CH_3C{\equiv}CCH_2CH_3$	−101	55.5	0.714
3-Methyl-1-butyne	$HC{\equiv}CCH(CH_3)_2$	−90	29	0.665
1-Hexyne	$HC{\equiv}C(CH_2)_3CH_3$	−132	71	0.715
2-Hexyne	$CH_3C{\equiv}C(CH_2)_2CH_3$	−92	84	0.731
3-Hexyne	$CH_3CH_2C{\equiv}CCH_2CH_3$	−101	81	0.725
1-Heptyne	$HC{\equiv}C(CH_2)_4CH_3$	−81	100	0.733
1-Octyne	$HC{\equiv}C(CH_2)_5CH_3$	−80	127	0.747
1-Nonyne	$HC{\equiv}C(CH_2)_6CH_3$	−50	151	0.757
1-Decyne	$HC{\equiv}C(CH_2)_7CH_3$	−44	174	0.766

Physical Properties of Cyclic Saturated Alkanes

Name	mp (°C)	bp (°C)	Density (g/mL)
Cyclopropane	−128	−33	
Cyclobutane	−80	−12	
Cyclopentane	−94	50	0.751
Cyclohexane	6.5	81	0.779
Cycloheptane	−12	118	0.811
Cyclooctane	14	149	0.834
Methylcyclopentane	−142	72	0.749
Methylcyclohexane	−126	100	0.769
cis-1,2-Dimethylcyclopentane	−62	99	0.772
trans-1,2-Dimethylcyclopentane	−120	92	0.750

Physical Properties of Ethers

Name	Structure	mp (°C)	bp (°C)	Density (g/mL)
Dimethyl ether	CH_3OCH_3	−141	−24.8	
Diethyl ether	$CH_3CH_2OCH_2CH_3$	−116	34.6	0.706
Dipropyl ether	$CH_3(CH_2)_2O(CH_2)_2CH_3$	−123	88	0.736
Diisopropyl ether	$(CH_3)_2CHOCH(CH_3)_2$	−86	69	0.725
Dibutyl ether	$CH_3(CH_2)_3O(CH_2)_3CH_3$	−98	142	0.764
Divinyl ether	$CH_2{=}CHOCH{=}CH_2$		35	
Diallyl ether	$CH_2{=}CHCH_2OCH_2CH{=}CH_2$		94	0.830
Tetrahydrofuran	(five-membered ring with O)	−108	66	0.889
Dioxane	(six-membered ring with two O)	12	101	1.034

Physical Properties of Alcohols

Name	Structure	mp (°C)	bp (°C)	Solubility (g/100 g H_2O at 25 °C)
Methanol	CH_3OH	−97.8	64	∞
Ethanol	CH_3CH_2OH	−114.7	78	∞
1-Propanol	$CH_3(CH_2)_2OH$	−127	97.4	∞
1-Butanol	$CH_3(CH_2)_3OH$	−90	118	7.9
1-Pentanol	$CH_3(CH_2)_4OH$	−78	138	2.3
1-Hexanol	$CH_3(CH_2)_5OH$	−52	157	0.6
1-Heptanol	$CH_3(CH_2)_6OH$	−36	176	0.2
1-Octanol	$CH_3(CH_2)_7OH$	−15	196	0.05
2-Propanol	$CH_3CHOHCH_3$	−89.5	82	∞
2-Butanol	$CH_3CHOHCH_2CH_3$	−115	99.5	12.5
2-Methyl-1-propanol	$(CH_3)_2CHCH_2OH$	−108	108	10.0
2-Methyl-2-propanol	$(CH_3)_3COH$	25.5	83	∞
3-Methyl-1-butanol	$(CH_3)_2CH(CH_2)_2OH$	−117	130	2
2-Methyl-2-butanol	$(CH_3)_2COHCH_2CH_3$	−12	102	12.5
2,2-Dimethyl-1-propanol	$(CH_3)_3CCH_2OH$	55	114	∞
Allyl alcohol	$CH_2{=}CHCH_2OH$	−129	97	∞
Cyclopentanol	C_5H_9OH	−19	140	s. sol.
Cyclohexanol	$C_6H_{11}OH$	24	161	s. sol.
Benzyl alcohol	$C_6H_5CH_2OH$	−15	205	4

Physical Properties of Alkyl Halides

Name	bp (°C)			
	Fluoride	*Chloride*	*Bromide*	*Iodide*
Methyl	−78.4	−24.2	3.6	42.4
Ethyl	−37.7	12.3	38.4	72.3
Propyl	−2.5	46.6	71.0	102.5
Isopropyl	−9.4	34.8	59.4	89.5
Butyl	32.5	78.4	100	130.5
Isobutyl		68.8	90	120
sec-Butyl		68.3	91.2	120.0
tert-Butyl		50.2	73.1	dec.
Pentyl	62.8	108	130	157.0
Hexyl	92	133	154	179

Physical Properties of Amines

Name	Structure	mp (°C)	bp (°C)	Solubility (g/100 g H_2O at 25 °C)
Primary Amines				
Methylamine	CH_3NH_2	−93	−6.3	v. sol.
Ethylamine	$CH_3CH_2NH_2$	−81	17	∞
Propylamine	$CH_3(CH_2)_2NH_2$	−83	48	∞
Isopropylamine	$(CH_3)_2CHNH_2$	−95	33	∞
Butylamine	$CH_3(CH_2)_3NH_2$	−49	78	v. sol.
Isobutylamine	$(CH_3)_2CHCH_2NH_2$	−85	68	∞
sec-Butylamine	$CH_3CH_2CH(CH_3)NH_2$	−72	63	∞
tert-Butylamine	$(CH_3)_3CNH_2$	−67	46	∞
Cyclohexylamine	$C_6H_{11}NH_2$	−18	134	s. sol.
Secondary Amines				
Dimethylamine	$(CH_3)_2NH$	−93	7.4	v. sol.
Diethylamine	$(CH_3CH_2)_2NH$	−93	55	10.0
Dipropylamine	$(CH_3CH_2CH_2)_2NH$	−50	110	10.0
Dibutylamine	$(CH_3CH_2CH_2CH_2)_2NH$	−62	159	s. sol.
Tertiary Amines				
Trimethylamine	$(CH_3)_3N$	−115	2.9	91
Triethylamine	$(CH_3CH_2)_3N$	−114	89	14
Tripropylamine	$(CH_3CH_2CH_2)_3N$	−93	157	s. sol.

Physical Properties of Benzene and Substituted Benzenes

Name	Structure	mp (°C)	bp (°C)	Solubility (g/100 g H_2O at 25 °C)
Aniline	$C_6H_5NH_2$	−6	184	3.7
Benzene	C_6H_6	5.5	80.1	s. sol.
Benzaldehyde	C_6H_5CHO	−26	178	s. sol.
Benzamide	$C_6H_5CONH_2$	132	290	s. sol.
Benzoic acid	C_6H_5COOH	122	249	0.34
Bromobenzene	C_6H_5Br	−30.8	156	insol.
Chlorobenzene	C_6H_5Cl	−45.6	132	insol.
Nitrobenzene	$C_6H_5NO_2$	5.7	210.8	s. sol.
Phenol	C_6H_5OH	43	182	s. sol.
Styrene	$C_6H_5CH{=}CH_2$	−30.6	145.2	insol.
Toluene	$C_6H_5CH_3$	−95	110.6	insol.

Physical Properties of Carboxylic Acids

Name	Structure	mp (°C)	bp (°C)	Solubility (g/100 g H_2O at 25 °C)
Formic acid	$HCOOH$	8.4	101	∞
Acetic acid	CH_3COOH	16.6	118	∞
Propionic acid	CH_3CH_2COOH	−21	141	∞
Butanoic acid	$CH_3(CH_2)_2COOH$	−5	162	∞
Pentanoic acid	$CH_3(CH_2)_3COOH$	−34	186	4.97
Hexanoic acid	$CH_3(CH_2)_4COOH$	−4	202	0.97
Heptanoic acid	$CH_3(CH_2)_5COOH$	−8	223	0.24
Octanoic acid	$CH_3(CH_2)_6COOH$	17	237	0.068
Nonanoic acid	$CH_3(CH_2)_7COOH$	15	255	0.026
Decanoic acid	$CH_3(CH_2)_8COOH$	32	270	0.015

Physical Properties of Dicarboxylic Acids

Name	Structure	mp (°C)	Solubility (g/100 g H_2O at 25 °C)
Oxalic acid	$HOOCCOOH$	189	s
Malonic acid	$HOOCCH_2COOH$	136	v. sol.
Succinic acid	$HOOC(CH_2)_2COOH$	185	s. sol.
Glutaric acid	$HOOC(CH_2)_3COOH$	98	v. sol.
Adipic acid	$HOOC(CH_2)_4COOH$	151	s. sol.
Pimelic acid	$HOOC(CH_2)_5COOH$	106	s. sol.
Phthalic acid	1,2-$C_6H_4(COOH)_2$	231	s. sol.
Maleic acid	*cis*-$HOOCCH{=}CHCOOH$	130.5	v. sol.
Fumaric acid	*trans*-$HOOCCH{=}CHCOOH$	302	s. sol.

Physical Properties of Acyl Chlorides and Acid Anhydrides

Name	Structure	mp (°C)	bp (°C)
Acetyl chloride	CH_3COCl	−112	51
Propionyl chloride	CH_3CH_2COCl	−94	80
Butyryl chloride	$CH_3(CH_2)_2COCl$	−89	102
Valeryl chloride	$CH_3(CH_2)_3COCl$	−110	128
Acetic anhydride	$CH_3(CO)O(CO)CH_3$	−73	140
Succinic anhydride	(ring structure: O=C–O–C=O, five-membered)	120	261

Physical Properties of Esters

Name	Structure	mp (°C)	bp (°C)
Methyl formate	$HCOOCH_3$	−100	32
Ethyl formate	$HCOOCH_2CH_3$	−80	54
Methyl acetate	CH_3COOCH_3	−98	57.5
Ethyl acetate	$CH_3COOCH_2CH_3$	−84	77
Propyl acetate	$CH_3COO(CH_2)_2CH_3$	−92	102
Methyl propionate	$CH_3CH_2COOCH_3$	−87.5	80
Ethyl propionate	$CH_3CH_2COOCH_2CH_3$	−74	99
Methyl butyrate	$CH_3CH_2CH_2COOCH_3$	−84.8	102.3
Ethyl butyrate	$CH_3CH_2CH_2COOCH_2CH_3$	−93	121

Physical Properties of Amides

Name	Structure	mp (°C)	bp (°C)
Formamide	$HCONH_2$	3	200 d*
Acetamide	CH_3CONH_2	82	221
Propanamide	$CH_3CH_2CONH_2$	80	213
Butanamide	$CH_3(CH_2)_2CONH_2$	116	216
Pentanamide	$CH_3(CH_2)_3CONH_2$	106	232

*d means the substance decomposes.

Physical Properties of Aldehydes

Name	Structure	mp (°C)	bp (°C)	Solubility (g/100 g H_2O at 25 °C)
Formaldehyde	$HCHO$	−92	−21	v. sol.
Acetaldehyde	CH_3CHO	−121	21	∞
Propionaldehyde	CH_3CH_2CHO	−81	49	16
Butyraldehyde	$CH_3(CH_2)_2CHO$	−96	75	7
Pentanal	$CH_3(CH_2)_3CHO$	−92	103	s. sol.
Hexanal	$CH_3(CH_2)_4CHO$	−56	131	s. sol.
Heptanal	$CH_3(CH_2)_5CHO$	−43	153	0.1
Octanal	$CH_3(CH_2)_6CHO$		171	insol.
Nonanal	$CH_3(CH_2)_7CHO$		192	insol.
Decanal	$CH_3(CH_2)_8CHO$	−5	209	insol.
Benzaldehyde	C_6H_5CHO	−26	178	0.3

Physical Properties of Ketones

Name	Structure	mp (°C)	bp (°C)	Solubility (g/100 g H_2O at 25 °C)
Acetone	CH_3COCH_3	−95	56	∞
2-Butanone	$CH_3COCH_2CH_3$	−86	80	25.6
2-Pentanone	$CH_3CO(CH_2)_2CH_3$	−78	102	5.5
2-Hexanone	$CH_3CO(CH_2)_3CH_3$	−57	127	1.6
2-Heptanone	$CH_3\ CO(CH_2)_4CH_3$	−36	151	0.4
2-Octanone	$CH_3CO(CH_2)_5CH_3$	−16	173	insol.
2-Nonanone	$CH_3CO(CH_2)_6CH_3$	−7	195	insol.
2-Decanone	$CH_3CO(CH_2)_7CH_3$	14	210	insol.
3-Pentanone	$CH_3CH_2COCH_2CH_3$	−40	102	4.8
3-Hexanone	$CH_3CH_2CO(CH_2)_2CH_3$		123	1.5
3-Heptanone	$CH_3CH_2CO(CH_2)_3CH_3$	−39	149	0.3
Acetophenone	$CH_3COC_6H_5$	19	202	insol.
Propiophenone	$CH_3CH_2COC_6H_5$	18	218	insol.

Appendix II

pK_a Values

Compound	pK_a
$CH_3C{\equiv}\overset{+}{N}H$	−10.1
HI	−10
HBr	−9
$CH_3CH{=}\overset{+}{O}H$	−8
$CH_3C({=}\overset{+}{O}H)CH_3$	−7.3
HCl	−7
$C_6H_5SO_3H$ (benzenesulfonic acid)	−6.5
$CH_3C({=}\overset{+}{O}H)OCH_3$	−6.5
$CH_3C({=}\overset{+}{O}H)OH$	−6.1
H_2SO_4	−5
protonated pyrrole (2H-pyrrolium, CH_2 ring carbon, $\overset{+}{N}H$)	−3.8
$CH_3CH_2\overset{+}{O}(H)CH_2CH_3$	−3.6
$CH_3CH_2\overset{+}{O}H_2$	−2.4
$CH_3\overset{+}{O}H_2$	−2.5
H_3O^+	−1.7
HNO_3	−1.3
CH_3SO_3H	−1.2
$CH_3C({=}\overset{+}{O}H)NH_2$	0.0
F_3CCOOH	0.2
Cl_3CCOOH	0.64
protonated pyridine N-oxide ($C_5H_5\overset{+}{N}{-}OH$)	0.79
$O_2N{-}C_6H_4{-}\overset{+}{N}H_3$ (para)	1.0
protonated pyrimidine ($\overset{+}{N}H$)	1.0
$Cl_2CHCOOH$	1.3
HSO_4^-	2.0
H_3PO_4	2.1
protonated purine ($H\overset{+}{N}$)	2.5
FCH_2COOH	2.7
$ClCH_2COOH$	2.8
$BrCH_2COOH$	2.9
ICH_2COOH	3.2
HF	3.2
HNO_2	3.4
$O_2N{-}C_6H_4{-}COOH$ (para)	3.4
$HCOOH$	3.8
$Br{-}C_6H_4{-}\overset{+}{N}H_3$ (para)	3.9
$Br{-}C_6H_4{-}COOH$ (para)	4.0
C_6H_5COOH	4.2
$CH_3{-}C_6H_4{-}COOH$ (para)	4.3
$CH_3O{-}C_6H_4{-}COOH$ (para)	4.5
$C_6H_5\overset{+}{N}H_3$	4.6
CH_3COOH	4.8
quinolinium ($\overset{+}{N}H$)	4.9
$CH_3{-}C_6H_4{-}\overset{+}{N}H_3$ (para)	5.1
pyridinium ($\overset{+}{N}H$)	5.2
$CH_3O{-}C_6H_4{-}\overset{+}{N}H_3$ (para)	5.3
$CH_3C(CH_3){=}\overset{+}{N}HCH_3$	5.5
$CH_3C(O)CH_2CH(O)$	5.9
$HO\overset{+}{N}H_3$	6.0
H_2CO_3	6.4
imidazolium ($H\overset{+}{N}$, NH)	6.8
H_2S	7.0
$O_2N{-}C_6H_4{-}OH$ (para)	7.1
$H_2PO_4^-$	7.2
C_6H_5SH	7.8

[a]pK_a values are for the red H in each structure

pK_a Values (continued)

Compound	pK_a	Compound	pK_a	Compound	pK_a
Aziridinium (protonated aziridine, N^+ with H, H)	8.0	Cyclohexyl–$\overset{+}{N}H_3$	10.7	CH_3CHO (acetaldehyde)	17
$H_2N\overset{+}{N}H_3$	8.1	$(CH_3)_2\overset{+}{N}H_2$	10.7	$(CH_3)_3COH$	18
$CH_3C(=O)OOH$	8.2	Piperidinium (protonated, N^+ with H, H)	11.1	$CH_3C(=O)CH_3$	20
$CH_3CH_2NO_2$	8.6	$CH_3CH_2\overset{+}{N}H_3$	11.0	$CH_3C(=O)OCH_2CH_3$	24.5
$CH_3C(=O)CH_2C(=O)CH_3$	8.9	Pyrrolidinium (protonated, N^+ with H, H)	11.3	$HC{\equiv}CH$	25
$HC{\equiv}N$	9.1	HOOH	11.6	$CH_3C{\equiv}N$	25
Morpholinium (protonated, N^+ with H, H)	9.3	HPO_4^{2-}	12.3	$CH_3C(=O)N(CH_3)_2$	30
Cl–C_6H_4–OH	9.4	CF_3CH_2OH	12.4	NH_3	36
$\overset{+}{N}H_4$	9.4	$CH_3CH_2OC(=O)CH_2C(=O)OCH_2CH_3$	13.3	Pyrrolidine (N–H)	36
$HOCH_2CH_2\overset{+}{N}H_3$	9.5	$HC{\equiv}CCH_2OH$	13.5	CH_3NH_2	40
$H_3\overset{+}{N}CH_2C(=O)O^-$	9.8	$H_2NC(=O)NH_2$	13.7	C_6H_5–CH_3	41
C_6H_5–OH	10.0	$CH_3\overset{+}{N}(CH_3)(CH_3)CH_2CH_2OH$	13.9	Benzene	43
CH_3–C_6H_4–OH	10.2	Imidazole (N, NH)	14.4	$CH_2{=}CHCH_3$	43
HCO_3^-	10.2	CH_3OH	15.5	$CH_2{=}CH_2$	44
CH_3NO_2	10.2	H_2O	15.7	Cyclopropane	46
H_2N–C_6H_4–OH	10.3	CH_3CH_2OH	16.0	CH_4	60
CH_3CH_2SH	10.5	$CH_3C(=O)NH_2$	16	CH_3CH_3	> 60
$(CH_3)_3\overset{+}{N}H$	10.6	C_6H_5–$C(=O)CH_3$	16.0		
$CH_3C(=O)CH_2C(=O)OCH_2CH_3$	10.7	Pyrrole (N–H)	~17		
$CH_3\overset{+}{N}H_3$	10.7				

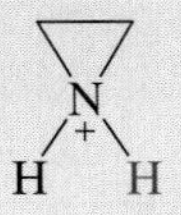

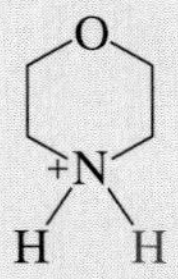

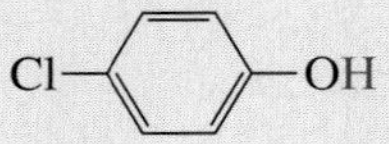

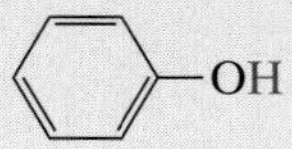

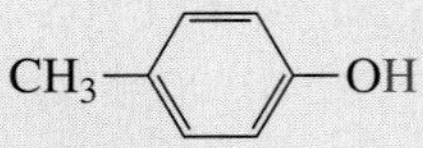

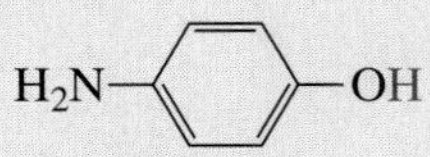

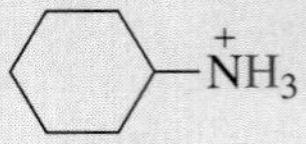

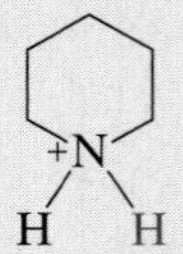

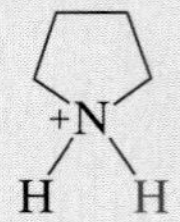

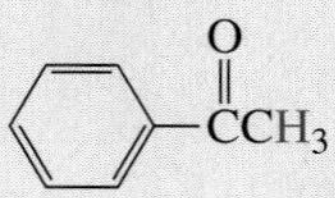

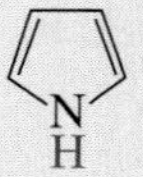

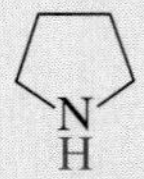

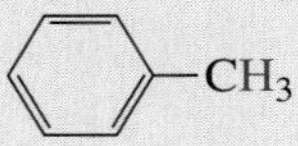

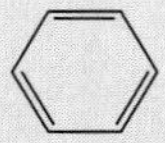

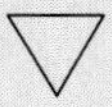

Appendix III

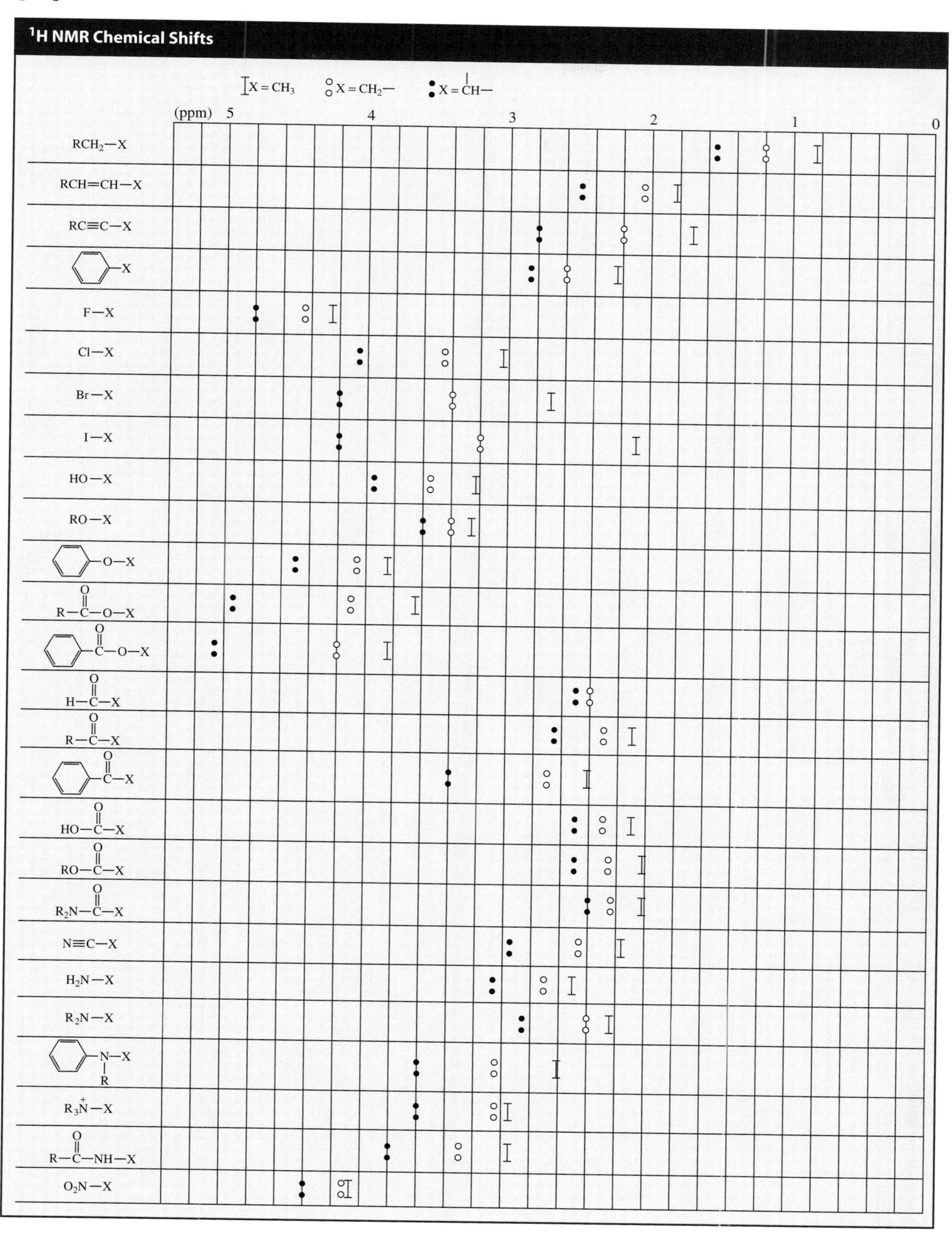

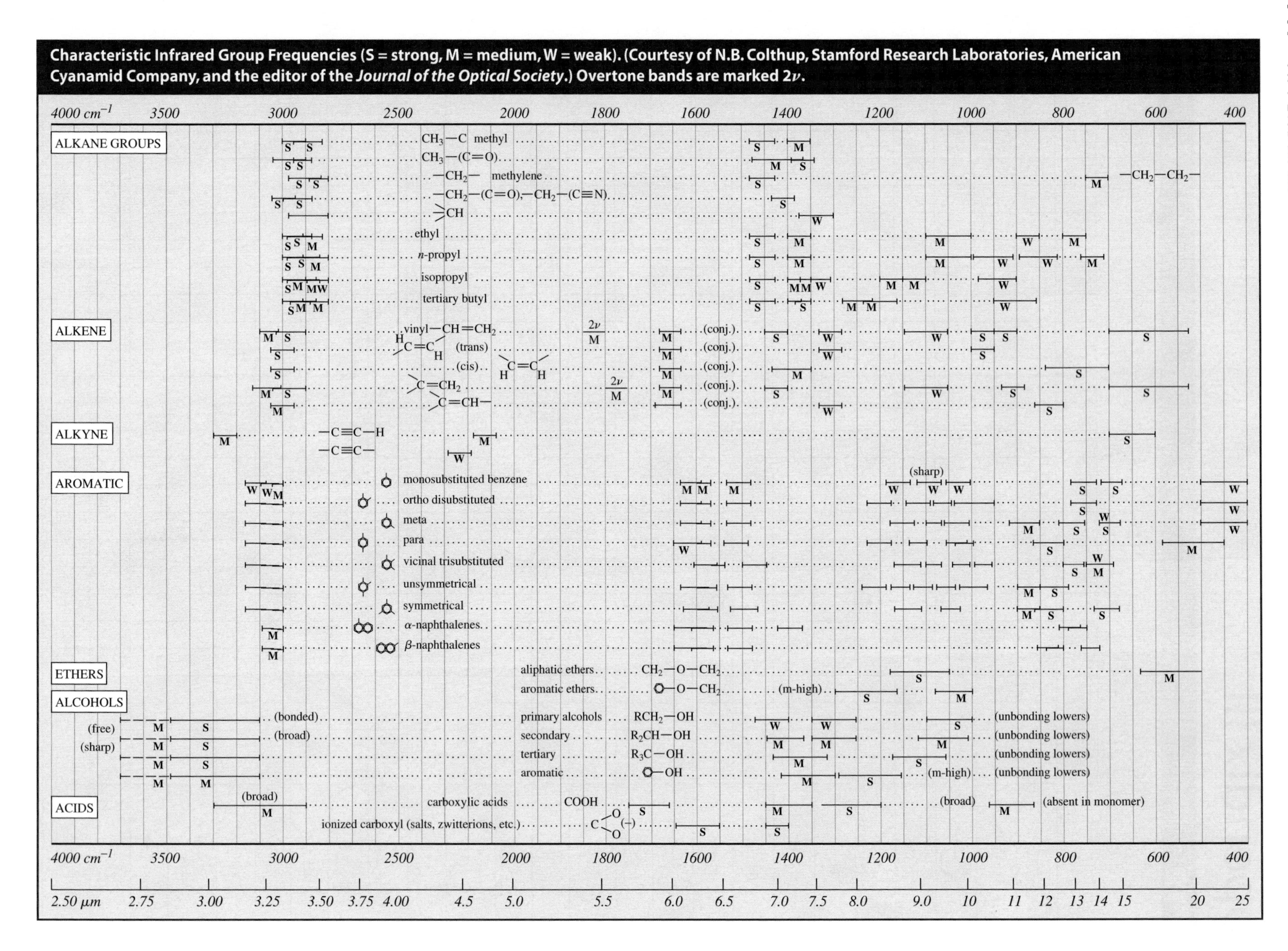
Characteristic Infrared Group Frequencies (S = strong, M = medium, W = weak). (Courtesy of N.B. Colthup, Stamford Research Laboratories, American Cyanamid Company, and the editor of the Journal of the Optical Society.) Overtone bands are marked 2ν.
4000 cm⁻¹
3500
3000
2500
2000
1800
1600
1400
1200
1000
800
600
400
ALKANE GROUPS
CH₃—C methyl
CH₃—(C=O)
—CH₂— methylene
—CH₂—(C=O),—CH₂—(C≡N)
CH
—CH₂—CH₂—
ethyl
n-propyl
isopropyl
tertiary butyl
ALKENE
vinyl—CH=CH₂
(trans)
(cis)
C=CH₂
C=CH—
(conj.)
2ν
ALKYNE
—C≡C—H
—C≡C—
AROMATIC
monosubstituted benzene
ortho disubstituted
meta
para
vicinal trisubstituted
unsymmetrical
symmetrical
α-naphthalenes
β-naphthalenes
(sharp)
ETHERS
aliphatic ethers
CH₂—O—CH₂
aromatic ethers
(m-high)
ALCOHOLS
(free)
(sharp)
(bonded)
(broad)
primary alcohols
RCH₂—OH
secondary
R₂CH—OH
tertiary
R₃C—OH
aromatic
—OH
(unbonding lowers)
(m-high)
ACIDS
(broad)
carboxylic acids
COOH
ionized carboxyl (salts, zwitterions, etc.)
(absent in monomer)
2.50 μm
2.75
3.00
3.25
3.50
3.75
4.00
4.5
5.0
5.5
6.0
6.5
7.0
7.5
8.0
9.0
10
11
12
13
14
15
20
25

Characteristic Infrared Group Frequencies (continued)

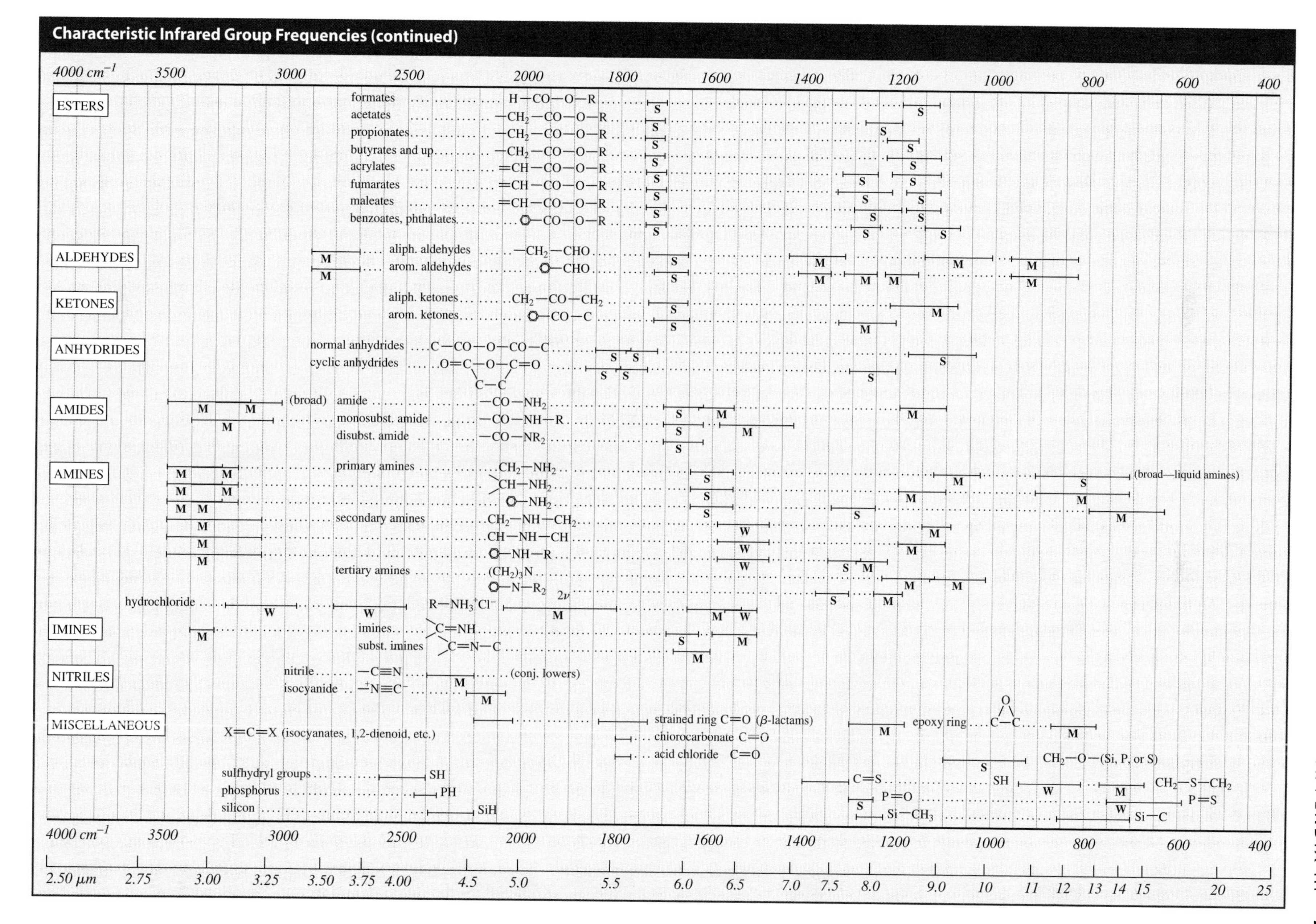

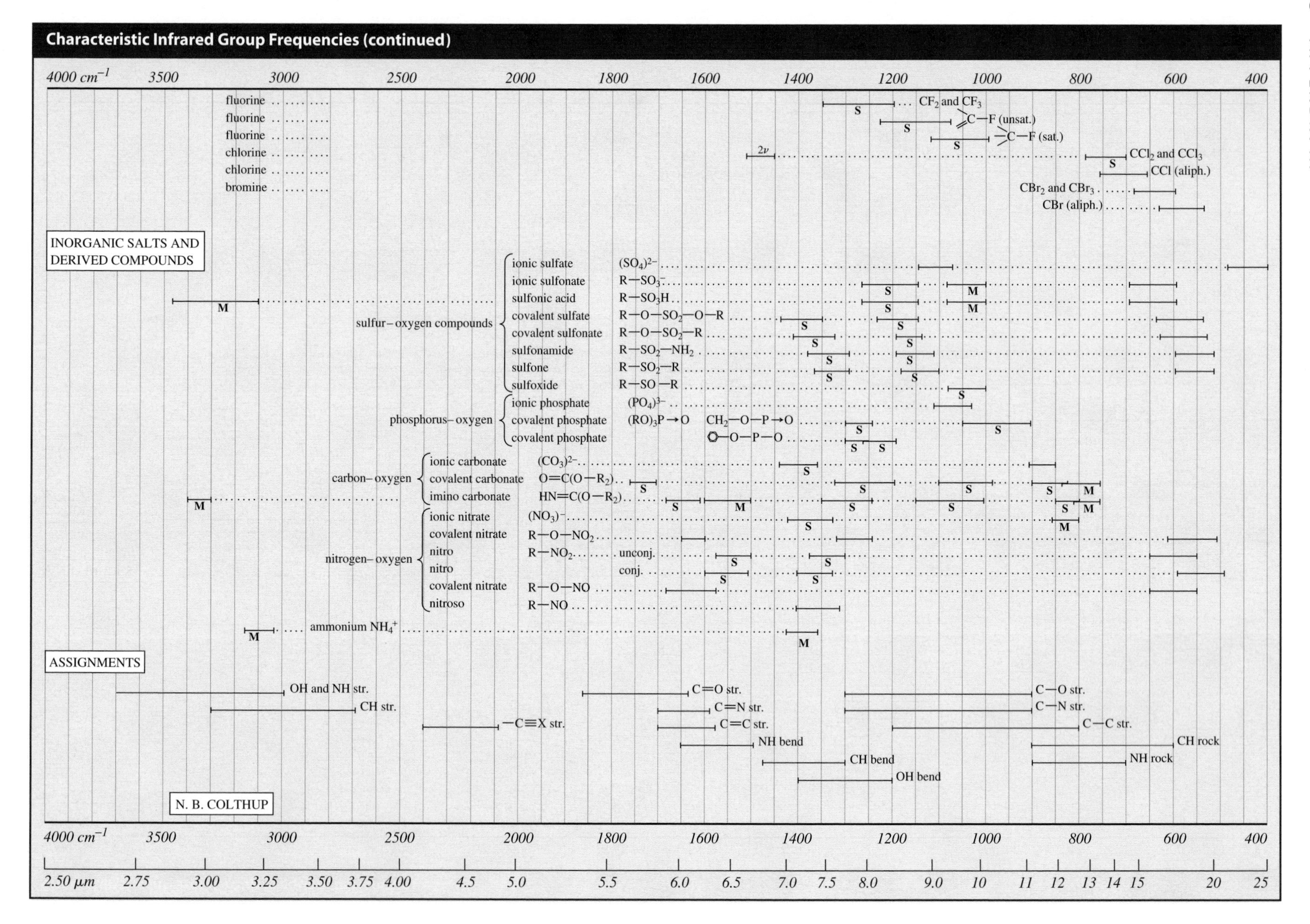
Characteristic Infrared Group Frequencies (continued)
4000 cm^{-1}
3500
3000
2500
2000
1800
1600
1400
1200
1000
800
600
400
fluorine
fluorine
fluorine
chlorine
chlorine
bromine
CF_2 and CF_3
C—F (unsat.)
C—F (sat.)
2ν
CCl_2 and CCl_3
CCl (aliph.)
CBr_2 and CBr_3
CBr (aliph.)
S
M
INORGANIC SALTS AND DERIVED COMPOUNDS
sulfur–oxygen compounds
ionic sulfate
ionic sulfonate
sulfonic acid
covalent sulfate
covalent sulfonate
sulfonamide
sulfone
sulfoxide
$(SO_4)^{2-}$
$R—SO_3^-$
$R—SO_3H$
$R—O—SO_2—O—R$
$R—O—SO_2—R$
$R—SO_2—NH_2$
$R—SO_2—R$
R—SO—R
phosphorus–oxygen
ionic phosphate
covalent phosphate
covalent phosphate
$(PO_4)^{3-}$
$(RO)_3P\rightarrow O$
$CH_2—O—P\rightarrow O$
⌬—O—P—O
carbon–oxygen
ionic carbonate
covalent carbonate
imino carbonate
$(CO_3)^{2-}$
$O{=}C(O—R_2)$
$HN{=}C(O—R_2)$
nitrogen–oxygen
ionic nitrate
covalent nitrate
nitro
nitro
covalent nitrate
nitroso
$(NO_3)^-$
$R—O—NO_2$
$R—NO_2$
unconj.
conj.
R—O—NO
R—NO
ammonium NH_4^+
ASSIGNMENTS
OH and NH str.
CH str.
—C≡X str.
C=O str.
C=N str.
C=C str.
NH bend
CH bend
OH bend
C—O str.
C—N str.
C—C str.
CH rock
NH rock
N. B. COLTHUP
2.50 μm
2.75
3.00
3.25
3.50
3.75
4.00
4.5
5.0
5.5
6.0
6.5
7.0
7.5
8.0
9.0
10
11
12
13
14
15
20
25

Answers to Selected Problems

CHAPTER 1

1. mn = 16 = 8 p + 8n; mn = 17 = 8 p + 9 n; mn = 18 = 8 p + 10 n **2. a.** 4 **b.** 5 **c.** 6 **d.** 7 **3.** 1 **4. a.** N = 2 core, 5 valence; P = 10 core, 5 valence **b.** O = 2 core, 6 valence; S = 10 core, 6 valence **5.** 7 **6. a.** Cl—CH_3 **b.** H—OH **c.** H—F **d.** Cl—CH_3 **7. a.** KCl **b.** Cl_2

8. a. $\overset{\delta-}{HO}—\overset{\delta+}{H}$ **b.** $H_3\overset{\delta+}{C}—\overset{\delta-}{NH_2}$ **c.** $\overset{\delta-}{HO}—\overset{\delta+}{Br}$ **d.** $\overset{\delta+}{I}—\overset{\delta-}{Cl}$

9. a. LiH and HF **b.** HF **10. a.** oxygen **b.** oxygen **c.** oxygen **d.** hydrogen

11. a. $CH_3—\overset{+}{\ddot{O}}(H)—CH_3$ **b.** $H—\overset{-}{\ddot{C}}(H)—H$ **c.** $CH_3—\overset{+}{N}(CH_3)_2—CH_3$ **d.** $H—\overset{+}{N}H_2—\overset{-}{B}H_2—H$

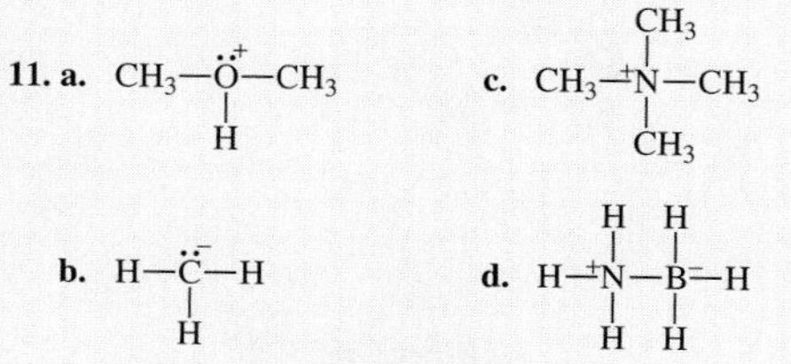

12. a. $H:\overset{HH}{\underset{HH}{C:N}}\overset{+}{:}H$ **b.** $H:\underset{HH}{\overset{H}{C}:\ddot{C}}:H$ **c.** Na^+ $:\ddot{O}:H$ **d.** $H:\overset{H}{\underset{H}{\ddot{N}}}\overset{+}{:}H$ $:\ddot{Cl}:^-$

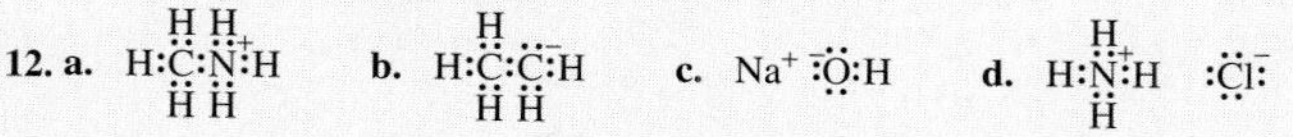

13. a. $CH_3CH_2\ddot{N}H_2$ **b.** $CH_3\ddot{N}HCH_3$ **c.** $CH_3CH_2\ddot{O}H$ **d.** $CH_3\ddot{O}CH_3$ **e.** $CH_3CH_2\ddot{Cl}:$ **f.** $H\ddot{O}\ddot{N}H_2$

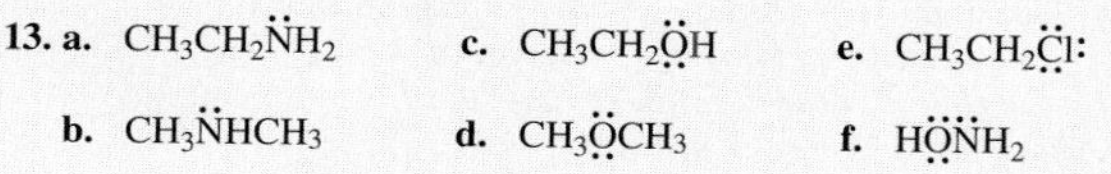

14. a. $CH_3CH_2CH_2Cl$ **b.** $CH_3CH_2C{\equiv}N$ **c.** $CH_3\overset{O}{\overset{\|}{C}}OCH_2CH_3$ **d.** $CH_3CH_2\overset{O}{\overset{\|}{C}}N(CH_3)CH_2CH_3$

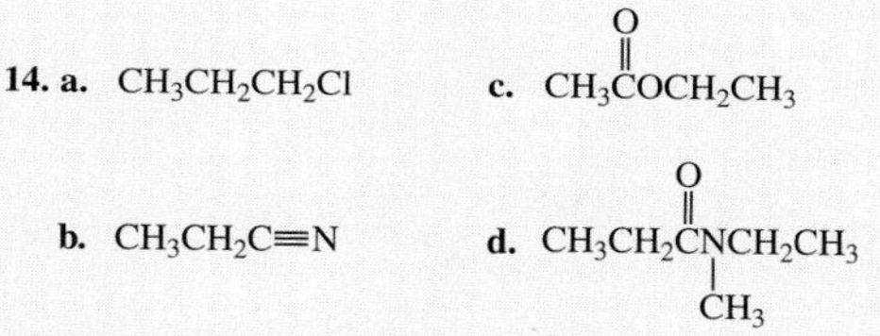

15. a. Cl **b.** O **c.** N **d.** C and H **16.** C_2H_7, C_3H_9 **19.** 2 C—C bonds from $sp^3—sp^3$ overlap; 8 C—H bonds from $sp^3—s$ overlap **21.** greater than 104.5° and less than 109.5° **22.** the hydrogens **23.** Water is the most polar. Methane is the least polar. **24.** ~ 107.3° **25. a.** relative lengths: $Br_2 > Cl_2$; relative strengths: $Cl_2 > Br_2$ **b.** relative lengths: HBr > HCl > HF; relative strengths: HF > HCl > HBr **26. a. 1.** C—Br **2.** C—C **3.** H—Cl **b. 1.** C—Cl **2.** C—H **3.** H—H **27.** σ bond

28. a. $CH_3CH(CH_3)CH{=}CHCH_2C{\equiv}CCH_3$ (labels: sp^3, sp^2, sp, sp^3, sp^3)

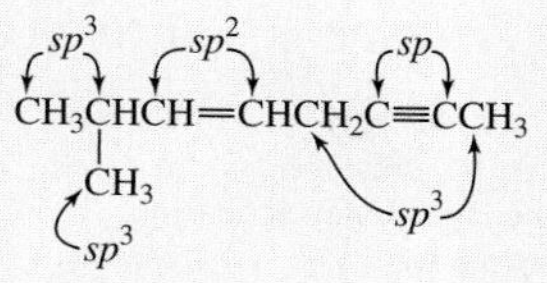

29. a. 109.5° **b.** 107.3° **c.** 109.5° **d.** 104.5°

CHAPTER 2

1. a. 1. $^+NH_4$ **2.** HCl **3.** H_2O **4.** H_3O^+ **b. 1.** $^-NH_2$ **2.** Br^- **3.** NO_3^- **4.** HO^- **3. a.** compound with $pK_a = 5.2$ **b.** compound with dissociation constant = 3.4×10^{-3} **4.** $K_a = 1.51\ 10^{-5}$; weaker **6. a.** basic **b.** acidic **c.** basic **7. a.** CH_3COO^- **b.** $^-NH_2$ **c.** H_2O

8. CH_3NH^- > CH_3O^- > CH_3NH_2 > $CH_3\overset{O}{\overset{\|}{C}}O^-$ > CH_3OH

9. CH_3O^- + $CH_3\overset{+}{N}H_3$

11. a. HBr **b.** $CH_3CH_2CH_2\overset{+}{O}H_2$ **c.** structure on the right **12. a.** F^- **b.** I^- **13. a.** oxygen **b.** H_2S **c.** CH_3SH **d.** $CH_3C({=}O)SH$ **14. a.** HO^- **b.** NH_3 **c.** CH_3O^- **d.** CH_3O^- **15. a.** CH_3COO^- **b.** $CH_3CH_2\overset{+}{N}H_3$ **c.** H_2O **d.** Br^-

e. $^+NH_4$ **f.** HC≡N **g.** NO_2^- **h.** NO_3^- **16. a. 1.** neutral **2.** neutral **3.** charged **4.** charged **5.** charged **6.** charged **b. 1.** charged **2.** charged **3.** charged **4.** charged **5.** neutral **6.** neutral **c. 1.** neutral **2.** neutral **3.** neutral **4.** neutral **5.** neutral **6.** neutral

17. a. 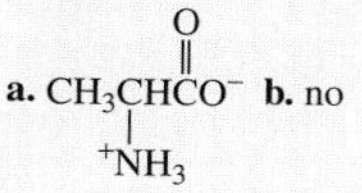 $CH_3CH(\overset{+}{N}H_3)\overset{O}{\overset{\|}{C}}O^-$ **b.** no

CHAPTER 3

1. a. *n*-propyl alcohol or propyl alcohol **b.** dimethyl ether **c.** *n*-propylamine or propylamine

3. a. $CH_3CH(CH_3)OH$ **b.** $CH_3CH(CH_3)CH_2CH_2F$ **c.** $CH_3CH_2OCH_2CH_2CH_3$ **d.** $CH_3CH_2CH(CH_3)I$ **e.** $CH_3C(CH_3)_2NH_2$ **f.** $CH_3CH_2CH_2CH_2CH_2CH_2CH_2CH_2Br$

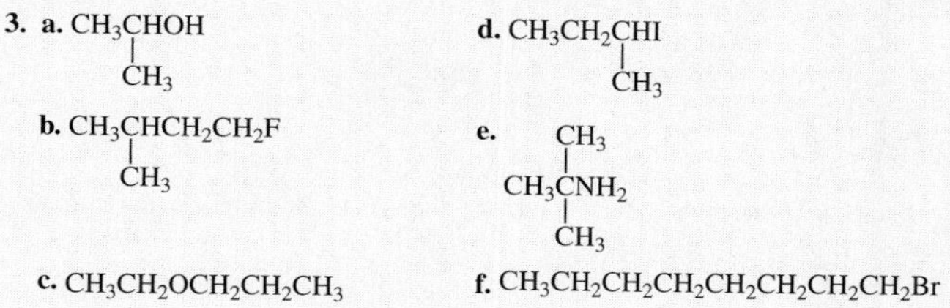

4. a. ethyl methyl ether **b.** methyl propyl ether **c.** *sec*-butylamine **d.** butyl alcohol **e.** isobutyl bromide **f.** *sec*-butyl chloride

5. a. $CH_3CH(CH_3)CH_2CH_3$ 2-methylbutane **b.** $CH_3CH_2CH_2CH_2CH_3$ pentane **c.** $CH_3C(CH_3)_2CH_3$ 2,2-dimethylpropane

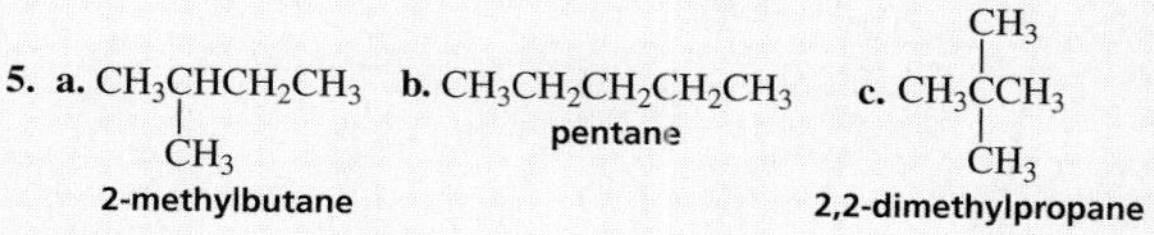

6. a. $CH_3CH(CH_3)CH(CH_3)CH_2CH_2CH_3$ **b.** $CH_3CH(CH_3)CH_2C(CH_3)(CH(CH_3)_2)—CH(CH_3)CH_2CH_3$ **c.** $CH_3C(CH_3)_2CH_2CH(CH_2CH_2CH_3)CH_2CH_2CH_2CH_3$ **d.** $CH_3CH(CH_3)CH_2CH(CH_2CH(CH_3)_2)CH(CH_3)CH_2CH_2CH_3$

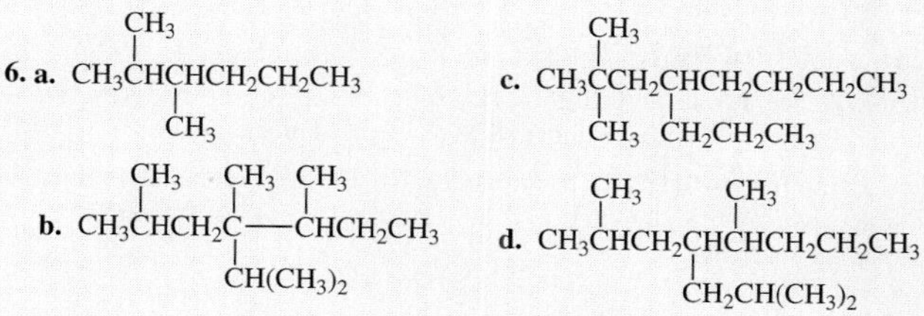

8. a. 2,2,4-trimethylhexane **b.** 2,2-dimethylbutane **c.** 3,3-diethylhexane **d.** 2,5-dimethylheptane **e.** 4-isopropyloctane **f.** 4-ethyl-2,2,3-trimethylhexane

10. a. (skeletal structure, OH) **b.** (skeletal structure) **c.** (skeletal structure, Br) **d.** (skeletal structure, O)

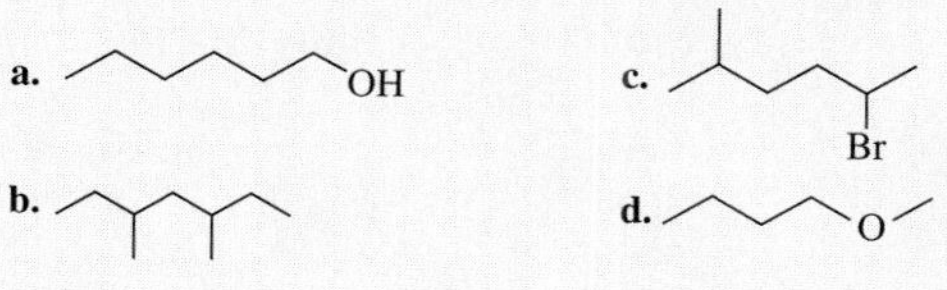

11. a. $C_{10}H_{20}O$ **b.** $C_{10}H_{20}O_2$ **12. a.** 1-ethyl-2-methylcyclopentane **b.** ethylcyclobutane **c.** 3,6-dimethyldecane **d.** 5-isopropylnonane **13. a.** same **b.** same **14. a.** *sec*-butyl chloride, 2-chlorobutane **b.** isohexyl chloride, 1-chloro-4-methylpentane **c.** cyclohexyl bromide, bromocyclohexane **d.** isopropyl fluoride, 2-fluoropropane **15. a.** a tertiary alkyl bromide **b.** a tertiary alcohol **c.** a primary amine **16. a.** methylpropylamine; secondary **b.** trimethylamine; tertiary **c.** diethylamine; secondary **d.** butyldimethylamine; tertiary

17. a. CH_2Cl (on cyclohexane) **b.** Cl CH_3 (on cyclohexane)

c. CH_3 / Cl (cyclohexane) CH_3 / Cl (cyclohexane) CH_3 / Cl (cyclohexane)

18. a. ~ 104.5° **b.** ~ 107.3° **c.** ~ 104.5° **19.** pentane **20. a.** O—H hydrogen bond is longer. **b.** O—H covalent bond is stronger. **21. a.** 1, 4, and 5 **b.** 1, 2, 4, 5, and 6

22. HO–(OH)–OH > (OH)–OH > (OH) > (NH_2) > (chain) > (branched chain)

24. even number

25. a. $HOCH_2CH_2CH_2OH > CH_3CH_2CH_2OH > CH_3CH_2CH_2CH_2OH > CH_3CH_2CH_2CH_2Cl$

b. NH_2 (cyclopentane) > OH (cyclopentane) > CH_3 (cyclopentane)

26. ethanol **27.** hexethal **31.** isopropylcyclohexane **32. a.** cis **b.** cis **c.** trans **d.** trans

CHAPTER 4

1. a. C_5H_8 **b.** C_4H_6 **c.** $C_{10}H_{16}$ **2. a.** 3 **b.** 4 **c.** 1 **4. a.** 4-methyl-2-pentene **b.** 2-chloro-3,4-dimethyl-3-hexene **c.** 1-bromo-4-methyl-3-hexene **d.** 1,5-dimethylcyclohexene **5. a.** 5 **b.** 4 **c.** 4 **d.** 6 **6. a.** 1 and 3

7. a. —I > —Br > —OH > $—CH_3$
b. —OH > $—CH_2Cl$ > $—CH{=}CH_2$ > $—CH_2CH_2OH$

9. a. *Z* **b.** *E*

10. H_3C(H)C=C($CHCH_3$ with CH_3)($CH_2CH_2CH_2CH_3$)

11. a. middle one **b.** one on the right **12.** *cis*-3,4-dimethyl-3-hexene > *trans*-3-hexene > *cis*-3-hexene > *cis*-2,5-dimethyl-3-hexene **14.** nucleophiles: H^-, CH_3O^-, $CH_3C{\equiv}CH$, NH_3; electrophiles: $CH_3\overset{+}{C}HCH_3$ **18. a.** 2 **b.** B **c.** 3 **d.** the first one **e.** products **f.** C to D **g.** B **h.** Yes, since the products are more stable than the reactants.

CHAPTER 5

2. ethyl cation

3. $CH_3CH_2\overset{+}{C}(CH_3)CH_3$ > $CH_3CH_2\overset{+}{C}HCH_3$ > $CH_3CH_2CH_2\overset{+}{C}H_2$

4. a. $CH_3CH_2CH(Br)CH_3$ **c.** CH_3, Br (cyclopentane) **e.** CH_3, Br (cyclohexane)

b. $CH_3CH_2C(CH_3)(Br)CH_3$ **d.** $CH_3C(CH_3)(Br)CH_2CH_2CH_3$ **f.** $CH_3CH_2CH(Br)CH_3$

5. a. $CH_2{=}C(CH_3)CH_3$ **c.** (cyclohexyl)–$C(CH_3){=}CH_2$

b. (cyclohexyl)–$CH_2CH{=}CH_2$ **d.** (cyclohexylidene)$=CHCH_3$ or (cyclohexenyl)–CH_2CH_3

6. a. $CH_3CH_2CH_2CH(OH)CH_3$ **b.** (cyclohexyl)–OH

c. $CH_3CH_2CH_2CH_2CH(OH)CH_3$ and $CH_3CH_2CH_2CH(OH)CH_2CH_3$

d. (cyclohexane with CH_3 and OH on same carbon)

9. a. H_2O is the solvent: $CH_3CH_2CH(OH)CH_3$
major

b. CH_3OH is the solvent: $CH_3CH_2CH(OCH_3)CH_3$
major

11. C_nH_{2n-4} **12.** $C_{14}H_{20}$

13. a. $ClCH_2CH_2C{\equiv}CCH_2CH_3$

b. $CH_3C{\equiv}CCH(Br)CH_3$ **c.** $HC{\equiv}CCH_2C(CH_3)_2CH_3$

14. a. 4-methyl-1-pentyne **b.** 2-hexyne **16. a.** 5-bromo-2-pentyne **b.** 6-bromo-2-chloro-4-octyne **c.** 5-methyl-3-heptyne **d.** 3-ethyl-1-hexyne **17. a.** $sp^2–sp^2$ **b.** $sp^2–sp^3$ **c.** $sp–sp^2$ **d.** $sp–sp^3$ **e.** $sp–sp$ **f.** $sp^2–sp^2$ **g.** $sp^2–sp^3$ **h.** $sp–sp^3$ **i.** $sp^2–sp$

18. a. $CH_2{=}C(Br)CH_3$ **b.** $CH_3C(Br)_2CH_3$ **c.** $CH_3CH_2C(Br)_2CH_3$

d. $CH_3C(Br)_2CH_2CH_2CH_3$ + $CH_3CH_2C(Br)_2CH_2CH_3$

19. $CH_3CH_2C({=}O)CH_2CH_2CH_2CH_3$ and $CH_3CH_2CH_2C({=}O)CH_2CH_2CH_3$

20. a. $CH_3C{\equiv}CH$ **b.** $CH_3CH_2C{\equiv}CCH_2CH_3$ **c.** $HC{\equiv}C$–(cyclohexyl)

21. $CH_3CH{=}C(OH)CH_2CH_2CH(CH_3)_2$ and $CH_3CH_2C(OH){=}CHCH_2CH(CH_3)_2$

22. a. 2-butyne + H_2 + Lindlar cat. **b.** 1-hexyne + H_2 + Lindlar cat. **23. a.** 1-butene, *cis*-2-butene, *trans*-2-butene **b.** 1-pentene, *cis*-2-pentene, or *trans*-2-pentene **c.** 1-methylcyclopentene, 3-methylcyclopentene, 4-methylcyclopentene, methylenecyclopentene **24.** The carbanion that would be formed is a stronger base than the amide ion.

25. a. $CH_3CH_2CH_2\bar{C}H_2 > CH_3CH_2CH{=}\bar{C}H > CH_3CH_2C{\equiv}\bar{C}$
b. $^-NH_2 > CH_3C{\equiv}C^- > CH_3CH_2O^- > F^-$

26. a. $CH_3\overset{+}{C}H_2$ **b.** $H_2C{=}\overset{+}{C}H$

30. a. $CH_2{=}CHCl$ **b.** $CH_2{=}C(CH_3)C({=}O)OCH_3$ **c.** $CF_2{=}CF_2$

32. beach balls

CHAPTER 6

1. a. $CH_3CH_2CH_2OH$ $CH_3CH(OH)CH_3$ $CH_3CH_2OCH_3$ **b.** 7

3. a. F, G, J, L, N, P, Q, R, S, Z **b.** A, C, D, H, I, M, O, T, U, V, W, X, Y **4.** a, c, and f **6.** a, c, and f

7. a. 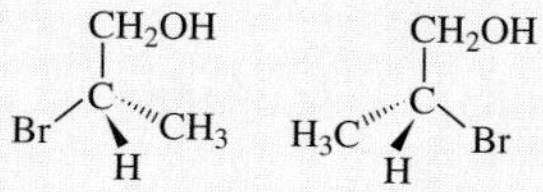 (structures: CH_2OH, Br, CH_3, H on C — pair of enantiomers)

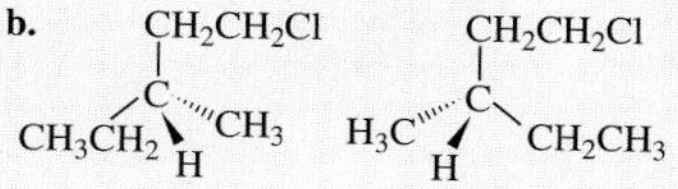

b. (structures: CH_2CH_2Cl, CH_3CH_2, CH_3, H)

c. (structures: $CH(CH_3)_2$, HO, CH_3, H)

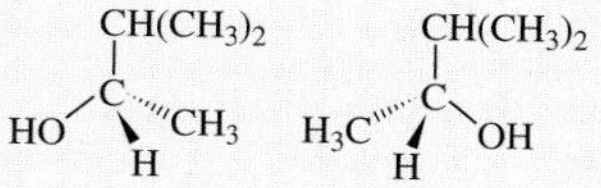

9. a, b, and c

10.

a. 1 $-CH_2OH$ 3 $-CH_3$ 2 $-CH_2CH_2OH$ 4 $-H$

b. 2 $-CH{=}O$ 1 $-OH$ 4 $-CH_3$ 3 $-CH_2OH$

c. 2 $-CH(CH_3)_2$ 3 $-CH_2CH_2Br$ 1 $-Cl$ 4 $-CH_2CH_2CH_2Br$

d. 2 $-CH{=}CH_2$ 3 $-CH_2CH_3$ 1 $-C{\equiv}CH$ 4 $-CH_3$

11. a. *R* **b.** *R* **12. a.** (*R*)-2-bromobutane **b.** (*R*)-1,3-dichlorobutane **13. a.** enantiomers **b.** enantiomers **15. a.** levorotatory **b.** dextrorotatory **16. a.** *S* **b.** *R* **c.** *R* **d.** *S* **17.** +168 **18. a.** −24 **b.** 0 **19.** From the data given, you cannot determine the configuration. **20. a.** enantiomers **b.** identical compounds (Therefore, they are not isomers.) **c.** diastereomers **21. a.** 8 **b.** $2^8 = 256$ **23. a.** diastereomers **b.** enantiomers **c.** identical **d.** constitutional isomers **24.** B **26.** the one on the left

28. a. 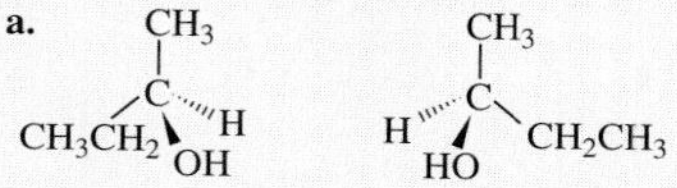 (structures: CH_3, CH_3CH_2, H, OH)

b. 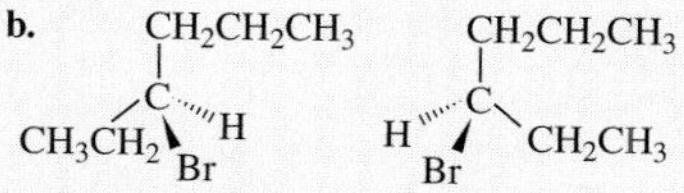 (structures: $CH_2CH_2CH_3$, CH_3CH_2, H, Br)

c. 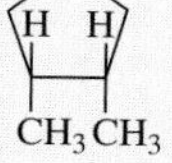 (cyclopentane with H, H and CH_3 CH_3)

29. a. (*R*)-malate and (*S*)-malate **b.** (*R*)-malate and (*S*)-malate

CHAPTER 7

1. a, b, e, f, and g have delocalized electrons.

4. a. 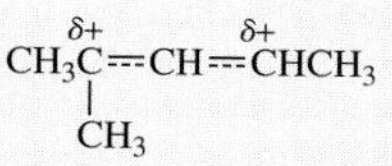 $CH_3\overset{\delta+}{C}{=}CH{=}\overset{\delta+}{C}HCH_3$ (with CH_3 on C) **b.** (cyclohexenone with δ+, δ−) **c.** $CH_3\overset{\delta+}{C}H{=}CH{=}\overset{\delta+}{C}HCH_3$

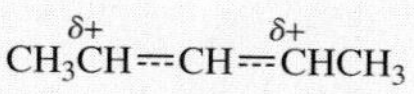

5. a. All have the same length. **b.** 2/3 of a negative charge

6. 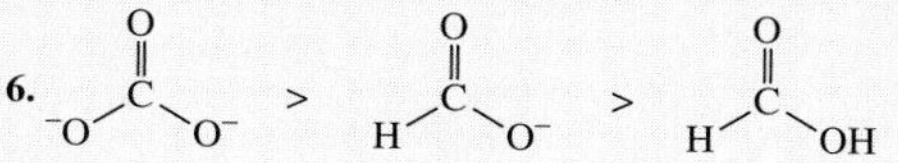 $^-O-C(=O)-O^-$ > $H-C(=O)-O^-$ > $H-C(=O)-OH$ **7.** 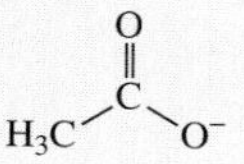 $H_3C-C(=O)-O^-$

8. $\dot{C}H_2-CH{=}CH-CH{=}CH_2 \longleftrightarrow CH_2{=}CH-\dot{C}H-CH{=}CH_2 \longleftrightarrow CH_2{=}CH-CH{=}CH-\dot{C}H_2$

9. 6 **10. a.** $CH_3CH{=}CH\overset{+}{C}HCH_3$ **b.** 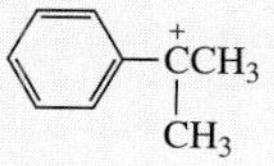 $C_6H_5-\overset{+}{C}CH_3$ (with CH_3)

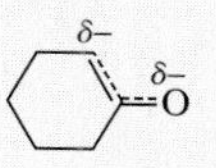

11. a. 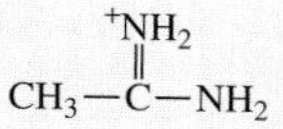 $CH_3-\overset{^+NH_2}{C}-NH_2$ **b.** 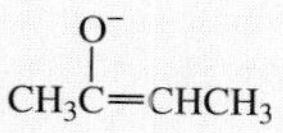 $CH_3\overset{O^-}{C}{=}CHCH_3$

13. a. 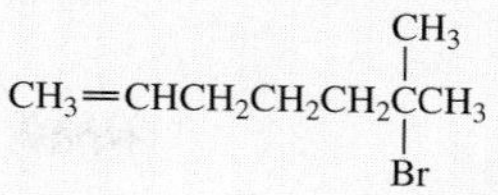 $CH_3{=}CHCH_2CH_2CH_2CCH_3$ (with CH_3 and Br on C) **b.** 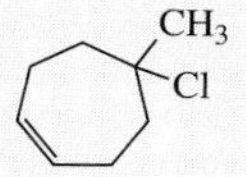 (1-chloro-1-methylcycloheptene structure: CH_3, Cl)

14. 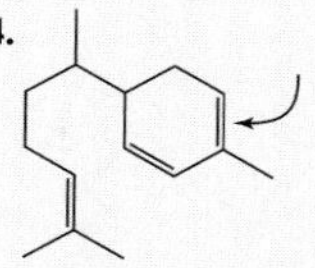 (structure)

15. 2,4-heptadiene

16. a. 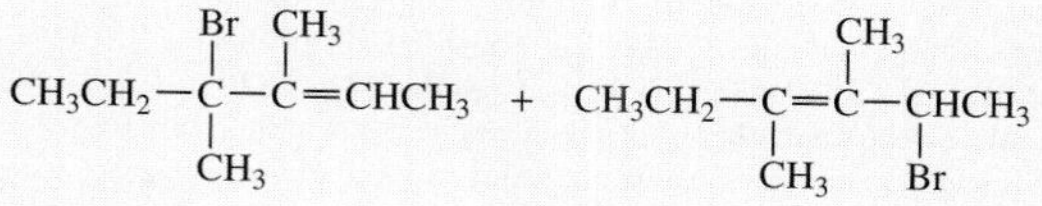 $CH_3CH_2-C(Br)(CH_3)-C(CH_3){=}CHCH_3$ + $CH_3CH_2-C(CH_3){=}C(CH_3)-CH(Br)CH_3$

b. (cyclohexene with Br, CH_3, CH_3) + (cyclohexene with CH_3, CH_3, Br)

17. a. $CH_3CH{=}CHOH$ **b.** 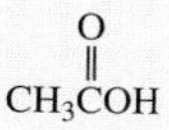 $CH_3\overset{O}{\overset{\|}{C}}OH$ **c.** $CH_3CH{=}CHOH$ **d.** $CH_3CH{=}CH\overset{+}{N}H_3$

18. a. ethylamine **b.** ethoxide ion **c.** ethoxide ion

19. 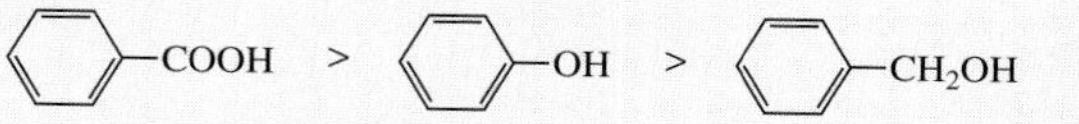 C_6H_5-COOH > C_6H_5-OH > $C_6H_5-CH_2OH$

20. 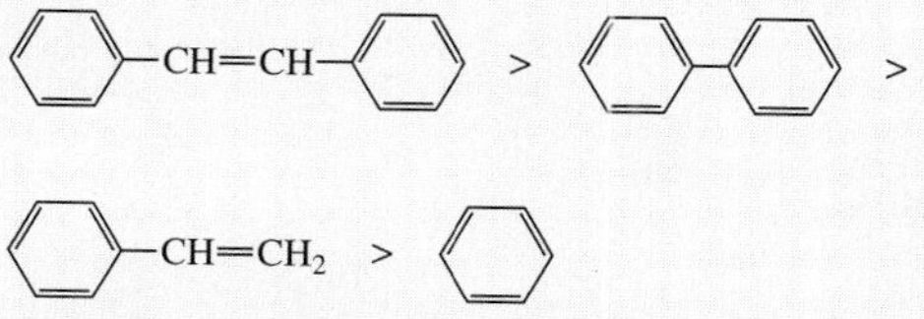 $C_6H_5-CH{=}CH-C_6H_5$ > $C_6H_5-C_6H_5$ > $C_6H_5-CH{=}CH_2$ > C_6H_6

21. a. The compound on the right is blue. **b.** They will be the same color.

CHAPTER 8

3. the cycloheptatrienyl cation **4.** only b **7. a.** The nitrogen donates electrons by resonance into the ring. **b.** The nitrogen is the most electronegative atom in the molecule. **c.** The nitrogen atom withdraws electrons inductively from the ring.

8. a. 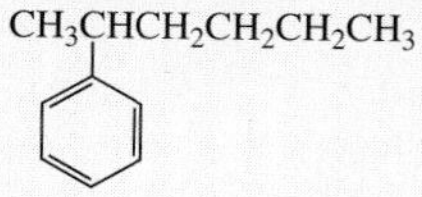 $CH_3CHCH_2CH_2CH_2CH_3$ (with C_6H_5 on C-2) **b.** 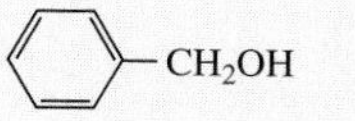 $C_6H_5-CH_2OH$ **c.** $CH_3CH_2CHCH_2CH_3$ (with $CH_2C_6H_5$)

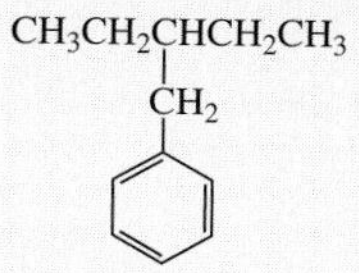

11. a. 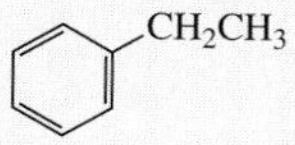 $C_6H_5-CH_2CH_3$ **b.** 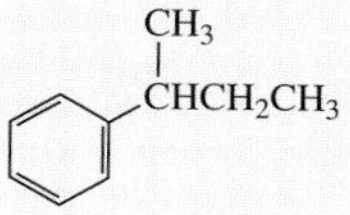 $C_6H_5-CHCH_2CH_3$ (with CH_3) **c.** $C_6H_5-CH_2CH{=}CH_2$

12. a. 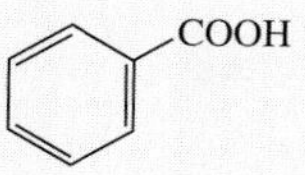 C_6H_5-COOH **b.** 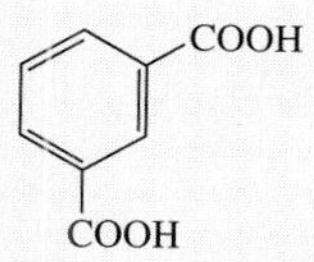 (benzene-1,3-dicarboxylic acid: COOH, COOH)

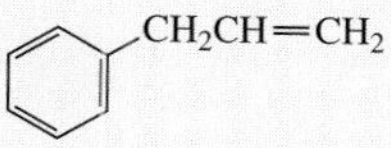

14. a. 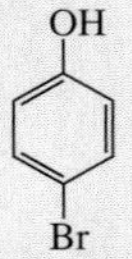 OH / Br (*para*-bromophenol ring) **c.** 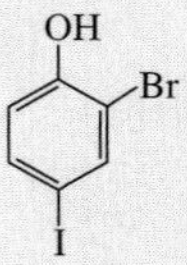 OH, Br, I (ring)

b. 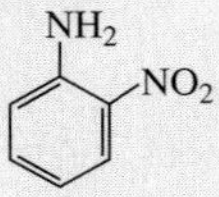 NH_2, NO_2 (ring) **d.** 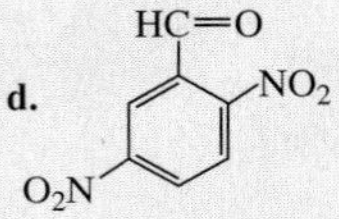 HC=O, NO_2, O_2N (ring)

15. a. *meta*-ethylphenol or 3-ethylphenol **b.** *meta*-bromochlorobenzene or 1-bromo-3-chlorobenzene **c.** *para*-bromobenzaldehyde or 4-bromobenzaldehyde **d.** *ortho*-ethyltoluene or 2-ethyltoluene **16. a.** 1,3,5-tribromobenzene **b.** *meta*-nitrophenol or 3-nitrophenol **c.** *para*-bromotoluene or 4-bromotoluene **d.** *ortho*-dichlorobenzene or 1,2-dichlorobenzene **17. a.** donates electrons by resonance and withdraws electrons inductively **b.** donates electrons inductively **c.** withdraws electrons by resonance and withdraws electrons inductively **d.** donates electrons by resonance and withdraws electrons inductively **e.** donates electrons by resonance and withdraws electrons inductively **f.** withdraws electrons inductively
18. a. phenol > toluene > benzene > bromobenzene > nitrobenzene
b. toluene > chloromethylbenzene > dichloromethylbenzene > difluoromethylbenzene

20. a. 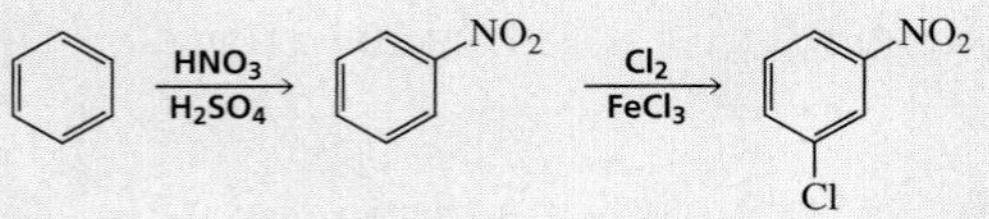 benzene $\xrightarrow[H_2SO_4]{HNO_3}$ NO_2-benzene $\xrightarrow[FeCl_3]{Cl_2}$ (NO_2, Cl ring)

b. 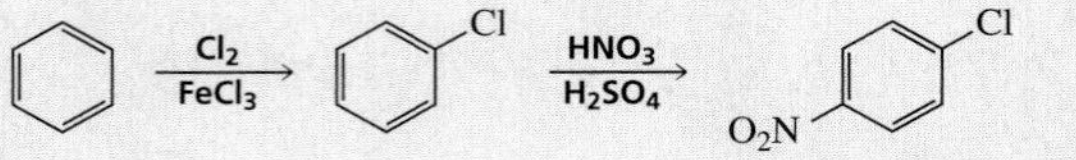 benzene $\xrightarrow[FeCl_3]{Cl_2}$ Cl-benzene $\xrightarrow[H_2SO_4]{HNO_3}$ (Cl, O_2N ring)

c. 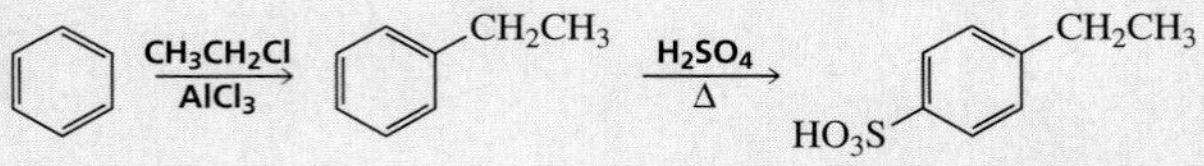 benzene $\xrightarrow[AlCl_3]{CH_3CH_2Cl}$ CH_2CH_3-benzene $\xrightarrow[\Delta]{H_2SO_4}$ (CH_2CH_3, HO_3S ring)

22.  a. $ClCH_2\overset{O}{\overset{\|}{C}}OH$ **b.** $H_3\overset{+}{N}CH_2\overset{O}{\overset{\|}{C}}OH$ **c.** $FCH_2\overset{O}{\overset{\|}{C}}OH$ **d.** $H\overset{O}{\overset{\|}{C}}OH$

CHAPTER 9

1. a. It is tripled. **b.** It is half the original rate.

2. 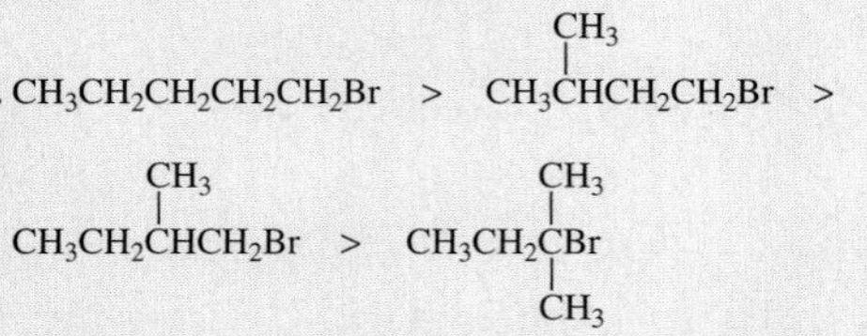 $CH_3CH_2CH_2CH_2CH_2Br$ > $CH_3\underset{}{\overset{CH_3}{\overset{|}{C}}}HCH_2CH_2Br$ > $CH_3CH_2\overset{CH_3}{\overset{|}{C}}HCH_2Br$ > $CH_3CH_2\underset{CH_3}{\underset{|}{\overset{CH_3}{\overset{|}{C}}}}Br$

3. b. (*S*)-2-butanol **c.** (*R*)-3-hexanol **d.** 3-pentanol

5. a. $CH_3CH_2Br + HO^-$ **c.** $CH_3CH_2Br + I^-$
b. 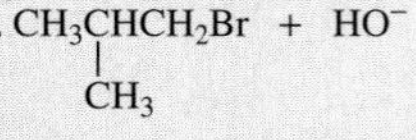 $CH_3\underset{CH_3}{\underset{|}{C}}HCH_2Br + HO^-$ **d.** $CH_3CH_2CH_2I + HO^-$

6. a. $CH_3CH_2CH_2OCH_2CH_3$ **c.** $(CH_3)_3\overset{+}{N}CH_2CH_3$
b. $CH_3C{\equiv}CCH_2CH_3$ **d.** $CH_3CH_2SCH_2CH_3$

7. 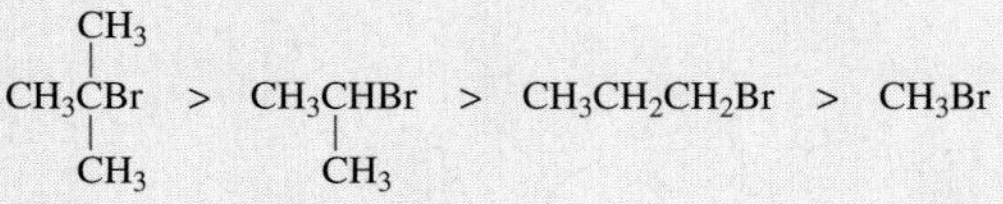 $CH_3\underset{CH_3}{\underset{|}{\overset{CH_3}{\overset{|}{C}}}}Br$ > $CH_3\underset{CH_3}{\underset{|}{C}}HBr$ > $CH_3CH_2CH_2Br$ > CH_3Br **8. a.** 2 **b.** 1

9. 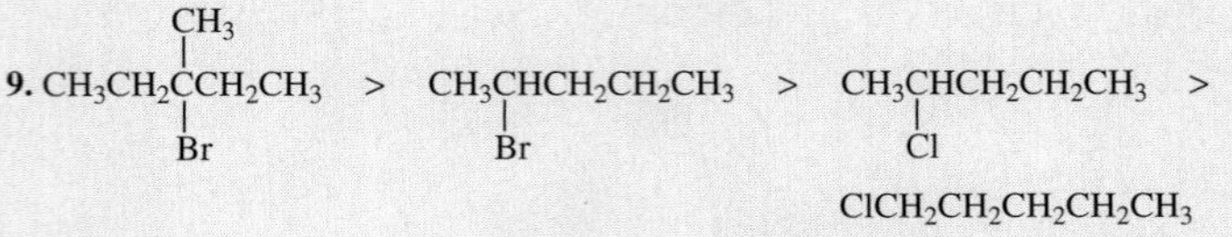 $CH_3CH_2\underset{Br}{\underset{|}{\overset{CH_3}{\overset{|}{C}}}}CH_2CH_3$ > $CH_3\underset{Br}{\underset{|}{C}}HCH_2CH_2CH_3$ > $CH_3\underset{Cl}{\underset{|}{C}}HCH_2CH_2CH_3$ > $ClCH_2CH_2CH_2CH_2CH_3$

10. a. $CH_3CH_2CH_2Br$ **c.** 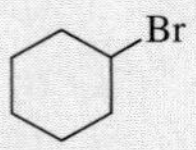 bromocyclohexane (Br)

b. 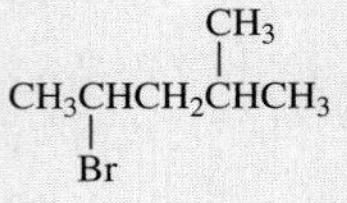 $CH_3\underset{Br}{\underset{|}{C}}HCH_2\overset{CH_3}{\overset{|}{C}}HCH_3$ **d.** 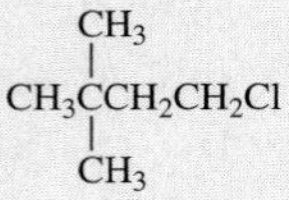 $CH_3\underset{CH_3}{\underset{|}{\overset{CH_3}{\overset{|}{C}}}}CH_2CH_2Cl$

11. a. 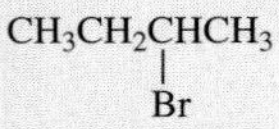 $CH_3CH_2\underset{Br}{\underset{|}{C}}HCH_3$ **c.** bromocyclohexane (Br)

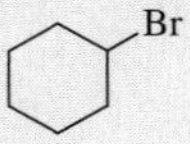

b. 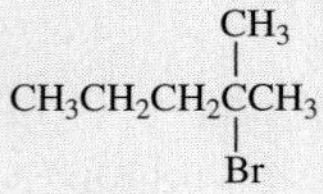 $CH_3CH_2CH_2\underset{Br}{\underset{|}{\overset{CH_3}{\overset{|}{C}}}}CH_3$ **d.** equally unreactive

13. a. 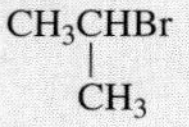 $CH_3\underset{CH_3}{\underset{|}{C}}HBr$ **b.** 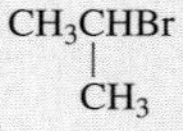 $CH_3\underset{CH_3}{\underset{|}{C}}HBr$

14. a. 2 is more reactive than 1. **b.** 2 is more reactive than 1. **c.** 2 is more reactive than 1. **d.** 1 is more reactive than 2.

15. b. 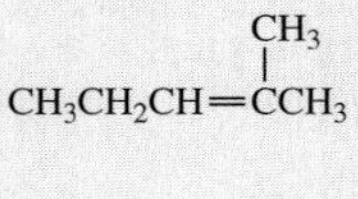 $CH_3CH_2CH{=}\overset{CH_3}{\overset{|}{C}}CH_3$ **d.** $CH_3\underset{CH_3}{\underset{|}{\overset{CH_3}{\overset{|}{C}}}}CH{=}CH_2$

c. $CH_3CH{=}CH\underset{CH_3}{\underset{|}{C}}HCH_3$

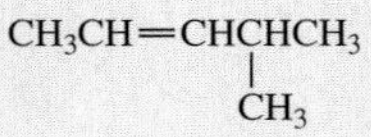

16. 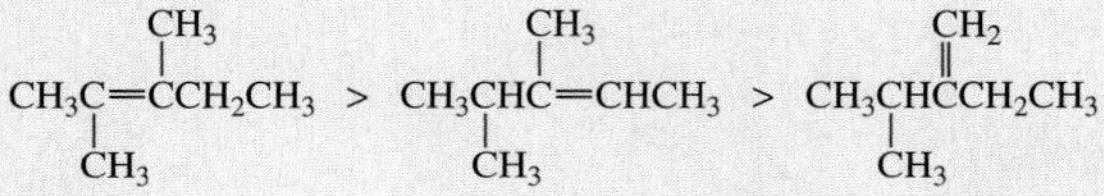 $CH_3\underset{CH_3}{\underset{|}{C}}{=}\overset{CH_3}{\overset{|}{C}}CH_2CH_3$ > $CH_3\underset{CH_3}{\underset{|}{C}}H\overset{CH_3}{\overset{|}{C}}{=}CHCH_3$ > $CH_3\underset{CH_3}{\underset{|}{C}}H\overset{CH_2}{\overset{\|}{C}}CH_2CH_3$

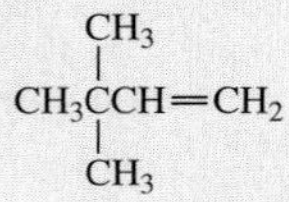

18. a. 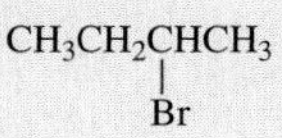 $CH_3CH_2\underset{Br}{\underset{|}{C}}HCH_3$ **c.** bromocyclohexane (Br)

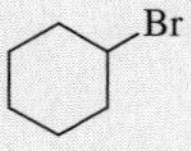

b. 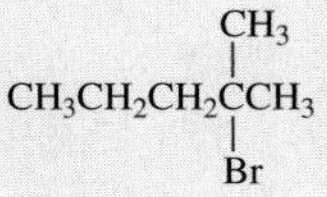 $CH_3CH_2CH_2\underset{Br}{\underset{|}{\overset{CH_3}{\overset{|}{C}}}}CH_3$ **d.** 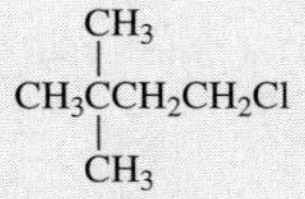 $CH_3\underset{CH_3}{\underset{|}{\overset{CH_3}{\overset{|}{C}}}}CH_2CH_2Cl$

19. a. 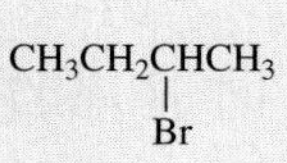 $CH_3CH_2\underset{Br}{\underset{|}{C}}HCH_3$ **c.** bromocyclohexane (Br)

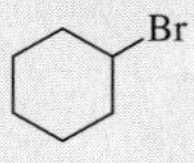

b. 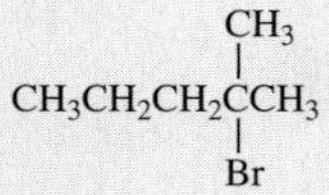 $CH_3CH_2CH_2\underset{Br}{\underset{|}{\overset{CH_3}{\overset{|}{C}}}}CH_3$

22. a. primarily substitution **b.** substitution and elimination **c.** substitution and elimination **d.** only elimination **24. a.** no reaction **b.** no reaction **c.** substitution and elimination **d.** substitution and elimination **25.** a and b

CHAPTER 10

2. a. 1-pentanol **primary b.** 4-methylcyclohexanol **secondary c.** 5-chloro-2-methyl-2-pentanol **tertiary d.** 2-ethyl-1-pentanol **primary e.** 5-methyl-3-hexanol **secondary f.** 2,6-dimethyl-4-octanol **secondary**

3. 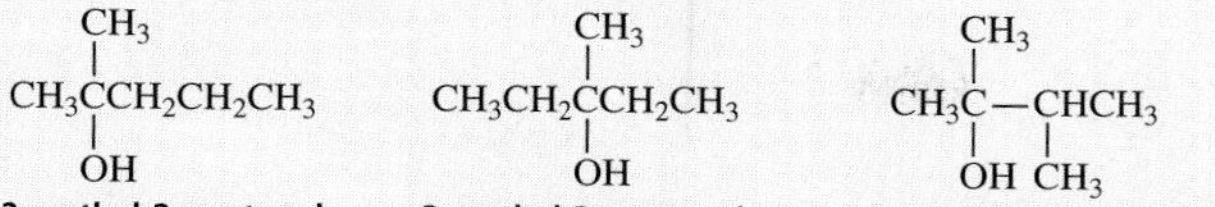 $CH_3C(CH_3)(OH)CH_2CH_2CH_3$ 2-methyl-2-pentanol; $CH_3CH_2C(CH_3)(OH)CH_2CH_3$ 3-methyl-3-pentanol; $CH_3C(CH_3)(OH)-CH(CH_3)CH_3$ 2,3-dimethyl-2-butanol

4. Their nucleophilic ability results from the lone pair.

6. a. 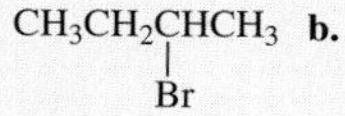 $CH_3CH_2CH(Br)CH_3$ **b.** 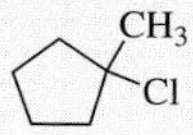 1-chloro-1-methylcyclopentane (cyclopentane with CH_3 and Cl)

8. 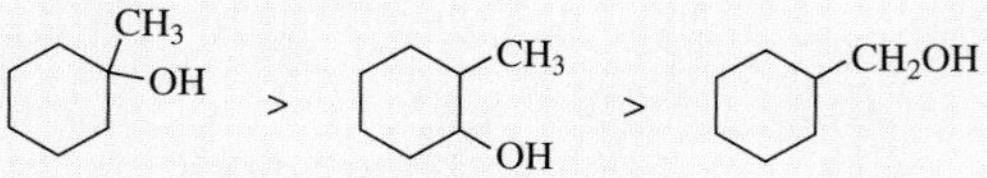 1-methylcyclohexanol (cyclohexane with CH_3 and OH) > 2-methylcyclohexanol (cyclohexane with CH_3, OH) > cyclohexylmethanol (cyclohexane–CH_2OH)

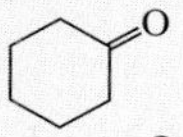

10. a. 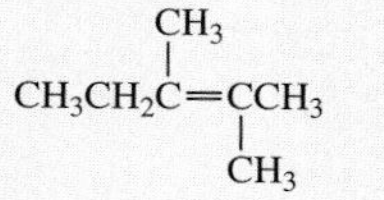 $CH_3CH_2C(CH_3){=}C(CH_3)CH_3$ **b.**  cyclohexene

11. a. 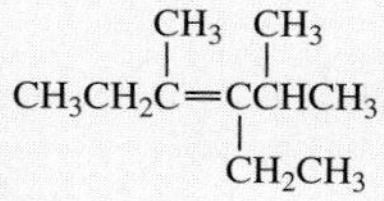 $CH_3CH_2C(CH_2CH_3){=}C(CH_3)CHCH_3$ with CH_3 on C-2 (CH₃ CH₃ above) **b.** 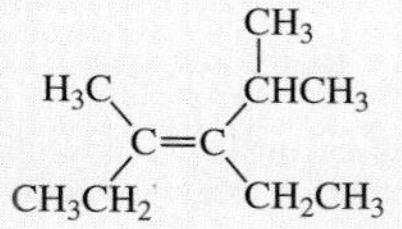 $(H_3C)(CH_3CH_2)C{=}C(CH(CH_3)CH_3)(CH_2CH_3)$

12. a. 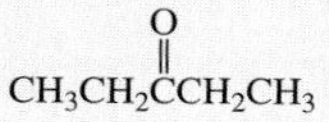 $CH_3CH_2\overset{O}{\overset{\|}{C}}CH_2CH_3$ **b.** 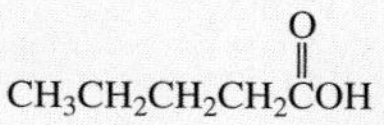 $CH_3CH_2CH_2CH_2\overset{O}{\overset{\|}{C}}OH$ **c.** cyclohexanone **d.** 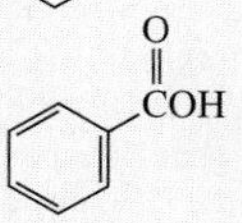 benzoic acid ($C_6H_5\overset{O}{\overset{\|}{C}}OH$)

13. a. 2-butanol **b.** cyclohexanol **c.** 1-butanol **15. a. 1.** methoxyethane **2.** ethoxyethane **3.** 4-methoxyoctane **4.** 1-propoxybutane **b.** no **c. 1.** ethyl methyl ether **2.** diethyl ether **3.** no common name **4.** butyl propyl ether

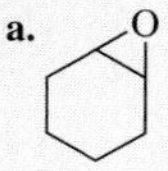

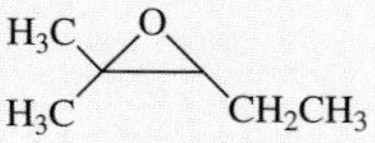

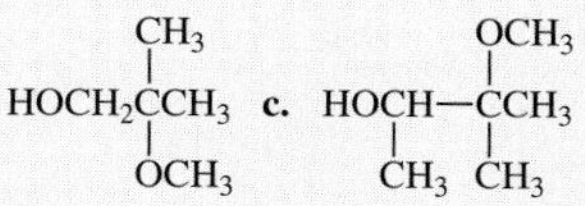

17. a. 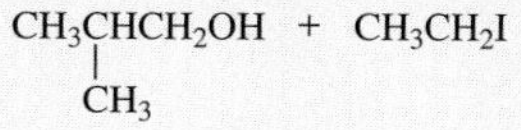 $CH_3CH(CH_3)CH_2OH + CH_3CH_2I$ **b.** $HOCH_2CH_2CH_2CH_2CH_2I$ **c.** 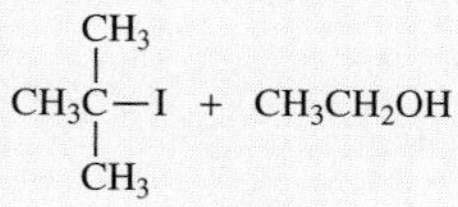 $CH_3C(CH_3)_2{-}I + CH_3CH_2OH$ **d.** 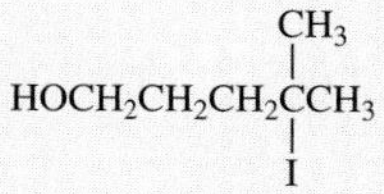 $HOCH_2CH_2CH_2C(CH_3)(I)CH_3$

18. a. cyclohexene oxide (epoxide fused to cyclohexane) **b.** epoxide with H_3C, H_3C on one carbon and CH_2CH_3 on the other

19. a. $HOCH_2C(CH_3)(OCH_3)CH_3$ **b.** $CH_3OCH_2C(OH)(CH_3)CH_3$ **c.** $HOCH(CH_3)-C(OCH_3)(CH_3)CH_3$ **d.** 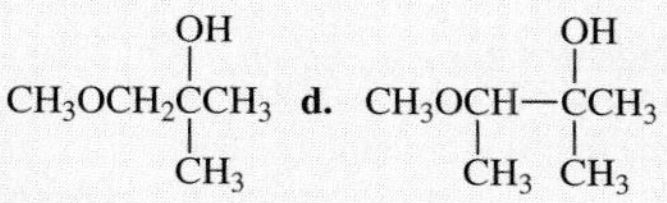 $CH_3OCH(CH_3)-C(OH)(CH_3)CH_3$

20. noncyclic ether

21. 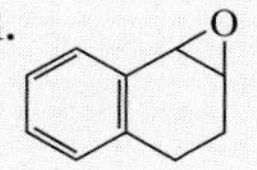 epoxide of 1,2-dihydronaphthalene (O bridging the ring fused to benzene)

CHAPTER 11

1. a. propanamide, propionamide **b.** isobutyl butanoate, isobutyl butyrate **c.** potassium butanoate, potassium butyrate **d.** pentanoyl chloride, valeryl chloride **e.** *N,N*-dimethylhexanamide **f.** *N*-ethyl-2-methylbutanamide

2. 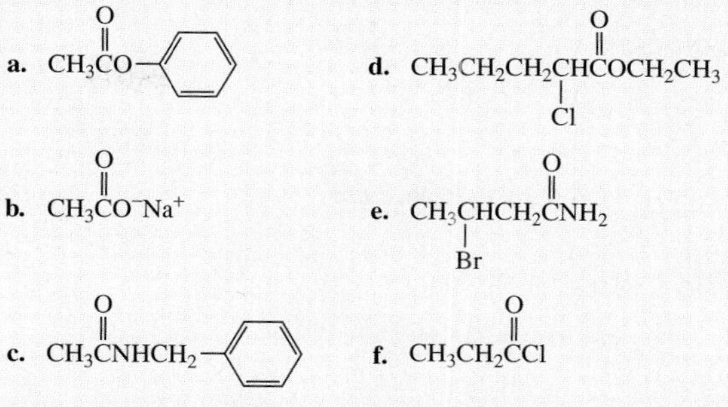 a. $CH_3\overset{O}{\overset{\|}{C}}O-C_6H_5$ **b.** $CH_3\overset{O}{\overset{\|}{C}}O^-Na^+$ **c.** $CH_3\overset{O}{\overset{\|}{C}}NHCH_2-C_6H_5$ **d.** $CH_3CH_2CH_2CH(Cl)\overset{O}{\overset{\|}{C}}OCH_2CH_3$ **e.** $CH_3CH(Br)CH_2\overset{O}{\overset{\|}{C}}NH_2$ **f.** $CH_3CH_2\overset{O}{\overset{\|}{C}}Cl$

3. The carbon–oxygen bond in an alcohol. **4.** The bond between oxygen and the methyl group is the longest; the bond between carbon and the carbonyl oxygen is the shortest. **6. a.** a new carboxylic acid derivative **b.** no reaction **c.** a mixture of two carboxylic acid derivatives **7. a.** acetate ion **b.** no reaction **10. a.** $CH_3CH_2CH_2OH$ **b.** $CH_3CH_2NH_2$ **c.** $(CH_3)_2NH$ **d.** 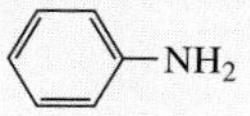 $C_6H_5-NH_2$ **e.** H_2O **f.** 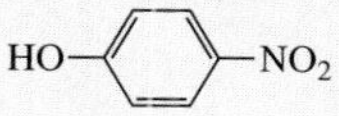 $HO-C_6H_4-NO_2$ **13.** phenyl acetate

14. a. 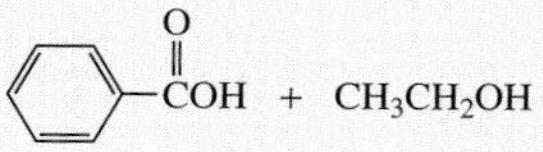 $C_6H_5-\overset{O}{\overset{\|}{C}}OH + CH_3CH_2OH$ **b.** 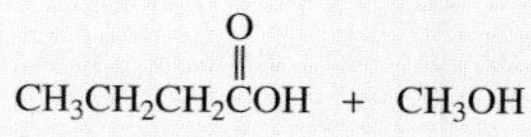 $CH_3CH_2CH_2\overset{O}{\overset{\|}{C}}OH + CH_3OH$

15. 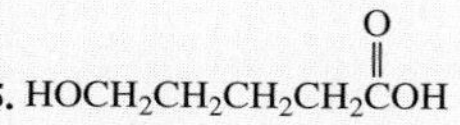 $HOCH_2CH_2CH_2CH_2\overset{O}{\overset{\|}{C}}OH$

16. 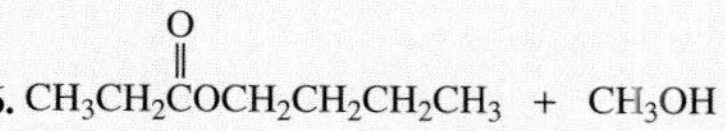 $CH_3CH_2\overset{O}{\overset{\|}{C}}OCH_2CH_2CH_2CH_3 + CH_3OH$

19. a. 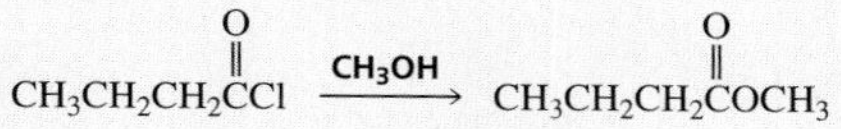 $CH_3CH_2CH_2\overset{O}{\overset{\|}{C}}Cl \xrightarrow{CH_3OH} CH_3CH_2CH_2\overset{O}{\overset{\|}{C}}OCH_3$

b. 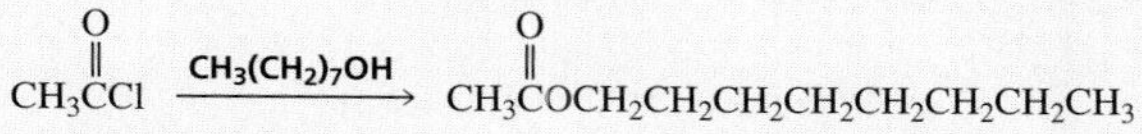 $CH_3\overset{O}{\overset{\|}{C}}Cl \xrightarrow{CH_3(CH_2)_7OH} CH_3\overset{O}{\overset{\|}{C}}OCH_2CH_2CH_2CH_2CH_2CH_2CH_2CH_3$

22. a. 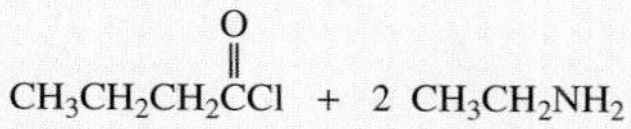 $CH_3CH_2CH_2\overset{O}{\overset{\|}{C}}Cl + 2\ CH_3CH_2NH_2$ **b.** 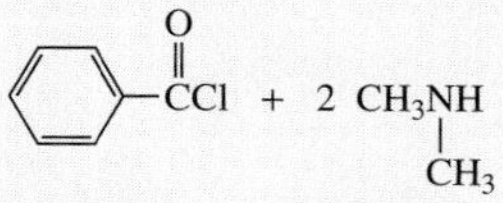 $C_6H_5-\overset{O}{\overset{\|}{C}}Cl + 2\ CH_3NHCH_3$

23. 2 and 4 **24. a. 1.** $SOCl_2$ **2.** 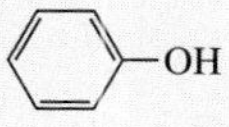 C_6H_5-OH **b. 1.** $SOCl_2$ **2.** 2 $CH_3CH_2NH_2$

25. a. $CH_3CH_2CH_2Br$ **b.** 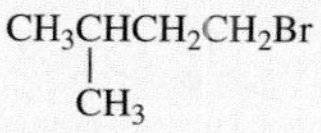 $CH_3CH(CH_3)CH_2CH_2Br$

CHAPTER 12

1. If the ketone functional groups were anywhere else in these compounds, they would not be ketones and would not have the "one" suffix. **2. a.** 4-heptanone **b.** 4-phenylbutanal **3. a.** 3-methylpentanal, β-methylvaleraldehyde **b.** 2-methyl-4-heptanone, isobutyl propyl ketone **c.** 4-ethylhexanal, γ-ethylcaproaldehyde **d.** 6-methyl-3-heptanone, ethyl isopentyl ketone **4. a.** 2-heptanone **b.** 5-methyl-3-hexanone **5. a.** $CH_3CH_3 + HO^-$ **b.** $CH_3CH_3 + CH_3O^-$ **c.** $CH_3CH_3 + CH_3\bar{N}H$

6. a. $CH_3CH_2CH_2CH_2CH(OH)CH_3$ **b.** $CH_3CH_2CH_2C(OH)(CH_3)CH_3$ **c.** 1-methylcyclohexanol (HO, CH_3 on cyclohexane)

7. $CH_3C(=O)CH_2CH_3$ + $CH_3CH_2CH_2MgBr$ and $CH_3CH_2C(=O)CH_2CH_2CH_3$ + CH_3MgBr

9. $CH_3CH(OH)CH_3$ and $CH_3CH_2CH(OH)CH_2CH_3$ **10.** A and C

11. a. $CH_3CH(CH_3)CH_2OH$ **b.** cyclohexane–OH **c.** benzene–CH_2OH **d.** benzene–$CH(OH)CH_3$

12. a. benzene–$C(=O)NHCH_3$ **b.** $CH_3C(=O)NH_2$ **c.** $CH_3C(=O)NHCH_2CH_3$ **d.** $CH_3C(=O)N(CH_2CH_3)_2$

13. a. cyclopentane$=NCH_2CH_3$ + H_2O **b.** cyclopentene–$N(CH_2CH_3)_2$ + H_2O **c.** $(CH_3CH_2)_2C{=}N(CH_2)_5CH_3$ + H_2O **d.** $(CH_3CH_2)_2C{=}N$–cyclohexane + H_2O

14. Electron-withdrawing groups decrease the stability of the aldehyde and increase the stability of the hydrate.

15. O_2N–benzene–$C(=O)$–benzene–NO_2 **17. a.** hemiacetals: 7 **b.** acetals: 2, 3 **c.** hemiketals: 1, 8 **d.** ketals: 5 **e.** hydrates: 4, 6

18. a. (bicyclic ketone with Br) **b.** (bicyclic alkene with OH, CH_3)

CHAPTER 13

2. a. the ketone **b.** the ester **3.** The electrons left behind when the proton is removed are readily delocalized onto two oxygen atoms. **4.** They have a hydrogen bonded to the nitrogen, which is more acidic than the hydrogen attached to the α-carbon. **5.** It is easier for hydroxide ion to attack the reactive carbonyl group than to remove a hydrogen from an α-carbon.

6. a. $CH_3CH{=}C(OH)CH_2CH_3$ **b.** benzene–$C(OH){=}CH_2$ **c.** 1-cyclohexenol

7. 3-hydroxy-2-cyclohexenone (more stable) and 3-hydroxy-3-cyclohexenone

more stable

9. a. $CH_3CH_2C(=O)CH(CH_2CH_3)CH_3$ **b.** 2-ethylcyclohexanone

11. a. $CH_3CH_2CH_2CHO$ **b.** $CH_3C(=O)CH_3$ **c.** cyclohexane–CH_2CHO **d.** $CH_3CH_2C(=O)CH_2CH_3$

12. (cyclohexylidene cyclohexanone)

14. a. $CH_3CH_2CH_2C(=O)CH(CH_2CH_3)C(=O)OCH_3$ **b.** $CH_3CH(CH_3)CH_2C(=O)CH(CH(CH_3)CH_3)C(=O)OCH_2CH_3$

15. A, B, and D **16.** $CH_3CH_2CH_2CH_2C(=O)OCH_3$ + CH_3O^-

17. A and D **18. a.** methyl bromide **b.** benzyl bromide **c.** isobutyl bromide **20. a.** ethyl bromide **b.** pentyl bromide **c.** benzyl bromide **22.** 7 **23. a.** 3 **b.** 7

CHAPTER 14

1. 2, 3, 5 **2.** m/z = 57 **4.** 1-bromopropane **6.** 2-pentanone **8.** C_6H_{14} **9. a.** IR **b.** UV **10. a.** 2000 cm^{-1} **b.** 8 μm **c.** 2 μm **11. a.** C≡C stretch **b.** C—H stretch **c.** C≡N stretch **d.** C=O stretch **13. a.** the carbon–oxygen stretch of phenol **b.** the carbon–oxygen double-bond stretch of a ketone **c.** the C—O stretch **14.** one bonded to an sp^3 carbon **15.** C=O **17. a.** a ketone **b.** a tertiary amine **20. a.** 2 **b.** 1 **c.** 4 **d.** 3 **e.** 3 **f.** 3 **21.** **A** = 2 signals, **B** = 1 signal, **C** = 3 signals **22. a.** 2 ppm **b.** 2 ppm **23.** 1.5 ppm **24.** upfield from the TMS peak **26.** first spectrum = 1-iodopropane

27. a. $CH_3CH(Br)CHBr$ **b.** $CH_3CH(CH_3)OCH_3$ **c.** $CH_3CH_2CH(Cl)CH_3$

28. a. $CH_3CH_2CH(Cl)CH_3$ **b.** $CH_3CH_2CH_2Cl$ **31.** B

32. a. $CH_3CH(Cl)C(=O)OH$ **b.** $ClCH_2CH_2C(=O)OH$

37. CH_3O–benzene–CH_3 **38. a. 1.** 3 **2.** 3 **3.** 4 **4.** 3 **5.** 2 **6.** 3

40. $CH_3CH_2CH_2CH_2CH_2C(=O)CH_2CH_2CH_2CH_2CH_3$

CHAPTER 15

1. D-Ribose is an aldopentose. D-Sedoheptulose is a ketoheptose. D-Mannose is an aldohexose. **3. a.** D-ribose **b.** L-talose **4.** D-psicose **5. a.** 16 **b.** 32 **c.** none **6.** D-tagatose, D-galactose, and D-talose **8. a.** D-iditol **b.** D-iditol and D-gulitol **9.** L-galactose **10.** D-tagatose **11. a.** L-gulose **b.** L-gularic acid **c.** D-allose and L-allose, D-altrose and D-talose, L-altrose and L-talose, D-galactose and L-galactose **12. a.** D-gulose and D-idose **b.** L-xylose and L-lyxose **14. b.** C-2, C-3, and C-4 **c.** C-1 and C-3 **15. b.** methyl α-D-galactoside; nonreducing

CHAPTER 16

1. (*S*)-alanine **3.** isoleucine

6. b. 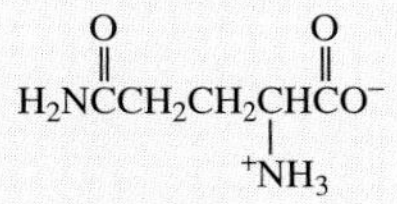 $H_2N\overset{O}{\overset{\|}{C}}CH_2CH_2\underset{^+NH_3}{\underset{|}{C}}H\overset{O}{\overset{\|}{C}}O^-$ **c.** $NH_2\overset{^+NH_2}{\overset{\|}{C}}NHCH_2CH_2CH_2\underset{^+NH_3}{\underset{|}{C}}H\overset{O}{\overset{\|}{C}}O^-$

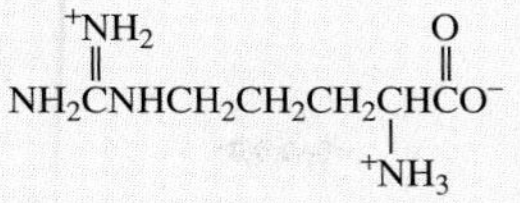

8. a. 5.43 **b.** 10.76 **c.** 5.68 **d.** 2.98 **9. a.** Asp **b.** Arg **10.** 2-methylpropanal **14.** A-G-M A-M-G M-G-A M-A-G G-A-M G-M-A **16. a.** glutamate, cysteine, and glycine **17.** Leu-Val and Val-Val **18.** 5.8% **20.** Gly-Arg-Trp-Ala-Glu-Leu-Met-Pro-Val-Asp **21. a.** His-Lys, Leu-Val-Glu-Pro-Arg, Ala-Gly-Ala **b.** Leu-Gly-Ser-Met-Phe-Pro-Tyr, Gly-Val **23.** Leu-Tyr-Lys-Arg-Met-Phe-Arg-Ser **25. a.** cigar-shaped protein **b.** subunit of a hexamer

CHAPTER 17

2. 2 and **3** **3.** Only **2** can form an imine with the substrate.

4. 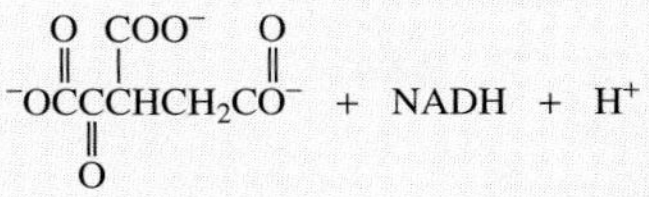 $^-O\overset{O}{\overset{\|}{C}}\underset{O}{\underset{\|}{C}}\overset{COO^-}{\overset{|}{C}}HCH_2\overset{O}{\overset{\|}{C}}O^-$ + NADH + H^+ **5.** 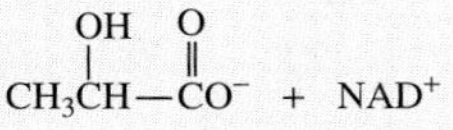 $CH_3\overset{OH}{\overset{|}{C}}H-\overset{O}{\overset{\|}{C}}O^-$ + NAD^+

6. a. 7 **b.** 3 and a separate group of 2

7. 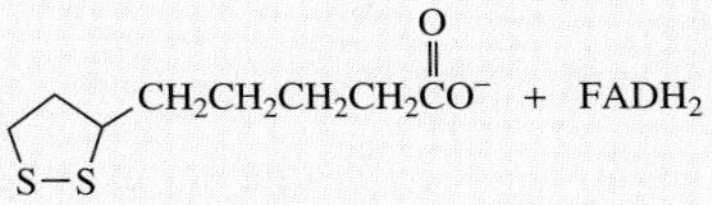 (ring with S–S) $CH_2CH_2CH_2CH_2\overset{O}{\overset{\|}{C}}O^-$ + $FADH_2$

8. a. 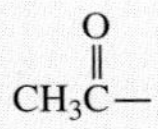 $CH_3\overset{O}{\overset{\|}{C}}-$ **b.** 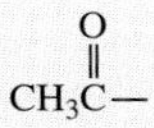 $CH_3\overset{O}{\overset{\|}{C}}-$ **9. a.** 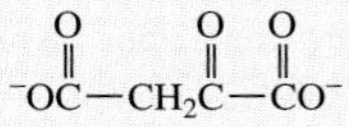 $^-O\overset{O}{\overset{\|}{C}}-CH_2\overset{O}{\overset{\|}{C}}-\overset{O}{\overset{\|}{C}}O^-$ **10.** 9

12. a. 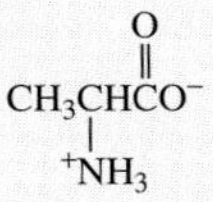 $CH_3\underset{^+NH_3}{\underset{|}{C}}H\overset{O}{\overset{\|}{C}}O^-$ **b.** 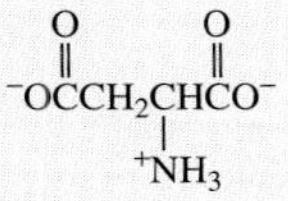 $^-O\overset{O}{\overset{\|}{C}}CH_2\underset{^+NH_3}{\underset{|}{C}}H\overset{O}{\overset{\|}{C}}O^-$

13. 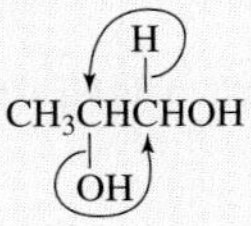 $CH_3\underset{OH}{\underset{|}{C}}H\overset{H}{\overset{|}{C}}HOH$ **14.** It is obtained by adding four hydrogens to folate.

15. 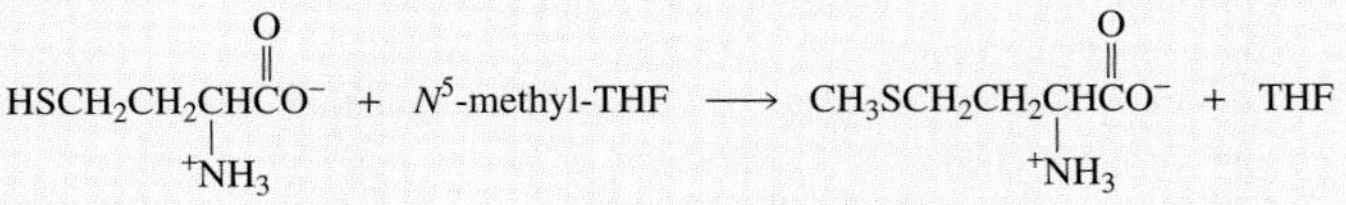 $HSCH_2CH_2\underset{^+NH_3}{\underset{|}{C}}H\overset{O}{\overset{\|}{C}}O^-$ + N^5-methyl-THF $\longrightarrow$ $CH_3SCH_2CH_2\underset{^+NH_3}{\underset{|}{C}}H\overset{O}{\overset{\|}{C}}O^-$ + THF

16. the methylene group of N^5,N^{10}-methylene-THF

CHAPTER 18

4. The β-carbon has a partial positive charge. **5.** eight **6.** seven **9.** two **10.** a ketone **11.** thiamine pyrophosphate **12.** an aldehyde **14.** pyridoxal phosphate **15.** pyruvate **16.** a secondary alcohol **17.** two **18.** citrate and isocitrate **19.** 11

CHAPTER 19

2. glyceryl tripalmitate

3. 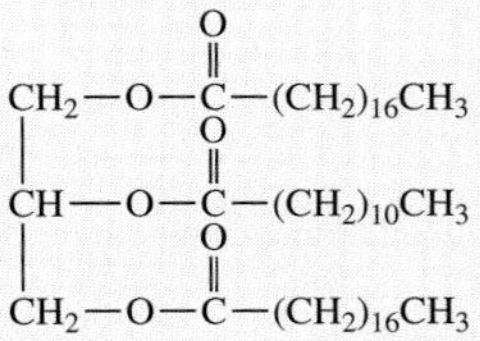
$$\begin{array}{l} CH_2-O-\overset{O}{\overset{\|}{C}}-(CH_2)_{16}CH_3 \\ CH-O-\overset{O}{\overset{\|}{C}}-(CH_2)_{10}CH_3 \\ CH_2-O-\overset{O}{\overset{\|}{C}}-(CH_2)_{16}CH_3 \end{array}$$

4. 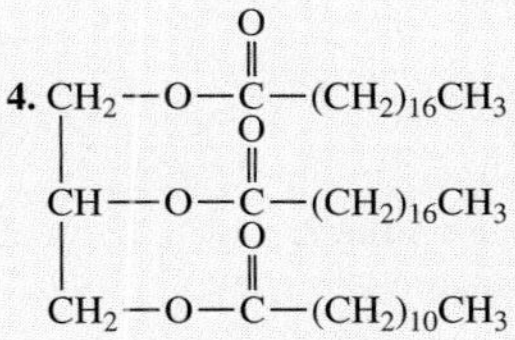
$$\begin{array}{l} CH_2-O-\overset{O}{\overset{\|}{C}}-(CH_2)_{16}CH_3 \\ CH-O-\overset{O}{\overset{\|}{C}}-(CH_2)_{16}CH_3 \\ CH_2-O-\overset{O}{\overset{\|}{C}}-(CH_2)_{10}CH_3 \end{array}$$

6. integral proteins have a higher percentage of nonpolar amino acids. **7.** The bacteria could synthesize phosphoglycerides with more saturated fatty acids. **9. b.** a monoterpene **11.** two ketone groups, a double bond, an aldehyde group, a primary alcohol, a secondary alcohol

CHAPTER 20

2. a. 3′—C—C—T—G—T—T—A—G—A—C—G—5′

b. guanine **5.** Met-Asp-Pro-Val-Ile-Lys-His **6.** Met-Asp-Pro-Leu-Leu-Asn

8. 5′—G—C—A—T—G—G—A—C—C—C—C—G—T—T—A—T—T—A—A—A—C—A—C—3′

9. a

Glossary

absorption band a peak in a spectrum that occurs as a result of the absorption of energy.

acetal $R-\overset{\displaystyle OR}{\underset{\displaystyle OR}{C}}-H$

acetoacetic ester synthesis synthesis of a methyl ketone, using ethyl acetoacetate as the starting material.

acetylide ion $RC{\equiv}C^-$

achiral (optically inactive) an achiral molecule is identical to (that is, superimposable upon) its mirror image.

acid a substance that donates a proton.

acid anhydride

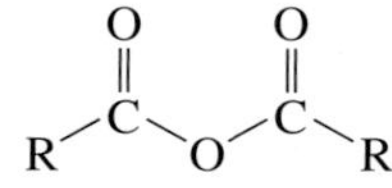

acid–base reaction a reaction in which an acid donates a proton to a base or accepts a share in a base's electrons.

acid catalyst a catalyst that increases the rate of a reaction by donating a proton.

acid-catalyzed reaction a reaction catalyzed by an acid.

acid dissociation constant a measure of the degree to which an acid dissociates in solution.

activating substituent a substituent that increases the reactivity of an aromatic ring. Electron-donating substituents activate aromatic rings toward electrophilic attack, and electron-withdrawing substituents activate aromatic rings toward nucleophilic attack.

active site a pocket or cleft in an enzyme where the substrate is bound.

acyl chloride

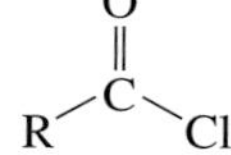

acyl group a carbonyl group bonded to an alkyl group or to an aryl group.

1,2-addition addition to the 1- and 2-positions of a conjugated system.

1,4-addition addition to the 1- and 4-positions of a conjugated system.

addition reaction a reaction in which atoms or groups are added to the reactant.

alcohol a compound with an OH group in place of one of the hydrogens of an alkane: ROH.

alcoholysis reaction with an alcohol.

aldaric acid a dicarboxylic acid with an OH group bonded to each carbon. Obtained by oxidizing the aldehyde and primary alcohol groups of an aldose.

aldehyde

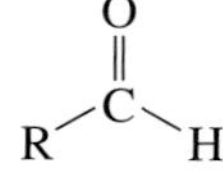

alditol a compound with an OH group bonded to each carbon. Obtained by reducing an aldose or a ketose.

aldol addition a reaction between two molecules of an aldehyde (or two molecules of a ketone) that connects the α-carbon of one with the carbonyl carbon of the other.

aldol condensation an aldol addition followed by the elimination of water.

aldonic acid a carboxylic acid with an OH group bonded to each carbon. Obtained by oxidizing the aldehyde group of an aldose.

aldose a polyhydroxyaldehyde.

aliphatic a nonaromatic organic compound.

alkaloid a natural product, with one or more nitrogen heteroatoms, found in the leaves, bark, or seeds of plants.

alkane a hydrocarbon that contains only single bonds.

alkene a hydrocarbon that contains a double bond.

alkyl halide a compound with a halogen in place of one of the hydrogens of an alkane.

alkyl substituent (alkyl group) a substituent formed by removing a hydrogen from an alkane.

alkyne a hydrocarbon that contains a triple bond.

allyl group $CH_2{=}CHCH_2{-}$

allylic carbon an sp^3 carbon adjacent to a vinylic carbon.

allylic cation a species with a positive charge on an allylic carbon.

amide

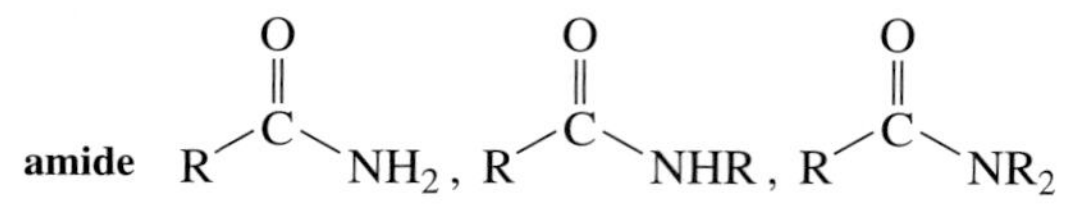

amine a compound with alkyl groups in place of one or more of the hydrogens of ammonia: RNH_2, R_2NH, R_3N.

amino acid an α-aminocarboxylic acid. Naturally occurring amino acids have the L configuration.

amino acid analyzer an instrument that automates the ion-exchange separation of amino acids.

amino acid residue a monomeric unit of a peptide or protein.

amino acid side chain the substituent attached to the α-carbon of an amino acid.

aminolysis reaction with an amine.

amino sugar a sugar in which one of the OH groups is replaced by an NH_2 group.

anabolic reaction a reaction that a living organism carries out in order to synthesize complex molecules from simple precursor molecules.

anabolism reactions that living organisms carry out in order to synthesize complex molecules from simple precursor molecules.

angle strain the strain introduced into a molecule as a result of its bond angles being distorted from their ideal values.

angstrom unit of length; 100 picometers = 10^{-8} cm = 1 angstrom.

anomeric carbon the carbon in a cyclic sugar that is the carbonyl carbon in the open-chain form.

anomers two cyclic sugars that differ in configuration only at the carbon that is the carbonyl carbon in the open-chain form.

antibiotic a compound that interferes with the growth of a microorganism.

antibodies compounds that recognize foreign particles in the body.

anticodon the three bases at the bottom of the middle loop in tRNA.

antigens compounds that can generate a response from the immune system.

antiviral drug a drug that interferes with DNA or RNA synthesis in order to prevent a virus from replicating.

applied magnetic field the externally applied magnetic field.

aprotic solvent a solvent that does not have a hydrogen bonded to an oxygen or to a nitrogen.

arene oxide an aromatic compound that has had one of its double bonds converted to an epoxide.

aromatic a cyclic and planar compound with an uninterrupted ring of *p* orbital-bearing atoms containing an odd number of pairs of π electrons.

aryl group a benzene or a substituted-benzene group.

asymmetric center a carbon bonded to four different atoms or groups.

atomic number the number of protons (or electrons) that the neutral atom has.

atomic orbital an orbital associated with an atom.

atomic weight the average mass of the atoms in the naturally occurring element.

autoradiograph the exposed photographic plate obtained in autoradiography.

axial bond a bond of the chair conformation of cyclohexane that is perpendicular to the plane in which the chair is drawn (an up–down bond).

back-side attack nucleophilic attack on the side of the carbon opposite the side bonded to the leaving group.
bactericidal drug a drug that kills bacteria.
bacteriostatic drug a drug that inhibits the further growth of bacteria.
base[1] a substance that accepts a proton.
base[2] a heterocyclic compound (a purine or a pyrimidine) in DNA and RNA.
base catalyst a catalyst that increases the rate of a reaction by removing a proton.
basicity the tendency of a compound to share its electrons with a proton.
bending vibration a vibration that does not occur along the line of the bond. It results in changing bond angles.

benzyl group C_6H_5—CH_2—

benzylic carbon an sp^3 carbon bonded to a benzene ring.
benzylic cation a compound with a positive charge on a benzylic carbon.
bimolecular reaction (second-order reaction) a reaction whose rate depends on the concentration of two reactants.
biochemistry (biological chemistry) the chemistry of biological systems.
biodegradable polymer a polymer that can be broken into small segments by an enzyme-catalyzed reaction.
bioorganic compound an organic compound found in biological systems.
biopolymer a polymer that is synthesized in nature.
biosynthesis synthesis in a biological system.
biotin the coenzyme required by enzymes that catalyze carboxylation of a carbon adjacent to an ester or a keto group.
blind screen (random screen) the search for a pharmacologically active compound without any information about which chemical structures might show activity.
boiling point the temperature at which the vapor pressure equals the atmospheric pressure.
bond an attractive force between two atoms.
bond dissociation energy the energy required to break a bond, or the amount of energy released when a bond is formed.
bond length the internuclear distance between two atoms at minimum energy (maximum stability).
bond strength the energy required to break a bond.
brand name name that identifies a commercial product and distinguishes it from other products. It can be used only by the owner of the registered trademark.
buffer a weak acid and its conjugate base.
carbanion a compound containing a negatively charged carbon.
carbocation a species containing a positively charged carbon.
carbohydrate a sugar or a saccharide. Naturally occurring carbohydrates have the D configuration.
α-carbon a carbon bonded to a carbonyl carbon or to a leaving group.
β-carbon a carbon adjacent to an α-carbon carbonyl.
carbonyl carbon the carbon of a carbonyl group.
carbonyl compound a compound that contains a carbonyl group.
carbonyl group a carbon doubly bonded to an oxygen.
carbonyl oxygen the oxygen of a carbonyl group.
carboxyl group COOH

carboxylic acid 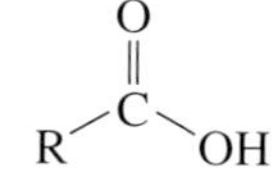

carboxylic acid derivative a compound that is hydrolyzed to a carboxylic acid.
carboxyl oxygen the single-bonded oxygen of a carboxlic acid or an ester.
carotenoid a class of compounds (a tetraterpene) responsible for the red and orange colors of fruits, vegetables, and fall leaves.
catabolic reaction a reaction that a living organism carries out in order to break down complex molecules into simple molecules and energy.
catabolism reactions that living organisms carry out in order to break down complex molecules into simple molecules and energy.
catalyst a species that increases the rate at which a reaction occurs without being consumed in the reaction. Because it does not change the equilibrium constant of the reaction, it does not change the amount of product that is formed.
catalytic hydrogenation the addition of hydrogen to a double or a triple bond with the aid of a metal catalyst.
cation-exchange resin a negatively charged resin used in ion-exchange chromatography.
chain-growth polymer a polymer made by adding monomers to the growing end of a chain.
chain reaction a reaction in which propagating steps are repeated over and over.
chair conformation the conformation of cyclohexane that roughly resembles a chair. It is the most stable conformation of cyclohexane.
chemically equivalent protons protons with the same connectivity relationship to the rest of the molecule.
chemical shift the location of a signal in an NMR spectrum. It is measured downfield from a reference compound (most often, TMS).
chiral (optically active) a chiral molecule has a nonsuperimposable mirror image.
cholesterol a steroid that is the precursor of all other animal steroids.
chromatography a separation technique in which the mixture to be separated is dissolved in a solvent and the solvent is passed through a column packed with an absorbent stationary phase.
cis fused two cyclohexane rings fused together such that if the second ring were considered to be two substituents of the first ring, one substituent would be in an axial position and the other would be in an equatorial position.
cis isomer the isomer with identical substituents on the same side of the double bond or on the same side of a cyclic structure.
cis–trans isomers geometric isomers.
citric acid cycle (Krebs cycle) a series of reactions that converts the acetyl group of acetyl-CoA into two molecules of CO_2.
Claisen condensation a reaction between two molecules of an ester that connects the α-carbon of one with the carbonyl carbon of the other and eliminates an alkoxide ion.
codon a sequence of three bases in mRNA that specifies the amino acid to be incorporated into a protein.
coenzyme a cofactor that is an organic molecule.
coenzyme A a thiol used by biological organisms to form thioesters.
coenzyme B_{12} the coenzyme required by enzymes that catalyze certain rearrangement reactions.
cofactor an organic molecule or a metal ion that certain enzymes need to catalyze a reaction.
common name nonsystematic nomenclature.
complex carbohydrate a carbohydrate containing two or more sugar molecules linked together.
condensation reaction a reaction combining two molecules while removing a small molecule (usually water or an alcohol).
configuration the three-dimensional structure of a particular atom in a compound. The configuration is designated by *R* or *S*.
conformation the three-dimensional shape of a molecule at a given instant that can change as a result of rotations about σ bonds.
conformers different conformations of a molecule.
conjugate acid a species accepts a proton to form its conjugate acid.
conjugate addition 1,4-addition.
conjugate base a species loses a proton to form its conjugate base.
conjugated diene a hydrocarbon with two conjugated double bonds.
conjugated double bonds double bonds separated by one single bond.
constitutional isomers molecules that have the same molecular formula but differ in the way their atoms are connected.
core electrons electrons in inner shells.
coupled protons protons that split each other. Coupled protons have the same coupling constant.
coupling constant the distance (in hertz) between two adjacent peaks of a split NMR signal.
covalent bond a bond created as a result of sharing electrons.
crown ether a cyclic molecule that contains several ether linkages.
crown–guest complex the complex formed when a crown ether binds a substrate.
C-terminal amino acid the terminal amino acid of a peptide (or protein) that has a free carboxyl group.
cycloalkane an alkane with its carbon chain arranged in a closed ring.
deactivating substituent a substituent that decreases the reactivity of an aromatic ring. Electron-withdrawing substituents deactivate aromatic rings toward electrophilic attack, and electron-donating substituents deactivate aromatic rings toward nucleophilic attack.

deamination loss of ammonia.
decarboxylation loss of carbon dioxide.
dehydration loss of water.
dehydrogenase an enzyme that carries out an oxidation reaction by removing hydrogen from the substrate.
delocalization energy (resonance energy) the extra stability a compound achieves as a result of having delocalized electrons.
delocalized electrons electrons that are shared by more than two atoms.
denaturation destruction of the highly organized tertiary structure of a protein.
deoxyribonucleic acid (DNA) a polymer of deoxyribonucleotides.
deoxyribonucleotide a nucleotide in which the sugar component is D-2-deoxyribose.
deoxy sugar a sugar in which one of the OH groups has been replaced by an H.
dextrorotatory the enantiomer that rotates polarized light in a clockwise direction.
diastereomer a configurational stereoisomer that is not an enantiomer.
diene a hydrocarbon with two double bonds.
dinucleotide two nucleotides linked by phosphodiester bonds.
dipeptide two amino acids linked by an amide bond.
dipole–dipole interaction an interaction between the dipole of one molecule and the dipole of another.
disaccharide a compound containing two sugar molecules linked together.
disulfide R—S—S—R
disulfide bridge a disulfide (—S—S—) bond in a peptide or protein.
DNA (deoxyribonucleic acid) a polymer of deoxyribonucleotides.
donation of electrons by resonance donation of electrons through *p* orbital overlap with neighboring π bonds.
double bond a σ bond and a π bond between two atoms.
double helix the secondary structure of DNA.
doublet an NMR signal split into two peaks.
downfield at a higher frequency in an NMR spectrum.
drug a compound that reacts with a biological molecule, triggering a physiological effect.
drug resistance biological resistance to a particular drug.
eclipsed conformer a conformer in which the bonds on adjacent carbons are aligned as viewed looking down the carbon–carbon bond.
***E* conformation** the conformation of a carboxylic acid or carboxylic acid derivative in which the carbonyl oxygen and the substituent bonded to the carboxyl oxygen or nitrogen are on opposite sides of the single bond.
Edman's reagent phenyl isothiocyanate. A reagent used to determine the N-terminal amino acid of a polypeptide.
effective magnetic field the magnetic field that a proton "senses" through the surrounding cloud of electrons.
***E* isomer** the isomer with the high-priority groups on opposite sides of the double bond.
electronegative element an element that readily acquires an electron.
electronegativity tendency of an atom to pull electrons toward itself.
electronic configuration description of the orbitals that the electrons in an atom occupy.
electrophile an electron-deficient atom or molecule.
electrophilic addition reaction an addition reaction in which the first species that adds to the reactant is an electrophile.
electrophilic aromatic substitution a reaction in which an electrophile substitutes for a hydrogen of an aromatic ring.
electrophoresis a technique that separates amino acids on the basis of their pI values.
electrostatic attraction attractive force between opposite charges.
electrostatic potential map a model that shows the charge distribution in a species.
elimination reaction a reaction that involves the elimination of atoms (or molecules) from the reactant.
enamine an α,β-unsaturated tertiary amine.
enantiomers nonsuperimposable mirror-image molecules.
enkephalins pentapeptides synthesized by the body to control pain.
enol an α,β-unsaturated alcohol.
enolization keto–enol interconversion.
enzyme a protein that is a catalyst.
epimers monosaccharides that differ in configuration at only one carbon.
epoxide an ether in which the oxygen is incorporated into a three-membered ring.
epoxy resin substance formed by mixing a low-molecular-weight prepolymer with a compound that forms a cross-linked polymer.
equatorial bond a bond of the chair conformer of cyclohexane that juts out from the ring in approximately the same plane that contains the chair.
equilibrium constant the ratio of products to reactants at equilibrium or the ratio of the rate constants for the forward and reverse reactions.
E1 reaction a first-order elimination reaction.
E2 reaction a second-order elimination reaction.
essential amino acid an amino acid that humans must obtain from their diet because they cannot synthesize it at all or cannot synthesize it in adequate amounts.
ester 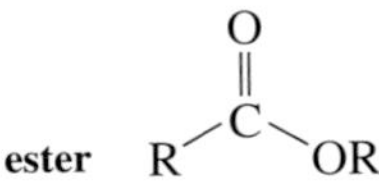

ether a compound containing an oxygen bonded to two carbons (ROR).
fat a triester of glycerol that exists as a solid at room temperature.
fatty acid a long-chain carboxylic acid.
favorable reaction a reaction in which the concentration of products is greater than the concentration of reactants at equilibrium.
fibrous protein a water-insoluble protein in which the polypeptide chains are arranged in bundles.
fingerprint region the right-hand third of an IR spectrum where the absorption bands are characteristic of the compound as a whole.
Fischer esterification reaction the reaction of a carboxylic acid with alcohol in the presence of an acid catalyst to form an ester.
Fischer projection a method of representing the spatial arrangement of groups bonded to a chirality center. The chirality center is the point of intersection of two perpendicular lines; the horizontal lines represent bonds that project out of the plane of the paper toward the viewer, and the vertical lines represent bonds that point back from the plane of the paper away from the viewer.
flavin adenine dinucleotide (FAD) a coenzyme required in certain oxidation reactions. It is reduced to $FADH_2$, which can act as a reducing agent in another reaction.
formal charge the number of valence electrons – (the number of nonbonding electrons +1/2 the number of bonding electrons).
free energy of activation ($\Delta G^{\ddagger}$) the true energy barrier to a reaction.
free radical a species with an unpaired electron.
frequency the velocity of a wave divided by its wavelength (in units of cycles/s).
Friedel–Crafts acylation an electrophilic substitution reaction that puts an acyl group on a benzene ring.
functional group the center of reactivity in a molecule.
functional group interconversion the conversion of one functional group into another functional group.
functional group region the left-hand two-thirds of an IR spectrum where most functional groups show absorption bands.
furanose a five-membered-ring sugar.
furanoside a five-membered-ring glycoside.
gene a segment of DNA.
generic name a commercially nonrestricted name for a drug.
genetic code the amino acid specified by each three-base sequence of mRNA.
geometric isomers cis–trans (or *E,Z*) isomers.
globular protein a water-soluble protein that tends to have a roughly spherical shape.
gluconeogenesis the synthesis of D-glucose from pyruvate.
glycolysis (glycolytic pathway) the series of reactions that converts D-glucose into two molecules of pyruvate.
glycoprotein a protein that is covalently bonded to a polysaccharide.
glycoside the acetal of a sugar.
glycosidic bond the bond between the anomeric carbon and the alcohol in a glycoside.
α-1,4'-glycosidic linkage a glycosidic linkage between the C-1 oxygen of one sugar and the C-4 of a second sugar with the oxygen atom of the glycosidic linkage in the axial position.

α-1,6'-glycosidic linkage a glycosidic linkage between the C-1 oxygen of one sugar and the C-6 of a second sugar with the oxygen atom of the glycosidic linkage in the axial position.
β-1,4'-glycosidic linkage a glycosidic linkage between the C-1 oxygen of one sugar and the C-4 of a second sugar with the oxygen atom of the glycosidic linkage in the equatorial position.
Grignard reagent the compound that results when magnesium is inserted between the carbon and halogen of an alkyl halide (RMgBr, RMgCl).
halogenation reaction with halogen (Br_2, Cl_2, I_2).
α-helix the backbone of a polypeptide coiled in a right-handed spiral with hydrogen bonding occurring within the helix.
hemiacetal $R{-}C(OH)(OR){-}H$
hemiketal $R{-}C(OH)(OR){-}R$
heptose a monosaccharide with seven carbons.
heteroatom an atom other than carbon or hydrogen.
heterocyclic compound (heterocycle) a cyclic compound in which one or more of the atoms of the ring are heteroatoms.
hexose a monosaccharide with six carbons.
hormone an organic compound synthesized in a gland and delivered by the bloodstream to its target tissue.
human genome the total DNA of a human cell.
hybrid orbital an orbital formed by mixing (hybridizing) orbitals.
hydrate $R{-}C(OH)_2{-}R$ (H)
hydrated water has been added to a compound.
hydration addition of water to a compound.
hydride ion a negatively charged hydrogen (a hydrogen atom with an extra electron).
hydrocarbon a compound that contains only carbon and hydrogen.
α-hydrogen usually, a hydrogen bonded to the carbon adjacent to a carbonyl carbon.
hydrogenation addition of hydrogen.
hydrogen bond an unusually strong dipole–dipole attraction (5 kcal/mol) between a hydrogen bonded to O, N, or F and the nonbonding electrons of an O, N, or F of another molecule.
hydrogen ion (proton) a positively charged hydrogen (a hydrogen atom without its electron).
hydrolysis reaction with water.
hydrophobic interactions interactions between nonpolar groups. These interactions increase stability by decreasing the amount of structured water (increasing entropy).
imine $R_2C{=}NR$
induced-dipole–induced-dipole interaction an interaction between a temporary dipole in one molecule and the dipole the temporary dipole induces in another molecule.
induced-fit model a model that describes the specificity of an enzyme for its substrate: The shape of the active site does not become completely complementary to the shape of the substrate until after the enzyme binds the substrate.
inductive electron donation donation of electrons through σ bond(s).
inductive electron withdrawal withdrawal of electrons through σ bond(s).
infrared radiation electromagnetic radiation familiar to us as heat.
infrared spectroscopy uses infrared energy to provide knowledge of the functional groups in a compound.
infrared (IR) spectrum a plot of percent transmission versus wave number (or wavelength) of infrared radiation.
initiation step the step in which radicals are created, or the step in which the radical needed for the first propagation step is created.
intermediate a species that is formed during a reaction and that is not the final product of the reaction.
intermolecular reaction a reaction that takes place between two molecules.
internal alkyne an alkyne with the triple bond not at the end of the carbon chain.
intramolecular reaction a reaction that takes place within a molecule.
inversion of configuration turning the configuration of a carbon inside out like an umbrella in a windstorm, so that the resulting product has a configuration opposite that of the reactant.
ion–dipole interaction the interaction between an ion and the dipole of a molecule.
ion-exchange chromatography a technique that uses a column packed with an insoluble resin to separate compounds on the basis of their charges and polarities.
ionic bond a bond formed through the attraction of two ions of opposite charges.
isoelectric point (pI) the pH at which there is no net charge on an amino acid.
isolated diene a hydrocarbon containing two isolated double bonds.
isolated double bonds double bonds separated by more than one single bond.
isomers nonidentical compounds with the same molecular formula.
isotopes atoms with the same number of protons but different numbers of neutrons.
IUPAC nomenclature systematic nomenclature of chemical compounds.
Kekulé structure a model that represents the bonds between atoms as lines.
ketal $R{-}C(OR)_2{-}R$
keto–enol tautomerism (keto–enol interconversion) interconversion of keto and enol tautomers.
keto–enol tautomers a ketone and its isomeric α,β-unsaturated alcohol.
β-keto ester an ester with a second carbonyl group at the β-position.
ketone $R{-}C({=}O){-}R$
ketose a polyhydroxyketone.
Kiliani–Fischer synthesis a method used to increase the number of carbons in an aldose by one, resulting in the formation of a pair of C-2 epimers.
λ_{max} the wavelength at which there is maximum UV or Vis absorbance.
lead compound the prototype in a search for other biologically active compounds.
leaving group the group that is displaced in a nucleophilic substitution reaction.
levorotatory the enantiomer that rotates polarized light in a counterclockwise direction.
Lewis acid a substance that accepts an electron pair.
Lewis base a substance that donates an electron pair.
Lewis structure a model that represents the bonds between atoms as lines or dots and the valence electrons as dots.
ligation sharing of nonbonding electrons with a metal ion.
lipid a water-insoluble compound found in a living system.
lipid bilayer two layers of phosphoacylglycerols arranged so that their polar heads are on the outside and their nonpolar fatty acid chains are on the inside.
localized electrons electrons that are restricted to a particular locality.
lone-pair electrons (nonbonding electrons) valence electrons not used in bonding.
magnetic resonance imaging (MRI) NMR used in medicine. The difference in the way water is bound in different tissues produces a variation in signal between organs as well as between healthy and diseased tissue.
major groove the wider and deeper of the two alternating grooves in DNA.
malonic ester synthesis the synthesis of a carboxylic acid, using diethyl malonate as the starting material.
mass number the number of protons plus the number of neutrons in an atom.
mechanism of a reaction a description of the step-by-step process by which reactants are changed into products.
melting point the temperature at which a solid becomes a liquid.
membrane the material that surrounds a cell in order to isolate its contents.
meso compound a compound that contains chirality centers and a plane of symmetry.

metabolism reactions that living organisms carry out in order to obtain the energy and to synthesize the compounds they require.
meta director a substituent that directs an incoming substituent meta to an existing substituent.
methine hydrogen a tertiary hydrogen.
methylene group a CH_2 group.
micelle a spherical aggregation of molecules, each with a long hydrophobic tail and a polar head, arranged so that the polar head points to the outside of the sphere.
minor groove the narrower and more shallow of the two alternating grooves in DNA.
mixed triglyceride a triacylglycerol in which the fatty-acid components are different.
molecular modification changing the structure of a lead compound.
molecular recognition the recognition of one molecule by another as a result of specific interactions; for example, the specificity of an enzyme for its substrate.
molecular weight the average weighted mass of the atoms in a molecule.
monomer a repeating unit in a polymer.
monosaccharide (simple carbohydrate) a single sugar molecule.
monoterpene a terpene that contains 10 carbons.
MRI scanner an NMR spectrometer used in medicine for whole-body NMR.
multiplet an NMR signal split into more than seven peaks.
multiplicity the number of peaks in an NMR signal.
multistep synthesis preparation of a compound by a route that requires several steps.
mutarotation a slow change in optical rotation to an equilibrium value.
***N* + 1 rule** an ^1H NMR signal for a hydrogen with *N* equivalent hydrogens bonded to an adjacent carbon is split into $N + 1$ peaks. A ^{13}C NMR signal for a carbon bonded to *N* hydrogens is split into $N + 1$ peaks.
natural-abundance atomic weight the average mass of the atoms in the naturally occurring element.
natural product a product synthesized in nature.
nicotinamide adenine dinucleotide (NAD^+) a coenzyme required in certain oxidation reactions. It is reduced to NADH, which can act as a reducing agent in another reaction.
nitration substitution of a nitro group (NO_2) for a hydrogen of a benzene ring.
nitrile a compound that contains a carbon–nitrogen triple bond ($RC{\equiv}N$).
NMR spectroscopy the absorption of electromagnetic radiation to determine the structural features of an organic compound. In the case of NMR spectroscopy, it determines the carbon–hydrogen framework.
nonbonding electrons (lone-pair electrons) valence electrons not used in bonding.
nonpolar covalent bond a bond formed between two atoms that share the bonding electrons equally.
nonpolar molecule a molecule with no charge or partial charge on any of its atoms.
nonreducing sugar a sugar that cannot be oxidized by reagents such as Ag^+ and Cu^+. Nonreducing sugars are not in equilibrium with the open-chain aldose or ketose.
***N*-terminal amino acid** the terminal amino acid of a peptide (or protein) that has a free amino group.
nucleic acid the two kinds of nucleic acid are DNA and RNA.
nucleophile an electron-rich atom or molecule.
nucleophilic acyl substitution reaction a reaction in which a group bonded to an acyl or aryl group is substituted by another group.
nucleophilic addition reaction a reaction that involves the addition of a nucleophile to a reagent.
nucleophilicity a measure of how readily an atom or a molecule with a pair of nonbonding electrons attacks an atom.
nucleophilic substitution reaction a reaction in which a nucleophile substitutes for an atom or a group.
nucleoside a heterocyclic base (a purine or a pyrimidine) bonded to the anomeric carbon of a sugar (D-ribose or D-2-deoxyribose).
nucleotide a heterocycle attached in the β-position to a phosphorylated ribose or deoxyribose.
observed rotation the amount of rotation observed in a polarimeter.
octet rule states that an atom will give up, accept, or share electrons in order to achieve a filled shell. Because a filled second shell contains eight electrons, this is known as the octet rule.
oil a triester of glycerol that exists as a liquid at room temperature.
oligonucleotide 3 to 10 nucleotides linked by phosphodiester bonds.
oligopeptide 3 to 10 amino acids linked by amide bonds.
oligosaccharide 3 to 10 sugar molecules linked by glycosidic bonds.
operating frequency the frequency at which an NMR spectrometer operates.
optically active rotates the plane of polarized light.
optically inactive does not rotate the plane of polarized light.
orbital the volume of space around the nucleus in which an electron is most likely to be found.
orbital hybridization mixing of orbitals.
organic compound a compound that contains carbon.
organic synthesis preparation of organic compounds from other organic compounds.
organometallic compound a compound containing a carbon–metal bond.
orphan drugs drugs for diseases or conditions that affect fewer than 200,000 people.
ortho-para-director a substituent that directs an incoming substituent ortho and para to an existing substituent.
oxidation reaction a reaction in which the number of C—H bonds decreases.
oxidation–reduction reaction (redox reaction) a reaction that involves the transfer of electrons from one species to another.
oxidative cleavage an oxidation reaction that cuts the reactant into two or more pieces.
packing the fitting of individual molecules into a frozen crystal lattice.
paraffin an alkane.
parent hydrocarbon the longest continuous carbon chain in a molecule.
partial hydrolysis a technique that hydrolyzes only some of the peptide bonds in a polypeptide.
pentose a monosaccharide with five carbons.
peptide polymer of amino acids linked together by amide bonds. A peptide contains fewer amino acid residues than a protein does.
peptide bond the amide bond that links the amino acids in a peptide or protein.
peroxyacid a carboxylic acid with an OOH group instead of an OH group.
perspective formula a method of representing the spatial arrangement of groups bonded to a chirality center. Two bonds are drawn in the plane of the paper; a solid wedge is used to depict a bond that projects out of the plane of the paper toward the viewer, and a hatched wedge is used to represent a bond that projects back from the plane of the paper away from the viewer.
pH the pH scale is used to describe the acidity of a solution ($pH = -\log[H^+]$).
phenyl group [structure: benzene ring with one bond]
pheromone a compound secreted by an animal that stimulates a physiological or behavioral response from a member of the same species.
phosphoacylglycerol (phosphoglyceride) a compound formed when two OH groups of glycerol form esters with fatty acids and the terminal OH group forms a phosphate ester.
phosphoanhydride bond the bond holding two phosphoric acid molecules together.
phospholipid a lipid that contains a phosphate group.
phosphoryl transfer reaction the transfer of a phosphate group from one compound to another.
pi (π) bond a bond formed as a result of side-to-side overlap of *p* orbitals.
pK_a describes the tendency of a compound to lose a proton ($pK_a = -\log K_a$, where K_a is the acid dissociation constant).
plane polarized light light that oscillates only in one plane.
plane of symmetry an imaginary plane that bisects a molecule into mirror images.
β-pleated sheet the backbone of a polypeptide that is extended in a zigzag structure with hydrogen bonding between neighboring chains.
polar covalent bond a covalent bond between atoms with different electronegativities.
polarimeter an instrument that measures the rotation of polarized light.
polymer a large molecule made by linking monomers together.

polymerization the process of linking up monomers to form a polymer.
polynucleotide many nucleotides linked by phosphodiester bonds.
polypeptide many amino acids linked by amide bonds.
polysaccharide a compound containing more than 10 sugar molecules linked together.
polyunsaturated fatty acid a fatty acid with more than one double bond.
porphyrin ring system consists of four pyrrole rings joined by one-carbon bridges.
primary alcohol an alcohol in which the OH group is bonded to a primary carbon.
primary alkyl halide an alkyl halide in which the halogen is bonded to a primary carbon.
primary alkyl radical a radical with the unpaired electron on a primary carbon.
primary amine an amine with one alkyl group bonded to the nitrogen.
primary carbocation a carbocation with the positive charge on a primary carbon.
primary carbon a carbon bonded to only one other carbon.
primary hydrogen a hydrogen bonded to a primary carbon.
primary structure (of a nucleic acid) the sequence of bases in a nucleic acid.
primary structure (of a protein) the sequence of amino acids in a protein.
propagating site the reactive end of a chain-growth polymer.
propagation step in the first of a pair of propagation steps, a radical (or an electrophile or a nucleophile) reacts to produce another radical (or an electrophile or a nucleophile) that reacts in the second step to produce the radical (or the electrophile or the nucleophile) that was the reactant in the first propagation step.
protein a polymer containing 40 to 4000 amino acids linked by amide bonds.
proton a positively charged hydrogen (hydrogen ion); a positively charged particle in an atomic nucleus.
proton-coupled ^{13}C NMR spectrum a ^{13}C NMR spectrum in which each signal is split by the hydrogens bonded to the C that produced the spectrum.
pyranose a six-membered-ring sugar.
pyranoside a six-membered-ring glycoside.
pyridoxal phosphate the coenzyme required by enzymes that catalyze certain transformations of amino acids.
quartet an NMR signal split into four peaks.
quaternary structure a description of the way the individual polypeptide chains of a protein are arranged with respect to each other.
racemic mixture (racemate, racemic modification) a mixture of equal amounts of a pair of enantiomers.
radical an atom or a molecule with an unpaired electron.
radical chain reaction a reaction in which radicals are formed and react in repeating propagating steps.
radical inhibitor a compound that traps radicals.
radical substitution reaction a substitution reaction that has a radical intermediate.
random screen (blind screen) the search for a pharmacologically active compound without any information about what chemical structures might show activity.
rate constant a measure of how easy or difficult it is to reach the transition state of a reaction (to get over the energy barrier to the reaction).
rate law the relationship between the rate of a reaction and the concentration of the reactants.
rate-determining step (rate-limiting step) the step in a reaction that has the transition state with the highest energy.
***R* configuration** after assigning relative priorities to the four groups bonded to a chirality center, if the lowest priority group is on a vertical axis in a Fischer projection (or pointing away from the viewer in a perspective formula), an arrow drawn from the highest priority group to the next-highest-priority group goes in a clockwise direction.
reaction coordinate diagram describes the energy changes that take place during the course of a reaction.
reactivity–selectivity principle states that the greater the reactivity of a species, the less selective it will be.
receptor a site on a cell at which a drug binds in order to exert its physiological effect.
reducing sugar a sugar that can be oxidized by reagents such as Ag^+ or Br_2. Reducing sugars are in equilibrium with the open-chain aldose or ketose.
reduction reaction a reaction in which the number of C—H bonds increases.
reference compound a compound added to a sample whose NMR spectrum is to be taken. The positions of the signals in the NMR spectrum are measured from the position of the signal given by the reference compound.
regioselective reaction a reaction that leads to the preferential formation of one constitutional isomer over another.
replication the synthesis of identical copies of DNA.
resonance a compound with delocalized electrons is said to have resonance.
resonance contributor a structure with localized electrons that approximates the true structure of a compound with delocalized electrons.
resonance electron donation donation of electrons through p orbital overlap with neighboring π bonds.
resonance electron withdrawal withdrawal of electrons through p orbital overlap with neighboring π bonds.
resonance hybrid the actual structure of a compound with delocalized electrons; it is represented by two or more structures with localized electrons.
resonance stabilization the extra stability associated with a compound as a result of its having delocalized electrons.
restriction endonuclease an enzyme that cleaves DNA at a specific base sequence.
restriction fragment a fragment that is formed when DNA is cleaved by a restriction endonuclease.
ribonucleic acid (RNA) a polymer of ribonucleotides.
ribonucleotide a nucleotide in which the sugar component is D-ribose.
RNA (ribonucleic acid) a polymer of ribonucleotides.
saponification hydrolysis of an ester (such as a fat) under basic conditions.
saturated hydrocarbon a hydrocarbon that is completely saturated (that is, contains no double or triple bonds) with hydrogen.
***S* configuration** after assigning relative priorities to the four groups bonded to a chirality center, if the lowest priority group is on a vertical axis in a Fischer projection (or pointing away from the viewer in a perspective formula), an arrow drawn from the highest priority group to the next-highest priority group goes in a counterclockwise direction.
secondary alcohol an alcohol in which the OH group is bonded to a secondary carbon.
secondary alkyl halide an alkyl halide in which the halogen is bonded to a secondary carbon.
secondary alkyl radical a radical with the unpaired electron on a secondary carbon.
secondary amine an amine with two alkyl groups bonded to the nitrogen.
secondary carbocation a carbocation with the positive charge on a secondary carbon.
secondary carbon a carbon bonded to two other carbons.
secondary hydrogen a hydrogen bonded to a secondary carbon.
secondary structure a description of the conformation of the backbone of a protein.
semiconservative replication the mode of replication that results in a daughter molecule of DNA having one of the original DNA strands plus a newly synthesized strand.
sense strand the strand in DNA that is not read during transcription; it has the same sequence of bases as the synthesized mRNA strand (with a U, T difference).
separated charges a positive and a negative charge that can be neutralized by the movement of electrons.
sesquiterpene a terpene that contains 15 carbons.
shielding phenomenon caused by electron donation to the environment of a proton. The electrons shield the proton from the full effect of the applied magnetic field. The more a proton is shielded, the farther to the right its signal appears in an NMR spectrum.
sigma (σ) bond a bond with a cylindrically symmetrical distribution of electrons.
simple carbohydrate (monosaccharide) a single sugar molecule.
simple triglyceride a triacylglycerol in which the fatty acid components are the same.
single bond a σ bond.
singlet an unsplit NMR signal.
skeletal structure shows the carbon–carbon bonds as lines and does not show the carbon–hydrogen bonds.
S_N1 reaction a unimolecular nucleophilic substitution reaction.

S_N2 reaction a bimolecular nucleophilic substitution reaction.
soap a sodium or potassium salt of a fatty acid.
solvation the interaction between a solvent and another molecule (or ion).
specific rotation the amount of rotation that will be caused by a compound with a concentration of 1.0 g/mL in a sample tube 1.0 dm long.
spectroscopy study of the interaction of matter and electromagnetic radiation.
sphingolipid a lipid that contains sphingosine.
α-spin state nuclei in this spin state have their magnetic moments oriented in the same direction as the applied magnetic field.
β-spin state nuclei in this spin state have their magnetic moments oriented opposite the direction of the applied magnetic field.
squalene a triterpene that is a precursor of steroid molecules.
staggered conformer a conformer in which the bonds on one carbon bisect the bond angle on the adjacent carbon when viewed looking down the carbon–carbon bond.
stereochemistry the field of chemistry that deals with the structures of molecules in three dimensions.
stereoisomers isomers that differ in the way their atoms are arranged in space.
steric effects effects due to the fact that groups occupy a certain volume of space.
steric hindrance refers to bulky groups at the site of a reaction that make it difficult for the reactants to approach each other.
steric strain (van der Waals strain, van der Waals repulsion) the repulsion between the electron cloud of an atom or a group of atoms and the electron cloud of another atom or group of atoms.
steroid a class of compounds that contains a steroid ring system.
straight-chain alkane an alkane in which the carbons form a continuous chain with no branches.
stretching frequency the frequency at which a stretching vibration occurs.
substitution reaction a reaction in which an atom or group substitutes for another atom or group.
substrate the reactant of an enzyme-catalyzed reaction.
subunit an individual chain of an oligomer.
sulfonation substitution of a hydrogen of a benzene ring by a sulfonic acid group ($-SO_3H$).
synthetic polymer a polymer that is not synthesized in nature.
systematic nomenclature nomenclature based on structure.
tautomerism interconversion of tautomers.
tautomers rapidly equilibrating isomers that differ in the location of their bonding electrons.
template strand (antisense strand) the strand in DNA that is read during transcription.
terminal alkyne an alkyne with the triple bond at the end of the carbon chain.
termination step when two radicals combine to produce a molecule in which all the electrons are paired.
terpene a lipid, isolated from a plant, that contains carbon atoms in multiples of five.
tertiary alcohol an alcohol in which the OH group is bonded to a tertiary carbon.
tertiary alkyl halide an alkyl halide in which the halogen is bonded to a tertiary carbon.
tertiary alkyl radical a radical with the unpaired electron on a tertiary carbon.
tertiary amine an amine with three alkyl groups bonded to the nitrogen.
tertiary carbocation a carbocation with the positive charge on a tertiary carbon.
tertiary carbon a carbon bonded to three other carbons.
tertiary hydrogen a hydrogen bonded to a tertiary carbon.
tertiary structure a description of the three-dimensional arrangement of all the atoms in a protein.
tetrahedral bond angle the bond angle (109.5°) formed by adjacent bonds of an sp^3 carbon.
tetrahedral carbon an sp^3 carbon; a carbon that forms covalent bonds by using four sp^3 hybridized orbitals.
tetrahedral intermediate the intermediate formed in a nucleophilic acyl substitution reaction.
tetrahydrofolate (THF) the coenzyme required by enzymes that catalyze reactions that donate a group containing a single carbon to their substrates.
tetraterpene a terpene that contains 40 carbons.
tetrose a monosaccharide with four carbons.
thiamine pyrophosphate (TPP) the coenzyme required by enzymes, which catalyze a reaction that transfers a two-carbon fragment to a substrate.
thioester the sulfur analog of an ester. $R-C(=O)-SR$
thiol the sulfur analog of an alcohol (RSH).
trademark a registered name, symbol, or picture.
transamination a reaction in which an amino group is transferred from one compound to another.
transcription the synthesis of mRNA from a DNA blueprint.
transesterification reaction the reaction of an ester with an alcohol to form a different ester.
trans fused two cyclohexane rings fused together such that if the second ring were considered to be two substituents of the first ring, both substituents would be in equatorial positions.
trans isomer the isomer with identical substituents on opposite sides of the double bond or on opposite sides of a cyclic structure.
transition state the highest point on a hill in a reaction coordinate diagram. In the transition state, bonds in the reactant that will break are partially broken and bonds in the product that will form are partially formed.
translation the synthesis of a protein from an mRNA blueprint.
triglyceride the compound formed when the three OH groups of glycerol are esterified with fatty acids.
triose a monosaccharide with three carbons.
tripeptide three amino acids linked by amide bonds.
triple bond a σ bond plus two π bonds.
triplet an NMR signal split into three peaks.
triterpene a terpene that contains 30 carbons.
ultraviolet light electromagnetic radiation with wavelengths ranging from 180 to 400 nm.
upfield toward lower frequency in an NMR spectrum.
unimolecular reaction (first-order reaction) a reaction whose rate depends on the concentration of one reactant.
unsaturated hydrocarbon a hydrocarbon that contains one or more double or triple bonds.
UV/Vis spectroscopy the absorption of electromagnetic radiation in the ultraviolet and visible regions of the spectrum; used to determine information about conjugated systems.
valence electron an electron in an unfilled shell.
van der Waals forces induced-dipole–induced-dipole interactions.
vinyl group $CH_2{=}CH-$
vinylic carbon a carbon in a carbon–carbon double bond.
vinylic cation a compound with a positive charge on a vinylic carbon.
vinylic radical a compound with an unpaired electron on a vinylic carbon.
visible light electromagnetic radiation with wavelengths ranging from 400 to 780 nm.
vitamin a substance needed in small amounts for normal body function that the body cannot synthesize at all or cannot synthesize in adequate amounts.
vitamin KH_2 the coenzyme required by the enzyme that catalyzes the carboxylation of glutamate side chains.
wavelength distance from any point on one wave to the corresponding point on the next wave (usually in units of μm or nm).
wavenumber the number of waves in 1 cm.
wax an ester formed from a long-chain carboxylic acid and a long-chain alcohol.
withdraw electrons by resonance withdrawal of electrons through *p* orbital overlap with neighboring bonds.
***Z* isomer** the isomer with the high-priority groups on the same side of the double bond.
zwitterion a compound with a negative charge and a positive charge on nonadjacent atoms.

Photo Credits

CHAPTER 1 p. 1 Science & Society Picture Library\The Image Works p. 2 Stephane Alix\USDA/APHIS Animal And Plant Health Inspection Service p. 5 (top) AP Wide World Photos p. 5 (bottom) Photograph of the Albert Einstein Memorial Sculpture, Robert Berks, sculptor, at the National Academy of Sciences. Credit: National Academy of Sciences, © 1978 Robert Berks p. 6 Courtesy of the Bancroft Library, University of California, Berkeley. p. 7 Blaz Kure\Shutterstock p. 17 Joe Mc Nally\Joe McNally Photography.

CHAPTER 2 p. 33 Edgar Fahs Smith Memorial Collection p. 35 (left) © 1994 NYC Parks Photo Archive, Fundamental Photographs, NYC p. 35 (right) © 1994 Kristen Brochmann, Fundamental Photographs, NYC p. 46 Richard Megna\Fundamental Photographs, NYC.

CHAPTER 3 p. 63 Photo by Gen. Stab. Lit. Anst., courtesy AIP Emilio Segre Visual Archives, W.F. Meggers Gallery of Nobel Laureates and Weber Collections p. 69 Simon Fraser/Science Photo Library\Photo Researchers, Inc. p. 71 © & Courtesy of Dr. John D. Roberts p. 79 The University of Texas Southwestern Medical Center at Dallas.

CHAPTER 4 p. 85 Courtesy of www.istockphoto.com.

CHAPTER 5 p. 112 Edgar Fahs Smith Collection, University of Pennsylvania Library, Van Pelt Libray. p. 133 Renn Sminkey\CDV LLC, Creative Digital Visions p. 136 Philippe Plailly/Science Photo Library\Photo Researchers, Inc. p. 137 (top) NASA/Goddard Space Flight Center p. 137 (bottom) NASA/Goddard Space Flight Center.

CHAPTER 6 p. 152 Photos.com. p. 153 Diane Schiumo\Fundamental Photographs, NYC p. 159 (top) Getty Images Inc. - Hulton Archive Photos p. 159 (bottom) © Dr. Jeremy Burgess/Science Photo Library/Photo Researchers, Inc. p. 160 (top) Photos.com. p. 160 (bottom) U.S. Food and Drug Administration p. 163 © Bettmann/CORBIS All Rights Reserved p. 164 © & Courtesy of Dr. John D. Roberts.

CHAPTER 7 p. 171 Corbis/Bettmann p. 191 Elena Elisseeva\Shutterstock.

CHAPTER 8 p. 197 Sheila Terry\Photo Researchers, Inc. p. 200 (left) Xuanlu Wang\Shutterstock p. 200 (right) Mark Seggie, Resin[sistem]Design p. 204 Stan Shebs/Wikimedia\Wikemedia Commons p. 210 Edgar Fahs Smith Collection, University of Pennsylvania Library p. 211 Edgar Fahs Smith Collection, University of Pennsylvania Library.

CHAPTER 9 p. 230 Marevision \Photolibrary.com p. 231 © & Courtesy of Dr. John D. Roberts p. 246 Butlerov Institute of Chemistry p. 255 Edgar Fahs Smith Collection/University of Pennsylvania Library.

CHAPTER 10 p. 276 (left) Corbis/Bettmann p. 276 (right) Boston Medical Library in the Francis A. Countway Library of Medicine p. 285 Getty Images Inc. - Hulton Archive Photos.

CHAPTER 11 p. 298 Photos.com. p. 311 Corbis/Bettmann.

CHAPTER 12 p. 328 © The Nobel Foundation 1912.

CHAPTER 13 p. 358 Reproduced courtesy of the Library and Information Centre of The Royal Society of Chemistry.

CHAPTER 14 p. 387 Courtesy of the Library of Congress p. 407 (left) Scott Camazine\Photo Researchers, Inc. p. 407 (right) Simon Fraser/Science Photo Library\Photo Researchers, Inc.

CHAPTER 15 p. 428 Brown Brothers p. 436 (left) Dr. Paula Bruice p. 436 (right) Dr. Paula Bruice p. 439 (top) Visuals Unlimited p. 439 (bottom) Justin Black\Shutterstock p. 440 Dr. Jeremy Burgess/Science Photo Library\Photo Researchers, Inc. p. 443 Science Photo Library\Photo Researchers, Inc.

CHAPTER 16 p. 466 AP Wide World Photos p. 471 © Bettmann/CORBIS. p. 472 © Bettmann/CORBIS.

CHAPTER 17 p. 481 (top) Corbis/Bettmann p. 481 (bottom) Edgar Fahs Smith Collection, University of Pennsylvania Library p. 490 UPI/Corbis-Bettmann\Corbis/Bettmann p. 494 Ultimathule\Shutterstock.

CHAPTER 18 p. 497 Erich Lessing\Art Resource, N.Y. p. 499 Corbis/Bettmann.

CHAPTER 19 p. 516 Dr. Paula Bruice p. 517 (top) Andre Nantel\Shutterstock p. 517 (bottom) Arvedphoto\istockphoto.com p. 519 AtWaG\istockphoto.com p. 520 (top) Courtesy Procter & Gamble Company p. 520 (bottom) Brett Atkins\Shutterstock p. 522 (top) age fotostock\SuperStock, Inc. p. 522 (bottom) Courtesy of www.istockphoto.com p. 523 Corbis/Bettmann.

CHAPTER 20 p. 534 (left) National Cancer Institute p. 534 (right) By courtesy of the National Portrait Gallery, London. p. 539 Science Photo Library\Photo Researchers, Inc. p. 542 AP Wide World Photos p. 544 (top) Dr. Paula Bruice p. 544 (bottom left) Dr. Gopal Murti/Science Photo Library\Custom Medical Stock Photo, Inc. p. 544 (bottom right) OMIKRON\Photo Researchers, Inc. p. 547 (right) Getty Images, Inc. p. 548 Nancy Cosgrove Mullis p. 549 (top) George Poinar p. 549 (bottom) Pioneer Hi-Bred, International, Inc.

CHAPTER 21 p. 555 Ralf Gosch\istockphoto.com p. 556 Gregory G. Dimijian\Photo Researchers, Inc. p. 558 AP Wide World Photos p. 559 © Bettmann/CORBIS All Rights Reserved p. 560 Hoffmann-La Roche, Inc. p. 562 AP Wide World Photos.

Index

D

Periodic Table of the Elements

	Main groups 1A[a] 1	2A 2	3B 3	4B 4	5B 5	6B 6	7B 7	8B 8	8B 9	8B 10	1B 11	2B 12	Main groups 3A 13	4A 14	5A 15	6A 16	7A 17	8A 18
1	1 **H** 1.00794																	2 **He** 4.002602
2	3 **Li** 6.941	4 **Be** 9.012182											5 **B** 10.811	6 **C** 12.0107	7 **N** 14.0067	8 **O** 15.9994	9 **F** 18.998403	10 **Ne** 20.1797
3	11 **Na** 22.989770	12 **Mg** 24.3050											13 **Al** 26.981538	14 **Si** 28.0855	15 **P** 30.973761	16 **S** 32.065	17 **Cl** 35.453	18 **Ar** 39.948
4	19 **K** 39.0983	20 **Ca** 40.078	21 **Sc** 44.955910	22 **Ti** 47.867	23 **V** 50.9415	24 **Cr** 51.9961	25 **Mn** 54.938049	26 **Fe** 55.845	27 **Co** 58.933200	28 **Ni** 58.6934	29 **Cu** 63.546	30 **Zn** 65.39	31 **Ga** 69.723	32 **Ge** 72.64	33 **As** 74.92160	34 **Se** 78.96	35 **Br** 79.904	36 **Kr** 83.80
5	37 **Rb** 85.4678	38 **Sr** 87.62	39 **Y** 88.90585	40 **Zr** 91.224	41 **Nb** 92.90638	42 **Mo** 95.94	43 **Tc** [98]	44 **Ru** 101.07	45 **Rh** 102.90550	46 **Pd** 106.42	47 **Ag** 107.8682	48 **Cd** 112.411	49 **In** 114.818	50 **Sn** 118.710	51 **Sb** 121.760	52 **Te** 127.60	53 **I** 126.90447	54 **Xe** 131.293
6	55 **Cs** 132.90545	56 **Ba** 137.327	71 **Lu** 174.967	72 **Hf** 178.49	73 **Ta** 180.9479	74 **W** 183.84	75 **Re** 186.207	76 **Os** 190.23	77 **Ir** 192.217	78 **Pt** 195.078	79 **Au** 196.96655	80 **Hg** 200.59	81 **Tl** 204.3833	82 **Pb** 207.2	83 **Bi** 208.98038	84 **Po** [208.98]	85 **At** [209.99]	86 **Rn** [222.02]
7	87 **Fr** [223.02]	88 **Ra** [226.03]	103 **Lr** [262.11]	104 **Rf** [261.11]	105 **Db** [262.11]	106 **Sg** [266.12]	107 **Bh** [264.12]	108 **Hs** [269.13]	109 **Mt** [268.14]	110 [271.15]	111 [272.15]	112 [277]		114 [285]		116 [289]		

Transition metals: groups 3B–2B (3–12).

*Lanthanide series	57 ***La** 138.9055	58 **Ce** 140.116	59 **Pr** 140.90765	60 **Nd** 144.24	61 **Pm** [145]	62 **Sm** 150.36	63 **Eu** 151.964	64 **Gd** 157.25	65 **Tb** 158.92534	66 **Dy** 162.50	67 **Ho** 164.93032	68 **Er** 167.259	69 **Tm** 168.93421	70 **Yb** 173.04
†Actinide series	89 **†Ac** [227.03]	90 **Th** 232.0381	91 **Pa** 231.03588	92 **U** 238.02891	93 **Np** [237.05]	94 **Pu** [244.06]	95 **Am** [243.06]	96 **Cm** [247.07]	97 **Bk** [247.07]	98 **Cf** [251.08]	99 **Es** [252.08]	100 **Fm** [257.10]	101 **Md** [258.10]	102 **No** [259.10]

[a] The labels on top (1A, 2A, etc.) are common American usage. The labels below these (1, 2, etc.) are those recommended by the International Union of Pure and Applied Chemistry.

The names and symbols for elements 110 and above have not yet been decided.

Atomic weights in brackets are the masses of the longest-lived or most important isotope of radioactive elements.

Further information is available at *http://www.shef.ac.uk/chemistry/web-elements/*.

The production of element 116 was reported in May 1999 by scientists at Lawrence Berkeley National Laboratory.

Common Functional Groups

Alkane	RCH_3
Alkene	$>C{=}C<$ (internal), $>C{=}CH_2$ (terminal)
Alkyne	$RC{\equiv}CR$ (internal), $RC{\equiv}CH$ (terminal)
Nitrile	$RC{\equiv}N$
Ether	$R{-}O{-}R$
Thiol	$RCH_2{-}SH$
Disulfide	$R{-}S{-}S{-}R$
Epoxide	three-membered ring containing O
Aniline	$C_6H_5{-}NH_2$
Phenol	$C_6H_5{-}OH$
Carboxylic acid	$RC(=O)OH$
Acyl chloride	$RC(=O)Cl$
Acid anhydride	$RC(=O)OC(=O)R$
Ester	$RC(=O)OR$
Amide	$RC(=O)NH_2$, $RC(=O)NHR$, $RC(=O)NR_2$
Aldehyde	$RC(=O)H$
Ketone	$RC(=O)R$

	primary	secondary	tertiary
Alkyl halide	$R{-}CH_2{-}X$ X = F, Cl, Br, or I	$R_2CH{-}X$	$R_3C{-}X$
Alcohol	$R{-}CH_2{-}OH$	$R_2CH{-}OH$	$R_3C{-}OH$
Amine	$R{-}NH_2$	R_2NH	R_3N

Approximate pK_a Values **See Appendix II for more detailed information.**

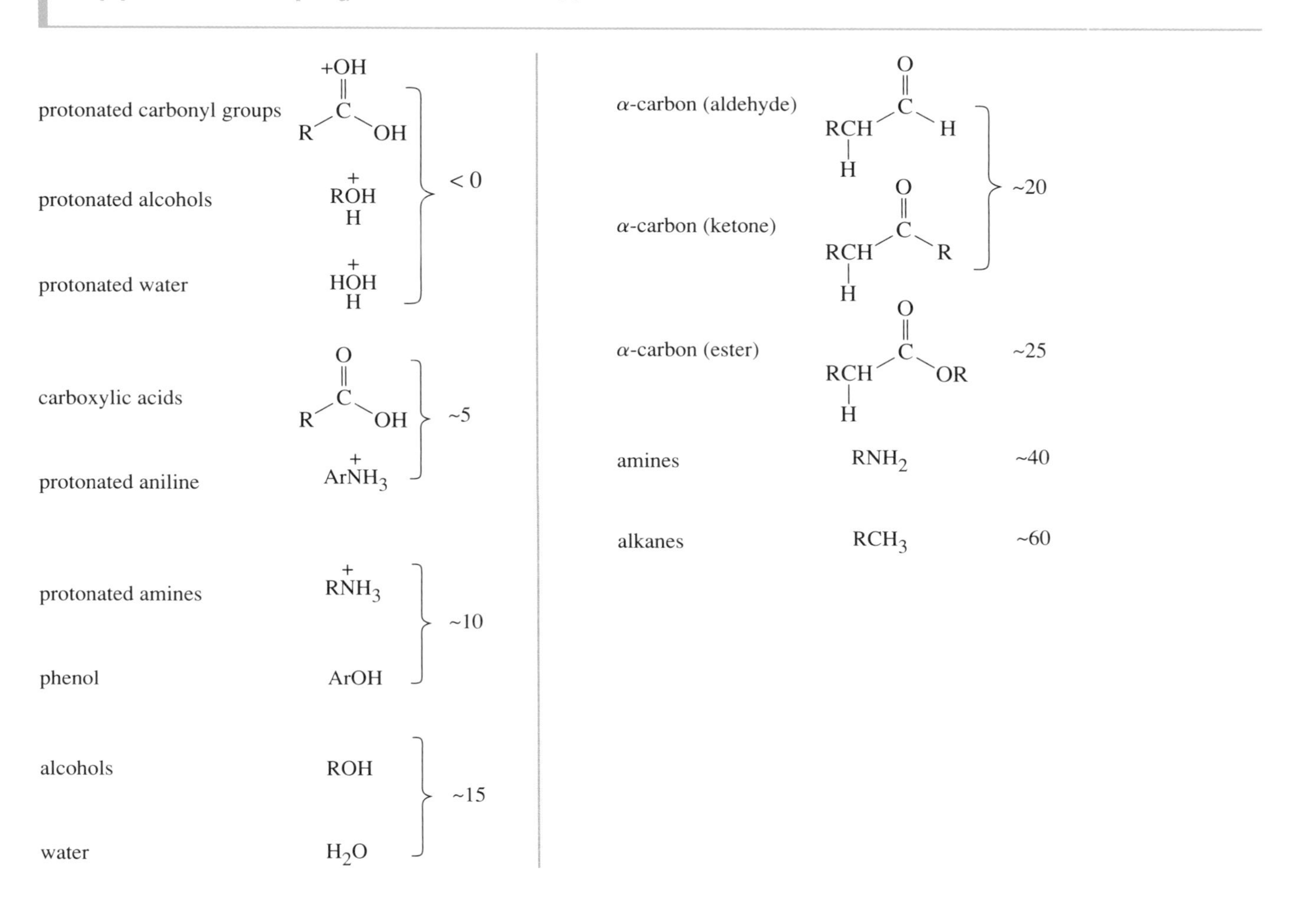

Common Symbols and Abbreviations

Ar	phenyl group or substituted phenyl group	K_a	acid dissociation constant
[α]	specific rotation	K_{eq}	equilibrium constant
B_0	applied magnetic field	NMR	nuclear magnetic resonance
DCC	dicyclohexylcarbodiimide	PCC	pyridinium chlorochromate
δ	chemical shift	pH	measure of the acidity of a solution ($= -\log [H^+]$)
δ	partial	pI	isoelectric point
Δ	heat	pK_a	measure of the strength of an acid ($= -\log K_a$)
E	entgegen (opposite sides in *E,Z* nomenclature)	R	alkyl group; group derived from a hydrocarbon
$\Delta G^{\ddagger}$	free energy of activation	*R,S*	configuration about an asymmetric center
H_2CrO_4	chromic acid	TMS	tetramethylsilane, $(CH_3)_4Si$
IR	infrared	UV/Vis	ultraviolet/visible
k	rate constant	*Z*	zusammen (same side in *E,Z* nomenclature)